Resolution of Curve and Surface Singularities

T0225002

Algebras and Applications

Volume 4

Editors:

F. Van Oystaeyen
University of Antwerp, UIA, Wilrijk, Belgium

A. Verschoren
University of Antwerp, RUCA, Antwerp, Belgium

Advisory Board:

M. Artin
Massachusetts Institute of Technology
Cambridge, MA, USA

A. Bondal
Moscow State University, Moscow, Russia

I. Reiten
Norwegian University of Science and Technology
Trondheim, Norway

The theory of rings, algebras and their representations has evolved into a well-defined subdiscipline of general algebra, combining its proper methodology with that of other disciplines and thus leading to a wide variety of applications ranging from algebraic geometry and number theory to theoretical physics and robotics.

Due to this, many recent results in these domains were dispersed in the literature, making it very hard for researchers to keep track of recent developments.

In order to remedy this, *Algebras and Applications* aims to publish carefully refereed monographs containing up-to-date information about progress in the field of algebras and their representations, their classical impact on geometry and algebraic topology and applications in related domains, such as physics or discrete mathematics.

Particular emphasis will thus be put on the state-of-the-art topics including rings of differential operators, Lie algebras and super-algebras, groups rings and algebras, C* algebras, Hopf algebras and quantum groups, as well as their applications.

Resolution of Curve and Surface Singularities

in Characteristic Zero

by

K. Kiyek

Department of Mathematics,
University of Paderborn, Paderborn, Germany

and

J.L. Vicente

Departamento de Algebra,
Universidad de Sevilla, Sevilla, Spain

KLUWER ACADEMIC PUBLISHERS

DORDRECHT / BOSTON / LONDON

A C.I.P. Catalogue record for this book is available from the Library of Congress.

ISBN   978-90-481-6573-5 (PB)
ISBN   978-1-4020-2029-2 (e-book)

Published by Kluwer Academic Publishers,
P.O. Box 17, 3300 AA Dordrecht, The Netherlands.

Sold and distributed in North, Central and South America
by Kluwer Academic Publishers,
101 Philip Drive, Norwell, MA 02061, U.S.A.

In all other countries, sold and distributed
by Kluwer Academic Publishers,
P.O. Box 322, 3300 AH Dordrecht, The Netherlands.

Printed on acid-free paper

All Rights Reserved
© 2004 Kluwer Academic Publishers
Softcover reprint of the hardcover 1st edition 2004
No part of this work may be reproduced, stored in a retrieval system, or transmitted
in any form or by any means, electronic, mechanical, photocopying, microfilming, recording
or otherwise, without written permission from the Publisher, with the exception
of any material supplied specifically for the purpose of being entered
and executed on a computer system, for exclusive use by the purchaser of the work.

Contents

Preface

The Curves

The Point of View of Max Noether

Probably the oldest references to the problem of resolution of singularities are found in Max Noether's works on plane curves [cf. [148], [149]]. And probably the origin of the problem was to have a formula to compute the genus of a plane curve. The genus is the most useful birational invariant of a curve in classical projective geometry. It was long known that, for a plane curve of degree n having m *ordinary singular points*[1] with respective multiplicities r_i, $i \in \{1, \ldots, m\}$, the genus p of the curve is given by the formula

$$p = \frac{(n-1)(n-2)}{2} - \frac{1}{2} \sum r_i(r_i - 1).$$

Of course, the problem now arises: how to compute the genus of a plane curve having some *non-ordinary* singularities. This leads to the natural question: can we birationally transform any (singular) plane curve into another one having only ordinary singularities? The answer is positive. Let us give a flavor (without proofs) on how Noether did it[2].

To solve the problem, it is enough to consider a special kind of Cremona transformations, namely quadratic transformations of the projective plane. Let Σ be a linear system of conics with three non-collinear base points $\Gamma = \{A_0, A_1, A_2\}$, and take a projective frame of the type $\{A_0, A_1, A_2; U\}$. Let us consider the three conics of Σ having equations

$$0 = \varphi_0 = x_1 x_2, \quad 0 = \varphi_1 = x_0 x_2, \quad 0 = \varphi_2 = x_0 x_1$$

and the map

$$T : \mathbb{P}^2_{\mathbb{C}} \setminus \Gamma \longrightarrow \mathbb{P}^2_{\mathbb{C}}$$

given by the equations

$$\varrho x'_0 = x_1 x_2, \quad \varrho x'_1 = x_0 x_2, \quad \varrho x'_2 = x_0 x_1;$$

then T is called a *quadratic transformation* associated with Γ. It has nice properties:

1. The conics of Σ are mapped onto straight lines.

2. T is self-inverse outside $A_0 A_1 \cup A_0 A_2 \cup A_1 A_2$, hence it is birational.

---

[1] An ordinary singular point is a point of multiplicity r with r different tangents.
[2] For details see [190], Th. 7.4., p. 80 or [69], Ch. 7, Th. 2.

3. The points A_0, A_1, A_2 have no image: they are the *fundamental points* of the correspondence.

4. The reference lines (with the reference points removed) contract to points.

The algorithm is to apply a sequence of quadratic transformations to the given curve C of degree n (i.e. to C and its successive transforms) in such a way that, if $P \in C$ is a singular r-fold point, then we take:

1. $A_0 = P$.

2. The line $A_0 A_i$, $i = 1, 2$, cuts C at exactly $n - r$ pairwise different points other than A_0, A_i.

3. The line $A_1 A_2$ cuts C at exactly n pairwise different points other than A_1, A_2

[by an appropriate choice of A_0, A_1, A_2 this can always be achieved; on says that then C is in excellent position with respect to the points A_0, A_1, A_2]. In this situation, P "blows up" to a finite number of points on the opposite side of the reference triangle, which are "less complicated" than P. A finite number of steps in this process allows us to reach the desired result.

The First Resolution of Singularities of Curves

The first general proof of resolution of singularities for projective curves over a ground field of arbitrary characteristic was performed by Zariski-Muhly [cf. [143]]. Rather than faithfully comment the arguments of Zariski-Muhly, we are going to think about what a resolution of singularities is, and then we shall deal with the curve case in the spirit of these authors, but borrowing from modern algebraic language.

The Noether's theorem of the preceding section leads us to the natural question of whether it would be possible to birationally transform a plane curve into another one which is free from singularities. This cannot be done in general. However, the proper question is not that one; it is rather the following: given an irreducible curve $C \subset \mathbb{P}_k^n$, does there exist another irreducible curve $C' \subset \mathbb{P}_k^{n'}$, birationally equivalent to C and free from singularities? The answer to this question is affirmative, as was shown by Zariski-Muhly.

Resolution of singularities of curves *à la Zariski-Muhly* is achieved in the following way. Let k be an algebraically closed field of arbitrary characteristic (the ground field), let $C \subset \mathbb{P}_k^n$ be an irreducible curve whose ideal in $k[X_0, X_1, \ldots, X_n] = k[\mathbf{X}]$ is $\mathfrak{p}$, and let us denote by

$$A(C) = k[\mathbf{X}]/\mathfrak{p} = k[\mathbf{x}] \quad \text{where } x_i := X_i + \mathfrak{p} \text{ for } i \in \{0, \ldots, n\}$$

the ring of homogeneous coordinates on C. By Noether normalization lemma, we may always assume that

$$A(C) = k[x_0, x_1][x_2, \ldots, x_n],$$

where $\{x_0, x_1, \ldots, x_n\}$ are homogeneous of degree 1, $A(C)$ is integral over $k[x_0, x_1]$, the field $K := k(x_1/x_0)$ is a field of rational functions in one variable over k, and the field $F = k(x_1/x_0, \ldots, x_n/x_0)$ is a finite separable extension of K. Note that F is the subfield consisting of elements of degree 0 of the graded ring of fractions of $Q_+(A)$ (which consists of all quotients z/w where z, $w \in A(C)$ are homogeneous). Now it is not too difficult to show that the integral closure $\overline{A}(C)$ of $A(C)$ in $Q_+(A)$ is a finitely generated $k[x_0, x_1]$-module.

However, there is something far more important, namely that $\overline{A(C)}$ has a natural structure of a graded ring and of a graded $A(C)$-module: it is a homogeneous subring of $Q_+(A)$. Then there exists a finite set of elements $\{z_0, z_1, \ldots, z_{n'}\}$ in $\overline{A(C)}$, homogeneous of degree 1, such that $\overline{A(C)} = k[z_0, z_1, \ldots, z_{n'}]$. Then $(z_0, z_1, \ldots, z_{n'})$ are the homogeneous coordinates of the generic point of an irreducible curve $C' \subset \mathbb{P}_k^{n'}$, which is birationally equivalent to C and is nonsingular. The smoothness comes from the fact that, $A(C')$ being integrally closed, the singular locus of C' has codimension 2, that is, it is empty[3]. This proves resolution of singularities for curves.

It turns out that, for many purposes, this resolution process is not satisfactory. It is better to have a finite algorithm, i.e., a finite sequence of birational transformations which, when applied to an irreducible curve $C \subset \mathbb{P}_k^n$, produces a nonsingular curve C'. This will be the main theme of the next section.

The Controlled Process

The above resolution of singularities can be considerably refined. In general, we seek not only C' but a finite sequence of birational transformations that, when successively applied to C, produces C'. These transformations must be as simple as possible. If we take this point of view, we need a control of the process of applying transformations that, on the one side, guarantees us the finiteness of the process and, on the other side, allows us to measure the "improvement" at each step. One must never forget this: if we transform, we need a control measure. Overlooking this obvious fact, has driven several early authors to false proofs of resolution of singularities (cf., for instance, a historic case, [198]).

The transformations which work are called *blowing ups*. We will briefly describe a special case: the blowing up of a linear variety in $\mathbb{P}_k^n$. This will be enough for our purposes, since a curve has a finite number of singular points, and we will blow up these ones.

Let us write $n = r + s + 1$, $n, r, s \in \mathbb{N}_0$, $n \geq 2$, consider the projective space $\mathbb{P}_k^n$ and a linear subvariety L of dimension r; then the radiation P_k^n/L is a projective space of dimension s [cf. appendix A, section 12]. We also consider the map

$$f: \mathbb{P}_k^n \longrightarrow \mathbb{P}_k^n/L$$

that, to each point $P \in \mathbb{P}_k^n$, $P \notin L$, associates the point $(P + L)/L$ of $\mathbb{P}_k^n/L$. Let $\Gamma \subset \mathbb{P}_k^n \times (\mathbb{P}_k^n/L)$ be the graph of f, and let $\overline{\Gamma}$ be its Zariski closure in the product

---

[3]This always happens: the singular locus of a projective variety whose coordinate ring is integrally closed has always codimension 2 at least.

space. The first projection $\pi \colon \overline{\Gamma} \to \mathbb{P}_k^n$ is called the *blowing up* of L in $\mathbb{P}_k^n$, or the blowing up of $\mathbb{P}_k^n$ with *center L*. It is a straightforward matter to get a set of bi-homogeneous equations for $\overline{\Gamma}$.

Let V be a projective variety in $\mathbb{P}_k^n$ containing L; then the strict transform of V by π is the Zariski closure of $f^{-1}(V \setminus L)$. If C is a curve and $L = P$ is one of its points, then it is evident that the strict transform C_1 of C is isomorphic to C outside P and that $\pi^{-1}(C)$ is a finite set of points, namely one for every tangent line of C at P.

We now apply to C the following algorithm. We take a singular point $P_0 \in C_0 := C$ and blow up P_0. If the strict transform C_1 of C has some singular points, we pick one of them, do the same, and so on. In this way, we get a sequence of blowing ups and strict transforms

$$\cdots \to C_m \xrightarrow{\pi_n} C_{m-1} \xrightarrow{\pi_{m-1}} \cdots \xrightarrow{\pi_2} C_1 \xrightarrow{\pi_1} C_0,$$

and the task is to prove that this sequence is, at any rate, finite[4].

The proof of this fact can be done in two ways. One is purely algebraic, and the idea is very simple. If we start with $P_0 \in C_0$, then we take on C_1 any point P_1 such that $\pi_1(C_1) = P_0$, and so on. We may call this "to follow the track" of P_0. Passing to local rings, and given the finiteness of the integral closure, this "following the track" ends in a regular local ring, so at a smooth point. Note that, "to follow the track" is enough to prove what we wanted, because, when dealing with one singular point, the others are unaltered, and the total number of simple points is finite.

The second way to deal with the proof is analytic and works only for an algebraically closed field of characteristic zero, essentially $\mathbb{C}$. The method is the same: "to follow the track". However, this time we use complex-analytic parametric equations instead of local rings, but with the same aim. To simplify things, we will assume that C is analytically irreducible at P_0, i.e., it has only one branch. Since we are dealing with one point at each step, we may go affine. If P_0 is the origin of $\mathbb{C}^n$, then we may describe a neighborhood of P_0 by a set of analytic parametric equations

$$x_1 = t^p, \; x_2 = \sum_{i>p} a_{2i} t^i, \ldots, x_n = \sum_{i>p} a_{ni} t^i$$

where p is the multiplicity of P_0 in C_0. The parametric equations of C_1 in a neighborhood of P_1 are

$$x_1 = t^p, \; x_2 = \sum_{i>p} a_{2i} t^{i-p}, \ldots, x_n = \sum_{i>p} a_{ni} t^{i-p},$$

and we see that, after a finite number of steps, the multiplicity must drop. This proves everything. In the case that there are several branches at P_0 (always a

---

[4] Remember that it continues one step further only if the last curve has a singular point.

finite number!), the same process has a step at which either multiplicity drops or two branches separate (hence multiplicity also drops!), so we continue with two sequences with less branches and smaller multiplicities.

This fact is a major issue when we solve singularities of surfaces. The complete neighborhood of a singularity of a complex curve can be analytically parametrized. *This is in fact false in the case of surfaces.* Hence, we can apply these methods only in some very special cases in which analytic parametric equations exist.

The Surfaces

Jung Methods

When we dealt before with resolution of singularities of curves, we explained how to take advantage of the two key facts: local parametrization and the finiteness of the singular locus. Now that we are dealing properly with the case of surfaces, we find that none of these facts is true.

Let us deal first with the difficulties. It is enough to see them in very easy cases, for instance, surfaces embedded in dimension 3. Let $S : z^2 - 2yz + x^2 = 0$ be a surface with a singularity in $(0, 0, 0)$. By solving the second-degree equation in z we get the two solutions

$$f_1 = y \left(1 + \sqrt{1 - \frac{x^2}{y^2}} \right), \quad f_2 = y \left(1 - \sqrt{1 - \frac{x^2}{y^2}} \right)$$

which do not make sense, even as meromorphic functions. However, these "expressions" converge for $|x| < |y|$; this gives us a region around $(0,0) \in \mathbb{C}^2$, which is "one half" of a neighborhood, something of the kind Walker [cf. [189]] called *a wedge*. The world would be perfect if we could find, for every (embedded) surface (around a singular point, say $(0,0,0)$), a Puiseux expansion

$$z = \sum_{(i,j) \in \mathbb{N}_0^2, i+j > 0} \alpha_{ij} x^{r_i/q} y^{s_j/q},$$

but this is not true.

Nevertheless, Jung [cf. [102]] did something in 1908. He proved (essentially in the complex-analytic case) that, if a neighborhood of a singular point (say $O = (0, 0, 0)$) of a surface can be projected onto a plane in such a way that the discriminant locus is a normal crossing divisor[5], then the complete neighborhood of O can be represented by conjugate Puiseux power series in two variables, of a special type called quasi-ordinary, which we are not to describe here. For instance, around the origin, the surface $z^4 - 2xz^2 + x^2 - x^2y = 0$ has a discriminant $D = -256 x^6 (-1 + y) y^2$, hence, by Jung, the complete neighborhood of $(0, 0, 0)$ is described by the Puiseux power series $z = \sqrt{x}\sqrt{1 - \sqrt{y}}$.

---

[5]i.e., the set of zeroes of $x^a y^b U(x, y)$, with U a unit in the ring of convergent power series

There are several proofs of Jung's theorem besides Jung's own one, which is of a topological nature. Abhyankar felt the need of having a purely algebraic proof, which he gave in 1955 [cf. [1]] making use of very sophisticated Galois theory. The authors of this book gave a third one by using modern techniques of ramification theory [cf. [108]]. Finally, there are hints for the existence of a purely combinatorial way of proving it [see, for instance, [175] for the two-dimensional case].

Now that we have introduced local parametrization of a quasi-ordinary (or Jungian) surface singularity, the obvious question is the following: how can we apply this to obtain a full resolution of singularities of a surface? We have two problems:

1. Given a projection $\pi : X \to S$ where X and S are surfaces and S is smooth, is there a morphism $\varphi : S' \to S$ such that π *pulls back* to a projection $\pi' : X' \to S'$, with S' smooth and such that the discriminant locus of π' is a normal crossing divisor?

2. Given a projection $\pi : X \to S$ having a normal crossing divisor as its discriminant locus, how can X be desingularized?

For a general reference, see also [50].

The answer to the first problem is a pure technicality. The morphism φ is, essentially, the embedded resolution of the discriminant locus D_π of π, that is, a finite sequence of blowing ups transforming D_π into a normal crossing divisor.

Let us deal with the second problem. The first step to desingularization is to pass to the derived normal model. With the information we are given on Puiseux power series, the reader should be able to convince himself, after a little thought, of the following fact: the derived normal model of a surface projected onto a smooth one, with discriminant locus a normal crossing divisor, is a monomial variety. Now, these varieties can be desingularized by blowing up closed points, as it has been proved by Hirzebruch. This implies global desingularization because the derived normal model of X has only a finite number of singularities which are closed points, and local desingularization of each of these (as monomial varieties) does not affect the others. There are a lot of technicalities inside the process; essentially one must study Hirzebruch-Jung singularity-type singularities and Hirzebruch continued fraction expansions for a monomial variety.

The Zariski Method

Oscar Zariski is, by far, the most important author in the field of resolution of singularities before Hironaka. In fact, Hironaka took some inspiration from Zariski, as well as did Abhyankar, who has been the leader of the movement towards "the positive characteristic", i.e., the extension of the theorem of resolution of singularities to ground fields of positive characteristic.

To understand the philosophy behind Zariski's papers on resolution of singularities, one needs to say a few words on the ambient he lived in. This ambient was dominated by the ancient italian geometry, and its lack of rigor for proofs is very well known. To do justice to the italian geometers, one must say that most of the results they pretended to have proved were finally true, once rigor was introduced much later. But, for Zariski, the situation was completely unclear. In fact, at that time, confusion was the name of the game, and some major flaws of the italian geometry had been found. This explains the almost obsessive search for rigor in proofs he had. He sought rigor by making proofs in algebraic or arithmetical ways. He knew quite a lot of the so called "modern algebra", and he was the first one who saw that the algebraic techniques could be applied to give rigorous foundations to italian geometry. This is better explained by using his own words. In a review he wrote for Math. Reviews, on some paper that gave an incorrect proof of resolution of singularities [cf. [199]], he stated: *The geometric language, when it is not based on a carefully prepared algebraic basis, is never explicit or convincing in algebraic geometry. On this ground alone the author's 'proof' could be dismissed as incomplete, for in scientific work it is right to hold every author guilty until he proves himself innocent.*

Zariski's first papers on resolution of singularities dealt with the surface case. Resolution of singularities of surfaces was already known at this time. There were several (correct!) proofs by italian geometers, namely by Albanese, Beppo Levi and Chisini [see [17], [124] and [43]]. There was also Walker's proof [cf. [189]] which deserves a little comment. Walker's proof is analytical and is based upon Jung's ideas, too. He shows first that the general problem of resolution of singularities of an algebraic surface can be reduced to the case in which there is a projection onto a smooth surface whose discriminant locus is a normal crossing divisor (see above). Then, there exist Jung's parametrizations of the neighborhoods of singular points. He calls these parametrized neighborhoods *wedges*. Now, he introduces the concept of *resolving system* for a set of wedges, which is nothing else but a set of Cremona transformations essentially solving the singularities of the branches (of dimension 1) contained in the wedges. The very definition of resolving system (which is too technical to be given here) implies that all these local resolutions automatically glue together, so the only problem is to find explicitly a resolving system for each set of wedges, which is done in the main part of his paper.

However, Zariski wanted an algebraic proof for the resolution of singularities of surfaces mainly because, by using algebra, he believed he could adapt the italian proofs. There are several proofs of resolution of singularities by Zariski. We are going to comment here, with some detail, only the key idea of the one included in this book [cf. [192]], of course translated to modern techniques.

The very first (and rough) idea is to use a technique similar to Zariski-Muhly [cf. [143]] resolution of singularities for curves, taking into account that the construction of the integral closure does not depend on dimension. But two difficulties immediately arise:

1. A variety of dimension 2 with integrally closed coordinate ring is not smooth (as curves were!), but its singular locus might have a finite number of points.

2. If we consider the process of reaching the integral closure by means of blowing

ups, we find that these two operations are not compatible in the following sense: the blowing up of a normal surface usually is not normal. For instance, (and very roughly), if we blow up the origin of the surface $z^3 - xy = 0$, then we have a point in the exceptional divisor whose local ring is determined by the equation

$$\left(\frac{z}{x}\right)^3 - \frac{1}{x} \cdot \frac{y}{x} = 0,$$

so the strict transform is clearly out of the integral closure.

The solution is achieved by a sequence of blowing ups and normalizations (that is, passage to the integral closure). The huge difficulty here is the control of the process. One must keep in mind that normalization means elimination of the singular curves of the surface.

The obvious possibility of control is to have something covering (as an umbrella) the process of blowing up and normalizing. There is something which behaves very well with respect to both processes: valuations. If we start with a normal variety and a singular point, and if we blow up and normalize, we end up with a finite number of points projecting onto the initial one. If we continue like that, we must show that (globally!) there are only a finite number of steps with non-empty singular loci. But a sequence of points and transforms is covered by a valuation, which can be uniformized (i.e., there is a projective model at which the center of the valuation is smooth). This is the key for the finiteness of the process.

The proofs are highly technical, and require a great deal of valuation theory, and ideal theory of two-dimensional regular local rings.

Just in a few words, and to end our preface: three key things are proved in this book:

1. Resolution of singularities of curves (over an algebraically closed field of arbitrary characteristic).

2. Resolution of surface singularities à la *Jung* (over an algebraically closed field of characteristic zero)

3. Resolution of surface singularities by blowing up and normalizing, i.e., resolution á la *Zariski* (over an algebraically closed field of characteristic zero).

Prerequisites

The book is self-contained, except for very common and elementary notions of Commutative Algebra. With regard to notions and results from Commutative Algebra, in this book we rely mostly on Eisenbud's "Commutative Algebra" [63]; to make life for the reader more easy, we gave a lot of references to [63] which the more experienced reader naturally may skip.

There are two appendices: We work inside the frame of classical Algebraic Geometry; no schemes are around. Thus, we felt the need to introduce those notions which we could not found readily in the existing literature; in particular, we needed more results on blowing up points of a projective variety than one usually finds in textbooks. This is the justification for the presence of appendix A. In appendix B we collect a lot of miscellaneous results which we needed in this book, and which mostly are not readily available in existing textbooks. In particular, at the beginning of this appendix, we prove the results on totally ordered abelian groups which are needed when dealing with valuations.

Main Sources

For the results in classical valuation theory in chapter I, we used mainly the books of Zariski and Samuel [204] and Bourbaki [34], ch. 5; for the proof of Abhyankar's results on valuations dominating a local domain we used the presentation given by Vaquié [181]. The results on one-dimensional local Cohen-Macaulay rings in chapter II have their origin in the lecture notes of Matlis [138] and a paper of Lipman [128] which appeared nearly at the same time. For the theory of quasi-ordinary singularities our main source was the thesis of Lipman [126]; the algorithm in section 5 of chapter V comes from the thesis of Herrera [87]. With regard to toric varieties, there are the three books of Ewald [68], Fulton [70] and Oda [157]. We develop in chapter VI the theory only as far as we need it; in particular, we restrict our considerations mainly to the case of dimension 2, and follow to a certain extent the presentation in Oda's book. For the theory of two-dimensional regular local rings, we rely on appendix 4 in the book of Zariski and Samuel [204], the article of Huneke [98], the two papers [131] and [133] of Lipman, and Northcott's paper [152]. The proof of the uniformization theorem which we present in chapter VIII is based on Abhyankar's book [3]; the proof that repeatedly blowing up and normalizing leads to a regular surface is, in principle, Zariski's own proof as presented by Lipman in a far more general context in his paper [130]. With regard to appendix A in which we prepare the necessary background from Algebraic Geometry, our main sources where the books of Brodmann [36], Hartshorne [82] and Shafarevich [174].

Acknowledgements

We thank all persons who helped us in on way or other. The Spanish research authorities, both at national and andalusian level, helped by financing this project during several years. The staff of the Algebra Department of the University of Sevilla (and some from Granada, too!) gave us invaluable help. In particular, J. L. Vicente was partially supported by BFM2001-3207 and FEDER. The University of Paderborn gave us financial support. The secretaries at various times of the first author, Mrs. Borchert, Mrs. Schapkow and Mrs. Kube, typed numerous versions of the manuscript. Kluwer and, especially, the editor of this collection must be deeply thanked for their quick understanding and gentleness.
To end, we quote Zariski's own words [cf. [192]]: *The length of this paper is not entirely due to the lengthiness of the proof. It is due in part to the fact that we treat the entire problem "ab initio". We also wished to avoid, as much as possible, references to scattered sources, especially when we are not certain whether the reference would exactly cover the topic under consideration.*

Note to the Reader

Definitions, lemmas, propositions, theorems ect. are numbered consecutively within each section. References to such an item in the same chapter are made by (m.n) where m is the section number and n indicates the position within the section. References to such an item in another chapter (resp. in one of the two appendices) are made in the form R(m.n) (resp. A(m.n) resp. B(m.n)) where R is the (capital roman) chapter number (resp. A is appendix A resp. B is appendix B), and (m.n) are as before. References to the bibliography at the end of this book are indicated in the text by numbers between square brackets.

We listed in the bibliography most of the books which treat (partial) aspects of the subjects of this book. Also, all the papers which we used more or less explicitly are listed there.

For an extensive bibliography on resolution of singularities the reader should refer to the article of Hauser [85]; we added to our bibliography some newer items which are not yet listed in Hauser's paper.

It is planned to provide—in the course of time—additional material concerning the subjects dealt with in this book. In particular, a chapter on algebroid curves seems to be a nice supplement.

This additional material can be found on the homepages of the authors, namely

K. Kiyek: http://math-www.uni-paderborn.de/~karlh/Singularities

J. L. Vicente: http://www.us.es/da

Terminology

We use the terminology of commutative algebra.

A ring R is, unless stated otherwise, a commutative ring with unit element, and R-modules are unitary R-modules. The ring R is called reduced if 0 is the only nilpotent element. The radical of an ideal $\mathfrak{a}$ of a ring R is denoted by $\mathrm{rad}(\mathfrak{a})$; we set $R_{\mathrm{red}} := R/\mathrm{rad}(\{0\})$. Note that R_{red} is reduced. An ideal is called a nil ideal if every element of it is nilpotent.

Let R be a ring; an R-algebra is a commutative ring together with a ring homomorphism $R \to S$.

Let R be a ring; then $\mathrm{Spec}(R)$ is the set of prime ideals of R, endowed with the Zariski topology, and $\mathrm{Max}(R)$ is the set of maximal ideals of R. For every ideal $\mathfrak{a}$ of R we denote by $V(\mathfrak{a})$ the subset of $\mathrm{Spec}(R)$ consisting of all prime ideals of R which contain $\mathfrak{a}$, and for $f \in R$ we set $D(f) := \{\mathfrak{p} \in \mathrm{Spec}(R) \mid f \notin \mathfrak{p}\}$.

Let R be a ring and M an R-module; then $\ell_R(M)$ is the length of M. An ideal $\mathfrak{a}$ of R is said to have finite colength if the R-module $R/\mathfrak{a}$ has finite length.

A ring R it called quasilocal if it has only one maximal ideal $\mathfrak{m}$; sometimes we write $\mathfrak{m}(R)$ for this ideal $\mathfrak{m}$. If, in addition, R is noetherian, then R is called local. The field $R/\mathfrak{m}$ is called the residue field of R. A ring is called quasisemilocal if it has finitely many maximal ideals; if it is, in addition, noetherian, then it is called semilocal.

A homomorphism $R \to S$ of rings is called finite if S is a finitely generated R-module, and it is called of finite type if there exists a surjective R-algebra homomorphism $R[T_1, \ldots, T_n] \to S$ where $R[T_1, \ldots, T_n]$ is the polynomial ring over R in n indeterminates $T_1, \ldots, T_n$.

Let S be a ring, and let R be a subring of S. Let M be a submodule of the R-module S, and let U, V be subsets of S. Then we set

$$(M : U)_V := \{v \in V \mid vU \subset M\};$$

$(M : U)_V$ is a submodule of the R-module S.

Let R be a ring, and let $m, n \in \mathbb{N}$. Then $M(m, n; R)$ denotes the additive group of (m, n)-matrices with entries in R, and $M(m; R)$ is the ring of m-rowed square matrices with entries in R. I_m denotes the m-rowed identity matrix of $M(m; R)$. For $A \in M(m, n; R)$ we denote by ${}^t A$ the transpose of A.

If $k \subset K$ are fields, then $\mathrm{tr.\,d}_k(K)$ denotes the transcendence degree of K over k.

A normal extension of fields must not be finite; a Galois extension of fields is always a finite Galois extension.

Let a be a real number; then $\lfloor a \rfloor$ is the largest integer n with $n \le a$, and $\lceil a \rceil$ is the smallest integer n with $a \le n$.

Chapter I

Valuation Theory

INTRODUCTION: In this chapter we study valuation rings and valuations. We start with the concept of a Manis valuation of a ring K; Manis valuations behave nicely if K is a ring having large Jacobson radical. We show that a Manis valuation of K gives rise to a Manis valuation ring of K and that, conversely, every Manis valuation ring of K determines a Manis valuation of K. Also, for discrete Manis valuations we prove an approximation theorem.

From now on let K be a field; in this case a Manis valuation resp. a Manis valuation ring of K is a valuation resp. a valuation ring of K. In section 3 we introduce, among other things, the notion of composite valuations; here the reader is supposed to have some knowledge of ordered groups; we refer to section 1 of appendix B where results on ordered groups are proved. In section 4 we prove the approximation theorem for independent valuations of a field.

After some generalities in section 5 on extensions of valuations we study in section 6 extensions of valuations of a field K to an algebraic overfield L of K and consider, in particular, the case that L is a finite extension of K. Extensions of discrete valuations to finite extensions of K are the subject of section 7. Ramification theory of valuations is needed for the proof of the theorem of uniformization in section 6 of chapter VIII; before reading this section the reader should study section 7 of appendix B. In section 9 we consider extensions of valuations of a field K to non-algebraic overfields. We use these results for getting results on valuations of algebraic function fields in section 10 and for proving Abhyankar's result on valuations dominating a local domain in section 11.

1 Marot Rings

(1.0) NOTATION: Let A be a ring. We use without further explanations the following notations. The intersection of all maximal ideals of A is called the Jacobson radical of A. The intersection of all prime ideals of A is called the nilradical of A; it is also the set of nilpotent elements of A [cf. [63], Cor. 2.12], hence it is a nil ideal. If S is multiplicatively closed in A, and if $\mathfrak{n}$ is the nilradical of A, then $S^{-1}\mathfrak{n}$ is the nilradical of $S^{-1}A$.

The set of regular elements of A is denoted by $\operatorname{Reg}(A)$, and an ideal of A is called regular if it contains regular elements. The group of units of A is denoted by $U(A)$ or $A^\times$. The ring of quotients of A is denoted by $Q(A)$; hence we have $Q(A) = \operatorname{Reg}(A)^{-1}A$, and we have an embedding $A \hookrightarrow Q(A)$. Note that the units of $Q(A)$ are just the regular elements of $Q(A)$. An A-submodule of $Q(A)$ is called regular if it contains a regular element of $Q(A)$; this is the case iff it contains a regular element of A.

(1.1) CONDUCTOR: Let B be an overring of a ring A; then

$$(A : B)_B = \{b \in B \mid Bb \subset A\}$$

is an ideal of the ring A and of the ring B; it is the largest ideal of A which is also an ideal of B. This ideal is called the conductor of A in B.

1.1 Marot Rings

(1.2) DEFINITION: A ring is called a Marot ring if each of its regular ideals is generated by its regular elements.

(1.3) Proposition: *Let A be a ring. The following statements are equivalent:*
(1) A is a Marot ring.
(2) For every two elements a, $b \in A$ such that $a \in \operatorname{Reg}(A)$, the ideal $Aa + Ab$ admits a finite system of regular elements as generators.
(3) Every regular submodule of the A-module $Q(A)$ admits a system of generators contained in the set $\operatorname{Reg}(Q(A))$.
Assume that A is a Marot ring. Then every finitely generated regular submodule M of the A-module $Q(A)$ admits a finite system of generators which is contained in the set $\operatorname{Reg}(Q(A))$.

Proof (1) $\Rightarrow$ (2): We set $\mathfrak{a} := Aa + Ab$; the ideal $\mathfrak{a}$ is generated by its regular elements, hence we can write $a = r_1 c_1 + \cdots + r_p c_p$ and $b = s_1 d_1 + \cdots + s_q d_q$ where $r_1, \ldots, r_p$ and $s_1, \ldots, s_q$ are elements of A and $c_1, \ldots, c_p$ and $d_1, \ldots, d_q$ are regular elements of $\mathfrak{a}$. Then $\{c_1, \ldots, c_p, d_1, \ldots, d_q\}$ is a finite system of generators of $\mathfrak{a}$ consisting of regular elements.

(2) $\Rightarrow$ (3): Let M be a regular submodule of the A-module $Q(A)$, and let $a \in M$ be a regular element of A. Let $x \in M$. We can choose $s \in \operatorname{Reg}(A)$ such that $sx \in A$; let $G'_x \subset A$ be a system of generators for the ideal $Asa + Asx \subset A$ consisting of regular elements [note that $sa \in \operatorname{Reg}(A)$]. Then $G_x := s^{-1}G'_x$ is a system of generators for the A-module $Aa + Ax \subset M$ consisting of regular elements of $Q(A)$, and x is an element of the A-module generated by G_x. The union of the sets G_x, $x \in M$, is a system of generators of M contained in $\operatorname{Reg}(Q(A))$.

(3) $\Rightarrow$ (1): This is clear.

Now we assume that A is a Marot ring, and that M is a finitely generated regular submodule of the A-module $Q(A)$. By adapting the proof of (1) $\Rightarrow$ (2), it is easy to see that M has a finite system of generators contained in $\mathrm{Reg}(Q(A))$.

(1.4) Corollary: *Let A have the following property: For every two elements a, $b \in A$ with $a \in \mathrm{Reg}(A)$ there exists $c \in A$ such that $b + ca$ is regular. Then A is a Marot ring.*

Proof: Note that the ideal $Aa + Ab$ can be generated by the regular elements a and $b + ca$.

(1.5) Corollary: (1) *If A is a Marot ring, then every subring of $Q(A)$ containing A is a Marot ring.*
(2) *If every finitely generated regular ideal of A is principal, then A is a Marot ring.*

1.2 Large Quotient Ring

(1.6) REMARK: Let A be a ring, and let $\mathfrak{p}$ be a prime ideal of A. It is easy to see that the set

$$A_{[\mathfrak{p}]} := \{x \in Q(A) \mid sx \in A \text{ for some } s \in A \setminus \mathfrak{p}\}$$

is a subring of $Q(A)$; this subring is called the large quotient ring of A with respect to $\mathfrak{p}$. Note that $A_{[\mathfrak{p}]} = A_{\mathfrak{p}}$ if A is a domain.

Properties of the large quotient ring are listed in [118] and [96]. In case A is a Marot ring, the ring $A_{[\mathfrak{p}]}$ is the localization of A with respect to a multiplicatively closed system, as the next result shows; hence in this case all the usual properties concerning localization can be applied.

(1.7) Proposition: *Let A be a Marot ring, and let $\mathfrak{p}$ be a prime ideal of A. We set $S := \mathrm{Reg}(A) \cap (A \setminus \mathfrak{p})$. Then S is a multiplicatively closed system in A, and we have $A_{[\mathfrak{p}]} = S^{-1}A$. The ideal $\mathfrak{q} := \mathfrak{p}A_{[\mathfrak{p}]}$ is a prime ideal of $A_{[\mathfrak{p}]}$, and we have $A_{\mathfrak{p}} = (A_{[\mathfrak{p}]})_{\mathfrak{q}}$.*

Proof: It is clear that S is a multiplicatively closed system in A, and that $S^{-1}A \subset A_{[\mathfrak{p}]}$. Let $x \in A_{[\mathfrak{p}]}$, and choose $s \in A \setminus \mathfrak{p}$ with $sx \in A$. There exists a regular element $a \in A$ with $ax \in A$. We set $\mathfrak{a} := Aa + As$; then $\mathfrak{a}$ is a regular ideal of A, and we have $\mathfrak{a}x \subset A$ and $\mathfrak{a} \not\subset \mathfrak{p}$. If $\mathfrak{p}$ contains only zero-divisors, then we set $t := a$, and if $\mathfrak{p}$ contains regular elements, then there exists a regular element $t \in \mathfrak{a}$ which does not lie in $\mathfrak{p}$ [since A is a Marot ring]. In both cases we have $tx \in A$, hence we have shown that $x \in S^{-1}A$.
Since S does not meet $\mathfrak{p}$, the extended ideal $\mathfrak{q} = S^{-1}\mathfrak{p}$ is a prime ideal of $A_{[\mathfrak{p}]}$. Every element of $A \setminus \mathfrak{p}$ is invertible in $(A_{[\mathfrak{p}]})_{\mathfrak{q}}$, hence we have a canonical homomorphism $A_{\mathfrak{p}} \to (A_{[\mathfrak{p}]})_{\mathfrak{q}}$ of rings, and it is easy to check that it is bijective.

1.3   Rings With Large Jacobson Radical

(1.8) Lemma: *Let A be a ring; assume that every prime ideal of A is maximal, and that the Jacobson radical of A is the zero ideal. Then, for every $a \in A$, there exists $b \in A$ such that $ab = 0$ and that $a + b$ is a unit of A.*

Proof: If $a \in U(A)$, then we may choose $b = 0$. If $a \notin U(A)$, then $Aa \neq A$ and the set Ω of those maximal ideals of A which contain a is not empty. The set $S := A \setminus (\bigcup_{\mathfrak{m} \in \Omega} \mathfrak{m})$ is multiplicatively closed. The Jacobson radical $\mathfrak{r}(A)$ and the nilradical $\mathfrak{n}(A)$ of A are equal—namely equal to the zero ideal of A—; likewise, the Jacobson radical $\mathfrak{r}(S^{-1}A)$ and the nilradical $\mathfrak{n}(S^{-1}A)$ of $S^{-1}A$ are equal. Since $a/1 \in \mathfrak{r}(S^{-1}A)$ and $\mathfrak{n}(S^{-1}A) = S^{-1}\mathfrak{n}(A) = \{0\}$, we have $a/1 = 0$ in $S^{-1}A$, hence there exists $b \in S$ such that $ba = 0$. Let $\mathfrak{m}$ be any maximal ideal of A. If $a \notin \mathfrak{m}$, then $b \in \mathfrak{m}$ and if $a \in \mathfrak{m}$, then $b \notin \mathfrak{m}$, hence $a + b \notin \mathfrak{m}$, and therefore $a + b \in U(A)$.

(1.9) Proposition: *Let A be a ring, and let $\mathfrak{r}$ be its Jacobson radical. The following statements are equivalent:*
(1) *Every prime ideal of A containing $\mathfrak{r}$ is maximal.*
(2) *For each $a \in A$ there exists $b \in A$ such that, for all $d \in A$ and all units u of A, $a + ub$ and $1 + dab$ are units of A.*
(3) *For each $a \in A$ there exists $b \in A$ such that $ab \in \mathfrak{r}$ and that $a + b \in U(A)$.*

Proof (1) $\Rightarrow$ (2): Let $\varphi\colon A \to A/\mathfrak{r}$ be the canonical homomorphism. By assumption, every prime ideal of $A/\mathfrak{r}$ is maximal and the Jacobson radical of $A/\mathfrak{r}$ is the zero ideal. By (1.8) there exists $b \in A$ such that $\varphi(a)\varphi(b) = 0$ and that $\varphi(a)+\varphi(b)$ is a unit of $A/\mathfrak{r}$. Thus, we have $ab \in \mathfrak{r}$ and there exist $c \in A$, $f \in \mathfrak{r}$ such that $c(a + b) = 1 + f$. Now, $1 + f$ is a unit of A [cf. [63], Ex. 4.7], hence $a + b \in U(A)$. Since $ab \in \mathfrak{r}$, we have $1 + dab \in U(A)$ for every $d \in A$ [cf. [63], Ex. 4.7]. Let $u \in U(A)$. For every maximal ideal $\mathfrak{m}$ of A we have $a + b \notin \mathfrak{m}$ and $ab \in \mathfrak{m}$, hence either $a \in \mathfrak{m}$ and $b \notin \mathfrak{m}$, whence $ub \notin \mathfrak{m}$ and therefore $a + ub \notin \mathfrak{m}$, or $a \notin \mathfrak{m}$ and $b \in \mathfrak{m}$, whence $ub \in \mathfrak{m}$ and again $a + ub \notin \mathfrak{m}$. Therefore we have $a + ub \in U(A)$.
(2) $\Rightarrow$ (3): By assumption, there exists $b \in A$ such that $a + b$ is a unit and such that, for every $d \in A$, the element $1 + dab$ is a unit; hence we have $ab \in \mathfrak{r}$.
(3) $\Rightarrow$ (1): Let $\mathfrak{p}$ be a prime ideal of A containing $\mathfrak{r}$ and let $\mathfrak{m}$ a maximal ideal of A containing $\mathfrak{p}$. Suppose that $\mathfrak{m} \neq \mathfrak{p}$, and let $a \in \mathfrak{m} \setminus \mathfrak{p}$. By assumption, there exists $b \in A$ such that $ab \in \mathfrak{r}$ and that $a + b \in U(A)$. Now we have $b \in \mathfrak{p}$ since $ab \in \mathfrak{r} \subset \mathfrak{p}$; thus, we have $a + b \in \mathfrak{m}$, in contradiction with the fact that $a + b$ is a unit.

(1.10) DEFINITION: A ring is said to have large Jacobson radical if it satisfies the conditions of (1.9).

(1.11) REMARK: (1) A ring in which every prime ideal is maximal is a ring with large Jacobson radical.
(2) A quasisemilocal ring is a ring with large Jacobson radical.

(1.12) Proposition: *If the ring of quotients of A is a ring having large Jacobson radical, then A is a Marot ring.*

Proof: Let $a \in \text{Reg}(A)$ and $b \in A$. By (1.9) we may choose $c \in Q(A)$ such that $b + uc \in \text{Reg}(Q(A))$ for every $u \in \text{Reg}(Q(A))$. We write $c = zr^{-1}$ with $z \in A$, $r \in \text{Reg}(A)$. Now we have $b + (ar)c = b + az \in \text{Reg}(A)$, hence A is a Marot ring by (1.4).

(1.13) Corollary: *A noetherian ring is a Marot ring.*

Proof: The ring of quotients of a noetherian ring A is a semilocal ring since the set of regular elements of A is the complement of the union of the finitely many prime ideals in $\text{Ass}(A)$ [cf. [63], Th. 3.1], hence it is a ring having large Jacobson radical [cf. (1.11)(2)].

2 Manis Valuation Rings

(2.0) NOTATION: *In this section K always denotes a ring having large Jacobson radical and which is its own ring of quotients. Every subring of K having K as ring of quotients is a Marot ring* [cf. (1.12)].

2.1 Manis Valuation Rings

(2.1) Lemma: *Let V be a subring of K with the following property: The product of elements in $K \setminus V$ lies in $K \setminus V$. (By abuse of language, in this section such a set shall be called multiplicatively closed.) Then V is integrally closed in K, and K is the ring of quotients of V.*

Proof: There is nothing to show if $V = K$. Therefore, we treat the case that $V \neq K$.
(1) Let $a, b \in K$. If $ab \in V$, then $a \in V$ or $b \in V$ since $K \setminus V$ is a multiplicatively closed set. In particular, for every $a \in \text{Reg}(K)$ we have $a \in V$ or $a^{-1} \in V$.
(2) Let $t \in K \setminus V$, and suppose that t is integral over V. Let $F = T^n + a_1 T^{n-1} + \cdots + a_n \in V[T]$ be a monic polynomial of least degree n such that $F(t) = 0$. Note that $n \geq 2$ since $t \notin V$. Now $t(t^{n-1} + a_1 t^{n-2} + \cdots + a_{n-1}) = -a_n \in V$ implies that $t^{n-1} + a_1 t^{n-2} + \cdots + a_{n-1} \in V$, contradicting the choice of n. Hence V is integrally closed in K.
(3) Since K is its own ring of quotients, the ring of quotients $Q(V)$ of V can be considered as a subring of K.
(a) The Jacobson radical $\mathfrak{r}(K)$ of K is contained in $Q(V)$. Indeed, let $a \in \mathfrak{r}(K)$; then $b := 1 + a$ is a unit of K, hence $b \in V$ or $b^{-1} \in V$ [cf. (1)]. If $b \in V$, then $a \in V \subset Q(V)$, and if $b^{-1} \in V$, then $a = (1 - b^{-1})(b^{-1})^{-1} \in Q(V)$.
(b) Let $a \in K$. By (1.9) there exists $c \in K$ such that $b := a + c$ is a unit of K and that $ac \in \mathfrak{r}(K)$. By (a) there exist elements $x \in V$, $r \in \text{Reg}(V)$ with $ac = x/r$. Now $(ar)^2 - br(ar) + xr = 0$, hence $(ab^{-1}r)^2 - r(ab^{-1}r) + xb^{-2}r = 0$ since b is a

unit of K. If $b \in V$, then ar is integral over V, hence $ar \in V$ by (2), and therefore we have $a \in Q(V)$. If $b^{-1} \in V$, then $ab^{-1}r \in V$, and therefore we have $a \in Q(V)$.

(2.2) Theorem: Let $V \neq K$ be a subring of K having K as its ring of quotients. The following statements are equivalent:

(1) There exists a prime ideal $\mathfrak{p}$ of V such that $\mathfrak{p}A = A$ for every subring A of K properly containing V.

(2) $K \setminus V$ is a multiplicatively closed set.

(3) For every regular element $x \in K$ we have either $x \in V$ or $x^{-1} \in V$.

(4) The set of regular cyclic V-submodules of K is totally ordered by inclusion.

(5) The set of regular V-submodules of K is totally ordered by inclusion.

Let these conditions be satisfied. Then the regular non-units of V are contained in a unique maximal ideal $\mathfrak{m}$ of V, and for every subring A of K properly containing V we have $\mathfrak{m}A = A$.

Proof (1) $\Rightarrow$ (2): Let $t, t' \in K \setminus V$, and suppose that $tt' \in V$. Now V is a proper subring of $V[t]$ and of $V[t']$. Since $\mathfrak{p}V[t] = V[t]$ by assumption, there exist $m \in \mathbb{N}$ and $z_0, \ldots, z_m \in \mathfrak{p}$ with $1 = z_0 + z_1 t + \cdots + z_m t^m$, hence $1 - z_0 = z_1 t + \cdots + z_m t^m \in (V \setminus \mathfrak{p}) \cap \mathfrak{p}V[t]$. We choose $m \in \mathbb{N}$ minimal with the property that there exist $z_1, \ldots, z_m \in \mathfrak{p}$ such that $a := z_1 t + \cdots + z_m t^m \in (V \setminus \mathfrak{p}) \cap \mathfrak{p}V[t]$. Likewise, we choose $n \in \mathbb{N}$ minimal with the property that there exist $w_1, \ldots, w_n \in \mathfrak{p}$ such that $a' := w_1 t' + \cdots + w_n t'^n \in (V \setminus \mathfrak{p}) \cap \mathfrak{p}V[t']$. We may assume that $m \geq n$. Since $a' z_m t^m = w_1 z_m (tt') t^{m-1} + w_2 z_m (tt')^2 t^{m-2} + \cdots + w_n z_m (tt')^n t^{m-n}$, we see that

$$a'a = a' z_1 t + \cdots + a' z_{m-1} t^{m-1} + a' z_m t^m$$
$$= a' z_1 t + \cdots + a' z_{m-n-1} t^{m-n-1}$$
$$\quad + \left(a' z_{m-n} + w_n z_m (tt')^n \right) t^{m-n} + \cdots + \left(a' z_{m-1} + w_1 z_m (tt') \right) t^{m-1}.$$

Now $a' z_1, \ldots, a' z_{m-n-1}$ and $a' z_{m-n} + w_n z_m (tt')^n, \ldots, a' z_{m-1} + w_1 z_m (tt')$ are elements of $\mathfrak{p}$; on the other hand, we have $a'a \in V \setminus \mathfrak{p}$ since $\mathfrak{p}$ is a prime ideal, contradicting the choice of m.

(2) $\Rightarrow$ (3): Let $x \in K$ be regular. Since $1 = xx^{-1} \in V$, we see that, if $x \notin V$, then $x^{-1} \in V$.

(3) $\Rightarrow$ (4): A regular cyclic V-submodule of K has the form Vc for some regular $c \in K$. Let a, b be regular elements of K. If $Va \not\subset Vb$, then $b^{-1}a \notin V$, hence $a^{-1}b \in V$ and we have $Vb \subset Va$.

(4) $\Rightarrow$ (5): Let M, N be regular V-submodules of K such that $M \not\subset N$. The ring V is a Marot ring, hence the regular V-modules M and N are generated by their regular elements [cf. (1.3)], and therefore M contains a regular element a with $a \notin N$. Then we have $Va \not\subset Vb$ for every regular $b \in N$, hence $Vb \subset Va$ for every regular $b \in N$, and we see that $N \subset Va \subset M$.

(5) $\Rightarrow$ (1): The set of regular ideals of V which are different from V is not empty since V contains regular non-units as $V \neq K$, and, by assumption, it is totally ordered by inclusion. Therefore the union of all these ideals is an ideal of V, say

m. This ideal is a maximal ideal of V [let $x \in V \setminus \mathfrak{m}$; the ideal $\mathfrak{m} + Vx$ is a regular ideal of V properly containing $\mathfrak{m}$, hence $\mathfrak{m} + Vx = V$]. Let B be a subring of K containing V and such that $\mathfrak{m}B \neq B$. It is enough to show that $B = V$. The V-module B is generated by regular elements since V is a Marot ring; we show that every regular element of B lies in V. Let $x \in B$ be regular, and choose y, $r \in \mathrm{Reg}(V)$ with $x = y/r$. If $Vy \subset Vr$, then $x \in V$. If $Vy \not\subset Vr$, then $Vr \subset Vy$ and $x^{-1} \in V$, but $x^{-1} \notin \mathfrak{m}$ since otherwise $1 = xx^{-1} \in \mathfrak{m}B$. Therefore, the regular element x^{-1} of V is a unit of V, hence $(x^{-1})^{-1} = x \in V$.

(2.3) DEFINITION: A subring $V \neq K$ of K having K as ring of quotients is called a Manis valuation ring of K if it satisfies the conditions of (2.2). The maximal ideal $\mathfrak{m}$ of V containing all regular non-units of V is called the regular maximal ideal of V. [Note that a subring V of K satisfying only (2) in (2.2) already has K as ring of quotients by (2.1).]

(2.4) Proposition: *Let V be a Manis valuation ring of K. Then:*
(1) V is integrally closed, and every proper subring of K containing V is a Manis valuation ring of K.
(2) Every finitely generated regular V-submodule of K is cyclic.
(3) Let $\mathfrak{m}$ be the regular maximal ideal of V. Then:
(a) The conductor $(V : K)_K$ of V in K is a prime ideal of V and of K; the ideal $(V : K)_K$ is the intersection of all regular ideals of V, and $(V : K)_K \subset \mathfrak{m}$.
(b) Let $x \in K$. If $x \notin (V : K)_K$, then the intersection $M(x)$ of all regular V-submodules of K containing x is a regular cyclic V-submodule of K, and for every regular $z \in K$ we have $M(x) = Vz$ iff $xz^{-1} \in V \setminus \mathfrak{m}$.

Proof: (1) This follows from (2.1) and (2.2)(3).
(2) A finitely generated regular V-submodule of K has a finite system of regular elements of K as generators [cf. (1.3)]; now one may use (2.2)(4).
(3)(a) We set $\mathfrak{p} := (V : K)_K$. We have $\mathfrak{p} \neq V$ since $V \neq K$. Let $x, y \in V \setminus \mathfrak{p}$. There exist $s, t \in K \setminus V$ with $sx \notin V$, $ty \notin V$, hence $xyst \notin V$ [since $K \setminus V$ is multiplicatively closed], and therefore $xy \in V \setminus \mathfrak{p}$; thus, we have shown that $\mathfrak{p}$ is a prime ideal of V. In the same way we can show that $\mathfrak{p}$ is a prime ideal of K.
Let $a \in \mathfrak{p}$. For every regular $t \in V$ we have $a \cdot 1/t \in V$, hence $a \in Vt$. This implies that $\mathfrak{p} \subset Vt$, and therefore $\mathfrak{p}$ is contained in the intersection of all regular ideals of V. Conversely, let $a \in V$ be contained in the intersection of all regular ideals of V. Let $x \in K$, and choose $y \in V$, $r \in \mathrm{Reg}(V)$ with $x = yr^{-1}$. Then $a \in Vr$, hence $ax = a(yr^{-1}) \in V$, and therefore $a \in \mathfrak{p}$. This characterization implies that $\mathfrak{p} \subset \mathfrak{m}$.
(b) Let $x \in K$, $x \notin \mathfrak{p}$. By (a) there exists a regular element $a \in V$ such that $x \notin Va$. For every regular element $b \in K$ satisfying $x \in Vb$ we have $Va \subset Vb$ [cf. (2.2)], hence $Va \subset M(x)$, and $M(x)$ is a regular V-submodule of K. The regular V-submodule $Va + Vx$ of K is cyclic by (2); it contains x and therefore it contains $M(x)$, hence $M(x)$ is equal to $Va + Vx$. Thus, we have shown that the V-module $M(x)$ is generated by a regular element of K.

Let $M(x) = Vz$ for some regular $z \in K$. Then $xz^{-1} \in V$, and we have to show that $xz^{-1} \notin \mathfrak{m}$. Suppose, on the contrary, that $xz^{-1} \in \mathfrak{m}$. Since $xz^{-1} \notin \mathfrak{p}$ as $x \notin \mathfrak{p}$ and $\mathfrak{p}$ is a prime ideal of K, by the first part of the proof there exists a regular $b \in V$ such that $M(xz^{-1}) = Vb$ and that $Vb \neq V$ [since $xz^{-1} \in \mathfrak{m}$, and since $\mathfrak{m}$ contains regular elements]; therefore we have $b \in \mathfrak{m}$. Now x is an element of the V-module Vzb which is a proper regular V-submodule of $Vz = M(x)$, contradicting the choice of $M(x)$. Conversely, let z be a regular element of K such that $xz^{-1} \in V \setminus \mathfrak{m}$. Since $x \in Vz$, we have $M(x) \subset Vz$. By the first part of the proof we know that $M(x) = Vw$ for some regular element w of K, and by the second part of the proof we know that $xw^{-1} \in V \setminus \mathfrak{m}$. Suppose that $M(x) \neq Vz$; then $wz^{-1} \in \mathfrak{m}$, and therefore $xz^{-1} = (xw^{-1})(wz^{-1}) \in \mathfrak{m}$, contradicting $xz^{-1} \notin \mathfrak{m}$.

(2.5) Proposition: *Let R be a subring of K. For every $x \in K$ which is not integral over R there exists a Manis valuation ring of K containing R and such that x is not integral over V.*

Proof: Let $\mathcal{A}$ be the set of subrings A of K which contain R, and with the property that x is not integral over R. $\mathcal{A}$ is not empty [since $R \in \mathcal{A}$]; we show that $\mathcal{A}$ is inductive. Let $(A_i)_{i \in I}$ be a chain in $\mathcal{A}$, and set $A := \bigcup_{i \in I} A_i$. Then A is a subring of K which contains R. Suppose that x is integral over A, and let $F = T^n + a_1 T^{n-1} + \cdots + a_n \in A[T]$ be an equation of integral dependence of x over A. Since $\{a_1, \ldots, a_n\}$ is a finite subset of A, there exists $j \in I$ such that $a_1, \ldots, a_n$ lie in A_j; this implies that x is integral over A_j, contradicting $A_j \in \mathcal{A}$. Therefore $A \in \mathcal{A}$ is an upper bound for $(A_i)_{i \in I}$.

Let V be a maximal element of $\mathcal{A}$ [Zorn's lemma]. Then we have $V \neq K$ since $x \notin V$. We show that $K \setminus V$ is multiplicatively closed. Let $s, t \in K \setminus V$, and suppose that $a := st \in V$. Since $B := V[s]$, $C := V[t]$ are subrings of K which contain R and are not equal to V, x is integral over B and C [by maximality of V]. Therefore there exist polynomials $F, G \in V[X, Y]$, monic with respect to X, and such that $F(x, s) = 0$, $G(x, t) = 0$. We write

$$F = F_0(X) + F_1(X)Y + \cdots + F_m(X)Y^m, \quad G = G_0(X) + G_1(X)Y + \cdots + G_n(X)Y^n$$

where $F_0, \ldots, F_m, G_0, \ldots, G_n \in V[X]$, F_0, G_0 are monic and $\deg(F_i) < \deg(F_0)$ for every $i \in \{1, \ldots, m\}$, $\deg(G_j) < \deg(G_0)$ for every $j \in \{1, \ldots, n\}$. We choose F and G in such a way that m and n are minimal; note that $m, n \geq 1$. We may assume that $m \geq n$. Then we have

$$F_m(x)s^m(G_1(x)t + \cdots + G_n(x)t^n) = aF_m(x)G_1(x)s^{m-1} + \cdots + a^n F_m(x)G_n(x)s^{m-n},$$

and therefore, since $G_0(x) = -(G_1(x)t + \cdots + G_n(x)t^n)$, it follows that

$$-F_0(x)G_0(x) = G_0(x)F_1(x)s + \cdots + G_0(x)F_{m-1}(x)s^{m-1}$$
$$- (aF_m(x)G_1(x)s^{m-1} + \cdots + a^n F_m(x)G_n(x)s^{m-n}).$$

Thus, we have found a polynomial $H(X,Y) \in V[X,Y]$,

$$H(X,Y) = H_0(X) + H_1(X)Y + \cdots + H_{m-1}(X)Y^{m-1} \text{ with } H_0, \ldots, H_{m-1} \in V[X]$$

where $H_0 = F_0 G_0$, $H_i = G_0 F_i$ for $i \in \{1, \ldots, m-n-1\}$, and $H_i = G_0 F_i - a^{m-i} F_m G_{m-i}$ for $i \in \{m-n, \ldots, m-1\}$. Note that $H_0 \in V[X]$ is monic, and that $\deg(H_i) < \deg(F_0) + \deg(G_0) = \deg(F_0 G_0)$ for every $i \in \{1, \ldots, m-1\}$. We have $H(x, s) = 0$, contradicting the choice of m. Therefore it follows that $st \in K \setminus V$. Now V is a proper subring of K since $x \notin V$, and $Q(V) = K$ by (2.1); it follows that V is a Manis valuation ring of K [cf. (2.2)].

2.2   Manis Valuations

(2.6) We now introduce the notion of a Manis valuation of K, and we show that a Manis valuation of K can be used to define a Manis valuation ring of K, and, conversely, that a Manis valuation ring of K gives rise to a Manis valuation of K.

(2.7) NOTATION: For the rest of this section let $(\Gamma, <)$ be a totally ordered abelian group [cf. appendix B, section 1]. We adjoin to Γ an element ∞, and we set $\Gamma_\infty := \Gamma \cup \{\infty\}$. We make Γ_∞ into a totally ordered monoid containing Γ as an ordered submonoid by defining $\gamma < \infty$ for every $\gamma \in \Gamma$ and $\gamma + \infty = \infty$ for every $\gamma \in \Gamma_\infty$.

(2.8) DEFINITION: A *surjective* map $v: K \to \Gamma_\infty$ such that $v(1) = 0$, $v(0) = \infty$, and

$$v(ab) = v(a) + v(b), \quad v(a+b) \geq \min(\{v(a), v(b)\}) \quad \text{for all } a, b \in K, \quad (*)$$

is called a Manis valuation of K; the inequality in $(*)$ is called the strong triangle inequality.
Let $v: K \to \Gamma_\infty$ be a Manis valuation of K. If $v(K) = \{v(1), \infty\}$ [hence $\Gamma = \{v(1)\}$], then v is called a trivial Manis valuation, and if $\Gamma = \mathbb{Z}$, then v is called a discrete Manis valuation.

(2.9) REMARK: A Manis valuation $v: K \to \Gamma_\infty$ has the following simple properties: Let $x \in K$, and assume that $x^n = 1$ for some $n \in \mathbb{N}$; then we have $v(x) = 0$ since $0 = v(1) = nv(x)$. In particular, we have $v(-x) = v(x)$ for every $x \in K$. For every regular element $x \in K$ we have $\infty \neq v(x) = -v(x^{-1})$ since $0 = v(1) = v(xx^{-1})$. Note that $v(a1_K) \geq 0$ for every $a \in \mathbb{Z}$.

(2.10) Proposition: Let $v: K \to \Gamma_\infty$ be a Manis valuation of K. Then:
(1) For elements $x_1, \ldots, x_h \in K$ we have

$$v(x_1 + \cdots + x_h) \geq \min(\{v(x_i) \mid i \in \{1, \ldots, h\}\}); \quad (*)$$

if the minimum is reached by exactly one x_k, then this is an equality.

(2) If x, $y \in K$ and $v(x) \neq v(y)$, then $v(x+y) = \min(\{v(x), v(y)\})$.

(3) Let $h \geq 2$ and $x_1, \ldots, x_h \in K$ be such that $x_1 + \cdots + x_h = 0$; then, for at least two j, $k \in \{1, \ldots, h\}$ with $j \neq k$, we have $v(x_j) = v(x_k) = \min(\{v(x_i) \mid i \in \{1, \ldots, h\}\})$.

Proof: (1) We can easily prove (∗) by induction.

Now let us assume that there is only one $k \in \{1, \ldots, h\}$ such that $v(x_k) = \min(\{v(x_i) \mid i \in \{1, \ldots, h\}\})$, set $y := \sum_{i \neq k} x_i$ and $z := \sum_i x_i$. Then, by (∗), we have $v(y) > v(x_k)$ and $v(z) \geq v(x_k)$. Suppose that $v(z) > v(x_k)$; from $x_k = z - y$ we get $v(x_k) \geq \min(\{v(z), v(y)\}) > v(x_k)$ which is absurd, hence we have $v(z) = v(x_k)$.

(2) and (3) follow immediately from (1).

(2.11) NOTATION: Let $v \colon K \to \Gamma_\infty$ be a Manis valuation of K.

(1) $A_v := \{x \in K \mid v(x) \geq 0\}$ is a subring of K; it is called *the ring of* v.

(2) $\mathfrak{p}_v := \{x \in K \mid v(x) > 0\}$ is a prime ideal of A_v; it is called *the prime ideal of* v. Note that $\mathfrak{n}_v := v^{-1}(\{\infty\})$ is a prime ideal of A_v and of K; it is called *the kernel of* v. Clearly we have $v^{-1}(\{\infty\}) \subset (A_v : K)_K$.

(2.12) Proposition: *Let* $v \colon K \to \Gamma_\infty$ *be a Manis valuation of* K, *and let* A_v *be the ring of* v. *Assume that* $A_v \neq K$. *Then* A_v *is a Manis valuation ring of* K, $\mathfrak{p}_v$ *is its regular maximal ideal,* $(A_v : K)_K = \mathfrak{n}_v$ *is the kernel of* v, *and we have* $v(\operatorname{Reg}(K)) = \Gamma$.

Proof: (1) It is clear that $K \setminus A_v$ is multiplicatively closed, hence K is the ring of quotients of A_v [cf. (2.1)], and A_v is a Manis valuation ring of K [cf. (2.2)].

(2) Let $\mathfrak{m}$ be the regular maximal ideal of A_v. Every regular element $x \in \mathfrak{m}$ is a non-unit of A_v, hence $v(x) > 0$, and therefore $\mathfrak{m} \subset \mathfrak{p}_v$ [since A_v is a Marot ring]. Conversely, let $x \in \mathfrak{p}_v$ be regular. Then x is a non-unit of A_v, hence $x \in \mathfrak{m}$ and therefore $\mathfrak{p}_v \subset \mathfrak{m}$. Now we have shown that $\mathfrak{m} = \mathfrak{p}_v$.

(3) We have $\mathfrak{n}_v \subset (A_v : K)_K$. Suppose that there exists $x \in (A_v : K)_K$ with $v(x) < \infty$. Since v is surjective, there exists $y \in K$ with $v(y) = -v(x)$. Since $A_v \neq K$, there exists $z \in K$ with $v(z) < 0$. Now we have $x(yz) \in A_v$ but $v(xyz) < 0$, contrary to the definition of A_v.

(4) Let $x \in K$ and $v(x) \neq \infty$. Then $x \notin \mathfrak{n}_v = (A_v : K)_K$, and there exists $a \in K$ with $ax \notin A_v$. We choose a regular $b \in A_v$ with $ba \in A_v$ [note that $Q(A_v) = K$]. Now $v(x) < -v(a) \leq v(b)$. The regular A_v-module $A_v x + A_v b$ is cyclic [cf. (2.4)(2)], hence we have $A_v x + A_v b = A_v y$ for some regular $y \in K$. This yields $y \in A_v x + A_v b$, hence $v(y) \geq \min(\{v(x), v(b)\}) = v(x)$, and $x \in A_v y$, hence $v(y) \leq v(x)$. Thus, we have shown that $v(y) = v(x)$; since y is regular, it follows that we have $v(\operatorname{Reg}(K)) = \Gamma$.

(2.13) Proposition: *Every Manis valuation ring of* K *is the ring of a Manis valuation of* K.

Proof: Let V be a Manis valuation ring of K, and let $\mathfrak{m}$ be the regular maximal ideal of V.

Let Δ be the group of regular cyclic V-submodules of K (written additively); its zero element is V. We define $\Pi := \{Va \mid a \in V \text{ regular}\}$. Then we have $\Pi + \Pi \subset \Pi$, $\Pi \cap (-\Pi) = \{V\}$, and $\Pi \cup (-\Pi) = \Delta$. We define $Va \leq Vb$ if $Va^{-1}b \subset \Pi$; then $(\Delta, <)$ is a totally ordered abelian group. We define a map $w: \mathrm{Reg}(K) \to \Delta$ by $w(a) := Va$. Now $w(ab) = w(a) + w(b)$ for all a, $b \in \mathrm{Reg}(K)$, and if a, b and $a + b \in \mathrm{Reg}(K)$, then $w(a + b) \geq \min(\{w(a), w(b)\})$ [since $Va \subset Vb$ or $Vb \subset Va$]. Let us now define a Manis valuation $v: K \to \Delta_\infty$. Let $x \in K$. If $x \in (V : K)_K$, then we define $v(x) := \infty$. If $x \notin (V : K)_K$, then the intersection $M(x)$ of all regular V-submodules of K containing x is a regular cyclic V-submodule Va by $(2.4)(3)(b)$, and we define $v(x) := w(a) = Va$. Clearly we have $v(x) \geq 0$ iff $x \in V$. It is clear that $v(1) = 0$ and $v(0) = \infty$. Let x, $y \in K$. If $x \in (V : K)_K$ or $y \in (V : K)_K$, then $xy \in (V : K)_K$ and $v(xy) = v(x) + v(y)$; if $x \notin (V : K)_K$ and $y \notin (V : K)_K$, then $xy \notin (V : K)_K$ since $(V : K)_K$ is a prime ideal of K [cf. $(2.4)(3)(a)$], and there exist regular elements a, $b \in K$ such that $M(x) = Va$, $M(y) = Vb$, hence we have $v(x) = w(a)$, $v(y) = w(b)$. By $(2.4)(3)(b)$ we know that xa^{-1} and $yb^{-1} \in V \setminus \mathfrak{m}$, hence we have $xy(ab)^{-1} \in V \setminus \mathfrak{m}$ since $\mathfrak{m}$ is a prime ideal of V, and therefore $M(xy) = Vab$ [cf. $(2.4)(3)(b)$]. This implies that $v(xy) = w(ab) = w(a) + w(b) = v(x) + v(y)$.

We now show that the strong triangle inequality holds. If $x \in (V : K)_K$ and $y \in (V : K)_K$, then $x + y \in (V : K)_K$, hence $v(x + y) = \infty = \min(\{v(x), v(y)\})$. If $x \in (V : K)_K$ and $y \notin (V : K)_K$, then $x + y \notin (V : K)_K$. The regular V-submodule $M(x + y)$ of K contains $(V : K)_K$ [cf. $(2.4)(3)(a)$], and therefore $x \in (V : K)_K \subset M(x + y)$. This implies that $y = (y + x) - x \in M(x + y)$, and therefore $M(y) \subset M(x + y)$. We know that $x \in (V : K)_K \subset M(y)$, hence $x + y \in M(y)$, and therefore $M(x + y) \subset M(y)$. Thus, we have shown that $M(x + y) = M(y)$, and this implies that $v(x + y) = v(y) = \min(\{v(x), v(y)\})$. If $x \notin (V : K)_K$ and $y \in (V : K)_K$, we again see, by interchanging the roles of x and y, that $v(x + y) = v(x) = \min(\{v(x), v(y)\})$. Now we consider the case that $x \notin (V : K)_K$ and $y \notin (V : K)_K$. There exist regular elements a, $b \in K$ such that $M(x) = Va$ and $M(y) = Vb$, hence we have $v(x) = w(a)$, $v(y) = w(b)$. We know that $Va \subset Vb$ or $Vb \subset Va$. If $Va \subset Vb$, then $x + y \in Vb$, hence $M(x + y) \subset Vb$, and therefore $v(x + y) \geq w(b) = \min(\{w(a), w(b)\}) = \min(\{v(x), v(y)\})$. Likewise, if $Vb \subset Va$, we see that $v(x + y) \geq \min(\{v(x), v(y)\})$.

(2.14) MANIS VALUATION RINGS AND MANIS VALUATIONS: Let $v: K \to \Gamma_\infty$ be a Manis valuation of K, and let $V := A_v$ be the ring of v. We assume that $V \neq K$. Then V is a Manis valuation ring of K by (2.12).

(1) Let $\Delta := \{Va \mid a \in \mathrm{Reg}(K)\}$ be the totally ordered group of regular cyclic V-submodules of K [cf. proof of (2.13)]. Then

$$Va \mapsto v(a) : \Delta \to \Gamma \qquad (*)$$

is an isomorphism of ordered groups. In fact, let a, $b \in \mathrm{Reg}(K)$; if $Va = Vb$, then ab^{-1} is a unit of V, and therefore the map in $(*)$ is well-defined. Moreover, since

$(Va)(Vb) = Vab$ for $a, b \in \text{Reg}(K)$, the map in $(*)$ is a homomorphism which is surjective by (2.12). Let $a \in \text{Reg}(K)$ and $v(a) = 0$. Then a is a unit of V, hence $Va = V$, and the homomorphism in $(*)$ is injective. Lastly, if $a, b \in \text{Reg}(K)$ and $Va \subset Vb$, then $a = zb$ with $z \in V$, hence $v(a) \geq v(b)$. Therefore the map in $(*)$ is an isomorphism of ordered groups.
(2) By (1) we may identify the group $\Gamma = v(\text{Reg}(K))$ [cf. (2.12)] with the ordered group $\{Va \mid a \in \text{Reg}(K)\}$ of regular cyclic V-submodules of K.

An important class of Manis valuation rings is characterized in the following proposition.

(2.15) Proposition: *Let V be a Manis valuation ring of K, and let $\mathfrak{m}$ be the regular maximal ideal of V. The following statements are equivalent:*
(1) *V is the ring of a discrete Manis valuation $v \colon K \to \mathbb{Z}_\infty$.*
(2) *Every regular ideal of V is finitely generated.*
(3) *$\mathfrak{m}$ is a finitely generated ideal, and $\mathfrak{m}$ is the only regular prime ideal of V.*
Assume that these properties are satisfied. Then:
(a) *Every regular ideal of V is a principal ideal.*
(b) *Let $\mathfrak{m} = Vp$. Then p is a regular element, we have $K = V[p^{-1}]$, and every regular element x of K has a unique representation $x = ap^k$ where $a \in U(V)$ and $k \in \mathbb{Z}$. Moreover, $\{Vp^h \mid h \in \mathbb{Z}\}$ is the set of regular V-submodules of K.*
(c) *We have $(V : K)_K = \bigcap_{n \in \mathbb{N}} \mathfrak{m}^n$.*
(d) *Let A be a subring of K containing V. Then we have $A = V$ or $A = K$.*

Proof (1) $\Rightarrow$ (2): Let $\mathfrak{a}$ be a regular ideal of V. Then $\text{Reg}(V) \cap \mathfrak{a}$ is not empty; let n be the smallest integer in the set $\{v(x) \mid x \in \text{Reg}(V) \cap \mathfrak{a}\}$, and choose a regular element $a \in \mathfrak{a}$ with $v(a) = n$. Let $y \in \mathfrak{a} \cap \text{Reg}(V)$; then we have $v(y/a) \geq 0$ and therefore $y \in Va$, hence $\mathfrak{a} = Va$ [since V is a Marot ring].
(2) $\Rightarrow$ (3): Every finitely generated regular ideal of V is a principal ideal [cf. (2.4)(2)]. Let $\mathfrak{m} = Vp$. Suppose that $\mathfrak{p}$ is a regular prime ideal of V and $\mathfrak{p} \neq \mathfrak{m}$. Then $\mathfrak{p} \subset \mathfrak{m}$ [cf. (2.2)(5)], and $\mathfrak{p}$ is a principal ideal since it is finitely generated by hypothesis [cf. (2.4)(2)], say $\mathfrak{p} = Vx$, where x is a regular element of V. We have $x = yp$ for some $y \in V$, and since $p \notin \mathfrak{p}$, we must have $y \in \mathfrak{p}$ [since $\mathfrak{p}$ is a prime ideal], and therefore $y = zx$ for some $z \in V$. Now $x = yp = xzp$, hence $1 = zp$ since x is a regular element of V, and therefore $\mathfrak{m} = V$, contradicting $\mathfrak{m} \neq V$.
(3) $\Rightarrow$ (1): $\mathfrak{m}$ is a principal ideal by (2.4)(2), say $\mathfrak{m} = Vp$. The ring $W := V[p^{-1}]$ is a subring of K properly containing V. Suppose that $W \neq K$. Then W is a Manis valuation ring of K and $W = Wp$ [cf. (2.4) and (2.2)]; let $\mathfrak{n}$ be the regular maximal ideal of W. The ideal $V \cap \mathfrak{n}$ is a regular prime ideal of V different from $\mathfrak{m}$ [since $(\mathfrak{n} \cap V)W \subset \mathfrak{n}$], contradicting the hypothesis that V has only one regular prime ideal. Therefore we have $V[p^{-1}] = K$. Set $\mathfrak{a} := \bigcap_{h \in \mathbb{N}} \mathfrak{m}^h$; then $(V : K)_K \subset \mathfrak{a}$ by (2.4)(3)(a). Since $\mathfrak{a} \subset \mathfrak{m}^n = Vp^n$ for every $n \in \mathbb{N}$, we have $(V : \mathfrak{a})_K \supset (V : \mathfrak{m}^n)_K = Vp^{-n}$ for every $n \in \mathbb{N}$, and therefore $(V : \mathfrak{a})_K = K$. This implies that $\mathfrak{a} \subset (V : K)_K$. Thus, we have shown that $(V : K)_K = \mathfrak{a}$, and therefore $\mathfrak{a} = \bigcap_{h \in \mathbb{N}} \mathfrak{m}^h$ does not contain regular elements.

Let $x \in \mathrm{Reg}(V)$. There exists $n \in \mathbb{N}_0$ such that $x \in \mathfrak{m}^n$, $x \notin \mathfrak{m}^{n+1}$, and therefore $x = ap^n$ where $a \in V$ is regular and not contained in $\mathfrak{m}$, hence a is a unit of V. It is easy to see that such a representation is unique. Therefore, every regular element x of K has a unique representation $x = ap^k$ where $a \in U(V)$ and $k \in \mathbb{Z}$. Let M be a regular V-submodule of K different from K. Since $K = V[p^{-1}]$, there exists $h \in \mathbb{Z}$ with $Vp^h \subset M$, $Vp^{h-1} \subsetneq M$. Let $x \in M$ be regular. We may write $x = ap^k$ where $a \in U(V)$ and $k \in \mathbb{Z}$. Then $p^k \in M$, hence $k \geq h$, and $x = ap^{k-h}p^h \in Vp^h$, whence $M \subset Vp^h$ since V is a Marot ring. Thus, we have shown that $M = Vp^h$.

Let $x \in K$. If $x \in (V : K)_K$, then we set $v(x) := \infty$. If $x \notin (V : K)_K$, then the intersection $M(x)$ of all regular V-submodules of K containing x has the form Vp^h where $h \in \mathbb{Z}$ is determined uniquely. In this case we set $v(x) := h$. It is immediate that $v \colon K \to \mathbb{Z}_\infty$ is a Manis valuation of K, and that $V = A_v$. This ends the proof of the equivalence of the properties (1)–(3).

Now assume that V satisfies the properties (1)–(3). Then (a) follows from (2) and (2.4)(2), and (b)–(c) were shown while proving the implication (3) $\Rightarrow$ (1). We show that (d) holds. Let A be a subring of K containing V such that $V \neq A$. Then there exists a regular element $x \in A \setminus V$, and we have $x = ap^k$ where $a \in U(V)$ and $k \in \mathbb{Z}$. Since $x \notin V$, we must have $k < 0$, and $K = V[p^k] \subset A$, hence $A = K$.

(2.16) DEFINITION: A Manis valuation ring V of K satisfying the conditions stated in (2.15) is called a discrete Manis valuation ring of K, and the Manis valuation $v \colon K \to \mathbb{Z}_\infty$ constructed in (2.15) is called the Manis valuation of K defined by V. Any generator of the maximal ideal of V is called a uniformizing parameter for V.

(2.17) Proposition: Let $\mathfrak{d}$ be a maximal ideal of K, set $\overline{K} := K/\mathfrak{d}$, and let $\varphi \colon K \to \overline{K}$ be the canonical homomorphism. Then:
(1) The map $V \mapsto V/\mathfrak{d}$ is a bijective map between the set of Manis valuation rings of K with conductor $\mathfrak{d}$ and the set of Manis valuation rings of $\overline{K}$, and such a V is a discrete Manis valuation ring of K iff $V/\mathfrak{d}$ is a discrete Manis valuation ring of $\overline{K}$.
(2) Let V be a Manis valuation ring of K with conductor $\mathfrak{d}$. For every Manis valuation $v \colon K \to \Gamma_\infty$ of K with ring V there exists a unique Manis valuation $\overline{v} \colon \overline{K} \to \Gamma_\infty$ with $\overline{v} \circ \varphi = v$ which has $V/\mathfrak{d}$ as Manis valuation ring.

Proof: (a) Let V be a Manis valuation ring of K with conductor $\mathfrak{d}$, and set $\overline{V} := V/\mathfrak{d}$. Now $\overline{V}$ is a proper subring of $\overline{K}$, and $\overline{K} \setminus \overline{V}$ is multiplicatively closed since $K \setminus V$ has this property. Then $\overline{V}$ has $\overline{K}$ as ring of quotients, and it is a Manis valuation ring of $\overline{K}$ [cf. (2.1) and (2.2)]. Let $v \colon K \to \Gamma_\infty$ be a Manis valuation of K with ring V. It is easy to see that there exists a unique Manis valuation $\overline{v} \colon \overline{K} \to \Gamma_\infty$ with $\overline{v} \circ \varphi = v$; then $\overline{V}$ is the ring of $\overline{v}$.
(b) Let $\overline{V}$ be a Manis valuation ring of $\overline{K}$, and let V be the preimage of $\overline{V}$ in K. Then $\mathfrak{d}$ is an ideal of V, hence $\mathfrak{d}$ is the conductor of V in K [since $\mathfrak{d}$ is a maximal ideal of K], and we have $V/\mathfrak{d} = \overline{V}$. Since $\overline{K} \setminus \overline{V}$ is multiplicatively closed, the

same is true for $K \setminus V$, hence V is a Manis valuation ring of K [cf. (2.2)]. Let $\overline{v}: \overline{V} \to \overline{K}$ be a Manis valuation of $\overline{K}$ with ring $\overline{V}$, then $v := \overline{v} \circ \varphi$ is a Manis valuation of K with ring V.

(c) Now we have shown the assertions in (1) and (2).

2.3 The Approximation Theorem For Discrete Manis Valuations

(2.18) **Lemma:** Let $v_1: K \to \Gamma_{1\infty}, \ldots, v_n: K \to \Gamma_{n\infty}$ be Manis valuations of K. For every $x \in K$ there exists a polynomial $F \in \mathbb{Z}[T]$ of degree $k \geq 2$ having the form

$$F = 1 + a_1 T + \cdots + a_{k-1} T^{k-1} + T^k$$

and such that, for every $i \in \{1, \ldots, n\}$,

$$v_i(F(x)) = 0 \qquad \qquad \text{if} \quad v_i(x) \geq 0,$$
$$v_i(F(x)) < v_i(x) \qquad \text{if} \quad v_i(x) < 0.$$

In particular, we have $v_i(F(x)) \leq 0$ for every $i \in \{1, \ldots, n\}$.

Proof: We define $\mathcal{F}$ as the set of polynomials $F \in \mathbb{Z}[T]$ of the form

$$F = 1 + a_1 T + \cdots + a_{k-1} T^{k-1} + T^k \quad \text{where } k \in \mathbb{N}, \ k \geq 2.$$

Let $i \in \{1, \ldots, n\}$. If there exists a polynomial $G \in \mathcal{F}$ with $v_i(G(x)) > 0$, then let $F_i \in \mathcal{F}$ be such a polynomial; otherwise we set $F_i := 1$. We set $F := 1 + T^2 F_1 \cdots F_n \in \mathbb{Z}[T]$. It is clear that $F \in \mathcal{F}$; we set $k := \deg(F)$. Let $i \in \{1, \ldots, n\}$. First, we consider the case $v_i(x) \geq 0$; then we have $v_i(F(x)) \geq 0$. Suppose that $v_i(F(x)) > 0$; then $v_i(F_i(x)) > 0$ [by definition of F_i], hence $v_i(x^2 F_1(x) \cdots F_n(x)) > 0$, and therefore $v_i(F(x)) = 0$ [cf. (2.10)(2)]; this contradiction shows that $v_i(F(x)) = 0$. Now, we consider the case $v_i(x) < 0$. Then $v_i(F(x)) = v_i(x^k)$ [cf. (2.10)(1)], and $v_i(x^k) = k v_i(x) = (k-1)v_i(x) + v_i(x) < v_i(x)$ [since $k \geq 2$].

(2.19) **Proposition:** Let $v_1: K \to \Gamma_{1\infty}, \ldots, v_n: K \to \Gamma_{n\infty}$ be Manis valuations of K, and set $R := A_{v_1} \cap \cdots \cap A_{v_n}$ and $\mathfrak{m}_i := R \cap \mathfrak{p}_{v_i}$ for every $i \in \{1, \ldots, n\}$. We assume that $A_{v_i} \not\subset A_{v_j}$ for all $i, j \in \{1, \ldots, n\}$, $i \neq j$. Then $\{\mathfrak{m}_1, \ldots, \mathfrak{m}_n\}$ is the set of regular maximal ideals of R, they are all different, K is the ring of quotients of R, and we have $R_{[\mathfrak{m}_i]} = A_{v_i}$ and $\mathfrak{m}_i A_{v_i} = \mathfrak{p}_{v_i}$ for every $i \in \{1, \ldots, n\}$.

Proof: Let $\mathfrak{r}$ be the Jacobson radical of K, and set $\mathfrak{q}_i := \{x \in R \mid v_i(x) = \infty\}$ for every $i \in \{1, \ldots, n\}$. Clearly $\mathfrak{q}_i$ is a prime ideal of R.

(1) Let $i \in \{1, \ldots, n\}$, and assume that $\mathfrak{r} \not\subset \mathfrak{q}_i$. We show: for every $\gamma \in \Gamma_i$ there exists a regular $x \in \mathfrak{m}_i$ with $v_i(x) > \gamma$.

For the proof of this assertion we may assume that $\gamma > 0$. We choose $a \in \mathfrak{r}$ and $a \notin \mathfrak{q}_i$. Since v_i is surjective, there exists $b \in K$ such that $v_i(ab) = -\gamma$. We choose with respect to ab a polynomial $F \in \mathbb{Z}[T]$ as in (2.18); then we have $v_j(F(ab)) \leq 0$ for every $j \in \{1, \dots, n\}$ and $v_i(F(ab)) < -\gamma$. Now $F(ab)$ is of the form $1 + da$ for some $d \in K$, and since $a \in \mathfrak{r}$, the element $1 + da$ is a unit of K. The inverse $(1 + da)^{-1}$ of $1 + da$ lies in R, and we have $v_i((1 + da)^{-1}) > \gamma$.

(2) Let $a \in K$, and assume that $v_i(a) \neq \infty$ for every $\{1, \dots, n\}$. For every $t \in K$ with $ta \in \mathfrak{r}$ there exists a regular $u \in R$ with $v_i(ut) > v_i(a)$ for every $i \in \{1, \dots, n\}$. Indeed, set $J := \{j \in \{1, \dots, n\} \mid v_j(t) \neq \infty\}$. First, we consider the case $J \neq \emptyset$. For every $j \in J$ we have $\mathfrak{r} \not\subset \mathfrak{q}_j$ since $ta \in \mathfrak{r}$. Let $j \in J$; then there exists a regular element $t_j \in \mathfrak{m}_j$ with $v_j(t_j) > v_j(a) - v_j(t)$ by (1), i.e., $v_j(tt_j) > v_j(a)$. We set $u := \prod_{j \in J} t_j$; u is a regular element of R. Since $v_i(t) = \infty$ for every $i \in \{1, \dots, n\} \setminus J$, it follows that $v_i(ut) > v_i(a)$ for every $i \in \{1, \dots, n\}$.
Now we consider the case $J = \emptyset$. Then $v_i(t) = \infty$ for every $i \in \{1, \dots, n\}$, and we choose $u = 1$.

(3) For every $a \in K$ there exists a regular element $b \in R$ with $ab \in R$ and with the following property: for every $i \in \{1, \dots, n\}$ with $v_i(a) \geq 0$ we have $v_i(b) = 0$. To see this, we choose with respect to a a polynomial $F \in \mathbb{Z}[T]$ as in (2.18) and set $y := F(a)$. Then we have $v_i(y) \leq 0$ for every $i \in \{1, \dots, n\}$. Since K has large Jacobson radical, there exists $d \in K$ such that $y + ud$ is a regular element of K for every regular element u of K, and $1 + zyd$ is a regular element of K for every $z \in K$ [cf. (1.9)]. In particular, $y + d$ is a regular element of K and $yd \in \mathfrak{r}$ [cf. [63], Ex. 4.7]. By (2) there exists a regular element $r \in R$ with $v_i(rd) > v_i(y)$ for every $i \in \{1, \dots, n\}$. Therefore we have $v_i(y + rd) = v_i(y) \leq 0$ for every $i \in \{1, \dots, n\}$. Since $y + rd \in \mathrm{Reg}(K)$, we may set $b := (y + rd)^{-1}$; it is easy to check that b has the desired properties.

(4) For every $i \in \{1, \dots, n\}$ we have $A_{v_i} = R_{[\mathfrak{m}_i]}$, hence we have $K = Q(R)$. Indeed, it is clear that $R_{[\mathfrak{m}_i]} \subset A_{v_i}$. Let $a \in A_{v_i}$, and let b be chosen as in (3). Then $b \in R \setminus \mathfrak{m}_i$, $ab \in R$, and $a = ab/b \in R_{[\mathfrak{m}_i]}$. In particular, if $v_i(a) > 0$, then we have $ab \in \mathfrak{m}_i$, hence $\mathfrak{m}_i A_{v_i} = \mathfrak{p}_{v_i}$. Since $Q(A_{v_i}) = K$, it follows that $Q(R) = K$.

(5) $\{\mathfrak{m}_1, \dots, \mathfrak{m}_n\}$ is the set of pairwise different regular maximal ideals of R. Indeed, let $\mathfrak{n}$ be a regular maximal ideal of R and let $r \in \mathfrak{n}$ be a regular element. Suppose that $\mathfrak{n} \not\subset \mathfrak{m}_1 \cup \dots \cup \mathfrak{m}_n$. Then there exists an element $a \in \mathfrak{n}$ such that $v_i(a) = 0$ for every $i \in \{1, \dots, n\}$. Since K has large Jacobson radical, there exists $t \in K$ with $a + xt \in \mathrm{Reg}(K)$ for every $x \in \mathrm{Reg}(K)$, and with $1 + yat \in \mathrm{Reg}(K)$ for every $y \in K$ [cf. (1.9)]; in particular, $at \in \mathfrak{r}$. By (2) there exists $u \in \mathrm{Reg}(K)$ such that $v_i(ut) > 0$ for every $i \in \{1, \dots, n\}$. Now $a + urt$ lies in $\mathrm{Reg}(K)$ since $ur \in \mathrm{Reg}(K)$, and therefore $a + urt$ is a unit of R since $v_i(a + urt) = v_i(a) = 0$ for every $i \in \{1, \dots, n\}$. On the other hand, since a and r lie in $\mathfrak{n}$ and ut lies in R, it follows that $a + urt \in \mathfrak{n}$. This contradiction shows that $\mathfrak{n} \subset \mathfrak{m}_1 \cup \dots \cup \mathfrak{m}_n$, hence that $\mathfrak{n} \subset \mathfrak{m}_i$ for some $i \in \{1, \dots, n\}$ [cf. [63], Lemma 3.3], and therefore that $\mathfrak{n} = \mathfrak{m}_i$.
Let $i, j \in \{1, \dots, n\}$, $i \neq j$; then we have $\mathfrak{m}_i \not\subset \mathfrak{m}_j$ since $R_{[\mathfrak{m}_j]} = A_{v_j} \not\subset A_{v_i} = R_{[\mathfrak{m}_i]}$ by assumption. Let $i \in \{1, \dots, n\}$; suppose that $\mathfrak{m}_i$ is not a maximal ideal of R. Then there exists a regular maximal ideal $\mathfrak{n}$ of R containing $\mathfrak{m}_i$, and, by what we have just shown, we have $\mathfrak{n} \subset \mathfrak{m}_j$ for some $j \in \{1, \dots, n\}$, i.e., we have $\mathfrak{m}_i \subset \mathfrak{m}_j$.

Therefore we have shown that $\{\mathfrak{m}_1, \ldots, \mathfrak{m}_n\}$ is the set of pairwise different maximal regular ideals of R.

(2.20) Theorem: [Approximation theorem for discrete Manis valuations] *Let $V_1, \ldots, V_n$ be pairwise different discrete Manis valuation rings of K. For every $i \in \{1, \ldots, n\}$ let $v_i \colon K \to \mathbb{Z}_\infty$ be the Manis valuation of K defined by V_i, and let $\mathfrak{p}_{v_i}$ be the regular maximal ideal of V_i. We set $R := V_1 \cap \cdots \cap V_n$ and $\mathfrak{m}_i := R \cap \mathfrak{p}_{v_i}$ for every $i \in \{1, \ldots, n\}$. Then:*

(1) *The prime ideals $\mathfrak{m}_1, \ldots, \mathfrak{m}_n$ are regular maximal ideals of R which are pairwise different, and they exhaust the set of regular prime ideals of R. Moreover, for every $i \in \{1, \ldots, n\}$ we have $R_{[\mathfrak{m}_i]} = V_i$ and $\mathfrak{m}_i V_i = \mathfrak{p}_{v_i}$.*

(2) *For any $a_1, \ldots, a_n \in K$ and $m_1, \ldots, m_n \in \mathbb{Z}$ there exists $a \in K$ such that*

$$v_i(a - a_i) \geq m_i \quad \text{for every } i \in \{1, \ldots, n\}.$$

(3) *For any $m_1, \ldots, m_n \in \mathbb{Z}$ there exists $a \in K$ such that*

$$v_i(a) = m_i \quad \text{for every } i \in \{1, \ldots, n\}.$$

Proof: The assumption on $V_1, \ldots, V_n$ implies that $V_i \not\subset V_j$ for $i, j \in \{1, \ldots, n\}$, $i \neq j$ [cf. (2.15)].

(1) By (2.19) we know that $\{\mathfrak{m}_1, \ldots, \mathfrak{m}_n\}$ is the set of regular maximal ideals of R, that they are pairwise different, and that $V_i = R_{[\mathfrak{m}_i]}$ and $\mathfrak{m}_i V_i = \mathfrak{p}_{v_i}$ for every $i \in \{1, \ldots, n\}$. Let $\mathfrak{p}$ be a regular prime ideal of R. There exists $i \in \{1, \ldots, n\}$ such that $\mathfrak{p} \subset \mathfrak{m}_i$, hence $\mathfrak{p} V_i = \mathfrak{m}_i V_i = \mathfrak{p}_{v_i}$ [since $\mathfrak{p} V_i$ is a regular prime ideal of $V_i = R_{[\mathfrak{m}_i]}$ and since V_i has only one regular prime ideal by (2.15)]. Let $x \in \mathfrak{m}_i$. Then there exist $y \in \mathfrak{p}$ and a regular element $s \in R \setminus \mathfrak{m}_i$ such that $x = y/s$, hence that $sx = y \in \mathfrak{p}$, and that therefore $x \in \mathfrak{p}$.

(2) Let $a_1, \ldots, a_n \in K$. Since $Q(R) = K$ by (2.19), there exist elements $r_1, \ldots, r_n \in R$ and a regular element $s \in R$ such that $a_1 = r_1/s, \ldots, a_n = r_n/s$. Since s is regular, we have $v_i(s) \neq \infty$ for $i \in \{1, \ldots, n\}$. Set $m_i' := m_i + v_i(s)$ for every $i \in \{1, \ldots, n\}$. The existence of $a \in K$ satisfying (2) follows from the existence of $b \in K$ satisfying $v_i(b - r_i) \geq m_i'$ for every $i \in \{1, \ldots, n\}$. To prove (2) we may assume, therefore, that $a_1, \ldots, a_n \in R$; furthermore, we may assume that $m_1, \ldots, m_n \in \mathbb{N}$.

Let $i \in \{1, \ldots, n\}$ and set $\mathfrak{q}_i := \{x \in R \mid v_i(x) \geq m_i\}$. It is clear that $\mathfrak{q}_i$ is an ideal of R and that $\mathfrak{m}_i^{m_i} \subset \mathfrak{q}_i \subset \mathfrak{m}_i$ [since the ideal $\mathfrak{m}_i^{m_i}$ is generated by all products $x_1 \cdots x_{m_i}$ where $x_1, \ldots, x_{m_i} \in \mathfrak{m}_i$]. Now we apply B(10.1): the ideals $\mathfrak{m}_1^{m_1}, \ldots, \mathfrak{m}_n^{m_n}$ are pairwise comaximal, hence the ideals $\mathfrak{q}_1, \ldots, \mathfrak{q}_n$ are pairwise comaximal, and therefore there exists $a \in R$ such that $a \equiv a_i \pmod{\mathfrak{q}_i}$ for every $i \in \{1, \ldots, n\}$, hence we have $v_i(a - a_i) \geq m_i$ for every $i \in \{1, \ldots, n\}$.

(3) For every $i \in \{1, \ldots, n\}$ there exists $a_i \in K$ such that $v_i(a_i) = m_i$. By (2) there exists $a \in K$ such that $v_i(a - a_i) \geq m_i + 1$ for every $i \in \{1, \ldots, n\}$. Now we have $v_i(a) = v_i(a_i)$ for every $i \in \{1, \ldots, n\}$ [cf. (2.10)(2)].

(2.21) Corollary: *We assume, furthermore, that* $(V_1 : K)_K, \ldots, (V_n : K)_K$ *is the set of prime ideals of* K. *Then:*
(1) *Let* $x \in K$; *then* x *is regular iff* $v_i(x) < \infty$ *for every* $i \in \{1, \ldots, n\}$.
(2) *Let* $m_1, \ldots, m_n \in \mathbb{Z}$. *Then there exists a regular* $x \in K$ *with*

$$v_i(x) = m_i \quad \text{for every } i \in \{1, \ldots, n\}.$$

(3) *Every regular ideal of* R *is a principal ideal.*

Proof: (1) Let $x \in K$; if x is regular, then $v_i(x) < \infty$ for every $i \in \{1, \ldots, n\}$. Conversely, assume that $v_i(x) < \infty$ for every $i \in \{1, \ldots, n\}$, i.e., that $x \notin (V_i : K)_K$ for every $i \in \{1, \ldots, n\}$ [cf. (2.12)]. Then x is a unit of K [since, by assumption, the ideals $(V_1 : K)_K, \ldots, (V_n : K)_K$ are all the prime ideals of K].
(2) follows from (1) and (2.20)(3).
(3) Let $i \in \{1, \ldots, n\}$. By (2) there exists a regular $x_i \in K$ such that $v_i(x_i) = 1$, $v_j(x_i) = 0$ for every $j \in \{1, \ldots, n\}$, $j \neq i$, and therefore we have $Rx_i \subset \mathfrak{m}_i$. Let $y \in \mathfrak{m}_i$. Then $v_i(y) \geq 1$, hence $v_j(y/x_i) \geq 0$ for every $j \in \{1, \ldots, n\}$, and we have shown that $y \in Rx_i$, hence that $\mathfrak{m}_i = Rx_i$.
Let $\mathfrak{a} \subset R$ be a regular ideal, and, for every $i \in \{1, \ldots, n\}$, set $m_i := \min(\{v_i(a) \mid a \in \mathfrak{a}\})$ [$m_i \in \mathbb{N}_0$ since $\mathfrak{a}$ is a regular ideal]. First, we consider the case $m_1 = \cdots = m_n = 0$. Then $\mathfrak{a} \not\subset \mathfrak{m}_j$ for every $j \in \{1, \ldots, n\}$, hence $\mathfrak{a} = R$ [since $\mathfrak{m}_1, \ldots, \mathfrak{m}_n$ is the set of maximal ideals of R]. In the general case we apply this argument to the regular ideal $\mathfrak{a}x_1^{-m_1} \cdots x_n^{-m_n}$ of R.

3   Valuation Rings and Valuations

(3.0) NOTATION: In this section K always denotes a field.

3.1   Valuation Rings

(3.1) DEFINITION: A subring V of K is called a valuation ring of K if it is a Manis valuation ring of K or if $V = K$.

(3.2) NOTATION: Let V be a valuation ring of K. Then V is integrally closed and has K as its field of quotients [cf. (2.1)]. Furthermore, it is quasilocal. Let $\mathfrak{m}$ be the maximal ideal of V; the field $V/\mathfrak{m}$ is called the residue field of the valuation ring V.

(3.3) NOTATION: Let V be a valuation ring of K, and let $\mathfrak{m}$ be the maximal ideal of V. Let A be a subring of K with $A \subset V$. The prime ideal $\mathfrak{m} \cap A$ is called the center of V in A. In particular, if A is quasilocal and $\mathfrak{m} \cap A$ is the maximal ideal of A, then we say that V dominates A.
Let V dominate the quasilocal ring A. The residue field $A/(\mathfrak{m} \cap A)$ of A is a subfield of the residue field $V/\mathfrak{m}$ of V. The valuation ring V is said to be of the first kind (resp. of the second kind) with respect to A if the residue field $V/\mathfrak{m}$ of V is algebraic (resp. not algebraic) over the residue field $A/(\mathfrak{m} \cap A)$ of A.

(3.4) Proposition: *Let A be a subring of K. The set of valuation rings of K containing A is not empty, and its intersection is the integral closure of A in K.*

Proof: This follows from (2.4) and (2.5).

(3.5) Theorem: [Existence theorem] *Let A be a subring of K, and let $\mathfrak{p}$ be a non-zero prime ideal of A. Then there exists a valuation ring $V \neq K$ of K which has center $\mathfrak{p}$ in A.*

Proof: By replacing A by $A_\mathfrak{p}$ we may assume that A is quasilocal with maximal ideal $\mathfrak{m} \neq \{0\}$. The set $\mathcal{B}$ of subrings B of K with $A \subset B$ and $\mathfrak{m}B \neq B$ is not empty [since $A \in \mathcal{B}$] and it is ordered by inclusion. Let $(B_i)_{i \in I}$ be a chain in $\mathcal{B}$, and set $B := \bigcup_{i \in I} B_i$; then B is a subring of K. Suppose that $\mathfrak{m}B = B$. Then there exist $x_1, \ldots, x_n \in \mathfrak{m}$ and $b_1, \ldots, b_n \in B$ with $1 = x_1 b_1 + \cdots + x_n b_n$. Since $(B_i)_{i \in I}$ is a chain, there exists $j \in I$ with $b_1, \ldots, b_n \in B_j$, hence we get $\mathfrak{m}B_j = B_j$, contradicting $B_j \in \mathcal{B}$. Therefore $B \in \mathcal{B}$ is an upper bound of $(B_i)_{i \in I}$. Let V be a maximal element of $\mathcal{B}$ [Zorn's lemma]; then we have $\mathfrak{m}V \neq V$, and therefore there exists a maximal ideal $\mathfrak{n}$ of V containing $\mathfrak{m}V$; for every subring C of K properly containing V we have $\mathfrak{m}C = C$ by maximality of V, hence $\mathfrak{n}C = C$, and therefore V is a valuation ring of K [cf. (2.2)], and $V \neq K$ since $\mathfrak{m}V \neq V$.

(3.6) Corollary: *Let A be a subring of K, and let $\mathfrak{a}$ be a proper non-zero ideal of A. Then there exists a valuation ring V of K with $V \supset A$ and $\mathfrak{a}V \neq V$.*

Proof: Since $\mathfrak{a}$ is a proper non-zero ideal of A, there exists a non-zero maximal ideal $\mathfrak{n}$ of A containing $\mathfrak{a}$. By (3.5) there exists a valuation ring V of K which dominates $A_\mathfrak{n}$.

(3.7) Proposition: *Let V be a valuation ring of K, and let M be a finitely generated torsion-free V-module. Then M is a free V-module.*

Proof: Let $\{x_1, \ldots, x_n\}$ be an irredundant system of generators of M. Suppose that there exist elements $a_1, \ldots, a_n$ of V which are not all zero, such that $a_1 x_1 + \cdots + a_n x_n = 0$. The ideal of V generated by $a_1, \ldots, a_n$ is a principal non-zero ideal [cf. (2.4)]. By relabelling, we may assume that this ideal is generated by a_1. Then we have $a_1(x_1 + (a_2/a_1)x_2 + \cdots + (a_n/a_1)x_n) = 0$, and since M is torsion-free, we have $x_1 + (a_2/a_1)x_2 + \cdots + (a_n/a_1)x_n = 0$, hence $x_1 = -((a_2/a_1)x_2 + \cdots + (a_n/a_1)x_n)$, and therefore $x_2, \ldots, x_n$ also generate M, contradicting the assumption that $\{x_1, \ldots, x_n\}$ is an irredundant system of generators of M. Thus, $\{x_1, \ldots, x_n\}$ is a free system of generators of M.

3.2   Subrings and Overrings of Valuation Rings

(3.8) Proposition: *Let V be a valuation ring of K. Then:*
(1) Let W be a subring of K containing V. Then W is a valuation ring of K, and its maximal ideal is a prime ideal of V.

(2) *The map $\mathfrak{p} \mapsto V_{\mathfrak{p}}$ is a bijective inclusion-reversing map from the set of prime ideals of V to the set of valuation rings of K containing V. The inverse map maps such a valuation ring of K containing V to its maximal ideal.*
(3) *The set of subrings of K containing V is totally ordered by inclusion.*

Proof: (1) The first statement follows from (2.4)(1). Let $\mathfrak{n}$ be the maximal ideal of W. If $\mathfrak{n} = \{0\}$, then $W = K$ and $\mathfrak{n}$ is a prime ideal of V. In the other case, let $x \in \mathfrak{n}$ be non-zero, and suppose that $x \notin V$. Then $x^{-1} \in V$, hence $1 = xx^{-1} \in \mathfrak{n}$, contradicting $\mathfrak{n} \neq W$. Therefore $\mathfrak{n}$ is an ideal of V; clearly it is a prime ideal of V.
(2) Let $\mathfrak{p}$ be a prime ideal of V, and set $W := V_{\mathfrak{p}}$. Then W is a subring of K, hence a valuation ring of K, and its maximal ideal $\mathfrak{p}V_{\mathfrak{p}}$ is contained in V by (1), hence $\mathfrak{p} = \mathfrak{p}V_{\mathfrak{p}} \cap V = \mathfrak{p}V_{\mathfrak{p}}$. Conversely, let W be a valuation ring of K containing V, and let $\mathfrak{n}$ be its maximal ideal. We show that $V_{\mathfrak{n}} = W$. Since every element in $V \setminus \mathfrak{n}$ is a unit of W, we have $V_{\mathfrak{n}} \subset W$. Let $z \in W$. If $z \in V$, then $z \in V_{\mathfrak{n}}$. If $z \notin V$, then $w := 1/z \in V$, and since $z \notin \mathfrak{n}$ [as $z \notin V$], z is a unit of W, hence w is a unit of W and therefore $w \notin \mathfrak{n}$, hence $z = 1/w \in V_{\mathfrak{n}}$.
(3) The set of prime ideals of V is totally ordered by inclusion [cf. (2.2)]; the assertion follows from (2).

(3.9) DEFINITION: Let V be a valuation ring of K. If V has only finitely many non-zero prime ideals, then the number of those prime ideals is called the rank of the valuation ring V, and it is denoted by rank(V). In the other case, we say that V has infinite rank, and we set rank$(V) = \infty$.

(3.10) VALUATION RINGS OF RANK > 1: Let V be a valuation ring of K with rank$(V) > 1$. Then V is not noetherian.
In fact, suppose that V is noetherian. Let $\mathfrak{p} \supsetneq \mathfrak{q} \neq \{0\}$ be prime ideals of V [remember that rank$(V) > 1$]. Since $\mathfrak{p}$ and $\mathfrak{q}$ are finitely generated, these ideals are principal ideals $\mathfrak{p} = Vs$, $\mathfrak{q} = Vt$ [cf. (2.4)], hence s and t are prime elements, hence irreducible, and since $t = us$ with $u \in V$, u must be a unit of V, hence $\mathfrak{p} = \mathfrak{q}$, contrary to our assumption.

(3.11) VALUATION RINGS OF RANK 1: Let V be a valuation ring of K of rank 1. Then the maximal ideal $\mathfrak{m}$ of V is the only non-zero prime ideal of V; moreover, by (3.8), K is the only subring of V which properly contains V. We have by (2.15): V is noetherian iff $\mathfrak{m}$ is principal. Moreover, in this case V is a principal ideal domain with only one non-zero prime ideal, and it is a one-dimensional regular local ring.

(3.12) Proposition: *Let V be a valuation ring of K with maximal ideal $\mathfrak{m}$, and let $\varphi : V \to V/\mathfrak{m} =: \kappa$ be the canonical homomorphism. Then:*
(1) *The map $W \mapsto \varphi(W)$ is a bijective inclusion-preserving map from the set of valuation rings of K contained in V to the set of valuation rings of κ.*

(2) *Let $T \subset V$ be a quasilocal subring of K having K as field of quotients. The map in (1) induces a bijective map from the set of valuation rings of K contained in V and dominating T to the set of valuation rings of κ dominating $\varphi(T)$.*

Proof: (1) Let W be a valuation ring of K contained in V with maximal ideal $\mathfrak{n}$. The maximal ideal $\mathfrak{m}$ of V is a prime ideal of W and $W_{\mathfrak{m}} = V$ [cf. (3.8)]. Let $z \in V \setminus \mathfrak{m}$. If $z \in W$, then $\varphi(z) \in \varphi(W)$. If $z \notin W$, then $1/z \in W$, hence $\varphi(1/z) = 1/\varphi(z) \in \varphi(W)$. Therefore $\varphi(W)$ is a valuation ring of κ. Conversely, let $\overline{W}$ be a valuation ring of κ, and set $W := \varphi^{-1}(\overline{W})$. Then W is a subring of V, and $\mathfrak{m}$ is an ideal of W. Let $z \in K^{\times}$. If $z \notin V$, then $1/z \in \mathfrak{m} \subset W$. If $z \in \mathfrak{m}$, then $z \in W$. If $z \in V \setminus \mathfrak{m}$, then z is a unit of V, hence $\varphi(z) \neq 0$. If $\varphi(z) \in \overline{W}$, then there exists $z' \in W$ with $z - z' \in \mathfrak{m} \subset W$, hence $z \in W$. If $\varphi(z) \notin \overline{W}$, then $1/\varphi(z) \in \overline{W}$, hence there exists $z' \in W$ with $z' - 1/z \in \mathfrak{m}$, hence $1/z \in W$. Therefore W is a valuation ring of K.
(2) Let $\mathfrak{p}$ be the maximal ideal of T. Then $\mathfrak{m} \cap T \subset \mathfrak{p}$, hence $\varphi(T) \cong T/(\mathfrak{m} \cap T)$ is a quasilocal subring of κ. It is easy to check that, if W is a valuation ring of K which is contained in V and which dominates T, then $\varphi(W)$ dominates $\varphi(T)$. Conversely, if $\overline{W}$ is a valuation ring of κ which dominates $\varphi(T)$, then $W := \varphi^{-1}(\overline{W})$ is a valuation ring of K which dominates T.

3.3   Valuations

(3.13) NOTATION: For the rest of this section let $(\Gamma, <)$ be a totally ordered abelian group.

(3.14) NOTATION: A map $v\colon K \to \Gamma_{\infty}$ is called a valuation of K if it satisfies the conditions stated in (2.8), but it is not necessarily surjective. Note that $v(K^{\times})$ is an ordered subgroup of Γ; it is called the value group Γ_v of v. Therefore $v\colon K \to v(K^{\times})_{\infty}$ is a Manis valuation of K. In particular, v is called a trivial valuation of K if $v(K^{\times}) = \{v(1)\} = \{0_{\Gamma}\}$.

(3.15) NOTATION: Let $v\colon K \to \Gamma_{\infty}$ be a valuation of K. Clearly the set $A_v := \{x \in K \mid v(x) \geq 0\}$ is a valuation ring of K, the ring of v, and $\mathfrak{p}_v := \{x \in A_v \mid v(x) > 0\}$ is the maximal ideal of the quasilocal ring A_v [cf. (2.12)]. The field $\kappa_v := A_v/\mathfrak{p}_v$ is called the residue field of v. The rank of A_v is said to be the rank of v, and we denote it by $\mathrm{rank}(v)$. The rational rank $\mathrm{rat.\,rank}(\Gamma_v)$ of Γ_v [cf. B(1.29)] is called the rational rank of v; it is denoted by $\mathrm{rat.\,rank}(v)$.

(3.16) REMARK: Let v be a valuation of K with value group Γ. If Γ is finitely generated, then the $\mathbb{Z}$-modules Γ and $\mathbb{Z}^d$ with $d := \mathrm{rat.\,rank}(v)$ are isomorphic [cf. [32], Ch. VII, § 4, Cor. 2 to Th. 2, and note that Γ is torsion-free].

(3.17) NOTATION: Valuations v and w of K are called equivalent if $A_v = A_w$.

(3.18) NOTATION: Let v be a valuation of K, and let A be a subring of K contained in A_v. The center of A_v in A is called the center of v in A, and if A is quasilocal and A_v dominates A, then we say that v dominates A.

(3.19) REMARK: (1) Let $v\colon K^\times \to \Gamma$ be a valuation of K, and let V be the valuation ring of v. Let Δ be defined as in (2.14); the map

$$Va \mapsto v(a) : \Delta \to v(K^\times) \qquad (*)$$

in an isomorphism of ordered groups. Therefore we always may identify the value group $v(K^\times)$ of v with the totally ordered group $\{Va \mid a \in K^\times\}$ of non-zero cyclic V-submodules of K.
(2) Let V be a valuation ring of K. If $V \neq K$, then we constructed in (2.13) a valuation v of K such that $V = A_v$; this valuation shall be called the canonical valuation of K defined by V. If $V = K$, then V is the ring of the trivial valuation of K, which also shall be called the canonical valuation of K defined by V.

(3.20) REMARK: Let A be a subring of K having K as field of quotients, and let $v\colon A \to \Gamma_\infty$ be a map with $v(1) = 0$, $v(x) = \infty$ iff $x = 0$, and with

$$v(a + b) \geq \min(\{v(a), v(b)\}), \quad v(ab) = v(a) + v(b) \quad \text{for all } a, b \in A.$$

Then v admits a unique extension $w\colon K \to \Gamma_\infty$ such that w is a valuation of K. In fact, setting $w(a/b) := v(a) - v(b)$ for $a, b \in A$, $b \neq 0$, gives a well-defined map $w\colon K \to \Gamma_\omega$ since, from $a/b = a'/b'$ for $a, a' \in A$, $b, b' \in A \setminus \{0\}$, we get $ab' = a'b$, hence $v(a) + v(b') = v(a') + v(b)$. It is clear that $w|K^\times\colon K^\times \to \Gamma$ is a homomorphism; since elements $x, x' \in K$ can be written in the form $x = a/b$, $x' = a'/b$ with $a, a', b \in A$, $b \neq 0$, we have $w(x + x') = w((a + a')/b) \geq \min(\{w(x), w(x')\})$, hence w is a valuation of K. Let w' be a valuation of K with $w'|A = v$. For $x = a/b$ with $a, b \in A$ and $b \neq 0$ we have $w'(x) = w'(a) - w'(b) = v(a) - v(b) = w(a/b) = w(x)$, hence we have $w' = w$.

(3.21) **Proposition:** *Let $v\colon K \to \Gamma_\infty$ be a valuation of K, and let V be the ring of v. For every $\gamma \in \Gamma$ with $\gamma \geq 0$ the sets $\mathfrak{a}_\gamma$ (resp. $\mathfrak{a}_\gamma^+$) of elements of V with $v(x) \geq \gamma$ (resp. $v(x) > \gamma$) are ideals of V, and every non-zero ideal of V contains an ideal $\mathfrak{a}_\gamma$ for some $\gamma \in \Gamma$ with $\gamma \geq 0$.*

Proof: Clearly $\mathfrak{a}_\gamma$ (resp. $\mathfrak{a}_\gamma^+$) is an ideal of V. Let $\mathfrak{a} \neq \{0\}$ be an ideal of V, and let $x \in \mathfrak{a}$ be a non-zero element. Set $\gamma := v(x)$. Then $\gamma \geq 0$ and $\mathfrak{a}_\gamma \subset \mathfrak{a}$ [let $y \in \mathfrak{a}_\gamma$, then $y = zx$ with $z := y/x \in V$].

(3.22) V-SUBMODULES OF K AND MAJOR SUBSETS OF Γ: Let v be a valuation of K, let V be the ring of v, and let Γ be the value group of v.
(1) For any V-submodule N of K we set $\varphi(N) := \{v(z) \mid z \in N \setminus \{0\}\}$. Then $\varphi(N)$ is a major subset of Γ: let $z \in N$ be a non-zero element, and let $\gamma \in \Gamma$ with

$\gamma \geq v(z)$. There exists $x \in K^\times$ with $v(x) = \gamma$, hence $v(x/z) \geq 0$, and we have $x = (x/z)z \in N$ [since $x/z \in V$], hence $\gamma \in \varphi(N)$.

(2) For any major subset M of Γ we set $\psi(M) := \{z \in K \mid v(z) \in M \cup \{\infty\}\}$. Then $\psi(M)$ is a V-submodule of K: Let x, $y \in \psi(M)$, then from $v(x + y) \geq \min(\{v(x), v(y)\})$ we see that $x + y \in \psi(M)$, and if $z \in V$, we have $v(xz) \geq v(x)$, hence $xz \in \psi(M)$.

(3.23) Proposition: *Using the notations of (3.22) we have:*
(1) The map $N \mapsto \varphi(N)$ is an inclusion-preserving bijective map from the set of V-submodules of K to the set of major subsets of Γ.
(2) The map $\mathfrak{a} \mapsto \varphi(\mathfrak{a})$ is an inclusion-preserving bijective map from the set of ideals of V to the set of major subsets of Γ_+, and $\mathfrak{p}$ is a non-zero prime ideal of V iff $\Gamma \setminus \varphi(\mathfrak{p})$ is an isolated subgroup of Γ.
(3) The map $\mathfrak{p} \mapsto v(V_{\mathfrak{p}}^\times)$ is an inclusion-reversing bijective map from the set of non-zero prime ideals of V to the set of isolated subgroups of Γ.
(4) The map $W \mapsto v(W^\times)$ is an inclusion-preserving bijective map from the set of subrings of K properly containing V to the set of isolated subgroups of Γ, and its inverse is the map $\Delta \mapsto v^{-1}(\Gamma_+ \cup \Delta \cup \{\infty\})$.

Proof: (1) It is enough to show that

$$\varphi(\psi(M)) = M \quad \text{for every major subset } M \text{ of } \Gamma,$$
$$\psi(\varphi(N)) = N \quad \text{for every } V\text{-submodule } N \text{ of } K.$$

Let M be a major subset of Γ. It is clear that $M \subset \varphi(\psi(M))$. Conversely, let $\gamma \in \varphi(\psi(M))$. Then there exists $x \in \psi(M)$ with $v(x) = \gamma$, hence $\gamma = v(x) \in M$.
Let N be a V-submodule of K. It is clear that $N \subset \psi(\varphi(N))$. Conversely, let $x \in \psi(\varphi(N))$, $x \neq 0$. Then we have $v(x) \in \varphi(N)$, hence there exists $z \in N$ with $v(x) = v(z)$, hence $x = (x/z)z \in N$ since $x/z \in V$.
(2) The first assertion follows from (1) since $\varphi(V) = \Gamma_+$. Let $\mathfrak{p}$ be a non-zero prime ideal of V, and set $\Delta := \Gamma \setminus \varphi(\mathfrak{p})$; note that $\Delta \neq \Gamma$. Then Δ is a segment: Let $\delta \in \Delta$ be positive and $\gamma \in \Gamma$ with $0 \leq \gamma \leq \delta$. Then we have $\gamma \in \Delta$ since $\varphi(\mathfrak{p})$ is a major subset of Γ. Also, for $\delta_1, \delta_2 \in \Delta$, we have $\delta_1 + \delta_2 \in \Delta$ if not both of these elements are positive, and if these elements lie in Δ_+, then their sum also lies in Δ since $\mathfrak{p}$ is a prime ideal. Therefore Δ is an isolated subgroup of Γ [cf. B(1.5)(2)]. Conversely, let Δ be an isolated subgroup of Γ, and set $M := \Gamma \setminus \Delta$. Then M is a major subset of Γ, and it consists of strictly positive elements [if an element $\gamma < 0$ would belong to M, then $0 \in M$]. Set $\mathfrak{p} := \psi(M)$; then $\mathfrak{p} \neq V$ is a non-zero ideal of V, and if two elements of V do not lie in $\mathfrak{p}$, then their product also does not lie in $\mathfrak{p}$ since Δ is a subgroup of Γ, hence $\mathfrak{p}$ is a prime ideal of V.
(3) Let $\mathfrak{p}$ be a non-zero prime ideal of V; we set $\Delta := \Gamma \setminus \varphi(\mathfrak{p})$. Using (3.8) and (2), it is enough to show that $\Delta = v(V_{\mathfrak{p}}^\times)$. Let $x \in V_{\mathfrak{p}}^\times$; then $x = y/z$ with y, $z \in V \setminus \mathfrak{p}$, hence $v(y)$, $v(z) \in \Delta$, and therefore $v(x) \in \Delta$. Conversely, let $\gamma \in \Delta$, and choose $x \in K^\times$ with $v(x) = \gamma$. We write $x = y/z$ with y, $z \in V$. We have

$v(x) = v(y) - v(z) \geq v(y) \geq 0$, and therefore $v(y) \in \Delta$, hence $y \notin \mathfrak{p}$, and since $v(1/x) = -\gamma \in \Delta$, we conclude that also $z \notin \mathfrak{p}$, hence $x \in V_{\mathfrak{p}}^{\times}$.
(4) This follows from (3) and (3.8)(2).

(3.24) Corollary: *The rank of v (and of V) is equal to the rank of Γ.*

Proof: This follows immediately from (3) and (3.8).

3.4 Composite Valuations

(3.25) REMARK: Let v be a valuation of K with valuation ring V and value group Γ. We know that the valuation rings V' of K which contain V correspond uniquely to the set of isolated subgroups of Γ [cf. (3.23)(4)], and that V determines a valuation ring of the residue field of such a valuation ring V' [cf. (3.12)]. We study this situation in more detail.

(3.26) Proposition: *Let v be a valuation of K, let Γ be the value group of v, and let V be the valuation ring of v. Let v' be a valuation of K, let Δ be the value group of v', and let V' be the valuation ring of v'. We assume that $V \subset V'$, and we set $\Gamma' := v(V'^{\times})$ which is the isolated subgroup of Γ corresponding to V'. Then:*
(1) There exists a uniquely determined homomorphism $\omega: \Gamma \to \Delta$ of ordered groups with $\omega \circ v = v'$; ω is surjective and its kernel is Γ'.
(2) Let $\varphi: V' \to V'/\mathfrak{p}_{v'} = \kappa_{v'}$ be the canonical homomorphism, and set $\overline{V} := \varphi(V)$. Let $\overline{v}$ be a valuation of $\kappa_{v'}$ with valuation ring $\overline{V}$ and value group $\overline{\Gamma}$. There exists a uniquely determined homomorphism of ordered groups $\overline{\omega}: \overline{\Gamma} \to \Gamma'$ with $\overline{\omega} \circ \overline{v} \circ \varphi | V'^{\times} = v | V'^{\times}$, and $\overline{\omega}$ is an isomorphism.

Proof: (1) We have an exact sequence

$$0 \to V^{\times} \to V'^{\times} \xrightarrow{v} v(V'^{\times}) \to 0$$

of abelian groups, and we have a commutative diagram with exact rows

$$
\begin{array}{ccccccccc}
0 & \to & V^{\times} & \to & K^{\times} & \xrightarrow{v} & \Gamma & \to & 0 \\
 & & \downarrow & & \downarrow & & & & \\
0 & \to & V'^{\times} & \to & K^{\times} & \xrightarrow{v'} & \Delta & \to & 0
\end{array}
.
$$

Therefore there exists a unique homomorphism of groups $\omega: \Gamma \to \Delta$ with $\omega \circ v = v'$; ω is surjective and $v(V'^{\times})$ is its kernel. Moreover, if $a \in K^{\times}$ lies in V, then $a \in V'$, hence ω is a homomorphism of ordered groups.
(2) For any $a \in V'$ we write $\overline{a}$ instead of $\varphi(a)$. Let $\overline{\gamma} \in \overline{\Gamma}$, and choose $a \in V'^{\times}$ with $\overline{v}(\overline{a}) = \overline{\gamma}$. Note that, if $\overline{\gamma} \geq 0$, then $a \in V$. Let also $b \in V'^{\times}$ with $\overline{v}(\overline{b}) = \overline{\gamma}$. Then $\overline{a}^{-1}\overline{b}$ is a unit of $\overline{V}$, hence there exists $z \in V \setminus \mathfrak{p}_v$ with $a^{-1}b - z \in \mathfrak{p}_{v'}$, hence $a^{-1}b \in V \setminus \mathfrak{p}_v$, and therefore $v(a) = v(b)$. We set $\overline{\omega}(\overline{\gamma}) := v(a) \in \Gamma'$; we have $\overline{v}(\overline{\gamma}) \geq 0$ if $\overline{\gamma} \geq 0$. It is easy to check that $\overline{\omega}: \overline{\Gamma} \to \Gamma'$ is an isomorphism of ordered groups; we have $\overline{\omega} \circ \overline{v} \circ \varphi | V'^{\times} = v | V'^{\times}$ by construction. Uniqueness of $\overline{\omega}$ is clear.

(3.27) NOTATION: In the situation of (3.26) we say that the valuation v is composite with valuations v' and $\overline{v}$, and we write $v = v' \circ \overline{v}$. We also say that v is decomposed, and that $v = v' \circ \overline{v}$ is a decomposition of v. We can assume that $\Gamma_{\overline{v}} = \Gamma'$, i.e., that $\overline{v} \circ \varphi | V'^{\times} = v | V'^{\times}$. Note that we have $\mathrm{rank}(v) = \mathrm{rank}(\overline{v}) + \mathrm{rank}(v')$. A decomposition $v = v' \circ \overline{v}$ is called non-trivial if $A_v \subsetneqq A_{v'} \subsetneqq K$.

3.5   Discrete Valuations

(3.28) NOTATION: Let v be a valuation of K with valuation ring V and value group Γ. Then v is said to be a discrete valuation of rank h if Γ is a discrete group of rank h [cf. B(1.23) for a definition]. In particular, if $h = 1$, then v is called a discrete valuation and V is called a discrete valuation ring. Note that in this case the value group Γ can be identified with $\mathbb{Z}$. Furthermore, a discrete Manis valuation ring which is a domain is a discrete valuation ring.

(3.29) Proposition: *Let V be a ring. The following statements are equivalent:*
(1) V is a discrete valuation ring.
(2) V is a one-dimensional local ring, and its maximal ideal $\mathfrak{m}$ is generated by a non-nilpotent element.
(3) V is a one-dimensional local integral domain which is integrally closed.
If these conditions are satisfied, then V is a one-dimensional regular local ring, hence a principal ideal domain with only one non-zero prime ideal.

Proof (1) $\Rightarrow$ (2): This follows immediately from (2.15).
(2) $\Rightarrow$ (3): Let $\mathfrak{m} = Vt$. By Krull's intersection theorem [cf. [63], Cor. 5.4] we have $\bigcap \mathfrak{m}^n = \{0\}$, and since t is not nilpotent, we have $\mathfrak{m}^n \neq \{0\}$ for every $n \in \mathbb{N}_0$. Let $x \in V$ be a non-zero element; then $x = ut^n$ where $u \in V$ is a unit and $n \in \mathbb{N}_0$. The element t is a regular element of V. In fact, suppose that there exists a non-zero $y \in V$ with $yt = 0$, then we write $y = ut^n$ where u is a unit of V and $n \in \mathbb{N}_0$, hence $t^{n+1} = 0$, and t would be nilpotent, contrary to our assumption. Therefore every non-zero $x \in V$ has a unique representation $x = ut^n$ where $u \in V$ is a unit and $n \in \mathbb{N}_0$; we set $v(x) := n$. In particular, V is an integral domain which is factorial, hence it is integrally closed [cf. [63], Prop. 4.10], and $\mathfrak{m}$ is the only non-zero prime ideal of V.
(3) $\Rightarrow$ (1): Let $\mathfrak{m}$ be the maximal ideal of V. Then $\mathfrak{m}$ is not the zero ideal [since $\dim(V) = 1$]. Let $a \in \mathfrak{m}$ be non-zero. We have $\mathrm{rad}(Va) = \mathfrak{m}$ [since $\mathfrak{m}$ is the only non-zero prime ideal of V], hence there exists $n \in \mathbb{N}$ with $\mathfrak{m}^n \subset Va$, $\mathfrak{m}^{n-1} \not\subset Va$. We chose $b \in \mathfrak{m}^{n-1}$ with $b \notin Va$. Set $x := a/b \in Q(V)$. The element x^{-1} lies not in V, hence is not integral over V and therefore we have $\mathfrak{m}x^{-1} \not\subset \mathfrak{m}$ [cf. [63], Cor. 4.6]. On the other hand, we have $\mathfrak{m}x^{-1} = \mathfrak{m}(1/a)b \subset \mathfrak{m}^n(1/a) \subset V$, and therefore we get $\mathfrak{m}x^{-1} = V$, hence we have $\mathfrak{m} = Vx$. Now it not difficult to show that every non-zero $y \in V$ has a unique representation $y = ux^n$ with a unit u of V and $n \in \mathbb{N}_0$, and that $v \colon V \to \mathbb{N}_0$, defined by $v(y) = n$, gives rise to a discrete valuation of $Q(V)$ having V as its ring.

(3.30) SOME REMARKS ON FACTORIAL DOMAINS: Let R be a factorial domain with field of quotients K, and let $\mathbb{P}$ be a set of representatives for the irreducible elements of R.

(1) Remember that a domain is factorial iff every non-zero element admits a representation as a product of irreducible elements and every irreducible element $p \in R$ is a prime element, i.e., Rp is a prime ideal.

(2) The set $\{Rp \mid p \in \mathbb{P}\}$ is the set of prime ideals of height 1 of R [note that every non-zero prime ideal of R contains an irreducible element].

(3) Let $M \subset R$ be multiplicatively closed with $0 \notin M$, and set $S := M^{-1}R$.

(a) The ring S is factorial, and $\{p \in \mathbb{P} \mid Rp \cap M = \emptyset\}$ is a set of representatives for the irreducible elements of S.

Proof: If $p \in R$ is irreducible, then Rp is a prime ideal of R, hence pS is either the unit ideal—and this is the case iff $Rp \cap M \neq \emptyset$—or it is a prime ideal of S, and in the latter case p is irreducible in S. Therefore every non-zero element of S admits a representation as a product of irreducible elements of S. Let $q \in S$ be irreducible. Then we have $q = a/m$ with $a \in R$ and $m \in M$. Let $a = up_1 \cdots p_h$ where $u \in R$ is a unit and $p_1, \ldots, p_h \in \mathbb{P}$. Now a is irreducible in S, hence exactly one of the elements $p_1, \ldots, p_h$ is not a unit of S, hence $aS = pS$ for some $p \in \mathbb{P}$, and therefore $aS = qS$ is a prime ideal of S. This implies, by (1), that S is factorial.

(b) Let $p_1, \ldots, p_h \in R$ be irreducible elements with $p_i S \neq S$ for $i \in \{1, \ldots, h\}$. Then we have $(p_1 \cdots p_h)S \cap R = R(p_1 \cdots p_h)$.

Proof: Let $x \in R$, $m \in M$ with $p_1 \cdots p_h(x/m) =: r \in R$. We have $p_i \nmid m$ for every $i \in \{1, \ldots, h\}$, and therefore $p_1 \cdots p_h \mid r$, hence $x = mr'$ for some $r' \in R$, which means that $x/m \in R$.

(4)(a) Let $\mathfrak{p} = Rp$ be a prime ideal of R of height 1 [cf. (2)]. Then $R_\mathfrak{p}$ is a discrete valuation ring.

Proof: The ring $R_\mathfrak{p}$ is factorial by (3)(a), and $\{p\}$ is a set of representatives for the irreducible elements of $R_\mathfrak{p}$. Let $x \in K^\times$. Then x has a unique representation $x = up^n$ where $n \in \mathbb{Z}$ and $u \in R_\mathfrak{p}$ is a unit. We set $v_p(x) := n$. Then $v_p : K^\times \to \mathbb{Z}$ is a discrete valuation of K, and $R_\mathfrak{p} = \{x \in K^\times \mid v_p(x) \geq 0\} \cup \{0\} = A_{v_p}$ is the ring of the valuation v_p, hence $R_\mathfrak{p}$ is a discrete valuation ring of K [cf. (3.29)]. The discrete valuation v_p of K is called the valuation associated with p. [It is clear that v_p is independent of the choice of p in its equivalence class.]

(b) Let x, y be non-zero elements of R. If $v_p(x) = v_p(y)$ for every $p \in \mathbb{P}$, then we have $Rx = Ry$.

In fact, the assumption means that every $p \in \mathbb{P}$ occurs in a product decomposition of x and y with the same multiplicity, hence x and y are associated.

(4) If p and $q \in \mathbb{P}$ are different, then v_p and v_q are not equivalent since $v_p(p) = 1 \neq 0 = v_q(p)$.

(5) Now we assume, in particular, that $\dim(R) = 1$. In this case the only valuation rings $V \neq K$ of K which contain R are the discrete valuation rings V_p with $p \in \mathbb{P}$, since for such a valuation ring V we have $\mathfrak{m}(V) \cap R = Rp$ for some $p \in \mathbb{P}$, hence $V_p \subset V$ and therefore $V = V_p$. In particular, R is the intersection of all the discrete valuation rings V_p with $p \in \mathbb{P}$ [cf. (3.4)].

(3.31) THE VALUATIONS OF THE RATIONAL NUMBER FIELD: Let $\mathbb{P}$ be the set of prime numbers. Every valuation ring $V \neq \mathbb{Q}$ of $\mathbb{Q}$ contains $\mathbb{Z}$, hence $V = V_p$ for some prime number p [cf. (3.30)(5)]. Therefore $\{V_p \mid p \in \mathbb{P}\}$ is the set of all valuation rings of $\mathbb{Q}$ different from $\mathbb{Q}$, and $\mathbb{Z}$ is the intersection of all these discrete valuation rings.

(3.32) THE VALUATIONS OF AN ALGEBRAIC NUMBER FIELD: Let K be an extension of $\mathbb{Q}$ of finite degree; such a field is called an algebraic number field. Let R be the integral closure of $\mathbb{Z}$ in K; then R is one-dimensional [cf. [63], Cor. 4.17], integrally closed and noetherian [cf. III(4.5)(2)]; a ring with these properties is called a Dedekind ring. Let $\mathfrak{p}$ be a non-zero prime ideal of R. Then $R_\mathfrak{p}$ is a one-dimensional local integrally closed [cf. [63], Prop. 4.13] domain, hence a discrete valuation ring [cf. (3.29)]. Every valuation ring $V \neq K$ of K contains the integers, hence also R [since it is integrally closed]; since $\mathfrak{m}(V) \cap R$ is a non-zero prime ideal $\mathfrak{p}$ of R, we have $R_\mathfrak{p} \subset V$, hence $V = R_\mathfrak{p}$. Therefore $\{R_\mathfrak{p} \mid \mathfrak{p}$ non-zero prime ideal of $R\}$ is the set of all valuation rings of K different from K; they are all discrete valuation rings, and their intersection is R.

(3.33) EXAMPLE: Let v be a valuation of K, and let $v = w \circ \overline{v}$ be a decomposition of v. We assume that w and $\overline{v}$ are discrete of rank 1. Then we may assume that $\Gamma_w = \Gamma_{\overline{v}} = \mathbb{Z}$, and that $\Gamma_v = \mathbb{Z} \times \mathbb{Z}$, ordered lexicographically. It is easy to check that for $z \in A_w$, $z \notin \mathfrak{p}_w$, we have $v(z) = (w(z), \overline{v}(\overline{z}))$ where $\overline{z}$ is the image of z in κ_w.

3.6   Existence of Valuations of the Second Kind

(3.34) Proposition: *Let R be a quasilocal integrally closed proper subring of K having field of quotients K, and let $\mathfrak{m}$ be the maximal ideal of R. Then:*
(1) Let $x \in K^\times$ with $x \notin R$ and $x^{-1} \notin R$, and set $S := R[x]$. Then $\mathfrak{q} := \mathfrak{m}S$ is a prime ideal of S, and the image $\overline{x}$ of x in $S/\mathfrak{q}$ is transcendental over the residue field $R/\mathfrak{m}$ of R.
(2) If R is not a valuation ring of K, then there exists a valuation v of K which is of the second kind with respect to R.

Proof: (1)(a) Let $n \in \mathbb{N}$, and let $a_0, \ldots, a_n \in R$ not all belong to $\mathfrak{m}$. Then we have $a_0 + a_1 x + \cdots + a_n x^n \neq 0$.
Assume the contrary, i.e., that there exist $n \in \mathbb{N}$ and $a_0, \ldots, a_n \in R$ not all belonging to $\mathfrak{m}$ such that $a_0 + a_1 x + \cdots + a_n x^n = 0$. Let v be a valuation of K having center $\mathfrak{m}$ in R [cf. (3.5)]. First, we consider the case that $v(x) \geq 0$.
We choose $j \in \{0, \ldots, n\}$ with $a_n, \ldots, a_{j+1} \in \mathfrak{m}$, $a_j \notin \mathfrak{m}$. If $j = n$, then, upon division by a_n, we see that x would be integral over R, hence $x \in R$ [since R is integrally closed], contradicting our assumption. We cannot have $j = 0$, because $a_1, \ldots, a_n \in \mathfrak{m}$ and $v(x) \geq 0$ imply that $v(a_0) > 0$. Therefore we may assume that $0 < j < n$. We set

$$r := a_j + a_{j+1} x + \cdots + a_n x^{n-j}, \quad s := a_{j-1} + a_{j-2} \cdot 1/x + \cdots + a_0 \cdot (1/x)^{j-1};$$

note that $rx + s = 0$. Let w be any valuation of K with $R \subset A_w$. If $w(x) \geq 0$, then $w(r) \geq 0$, and also $w(s) = w(-rx) \geq 0$. If $w(x) < 0$, then $w(s) \geq 0$, and since $r + s/x = 0$, we have $w(r) > 0$. In both cases we have $w(r) \geq 0$ and $w(s) \geq 0$; since R is the intersection of all valuation rings of K containing R [cf. (3.4)], we see that $r, s \in R$. Since $v(x) \geq 0$, $a_{j+1}, \ldots, a_n \in \mathfrak{m}$ and a_j is a unit of R, we have $v(r) = 0$, hence r is a unit of R, and $x = -s/r \in R$, again contrary to our assumption.

Now we consider the case that $v(x) \leq 0$. Then we have $a_n + a_{n-1}x^{-1} + \cdots + a_0 x^{-n} = 0$, and, by a similar argument, we would get that $x^{-1} \in R$, again contradicting our assumption.

(b) The canonical homomorphism of rings $\alpha \colon R \to R/\mathfrak{m}$ admits a unique extension as a homomorphism of rings

$$\alpha' \colon R[x] \to (R/\mathfrak{m})[X]$$

[polynomial ring over $R/\mathfrak{m}$ in an indeterminate X] with $\alpha'(x) = X$ by defining

$$\alpha'(a_0 + a_1 x + \cdots + a_n x^n) = \alpha(a_0) + \alpha(a_1)X + \cdots + \alpha(a_n)X^n$$

[$a_0, \ldots, a_n \in R$]: α' is well-defined since $a_0 + a_1 x + \cdots + a_n x^n \neq 0$ if not all $a_0, \ldots, a_n$ lie in $\mathfrak{m}$ [cf. (a)]. In particular, α' is surjective and $\mathfrak{q}$ is its kernel, hence $\mathfrak{q}$ is a prime ideal of S with $\mathfrak{q} \cap R = \mathfrak{m}$, and since $S/\mathfrak{q} \cong R/\mathfrak{m}[X]$, we see that $\overline{x}$ is transcendental over $R/\mathfrak{m}$.

(2) Since R is not a valuation ring, there exists $x \in K$ with $x \notin R$ and $1/x \notin R$. There exists a valuation v of K having center $\mathfrak{m}R[x]$ in $R[x]$ [cf. (3.5)]. Then the residue field κ_v of v can be considered as an extension of the quotient field $(R/\mathfrak{m})(\overline{x})$ of $R[x]/\mathfrak{m}R[x]$ which is the field of rational function in the variable $\overline{x}$ over $R/\mathfrak{m}$, hence v is of the second kind with respect to R.

4 The Approximation Theorem For Independent Valuations

(4.0) NOTATION: In this section K always denotes a field.

(4.1) **Proposition:** Let $v_1, \ldots, v_n$ be valuations of K, and set $A_i := A_{v_i}$ for $i \in \{1, \ldots, n\}$. We assume that $A_i \not\subset A_j$ for $i, j \in \{1, \ldots, n\}$, $i \neq j$. Then:
(1) Let $a_i \in A_i$ for $i \in \{1, \ldots, n\}$. Then there exists $a \in A_1 \cap \cdots \cap A_n$ with $a \equiv a_i$ (mod $\mathfrak{m}(A_i)$) for $i \in \{1, \ldots, n\}$.
(2) Let $i \in \{1, \ldots, n\}$; there exists $x_i \in K$ with $v_i(x_i) = 0$, $v_j(x_i) > 0$ for all $j \in \{1, \ldots, n\}$, $j \neq i$.

Proof: (1) We set $R := A_1 \cap \cdots \cap A_n$, $\mathfrak{m}_i := \mathfrak{m}(A_i) \cap R$. Then $\{\mathfrak{m}_1, \ldots, \mathfrak{m}_n\}$ is the set of maximal ideals of R, we have $R_{\mathfrak{m}_i} = A_i$ and $\mathfrak{m}_i A_i = \mathfrak{m}(A_i)$ [cf. (2.19)]. Since $A_i/\mathfrak{m}(A_i) = R_{\mathfrak{m}_i}/\mathfrak{m}_i R_{\mathfrak{m}_i} = R/\mathfrak{m}_i$, we may assume that $a_1, \ldots, a_n \in R$. Now the assertion follows from the Chinese remainder theorem B(10.1).
(2) This follows from (1) by choosing $a_i = 1$, $a_j = 0$ for $j \in \{1, \ldots, n\}$, $j \neq i$.

(4.2) DEFINITION: (1) Valuation rings V, W of K are called independent if K is the smallest subring of K which contains V and W; otherwise they are called dependent.

(2) Valuations v and w of K are called independent if the valuation rings A_v and A_w are independent, otherwise they are called dependent.

(4.3) DEPENDENCY AND DECOMPOSITION: (1) Let v and w be non-equivalent non-trivial valuations of K with $A_v \not\subset A_w$ and $A_w \not\subset A_v$. If they are dependent, then there exists a subring $A \neq K$ of K with $A_v \subset A$ and $A_w \subset A$, hence there exists a valuation ring $V' \neq K$ of K with $A \subset V'$ [cf. (3.5)]. Let v' be a valuation of K defined by V'. Then we have non-trivial decompositions $v = v' \circ \overline{v}$ and $w = v' \circ \overline{w}$.

(2) Let v and w be valuations of K which admit non-trivial decompositions $v = v' \circ \overline{v}$ and $w = v' \circ \overline{w}$. Then v and w are dependent.

(3) Note that two different valuation rings of K of rank 1 are always independent.

(4.4) Theorem: [Approximation theorem for independent valuations] *Let v_1, $\ldots, v_n$ be pairwise independent valuations of K. Let $x_1, \ldots, x_n \in K$ and, for $i \in \{1, \ldots, n\}$, let $\gamma_i \in \Gamma_{v_i}$. Then there exists $x \in K$ with $v_i(x - x_i) \geq \gamma_i$ for every $i \in \{1, \ldots, n\}$.*

Proof: If v_i is a trivial valuation for some $i \in \{1, \ldots, n\}$, then we have $\gamma_i = 0$, and $v_i(x - x_i) \geq \gamma_i$ is true for every $x \in K$. Therefore we may assume that none of the valuations $v_1, \ldots, v_n$ is trivial. In this case we have $A_{v_i} \not\subset A_{v_j}$ for $i, j \in \{1, \ldots, n\}$, $i \neq j$.

Just as in the proof of (2.20) we may assume that $\gamma_i > 0$ and that $x_i \in R := A_{v_1} \cap \cdots \cap A_{v_n}$ for $i \in \{1, \ldots, n\}$. Let $i \in \{1, \ldots, n\}$. We set $\mathfrak{a}_i := \{z \in A_{v_i} \mid v_i(z) \geq \gamma_i\}$ and $\mathfrak{q}_i := R \cap \mathfrak{a}_i$. Now $\mathrm{rad}(\mathfrak{a}_i)$ is a prime ideal of A_{v_i} [let x, $y \in A_{v_i}$ with $xy \in \mathrm{rad}(\mathfrak{a}_i)$, hence with $(xy)^m \in \mathfrak{a}_i$ for some $m \in \mathbb{N}$, and if we assume, as we may, that $v_i(x) \geq v_i(y)$, then we have $v_i(x^{2m}) \geq v_i(x^m y^m)$, hence $x^{2m} \in \mathfrak{a}_i$, hence $x \in \mathrm{rad}(\mathfrak{a}_i)$], and therefore $\mathfrak{p}_i := \mathrm{rad}(\mathfrak{q}_i) = \mathrm{rad}(\mathfrak{a}_i) \cap R$ is a prime ideal of R. It is enough to show that the ideals $\mathfrak{q}_1, \ldots, \mathfrak{q}_n$ are pairwise comaximal; then the result follows from the Chinese remainder theorem. Let $i \in \{1, \ldots, n\}$; since $\mathfrak{m}_1, \ldots, \mathfrak{m}_n$ are all the maximal ideals of R [cf. (2.19)], and since $\mathfrak{q}_i \subset \mathfrak{m}_i$, it is enough to show that $\mathfrak{q}_i \not\subset \mathfrak{m}_j$ for every $j \in \{1, \ldots, n\}$, $j \neq i$. Thus, suppose that $\mathfrak{q}_i \subset \mathfrak{m}_j$ for some $j \in \{1, \ldots, n\}$ with $j \neq i$. Now $\mathfrak{p}_i$ is a prime ideal of R which is contained in $\mathfrak{m}_i$ and $\mathfrak{m}_j$, hence $A_{v_i} = R_{\mathfrak{m}_i}$ and $A_{v_j} = R_{\mathfrak{m}_j}$ are contained in $R_{\mathfrak{p}_i}$. Since $\gamma_i > 0$, we have $\mathfrak{a}_i \neq \{0\}$, and as $\mathfrak{q}_i A_{v_i} = \mathfrak{a}_i$, we have $\mathfrak{q}_i \neq \{0\}$, hence $\mathfrak{p}_i \neq \{0\}$, and therefore $R_{\mathfrak{p}_i} \neq K$, contradicting our assumption that v_i and v_j are independent.

(4.5) Corollary: *Under the assumptions of (4.4) we have: For $i \in \{1, \ldots, n\}$ let $\gamma_i \in \Gamma_{v_i}$. Then there exists $x \in K$ with $v_i(x) = \gamma_i$ for every $i \in \{1, \ldots, n\}$.*

Proof: We may assume that $A_{v_i} \neq K$ for $i \in \{1, \ldots, n\}$. Let $i \in \{1, \ldots, n\}$, and choose $x_i \in K$ with $v_i(x_i) = \gamma_i$ and $\delta_i \in \Gamma_{v_i}$ with $\gamma_i < \delta_i$. By (4.4), applied to the elements x_i and the elements δ_i, there exists $x \in K$ with $v_i(x - x_i) > v_i(x_i)$ for $i \in \{1, \ldots, n\}$. From $x = x_i + (x - x_i)$ we now get $v_i(x) = \gamma_i$ for every $i \in \{1, \ldots, n\}$ [cf. (2.10)].

5 Extensions of Valuations

5.1 Existence of Extensions

(5.1) NOTATION: Let L/K be an extension of fields. and let w be a valuation of L. Then the restriction $v := w|K$ of w to K is a valuation of K. Let W be the valuation ring of w, and let $\Gamma_w := w(L^\times)$ be the value group of w. Then $V := W \cap K$ is the valuation ring of v, and the value group $\Gamma_v := v(K^\times)$ of v is a subgroup of Γ_w.

In this situation we often say: The valuation w is an extension to L of the valuation v. By abuse of language, we also say that W is an extension of V to L.

(5.2) REMARK: Let L/K be an extension of fields, let W be a valuation ring of L, and let V be a valuation ring of K. Then W dominates V iff $W \cap K = V$.

In fact, if W dominates V, then $V \subset W \cap K$, and, for any $x \in K \setminus V$, x^{-1} lies in the maximal ideal of V, hence in the maximal ideal of W, and therefore $x \notin W$. This shows that $W \cap K = V$. Conversely, if $V = W \cap K$, then we have $V \subset W$, and if $x \in K$ lies in the maximal ideal of V, then we have $x^{-1} \notin V$, hence $x^{-1} \notin W$, and therefore x lies in the maximal ideal of W. This shows that W dominates V. In particular, this shows that if $V = W \cap K$, then the residue field of v can be considered as a subfield of the residue field of W.

(5.3) REMARK: Let L/K be an extension of fields, let W be a valuation ring of L, and set $V := W \cap K$. We set $\Gamma := \{Va \mid a \in K^\times\}$ and $\Delta := \{Wb \mid b \in L^\times\}$, considered as totally ordered groups [cf. proof of (2.13)]. Now (5.2) easily yields: the map $Va \mapsto Wa : \Gamma \to \Delta$ is an injective homomorphism of ordered groups. Therefore the canonical valuation of L defined by W is an extension of the canonical valuation of K defined by V.

(5.4) Proposition: *Let L/K be an extension of fields, and let v be a valuation of K. Then v admits at least one extension as a valuation of L.*

Proof: The valuation ring V of v is a quasilocal subring of L. By (3.5) there exists a valuation ring W of L which dominates V, hence $V = W \cap K$ by (5.2). Let $\Gamma \to \{Va \mid a \in K^\times\}$ be the inverse of the isomorphism in (3.19); then we have an injective homomorphism of ordered groups $\Gamma \to \{Wa \mid a \in L^\times\}$; the canonical valuation w of L defined by W is an extension of v to L.

5.2   Reduced Ramification Index and Residue Degree

(5.5) NOTATION: Let L/K be an extension of fields, let v be a valuation of K, let V be the valuation ring of v, and let w be a valuation of L with valuation ring W which is an extension of v. Let Γ_w (resp. Γ_v) be the value group of w (resp. of v), and let κ_w (resp. κ_v) be the residue field of w (resp. of v). Then

$$e(W/V) = e(w/v) := (\Gamma_w : \Gamma_v),$$

the index of Γ_v in Γ_w, is called the reduced ramification index of w over v, and

$$f(W/V) = f(w/v) := [\kappa_w : \kappa_v]$$

is called the residue degree of w over v.

Note that Γ_w/Γ_v must not be a finite group, and that κ_w must not be an algebraic extension of κ_v, and if κ_w is an algebraic extension of κ_v, it must not be a finite extension of κ_v. We shall show in (6.10) that $e(w/v)$ and $f(w/v)$ are finite if L is a finite extension of K.

(5.6) REMARK: Let $K \subset L \subset L'$ be a tower of field extensions, let w' be a valuation of L', set $w := w'|L$ and $v := w|K$. Then we have

$$e(w'/v) = e(w'/w)e(w/v), \quad f(w'/v) = f(w'/w)f(w/v),$$

and, in particular, $e(w'/v)$ (resp. $f(w'/v)$) is finite iff $e(w'/w)$ and $e(w/v)$ (resp. $f(w'/w)$ and $f(w/v)$) are finite.

These assertions follow immediately from the multiplicative behavior of degrees in a tower of field extensions and of indices in a tower of groups.

5.3   Extension of Composite Valuations

(5.7) Proposition: *Let L/K be an extension of fields, let v be a valuation of K, and let w be a valuation of L which is an extension of v. Then:*
(1) *Let $v = v' \circ \overline{v}$ be a decomposition of v where v' is a valuation of K with $A_v \subset A_{v'}$, and $\overline{v}$ is a valuation of $\kappa_{v'}$ with valuation ring $A_v/\mathfrak{p}_{v'}$. Then there exist an extension w' of v' to L and an extension $\overline{w}$ of $\overline{v}$ to $\kappa_{w'}$ such that $w = w' \circ \overline{w}$.*
(2) *Let $w = w' \circ \overline{w}$ be a decomposition of w where w' is a valuation of L with $A_w \subset A_{w'}$ and $\overline{w}$ is a valuation of $\kappa_{w'}$. We set $v' := w'|K$; then we have $A_v \subset A_{v'}$, $\kappa_{w'}$ is an extension field of $\kappa_{v'}$, and setting $\overline{v} := \overline{w}|\kappa_{v'}$, we have $v = v' \circ \overline{v}$.*
(3) *If, in the situation of (2), we set $w(A_{w'}^\times) =: \Delta'$, $v(A_{v'}^\times) =: \Gamma'$, then Δ' (resp. Γ') is an isolated subgroup of Γ_w (resp. of Γ_v), and $\Delta' \cap \Gamma_v = \Gamma'$.*

Proof: (1) We may consider v' to be the canonical valuation of K defined by the valuation ring of v' and $\overline{v}$ to be the canonical valuation of $\kappa_{v'}$ defined by the valuation ring $A_v/\mathfrak{p}_{v'}$. Let $\Gamma' := v(A_{v'}^\times)$ be the isolated subgroup of $\Gamma := \Gamma_v$ corresponding to v' [cf. (3.26)], and choose an isolated subgroup Δ' of $\Delta := \Gamma_w$ with $\Delta' \cap \Gamma = \Gamma'$ [cf. B(1.11)]. Let w' be the canonical valuation of L defined by the valuation ring $w^{-1}(\Delta_+ \cup \Delta' \cup \{\infty\})$ corresponding to Δ'; note that $A_w \subset A_{w'}$. Since $A_{v'} = v^{-1}(\Gamma_+ \cup \Gamma' \cup \{\infty\})$ [cf. (3.23)], we have $A_{w'} \cap K = A_{v'}$, hence w' is an extension of v' [cf. (5.3)]. Let $\varphi \colon A_{v'} \to A_{v'}/\mathfrak{p}_{v'} = \kappa_{v'}$, $\psi \colon A_{w'} \to A_{w'}/\mathfrak{p}_{w'} = \kappa_{w'}$ be the canonical homomorphisms. Since $A_{w'}$ dominates $A_{v'}$ [cf. (5.2)], we have $\psi|A_{v'} = \varphi$. Since $A_w \cap A_{v'} = A_w \cap (A_{w'} \cap K) = A_w \cap K = A_v$, we see that the canonical valuation $\overline{w}$ of $\kappa_{w'}$ defined by $\varphi(A_w)$ is an extension of $\overline{v}$.

(2) Again let $\varphi \colon A_{v'} \to A_{v'}/\mathfrak{p}_{v'} = \kappa_{v'}$, $\psi \colon A_{w'} \to A_{w'}/\mathfrak{p}_{w'} = \kappa_{w'}$ be the canonical homomorphisms. We have $\psi|A_{w'} = \varphi$, $A_w = \psi^{-1}(A_{\overline{w}})$, and since $A_{\overline{w}} \cap \kappa_{v'} = A_{\overline{v}}$, we have $A_v = \varphi^{-1}(A_{\overline{v}})$, hence $v = v' \circ \overline{v}$.

(3) This follows immediately from the considerations above.

6 Extending Valuations to Algebraic Overfields

6.1 Some General Results

(6.1) **Proposition:** Let K be a field, and let L be an algebraic extension of K. Let v be a valuation of K with valuation ring V, and let B be the integral closure of V in L. Then:

(1) The map $\mathfrak{n} \mapsto B_\mathfrak{n}$ is a bijective inclusion-reversing map from the set of maximal ideals of B to the set of valuation rings W of L with $W \cap K = V$, the inverse map being $W \mapsto \mathfrak{m}(W) \cap B$.

(2) Every valuation of L extending v is equivalent to a valuation of L defined by a ring $B_\mathfrak{n}$ where $\mathfrak{n}$ is a maximal ideal of B.

(3) B is the intersection of all the valuation rings of L which are extensions of V.

Proof: (1) Let W be a valuation ring of L with $W \cap K = V$. Since W is integrally closed [cf. (3.2)], it is clear that $B \subset W$. We set $\mathfrak{n} := \mathfrak{m}(W) \cap B$. Since W dominates V [cf. (5.2)], we have $\mathfrak{n} \cap V = \mathfrak{m}(W) \cap V = \mathfrak{m}(V)$, hence $\mathfrak{n}$ is a maximal ideal of B [cf. [63], Cor. 4.17]. Let $x \in W$ be a non-zero element, and let $a_0 T^n + a_1 T^{n-1} + \cdots + a_n$ with $a_0 = 1$ be the minimal polynomial of x over K. We choose $j \in \{0, \ldots, n\}$ with $v(a_j) \leq v(a_i)$ for $i \in \{0, \ldots, n\}$, and we set $b_i := a_i/a_j$, $i \in \{0, \ldots, n\}$. The elements $b_0, \ldots, b_n$ lie in $V \subset B_\mathfrak{n}$, and $B_\mathfrak{n}$ is integrally closed in L by [63], Prop. 4.13. We have $b_j = 1$; by the first part of the proof of (3.34) we get $x \in B_\mathfrak{n}$ or $1/x \in B_\mathfrak{n}$. Suppose that $x \notin B_\mathfrak{n}$. Then $1/x \in B_\mathfrak{n}$ is not a unit of $B_\mathfrak{n}$, hence $1/x \in \mathfrak{n}B_\mathfrak{n} \subset \mathfrak{m}(W)$ which implies that $x \notin W$, contrary to our choice of x. Therefore we have $x \in B_\mathfrak{n}$, hence we have $B_\mathfrak{n} = W$. Conversely, let $\mathfrak{n}$ be a maximal ideal of B. Then we have $\mathfrak{n} \cap V = \mathfrak{m}(V)$. Let W be a valuation ring of L which dominates $B_\mathfrak{n}$ [cf. (3.5)]. Then W dominates V, hence $W \cap K = V$ [cf. (5.2)], and therefore $B_\mathfrak{n} = W$ by the first part of the proof.

(2) follows from (1) and (3) follows from (1) and B(2.5).

(6.2) **Corollary:** If $V = K$, then $W = L$ is the only extension of V to L.

(6.3) **Corollary:** (1) Every valuation ring of L which contains B also contains the valuation ring of an extension of v to L.

(2) *Let w, w' be non-equivalent extensions of v to L. Then we have $A_w \not\subset A_{w'}$ and $A_{w'} \not\subset A_w$.*
(3) *Let $\mathfrak{n}$ be a maximal ideal of B, and set $W := B_\mathfrak{n}$, an extension of V to L. We have $B/\mathfrak{n} = W/\mathfrak{m}(W)$.*

Proof: (1) Let W be such a valuation ring, and set $\mathfrak{p} := B \cap \mathfrak{m}(W)$. Let $\mathfrak{n}$ be a maximal ideal of B with $\mathfrak{p} \subset \mathfrak{n}$. Then we have $B_\mathfrak{n} \subset W$.
(2) We have $A_w = B_\mathfrak{n}$, $A_{w'} = B_{\mathfrak{n}'}$ where $\mathfrak{n} \neq \mathfrak{n}'$ are maximal ideals of B.
(3) This is clear.

(6.4) Corollary: *Let L/K be a finite extension of fields. Let v be a valuation of K, and let B be the integral closure of A_v in L. Then B is quasisemilocal, and every extension w of v to L is equivalent to a valuation defined by a valuation ring of the form $B_\mathfrak{n}$ where $\mathfrak{n}$ is a maximal ideal of B. In particular, the non-empty set of pairwise non-equivalent extensions of v to L has at most $[L:K]_{\mathrm{sep}}$ elements.*

Proof: This follows immediately by applying B(7.3).

(6.5) Definition: Let L/K be a finite extension of fields, and let v be a valuation of K. A system of extensions $\{w_1, \ldots, w_g\}$ of v to L is called a complete set of extensions of v to L if the valuations $w_1, \ldots, w_g$ are pairwise non-equivalent, and if every extension of v to L is equivalent to a valuation w_i for some $i \in \{1, \ldots, g\}$.

(6.6) Corollary: *Let L/K be a finite extension of fields, and let L' be a subfield of L containing K. Let v be a valuation of K. If a complete set of extensions of v to L is pairwise independent, then any complete set of extensions of v to L' is also pairwise independent.*

Proof: (1) Let B be the integral closure of A_v in L. We show: A complete set of extensions of v to L is pairwise independent iff every non-zero prime ideal of B is contained in exactly one maximal ideal of B. In fact, assume that a complete set of extensions of v to L is pairwise independent. Let $\mathfrak{p}$ be a non-zero prime ideal of B which is contained in two different maximal ideals $\mathfrak{n}_1$, $\mathfrak{n}_2$ of B. Then $B_{\mathfrak{n}_1}$ and $B_{\mathfrak{n}_2}$ are contained in $B_\mathfrak{p} \neq L$, hence the valuations of L defined by $B_{\mathfrak{n}_1}$ and $B_{\mathfrak{n}_2}$ are dependent, contradicting our assumption. Now assume that every non-zero prime ideal of B is contained in only one maximal ideal of B. Let $\mathfrak{n}_1 \neq \mathfrak{n}_2$ be maximal ideals of B, and suppose that there exists a subring $C \neq L$ with $B_{\mathfrak{n}_i} \subset C$ for $i = 1, 2$. Let $W \neq L$ be a valuation ring of L with $W \supset C$ [cf. (3.5)]. Then we have $\{0\} \neq (\mathfrak{m}(W) \cap B) \subset \mathfrak{n}_1 \cap \mathfrak{n}_2$ [cf. (3.8)], contradicting our assumption.
(2) Let B' be the integral closure of V in L'; then $B' = B \cap L'$. Let $\mathfrak{p}'$ be a non-zero prime ideal of B', and suppose that there exist two different maximal ideals $\mathfrak{n}'_1$, $\mathfrak{n}'_2$ of B' with $\mathfrak{p}' \subset \mathfrak{n}'_1 \cap \mathfrak{n}'_2$. Using [63], Prop. 4.15, we get the existence of prime ideals $\mathfrak{p}$, $\mathfrak{n}_1$ and $\mathfrak{n}_2$ of B with $\mathfrak{p} \subset \mathfrak{n}_1 \cap \mathfrak{n}_2$, $\mathfrak{p} \cap B' = \mathfrak{p}'$ and $\mathfrak{n}_i \cap B' = \mathfrak{n}'_i$ for $i = 1, 2$. Since $\mathfrak{n}_1$ and $\mathfrak{n}_2$ are maximal ideals of B [cf. [63], Cor. 4.17], this contradicts the condition shown in (1) to be equivalent to the hypothesis of the corollary.

(6.7) REMARK: Let K be a field, and let L be a finite normal extension of K; let $G := \mathrm{Gal}(L/K)$ be the group of K-automorphisms of L. Remember that $\mathrm{Card}(G) = [L:K]_{\mathrm{sep}}$, the degree of separability of L/K [cf. B(7.3)]. Let v be a valuation of K, and let w be an extension of v to L. For every $\sigma \in G$ the map $x \mapsto w(\sigma^{-1}(x)) : L^\times \to \Gamma_w$ is a valuation of L with value group Γ_w and valuation ring $\sigma(A_w)$; it shall be denoted by w^σ. Clearly the residue fields of w and w^σ are isomorphic; note that w^σ also is an extension of v. We say that the valuations w and w^σ are conjugate.
Note that, by (6.3)(2), if $A_{w^\sigma} \neq A_w$, then we have $A_{w^\sigma} \not\subset A_w$ and $A_w \not\subset A_{w^\sigma}$.

(6.8) Corollary: *Let K be a field, and let L be a finite normal extension of K. Let v be a valuation of K, and let w_1, w_2 be extensions of v to L. Then w_2 is equivalent to a valuation w' of L such that w' and w_1 are conjugate valuations.*

Proof: Let B be the integral closure of A_v in L, and set $\mathfrak{q}_i := B \cap \mathfrak{p}_{w_i}$, $i \in \{1,2\}$. The ideals $\mathfrak{q}_1$, $\mathfrak{q}_2$ are maximal ideals of B, hence there exists $\sigma \in \mathrm{Gal}(L/K)$ with $\sigma(\mathfrak{q}_1) = \mathfrak{q}_2$ [cf. [63], Prop. 13.10], and since $A_{w_i} = B_{\mathfrak{q}_i}$ for $i \in \{1,2\}$ by (6.1), we have $\sigma(A_{w_1}) = A_{w_2}$, hence w_2 and w_1^σ are equivalent valuations.

6.2 The Formula $ef \leq n$

(6.9) Proposition: *Let L/K be a finite extension of fields, let v be a valuation of K, and let $w_1, \ldots, w_h$ be extensions of v to L. If the valuations $w_1, \ldots, w_h$ are pairwise independent, then we have*

$$e(w_1/v)f(w_1/v) + \cdots + e(w_h/v)f(w_h/v) \leq [L:K].$$

Proof: For $i \in \{1, \ldots, h\}$ we choose natural integers e_i, f_i with $e_i \leq e(w_i/v)$, $f_i \leq f(w_i/v)$.
(1) Let $i \in \{1, \ldots, h\}$. We choose elements $\gamma_{i1}, \ldots, \gamma_{ie_i} \in \Gamma_{w_i}$ which are pairwise incongruent modulo Γ_v, and elements $z_{i1}, \ldots, z_{if_i} \in A_{w_i}$ whose images in κ_{w_i} are linearly independent over κ_v. We set $\gamma_i := \max(\{\gamma_{i1}, \ldots, \gamma_{ie_i}\}) \in \Gamma_{w_i}$.
(2) Let $i \in \{1, \ldots, h\}$. By (4.5), for every $s \in \{1, \ldots, e_i\}$ there exists $x_{is} \in L$ with

$$w_i(x_{is}) = \gamma_{is}, \ w_j(x_{is}) > \gamma_j \text{ for } j \in \{1, \ldots, h\}, j \neq i,$$

and, by (4.4), for every $t \in \{1, \ldots, f_i\}$ there exists $y_{it} \in L$ with

$$w_i(y_{it} - z_{it}) > 0, \ w_j(y_{it}) > 0 \text{ for } j \in \{1, \ldots, h\}, j \neq i.$$

(3) We show: The elements $x_{is}y_{it}$, $i \in \{1, \ldots, h\}$, $s \in \{1, \ldots, e_i\}$, $t \in \{1, \ldots, f_i\}$, are linearly independent over K.
In fact, suppose that there exists a relation

$$\sum_{i=1}^{h} \sum_{s=1}^{e_i} \sum_{t=1}^{f_i} a_{ist} x_{is} y_{it} = 0 \tag{$*$}$$

with elements $a_{ist} \in K$ which are not all 0. We may assume that all the elements a_{ist} lie in A_v, and that $a_{111} = 1$.

(a) Let $i \in \{1, \ldots, h\}$. Since $y_{it} \equiv z_{it} \pmod{\mathfrak{m}(A_{w_i})}$ for $t \in \{1, \ldots, f_i\}$, the images of the elements $y_{i1}, \ldots, y_{if_i}$ in κ_{w_i} are linearly independent over κ_v.

(b) The w_1-value of any element $\tilde{y} := b_1 y_{11} + \cdots + b_{f_1} y_{1f_1}$ where $b_1, \ldots, b_{f_1} \in A_v$ are not all 0, lies in Γ_v.

In fact, we choose $t' \in \{1, \ldots, f_1\}$ with $v(b_{t'}) \leq v(b_t)$ for $t \in \{1, \ldots, f_1\}$; then we can write

$$\tilde{y} = b_1 y_{11} + \cdots + b_{f_1} y_{1f_1} = b_{t'}(c_1 y_{11} + \cdots + c_{f_1} y_{1f_1})$$

with $c_1, \ldots, c_{f_1} \in A_v$ and $c_{t'} = 1$. From (a) we get $w_1(c_1 y_{11} + \cdots + c_{f_1} y_{1f_1}) = 0$, hence $w_1(\tilde{y}) = v(b_{t'}) \in \Gamma_v$.

(c) We consider the elements

$$x_s := (a_{1s1} y_{11} + \cdots + a_{1sf_1} y_{1f_1}) x_{1s} \quad \text{for } s \in \{1, \ldots, e_1\}.$$

If $x_s \neq 0$, then we have $w_1(x_s) \in \gamma_{1s} + \Gamma_v$ by (b), hence the w_1-values of those elements $x_1, \ldots, x_{e_1}$ which are different from 0 are pairwise different, and therefore we have [cf. (2.10)]

$$w_1(x_1 + \cdots + x_{e_1}) = \min(\{w_1(x_1), \ldots, w_1(x_{e_1})\}).$$

Since $a_{111} = 1$, we have by (a) $w_1(a_{111} y_{11} + \cdots + a_{11f_1} y_{1f_1}) = 0$, hence $w_1(x_1) = w_1(x_{11}) = \gamma_{11}$, and therefore

$$w_1\left(\sum_{s=1}^{e_1} \sum_{t=1}^{f_1} a_{1st} x_{1s} y_{1t}\right) = w_1(x_1 + \cdots + x_{e_1}) \leq \gamma_{11} \leq \gamma_1. \tag{$\dagger$}$$

(d) We have, by the choice made in (2),

$$w_1\left(\sum_{i=2}^{h} \sum_{s=1}^{e_i} \sum_{t=1}^{f_i} a_{ist} x_{is} y_{it}\right) > \gamma_1. \tag{$\dagger\dagger$}$$

From ($\dagger$) and ($\dagger\dagger$) we get

$$w_1\left(\sum_{i=1}^{h} \sum_{s=1}^{e_i} \sum_{t=1}^{f_i} a_{ist} x_{is} y_{it}\right) \leq \gamma_1,$$

contradicting (∗) in (3). Therefore we have $e_1 f_1 + \cdots + e_h f_h \leq [L : K]$. Since $e_1, \ldots, e_h, f_1, \ldots, f_h$ where arbitrary natural integers with $e_i \leq e(w_i/v)$, $f_i \leq f(w_i/v)$ for $i \in \{1, \ldots, h\}$, the inequality of the proposition follows.

(6.10) Corollary: Let L/K be a finite extension of fields, let v be a valuation of K, and let w be an extension of v to L. Then we have $e(w/v) f(w/v) \leq [L : K]$.

Proof: The proof of (6.9) works also if we consider only one extension w of v to L; in this case the approximation theorem and its corollary are not needed.

(6.11) Corollary: *Let L/K be an algebraic extension of fields, let v be a valuation of K, and let w be a valuation of L which is an extension of v. Then:*
(1) *The residue field κ_w of w is an algebraic extension of the residue field κ_v of v, and the quotient group Γ_w/Γ_v is a torsion group.*
(2) *Γ_w and Γ_v have the same rational rank.*
(3) *Γ_w and Γ_v have the same rank.*
(4) *If, moreover, L is a finite extension of K, then Γ_w is discrete iff Γ_v is discrete.*

Proof: (1) We write $L = \bigcup_{i \in I} L_i$ where $(L_i)_{i \in I}$ is the set of finite extensions of K contained in L. For every $i \in I$ we set $\Gamma_i := w(L_i^{\times})$, and we let κ_i be the residue field of $w|L_i$. Clearly we have $\Gamma_w = \bigcup \Gamma_i$, $\kappa_w = \bigcup \kappa_i$. Since Γ_i/Γ_v is a torsion group and κ_i is a finite extension of κ for every $i \in I$ [cf. (6.10)], the assertions follow immediately.
(2) Tensoring the exact sequence of $\mathbb{Z}$-modules [note that $\mathbb{Q}$ is a flat $\mathbb{Z}$-module]

$$0 \to \Gamma_v \to \Gamma_w \to \Gamma_w/\Gamma_v \to 0$$

with $\mathbb{Q}$ and noting that $\Gamma_w/\Gamma_v \otimes_{\mathbb{Z}} \mathbb{Q} = \{0\}$ by (1), proves the assertion.
(3) Γ_w/Γ_v is a torsion group by (1), and the result follows from B(1.11)(2).
(4) In this case the factor group Γ_w/Γ_v is finite by (6.10), and the result follows from B(1.27).

(6.12) Corollary: *Let L/K be an algebraic extension of fields. In the situation of (5.7)(1), there exists only one decomposition $w = w' \circ \overline{w}$. Moreover, we have $e(w/v) = e(w'/v')e(\overline{w}/\overline{v})$, $f(w/v) = f(\overline{w}/\overline{v})$.*

Proof: The first part of the assertion follows from B(1.11). Clearly we have $f(w/v) = f(\overline{w} \mid \overline{v})$. We set $\Delta := \Gamma_w$, $\Gamma := \Gamma_v$, and define Δ', Γ' as in (5.7). Note that $e(w/v) = (\Delta : \Gamma)$, and, by (3.26), $e(\overline{w}/\overline{v}) = (\Delta' : \Gamma')$ and $e(w'/v') = (\Delta/\Delta' : \Gamma/\Gamma')$. The factor groups $(\Delta/\Delta')/(\Gamma/\Gamma')$ and $(\Delta/\Gamma)/(\Delta'/\Gamma')$ are isomorphic; now the assertion follows since $(\Delta : \Gamma) = (\Delta/\Gamma : \Delta'/\Gamma') \cdot (\Delta' : \Gamma')$.

(6.13) Notation: *Let L/K be an algebraic extension of fields, let v be a valuation of K, and let w be an extension of v to L. Let V (resp. W) be the valuation ring of v (resp. w). We set*

$$r(W/V) = \overset{\bullet}{r}(w/v) := e(w/v)[\kappa_w : \kappa_v]_{\mathrm{ins}};$$

$r(w/v)$ is called the ramification index of w over v [cf., e.g., [204], Ch. II, §5, (3), for the notion of degree of inseparability of algebraic field extensions].

(6.14) REMARK: Let $K \subset L \subset L'$ be a tower of field extension with L' algebraic over K. Let w' be a valuation of L', set $w := w'|L$ and $v := w'|K$. Then we have

$$r(w'/v) = r(w'/w)r(w/v)$$

[by (5.6) and by the multiplicative behavior of the degrees of inseparability in a tower of field extensions].

6.3 The Formula $\sum e_i f_i \leq n$

(6.15) A PARTICULAR DECOMPOSITION OF A VALUATION: (1) The results of this paragraph should be stated better in terms of valuation rings instead of in terms of valuations. This would make the statements a little bit clumsy. Therefore, when we speak in this paragraph of a valuation of a field K, we always mean the canonical valuation of K which is associated with the valuation ring of the valuation [cf. (2.13)].

(2) Let L/K be a finite extension of fields.

(a) For every non-trivial valuation v of K we denote by $E(v)$ the set of all extensions of v to L [note that, according to our agreement in (1), two different extensions of v to L are not equivalent]. We have $1 \leq \text{Card}(E(v)) \leq [L : K]$ [cf. (5.4) and (6.4)].

(b) For valuations v, v' of K we set $v' < v$ if $A_v \subsetneq A_{v'}$; clearly this gives a total ordering on the set of valuations of K. For a valuation v of K with $\text{rank}(v) > 1$ we denote by $I(v)$ the set of all non-trivial valuations v' of K with $v' < v$; note that $I(v)$ is not empty.

(c) Let v, v' be non-trivial valuations of K with $v' \leq v$. We define a map $\varphi_{v,v'} : E(v) \to E(v')$ in the following way: First, we consider the case $v' < v$. Let $w \in E(v)$. Then there exists a unique extension w' of v' to L with $w' < w$ [cf. (5.7)(3) and B(1.11)]; we set $w' =: \varphi_{v,v'}(w)$. From (5.7)(2) we also get that $\varphi_{v,v'}$ is surjective. Second, if $v' = v$, then we set $\varphi_{v,v} := \text{id}_{E(v)}$. It is immediate that for three non-trivial valuations $v'' \leq v' \leq v$ of K we have $\varphi_{v,v''} = \varphi_{v',v''} \circ \varphi_{v,v'}$.

(d) Let v be a valuation of K with $\text{rank}(v) > 1$ and let $w \in E(v)$; note that $\text{rank}(w) > 1$ [cf. (6.11)]. We define a map $\psi_w : I(v) \to I(w)$ by $\psi_w(v') = \varphi_{v,v'}(w)$ for every $v' \in I(v)$. The map ψ_w is order-preserving and bijective. In fact, for every $v' \in I(v)$ we have $\psi_w(v')|K = v'$, hence ψ_w is injective. Let $w' \in I(w)$; then $v' := w'|K$ is non-trivial and satisfies $v' < v$, hence $\varphi_{v,v'}(w) = w'$, and therefore ψ_w is surjective. Let $v'' < v'$ be in $I(v)$, and set $w' := \varphi_{v,v'}(w)$, $w'' := \varphi_{v,v''}(w)$. Then we have $w'' = \varphi_{v',v''}(\varphi_{v,v'}(w)) = \varphi_{v',v''}(w')$, hence $w'' < w'$.

(3) Let L/K be a finite extension of fields, and let v be a non-trivial valuation of K. Then there exist $w_1 \neq w_2$ in $E(v)$ which are dependent iff $\text{rank}(v) > 1$ and there exists $v' \in I(v)$ with $\text{Card}(E(v')) < \text{Card}(E(v))$.

In fact, assume that there exist $w_1 \neq w_2$ in $E(v)$ which are dependent. Then we have $\text{rank}(v) > 1$, and there exists a non-trivial valuation w' of L with $w' < w_1$ and with $w' < w_2$. We set $v' := w'|K$. Then we have $\varphi_{v,v'}(w_1) = \varphi_{v,v'}(w_2) = w'$, hence $\text{Card}(E(v')) < \text{Card}(E(v))$. Conversely, assume that $\text{rank}(v) > 1$, and let

$v' \in I(v)$ be as stated. Since $\varphi_{v,v'}$ is surjective, there exist $w_1 \neq w_2$ in $E(v)$ with $\varphi_{v,v'}(w_1) = \varphi_{v,v'}(w_2) =: w'$, hence we have $w' < w_i$ for $i \in \{1, 2\}$, hence w_1, w_2 are not independent [cf. (4.3)].

(4) Let L/K be a finite extension of fields, and let v be a non-trivial valuation of K. If there exist $w_1 \neq w_2$ in $E(v)$ which are dependent, then there exists a decomposition $v = v' \circ \overline{v}$ with the following properties:

(i) We have $\mathrm{Card}(E(v')) < \mathrm{Card}(E(v))$.

(ii) For every $w' \in E(v')$, any two extensions of the valuation $\overline{v}$ of $\kappa_{v'}$ to $\kappa_{w'}$ are independent.

Proof: (a) We construct v' in the following way: By (3) the set $I_1(v)$ of valuations $\widetilde{v} \in I(v)$ with $\mathrm{Card}(E(\widetilde{v})) < \mathrm{Card}(E(v))$ is not empty. We choose $v_0 \in I_1(v)$ such that $\mathrm{Card}(E(v_0)) \geq \mathrm{Card}(E(\widetilde{v}))$ for every $\widetilde{v} \in I_1(v)$, and we set $I_0(v) := \{\widetilde{v} \in I_1(v) \mid v_0 \leq \widetilde{v}\}$. For every $\widetilde{v} \in I_0(v)$ the map $\varphi_{\widetilde{v},v_0}: E(\widetilde{v}) \to E(v_0)$ is bijective. In fact, it is surjective by (2)(c), hence we have $\mathrm{Card}(E(\widetilde{v})) \geq \mathrm{Card}(E(v_0))$, and therefore the two sets $E(\widetilde{v})$, $E(v_0)$ have the same number of elements by our choice of v_0, hence $\varphi_{\widetilde{v},v_0}$ is also injective.

The intersection V' of all the valuation rings $A_{\widetilde{v}}$ with $\widetilde{v} \in I_0(v)$ is itself a valuation ring of K [since $A_v \subset A_{\widetilde{v}}$ for $\widetilde{v} \in I_0(v)$]; let v' be the valuation of K defined by V'. Note that $v_0 \leq \widetilde{v} \leq v' \leq v$ for every valuation $\widetilde{v} \in I_0(v)$.

Let $w' \in E(v')$ and set $w_0 := \varphi_{v',v_0}(w') \in E(v_0)$. Let W' be the intersection of all the rings $A_{\psi_{w'}(\widetilde{v})}$ with $\widetilde{v} \in I_0(v)$. We have $A_{\psi_{w'}(\widetilde{v})} \cap K = A_{\widetilde{v}}$ for every $\widetilde{v} \in I_0(v)$, hence we have $W' \cap K = V'$. On the other hand, we have $\psi_{w'}(\widetilde{v}) < w'$ for every $\widetilde{v} \in I_0(v)$ with $\widetilde{v} < v'$, hence $A_{w'} \subset W'$, and therefore $A_{w'} = W'$ [cf. (6.3)(2)].

Let w'' be another extension of v' to L with $w_0 = \varphi_{v',v_0}(w'')$. For every $\widetilde{v} \in I_0(v)$ we have $w_0 = \varphi_{v',v_0}(w') = \varphi_{\widetilde{v},v_0}(\varphi_{v',\widetilde{v}}(w')) = \varphi_{\widetilde{v},v_0}(\varphi_{v',\widetilde{v}}(w''))$, hence $\varphi_{v',\widetilde{v}}(w') = \varphi_{v',\widetilde{v}}(w'')$ since $\varphi_{\widetilde{v},v_0}$ is injective. Therefore we have $\psi_{w'}(I_0(v)) = \psi_{w''}(I_0(v))$, hence $w' = w''$ by what we have shown above. From this we get: The valuation v' has only one extension w' to L with $w_0 \leq w'$, hence v' and v_0 have the same number of extensions to L. In particular, we have $\mathrm{Card}(E(v')) < \mathrm{Card}(E(v))$.

(b) Now we have a decomposition $v = v' \circ \overline{v}$ of v. Let w' be an extension of v' to L, and let $\overline{w}_1$, $\overline{w}_2$ be different extensions to $\kappa_{w'}$ of the valuation $\overline{v}$ of $\kappa_{v'}$. Suppose that $\overline{w}_1$ and $\overline{w}_2$ are dependent, and let $\overline{w}$ be a non-trivial valuation of $\kappa_{w'}$ with $\overline{w} < \overline{w}_i$ for $i \in \{1, 2\}$. We set $w_i := w' \circ \overline{w}_i \in E(v)$ for $i \in \{1, 2\}$ and $w := w' \circ \overline{w}$. We set $\nu := w|K$; then we have $v' < \nu < v$, hence $E(v)$ and $E(\nu)$ have the same number of elements, and therefore $\varphi_{v,\nu}$ is bijective. On the other hand, we have $\varphi_{v,\nu}(w_1) = \varphi_{v,\nu}(w_2) = w$, hence $\varphi_{v,\nu}$ is not bijective. This contradiction shows that $\overline{w}_1$ and $\overline{w}_2$ are independent.

(6.16) Theorem: Let L/K be a finite extension of fields, let v be a valuation of K, and let $\{w_1, \ldots, w_g\}$ be a complete set of extensions of v to L. Then we have

$$e(w_1/v)f(w_1/v) + \cdots + e(w_g/v)f(w_g/v) \leq [L:K].$$

Proof: We prove the assertion by induction on the number g of elements of a complete set of extensions of v to L. If $g = 1$, then the assertion follows from

(6.10). Let $g > 1$, and assume that the assertion has been proved for all valuations of K which have a complete set of extensions to L consisting of at most $g - 1$ elements. Now let v be a valuation of K whose complete set of extensions to L has g elements.

(a) If the elements of a complete set of extensions of v to L are pairwise independent, then the result follows from (6.9).

(b) In the other case, there exists a decomposition $v = v' \circ \overline{v}$ as in (6.15)(4). A complete set of extensions of v' to L has the form $\{w_1', \ldots, w_h'\}$ with $h < g$; then we have by induction, applied to the extensions of v' to L,

$$e(w_1'/v')f(w_1'/v') + \cdots + e(w_h'/v')f(w_h'/v') \leq [L : K]. \tag{*}$$

For every $i \in \{1, \ldots, h\}$ let $\{\overline{w}_{i1}, \ldots, \overline{w}_{ig_i}\}$ be a complete set of extensions of the valuation $\overline{v}$ of $\kappa_{v'}$ to the finite extension $\kappa_{w_i'}$; the g_i valuations $\overline{w}_{i1}, \ldots, \overline{w}_{ig_i}$ are pairwise independent [cf. (6.15)(4)]. Then we have, by (6.9),

$$e(\overline{w}_{i1}/\overline{v})f(\overline{w}_{i1}/\overline{v}) + \cdots + e(\overline{w}_{ig_i}/\overline{v})f(\overline{w}_{ig_i}/\overline{v}) \leq [\kappa_{w_i'} : \kappa_{v'}] \quad \text{for } i \in \{1, \ldots, h\}.$$

We multiply the i-th equation by $e(w_i'/v')$, and add all the equations; we have by definition $[\kappa_{w_i'} : \kappa_{v'}] = f(w_i'/v')$, hence we obtain by $(*)$

$$\sum_{i=1}^{h} \sum_{j=1}^{g_i} e(w_i'/v')e(\overline{w}_{ij}/\overline{v})f(\overline{w}_{ij}/\overline{v}) \leq \sum_{i=1}^{h} e(w_i'/v')f(w_i'/v') \leq [L : K].$$

From (5.7)(3) we see that the set $\{w_i' \circ \overline{w}_{ij} \mid i \in \{1, \ldots, h\}, j \in \{1, \ldots, g_i\}\}$ is a complete set of extensions of v to L. Since $e(w_i'/v')e(\overline{w}_{ij}/\overline{v}) = e(w_i' \circ \overline{w}_{ij}/v)$ and $f(\overline{w}_{ij}/\overline{v}) = f(w_i' \circ \overline{w}_{ij}/v)$, we have proved the theorem.

(6.17) Theorem: *Let K be a field, let V be a valuation ring of K, let L be a finite Galois extension of K, let G be the Galois group of L over K, and let $W_1, \ldots, W_g$ be the extensions of V to L. Then $\{W_1, \ldots, W_g\} = \{\sigma(W_1) \mid \sigma \in G\}$, hence $e(W_i/V) =: e$, $f(W_i/V) =: f$ for every $i \in \{1, \ldots, g\}$, and $efg \leq [L : K]$.*

Proof: Set $B := W_1 \cap \cdots \cap W_g$, let $\mathfrak{m}$ be the maximal ideal of V, and, for every $i \in \{1, \ldots, g\}$, set $\mathfrak{m}_i := B \cap \mathfrak{m}(W_i)$. Then B is the integral closure of V in L, and we have $W_i = B_{\mathfrak{n}_i}$ for $i \in \{1, \ldots, g\}$ [cf. (6.1)]. Moreover, B is invariant under the operation of the Galois group G, and we have $\{\sigma(\mathfrak{m}_1) \mid \sigma \in G\} = \{\mathfrak{m}_1, \ldots, \mathfrak{m}_g\}$ by [63], Prop. 13.10. The assertions of the theorem follow immediately.

6.4   The Formula $\sum e_i f_i = n$

(6.18) NOTATION: Let L/K be a finite extension of fields, let v be a nontrivial valuation of K, and let w be an extension of v to L. Then $\varepsilon(\Gamma_w, \Gamma_v)$ [cf. B(1.14)] is called the initial ramification index of w over v and it is denoted by $\varepsilon(w/v)$. Note that $\varepsilon(w/v)$ divides $e(w/v)$ by B(1.16), hence $\varepsilon(w/v) \leq e(w/v)$.

(6.19) Proposition: *Under the assumptions of* (6.18) *we have*

$$\dim_{\kappa_v}(A_w/\mathfrak{p}_v A_w) = \varepsilon(w/v) f(w/v).$$

Proof: The ideals of A_w containing $\mathfrak{p}_v A_w$ correspond uniquely to the major subsets of Γ_w containing only strictly positive elements and containing the strictly positive elements of Γ_v [cf. (3.23)]. Since the ideals of A_w are totally ordered by inclusion [cf. (2.2)], we have a composition series of the A_w-module $A_w/\mathfrak{p}_v A_w$ of length $\varepsilon(w/v)$, hence $\ell_{A_w}(A_w/\mathfrak{p}_v A_w) = \varepsilon(w/v)$. Now every simple A_w-module is isomorphic to $\kappa_w = A_w/\mathfrak{p}_w$, and we have $\dim_{\kappa_v}(\kappa_w) = f(w/v)$.

(6.20) Proposition: *Let L/K be a finite extension of fields, let v be a nontrivial valuation of K, and let $\{w_1, \ldots, w_g\}$ be a complete set of extensions of v to L. Let B be the integral closure of A_v in L. Then we have*

$$\varepsilon(w_1/v) f(w_1/v) + \cdots + \varepsilon(w_g/v) f(w_g/v) = \dim_{\kappa_v}(B/\mathfrak{p}_v B).$$

Proof: By (6.1) we have $B = A_{w_1} \cap \cdots \cap A_{w_g}$, and setting $\mathfrak{n}_i := \mathfrak{p}_{w_i} \cap B$, $\{\mathfrak{n}_1, \ldots, \mathfrak{n}_g\}$ is the set of pairwise different maximal ideals of B, and we have $B_{\mathfrak{n}_i} = A_{w_i}$ for $i \in \{1, \ldots, g\}$. We have $\mathfrak{p}_v B = \mathfrak{q}_1 \cap \cdots \cap \mathfrak{q}_g$ where $\mathfrak{q}_i$ is $\mathfrak{n}_i$-primary [cf. B(10.21)]. Let $i \in \{1, \ldots, g\}$. The ring $B/\mathfrak{q}_i$ is quasilocal with maximal ideal $\mathfrak{n}_i/\mathfrak{q}_i$; therefore we have $B/\mathfrak{q}_i = (B/\mathfrak{q}_i)_{\mathfrak{n}_i/\mathfrak{q}_i} \cong B_{\mathfrak{n}_i}/\mathfrak{q}_i B_{\mathfrak{n}_i} = A_{w_i}/\mathfrak{p}_v A_{w_i}$. Since $B/\mathfrak{p}_v B \to B/\mathfrak{q}_1 \times \cdots \times B/\mathfrak{q}_g$ is an isomorphism by the Chinese remainder theorem [cf. B(10.1)], the assertion follows from (6.19).

(6.21) Corollary: *We have*

$$\dim_{\kappa_v}(B/\mathfrak{p}_v B) \leq \sum_{i=1}^{g} e(w_i/v) f(w_i/v) \leq [L : K].$$

Proof: The first inequality follows from (6.20) and B(1.16), the second inequality follows from (6.16).

(6.22) Theorem: *Let L/K be a finite extension of fields, let v be a nontrivial valuation of K, and let $\{w_1, \ldots, w_g\}$ be a complete set of extensions of v to L. We set $B := A_{w_1} \cap \cdots \cap A_{w_g}$. The following statements are equivalent:*
(1) B *is a finitely generated A_v-module.*
(2) B *is a free A_v-module.*
(3) *We have* $\dim_{\kappa_v}(B/\mathfrak{p}_v B) = [L : K]$.
(4) *We have*

$$\sum_{i=1}^{g} e(w_i/v) f(w_i/v) = [L : K] \text{ and } \varepsilon(w_i/v) = e(w_i/v) \text{ for every } i \in \{1, \ldots, g\}.$$

Proof: We set $n := [L : K]$. The equivalence of (1) and (2) follows from (3.7). Now we assume that B is a free A_v-module. Then B is a free A_v-module of rank n [as B has L as field of quotients], hence (2) implies (3). The equivalence of (3) and (4) follows from (6.18) and (6.21). We have to show that (3) implies (2). Let M be any A_v-module; the κ_v-vector space $M/\mathfrak{p}_v M$ shall be denoted by $\overline{M}$. Now $\overline{B}$ is a κ_v-vector space of dimension n. Let $x_1, \ldots, x_n$ be elements in B such that their images in $\overline{B}$ are a κ_v-basis of $\overline{B}$, and set $M := A_v x_1 + \cdots + A_v x_n \subset B$; M is a finitely generated torsion-free A_v-module, hence a free A_v-module by (3.7), and clearly $\{x_1, \ldots, x_n\}$ is an A_v-basis of M. We show that $B = M$. Thus, let $y \in B$ and set $N := M + A_v y$; N is also a finitely generated torsion-free $A_v$module, hence a free A_v-module of rank n. The inclusions $M \subset N \subset B$ induce κ_v-linear maps $\overline{M} \to \overline{N} \to \overline{B}$; since $\dim_{\kappa_v}(\overline{M}) = \dim_{\kappa_v}(\overline{N}) = n$ and $\overline{M} \to \overline{B}$ is surjective, the map $\overline{M} \to \overline{N}$ is also surjective, hence $N = M + \mathfrak{p}_v N$, and therefore $M = N$ by Nakayama's lemma [cf. [63], Cor. 4.8], hence $y \in M$ and $M = B$. Thus, we have shown that B is a free A_v-module.

(6.23) Corollary: *Let L/K be a finite separable extension of fields. Let v be a discrete valuation of K of rank 1, and let $\{w_1, \ldots, w_g\}$ be a complete set of extensions of v to L. Then we have*

$$e(w_1/v) f(w_1/v) + \cdots + e(w_g/v) f(w_g/v) = [L : K].$$

Proof: B is a finitely generated A_v-module by III(4.5)(2) and (3.29), and we have $\varepsilon(w_i/v) = e(w_i/v)$ for every $i \in \{1, \ldots, g\}$ [cf. B(1.16)].

7   Extensions of Discrete Valuations

7.1   Intersections of Discrete Valuation Rings

(7.1) In this subsection we consider a finite intersection of discrete valuation rings of a field K; the results in (2.20) and (2.21) can be applied and yield

(7.2) Theorem: *Let $V_1, \ldots, V_n$ be pairwise different discrete valuation rings of a field K. We set $R := V_1 \cap \cdots \cap V_n$ and $\mathfrak{m}_i := R \cap \mathfrak{m}(V_i)$ for every $i \in \{1, \ldots, n\}$. Then $\mathfrak{n}_1, \ldots, \mathfrak{n}_n$ are the maximal ideals of R, and R is a one-dimensional semilocal principal integrally closed domain. Moreover, $V_1, \ldots, V_n$ are the only proper valuation rings of K which contain R.*

Proof: All these results except the last one follow at once from the statements alluded to above. The last statement follows from (6.3) and (2.15).

7.2   Extensions of Discrete Valuations

(7.3) REMARK: Let L/K be a finite extension of fields, let V be a discrete valuation ring of K, and let W be an extension of V to L; then W is a discrete valuation ring by (6.11), and $e(W/V)$ is a positive integer by (6.10). Let $w\colon L \to \mathbb{Z}_\infty$ be the valuation of L defined by W, and let $v\colon K^\times \to \mathbb{Z}$ denote the valuation of K defined by V [cf. (2.13)]. Then we have

$$w(z) = e(W/V)v(z) \quad \text{for every } z \in K^\times.$$

Proof: We choose $t \in K$ with $v(t) = 1$. Each $z \in K^\times$ has a unique representation $z = at^n$ where $a \in V$ is a unit and $n \in \mathbb{Z}$, hence $w(z) = nw(t)$ [since a is a unit of W], and $(w(L^\times) : w(K^\times)) = (\mathbb{Z} : w(t)\mathbb{Z})$, hence $e(W/V) = w(t)$ by definition. Therefore we have $w(z) = nw(t) = v(z)e(W/V)$.

(7.4) Proposition: *Let L/K be a finite extension of fields, let V be a discrete valuation ring of K, and let B be the integral closure of V in L. Then:*
(1) B is a semilocal principal ideal domain.
(2) Let $\mathfrak{n}_1 = Bt_1, \ldots, \mathfrak{n}_g = Bt_g$ be the maximal ideals of B; then $W_1 := B_{\mathfrak{n}_1}, \ldots, W_g := B_{\mathfrak{n}_g}$ are the extensions of V to L, $W_1, \ldots, W_g$ are discrete valuation rings, and $B = W_1 \cap \cdots \cap W_g$.
(3) Let t be a generator of the maximal ideal $\mathfrak{m}$ of V. Then we have $Bt = Bt_1^{e(W_1/V)} \cdots t_g^{e(W_g/V)}$, hence $\mathfrak{m}B = \mathfrak{n}_1^{e(W_1/V)} \cdots \mathfrak{n}_g^{e(W_g/V)}$.
(4) If, in particular, L is a Galois extension of K, then we have $e := e(W_1/V) = \cdots = e(W_g/V)$, $f := f(W_1/V) = \cdots = f(W_g/V)$ and $efg = [L : K]$.

Proof: (1) and (2) follow from (6.1), (6.11) and (7.2).
(3) Let v be the valuation of K defined by V, and, for $i \in \{1, \ldots, g\}$, set $e_i := e(W_i/V)$ and let w_i be the valuation of L defined by W_i; note that, by (7.3), we have $w_i(t) = e_i v(t)$. Therefore $t/t_1^{e_1} \cdots t_g^{e_g}$ is a unit of B, and this yields $Bt = Bt_1^{e_1} \cdots t_g^{e_g}$.
(4) follows from (6.17) and (6.23).

7.3   Some Classes of Extensions

(7.5) In this subsection we show how in special cases we can calculate the reduced ramification indices e_i and the relative degrees f_i.

(7.6) EXAMPLE: Let $m \in \mathbb{N}$, $m \geq 2$, and let K be a field such that m is not divisible by the characteristic of K.
(1) Let $F = X^m - 1 \in K[X]$. For the discriminant $\operatorname{dis}(F)$ of F we have

$$\operatorname{dis}(F) = (-1)^{m(m-1)/2} m^m \cdot 1_K.$$

Proof: Since the characteristic of K does not divide m, there exists a finite extension L of K containing a primitive m-th root of unity ζ. By B(10.22)(1)(**) we have

$$\operatorname{dis}(F) = \prod_{1 \leq i < j \leq m} (\zeta^j - \zeta^i)^2 = (-1)^{m(m-1)/2} \prod_{i \neq j} (\zeta^i - \zeta^j). \qquad (*)$$

Now $X^m - 1 = \prod_{i=1}^{m}(X - \zeta^i)$, hence [set $X := 0$]

$$(-1)^{m-1} = \prod_{i=1}^{m} \zeta^i. \qquad (**)$$

By taking the derivative of F we get

$$mX^{m-1} = \sum_{k=1}^{m} \prod_{\substack{j=1 \\ j \neq k}}^{m} (X - \zeta^j),$$

and therefore we get

$$m\zeta^{i(m-1)} = \prod_{\substack{j=1 \\ j \neq i}}^{m} (\zeta^i - \zeta^j) \quad \text{for every } i \in \{1, \ldots, m\}.$$

This equation and $(**)$ together with $(*)$ prove the assertion.
(2) Let $z \in K^\times$ and $F := X^m - z \in K[X]$. Then we have

$$\text{dis}(F) = (-1)^{m(m-1)/2} m^m z^{m-1}.$$

In fact, there exists a finite extension L of K containing a primitive m-th root of unity ζ and an element $y \in L$ such that $y^m = z$. By (1) and B(10.22)(1)$(**)$ we have

$$\text{dis}(F) = \prod_{1 \leq i < j \leq m} (y\zeta^j - y\zeta^i)^2 = (-1)^{m(m-1)/2} m^m z^{m-1}.$$

(3) If the characteristic of K divides m, the formulae in (1) and (2) are true also, because in this case $\text{dis}(F) = 0$ [F has zeroes of multiplicity > 1].

(7.7) Proposition: *Let A be an integrally closed domain with field of quotients K, let L be a finite separable extension of K of degree n and let B be the integral closure of A in L. Assume that there exists a primitive element $x \in B$ for L/K such that the discriminant of the minimal polynomial F of x over K is a unit of A. Then F has coefficients in A, B is a free A-module having $\{1, x, \ldots, x^{n-1}\}$ as a free system of generators, and the canonical A-algebra homomorphism $A[X] \to B$, defined by mapping X to x, is surjective with kernel (F).*

Proof: We have $F \in A[X]$ by III(2.12)(2). By chapter III, (4.5)(4) and (4.3), we have

$$D_{L/K}(1, x, \ldots, x^{n-1})B \subset A[x] \subset B,$$

and since $D_{L/K}(1, x, \ldots, x^{n-1}) = \text{dis}(F)$ [cf. B(10.22)(1)] is a unit of A, we have $B = A[x]$. The last assertion is clear.

(7.8) Lemma: *Let A be a quasilocal ring with maximal ideal $\mathfrak{m}$ and residue field $k = A/\mathfrak{m}$. Let $F \in A[X]$ be monic of degree n, and set $A_F := A[X]/(F)$. Let $\overline{F}$ be the image of F in $k[X]$, obtained by reducing the coefficients of F modulo $\mathfrak{m}$, and let $\overline{F}_1, \ldots, \overline{F}_g \in k[X]$ be irreducible monic pairwise different polynomials such that $\overline{F} = \overline{F}_1^{e_1} \cdots \overline{F}_g^{e_g}$ is the factorization of $\overline{F}$ in $k[X]$. For every $i \in \{1, \ldots, g\}$ let $F_i \in A[X]$ be a monic polynomial having $\overline{F}_i$ as image in $k[X]$, and set $\mathfrak{n}_i := \mathfrak{m}A_F + F_i A_F$. Then A_F is quasisemilocal with maximal ideals $\mathfrak{n}_1, \ldots, \mathfrak{n}_g$. For $i \in \{1, \ldots, g\}$ we have $A_F/\mathfrak{n}_i \cong k[X]/(\overline{F}_i)$ [as k-algebras].*

Proof: Let x be the image of X in A_F; then A_F is a free A-module having $\{1, x, \ldots, x^{n-1}\}$ as a free set of generators. Let $\pi \colon A_F \to \overline{A}_F := A_F/\mathfrak{m}A_F$ be the canonical homomorphism; note that $\overline{A}_F \cong A[X]/(\mathfrak{m}A[X]+FA[X]) \cong k[X]/(\overline{F})$ [as k-algebras]. For $i \in \{1, \ldots, g\}$ let $\overline{\mathfrak{n}}_i'$ be the ideal of $k[X]/(\overline{F})$ which is generated by the image of $\overline{F}_i$, and let $\overline{\mathfrak{n}}_i$ be the ideal of $\overline{A}_F$ corresponding to $\overline{\mathfrak{n}}_i'$. Now $\{\overline{\mathfrak{n}}_1', \ldots, \overline{\mathfrak{n}}_g'\}$ is the set of maximal ideals of $k[X]/(\overline{F})$, hence $\{\overline{\mathfrak{n}}_1, \ldots, \overline{\mathfrak{n}}_g\}$ is the set of maximal ideals of $\overline{A}_F$, and $\overline{A}_F/\overline{\mathfrak{n}}_i \cong (k[X]/(\overline{F}))/\overline{\mathfrak{n}}_i' \cong k[X]/(\overline{F}_i)$. Let $i \in \{1, \ldots, g\}$; we have $\mathfrak{n}_i = \pi^{-1}(\overline{\mathfrak{n}}_i)$, hence the ideal $\mathfrak{n}_i$ is maximal. Let $\mathfrak{n}$ be a maximal ideal of A_F. We have to show that $\mathfrak{n}$ is one of the ideals $\mathfrak{n}_1, \ldots, \mathfrak{n}_g$. For this, it is enough to show that $\mathfrak{m}A_F \subset \mathfrak{n}$. Suppose, therefore, that $\mathfrak{m}A_F \not\subset \mathfrak{n}$; then we have $\mathfrak{n} + \mathfrak{m}A_F = A_F$, hence $\mathfrak{n} = A_F$ by Nakayama's lemma [cf. [63], Cor. 4.8] which is absurd. The last assertion is clear.

(7.9) Proposition: [E. E. Kummer] *Let V be a discrete valuation ring with field of quotients K, maximal ideal $\mathfrak{m}$ and residue field k. Let L be a finite separable extension of K, and let B be the integral closure of V in L. Assume that there exists a primitive element $x \in B$ for L/K such that the discriminant of the minimal polynomial F of x over K is a unit of V. Let $\overline{F} \in k[X]$ be the image of F, obtained by reducing the coefficients of F modulo $\mathfrak{m}$, and let $\overline{F} = \overline{F}_1^{e_1} \cdots \overline{F}_g^{e_g}$ be the factorization of $\overline{F}$ in $k[X]$. Then there are exactly g extension $W_1, \ldots, W_g$ of V to L, and we have*

$$e(W_i/V) = e_i, \quad f(W_i/V) = \deg(\overline{F}_i) \quad \text{for } i \in \{1, \ldots, g\}.$$

Proof: Since $B \cong V[X]/(F)$ [cf. (7.7)], we can apply the results of (7.8) to our situation. For $i \in \{1, \ldots, g\}$ we choose a monic polynomial $F_i \in V[X]$ such that $\overline{F}_i$ is the image of F_i in $k[X]$; then $\{\mathfrak{n}_1 = \mathfrak{m}B + F_1 B, \ldots, \mathfrak{n}_g = \mathfrak{m}B + F_g B\}$ is the set of maximal ideals of B, hence $W_1 := B_{\mathfrak{n}_1}, \ldots, W_g := B_{\mathfrak{n}_g}$ are the extensions of V to L [cf. (7.4)]. We set $\overline{B} := B/\mathfrak{m}B$ and, for $i \in \{1, \ldots, g\}$, $e_i' := e(W_i/V)$ and $f_i := f(W_i/V) = [B/\mathfrak{n}_i : V/\mathfrak{m}]$ [cf. (7.4)]. Note that $B/\mathfrak{n}_i \cong \overline{B}/\mathfrak{n}_i\overline{B} \cong k[X]/(\overline{F}_i)$, hence that $f_i = \deg(\overline{F}_i)$.

We have $\mathfrak{m}B = \mathfrak{n}_1^{e_1'} \cdots \mathfrak{n}_g^{e_g'}$ [cf. (7.4)(3)], and $\dim_k(\overline{B}) = \sum e_i' f_i$ [cf. (6.22)]. Since the ideals $\mathfrak{n}_1, \ldots, \mathfrak{n}_g$ are pairwise comaximal, we get by the Chinese remainder theorem [cf. B(10.1)]

$$\mathfrak{m}B = \mathfrak{n}_1^{e_1'} \cdots \mathfrak{n}_g^{e_g'} = \mathfrak{n}_1^{e_1'} \cap \cdots \cap \mathfrak{n}_g^{e_g'}, \quad \overline{B} \cong B/\mathfrak{n}_1^{e_1'} \times \cdots \times B/\mathfrak{n}_g^{e_g'},$$

hence $\{0\} = \overline{\mathfrak{n}}_1^{e_1'}\overline{B} \cap \cdots \cap \overline{\mathfrak{n}}_g^{e_g'}\overline{B}$. Now $\overline{B} \cong k[X]/(\overline{F})$, and, for $i \in \{1,\dots,g\}$, the maximal ideal $\overline{\mathfrak{n}}_i$ of $\overline{B}$ corresponds to the maximal ideal $\overline{F}_i(k[X]/(\overline{F}))$ of $k[X]/(\overline{F})$. Therefore we have $\{0\} = \overline{F}_1^{e_1'}(k[X]/(\overline{F})) \cap \cdots \cap \overline{F}_g^{e_g'}(k[X]/(\overline{F})) = (\overline{F}_1^{e_1'} \cdots \overline{F}_g^{e_g'})k[X]/(\overline{F})$, hence $\overline{F}$ divides $\overline{F}_1^{e_1'} \cdots \overline{F}_g^{e_g'}$, and therefore we have $e_i' \geq e_i$ for $i \in \{1,\dots,g\}$. On the other hand we have, again by B(10.1), $k[X]/(\overline{F}) \cong k[X]/(\overline{F}_1^{e_1}) \times \cdots \times k[X]/(\overline{F}_g^{e_g})$, hence $\dim_k(k[X]/(\overline{F})) = \sum e_i f_i$. Therefore we have $e_i' = e_i$ for $i \in \{1,\dots,g\}$.

(7.10) REMARK: We can replace in (7.7) and (7.9) the condition concerning the discriminant by the condition that $\{1, x, \dots, x^{n-1}\}$ is a basis of B.

(7.11) Corollary: *If $\overline{F}$ is a separable polynomial, then we have*

$$e(W_i/V) = 1 \quad for \ i \in \{1,\dots,g\}.$$

(7.12) Proposition: *Let V be a discrete valuation ring with maximal ideal $\mathfrak{m}$ and field of quotients K, and let $F = X^n + a_1 X^{n-1} + \cdots + a_n \in V[X]$ with $a_1, \dots, a_n \in \mathfrak{m}$, $a_n \notin \mathfrak{m}^2$. Then F is irreducible in $K[X]$, and setting $L := K[X]/(F)$, $W := V[X]/(F)$, then W is the only extension of V to L, and we have*

$$e(W/V) = [L:K] = n, \ f(W/V) = 1.$$

Proof: V is a principal ideal domain, hence F is irreducible in $V[X]$ by Eisenstein's criterion, and therefore F also is irreducible in $K[X]$, and we have $FK[X] \cap V[X] = FV[X]$. Let x be the image of X in L. Then $L = K(x)$, we can consider $W = V[x]$ as a subring of L containing V, and W is integral over V, hence $\dim(V) = \dim(W)$ [cf. [63], Prop. 9.2], and therefore W is a one-dimensional domain since $\dim(V) = 1$ by (3.29). We have $\overline{F} = X^n$, hence W is a local ring with maximal ideal $\mathfrak{n} := \mathfrak{m}W + Wx$ [cf. (7.8)]. Since $a_n \in \mathfrak{m} \setminus \mathfrak{m}^2$, $a_n =: t$ is a uniformizing parameter for V, hence $\mathfrak{m} = Vt$, and from $-t = x^n + a_1 x^{n-1} + \cdots + a_{n-1}x$ we get $t \in Wx$, hence $\mathfrak{n} = Wx$. Therefore W is a discrete valuation ring [cf. (3.29)], hence W is the only extension of V to L [cf. (7.4)], and from Kummer's theorem [cf. (7.9)] we get our assertion.

(7.13) NOTATION: A polynomial as in (7.12) is called an Eisenstein polynomial.

7.4 Quadratic Number Fields

We apply (7.9) and (7.12) in order to determine the decomposition of primes in quadratic number fields K.

(7.14) GENERATION: Let $K = \mathbb{Q}(\theta)$ be an extension of $\mathbb{Q}$ of degree 2, and let $T^2 + aT + b$ be the minimal polynomial of θ over $\mathbb{Q}$. Then $\theta' := \theta + a/2$ is also a primitive element of $K/\mathbb{Q}$, and $T^2 + (4b - a^2)/4$ is the minimal polynomial for

θ' over $\mathbb{Q}$. Writing $(a^2 - 4b)/4 = (n_1/n_2)^2 m$ with n_1, $n_2 \in \mathbb{N}$ and $m \in \mathbb{Z}$ being squarefree and $\neq 1$, we see that $K = \mathbb{Q}(\sqrt{m})$ and that $T^2 - m$ is the minimal polynomial of $\sqrt{m}$ over $\mathbb{Q}$. Note that K is a Galois extension of $\mathbb{Q}$.

(7.15) INTEGRAL BASIS: Let R be the integral closure of $\mathbb{Z}$ in K. Note that $\{1, \sqrt{m}\}$ is a basis of the $\mathbb{Q}$-vector space K.
(1) Let $\alpha \in K$, and write $\alpha = (a + b\sqrt{m})/2$ with a, $b \in \mathbb{Q}$. We show: $\alpha \in R$ iff $a \equiv b \pmod{2}$ in case $m \equiv 1 \pmod 4$, and $a \equiv b \equiv 0 \pmod 2$ in case $m \equiv 2, 3 \pmod 4$.
Proof: If $b = 0$, then $\alpha \in R$ iff $a \equiv 0 \pmod 2$. Now we consider the case that $b \neq 0$. The minimal polynomial F of α over $\mathbb{Q}$ is $F = T^2 - aT + (a^2 - b^2 m)/4$, and we have $\alpha \in R$ if the coefficients of this polynomial lie in $\mathbb{Z}$ [cf. III(2.12)(2)]. If a and b satisfy the conditions above, then $F \in \mathbb{Z}[T]$. Conversely, assume that $F \in \mathbb{Z}[T]$. Then $a \in \mathbb{Z}$ and $(a^2 - b^2 m)/4 \in \mathbb{Z}$, hence $a^2 - 4(a^2 - b^2 m)/4 = b^2 m \in \mathbb{Z}$. Since m is squarefree, we have $b \in \mathbb{Z}$, and therefore $a^2 \equiv b^2 m \pmod 4$. If $m \equiv 1 \pmod 4$, then this condition is equivalent to $a \equiv b \pmod 4$, and if $m \equiv 2, 3 \pmod 4$, then this condition is equivalent to $a \equiv b \equiv 0 \pmod 4$.
(2) Now it is easy to check that (1) yields: $\{1, \sqrt{m}\}$ in case $m \equiv 2, 3 \pmod 4$ and $\{1, (1 + \sqrt{m})/2\}$ in case $m \equiv 1 \pmod 4$ is a $\mathbb{Z}$-basis of R; in the first case $\Delta := 4m$ and in the second case $\Delta := m$ generate $\Delta_{R/\mathbb{Z}}$ [cf. III(2.13) and B(10.22)].

(7.16) EXTENSION OF VALUATIONS: (1) Let p be a prime number. We set $V_p := \mathbb{Z}_{\mathbb{Z}p}$ and $R_p := R_{\mathbb{Z}p}$; then R_p is the integral closure of V_p in K [cf. [63], Prop. 4.13]. Let $\mathfrak{p}$ be a prime ideal of R lying over $\mathbb{Z}p$. Then $R_{\mathfrak{p}} = (R_p)_{\mathfrak{p}R_p}$ is a discrete valuation ring which is an extension of V_p to K, and $[R/\mathfrak{p} : \mathbb{Z}/\mathbb{Z}p] = f(R_{\mathfrak{p}}/V_p)$ is called the degree of $\mathfrak{p}$.
We leave it to the reader to prove that from a decomposition of p in R_p we get a corresponding decomposition of p in R [note that $pR_q = R_q$ for distinct primes p and q of $\mathbb{Z}$].
(2) Let p be an odd prime; then $\{1, \sqrt{m}\}$ is a V_p-basis of R_p [since 2 is a unit of V_p], and in $\mathbb{F}_p[T]$ we have

$$
T^2 - m = \begin{cases} T^2 & \text{if } p \mid m, \\ T^2 - m & \text{if } m \text{ is not a square modulo } p, \\ (T - a)(T + a) & \text{if } m \equiv a^2 \pmod{m} \text{ for some } a \in \mathbb{Z} \text{ with } p \nmid a. \end{cases}
$$

By Kummer's theorem (7.9) and (7.10) we get $pR = (p, \sqrt{m})^2$ in the first case, pR is a prime ideal of R in the second case, and $pR = (p, \sqrt{m} - a)(p, \sqrt{m} + a)$, a product of distinct prime ideals, in the third case.
(3) Now we consider the case $p = 2$. Let $m \equiv 2 \pmod 4$. According to (7.12) we have $pR = (2, \sqrt{m})^2$. Let $m \equiv 3 \pmod 4$; we have $T^2 - m = (T - 1)^2 \pmod 2$, hence $pR = (2, \sqrt{m} - 1)^2$. Now let $m \equiv 1 \pmod 4$. The minimal polynomial F of $(1 + \sqrt{m})/2$ over $\mathbb{Q}$ is $F = T^2 - T + (1 - m)/4$. If $m \equiv 1 \pmod 8$, then $F = T(T - 1) \pmod 2$, hence $2R = (2, (\sqrt{m} + 1)/2)(2, (\sqrt{m} - 1)/2)$, and if $m \equiv 5 \pmod 8$, then $F = T^2 + T + 1 \pmod 2$ is irreducible, hence $2R$ is a prime ideal.

(7.17) Theorem: *The decomposition of the prime p of $\mathbb{Z}$ in the integral closure R of $\mathbb{Z}$ in $\mathbb{Q}(\sqrt{m})$ is as follows:*
(1) *A prime p which divides Δ is the square of a prime ideal of R.*
(2) *An odd prime p which does not divide Δ decomposes as the product of two distinct conjugate prime ideals of degree 1 or becomes a prime ideal of degree 2 according as Δ is a quadratic residue or a quadratic nonresidue modulo p.*
(3) *If $p = 2$, when $m \equiv 1$ (mod 4), then pR is the product of two distinct conjugate prime ideals or is itself a prime ideal according as $m \equiv 1$ or $\equiv 5$ (mod 8).*

8 Ramification Theory of Valuations

8.1 Generalities

(8.1) In this section we consider the following situation: Let K be a field, let L be a finite Galois extension of K of degree n, let $G := \mathrm{Gal}(L/K)$ be the Galois group of L/K, let v be a valuation of K, and let w be an extension of v to L. Let κ be the residue field and Γ the value group of v, let V be the valuation ring of v and $\mathfrak{m}$ its maximal ideal. Let λ be the residue field and Δ the value group of w, let W be the valuation ring of w and $\mathfrak{n}$ its maximal ideal. We set

$$e := e(w/v) = (\Delta : \Gamma), \quad f := f(w/v) = [\lambda : \kappa].$$

Let $\{w_1 = w, w_2, \ldots, w_g\}$ be a complete set of extensions of v to L. Then we have $e(w_i/v) = e$, $f(w_i/v) = f$ for $i \in \{1, \ldots, g\}$, and $efg \le n$ [cf. (6.17)].

(8.2) REMARK: In the following, we use the results of section 7 of appendix B; the reader is advised to recall the contents of that section.

(8.3) INERTIA GROUP AND DECOMPOSITION GROUP: In the sense of section 7 of appendix B, W is a quasilocal ring which lies over V [cf. (6.1)], hence the decomposition group $G_Z := G_Z(W/V) =: G_Z(w/v)$ and the inertia group $G_T := G_T(W/V) =: G_T(w/v)$ are defined. Remember that

$$G_Z = \{\sigma \in G \mid \sigma(W) = W\}, \quad G_T = \{\sigma \in G_Z \mid \sigma(x) - x \in \mathfrak{n} \text{ for every } x \in W\};$$

G_Z is a subgroup of G, and G_T is a normal subgroup of G_Z [cf. B(7.8)]. We denote the fixed field of G_Z (resp. of G_T) by K_Z (resp. by K_T), and we set $v_Z := w|K_Z$, $v_T := w|K_T$. Let $V_Z := A_{v_Z}$ be the valuation ring, κ_Z be the residue field and Γ_Z be the value group of v_Z, and let $V_T := A_{v_T}$ be the valuation ring, κ_T be the residue field and Γ_T be the value group of v_T.
(1) W is the only extension of V_Z to L, $\mathfrak{m}V_Z$ is the maximal ideal of V_Z and $\kappa_Z = \kappa$ [cf. B(7.7)].
(2) λ/κ is a finite normal extension, $\mathfrak{m}V_T$ is the maximal ideal of V_T, κ_T is the separable algebraic closure of κ in λ, λ/κ_T is purely inseparable, and $[K_T : K_Z] = [\kappa_T : \kappa_Z] = [\lambda : \kappa]_{\mathrm{sep}}$ [cf. B(7.11)].

(3) From B(7.7) we get

$$[K_Z : K] = (G : G_Z) = g.$$

(4) Let p be the characteristic index of κ, i.e., $p = 1$ if κ has characteristic 0, and otherwise p is the characteristic of κ. We set $f_0 := [\lambda : \kappa]_{\text{sep}}$, $p^s := [\lambda : \kappa]_{\text{ins}}$ with $s \in \mathbb{N}_0$, hence $f = f_0 p^s$. Note that by (2) we have

$$f_0 = [\kappa_T : \kappa_Z] = [K_T : K_Z] = (G_Z : G_T).$$

In particular, if $p = 1$, then we have $f = f_0$ and $n = fg\,\text{ord}(G_T)$ [since $(G : G_T) = (G : G_Z)(G_Z : G_T) = fg$].

(5) We have $G_Z = \{\sigma \in G \mid w = w^\sigma\}$.

Proof: Let $\sigma \in G_Z$; then we have $\sigma(W) = W$, hence w^σ and w are equivalent. We identify the value group of w^σ with Δ. Then we have an automorphism $\varphi_\sigma := w^\sigma \circ w^{-1} : \Delta \to \Delta$ of ordered groups, and for $h \in \mathbb{N}$ we have $(\varphi_\sigma)^h = w^{\sigma^h} \circ w^{-1}$. This means that φ_σ has finite order, and therefore $\varphi^\sigma = \text{id}_\Delta$ [cf. the following lemma], hence $w^\sigma = w$. Conversely, let $\sigma \in G$, and assume that $w^\sigma = w$; then we have $\sigma(W) = W$, hence $\sigma \in G_Z$.

(8.4) Lemma: *Let Δ be a totally ordered abelian group, and let φ be an order-preserving automorphism of Δ of finite order. Then we have $\varphi = \text{id}_\Delta$.*

Proof: We choose $h \in \mathbb{N}$ with $\varphi^h = \text{id}_\Delta$. Suppose that $\varphi \neq \text{id}_\Delta$, and choose $\delta \in \Delta$ with $\varphi(\delta) \neq \delta$, say $\varphi(\delta) > \delta$. Then we get $\delta = \varphi^h(\delta) > \delta$ which is absurd. Therefore we have $\varphi = \text{id}_\Delta$.

(8.5) NOTATION: (1) Assume that $p > 1$. Remember that a finite group H is called a p-group if the order of every element of H is a power of p. Moreover, every finite abelian group A has a unique decomposition $A = B \oplus C$ where C is a p-group and no element of B has an order which is a positive power of p; the group C is the p-subgroup of A.

(2) If $p = 1$, then we say that a finite group is a p-group if it is trivial.

8.2   The Value Groups Γ, Γ_Z and Γ_T

The following simple result shall be used at several places.

(8.6) Lemma: *Let $x \in K_Z$, and assume that for every extension v' of v to K_Z which is not equivalent to v_Z we have $v'(x) > 0$ (resp. $v'(x) = 0$). Then, for any conjugate $y \neq x$ of x over K, we have $w(y) > 0$ (resp. $w(y) = 0$).*

Proof: We choose $\sigma \in G$ with $x = \sigma(y)$; since $x \neq y$, we have $\sigma \notin G_Z$. We have $w(y) = w^\sigma(x)$. Since w and w^σ are not equivalent [since $\sigma \notin G_Z$, cf. (8.3)(5)], $w^\sigma|K_Z$ and v_Z are not equivalent [since W is the only extension of V_Z to L, cf. (8.3)(1)], and therefore we have by assumption $w(y) = w^\sigma(x) > 0$ (resp. $w(y) = w^\sigma(x) = 0$).

(8.7) Proposition: *We have* $\Gamma = \Gamma_Z = \Gamma_T$.

Proof: (1) We have $[K_T : K_Z] = [\kappa_T : \kappa_Z]$ [cf. (8.3)(2)] and $(\Gamma_T : \Gamma_Z)[\kappa_T : \kappa_Z] \leq [K_T : K_Z]$ [cf. (6.10)], hence $(\Gamma_T : \Gamma_Z) = 1$, hence $\Gamma_T = \Gamma_Z$.
(2) We show now that $\Gamma_Z = \Gamma$.
(a) First, we assume that a complete set of extensions of v to L is pairwise independent. Then a complete set of extensions of v to K_Z is also pairwise independent [cf. (6.6)]. It is enough to show that every positive element of Γ_Z lies in Γ. Thus, let $\gamma \in \Gamma_Z$ be positive; by (4.5) there exists $x \in K_Z$ with $v_Z(x) = \gamma$ and $\omega(x) = 0$ for every other extension ω of v to K_Z which is not equivalent to v_Z. We set $m := [K_Z : K]$, and let $x_1 = x, x_2, \ldots, x_m$ be the conjugates of x over K. By (8.6) we have $w(x_2) = \cdots = w(x_m) = 0$, hence $v(N_{K_Z/K}(x)) = w(N_{K_Z/K}(x)) = w(x_1) + w(x_2) + \cdots + w(x_m) = v_Z(x_1) = \gamma \in \Gamma$.
(b) In the general case, we prove the theorem by induction on the number g of elements of a complete set of extensions of v to L. If $g = 1$, then we have $K = K_Z$ [cf. (8.3)(3)], hence $\Gamma = \Gamma_Z$. Now we assume that $g > 1$, and that the theorem has been proved for all valuations of K having a complete set of extension to L with less than g elements. Let v be a valuation of K having a complete set of extensions to L consisting of g elements. We choose a decomposition $v = v' \circ \overline{v}$ as in (6.16)(4). Since w is an extension of v, we have $w = w' \circ \overline{w}$ where w' is a uniquely determined extension of v' to L, and $\overline{w}$ is an extension of $\overline{v}$ to $\kappa_{w'}$ [cf. (6.12)]. We set $G_{Z'} := G_Z(w'/v')$, $G_{T'} := G_T(w'/v')$. We show that we have the inclusions

$$G_{Z'} \supset G_Z \supset G_T \supset G_{T'}.$$

Let $\sigma \in G_Z$; then we have $w^\sigma = w$ by (8.3)(5). On the other hand, w' is the only extension of v' to L such that w is composed with w'. Therefore we have $w'^\sigma = w'$, hence $\sigma \in G_{Z'}$. We have $G_Z \supset G_T$ by definition, and $G_T \supset G_{T'}$ by B(7.10).

We denote by $K_{Z'}$ (resp. by $K_{T'}$) the fixed field of $G_{Z'}$ (resp. of $G_{T'}$); $K_{Z'}$ is the decomposition field of w'/v', and $K_{T'}$ is the inertia field of w'/v'. By what we have shown above, we have

$$K \subset K_{Z'} \subset K_Z \subset K_T \subset K_{T'} \subset L. \tag{*}$$

We denote by $v_{Z'}$, v_Z, v_T, $v_{T'}$ the restrictions of w to $K_{Z'}$, K_Z, K_T, $K_{T'}$, respectively, and by $v'_{Z'}$, v'_Z, v'_T, $v'_{T'}$ the restrictions of w' to $K_{Z'}$, K_Z, K_T, $K_{T'}$, respectively. The value groups of the valuations $v_{Z'}, \ldots, v_{T'}$, $v'_{Z'}, \ldots, v'_{T'}$ shall be denoted by $\Gamma_{Z'}, \ldots, \Gamma_{T'}$, $\Gamma'_{Z'}, \ldots, \Gamma'_{T'}$. We set $\Gamma' := \Gamma_{v'}$. By induction, applied to w'/v', we have $\Gamma' = \Gamma'_{Z'} = \Gamma'_{T'}$. From (*) we therefore get $\Gamma' = \Gamma'_Z = \Gamma'_T$. Therefore we have

$$\Gamma' = \Gamma'_{Z'} = \Gamma'_Z. \tag{**}$$

The decomposition $w = w' \circ \overline{w}$ of w gives rise to a decomposition of $w|K_{Z'} = v_{Z'}$ as $v_{Z'} = v'_{Z'} \circ \overline{v}_{Z'}$ [cf. (5.7)]; here $v'_{Z'} = w'|K_{Z'}$ and $\overline{v}_{Z'} = \overline{w}|\kappa_{v'_{Z'}}$. From (8.3)(1) we get $\kappa_{v'_{Z'}} = \kappa_{v'}$. On the other hand, $\overline{v}_{Z'}$ is an extension of the valuation $\overline{v}$ of $\kappa_{v'}$ to $\kappa_{v'_{Z'}}$. Therefore we have $\overline{v}_{Z'} = \overline{v}$, hence $\Gamma_{\overline{v}} = \Gamma_{\overline{v}_{Z'}}$. Applying (3.26) yields: $\Gamma_{\overline{v}_{Z'}}$

is an isolated subgroup of $\Gamma_{Z'}$ and $\Gamma'_{Z'} = \Gamma_{Z'}/\Gamma_{\bar{v}_{Z'}}$, $\Gamma_{\bar{v}}$ is an isolated subgroup of Γ and $\Gamma' = \Gamma/\Gamma_{\bar{v}}$. Since $\Gamma' = \Gamma'_{Z'}$ by (**), we get $\Gamma = \Gamma_{Z'}$. It is therefore enough to prove that $\Gamma_Z = \Gamma_{Z'}$.

We may replace K by $K_{Z'}$; then w' is the only extension of v' to L, and the valuation $\bar{v}$ has a complete set of extensions to L which are pairwise independent. We consider the decompositions $v = v' \circ \bar{v}$ and $v_Z = v'_Z \circ \bar{v}_Z$ [$\bar{v}_Z$ is the restriction of $\bar{w}$ to $\kappa_{v'_Z}$]. We set $\Delta := \Gamma_{\bar{v}}$ and $\Delta_Z := \Gamma_{\bar{v}_Z}$; then we have $\Delta = \Delta_Z \cap \Gamma$ [cf. (5.7)]. Applying (3.26) yields: Δ is an isolated subgroup of Γ and $\Gamma' = \Gamma/\Delta$, Δ_Z is an isolated subgroup of Γ_Z and $\Gamma'_Z = \Gamma_Z/\Delta_Z$. From $\Gamma' = \Gamma'_Z$ [cf. (**)] we get $\Gamma/\Delta = \Gamma_Z/\Delta_Z$. We show that $\Delta = \Delta_Z$; then we have $\Gamma = \Gamma_Z$. It is enough to show that $\Delta_Z \subset \Gamma$. A complete set of extensions of $\bar{v}$ to the residue field $\kappa_{w'}$ of w' is pairwise independent, hence a complete set of extensions of $\bar{v}$ to $\kappa_{v'}$ is also pairwise independent [cf. (6.6)]. Let $\gamma \in \Delta_Z$ be positive. By (4.5) there exists $\bar{x} \in \kappa_{v'_Z}$ such that $\bar{v}_Z(\bar{x}) = \gamma$ and that $\bar{\nu}(\bar{x}) = 0$ for all the other extensions $\bar{\nu}$ of $\bar{v}$ to $\kappa_{v'_Z}$. Let $x \in K_Z$ be an element having v'_Z-residue $\bar{x}$; then we have $v_Z(x) = \gamma$ and $\nu(x) = 0$ for all the other extensions ν of v to K_Z [cf. (3.27)]. Therefore we have, just as in (a), $v(N_{K_Z/K}(x)) = v_Z(x) = \gamma$, hence $\gamma \in \Gamma$.

8.3 The Ramification Group

(8.8) Proposition: Let $\psi: W \to \lambda$ be the canonical homomorphism. Then:
(1) For every $x \in L^\times$ and every $\sigma \in G_T$ the element $\sigma(x)/x$ is a unit of W.
(2) The map

$$(x, \sigma) \mapsto \langle x, \sigma \rangle := \psi\left(\frac{\sigma(x)}{x}\right) : L^\times \times G_T \to \lambda^\times$$

is a pairing of groups.
(3) For every $x \in W^\times$ and every $\sigma \in G_T$ we have $\langle x, \sigma \rangle = 1_\kappa$.

Proof: (1) For $\sigma \in G_T$ we have $w^{\sigma^{-1}} = w$ by (8.3)(5) [note that $G_T \subset G_Z$], hence $w(\sigma(x)) = w(x)$ for $x \in L^\times$, and therefore $\sigma(x)/x$ is a unit of W.
(3) Let $x \in W$ be a unit. We have $\sigma(x) - x =: y \in \mathfrak{n}$ [since $\sigma \in G_T$], hence $\sigma(x)/x = 1 + y/x \in 1 + \mathfrak{n}$, and therefore we have $\langle x, \sigma \rangle = \psi(1) = 1_\kappa$.
(2) Let $x, y \in L^\times$ and $\sigma \in G_T$; we have $\sigma(xy) = \sigma(x)\sigma(y)$, hence $\langle xy, \sigma \rangle = \langle x, \sigma \rangle \langle y, \sigma \rangle$.
Let $x \in L^\times$, $\sigma, \tau \in G_T$. We have

$$\frac{\sigma\tau(x)}{x} = \frac{\sigma(\tau(x))}{\sigma(x)} \frac{\sigma(x)}{x} = \sigma\left(\frac{\tau(x)}{x}\right) \frac{\sigma(x)}{x};$$

since $\sigma(\tau(x)/x)/(\tau(x)/x)$ is a unit of W by (1), we have $\psi(\sigma(\tau(x)/x)) = \psi(\tau(x)/x)$ by (3), hence $\langle x, \sigma\tau \rangle = \langle x, \sigma \rangle \langle x, \tau \rangle$.

(8.9) NOTATION: For every $x \in L^\times$ we have a homomorphism of groups

$$\sigma \mapsto \langle x, \sigma \rangle : G_T \to \lambda^\times;$$

let $\alpha: G_T \to \mathrm{Hom}(L^\times, \lambda^\times)$ with $\alpha(\sigma)(x) := \langle x, \sigma \rangle$ for every $x \in L^\times$, $\sigma \in G_T$, be the induced homomorphism. Likewise, for every $\sigma \in G_T$ we have a homomorphism of groups

$$x \mapsto \langle x, \sigma \rangle : L^\times \to \lambda^\times;$$

let $\beta: L^\times \to \mathrm{Hom}(G_T, \lambda^\times)$ with $\beta(x)(\sigma) = \langle x, \sigma \rangle$ for every $x \in L^\times$, $\sigma \in G_T$, be the induced homomorphism.

(8.10) THE KERNELS OF α AND β: (1) The kernel

$$G_V := \{\sigma \in G_T \mid \langle x, \sigma \rangle = 1 \text{ for all } x \in L^\times\}$$

of α is a normal subgroup of G_T; it is called the ramification group $G_V = G_V(W/V) = G_V(w/v)$ of W over V ["Verzweigungsgruppe" in german]. We have

$$G_V = \{\sigma \in G_T \mid w(\sigma(x) - x) > w(x) \text{ for every } x \in L^\times\}.$$

In fact, let $\sigma \in G_T$. For $x \in L^\times$ we have $\langle x, \sigma \rangle = 1$ iff $w(\sigma(x)/x - 1) > 0$, hence iff $w(\sigma(x) - x) > w(x)$.
(2) We have $L_T^\times \subset \mathrm{ker}(\beta)$ and $\{x \in L^\times \mid w(x) \in \Gamma\} \subset \mathrm{ker}(\beta)$.
In fact, let $x \in L_T^\times$; for every $\sigma \in G_T$ we have $\sigma(x) = x$, hence $\langle x, \sigma \rangle = 1$. Therefore we have $L_T^\times \subset \mathrm{ker}(\beta)$.
For every unit $x \in W$ we also have $x \in \mathrm{ker}(\beta)$ [cf. (8.8)(3)]. Let $x \in L^\times$ with $w(x) \in \Gamma$. We choose $y \in K^\times$ with $v(y) = w(x)$; then $x/y =: z$ is a unit of W. Since y and z lie in $\mathrm{ker}(\beta)$, also x lies in $\mathrm{ker}(\beta)$.

(8.11) ANOTHER PAIRING: For $\sigma \in G_T$ we have $\langle x, \sigma \rangle = 1$ if $x \in W$ is a unit [cf. (8.8)(3)]. Let $\delta \in \Delta$, and choose $x, y \in L^\times$ with $w(x) = w(y) = \delta$; then we have $\langle x, \sigma \rangle = \langle y, \sigma \rangle$, and by defining $\langle \delta, \sigma \rangle := \langle x, \sigma \rangle$, we have a pairing of groups

$$\Delta \times G_T \to \lambda^\times.$$

By (1) and (2) in (8.10) we get an induced pairing of groups

$$\Delta/\Gamma \times G_T/G_V \to \lambda^\times.$$

Let
$$\overline{\alpha}: G_T/G_V \to \mathrm{Hom}(\Delta/\Gamma, \lambda^\times), \quad \overline{\beta}: \Delta/\Gamma \to \mathrm{Hom}(G_T/G_V, \lambda^\times) \qquad (*)$$

be the induced homomorphisms of groups; $\overline{\alpha}$ is injective.

(8.12) CHARACTER GROUP OF A FINITE ABELIAN GROUP: Let $n \in \mathbb{N}$ and let H be a finite abelian group of order n.
(1) The abelian group $\widehat{H} := \mathrm{Hom}(H, \mathbb{Z}) = \mathrm{Hom}(H, \mathbb{Z}/\mathbb{Z}n)$ is called the group of characters of H. The canonical pairing

$$(h, \chi) \mapsto \chi(h) : H \times \widehat{H} \to \mathbb{Z}/\mathbb{Z}n$$

is faithful; in particular, H and $\widehat{H}$ are isomorphic [cf. [117], Ch. I, § 1, Th. 10].
(2) Let k be a field, and set $U_n(k) := \{x \in k^\times \mid x^n = 1\}$; $U_n(k) = \langle \zeta \rangle$ is a finite cyclic group of order m where $m \mid n$ [cf. [117], Ch. VIII, § 3], and there exists an injective homomorphism $U_n(k) \to \mathbb{Z}/\mathbb{Z}n$ by mapping ζ to the image of $n/m \in \mathbb{Z}$ in $\mathbb{Z}/\mathbb{Z}n$ which induces an injective homomorphism $\mathrm{Hom}(H, U_n(k)) \to \mathrm{Hom}(H, \mathbb{Z}/\mathbb{Z}n) = \widehat{H}$.
(3) Now we have $\mathrm{Hom}(H, k^\times) = \mathrm{Hom}(H, U_n(k))$, hence $\mathrm{Hom}(H, k^\times)$ can be considered as a subgroup of the group $\widehat{H}$ of characters of H. If k has positive characteristic p, and if $H = A \oplus B$ where B is the p-subgroup of H, then we have $\mathrm{Hom}(H, k^\times) = \mathrm{Hom}(A, k^\times)$ since $U_p(k) = \{1\}$.

(8.13) NOTATION: Δ/Γ is a finite abelian group of order e; we write

$$\Delta/\Gamma = A \oplus B$$

where B is the p-subgroup of Δ/Γ [$B = \{0\}$ if $p = 1$]. We set $\mathrm{ord}(A) =: e_0$ and $\mathrm{ord}(B) =: p^h$; then we have $e = e_0 p^h$, and $p \nmid e_0$ if $p > 1$. These notations shall be kept until the end of this section.

(8.14) NOTATION: We set $e_0' := (G_T : G_V) = [K_V : K_T]$. We show: If $p > 1$, then $p \nmid e_0'$.
In fact, the map $a \mapsto pa : A \to A$ is an isomorphism and $U_p(\lambda) = \{1\}$. Therefore p does not divide the order of $\mathrm{Hom}(A, \lambda^\times)$. Since $\overline{\alpha}: G_T/G_V \to \mathrm{Hom}(\Delta/\Gamma, \lambda^\times) = \mathrm{Hom}(A, \lambda^\times)$ is injective [cf. (8.11)], we get the assertion.

(8.15) **Proposition:** *The group G_V is a p-group.*

Proof: It is enough to show the following: If q is a prime number different from p (if $p = 1$, this means that q is any prime number), then there exists no element of G_V having order equal to q. Suppose that there exists $\sigma \in G_V$ with $\mathrm{ord}(\sigma) = q$. Let $M = \{x \in L \mid \sigma(x) = x\}$ be the fixed field of $\langle \sigma \rangle$; L/M is a cyclic Galois extension of degree q and $\mathrm{Gal}(L/M) = \langle \sigma \rangle$. Let $z \in L$ be a primitive element for L/M, and let $T^q + a_1 T^{q-1} + \cdots + a_q \in M[T]$ be the minimal polynomial of z over M. Since we can replace z by $z - a_1/q$ with $a_1 \in M$ [note that $q 1_M \neq 0$ since $q \neq p$], we may assume that z is a primitive element for L/M with $\mathrm{Tr}_{L/M}(z) = 0$. For $i \in \{0, \ldots, q-1\}$ we have $\psi(\sigma^i(z)/z) = 1_\kappa$ [since $\sigma^i \in G_V$], hence $\psi(\sigma^{q-1}(z)/z + \cdots + \sigma(z)/z + 1_\kappa) = q 1_\kappa \neq 0$, contradicting $\mathrm{Tr}_{L/M}(z) = \sigma^{q-1}(z) + \cdots + \sigma(z) + z = 0$.

(8.16) **Corollary:** *If κ has characteristic 0, then $G_V = \{1\}$, $\overline{\alpha}$ is an isomorphism, and the groups G_T and Δ/Γ are isomorphic. In particular, G_T is abelian, and we have $efg = n$.*

Proof: We have $G_V = \{1\}$ by (8.15), hence, by (8.11), G_T is isomorphic to a subgroup of $\mathrm{Hom}(\Delta/\Gamma, \lambda^\times) = \mathrm{Hom}(\Delta/\Gamma, U_e(\lambda^\times)) \subset \mathrm{Hom}(\Delta/\Gamma, \mathbb{Z}/\mathbb{Z}e)$ [cf. (8.12)(2)], hence to a subgroup of a group of order e [cf. (8.12)(1)], and therefore $\mathrm{ord}(G_T) \leq e$. On the other hand, we have $n = fg \, \mathrm{ord}(G_T) \leq efg$ by (8.3)(4), hence $n = efg$ [cf. (8.1)] and $\mathrm{ord}(G_T) = e$. This implies all the assertions.

(8.17) Lemma: *The homomorphism* $\overline{\beta}\colon \Delta/\Gamma \to \mathrm{Hom}(G_T/G_V, \lambda^\times)$ *has kernel B.*

Proof: Since G_V is a p-group by (8.15), we have $\mathrm{ord}(G_V) = p^m$ for some $m \in \mathbb{N}_0$. Let $\overline{\delta} \in \Delta/\Gamma$. If $\overline{\delta} \in B$, then $\overline{\beta}(\overline{\delta})$ is the neutral element of $\mathrm{Hom}(G_T/G_V, \lambda^\times)$ [since the order of $\overline{\beta}(\overline{\delta})$ is a power of p]. Conversely, if $\overline{\beta}(\overline{\delta})$ is the neutral element of $\mathrm{Hom}(G_T/G_V, \lambda^\times)$, then we show that $p^m\overline{\delta} = 0$, hence that $\overline{\delta} \in B$.
Thus, let $\delta \in \Delta$ be a preimage of $\overline{\delta}$; it is enough to show that $p^m\delta \in \Gamma$. We choose $x \in L^\times$ with $w(x) = \delta$; we have $\langle x, \sigma \rangle = 1$ for every $\sigma \in G_T$ [since $\overline{\beta}(\overline{\delta})$ is the neutral element of $\mathrm{Hom}(G_T/G_V, \lambda^\times)$]. For every $\sigma \in G_T$ there exists therefore an element $z_\sigma \in \mathfrak{n}$ with $\sigma(x) = x(1 + z_\sigma)$, hence we get

$$y := \mathrm{N}_{L/K_V}(x) = \prod_{\tau \in G_V} \tau(x) = x^{p^m}(1 + z) \quad \text{with } z \in \mathfrak{n},$$

hence $w(y) = p^m w(x)$.
Let $\sigma \in G_T$. Remember that G_V is a normal subgroup of G_T. Therefore, we have $\mathrm{N}_{L/K_V}(\sigma(x)) = \sigma(\mathrm{N}_{L/K_V}(x)) = \sigma(y)$. On the other hand, taking the norm to K_V on both sides of the equation $\sigma(x) = x(1 + z_\sigma)$ yields $\sigma(y) = y(1 + u_\sigma)$ with $u_\sigma \in \mathfrak{n}$. From this result we see: The e_0' conjugates $y_1 = y, y_2, \ldots, y_{e_0'}$ of $y \in K_V$ over K_T can be written in the form $y_i = y(1 + u_i)$ with $u_i \in \mathfrak{n}$ for $i \in \{1, \ldots, e_0'\}$, hence

$$\mathrm{Tr}_{K_V/K_T}(y) = y(e_0' + u) \quad \text{with } u := u_1 + \cdots + u_{e_0'} \in \mathfrak{n}.$$

If $p = 1$, then e_0' is a unit of W, and if $p > 1$, then p does not divide e_0' [cf. (8.14)], hence e_0' is also a unit of W, and therefore in both cases $e_0' + u$ is a unit of W. Thus, we have $p^m\delta = p^m w(x) = w(y) = w(\mathrm{Tr}_{K_V/K_T}(y)) \in \Gamma_T$, and since $\Gamma_T = \Gamma$ [cf. (8.7)], the assertion has been proved.

(8.18) From (8.17) we see that $\overline{\beta}|A\colon A \to \mathrm{Hom}(G_T/G_V, \lambda^\times)$ is injective, hence the pairing of groups

$$A \times G_T/G_V \to \lambda^\times \tag{$*$}$$

is faithful. In particular, G_T/G_V can be considered as a subgroup of the abelian group $\mathrm{Hom}(A, \lambda^\times)$, and therefore it is an abelian group. Since $\mathrm{ord}(A) = e_0$ and $\mathrm{ord}(G_T/G_V) = e_0'$, we may replace $\lambda^\times$ in $(*)$ either by $U_{e_0'}(\lambda)$ or by $U_{e_0}(\lambda)$. Both groups are cyclic; from [117], Ch. I, § 1, Th. 10, we now get:

(8.19) Theorem: *The groups A and G_T/G_V are isomorphic, hence, in particular, G_T/G_V is abelian, and we have $e_0' = e_0$.*

(8.20) Corollary: *We have $n = efgp^t$ with $t \in \mathbb{N}_0$.*

Proof: We have $n = (G : G_Z)(G_Z : G_T)(G_T : G_V)(G_V : (1)) = gf_0e_0'p^m = g(fp^{-s})(ep^{-h})p^m = efgp^{m-h-s}$ [cf. (8.3)(4) and (8.13)]. Since $efg \leq n$ [cf. (8.1)], we have $t := m - h - s \geq 0$, and the assertion follows.

9 Extending Valuations to Non-Algebraic Overfields

(9.1) THE GAUSS VALUATION: Let K be a field, let v be a valuation of K with value group Γ, and set $L := K(x_1, \ldots, x_r)$ where $x_1, \ldots, x_r$ are algebraically independent over K. Let Δ be a totally ordered abelian group which contains Γ as an ordered subgroup, and let $(\delta_1, \ldots, \delta_r) \in \Delta^r$. We define a map $w \colon K[x_1, \ldots, x_r] \to \Delta$ by

$$w\left(\sum a_j x^j\right) = \min(\{v(a_j) + j_1\delta_1 + \cdots + j_r\delta_r \mid j = (j_1, \ldots, j_r) \in \mathbb{N}_0^r\})$$

for every polynomial $\sum a_j x^j \in K[x_1, \ldots, x_r]$ [we set $x^j = x_1^{j_1} \cdots x_r^{j_r}$ for $j = (j_1, \ldots, j_r) \in \mathbb{N}_0^r$]. Then it is easy to check that for $f, g \in K[x_1, \ldots, x_r]$ we have

$$w(fg) = w(f) + w(g), \quad w(f + g) \geq \min(\{w(f), w(g)\}).$$

The unique extension of w to a valuation of L [cf. (3.20)] shall be again denoted by w. Note that $w(L^\times) = \Gamma + \mathbb{Z}\delta_1 + \cdots + \mathbb{Z}\delta_r$. The valuation w shall be called the Gauß valuation of L defined by v and the r-tuple $(\delta_1, \ldots, \delta_r) \in \Delta^r$.

(9.2) Proposition: Let K be a field, and let v be a valuation of K with value group Γ and residue field κ. Let $L := K(x)$ be a purely transcendental extension of K of transcendence degree 1. Then:
(1) Let Δ be a totally ordered abelian group containing Γ as an ordered subgroup, and let $\delta \in \Delta$ be such that the image of δ in Δ/Γ is not a torsion element. Then there exists a unique valuation w of L extending v with values in Δ and with $w(x) = \delta$. We have $w(L^\times) = \Gamma \oplus \mathbb{Z}\delta$ and $\kappa_w = \kappa$.
(2) There exists a unique valuation w of L extending v with $w(x) = 0$ and such that the image $\bar{x}$ of x in the residue field κ_w of w is transcendental over κ. We have $w(L^\times) = \Gamma$ and $\kappa_w = \kappa(\bar{x})$.

Proof: (1) Let ω be any valuation of L which extends v, and with $\omega(x) = \delta$. For every polynomial $a_0 + a_1 x + \cdots + a_n x^n \in K[x]$ we have

$$\omega(a_0 + a_1 x + \cdots + a_n x^n) \geq \min(\{v(a_j) + j\delta \mid j \in \{0, \ldots, n\}\}). \qquad (*)$$

The assumption on δ implies that those of the values $v(a_j) + j\delta$, $j \in \{0, \ldots, n\}$, which are finite, are pairwise different; therefore we have equality in $(*)$ [cf. (2.10)], hence $\omega = w$ where w is the Gauß valuation of L defined by v and δ. Let $f = a_0 + a_1 x + \cdots + a_n x^n \in K[x]$ be a non-zero polynomial; as we just showed, there exists $i \in \{0, \ldots, n\}$ with

$$w(a_i x^i) < w(a_j x^j) \quad \text{for } j \in \{0, \ldots, n\}, \, j \neq i.$$

Then we have

$$f = a_i x^i (1 + r) \quad \text{with } r := \frac{1}{a_i x^i} \sum_{j \neq i} a_j x^j \in K(x) \text{ and } w(r) > 0.$$

Therefore every non-zero element $h \in L = K(x)$ can be written in the form $h = x^n a(1 + r_1)/(1 + r_2)$ with $n \in \mathbb{Z}$, $a \in K^\times$, r_1, $r_2 \in K(x)$ and $w(r_1) > 0$, $w(r_2) > 0$. From this we see that $w(h) = 0$ iff $n\delta + v(a) = 0$, hence iff $n = 0$ and $v(a) = 0$ [since the image of δ in Δ/Γ is not a torsion element]. Therefore κ is the residue field of w, and $w(L^\times) = \Gamma \oplus \mathbb{Z}\delta$.

(2) Let ω be a valuation of L which extends v, and with $\omega(x) = 0$. For every polynomial $a_0 + a_1 x + \cdots + a_n x^n \in K[x]$ we have

$$\omega(a_0 + a_1 x + \cdots + a_n x^n) \geq \min(\{v(a_j) \mid j \in \{0, \ldots, n\}\}). \qquad (**)$$

We show that in $(**)$ we have equality iff the image $\overline{x}$ of x in the residue field of ω is transcendental over κ. To prove this, it is enough to consider polynomials $a_0 + a_1 x + \cdots + a_n x^n$ which satisfy $\min(\{v(a_j) \mid j \in \{0, \ldots, n\}\}) = 0$ [multiply an arbitrary non-zero polynomial $a_0 + a_1 x + \cdots + a_n x^n \in K[x]$ by the inverse $1/b$ of an element $b \in K^\times$ which satisfies $v(b) = \min(\{v(a_j) \mid j \in \{0, \ldots, n\}\})$]. Let $f = a_0 + a_1 x + \cdots + a_n x^n \in K[x]$ with $0 = \min(\{v(a_j) \mid j \in \{0, \ldots, n\}\})$ and with $\omega(f) > 0$. Then we get $\overline{a}_0 + \overline{a}_1 \overline{x} + \cdots + \overline{a}_n \overline{x}^n = 0$ where $\overline{a}_j$ is the image of a_j in κ, $j \in \{0, \ldots, n\}$. This shows that we have equality in $(**)$ iff $\overline{x}$ is transcendental over κ, and in this case we have $\omega = w$ where w is the Gauß valuation of L defined by v and 0. Conversely, the Gauß valuation of L defined by v and 0 has Γ as value group and $\kappa(\overline{x})$ as residue field.

(9.3) Proposition: *Let K be a field, let v be a valuation of K, and let Γ_v be the value group of v. Let L be an overfield of K, let w be a valuation of L which extends v, let W be the valuation ring and Γ_w be the value group of w. Let r, $s \in \mathbb{N}_0$, and assume that there exist elements $x_1, \ldots, x_r$ of W such that the images $\overline{x}_1, \ldots, \overline{x}_r$ in the residue field κ_w of w are algebraically independent over the residue field κ_v of v, and elements $y_1, \ldots, y_s$ of L such that the images of the elements $w(y_1), \ldots, w(y_s)$ in Γ_w/Γ_v are free over $\mathbb{Z}$. Then the $r + s$ elements $x_1, \ldots, x_r, y_1, \ldots, y_s$ of L are algebraically independent over K, and setting $L' := K(x_1, \ldots, x_r, y_1, \ldots, y_s)$ and $w' := w|L'$, we have $\Gamma_{w'} = w(L'^\times) = \Gamma_v \oplus (\mathbb{Z}w(y_1) + \cdots + \mathbb{Z}w(y_s))$, and $\kappa_v(\overline{x}_1, \ldots, \overline{x}_r)$ is the residue field $\kappa_{w'}$ of w', hence a purely transcendental extension of κ_v of transcendence degree r.*
Moreover, for every $f = \sum c_{ij} x^i y^j \in K[x_1, \ldots, x_r, y_1, \ldots, y_s]$ we have

$$w'(f) = \min(\{v(c_{ij}) + j_1 w(y_1) + \cdots + j_s w(y_s) \mid i \in \mathbb{N}_0^r, j \in \mathbb{N}_0^s\}).$$

[Here $x^i y^j := x_1^{i_1} \cdots x_r^{i_r} y_1^{j_1} \cdots y_s^{j_s}$ for $i = (i_1, \ldots, i_r) \in \mathbb{N}_0^r$, $j = (j_1, \ldots, j_s) \in \mathbb{N}_0^s$.]
Proof: We prove the theorem by induction on $r + s$. There is nothing to show if $r + s = 0$. Let $r + s > 0$, and assume that the theorem holds for all pairs (r', s') with $0 \leq r' \leq r$, $0 \leq s' \leq s$ and $r' + s' < r + s$. We set $K' := K(x_1, \ldots, x_{r'}, y_1, \ldots, y_{s'})$ and $v' := w|K'$; note that now $\Gamma_{v'} = \Gamma_v \oplus (\mathbb{Z}w(y_1) + \cdots + \mathbb{Z}w(y_{s'}))$, and $\kappa_{v'} = \kappa_v(\overline{x}_1, \ldots, \overline{x}_{r'})$. Replacing K by K' and v by v', it is therefore enough to prove the theorem for the two pairs $(r, s) = (1, 0)$ and $(r, s) = (0, 1)$.

(a) Let $r = 1$, $s = 0$: In this case we have $L = K(x)$, and the image of x in the residue field of w is transcendental over the residue field of v. Therefore, by (6.11), the element x is transcendental over K.

(b) Let $r = 0$ and $s = 1$: In this case we have $L = K(y)$, and the image of $w(y)$ in $w(L^\times)/v(K^\times)$ is not a torsion element. Therefore, by (6.11), the element y is transcendental over K.

In both cases the result follows immediately from (9.2).

The case of a general field extension L/K is covered by the following theorem.

(9.4) Theorem: *Let K be a field, let v be a valuation of K, let Γ_v be the value group and κ_v the residue field of v. Let L be an overfield of K, let w be a valuation of L which extends v, let Γ_w be the value group and κ_w be the residue field of w. Then:*

(1) We have

$$\operatorname{tr.d}_{\kappa_v}(\kappa_w) + \operatorname{rat.rank}(\Gamma_w/\Gamma_v) \leq \operatorname{tr.d}_K(L). \tag{*}$$

If, moreover, L is a finitely generated extension of K, and if we have equality in (*), then Γ_w/Γ_v is a finitely generated $\mathbb{Z}$-module, and κ_w is a finitely generated extension of κ_v.

(2) We have

$$\operatorname{tr.d}_{\kappa_v}(\kappa_w) + \operatorname{rank}(\Gamma_w) \leq \operatorname{tr.d}_K(L) + \operatorname{rank}(\Gamma_v). \tag{**}$$

If, moreover, L is a finitely generated extension of K, if Γ_v is discrete of finite rank and if we have equality in (**), then Γ_w is discrete of rank $\operatorname{rank}(\Gamma_v) + \operatorname{rat.rank}(\Gamma_w/\Gamma_v)$, and κ_w is a finitely generated extension of κ_v.

Proof: (1) Let $r, s \in \mathbb{N}_0$ with $r \leq \operatorname{tr.d}_{\kappa_v}(\kappa_w)$ and $s \leq \operatorname{rat.rank}(\Gamma_w/\Gamma_v)$; by (9.3) we get $r + s \leq \operatorname{tr.d}_K(L)$. Therefore (*) has been proved. Now we assume that L is a finitely generated extension of K—hence $d := \operatorname{tr.d}_K(L)$ is finite—and that we have equality in (*). Setting $r := \operatorname{tr.d}_{\kappa_v}(\kappa_w)$ and $s := \operatorname{rat.rank}(\Gamma_w/\Gamma_v)$, we have $r + s = d$. Then there exist elements $x_1, \ldots, x_r$ in the valuation ring of w such that their images $\overline{x}_1, \ldots, \overline{x}_r$ in κ_w are algebraically independent over κ_v, and elements $y_1, \ldots, y_s$ in $L^\times$ such that the images of the elements $w(y_1), \ldots, w(y_s)$ in Γ_w/Γ_v are free over $\mathbb{Z}$. We set $L' := K(x_1, \ldots, x_r, y_1, \ldots, y_s)$ and $w' := w|L'$. By (9.3) we have $\kappa_{w'} = \kappa_v(\overline{x}_1, \ldots, \overline{x}_r)$, $\Gamma_{w'} = \Gamma_v \oplus (\mathbb{Z}w(y_1) + \cdots + \mathbb{Z}w(y_s))$, and the $r + s = d$ elements $x_1, \ldots, x_r, y_1, \ldots, y_s$ are algebraically independent over K. Therefore $\{x_1, \ldots, x_r, y_1, \ldots, y_s\}$ is a transcendence basis of L over K, hence L, being a finitely generated extension of K, is a finite extension of L'. This implies, by (6.10), that κ_w is a finite extension of $\kappa_{w'}$ and that $\Gamma_w/\Gamma_{w'}$ is a finite group. Therefore κ_w is a finitely generated extension of κ_v. The $\mathbb{Z}$-module $\Gamma_{w'}/\Gamma_v$ is finitely generated, and from the exact sequence

$$0 \to \Gamma_{w'}/\Gamma_v \to \Gamma_w/\Gamma_v \to \Gamma_w/\Gamma_{w'} \to 0$$

we see that Γ_w/Γ_v is a finitely generated $\mathbb{Z}$-module.

(2) By B(1.30)(2) we have

$$\text{rank}(\Gamma_w) \leq \text{rank}(\Gamma_v) + \text{rat.} \, \text{rank}(\Gamma_w/\Gamma_v), \qquad (\dagger)$$

and by (1) above we have

$$\text{tr.} \, d_{\kappa_v}(\kappa_w) + \text{rat.} \, \text{rank}(\Gamma_w/\Gamma_v) \leq \text{tr.} \, d_K(L), \qquad (\dagger\dagger)$$

whence $(**)$ holds [this is clear if Γ_w/Γ_v has finite rational rank; in the other case $\text{tr.} \, d_K(L)$ is not finite, and therefore $(**)$ holds trivially]. Now we assume that L is a finitely generated extension of K, hence that $\text{tr.} \, d_K(L)$ is finite, and that Γ_v is discrete of rank h—hence that $\text{rank}(\Gamma_v) = \text{rat.} \, \text{rank}(\Gamma_v) = h$ is finite [cf. B(1.25)]. By $(\dagger\dagger)$ we get that $\text{rat.} \, \text{rank}(\Gamma_w/\Gamma_v)$ is finite. Furthermore, assume that we have equality in $(**)$. Then we must have also equality in $(\dagger)$ and $(\dagger\dagger)$, and this means that, in particular, we have equality in $(*)$, hence, by (1), that κ_w is a finitely generated extension of κ_v, and that Γ_w/Γ_v is a finitely generated $\mathbb{Z}$-module. We set $r := \text{tr.} \, d_{\kappa_v}(\kappa_w)$ and $s := \text{rat.} \, \text{rank}(\Gamma_w/\Gamma_v)$. Then there exists a transcendence basis $\{x_1, \ldots, x_r, y_1, \ldots, y_s\}$ of L over K such that, setting $L' := K(x_1, \ldots, x_r, y_1, \ldots, y_s)$ and $w' := w|L'$, we have $w'(L'^\times) = \Gamma_{w'} = \Gamma_v \oplus (\mathbb{Z}w(y_1) + \cdots + \mathbb{Z}w(y_s))$, and $\{w(y_1), \ldots, w(y_s)\}$ is a $\mathbb{Z}$-free set [cf. proof of (1)]. From $\text{rank}(\Gamma_w) = \text{rank}(\Gamma_v) + \text{rat.} \, \text{rank}(\Gamma_w/\Gamma_v)$ [cf. $(\dagger)$] and $\text{rat.} \, \text{rank}(\Gamma_w/\Gamma_v) = \text{rat.} \, \text{rank}(\Gamma_w) - \text{rat.} \, \text{rank}(\Gamma_v)$ [cf. B(1.30)] we get $\text{rank}(\Gamma_w) = \text{rat.} \, \text{rank}(\Gamma_w)$, and since L is a finite extension of L', we have $\text{rank}(\Gamma_w) = \text{rank}(\Gamma_{w'})$ and $\text{rat.} \, \text{rank}(\Gamma_w) = \text{rat.} \, \text{rank}(\Gamma_{w'})$ by (6.11). Moreover, we have $\text{rat.} \, \text{rank}(\Gamma_w/\Gamma_v) = \text{rat.} \, \text{rank}(\Gamma_{w'}/\Gamma_v)$ by B(1.30), since $\Gamma_{w'}/\Gamma_w$ is a finite group by (6.10). Now $\Gamma_{w'}$ is a finitely generated free $\mathbb{Z}$-module, and we have $\text{rat.} \, \text{rank}(\Gamma_{w'}) = \text{rat.} \, \text{rank}(\Gamma_v) + \text{rat.} \, \text{rank}(\Gamma_{w'}/\Gamma_v) = \text{rank}(\Gamma_v) + \text{rat.} \, \text{rank}(\Gamma_w/\Gamma_v) = \text{rank}(\Gamma_w) = \text{rank}(\Gamma_{w'})$, hence $\text{rat.} \, \text{rank}(\Gamma_{w'}) = \text{rank}(\Gamma_{w'})$, and therefore $\Gamma_{w'}$ is discrete of rank $h + \text{rat.} \, \text{rank}(\Gamma_w/\Gamma_v)$ [cf. B(1.39)], and the same is true for Γ_w [cf. (6.11)].

10　Valuations of Algebraic Function Fields

(10.0) In this section k is an arbitrary field.

(10.1) DEFINITION: A finitely generated extension field K of k is called an algebraic function field over k. We set $r := \text{tr.} \, d_k(K)$; then K is said to be an algebraic function field in r variables over k.

For the theory of algebraic function fields of one variable we refer to the books of Chevalley [42] and Eichler [62].

(10.2) REMARK: In the following, when we consider a valuation v of an algebraic function field K over k, we tacitly assume that $k \subset A_v$, i.e., that $v|k$ is the trivial valuation; then k can be considered as a subfield of the residue field κ_v of v. The transcendence degree $\text{tr.} \, d_k(\kappa_v)$ is called the dimension $\dim(v)$ of v.

(10.3) Proposition: *Let $K = k(x)$ be the field of rational functions over k. Then:*

(1) *Let $\mathbb{P}$ be a set of representatives for the irreducible polynomials of $k[x]$. For every $p \in \mathbb{P}$ the local ring $V_p := k[x]_{(p)}$ is a discrete valuation ring. Let v_p be the valuation of K defined by v_p. The residue field of v_p is the field $k[x]/(p)$, hence an extension of k of degree $\deg(p)$. Let $f \in K^\times$, and let $f = \gamma \prod_{q \in \mathbb{P}} q^{a_q}$ with $\gamma \in k^\times$ and $a_q \in \mathbb{Z}$ for every $q \in \mathbb{P}$ be the factorization of f; then $v_p(f) = a_p$. If $p \neq p'$ are in $\mathbb{P}$, then v_p and $v_{p'}$ are not equivalent.*

(2) *We set $V_\infty := k[1/x]_{(1/x)}$; then V_∞ is a discrete valuation ring. Let v_∞ be the valuation of K defined by V_∞. The residue field of v_∞ is k. Let $f \in K^\times$, and write $f = g/h$ where $g, h \in k[x]$. Then $v_\infty(f) = \deg(h) - \deg(g)$. For any $p \in \mathbb{P}$ the valuations v_p and v_∞ are not equivalent.*

(3) *Every non-trivial valuation v of the function field K is either equivalent to v_p for some $p \in \mathbb{P}$ or equivalent to v_∞.*

Proof: $k[x]$ is a principal ideal domain, hence $\{V_p \mid p \in \mathbb{P}\}$ is the set of all valuation rings of K containing $k[x]$ [cf. (3.30)(5)], and $V_p/V_pp = k[x]/(p)$. Since p generates the maximal ideal of V_p, it is clear that $v_p(f) = a_p$. We have $v_p(p) = 1 \neq 0 = v_{p'}(p)$, hence v_p and $v_{p'}$ are not equivalent.

(2) The ring V_∞ is a discrete valuation ring by (1), and its residue field is k. Set $m := \deg(g)$, $n := \deg(h)$, and write $g = a_0 + a_1 x + \cdots + a_m x^m = x^m(a_m + a_{m-1}x^{-1} + \cdots + a_0 x^{-m})$, $h = b_0 + b_1 + \cdots + b_n x^n = x^{-n}(b_n + b_{n-1}x^{-1} + \cdots + b_0 x^{-n})$; by (1) we have $v_\infty(g) = -m$, $v_\infty(h) = -n$, hence $v_\infty(f) = n - m$. Let $p \in \mathbb{P}$. We have $v_p(p) = 1 \neq -\deg(p) = v_\infty(p)$, hence v_p and v_∞ are not equivalent.

(3) If the ring A_v of v contains $k[x]$, then we have $A_v = V_p$ for some $p \in \mathbb{P}$ [cf. (3.30)(5)]. If the ring A_v of v does not contain $k[x]$, then $v(x) < 0$, $A_v \supset k[1/x]$, and the center of v in $k[1/x]$ is the prime ideal generated by $1/x$. Just as before we get that v and v_∞ are equivalent.

(10.4) Corollary: *Let K be a function field in one variable over k. There exist non-trivial valuations of K, every such valuation of K is discrete of rank 1, and its residue field is a finite extension of k.*

Proof: We chose $x \in K$ which is transcendental over k. Then K is a finite extension of $k(x)$, and the result follows from (10.3) and (6.10).

(10.5) REMARK: In (10.4) we have classified the valuations of a field of algebraic functions in one variable. We shall classify valuations of a field of algebraic functions in two variables in section 5 of chapter VIII.

Applying the results of (9.4) we get

(10.6) Theorem: *Let K be an algebraic function field in r variables over k, and let v be a valuation of K. Then:*

(1) *We have*

$$\dim(v) + \mathrm{rat.\,rank}(v) \leq r; \tag{$*$}$$

if we have equality in $(*)$, *then* κ_v *is an algebraic function field in* $r - \mathrm{rat.\,rank}(\Gamma_v)$ *variables over* k, *and* Γ_v *is a finitely generated free* $\mathbb{Z}$-*module, minimally generated by* $r - \dim(v)$ *elements.*

(2) *We have*

$$\dim(v) + \mathrm{rank}(\Gamma_v) \leq r; \tag{$**$}$$

if we have equality in $(**)$, *then* $\mathrm{rat.\,rank}(v) = \mathrm{rank}(v)$, v *is discrete, and* κ_v *is an algebraic function field in* $r - \mathrm{rank}(v)$ *variables over* k.

(10.7) Proposition: *Let* K *be an algebraic function field in* $r \geq 1$ *variables over* k, *and let* v *be a valuation of* K *with* $\dim(v) = r-1$. *Then* v *is a discrete valuation of rank 1, and* κ_v *is an algebraic function field in* $r - 1$ *variables over* k. *Moreover, the valuation ring* A_v *of* v *is a localization with respect to a prime ideal of height 1 of an integrally closed* k-*subalgebra* R *of* K *of finite type with* $Q(R) = K$.

Proof: The first assertions follow from (10.6). We prove the last assertion. We choose elements $x_1, \ldots, x_{r-1} \in A_v$ whose v-residues in κ_v are algebraically independent over k. Then $x_1, \ldots, x_{r-1}$ are algebraically independent over k by (9.3). Let $x_r \in K$ be such that $\{x_1, \ldots, x_r\}$ is a transcendence basis of K over k. This property holds also for $\{x_1, \ldots, x_{r-1}, 1/x_r\}$, hence we may assume that $x_r \in A_v$. We set $R' := k[x_1, \ldots, x_r]$, and we let R be the integral closure of R' in K which is a k-algebra of finite type [cf. B(3.6)] contained in A_v [cf. (3.4)] and with $Q(R) = K$. Note that K is a finite extension of $Q(R')$. Set $\mathfrak{p}' := \mathfrak{p}_v \cap R'$, $\mathfrak{p} := \mathfrak{p}_v \cap R$, and $v' = v|k(x_1, \ldots, x_r)$. The images in $R'/\mathfrak{p}'$ of the elements $x_1, \ldots, x_{r-1}$ are algebraically independent over k, and since $\mathrm{tr.\,d}_k(\kappa_{v'}) = \mathrm{tr.\,d}_k(\kappa_v)$ [cf. (6.10)], we have $\mathrm{tr.\,d}_k(Q(R'/\mathfrak{p}')) = r - 1$, hence $\mathfrak{p}'$ is a prime ideal of R' of height 1 [cf. [63], Cor. 13.4]. Therefore $\mathfrak{p}$ is a prime ideal of R of height 1 [cf. [63], Prop. 9.2]. Now $R_{\mathfrak{p}}$ is a discrete valuation ring of K [cf. B(10.5)] contained in A_v, hence we have $R_{\mathfrak{p}} = A_v$ [cf. (2.15)].

(10.8) Corollary: *Let* K *be an algebraic function field in* $r \geq 1$ *variables over* k, *let* R *be subring of* K *which has* K *as field of quotients and is a* k-*algebra of finite type, and let* $\mathfrak{p}$ *be a prime ideal of* R *of height 1. The set of pairwise non-equivalent* $(r-1)$-*dimensional valuations of* K *which have center* $\mathfrak{p}$ *in* R *is finite and not empty. For every such valuation, its residue field is an algebraic function field in* $r - 1$ *variables over* k, *and it is a finite extension of the quotient field of* $R/\mathfrak{p}$.

Proof: Let $R = k[t_1, \ldots, t_n]$. We have $\dim(R/\mathfrak{p}) = \mathrm{tr.\,d}_k(Q(R/\mathfrak{p})) = r - 1$ by [63], Cor. 13.4; we label the elements $t_1, \ldots, t_n$ in such a way that the images of $t_1, \ldots, t_{r-1}$ in $R/\mathfrak{p}$ are algebraically independent over k. Then the elements $t_1, \ldots, t_{r-1}$ are algebraically independent over k [cf. (9.3)], and we may assume that $\{t_1, \ldots, t_r\}$ is a transcendence basis of K over k. Set $R_0 := k[t_1, \ldots, t_r]$, a polynomial ring, and $\mathfrak{p}_0 := \mathfrak{p} \cap R_0$. We have $r - 1 = \operatorname{tr.d}_k(Q(R/\mathfrak{p})) \geq \operatorname{tr.d}_k(Q(R_0/\mathfrak{p}_0)) \geq r - 1$ [since the images of $t_1, \ldots, t_{r-1}$ are algebraically independent over k], hence $\dim(R_0/\mathfrak{p}_0) = r - 1$ and therefore $\operatorname{ht}(\mathfrak{p}_0) = 1$ [cf. [63], Cor. 13.4]. Now R_0 is integrally closed in $k(t_1, \ldots, t_r)$; therefore $V := (R_0)_{\mathfrak{p}_0}$ is a discrete valuation ring of $k(t_1, \ldots, t_r)$ [cf. B(10.5)]. Any $(r - 1)$-dimensional valuation of K is discrete of rank 1 [cf. (10.7)], and if it has center $\mathfrak{p}$ in R, it is an extension of V. Now K is a finite extension of $k(t_1, \ldots, t_r)$, hence the set of such valuations is finite and not empty [cf. (7.4)]. Furthermore, the residue field of such a valuation of K is an algebraic function field in $r - 1$ variables over k [cf. (10.6)], and since $\operatorname{tr.d}_k(Q(R/\mathfrak{p})) = r - 1$, the last assertion also has been proved.

(10.9) Proposition: *Let K be an algebraic function field in $r \geq 1$ variables over k, and let R be a subring of K which has K as field of quotients and is a k-algebra of finite type. Then:*
(1) *If $\mathfrak{m}$ is a maximal ideal of R, then there exists a zero-dimensional valuation v of K with $R \subset A_v$ and with $\mathfrak{m} = R \cap \mathfrak{p}_v$.*
(2) *Let v be a zero-dimensional valuation of K with $R \subset A_v$. Then $\mathfrak{m} := \mathfrak{p}_v \cap R$ is a maximal ideal of R.*

Proof: (1) The assertion in case $r = 1$ follows from (10.8) [since $\dim(R) = 1$ by [63], Th. A on p. 286]. Let $r > 1$, and assume that the assertion is true for algebraic function fields in $r - 1$ variables over k. Let $\mathfrak{p} \subset \mathfrak{m}$ be a prime ideal of R of height 1, and let w be an $(r - 1)$-dimensional valuation of K having center $\mathfrak{p}$ in R [cf. (10.8)]; then κ_w is an algebraic function field in $r - 1$ variables over k, and it is a finite extension of the field of quotients of $R/\mathfrak{p}$. There exists a subring S of κ_w containing $R/\mathfrak{p}$ which is a finitely generated $R/\mathfrak{p}$-module and has κ_w as field of quotients [obtained by adjoining to $R/\mathfrak{p}$ a basis of κ_w over $Q(R/\mathfrak{p})$ consisting of elements which are integral over $R/\mathfrak{p}$]. Let $\mathfrak{n}$ be a maximal ideal of S lying over $\mathfrak{m}/\mathfrak{p}$ [cf. [63], Prop. 4.15 and Cor. 4.17]. By induction, there exists a zero-dimensional valuation $\overline{v}$ of κ_w having center $\mathfrak{n}$ in S. Set $v := w \circ \overline{v}$; then v is zero-dimensional and has center $\mathfrak{m}$ in R [cf. (3.12) and (3.26)].
(2) We have $\dim(R/\mathfrak{m}) = \operatorname{tr.d}_k(Q(R/\mathfrak{m}))$ and $\operatorname{tr.d}_k(\kappa_v) = 0$, hence $\dim(R/\mathfrak{m}) = 0$, and therefore $\mathfrak{m}$ is a maximal ideal of R [cf. [63], Cor. 9.1].

11   Valuations Dominating a Local Domain

(11.0) NOTATION: In this section R is a local domain with maximal ideal $\mathfrak{m}$, field of quotients K and residue field $\kappa := R/\mathfrak{m}$. Let v be a valuation of K having center $\mathfrak{m}$ in R, let V be the valuation ring of v, let $\Gamma := v(K^\times)$ be the value-group of v, and let κ_v be the residue field of V. We assume in this section that $V \neq K$. Since V dominates R, we can consider κ_v as an overfield of κ; we define $\operatorname{tr.d}(v) := \operatorname{tr.d}_\kappa(\kappa_v)$. The set $\Sigma := v(R \setminus \{0\})$ is a subsemigroup of Γ, and Γ is generated by Σ [since K is the field of quotients of R]. The element

$\alpha := v(\mathfrak{m}) := \min(\{v(x) \mid x \in \mathfrak{m}\})$ is the smallest positive element in Σ [since $\mathfrak{m}$ is finitely generated, $v(\mathfrak{m})$ is well-defined].

(11.1) Proposition: *The semigroup Σ is a well-ordered set.*

Proof: We have to show that every non-empty subset S of Σ has a smallest element. We set $S' := v^{-1}(S) \cap R$; let $\mathfrak{a}$ be the ideal of R generated by S'. Let $\{x_1, \ldots, x_n\} \subset S'$ be a system of generators of $\mathfrak{a}$, and set $\gamma := \min(\{v(x_1), \ldots, v(x_n)\})$. Then we have $\gamma \le v(x)$ for every $x \in \mathfrak{a}$ [since v is non-negative on R], and therefore $\gamma \in S$ is the smallest element of S [since $S \subset \{v(x) \mid x \in \mathfrak{a}\}$].

(11.2) REMARK: For every $\gamma \in \Sigma$ we define ideals $\mathfrak{a}_\gamma$ and $\mathfrak{a}_\gamma^+$ of R by setting

$$\mathfrak{a}_\gamma := \{x \in R \mid v(x) \ge \gamma\}, \; \mathfrak{a}_\gamma^+ := \{x \in R \mid v(x) > \gamma\}.$$

Clearly $\mathfrak{a}_\gamma$ and $\mathfrak{a}_\gamma^+$ are v-ideals [cf. B(6.11) for the definition of a v-ideal]. (Moreover, for every v-ideal $\mathfrak{a}$ of R, $\mathfrak{a} \ne \{0\}$, we have $\mathfrak{a} = \mathfrak{a}_{v(\mathfrak{a})}$.) The set $\{\delta \in \Sigma \mid \delta > \gamma\}$ is not empty, hence has a smallest element γ^+ by (11.1), and therefore we have $\mathfrak{a}_\gamma^+ = \mathfrak{a}_{\gamma^+}$. Note that $\mathfrak{a}_\gamma^+ \subsetneqq \mathfrak{a}_\gamma$ since there exist elements $x \in R$ with $v(x) = \gamma$.

(11.3) THE ASSOCIATED GRADED RING: We consider the graded abelian group

$$\mathrm{gr}_v(R) := \bigoplus_{\gamma \in \Sigma} \mathfrak{a}_\gamma / \mathfrak{a}_\gamma^+.$$

Let $\gamma, \delta \in \Sigma$; then we have $\mathfrak{a}_\gamma \mathfrak{a}_\delta \subset \mathfrak{a}_{\gamma+\delta}$, $\mathfrak{a}_\gamma^+ \mathfrak{a}_\delta^+ \subset \mathfrak{a}_{\gamma+\delta}^+$. Let $x \in \mathfrak{a}_\gamma$, $y \in \mathfrak{a}_\delta$; we define

$$(x + \mathfrak{a}_\gamma^+)(y + \mathfrak{a}_\delta^+) := xy + \mathfrak{a}_{\gamma+\delta}^+;$$

it is easily seen that this gives a well-defined product of two homogeneous elements of $\mathrm{gr}_v(R)$, and we define a product of two arbitrary elements of $\mathrm{gr}_v(R)$ by using the distributive law. Therefore $\mathrm{gr}_v(R)$ with this multiplication is a Γ-graded ring of type Σ.

(11.4) DEFINITION: The semigroup Σ is called archimedean if, for any γ and $\delta \in \Sigma$ with $\gamma \ne 0$, there exists an integer n with $n\gamma > \delta$.

(11.5) REMARK: Note that Σ is archimedean if Γ is archimedean, but, in general, the condition for Σ to be archimedean is weaker, as the following example shows.

(11.6) EXAMPLE: Let R be a two-dimensional regular local ring with maximal ideal $\mathfrak{m} = Rx + Ry$, field of quotients K and residue field κ. Then $\mathrm{gr}_\mathfrak{m}(R) = \kappa[\overline{x}, \overline{y}]$ [we use the notations introduced in VII(2.0)]. Let $p \in \mathbb{P}_R$ be the homogeneous prime ideal in $\mathrm{gr}_\mathfrak{m}[\overline{x}, \overline{y}]$ which is generated by $\overline{y}$, set $\Gamma := \mathbb{Z} \oplus \mathbb{Z}$, ordered lexicographically, and let $\nu_p \colon K^\times \to \Gamma$ be the valuation of K as defined in VII(7.5). The value group of ν_p is Γ, and we have $\nu_p(x) = (1,0)$, $\nu_p(y) = (1,1)$, hence $\Sigma := \nu_p(R \setminus \{0\}) = \{(m,n) \in \mathbb{Z}^2 \mid m \ge n \ge 0\}$. Clearly Σ is archimedean, but Γ is not archimedean [cf. B(1.19)].

(11.7) Proposition: *The following statements are equivalent:*
(1) Σ *is an archimedean semigroup.*
(2) *For every isolated subgroup Δ of Γ we have $\Sigma \cap \Delta = \{0\}$.*
(3) *Every non-zero v-ideal of R contains a power of $\mathfrak{m}$.*

Proof (1) $\Rightarrow$ (2): Let Δ be an isolated subgroup of Γ and suppose that $\Delta \cap \Sigma \neq \{0\}$. Let γ be the smallest element of the non-empty set $(\Delta \cap \Sigma) \setminus \{0\}$ [cf. (11.1)]. Let $\delta \in \Sigma$; then there exists $n \in \mathbb{Z}$ with $n\gamma > \delta$, hence $\delta \in \Delta$ [since Δ is a segment] which implies that $\Sigma \subset \Delta$, hence that $\Delta = \Gamma$ [since Γ is generated by Σ], in contradiction with $\Delta \neq \Gamma$. Therefore we have $\Gamma \cap \Sigma = \{0\}$.
(2) $\Rightarrow$ (1): Let $\gamma \in \Sigma$ be a non-zero element. The subset Δ of Γ consisting of all elements δ of Γ such that there exists an integer n (depending on δ) with $n\gamma > \delta$ clearly is a segment; since the sum of two positive elements of Δ lies in Δ, Δ is a subgroup of Γ [cf. B(1.5)(3)]. This subgroup must be Γ, since otherwise it would be an isolated subgroup of Γ having non-zero intersection with Σ. Therefore, for $\delta \in \Sigma$, there exists $n \in \mathbb{Z}$ with $n\gamma > \delta$, hence Σ is archimedean.
(1) $\Leftrightarrow$ (3): Assume that Σ is archimedean. Let $\mathfrak{a}$ be a non-zero v-ideal of R, set $\beta := v(\mathfrak{a})$, and choose $n \in \mathbb{N}$ with $n\alpha > \beta$. Then we have $\mathfrak{m}^n \subset \mathfrak{a}$. Conversely, assume that every non-zero v-ideal of R contains a power of $\mathfrak{m}$. Let $0 \neq \gamma, \delta \in \Sigma$. We have $\alpha \leq \gamma$. We choose $n \in \mathbb{N}$ with $\mathfrak{m}^n \subsetneqq \mathfrak{a}_\delta$; then we have $n\gamma \geq n\alpha > \delta$ [since $\mathfrak{a}_\delta$ is a v-ideal by (11.3)].

(11.8) Proposition: *We assume that Σ is archimedean. Let $p \in \mathbb{N}$. For every $n \in \mathbb{N}$ with $n \leq \mathrm{rat.\,rank}(v)$ there exists $m \in \mathbb{N}$ such that setting $\Sigma_p := \{\beta \in \Sigma \mid \beta < pm\alpha\}$, we have*

$$\ell_R(R/\mathfrak{m}^{pm}) \geq \ell_R(R/\mathfrak{a}_{pm\alpha}) \geq \mathrm{Card}(\Sigma_p) \geq \binom{p+n}{n}.$$

Proof: Let $n \in \mathbb{N}$ with $n \leq \mathrm{rat.\,rank}(v)$. Since the image of Σ in $\Gamma \otimes_{\mathbb{Z}} \mathbb{Q}$ generates this $\mathbb{Q}$-vector space, there exist elements $\gamma_1, \ldots, \gamma_n \in \Sigma$ which generate a free $\mathbb{Z}$-submodule Γ' of Γ of rank n. Since Σ is archimedean, there exists $m \in \mathbb{N}$ with $\gamma_i < m\alpha$ for every $i \in \{1, \ldots, n\}$. We have $\mathfrak{m}^{pm} \subset \mathfrak{a}_{pm\alpha}$. Furthermore, setting

$$\Gamma'_p := \{l_1\gamma_1 + \cdots + l_n\gamma_n \mid l = (l_1, \ldots, l_n) \in \mathbb{N}_0^n, |l| := l_1 + \cdots + l_n \leq p\},$$

we have $\Gamma'_p \subset \Sigma_p$ and $\mathrm{Card}(\Gamma'_p) = \binom{p+n}{n}$, and therefore we have [cf. (11.3)]

$$\ell(R/\mathfrak{m}^{pm}) \geq \ell_R(R/\mathfrak{a}_{pm\alpha}) = \sum_{\beta \in \Sigma_p} \ell_R(\mathfrak{a}_\beta / \mathfrak{a}_\beta^+) \geq \mathrm{Card}(\Sigma_p) \geq \mathrm{Card}(\Gamma'_p) = \binom{p+n}{n}.$$

(11.9) Theorem: [S. S. Abhyankar] *Let R be a local domain, and let v be a valuation of the field of quotients of R which dominates R. We set $d := \mathrm{rat.\,rank}(v)$. Then:*

(1) *We have the inequality*

$$\mathrm{rat.\,rank}(v) + \mathrm{tr.\,d}(v) \leq \dim(R). \tag{$*$}$$

(2) *If in $(*)$ we have equality, then the group Γ is isomorphic to $\mathbb{Z}^d$, and κ_v is a finitely generated extension of κ.*
(3) *If $\mathrm{rank}(v) + \mathrm{tr.\,d}(v) = \dim(R)$, then we have $\mathrm{rank}(v) = \mathrm{rat.\,rank}(v)$, equality holds in $(*)$, and Γ is discrete of rank d.*

Before proving Abhyankar's theorem, we need three auxiliary results.

(11.10) Lemma: *Let A be a noetherian integral domain, and let x be an element of the field of quotients of A. Then we have*

$$\dim(A[\,x\,]) \leq \dim(A). \tag{$*$}$$

Proof: We write $x = a/b$ with a, $b \in A$, $b \neq 0$. The surjective A-algebra homomorphism $A[\,T\,] \to A[\,x\,]$ mapping T to x has non-zero kernel $\mathfrak{p}$ since $bT - a$ is mapped to 0, and we have $A[\,x\,] \cong A[\,T\,]/\mathfrak{p}$. Now $(*)$ is trivially true if the Krull dimension of A is not finite. In the other case we have

$$\dim(A[\,x\,]) = \dim(A[\,T\,]/\mathfrak{p}) \leq \dim(A[\,T\,]) - 1 = \dim(A)$$

[since $\mathfrak{p} \neq (0)$ and $\dim(A[\,T\,]) = \dim(A) + 1$, cf. [63], Cor. 10.13].

(11.11) Lemma: *Let A be a local domain with maximal ideal $\mathfrak{m}$, let x be an element of the field of quotients of A, let $\mathfrak{q}$ be a prime ideal of $A[\,x\,]$ with $\mathfrak{q} \cap A = \mathfrak{m}$, and set $B := A[\,x\,]_{\mathfrak{q}}$, $\mathfrak{n} := \mathfrak{q}B$. The local ring B dominates A; if the image $\overline{x} \in B/\mathfrak{n}$ of x is transcendental over $A/\mathfrak{m}$, then we have*

$$\dim(B) \leq \dim(A[\,x\,]) - 1.$$

Proof: It is enough to show that $\mathfrak{q}$ is not a maximal ideal of $A[\,x\,]$. If $\mathfrak{q}$ were a maximal ideal of $A[\,x\,]$, then $A[\,x\,]/\mathfrak{q} = \kappa[\overline{x}]$ would be a field, hence $\overline{x}$ would be algebraic over κ, in contradiction with our choice of x.

(11.12) Lemma: *Let R and V be as in (11.0). Then $\mathrm{tr.\,d}(v) =: h$ is finite, and if $h \geq 1$, then there exists a local domain R_h which dominates R and is dominated by V such that $\dim(R) \geq \dim(R_h) + h$, that κ_v is algebraic over the residue field κ_h of R_h, and that κ_h is a purely transcendental extension of κ with $\mathrm{tr.\,d}_\kappa(\kappa_h) = h$.*

Proof: There is nothing to show if $\mathrm{tr.\,d}(v) = 0$. Now let κ_v be transcendental over κ, and choose an element $x \in V$ whose image in κ_v is transcendental over κ; we have $R[\,x\,] \subset V$. Let $\mathfrak{q}$ be the center of V in $R[\,x\,]$, and set $R_1 := R[\,x\,]_{\mathfrak{q}}$, $\mathfrak{m}_1 := \mathfrak{q}R_1$. Then we have by (11.10) and (11.11)

$$\dim(R_1) \leq \dim(R[\,x\,]) - 1 \leq \dim(R) - 1.$$

Note that R_1 dominates R, that V dominates R_1, and setting $\kappa_1 := R_1/\mathfrak{m}_1$, that we have $\kappa_1 = \kappa(\overline{x})$, hence that $\mathrm{tr.}\,\mathrm{d}_\kappa(\kappa_1) = 1$. If κ_v is transcendental over κ_1, we can repeat this construction. Thus, we get a finite sequence $R =: R_0 \subset R_1 \subset \cdots \subset R_h$ of local domains with strictly decreasing dimensions which are dominated by V [with $h \geq 1$], and denoting by κ_i the residue field of R_i, $i \in \{1,\dots,h\}$, we have $\kappa =: \kappa_0 \subset \kappa_1 \subset \cdots \subset \kappa_h \subset \kappa_v$, κ_i/κ_{i-1} is a purely transcendental extension of transcendence degree 1 for $i \in \{1,\dots,h\}$, hence κ_h/κ is a purely transcendental extension of transcendence degree h, and κ_v is algebraic over κ_h. Clearly we have $\dim(R_h) \leq \dim(R) - h$.

Proof of Abhyankar's theorem:

(1)(a) We show: If Σ is archimedean and $\mathrm{tr.}\,\mathrm{d}(v) = 0$, then $\mathrm{rat.}\,\mathrm{rank}(v) \leq \dim(R)$. It is enough to show the following: For every $n \in \mathbb{N}_0$ with $n \leq \mathrm{rat.}\,\mathrm{rank}(v)$ we have $n \leq \dim(R)$.
Thus, let $n \in \mathbb{N}_0$ with $n \leq \mathrm{rat.}\,\mathrm{rank}(v)$. By (11.8) there exists $m \in \mathbb{N}$ such that

$$\ell_R(R/\mathfrak{m}^{pm}) \geq \binom{p+n}{n} \quad \text{for every } p \in \mathbb{N}.$$

Now $\ell_R(R/\mathfrak{m}^{pm})$ is for all large p a polynomial in p with rational coefficients of degree $\dim(R)$ [cf. [63], Th. 12.4], and therefore we have $\dim(R) \geq n$.
(b) We show that

$$\mathrm{rat.}\,\mathrm{rank}(v) + \mathrm{tr.}\,\mathrm{d}(v) \leq \dim(R). \tag{$*$}$$

We prove this by induction on $\dim(R)$. Let us consider the case $\dim(R) = 1$. Then Σ is archimedean and $\mathrm{tr.}\,\mathrm{d}(v) = 0$. In fact, suppose that Σ is not archimedean. Then Γ has an isolated subgroup Δ with $\Delta \cap \Sigma \neq \{0\}$ [cf. (11.7)]. Let $\mathfrak{q} := \{x \in V \mid v(x) \notin \Delta\}$ be the prime ideal of V corresponding to Δ [cf. (3.23)(2)]. Then $\mathfrak{p} := \mathfrak{q} \cap R$ is different from $\{0\}$ [since $\mathfrak{q} \neq \{0\}$]. There exists $x \in R$ with $v(x) \in \Sigma \cap \Delta$, $v(x) \neq 0$, hence $x \in \mathfrak{m}$ and $x \notin \mathfrak{p}$, which means that $\mathfrak{p} \neq \mathfrak{m}$, contradicting $\dim(R) = 1$. Moreover, using (11.12), we have $\mathrm{tr.}\,\mathrm{d}(v) = 0$. From (a) we see that $(*)$ holds for R.
Now let $n > 1$, and assume that $(*)$ holds for all pairs (R', v') where R' is a local domain, v' is a non-trivial valuation of the field of quotients of R' dominating R', and $\dim(R') < n$.
Let $\dim(R) = n$. If $h := \mathrm{tr.}\,\mathrm{d}(v) \geq 1$, then we can apply our induction assumption to (R_h, v) [cf. (11.12)], hence $\mathrm{rat.}\,\mathrm{rank}(v) \leq \dim(R_h) \leq \dim(R) - h$, and therefore we get $\mathrm{rat.}\,\mathrm{rank}(v) + \mathrm{tr.}\,\mathrm{d}(v) \leq \dim(R)$. If $h = 0$ and Σ is archimedean, then the result follows from (a). If $h = 0$ and Σ is not archimedean, then we can choose an isolated subgroup Δ of Γ with $\Delta \cap \Sigma \neq \{0\}$; we define $\mathfrak{q}$ and $\mathfrak{p}$ as above, i.e., $\mathfrak{q}$ is the prime ideal of V corresponding to Δ and $\mathfrak{p} := \mathfrak{q} \cap R$. Then we have $\{0\} \subsetneq \mathfrak{p} \subsetneq \mathfrak{m}$. In particular, the local rings $S := R_\mathfrak{p}$ and $\overline{R} := R/\mathfrak{p}$ have dimension strictly less than n. We set $W := V_\mathfrak{q}$, $\overline{V} := V/\mathfrak{q}$. Now W is a valuation ring of K with maximal ideal $\mathfrak{q}$ [cf. (3.8)] and $v(W^\times) = \Delta$ [cf. (3.23)], and $\overline{V}$ is a valuation ring of the residue field $\overline{K} = W/\mathfrak{q}$ of W [cf. (3.26)]. The valuation v is composite

with valuations w and $\overline{v}$ where w is a valuation of K defined by the valuation ring W, and $\overline{v}$ is a valuation of $\overline{K}$ defined by the valuation ring $\overline{V}$ [cf. (3.27)]. We may assume that w is a valuation with value-group Γ/Δ, and that $\overline{v}$ is a valuation with value-group Δ [cf. (3.26)]. Note that W dominates S. By induction, we have

$$\operatorname{rat.rank}(\Gamma/\Delta) + \operatorname{tr.d}(w) \leq \dim(S) \tag{i}$$

where $\operatorname{tr.d}(w)$ is the transcendence degree of the residue field $\overline{K}$ of W over the residue field of S.

Let $\widetilde{K}$ be the field of quotients of $\overline{R}$; note that $\widetilde{K}$ is a subfield of $\overline{K}$. Let $\widetilde{v}$ be the restriction of $\overline{v}$ to $\widetilde{K}$; $\widetilde{v}$ dominates $\overline{R}$. Note that $\widetilde{K}$ also is the residue field of S, hence that

$$\operatorname{tr.d}(w) = \operatorname{tr.d}_{\widetilde{K}}(\overline{K}).$$

The value group $\widetilde{\Delta}$ of $\widetilde{v}$ is a subgroup of Δ, and the residue field of $\overline{R}$ is κ; by induction, we have

$$\operatorname{rat.rank}(\widetilde{\Delta}) + \operatorname{tr.d}(\widetilde{v}) \leq \dim(\overline{R}) \tag{ii}$$

where $\operatorname{tr.d}(\widetilde{v})$ is the transcendence degree of the residue field of $\widetilde{v}$ over κ. Furthermore, by (9.4) we have

$$\operatorname{rat.rank}(\Delta/\widetilde{\Delta}) + \operatorname{tr.d}_{\kappa_{\widetilde{v}}}(\kappa_{\overline{v}}) \leq \operatorname{tr.d}_{\widetilde{K}}(\overline{K}). \tag{iii}$$

Adding inequalities (ii) and (iii) we get, using B(1.30)(1),

$$\operatorname{rat.rank}(\Delta) + \operatorname{tr.d}(v) \leq \operatorname{tr.d}_{\widetilde{K}}(\overline{K}) + \dim(\overline{R}),$$

and adding (i) to this inequality, using (1.30)(1) and that $\operatorname{tr.d}_{\widetilde{K}}(\overline{K}) = \operatorname{tr.d}(w)$, we get

$$\operatorname{rat.rank}(\Gamma) + \operatorname{tr.d}(v) \leq \dim(S) + \dim(\overline{R}) \leq \dim(R).$$

Therefore we have proved the inequality $(*)$ in Abhyankar's theorem.
(2) We now assume that

$$\operatorname{rat.rank}(v) + \operatorname{tr.d}(v) = \dim(R). \tag{iv}$$

(a) We show: If Σ is archimedean and $\operatorname{rat.rank}(v) = \dim(R) =: n$, then Γ is isomorphic to $\mathbb{Z}^n$ and κ_v is a finite extension of κ.
We choose elements $\gamma_1, \ldots, \gamma_n \in \Sigma$ which generate a free $\mathbb{Z}$-submodule Γ' of Γ of rank n, and we define m and for $p \in \mathbb{N}$ the sets Σ_p and Γ'_p as in the proof of (11.8). There exists $q \in \mathbb{N}$ with

$$\ell_R(R/\mathfrak{m}^{2pm}) \leq q \binom{p+n}{n} \quad \text{for all } p \in \mathbb{N}$$

[since $\ell_R(R/\mathfrak{m}^{2pm})$ is for all large enough p a polynomial in p with rational coefficients of degree $n = \dim(R)$, cf. [63], Th. 12.4].
(α) We show that $\Gamma \cong \mathbb{Z}^n$ [as a group].

It is enough to show that Γ/Γ' is a finite group, since then Γ is a finitely generated $\mathbb{Z}$-module, hence we get $\Gamma \cong \mathbb{Z}^n$ by (3.16).

Since Γ/Γ' is a torsion group, for every $\beta \in \Sigma$ there exists $h_\beta \in \mathbb{N}$ with $h_\beta\beta \in \Gamma'$, hence $-\beta \equiv (h_\beta - 1)\beta \pmod{\Gamma'}$; since Γ is generated by Σ, it follows that for every $\gamma \in \Gamma$ there exists $\beta \in \Sigma$ with $\gamma \equiv \beta \pmod{\Gamma'}$. Therefore there exists a system of representatives for Γ/Γ' consisting of elements of Σ, hence we can write

$$\Gamma = \bigcup_{i \in I}(\beta_i + \Gamma') \quad \text{with } \beta_i \in \Sigma \text{ for every } i \in I$$

[disjoint union]. Let $J \subset I$ be a finite subset. Then there exists $p^* \in \mathbb{N}$ with $\beta_j < p^* m\alpha$ for every $j \in J$. Let $j \in J$; for every $p \geq p^*$ we have $\beta_j + \Gamma'_p \subset \Sigma_{2p}$, and therefore [cf. (11.8)]

$$\text{Card}(J)\binom{p+n}{n} \leq \text{Card}(\Sigma_{2p}) \leq \ell_R(R/\mathfrak{m}^{2pm}) \leq q\binom{p+n}{n} \quad \text{for all } p \geq p^*,$$

hence we have $\text{Card}(J) \leq q$. Therefore I is a finite set.

(β) We show that κ_v is a finite extension of κ.

First, we show that

$$\ell_R(\mathfrak{a}_\beta/\mathfrak{a}_\beta^+) \leq q \quad \text{for every } \beta \in \Sigma. \tag{$*$}$$

In fact, suppose that there exists $\gamma \in \Sigma$ with $\ell_R(\mathfrak{a}_\gamma/\mathfrak{a}_\gamma^+) \geq q + 1$. We choose elements $y_1, \ldots, y_n \in R$ with $v(y_i) = \gamma_i$ for $i \in \{1, \ldots, n\}$. Let $l := (l_1, \ldots, l_n) \in \mathbb{N}_0^n$, and set $\beta := \gamma + l_1 v(y_1) + \cdots + l_n v(y_n)$. Then $\mathfrak{a}_\beta = y^l \mathfrak{a}_\gamma$, $\mathfrak{a}_\beta^+ = y^l \mathfrak{a}_\gamma^+$, and the R-modules $\mathfrak{a}_\beta/\mathfrak{a}_\beta^+$ and $\mathfrak{a}_\gamma/\mathfrak{a}_\gamma^+$ are isomorphic. Therefore we have $\ell_R(\mathfrak{a}_\beta/\mathfrak{a}_\beta^+) \geq q+1$ for every $\beta \in \gamma + \Gamma'_+$. We choose $p^* \in \mathbb{N}$ with $\gamma < p^* m\alpha$. For every $p \geq p^*$ we have $\gamma + \Gamma'_p \subset \Sigma_{2p}$, hence

$$q\binom{p+n}{n} \geq \ell_R(R/\mathfrak{m}^{2pm}) \geq \ell_R(R/\mathfrak{a}_{2pm\alpha}) = \sum_{\beta \in \Sigma_{2p}} \ell_R(\mathfrak{a}_\beta/\mathfrak{a}_\beta^+)$$

$$\geq \sum_{\beta \in \Gamma'_p} \ell_R(\mathfrak{a}_{\beta+\gamma}/\mathfrak{a}_{\beta+\gamma}^+) \geq (q+1)\binom{p+n}{n},$$

hence $q \geq q + 1$ which is absurd. Therefore ($*$) holds.

Second, we show that κ_v is a finite extension of κ. Let $t \leq [\kappa_v : \kappa]$ be a positive integer, and choose elements $x_1, \ldots, x_t$ in V such that their images $\overline{x}_1, \ldots, \overline{x}_t$ in κ_v are linearly independent over κ. For $i \in \{1, \ldots, t\}$ let us write $x_i = a_i/b$ with a_i and $b \in R$, $b \neq 0$. We set $\beta := v(b)$. The elements $a_1 = x_1 b, \ldots, a_t = x_t b$ lie in $\mathfrak{a}_\beta$ [since $v(x_1) = \cdots = v(x_t) = 0$]. Let $\overline{a}_1, \ldots, \overline{a}_t$ be the images of these elements in the κ-vector space $\mathfrak{a}_\beta/\mathfrak{a}_\beta^+$ [note $\mathfrak{a}_\beta/\mathfrak{a}_\beta^+$ is annihilated by $\mathfrak{m}$]. These elements are linearly independent over κ. In fact, suppose that there exist elements $u_1, \ldots, u_t \in R$ with $v(u_1 a_1 + \cdots + u_t a_t) > \beta$; then we have $v(u_1 x_1 + \cdots + u_t x_t) > 0$, hence $\overline{u}_1 \overline{x}_1 + \cdots + \overline{u}_x \overline{x}_t = 0$, and therefore $\overline{u}_1 = \cdots = \overline{u}_t = 0$ [where $\overline{u}_i$ is the image of u_i in κ, $i \in \{1, \ldots, t\}$]. This means that $t \leq \ell_R(\mathfrak{a}_\beta / \mathfrak{a}_\beta^+)$, hence that $t \leq q$ by what we have shown above, and therefore we find that $[\kappa_v : \kappa] \leq q$.

(b) We prove the assertions in (2) by induction on $\dim(R)$.

If $\dim(R) = 1$, then Σ is archimedean and $\mathrm{tr. d}(v) = 0$ [cf. the claim in (b) of part (1)], hence the assertions in (2) follow from (a).

Now let $n > 1$, and assume that the assertions in (2) hold for all pairs (R', v') where R' is a local domain, v' is a valuation of the field of quotients of R' dominating R', $1 \leq \dim(R') < n$ and $\mathrm{rat. rank}(v') + \mathrm{tr. d}(v') = \dim(R')$.

Let $\dim(R) = n$; by assumption, we have $\mathrm{rat. rank}(v) + \mathrm{tr. d}(v) = \dim(R)$, i.e., (iv) holds for (R, v). First, we consider the case $h := \mathrm{tr. d}(v) \geq 1$. Then we can apply (11.12); for the ring R_h we find, using $(*)$ in (1), that

$$\dim(R_h) \geq \mathrm{rat. rank}(v) = \mathrm{rat. rank}(v) + \mathrm{tr. d}_\kappa(\kappa_v) - h$$
$$= \dim(R) - h \geq \dim(R_h),$$

hence $\dim(R_h) = \mathrm{rat. rank}(v)$, and therefore $\mathrm{tr. d}_{\kappa_h}(\kappa) = 0$ by $(*)$ in (1), applied to R_h. Therefore we have $\dim(R_h) = \mathrm{rat. rank}(v) + \mathrm{tr. d}(v)$. Now we can apply our induction assumption to (R_h, v) since $\dim(R_h) < \dim(R) = n$: we get that κ_v is a finite extension of κ_h [since it is a finitely generated extension by induction], hence that κ_v is a finitely generated extension of κ, and that $\Gamma \cong \mathbb{Z}^d$. Second, we consider the case $h = 0$. If Σ is archimedean, then the result follows from (a). If Σ is not archimedean, then we use the notations in part (b) of the proof of (1). If one of the inequalities (i), (ii) and (iii) would be a strict inequality, then, by the proof of $(*)$ in part (b) of (1), we would get strict inequality in $(*)$. Therefore (i), (ii) and (iii) are equalities. By induction, using equality in (i) and (ii), we get that Γ/Δ and $\widetilde{\Delta}$ are finitely generated $\mathbb{Z}$-modules, and that $\overline{K}/\widetilde{K}$ and $\kappa_{\widetilde{v}}/\kappa$ are finitely generated extensions. Furthermore, by equality in (iii), we get from (9.4)(1), using the fact that $\overline{K}/\widetilde{K}$ is a finitely generated extension, that the extension $\kappa_{\overline{v}}/\kappa_{\widetilde{v}}$ is finitely generated and that $\Delta/\widetilde{\Delta}$ is a finitely generated $\mathbb{Z}$-module. Since $\Delta/\widetilde{\Delta}$ and $\widetilde{\Delta}$ are finitely generated $\mathbb{Z}$-modules, Δ is also a finitely generated $\mathbb{Z}$-module, and since Γ/Δ and Δ are finitely generated $\mathbb{Z}$-modules, Γ is a finitely generated $\mathbb{Z}$-module, and therefore also a free $\mathbb{Z}$-module of finite rank d [cf. (3.16)].

This ends the proof of the assertions in (2) of Abhyankar's theorem.

(3) Now the assumption is

$$\mathrm{rank}(v) + \mathrm{tr. d}(v) = \dim(R).$$

From $\mathrm{rank}(v) \leq \mathrm{rat. rank}(v)$ [cf. B(1.30)(2)] we get $\mathrm{rank}(v) = \mathrm{rat. rank}(v)$ by (1), and $\mathrm{rat. rank}(v) + \mathrm{tr. d}(v) = \dim(R)$, hence Γ and $\mathbb{Z}^d$ are isomorphic by (2) [as groups]. From B(1.39) it follows that Γ is discrete of rank d.

Chapter II

One-Dimensional Semilocal Cohen-Macaulay Rings

INTRODUCTION: The local ring of a point on a curve is a one-dimensional local Cohen-Macaulay ring; in this chapter we study this class of rings. After proving some results on transversal elements in section 1, our main interest in section 2 is the integral closure of a one-dimensional local Cohen-Macaulay ring; we use Manis valuations in describing the integral closure. In section 3 we give necessary and sufficient conditions in order to ensure that the completion of a one-dimensional local Cohen-Macaulay ring which is a domain (resp. has no nilpotent elements) again is a domain (resp. has no nilpotent elements). Here the reader is supposed to be acquainted with the notion of the completion of a local ring and its properties.

In section 4 we introduce in the context of this chapter that technique which shall be used over and over again: Blowing up. After defining the notion of the blow-up of a one-dimensional semilocal Cohen-Macaulay ring, we show that by a sequence of blow-ups we can reach the integral closure of such a ring, and prove some interesting results on stable ideals. The notion of infinitely near rings, which shall play a decisive role in section 1 of chapter VIII, is introduced in section 5, and some simple properties are proved.

1 Transversal Elements

1.1 Adic topologies

(1.1) In this subsection we collect for easy reference some results on adic topologies.

(1.2) ADIC TOPOLOGIES: Let A be a ring, let $\mathfrak{a}$ be an ideal of A, and let M be an A-module.

(1) The $\mathfrak{a}$-adic topology on A is the topology such that, for every $a \in A$, the family $(a + \mathfrak{a}^n)_{n \in \mathbb{N}_0}$ is a basis for the neighborhoods of a. With this topology, A becomes a topological ring. Note that A is a Hausdorff space in the $\mathfrak{a}$-adic topology iff $\bigcap_{n=0}^{\infty} \mathfrak{a}^n = \{0\}$.

(2) The $\mathfrak{a}M$-adic topology on M is the topology such that, for every $x \in M$, the family $(x + \mathfrak{a}^n M)_{n \in \mathbb{N}_0}$ is a basis for the neighborhoods of x. With this topology, M becomes a topological module over the topological ring A. Note that M is a Hausdorff space in the $\mathfrak{a}M$-adic topology iff $\bigcap_{n=0}^{\infty} \mathfrak{a}^n M = \{0\}$. A submodule N of M is open iff there exists $n \in \mathbb{N}$ with $\mathfrak{a}^n M \subset N$.

(1.3) Lemma: *Let A be a ring, $\mathfrak{a}$ an ideal of A and M an A-module, endowed with the $\mathfrak{a}M$-adic topology. Then:*

(1) *For every subset $X \subset M$ its closure $\overline{X}$ is equal to $\bigcap_{n=0}^{\infty}(X + \mathfrak{a}^n M)$.*

(2) *Every open submodule of M is closed.*

Proof: (1) Let $x \in \overline{X}$ and $n \in \mathbb{N}_0$; there exists an element $x_n \in X$ such that $x_n \in x + \mathfrak{a}^n M$, hence we have $x \in x_n + \mathfrak{a}^n M \subset X + \mathfrak{a}^n M$, hence $x \in \bigcap_{n=0}^{\infty}(X + \mathfrak{a}^n M)$. Conversely, if $x \in \bigcap_{n=0}^{\infty}(X + \mathfrak{a}^n M)$, then there exists, for every $n \in \mathbb{N}_0$, an element x_n of X such that $x \in x_n + \mathfrak{a}^n M$, whence $x_n \in x + \mathfrak{a}^n M$, and therefore $x \in \overline{X}$.

(2) If N is an open submodule of M, then we have $\mathfrak{a}^n M \subset N$ for some $n \in \mathbb{N}$.

(1.4) Let A be a noetherian ring, $\mathfrak{a}$ an ideal of A which is contained in the Jacobson radical of A, M a finitely generated A-module and N a submodule of M. Then $\bigcap_{n=0}^{\infty}(N + \mathfrak{a}^n M) = N$ since we have $\bigcap_{n=0}^{\infty} \mathfrak{a}^n(M/N) = \{0\}$ by Krull's intersection theorem [cf. [63], Cor. 5.4], and therefore N is closed in the $\mathfrak{a}M$-adic topology of M [cf. (1.3)].

(1.5) NOTATION: In this chapter we are mainly interested in one-dimensional noetherian Cohen-Macaulay rings (CM-rings). The general definition [cf., e.g., [63], definition before Prop. 18.8] says in this case: A one-dimensional noetherian ring R is a CM-ring iff every maximal ideal of R contains a regular element of R.

(1.6) OPEN IDEALS: Let R be a semilocal ring of dimension d, and let $\mathfrak{r}$ be the Jacobson radical of R. R always will be endowed with the $\mathfrak{r}$-adic topology which is called the natural topology of R. Thus, an ideal $\mathfrak{a}$ of R is open iff it contains a power of $\mathfrak{r}$. Note that, for every $n \in \mathbb{N}_0$, $R/\mathfrak{r}^n$ is an artinian ring [cf. [63], Cor. 9.1].

(1) An ideal $\mathfrak{a}$ of R is open iff $R/\mathfrak{a}$ is an R-module of finite length.

In fact, if $\mathfrak{a}$ is an open ideal of R, then $R/\mathfrak{a}$ is an artinian ring [because $R/\mathfrak{a}$ is a homomorphic image of $R/\mathfrak{r}^n$ for some $n \in \mathbb{N}$]. Conversely, assume that $\mathfrak{a}$ is an ideal of R such that $R/\mathfrak{a}$ is an artinian ring. Then $\mathfrak{r}^n + \mathfrak{a} = \mathfrak{r}^{n+1} + \mathfrak{a}$ for all sufficiently large integers n; on the other hand, $\mathfrak{a}$ is a closed set of R since R is semilocal [cf. (1.4)], and is therefore the intersection of the ideals $\mathfrak{r}^m + \mathfrak{a}$, $m \in \mathbb{N}$ [cf. (1.3)]. This implies that $\mathfrak{a}$ contains a power of $\mathfrak{r}$.

(2) Now assume that R is a one-dimensional semilocal CM-ring. An ideal $\mathfrak{a}$ of R is open iff it contains a regular element.

Indeed, if $\mathfrak{a}$ is open, then $\mathfrak{a}$ contains a power of $\mathfrak{r}$, hence contains a regular element of R [since R is a CM-ring]. Conversely, let $r \in \mathfrak{a}$ be a regular element. It is enough to consider the case where r is a non-unit. Let $Rr = \mathfrak{q}_1 \cap \cdots \cap \mathfrak{q}_k$ be an irredundant primary decomposition of Rr. Then $\mathfrak{n}_1 := \operatorname{rad}(\mathfrak{q}_1), \ldots, \mathfrak{n}_k := \operatorname{rad}(\mathfrak{q}_k)$ are maximal ideals of R, and Rr contains a power of $\operatorname{rad}(Rr) = \mathfrak{n}_1 \cap \cdots \cap \mathfrak{n}_k = \mathfrak{n}_1 \cdots \mathfrak{n}_k$ [cf. B(10.1) and [63], Ex. 1.13], hence a fortiori a power of $\mathfrak{r}$. Since $Rr \subset \mathfrak{a}$, we have proved the assertion.

1.2 The Hilbert Polynomial

(1.7) NOTATION: Let $S := \bigoplus_{n \geq 0} S_n$ be a graded ring; assume that S_0 is an artinian ring, and that S as S_0-algebra is generated by finitely many elements in S_1. Let $M = \bigoplus_{n \in \mathbb{Z}} M_n$ be a finitely generated graded S-module. Then, for every $n \in \mathbb{Z}$, M_n is a finitely generated S_0-module [cf. B(4.32)], hence has finite length. By [63], Prop. 12.2, there exists a uniquely determined polynomial $P(M) \in \mathbb{Q}[X]$ such that

$$\ell_{S_0}(M_n) = P(M)(n) \quad \text{for all sufficiently large integers } n;$$

$P(M)$ is called the Hilbert polynomial of M.

(1.8) HILBERT POLYNOMIAL AND MULTIPLICITY: Let R be a semilocal ring of dimension d, let $\mathfrak{r}$ be the Jacobson radical of R, and let $\{\mathfrak{m}_1, \ldots, \mathfrak{m}_h\}$ be the set of maximal ideals of R. For every $i \in \{1, \ldots, h\}$ set $R_i := R_{\mathfrak{m}_i}$; then we have $\dim(R_i) \leq d$ with equality for at least one $i \in \{1, \ldots, h\}$.

Let $\mathfrak{a}$ be an open ideal of R. For every $n \in \mathbb{N}_0$ the R-module $\mathfrak{a}^n / \mathfrak{a}^{n+1}$ has finite length [cf. (1.6)(1)]. We set

$$H(\mathfrak{a}, R)(n) := \ell_R(R/\mathfrak{a}^n) = \sum_{\nu=0}^{n-1} \ell_R(\mathfrak{a}^\nu/\mathfrak{a}^{\nu+1}) \quad \text{for every } n \in \mathbb{N}_0.$$

The function $H(\mathfrak{a}, R) \colon \mathbb{N}_0 \to \mathbb{N}_0$ is called the Hilbert function of $\mathfrak{a}$.

(1) *Let $\mathfrak{a}$ be a proper open ideal of R. Then there exists a non-zero polynomial $P(\mathfrak{a}, R) \in \mathbb{Q}[X]$ of degree $\max(\{\operatorname{ht}(\mathfrak{m}_i) \mid i \in \{1, \ldots, h\}, \mathfrak{a} \subset \mathfrak{m}_i\})$ such that*

$$H(\mathfrak{a}, R)(n) = P(\mathfrak{a}, R)(n) \quad \text{for all sufficiently large integers } n.$$

Proof: If R is local, this is shown in [63], Th. 12.4. Now assume that R is semilocal. Set $\mathfrak{a}_i := \mathfrak{a}R_i$ for every $i \in \{1, \ldots, h\}$. By B(10.9) we have

$$\ell_R(\mathfrak{a}^n/\mathfrak{a}^{n+1}) = \sum_{i=1}^{h} \ell_{R_i}(\mathfrak{a}_i^n/\mathfrak{a}_i^{n+1}) \quad \text{for every } n \in \mathbb{N}_0,$$

and, for $i \in \{1, \ldots, h\}$, $\mathfrak{a}_i$ is $\mathfrak{m}_i R_i$-primary if $\mathfrak{a} \subset \mathfrak{m}_i$ and $\mathfrak{a}_i = R_i$ otherwise, hence

$$H(\mathfrak{a}, R) = \sum_{i=1}^{h} H(\mathfrak{a}_i, R_i),$$

and the assertion follows from the corresponding assertion in the local case.
The polynomial $P(\mathfrak{a}, R)$ is called the Hilbert polynomial of $\mathfrak{a}$.
(2) Let $\mathfrak{a}$ be an open ideal of R that is contained in a maximal ideal of R of height d. By [63], Ex. 1.21, the polynomial $P(\mathfrak{a}, R) \in \mathbb{Q}[X]$ has the form

$$P(\mathfrak{a}, R)(X) = e(\mathfrak{a}, R)\frac{X^d}{d!} + \cdots \quad [\text{terms of lower degree}]$$

where $e(\mathfrak{a}, R)$ is a positive integer which is called the multiplicity of $\mathfrak{a}$ in R; in particular, $e(\mathfrak{r}, R) =: e(R)$ is called the multiplicity of R. Assume that $\dim(R) = \dim(R_{\mathfrak{m}_i})$ for every $i \in \{1, \ldots, h\}$. Then we have

$$e(R) = \sum_{i=1}^{h} e(R_i).$$

(3) Assume that R is a one-dimensional CM-ring; then R_i is a one-dimensional local CM-ring for every $i \in \{1, \ldots, h\}$. Let $\mathfrak{a}$ be a proper open ideal of R. Then $\mathfrak{a}R_i$ is an $\mathfrak{m}_i R_i$-primary ideal for at least one $i \in \{1, \ldots, h\}$, hence the Hilbert polynomial $P(\mathfrak{a}, R)$ is a polynomial of degree 1, i.e.,

$$P(\mathfrak{a}, R)(X) = \rho(\mathfrak{a})X - \nu(\mathfrak{a}), \quad \rho(\mathfrak{a}) \in \mathbb{N}, \ \nu(\mathfrak{a}) \in \mathbb{Z},$$

where we have defined $\rho(\mathfrak{a}) := e(\mathfrak{a}, R)$ [to simplify notation and to cover also the case that $\mathfrak{a} = R$, cf. below]. It shall be shown later that $\nu(\mathfrak{a}) \in \mathbb{N}_0$ [cf. (4.11) below].
If $\mathfrak{a} = R$, then $R/\mathfrak{a}^n = \{0\}$ for every $n \in \mathbb{N}_0$, and we define $\rho(\mathfrak{a}) = \nu(\mathfrak{a}) = 0$.
For all sufficiently large integers n we have $\rho(\mathfrak{a}) = P(\mathfrak{a}, R)(n+1) - P(\mathfrak{a}, R)(n) = \ell_R(\mathfrak{a}^n/\mathfrak{a}^{n+1})$, hence we have $\ell_R(\mathfrak{a}^n/\mathfrak{a}^{n+1}) = \ell_R(\mathfrak{a}^m/\mathfrak{a}^{m+1})$ for all sufficiently large integers m, n.

1.3 Transversal Elements

(1.9) DEFINITION: Let R be a one-dimensional semilocal CM-ring, and let $\mathfrak{a}$ be an open ideal of R. Let $t \in \mathbb{N}$; an element $a \in \mathfrak{a}^t$ such that $a\mathfrak{a}^n = \mathfrak{a}^{n+t}$ for all sufficiently large integers n is called a transversal element for $\mathfrak{a}$ of order t. If $t = 1$, then a is called $\mathfrak{a}$-transversal; if R is local and $\mathfrak{m}$ is its maximal ideal, then an $\mathfrak{m}$-transversal element is called a transversal element of R.

To show the existence of tranversal elements of some order, we need the following four results. We consider graded rings $A = \bigoplus_{n \geq 0} A_n$, and we set $A_+ := \bigoplus_{n > 0} A_n$.

(1.10) Lemma: *Every homogeneous ideal $\mathfrak{M} \neq A$ of A which is maximal in the set of proper homogeneous ideals of A, has the form $\mathfrak{M} = \mathfrak{M}_0 + A_+$ where $\mathfrak{M}_0$ is a maximal ideal of A_0; in particular, we have $\mathfrak{M} \supset A_+$.*

Proof: Let $\mathfrak{M} \neq A$ be a homogeneous ideal of A. We have $\mathfrak{M} \cap A_0 \neq A_0$, hence $\mathfrak{M} \subset (\mathfrak{M} \cap A_0) + A_+ \neq A$. Now we assume that $\mathfrak{M}$ is, in addition, a maximal element in the set of proper homogeneous ideals of A. Then $\mathfrak{M} = (\mathfrak{M} \cap A_0) + A_+$, and $\mathfrak{M}_0 := \mathfrak{M} \cap A_0$ is a maximal ideal of A_0.

(1.11) Lemma: *The following statements are equivalent:*
(1) A is an artinian ring.
(2) For every $n \in \mathbb{N}_0$ the A_0-module A_n has finite length, and $A_n = \{0\}$ for all large enough integers n.

Proof (1) $\Rightarrow$ (2): Since A is an artinian ring, the ring $A_0 \cong A/A_+$ is also artinian. Since A is an artinian ring, it is also a noetherian ring [cf. [63], Th. 2.14], hence we have $A = A_0[x_1, \ldots, x_p]$ with homogeneous elements $x_1, \ldots, x_p$ of positive degree [cf. [63], Ex. 1.4]; set $k_i := \deg(x_i)$ for $i \in \{1, \ldots, p\}$. Let $n \in \mathbb{N}_0$. Now A_n is a finitely generated A_0-module, hence satisfies both chain conditions, and therefore has finite length. Since A is artinian, the set of prime ideals $\mathrm{Ass}(A) = \{\mathfrak{M}_1, \ldots, \mathfrak{M}_s\}$ of A consists of maximal ideals [cf. [63], Th. 2.14], and each of them is homogeneous [cf. [63], Ex. 3.5], hence $\mathfrak{M}_i \supset A_+$ for $i \in \{1, \ldots, s\}$ [cf. (1.10)]. Since the ideals $\mathfrak{M}_1, \ldots, \mathfrak{M}_s$ are pairwise comaximal, we have $\mathfrak{M}_1 \cdots \mathfrak{M}_s = \mathfrak{M}_1 \cap \cdots \cap \mathfrak{M}_s =: \mathfrak{N} \supset A_+$ [cf. B(10.1)], and $\mathfrak{N}$ is the nilradical of A, hence it is nilpotent since A is noetherian [cf. [63], Cor. 2.12 and Ex. 1.13]. We choose $h \in \mathbb{N}$ with $\mathfrak{N}^h = \{0\}$; then we have $A_+^h = \{0\}$. This implies that $x_1^{i_1} \cdots x_p^{i_p} = 0$ for all $(i_1, \ldots, i_p) \in \mathbb{N}_0^p$ with $i_1 + \cdots + i_p \geq h$. We set $k := \max(\{k_1, \ldots, k_p\})$. Let $n \geq kh$. Then the A_0-module A_n is generated by the power products $x_1^{i_1} \cdots x_p^{i_p}$ with $i_1 k_1 + \cdots + i_p k_p = n \geq kh$; for any such p-tuple $(i_1, \ldots, i_p)$ we have $i_1 + \cdots + i_p \geq i_1(k_1/k) + \cdots + i_p(k_p/k) = n/k \geq h$, hence $x_1^{i_1} \cdots x_p^{i_p} = 0$, and therefore we have $A_n = \{0\}$.

(2) $\Rightarrow$ (1): The assumptions imply that A_0 is an artinian ring and that A is an A_0-module of finite length, hence that A is an artinian A_0-module [cf. [63], Th. 2.13], and therefore A is an artinian ring.

(1.12) Lemma: *Let A be noetherian with A_0 artinian, and let M be a finitely generated graded A-module. Then we have $M_n = \{0\}$ for all sufficiently large integers n iff every prime ideal in $\mathrm{Ass}(M)$ contains the ideal A_+.*

Proof: First, we remark that A as A_0-algebra is generated by finitely many homogeneous elements [cf. [63], Ex. 1.4], hence, for every $n \in \mathbb{Z}$, M_n is a finitely generated A_0-module [cf. B(4.32)] and therefore of finite length, and $M_n = \{0\}$ for all sufficiently small integers n.

Assume that $M_n = \{0\}$ for all sufficiently large integers n. Then the A_0-module M has finite length, hence satisfies both chain conditions, and therefore the A-module M satisfies both chain conditions, hence has finite length. This implies that every prime ideal in $\mathrm{Ass}(M)$ is a maximal [cf. [63], Th. 2.14] and homogeneous [cf. [63], Ex. 3.5] ideal, hence it contains A_+ [cf. (1.10)]. Conversely, assume that every prime ideal in $\mathrm{Ass}(M)$ contains A_+. Let $\mathfrak{P} \in \mathrm{Ass}(M)$; since $A/A_+ \cong A_0$ is an artinian ring by assumption, $\mathfrak{P}/A_+$ is a maximal ideal of A/A_+ [cf. [63], Th. 2.14], hence $\mathfrak{P}$ is a maximal ideal of A. This implies that M is an A-module of finite length [cf. [63], Cor. 2.17 and Th. 3.1], and setting $\mathfrak{A} := \mathrm{Ann}_A(M)$, $A/\mathfrak{A}$ is a graded artinian ring [cf. [63], Cor. 2.17 and note that $\mathfrak{A}$ is a homogeneous ideal of A, cf. [63], Ex. 3.5]. Now there exists $n_0 \in \mathbb{N}$ with $(A/\mathfrak{A})_n = \{0\}$ for every $n \geq n_0$ [cf. (1.11)]. The $A/\mathfrak{A}$-module M is finitely generated; let $\{x_1, \ldots, x_p\}$ be a homogeneous system of generators of M, and set $k := \max(\{\deg(x_1), \ldots, \deg(x_p)\})$. Then we have $M_n = \{0\}$ for every $n \geq n_0 + k$.

(1.13) Lemma: *Let A be noetherian with A_0 artinian, and let M be a finitely generated graded A-module. Let $\mathfrak{P}_1, \ldots, \mathfrak{P}_h$ be those prime ideals in $\mathrm{Ass}(M)$ which do not contain A_+. Let $t \in \mathbb{N}$, $\alpha \in A_t$, and $\varphi\colon M \to M$ be multiplication by α. Then $\varphi_n\colon M_n \to M_{n+t}$ is injective for all large enough n iff α is not contained in any of the prime ideals $\mathfrak{P}_1, \ldots, \mathfrak{P}_h$.*

Proof: The kernel $\ker(\varphi)$ of φ is a finitely generated graded submodule of M. On the one hand, for any homogeneous prime ideal $\mathfrak{P}$ of A we have $\mathfrak{P} \in \mathrm{Ass}(\ker(\varphi))$ iff $\mathfrak{P} = \mathrm{Ann}_A(A\beta)$ for some homogeneous $\beta \in \ker(\varphi)$, hence iff $\mathfrak{P} \in \mathrm{Ass}(M)$ and $\alpha \in \mathfrak{P}$. On the other hand, by (1.12), $\ker(\varphi)_n = \{0\}$ for all sufficiently large n iff every prime ideal in $\mathrm{Ass}(\ker(\varphi))$ contains A_+. Since φ_n is injective for all large enough n iff $\ker(\varphi)_n = \{0\}$ for all large enough n, we have proved the assertion.

(1.14) REMARK: Let R be a one-dimensional semilocal CM-ring, and let $\mathfrak{r}$ be its Jacobson radical. Let $\mathfrak{a}$ be an open ideal of R.
(1) We set $S = \bigoplus_{n \geq 0} S_n$ with $S_n := \mathfrak{a}^n/\mathfrak{r}\mathfrak{a}^n$ for every $n \in \mathbb{N}_0$ [direct sum of abelian groups]. Let $\alpha \in \mathfrak{a}^m/\mathfrak{r}\mathfrak{a}^m$, $\beta \in \mathfrak{a}^n/\mathfrak{r}\mathfrak{a}^n$, and choose $a \in \mathfrak{a}^m$ with $\alpha = a + \mathfrak{r}\mathfrak{a}^m$, and $b \in \mathfrak{a}^n$ with $\beta := b + \mathfrak{r}\mathfrak{a}^n$. Then $\alpha\beta := ab + \mathfrak{r}\mathfrak{a}^{m+n}$ is easily seen to be independent of the choice of the representatives a of α and b of β, and therefore we can define a product on S of homogeneous elements of S which can be extended by the distributive law to make S, endowed with this product, into a graded ring. Then $S_0 = R/\mathfrak{r}$ is an artinian ring, and the S_0-algebra S is generated by finitely many elements of S_1.
(2) By (1.7) and [63], Ex. 1.21, there exists a polynomial $Q(\mathfrak{a}, R) \in \mathbb{Q}[X]$ such that

$$Q(\mathfrak{a}, R)(n) = \sum_{i=0}^{n-1} \ell_{S_0}(\mathfrak{a}^i/\mathfrak{r}\mathfrak{a}^i) \quad \text{for all sufficiently large integers } n.$$

We show that $\deg(Q(\mathfrak{a}, R)) = 1$. By the argument used in (1.8)(1), it is enough to show the following: If R is local, and if $\mathfrak{a}$ is an open ideal of R, then $Q(\mathfrak{a}, R)$ is

a polynomial of degree 1. This is clear if $\mathfrak{a} = R$ [then we have $Q(\mathfrak{a}, R)(X) = X$].
Let $\mathfrak{m}$ be the maximal ideal of R, and let $\mathfrak{a}$ be a proper ideal of R; then $\mathfrak{a}^n \neq \mathfrak{m}\mathfrak{a}^n$
for every $n \in \mathbb{N}_0$ [by Nakayama's lemma, cf. [63], Cor. 4.8], hence $Q(\mathfrak{a}, R)$ is a
polynomial of degree 1 at least, and since $\mathfrak{a} \subset \mathfrak{m}$, the estimate $\ell_{R/\mathfrak{m}}(\mathfrak{a}^n/\mathfrak{a}^{n+1}) \geq$
$\ell_{R/\mathfrak{m}}(\mathfrak{a}^n/\mathfrak{m}\mathfrak{a}^n)$ for every $n \in \mathbb{N}_0$ shows that $\deg(Q(\mathfrak{a}, R)) = 1$ [since the Hilbert
polynomial $P(\mathfrak{a}, R)$ has degree 1 by (1.8)(1)].
(3) For the highest coefficient $\sigma(\mathfrak{a})$ of $Q(\mathfrak{a}, R)$ we have $\sigma(\mathfrak{a}) = Q(\mathfrak{a}, R)(n + 1) -$
$Q(\mathfrak{a}, R)(n) = \ell_{S_0}(S_n)$ for all sufficiently large n, and therefore we have $\ell_{S_0}(S_n) =$
$\ell_{S_0}(S_m)$ for all sufficiently large integers m, n.

(1.15) Proposition: *Let R be a one-dimensional semilocal CM-ring, let $\mathfrak{r}$ be its
Jacobson radical, and let $\mathfrak{a}$ be an open ideal of R. We set $S = \bigoplus_{n \geq 0} S_n$ with
$S_n := \mathfrak{a}^n/\mathfrak{r}\mathfrak{a}^n$ for every $n \in \mathbb{N}_0$, and we let $\mathfrak{P}_1, \ldots, \mathfrak{P}_h$ be those prime ideals
in $\mathrm{Ass}(S)$ which do not contain $S_+ := \bigoplus_{n > 0} S_n$. Let $t \in \mathbb{N}$, $a \in \mathfrak{a}^t$, and set
$\alpha := a + \mathfrak{r}\mathfrak{a}^t$. Then a is transversal for $\mathfrak{a}$ of order t iff α is not contained in any of
the prime ideals $\mathfrak{P}_1, \ldots, \mathfrak{P}_h$. In particular, there exist transversal elements for $\mathfrak{a}$
of some order, and every such element is a regular element of R.*

Proof: Let $\varphi \colon S \to S$ be multiplication by α; then α is not contained in any of
the prime ideals $\mathfrak{P}_1, \ldots, \mathfrak{P}_h$ iff $\varphi_n \colon S_n \to S_{n+t}$ is injective for all sufficiently large
integers n [cf. (1.13)], hence iff $\ell_{S_0}(S_n) = \ell_{S_0}(\alpha S_n)$ for all sufficiently large integers
n. Since we have $\ell_{S_0}(S_n) = \ell_{S_0}(S_m)$ for all sufficiently large integers m, n [cf.
(1.14)(3)], the latter condition is equivalent to $\alpha S_n = S_{n+t}$ for all sufficiently large
integers n, and this condition is equivalent to $\mathfrak{a}^{n+t} = a\mathfrak{a}^n + \mathfrak{r}\mathfrak{a}^{n+t}$ for all sufficiently
large integers n, which in turn is equivalent to $\mathfrak{a}^{n+t} = a\mathfrak{a}^n$ for all sufficiently large
integers n by Nakayama's lemma [cf. [63], Cor. 4.8]. Now we have proved the first
part of the assertion.
There exist $t \in \mathbb{N}$ and a homogeneous element $\alpha \in S$ of positive degree t which
is not contained in any of the prime ideals $\mathfrak{P}_1 \ldots, \mathfrak{P}_h$ [cf. [63], Lemma 3.3]. We
choose $a \in \mathfrak{a}^t$ such that $\alpha = a + \mathfrak{r}\mathfrak{a}^t$; then a is transversal for $\mathfrak{a}$ of order t by what
we have just shown, and clearly a is regular [since $\mathfrak{a}$ contains regular elements].

(1.16) REMARK: Let R be a one-dimensional semilocal CM-ring, and let $\mathfrak{a}$ be an
open ideal of R.
(1) Let s, $t \in \mathbb{N}$, $a \in \mathfrak{a}^s$ transversal for $\mathfrak{a}$ of order s, $b \in \mathfrak{a}^t$ transversal for $\mathfrak{a}$ of
order t; then, clearly, $ab \in \mathfrak{a}^{s+t}$ is transversal for $\mathfrak{a}$ of order $s + t$.
(2) $\{t \in \mathbb{N} \mid \text{there exists } a \in \mathfrak{a}^t \text{ transversal for } \mathfrak{a} \text{ of order } t\} \cup \{0\}$ is a subsemigroup
of $\mathbb{N}_0$ by (1), and it contains positive integers by (1.15).

We can prove the existence of transversal elements of any given order under additional hypotheses. We need the following lemma.

(1.17) Lemma: *Let A be an artinian ring, and assume that $A/\mathfrak{m}$ is an infinite
field for every maximal ideal $\mathfrak{m}$ of A. Let M be a finitely generated A-module,
and let $N_1, \ldots, N_h$ be proper submodules of M. Then M is not the union of the
modules $N_1, \ldots, N_h$.*

Proof: The case $A = \{0\}$ is clear; now we assume that $A \neq \{0\}$. We argue by induction on $l := \ell_A(A)$. Let $\ell_A(A) = 1$. Then A is an infinite field, and the assertion is an easy exercise in linear algebra. Now let $l > 1$, and assume that the assertion is true for all artinian rings with length smaller than l. Let A be an artinian ring with $\ell_A(A) = l$, and let $\mathfrak{r}$ be the Jacobson radical of A. If $\mathfrak{r} = \{0\}$, then $A = k_1 \oplus \cdots \oplus k_t$ is a direct sum of infinite fields [cf. B(10.1)]; using the case $t = 1$ it is easy to prove the assertion. Now we consider the case $\mathfrak{r} \neq \{0\}$; we set $\overline{A} := A/\mathfrak{r}$ and $\overline{N}_i := (N_i + \mathfrak{r}M)/\mathfrak{r}M$ for every $i \in \{1,\ldots,h\}$. Note that, for every $i \in \{1,\ldots,h\}$, we have $N_i + \mathfrak{r}M \neq M$ by Nakayama's lemma [cf. [63], Cor. 4.8], hence $\overline{N}_1,\ldots,\overline{N}_h$ are proper submodules of $\overline{M}$. Since $\overline{A}/\overline{\mathfrak{m}}$ is an infinite field for every maximal ideal $\overline{\mathfrak{m}}$ of $\overline{A}$, and since $\ell_{\overline{A}}(\overline{A}) < \ell_A(A)$ [as $\mathfrak{r} \neq \{0\}$], we can apply the induction assumption to $\overline{A}$ and the $\overline{A}$-module $\overline{M}$ and its proper submodules $\overline{N}_1,\ldots,\overline{N}_h$. Thus, there exists an element $\overline{y} \in \overline{M}$ which does not lie in any of the $\overline{A}$-modules $\overline{N}_1,\ldots,\overline{N}_h$. Let $y \in M$ be a preimage of $\overline{y}$; then y does not lie in any of the A-modules $N_1,\ldots,N_h$.

(1.18) **Proposition:** *Let R be a one-dimensional semilocal CM-ring, and assume that $R/\mathfrak{m}$ is an infinite field for every maximal ideal $\mathfrak{m}$ of R. Let $\mathfrak{a}$ be an open ideal of R. For any $t \in \mathbb{N}$ there exists an element in $\mathfrak{a}^t$ which is transversal for $\mathfrak{a}$ of order t.*

Proof: Let $\mathfrak{r}$ be the Jacobson radical of R, and define S as in (1.15); the maximal ideals of the artinian ring $S_0 = R/\mathfrak{r}$ are the ideals $\mathfrak{m}/\mathfrak{r}$ where $\mathfrak{m}$ is a maximal ideal of R. Therefore S_0 satisfies the hypothesis in (1.17). Let $\mathfrak{P}_1,\ldots,\mathfrak{P}_h$ be those prime ideals in $\mathrm{Ass}(S)$ which do not contain $S_+ := \bigoplus_{n>0} S_n$. Since the ideal S_+ is generated by the elements of S_1, the S_0-modules $S_1 \cap \mathfrak{P}_1,\ldots,S_1 \cap \mathfrak{P}_h$ are proper submodules of S_1, and therefore [cf. (1.17)] there exists $\alpha \in S_1$ which is not contained in any of the ideals $\mathfrak{P}_1,\ldots,\mathfrak{P}_h$. We choose $a \in \mathfrak{a}$ with $\alpha = a + \mathfrak{r}\mathfrak{a}$. Then a is tranversal for $\mathfrak{a}$ of order 1 by (1.15), hence a^t is transversal for $\mathfrak{a}$ of order t by (1.16).

2 Integral Closure of One-Dimensional Semilocal Cohen-Macaulay Rings

2.1 Invertible Modules

(2.1) REMARK: Let A be a ring and K its ring of quotients.
(1) Let M, N be A-submodules of K. Remember that MN is the A-submodule of K which is generated by all products xy with $x \in M$ and $y \in N$ [cf. B(6.0)].
(2) Let M be an A-submodule of K. Then we define

$$M^{-1} := (A : M)_K = \{x \in K \mid xM \subset A\};$$

M^{-1} is an A-submodule of K, and $M(A : M)_K \subset A$. If M, N are A-submodules

of K with $N \subset M$, then $(A : M)_K \subset (A : N)_K$. In particular, let $\mathfrak{a}$ be an ideal of A. Then we have $A = (A : A)_K \subset (A : \mathfrak{a})_K$, hence we have $\mathfrak{a} \subset \mathfrak{a}(A : \mathfrak{a})_K \subset A$.
(3) An A-submodule M of K is called invertible if $MN = A$ for some A-submodule N of K. An invertible A-submodule of K is a regular submodule.

(2.2) Proposition: Let A be a ring and K its ring of quotients. Then:
(1) If M is an invertible A-submodule of K, then $(A : M)_K$ is the only A-submodule N of K satisfying $MN = A$.
(2) Every invertible A-submodule of K is finitely generated.
(3) If, in particular, A is quasisemilocal, then every invertible A-submodule of K is cyclic.

Proof: (1) Let N be an A-submodule of K satisfying $MN = A$. Clearly N is contained in $(A : M)_K$, hence we have $A = MN \subset M(A : M)_K \subset A$, and therefore $A = M(A : M)_K$. Multiplying this equation by N gives $N = (A : M)_K$.
(2) Let M be an invertible A-submodule of K. Since $M(A : M)_K = A$ by (1), there exist elements $x_1, \ldots, x_n \in M$ and $y_1, \ldots, y_n \in (A : M)_K$ such that $x_1 y_1 + \cdots + x_n y_n = 1$. Let $x \in M$; then $x = (xy_1)x_1 + \cdots + (xy_n)x_n$. Thus, we have $M = Ax_1 + \cdots + Ax_n$ since $xy_1, \ldots, xy_n \in A$.
(3) Let $\mathfrak{m}_1, \ldots, \mathfrak{m}_h$ be the maximal ideals of A and let $i \in \{1, \ldots, h\}$. Let M be an invertible A-submodule of K. Since $A = M(A : M)_K \not\subset \mathfrak{m}_i$, there exist $x_i \in M$ and $y_i \in (A : M)_K$ such that $x_i y_i \notin \mathfrak{m}_i$, and since $\mathfrak{m}_i$ does not contain the intersection of the other maximal ideals of A, there exists $z_i \in A \setminus \mathfrak{m}_i$ which lies in all the other maximal ideals of A. We set $y := z_1 y_1 + \cdots + z_h y_h$. Then we have $y \in (A : M)_K$, hence yM is an ideal of A. We show that yM is not contained in a maximal ideal of A. Indeed, suppose that $yM \subset \mathfrak{m}_i$ for some $i \in \{1, \ldots, h\}$; then we have $yx_i \in \mathfrak{m}_i$. We consider $yx_i = (z_1 y_1 + \cdots + z_h y_h)x_i \in \mathfrak{m}_i$; by our choice we have $z_i(x_i y_i) \notin \mathfrak{m}_i$ and $z_j(y_j x_i) \in \mathfrak{m}_i$ for every $j \in \{1, \ldots, h\}$, $j \neq i$, hence we get $z_i(x_i y_i) \in \mathfrak{m}_i$. This contradiction shows that $yM = A$. In particular, y is a regular element of K and $M = y^{-1}A$.

2.2 The Integral Closure

(2.3) REMARK: Let R be a one-dimensional semilocal CM-ring and let Q be its ring of quotients. In this remark we collect at first some properties of R.
Let $\mathrm{Ass}(R) = \{\mathfrak{c}_1, \ldots, \mathfrak{c}_l\}$; note that $\mathrm{Ass}(R)$ is also the set of minimal prime ideals of R. Then $\Sigma := R \setminus (\mathfrak{c}_1 \cup \cdots \cup \mathfrak{c}_l)$ is the set $\mathrm{Reg}(R)$ of regular elements of R [cf. [63], Th. 3.1], $Q = \Sigma^{-1}R$, and $\mathfrak{d}_1 := \Sigma^{-1}\mathfrak{c}_1, \ldots, \mathfrak{d}_l := \Sigma^{-1}\mathfrak{c}_l$ are the prime ideals of Q. Each of them is a maximal ideal, and therefore Q is an artinian ring [cf. [63], Cor. 9.1]; moreover, the canonical homomorphism $Q \to Q/\mathfrak{d}_1 \times \cdots \times Q/\mathfrak{d}_l$ is surjective and has kernel $\mathfrak{d}_1 \cap \cdots \cap \mathfrak{d}_l$ [cf. B(10.1)(2)]. Let $i \in \{1, \ldots, l\}$; then we have $\mathfrak{d}_i \cap R = \mathfrak{c}_i$ [cf. [63], Prop. 2.2], $R/\mathfrak{c}_i$ is a one-dimensional semilocal domain, and it is easy to check that $Q/\mathfrak{d}_i$ is the field of quotients of $R/\mathfrak{c}_i$.

Now we specialize to the case of a one-dimensional *local* CM-ring R with maximal ideal $\mathfrak{m}$. Then we have:

(1) Let $r \in \mathrm{Reg}(R) \cap \mathrm{m}$; then $\mathrm{Ass}(R/Rr) = \{\mathrm{m}\}$, hence $\mathrm{rad}(Rr) = \mathrm{m}$. Therefore there exists $h \in \mathbb{N}$ such that $\mathrm{m}^h \subset Rr$ [cf. [63], Ex. 1.13]. In particular, the Rr-adic topology and the m-adic topology on R coincide.

(2) For every $r \in \mathrm{Reg}(R) \cap \mathrm{m}$ we have $Q = R_r = R[1/r]$.

In fact, let $x \in Q$; then we write $x = a/s$ for some $a \in R$, $s \in \mathrm{Reg}(R)$. If $s \notin \mathrm{m}$, then we have $x \in R$. If $s \in \mathrm{m}$, then we may choose $n \in \mathbb{N}$ such that $r^n \in \mathrm{m}^n \subset Rs$, hence $r^n = ts$ for some $t \in \mathrm{Reg}(R)$, and we have $x = ta/ts = ta/r^n \in R_r$.

(2.4) COMPLETION: Let R be a one-dimensional local CM-ring with maximal ideal m and ring of quotients Q. Let $\widehat{R}$ be the completion of R, and let $Q(\widehat{R})$ be the ring of quotients of $\widehat{R}$. Then we have $\dim(\widehat{R}) = 1$ [cf. [63], Cor. 10.12], and $\widehat{R}$ is a faithfully flat extension of R; in particular, we have $(\widehat{R}M) \cap Q = M$ for every R-submodule M of Q [cf. [34], Ch. I, §3, no. 5, Prop. 10] and $\mathrm{Reg}(R) \subset \mathrm{Reg}(\widehat{R})$. This implies that $\widehat{R}$ is a one-dimensional local CM-ring. Moreover, for every regular element $r \in \mathrm{m}$ we have $Q(\widehat{R}) = \widehat{R}[1/r]$ [cf. (2.3)], hence we have $\widehat{R}Q = Q(\widehat{R})$ [$\widehat{R}Q$ is the smallest subring of $Q(\widehat{R})$ containing $\widehat{R}$ and Q].

(2.5) Proposition: *Let R be a one-dimensional local integrally closed CM-ring. Then R is a discrete valuation ring.*

Proof: Let m be the maximal ideal of R, let Q be the ring of quotients of R, and define $\mathrm{m}' := (R : \mathrm{m})_Q$. We have $\mathrm{m} \subset \mathrm{mm}' \subset R$.

We show that $\mathrm{m}' \neq R$. Let $r \in \mathrm{m}$ be regular. By (2.3)(1) there exists $h \in \mathbb{N}$ with $\mathrm{m}^h \subset Rr$. We chose h minimal with this property. Then there exists $y \in \mathrm{m}^{h-1}$ with $y \notin Rr$, hence we have $y/r \notin R$. On the other hand, we have $y\mathrm{m} \subset Rr$, hence we have $y/r \in \mathrm{m}'$.

Suppose that $\mathrm{mm}' = \mathrm{m}$. Then every element of m' is integral over R [cf. [63], Cor. 4.6], hence $\mathrm{m}' = R$, contrary to what we have shown. Therefore we have $\mathrm{mm}' = R$, hence m is generated by one element [cf. (2.2)], and therefore R is a discrete valuation ring [cf. I(3.29)].

(2.6) Corollary: *Let R be a one-dimensional local CM-ring, and let S be the integral closure of R. Assume that S is a finitely generated R-module. Then S is a one-dimensional semilocal CM-ring, and for every maximal ideal n of S the localization S_n is a discrete valuation ring.*

Proof: Let Q be the ring of quotients of R. We have $\dim(S) = 1$ by [63], Prop. 9.2, and S is semilocal by B(3.8), hence S is a semilocal CM-ring. Let n be a maximal ideal of S. Since S is a noetherian ring, also $V := S_\mathrm{n}$ is noetherian [cf. [63], Cor. 2.3], hence V is a one-dimensional local CM-ring. Since S is integrally closed in Q, V is integrally closed in Q_n [cf. [63], Prop. 4.13]. Now Q_n is an artinian ring since Q is artinian [cf. (2.3)], hence it is its own ring of quotients, and therefore the ring of quotients $Q(V)$ of V is contained in Q_n, hence V is, a priori, integrally closed in $Q(V)$. By (2.5) we see that V is a discrete valuation ring.

(2.7) REMARK: Let A be a noetherian ring, and assume that the Jacobson radical of A contains a regular element r. If A is not reduced, then there exist subrings of the ring of quotients K of A containing A which are integral over A and not noetherian; in particular, the integral closure of A is not a finitely generated A-module.

Proof: Let $a \in A$ be a non-zero nilpotent element. We may assume that $a^2 = 0$. For any $n \in \mathbb{N}$ we define $B_n := A + A(a/r^n) \subset K$. Now B_n is a subring of K containing A, and $B_n \subset B_{n+1}$ for every $n \in \mathbb{N}$. Thus, $B := \bigcup B_n$ is a subring of K. Suppose that the increasing chain $\{B(a/r^n)\}_{n \in \mathbb{N}}$ of ideals of B becomes stationary, hence that there exists $m \in \mathbb{N}$ such that $a/r^{m+1} = b \cdot a/r^m$ for some $b \in B$. We may write $b = c + d \cdot a/r^k$ for some $c, d \in A$, $k \in \mathbb{N}$. Then we have $a/r^{m+1} = ca/r^m$, hence $a(1 - rc) = 0$. Now $1 - rc$ is a unit of A [cf. [63], Ex. 4.7], hence we have $a = 0$ which is in contradiction with the assumption $a \neq 0$. Therefore B is not a noetherian ring. Clearly B is integral over A. Now B is contained in the integral closure $\overline{A}$ of A; if $\overline{A}$ would be a finitely generated A-module, then also B would be a finitely generated A-module [cf. [63], Ex. 1.3].

2.3   Integral Closure and Manis Valuation Rings

(2.8) For the rest of this section let R be a one-dimensional semilocal CM-ring, let Q be the ring of quotients of R, and let S be the integral closure of R.

(2.9) DEFINITION: The Manis valuation rings of Q which contain R are called the Manis valuation rings belonging to R.
Note that, in case R is a domain, a Manis valuation ring belonging to R is a valuation ring of Q which is different from Q and contains R.

(2.10) Proposition: *Let R be a one-dimensional semilocal integral domain. Then:*
(1) The set of Manis valuation rings of Q belonging to R is finite and not empty; each such valuation ring is a discrete valuation ring.
(2) Let $V_1, \ldots, V_h$ be the Manis valuation rings of Q belonging to R; then the integral closure S of R is the intersection $V_1 \cap \cdots \cap V_h$, S is a principal ideal domain with h maximal ideals $\mathfrak{n}_1, \ldots, \mathfrak{n}_h$, and we have $V_i = S_{\mathfrak{n}_i}$ for $i \in \{1, \ldots, h\}$.

Proof: (a) S is a one-dimensional semilocal domain by the theorem of Krull-Akizuki [cf. [63], Th. 11.13]. Let $\mathfrak{n}_1, \ldots, \mathfrak{n}_h$ be the pairwise different maximal ideals of S. Then $S_{\mathfrak{n}_1}, \ldots, S_{\mathfrak{n}_h}$ are discrete valuation rings of Q [cf. I(3.29)] and we have $S = \bigcap_{i=1}^h S_{\mathfrak{n}_i}$ [cf. B(2.5)]; S is a principal ideal domain by I(7.2).
(b) Let $V \neq Q$ be a valuation ring of Q containing R; then S is contained in V [cf. I(3.4)], and therefore we have $V = S_{\mathfrak{n}_i}$ for some $i \in \{1, \ldots, h\}$ [cf. I(7.2)].

(2.11) Theorem: *Let R be a one-dimensional semilocal CM-ring. Then:*

(1) *The set of Manis valuation rings of Q belonging to R is finite and not empty; each such Manis valuation ring is a discrete Manis valuation ring.*

(2) *Let $\{V_1, \ldots, V_h\}$ be the set of pairwise different Manis valuation rings of Q belonging to R; then $\{(V_1 : Q)_Q, \ldots, (V_h : Q)_Q\}$ is the set of maximal ideals of Q.*

(3) *Let $\mathfrak{d}$ be a maximal ideal of Q. Then $Q/\mathfrak{d}$ is a field, $R/(\mathfrak{d} \cap R)$ is a one-dimensional semilocal domain with field of quotients $Q/\mathfrak{d}$, $V \mapsto V/\mathfrak{d}$ is a map from the set of Manis valuation rings of Q belonging to R and having conductor $\mathfrak{d}$ to the set of Manis valuation rings of $Q/\mathfrak{d}$ belonging to $R/(\mathfrak{d}\cap R)$, and it is bijective.*

(4) *The integral closure S of R in Q is the intersection $V_1 \cap \cdots \cap V_h$, S has h maximal ideals $\mathfrak{n}_1, \ldots, \mathfrak{n}_h$, each of them is regular, and every regular ideal of S is a principal ideal. Moreover, $\mathfrak{n}_j V_j$ is the regular maximal ideal of V_j and $V_j = S_{[\mathfrak{n}_j]}$ for every $j \in \{1, \ldots, h\}$.*

Proof: We use the notations introduced in (2.3).

Let V be a Manis valuation ring of Q belonging to R, and set $\mathfrak{d} := (V : Q)_Q$. Then $\mathfrak{d}$ is a prime ideal of Q [cf. I(2.4)(3)(a)], hence $\mathfrak{d} = \mathfrak{d}_i$ for some $i \in \{1, \ldots, l\}$, $Q/\mathfrak{d}$ is a field, and $V/\mathfrak{d} \neq Q/\mathfrak{d}$ is a valuation ring of $Q/\mathfrak{d}$ by I(2.17)(1), and therefore a discrete valuation ring by (2.10). This implies that V is a discrete Manis valuation ring of Q, again by I(2.17)(1).

Conversely, let $i \in \{1, \ldots, l\}$, and let $\overline{V} \neq Q/\mathfrak{d}_i$ be a Manis valuation ring of $Q/\mathfrak{d}_i$ which contains $R/\mathfrak{c}_i$; then $\overline{V}$ is a discrete valuation ring of $Q/\mathfrak{d}_i$ by (2.10). The preimage $V \subset Q$ of $\overline{V}$ is a discrete Manis valuation ring of Q [cf. I(2.17)(1)] containing R with conductor $\mathfrak{d}_i$, and there are only finitely many discrete Manis valuation rings belonging to R and having conductor $\mathfrak{d}_i$ [cf. (2.10)].

Now we have proved the assertions in (1)–(3). The assertion in (4) follows from chapter I, (2.5), (2.19), (2.21), and the fact that every maximal ideal of S lies over a maximal ideal of R [cf. [63], Cor. 4.17], hence it is a regular ideal.

(2.12) THE REDUCED CASE: We use the notations of (2.3). Let R be a *reduced* one-dimensional semilocal CM-ring. In this case we can describe the Manis valuation rings of Q which belong to R more precisely. Since R is reduced, we have $\mathfrak{c}_1 \cap \cdots \cap \mathfrak{c}_l = \{0\}$, $\mathfrak{d}_1 \cap \cdots \cap \mathfrak{d}_l = \{0\}$. Let $i \in \{1, \ldots, l\}$. Then the one-dimensional semilocal CM-domain $R/\mathfrak{c}_i$ has $K_i := Q/\mathfrak{d}_i$ as field of quotients; let S_i be the integral closure of $R/\mathfrak{c}_i$. Note that S_i is a one-dimensional semilocal domain [cf. (2.10)]. Let $\mathfrak{m}'_{i1}, \ldots, \mathfrak{m}'_{il_i}$ be the maximal ideals of S_i; then $\{W_{i1} := (S_i)_{\mathfrak{m}'_{i1}}, \ldots, W_{il_i} := (S_i)_{\mathfrak{m}'_{il_i}}\}$ is the set of Manis valuation rings of K_i belonging to $R/\mathfrak{c}_i$. For $j \in \{1, \ldots, l_i\}$ we denote by $w_{ij} : K_i \to \mathbb{Z}_\infty$ the valuation of K_i defined by W_{ij}.

Set $K := K_1 \times \cdots \times K_l$ and $\widetilde{S} := S_1 \times \cdots \times S_l$. We choose $i \in \{1, \ldots, l\}$ and $j \in \{1, \ldots, l_i\}$. We define a map $\widetilde{v}_{ij} : K \to \mathbb{Z}_\infty$ by $\widetilde{v}_{ij}((x_1, \ldots, x_l)) = w_{ij}(x_i)$; it is easy to check that $\widetilde{v}_{ij}$ is a discrete Manis valuation of K having $\widetilde{V}_{ij} := K_1 \times \cdots \times K_{i-1} \times W_{ij} \times K_{i+1} \times \cdots \times K_l$ as its valuation ring; $K_1 \times \cdots \times K_{i-1} \times \mathfrak{m}'_{ij} \times K_{i+1} \times \cdots \times K_l$ is the regular maximal ideal of $\widetilde{V}_{ij}$, $\widetilde{\mathfrak{m}}_{ij} := S_1 \times \cdots \times S_{i-1} \times \mathfrak{m}'_{ij} \times S_{i+1} \times \cdots \times S_l$ is the intersection of this ideal with $\widetilde{S}$, we have $\widetilde{S}_{[\widetilde{m}_{ij}]} = \widetilde{V}_{ij}$, and $K_1 \times \cdots \times K_{i-1} \times \{0\} \times K_{i+1} \times \cdots \times K_l$ is the kernel of $\widetilde{v}_{ij}$.

The canonical isomorphism $\varphi\colon Q \to K$ maps S onto $\widetilde{S}$ [cf. B(3.4)]. We set $V_{ij} := \varphi^{-1}(\widetilde{V}_{ij})$; V_{ij} is a discrete Manis valuation ring of Q containing R, the intersection $\mathfrak{m}_{ij}$ of the regular maximal ideal of V_{ij} with S is a maximal ideal of S, and $V_{ij} = S_{[\mathfrak{m}_{ij}]}$. Now $\{V_{ij} \mid i \in \{1,\ldots,l\}, j \in \{1,\ldots,l_i\}\}$ is the set of Manis valuation rings of Q belonging to R. In particular, we can state: S is a one-dimensional integrally closed reduced semilocal ring, and every ideal of S is a principal ideal.

3 One-Dimensional Analytically Unramified and Analytically Irreducible CM-Rings

(3.0) In this section R is a one-dimensional *local* CM-ring with maximal ideal $\mathfrak{m}$ and ring of quotients Q.

3.1 Two Length Formulae

(3.1) **Proposition:** (1) *Let M be a finitely generated R-submodule of Q containing a regular element of R. Then we have*

$$\ell_R(R/Rr) = \ell_R(M/rM) \quad \text{for every } r \in \mathrm{Reg}(R).$$

(2) *Let M be an R-submodule of Q. Then we have*

$$\ell_R(M/rM) \leq \ell_R(R/Rr) \quad \text{for every } r \in \mathrm{Reg}(R).$$

Proof: (1) For every regular element $a \in R$ the map $x \mapsto ax : Q \to Q$ is an isomorphism of R-modules; thus, we may assume that M is a regular ideal $\mathfrak{a}$ of R. Consider the following two exact sequences

$$0 \to \mathfrak{a}/r\mathfrak{a} \to R/r\mathfrak{a} \to R/\mathfrak{a} \to 0, \quad 0 \to Rr/r\mathfrak{a} \to R/r\mathfrak{a} \to R/Rr \to 0.$$

All the R-modules in these sequences have finite length by (1.6), hence we obtain

$$\ell_R(\mathfrak{a}/r\mathfrak{a}) + \ell_R(R/\mathfrak{a}) = \ell_R(Rr/r\mathfrak{a}) + \ell_R(R/Rr).$$

The R-modules $R/\mathfrak{a}$ and $Rr/r\mathfrak{a}$ are isomorphic, whence $\ell_R(\mathfrak{a}/r\mathfrak{a}) = \ell_R(R/Rr)$.
(2) Suppose that we have $\ell_R(M/rM) > \ell_R(R/Rr)$ for some $r \in \mathrm{Reg}(R)$. Then there exists a finitely generated submodule N of M such that $\ell((N+rM)/rM) > \ell_R(R/Rr)$. Now we get

$$\ell_R(N/rN) \geq \ell_R(N/(N \cap rM)) = \ell_R((N+rM)/rM) > \ell_R(R/Rr),$$

which is in contradiction with (1).

(3.2) Proposition: *Let $\mathfrak{a}$ be a regular ideal of R, and let $r \in \mathfrak{m}$ be regular. Then we have*

$$\ell_R(R/r\mathfrak{a}) = \ell_R(R/Rr) + \ell_R(R/\mathfrak{a}).$$

Proof: The R-modules $R/\mathfrak{a}$ and $Rr/r\mathfrak{a}$ are isomorphic, hence $\ell_R(R/\mathfrak{a}) = \ell_R(Rr/r\mathfrak{a})$. On the other hand, we have $\ell_R(R/r\mathfrak{a}) = \ell_R(R/Rr) + \ell_R(Rr/r\mathfrak{a})$. This implies the assertion.

3.2   Divisible Modules

(3.3) DEFINITION: Let A be a ring and M an A-module; for $a \in A$ let $a_M : M \to M$ be multiplication by a.
(1) M is said to be divisible if, for every $a \in \mathrm{Reg}(A)$, the map $a_M : M \to M$ is surjective.
(2) M is said to be torsion-free if, for every $a \in \mathrm{Reg}(A)$, the map $a_M : M \to M$ is injective.
(3) M is said to be a torsion module if $\{0\}$ is the only torsion-free submodule of M.
(4) M is said to be simple divisible if M is a non-zero torsion divisible module and admits no proper non-zero divisible submodules.

(3.4) DIVISIBLE MODULES: Let A be a ring and M an A-module.
(1) Let N be a submodule of M. Then M/N is a divisible module iff $M = N + aM$ for every $a \in \mathrm{Reg}(A)$. If M is divisible, then M/N is divisible, and if N and M/N are divisible, then M is divisible.
(2) A sum of divisible submodules of M is a divisible submodule of M, hence M contains a unique largest divisible submodule $d_A(M)$, namely the sum of all divisible submodules; $d_A(M)$ is called the divisible submodule of M. Clearly M is a divisible A-module iff $M = d_A(M)$; we have $d_A(M/d_A(M)) = \{0\}$ [cf. (1)].
(3) Let M be an A-module. We set

$$N := \bigcap_{a \in \mathrm{Reg}(A)} aM.$$

Then we have $d_A(M) \subset N$, and if M is torsion-free, then we have $d_A(M) = N$.
Proof: (a) Let $x \in d_A(M)$; for every $a \in \mathrm{Reg}(A)$ there exists $y_a \in d_A(M)$ such that $x = ay_a \in aM$, hence we have $d_A(M) \subset N$.
(b) Now we assume that M is torsion-free. We show that N is a divisible submodule of M, hence that $d_A(M) = N$. Let $x \in N$ and $b \in \mathrm{Reg}(A)$. For every $a \in \mathrm{Reg}(A)$ there exists $y_a \in M$ such that $x = bay_a$ [since $ba \in \mathrm{Reg}(A)$], and we have $ay_a = a'y_{a'}$ for all a, $a' \in \mathrm{Reg}(A)$ [since M is torsion-free]. We set $z := y_{1_A}$; then we have $z = ay_a$ for every $a \in \mathrm{Reg}(A)$, hence $z \in N$ and $x = bz$, i.e., $N = bN$.
(4) The divisible submodule $d_A(A)$ of A is the conductor $\mathfrak{f}_A$ of A in its ring of quotients $Q(A)$.

Indeed, for every $a \in \mathrm{Reg}(A)$ we have $a\mathfrak{f}_A = \mathfrak{f}_A$ since a is a unit of $Q(A)$ and $\mathfrak{f}_A$ is an ideal of $Q(A)$, hence we have $\mathfrak{f}_A \subset d_A(A)$. Let $x \in d_A(A)$. Take any $z \in Q(A)$; then $z = a/r$ for some $a \in A$ and $r \in \mathrm{Reg}(A)$. There exists $y \in d_A(A)$ such that $x = ry$. Now $xz = ry(a/r) = ya \in A$, and therefore $Q(A)x \subset A$, hence $x \in \mathfrak{f}_A$.

(3.5) DIVISIBLE R-MODULES: (1) An R-module M is divisible iff $M = \mathfrak{m}M$; in particular, $\{0\}$ is the only finitely generated divisible R-module.

Indeed, if M is a divisible R-module, then $M = rM$ for a regular $r \in \mathfrak{m}$, hence $M = \mathfrak{m}M$. Conversely, assume that $M = \mathfrak{m}M$; this implies that $M = \mathfrak{m}^h M$ for every $h \in \mathbb{N}$. Let $r \in \mathrm{Reg}(R)$. Then there exists $h \in \mathbb{N}$ with $\mathfrak{m}^h \subset Rr$ [cf. (2.3)(1)], and therefore we have $M = rM$, i.e., M is a divisible R-module. Now, if M is a finitely generated divisible R-module, then $M = \mathfrak{m}M$ implies $M = \{0\}$ by Nakayama's lemma [cf. [63], Cor. 4.8].

(2) Let M be an R-submodule of Q containing R. Then M/R is divisible iff $M = R + \mathfrak{m}M$ iff $M = R + rM$ for every $r \in \mathrm{Reg}(R)$ [cf. (1)]. Moreover, in this case M is a subring of Q. In fact, let $x \in M$; then we can write $x = y/r$ with $y \in R$ and $r \in \mathrm{Reg}(R)$. From $M = R + rM$ we get $xM = Rx + yM \subset M + M = M$, hence the product of two element of M lies in M.

(3) The R-module $\widehat{R}/R$ is divisible.

In fact, let $x \in \widehat{R}$ and $r \in \mathrm{Reg}(R) \cap \mathfrak{m}$. Since R is dense in $\widehat{R}$ [as a topological subspace], we may write $x = y + rz$ for some $y \in R$ and $z \in \widehat{R}$ [note that the $\widehat{\mathfrak{m}}$-adic topology and the $r\widehat{R}$-adic topology of $\widehat{R}$ are identical, cf. (2.3)]. Hence $\widehat{R}/R$ is a divisible R-module.

3.3 Compatible Extensions

(3.6) DEFINITION: A proper subring A of Q containing R is called a compatible extension of R if A/R is a divisible R-module.

(3.7) REMARK: Let M be a proper R-submodule of Q containing R. If M/R is a divisible R-module, then M is a compatible extension of R by (3.5)(2).

(3.8) REMARK: Let A be a compatible extension of R.
(1) The only regular maximal ideal of A is $\mathfrak{m}A$.
In fact, since $A \neq Q$, there exist regular elements in A which are not units, hence there exist regular maximal ideals of A. Let $\mathfrak{n}$ be such an ideal. Since it contains regular elements of R, we have $A = R + \mathfrak{n}$ [since A is a compatible extension of R], hence $A/\mathfrak{n} \cong R/(\mathfrak{n} \cap R)$, hence $\mathfrak{n} \cap R = \mathfrak{m}$ [since $A/\mathfrak{n}$ is a field], hence $\mathfrak{m}A \subset \mathfrak{n}$. We have $A = R + \mathfrak{m}A$ [since $\mathfrak{m}A$ contains regular elements of R]. Let $y \in \mathfrak{n}$. Then we can write $y = r + z$ with $r \in R$, $z \in \mathfrak{m}A$, hence $r = y - z \in R \cap \mathfrak{n} = \mathfrak{m}$, and therefore $y \in \mathfrak{m}A$, hence $\mathfrak{n} \subset \mathfrak{m}A$; thus, we have shown that $\mathfrak{m}A = \mathfrak{n}$.
(2) Since A is a torsion-free R-module, we have $d_R(A) = \bigcap \mathfrak{m}^n A$ [cf. (3.4)(3)], and since $\mathfrak{m}A$ contains all regular elements of A which are non-units [cf. (1)], we have

$d_A(A) = \bigcap m^n A$. Therefore we have $d_R(A) = d_A(A)$, and this is the conductor of A in Q by (3.4)(4).

(3.9) Proposition: *Assume that R is complete. Let $\mathcal{D}$ be the set of proper ideals of Q, and let $\mathcal{C}$ be the set of compatible extensions of R. Then the map $\mathfrak{d} \mapsto R + \mathfrak{d} : \mathcal{D} \to \mathcal{C}$ is an inclusion-preserving bijective map, $B \mapsto d_B(B) = \bigcap m^n B$ is the inverse map, and $\mathfrak{d}$ is the conductor of $R + \mathfrak{d}$ in Q.*

Proof: (1) Let B be a compatible extension of R, and set $\mathfrak{d} := \bigcap m^n B$. By (3.8)(2) $\mathfrak{d}$ is an ideal of Q. Let $r \in \operatorname{Reg}(R) \cap m$; then we have $B = R + rB$ [cf. (3.5)(2)]. Let $x \in B$. By induction, there exist elements $y_0, y_1, \ldots \in R$ and $z_1, z_2, \ldots \in B$ such that, for every $n \in \mathbb{N}_0$, $x = y_0 + y_1 r + \cdots + y_n r^n + z_{n+1} r^{n+1}$. Set $y := \sum_{\nu=0}^{\infty} y_\nu r^\nu \in R$ [note that R is complete]. Now we have $x - y \in m^n B$ for every $n \in \mathbb{N}$, hence we have $x - y \in \mathfrak{d}$, and therefore we have $B = R + \mathfrak{d}$.
(2) Let $\mathfrak{d} \in \mathcal{D}$ and set $B := R + \mathfrak{d}$. Then we have $B \neq Q$. Indeed, suppose that $B = Q$. Let $r \in m \cap \operatorname{Reg}(R)$. Then $1/r = x + d$ for some $x \in R$, $d \in \mathfrak{d}$, hence $rd = 1 - rx$ is a unit of R, which is in contradiction with the fact that d is a zero-divisor in Q. Thus, we have shown that B is a proper submodule of Q. For every $r \in \operatorname{Reg}(R)$ we have $R + rB = R + Rr + r\mathfrak{d} = R + \mathfrak{d} = B$, hence B/R is a divisible R-module, and therefore B is a compatible extension of R [cf. (3.7)]. Set $\mathfrak{d}' := \bigcap m^n B$; then $\mathfrak{d}'$ is the conductor of B in Q, and we have $B = R + \mathfrak{d}'$ [cf. (1)]. We have $\mathfrak{d} \subset \mathfrak{d}'$, and from $B/R = R/(R \cap \mathfrak{d}) = R/(R \cap \mathfrak{d}')$ we get $R \cap \mathfrak{d} = R \cap \mathfrak{d}'$, hence we have $\mathfrak{d} = \mathfrak{d}'$ since Q is a localization of R.

(3.10) REMARK: In the sequel, we study the set of maximal compatible extensions of a one-dimensional local CM-ring. As we just saw in (3.9), it is easier to work with a complete ring; therefore, the following result shall play a crucial role.

(3.11) Proposition: *Let $\mathcal{M}$ be the set of all R-submodules of Q containing R, and let $\mathcal{M}^*$ be the set of all $\widehat{R}$-submodules of $Q(\widehat{R})$ containing $\widehat{R}$. Then:*
(1) The maps

$$M \mapsto \widehat{R}M : \mathcal{M} \to \mathcal{M}^*, \quad M^* \mapsto M^* \cap Q : \mathcal{M}^* \to \mathcal{M}$$

are bijective, inclusion-preserving and mutually inverse. If $M^* \in \mathcal{M}^*$, then $M^* = \widehat{R} + (Q \cap M^*)$.
(2) The map $A \mapsto \widehat{R}A$ is a bijective inclusion-preserving map between the set of compatible extensions of R and the set of compatible extensions of $\widehat{R}$.

Proof: (1) By (2.4) we have $\widehat{R}M \cap Q = M$ for every R-submodule M of Q. Let $M^* \in \mathcal{M}^*$. It is enough to show that $M^* = \widehat{R} + (M^* \cap Q)$. Let $x \in M^*$, and choose $r \in \operatorname{Reg}(R)$ such that $rx =: y \in \widehat{R}$. The R-module $\widehat{R}/R$ is divisible [cf. (3.5)(3)], hence there exist $u \in R$, $v \in \widehat{R}$ such that $y = u + rv$. Now we have $x - v = u/r \in M^* \cap Q$, and therefore we have $x = v + u/r \in \widehat{R} + (M^* \cap Q)$.

(2) Let A be a compatible extension of R, hence $A = R + \mathfrak{m}A$; then we have $\widehat{R}A = \widehat{R} + \mathfrak{m}\widehat{R}A = \widehat{R} + \widehat{\mathfrak{m}}\widehat{R}A$, and $\widehat{R}A \neq Q(\widehat{R})$ by (1). Thus, $\widehat{R}A$ is a compatible extension of $\widehat{R}$. Conversely, assume that B is a compatible extension of $\widehat{R}$; then we have $B = \widehat{R}A$ where $A = B \cap Q$ is a proper R-submodule of Q containing R [cf. (1)]. Choose any $r \in \mathrm{Reg}(R)$; we show that $A = R + rA$. Let $x \in A$. Then $x = y + rz$ for some $y \in \widehat{R}$, $z \in B$ [because $B = \widehat{R} + rB$]. Since $\widehat{R}/R$ is a divisible R-module, we may write $y = y_1 + rz'$ for some $y_1 \in R$, $z' \in \widehat{R}$. Set $z_1 := z + z' \in B$. Now $x = y_1 + rz_1$, $z_1 = (x - y_1)/r \in Q$, hence $z_1 \in A$, and therefore $x \in R + rA$. Thus, we have shown that A is a compatible extension of R.

(3.12) Corollary: *We have the following:*
(1) The set of compatible extensions of R satisfies the ascending and the descending chain condition.
(2) The set of maximal compatible extensions of R is finite and not empty.

Proof: This follows from (3.11) and (3.9) since $Q(\widehat{R})$ is an artinian ring.

(3.13) Proposition: *Let A a compatible extension of R. Then:*
(1) Every R-submodule M of Q containing A is an A-module.
(2) Let $M \neq Q$ be an R-submodule of Q containing A. If the only compatible extension of R lying between A and M is A itself, then M is a finitely generated A-module, and M/A is an R-module of finite length.

Proof: (1) Let $a \in A$ and $x \in M$. We write $x = y/r$ for some $y \in R$, $r \in \mathrm{Reg}(R)$. Since $A = R + rA$ [cf. (3.5)(2)], we may write $a = s + rb$ for some $s \in R$ and $b \in A$. Now we have $ax = (s + rb)x = sx + by \in M + A \subset M$, and therefore M is an A-module.
(2) Let $r \in \mathrm{Reg}(R) \cap \mathfrak{m}$. Then M/rM is a finitely generated R-module by (3.1), hence $M = N + rM$ where $N = Rx_1 + \cdots + Rx_n$ is a finitely generated R-submodule of M. There exists $h \in \mathbb{N}$ such that $\mathfrak{m}^h x_i \subset R$ for every $i \in \{1, \ldots, n\}$, hence $\mathfrak{m}^h M \subset R + \mathfrak{m}^{h+1}M$. Set $B := R + \mathfrak{m}^h M$. Then we have $B \supset R + \mathfrak{m}B = R + \mathfrak{m}^{h+1}M \supset R + \mathfrak{m}^h M = B$, hence $B = R + \mathfrak{m}B$, and the R-module $B \neq Q$ is a compatible extension of R by (3.5)(2). Since $A = R + \mathfrak{m}^h A \subset R + \mathfrak{m}^h M = B \subset M$, we have $A = B$, and therefore we have $\mathfrak{m}^h M \subset A$. Now $M = N + rM = \cdots = N + r^h M$, and since $r^h M \subset A$, we have $M = Ax_1 + \cdots + Ax_n + A$. Thus, we have shown that M is a finitely generated A-module. Since the R-module M/A is annihilated by $\mathfrak{m}^h$ and since it is a homomorphic image of the finitely generated R-module $M/r^h M$, it has finite length.

(3.14) Theorem: *The R-module Q/R is artinian.*

Proof: By (3.12)(1) the set of compatible extensions of R satisfies both chain conditions. Therefore we can construct a chain $R = A_0 \subset A_1 \subset \cdots \subset A_n = Q$ with compatible extensions $A_1, \ldots, A_{n-1}$ of R, and there is no other compatible extension of R lying properly between two consecutive ones. Let $i \in \{0, \ldots, n-1\}$, and let M be an R-submodule of Q with $A_i \subset M \subsetneqq A_{i+1}$. Then M/A_i is an R-module of finite length by (3.13), hence A_{i+1}/A_i is an artinian R-module, and the claim follows easily.

(3.15) Proposition: *Let M be an artinian R-module. Then M is a noetherian R-module iff M has no divisible submodules except $\{0\}$.*

Proof (1) Assume that M is noetherian; then the divisible submodule $d_R(M)$ is noetherian. Now we have $d_R(M) = \{0\}$ by (3.5)(1).
(2) Assume that M has no divisible submodules except $\{0\}$. Suppose that M is not noetherian. Let M' be a minimal element in the non-empty set of submodules of M which are not noetherian; then $M' \neq \{0\}$. M' is not divisible by assumption; hence there exists $r \in \mathrm{Reg}(R)$ such that $rM' \subsetneqq M'$. Now, by the choice of M', rM' is a noetherian module, and $N := M'/rM'$ is not noetherian. We choose $h \in \mathbb{N}$ such that $\mathfrak{m}^h \subset Rr$ [cf. (2.3)(1)], and consider the decreasing chain $N \supset \mathfrak{m}N \supset \cdots \supset \mathfrak{m}^h N = \{0\}$ of submodules of N. Let $i \in \{0, \ldots, h-1\}$. The R-module $\mathfrak{m}^i N/\mathfrak{m}^{i+1} N$ is artinian, and it is annihilated by $\mathfrak{m}$, hence it is an artinian $R/\mathfrak{m}$-module; since $R/\mathfrak{m}$ is a field, $\mathfrak{m}^i N/\mathfrak{m}^{i+1} N$ is a noetherian $R/\mathfrak{m}$-module [cf. [21], Prop. 6.10], and therefore it is a noetherian R-module. This implies that also N is a noetherian R-module, which is in contradiction with the choice of N. This contradiction shows that M is noetherian.

3.4 Criteria for One-Dimensional Analytically Unramified and Analytically Irreducible CM-Rings

(3.16) DEFINITION: A local ring is called analytically irreducible (resp. analytically unramified) if its completion is a domain (resp. is reduced).

(3.17) Theorem: *The following statements are equivalent:*
(1) R is analytically irreducible.
(2) Q/R is a simple divisible R-module.
(3) Q is a field, and the integral closure V of R in Q is a discrete valuation ring and a finitely generated R-module.

Proof (1) $\Leftrightarrow$ (2): We know that Q/R is a simple divisible R-module iff $Q(\widehat{R})/\widehat{R}$ is a simple divisible $\widehat{R}$-module [cf. (3.11)], and that there is a bijective map between the set of divisible $\widehat{R}$-submodules of $Q(\widehat{R})/\widehat{R}$ and the set of ideals of $Q(\widehat{R})$ [cf. (3.9)]; in particular, $Q(\widehat{R})$ is a field iff $Q(\widehat{R})/\widehat{R}$ is a simple divisible $\widehat{R}$-module.
(2) $\Rightarrow$ (3): (a) By the equivalence of (1) and (2) we know that Q is a field.
(b) A proper R-submodule M of Q which contains R is noetherian. In fact, M/R is a submodule of the artinian module Q/R [cf. (3.14)], hence it is artinian, and has no proper divisible submodules except $\{0\}$ by assumption, hence it is noetherian by (3.15), and therefore M is a noetherian R-module.
(c) In particular, the integral closure V of R is a finitely generated R-module, hence V is an integrally closed noetherian domain, and we have $\dim(V) = 1$ [cf.

[63], Prop. 9.2]. We show that V is local, hence that V is a discrete valuation ring [cf. I(3.28)].

In fact, let $\mathfrak{n}$ be a maximal ideal of V. Then $V_{\mathfrak{n}}$ is a proper subring of Q containing R, hence it is a finitely generated R-module by (a), and therefore integral over R [cf. [63], Cor. 4.5]. It follows that $V_{\mathfrak{n}} \subset V$, and therefore we have $V = V_{\mathfrak{n}}$. This implies that every element of V which lies not in $\mathfrak{n}$ is a unit of V, hence V is local, and $\mathfrak{n}$ is its maximal ideal.

(3) $\Rightarrow$ (2): We show that every proper subring A of Q containing R is contained in V. In fact, assume that $A \not\subset V$, and choose $a \in A$, $a \notin V$. Then $1/a$ lies in the maximal ideal of V, hence we have $Q = V[a]$ [cf. (2.3)]. There exists $r \in \mathrm{Reg}(R)$ with $rV \subset R$, and therefore $Q = rV[a] \subset R[a] \subset A$, hence we have $A = Q$, contradicting the choice of A.

This result implies, in particular, the following: Let A/R be a divisible submodule of Q/R; then A is a subring of Q containing R [cf. (3.5)], hence we have $A \subset V$, and therefore $A = R$ [cf. (3.5)(1)]. Thus, we have shown that Q/R is a simple divisible R-module. This ends the proof.

As an easy application of this result we prove

(3.18) **Proposition:** *Let V be a complete discrete valuation ring, let K be its field of quotients, and let L be a finite extension of K. Then there exists a unique extension W of V to L, W is a finitely generated free V-module, and we have*

$$e(W/V)f(W/V) = [L:K].$$

Let v (resp. w) be the valuation of K (resp. of L) defined by V (resp. W). Then we have

$$v(\mathrm{N}_{L/K}(x)) = f(W/V)w(x) \quad \text{for every } x \in L.$$

Proof: We set $n := [L:K]$ and let T be the integral closure of V in L. Then T contains a basis $\{x_1, \ldots, x_n\}$ of the K-vector space L. The ring $S = V[x_1, \ldots, x_n]$ is a finitely generated V-module [cf. [63], Cor. 4.5], hence S is a complete semilocal domain [cf. B(3.8)], and therefore it is local [cf. B(3.5)]. Now T is the integral closure of S, hence it is a discrete valuation ring which is a finitely generated S-module [cf. (3.17)]. Therefore we have $W = T$, and W is a finitely generated free V-module [cf. [63], Ex. 4.1 for the first assertion and I(3.7) for the second one]. The assertion $e(W/V)f(W/V) = [L:K]$ now follows from I(6.22).

Let t be a generator of the maximal ideal of V, and let u be a generator of the maximal ideal of W. It is enough to show that $v(\mathrm{N}_{L/K}(u)) = f(W/V)$; this follows from [note that the norm of a unit of W is a unit of V]

$$0 = v(\mathrm{N}_{L/K}(u^{e(W/V)}/t)) = e(W/V)v(\mathrm{N}_{L/K}(u)) - [L:K]$$
$$= e(W/V)(v(\mathrm{N}_{L/K}(u)) - f(W/V)).$$

(3.19) **Theorem:** (1) *The map $A \mapsto$ (integral closure of A in Q) is a bijective inclusion-preserving map from the set of maximal compatible extensions of R to*

*the set of Manis valuation rings belonging to R; the inverse map is given by $V \mapsto A$
where $A/R = d_R(V/R)$.*
*(2) The map $V \mapsto V\widehat{R}$ is a bijective inclusion-preserving map from the set of Manis
valuation rings belonging to R to the set of Manis valuation rings belonging to $\widehat{R}$;
the inverse map is given by $W \mapsto W \cap Q$.*
*(3) Let S be the integral closure of R. Then $S\widehat{R}$ is the integral closure of $\widehat{R}$
in $Q(\widehat{R})$.*

Proof: (1)(a) Let A be a maximal compatible extension of R, and let V be the
integral closure of A in Q. Let $x \in Q$ be regular. We show that $A[x] \neq Q$ or
$A[x^{-1}] \neq Q$. In fact, suppose that $A[x] = Q$ and $A[x^{-1}] = Q$. Since $\mathfrak{m}Q = Q$,
we have relations

$$1 = a_0 + a_1 x + \cdots + a_m x^m, \; 1 = b_1 + b_1 x^{-1} + \cdots + b_n x^{-n} \qquad (*)$$

with $a_0, \ldots, a_m$ and $b_0, \ldots, b_n \in \mathfrak{m}A$. We may assume that the relations in $(*)$ are
of the smallest possible degrees m and n. Since $\mathfrak{m}A \neq A$ [cf. (3.8)(1)], we have
$m \geq 1$ and $n \geq 1$. Let, say, $n \leq m$. Multiplying the first equation in $(*)$ by $1 - b_0$
and the second equation by $a_m x^m$, we get

$$1 - b_0 = (1 - b_0)a_0 + \cdots + (1 - b_0)a_m x^m,$$
$$(1 - b_0)a_m x^m = a_m b_1 x^{m-1} + \cdots + a_m b_n x^{m-n},$$

hence we get a relation

$$1 = c_0 + \cdots + c_{m-1} x^{m-1} \quad \text{with } c_0, \ldots, c_{m-1} \in \mathfrak{m}A$$

which is in contradiction with the choice of m.
Let $x \in Q$ be regular; by what we have just shown, we may assume that $A[x] \neq Q$.
Then $A[x]$ is a finitely generated R-module by (3.13), hence x is integral over A
[cf. [63], Cor. 4.6], and therefore $x \in V$. Thus, we have shown that V is a Manis
valuation ring of Q, and therefore a Manis valuation ring of Q belonging to R.
Let B be defined by $B/R = d_R(V/R)$; then B is a compatible extension of R.
Since A/R is a divisible R-module, we have $B \supset A$, hence we have $B = A$ since A
is a maximal compatible extension of R.
(b) Let V be a Manis valuation ring of Q belonging to R, and let A be defined by
$A/R = d_R(V/R)$; then A is a compatible extension of R, and there is no further
compatible extension of R between A and V. The R-module V/A is artinian by
(3.14), and, by (3.13), V is a finitely generated A-module and V/A is a noetherian
R-module. Therefore V is integral over A, and we have $d_R(V/A) = \{0\}$ [cf. (3.15)].
(c) Thus, we have shown that the maps mentioned in the statement of (1) are
mutually inverse; clearly they are inclusion-preserving.
(2) Let V be a Manis valuation ring of Q belonging to R, let A be defined by
$A/R = d_R(V/R)$, and set $B := A\widehat{R}$. Then A is a maximal compatible extension
of R and V is its integral closure by (1), and B is a maximal compatible extension

of $\widehat{R}$ by (3.11). Let W be the integral closure of B in $Q(\widehat{R})$. Clearly we have $\widehat{R}V \subset W$. We set $U := Q \cap W$; then we have $V \subset U$ by (3.11). Since U is a finitely generated A-module [cf. (3.13)], it is integral over A [cf. [63], Cor. 4.5], hence $U \subset V$, and therefore we have $W = \widehat{R}V$ by (3.11).

(3) Let $V_1, \ldots, V_h$ be the Manis valuation rings belonging to R. Then $V_1\widehat{R}, \ldots, V_h\widehat{R}$ are the Manis valuation rings belonging to $\widehat{R}$ [cf. (2)], $S = V_1 \cap \cdots \cap V_h$ is the integral closure of R in Q by (2.10), $V_1\widehat{R} \cap \cdots \cap V_h\widehat{R}$ is the integral closure of $\widehat{R}$ in $Q(\widehat{R})$ by (2.11), and $S\widehat{R} = V_1\widehat{R} \cap \cdots \cap V_h\widehat{R}$ [cf. [34], Ch. I, § 3, no. 5, Prop. 10, and note that $\widehat{R}$ is a faithfully flat R-module].

(3.20) Proposition: *Let $A_1, \ldots, A_h$ be the maximal compatible extensions of R, and, for every $i \in \{1, \ldots, h\}$, let V_i be the integral closure of A_i. Set $S := V_1 \cap \cdots \cap V_h$, $T := A_1 \cap \cdots \cap A_h$. Then:*
(1) *The following sequences of R-modules*

$$0 \to T \to Q \to \bigoplus_{i=1}^{h} Q/A_i \to 0, \quad 0 \to S \to Q \to \bigoplus_{i=1}^{h} Q/V_i \to 0$$

are exact.
(2) *The R-module S/T has finite length; in particular, we have the estimate*

$$\ell_R(S/T) \leq \sum_{i=1}^{h} \ell_R(V_i/A_i).$$

Proof: (1) Since $\widehat{R}$ is a faithfully flat R-module, a sequence $M' \to M \to M''$ of R-modules is exact iff $M' \otimes_R \widehat{R} \to M \otimes_R \widehat{R} \to M'' \otimes_R \widehat{R}$ is exact. We have $S\widehat{R} = V_1\widehat{R} \cap \cdots \cap V_h\widehat{R}$ and $T\widehat{R} = A_1\widehat{R} \cap \cdots \cap A_h\widehat{R}$ by [34], Ch. I, § 3, no. 5, Prop. 10, $Q(\widehat{R}) = \widehat{R}Q$ by (2.4)(1), $V_1\widehat{R}, \ldots, V_h\widehat{R}$ are the Manis valuation rings belonging to $\widehat{R}$ by (3.19)(2), and $A_1\widehat{R}, \ldots, A_h\widehat{R}$ are the maximal compatible extensions of $\widehat{R}$ by (3.11). We may assume, therefore, that R is complete.
Let $\mathfrak{p}_1, \ldots, \mathfrak{p}_h$ be the maximal ideals of Q labelled in such a way that $A_i = R + \mathfrak{p}_i$ for every $i \in \{1, \ldots, h\}$ [cf. (3.9)]. Then, for any $i \in \{1, \ldots, h-1\}$, we have $(A_1 \cap \cdots \cap A_i) + A_{i+1} = Q$ since $(\mathfrak{p}_1 \cap \cdots \cap \mathfrak{p}_i) + \mathfrak{p}_{i+1} = Q$. The assertions in (1) are now a consequence of the lemma below, whose proof is simple and is left as an exercise.
(2) The R-module homomorphism $S \to \bigoplus_{i=1}^{h} V_i/A_i$, defined by $S \to V_i \to V_i/A_i$ for every $i \in \{1, \ldots, h\}$, clearly has T as its kernel. The R-modules $V_1/A_1, \ldots, V_h/A_h$ have finite length [cf. (3.13)], hence we get the estimate in (2).

(3.21) Lemma: *Let A be a ring, let M be an A-module, and let $N_1, \ldots, N_h$ be submodules of M with the following property: For every $j \in \{1, \ldots, h-1\}$ we have $(N_1 \cap \cdots \cap N_j) + N_{j+1} = M$. Then the canonical homomorphism $M \to M/N_1 \oplus \cdots \oplus M/N_h$ is surjective and has kernel $N_1 \cap \cdots \cap N_h$.*

(3.22) Theorem: *The following statements are equivalent:*

(1) *R is analytically unramified.*

(2) *The integral closure S of R in Q is a finitely generated R-module.*

Proof: Let $A_1, \ldots, A_h$ be the maximal compatible extensions of R, and set $T := A_1 \cap \cdots \cap A_h$. Then S/T is an R-module of finite length by (3.20); in particular, S/T is a noetherian R-module. Now, S is a noetherian R-module iff T is a noetherian R-module by [63], Ex. 1.3, and T is a noetherian R-module iff $T\widehat{R}$ is a noetherian $\widehat{R}$-module by [34], Ch. I, § 3, no. 5, Prop. 10 and no. 6, Prop. 11. Let $\{\mathfrak{p}_1, \ldots, \mathfrak{p}_h\}$ be the set of maximal ideals of $Q(\widehat{R})$; then $\mathfrak{n} := \mathfrak{p}_1 \cap \cdots \cap \mathfrak{p}_h$ is the nilradical of $Q(\widehat{R})$. Thus, we have to show that $\mathfrak{n} = \{0\}$ iff $T\widehat{R}$ is a noetherian $\widehat{R}$-module.

Let the prime ideals $\mathfrak{p}_1, \ldots, \mathfrak{p}_h$ be labelled in such a way that $B_1 := A_1\widehat{R} = \widehat{R} + \mathfrak{p}_1, \ldots, B_h := A_h\widehat{R} = \widehat{R} + \mathfrak{p}_h$ are the maximal compatible extensions of $\widehat{R}$ [cf. (3.9)]. Note that $T\widehat{R} = B_1 \cap \cdots \cap B_h$ by [34], Ch. I, § 3, no. 5, Prop. 10.

We assume that $\mathfrak{n} = \{0\}$. The canonical $\widehat{R}$-linear map $T\widehat{R} = B_1 \cap \cdots \cap B_h \to B_1/\mathfrak{p}_1 \oplus \cdots \oplus B_h/\mathfrak{p}_h$ has kernel $\mathfrak{n}$, hence $T\widehat{R}$ is a submodule of the $\widehat{R}$-module $B_1/\mathfrak{p}_1 \oplus \cdots \oplus B_h/\mathfrak{p}_h$. Now $B/\mathfrak{p}_i = \widehat{R}/(\widehat{R} \cap \mathfrak{p}_i)$ is a cyclic $\widehat{R}$-module for every $i \in \{1, \ldots, h\}$, hence $T\widehat{R}$ is a noetherian $\widehat{R}$-module.

We assume that $T\widehat{R}$ is a noetherian $\widehat{R}$-module. Since $\widehat{R} + \mathfrak{n} \subset T\widehat{R}$, we see that $(\widehat{R}+\mathfrak{n})/\widehat{R}$ is a divisible [cf. (3.9)] noetherian $\widehat{R}$-module, hence it is the zero module by (3.5)(1). Therefore $\mathfrak{n}$ is contained in $\widehat{R}$. Suppose that $\mathfrak{n}$ is not the zero ideal. Then there exists $x \in \mathfrak{n}$ with $x \neq 0$, $x^2 = 0$. Let $r \in \mathfrak{m}$ be regular. Since $\mathfrak{n}$ is an ideal in $\widehat{Q}$, we have in $\widehat{R}$ the ascending chain $((x/r^n)\widehat{R})_{n\in\mathbb{N}}$ of ideals. This chain becomes stationary, hence there exist $n \in \mathbb{N}$ and $s \in R$ with $sx = x/r$, hence $x(1 - rs) = 0$. Now $1 - sr$ is a unit of R since $r \in \mathfrak{m}$, hence we have $x = 0$, which contradicts our choice of x. Therefore we have $\mathfrak{n} = \{0\}$.

4　Blowing up Ideals

(4.0) NOTATION: In this section R always is a one-dimensional semilocal CM-ring, $\mathfrak{r}$ is its Jacobson radical, Q is its ring of quotients, and S is the integral closure of R in Q.

4.1　The Blow-up Ring R^a

(4.1) NOTATION: (1) For R-submodules M, N of S we define

$$M : N := (M : N)_S = \{x \in S \mid xN \subset M\}$$

[the notation $M : N$ instead of the more exact notation $(M : N)_S$ will be used only in this section].

(2) Let $\mathfrak{a}$ be an ideal of R, and let $n \in \mathbb{N}$; then, clearly, $\mathfrak{a}^n : \mathfrak{a}^n$ is a subring of S containing R, and we have $\mathfrak{a}^n : \mathfrak{a}^n \subset \mathfrak{a}^{n+1} : \mathfrak{a}^{n+1}$, showing that

$$R^{\mathfrak{a}} := \bigcup_{n \in \mathbb{N}} (\mathfrak{a}^n : \mathfrak{a}^n)$$

is a union of an increasing sequence of subrings of S, hence it is a subring of S containing R.

(4.2) Definition: Let $\mathfrak{a}$ be an open ideal of R. -
(1) The ring $R^{\mathfrak{a}}$ is called the blow-up of R with respect to $\mathfrak{a}$; we also say that we get $R^{\mathfrak{a}}$ by blowing up the ideal $\mathfrak{a}$.
(2) Let R be local with maximal ideal $\mathfrak{m}$. The ring $R^{\mathfrak{m}}$ is called the quadratic transform of R.

(4.3) Proposition: *Let $\mathfrak{a}$ be an open ideal of R. Then:*
(1) *$R^{\mathfrak{a}} = \{x/a \mid s \in \mathbb{N}, a \text{ transversal for } \mathfrak{a} \text{ of order } s, x \in \mathfrak{a}^s\}$.*
(2) *There exist $t \in \mathbb{N}$ and $a \in \mathfrak{a}^t$ transversal for $\mathfrak{a}$ of order t such that $R^{\mathfrak{a}} = \mathfrak{a}^t a^{-1}$.*
(3) *$R^{\mathfrak{a}}$ is a finitely generated R-module, and therefore is a one-dimensional semilocal CM-ring.*
(4) *For all sufficiently large integers n we have $R^{\mathfrak{a}} = \mathfrak{a}^n : \mathfrak{a}^n$ and $\mathfrak{a}^n = \mathfrak{a}^n R^{\mathfrak{a}}$.*
(5) *There exists a regular $r \in R^{\mathfrak{a}}$ such that $\mathfrak{a}R^{\mathfrak{a}} = R^{\mathfrak{a}}r$.*
(6) *If A is any subring of S containing R and $\mathfrak{a}A$ is a principal ideal in A, then we have $R^{\mathfrak{a}} \subset A$.*

Proof: (a) The set on the righthand side in (1) clearly is a subring of Q containing R [cf. (1.16)(2)]; call it B. Let $b \in \mathfrak{a}^t$ be a transversal element for $\mathfrak{a}$ of order t [for some $t \in \mathbb{N}$]; b is a regular element of R by (1.15). We choose $k \in \mathbb{N}$ so large that $b\mathfrak{a}^{kt} = \mathfrak{a}^{kt+t}$. Then we have $b^k \mathfrak{a}^{kt} = \mathfrak{a}^{2kt}$. We set $s := kt$ and $a := b^k$; then we have $a \in \mathfrak{a}^s$ and $a\mathfrak{a}^s = \mathfrak{a}^{2s}$, hence a is transversal for $\mathfrak{a}$ of order s. We set $C := \mathfrak{a}^s a^{-1} \subset Q$. Clearly we have $C \cdot C = C$, hence C is a subring of B containing R, and, moreover, it is a finitely generated R-module, hence C is integral over R [cf. [63], Cor 4.5]. Since $\mathfrak{a}^t b^{-1} \subset C$, we see that every element of $\mathfrak{a}^t b^{-1}$ is integral over R. Now, b was an arbitrary element of R being transversal for $\mathfrak{a}$ [of some order], hence B is integral over R, and a fortiori over C.
(b) Keeping the notation introduced in (a), we shall show that $B \subset C$—whence $B = C$. Let $y \in B$; then $y = z/c$ where c is transversal for $\mathfrak{a}$, say of order m, and where $z \in \mathfrak{a}^m$. Let us write $y = (zc^{s-1}/a^m)(a^m/c^s)$; clearly, zc^{s-1}/a^m and c^s/a^m are elements of C. Moreover, c^s/a^m is a unit of B, and since B is integral over C, c^s/a^m is a unit of C [otherwise, there would exist a maximal ideal $\mathfrak{n}$ of C containing c^s/a^m, and since there exist maximal ideals of B lying over $\mathfrak{n}$ by [63], Prop. 4.15 and Cor. 4.17, c^s/a^m would not be a unit of B], hence we have $y \in C$.
(c) We shall show that $R^{\mathfrak{a}} = \mathfrak{a}^s a^{-1}$ $[= C = B]$. Let $y \in R^{\mathfrak{a}}$, and choose $h \in \mathbb{N}$ such that $y \in \mathfrak{a}^{hs} : \mathfrak{a}^{hs}$; then $y \in \mathfrak{a}^{hs} a^{-h} = C^h = C$. Let $y \in \mathfrak{a}^s a^{-1}$. Then we have $y \in S$ [cf. (a)], and $y\mathfrak{a}^{2s} = ya\mathfrak{a}^s \subset \mathfrak{a}^{2s}$, hence we have $y \in \mathfrak{a}^{2s} : \mathfrak{a}^{2s} \subset R^{\mathfrak{a}}$.

(d) As R^a is a finitely generated R-module, (3) holds [cf. B(3.8) and [63], Prop. 9.2], the first statement in (4) holds, and from $\mathfrak{a}^s = aR^a$ we get $\mathfrak{a}^s = aR^a \subset \mathfrak{a}^s R^a = aR^a$, hence $\mathfrak{a}^s R^a = aR^a$ and $\mathfrak{a}^s R^a = \mathfrak{a}^s$, and therefore we have $\mathfrak{a}^n R^a = \mathfrak{a}^n$ for every integer $n \geq s$.

(e) The statement in (5) is equivalent to $\mathfrak{a}R^a$ being an invertible R^a-module [cf. (2.2)(3), and note that R^a is semilocal]. By (d) we have $\mathfrak{a}R^a \cdot \mathfrak{a}^{s-1} R^a = aR^a$, hence $\mathfrak{a}R^a$ is invertible.

(f) Now we prove (6). Assume that $\mathfrak{a}A = Aw$ for some $w \in A$. Then w is a regular element of A [because $\mathfrak{a}$ contains regular elements of R]. Let $y \in R^a$, and choose $n \in \mathbb{N}$ such that $y\mathfrak{a}^n \subset \mathfrak{a}^n$. Then we have $y\mathfrak{a}^n A \subset \mathfrak{a}^n A$, hence $yw^n \in w^n A$, and therefore we have $y \in A$ [since w^n is a regular element of A].

(4.4) REMARK: Let $\mathfrak{a}$ be an open ideal of R.

(1) In the course of proving (4.3), we have shown the following: There exist $s \in \mathbb{N}$ and $a \in \mathfrak{a}^s$ such that a is transversal for $\mathfrak{a}$ of order s and that $a\mathfrak{a}^s = \mathfrak{a}^{2s}$; in this case we have $R^a = \mathfrak{a}^s a^{-1} = R[\mathfrak{a}^s/a]$.

(2) Assume that there exists an $\mathfrak{a}$-transversal element a; then we have $aR^a = \mathfrak{a}R^a$. In fact, choose $n \in \mathbb{N}$ with $a\mathfrak{a}^n = \mathfrak{a}^{n+1}$. Then we have $a\mathfrak{a}^n R^a = \mathfrak{a}^{n+1} R^a$, and since $\mathfrak{a}R^a = R^a r$ for some regular $r \in R^a$ [cf. (4.3)(5)], we have $aR^a = rR^a = \mathfrak{a}R^a$. Moreover, we have $a^n \mathfrak{a}^n = \mathfrak{a}^{2n}$ and, by (1), we have $R^a = R[\mathfrak{a}^n/a^n] = R[\mathfrak{a}/a]$. In particular, we have $R^a = R[a_1/a, \ldots, a_h/a]$ for any system of generators $\{a_1, \ldots, a_h\}$ of $\mathfrak{a}$.

(3) Now we assume that $\mathfrak{a}$ is contained in the Jacobson radical of R, and that there exists an $\mathfrak{a}$-transversal element a; then we have $R^a = R[\mathfrak{a}/a]$ [cf. (2)]. Let $\mathcal{R}(\mathfrak{a}, R) = \bigoplus_{n \geq 0} \mathfrak{a}^n T^n$ be the Rees ring of R with respect to $\mathfrak{a}$ [cf. B(5.1)]; there exists a canonical isomorphism $\psi\colon \mathcal{R}(\mathfrak{a}, R)_{(aT)} \to R^a$ [cf. B(5.5)]. Let $\varphi\colon \mathcal{R}(\mathfrak{a}, R) \to \mathrm{gr}_\mathfrak{a}(R)$ be the canonical homomorphism, and let $J := \bigoplus_{n > 0} \mathfrak{a}^{n+1} T^n$ be its kernel [cf. B(5.2)]; we set $E := V_+(J) \subset \mathrm{Proj}(\mathcal{R}(\mathfrak{a}, R))$ [cf. B(5.7)]. Every homogeneous ideal $\mathfrak{A}$ of $\mathrm{gr}_\mathfrak{a}(R)$ which contains $a \bmod \mathfrak{a}^2$ contains a power of $(\mathrm{gr}_\mathfrak{a}(R))_+$ [since $\mathfrak{A}_n = (\mathrm{gr}_\mathfrak{a}(R))_n$ for every n with $a\mathfrak{a}^{n-1} = \mathfrak{a}^n$], hence we have $\mathrm{Proj}(\mathrm{gr}_\mathfrak{a}(R)) = D_+(a \bmod \mathfrak{a}^2)$.

(a) The element a is a regular element of R, and we have $\mathfrak{a}R^a = aR^a$ [cf. (2)]. Moreover, since R^a is integral over R, every maximal ideal of R^a lies over a maximal ideal of R, and over every maximal ideal of R there lies a maximal ideal of R^a [cf. [63], Cor. 4.17] and therefore the set $\mathrm{Max}(R^a)$ of maximal ideals of R^a is the set $V(\mathfrak{a}R^a)$ of prime ideals of R^a containing $\mathfrak{a}$, and there exists a bijective map $E \cap D_+(aT) \to \mathrm{Max}(R^a)$ [cf. B(5.7)(2)]. Let $\mathfrak{p}$ be a maximal ideal of R^a, and let $\mathfrak{q}$ be the homogeneous prime ideal of $\mathcal{R}(\mathfrak{a}, R)$ corresponding to $\mathfrak{p}$. Then there exists a canonical isomorphism $\mathcal{R}(\mathfrak{a}, R)_{(\mathfrak{q})} \to (R^a)_\mathfrak{p}$ [cf. B(5.6)(2)].

(b) Now we assume that R is local and that $\mathfrak{a} = \mathfrak{m}$, the maximal ideal of R. Since $\dim(\mathrm{gr}_\mathfrak{m}(R)) = 1$ [cf. [63], Ex. 13.8, 3)], and since the maximal ideal $(\mathrm{gr}_\mathfrak{m}(R))_+$ of $\mathrm{gr}_\mathfrak{m}(R)$ is its only irrelevant homogeneous prime ideal, we see the following: $\mathrm{Proj}(\mathrm{gr}_\mathfrak{m}(R))$ consists of all homogeneous prime ideals of $\mathrm{gr}_\mathfrak{m}(R)$ different from $(\mathrm{gr}_\mathfrak{m}(R))_+$, we have $E \subset D_+(aT)$, and there exists a bijective map

$\text{Proj}(\text{gr}_{\mathfrak{m}}(R)) \to \text{Max}(R^m)$.
For another description the reader should read VIII(1.3).

(4.5) Corollary: *Let T be a one-dimensional semilocal CM-ring with ring of quotients $Q(T)$, and let $\varphi\colon R \to T$ be a homomorphism such that $\varphi(\text{Reg}(R)) \subset \text{Reg}(T)$. Let $\mathfrak{a}$ be an open ideal of R. Then:*
(1) The ideal $\mathfrak{a}T$ is an open ideal of T, φ admits a unique extension to a homomorphism of rings $\varphi^{\mathfrak{a}}\colon R^{\mathfrak{a}} \to T^{\mathfrak{a}T}$, and the induced homomorphism $\varphi_T^{\mathfrak{a}}\colon T \otimes_R R^{\mathfrak{a}} \to T^{\mathfrak{a}T}$, defined by $\varphi_T^{\mathfrak{a}}(x \otimes y) = x\varphi^{\mathfrak{a}}(y)$ for all $x \in T$, $y \in R^{\mathfrak{a}}$, is surjective.
(2) If $\varphi\colon R \to T$ is a flat homomorphism, then $\varphi_T^{\mathfrak{a}}$ is an isomorphism (in this case, the condition $\varphi(\text{Reg}(R)) \subset \text{Reg}(T)$ holds automatically).

Proof: (1) Since $\mathfrak{a}$ is an open ideal of R, $\mathfrak{a}$ contains a regular element of R, hence $\mathfrak{a}T$ contains a regular element of T, and it is therefore an open ideal of T [cf. (1.6)(2)]. Clearly φ has a unique extension to a homomorphism $\psi\colon Q \to Q(T)$. If $\varphi^{\mathfrak{a}}$ exists, it must be the restriction of ψ to $R^{\mathfrak{a}}$. Choose s and a as in (4.4)(1); then we have $R^{\mathfrak{a}} = \mathfrak{a}^s a^{-1}$. Furthermore, since $\mathfrak{a}^s T = (\mathfrak{a}T)^s$, we have $a\mathfrak{a}^s T = \mathfrak{a}^{2s}T$, hence $T^{\mathfrak{a}T} = (\mathfrak{a}^s T)\psi(a)^{-1}$, and therefore $\psi(R^{\mathfrak{a}}) = \psi(\mathfrak{a})^s\psi(a)^{-1} \subset \mathfrak{a}^s\psi(a)^{-1}T = T^{\mathfrak{a}T}$; this implies that $\psi(R^{\mathfrak{a}})T = T^{\mathfrak{a}T}$. Thus, $\varphi^{\mathfrak{a}}$ exists, and $\varphi_T^{\mathfrak{a}}$ is surjective.
(2) The homomorphism $\psi\colon Q \to Q(T)$ induces a homomorphism $T \otimes_R Q \to Q(T)$ which is easily seen to be injective; furthermore, $T \otimes_R R^{\mathfrak{a}} \to T \otimes_R Q$ is injective [since φ is flat], hence $\varphi_T^{\mathfrak{a}}$ is injective, and therefore an isomorphism [since it is surjective by (1)].

(4.6) REMARK: Let $\mathfrak{a}$ be an open ideal of R. Corollary (4.5) can be used in the following three cases:
(1) T is the $\mathfrak{r}$-adic completion $\widehat{R}$ of R; then $\mathfrak{a}\widehat{R} = \widehat{\mathfrak{a}}$ and $\widehat{R}^{\widehat{\mathfrak{a}}} = R^{\mathfrak{a}}\widehat{R} \subset Q(\widehat{R})$.
(2) T is the ring of fractions $\Sigma^{-1}R$ of R with respect to a multiplicatively closed system Σ of R such that $\Sigma^{-1}R$ is a one-dimensional CM-ring [this is the case if Σ is the complement of a union of maximal ideals of R]. Then the $\Sigma^{-1}R$-algebras $(\Sigma^{-1}R)^{\Sigma^{-1}\mathfrak{a}}$ and $\Sigma^{-1}(R^{\mathfrak{a}})$ are canonically isomorphic.
(3) $T := R/\mathfrak{c}$, where $\mathfrak{c}$ is an ideal of R such that $\text{Ass}(R/\mathfrak{c})$ contains no maximal ideal. Then $R/\mathfrak{c}$ is a one-dimensional semilocal CM-ring, and $Q(R)/\mathfrak{c}Q(R)$ is the ring of quotients of $R/\mathfrak{c}$ [as one checks easily]. Let $\varphi\colon R \to R/\mathfrak{c}$ be the canonical homomorphism and $\psi\colon Q(R) \to Q(R/\mathfrak{c})$ its extension; since $(R/\mathfrak{c})^{\mathfrak{a}(R/\mathfrak{c})}$ is the smallest subring of $Q(R/\mathfrak{c})$ containing $R/\mathfrak{c}$ and $\psi(R^{\mathfrak{a}})$ [cf. (4.5)(1)], it is equal to the ring $R^{\mathfrak{a}}/(\mathfrak{c}Q(R) \cap R^{\mathfrak{a}})$.

4.2 Integral Closure

(4.7) NOTATION: Set $R_0 := R$, and, for every $i \in \mathbb{N}_0$, let R_{i+1} be the blow-up of R_i with respect to its Jacobson radical. We have an ascending chain

$$R = R_0 \subset R_1 \subset \cdots \subset S$$

of subrings of the integral closure S of R; the sequence $(R_i)_{i\geq 0}$ is called the blow-up sequence of R, or the sequence of blow-ups of R. Note that $R_{i+j} = (R_i)_j$ for all $i, j \in \mathbb{N}_0$ [$(R_i)_j$ is the j-th blow-up of R_i].

The following result allows us to construct the integral closure S of R "step by step"—at least in the case where S is a finitely generated R-module.

(4.8) **Proposition:** Let $(R_i)_{i\geq 0}$ be the blow-up sequence of R. Then we have

$$S = \bigcup_{i\geq 0} R_i.$$

Proof: Let $n \in \mathbb{N}$. We assume that for every one-dimensional semilocal CM-ring T and every $t \in \text{Reg}(T)$ such that $\ell_R(T/Tt) < n$, the following holds true: For every $b \in T$ such that b/t is integral over T, there exists $i \in \mathbb{N}$ such that $b/t \in T_i$, the i-th blow-up of T. [If $\ell_R(T/Tt) = 0$, then t is a unit of T, and the assumption is true.] Let $r \in \text{Reg}(R)$ be such that $1 \leq \ell_R(R/Rr) \leq n$, and let $a \in R$ be such that a/r is integral over R. Let $\mathfrak{m}$ be a maximal ideal of R such that $r \in \mathfrak{m}$, and let $Z^h + a_1 Z^{h-1} + \cdots + a_h \in R[Z]$ be an equation of integral dependence of a/r over R. Multiplying this equation by r^h shows that also $a \in \mathfrak{m}$. Note that $\mathfrak{r} = \mathfrak{m}\mathfrak{q}$ for some ideal $\mathfrak{q}$ of R. By (4.3)(5) there exists a regular element $w \in R_1$ such that $R_1 w = \mathfrak{r}R_1 = (\mathfrak{m}R_1)(\mathfrak{q}R_1)$. This implies that $\mathfrak{m}R_1$ is a principal ideal of R_1 [cf. (2.2)], say $\mathfrak{m}R_1 = R_1 v$ where $v \in \text{Reg}(R_1)$. Since R_1 is integral over R, the ideal $\mathfrak{m}R_1$ is a proper ideal of R_1 [cf. [63], Prop. 4.15], and therefore v is not a unit of R_1. Now, av^{-1} and rv^{-1} lie in R_1, and from (3.1) we get [note that R_1 is a finitely generated R-module by (4.3)(3)]

$$\ell_R(R/Rr) = \ell_R(R_1/rR_1) \geq \ell_{R_1}(R_1/rR_1) > \ell_{R_1}(R_1/R_1(rv^{-1})).$$

By the induction assumption, there exists $j \in \mathbb{N}$ such that $av^{-1}/rv^{-1} \in (R_1)_{j-1}$, and since $(R_1)_{j-1} = R_j$, it follows that $a/r \in R_j$.

4.3 Stable Ideals

(4.9) DEFINITION: An open ideal $\mathfrak{a}$ of R is called stable if $\mathfrak{a}R^{\mathfrak{a}} = \mathfrak{a}$.

(4.10) **Proposition:** Let $\mathfrak{a}$ be an open ideal of R. Then:
(1) There exists $n \in \mathbb{N}$ such that $\mathfrak{a}^n$ is stable.
(2) If $\mathfrak{a}^n$ is stable for some $n \in \mathbb{N}$, then $\mathfrak{a}^m$ is stable for every integer $m \geq n$.
(3) $\mathfrak{a}$ is a stable ideal of R iff $\mathfrak{a}R_\mathfrak{m}$ is a stable ideal of $R_\mathfrak{m}$ for every maximal ideal $\mathfrak{m}$ of R.

Proof: (1) Since $\mathfrak{a}R^{\mathfrak{a}} = \mathfrak{a}$ iff $R^{\mathfrak{a}} = (\mathfrak{a} : \mathfrak{a})_S$, and since $R^{\mathfrak{a}^n} = R^{\mathfrak{a}}$ for every $n \in \mathbb{N}$, we get the assertion from (4.3)(4).
(2) Let $n \in \mathbb{N}$, and assume that $\mathfrak{a}^n$ is stable; then $\mathfrak{a}^n R^{\mathfrak{a}} = \mathfrak{a}^n$, hence $\mathfrak{a}^m R^{\mathfrak{a}} = \mathfrak{a}^m$ for every integer $m \geq n$, and therefore $\mathfrak{a}^m$ is stable.

(3) Assume that $\mathfrak{a}$ is a stable ideal of R. Then we have $\mathfrak{a}R^{\mathfrak{a}} = \mathfrak{a}$. Let $\mathfrak{m}$ be a maximal ideal of R. We have $(R^{\mathfrak{a}})_{\mathfrak{m}} = R_{\mathfrak{m}}^{\mathfrak{a}R_{\mathfrak{m}}}$ by (4.6), from this we get $(\mathfrak{a}R_{\mathfrak{m}})R_{\mathfrak{m}}^{\mathfrak{a}R_{\mathfrak{m}}} = \mathfrak{a}R_{\mathfrak{m}}$, and therefore $\mathfrak{a}R_{\mathfrak{m}}$ is a stable ideal of $R_{\mathfrak{m}}$. Conversely, assume that $\mathfrak{a}R_{\mathfrak{m}}$ is a stable ideal of $R_{\mathfrak{m}}$ for every maximal ideal $\mathfrak{m}$ of R. We have to show that the inclusion $\mathfrak{a} \hookrightarrow \mathfrak{a}R^{\mathfrak{a}}$ is bijective. Since this R-linear map induces an $R_{\mathfrak{m}}$-linear isomorphism $\mathfrak{a}R_{\mathfrak{m}} \to (\mathfrak{a}R_{\mathfrak{m}})R_{\mathfrak{m}}^{\mathfrak{a}R_{\mathfrak{m}}} = (\mathfrak{a}R^{\mathfrak{a}})_{\mathfrak{m}}$ for every maximal ideal $\mathfrak{m}$ of R by assumption, we obtain $\mathfrak{a} = \mathfrak{a}R^{\mathfrak{a}}$ [cf. [63], Cor. 2.9].

(4.11) Theorem: *Let $\mathfrak{a}$ be an open ideal of R. Then we have*

$$\rho(\mathfrak{a}) = \ell_R(R^{\mathfrak{a}}/\mathfrak{a}R^{\mathfrak{a}}), \quad \nu(\mathfrak{a}) = \ell_R(R^{\mathfrak{a}}/R).$$

Furthermore, we have

$$H(\mathfrak{a}, R)(n) = \ell_R(R/\mathfrak{a}^n) \geq \rho(\mathfrak{a})n - \nu(\mathfrak{a}) = P(\mathfrak{a}, R)(n) \quad \text{for every } n \in \mathbb{N}, \qquad (*)$$

and, for $n \in \mathbb{N}$, equality holds iff $\mathfrak{a}^n$ is stable.

Proof: Let $r \in \mathrm{Reg}(R^{\mathfrak{a}})$ be such that $\mathfrak{a}R^{\mathfrak{a}} = R^{\mathfrak{a}}r$ [cf. (4.3)(5)]. Note that the R-modules $\mathfrak{a}^{\nu}R^{\mathfrak{a}}/\mathfrak{a}^{\nu+1}R^{\mathfrak{a}} = R^{\mathfrak{a}}r^{\nu}/R^{\mathfrak{a}}r^{\nu+1}$ and $R^{\mathfrak{a}}/R^{\mathfrak{a}}r$ are isomorphic for every $\nu \in \mathbb{N}_0$.
Let $n \in \mathbb{N}$; then we have

$$\ell_R(R^{\mathfrak{a}}/\mathfrak{a}^n R^{\mathfrak{a}}) = \sum_{\nu=0}^{n-1} \ell_R(\mathfrak{a}^{\nu}R^{\mathfrak{a}}/\mathfrak{a}^{\nu+1}R^{\mathfrak{a}}) = n \cdot \ell_R(R^{\mathfrak{a}}/\mathfrak{a}R^{\mathfrak{a}}).$$

From the exact sequences $0 \to R/\mathfrak{a}^n \to R^{\mathfrak{a}}/\mathfrak{a}^n \to R^{\mathfrak{a}}/R \to 0$, $0 \to \mathfrak{a}^n R^{\mathfrak{a}}/\mathfrak{a}^n \to R^{\mathfrak{a}}/\mathfrak{a}^n \to R^{\mathfrak{a}}/\mathfrak{a}^n R^{\mathfrak{a}} \to 0$ we get

$$\ell_R(R^{\mathfrak{a}}/R) + \ell_R(R/\mathfrak{a}^n) = \ell_R(R^{\mathfrak{a}}/\mathfrak{a}^n) = \ell_R(R^{\mathfrak{a}}/\mathfrak{a}^n R^{\mathfrak{a}}) + \ell_R(\mathfrak{a}^n R^{\mathfrak{a}}/\mathfrak{a}^n),$$

hence

$$\ell_R(R^{\mathfrak{a}}/R) + \ell_R(R/\mathfrak{a}^n) - n\ell_R(R^{\mathfrak{a}}/\mathfrak{a}R^{\mathfrak{a}}) = \ell_R(\mathfrak{a}^n R^{\mathfrak{a}}/\mathfrak{a}^n) \geq 0$$

with equality iff $\mathfrak{a}^n$ is stable.
By (4.10)(1) there exists $n_0 \in \mathbb{N}$ such that $\mathfrak{a}^{n_0}$ is stable, and this implies, by (4.10)(2), that

$$P(\mathfrak{a}, R)(X) = \ell_R(R^{\mathfrak{a}}/\mathfrak{a}R^{\mathfrak{a}})X - \ell_R(R^{\mathfrak{a}}/R).$$

Thus, we have proved the assertion.

(4.12) Corollary: *Assume that R is local with maximal ideal $\mathfrak{m}$. Then we have*

$$e(R) = \ell_R(R^{\mathfrak{m}}/\mathfrak{m}R^{\mathfrak{m}}).$$

(4.13) Corollary: *Assume that R is local, and that r is a regular element of the maximal ideal of R. Then we have $\ell_R(R/Rr) \geq e(R)$.*

Proof: For all large enough $n \in \mathbb{N}$ we have, using (3.2),

$$n\ell_R(R/Rr) = \ell_R(R/Rr^n) \geq \ell_R(R/\mathfrak{m}^n) = e(R)n - \nu(\mathfrak{m}) = P(\mathfrak{m}, R)(n),$$

hence $\ell_R(R/Rr) \geq e(R)$ [since $\nu(\mathfrak{m}) \geq 0$].

(4.14) Corollary: Let $n_0 \in \mathbb{N}$. Then the ideal $\mathfrak{a}^{n_0}$ is stable iff $\ell_R(\mathfrak{a}^n/\mathfrak{a}^{n+1}) = \rho(\mathfrak{a})$ for all integers $n \geq n_0$.

Proof: (1) Assume that $\mathfrak{a}^{n_0}$ is stable and that $n \geq n_0$. Then $\mathfrak{a}^n$ and $\mathfrak{a}^{n+1}$ are stable [cf. (4.10)(2)], and, by (4.11), $\ell_R(\mathfrak{a}^n/\mathfrak{a}^{n+1}) = \ell_R(R/\mathfrak{a}^{n+1}) - \ell_R(R/\mathfrak{a}^n) = \rho(\mathfrak{a})$.
(2) Conversely, assume that $\ell_R(\mathfrak{a}^n/\mathfrak{a}^{n+1}) = \rho(\mathfrak{a})$ for all integers $n \geq n_0$. Write

$$\ell_R(R/\mathfrak{a}^{n_0}) = \rho(\mathfrak{a})n_0 - \sigma \quad \text{for some } \sigma \in \mathbb{Z}.$$

Then, for all sufficiently large integers n, we have

$$P(\mathfrak{a}, R)(n) = H(\mathfrak{a}, R)(n) = \sum_{\nu=0}^{n-1} \ell_R(\mathfrak{a}^\nu/\mathfrak{a}^{\nu+1})$$

$$= (\rho(\mathfrak{a})n_0 - \sigma) + (n - n_0)\rho(\mathfrak{a}) = \rho(\mathfrak{a})n - \sigma,$$

hence $\sigma = \nu(\mathfrak{a})$ and $P(\mathfrak{a}, R)(n_0) = H(\mathfrak{a}, R)(n_0)$, and therefore $\mathfrak{a}^{n_0}$ is a stable ideal [cf. (4.11)].

(4.15) Corollary: Let $m \in \mathbb{N}$ and $r \in R$, and assume that $\mathfrak{a}^m R^{\mathfrak{a}} = rR^{\mathfrak{a}}$. Then we have $\rho(\mathfrak{a}^m) = m \cdot \rho(\mathfrak{a}) = \ell_R(R/Rr)$.

Proof: Clearly, we have $\rho(\mathfrak{a}^m) = m \cdot \rho(\mathfrak{a})$ and $R^{\mathfrak{a}^m} = R^{\mathfrak{a}}$, and we have [cf. (3.1)]

$$\rho(\mathfrak{a}^m) = \ell_R(R^{\mathfrak{a}}/\mathfrak{a}^m R^{\mathfrak{a}}) = \ell_R(R^{\mathfrak{a}}/rR^{\mathfrak{a}}) = \ell_R(R/Rr).$$

(4.16) TRANSVERSAL ELEMENTS AND STABILITY: Let $\mathfrak{a}$ be an open ideal of R. We prove some properties of $\mathfrak{a}$-transversal elements and of stable ideals.
(1) An element $a \in \mathfrak{a}$ is $\mathfrak{a}$-transversal iff $aR^{\mathfrak{a}} = \mathfrak{a}R^{\mathfrak{a}}$.
Indeed, let a be $\mathfrak{a}$-transversal; then we have $aR^{\mathfrak{a}} = \mathfrak{a}R^{\mathfrak{a}}$ by (4.4)(2). Conversely, assume that $aR^{\mathfrak{a}} = \mathfrak{a}R^{\mathfrak{a}}$. Since $\mathfrak{a}^n R^{\mathfrak{a}} = \mathfrak{a}^n$ for some $n \in \mathbb{N}$ [cf. (4.10)(1)], and since we have

$$a\mathfrak{a}^n = a\mathfrak{a}^n R^{\mathfrak{a}} = \mathfrak{a}^n(aR^{\mathfrak{a}}) = \mathfrak{a}^n(\mathfrak{a}R^{\mathfrak{a}}) = \mathfrak{a}^{n+1}R^{\mathfrak{a}} = \mathfrak{a}^{n+1}, \quad (*)$$

we have shown that a is $\mathfrak{a}$-transversal.
(2) Let a be $\mathfrak{a}$-transversal, and let $n \in \mathbb{N}$; then $a\mathfrak{a}^n = \mathfrak{a}^{n+1}$ iff $\mathfrak{a}^n$ is stable.
Indeed, we have $aR^{\mathfrak{a}} = \mathfrak{a}R^{\mathfrak{a}}$ since a is $\mathfrak{a}$-transversal [cf. (1)]. If $\mathfrak{a}^n$ is stable, then we have just shown that $a\mathfrak{a}^n = \mathfrak{a}^{n+1}$ [cf. (*)]. Conversely, assume that $a\mathfrak{a}^n = \mathfrak{a}^{n+1}$.

By induction we get $a^l a^n = a^{l+n}$ for all $l \in \mathbb{N}$, and, since a is regular [cf. (1.15)], we have $(a^{n+l} : a^{n+l})_S = (a^n : a^n)_S$ for all $l \in \mathbb{N}$, hence we have

$$R^a = \bigcup_{l \in \mathbb{N}} (a^{n+l} : a^{n+l})_S = (a^n : a^n)_S;$$

since $(a^n : a^n)_S = R^a$ is equivalent to $a^n R^a = a^n$, it follows that a^n is stable.

(3) Let $a \in \mathfrak{a}$; the following conditions are equivalent:
(a) $\mathfrak{a}^2 = a\mathfrak{a}$.
(b) $a \in \mathrm{Reg}(R)$ and $\mathfrak{a}a^{-1} = R^a$.
(c) $a \in \mathrm{Reg}(R)$ and $\mathfrak{a}a^{-1}$ is a subring of Q.
Indeed, if $\mathfrak{a}^2 = a\mathfrak{a}$, then a is $\mathfrak{a}$-transversal, hence a is regular [cf. (1.15)] and we have $\mathfrak{a}a^{-1} = R^a$ [cf. (4.4)(1)], i.e., (a) implies (b). It is clear that (b) implies (c). Assume that a is regular and that $\mathfrak{a}a^{-1}$ is a subring of $Q(R)$; then $(\mathfrak{a}a^{-1})^2 = a^{-1}\mathfrak{a}$, hence we have $\mathfrak{a}^2 = a\mathfrak{a}$, i.e., (c) implies (a).

(4) $\mathfrak{a}$ is stable iff there exists $a \in \mathfrak{a}$ satisfying the conditions in (3).
Indeed, we have $\mathfrak{a}R^a = R^a r$ for some regular element $r \in R^a$ by (4.3)(5). Assume that $\mathfrak{a}$ is stable; then $\mathfrak{a}R^a = \mathfrak{a}$, hence $r \in \mathfrak{a}$ and r is an $\mathfrak{a}$-transversal element [cf. (1)] such that $a\mathfrak{a} = \mathfrak{a}^2$ [cf. (2) with $n = 1$]. Conversely, let $a \in \mathfrak{a}$ be such that $a\mathfrak{a} = \mathfrak{a}^2$; then a is $\mathfrak{a}$-transversal and $\mathfrak{a}$ is stable [cf. (2)].

(5) Assume that $\mathfrak{a}$ is stable. Then every $\mathfrak{a}$-transversal element $a \in \mathfrak{a}$ satisfies the conditions in (3) [cf. (2)].

(6) Assume that R^a is a local ring. Then there exist $\mathfrak{a}$-transversal elements.
Indeed, let $\mathfrak{a} = Ra_1 + \cdots + Ra_h$, and choose a regular $r \in R^a$ such that $\mathfrak{a}R^a = R^a r$ [cf. (4.3)(5)]; then there exist $b_1, \ldots, b_h, c_1, \ldots, c_h \in R^a$ such that $a_i = b_i r$ for every $i \in \{1, \ldots, h\}$ and $r = a_1 c_1 + \cdots + a_h c_h$, hence $1 = c_1 b_1 + \cdots + c_h b_h$. Since R^a is local, there exists $i \in \{1, \ldots, h\}$ such that $c_i b_i$ is invertible in R^a, i.e., b_i is invertible in R^a, and $\mathfrak{a}R^a = R^a r = R^a a_i$. The assertion follows from (1).

(4.17) Proposition: Let $\mathfrak{a}$ be an open ideal of R and $a \in \mathfrak{a}$. Then a is $\mathfrak{a}$-transversal iff a is regular and $\rho(\mathfrak{a}) = \ell_R(R/Ra)$.

Proof: Assume that a is $\mathfrak{a}$-transversal. Then a is regular, and we have $aR^a = \mathfrak{a}R^a$ by (4.16)(1) and $\rho(\mathfrak{a}) = \ell_R(R/Ra)$ by (4.15). Conversely, assume that a is regular, and that $\rho(\mathfrak{a}) = \ell_R(R/Ra)$. By (3.1) we have

$$\rho(\mathfrak{a}) = \ell_R(R/Ra) = \ell_R(\mathfrak{a}^n/a\mathfrak{a}^n) \geq \ell_R(\mathfrak{a}^n/\mathfrak{a}^{n+1}) \quad \text{for every } n \in \mathbb{N}_0 \qquad (*)$$

with equality iff $a\mathfrak{a}^n = \mathfrak{a}^{n+1}$. By (4.10)(1) there exists $n_0 \in \mathbb{N}$ such that $\mathfrak{a}^{n_0}$ is stable, hence we obtain $a\mathfrak{a}^{n_0} = \mathfrak{a}^{n_0+1}$ [cf. (4.14)].

(4.18) Theorem: Let $\mathfrak{a}$ be an open ideal of R. Then we have

$$\ell_R(\mathfrak{a}^n/\mathfrak{a}^{n+1}) \leq \rho(\mathfrak{a}) \quad \text{for every } n \in \mathbb{N}_0,$$

and for $n \in \mathbb{N}$ equality holds iff $\mathfrak{a}^n$ is stable.

Proof: (1) We use the trick of adjoining an indeterminate [cf. B(10.4)]. Let $m_1, \ldots,$ m_h be the maximal ideals of R, and set $\Sigma := R[X] \setminus (m_1 R[X] \cup \cdots \cup m_h R[X])$, $R(X) := \Sigma^{-1} R[X]$. Then $R(X)$ is a semilocal ring, $m_1 R(X), \ldots, m_h R(X)$ are its maximal ideals, the residue fields $R(X)/m_i R(X) \cong (R/m_i)(X)$ are infinite fields for every $i \in \{1, \ldots, h\}$, $R \to R(X)$ is a faithfully flat extension, and maximal ideals of R generate maximal ideals of $R(X)$. Therefore we have $\ell_R(M) = \ell_{R(X)}(M \otimes_R R(X))$ for any R-module M of finite length [note that if P is a simple R-module, hence $P \cong R/m$ where m is a maximal ideal of R, then $P \otimes_R R(X) \cong R(X)/m R(X)$], and the map $a^n \otimes_R R(X) \to a^n R(X)$ is an isomorphism of $R(X)$-modules, in particular,

$$\ell_R(a^n/a^{n+1}) = \ell_{R(X)}(a^n R(X)/a^{n+1} R(X)).$$

Therefore we have $P(a, R) = P(aR(X), R(X))$, hence we obtain $\dim(R(X)) = 1$ [cf. (1.8)(1)], hence $R(X)$ is a one-dimensional semilocal CM-ring, and $aR(X)$ is an open ideal of $R(X)$. Let $n \in \mathbb{N}$; a^n is a stable ideal of R iff $a^n R(X)$ is a stable ideal of $R(X)$ by (4.11).
(2) By the results of (1), and by (1.18), it is enough to prove (4.18) under the additional assumption that there exist a-transversal elements in R.
Let a be an a-transversal element. Then we have $\rho(a) = \ell_R(R/Ra)$ by (4.17), hence we have $\rho(a) \geq \ell_R(a^n/a^{n+1})$ for every $n \in \mathbb{N}_0$ with equality iff $aa^n = a^{n+1}$ [cf. (*) in the proof of (4.17)]; the last part of the assertion follows from (4.16)(2).

(4.19) REMARK: Let A be a local ring with maximal ideal m and residue field κ. We call

$$\mathrm{emdim}(A) := \dim_\kappa(m/m^2)$$

the embedding dimension of A. Note that $\dim(A) \leq \mathrm{emdim}(A)$ by [63], Cor. 10.7, and that A is a regular local ring iff $\dim(A) = \mathrm{emdim}(A)$.

(4.20) Corollary: *Assume that R is local. We have $\mathrm{emdim}(R) \leq e(R)$ with equality iff the maximal ideal m of R is stable. In particular, R is a regular local ring iff $e(R) = 1$.*

Proof: We have $\rho(m) = e(m, R) = e(R)$, and the first assertion follows from (4.18) [with $a = m$ and $n = 1$]. This implies the second assertion.

(4.21) REMARK: This result is a particular case of a result in [88], Ch. I, Th. 6.8. In section 7 of the book just mentioned there are examples of local rings having multiplicity 1 without being regular.

5　Infinitely Near Rings

(5.0) In this section R always is a one-dimensional semilocal CM-ring, $\mathfrak{r}$ is its Jacobson radical, Q is its ring of quotients, and S is the integral closure of R in Q.

(5.1) BLOWING UP AND LOCALIZATION: We recall the following facts. Let $\Sigma \subset R$ be multiplicatively closed, and *assume that* $\dim(\Sigma^{-1}R) = 1$. Then $\Sigma^{-1}R$ is a one-dimensional CM-ring and $\Sigma^{-1}Q$ is the ring of quotients of $\Sigma^{-1}R$.

Let $(R_i)_{i\geq 0}$ be the sequence of blow-ups of R. The ideal $\Sigma^{-1}\mathfrak{r}$ is the Jacobson radical of $\Sigma^{-1}R$, and $\Sigma^{-1}(R_1)$ can be canonically identified with $(\Sigma^{-1}R)_1$ [cf. (4.6)]. Let $i \in \mathbb{N}_0$, and let $\mathfrak{r}_i$ be the Jacobson radical of R_i. Then $\Sigma^{-1}\mathfrak{r}_i$ is the Jacobson radical of $\Sigma^{-1}R_i$, and, by induction, $\Sigma^{-1}R_{i+1}$ can be canonically identified with $(\Sigma^{-1}R)_{i+1}$. Therefore, the sequence of blow-ups $(\Sigma^{-1}R_i)_{i\geq 0}$ of $\Sigma^{-1}R$ can be canonically identified with the sequence $((\Sigma^{-1}R)_i)_{i\geq 0}$; in particular, $\Sigma^{-1}S$ is the integral closure of $\Sigma^{-1}R$ in $\Sigma^{-1}Q$ [cf. (4.8)].

(5.2) NOTATION: (1) Let $(R_i)_{i\geq 0}$ be the sequence of blow-ups of R [cf. (4.7)]. A local ring $A = (R_i)_\mathfrak{n}$, where $i \in \mathbb{N}_0$ and $\mathfrak{n}$ is a maximal ideal of R_i, is said to lie in the i-th neighborhood of R; the set of all local rings in the i-th neighborhood of R shall be denoted by $\Delta_i(R)$. A local ring A is said to be infinitely near to R if $A \in \Delta_i(R)$ for some $i \in \mathbb{N}_0$, and we denote by $\Delta(R)$ the set of all rings which are infinitely near to R.

(2) Let $A \in \Delta(R)$; then there exists $i \in \mathbb{N}_0$ and a maximal ideal $\mathfrak{n}$ of R_i such that $A = (R_i)_\mathfrak{n}$. Now R_i is a finitely generated R-module, hence $R_{\mathfrak{n}(A)\cap R} \to A$ is quasifinite [cf. B(10.10)] and

$$f(A) := [A/\mathfrak{n}(A) : R/(\mathfrak{n}(A) \cap R)]$$

is finite.

(5.3) INFINITELY NEAR RINGS: We collect some simple properties of infinitely near rings. Let $(R_i)_{i\geq 0}$ be the sequence of blow-ups of R.

(1) A local ring is infinitely near to R iff it is infinitely near to $R_\mathfrak{m}$ for some maximal ideal $\mathfrak{m}$ of R.

In fact, let $A := (R_i)_\mathfrak{n}$ for some $i \in \mathbb{N}_0$ and some maximal ideal $\mathfrak{n}$ of R_i. Set $\mathfrak{m} := \mathfrak{n} \cap R$; the ideal $\mathfrak{m}$ is a maximal ideal of R [cf. [63], Cor. 4.17]. Then $(R_i)_\mathfrak{n}$ is the localization of $(R_i)_\mathfrak{m}$ with respect to the maximal ideal $\mathfrak{n} \cdot (R_i)_\mathfrak{m}$ of $(R_i)_\mathfrak{m}$; by (5.1) we have $(R_i)_\mathfrak{m} \cong (R_\mathfrak{m})_i$, hence A is infinitely near to $R_\mathfrak{m}$. In the same way we see that if a local ring is infinitely near to $R_\mathfrak{m}$ for some maximal ideal $\mathfrak{m}$ of R, then it is infinitely near to R.

(2) If A is a local ring infinitely near to R, and if B is a local ring infinitely near to A, then B is infinitely near to R.

In fact, choose $i \in \mathbb{N}_0$ and a maximal ideal $\mathfrak{n}$ of R_i such that $A = (R_i)_\mathfrak{n}$. Now B is infinitely near to R_i by (1), and therefore B is infinitely near to R.

(3) Let $\mathfrak{p}$ be a maximal ideal of R. Every local ring A which is infinitely near to $R^\mathfrak{p}$ is also infinitely near to R.

In fact, A is infinitely near to $(R^\mathfrak{p})_\mathfrak{n}$ for some maximal ideal $\mathfrak{n}$ of $R^\mathfrak{p}$ by (1). Set $\mathfrak{m} := R \cap \mathfrak{n}$. Note that $(R_\mathfrak{m})^{\mathfrak{p}R_\mathfrak{m}}$ and $(R^\mathfrak{p})_\mathfrak{m}$ are canonically isomorphic [cf. (5.1)], and that $(R_\mathfrak{m})^{\mathfrak{p}R_\mathfrak{m}}$ is equal to $R_\mathfrak{m}$ if $\mathfrak{p} \neq \mathfrak{m}$, and is equal to the quadratic transform of $R_\mathfrak{m}$ if $\mathfrak{p} = \mathfrak{m}$; hence $(R^\mathfrak{p})_\mathfrak{n}$, which is the localization of $(R^\mathfrak{p})_\mathfrak{m}$ with respect to the maximal ideal $\mathfrak{n} \cdot (R^p)_\mathfrak{m}$, is infinitely near to the local ring $R_\mathfrak{m}$. Now $(R^p)_\mathfrak{n}$ is infinitely near to R by (1), hence A is infinitely near to R by (2).

(5.4) REMARK: The integral closure S of R has only finitely many maximal ideals, and each of them is a regular ideal [cf. (2.11)]. Let $\mathfrak{n}_1, \ldots, \mathfrak{n}_h$ be the maximal ideals of S. Then $V_1 = S_{[\mathfrak{n}_1]}, \ldots, V_h = S_{[\mathfrak{n}_h]}$ are the Manis valuation rings of Q belonging to R, we have $S = V_1 \cap \cdots \cap V_h$, and, for every $j \in \{1, \ldots, h\}$, $\mathfrak{p}_j := \mathfrak{n}_j V_j$ is the regular maximal ideal of V_j [cf. (2.11)].
Let $(R_i)_{i \geq 0}$ be the sequence of blow-ups of R. We have $S = \bigcup_{i \geq 0} R_i$ [cf. (4.8)]; therefore there exists $i_0 \in \mathbb{N}$ such that, for every integer $i \geq i_0$, the ideals $\mathfrak{n}_1 \cap R_i, \ldots, \mathfrak{n}_h \cap R_i$ are pairwise different. Let $j \in \{1, \ldots, h\}$.
(1) Define $\Sigma_j := R_{i_0} \setminus (R_{i_0} \cap \mathfrak{n}_j)$. Every maximal ideal of S different from $\mathfrak{n}_j$ meets Σ_j, hence the only prime ideal of $\Sigma_j^{-1} S$ lying over the maximal ideal of the local ring $\Sigma_j^{-1} R_{i_0}$ is the ideal $\mathfrak{n}_j(\Sigma_j^{-1} S)$. Since every maximal ideal of $\Sigma_j^{-1} S$ must lie over the maximal ideal of $\Sigma_j^{-1} R_{i_0}$ [cf. [63], Cor. 4.17], the ring $\Sigma_j^{-1} S$ is a quasilocal ring with maximal ideal $\mathfrak{n}_j(\Sigma_j^{-1} S)$, hence we have $\Sigma_j^{-1} S = S_{\mathfrak{n}_j}$. The same argument shows also that $(R_l)_{\mathfrak{n}_j \cap R_l} = \Sigma_j^{-1} R_l$ for every $l \geq i_0$.
(2) The ring $\Sigma_j^{-1} S$ is the integral closure of $\Sigma_j^{-1} R_{i_0}$ in $\Sigma_j^{-1} Q$ [cf. [63], Prop. 4.13], and $\Sigma_j^{-1} Q$, being its own ring of quotients since it is artinian, is also the ring of quotients of $\Sigma_j^{-1} R_{i_0}$. Therefore the ring $S_{\mathfrak{n}_j} = (V_j)_{\mathfrak{p}_j}$ [cf. (1.7)], being quasilocal and the integral closure of a one-dimensional local CM-ring, is a discrete Manis valuation ring [cf. (2.11)(4)].
(3) For every $i \geq i_0$ we have $(R_{i+1})_{\mathfrak{n}_j \cap R_i} = (R_{i+1})_{\mathfrak{n}_j \cap R_{i+1}}$. By (5.1) we see that $(R_{i+1})_{\mathfrak{n}_j \cap R_i}$ is the blow-up of $(R_i)_{\mathfrak{n}_j \cap R_i}$, and therefore, by (4.8), we have $(V_j)_{\mathfrak{p}_j} = \bigcup_{i \geq i_0} (R_i)_{\mathfrak{n}_j \cap R_i}$.
(4) Assume that S is a finitely generated R-module. Then S is a noetherian ring, and therefore $(V_j)_{\mathfrak{p}_j}$ is a local ring whose maximal ideal is generated by one element, hence it is a one-dimensional regular local ring, and therefore it is a discrete valuation ring [cf. I(3.29)].

(5.5) BRANCH SEQUENCE: Let $(R_i)_{i \geq 0}$ be the sequence of blow-ups of R, let $\mathfrak{n}$ be a maximal ideal of S, set $V := S_{[\mathfrak{n}]}$, and let $\mathfrak{p}$ be the regular maximal ideal of the Manis valuation ring V. If, in particular, S is a finitely generated R-module, then $V_\mathfrak{p}$ is a discrete valuation ring [cf. (5.4)(4)].
(1) For every $i \in \mathbb{N}_0$ let $A_i = A_i(\mathfrak{n})$ be the local ring $(R_i)_{\mathfrak{n} \cap R_i}$. The sequence $(A_i)_{i \geq 0}$ is called the branch sequence of R along $\mathfrak{n}$ or along V. Note that, for all large enough i, the ring $V_\mathfrak{p}$ is the only Manis valuation ring belonging to the ring A_i. One often says: Blowing up separates the branches. This will get a geometric interpretation in chapter VIII, section 1.
(2) The sequence $(e(A_i), f(A_i))_{i \in \mathbb{N}_0}$ is called the multiplicity sequence of R along $\mathfrak{n}$ or along V. If, in particular, $f(A_i) = 1$ for every $i \in \mathbb{N}_0$, then the sequence $(e(A_i))_{i \geq 0}$ is called the multiplicity sequence of R along V.

In the rest of this section, we prove for later use some length formulae.

(5.6) Proposition: *Assume that S is a finitely generated R-module. Then S/R is an R-module of finite length, and we have*

$$\ell_R(S/R) = \sum_{A \in \Delta(R)} f(A) \cdot \ell_R(A^{\mathfrak{n}(A)}/A).$$

Proof: (1) Assume, at first, that R is local, and let $(R_i)_{i \geq 0}$ be the sequence of blow-ups of R. Since S is a finitely generated R-module, it is clear that S/R is an R-module of finite length [there exists a regular $r \in R$ which annihilates S/R], and therefore we have

$$\ell_R(S/R) = \sum_{i \geq 0} \ell_R(R_{i+1}/R_i).$$

Fix $i \in \mathbb{N}_0$, let $\mathfrak{m}_1, \ldots, \mathfrak{m}_h$ be the maximal ideals of R_i, and set $A_j = (R_i)_{\mathfrak{m}_j}$, $\mathfrak{n}_j = \mathfrak{m}_j A_j$ for $j \in \{1, \ldots, h\}$. Note that, for $j \in \{1, \ldots, h\}$, we have a canonical isomorphism $(R_{i+1}/R_i)_{\mathfrak{m}_j} = A_j^{\mathfrak{n}_j}/A_j$ [cf. (5.1)]; $A_j^{\mathfrak{n}_j}$ is the quadratic transform of the local ring A_j. Therefore we have [cf. B(10.10)]

$$\ell_R(R_{i+1}/R_i) = \sum_{j=1}^{h} f(A_j)\ell_{A_j}(A_j^{\mathfrak{n}_j}/A_j).$$

(2) Now assume that R is semilocal, and let $\mathfrak{m}_1, \ldots, \mathfrak{m}_h$ be its maximal ideals. Then we have [cf. B(10.9)]

$$\ell_R(R_{i+1}/R_i) = \sum_{j=1}^{h} \ell_{R_{\mathfrak{m}_j}}((R_{i+1}/R_i)_{\mathfrak{m}_j}) \quad \text{for every } i \in \mathbb{N}_0.$$

For every $j \in \{1, \ldots, h\}$ we know that $((R_i)_{\mathfrak{m}_j})_{i \geq 0}$ can be identified with the blow-up sequence of the local ring $R_{\mathfrak{m}_j}$ [cf. (5.1)]; now we get the assertion from (1).

(5.7) Proposition: *Assume that R is local with maximal ideal $\mathfrak{m}$, and let $r \in \mathfrak{m}$ be regular. Then we have*

$$\ell_R(R/Rr) = \sum_{A \in \Delta_1(R)} f(A)\ell_A(A/rA).$$

Proof: We have $\ell_R(R/Rr) = \ell_R(R_1/rR_1)$ by (3.1) and

$$\ell_R(R_1/rR_1) = \sum_{A \in \Delta_1(R)} f(A)\ell_A(A/rA)$$

by B(10.10)(2).

(5.8) Corollary: *Assume that R is local. Then we have*

$$e(R) \geq \sum_{A \in \Delta_1(R)} e(A);$$

in particular, we have $e(R) \geq e(A)$ for every local ring A in the first neighborhood of R.

Proof: Set $\mathfrak{m} := \mathfrak{m}(R)$. We have $e(R) = \ell_R(R^{\mathfrak{m}}/\mathfrak{m}R^{\mathfrak{m}})$ by (4.12) and $\mathfrak{m}R^{\mathfrak{m}} = R^{\mathfrak{m}}r$ for some regular $r \in R^{\mathfrak{m}}$ by (4.3), hence $\ell_R(R^{\mathfrak{m}}/R^{\mathfrak{m}}r) \geq \ell_{R^{\mathfrak{m}}}(R^{\mathfrak{m}}/R^{\mathfrak{m}}r) = \sum_{A \in \Delta_1(R)} \ell_A(A/Ar)$ by B(10.9), and we have $\ell_A(A/Ar) \geq e(A)$ for every $A \in \Delta_1(R)$ by (4.13).

(5.9) Corollary: *Let R be local, let $\mathfrak{n}$ be a maximal ideal of the integral closure S of R, and let $(A_i)_{i \geq 0}$ be the branch sequence of R along $\mathfrak{n}$. Then $(e(A_i))_{i \geq 0}$ is a decreasing sequence of positive integers. If, in particular, S is a finitely generated R-module, then 1 is the smallest element of this sequence.*

Proof: Only the last assertion needs a proof. In this case $S_{\mathfrak{n}}$ is a one-dimensional local ring, and since $S_{\mathfrak{n}}$ is a localization of the discrete Manis valuation ring $V = S_{[\mathfrak{n}]}$ with respect to its maximal ideal $\mathfrak{m}(V)$, the maximal ideal of $S_{\mathfrak{n}}$ is a principal ideal, generated by a regular element. Therefore $S_{\mathfrak{n}}$ is a discrete valuation ring by I(3.28).

Chapter III

Differential Modules and Ramification

INTRODUCTION: Differential modules and ramification is the subject of this chapter. With regard to the principal properties of differential modules, we refer to chapter 16 of Eisenbud's book [63], and add some more results in section 1. After dealing with norms and traces in section 2, we introduce the notion of formally unramified and unramified extensions of rings $R \to S$ in section 3 and study, in particular, ramification of pairs of local rings in subsection 3.3, ramification of pairs of quasilocal rings in section 5. Under additional hypotheses for a ring extension $R \to S$, the set of prime ideals of R which are unramified in S can be described as the closure in $\mathrm{Spec}(R)$ of an ideal, the Noether discriminant ideal, which is generated by discriminants; this is done in section 4. In the last section 6 we prove a theorem of Chevalley and a theorem of Zariski: Let k be a perfect field, and let A be a local k-algebra essentially of finite type. Then A is analytically irreducible and analytically normal. These two results shall play an important role when proving the uniformization theorem VIII(6.9): in the proof of it we use namely VIII(6.4) and VIII(6.3) which rely on Chevalley's result.

1 Introduction

(1.1) As a source for the main properties of differential modules, we use chapter 16 of Eisenbud's book [63]. Let $A \to B$ be a homomorphism of rings; $\Omega_{B/A}$ denotes the B-module of differentials of B/A, and $d_{B/A}\colon B \to \Omega_{B/A}$ is the differential map. The following easy results should be kept in mind.

(1.2) REMARK: (1) Let $\alpha\colon A \to B$ be a homomorphism of rings, let M be a B-module, and let $D\colon B \to M$ be an A-derivation of B with values in M.
(a) The map $D\colon B \to M$ is, in particular, an A-linear map of A-modules. Furthermore, we have $D(\alpha(a)1_B) = \alpha(a)D(1_B)$ and $D(\alpha(a)1_B) = \alpha(a)D(1_B) +$

101

$1_B D(\alpha(a))$ for every $a \in A$, hence we have

$$D(\alpha(a)) = 0 \quad \text{for every } a \in A.$$

(b) For every $n \in \mathbb{Z}$ we have $D(n1_B) = D(\alpha(n1_A)) = 0$.

(2) Let $\alpha: A \to B$ be a homomorphism of rings, and let $D: B \to B$ be an A-derivation. The following result is easily proved by induction:

$$D^n(bc) = \sum_{\nu=0}^{n} \binom{n}{\nu} D^\nu(b) D^{n-\nu}(c) \quad \text{for every } n \in \mathbb{N}_0 \text{ and all } b, c \in B.$$

(3) Let $\alpha: A \to B$ be a homomorphism of rings, let M be a B-module and let $D: B \to M$ be an A-derivation. For every $b \in B$ and $n \in \mathbb{N}$ we have

$$D(b^n) = nb^{n-1} D(b).$$

For every unit b of B we have $D(bb^{-1}) = 0$, hence we have $D(b^{-1}) = -b^{-2}D(b)$.

(1.3) EXAMPLE: Let $B = A[T_1, \ldots, T_n]$ be a polynomial ring over A. For every $i \in \{1, \ldots, n\}$ we have the partial derivative $D_i: B \to B$ with $D_i(T_j) = \delta_{ij}1_B$; if $n = 1$, then we often write $F' = D_1(F)$ for any $F \in B$, and call F' the derivative of F. The differential module $\Omega_{B/A}$ is a free B-module of rank n, generated by the elements $d_{B/A}(T_1), \ldots, d_{B/A}(T_n)$.

We need the following results on extending derivations.

(1.4) RING OF FRACTIONS: (1) Let $\alpha: A \to B$ be a homomorphism of rings, let S be a multiplicatively closed subset of A, let T be a multiplicatively closed subset of B, and assume that $\alpha(S) \subset T$. Then there exists a unique algebra-homomorphism $\alpha_{A,B}^{S,T}: S^{-1}A \to T^{-1}B$ such that $i_B^T \circ \alpha = \alpha_{A,B}^{S,T} \circ i_A^S$ [cf. B(2.3)].

(2) Let N be a B-module, and let $d: B \to N$ be an A-derivation. Then there exists a unique $S^{-1}A$-derivation $\tilde{d}: T^{-1}B \to T^{-1}N$ such that the following diagram

$$
\begin{array}{ccc}
B & \xrightarrow{\ d\ } & N \\
\Big\downarrow{\scriptstyle i_B^T} & & \Big\downarrow{\scriptstyle i_N^T} \\
T^{-1}B & \xrightarrow[\ \tilde{d}\ ]{} & T^{-1}N
\end{array}
$$

is commutative.

Proof [Existence]: Let $b, b' \in B$ and $t, t' \in T$ satisfy $b/t = b'/t'$ in $T^{-1}B$. Then we have

$$\frac{td(b) - bd(t)}{t^2} = \frac{t'd(b') - b'd(t')}{t'^2} \quad \text{in } T^{-1}N.$$

Indeed, there exists $t'' \in T$ satisfying $t''(bt' - tb') = 0$. Then we have

$$d(t'')(bt' - tb') + t''(t'd(b) + bd(t') - td(b') - b'd(t)) = 0,$$

whence, by multiplying with $1/tt'$, we get

$$d(t'')\left(\frac{b}{t} - \frac{b'}{t'}\right) + t''\left(\frac{td(b) - bd(t)}{t^2} - \frac{t'd(b') - b'd(t')}{t'^2}\right) = 0 \quad \text{in } T^{-1}N,$$

and therefore the claim has been proved.

This shows that there exists a well-defined map $\tilde{d} \colon T^{-1}B \to T^{-1}N$, defined by

$$\tilde{d}(b/t) = \frac{td(b) - bd(t)}{t^2} \quad \text{for all } b \in B, t \in T.$$

It is clear that $\tilde{d}$ is $S^{-1}A$-linear; moreover, for all b, $b' \in B$, t, $t' \in T$ we have

$$\tilde{d}\left(\frac{b}{t}\frac{b'}{t'}\right) = \frac{tt'd(bb') - bb'd(tt')}{(tt')^2} = \frac{b'}{t'}\tilde{d}\left(\frac{b}{t}\right) + \frac{b}{t}\tilde{d}\left(\frac{b'}{t'}\right),$$

hence $\tilde{d}$ is an $S^{-1}A$-derivation. Clearly, the above diagram is commutative.

[Uniqueness] Let $d' \colon T^{-1}B \to T^{-1}N$ be any $S^{-1}A$-derivation such that $d'(b/1) = d(b)/1$ for every $b \in B$. Let $b \in B$, $t \in T$. Then we have by (1.2)(3)

$$d'(b/t) = d'((b/1)(1/t)) = d'(b/1)/t - (b/1)d'(t/1)/t^2 = (d(b)t - bd(t))/t^2,$$

i.e., we have $d'(b/t) = \tilde{d}(b/t)$.

(3) In particular, let $A = \mathbb{Z}$, $S = \{1\}$, B a domain and $T = B \setminus \{0\}$. Every derivation $d \colon B \to B$ has a unique extension to a derivation $\tilde{d} \colon Q(B) \to Q(B)$ of the field of quotients $Q(B) = T^{-1}B$ of B [cf. (2)].

(1.5) SEPARABLE EXTENSION OF FIELDS: Let K be a field, and let $d \colon K \to K$ be a derivation. Let L be a finite separable extension of K. Then there exists a unique derivation $D \colon L \to L$ which extends d.

Proof: We have $L = K(x)$ by the theorem of the primitive element; let f be the minimal polynomial of x over K, and let f' be its derivative. For every polynomial $F = \sum a_i T^i \in K[T]$ we define $F^d := \sum d(a_i)T^i$; it is clear that $F \mapsto F^d \colon K[T] \to K[T]$ is a derivation. We define $D(x) := f^d(x)/f'(x)$. Let $y \in L$, and choose $F \in K[T]$ with $y = F(x)$; then we define $D(y) := F^d(x) + F'(x)D(x)$. The map D is well-defined: let F, $G \in K[T]$ with $F(x) = G(x)$; then we have $F - G = Hf$ for $H \in K[T]$, and therefore we find that $F^d(x) + F'(x)D(x) = G^d(x) + G'(x)D(x)$. It is immediate that D is a derivation of L which extends d, and since $0 = f(x) = f^d(x) + f'(x)D(x)$, it is clear that D is the unique extension of d to L.

(1.6) Let $A \to B$ be a homomorphism of rings. Assume that B is an A-algebra of finite type. By [63], Prop. 16.1 and Prop. 16.3, the module of differentials $\Omega_{B/A}$ is a finitely generated B-module. More precisely, if $B = A[x_1, \ldots, x_n]$, then we have $\Omega_{B/A} = B d_{B/A}(x_1) + \cdots + B d_{B/A}(x_n)$ where $d_{B/A} \colon B \to \Omega_{B/A}$ is the canonical derivation.

(1.7) NOTATION: Let $\alpha \colon A \to C$ be a homomorphism of rings; we say that C is essentially of finite type over A or that α is essentially of finite type if C is A-isomorphic to an A-algebra of the form $S^{-1}B$ where B is an A-algebra of finite type and $S \subset B$ is multiplicatively closed.

(1.8) Proposition: *Let A be a ring. Then:*
(1) If B is an A-algebra essentially of finite type and C is a B-algebra essentially of finite type, then C is an A-algebra essentially of finite type.
(2) If B is an A-algebra essentially of finite type and A' is an A-algebra, then $B' := B \otimes_A A'$ is an A'-algebra essentially of finite type.

Proof: (1) Let $B = S^{-1}B_1$ and $C = T^{-1}C_1$ where B_1 (resp. C_1) is an A-algebra (resp. a B-algebra) of finite type and S (resp. T) is multiplicatively closed in B_1 (resp. in C_1). Let us write $C_1 = B[X_1, \ldots, X_n]/\mathfrak{b}$ as a homomorphic image of a polynomial ring; now, since $B[X_1, \ldots, X_n] = S^{-1}(B_1[X_1, \ldots, X_n])$, the ideal $\mathfrak{b}$ is of the form $S^{-1}\mathfrak{b}_1$ where $\mathfrak{b}_1$ is an ideal in $B_1[X_1, \ldots, X_n]$. Therefore we have $C_1 = S^{-1}B_2$ where $B_2 = B_1[X_1, \ldots, X_n]/\mathfrak{b}_1$ is a B_1-algebra of finite type. Now B_2 is an A-algebra of finite type, and $C = T^{-1}(S^{-1}B_2)$ is an A-algebra essentially of finite type [since $T^{-1}(S^{-1}B_2)$ can be considered as a localization of B_2 with respect to a multiplicatively closed system, cf. [21], Ch. 3, Ex. 3].
(2) If $B = S^{-1}C$ where C is an A-algebra of finite type and $S \subset C$ is multiplicatively closed, then $B' = S^{-1}(C \otimes_A A')$, and $C \otimes_A A'$ is an A'-algebra of finite type.

(1.9) Proposition: *Let A be a ring. Let $B_1, \ldots, B_h$ be A-algebras which are essentially of finite type. Then $B_1 \otimes_A \cdots \otimes_A B_h$ and $B_1 \times \cdots \times B_h$ are A-algebras essentially of finite type.*

The proof is easy and is left to the reader.

(1.10) Proposition: *Let $A \to C$ be a homomorphism essentially of finite type. Then $\Omega_{C/A}$ is a finitely generated C-module.*

In fact, we have $C = \Sigma^{-1}B$ where B is an A-algebra B of finite type, and $\Sigma \subset B$ is multiplicatively closed. Now we have $\Omega_{C/A} = \Sigma^{-1}\Omega_{B/A}$, and since $\Omega_{B/A}$ is a finitely generated B-module by (1.6), also $\Omega_{C/A}$ is a finitely generated C-module.

2 Norms and Traces

(2.0) In this section A always denotes a ring, and $A[X]$ is the polynomial ring over A in an indeterminate X.

2.1 Some Linear Algebra

(2.1) In the first part of this section we collect some results belonging to linear algebra.

(2.2) LINEAR MAPS OF FREE MODULES: The proof of the following two results is left to the reader. Let $\varphi: M \to N$ be an A-linear map of free A-modules having the same rank n.

(1) Let $Z \in M(n; A)$ be the matrix of φ with respect to bases of M and N. Then φ is injective iff $\det(Z)$ is a regular element of A, and φ is surjective if $\det(Z)$ is a unit of A. Moreover, φ is surjective iff φ is bijective.

(2) The map φ is bijective iff, for every maximal ideal $\mathfrak{m}$ of A, the induced map $\varphi \otimes \mathrm{id}_{A/\mathfrak{m}}: M \otimes_A A/\mathfrak{m} \to N \otimes_A A/\mathfrak{m}$ is bijective.

(2.3) THE DUAL MODULE: Let M be an A-module. The A-module

$$M^* := \mathrm{Hom}_A(M, A)$$

is called the module dual to M; the map

$$\langle x^*, x \rangle \mapsto x^*(x) : M^* \times M \to A$$

is A-bilinear. If M is a free A-module of finite rank n, then M^* is a free A-module of rank n. In particular, if $\{e_1, \ldots, e_n\}$ is a basis of M, then there exists a unique basis $\{e_1^*, \ldots, e_n^*\}$ of M^* which is defined by $\langle e_i^*, e_j \rangle = \delta_{ij}$ for all $i, j \in \{1, \ldots, n\}$; it is called the basis of M^* dual to the basis $\{e_1, \ldots, e_n\}$.

(2.4) SYMMETRIC BILINEAR FORMS: Let M be an A-module, and let $\beta: M \times M \to A$ be a symmetric A-bilinear form. For every $y \in M$ the map $x \mapsto \beta(x, y) : M \to A$ is an element β_y of M^*; the A-linear map $y \mapsto \beta_y : M \to M^*$ induced by β is denoted by τ_β^M. Then we have $\langle \tau_\beta^M(y), x \rangle = \beta(x, y)$ for all $x, y \in M$. The map β is called non-degenerate (resp. a perfect pairing) if τ_β^M is injective (resp. bijective).

(1) Let $(x_i)_{1 \le i \le n}$ be a sequence in M. The matrix $(\beta(x_i, x_j)) \in M(n; A)$ is called the Gramian of the sequence $(x_i)_{1 \le i \le n}$ with respect to β, and $\det((\beta(x_i, x_j))) \in A$ is called the discriminant of the sequence $(x_i)_{1 \le i \le n}$ with respect to β.

(2) Let M be a free A-module of rank n, and let $\{z_1, \ldots, z_n\}$ be a basis of M.

(a) The matrix of τ_β^M with respect to the basis $\{z_1, \ldots, z_n\}$ of M and the dual basis $\{z_1^*, \ldots, z_n^*\}$ of M^* is $(\beta(z_i, z_j))$. Therefore τ_β^M is injective (resp. bijective) iff $\det((\beta(z_i, z_j)))$ is a regular element (resp. a unit) of A [cf. (2.2)(1)].

(b) Let $(x_i)_{1 \le i \le n}$, $(y_i)_{1 \le i \le n}$ be sequences in M. Let $S = (s_{ij})$, $T = (t_{ij}) \in M(n; A)$ be the matrices such that $x_i = \sum_{j=1}^n s_{ij} z_j$, $y_i = \sum_{j=1}^n t_{ij} z_j$ for every $i \in \{1, \ldots, n\}$. Then we have $(\beta(x_i, y_j))_{1 \le i, j \le n} = S \cdot (\beta(z_i, z_j))_{1 \le i, j \le n} \cdot {}^t T$, and, in particular, $\det((\beta(x_i, x_j))) = (\det(S))^2 \det((\beta(z_i, z_j)))$.

(c) Using the notations of (b) we get: If $\det((\beta(x_i, y_j)))$ is invertible in A, then $\{x_1, \ldots, x_n\}$ and $\{y_1, \ldots, y_n\}$ are bases of M, and β is a perfect pairing.

(d) Using the notations of (b) we get: let $\{x_1, \ldots, x_n\}$ be a basis of M and $\det(S)$ be a unit of A; then $\det((\beta(x_i, x_j)))$ is a regular element (resp. a unit) of A iff $\det((\beta(z_i, z_j)))$ is a regular element (resp. a unit) of A.

(2.5) DISCRIMINANT AND COMPLEMENTARY BASIS: Let M be a free A-module of rank n, and let $\beta\colon M \times M \to A$ be a symmetric A-bilinear form.

(1) *The following statements are equivalent:*

(i) *The discriminant of any basis of M is a unit of A.*

(ii) *β is a perfect pairing.*

(iii) *For every basis $\{x_1, \ldots, x_n\}$ of M there exists a basis $\{y_1, \ldots, y_n\}$ of M such that $\beta(x_i, y_j) = \delta_{ij}$ for all $i, j \in \{1, \ldots, n\}$, and there is only one such basis.*

Proof: The equivalence of (i) and (ii) follows from (2.4)(2)(d).

Now let $\{x_1, \ldots, x_n\}$ be any basis of M such that its discriminant is a unit of A. Let $C = (c_{ij}) \in M(n; A)$ be the inverse of $(\beta(x_i, x_j))$, and define $y_j := \sum_{k=1}^{n} c_{jk} x_k$ for every $j \in \{1, \ldots, n\}$. It is clear that $\beta(x_i, y_j) = \delta_{ij}$ for all $i, j \in \{1, \ldots, n\}$, and that $\{y_1, \ldots, y_n\}$ is a basis of M, hence (i) implies (iii). Moreover, it is clear that there is only one such basis $\{y_1, \ldots, y_n\}$.

Conversely, let $\{x_1, \ldots, x_n\}$ and $\{y_1, \ldots, y_n\}$ be bases of M with $(\beta(x_i, y_j)) = I_n$. Then β is a perfect pairing by (2.4)(2)(c), hence (iii) implies (ii).

(2) Assume that β is a perfect pairing, and let $\{x_1, \ldots, x_n\}$ be a basis of M. The unique basis $\{y_1, \ldots, y_n\}$ of M such that $\beta(x_i, y_j) = \delta_{ij}$ for all $i, j \in \{1, \ldots, n\}$ is called the *complementary basis* of $\{x_1, \ldots, x_n\}$ with respect to β. Note that the complementary basis of $\{y_1, \ldots, y_n\}$ is the basis $\{x_1, \ldots, x_n\}$, hence it is legitimate to call a pair of such bases *complementary bases*.

(3) Assume that β is a perfect pairing, let $\{x_1, \ldots, x_n\}$ be a basis of M, and let $y_1, \ldots, y_n \in M$. Then $\{y_1, \ldots, y_n\}$ is the complementary basis of $\{x_1, \ldots, x_n\}$ iff $\{\tau_\beta^M(y_1), \ldots, \tau_\beta^M(y_n)\}$ is the dual basis of $\{x_1, \ldots, x_n\}$ as follows immediately from $\langle \tau_\beta^M(y_i), x_j \rangle = \beta(x_j, y_i)$ for all $i, j \in \{1, \ldots, n\}$.

2.2 Determinant and Characteristic Polynomial

(2.6) DETERMINANT AND CHARACTERISTIC POLYNOMIAL. THE CASE OF A FREE MODULE: Let M be a free A-module of finite rank n, and let $\varphi \in \mathrm{End}_A(M) = \mathrm{Hom}_A(M, M)$. The linear map $\varphi\colon M \to M$ induces a linear map $\psi\colon \bigwedge^n M \to \bigwedge^n M$. The A-module $\bigwedge^n M$ is free of rank 1, hence it is isomorphic to the A-module A, and ψ induces an endomorphism of A which is multiplication by an element a; $\det(\varphi) := a$ is called the *determinant* of φ.

$M[X] := M \otimes_A A[X]$ is a free $A[X]$-module of rank n; the *characteristic polynomial* $\mathrm{Pc}(X; \varphi) \in A[X]$ of φ is the determinant of the $A[X]$-linear map

$$X \, \mathrm{id}_{M[X]} - \varphi \otimes \mathrm{id}_{A[X]} \in \mathrm{End}_{A[X]}(M[X]).$$

The characteristic polynomial $\mathrm{Pc}(X; \varphi)$ has the form

$$\mathrm{Pc}(X; \varphi) = X^n + a_1 X^{n-1} + \cdots + a_n \quad \text{with } a_1, \ldots, a_n \in A,$$

2 Norms and Traces

the element $\text{Tr}(\varphi) := -a_1$ is called the *trace of* φ, and we have $\det(\varphi) = (-1)^n a_n$. Let $\{x_1, \ldots, x_n\}$ be a basis of M, and let $Z \in M(n; A)$ be the matrix of φ with respect to this basis. Then we have

$$\text{Pc}(X; \varphi) = \det(I_n X - Z), \; \text{Tr}(\varphi) = \text{Tr}(Z), \; \det(\varphi) = \det(Z).$$

(1) The map $\varphi \mapsto \det(\varphi) : \text{End}_A(M) \to A$ has the following properties:
(a) $\det(\varphi \circ \psi) = \det(\varphi)\det(\psi) = \det(\psi \circ \varphi)$ for all $\varphi, \psi \in \text{End}_A(M)$.
(b) $\det(\text{id}_M) = 1_A$.
(c) $\det(a\varphi) = a^n \det(\varphi)$ for every $a \in A$ and $\varphi \in \text{End}_A(M)$.
(d) φ is injective iff $\det(\varphi)$ is a regular element of A. Furthermore, φ is bijective iff $\det(\varphi)$ is invertible in A; in this case we have $\det(\varphi^{-1}) = (\det(\varphi))^{-1}$.
(2) The map $\varphi \mapsto \text{Tr}(\varphi) : \text{End}_A(M) \to A$ is an A-linear map of A-modules.
(3) We have $\text{Pc}(\varphi; \varphi) = 0$ for every $\varphi \in \text{End}_A(M)$ [theorem of Cayley-Hamilton].

(2.7) DETERMINANT AND CHARACTERISTIC POLYNOMIAL. THE CASE OF A MODULE OVER A DOMAIN: Let A be an integral domain, K its field of quotients, M a finitely generated A-module, and $\varphi \in \text{End}_A(M)$. We want to define also in this situation $\det(\varphi)$, $\text{Tr}(\varphi)$ and the characteristic polynomial $\text{Pc}(X; \varphi)$ of φ.
(1) $M_K := M \otimes_A K$ is a finite-dimensional K-vector space, hence $\varphi_K := \varphi \otimes \text{id}_K$ in $\text{End}_K(M_K)$ has a well-defined determinant, trace and characteristic polynomial. We shall call $\det(\varphi) := \det(\varphi_K) \in K$ the determinant of φ, $\text{Tr}(\varphi) := \text{Tr}(\varphi_K) \in K$ the trace of φ and $\text{Pc}(X; \varphi_K) \in K[X]$ the characteristic polynomial of φ. Furthermore, we call the minimal polynomial $m(X; \varphi_K) \in K[X]$ of φ_K the minimal polynomial of φ, and denote it by $m(X; \varphi)$.
(2) If, in particular, M is a free A-module of finite rank, then the determinant $\det(\varphi)$ (resp. the trace $\text{Tr}(\varphi)$ resp. the characteristic polynomial $\text{Pc}(X; \varphi)$) defined in (2.6) coincides with the determinant (resp. the trace resp. the characteristic polynomial) defined in (1).

(2.8) **Proposition:** *Let A be an integral domain, K its field of quotients, $\overline{A}$ the integral closure of A, M a finitely generated A-module, and $\varphi \in \text{End}_A(M)$. Then we have $\text{Pc}(X; \varphi) \in \overline{A}[X]$. In particular, $\text{Tr}(\varphi)$ and $\det(\varphi)$ are integral over A, and the minimal polynomial $m(X; \varphi)$ of φ has its coefficients in $\overline{A}$.*

Proof: First, we show that $\det(\varphi)$ is integral over A. Set $M_K := M \otimes_A K$, $n := \dim_K(M_K)$ and $d := \det(\varphi \otimes \text{id}_K) = \det(\varphi)$. The case $n = 0$ being trivial, we may assume that $M_K \neq \{0\}$. The image N of $\bigwedge^n M$ in $\bigwedge^n M_K$ is a finitely generated A-module such that $KN = \bigwedge^n M_K$. $\varphi \otimes \text{id}_K$ induces a K-linear map of $\bigwedge^n M_K$ which is just multiplication by d. On the other hand, this map is the extension to KN of the A-linear map $\bigwedge^n M \to \bigwedge^n M$ induced by φ. Therefore we have $dN \subset N$, hence d is integral over A [cf. [63], Cor. 4.6].
Now $\overline{A}[X]$ is the integral closure of $A[X]$ in $K[X]$ [cf. [63], Ex. 4.17], and since $K[X]$ is integrally closed, $\overline{A}[X]$ is the integral closure of $A[X]$ in its field of quotients $K(X)$. Set $M_K[X] := M_K \otimes_K K[X]$. By definition, we have

$Pc(X; \varphi) = \det(X \operatorname{id}_{M_K[X]} - \varphi_K \otimes \operatorname{id}_{K[X]})$. We can calculate this determinant also in $M_K[X] \otimes_{K[X]} K(X)$. Set $M[X] := M \otimes_A A[X]$; then $M[X] \otimes_{A[X]} K(X)$ can be identified canonically with $M_K[X] \otimes_{K[X]} K(X)$, and, by definition,

$$\det(X \operatorname{id}_{M[X]} - \varphi \otimes \operatorname{id}_{A[X]}) = \det\big((X \operatorname{id}_{M[X]} - \varphi \otimes \operatorname{id}_{A[X]}) \otimes \operatorname{id}_{K(X)}\big)$$
$$= \det(X \operatorname{id}_{M_K[X]} - \varphi_K \otimes \operatorname{id}_{K[X]});$$

this determinant lies in $\overline{A}[X]$ by the first part of the proof. This implies that $\operatorname{Tr}(\varphi) \in \overline{A}$, and that the minimal polynomial of φ has coefficients in $\overline{A}$ by [63], Prop. 4.11, since it divides $Pc(X; \varphi)$.

(2.9) NORM AND TRACE. THE CASE OF A FREE ALGEBRA: Let B be an A-algebra and M a B-module which is a free A-module of finite rank n. Remember that, for $b \in B$, $b_M : M \to M$ is multiplication by b, cf. II(3.3).
(1) Let $b \in B$, and consider b_M as an element of $\operatorname{End}_A(M)$. The trace (resp. the determinant, the characteristic polynomial) of the map b_M is called the trace $\operatorname{Tr}_{M/A}(b)$ (resp. the norm $N_{M/A}(b)$, the characteristic polynomial $\operatorname{Pc}_{M/A}(X; b) \in A[X]$) of b with respect to M. For all $b, b' \in B$, $a \in A$ we have

$$\operatorname{Tr}_{M/A}(b + b') = \operatorname{Tr}_{M/A}(b) + \operatorname{Tr}_{M/A}(b'), \quad N_{M/A}(bb') = N_{M/A}(b) N_{M/A}(b'),$$
$$\operatorname{Tr}_{M/A}(ab) = a \operatorname{Tr}_{M/A}(b), \quad N_{M/A}(ab) = a^n N_{M/A}(b).$$

(2) Now we assume that B is a free A-module of finite rank. The map $b \mapsto b_B :$ $B \to \operatorname{End}_A(B)$ is an injective A-algebra homomorphism [injective, because B has a unit element]. Let $b \in B$; note that $\operatorname{Pc}_{B/A}(b; b) = Pc(b_B; b_B) = 0$.
(3) In particular, let $f \in A[T]$ be monic, set $B := A[T]/(f)$, and let t be the image of T in B. Then we have $\operatorname{Pc}_{B/A}(T; t) = f$.

(2.10) BASE CHANGE: Let B be an A-algebra which is a free A-module of finite rank, and let $A \to A'$ be a homomorphism of rings. The A'-algebra $B' := B \otimes_A A'$ is a free A'-module having the same rank as B, and we have, for every $b \in B$,

$$\operatorname{Tr}_{B'/A'}(b \otimes 1_{A'}) = \operatorname{Tr}_{B/A}(b) 1_{A'}, \quad N_{B'/A'}(b \otimes 1_{A'}) = N_{B/A}(b) 1_{A'},$$
$$\operatorname{Pc}_{B'/A'}(X; b \otimes 1_{A'}) = \operatorname{Pc}_{B/A}(X; b) 1_{A'} \text{ in } A'[X].$$

(2.11) PRODUCTS: Let $B_1, \ldots, B_h$ be A-algebras which are free A-modules of finite rank, and define $B := B_1 \times \cdots \times B_h$. Then B is an A-algebra which is a free A-module of finite rank, and for every $b = (b_1, \ldots, b_h) \in B_1 \times \cdots \times B_h$ we have

$$\operatorname{Tr}_{B/A}(b) = \sum_{i=1}^{h} \operatorname{Tr}_{B_i/A}(b_i), \quad N_{B/A}(b) = \prod_{i=1}^{h} N_{B_i/A}(b_i),$$

$$\operatorname{Pc}_{B/A}(X; b) = \prod_{i=1}^{h} \operatorname{Pc}_{B_i/A}(X; b_i) \text{ in } A[X].$$

(2.12) NORM AND TRACE. THE CASE OF AN ALGEBRA OVER A DOMAIN: Let A be an integral domain, K its field of quotients, $\overline{A}$ the integral closure of A, and let B be a finite A-algebra.

(1) The maps $\mathrm{Tr}_{B_K/K}\colon B_K \to K$ and $\mathrm{N}_{B_K/K}\colon B_K \to K$ are well-defined, since $B_K := B \otimes_A K$ is a finite-dimensional K-algebra.

(2) For every $b \in B$ we define

$$\mathrm{Tr}_{B/A}(b) := \mathrm{Tr}_{B_K/K}(b \otimes 1), \quad \mathrm{N}_{B/A}(b) := \mathrm{N}_{B_K/K}(b \otimes 1),$$
$$\mathrm{Pc}_{B/A}(X;b) := \mathrm{Pc}_{B_K/K}(X;b \otimes 1), \quad m(X;b) := m(X;b \otimes 1).$$

Let $b \in B$; then we have $\mathrm{Pc}_{B/A}(X;b) \in \overline{A}[X]$ by (2.8), hence $\mathrm{Tr}_{B/A}(b)$ and $\mathrm{N}_{B/A}(b)$ lie in $\overline{A}$. Moreover, we have $\mathrm{Pc}_{B/A}(b \otimes 1;b) = 0$ by (2.6)(3), and the minimal polynomial $m(X;b)$ has coefficients in $\overline{A}$ by [63], Prop. 4.11, since it divides the characteristic polynomial $\mathrm{Pc}_{B/A}(X;b)$.

(3) Now let B be an A-algebra which is a free A-module of finite rank. Then B can be considered as an A-subalgebra of B_K, and, by (2.10), the first three elements defined in (2) coincide with the elements defined in (2.9).

(4) In particular, assume that $A = K$, and let L be a finite extension of K; then $\mathrm{Tr}_{L/K}\colon L \to K$ and $\mathrm{N}_{L/K}\colon L \to K$ are the usual trace and norm maps for finite field extensions, and for $x \in L$ the polynomial $\mathrm{Pc}_{L/K}(X;x)$ is the field polynomial of x, and the polynomial $m(X;x) \in K[X]$ is the monic polynomial of smallest degree which hat x as a zero.

2.3 The Trace Form

(2.13) THE TRACE FORM: Let B be an A-algebra which is a free A-module of finite rank n.

(1) The map

$$(b,c) \mapsto \mathrm{Tr}_{B/A}(bc)\colon B \times B \to A$$

is a symmetric A-bilinear form; it is called the *trace form* of B (with respect to A).

(2) For any sequence $(x_i)_{1\leq i \leq n}$ in B the element

$$D_{B/A}(x_1,\ldots,x_n) := \det((\mathrm{Tr}_{B/A}(x_i x_j)))$$

is called the *discriminant* of the sequence $(x_i)_{1\leq i \leq n}$ [with respect to the trace form].

(3) Let $(x_i)_{1\leq i \leq n}$ be a sequence in B, let $T = (t_{ij}) \in M(n;A)$, and define the sequence $(y_i)_{1\leq i \leq n}$ in B by $y_i := \sum_{j=1}^n t_{ij}x_j$ for every $i \in \{1,\ldots,n\}$. Then we have

$$D_{B/A}(y_1,\ldots,y_n) = (\det(T))^2 D_{B/A}(x_1,\ldots,x_n)$$

[cf. (2.4)(2)(b)]. In particular, the discriminants of two bases of B differ by a factor which is the square of an invertible element of A; all the discriminants of bases of B generate the same principal ideal of A which is called the *discriminant ideal* $\Delta_{B/A}$ of B over A. The discriminant of any sequence of n elements of B lies in $\Delta_{B/A}$.

(4) Let $\{x_1, \ldots, x_n\}$ be a basis of B, and let $z \in B$; let $S = (s_{ij}) \in M(n; A)$ be the matrix defined by $zx_i = \sum_{j=1}^{n} s_{ij} x_j$ for every $i \in \{1, \ldots, n\}$. Now we have $N_{B/A}(z) = \det(S)$, hence $D_{B/A}(zx_1, \ldots, zx_n) = (N_{B/A}(z))^2 D_{B/A}(x_1, \ldots, x_n)$.

(5) The trace form of B is a perfect pairing iff $\Delta_{B/A} = A$ [cf. (2.5)(1)].

(6) Let $B_1, \ldots, B_h$ be free A-algebras of finite rank, and set $B := B_1 \times \cdots \times B_h$. The trace form of B is a perfect pairing iff, for every $i \in \{1, \ldots, h\}$, the trace form of B_i is a perfect pairing [cf. (2.11)].

(7) Let $A \to A'$ be a homomorphism of rings, and set $B' := B \otimes_A A'$. If $\{x_1, \ldots, x_n\}$ is a basis of the A-module B, then $\{x_1 \otimes 1, \ldots, x_n \otimes 1\}$ is a basis of the A'-module B', and $D_{B/A}(x_1, \ldots, x_n) \otimes 1_{A'} = D_{B'/A'}(x_1 \otimes 1_{A'}, \ldots, x_n \otimes 1_{A'})$. Therefore, the trace form of B' is a perfect pairing if the trace form of B is a perfect pairing.

(2.14) REMARK: Let B be an A-algebra.

(1) $B^* = \mathrm{Hom}_A(B, A)$ carries a canonical structure as B-module: For all $b \in B$, $b^* \in B^*$ let $bb^* \in B^*$ be the A-linear map $c \mapsto \langle b^*, cb \rangle : B \to A$.

(2) Now we assume that B is a free A-module of finite rank. Let $\tau^B : B \to B^*$ be the A-linear map induced by the trace form [cf. (2.4)], i.e., we have $\tau^B(b)(b') = \mathrm{Tr}_{B/A}(bb')$ for all $b, b' \in B$; τ^B is also B-linear. We have $\tau^B(b) = b \,\mathrm{Tr}_{B/A}$ for every $b \in B$.

(3) Again we assume that B is a free A-module of finite rank. Let $A \to A'$ be a homomorphism of rings. We set $B' := B \otimes_A A'$. The canonical map $\omega : \mathrm{Hom}_A(B, A) \otimes_A A' \to \mathrm{Hom}_{A'}(B \otimes_A A', A')$ is bijective [since B is a free A-module of finite rank], and the induced map $\omega \circ (\tau^B \otimes \mathrm{id}_{A'}) : B \otimes_A A' \to \mathrm{Hom}_{A'}(B \otimes_A A', A')$ is the map $\tau^{B'} : B' \to \mathrm{Hom}_{A'}(B', A')$. In particular, if τ^B is bijective, then $\tau^{B'}$ is bijective [this result was already stated in (2.13)(7)].

(2.15) Proposition: *Let B be an A-algebra which is a free A-module of finite rank, and let $\{x_1, \ldots, x_n\}$ be a basis of B. Then:*

(1) *The trace form of B is a perfect pairing iff $\mathrm{Tr}_{B/A}$ generates the B-module B^*. In this case the B-modules B and B^* are isomorphic.*

(2) *Let $\{x_1^*, \ldots, x_n^*\}$ be the dual basis of $\{x_1, \ldots, x_n\}$; then we have*

$$\mathrm{Tr}_{B/A} = \sum_{i=1}^{n} x_i x_i^*.$$

(3) *If the trace form of B is a perfect pairing, and if $\{y_1, \ldots, y_n\}$ is the complementary basis of $\{x_1, \ldots, x_n\}$, then $D_{B/A}(x_1, \ldots, x_n) D_{B/A}(y_1, \ldots, y_n) = 1$.*

Proof: (1) By definition [cf. (2.4)], τ^B is an isomorphism of B-modules iff the trace form of B is a perfect pairing. On the other hand, τ^B is surjective iff τ^B is bijective [cf. (2.2)(1)], and τ^B is surjective iff $\mathrm{Tr}_{B/A}$ generates the B-module B^* [cf. (2.14)(2)].

(2) Let $b \in B$, and let $S = (s_{ij}) \in M(n; A)$ be the matrix of b_B with respect to the basis $\{x_1, \ldots, x_n\}$. Now we have

$$\sum_{j=1}^{n} x_j x_j^*(b) = \sum_{j=1}^{n} \langle x_j^*, bx_j \rangle = \sum_{j=1}^{n} \sum_{i=1}^{n} s_{ij} \delta_{ij} = \mathrm{Tr}(b_B) = \mathrm{Tr}_{B/A}(b).$$

(3) Let $T = (t_{ij}) \in M(n; A)$ be the matrix defined by $y_i = \sum_{j=1}^{n} t_{ij} x_j$ for every $i \in \{1, \ldots, n\}$. For every $k \in \{1, \ldots, n\}$ we have

$$\delta_{ik} = \mathrm{Tr}_{B/A}(y_i x_k) = \sum_{j=1}^{n} t_{ij} \mathrm{Tr}_{B/A}(x_j x_k) \quad \text{for every } i \in \{1, \ldots, n\}.$$

Therefore we have $T \cdot (\mathrm{Tr}_{B/A}(x_j x_k)) = I_n$, and it follows that [cf. (2.13)(3)]

$$D_{B/A}(y_1, \ldots, y_n) \cdot D_{B/A}(x_1, \ldots, x_n) = \det(T)^2 (D_{B/A}(x_1, \ldots, x_n))^2 = 1.$$

(2.16) REMARK: Let B be a ring, let $f = X^n + a_{n-1} X^{n-1} + \cdots + a_0 \in B[X]$ be a monic polynomial, and let $x \in B$ be a zero of f. We can write

$$f(X) = (X - x)(b_0 + b_1 X + \cdots + b_{n-1} X^{n-1}) \quad \text{in } B[X]$$

with elements $b_0, \ldots, b_{n-1} \in B$ which can be calculated by the recursion formula

$$b_{n-1} = a_n = 1, \quad b_{n-i} = x b_{n-i+1} + a_{n-i+1} \quad \text{for every } i \in \{2, \ldots, n\}.$$

Note that $f'(x) = b_0 + b_1 x + \cdots + b_{n-1} x^{n-1}$. Furthermore, it is easy to check that

$$b_i = \sum_{j=i+1}^{n} a_j x^{j-i-1} \quad \text{for every } i \in \{0, \ldots, n-1\}. \tag{$*$}$$

(2.17) THE DUAL BASIS OF $\{1, x, \ldots, x^{n-1}\}$: Let $f = X^n + a_{n-1} X^{n-1} + \cdots + a_0 \in A[X]$ be a monic polynomial, set $B := A[X]/(f)$, and let x be the image of X in B. We know that B is a free A-module of rank n with basis $\{1, x, \ldots, x^{n-1}\}$. The element $x \in B$ is a zero of f, hence we may write [cf. (2.16)]

$$f(X) = (X - x)(b_0 + b_1 X + \cdots + b_{n-1} X^{n-1}) \quad \text{in } B[X].$$

(1) Let $b^* \in B^*$ be defined by

$$\langle b^*, x^{n-1} \rangle = 1, \quad \langle b^*, x^i \rangle = 0 \quad \text{for every } i \in \{0, \ldots, n-2\}.$$

Then $\{b_0 b^*, \ldots, b_{n-1} b^*\}$ is the dual basis of $\{1, x, \ldots, x^{n-1}\}$, hence $B^* = Bb^*$.
Proof: We have to show that

$$\langle b_j b^*, x^i \rangle = \delta_{ij} \quad \text{for all } i, j \in \{0, \ldots, n-1\};$$

note that $\langle b^*, x^i b_j \rangle = \langle b_j b^*, x^i \rangle$ for all $i, j \in \{0, \ldots, n-1\}$. Since $b_{n-1} = 1$, it is clear that $\langle b^*, x^i b_{n-1} \rangle = \delta_{i,n-1}$ for every $i \in \{0, \ldots, n-1\}$. Let $j \in \{1, \ldots, n-1\}$, and assume that we have already shown that

$$\langle b^*, x^i b_{n-j} \rangle = \delta_{i,n-j} \quad \text{for every } i \in \{0, \ldots, n-1\}. \tag{*}$$

For every $i \in \{0, \ldots, n-2\}$ we have by (2.16) and (*)

$$\langle b^*, x^i b_{n-j-1} \rangle = \langle b^*, x^{i+1} b_{n-j} + x^i a_{n-j} \rangle = \delta_{i+1,n-j}$$

and

$$\begin{aligned}
\langle b^*, x^{n-1} b_{n-j-1} \rangle &= \langle b^*, x^n b_{n-j} + x^{n-1} a_{n-j} \rangle \\
&= \langle b^*, (-a_0 - a_1 x - \cdots - a_{n-1} x^{n-1}) b_{n-j} + x^{n-1} a_{n-j} \rangle \\
&= -a_{n-j} + a_{n-j} = 0.
\end{aligned}$$

(2) Let b^* be as in (1); then we have $\mathrm{Tr}_{B/A} = f'(x) b^*$ [where f' is the derivative of f]. Indeed, by (2.15)(2) we have

$$\mathrm{Tr}_{B/A} = \sum_{i=0}^{n-1} x^i (b_i b^*) = (b_0 + b_1 x + \cdots + b_{n-1} x^{n-1}) b^* = f'(x) b^*.$$

(3) For every $b \in B$ we have

$$f'(x) b = \sum_{i=0}^{n-1} \mathrm{Tr}_{B/A}(b_i b) x^i.$$

Indeed, since $\{b_0 b^*, \ldots, b_{n-1} b^*\}$ is the dual basis of $\{1, x, \ldots, x^{n-1}\}$, by (2) we have

$$f'(x) b = \sum_{i=0}^{n-1} \langle b_i b^*, f'(x) b \rangle x^i = \sum_{i=0}^{n-1} \langle f'(x) b^*, b_i b \rangle x^i = \sum_{i=0}^{n-1} \mathrm{Tr}_{B/A}(b_i b) x^i.$$

(2.18) Proposition: Let $f = X^n + a_{n-1} X^{n-1} + \cdots + a_0 \in A[X]$ be a monic polynomial, set $B := A[X]/(f)$, and let x be the image of X in B. The following statements are equivalent:
(1) f and f' generate the unit ideal of $A[X]$.
(2) $f'(x)$ is a unit of B.
(3) The trace form of B is a perfect pairing.
(4) The module of differentials $\Omega_{B/A}$ is the zero module.
Let these conditions be satisfied, and let $b_0, \ldots, b_{n-1} \in B$ be defined as in (2.17). Then $\{1, x, \ldots, x^{n-1}\}$ and $\{b_0/f'(x), b_1/f'(x), \ldots, b_{n-1}/f'(x)\}$ are complementary bases.

Proof (1) $\Rightarrow$ (2): There exist $g, h \in A[X]$ such that $gf + hf' = 1$, hence $f'(x)$ is a unit of B.

(2) $\Rightarrow$ (1): There exists $h \in A[X]$ such that $h(x)f'(x) = 1$, hence there exists $g \in A[X]$ satisfying $gf + hf' = 1$.

(2) $\Rightarrow$ (3): By (2.17), (1) and (2), we see that $\mathrm{Tr}_{B/A}$ generates B^* as B-module, hence the trace form of B is non-degenerate by (2.15)(1).

(3) $\Rightarrow$ (2): Let b^* be defined as in (2.17)(1). There exists $b \in B$ with $b^* = b \,\mathrm{Tr}_{B/A}$ [cf. (2.15)(1)], hence $b^* = bf'(x)b^*$ by (2.17)(2), and evaluating both sides at x^{n-1} gives $1 = bf'(x)$, hence $f'(x)$ is a unit of B.

(2) $\Leftrightarrow$ (4): By [63], section 16.1, we have $\Omega_{B/A} = B/(f'(x))$ which immediately implies the assertion.

Assume that $f'(x)$ is a unit of B. Since we have [cf. (2.14)(2) and (2.17)(2)]

$$\tau^B(b_i/f'(x)) = b_i/f'(x) \cdot \mathrm{Tr}_{B/A} = b_i b^* \quad \text{for every } i \in \{0, \ldots, n-1\},$$

we find that $\{1, x, \ldots, x^{n-1}\}$ and $\{b_0/f'(x), b_1/f'(x), \ldots, b_{n-1}/f'(x)\}$ are complementary bases of the free A-module B [cf. (2.5)(3) and (2.17)(1)].

(2.19) DEFINITION: A monic polynomial $f \in A[X]$ is said to be separable if the conditions of (2.18) are satisfied.

In particular, if A is a field, then this is the usual definition of a separable polynomial [cf. also [63], proof of Lemma 16.15].

3 Formally Unramified and Unramified Extensions

3.1 The Branch Locus

(3.1) DEFINITION: Let $\alpha \colon A \to B$ be a homomorphism of rings, and let $\Omega_{B/A}$ be the module of differentials of B over A.

(1) B is called formally unramified over A if $\Omega_{B/A} = \{0\}$, and B is called unramified over A if B is essentially of finite type over A and formally unramified over A.

(2) A prime ideal $\mathfrak{q}$ of B is called unramified over A if $B_{\mathfrak{q}}$ is formally unramified over A.

(3) A prime ideal $\mathfrak{p}$ of A is called unramified in B if $B_{\mathfrak{p}}$ is formally unramified over $A_{\mathfrak{p}}$.

(3.2) BRANCH LOCUS: Let $\alpha \colon A \to B$ be a homomorphism of rings.

(1) Let $\mathfrak{q}$ be a prime ideal of B. If $\mathfrak{q}$ is not unramified over A, then $\mathfrak{q}$ is called ramified over A. Since $\Omega_{B_{\mathfrak{q}}/A} = (\Omega_{B/A})_{\mathfrak{q}}$ [cf. [63], Th. 16.9], we see that $\mathfrak{q}$ is ramified over A iff $\mathfrak{q} \in \mathrm{Supp}_B(\Omega_{B/A})$.

(2) The set of prime ideals of B which are ramified over A is called the branch locus of B over A, and it is denoted by $\mathrm{Branch}(B/A)$. By (1) we have

$$\mathrm{Branch}(B/A) = \mathrm{Supp}_B(\Omega_{B/A}).$$

(3) Let $\mathfrak{p}$ be a prime ideal of A. If $\mathfrak{p}$ is not unramified in B, then $\mathfrak{p}$ is called ramified in B. Since $\Omega_{B_\mathfrak{p}/A_\mathfrak{p}} = \Omega_{B_\mathfrak{p}/A} = (\Omega_{B/A})_\mathfrak{p}$ [cf. [63], Th. 16.9], we see that $\mathfrak{p}$ is ramified in B iff $\mathfrak{p} \in \mathrm{Supp}_A(\Omega_{B/A})$.

(4) The set of prime ideals of A which are ramified in B is called the branch locus of A in B, and it is denoted by $\mathrm{branch}(B/A)$. By (3) we have

$$\mathrm{branch}(B/A) = \mathrm{Supp}_A(\Omega_{B/A}).$$

(5) The following statements are equivalent:
(a) B is formally unramified over A.
(b) The branch locus $\mathrm{Branch}(B/A)$ of B over A is empty.
(b$'$) Every maximal ideal of B is unramified over A.
(c) The branch locus $\mathrm{branch}(B/A)$ of A in B is empty.
(c$'$) Every maximal ideal of A is unramified in B.
Proof: We have $\Omega_{B/A} = \{0\}$ iff $(\Omega_{B/A})_\mathfrak{n} = \{0\}$ for every maximal ideal $\mathfrak{n}$ of B [cf. [63], Lemma 2.8], hence (a) and (b$'$) are equivalent by (2). In the same way we see, using (4), that (a) and (c$'$) are equivalent. The equivalence of (b$'$) and (b) as well as the equivalence of (c$'$) and (c) follow from [63], Lemma 2.8.

(6) Let $\mathfrak{p}$ be a prime ideal of A. Then $\mathfrak{p}$ is ramified in B iff there exists a prime ideal $\mathfrak{q}$ of B such that $\alpha^{-1}(\mathfrak{q}) \subset \mathfrak{p}$ and $\Omega_{B_\mathfrak{q}/A} \neq \{0\}$.
Proof: Let $\mathfrak{p}$ be a prime ideal of A, set $S := A \setminus \mathfrak{p}$, $T := \alpha(S)$, let $\mathfrak{q}$ be a prime ideal of B which does not meet T [equivalently, $\alpha^{-1}(\mathfrak{q}) \subset \mathfrak{p}$], and let $\mathfrak{q}' := \mathfrak{q}(T^{-1}B)$ be the extended prime ideal. Then we have $B_\mathfrak{p} = T^{-1}B$ and $B_\mathfrak{q} = (T^{-1}B)_{\mathfrak{q}'}$, hence $T^{-1}\Omega_{B/A} = \Omega_{B_\mathfrak{p}/A}$ and $\Omega_{(T^{-1}B)_{\mathfrak{q}'}/A} = \Omega_{B_\mathfrak{q}/A}$ [cf. [63], Th. 16.9]. If $\mathfrak{p}$ is ramified in B, then we have $\Omega_{B_\mathfrak{p}/A} \neq \{0\}$, hence $\mathrm{Supp}_{B_\mathfrak{p}}(\Omega_{B_\mathfrak{p}/A}) \neq \emptyset$, and therefore there exists a prime ideal $\mathfrak{q}'$ of $T^{-1}B$ such that $(\Omega_{B_\mathfrak{p}/A})_{\mathfrak{q}'} \neq \{0\}$. The prime ideal $\mathfrak{q} := (i_B^T)^{-1}(\mathfrak{q}')$ of B does not meet T, and its extension to $T^{-1}B$ is $\mathfrak{q}'$. Conversely, if there exists a prime ideal $\mathfrak{q}$ of B such that $\alpha^{-1}(\mathfrak{q}) \subset \mathfrak{p}$ and $\Omega_{B_\mathfrak{q}/A} \neq \{0\}$, then we have $\Omega_{B_\mathfrak{p}/A} \neq \{0\}$, hence $\mathfrak{p}$ is ramified in B.

(3.3) BRANCH LOCUS FOR EXTENSIONS OF FINITE TYPE: Let $\alpha : A \to B$ be essentially of finite type. Then $\Omega_{B/A}$ is a finitely generated B-module [cf. (1.6)] and we have

$$\mathrm{Branch}(B/A) = V(\mathrm{Ann}_B(\Omega_{B/A})), \quad \mathrm{branch}(B/A) = V(\mathrm{Ann}_A(\Omega_{B/A})),$$

hence $\mathrm{Branch}(B/A)$ is a closed subset of $\mathrm{Spec}(B)$, and $\mathrm{branch}(B/A)$ is a closed subset of $\mathrm{Spec}(A)$.
Proof: The first claim follows easily from [63], Cor. 2.7. Let $\{w_1, \ldots, w_n\}$ be a system of generators of the B-module $\Omega_{B/A}$, and set $M := Aw_1 + \cdots + Aw_n \subset$

$\Omega_{B/A}$. Let $\mathfrak{p}$ be a prime ideal of A which is ramified in B. By (3.2)(6) there exists a prime ideal $\mathfrak{q}$ of B with $\alpha^{-1}(\mathfrak{q}) \subset \mathfrak{p}$ and with $(\Omega_{B/A})_{\mathfrak{q}} \neq \{0\}$. Since $BM = \Omega_{B/A}$, and since $(\Omega_{B/A})_{\mathfrak{q}}$ is a localization of $(BM)_{\mathfrak{p}} = B_{\mathfrak{p}}M_{\mathfrak{p}}$, we see that $M_{\mathfrak{p}} \neq \{0\}$. Conversely, let $\mathfrak{p}$ be a prime ideal of A such that $M_{\mathfrak{p}} \neq \{0\}$; then $\mathfrak{p}$ is ramified in B because $(\Omega_{B/A})_{\mathfrak{p}} \neq \{0\}$. This shows that $\mathrm{Supp}_A(M) = \mathrm{Supp}_A(\Omega_{B/A}) = \mathrm{branch}(B/A)$. Furthermore, we have $\mathfrak{a} := \mathrm{Ann}_A(M) \supset \mathrm{Ann}_A(\Omega_{B/A}) =: \mathfrak{a}'$. On the other hand, we have $\{0\} = \mathfrak{a}M = \mathfrak{a}BM = \mathfrak{a}\Omega_{B/A}$, hence $\mathfrak{a} \subset \mathfrak{a}'$, and therefore $\mathfrak{a} = \mathfrak{a}'$. Since M is a finitely generated A-module, we have $V(\mathrm{Ann}_A(M)) = \mathrm{Supp}_A(M)$ [cf. [63], Cor. 2.7], and therefore $\mathrm{Supp}_A(\Omega_{B/A}) = V(\mathrm{Ann}_A(\Omega_{B/A}))$.

3.2 Some Ramification Criteria

(3.4) Proposition: *Let $\alpha \colon A \to B$ be a homomorphism of rings. Then:*
(1) Let B be an A-algebra essentially of finite type; then B is unramified over A iff $B_{\mathfrak{n}}/\mathfrak{m}B_{\mathfrak{n}}$ is unramified over the field $A_{\mathfrak{m}}/\mathfrak{m}A_{\mathfrak{m}}$ for every maximal ideal $\mathfrak{n}$ of B where we have defined $\mathfrak{m} := \alpha^{-1}(\mathfrak{n})$.
(2) Let B be a finite A-algebra; then B is unramified over A iff $B/\mathfrak{m}B$ is unramified over the field $A/\mathfrak{m}$ for every maximal ideal $\mathfrak{m}$ of A.

Proof: (1) $\Omega_{B/A}$ is a finitely generated B-module [cf. (1.6)]. Let $\mathfrak{n}$ be a maximal ideal of B, and set $\mathfrak{m} := \alpha^{-1}(\mathfrak{n})$. The condition $(\Omega_{B/A})_{\mathfrak{n}} = \{0\}$ is equivalent to $(\Omega_{B/A})_{\mathfrak{n}}/\mathfrak{m}(\Omega_{B/A})_{\mathfrak{n}} = \{0\}$ by Nakayama's lemma [cf. [63], Cor. 4.8]. Therefore we have the following: $\Omega_{B/A} = \{0\}$ iff $(\Omega_{B/A})_{\mathfrak{n}}/\mathfrak{m}(\Omega_{B/A})_{\mathfrak{n}} = \{0\}$ for every maximal ideal $\mathfrak{n}$ of B where $\mathfrak{m} := \alpha^{-1}(\mathfrak{n})$ [cf. [63], Cor. 2.8]. Moreover, $B_{\mathfrak{n}}/\mathfrak{m}B_{\mathfrak{n}}$ is essentially of finite type over $A_{\mathfrak{m}}/\mathfrak{m}A_{\mathfrak{m}}$. Now, by [63], Th. 16.9 and Prop. 16.3, we have $(\Omega_{B/A})_{\mathfrak{n}}/\mathfrak{m}(\Omega_{B/A})_{\mathfrak{n}} = \Omega_{(B_{\mathfrak{n}}/\mathfrak{m}B_{\mathfrak{n}})/(A_{\mathfrak{m}}/\mathfrak{m}A_{\mathfrak{m}})}$.
(2) It is clear that the condition is necessary [cf. [63], Prop. 16.3]. Conversely, assume that $B/\mathfrak{m}B$ is unramified over the field $A/\mathfrak{m}$ for every maximal ideal $\mathfrak{m}$ of A. Let $\mathfrak{n}$ be a maximal ideal of B. Now B, being a finite A-algebra, is integral over $\alpha(A)$ [cf. [63], Cor. 4.5]. Set $\mathfrak{m} := \alpha^{-1}(\mathfrak{n})$; then $A/\mathfrak{m} \to B/\mathfrak{n}$ is an injective finite map, and therefore $\mathfrak{m}$ is a maximal ideal of A [cf. [63], Cor. 4.17]. Then, by assumption, we have $\{0\} = (\Omega_{(B/\mathfrak{m}B)/(A/\mathfrak{m})})_{\mathfrak{n}} = \Omega_{(B_{\mathfrak{n}}/\mathfrak{m}B_{\mathfrak{n}})/(A_{\mathfrak{m}}/\mathfrak{m}A_{\mathfrak{m}})}$, and by (1) we see that B is unramified over A.

(3.5) Proposition: *Let $\alpha \colon A \to B$, $\beta \colon B \to C$ be homomorphisms of rings. Then:*
(1) If B is formally unramified over A and C is formally unramified over B, then C is formally unramified over A.
(2) If C is formally unramified over A, then C also is formally unramified over B.
(3) The statements in (1) and (2) hold also with formally unramified replaced by unramified.

Proof: By [63], Prop. 16.2, the sequence

$$\Omega_{B/A} \otimes_B C \to \Omega_{C/A} \to \Omega_{C/B} \to 0$$

is exact.

(1) From $\Omega_{B/A} = \Omega_{C/B} = \{0\}$ we see that $\Omega_{C/A} = \{0\}$.

(2) If we have $\Omega_{C/A} = \{0\}$, then we have also $\Omega_{C/B} = \{0\}$.

(3) This follows from (1), (2) and (1.8).

(3.6) Proposition: Let $\alpha\colon A \to B$ be a homomorphism of rings, let $\mathfrak{b}$ be an ideal of B, and let $\mathfrak{a}$ be an ideal of A with $\mathfrak{a}B \subset \mathfrak{b}$. If B is formally unramified (resp. unramified) over A, then $B/\mathfrak{b}$ is formally unramified (resp. unramified) over A and over $A/\mathfrak{a}$.

Proof: Note that $B/\mathfrak{b}$ is formally unramified over A by [63], Prop. 16.3. Then B is formally unramified over $B/\mathfrak{a}B$, and $B/\mathfrak{a}B$ is formally unramified over $A/\mathfrak{a}$ by [63], Prop. 16.3, hence B is formally unramified over $A/\mathfrak{a}$ by (3.5)(1). The remaining statement follows from this and from (1.8).

(3.7) Proposition: Let $\alpha\colon A \to B$ and $\rho\colon A \to A'$ be homomorphisms of rings. Then:

(1) If B is formally unramified (resp. unramified) over A, then $B \otimes_A A'$ is formally unramified (resp. unramified) over A'.

(2) Assume that A' is a faithfully flat A-algebra. Then:

(a) If $B \otimes_A A'$ is formally unramified over A', then B is formally unramified over A.

(b) If α is essentially of finite type and if $B \otimes_A A'$ is unramified over A', then B is unramified over A.

Proof: Set $B' := B \otimes_A A'$. It is easy to check that the B'-modules $\Omega_{B'/A'}$ and $\Omega_{B/A} \otimes_B B'$ are isomorphic. The claim in (1) follows immediately [cf. (1.8)], and for the claims in (2) one may use [21], Ch. 3, Ex. 16, and (1.8).

(3.8) Proposition: Let A be a ring, let $\alpha_1\colon A \to B_1, \ldots, \alpha_h\colon A \to B_h$ be homomorphisms of rings, and set $B := B_1 \times \cdots \times B_h$, $C := B_1 \otimes_A \cdots \otimes_A B_h$. Then:

(1) B is formally unramified over A iff B_i is formally unramified over A for every $i \in \{1, \ldots, h\}$.

(2) If $B_1, \ldots, B_h$ are formally unramified over A, then C is formally unramified over A.

(3) If $B_1, \ldots, B_h$ are unramified over A, then B and C are unramified over A.

Proof: (1) follows from [63], Prop. 16.10, and (2) follows from [63], Prop. 16.5.

(3) The first statement follows from (1) and (1.9), and the second statement follows from (2) and (1.9).

(3.9) Proposition: Let $\alpha\colon A \to B$ be a homomorphism of rings, let S be a multiplicatively closed system in A, let T be a multiplicatively closed system in B, and assume that $\alpha(S) \subset T$. If B is formally unramified (resp. unramified) over A, then $T^{-1}B$ is formally unramified (resp. unramified) over $S^{-1}A$.

Proof: By [63], Th. 16.9, the $T^{-1}B$-modules $\Omega_{T^{-1}B/S^{-1}A}$ and $T^{-1}\Omega_{B/A}$ are isomorphic; the remaining statement is clear.

(3.10) Proposition: *Let k be a field, and let K be a finite extension of k. Then K is unramified over k iff K is a separable extension of k.*

Proof: Let $x \in K$, and let $f \in k[X]$ be the minimal polynomial for x over k. We have $\Omega_{k(x)/k} = k[x]/(f'(x))$ [cf. [63], section 16.1], hence $\Omega_{k(x)/k} = \{0\}$ iff x is separable over k [recall that, by definition, x is separable over k if its minimal polynomial over k is a separable polynomial].

Assume that K is separable over k. Then, by the theorem of the primitive element, we have $K = k[x]$ for some $x \in K$ which is separable over k, and therefore we have $\Omega_{K/k} = \{0\}$.

Assume that K is not separable over k. The set of elements of K which are separable over k form a field k_{sep}, the separable closure of k in K. Let k' be a maximal proper subfield of K containing k_{sep} [such subfields exist since we have $1 < [K : k_{\mathrm{sep}}] < \infty$]; K is a purely inseparable extension of k'. Let $x \in K \backslash k'$; then we have $K = k'[x]$, and the minimal polynomial $f \in k'[T]$ of x over k' has the form $T^{p^e} - a$ for some $e \in \mathbb{N}$ and $a \in k'$ [p is the characteristic of k]. Since $f' = 0$, we get that $\Omega_{K/k'} = K \neq \{0\}$, and from the exactness of $\Omega_{K/k} \to \Omega_{K/k'} \to \{0\}$ [cf. [63], Prop. 16.2] we get $\Omega_{K/k} \neq \{0\}$.

(3.11) Remark: Let k be a field, and let B be a finite k-algebra. We state some properties of B. First of all, B is an artinian ring [cf. [63], Th. 2.14]. Furthermore, B has only finitely many prime ideals $\mathfrak{n}_1, \ldots, \mathfrak{n}_r$ and each of them is maximal [cf. [63], Cor. 2.15], $\mathfrak{r} := \mathfrak{n}_1 \cdots \mathfrak{n}_r = \mathfrak{n}_1 \cap \cdots \cap \mathfrak{n}_r$ is the nilradical of B [cf. B(10.1)], $\mathfrak{r}^s = \{0\}$ for some $s \in \mathbb{N}$ [cf. [63], Ex. 1.13 and Cor. 2.12], $B \to B/\mathfrak{n}_1^s \times \cdots \times B/\mathfrak{n}_r^s$ is an isomorphism of k-algebras [cf. B(10.1)], and, for every $i \in \{1, \ldots, r\}$, we have $B_{\mathfrak{n}_i} \cong B/\mathfrak{n}_i^s$ [since $B_{\mathfrak{n}_i} = B_{\mathfrak{n}_i}/(\mathfrak{n}_1^s \cdots \mathfrak{n}_r^s B_{\mathfrak{n}_i}) = B_{\mathfrak{n}_i}/\mathfrak{n}_i^s B_{\mathfrak{n}_i} \cong B/\mathfrak{n}_i^s$ because $B/\mathfrak{n}_i^s$ is local, having maximal ideal $\mathfrak{n}_i/\mathfrak{n}_i^s$], hence $B_{\mathfrak{n}_i}$ is, in particular, a finite k-algebra.

Now assume, in addition, that B is a reduced ring. Then $\mathfrak{n}_1 \cap \cdots \cap \mathfrak{n}_r = \{0\}$, and B is k-isomorphic to the product of a finite number of fields, each of which is a finite extension of k.

(3.12) Lemma: *Let k be an algebraically closed field, and let B be a finite k-algebra. If B is unramified over k, then $\mathfrak{n} = \mathfrak{n}^2$ for every maximal ideal $\mathfrak{n}$ of B.*

Proof: Let $\mathfrak{n}$ be a maximal ideal of B. Now $B/\mathfrak{n}$ is a finite extension of k, hence $B/\mathfrak{n} = k$ [since k is algebraically closed] and $B = k 1_B \oplus \mathfrak{n}$ [direct sum of k-vector spaces]. Every $b \in B$ has a unique representation $b = c 1_B + y$ where $c \in k$ and $y \in \mathfrak{n}$. The map $b \mapsto y + \mathfrak{n}^2 : B \to \mathfrak{n}/\mathfrak{n}^2$ is easily seen to be a k-derivation $D : B \to \mathfrak{n}/\mathfrak{n}^2$. We have $\Omega_{B/k} = \{0\}$ by assumption, hence $D = 0$ [cf. [63], p. 384], and therefore $y \in \mathfrak{n}^2$ for every $y \in \mathfrak{n}$, i.e., we have $\mathfrak{n} = \mathfrak{n}^2$.

(3.13) Theorem: *Let k be a field, let $\overline{k}$ be an algebraic closure of k, and let B be a k-algebra essentially of finite type. The following statements are equivalent:*
(1) *B is unramified over k.*
(2) *B is k-isomorphic to the product of a finite number of fields each of which is a finite separable extension of k.*
(3) *$B \otimes_k \overline{k}$ is $\overline{k}$-isomorphic to the product of a finite number of copies of $\overline{k}$.*
(4) *B is a finite k-algebra, and, for every extension field k' of k, the k'-algebra $B \otimes_k k'$ is reduced.*
(5) *There exist finitely many monic separable polynomials $f_1, \ldots, f_r \in k[T]$ such that B is k-isomorphic to $k[T]/(f_1) \times \cdots \times k[T]/(f_r)$. If k is infinite, one may choose $r = 1$, i.e., there exists a monic separable polynomial $f \in k[T]$ such that $B \cong k[T]/(f)$.*

Proof (3) $\Leftrightarrow$ (1): Since $k \to \overline{k}$ is a faithfully flat extension, B is formally unramified over k iff $B \otimes_k \overline{k}$ is formally unramified over $\overline{k}$ [cf. (3.7)]; hence we may assume, in proving the equivalence of (1) and (3), that k is algebraically closed.
Assume that (3) holds, i.e., that $B \cong k^n$ [as k-algebras] for some $n \in \mathbb{N}$. A product of a finite number of k-algebras is formally unramified over k iff each factor is formally unramified over k [cf. (3.8)(1)]; since $\Omega_{k/k} = \{0\}$, we see that (3) implies (1).
Now assume that (1) holds. First, we consider the case where B is a *finite* k-algebra. By (3.11) B is isomorphic to a product of a finite number of finite local k-algebras; by (3.8)(1) it is enough to treat the case of a finite local k-algebra B which is unramified over k. Let B be such a k-algebra, and let $\mathfrak{n}$ be the maximal ideal of B. We have $\mathfrak{n}^2 = \mathfrak{n}$ by (3.12), hence $\mathfrak{n} = \{0\}$ by Nakayama's lemma [cf. [63], Cor. 4.8]; from $B = k \, 1_B \oplus \mathfrak{n}$ [as k-vector-spaces] we see that $B = k$.
Next, assume that B is a k-algebra of finite type. Let $\mathfrak{n}$ be a maximal ideal of B. By [63], Th. 4.19, $B/\mathfrak{n}$ is a finite extension of k, hence $B/\mathfrak{n} = k$ since k is algebraically closed. Let $h \in \mathbb{N}$, and consider the exact sequence $0 \to \mathfrak{n}^{h-1}/\mathfrak{n}^h \to B/\mathfrak{n}^h \to B/\mathfrak{n}^{h-1} \to 0$; since $\mathfrak{n}^{h-1}/\mathfrak{n}^h$ is a finitely generated B-module which is annihilated by $\mathfrak{n}$, we see that $\mathfrak{n}^{h-1}/\mathfrak{n}^h$ is a k-vector space of finite dimension [since $B/\mathfrak{n} = k$]. By induction it follows easily that $B/\mathfrak{n}^h$ is a finite k-algebra. Consider $B \supset \mathfrak{n} \supset \mathfrak{n}^2$. In particular, $B/\mathfrak{n}^2$ is a finite k-algebra, and since $B/\mathfrak{n}^2$ is a homomorphic image of the unramified k-algebra B, $B/\mathfrak{n}^2$ is unramified over k [cf. (3.6)]. We have $\mathfrak{n}/\mathfrak{n}^2 = (\mathfrak{n}/\mathfrak{n}^2)^2 = \{0\}$ by (3.12), hence $\mathfrak{n} = \mathfrak{n}^2$, and therefore we have $\mathfrak{n}B_{\mathfrak{n}} = \mathfrak{n}^2 B_{\mathfrak{n}}$ which implies $\mathfrak{n}B_{\mathfrak{n}} = \{0\}$ by Nakayama's lemma. Then $\mathfrak{n}$ is a minimal prime ideal of B. Thus, we have shown that every maximal ideal of the noetherian ring B is a minimal prime ideal of B, whence B is an artinian ring [cf. [63], Cor. 9.1]. Now B has only finitely many maximal ideals $\mathfrak{n}_1, \ldots, \mathfrak{n}_r$ [cf. [63], Th. 2.14], and there exists $h \in \mathbb{N}$ such that $\mathfrak{n}_1^h \cap \cdots \cap \mathfrak{n}_r^h = \{0\}$ [cf. [63], Ex. 1.13 and Cor. 2.12]. Therefore we have $B \cong B/\mathfrak{n}_1^h \times \cdots \times B/\mathfrak{n}_r^h$ [cf. B(10.1)], hence B is a finite k-algebra [because, as we have just seen, $B/\mathfrak{n}_i^h$ is a finite k-algebra for every $i \in \{1, \ldots, r\}$]. By the first part of the proof we get the result.
Now assume that B is essentially of finite type over k, hence that $B = S^{-1}C$ where C is a k-algebra of finite type and $S \subset C$ is multiplicatively closed. Since

$\{0\} = \Omega_{B/k} = S^{-1}\Omega_{C/k}$ [cf. [63], Th. 16.9], and since $\Omega_{C/k}$ is a finitely generated C-module, there exists $s \in S$ with $\{0\} = (\Omega_{C/k})_s = \Omega_{C_s/k}$ [cf. [63], Prop. 2.1]. The k-algebra C_s is of finite type [since $C_s = C[X]/(1 - sX)$, cf. [63], Ex. 2.2] and unramified over k, hence artinian by the part just proved, and therefore B as a localization of an artinian ring is artinian. Now we are in the situation considered above.

$(5) \Rightarrow (3)$: We may restrict to the case $B = k[T]/(f)$ where $f \in k[T]$ is a monic separable polynomial of positive degree r. In $\overline{k}[T]$ we may write $f = \prod_{i=1}^{r}(T - x_i)$ where $x_1, \ldots, x_r \in \overline{k}$ are pairwise different [since f is separable]. Then we have $B \otimes_k \overline{k} \cong \overline{k}[T]/(f) \cong \overline{k}^r$.

$(3) \Rightarrow (5)$: Since B may be considered as a k-subalgebra of $\overline{B} := B \otimes_k \overline{k}$, and this latter ring is reduced by assumption, we see that B is reduced. Since $\dim_k(B) = \dim_{\overline{k}}(B \otimes_k \overline{k})$, we see that B is a finite k-algebra, and therefore a product of a finite number of fields which are finite extensions of k [cf. (3.11)]. First, we consider the case where k is a finite field. Every finite extension K of k has the form $K = k[T]/(f)$ where $f \in k[T]$ is a monic irreducible separable polynomial. Now, let k be an infinite field, and set $r := \dim_k(B)$. Every $b \in B$ is a zero of its characteristic polynomial $\mathrm{Pc}_{B/k}(T; b) \in k[T]$ over k [cf. (2.9)(2)].

Assume for the moment that there exists $b \in B$ such that its characteristic polynomial $\mathrm{Pc}_{B/k}(T; b) \in k[T]$ is separable. Then $\mathrm{Pc}_{B/k}(T; b)$ is a product of pairwise different monic irreducible separable polynomials in $k[T]$; since minimal polynomial and characteristic polynomial of a linear map of a finite-dimensional vector space have the same irreducible factors, it follows that $\mathrm{Pc}_{B/k}(T; b)$ is also the minimal polynomial of b over k. The k-algebra homomorphism $k[T] \to B$ defined by mapping T to b therefore has kernel $(\mathrm{Pc}_{B/k}(T; b))$, hence induces an injective k-algebra homomorphism $k[T]/(\mathrm{Pc}_{B/k}(T; b)) \to B$. The k-vector spaces $k[T]/(\mathrm{Pc}_{B/k}(T; b))$ and B both have the same dimension r, hence this homomorphism is an isomorphism of k-algebras, hence we may choose $f = \mathrm{Pc}(X; b)$.

Now we show that there exists $b \in B$ such that its characteristic polynomial $\mathrm{Pc}_{B/k}(T; b)$ over k is separable. By assumption we have $\overline{B} \cong \overline{k}^r$. Let $b' \in \overline{B}$, and let $(\beta_1', \ldots, \beta_r')$ be the image of b' in $\overline{k}^r$; then $(T - \beta_1') \cdots (T - \beta_r') \in \overline{k}[T]$ is the characteristic polynomial of b' over $\overline{k}$; thus, there exists an element $b' \in \overline{B}$ such that its characteristic polynomial $\mathrm{Pc}_{\overline{B}/\overline{k}}(T; b')$ over $\overline{k}$ has pairwise different zeroes in $\overline{k}$, hence the discriminant of the polynomial $\mathrm{Pc}_{\overline{B}/\overline{k}}(T; b')$ is a non-zero element of $\overline{k}$.

Let $R := k[X_1, \ldots, X_r]$ be the polynomial ring over k, and set $C := B \otimes_k R$. Furthermore, set $\overline{R} := R \otimes_k \overline{k}$ and $\overline{C} := C \otimes_k \overline{k}$. Then we have $\overline{R} = \overline{k}[X_1, \ldots, X_r]$ and $\overline{C} = B \otimes_k \overline{R} = B \otimes_k (R \otimes_k \overline{k})$. We may consider B and C as k-subalgebras of $\overline{C}$; note that C is a free R-module of rank r and that $\overline{C}$ is a free $\overline{R}$-module of rank r. Let $\{w_1, \ldots, w_r\}$ be a basis of the k-algebra B; then $\{w_1, \ldots, w_r\}$ is a basis of the R-algebra C and a basis of the $\overline{R}$-algebra $\overline{C}$.

Let $p_1, \ldots, p_r \in \overline{R}$ and set $z := p_1 w_1 + \cdots + p_r w_r$; then we have

$$\mathrm{Pc}_{\overline{C}/\overline{R}}(T; z) = T^r + a_1 T^{r-1} + \cdots + a_r \quad \text{with } a_1, \ldots, a_r \in \overline{R}.$$

Let $\alpha_1, \ldots, \alpha_r \in \overline{k}$, and set $\widetilde{z} := p_1(\alpha_1, \ldots, \alpha_r)w_1 + \cdots + p_r(\alpha_1, \ldots, \alpha_r)w_r$; clearly we have

$$\mathrm{Pc}_{\overline{B}/\overline{k}}(T; \widetilde{z}) = T^r + a_1(\alpha_1, \ldots, \alpha_r)T^{r-1} + \cdots + a_r(\alpha_1, \ldots, \alpha_r),$$

and therefore we have $\mathrm{dis}_T(\mathrm{Pc}_{\overline{B}/\overline{k}}(T; \widetilde{z})) = \mathrm{dis}_T(\mathrm{Pc}_{\overline{C}/\overline{R}}(T; z))(\alpha_1, \ldots, \alpha_r)$ [cf. B(10.22)(1)(*)].

Set $c := \sum_{i=1}^r X_i w_i \in \overline{C}$ and $\widetilde{c} := \alpha_1 w_1 + \cdots + \alpha_r w_r$ with $\alpha_1, \ldots, \alpha_r \in \overline{k}$. Set $D := \mathrm{dis}_T(\mathrm{Pc}_{\overline{C}/\overline{R}}(T; c)) \in \overline{R}$. From the last paragraph we get $D(\alpha_1, \ldots, \alpha_r) = \mathrm{dis}_T(\mathrm{Pc}_{\overline{B}/\overline{k}}(T; \widetilde{c}))$. We choose $b' \in \overline{B}$ with $\mathrm{dis}_T(\mathrm{Pc}_{\overline{B}/\overline{k}}(T; b')) \neq 0$ [cf. above] and elements $\alpha_1, \ldots, \alpha_r \in \overline{k}$ with $b' = \alpha_1 w_1 + \cdots + \alpha_r w_r$; then we have $D(\alpha_1, \ldots \alpha_r) \neq 0$, hence D is not the zero polynomial. Since k is an infinite field, there exist $\delta_1, \ldots, \delta_r \in k$ such that $D(\delta_1, \ldots, \delta_r) \neq 0$. Set $b := \sum_{i=1}^r \delta_i w_i \in B$; the discriminant of the characteristic polynomial $\mathrm{Pc}_{B/k}(T; b) \in k[T]$ is $D(\delta_1, \ldots, \delta_r)$, hence $\mathrm{Pc}_{B/k}(T; b)$ is a separable polynomial [cf. B(10.22)(2)].

(2) $\Leftrightarrow$ (5): If K is a finite separable extension of k, then we have $K \cong k[T]/(f)$ where $f \in k[T]$ is a monic separable polynomial, hence (2) implies (5), since the additional statement in (5) was shown to hold while we proved the implication (3) $\Rightarrow$ (5). On the other hand, let $f \in k[T]$ be a separable polynomial which may be assumed to be monic; then $f = f_1 \cdots f_h$ where $f_1, \ldots, f_h$ are pairwise different monic irreducible polynomials in $k[T]$ which are separable. The ideals (f_i) and (f_j) of $k[T]$ are pairwise comaximal for all $i, j \in \{1, \ldots, h\}$, $i \neq j$, hence $k[T]/(f) \cong k[T]/(f_1) \times \cdots \times k[T]/(f_h)$ [cf. B(10.1)], and we have shown that (5) implies (2).

Now we have shown that the statements (1), (2), (3) and (5) are equivalent.

(4) $\Rightarrow$ (3): By assumption, $B \otimes_k \overline{k}$ is a finite reduced $\overline{k}$-algebra, hence by (3.11) it is isomorphic to a product of a finite number of fields which are finite extensions of $\overline{k}$; hence $B \otimes_k \overline{k}$ is isomorphic to a finite number of copies of $\overline{k}$.

(2) $\Rightarrow$ (4): It is enough to show the following. Let K be a finite separable extension of k; then $K \otimes_k k'$ is reduced for any extension field k' of k. Now, if K is a finite separable extension of k, then K is unramified over k [cf. (3.10)], hence the finite k'-algebra $K \otimes_k k'$ is unramified over k' [cf. (3.7)(1)], and by the equivalence of (1) and (2), $K \otimes_k k'$ is a reduced ring.

This ends the proof of (3.13).

3.3   Ramification for Local Rings and Applications

(3.14) **Corollary:** *Let* $\alpha: A \to B$ *be a homomorphism essentially of finite type, let* $\mathfrak{q}$ *be a prime ideal of* B, *and set* $\mathfrak{p} := \alpha^{-1}(\mathfrak{q})$. *The following statements are equivalent:*

(1) $\mathfrak{q}$ *is unramified over* A.

(2) $\mathfrak{p}B_{\mathfrak{q}}$ *is the maximal ideal* $\mathfrak{q}B_{\mathfrak{q}}$ *of* $B_{\mathfrak{q}}$, *and the field* $B_{\mathfrak{q}}/\mathfrak{q}B_{\mathfrak{q}}$ *is a finite separable extension of the field* $A_{\mathfrak{p}}/\mathfrak{p}A_{\mathfrak{p}}$.

Proof (1) $\Rightarrow$ (2): B is an A-algebra essentially of finite type, hence $\Omega_{B/A}$ is a finitely generated B-module [cf. (1.6)]. From $\Omega_{B_q/A} = \{0\}$ we deduce that there exists $f \in B \setminus q$ such that $\Omega_{B_f/A} = \{0\}$ [cf. [63], Prop. 2.1 and Th. 16.9]. Now B_f is essentially of finite type over B and B is essentially of finite type over A, hence B_f is also essentially of finite type over A [cf. (1.8)(1)]. Therefore, to begin with, we may assume that B is unramified over A. Since $C := (B/\mathfrak{p}B)_\mathfrak{p}$ is essentially of finite type over the field $k := (A/\mathfrak{p})_\mathfrak{p} = A_\mathfrak{p}/\mathfrak{p}A_\mathfrak{p}$ [cf. (1.8)(2)], C is unramified over this field [cf. (3.6), (3.9)]. Now C is a finite k-algebra by (3.13), hence $C = C_1 \times \cdots \times C_h$ where $C_1, \ldots, C_h$ are finite local k-algebras by (3.11). Let $i \in \{1, \ldots, h\}$; C_i is unramified over k by (3.8), and, since C_i is local, C_i is a finite separable extension field of k by (3.13) [note that a quasilocal ring is indecomposable]. Since C is a product of a finite number of fields which are finite separable extensions of k, every localization of C with respect to a prime ideal of C is a field which is a finite separable extension of k [cf. (3.11)]. Now we have $C_q = (B/\mathfrak{p}B)_q = B_q/\mathfrak{p}B_q$, hence we have $\mathfrak{p}B_q = qB_q$, and B_q/qB_q is a finite separable extension of $k = A_\mathfrak{p}/\mathfrak{p}A_\mathfrak{p}$.

(2) $\Rightarrow$ (1): Since the field B_q/qB_q is a finite separable extension of the field $A_\mathfrak{p}/\mathfrak{p}A_\mathfrak{p}$, we have $\Omega_{(B_q/qB_q)/(A_\mathfrak{p}/\mathfrak{p}A_\mathfrak{p})} = \{0\}$ by (3.10). We know that $qB_q = \mathfrak{p}B_q$, and that therefore $\Omega_{(B_q/\mathfrak{p}B_q)/(A_\mathfrak{p}/\mathfrak{p}A_\mathfrak{p})} = \Omega_{B_q/A_\mathfrak{p}}/\mathfrak{p}\Omega_{B_q/A_\mathfrak{p}} = \Omega_{B_q/A_\mathfrak{p}}/q\Omega_{B_q/A_\mathfrak{p}} = \{0\}$ [cf. [63], Prop. 16.3]. Furthermore, since $\Omega_{B/A}$ is a finitely generated B-module [cf. (1.6)], $\Omega_{B_q/A_\mathfrak{p}} = (\Omega_{B/A})_q$ is a finitely generated B_q-module; by Nakayama's lemma [cf. [63], Cor. 4.8] we have $\Omega_{B_q/A_\mathfrak{p}} = \Omega_{B_q/A} = \{0\}$, i.e., B_q is unramified over A.

(3.15) REMARK: Let $\alpha \colon R \to S$ be a homomorphism of quasilocal rings, let $\mathfrak{m}$ be the maximal ideal of R, $\mathfrak{n}$ be the maximal ideal of S, and assume that $\mathfrak{m}S \subset \mathfrak{n}$ [i.e., the homomorphism is local]. If α is essentially of finite type, the result above says that S is unramified over R iff $\mathfrak{m}S = \mathfrak{n}$ and $S/\mathfrak{n}$ is a finite separable extension of $R/\mathfrak{m}$; it is the classical criterion for unramifiedness in case we have extensions essentially of finite type. This condition sometimes is even taken as a definition for unramifiedness in the absence of any finiteness conditions. If we do this, then we can interpret the results in I(8.3), (1) and (2), in the following way: The extensions $V \to V_Z$ and $V \to V_T$ are unramified. Similarly, in the situation of appendix B, (7.7) and (7.11), we can say that A_Z and A_T are unramified over A.

(3.16) Proposition: *Let A be a quasilocal ring with maximal ideal $\mathfrak{m}$ and residue field k, and let B be a finite A-algebra. Assume that either k is an infinite field or that B is a quasilocal ring. Then B is unramified over A iff there exists a monic separable polynomial $F \in A[T]$ such that B is isomorphic to a homomorphic image of $A[T]/(F)$; one may choose F in such a way that $\deg(F) = \dim_k(B \otimes_A k)$.*

Proof: (1) The conditions stated in the proposition are sufficient even without assuming that B is quasilocal or that k is an infinite field. Indeed, let $F \in A[T]$ be a monic separable polynomial, set $C := A[T]/(F)$, let t be the image of T in C, and assume that B is a homomorphic image of C. Since F is a separable polynomial, C is unramified over A [cf. (2.18) and (2.19)], and B is unramified over A by (3.6).

(2) We show that the conditions stated in the proposition are necessary. Set $L := B \otimes_A k$ and $n := \dim_k(L)$; then L is a finite k-algebra and it is unramified over k [cf. (3.7)(1)]. By (3.13) L is k-isomorphic to the product of a finite number of fields $k_1, \ldots, k_r$, each of which is a finite separable extension of k, i.e., there exist $x_1, \ldots, x_r$ in some algebraic closure of k such that $k_i = k[x_i]$ is a separable extension of k for every $i \in \{1, \ldots, r\}$. For every $i \in \{1, \ldots, r\}$ let $f_i \in k[T]$ be the minimal polynomial of x_i over k; f_i is a monic irreducible separable polynomial.
(a) Assume that B is quasilocal. Then $L = B \otimes_A k = B/\mathfrak{m}B$ is quasilocal and $r = 1$.
(b) Assume that k is an infinite field. Since we have $k[x_i + a] = k[x_i]$ for every $a \in k$ and every $i \in \{1, \ldots, r\}$, there exist elements $a_1, \ldots, a_r \in k$ such that, f_i denoting the minimal polynomial of the element $x_i + a_i$ over k for $i \in \{1, \ldots, r\}$, the polynomials $f_1, \ldots, f_r$ are pairwise different. Hence we may assume, to begin with, that the polynomials $f_1, \ldots, f_r$ are pairwise different and therefore relatively prime. Set $f := f_1 \cdots f_r$. Then L is k-isomorphic to $k[T]/(f)$.
In both cases, L is k-isomorphic to $k[T]/(f)$ where $f \in k[T]$ is a monic separable polynomial of degree n, and the ring homomorphism $B \to k[T]/(f)$ is surjective and has kernel $\mathfrak{m}B$. Let x be the image of T in $k[T]/(f)$; then f is the minimal polynomial of x over k [since $f(T) = \mathrm{Pc}_{L/k}(T; x) \in k[T]$, and since $\mathrm{Pc}_{L/k}(T; x)$ has the same irreducible factors as the minimal polynomial of x over k]. Let $b \in B$ be a preimage of x. Then we have $B = A 1_B + Ab + \cdots + Ab^{n-1} + \mathfrak{m}B$, hence, by Nakayama's lemma [cf. [63], Cor. 4.8], the A-module B is generated by the elements $1, b, \ldots, b^{n-1}$. Therefore there exists a monic polynomial $F \in A[T]$ of degree n such that $F(b) = 0$, and B is a homomorphic image of $C := A[T]/(F)$. Let t be the image of T in C; then b is the image of t in B. Denote by $\overline{F} \in k[T]$ the canonical image of F; since $\overline{F}(x) = 0$, we see that $\overline{F} = f$ [because f is the minimal polynomial of x over k]. Now C is a finite A-module, hence C is quasisemilocal, and every maximal ideal of C contains $\mathfrak{m}C$ [cf. B(3.8)]. Let $\mathfrak{n}_1, \ldots, \mathfrak{n}_r$ be the maximal ideals of C; then $\mathfrak{n}'_1 := \mathfrak{n}_1/\mathfrak{m}C, \ldots, \mathfrak{n}'_r := \mathfrak{n}_r/\mathfrak{m}C$ are the maximal ideals of $C/\mathfrak{m}C = L$. Let $i \in \{1, \ldots, r\}$ and set $L_i := L_{\mathfrak{n}'_i}$. The L-modules $\Omega_{L/k}$ and $L/Lf'(x)$ are isomorphic by [63], section 16.1, hence the L_i-modules $\Omega_{L_i/k}$ and $L_i/L_i f'(x)$ are isomorphic. By assumption, L is unramified over k, hence L_i is unramified over k [cf. (3.8)], i.e., we have $f'(x) \notin \mathfrak{n}'_i$, and therefore we have $F'(t) \notin \mathfrak{n}_i$. This implies that $F'(t)$ is a unit of C, hence F is a separable polynomial in $A[T]$ [cf. (2.18)].

(3.17) REMARK: Let A and B be as in (3.16); the proof of (3.16) gives us some additional information which we shall need later on. If B is unramified over A, then we have $B = A[b]$, and there exists a monic polynomial $F \in A[T]$ of degree $\dim_k(B/\mathfrak{m}B)$ with the following properties: $F(b) = 0$, $F'(b)$ is a unit of B, $B/\mathfrak{m}B = k[T]/(\overline{F})$, where $\overline{F}$ is the canonical image of F in $k[T]$, and $\overline{F}$ is the minimal polynomial of $b + \mathfrak{m}B$ over k.

(3.18) Corollary: *Let A be a quasilocal integrally closed domain with maximal ideal $\mathfrak{m}$, field of quotients K and residue field k. If B is a finite torsion-free A-algebra which is unramified over A, then B is a free A-module, and we have $\dim_K(B \otimes_A K) = \dim_k(B \otimes_A k)$.*

Proof: (1) First, we assume that k is an infinite field. Set $L := B \otimes_A k = B/\mathfrak{m}B$, $m := \dim_K(B \otimes_A K)$ and $n := \dim_k(L)$. Since $K = S^{-1}A$ where S is the multiplicatively closed set of non-zero elements of A, we have $B \otimes_A K = S^{-1}B$ [cf. [63], Lemma. 2.4]; the canonical map $c \mapsto c \otimes 1 \colon B \to B \otimes_A K$ is injective [since B is a torsion-free A-module]. By (3.17) there exists $b \in B$ such that $B = A + Ab + \cdots + Ab^{n-1}$, $L = k[x]$ where x is the image of b in L, and such that the minimal polynomial of x over k has degree n. The elements $1 \otimes 1, b \otimes 1, \ldots, b^{n-1} \otimes 1$ generate the K-vector space $B \otimes_A K$, hence we have $m \leq n$. Let $g := \operatorname{Pc}_{B/A}(T; b) \in K[T]$ be the characteristic polynomial of b. Now g is a monic polynomial of degree m having coefficients in A and with $g(b \otimes 1) = 0$ [since A is integrally closed in K, cf. (2.12)(2)]. Since the canonical map $B \to B \otimes_A K$ is injective, we see that $g(b) = 0$. Then x is a zero of the canonical image of g in $L[T]$, hence $m \geq n$. Therefore we have $m = n$, and B is a free A-module having $\{1, b, \ldots, b^{n-1}\}$ as a basis since $\{1 \otimes 1, b \otimes 1, \ldots, b^{n-1} \otimes 1\}$ is a basis of the K-vector space $B \otimes_A K$.
(2) If k is a finite field, we use the trick of adjoining an indeterminate [cf. B(10.4)]. Set $A' := A[X]$ and $\mathfrak{m}' := \mathfrak{m}A[X]$. Then A' is integrally closed in $K[X]$ [cf. [63], Ex. 4.18], and since $K[X]$ is integrally closed in $K' := K(X)$, A' is integrally closed in its field of quotients K'. The ring $A(X) := A'_{\mathfrak{m}'}$ is a quasilocal ring which is integrally closed in its field of quotients K' [cf. [63], Prop. 4.13], and its residue field is isomorphic to $k(X)$, hence it is an infinite field. The extension $A \to A(X)$ is faithfully flat [cf. B(10.4)]. Now we set $C := B \otimes_A A(X)$; note that C is unramified over $A(X)$ iff B is unramified over A [cf. (3.7)], that C is a free $A(X)$-module of finite rank iff B is a free A-module of finite rank [cf. [34], Ch. I, § 3, no. 6, Prop. 11], and that C is a torsion-free $A(X)$-module. Hence the result follows from (1).

(3.19) Corollary: *Let A be an integrally closed domain with field of quotients K, let L be a finite extension of K, and let M be a finite extension of L. Let B be the integral closure of A in L, and let C be the integral closure of A in M. Assume, furthermore, that B is a finitely generated A-module, and that C is a finitely generated B-module. If C is unramified over A, then B is unramified over A.*

Proof: Let $\mathfrak{m}$ be a maximal ideal of A. It is enough to show that $\mathfrak{m}$ is unramified in B [cf. (3.2)(5)]. Now $A_\mathfrak{m}$ is integrally closed in K, $B_\mathfrak{m}$ is the integral closure of $A_\mathfrak{m}$ in L, and $C_\mathfrak{m}$ is the integral closure of $A_\mathfrak{m}$ in M [cf. [63], Prop. 4.13]; furthermore, $B_\mathfrak{m}$ is a finitely generated $A_\mathfrak{m}$-module, and $C_\mathfrak{m}$ is a finitely generated $B_\mathfrak{m}$-module. Since C is unramified over A by assumption, we see that $C_\mathfrak{m}$ is unramified over $A_\mathfrak{m}$ by (3.9). We may assume therefore, to begin with, that A is quasilocal with maximal ideal $\mathfrak{m}$.

Let $\mathfrak{n}$ be a maximal ideal of B; then we have $\mathfrak{n} \cap A = \mathfrak{m}$, and it is enough to show that $B_{\mathfrak{n}}/\mathfrak{m}B_{\mathfrak{n}}$ is unramified over $A/\mathfrak{m}$ [cf. (3.4)(1)].

Thus, let $\mathfrak{n}$ be a maximal ideal of B, set $B' := B_{\mathfrak{n}}$ and $C' = C_{\mathfrak{n}}$. By [63], Prop. 4.13, B' is integrally closed, and C' is a finitely generated B'-module; since C is unramified over A, then C is unramified over B [cf. (3.5)(2)], hence C' is unramified over B' [cf. (3.9)]. Now C' is a free B'-module by (3.18), hence we have $\mathfrak{m}B' = B' \cap (\mathfrak{m}B')C' = B' \cap \mathfrak{m}C'$ [cf. [34], Ch. I, §3, no. 5, Prop. 10]. Thus, we may consider $B'/\mathfrak{m}B'$ as a subalgebra of the $A/\mathfrak{m}$-algebra $C'/\mathfrak{m}C'$. Now $C'/\mathfrak{m}C'$ is unramified over the field $A/\mathfrak{m}$ by (3.6), and therefore also $B'/\mathfrak{m}B'$ is unramified over $A/\mathfrak{m}$ by (3.13).

3.4   Discrete Valuation Rings and Ramification

(3.20) Proposition: *Let K be a field, let V be a discrete valuation ring of K, let L be a finite extension of K, and let W be an extension of V to L. Then W is unramified over V iff $r(W/V) = 1$ and L is a separable extension of K.*

Proof: (1) Assume that W is unramified over V; then, in particular, the zero ideal and the maximal ideal of V are unramified in W [cf. (3.2)]. The zero ideal of V is unramified in W iff $L = (V \setminus \{0\})^{-1}W$ is unramified over $K = (V \setminus \{0\})^{-1}V$, hence iff L is separable over K [cf. (3.10)]. Let T be the integral closure of V in L; then T is a finitely generated V-module [cf. (4.5)(2) below], and W is a localization of T with respect to a maximal ideal $\mathfrak{n}$ of T which lies over the maximal ideal $\mathfrak{m}$ of V [cf. I(7.4)]. Now $\mathfrak{n}$ is unramified over V iff $\mathfrak{m}W = \mathfrak{n}W$ and $W/\mathfrak{n}W$ is a separable extension of $V/\mathfrak{m}$ [cf. (3.14)], hence iff $e(W/V) = 1$ [cf. I(7.3)] and $[W/\mathfrak{n}W : V/\mathfrak{m}]_{\text{ins}} = 1$. Since $r(W/V) = e(W/V)[W/\mathfrak{n}W : V/\mathfrak{m}]_{\text{ins}}$, we have shown that the conditions are necessary.

(2) Now we assume that L is a separable extension of K, and that $r(W/V) = 1$, i.e., that $e(W/V) = 1$ and that $[W/\mathfrak{n}W : V/\mathfrak{m}]_{\text{ins}} = 1$. We use the notations introduced in (1). First, note that T is a finite V-algebra, hence that W is essentially of finite type over V. Since $e(W/V) = 1$, we have $\mathfrak{m}T_{\mathfrak{n}} = \mathfrak{n}T_{\mathfrak{n}}$ [cf. I(7.4)], and since $[W/\mathfrak{n}W : V/\mathfrak{m}]_{\text{ins}} = 1$, the field $W/\mathfrak{n}W$ is a finite separable extension of the field $V/\mathfrak{m}$.

(3.21) Proposition: *Let A be an integrally closed noetherian domain with field of quotients K, let L be a finite separable extension of K, and let B be the integral closure of A in L. Let $\mathfrak{p}$ be a prime ideal of height 1 of A. Then $V := A_{\mathfrak{p}}$ is a discrete valuation ring; $\mathfrak{p}$ is unramified in B iff for every extension W of V to L we have $r(W/V) = 1$.*

Proof: V is a discrete valuation ring [cf. B(10.5)], and $\mathfrak{p}$ is unramified in B iff for every maximal ideal $\mathfrak{n}$ of $T := B_{\mathfrak{p}}$ the ring $T_{\mathfrak{n}}$ is unramified over V [cf. (3.2)(5)]. Now T is the integral closure of V in L [cf. [63], Prop. 4.13], and the localizations $T_{\mathfrak{n}}$ are just the extensions of V to L [cf. I(7.4)]; the assertion follows from (3.20).

4   Unramified Extensions and Discriminants

(4.0) In the sequel, we shall characterize unramified extensions of a ring A by using the trace form [cf. section 2], and we shall describe the branch locus branch(B/A) as the closure in $\mathrm{Spec}(A)$ of an ideal of A.

(4.1) TRACE FORM OVER A FIELD: Let k be a field, and let A be a finite k-algebra. The trace form of A induces a k-linear map $\tau^A\colon A \to A^* = \mathrm{Hom}_k(A, k)$ [cf. (2.14)]. We collect the following useful facts.
(1) The trace form of A is non-degenerate iff it is a perfect pairing.
In fact, the k-linear map $\tau^A\colon A \to A^*$ is injective iff it is bijective.
(2) Let k' be an extension of k, and set $A' := A \otimes_k k'$. Let $\{x_1, \ldots, x_n\}$ be a k-basis of A. Then the set $\{x_1 \otimes 1_{A'}, \ldots, x_n \otimes 1_{A'}\}$ is a k'-basis of A', and we have $D_{A/k}(x_1, \ldots, x_n) \otimes 1_{A'} = D_{A'/k'}(x_1 \otimes 1_{A'}, \ldots, x_n \otimes 1_{A'})$ [cf. (2.13)(7)]. Therefore the trace form of A is a perfect pairing iff the trace form of A' is a perfect pairing.

(4.2) Proposition: *Let k be a field, and let A be a finite k-algebra. The following statements are equivalent:*
(1) *A is unramified over k.*
(2) *The discriminant of any k-basis of A is non-zero.*
(3) *The trace form of A is a perfect pairing.*

Proof: By (3.7)(2) and (4.1) we may assume that k is algebraically closed.
(1) $\Rightarrow$ (2): The k-algebra A is k-isomorphic to k^r for some $r \in \mathbb{N}$ [cf. (3.13)]. Let $\{e_1, \ldots, e_r\}$ be the canonical basis of k^r; then we have $e_i e_j = e_i \delta_{ij}$ for all $i, j \in \{1, \ldots, r\}$, hence we have $D_{k^r/k}(e_1, \ldots, e_r) = 1$ by (2.13)(2) [note that $\mathrm{Tr}_{k^r/k}(e_i) = 1$ for $i \in \{1, \ldots, r\}$].
(2) $\Rightarrow$ (3): This follows from (2.4)(2)(a).
(3) $\Rightarrow$ (1): The k-algebra A is an artinian ring, and its Jacobson radical $\mathfrak{r}$ is nilpotent [cf. (3.11)]. As is well known, the trace of a nilpotent endomorphism of a finite-dimensional vector space is zero. Let $x \in \mathfrak{r}$; for any $y \in A$ we have $xy \in \mathfrak{r}$, hence $\mathrm{Tr}_{A/k}(xy) = 0$, and therefore $x = 0$ since $\tau^A\colon A \to A^*$ is bijective. Thus, we have shown that $\mathfrak{r} = \{0\}$, hence that A is reduced, and that A is therefore isomorphic to a product of a finite number of copies of k [cf. (3.11)]. Now A is unramified over k by (3.13).

(4.3) Corollary: *Let k be a field, and let K be a finite extension of k. The following statements are equivalent:*
(1) *The trace form of K is a perfect pairing.*
(2) *K is a separable extension of k.*
(3) *The k-linear map $\mathrm{Tr}_{K/k}\colon K \to k$ is not the zero map.*

Proof: The equivalence of (1) and (2) follows immediately from (4.2) and (3.10). We show that (1) and (3) are equivalent. If $\mathrm{Tr}_{K/k}$ is the zero map, then the trace form of K is the zero map. Conversely, assume that the trace form of K

is not a perfect pairing. Let $\{x_1, \ldots, x_n\}$ be a k-basis of K. By (2.5) we have $D_{K/k}(x_1, \ldots, x_n) = 0$, i.e., the columns of the matrix $(\mathrm{Tr}_{K/k}(x_i x_j)) \in M(n; k)$ are linearly dependent. Therefore there exist elements $c_1, \ldots, c_n \in k$, not all of them being zero, such that

$$\sum_{j=1}^{n} c_j \, \mathrm{Tr}_{K/k}(x_i x_j) = 0 \quad \text{for every } i \in \{1, \ldots, n\}.$$

The element $z := c_1 x_1 + \cdots + c_n x_n$ is a non-zero element of K. Now we have

$$\mathrm{Tr}_{K/k}(x_i z) = \sum_{j=1}^{n} c_j \, \mathrm{Tr}_{K/k}(x_i x_j) = 0 \quad \text{for every } i \in \{1, \ldots, n\}.$$

Let $y \in K$, and choose $b_1, \ldots, b_n \in k$ such that $y = b_1 x_1 + \cdots + b_n x_n$. Then we have $\mathrm{Tr}_{K/k}(yz) = \sum_{i=1}^{n} b_i \, \mathrm{Tr}_{K/k}(x_i z) = 0$, and therefore we have

$$\mathrm{Tr}_{K/k}(y') = \mathrm{Tr}_{K/k}((y'/z)z) = 0 \quad \text{for every } y' \in K,$$

hence $\mathrm{Tr}_{K/k}$ is the zero map.

(4.4) COMPLEMENTARY MODULE: Let R be an integrally closed domain, K its field of quotients, L a finite unramified K-algebra and S a subring of L containing R which is integral over R and admits L as ring of quotients. Then S contains a K-basis of L. The trace form of L over K is a perfect pairing by (4.2).
(1) Let M be an S-submodule of L. We define

$$M^* := \{z \in L \mid \mathrm{Tr}_{L/K}(zy) \in R \text{ for every } y \in M\};$$

M^* is an S-submodule of L [for any $z \in M^*$ and $y' \in M$ we have $yy' \in M$ for every $y \in S$, hence $\mathrm{Tr}_{L/K}(zy'y) \in R$]. The S-module M^* is called the complementary module of M in L. Obviously, if M_1 and M_2 are S-submodules of L with $M_1 \subset M_2$, then we have $M_2^* \subset M_1^*$. (Note that the dual module $\mathrm{Hom}_S(M, S)$ of M was also denoted by M^*; this should not give rise to confusion.)
(2) The S-module S^* is denoted by $C_{S/R}$. Since R is integrally closed, we see that $S \subset C_{S/R}$ [cf. (2.12)(2)], hence we have $C_{S/R}L = L$.

(4.5) PROPERTIES OF THE COMPLEMENTARY MODULE: We consider the situation of (4.4). Let M be an S-submodule of L, and assume that there exists a basis $\{x_1, \ldots, x_n\}$ of the K-vector space L contained in M. Then we have the following:
(1) If $\{y_1, \ldots, y_n\}$ is the complementary basis of $\{x_1, \ldots, x_n\}$ [cf. (2.5)(2)], then

$$M^* \subset Ry_1 + \cdots + Ry_n, \tag{$*$}$$

hence M^* is a finitely generated R-module if R is noetherian. Moreover, M is a free R-module with basis $\{x_1, \ldots, x_n\}$ iff M^* is a free R-module with basis $\{y_1, \ldots, y_n\}$; if this is the case, then we have $(M^*)^* = M$.

Proof: Let $z \in M^*$, and choose $c_1, \ldots, c_n \in K$ with $z = c_1 y_1 + \cdots + c_n y_n$. Then we have $c_i = \mathrm{Tr}_{L/K}(z x_i) \in R$ for every $i \in \{1, \ldots, n\}$, hence $(*)$ is proved.

Assume that M is a free R-module with basis $\{x_1, \ldots, x_n\}$. If $z \in R y_1 + \cdots + R y_n$, then $z = c_1 y_1 + \cdots + c_n y_n$ with $c_1, \ldots, c_n \in R$, hence $\mathrm{Tr}_{L/K}(z x_i) = c_i \in R$ for every $i \in \{1, \ldots, n\}$, and therefore $\mathrm{Tr}_{L/K}(zy) \in R$ for every $y \in M$. Thus, we have shown that $z \in M^*$. By $(*)$ we see that $\{y_1, \ldots, y_n\}$ is a basis of the R-module M^*.

Assume that M^* is a free R-module with basis $\{y_1, \ldots, y_n\}$. Since $\{x_1, \ldots, x_n\}$ is the complementary basis of $\{y_1, \ldots, y_n\}$ [cf. (2.5)(2)], we see that $(M^*)^* = R x_1 + \cdots + R x_n$ by what we have just shown, hence we have $(M^*)^* = M$.

(2) The result of (1) implies, in particular, the following: If R is noetherian, then S and $C_{S/R}$ are finitely generated R-modules, and S is, in particular, a noetherian ring.

(3) We have

$$D_{L/K}(x_1, \ldots, x_n) C_{S/R} \subset R x_1 + \cdots + R x_n.$$

Proof: Let $z \in C_{S/R}$, and write $z = c_1 x_1 + \cdots + c_n x_n$ with $c_1, \ldots, c_n \in K$. Then we have

$$\mathrm{Tr}_{L/K}(z x_i) = \sum_{j=1}^{n} c_j \, \mathrm{Tr}_{L/K}(x_i x_j) \quad \text{for every } i \in \{1, \ldots, n\}. \tag{$*$}$$

Now $\mathrm{Tr}_{L/K}(z x_i) \in R$ for every $i \in \{1, \ldots, n\}$, and $\mathrm{Tr}_{L/K}(x_i x_j) \in R$ for all i, $j \in \{1, \ldots, n\}$. We consider $(*)$ as a system of linear equations for $c_1, \ldots, c_n$; its determinant is $D_{L/K}(x_1, \ldots, x_n)$. By Cramer's rule we get $D_{L/K}(x_1, \ldots, x_n) c_i \in R$ for every $i \in \{1 \ldots, n\}$, hence $D_{L/K}(x_1, \ldots, x_n) C_{S/R} \subset R x_1 + \cdots + R x_n$.

(4) In particular, since $S \subset C_{S/R}$, we have

$$D_{L/K}(x_1, \ldots, x_n) S \subset R x_1 + \cdots + R x_n.$$

(5) Assume, in particular, that $f \in K[X]$ is a monic separable polynomial having coefficients in R, and that $L = K[X]/(f)$; L is a finite K-algebra, and it is unramified over K by (3.13). Let S be a subring of L containing R which is integral over R and contains the image x of X in L. Then we have

$$f'(x) S \subset f'(x) C_{S/R} \subset R[x].$$

Proof: Write $f = (X - x)(b_0 + b_1 X + \cdots + b_{n-1} X^{n-1})$ in $S[X]$ where $b_0, \ldots, b_{n-1} \in R[x]$; then the bases $\{b_0/f'(x), \ldots, b_{n-1}/f'(x)\}$ and $\{1, x, \ldots, x^{n-1}\}$ are complementary bases [cf. (2.18)]. The assertion follows from (1).

(6) Let $\Sigma \subset R$ be a multiplicatively closed subset with $0 \notin \Sigma$; note that the elements of Σ are regular in L. Set $R' := \Sigma^{-1} R$ and $S' := \Sigma^{-1} S$. Then R' is integrally closed and S' is an integral extension of R' admitting L as ring of quotients [cf. [63], Prop. 4.13]. Let M be an S-submodule of L. We set $M' := \Sigma^{-1} M$; then $M'^* = \{z' \in L \mid \mathrm{Tr}_{L/K}(z' M') \subset R'\}$. We have $S' M^* = \Sigma^{-1} M^* \subset M'^*$ with equality if M is a finitely generated R-module.

Proof: (a) Let $z' \in \Sigma^{-1} M^*$; then we can write $z' = z/t$ for some $z \in M^*$, $t \in \Sigma$. Let $y' \in M'$; then we can write $y' = y/u$ for some $y \in M$, $u \in \Sigma$. Now we have $\mathrm{Tr}_{L/K}(z'y') = (tu)^{-1} \mathrm{Tr}_{L/K}(zy) \in R'$, and this yields $z' \in M'^*$.

(b) Assume that M is a finitely generated R-module, say $M = Ry_1 + \cdots + Ry_m$ for some $y_1, \ldots, y_m \in M$. Let $w \in L$; then we have $w \in M^*$ iff $\mathrm{Tr}_{L/K}(wy_i) \in R$ for every $i \in \{1, \ldots, m\}$, and $w \in M'^*$ iff $\mathrm{Tr}_{L/K}(wy_i) \in R'$ for every $i \in \{1, \ldots, m\}$. Now, let $w \in M'^*$; then we can write $wy_i = z_i/t$ for every $i \in \{1, \ldots, m\}$ where $z_1, \ldots, z_m \in M$ and $t \in \Sigma$, hence we have $tw \in M^*$ and $w \in \Sigma^{-1} M$.

(4.6) Proposition: *Let A be a ring, and let B be an A-algebra which is a free A-module of finite rank. B is unramified over A iff the trace form of B is a perfect pairing.*

Proof: Note that B is unramified over A iff, for every maximal ideal $\mathfrak{m}$ of A, $B/\mathfrak{m}B$ is unramified over the field $A/\mathfrak{m}$ [cf. (3.4)]. On the other side, let $\{x_1, \ldots, x_n\}$ be a basis of B. Then $D_{B/A}(x_1, \ldots, x_n)$ is a unit of A iff $D_{B/A}(x_1, \ldots, x_n)$ does not lie in any maximal ideal of A, i.e., iff $D_{B/A}(x_1, \ldots, x_n) \otimes 1_{A/\mathfrak{m}} \neq 0$ for every maximal ideal $\mathfrak{m}$ of A. Using (2.13)(7) and (2.5)(1) we therefore see: The trace form of the A-algebra B is a perfect pairing iff, for every maximal ideal $\mathfrak{m}$ of A, the trace form of the $A/\mathfrak{m}$-algebra $B/\mathfrak{m}B$ is a perfect pairing. From (4.2) we get the result.

(4.7) Proposition: *Let A be an integrally closed noetherian domain, let K be its field of quotients, let L be a finite separable extension of K, and let M be the smallest Galois extension of K containing L. Let B be the integral closure of A in L, and let C be the integral closure of A in M. Then B is unramified over A iff C is unramified over A.*

Proof: First, observe that L is the quotient field of B and M is the quotient field of C. Furthermore, B and C are finitely generated A-modules [cf. (4.5)(2)].

If C is unramified over A, then B is unramified over A by (3.19). Hence it is enough to show the following: If B is unramified over A, then C is unramified over A.

Let E be the smallest subalgebra of the A-algebra M containing all A-algebras $\tau(B)$ where $\tau \in \mathrm{Gal}(M/K)$ [$\mathrm{Gal}(M/K)$ is the Galois group of M over K]. The smallest subfield of M containing all the fields $\tau(L)$, $\tau \in \mathrm{Gal}(M/K)$, is M itself, hence E admits M as field of quotients. Now $\tau(B)$ is unramified over A for every $\tau \in \mathrm{Gal}(M/K)$ [since $\tau(A) = A$ and the A-algebras B and $\tau(B)$ are isomorphic]. Since E is a homomorphic image of the tensor product over A of the A-algebras $\tau(B)$, $\tau \in \mathrm{Gal}(M/K)$, we see that E is unramified over A [cf. (3.6) and (3.8)(3)]. Since B is a finitely generated A-module, E is a finitely generated A-module. Thus, we have shown that E is integral over A and admits M as field of quotients. Now E is a subring of C; we show that $E = C$, and that therefore C is unramified over A.

Let $\mathfrak{m}$ be a maximal ideal of A; then $E_\mathfrak{m}$ is integral and unramified over $A_\mathfrak{m}$ [cf. (3.9)], and $C_\mathfrak{m}$ is the integral closure of $A_\mathfrak{m}$ in M [cf. [63], Prop. 4.13]. We prove that $E_\mathfrak{m}$ is integrally closed, hence that $E_\mathfrak{m} = C_\mathfrak{m}$; this being true for every maximal ideal of A, we have $E = C$ by B(2.5).

Thus, we may assume that A is local. Let $\{x_1, \ldots, x_n\}$ be a basis of the free A-module E [cf. (3.18)]. Since K is the quotient field of A and M is the quotient field of E, and since every element of M has the form z/c for some $z \in E$, $c \in A$, a basis of the A-module E is a basis of the K-vector space M. By (4.6) the trace form of the A-algebra E is a perfect pairing, hence $d := D_{E/A}(x_1, \ldots, x_n)$ is a unit of A [cf. (2.5)(1)]. Let $x \in M$ be integral over E; then there exist $c_1, \ldots, c_n \in K$ with $x = c_1 x_1 + \cdots + c_n x_n$. Since x is, a priori, integral over E, $E[x]$ is integral over E [cf. [63], Th. 4.2], and, by transitivity [cf. [21], Cor. 5.4], $E[x]$ is integral over A. By (2.12)(2) we have $\operatorname{Tr}_{M/K}(y) \in A$ for every $y \in E[x]$; in particular, we have $\operatorname{Tr}_{M/K}(xx_i) =: a_i \in A$ for every $i \in \{1, \ldots, n\}$ and $\operatorname{Tr}_{M/K}(x_i x_j) \in A$ for all $i, j \in \{1, \ldots, n\}$. Now we have

$$\sum_{j=1}^n \operatorname{Tr}_{M/K}(x_i x_j) c_j = a_i \quad \text{for every } i \in \{1, \ldots, n\}. \tag{$*$}$$

$(*)$ is a system of linear equations for $c_1, \ldots, c_n$ having determinant d. By Cramer's rule we have $c_j \in A$ for every $j \in \{1, \ldots, n\}$, hence we have $x \in E$.

(4.8) DEFINITION: Let A be a domain, and let K be its field of quotients. Let B be a finite A-algebra, and set $n := \dim_K(B \otimes_A K)$. The A-linear map $\operatorname{Tr}_{B/A} : B \to K$ [cf. (2.12)] induces an A-linear map

$$(x_1 \wedge \cdots \wedge x_n) \otimes (y_1 \wedge \cdots \wedge y_n) \mapsto \det((\operatorname{Tr}_{B/A}(x_i y_j))) : \bigwedge^n B \otimes_A \bigwedge^n B \to K;$$

its image $\mathfrak{n}_{B/A} \subset K$ is called the Noether discriminant of B over A.

Note that the A-module $\mathfrak{n}_{B/A}$ is finitely generated [since $\bigwedge^n B$ is a finitely generated A-module], and that it is generated by all the elements $\det((\operatorname{Tr}_{B/A}(x_i y_j)))$ where $x_1, \ldots, x_n, y_1, \ldots, y_n \in B$.

(4.9) REMARK: Let A be a domain, and let B be an A-algebra which is a free A-module of finite rank. Then we have $\Delta_{B/A} = \mathfrak{n}_{B/A}$.
Proof: It is clear that $\Delta_{B/A} \subset \mathfrak{n}_{B/A}$. Now we prove the other inclusion. Let $\{z_1, \ldots, z_n\}$ be a basis of the A-module B, and let $x_1, \ldots, x_n$ and $y_1, \ldots, y_n$ be elements of B. By (2.4)(2)(b) we see that $\det((\operatorname{Tr}_{B/A}(x_i y_j))) \in \Delta_{B/A}$, hence that $\mathfrak{n}_{B/A} \subset \Delta_{B/A}$. The proof of the claim is complete.

(4.10) PROPERTIES OF THE NOETHER DISCRIMINANT: Let A be a domain, let K be its field of quotients, and let B be a finite A-algebra; we set $n := \dim_K(B \otimes_A K)$.
(1) Assume that A is integrally closed. Then we have $\mathfrak{n}_{B/A} \subset A$ [cf. (2.12)(2)].

(2) Let S be a multiplicatively closed system in A not containing 0. Then we have

$$\mathfrak{n}_{S^{-1}B/S^{-1}A} = \mathfrak{n}_{B/A} \cdot S^{-1}A = S^{-1}(\mathfrak{n}_{B/A}). \qquad (*)$$

Indeed, since $B \otimes_A K = S^{-1}B \otimes_{S^{-1}A} K$, we have $\mathfrak{n}_{B/A} \subset \mathfrak{n}_{S^{-1}B/S^{-1}A}$ by (2.10), hence we have $\mathfrak{n}_{B/A} \cdot S^{-1}A \subset \mathfrak{n}_{S^{-1}B/S^{-1}A}$. On the other hand, we have $S^{-1} \bigwedge^n B = S^{-1}A \otimes_A \bigwedge^n B = \bigwedge^n S^{-1}A \otimes_A B = \bigwedge^n S^{-1}B$ [cf. [63], Prop. A2.2], showing that $\mathfrak{n}_{S^{-1}B/S^{-1}A} \subset \mathfrak{n}_{B/A} \cdot S^{-1}A$. Therefore, we have shown that the first equality sign in $(*)$ holds. Since $S^{-1}A$ is a subring of K containing A, it is easy to see that the second equality sign in $(*)$ holds.

(4.11) REMARK: In the following theorem we use the Noether discriminant in order to describe branch(B/A). One can define the Noether different $\mathfrak{N}_{B/A}$, an ideal of B, and one can show: If B is an A-algebra essentially of finite type, then Branch$(B/A) = V(\mathfrak{N}_{B/A})$ [cf. [23], Th. 2.5 and [114], Appendix G].

(4.12) Theorem: *Let A be an integrally closed domain with field of quotients K, and let B be a finite torsion-free A-algebra. Then we have*

$$V(\mathfrak{n}_{B/A}) = \mathrm{branch}(B/A).$$

Proof: $\mathfrak{n}_{B/A}$ is an ideal of A by (4.10)(1). Let $\mathfrak{p}$ be a prime ideal of A; then $A_{\mathfrak{p}}$ is integrally closed in K [cf. [63], Prop. 4.13], hence $\mathfrak{n}_{B_{\mathfrak{p}}/A_{\mathfrak{p}}} = (\mathfrak{n}_{B/A})_{\mathfrak{p}}$ is an ideal of $A_{\mathfrak{p}}$ [cf. (4.10)(2)].

(1) Let $\mathfrak{p} \in V(\mathfrak{n}_{B/A})$, i.e., $\mathfrak{p}$ is a prime ideal of A containing $\mathfrak{n}_{B/A}$. By (4.10)(2) we have $\mathfrak{n}_{B/A}A_{\mathfrak{p}} = \mathfrak{n}_{B_{\mathfrak{p}}/A_{\mathfrak{p}}}$, and we have $A_{\mathfrak{p}} \neq \mathfrak{n}_{B/A}A_{\mathfrak{p}}$ since $\mathfrak{p} \supset \mathfrak{n}_{B/A}$. Suppose that $\mathfrak{p} \notin \mathrm{branch}(B/A)$. Then $B_{\mathfrak{p}}$ is unramified over $A_{\mathfrak{p}}$, hence $B_{\mathfrak{p}}$ is a free $A_{\mathfrak{p}}$-module of finite rank by (3.18), and the trace form of the $A_{\mathfrak{p}}$-algebra $B_{\mathfrak{p}}$ is a perfect pairing by (4.6). By (2.13)(5) this means that $\Delta_{B_{\mathfrak{p}}/A_{\mathfrak{p}}} = A_{\mathfrak{p}}$, hence that $\mathfrak{n}_{B_{\mathfrak{p}}/A_{\mathfrak{p}}} = A_{\mathfrak{p}}$ by (4.9). This contradiction shows that $V(\mathfrak{n}_{B/A}) \subset \mathrm{branch}(B/A)$.

(2) Let $\mathfrak{p}$ be a prime ideal of A such that $\mathfrak{p} \notin V(\mathfrak{n}_{B/A})$. We have to show that $B_{\mathfrak{p}}$ is unramified over $A_{\mathfrak{p}}$. Set $n := \dim_K(B \otimes_A K)$. Now $\mathfrak{n}_{B_{\mathfrak{p}}/A_{\mathfrak{p}}} = A_{\mathfrak{p}}$, hence not all elements $\det((\mathrm{Tr}_{B_{\mathfrak{p}}/A_{\mathfrak{p}}}(x_i y_j)))$ where $x_1, \ldots, x_n$, $y_1, \ldots, y_n \in B_{\mathfrak{p}}$, lie in the maximal ideal $\mathfrak{p}A_{\mathfrak{p}}$ of $A_{\mathfrak{p}}$. Therefore there exist elements $x_1, \ldots, x_n$, $y_1, \ldots, y_n \in B_{\mathfrak{p}}$ such that $\det((\mathrm{Tr}_{B_{\mathfrak{p}}/A_{\mathfrak{p}}}(x_i y_j))) = \det((\mathrm{Tr}_{(B_{\mathfrak{p}} \otimes_{A_{\mathfrak{p}}} K)/K}(x_i y_j \otimes 1)))$ is a unit of $A_{\mathfrak{p}}$. The K-vector space $B \otimes_A K = B_{\mathfrak{p}} \otimes_{A_{\mathfrak{p}}} K$ has dimension n; by (2.4)(2)(c) we see that $\{y_1 \otimes 1, \ldots, y_n \otimes 1\}$ is a basis of the K-vector space $B_{\mathfrak{p}} \otimes_{A_{\mathfrak{p}}} K$. We show that $\{y_1, \ldots, y_n\}$ is a system of generators of the $A_{\mathfrak{p}}$-module $B_{\mathfrak{p}}$. Let $y \in B_{\mathfrak{p}}$; then there exist $c_1, \ldots, c_n \in K$ such that $y \otimes 1 = c_1(y_1 \otimes 1) + \cdots + c_n(y_n \otimes 1)$. For every $i \in \{1, \ldots, n\}$ we have

$$\mathrm{Tr}_{(B_{\mathfrak{p}} \otimes_{A_{\mathfrak{p}}} K)/K}((x_i \otimes 1)(y \otimes 1)) = \sum_{j=1}^{n} c_j \, \mathrm{Tr}_{(B_{\mathfrak{p}} \otimes_{A_{\mathfrak{p}}} K)/K}(x_i y_j \otimes 1).$$

Now we have $\mathrm{Tr}_{(B_{\mathfrak{p}} \otimes_{A_{\mathfrak{p}}} K)/K}(x_i y \otimes 1) \in A_{\mathfrak{p}}$ for every $i \in \{1, \ldots, n\}$ [cf. (2.12)(2)]. Using Cramer's rule we see that $c_j \in A_{\mathfrak{p}}$ for every $j \in \{1, \ldots, n\}$. It follows that

$y = \sum_{j=1}^{n} c_j y_j$ [the map $B_\mathfrak{p} \to B_\mathfrak{p} \otimes_{A_\mathfrak{p}} K$ is injective since $B_\mathfrak{p}$ is a torsion-free $A_\mathfrak{p}$-module and $K = S^{-1}A$ where $S = A \setminus \{0\}$]. Therefore $\{y_1, \ldots, y_n\}$ is a basis of the $A_\mathfrak{p}$-module $B_\mathfrak{p}$, and the trace form of the $A_\mathfrak{p}$-algebra $B_\mathfrak{p}$ is a perfect pairing by (2.4)(2)(c), hence $B_\mathfrak{p}$ is unramified over $A_\mathfrak{p}$ by (4.6).

5 Ramification For Quasilocal Rings

(5.1) Lemma: *Let k be a field, and let A be a quasilocal k-algebra with maximal ideal $\mathfrak{m}$ and residue field $\kappa := A/\mathfrak{m}$. We assume that $\mathfrak{m}$ is a nil ideal, and that κ is algebraic over k. Then:*
(1) Let $a \in A$; the kernel of the canonical k-algebra homomorphism $k[T] \to k[a]$ has the form f^r where $f \in k[T]$ is monic and irreducible, and where $r \in \mathbb{N}$.
(2) Let k_{sep} be the separable closure of k in κ, and let $q \le [k_{sep} : k]$ be a positive integer. Then there exists $a \in A$ with $\dim_k(k[a]) \ge q$. If $\mathfrak{m} \ne \{0\}$ or $k_{sep} \ne \kappa$, then there even exists $a \in A$ with $\dim_k(k[a]) > q$.

Proof: For every $a \in A$ we denote by $\bar{a}$ the image of a in κ.
(1) Let $f \in k[T]$ be the minimal polynomial of $\bar{a}$ over k. We have $f(a) \in \mathfrak{m}$, hence there exists a minimal $r \in \mathbb{N}$ with $f^r(a) = 0$ [since $\mathfrak{m}$ is a nil ideal]. Let $g \in k[T]$ with $g(a) = 0$. Then we have $g(\bar{a}) = 0$, hence we can write $g = h f^s$ with $s \in \mathbb{N}$, $h \in k[T]$ and $f \nmid h$, hence we have $h(\bar{a}) \ne 0$. Then $h(a)$ lies not in $\mathfrak{m}$, hence $h(a)$ is a unit of A, and from $0 = g(a) = h(a) f^s(a)$ we see that $s \ge r$ by the choice of r.
(2) We choose $b \in A$ with $\bar{b} \in k_{sep}$ such that the minimal polynomial $g \in k[T]$ of $\bar{b}$ over k has degree $\ge q$ [this is possible by the theorem of the primitive element for finite separable extensions]; moreover, g is a separable polynomial, hence its derivative g' is not the zero polynomial.
(a) First, we consider the case $\mathfrak{m} = \{0\}$ and $k_{sep} = \kappa$. Then we have $\kappa = A$; we set $a := b$, and we have $\dim_k(k[a]) \ge q$.
(b) Now assume that $\mathfrak{m} \ne \{0\}$ or $\kappa \ne k_{sep}$. If $g(b) \ne 0$, then, by (1), the minimal polynomial for b over k has the form g^r with $r > 1$, hence $\dim_k(k[b]) > q$, and we set $a := b$. We consider the case that $g(b) = 0$. If $k_{sep} \ne \kappa$, then we choose $c \in A$ with $\bar{c} \notin k_{sep}$; $k(\bar{b}, \bar{c})$ is a simple extension of k [since $\bar{b}$ is separable over k, cf. [188], p. 138], and $[k(\bar{b}, \bar{c}) : k] > [k(\bar{b}) : k] \ge q$. We choose $a \in A$ with $k(\bar{a}) = k(\bar{b}, \bar{c})$. Then we have $\dim_k(k[a]) \ge \dim_k(k[\bar{a}]) > q$. If $k_{sep} = \kappa$, then we have $\mathfrak{m} \ne \{0\}$. We choose $c \in \mathfrak{m}$ with $c \ne 0$. We have $g'(\bar{b}) \ne 0$, hence $g'(b)$ is a unit of A. We can write $g(T) = g(b) + g'(b)(T - b) + h(T)(T - b)^2$ with $h(T) \in k[T]$. Setting $a := b + c$ we have $g(a) = c(g'(b) + c h(a))$ [since $g(b) = 0$]; now $g'(b) + c h(a)$ is a unit of A [since $g'(b)$ is a unit of A and $c \in \mathfrak{m}$] and therefore we have $g(a) \ne 0$. Since $g(\bar{a}) = g(\bar{b}) = 0$, and since g is irreducible, g is the minimal polynomial for $\bar{a}$ over k; the minimal polynomial for a over k has, by (1), the form g^r with $r \in \mathbb{N}$, $r > 1$ [since $g(a) \ne 0$], and therefore we obtain $\dim_k(k[a]) > q$.

(5.2) Lemma: *Let k be an infinite field, and let $A_1, \ldots, A_m$ be quasilocal k-algebras. We set $A := A_1 \times \cdots \times A_m$. For $i \in \{1, \ldots, m\}$ let $\mathfrak{m}_i$ be the maximal* *ideal of A_i and let $\kappa_i := A_i/\mathfrak{m}_i$ be the residue field of A_i. We assume that the fields $\kappa_1, \ldots, \kappa_m$ are algebraic extensions of k, and that the ideals $\mathfrak{m}_1, \ldots, \mathfrak{m}_m$ are nil ideals. For $i \in \{1, \ldots, m\}$ let k_i be the separable closure of k in κ_i, and let $q_i \leq [k_i : k]$ be a positive integer. Then:*

(1) For every $a \in A$ we have $\dim_k(k[a]) < \infty$.

(2) There exists $a \in A$ with $\dim_k(k[a]) \geq q_1 + \cdots + q_m$. If there exists $i \in \{1, \ldots, m\}$ with $\mathfrak{m}_i \neq \{0\}$ or with $\kappa_i \neq k_i$, then there even exists $a \in A$ with $\dim_k(k[a]) > q_1 + \cdots + q_m$.

Proof: (1) Let $a = (a_1, \ldots, a_m) \in A = A_1 \times \cdots \times A_m$. For every $i \in \{1, \ldots, m\}$ let $g_i \in k[T]$ be the minimal polynomial of a_i over k, and set $g := \gcd(g_1, \ldots, g_m)$. The kernel of the canonical k-algebra homomorphism $k[T] \to k[a_1] \times \cdots \times k[a_m]$ is the intersection of the ideals $(g_1), \ldots, (g_m)$, hence the ideal (g), and therefore we have $\dim_k(k[a]) \leq \deg(g)$.

(2) Let $i \in \{1, \ldots, m\}$; for $a_i \in A_i$ we denote by $\bar{a}_i$ the image of a_i in κ_i. There exists $a_i \in A_i$ such that, letting g_i be the minimal polynomial of $\bar{a}_i$ over k, the minimal polynomial of a_i over k has the form $g_i^{r_i}$ with $r_i \in \mathbb{N}$, $\deg(g_i^{r_i}) \geq q_i$, and $\deg(g_i^{r_i}) > q_i$ if $\mathfrak{m}_i \neq \{0\}$ or if $\kappa_i \neq k_i$ [cf. (5.1)]. For every $c_i \in k$ we have $k[\bar{a}_i + c_i] = k[\bar{a}_i]$. Since k is an infinite field, we may assume, replacing $\bar{a}_i$ by $\bar{a}_i + c_i$ with $c_i \in k$, and denoting the minimal polynomial of $\bar{a}_i + c_i$ over k again by g_i, that the irreducible polynomials $g_1, \ldots, g_m$ are pairwise different, hence, in particular, pairwise coprime. Then the polynomials $g_1^{r_1}, \ldots, g_m^{r_m}$ are pairwise coprime. Setting $a := (a_1, \ldots, a_m)$, we have, by the Chinese remainder theorem [cf. B(10.1)], $k[a] = k[a_1] \times \cdots \times k[a_m]$. This proves the assertion.

(5.3) REMARK: The following theorem shall be used in the sequel only in the case where S is quasilocal; in this case the proof works even if the residue field of R is finite. Nevertheless, we shall prove the theorem in its full generality, making use of an extension of the ground field in case R has finite residue field.

(5.4) Theorem: [Krull] *Let R be a quasilocal integrally closed domain, let K be the field of quotients of R, and let L be a finite extension of K. Let S be a subring of L which contains R and is integral over R. Let $\mathfrak{m}$ be the maximal ideal of R, and let $\mathfrak{n}_1, \ldots, \mathfrak{n}_m$ be the finitely many maximal ideals of S. Then:*

(1) We have

$$[S/\mathfrak{n}_1 : R/\mathfrak{m}]_{sep} + \cdots + [S/\mathfrak{n}_m : R/\mathfrak{m}]_{sep} \leq [L : K]. \tag{$*$}$$

(2) The equality sign in $(*)$ holds iff there exists a K-basis $\{x_1, \ldots, x_n\}$ of L contained in S with $D_{L/K}(x_1, \ldots, x_n) \in R \setminus \mathfrak{m}$. If this is the case, then S is the integral closure of R in L, and it is a finitely generated free R-module.

Proof: First, we consider the case that $k := R/\mathfrak{m}$ is an infinite field. For $y \in S$ we denote by $\bar{y}$ the image of y in $S/\mathfrak{m}S$.

(1) Set $n := [L : K]$. Let S' be the integral closure of R in L. Now S' has at most n maximal ideals by B(7.3), and for every maximal ideal $\mathfrak{n}'$ of S' the field $S'/\mathfrak{n}'$

is an algebraic extension of k by B(7.4). The number m of maximal ideals of S is therefore $\leq n$. By B(10.21) we have a primary decomposition $\mathfrak{m}S = \mathfrak{q}_1 \cap \cdots \cap \mathfrak{q}_m$ where $\mathfrak{q}_i$ is $\mathfrak{n}_i$-primary for $i \in \{1, \ldots, m\}$. We set $A := S/\mathfrak{m}S$, $A_i := S/\mathfrak{q}_i$ for $i \in \{1, \ldots, m\}$. Note that A_i is quasilocal with maximal ideal $\mathfrak{n}_i/\mathfrak{q}_i$ which is a nil ideal, and has residue field $S/\mathfrak{n}_i$ which is algebraic over k. Then $A = A_1 \times \cdots \times A_m$ as k-algebras [by the Chinese remainder theorem]. For every $a \in A$ we have $\dim_k(k[a]) \leq n$ [cf. B(7.4)], hence we get $(*)$ by using the first part of (5.2)(2).

(2) First, we assume that we have equality in $(*)$. By the second part of (5.2)(2) we see that $\mathfrak{n}_i/\mathfrak{q}_i = \{0\}$, and that A_i is a field which is a finite separable extension of k for every $i \in \{1, \ldots, m\}$. Therefore A_i is unramified over k [cf. (4.3)], hence A is unramified over k [cf. (3.8)(1)], hence $A = k[a]$ [cf. (3.13)], and therefore $D_{A/k}(1, a, \ldots, a^{n-1}) \neq 0$ [cf. (4.2)]. Since $\{1, a, \ldots, a^{n-1}\}$ is a k-basis of A, there exists a unique monic polynomial $f \in k[T]$ of degree n with $f(a) = 0$; remember that $\mathrm{dis}(f) = D_{A/k}(1, a, \ldots, a^{n-1})$. We choose $x \in S$ with $\overline{x} = a$, and we let $F \in K[T]$ be the minimal polynomial of x over K; we have $\deg(F) \leq n$. Now $F \in R[T]$ since R is integrally closed [cf. (2.12)(2)], hence $\mathrm{dis}(F) \in R$. Let $\overline{F}$ be the polynomial which we get from F by reducing its coefficients modulo $\mathfrak{m}$. We have $\overline{F}(a) = 0$, hence $\overline{F} = f$ and $\deg(F) = n$. Using B(10.22)(3) we see that the image of $\mathrm{dis}(F)$ in k is $\mathrm{dis}(f)$, hence that $\mathrm{dis}(F) = D_{L/K}(1, x, \ldots, x^{n-1})$ is a unit of R.

Second, we assume that there exists a K-Basis $\{x_1, \ldots, x_n\}$ of L which is contained in S, and such that $D_{L/K}(x_1, \ldots, x_n)$ is a unit of R. Then $S = Rx_1 + \cdots + Rx_n$, and S is the integral closure of R in L. In fact, we have $Rx_1 + \cdots + Rx_n \subset S$. Let $x \in L$ lie not in the R-module $Rx_1 + \cdots + Rx_n$. Then we have $x = a_1 x_1 + \cdots + a_n x_n$ with $a_1, \ldots, a_n \in K$, and not all of these elements lie in R. By relabelling, we may assume that $a_1 \notin R$. We have $D_{L/K}(x, x_2, \ldots, x_n) = a_1^2 D_{L/K}(x_1, x_2, \ldots, x_n)$ [cf. (2.13)(3)], hence $D_{L/K}(x, x_2, \ldots, x_n) \notin R$ [since $D_{L/K}(x_1, x_2, \ldots, x_n)$ is a unit of R], and this shows that x is not integral over R. Therefore $S = Rx_1 + \cdots + Rx_n$ is the integral closure of R in L. In particular, $A = S/\mathfrak{m}S$ is a quasisemilocal k-algebra with $\dim_k(A) = n$, and $D_{A/k}(\overline{x}_1, \ldots, \overline{x}_n)$ is the image of $D_{S/R}(x_1, \ldots, x_n)$ [cf. (2.13)(7)], hence $D_{A/k}(\overline{x}_1, \ldots, \overline{x}_n) \neq 0$. This implies that A is unramified over k [cf. (4.2)], hence that A is a direct product of fields which are finite separable extensions of k [cf. (3.13)], hence that A_i is a field which is a finite separable extension of k for $i \in \{1, \ldots, m\}$ [cf. B(8.2)], and therefore we have $[A_i : k]_{\mathrm{sep}} = [A_i : k]$ for $i \in \{1, \ldots, m\}$, hence we have equality in $(*)$.

Now we consider the case that k is a finite field. We use the trick of adjoining an indeterminate [cf. B(10.4)]. We set $K' := K(X)$, $L' := L(X)$; note that $[L' : K'] = [L : K]$. We set $\Sigma := R[X] \setminus \mathfrak{m}R[X]$, $R' := \Sigma^{-1}R[X]$, $\mathfrak{m}' := \mathfrak{m}R'$, $S' := \Sigma^{-1}S[X]$ and $\mathfrak{n}'_i := \mathfrak{n}_i S'$. Then R' is integrally closed, and S' is integral over R'; moreover, S is integrally closed iff S' is integrally closed [cf. [63], Ex. 4.18 and Prop. 4.13]. Furthermore, R' is quasilocal with maximal ideal $\mathfrak{m}'$, and S' is quasisemilocal with maximal ideals $\mathfrak{n}'_1, \cdots, \mathfrak{n}'_m$: if $\mathfrak{N}$ is a prime ideal of $S[X]$ with $\mathfrak{N} \supset \mathfrak{n}_i S[X]$ for some $i \in \{1, \ldots, m\}$ and $\mathfrak{N} \cap \Sigma = \emptyset$, then $\mathfrak{N} \cap R[X] = \mathfrak{m}R[X]$, hence $\mathfrak{N} = \mathfrak{n}_i S[X]$ by [63], Cor. 4.18. Therefore S' is also the localization of

$S[X]$ with respect to the complement of $\mathfrak{n}_1 S[X] \cup \cdots \cup \mathfrak{n}_m S[X]$. Since $S \to S'$ is faithfully flat [cf. B(10.4)], we have $\mathfrak{m}S' = \mathfrak{q}_1' \cap \cdots \cap \mathfrak{q}_m'$ [cf. [34], Ch. I, §3, no. 5, Prop. 10] where, for $i \in \{1, \ldots, m\}$, $\mathfrak{q}_i' := \mathfrak{q}_i S'$ is $\mathfrak{n}_i'$-primary. We set $k' := R'/\mathfrak{m}'$; for $i \in \{1, \ldots, m\}$ we set $l_i := S/\mathfrak{n}_i$, $l_i' := S'/\mathfrak{n}_i'$, and denote by k_i the separable closure of k in l_i. Then $k' \cong k(X)$, and, for $i \in \{1, \ldots, m\}$, $l_i' \cong l_i(X)$ and $k_i(X)$ is the separable closure of k' in l_i'. Note that $[k_i : k] = [k_i' : k']$ for $i \in \{1, \ldots, m\}$. Setting $A' := S'/\mathfrak{m}'S$, $A_1' := S_1'/\mathfrak{q}_1', \ldots, A_m' := S'/\mathfrak{q}_m'$, we have $A' = A_1' \times \cdots \times A_m'$ as k'-algebras, and we have $A' = A \otimes_k k(X)$, $A_i' = A_i \otimes_k k(X)$ for $i \in \{1, \ldots, m\}$. The assertions of the theorem hold for R' and S' by (1), hence also for R and S.

(5.5) Corollary: *If in* (∗) *we have equality, then we have* $\mathfrak{m}S_{\mathfrak{n}_i} = \mathfrak{n}_i S_{\mathfrak{n}_i}$ *and* $S/\mathfrak{n}_i$ *is a finite separable extension of* $R/\mathfrak{m}$ *for every* $i \in \{1, \ldots, m\}$. *In particular, if* R *is local, then* S *is unramified over* R.

Proof: Using the notation of the proof of the theorem, we have, for $i \in \{1, \ldots, m\}$, $\mathfrak{n}_i = \mathfrak{q}_i$, hence $\mathfrak{m}S_{\mathfrak{n}_i} = \mathfrak{n}_i S_{\mathfrak{n}_i}$ and $S/\mathfrak{n}_i$ is a finite separable extension of $R/\mathfrak{m}$. If R is local, then $\mathfrak{n}_i$ is unramified over A [cf. (3.14)], hence S is unramified over R [cf. (3.2)(5)].

6 Integral Closure and Completion

(6.1) NOTATION: Let R be a semilocal ring with maximal ideals $\mathfrak{m}_1, \ldots, \mathfrak{m}_h$, and let $\mathfrak{r} = \mathfrak{m}_1 \cap \cdots \cap \mathfrak{m}_h$ be its Jacobson radical. Set $R_i := R_{\mathfrak{m}_i}$ for $i \in \{1, \ldots, h\}$. Let $\widehat{R}$ be the $\mathfrak{r}$-adic completion of R. Generalizing the definition in II(3.16), we say that R is analytically unramified if $\widehat{R}$ is reduced.
(1) We have $\widehat{R} = \widehat{R}_1 \times \cdots \times \widehat{R}_h$ [cf. B(8.4)(1)]; hence R is analytically unramified iff the local rings $R_1, \ldots, R_h$ are analytically unramified.
(2) Assume that R is analytically unramified, and let $\mathfrak{p}_1^*, \ldots, \mathfrak{p}_r^*$ be the minimal prime ideals of $\widehat{R}$. Then we have $Q(\widehat{R}) = Q(\widehat{R})/\mathfrak{p}_1^* Q(\widehat{R}) \times \cdots \times Q(\widehat{R})/\mathfrak{p}_r^* Q(\widehat{R})$ [cf. B(3.4)]; let $e_1, \ldots, e_r$ be the elementary idempotents of $Q(\widehat{R})$ corresponding to this decomposition. Clearly these idempotents are integral over $\widehat{R}$. If $\widehat{R}$ is integrally closed, then we have $e_1, \ldots, e_r \in \widehat{R}$, hence $\widehat{R} = \widehat{R}/\mathfrak{p}_1^* \times \cdots \times \widehat{R}/\mathfrak{p}_r^*$; as the rings $\widehat{R}/\mathfrak{p}_i^*$ are local [being complete semilocal integral domains] and $\widehat{R}$ is decomposable [cf. B(8.4)], we have $r = h$ and, after relabelling, $\mathfrak{p}_i^* \subset \mathfrak{m}_i\widehat{R}$, $\widehat{R}/\mathfrak{p}_i^* = \widehat{R}_i$ and $\widehat{R}_i$ is an integrally closed domain for $i \in \{1, \ldots, h\}$ [cf. B(8.2)]. Conversely, if $\widehat{R}_1, \ldots, \widehat{R}_h$ are domains which are integrally closed, then $\widehat{R}$ is integrally closed [cf. B(3.3)].

(6.2) DEFINITION: A local ring R is called analytically normal if it is an integrally closed domain and if its completion is an integrally closed domain.

(6.3) EXAMPLE: Let R be a regular local ring. Then R is analytically normal. In fact, the completion $\widehat{R}$ is regular [since $\dim(R) = \dim(\widehat{R})$ by [63], Cor. 10.12, and emdim(R) = emdim$(\widehat{R})$ by [63], Th. 7.1], and a regular ring is a factorial domain [cf. [63], Th. 19.19], hence, in particular, is integrally closed [cf. [63], Prop. 4.10].

(6.4) Proposition: *Let R be a semilocal ring, $\widehat{R}$ its completion, and $\mathfrak{a}$ an ideal of R. Let $\mathfrak{a}\widehat{R} = \mathfrak{q}_1^* \cap \cdots \cap \mathfrak{q}_h^*$ be an irredundant primary decomposition of $\mathfrak{a}\widehat{R}$, and set $\mathfrak{p}_i^* := \mathrm{rad}(\mathfrak{q}_i^*)$ for every $i \in \{1, \ldots, h\}$. Then $\mathfrak{a} = (\mathfrak{q}_1^* \cap R) \cap \cdots \cap (\mathfrak{q}_h^* \cap R)$ is a primary decomposition of $\mathfrak{a}$. Let $i \in \{1, \ldots, h\}$; then we have $\mathfrak{p}_i^* \cap R = \mathrm{rad}(\mathfrak{q}_i^* \cap R)$, and there exists a prime ideal $\mathfrak{p}_i \in \mathrm{Ass}(R/\mathfrak{a})$ with $\mathfrak{p}_i^* \cap R \subset \mathfrak{p}_i$.*

Proof: Let $i \in \{1, \ldots, h\}$. Obviously $\mathfrak{q}_i^* \cap R$ is a primary ideal of R and $\mathfrak{p}_i^* \cap R$ is its prime ideal. Since $\widehat{R}$ is a faithfully flat R-module [cf. [34], Ch. III, §3, no. 5, Prop. 9], we have $\mathfrak{a}\widehat{R} \cap R = \mathfrak{a}$ [cf. [34], Ch. I, §3, no. 5, Prop. 10], and therefore $\mathfrak{a} = (\mathfrak{q}_1^* \cap R) \cap \cdots \cap (\mathfrak{q}_h^* \cap R)$ is a primary decomposition of $\mathfrak{a}$. Since $\widehat{R}/\mathfrak{a}\widehat{R}$ is the completion of $R/\mathfrak{a}$ [cf. [63], Lemma 7.15], every regular element of $R/\mathfrak{a}$ remains a regular element in $\widehat{R}/\mathfrak{a}\widehat{R}$ [note that an element a of a ring A is regular if the multiplication map $a_A : A \to A$ is injective]. Let $i \in \{1, \ldots, h\}$ and $x \in \mathfrak{p}_i^* \cap R$. The image of x in $\widehat{R}/\mathfrak{a}\widehat{R}$ is a zero-divisor [cf. [63], Th. 3.1], hence the image of x in $R/\mathfrak{a}$ is a zero-divisor, and therefore x is contained in the union of the prime ideals in $\mathrm{Ass}(R/\mathfrak{a})$. This implies that $\mathfrak{p}_i^* \cap R$ is contained in a prime ideal $\mathfrak{p} \in \mathrm{Ass}(R/\mathfrak{a})$ [cf. [63], Lemma 3.3].

(6.5) Corollary: *If, in (6.4), $\mathfrak{a}$ is a prime ideal, then $\mathfrak{a} = \mathfrak{q}_i^* \cap R = \mathfrak{p}_i^* \cap R$ for every $i \in \{1, \ldots, h\}$.*

Proof: We have $\mathfrak{a} \subset \mathfrak{q}_i^* \cap R \subset \mathfrak{p}_i^* \cap R \subset \mathfrak{a}$ [since $\{\mathfrak{a}\} = \mathrm{Ass}(R/\mathfrak{a})$].

(6.6) Proposition: *Let R be a semilocal domain, $\widehat{R}$ its completion, and let $\mathfrak{p}$ be a prime ideal of R such that $R_{\mathfrak{p}}$ is a discrete valuation ring and that $R/\mathfrak{p}$ is analytically unramified. Then an irredundant primary decomposition of $\mathfrak{p}\widehat{R}$ has the form $\mathfrak{p}\widehat{R} = \mathfrak{p}_1^* \cap \cdots \cap \mathfrak{p}_h^*$ where $\mathfrak{p}_1^*, \ldots, \mathfrak{p}_h^*$ are prime ideals of $\widehat{R}$. Moreover, for every $i \in \{1, \ldots, h\}$ we have:*
(a) the ring $\widehat{R}_{\mathfrak{p}_i^*}$ is a discrete valuation ring,
(b) for every $x \in R$ with $xR_{\mathfrak{p}} = \mathfrak{p}R_{\mathfrak{p}}$ we have $x\widehat{R}_{\mathfrak{p}_i^*} = \mathfrak{p}_i^*\widehat{R}_{\mathfrak{p}_i^*}$.

Proof: Set $\Sigma := R \setminus \mathfrak{p}$; then we have $\Sigma^{-1}R = R_{\mathfrak{p}}$. Since $\widehat{R}/\mathfrak{p}\widehat{R}$ is reduced, the ideal $\mathfrak{p}\widehat{R}$ admits an irredundant intersection as stated, and $\mathfrak{p}_i^* \cap R = \mathfrak{p}$ for every $i \in \{1, \ldots, h\}$ [cf. (6.5)]. This implies, in particular, that $\Sigma \cap \mathfrak{p}_i^* = \emptyset$ for every $i \in \{1, \ldots, h\}$. Since $R_{\mathfrak{p}}$ is a discrete valuation ring, there exists $x \in R$ with $xR_{\mathfrak{p}} = \mathfrak{p}R_{\mathfrak{p}}$. Using the commutative diagram

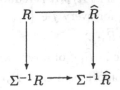

yields $x(\Sigma^{-1}\widehat{R}) = \mathfrak{p}(\Sigma^{-1}\widehat{R})$. On the other hand, $\mathfrak{p}(\Sigma^{-1}\widehat{R}) = \Sigma^{-1}\mathfrak{p}_1^* \cap \cdots \cap \Sigma^{-1}\mathfrak{p}_h^*$ is an irredundant primary decomposition of $\mathfrak{p}(\Sigma^{-1}\widehat{R})$ [cf. [63], Th. 3.10]. Therefore, by localizing, we get $x\widehat{R}_{\mathfrak{p}_i^*} = \mathfrak{p}_i^*\widehat{R}_{\mathfrak{p}_i^*}$ for every $i \in \{1,\ldots,h\}$. Now x is a regular element of $\widehat{R}_{\mathfrak{p}_i^*}$ since x is a regular element of $\widehat{R}$, hence $\widehat{R}_{\mathfrak{p}_i^*}$ is a local ring whose maximal ideal is generated by a regular element, and therefore $\widehat{R}_{\mathfrak{p}_i^*}$ is a discrete valuation ring [cf. I(3.29)].

(6.7) Let R be a ring, let $Q(R)$ be its ring of quotients, and let s, $t \in R$ be such that t is not a zero-divisor in R and that $(Rt : Rs)_R = Rt$. Let $z \in Q(R)$; if $sz \in R$ and $tz \in R$, then we have $z \in R$.
Proof: Since $szt \in Rt$, we have $zt \in (Rt : Rs)_R = Rt$, hence $zt = z't$ for some $z' \in R$. Since t is not a zero-divisor of R, we have $z = z' \in R$.

(6.8) Lemma: *Let R be an integrally closed semilocal domain, $\widehat{R}$ its completion, and $Q(\widehat{R})$ the ring of quotients of $\widehat{R}$. Assume that there exists a non-unit $t \in R\setminus\{0\}$ such that $R/\mathfrak{p}$ is analytically unramified for every prime ideal $\mathfrak{p} \in \mathrm{Ass}(R/Rt)$. Let $z \in Q(\widehat{R})$ be an element with $tz \in \widehat{R}$. If z is integral over $\widehat{R}$, then we have $z \in \widehat{R}$.*

Proof: (1) Let $\mathrm{Ass}(R/Rt) = \{\mathfrak{p}_1,\ldots,\mathfrak{p}_h\}$ and set $\Sigma := R\setminus(\mathfrak{p}_1 \cup \cdots \cup \mathfrak{p}_h)$. We have $\mathrm{ht}(\mathfrak{p}_i) = 1$ for every $i \in \{1,\ldots,h\}$ [cf. [63], Th. 11.2]. Now $\Sigma^{-1}R$ is a noetherian integrally closed domain [cf. [63], Prop. 4.13], and $\Sigma^{-1}\mathfrak{p}_1,\ldots,\Sigma^{-1}\mathfrak{p}_h$ is its set of prime ideals of height 1 which is also the set of maximal ideals of $\Sigma^{-1}R$. Therefore we have $\Sigma^{-1}R = R_{\mathfrak{p}_1} \cap \cdots \cap R_{\mathfrak{p}_h}$ [note that $\Sigma^{-1}R_{\Sigma^{-1}\mathfrak{p}_i} = R_{\mathfrak{p}_i}$ for $i \in \{1,\ldots,h\}$] and the rings $R_{\mathfrak{p}_1},\ldots,R_{\mathfrak{p}_h}$ are discrete valuation rings [cf. B(10.5)], hence $\Sigma^{-1}R$ is a semilocal principal ideal domain having $\Sigma^{-1}\mathfrak{p}_1,\ldots,\Sigma^{-1}\mathfrak{p}_h$ as set of non-zero prime ideals [cf. I(7.2)], and therefore there exist elements $x_1',\ldots,x_h' \in \Sigma^{-1}R$ with $(\Sigma^{-1}R)x_i' = \mathfrak{p}_i(\Sigma^{-1}R)$ for $i \in \{1,\ldots,h\}$. For $i \in \{1,\ldots,h\}$ we can write $x_i' = x_i/s$ with $x_1,\ldots,x_h \in R$ and $s \in \Sigma$. Let, for $i \in \{1,\ldots,h\}$, v_i be the valuation of $Q(R)$ defined by $R_{\mathfrak{p}_i}$. Since $s \in \Sigma$, we have $v_1(s) = \cdots = v_h(s) = 0$, hence we have $v_l(x_i) = \delta_{li}$ for all $l, i \in \{1,\ldots,h\}$.
For $i \in \{1,\ldots,h\}$ we set $e_i := v_i(t)$. We show that $x_1^{e_1} \cdots x_h^{e_h}/t \in R$. By [63], Cor. 11.4, it is enough to show that $x_1^{e_1} \cdots x_h^{e_h}/t \in R_{\mathfrak{p}}$ for every prime ideal $\mathfrak{p}$ of R of height 1. Thus, let $\mathfrak{p}$ be a such a prime ideal. If $\mathfrak{p} \in \mathrm{Ass}(R/Rt)$, then clearly $x_1^{e_1} \cdots x_h^{e_h}/t \in R_{\mathfrak{p}}$, and if $\mathfrak{p} \notin \mathrm{Ass}(R/Rt)$, then $t \notin \mathfrak{p}$, hence again $x_1^{e_1} \cdots x_h^{e_h}/t \in R_{\mathfrak{p}}$. Now we can write $x_1^{e_1} \cdots x_h^{e_h} = ts$ with $s \in R$ and $v_i(s) = 0$ for $i \in \{1,\ldots,h\}$, which implies that $s \in \Sigma$.
(2) Let $i \in \{1,\ldots,h\}$. From (6.6) we get an irredundant primary decomposition $\widehat{R}\mathfrak{p}_i = \mathfrak{p}_{i1}^* \cap \cdots \cap \mathfrak{p}_{ik(i)}^*$ where $\mathfrak{p}_{i1}^*,\ldots,\mathfrak{p}_{ik(i)}^*$ are prime ideals of $\widehat{R}$ and $\mathfrak{p}_{ij}^* \cap R = \mathfrak{p}_i$ for every $j \in \{1,\ldots,k(i)\}$ [note that $\widehat{R}/\mathfrak{p}_i\widehat{R}$ is reduced by assumption]. Moreover, $\widehat{R}_{\mathfrak{p}_{ij}^*}$ is a discrete valuation ring for every $j \in \{1,\ldots,k(i)\}$; let w_{ij} be the discrete valuation of $Q(\widehat{R}_{\mathfrak{p}_i^*})$ defined by $\widehat{R}_{\mathfrak{p}_{ij}^*}$.
Let $j \in \{1,\ldots,k(i)\}$. The image in $\widehat{R}_{\mathfrak{p}_{ij}^*}$ of a regular element of $\widehat{R}$ is a regular element, hence the canonical homomorphism $\widehat{R} \to \widehat{R}_{\mathfrak{p}_{ij}^*}$ extends to a homomorphism

$\varphi_{ij} \colon Q(\widehat{R}) \to Q(\widehat{R}_{\mathfrak{p}^*_{ij}})$ of rings of quotients. The map $\omega_{ij} := w_{ij} \circ \varphi_{ij} \colon Q(\widehat{R}) \to \mathbb{Z}_\infty$ clearly is a Manis valuation of $Q(\widehat{R})$ [cf. I(2.8), and note that $Q(\widehat{R})$ is a ring having large Jacobson radical, cf. I(1.11)(2)].

(3) Let $i \in \{1, \ldots, h\}$, $j \in \{1, \ldots, k(i)\}$. Since $x_i R_{\mathfrak{p}_i} = \mathfrak{p}_i R_{\mathfrak{p}_i}$, we have $\omega_{ij}(x_i) = 1$ for every $j \in \{1, \ldots, k(i)\}$ by (6.6), and for $l \in \{1, \ldots, h\}$, $l \neq i$, we have $x_i \notin \mathfrak{p}_l$, hence $x_i \notin \mathfrak{p}^*_{lj}$, and therefore we have $\omega_{lj}(x_i) = 0$ for every $j \in \{1, \ldots, k(l)\}$. From this we get, by (1), that $\omega_{ij}(t) = e_i$ for all $i \in \{1, \ldots, h\}$ and $j \in \{1, \ldots, k(i)\}$.

(4) Let $z \in Q(\widehat{R})$ be integral over R with $tz \in \widehat{R}$. We have to show that $z \in \widehat{R}$. Now t is a regular element of $\widehat{R}$. For every $s \in \Sigma$ we have $(Rt : Rs)_R = Rt$ [cf. [63], Th. 3.1], hence $(\widehat{Rt} : \widehat{Rs})_{\widehat{R}} = \widehat{R}t$ [note that $(Rt : Rs)_R \cong Rs/(Rs \cap Rt)$, and use [34], Ch. I, § 3, no. 5, Prop. 10]. We know that $x_1^{e_1} \cdots x_h^{e_h} = ts$ with $s \in \Sigma$ [cf. (1)]; by (6.7) it is enough to show that $(s'z)t \in \widehat{R}x_1^{e_1} \cdots x_h^{e_h}$ for some $s' \in \Sigma$.

(a) Let $i \in \{1, \ldots, h\}$, $j \in \{1, \ldots, k(i)\}$. Now $z \in Q(\widehat{R})$ is integral over $\widehat{R}$ by assumption, hence $\varphi_{ij}(z) \in Q(\widehat{R}_{\mathfrak{p}^*_{ij}})$ is integral over $\varphi_{ij}(\widehat{R}) \subset \widehat{R}_{\mathfrak{p}^*_{ij}}$, and therefore we have $\varphi_{ij}(z) \in \widehat{R}_{\mathfrak{p}^*_{ij}}$ [note that a discrete valuation ring is integrally closed]. This implies that $\omega_{ij}(z) \geq 0$.

(b) Since $tz \in \widehat{R}$, we can write $tz = x_1^{f_1} \cdots x_h^{f_h} y$ with $y \in \widehat{R}$ and $(f_1, \ldots, f_h) \in \mathbb{N}_0^h$. We consider the case that there exists $i \in \{1, \ldots, h\}$ with $f_i < e_i$. For every $j \in \{1, \ldots, k(i)\}$ we have

$$f_i + \omega_{ij}(y) = \omega_{ij}(x_1^{f_1} \cdots x_h^{f_h} y) = \omega_{ij}(tz) = e_i + \omega_{ij}(z), \qquad (*)$$

hence $\omega_{ij}(y) \geq 1$ [cf. (a)], i.e., we have $\varphi_{ij}(y) \in \mathfrak{p}^*_{ij} \widehat{R}_{\mathfrak{p}^*_{ij}}$, and therefore we have $y_i \in \mathfrak{p}^*_{ij}$ [since $\varphi_{ij}^{-1}(\mathfrak{p}^*_{ij}\widehat{R}_{\mathfrak{p}^*_{ij}}) = \mathfrak{p}^*_{ij}$]. We have, on the one hand, $x_i(\Sigma^{-1}R) = \mathfrak{p}_i(\Sigma^{-1}R)$, hence $x_i(\Sigma^{-1}\widehat{R}) = \mathfrak{p}_i(\Sigma^{-1}\widehat{R})$, and, on the other hand, $\mathfrak{p}_i(\Sigma^{-1}\widehat{R}) = \Sigma^{-1}\mathfrak{p}^*_{i1} \cap \cdots \cap \Sigma^{-1}\mathfrak{p}^*_{i,k(i)}$ [cf. [63], Th. 3.10]. Therefore there exist $s' \in \Sigma$, $y' \in \widehat{R}$ with $s'y = x_i y'$, hence we have

$$s'zt = x_1^{f_1} \cdots x_{i-1}^{f_{i-1}} x_i^{f_i+1} x_{i+1}^{f_{i+1}} \cdots x_h^{f_h} y'.$$

Note that $\omega_{ij}(s') = 0$ for all $j \in \{1, \ldots, k(i)\}$. If $f_i + 1 < e_i$, we may continue the procedure just described until we arrive at a representation

$$s''zt = x_1^{f_1} \cdots x_{i-1}^{f_{i-1}} x_i^{f} x_{i+1}^{f_{i+1}} \cdots x_h^{f_h} y''$$

with $f \geq e_i$ and with elements $s'' \in \Sigma$ and $y'' \in \widehat{R}$. Therefore we may assume that there exist $\widetilde{s} \in \Sigma$, $\widetilde{y} \in \widehat{R}$ and natural integers $f_1, \ldots, f_h$ with $f_i \geq e_i$ for every $i \in \{1, \ldots, n\}$ such that $tz\widetilde{s} = x_1^{f_1} \cdots x_h^{f_h} \widetilde{y}$.

(6.9) Theorem: Let R be an integrally closed local domain, K its field of quotients, L a finite separable extension of K, and S the integral closure of R in L. The ring S is semilocal. Assume that $S/\mathfrak{p}$ is analytically unramified for every prime ideal $\mathfrak{p}$ of S of height 1. If R is analytically normal, then S is analytically *unramified, the completion $\widehat{S}$ of S is a finite torsion-free $\widehat{R}$-algebra which is integrally closed in its ring of quotients, and it is the integral closure of $\widehat{R}$ in $Q(\widehat{S})$. Furthermore, the local rings $S_{\mathfrak{n}}$, $\mathfrak{n}$ maximal ideal of S, are analytically normal. Moreover, we have $\dim_{Q(\widehat{R})}(Q(\widehat{S})) = [L : K]$.*

Proof: (1) First, note that S is a finitely generated R-module by (4.5)(2), hence it is semilocal, and its natural topology and the $\mathfrak{m}S$-adic topology [$\mathfrak{m}$ being the maximal ideal of R] are identical [cf. B(3.8)]. We choose $x \in S$ such that $L = K(x)$; the minimal polynomial f of x over K has coefficients in R [cf. (2.12)(2)]. With $d := D_{L/K}(1, x, \ldots, x^{n-1})$ [$n := [L : K]$] we have $dS \subset R + Rx + \cdots + Rx^{n-1} = R[x] \subset S$ [cf. (4.5)(3)]. From $d\widehat{S} \subset \widehat{R}[x] \subset \widehat{S}$ we see that $\widehat{S}$ and $\widehat{R}[x]$ have the same ring of quotients, hence that $\widehat{S}$ and $\widehat{R}[x]$ have the same integral closure in $Q(\widehat{S})$, say T [note that d is a regular element in $\widehat{S}$ since $S \to \widehat{S}$ is flat, cf. [34], Ch. III, § 3, no. 5, Prop. 9, and that $\widehat{S}$ is a finitely generated $\widehat{R}$-module, hence is integral over $\widehat{R}$ by [63], Cor. 4.5]. Clearly $f \in Q(\widehat{R})[X]$ is a separable polynomial, hence $Q(\widehat{R})[x] = Q(\widehat{R})[X]/(f)$ is an unramified $Q(\widehat{R})$-algebra [cf. (3.13)]. Now we have $dT \subset \widehat{R}[x]$ by (4.5)(3) [because $\widehat{R}$ is an integrally closed domain by hypothesis], hence we have $dT \subset \widehat{S}$. Since S is an integrally closed noetherian domain, every prime ideal in $\mathrm{Ass}(S/Sd)$ has height 1 [cf. [63], Th. 11.2]; the assumption on S shows that we can apply (6.8), and we get $T = \widehat{S}$.

(2) Let $y \in \widehat{S}$ be a nilpotent element, and let $r \in \mathfrak{m}$ be a regular element. Then $y/r^n \in Q(\widehat{S})$ is integral over $\widehat{S}$ for every $n \in \mathbb{N}$, hence we have $y/r^n \in \widehat{S}$ and $y \in \mathfrak{m}^n \widehat{S}$. Since $\widehat{S}$ is a finitely generated $\widehat{R}$-module, we get $y = 0$ by Krull's intersection theorem [cf. [63], Cor. 5.4]. Thus, $\widehat{S}$ is a reduced ring, and $S_{\mathfrak{n}}$ is analytically normal for every maximal ideal $\mathfrak{n}$ of S [cf. (6.1)(2)].

(3) Since $R[x]$ is a free R-module of rank n, we see that $\widehat{R}[x] = R[x] \otimes_R \widehat{R}$ is a free $\widehat{R}$-module of rank n. Since $Q(\widehat{R}[x]) = \widehat{R}[x] \otimes_{\widehat{R}} Q(\widehat{R})$, we get $\dim_{Q(\widehat{R})}(Q(\widehat{S})) = n$ since $Q(\widehat{S}) = Q(\widehat{R}[x])$.

(6.10) Corollary: *We have $\mathfrak{n}_{S/R}\widehat{R} = \mathfrak{n}_{\widehat{S}/\widehat{R}}$ and $\mathrm{branch}(\widehat{S}/\widehat{R}) = V(\mathfrak{n}_{\widehat{S}/\widehat{R}})$.*

Proof: We set $L' := Q(\widehat{S})$, $K' := Q(\widehat{R})$. Clearly we have $\widehat{S} \otimes_{\widehat{R}} K' = L'$. From (2.10) we get $\mathfrak{n}_{S/R} \subset \mathfrak{n}_{\widehat{S}/\widehat{R}}$, hence $\mathfrak{n}_{S/R}\widehat{R} \subset \mathfrak{n}_{\widehat{S}/\widehat{R}}$. On the other hand, we have $\widehat{R} \otimes_R \bigwedge^n S = \bigwedge^n(\widehat{R} \otimes_R S) = \bigwedge^n \widehat{S}$ [cf. [63], Prop. A2.2], showing that $\mathfrak{n}_{\widehat{S}/\widehat{R}} \subset \mathfrak{n}_{S/R}\widehat{R}$. The last assertion follows from (4.12).

(6.11) REMARK: The following two results in (6.12) and (6.13) are needed for the proof of theorem (6.14). In (6.12) we mention only those results on homogenization which are needed in the sequel. For the notion of separable generated fields of algebraic functions and p-bases the reader should consult appendix A1 of Eisenbud's book [63].

(6.12) HOMOGENIZATION: Let k be a field, and let $R = k[x_1, \ldots, x_n]$ be an integral k-algebra of finite type. Let K be the field of quotients of R. Set $r := \operatorname{tr.d}_k(K) = \dim(R)$; note that $\operatorname{ht}(\mathfrak{m}) = r$ for every maximal ideal $\mathfrak{m}$ of R [cf. [63], Th. A on p. 286].

(1) Let y_0 be transcendental over K, set $y_i := x_i y_0$ for $i \in \{1, \ldots, n\}$ and $S := {}^hR := k[y_0, \ldots, y_n]$; note that $\dim(S) = \operatorname{tr.d}_k(Q(S)) = r + 1$ [cf. [63], Th. A on p. 286]. Let $p \in \mathbb{N}_0$, and let S_p be the k-vector space generated by all power products $y_0^{i_0} y_1^{i_1} \cdots y_n^{i_n}$ with $i_0 + \cdots + i_n = p$. Then the sum $S = \sum S_p$ is direct: In fact, if $\sum_p \sum_{i_0 + \cdots + i_n = p} c_{i_0, \ldots, i_n}^{(p)} y_0^{i_0} \cdots y_n^{i_n} = 0$ with $c_{i_0, \ldots, i_n}^{(p)} \in k$, then we get $\sum_p y_0^p \sum_{i_0 + \cdots + i_n = p} c_{i_0, \ldots, i_n}^{(p)} x_1^{i_1} \cdots x_n^{i_n} = 0$, hence $\sum_{i_0 + \cdots + i_n = p} c_{i_0, \ldots, i_n}^{(p)} x_1^{i_1} \cdots x_n^{i_n} = 0$ [since y_0 is transcendental over K], hence $\sum_{i_0 + \cdots + i_n = p} c_{i_0, \ldots, i_n}^{(p)} y_0^{i_0} y_1^{i_1} \cdots y_n^{i_n} = 0$ for every $p \in \mathbb{N}_0$. We set $S_+ := \sum_{p > 0} S_p$; S_+ is a homogeneous prime ideal of S.

(2) For any $g \in S$ we set ${}^ag := g(1, x_1, \ldots, x_r) \in R$; note that $g \mapsto {}^ag : S \to R$ is a surjective homomorphism of rings. For every homogeneous ideal $\mathfrak{b}$ of S we denote by ${}^a\mathfrak{b}$ the image of $\mathfrak{b}$ in R under this homomorphism, and for every ideal $\mathfrak{a}$ of R we denote by ${}^h\mathfrak{a}$ the homogeneous ideal of S which is generated by all homogeneous elements $g \in S$ with ${}^ag \in \mathfrak{a}$. Then:

(a) For every ideal $\mathfrak{a}$ of R we have ${}^a({}^h\mathfrak{a}) = \mathfrak{a}$.

(b) Let $\mathfrak{a}, \mathfrak{b}$ be ideals of R. If $\mathfrak{a} \subsetneq \mathfrak{b}$, then we have ${}^h\mathfrak{a} \subsetneq {}^h\mathfrak{b}$.

(c) If $\mathfrak{p}$ is a prime ideal of R, then ${}^h\mathfrak{p}$ is a prime ideal of S.

(d) If $\mathfrak{m}$ is a maximal ideal of R, then we have $\operatorname{ht}({}^h\mathfrak{m}) = r$.

Proof: (a) Clearly we have ${}^a({}^h\mathfrak{a}) \subset \mathfrak{a}$. Let $f \in \mathfrak{a}$; then there exists a homogeneous $g \in S$ with ${}^ag = f$, hence we have $g \in {}^h\mathfrak{a}$ and therefore we get $f \in {}^a({}^h\mathfrak{a})$.

(b) This follows immediately from the definition.

(c) Let $g_1, g_2 \in S$ be homogeneous with $g_1 g_2 \in {}^h\mathfrak{p}$. Then ${}^a(g_1 g_2) = {}^ag_1 {}^ag_2 \in \mathfrak{p}$, and therefore g_1 or g_2 lies in ${}^h\mathfrak{p}$, hence ${}^h\mathfrak{p}$ is a prime ideal [cf. B(4.1)].

Let $\{0\} =: \mathfrak{p}_0 \subset \mathfrak{p}_1 \subset \cdots \subset \mathfrak{p}_r := \mathfrak{m}$ be a strictly increasing chain of prime ideals of R. Then $\{0\} \subset {}^h\mathfrak{p}_1 \subset \cdots \subset {}^h\mathfrak{p}_r = {}^h\mathfrak{m} \subset S_+$ is a strictly increasing chain of prime ideals of S, and since $\dim(S) = r + 1$, we see that $\operatorname{ht}({}^h\mathfrak{m}) = r$.

(6.13) Lemma: *Let k be a field, let R be an integral k-algebra of finite type, and let $\mathfrak{m}$ be a maximal ideal of R. If $K := Q(R)$ is separably generated over k, then there exists a transcendence basis $\{z_1, \ldots z_r\}$ of K/k contained in $\mathfrak{m}$ such that*

(i) K is a finite separable extension of $k(z_1, \ldots, z_r)$,

(ii) R is a finitely generated $k[z_1, \ldots, z_r]$-module,

(iii) $\mathfrak{m} \cap k[z_1, \ldots, z_r] = (z_1, \ldots, z_r)$.

Proof: If k has characteristic 0, this is just Noether normalization [cf. [63], Th. 13.3]. We consider the case that k has positive characteristic p. We write $R = k[x_1, \ldots, x_n]$. We adjoin a transcendental element y_0 to K, and set $L := K(y_0)$, $y_i := x_i y_0$ for $i \in \{1, \ldots, n\}$, $S := k[y_0, \ldots, y_n]$; note that S is a graded ring by (6.12), and that L is separably generated over k. It is enough to show: There exist r homogeneous elements $\zeta_1, \ldots, \zeta_r$ of S such that, setting $\zeta_0 := y_0$, we have the following: $\{\zeta_0, \zeta_1, \ldots, \zeta_r\}$ is a separating transcendence

basis of L/k, $\zeta_1, \ldots, \zeta_r \in {}^h\mathfrak{m}$, and S is integral over $k[\zeta_0, \ldots, \zeta_r]$. In fact, if this is proved, then setting $z_i := {}^a\zeta_i$ for $i \in \{1, \ldots r\}$, we have: $z_1, \ldots, z_r$ lie in $\mathfrak{m}$, hence $\mathfrak{m} \cap k[z_1, \ldots, z_r] = (z_1, \ldots, z_r)$, R is integral over $k[z_1, \ldots, z_r]$ [since $g \mapsto {}^a g : S \to R$ is a surjective homomorphism] and K is a finite separable extension of $k(z_1, \ldots, z_r)$ [note that $z_i = \zeta_i/y_0^{m_i}$ with $\deg(\zeta_i) =: m_i$ for $i \in \{1, \ldots, r\}$, $k(\zeta_0, \zeta_1, \ldots, \zeta_r) = k(y_0, z_1, \ldots, z_r)$, and since $L = K(y_0)$ is a finite separable extension of $k(z_1, \ldots, z_r)(y_0)$, also K is a finite separable extension of $k(z_1, \ldots, z_r)$].
Now $S\zeta_0$ is a prime ideal of S, $\mathfrak{P}_{0,1} := S\zeta_0$ is not contained in ${}^h\mathfrak{m}$, we have $\mathrm{ht}(S\zeta_0) = 1$ [by Krull's principal ideal theorem, cf. [63], Th. 10.2] and $\{\zeta_0\}$ is a p-independent set over k [note that $kL^p = kK^p(\zeta_0^p)$].
We construct, for $j \in \{2, \ldots, r+1\}$, homogeneous elements $\zeta_1, \ldots, \zeta_{j-1}$ of ${}^h\mathfrak{m}$ such that, setting $\mathfrak{m}_{j-1} := S\zeta_0 + \cdots + S\zeta_{j-1}$, we have: $\mathrm{ht}(\mathfrak{m}_{j-1}) = j$, ${}^h\mathfrak{m}$ does not contain any of the isolated prime ideals $\mathfrak{P}_{j-1,i}$, $i \in \{1, \ldots, h_{j-1}\}$, of $\mathfrak{m}_{j-1}$, and $\{\zeta_0, \ldots, \zeta_{j-1}\} \subset L$ is a p-independent set over k.
Assume that $j \in \{1, \ldots, r\}$, and that $\zeta_0, \zeta_1, \ldots, \zeta_{j-1}$ have already been constructed. We choose homogeneous elements $u, v \in S$ with $u \notin \bigcup_i \mathfrak{P}_{j-1,i}$, $u \in {}^h\mathfrak{m}$ [by prime avoidance, cf. [63], Lemma 3.3], $v \in {}^h\mathfrak{m}$, $v \in \bigcap_i \mathfrak{P}_{j-1,i}$, and $v \notin kL^p(\zeta_0, \zeta_1, \ldots, \zeta_{j-1})$ [the existence of v follows from the fact that the elements of every non-zero ideal of S generate L over k,]. We write $\deg(u) = p^a s$, $\deg(v) = p^b t$ with $a, b \in \mathbb{N}_0$, $p \nmid s$ and $p \nmid t$. We set

$$\zeta_j := \begin{cases} u^{tp^{b-a}} + v^s & \text{if } a < b, \\ u^{tp} + \zeta_0^{p^{a+1-b}} v^s & \text{if } a \geq b. \end{cases}$$

Then $\zeta_0, \ldots, \zeta_j$ satisfy also the above conditions. In fact, ζ_j is homogeneous. With $L_{j-1} := kL^p(\zeta_0, \ldots, \zeta_{j-1})$ we have $[L_{j-1}[v] : L_{j-1}] = p$ [cf. [63], p. 559, first paragraph after the definition], hence $[L_{j-1}[v^s] : L_{j-1}] = p$ since $p \nmid s$, and therefore $\{\zeta_0, \ldots, \zeta_j\} \subset L$ is a p-independent set over k. We have $\mathrm{ht}(\mathfrak{m}_j) = j+1$ [by Krull's principal ideal theorem, cf. [63], Th. 10.2, since $\zeta_j \notin \bigcup_i \mathfrak{P}_{j-1,i}$]. We have $\mathrm{rad}(\mathfrak{m}_{j-1}) = \bigcap \mathfrak{P}_{j-1,i}$, $\mathrm{rad}(\mathfrak{m}_j) = \bigcap \mathfrak{P}_{j,i}$. Since $\mathfrak{m}_{j-1} \subset \mathfrak{m}_j$, every ideal $\mathfrak{P}_{j,i}$, $i \in \{1, \ldots, h_j\}$, contains an ideal $\mathfrak{P}_{j-1,i'}$ for some $i' \in \{1, \ldots, h_{j-1}\}$. Therefore none of the ideals $\mathfrak{P}_{j,i}$ is contained in ${}^h\mathfrak{m}$. We can continue the construction, and find elements $\zeta_0, \zeta_1, \ldots, \zeta_r$ as desired.
Since $\mathrm{ht}(\mathfrak{m}_r) = r+1 = \dim(S)$, $\mathfrak{m}_r \subset S_+$ and $\mathfrak{m}_r$ is homogeneous, we have $\mathrm{rad}(\mathfrak{m}_r) = S_+$, and therefore S is integral over $k[\zeta_0, \ldots, \zeta_r]$ [cf. B(4.34)], hence $\{\zeta_0, \ldots, \zeta_r\}$ is a transcendence basis of L/k. Since $\{\zeta_0, \ldots, \zeta_r\}$ is a p-independent set of L/k, it must be a p-basis of L/k [otherwise we could enlarge this set, in contradiction with the fact that in our case a p-basis is a transcendence basis, cf. [63], Cor A.1.5], and therefore $\{\zeta_0, \ldots, \zeta_r\}$ is a separating transcendence basis of L/k, again by [63], Cor. A.1.5.

(6.14) **Theorem:** *Let k be a perfect field, and let A be an integral local k-algebra essentially of finite type. Then:*
(1) [Chevalley] *The k-algebra A is analytically unramified.*
(2) [Zariski] *The k-algebra A is analytically normal.*

Proof: (a) We show (1) and (2) by induction on $r := \dim(A)$. The case $r = 0$ is trivial. Let $r > 0$, and assume that (1) and (2) hold for every integral local k-algebra essentially of finite type having dimension $< r$. By (6.1) this means, in particular: If S is a semilocal domain of dimension $< r$, and if, for every maximal ideal $\mathfrak{n}$ of S, $S_\mathfrak{n}$ is an integral local k-algebra essentially of finite type, then S is analytically unramified.

(b) Let A be an r-dimensional integral local k-algebra essentially of finite type with maximal ideal $\mathfrak{m}$ and field of quotients L. We show that there exists a subfield k' of L such that L is separably generated over k' and that A is the localization of a k'-algebra of finite type with respect to a *maximal* ideal. There exist elements $x_1, \ldots, x_n$ in A such that setting $T_0 := k[x_1, \ldots, x_n]$, A is a localization of T_0 with respect to the prime ideal $\mathfrak{m} \cap T_0$ of T_0. Set $s := \operatorname{tr.d}_k(L)$. Since $\operatorname{ht}(\mathfrak{m} \cap T_0) = r$, we have $\dim(T_0/(\mathfrak{m} \cap T_0)) = s - r$ [cf. [63], Th. A on p. 286 and Cor. 13.2]. Since L is separably generated over k [theorem of F. K. Schmidt, cf. [63], Cor. A1.7], we can apply (6.13) to T_0 (and an arbitrary maximal ideal of T_0) and get: There exist $u_1, \ldots, u_s \in T_0$ such that, setting $R_0 := k[u_1, \ldots, u_s]$, T_0 is a finitely generated R_0-module, and L is a finite separable extension of $K_0 := Q(R_0)$. Set $\overline{R} := R_0/(\mathfrak{m} \cap R_0) = k[\overline{u}_1, \ldots, \overline{u}_s]$ where $\overline{u}_i$ is the image of u_i in $\overline{R}_0$ for $i \in \{1, \ldots, s\}$. We have $\dim(\overline{R}_0) = \dim(T_0/(\mathfrak{m} \cap T_0))$ since $T_0/(\mathfrak{m} \cap T_0)$ is integral over $\overline{R}_0$ [cf. [63], Prop. 9.2], hence $\operatorname{tr.d}_k(Q(\overline{R}_0)) = s - r$ [cf. [63], Th. A on p. 286]. By relabelling, we may assume that $\{\overline{u}_1, \ldots, \overline{u}_{s-r}\}$ is a transcendence basis of $Q(\overline{R}_0)$ over k. We set $\Sigma := k[u_1, \ldots, u_{s-r}] \setminus \{0\}$. Note that $\Sigma \cap (\mathfrak{m} \cap R_0) = \emptyset$. We set $k' := \Sigma^{-1} k[u_1, \ldots, u_{s-r}] = k(u_1, \ldots, u_{s-r})$. Then $R' := \Sigma^{-1} R_0 = k'[u_{s-r+1}, \ldots, u_s]$, and since $R'/(\mathfrak{m} \cap R') = k'[\overline{u}_{s-r+1}, \ldots, \overline{u}_s]$, the ideal $\mathfrak{m} \cap R'$ is a maximal ideal of R' [the integral affine k'-algebra $R'' := R'/(\mathfrak{m} \cap R')$ is a field since $\operatorname{tr.d}_{k'}(Q(R'')) = 0$, hence $\dim(R'') = 0$, hence R'' is artinian]. We set $T := \Sigma^{-1} T_0$; then T is a k'-algebra of finite type, it is integral over R' [cf. [63], Prop. 4.13], $\mathfrak{m} \cap T$ is a maximal ideal of T [cf. [63], Cor. 4.17], and $A = T_{\mathfrak{m} \cap T}$. Note that K_0 is a purely transcendental extension of k', hence L is separably generated over k', and that $\operatorname{tr.d}_{k'}(L) = r$.

(c) We apply (6.13) to the k'-algebra T and the maximal ideal $\mathfrak{m} \cap T$ of T. This yields the existence of elements $z_1, \ldots, z_r \in T$ such that, setting $R := k'[z_1, \ldots, z_r]$, we have $\mathfrak{m} \cap R = Rz_1 + \cdots + Rz_r$, a maximal ideal of R, L is a finite separable extension of $K = Q(R)$ and T is a finitely generated R-module. Considering a primary decomposition of $(\mathfrak{m} \cap R)T$, we easily see that $\mathfrak{q} := (\mathfrak{m} \cap R)A$ is an $\mathfrak{m}$-primary ideal of A, hence that there exists $l \in \mathbb{N}$ with $\mathfrak{m}^l \subset \mathfrak{q}$.

(d) Let S be the integral closure of R in L; note that $T \subset S$. Then S is a finitely generated R-module by (4.5)(2), and $\dim(S) = r$ [cf. [63], Prop. 9.2], hence $\dim(S_\mathfrak{n}) = r$ for every maximal ideal of S [cf. [63], Th. A on p. 286]. Moreover, S is a k'-algebra of finite type, and k' is a k-algebra essentially of finite type, hence S is a k-algebra essentially of finite type [cf. (1.8)]. Let $\mathfrak{n}$ be a maximal ideal of S. Then $\mathfrak{n} \cap R$ is a maximal ideal of R [cf. [63], Cor. 4.17], and since $R' := R_{\mathfrak{n} \cap R}$ is a regular local ring [cf. [63], Cor. 19.14], it is analytically normal by (6.3). Set $S' := S_{\mathfrak{n} \cap R}$; S' is the integral closure of R' in L [cf. [63], Prop. 4.13], and since it is a finitely generated R'-module, it is semilocal [cf. B(3.8)]. Let $\mathfrak{n}'$ be any maximal ideal of S'. Then $\mathfrak{n}' \cap S$ is a maximal ideal of S by [63], Cor. 4.14 [since $(\mathfrak{n}' \cap S) \cap R = (\mathfrak{n}' \cap R') \cap R = \mathfrak{n} \cap R$], and we have $S_{\mathfrak{n}' \cap S} = S'_{\mathfrak{n}'}$, hence $S'_{\mathfrak{n}'}$ is an r-dimensional integral integrally closed local k-algebra which is essentially of finite type. Let $\mathfrak{p}'$ be a prime ideal of S' with $\mathrm{ht}(\mathfrak{p}') = 1$, and let $\mathfrak{n}'$ be a maximal ideal of S' containing $\mathfrak{p}'$. Now $\dim(S'_{\mathfrak{n}'}/\mathfrak{p}'S'_{\mathfrak{n}'}) < r$, and since $S'_{\mathfrak{n}'}/\mathfrak{p}'S'_{\mathfrak{n}'}$ is an integral local k-algebra essentially of finite type, it is analytically unramified by our induction hypothesis. Since $S'_{\mathfrak{n}'}/\mathfrak{p}'S'_{\mathfrak{n}'} \cong (S'/\mathfrak{p}')_{\mathfrak{n}'/\mathfrak{p}'}$, we get from (6.1) and (6.9): S' is analytically unramified, and $S'_{\mathfrak{n}'}$ is analytically normal for every maximal ideal $\mathfrak{n}'$ of S', hence, in particular, we have $\dim(S_{\mathfrak{n}}) = r$ and $S_{\mathfrak{n}}$ is analytically normal.

(e) We set $B := A[S]$, the smallest subring of L containing A and S. Since $S = R[t_1, \ldots, t_h]$ with elements $t_1, \ldots, t_h$ which are integral over R, we see that $B = A[t_1, \ldots, t_h]$, hence that B is a finitely generated A-module, and therefore we have $\dim(B) = r$ [cf. [63], Prop. 9.2]. Moreover, B is a semilocal domain and a power $\mathfrak{r}^e$ of the Jacobson radical $\mathfrak{r}$ of B is contained in $\mathfrak{m}B$ [cf. B(3.8)]. Let $\mathfrak{n}$ be a maximal ideal of B. Then $\mathfrak{n}^e B_{\mathfrak{n}} \subset \mathfrak{m}B_{\mathfrak{n}} \subset \mathfrak{n}B_{\mathfrak{n}}$, hence $\mathfrak{n}^{le} B_{\mathfrak{n}} \subset \mathfrak{m}^l B_{\mathfrak{n}} \subset \mathfrak{q}B_{\mathfrak{n}} \subset \mathfrak{n}B_{\mathfrak{n}}$, and therefore $\mathfrak{q}B_{\mathfrak{n}}$ is $\mathfrak{n}B_{\mathfrak{n}}$-primary. The elements $t_1, \ldots, t_h$ are, a priori, integral over T; setting $\widetilde{T} := T[t_1, \ldots, t_h]$, we see that $\widetilde{T}$ is integral over T, $\dim(\widetilde{T}) = r$ and that $B = (\widetilde{T})_{\mathfrak{m} \cap T}$. We have $\mathfrak{n} \cap A = \mathfrak{m}$, hence $\mathfrak{n} \cap \widetilde{T}$ lies over the maximal ideal $\mathfrak{m} \cap T$ of T and it is therefore maximal, hence $\mathrm{ht}(\mathfrak{n} \cap \widetilde{T}) = r$, and since $B_{\mathfrak{n}} = \widetilde{T}_{\mathfrak{n} \cap \widetilde{T}}$, we get $\dim(B_{\mathfrak{n}}) = r$. Now $(\mathfrak{n} \cap S) \cap R = (\mathfrak{n} \cap A) \cap R = \mathfrak{m} \cap R$, hence $\mathfrak{n} \cap S$ is a maximal ideal of S, and $(\mathfrak{m} \cap R)S \subset \mathfrak{n} \cap S$, hence $(\mathfrak{m} \cap R)B_{\mathfrak{n}} = \mathfrak{q}B_{\mathfrak{n}} \subset (\mathfrak{n} \cap S)B_{\mathfrak{n}} \subset \mathfrak{n}B_{\mathfrak{n}}$, and therefore $(\mathfrak{n} \cap S)B_{\mathfrak{n}}$ is an $\mathfrak{n}B_{\mathfrak{n}}$-primary ideal of $B_{\mathfrak{n}}$. Note that $S_{\mathfrak{n} \cap S}$ is analytically normal by (c). We consider the local homomorphism $S_{\mathfrak{n} \cap S} \to B_{\mathfrak{n}}$; note that $(\mathfrak{n} \cap S)B_{\mathfrak{n}}$, the ideal of $B_{\mathfrak{n}}$ generated by the maximal ideal $(\mathfrak{n} \cap S)S_{\mathfrak{n} \cap S}$ of $S_{\mathfrak{n} \cap S}$, is an $\mathfrak{n}B_{\mathfrak{n}}$-primary ideal of $B_{\mathfrak{n}}$. From B(9.6) we therefore get $S_{\mathfrak{n} \cap S} = B_{\mathfrak{n}}$. This being true for every maximal ideal $\mathfrak{n}$ of B, we see that B is analytically unramified [cf. (c) and (6.1)(1)]. Since the completion of A can be considered as a subring of the completion of B with respect to its Jacobson radical, we see that A is analytically unramified. Thus, we have shown (1).

From $S_{\mathfrak{n} \cap S} = B_{\mathfrak{n}}$ we see that $B_{\mathfrak{n}}$ is integrally closed; this being true for every maximal ideal of B, we get that B is integrally closed [cf. B(2.5)]. Therefore B is the integral closure of A. Hence, if A is integrally closed, we have $B = A$, hence B is local with maximal ideal $\mathfrak{n}$, and since $B = S_{\mathfrak{n} \cap S}$, B is analytically normal by (c). This proves (2), and the theorem.

(6.15) Corollary: *Let k be a perfect field, let A be an integral local k-algebra essentially of finite type, and let $\widehat{A}$ be its completion. If $\widehat{A}$ is integrally closed, then A is integrally closed.*

Proof: $\widehat{A}$ is a domain and $\widehat{A} \cap Q(A) = A$ [by [34], Ch. I, § 3, no. 5, Prop. 10 since $\widehat{A}$ is a faithfully flat A-module]; therefore A is integrally closed.

(6.16) Remark: For more general results concerning the completion of a local domain the reader should consult chapters 14 and 15 of [116].

Chapter IV

Formal and Convergent Power Series Rings

INTRODUCTION: In this chapter we prepare the ground for the proof of the Jung-Abhyankar theorem in section 2 and the study of quasiordinary power series in section 4 of chapter V. We assume that the reader is acquainted with the notion of power series over a field; in section 1, for the convenience of the reader, we give some background, introduce convergent power series over the real and complex numbers in section 2, and prove Weierstraß division and preparation theorem in section 3. The category of formal (resp. analytic) algebras—these are homomorphic images of formal (resp. convergent) rings of power series—is treated in section 4. The last section considers, in particular, the completion of an analytic algebra; these results are needed to prove the convergent case of the Jung-Abhyankar theorem.

1 Formal Power Series Rings

(1.0) In this section R is a ring.

(1.1) THE RING OF FORMAL POWER SERIES: (1) The ring of formal power series in indeterminates $T_1, \ldots, T_n$ with coefficients in R shall be denoted by $R[\![T_1, \ldots, T_n]\!]$ [for a definition of this ring and some simple properties which shall be used in the following without proof the reader should refer to, e.g., [32], Ch. V, §4, no. 1, or [204], Vol. II, Ch. VII, §1, and also [63], Ch. 7]. Usually a power series will be written in the form

$$F = \sum_{(i_1,\ldots,i_n)\in\mathbb{N}_0^n} F_{(i_1,\ldots,i_n)} T_1^{i_1} \cdots T_n^{i_n}, \quad F_{(i_1,\ldots,i_n)} \in R \text{ for all } (i_1,\ldots,i_n) \in \mathbb{N}_0^n,$$

and we call the support of F the set

$$\mathrm{Supp}(F) = \{(i_1,\ldots,i_n) \in \mathbb{N}_0^n \mid F_{(i_1,\ldots,i_n)} \neq 0\}.$$

143

Of course, polynomials are exactly power series with finite support.

(2) Sometimes we will use the representation of a power series as a sum of forms:

$$F = \sum_{h=0}^{\infty} F_h, \quad F_h = \sum_{\substack{(i_1,\ldots,i_n)\in\mathbb{N}_0^n \\ i_1+\cdots+i_n=h}} F_{(i_1,\ldots,i_n)} T_1^{i_1} \cdots T_n^{i_n},$$

and we will call F_h the form of degree h of F. If $F \neq 0$, we will call the *initial form* of F the form of F of smallest degree; this degree is the *order of F* and will be denoted $o(F)$. If $F = 0$, then we define $o(F) = \infty$ [with the usual rules for addition and order in $\mathbb{N}_\infty := \mathbb{N}_0 \cup \{\infty\}$, cf. I(2.7)]. The order function has the following properties:

$$o(F+G) \geq \min(\{o(F), o(G)\}), \quad o(FG) \geq o(F) + o(G).$$

A power series F is a unit of $R[\![T_1,\ldots,T_n]\!]$ iff F_0 is a unit of R. This F_0, whether it is zero or not, will be called the constant term of F.

Let $\mathfrak{m} = (T_1,\ldots,T_n)$ be the ideal of $R[\![T_1,\ldots,T_n]\!]$ consisting of the power series with zero constant term; the intersection of the powers of $\mathfrak{m}$ is the zero ideal. Note that for $F \in R[\![T_1,\ldots,T_n]\!]$, $F \neq 0$, we have $o(F) = n$ iff $F \in \mathfrak{m}^n$ but $F \notin \mathfrak{m}^{n+1}$. In particular, if $R = k$ is a field, then $k[\![T_1,\ldots,T_n]\!]$ is quasilocal with maximal ideal $\mathfrak{m}$ and residue field k.

If R is a domain, then the initial form of the product of two power series is the product of the corresponding initial forms, so $R[\![T_1,\ldots,T_n]\!]$ is also a domain, and the order of the product is the sum of the orders of the factors.

(3) The ring $R[\![T_1,\ldots,T_n]\!]$ is complete for the $\mathfrak{m}$-adic topology. Moreover, it is the completion of $R[T_1,\ldots,T_n]$ for the $\mathfrak{n} := \mathfrak{m} \cap R[T_1,\ldots,T_n]$-adic topology—note that $\mathfrak{n} = R[T_1,\ldots,T_n]T_1 + \cdots + R[T_1,\ldots,T_n]T_n$—and it is also the completion of $R[T_1,\ldots,T_n]_\mathfrak{n}$ for the $\mathfrak{n}R[T_1,\ldots,T_n]_\mathfrak{n}$-adic topology.

(1.2) SUBSTITUTION: Consider the ring of formal power series $R[\![U_1,\ldots,U_m]\!]$, and let $\mathfrak{m}$ be the ideal of this ring generated by $U_1,\ldots,U_m$. Let A be an R-algebra, $\mathfrak{a}$ an ideal of A, and assume that A is Hausdorff and complete in its $\mathfrak{a}$-adic topology. We choose elements $x_1,\ldots,x_m \in \mathfrak{a}$. Let $G \in R[\![U_1,\ldots,U_m]\!]$, and for every $h \in \mathbb{N}_0$ we consider the form G_h of degree h of G and the element $G_h(x_1,\ldots,x_m) \in A$. It is clear that $G_h(x_1,\ldots,x_m) \in \mathfrak{a}^h$, hence that the sum

$$\sum_{h=0}^{\infty} G_h(x_1,\ldots,x_m) =: G(x_1,\ldots,x_m)$$

converges for the $\mathfrak{a}$-adic topology of A.

The map

$$G \mapsto G(x_1,\ldots,x_m) : R[\![U_1,\ldots,U_m]\!] \to A$$

is an R-algebra homomorphism which is obviously continuous, for $\mathfrak{m}^h$ is mapped into $\mathfrak{a}^h$ for every integer $h \geq 0$. Since $R[U_1,\ldots,U_m]$ is dense in $R[\![U_1,\ldots,U_m]\!]$, there exists only one continuous R-algebra homomorphism $\varphi\colon R[\![U_1,\ldots,U_m]\!] \to A$ with $\varphi(U_i) = x_i$ for every $i \in \{1,\ldots,m\}$. This homomorphism is called a substitution. In particular, let $A = R[\![T_1,\ldots,T_n]\!]$ and the x_i be power series of positive order; then the homomorphism will be called a power series substitution. We get the constant term of a power series F by choosing $x_1 = \cdots = x_n = 0$: $F(0) := F(0,\ldots,0) = F_0$.

(1.3) PARTIAL DERIVATIVES: We set $R_n := R[\![T_1,\ldots,T_n]\!]$; let $\mathfrak{m}$ be the ideal of R_n generated by $T_1,\ldots,T_n$.
(1) Let $j \in \{1,\ldots,n\}$. For every

$$F = \sum_{(i_1,\ldots,i_n)\in\mathbb{N}_0^n} \gamma_{i_1,\ldots,i_n} T_1^{i_1}\cdots T_n^{i_n} \in R_n$$

set

$$D_j(F) := \sum_{(i_1,\ldots,i_n)\in\mathbb{N}_0^n} i_j\gamma_{i_1,\ldots,i_n} T_1^{i_1}\cdots T_{j-1}^{i_{j-1}} T_j^{i_j-1} T_{j+1}^{i_{j+1}}\cdots T_n^{i_n}.$$

It is easy to check that $D_j\colon R_n \to R_n$ is an R-derivation of R_n, and that $D_iD_j = D_jD_i$ for all $i, j \in \{1,\ldots,n\}$.
Let $j \in \{1,\ldots,n\}$; since $D_j(\mathfrak{m}^h) \subset \mathfrak{m}^{h-1}$ for every $h \in \mathbb{N}$, we see that D_j is a continuous map.
To emphasize that we are dealing with the indeterminates $T_1,\ldots,T_n$, we often write $\partial/\partial T_j$ instead of D_j; for $F \in R_n$ the power series $\partial/\partial T_j(F)$ is called the j-th partial derivative of F.
(2) For every $F \in R_n$ we have [we use the notation of (1)]

$$i_1!\cdots i_n!\gamma_{i_1,\ldots,i_n} = D_1^{i_1}\cdots D_n^{i_n}(F)(0) \quad \text{for all } (i_1,\ldots,i_n) \in \mathbb{N}_0^n.$$

Assume, in particular, that R is a $\mathbb{Q}$-algebra; then we have the Taylor expansion

$$F = \sum_{(i_1,\ldots,i_n)\in\mathbb{N}_0^n} \frac{D_1^{i_1}\cdots D_n^{i_n}(F)(0)}{i_1!\cdots i_n!} T_1^{i_1}\cdots T_n^{i_n}.$$

(1.4) JACOBIAN: Let $F^{(1)},\ldots,F^{(m)} \in R[\![T_1,\ldots,T_n]\!]$ be power series with constant term equal to zero. Then the (m,n)-matrix with elements in R

$$\left(\frac{\partial F^{(i)}}{\partial T_j}(0)\right)_{1\leq i\leq m, 1\leq j\leq n} \in M(m,n;R)$$

is called the Jacobian matrix of $F^{(1)},\ldots,F^{(m)}$. We consider the substitution homomorphism $\varphi\colon R[\![U_1,\ldots,U_m]\!] \to R[\![T_1,\ldots,T_n]\!]$ defined by $\varphi(U_i) := F^{(i)}$ for $i \in \{1,\ldots,m\}$; this substitution homomorphism is called the substitution homomorphism defined by $F^{(1)},\ldots,F^{(m)}$. The Jacobian matrix of $F^{(1)},\ldots,F^{(m)}$ is called also the Jacobian J_φ of φ.

We consider the special case $m = n$ and $F^{(i)} = \gamma_{i1}T_1 + \cdots + \gamma_{in}T_n$ for $i \in \{1,\ldots,n\}$ where $C = (\gamma_{ij}) \in M(n;R)$ is an invertible matrix. Then evidently the substitution homomorphism $\varphi\colon R[\![T_1,\ldots,T_n]\!] \to R[\![T_1,\ldots,T_n]\!]$ defined by $F^{(1)},\ldots,F^{(n)}$ is an automorphism with Jacobian $J_\varphi = C$; φ is called a linear change of variables.

2   Convergent Power Series Rings

(2.0) In this section we will assume that the ground ring R is one of the fields $\mathbb{R}$ or $\mathbb{C}$ which we will denote by k.

(2.1) DEFINITION: The polydisc $K_\rho \subset \mathbb{R}^n$ or $K_\rho \subset \mathbb{C}^n$ of radius $\rho > 0$ is the set

$$K_\rho = \{(s_1,\ldots,s_n) \in k^n \mid |s_i| < \rho \text{ for } i = 1,\ldots,n\},$$

where $|\;|$ means absolute value or modulus. A power series

$$F = \sum_{(i_1,\ldots,i_n)\in\mathbb{N}_0^n} \gamma_{i_1,\ldots,i_n} T_1^{i_1}\cdots T_n^{i_n}$$

is convergent when there is a polydisc K_ρ of radius ρ and an $M \in \mathbb{R}_+$ such that

$$|\gamma_{i_1,\ldots,i_n}| < \frac{M}{\rho^{i_1+\cdots+i_n}} \quad \text{for all } (i_1,\ldots,i_n) \in \mathbb{N}_0^n.$$

In this case, we will often say that F is convergent on the polydisc K_ρ, removing any reference to M.

(2.2) NOTATION: Let $t = (t_1,\ldots,t_n) \in \mathbb{R}_+^n$ be an n-tuple of positive real numbers $[\mathbb{R}_+ := \{t \in \mathbb{R} \mid t > 0\}]$. Let

$$F = \sum_{(i_1,\ldots,i_n)\in\mathbb{N}_0^n} \gamma_{i_1,\ldots,i_n} T_1^{i_1}\cdots T_n^{i_n} = \sum_{i=0}^{\infty} F_i \in k[\![T_1,\ldots,T_n]\!]$$

be a formal power series. We define

$$\|F\|_t := \sum_{(i_1,\ldots,i_n)\in\mathbb{N}_0^n} |\gamma_{i_1,\ldots,i_n}|\,t_1^{i_1}\cdots t_n^{i_n}$$

$[\|F\|_t = \infty$ is admissible$]$. It is clear that $\|F\|_t = \sum_{i=0}^{\infty} \|F_i\|_t$.

(2.3) SIMPLE PROPERTIES OF $\|F\|_t$: (1) Let $F \in k[\![T_1,\ldots,T_n]\!]$; then F is convergent iff there exists $t \in \mathbb{R}_+^n$ with $\|F\|_t < \infty$.
(2) Let $F, G \in k[\![T_1,\ldots,T_n]\!]$ and $\gamma \in k$. Then we have $\|F\|_t = 0$ iff $F = 0$, $\|\gamma F\|_t = |\gamma|\,\|F\|_t$ and $\|F + G\|_t \leq \|F\|_t + \|G\|_t$.

(3) Let $F, G \in k[\![T_1, \ldots, T_n]\!]$. Then we have $\|FG\|_t \leq \|F\|_t \|G\|_t$.
Proof: We write $F = \sum_{i=0}^{\infty} F_i$, $G = \sum_{i=0}^{\infty} G_i$. Then we have $FG =: H = \sum_{l=0}^{\infty} H_l$
and $H_l = \sum_{i+j=l} F_i G_j$ for every $l \in \mathbb{N}_0$. Now we obtain

$$\|FG\|_t = \sum_{l=0}^{\infty} \|H_l\|_t \leq \sum_{l=0}^{\infty} \sum_{i+j=l} \|F_i\|_t \|G_j\|_t$$

$$= \sum_{i=0}^{\infty} \|F_i\|_t \sum_{j=0}^{\infty} \|G_j\|_t = \|F\|_t \|G\|_t.$$

(2.4) PROPERTIES OF CONVERGENT POWER SERIES: We list some properties of
convergent power series whose proofs are straightforward.
(1) If F, G are convergent, then they have a common convergence polydisc.
(2) If F, G are convergent on a polydisc K and α, $\beta \in k$, then $\alpha F + \beta G$ is
convergent on K.
(3) If F, G are convergent on a polydisc K, then FG is convergent on K.
(4) If F is convergent on K and such that $F_0 \neq 0$, then the inverse of F in
$k[\![T_1, \ldots, T_n]\!]$ is convergent on some polydisc $K' \subset K$.
(5) The convergent power series form a subring $k[\langle T_1, \ldots, T_n \rangle]$ of $k[\![T_1, \ldots, T_n]\!]$
which is quasilocal with maximal ideal generated by $T_1, \ldots, T_n$ and residue field
k, and it contains the polynomials. Hence the completion of $k[\langle T_1, \ldots, T_n \rangle]$ is
$k[\![T_1, \ldots, T_n]\!]$.
(6) Substitution of a convergent power series into a convergent power series gives
a convergent power series.
(7) The partial derivative of a convergent power series is a convergent power series.
(8) Just as in (1.4) one may define the Jacobian of a finite set $F^{(1)}, \ldots, F^{(m)}$
of convergent power series in $k[\langle X_1, \ldots, X_n \rangle]$ having constant term 0, and the
Jacobian of the homomorphism $\varphi: k[\langle U_1, \ldots, U_m \rangle] \to k[\langle X_1, \ldots, X_n \rangle]$ defined by
$\varphi(U_i) = F^{(i)}$ for $i \in \{1, \ldots, m\}$.
We consider the special case $m = n$ and $F^{(i)} = \gamma_{i1} X_1 + \cdots + \gamma_{in} X_n$ for $i \in$
$\{1, \ldots, n\}$ where $C = (\gamma_{ij}) \in M(n; k)$ is an invertible matrix. Then evidently
the substitution homomorphism $\varphi: k[\langle X_1, \ldots, X_n \rangle] \to k[\langle X_1, \ldots, X_n \rangle]$ defined by
$F^{(1)}, \ldots, F^{(n)}$ is an automorphism with Jacobian $J_\varphi = C$; φ is called a linear
change of variables.

3   Weierstraß Preparation Theorem

3.1   Weierstraß Division Theorem

(3.1) NOTATION: Let R be a quasilocal ring, $\mathfrak{m}$ its maximal ideal and $k := R/\mathfrak{m}$
the residue field. When we consider polynomials over R, we will denote by $^-$ the
fact of taking residues modulo $\mathfrak{m}$. In other words, if $f \in R[Z]$, the polynomial
ring over R, then $\overline{f} \in k[Z]$ will denote the polynomial whose coefficients are
those of f taken modulo $\mathfrak{m}$. More generally, let $r \in \mathbb{N}_0$ and $\omega_r: R \to R/\mathfrak{m}^{r+1}$

be the canonical homomorphism, and extend it to an R-algebra homomorphism $R[Z] \to (R/\mathfrak{m}^{r+1})[Z]$; these maps will also be denoted by ω_r. Note that we have, in particular, $\overline{f} = \omega_0(f)$ for every $f \in R[Z]$.

Let $R[\![Z]\!]$ be the ring of formal power series over R in an indeterminate Z; for any $f \in R[\![Z]\!]$ we write $f = \sum_{i=0}^{\infty} f_i Z^i$ with $f_i \in R$ for every $i \in \mathbb{N}_0$. For every $r \in \mathbb{N}_0$ we extend the R-algebra homomorphism $\omega_r \colon R \to R/\mathfrak{m}^{r+1}$ to an R-algebra homomorphism $R[\![Z]\!] \to (R/\mathfrak{m}^{r+1})[\![Z]\!]$ which will be denoted also by ω_r; thus, we have

$$\omega_r(f) = \sum_{i=0}^{\infty} \omega_r(f_i) Z^i \in (R/\mathfrak{m}^r)[\![Z]\!] \quad \text{for every } f = \sum_{i=0}^{\infty} f_i Z^i \in R[\![Z]\!].$$

(3.2) Definition: Let R be a quasilocal ring with residue field k, and let $m \in \mathbb{N}_0$. A formal power series $f \in R[\![Z]\!]$ is called regular of order m if the power series $\omega_0(f) = \overline{f} \in k[\![Z]\!]$ [cf. (3.1)] has order m, i.e., if we have $\omega_0(f) = \widetilde{f}_1 Z^m$ where $\widetilde{f}_1 \in k[\![Z]\!]$ is a unit. In case $\overline{f} = 0$, we will say that f is not regular.

(3.3) Theorem: [abstract Weierstraß division theorem] *Let R be a quasilocal complete Hausdorff ring. Let $f \in R[\![Z]\!]$ be regular of order $m > 0$. Then, for every $g = \sum_{j=0}^{\infty} g_j Z^j \in R[\![Z]\!]$, there exist a polynomial $h_0 \in R[Z]$ and a power series $h_1 \in R[\![Z]\!]$ such that*

$$g = h_0 + h_1 f$$

and $h_0 = 0$ or $\deg(h_0) \leq m - 1$. Moreover, h_0 and h_1 are determined uniquely by f and g.

Proof: Let $\mathfrak{m}$ be the maximal ideal of R and k its residue field.
[Existence] First, we consider the case where $f = Z^m - p$, and $p \in R[\![Z]\!]$ has all its coefficients in the maximal ideal $\mathfrak{m}$ of R. We define recursively a sequence $(h_0^{(l)})_{l \in \mathbb{N}_0}$ of polynomials in $R[Z]$ and a sequence $(h_1^{(l)})_{l \in \mathbb{N}_0}$ of power series in $R[\![Z]\!]$ in the following way.
We define

$$h_0^{(0)} := \sum_{j=0}^{m-1} g_j Z^j, \quad h_1^{(0)} := \sum_{j=m}^{\infty} g_j Z^{j-m};$$

one sees that

$$g = h_0^{(0)} + h_1^{(0)} Z^m$$

and that $h_0^{(0)} = 0$ or that $h_0^{(0)} \in R[Z]$ is a polynomial of degree $m - 1$, at most. We now define $h_0^{(1)} \in R[Z]$, $h_1^{(1)} \in R[\![Z]\!]$ as follows:

$$(h_0^{(1)})_i := \begin{cases} (h_1^{(0)} p)_i & \text{if } i < m, \\ 0 & \text{otherwise,} \end{cases} \quad (h_1^{(1)})_i := (h_1^{(0)} p)_{i+m} \quad \text{for every } i \geq 0.$$

This shows that

$$h_1^{(0)} p = h_0^{(1)} + h_1^{(1)} Z^m = h_1^{(0)}(Z^m - f)$$

and that $h_0^{(1)} = 0$ or that $\deg(h_0^{(1)}) \leq m - 1$; moreover, the coefficients of $h_0^{(1)}$ and $h_1^{(1)}$ lie in $\mathfrak{m}$, and we have

$$g = (h_0^{(0)} + h_0^{(1)}) + h_1^{(0)} f + h_1^{(1)} Z^m.$$

The general step of the recursion is the following. Let us assume that $l \geq 1$ and that, for every $j \in \{0, \ldots, l\}$, we have constructed a polynomial $h_0^{(j)} \in R[Z]$ and a power series $h_1^{(j)} \in R[\![Z]\!]$, both having coefficients in $\mathfrak{m}^j$, such that

$$g = \sum_{j=0}^{l} h_0^{(j)} + \left(\sum_{j=0}^{l-1} h_1^{(j)} \right) f + h_1^{(l)} Z^m, \qquad (*)$$

and that $h_0^{(j)} = 0$ or that $\deg(h_0^{(j)}) \leq m - 1$. We construct a polynomial $h_0^{(l+1)} \in R[Z]$ and a power series $h_1^{(l+1)} \in R[\![Z]\!]$ in the following way:

$$(h_0^{(l+1)})_i := \begin{cases} (h_1^{(l)} p)_i & \text{if } i < m, \\ 0 & \text{otherwise,} \end{cases} \qquad (h_1^{(l+1)})_i := (h_1^{(l)} p)_{i+m} \quad \text{for every } i \geq 0.$$

This shows that

$$h_1^{(l)} p = h_0^{(l+1)} + h_1^{(l+1)} Z^m = h_1^{(l)} (Z^m - f),$$

hence that $(*)$ holds for $l + 1$ also, and that the coefficients of $h_0^{(l+1)}$ and $h_1^{(l+1)}$ lie in $\mathfrak{m}^{l+1}$. Moreover, we have $h_0^{(l+1)} = 0$ or $\deg(h_0^{(l+1)}) \leq m - 1$. Now, these conditions mean that the two sequences $\left(\sum_{j=0}^{l} (h_0^{(j)})_i \right)_{l \in \mathbb{N}_0}$, $\left(\sum_{j=0}^{l} (h_1^{(j)})_i \right)_{l \in \mathbb{N}_0}$ are convergent sequences in R for every $i \in \mathbb{N}_0$, and we will write

$$\lim_{l \to \infty} \left(\sum_{j=0}^{l} (h_0^{(j)})_i \right) =: h_{0,i}, \quad \lim_{l \to \infty} \left(\sum_{j=0}^{l} (h_1^{(j)})_i \right) =: h_{1,i} \quad \text{for every } i \in \mathbb{N}_0,$$

and define the formal power series

$$h_0 := \sum_{i=0}^{\infty} h_{0,i} Z^i, \quad h_1 := \sum_{i=0}^{\infty} h_{1,i} Z^i \in R[\![Z]\!].$$

In addition, we have $\lim_{l \to \infty} (h_1^{(l)})_i = 0$ for every $i \in \mathbb{N}_0$, hence we have [cf. $(*)$]

$$g = h_0 + h_1 f$$

and $h_0 = 0$ or $h_0 \in R[Z]$ is a polynomial of degree $m - 1$, at most. This proves the existence part in the theorem, if f has the special form mentioned above. In the general case, we write $f = f^{(1)} + Z^m f^{(2)}$ where

$$f^{(1)} := \sum_{j=0}^{m-1} f_j Z^j, \quad f^{(2)} := \sum_{j=m}^{\infty} f_j Z^{j-m};$$

$f^{(2)}$ is invertible in $R[Z]$ [cf. (1.1)(2), and note that $f_m \notin \mathfrak{m}$ since $\omega_0(f_m) \neq 0$] and $f^{(1)}f^{(2)-1}$ has all its coefficients in the maximal ideal $\mathfrak{m}$ of R [since $f_0, \ldots, f_{m-1}$ lie in $\mathfrak{m}$]. Now we have $f = \left(Z^m - (-f^{(1)}f^{(2)-1})\right)f^{(2)}$, hence, by what we have just shown, for any $g \in R[Z]$ there exist $h_0 \in R[Z]$, $h_0 = 0$ or of degree $m-1$ at most, and $h_1' \in R[Z]$ such that $g = h_0 + h_1' f f^{(2)-1} = h_0 + h_1 f$ with $h_1 := h_1' f^{(2)-1}$.

[Uniqueness] What we need to prove is that, if $g = 0$, then $h_0 = h_1 = 0$. Again, we may assume that $f = Z^m - p$ where $p \in R[Z]$ has its coefficients in the maximal ideal of R, i.e., that we have $\omega_0(p) = 0$.

Then, from $0 = h_0 + h_1 f$ we get $\omega_0(h_0) = -\omega_0(h_1)Z^m$, hence $\omega_0(h_0) = \omega_0(h_1) = 0$ [since $\omega_0(h_0) \in k[Z]$ has degree $m-1$ at most], hence all coefficients of h_0 and h_1 belong to $\mathfrak{m}$. Assume that $r \geq 1$ is an integer, and that we have proved that all the coefficients of h_0, h_1 belong to $\mathfrak{m}^r$. Again, we have $\omega_r(h_0) = -\omega_r(h_1)Z^m$, hence $\omega_r(h_0) = \omega_r(h_1) = 0$, so all the coefficients of h_0 and h_1 belong to $\mathfrak{m}^{r+1}$. Hence the coefficients of h_0 and h_1 belong to the intersection of all powers of the maximal ideal $\mathfrak{m}$, and this intersection is $\{0\}$ since R is Hausdorff by assumption. Therefore, we have shown that $h_0 = h_1 = 0$.

This ends the proof of the theorem.

(3.4) DEFINITION: Let R be a quasilocal ring having maximal ideal $\mathfrak{m}$. Let $m \in \mathbb{N}$; a polynomial

$$p = Z^m + a_1 Z^{m-1} + \cdots + a_m \in R[Z]$$

such that $a_i \in \mathfrak{m}$ for every $i \in \{1, \ldots, m\}$ will be called a Weierstraß polynomial. Note that the discriminant $\mathrm{dis}(F)$ of F is not a unit of R if $m > 1$ [cf. B(10.22)(1)(*)].

(3.5) NOTATION: In the following, we work with rings of formal power series $k[\![X_1, \ldots, X_n]\!]$ over an arbitrary field k, and also with rings of convergent power series $k[\langle X_1, \ldots, X_n \rangle]$ where $k = \mathbb{R}$ or $k = \mathbb{C}$. To avoid repetition, in the sequel we denote by $k[\{X_1, \ldots, X_n\}]$ one of the rings $k[\![X_1, \ldots, X_n]\!]$, $k[\langle X_1, \ldots, X_n \rangle]$ where in the latter case we have $k = \mathbb{R}$ or $k = \mathbb{C}$. Note that $k[\{X_1, \ldots, X_n\}]$ is a quasilocal domain with maximal ideal $\mathfrak{m}$ for which $X_1, \ldots, X_n$ is a minimal system of generators, and k is its residue field [cf. (1.1)(2) and (2.4)(4)]; furthermore, $k[\![X_1, \ldots, X_n]\!]$ is complete with respect to its natural topology, and it is the completion of the quasilocal ring $k[\langle X_1, \ldots, X_n \rangle]$. The field of quotients of $k[\![X_1, \ldots, X_n]\!]$ (resp. $k[\langle X_1, \ldots, X_n \rangle]$ resp. $k[\{X_1, \ldots, X_n\}]$) will be denoted by $k((X_1, \ldots, X_n))$ (resp. $k(\langle X_1, \ldots, x_n \rangle)$ resp. $k(\{X_1, \ldots, X_n\})$).

(3.6) NOTATION: A power series $F \in k[\{X_1, \ldots, X_{n+1}\}]$ is called X_{n+1}-regular of order m if F, considered as an element of $R_n[\![X_{n+1}]\!]$ where $R_n = k[\{X_1, \ldots, X_n\}]$, is regular of order m.

(3.7) Theorem: [Weierstraß division theorem, analytic case] *Let f be a convergent power series in $k[\langle X_1, \ldots, X_{n+1} \rangle]$, X_{n+1}-regular of order $m > 0$. Then, for* every $g \in k[\langle X_1, \ldots, X_{n+1} \rangle]$, there exist a polynomial $h_0 \in k[\langle X_1, \ldots, X_n \rangle][X_{n+1}]$ and a power series $h_1 \in k[\langle X_1, \ldots, X_{n+1} \rangle]$ such that

$$g = h_0 + h_1 f, \qquad\qquad (*)$$

and that $h_0 = 0$ or that $\deg(h_0) \leq m - 1$. Moreover, h_0 and h_1 are determined uniquely by f and g.

Proof: We shall use the notation of the proof of (3.3) [where Z must be replaced by X_{n+1}]. First, we consider the case $f = X_{n+1}^m - p$ where $p = \sum_{i=0}^{\infty} p_i X_{n+1}^i$, and $p_i \in k[\langle X_1, \ldots, X_n \rangle]$ has constant term zero for every $i \in \mathbb{N}_0$.

(1) For any $F \in k[\langle X_1, \ldots, X_n \rangle]$ and $(s, \ldots, s) \in \mathbb{R}_+^n$ write $\|F\|_{(s,\ldots,s)} =: \overline{F}(s)$, and for any $G \in k[\langle X_1, \ldots, X_{n+1} \rangle]$ and $(s, \ldots, s, t) \in \mathbb{R}_+^{n+1}$ write $\|G\|_{(s,\ldots,s,t)} := \overline{G}(s,t)$. In particular, writing $G = \sum_{i \geq 0} G_i X_{n+1}^i$ with $G_i \in k[\langle X_1, \ldots, X_n \rangle]$ for $i \geq 0$, we have $\overline{G}(s,t) = \sum_{i \geq 0} \overline{G_i}(s) t^i$.

(2) Let $F = \sum_{(i_1, \ldots, i_n) \in \mathbb{N}_0^n} \gamma_{i_1, \ldots, i_n} X_1^{i_1} \cdots X_n^{i_n} \in k[\langle X_1, \ldots, X_n \rangle]$ have constant term $\gamma_{0,\ldots,0} = 0$. Let s be a positive real number such that $\overline{F}(s) < \infty$. For every positive real number t such that $t < s$ we have

$$\overline{F}(t) = \sum_{(i_1, \ldots, i_n) \in \mathbb{N}_0^n} |\gamma_{i_1, \ldots, i_n}| \left(\frac{t}{s}\right)^{i_1} \cdots \left(\frac{t}{s}\right)^{i_n} s^{i_1} \cdots s^{i_n} \leq \frac{t}{s} \overline{F}(s).$$

(3) Since p and g are convergent, there exists $s \in \mathbb{R}_+$ such that $\overline{p}(s,s) < \infty$ and $\overline{g}(s,s) < \infty$. We choose $\gamma \in k$ such that $|\gamma| > 1/s$. Consider the linear change of variables defined by $\varphi(X_i) = X_i/\gamma$ for every $i \in \{1, \ldots, n+1\}$. Then we have $\overline{\varphi(p)}(1,1) < \infty$ and $\overline{\varphi(g)}(1,1) < \infty$. We may assume, therefore, that $\overline{p}(1,1) < \infty$ and $\overline{g}(1,1) < \infty$. Choose a positive real number $u < 1$, and then choose a positive real number t such that $t < 1$ and $t\,\overline{p}(1,1) < u^m(1-u)$. Writing $g = \sum_{i \geq 0} g_i X_{n+1}^i$ with $g_i \in k[\langle X_1, \ldots, X_n \rangle]$ for $i \geq 0$, it is clear that [cf. (1)]

$$\overline{g_i}(t) \leq \frac{\overline{g}(t,u)}{u^i} \quad \text{for every } i \in \mathbb{N}_0.$$

Moreover, by (2) we have

$$\overline{p_i}(t) \leq t\,\overline{p_i}(1) \leq t\,\overline{p}(1,1) \quad \text{for every } i \in \mathbb{N}_0.$$

Let us write $\overline{(h_0^{(j)})_i}(t) =: \overline{h}_{0,i}^{(j)}(t)$ for all $i, j \in \mathbb{N}_0$, and define $\overline{h}_{1,i}^{(j)}(t)$ in the same way. We show by induction on j: For every $i \in \mathbb{N}_0$ we have

$$\overline{h}_{0,i}^{(j)}(t) \leq \frac{\overline{g}(t,u)}{u^i} \left(\frac{t\,\overline{p}(1,1)}{u^m(1-u)}\right)^j, \quad \overline{h}_{1,i}^{(j)}(t) \leq \frac{\overline{g}(t,u)}{u^{i+m}} \left(\frac{t\,\overline{p}(1,1)}{u^m(1-u)}\right)^j.$$

It is clear that both formulae hold for $j = 0$. Assume that $j \geq 0$ and that the formulae hold for j. We show that the second formula holds for $j + 1$. We have

$$\overline{h}_{1,i}^{(j+1)}(t) \leq \sum_{l=0}^{i+m} \overline{p}_l(t) \overline{h}_{1,i+m-l}^{(j)}(t)$$

$$\leq t\,\overline{p}(1,1) \frac{\overline{g}(t,u)}{u^{m+i+m}} \left(\frac{t\,\overline{p}(1,1)}{u^m(1-u)} \right)^j \cdot \sum_{l=0}^{i+m} u^l$$

$$= \frac{\overline{g}(t,u)}{u^{i+m}} \left(\frac{t\,\overline{p}(1,1)}{u^m(1-u)} \right)^{j+1} \cdot (1-u) \cdot \sum_{l=0}^{i+m} u^l$$

$$\leq \frac{\overline{g}(t,u)}{u^{i+m}} \left(\frac{t\,\overline{p}(1,1)}{u^m(1-u)} \right)^{j+1}$$

since

$$(1-u) \cdot \sum_{l=0}^{i+m} u^l \leq (1-u) \cdot \sum_{l=0}^{\infty} u^l = 1.$$

The proof for the first formula is similar.

The general case can be handled as we did this in the proof of (3.3). The uniqueness assertion also follows from the corresponding statement in the abstract case.

(3.8) REMARK: We have $k[\![X_1, \ldots, X_{n+1}]\!] = k[\![X_1, \ldots, X_n]\!][\![X_{n+1}]\!]$, and since $k[\![X_1, \ldots, X_n]\!]$ is a quasilocal complete Hausdorff ring, the Weierstraß division theorem holds for $k[\![X_1, \ldots, X_n]\!][\![X_{n+1}]\!]$. We state the Weierstraß division theorem for $k[\{X_1, \ldots, X_{n+1}\}]$ usually in the following form: Let $F \in k[\{X_1, \ldots, X_{n+1}\}]$ be X_{n+1}-regular of order $m > 0$. Then, for every $G \in k[\{X_1, \ldots, X_{n+1}\}]$, there exist uniquely determined $H \in k[\{X_1, \ldots, X_{n+1}\}]$ and $A_1, \ldots, A_m \in k[\{X_1, \ldots, X_n\}]$ such that

$$G = HF + A_1 X_{n+1}^{m-1} + A_2 X_{n+1}^{m-2} + \cdots + A_m.$$

3.2 Weierstraß Preparation Theorem and Applications

(3.9) Theorem: [Weierstraß preparation theorem] *Let* $F \in k[\{X_1, \ldots, X_{n+1}\}]$ *be* X_{n+1}-*regular of order* $m > 0$. *Then* F *can be written, in a unique way, as* $F = U \cdot P$ *where* $U \in k[\{X_1, \ldots, X_{n+1}\}]$ *is a unit and* $P \in k[\{X_1, \ldots, X_n\}][X_{n+1}]$ *is a Weierstraß polynomial of degree* m.

Proof [Existence]: Set $R_n := k[\![X_1, \ldots, X_n]\!]$, and let $\mathfrak{m}$ be the maximal ideal of R_n. By Weierstraß division theorem [cf. (3.8)] we find a unique expression

$$X_{n+1}^m = H_0 + H_1 F, \quad H_0 \in R_n[X_{n+1}], \quad H_1 \in k[\{X_1, \ldots, X_{n+1}\}]$$

where $H_0 = 0$ or $\deg(H_0) \leq m - 1$. Taking residues modulo $\mathfrak{m}$ we see that $X_{n+1}^m = \overline{H}_0 + \overline{H}_1 \overline{F}$ and $\overline{F} = \widetilde{F}_1 \cdot X_{n+1}^m$ where $\widetilde{F}_1$ is a unit of $k[\![X_{n+1}]\!]$. This shows that $\overline{H}_0 = 0$ and that $\overline{H}_1 \widetilde{F}_1 = 1$, hence that $\overline{H}_1$ is a unit of $k[\![X_{n+1}]\!]$.

Therefore H_1 is a unit of $k[\{X_1, \ldots, X_{n+1}\}]$ [cf. (1.1)(2) resp. (2.4)(4)], and our result follows by taking $U := H_1^{-1}$, $P := X_{n+1}^m - H_0$.

[Uniqueness] Let $F = U \cdot P$ be a representation of F as stated. We set $H_1 := U^{-1}$, $H_0 := X_{n+1}^m - P$. Then we have $F = H_0 + H_1 F$, which shows [cf. (3.8) and note that $H_0 = 0$ or $\deg(H_0) \leq m - 1$] that U and P are determined uniquely.

(3.10) DEFINITION: In the situation of (3.9), the Weierstraß polynomial P is called the Weierstraß polynomial associated with F.

We add some consequences of Weierstraß division theorem and Weierstraß preparation theorem. We write $R_n = k[\{X_1, \ldots, X_n\}]$; note that $k[\langle X_1, \ldots, X_{n+1}\rangle]$ is a subring of $k[\langle X_1, \ldots, X_n\rangle][[X_{n+1}]]$, the power series ring over $k[\langle X_1, \ldots, X_n\rangle]$.

(3.11) REMARK: (1) Let $F \in R_n[X_{n+1}]$ be a Weierstraß polynomial of degree m, and let $G \in R_n[X_{n+1}]$ be a polynomial. The equation

$$G = HF + A_1 X_{n+1}^{m-1} + \cdots + A_m$$

[cf. (3.8)] is the division identity in $R_n[X_{n+1}]$, as follows from the uniqueness part of (3.8), hence, in particular, $H \in R_n[X_{n+1}]$ is a polynomial.

(2) Let $F \in R_n[X_{n+1}]$ be X_{n+1}-regular of order $m > 0$. Then in $F = UP$ [cf. (3.9)] the unit $U \in k[\{X_1, \ldots, X_{n+1}\}]$ is a polynomial in $R_n[X_{n+1}]$ of degree $\deg(F) - \deg(P)$.

(3.12) REMARK: Let $P \in R_n[X_{n+1}]$ be a Weierstraß polynomial, and assume that $P = F_1 \cdots F_h$ where $F_1, \ldots, F_h \in k[\{X_1, \ldots, X_{n+1}\}]$ are non-units. Clearly, the power series $F_1, \ldots, F_h$ are X_{n+1}-regular of positive orders, hence there exist units $U_1, \ldots, U_h$ in $k[\{X_1, \ldots, X_{n+1}\}]$ and Weierstraß polynomials $P_1, \ldots, P_h$ in $R_n[X_{n+1}]$ such that $F_i = U_i P_i$ for every $i \in \{1, \ldots, h\}$ [cf. (3.9)]. Since $P_1 \cdots P_h$ is a Weierstraß polynomial in $R_n[X_{n+1}]$, by (3.9) we get $P = P_1 \cdots P_h$ and $U_1 \cdots U_h = 1$.

(3.13) Corollary: *Let $F \in R_n[X_{n+1}]$ be a Weierstraß polynomial which is irreducible. Then F is also irreducible in $k[\{X_1, \ldots, X_{n+1}\}]$.*

Proof: Indeed, suppose that $F = F_1 F_2$ where $F_1, F_2 \in k[\{X_1, \ldots, X_{n+1}\}]$ are non-units. Since F is X_{n+1}-regular of positive order, also F_1 and F_2 are X_{n+1}-regular of positive orders. By (3.9) we have $F_i = U_i P_i$, $i = 1, 2$, where U_1, U_2 are units of $k[\{X_1, \ldots, X_{n+1}\}]$ and $P_1, P_2 \in R_n[X_{n+1}]$ are Weierstraß polynomials. By (3.12) we see that $F = P_1 P_2$, which contradicts the irreducibility of F.

(3.14) REMARK: Let

$$P = X_{n+1}^m + A_1 X_{n+1}^{m-1} + \cdots + A_m \in R_n[X_{n+1}]$$

be a Weierstraß polynomial, i.e., $A_1, \ldots, A_m \in R_n$ have positive orders.

(1) We show that

$$k[\{X_1,\dots,X_{n+1}\}]\,P \cap R_n[X_{n+1}] = R_n[X_{n+1}]\,P.$$

Indeed, let $F \in k[\{X_1,\dots,X_{n+1}\}]\,P \cap R_n[X_{n+1}]$, i.e., there exists a formal resp. convergent power series $H \in k[\{X_1,\dots,X_{n+1}\}]$ such that $F = HP$. We choose $q \in \mathbb{N}$ such that the degree of F as a polynomial in X_{n+1} is at most $m+q$. Let us write

$$H = \sum_{i=0}^{\infty} H_i X_{n+1}^i \quad \text{with } H_i \in R_n \quad \text{for every } i \in \mathbb{N}_0.$$

Then, for every $i \in \mathbb{N}$ such that $i > q$, we have

$$H_i + H_{i+1}A_1 + \cdots + H_{i+m}A_m = 0. \tag{*}$$

Since $A_1,\dots,A_m$ lie in the maximal ideal $\mathfrak{m}$ of R_n, we see that $H_i \in \mathfrak{m}$ for every $i \in \mathbb{N}$ such that $i > q$. Applying (*) again we conclude that $H_i \in \mathfrak{m}^2$ for every $i > q$. By induction, we see that $H_i \in \bigcap_{j\in\mathbb{N}} \mathfrak{m}^j$ for every $i > q$, hence that $H_i = 0$ for every $i > q$, and that therefore $H \in R_n[X_{n+1}]$.

(2) Now $R_n[X_{n+1}]/(P)$ can be considered as a subring of $k[\{X_1,\dots,X_{n+1}\}]/(P)$ by (1). Take any $G \in k[\{X_1,\dots,X_{n+1}\}]$. By Weierstraß division theorem there exists $H \in k[\{X_1,\dots,X_{n+1}\}]$ such that $G - HP \in R_n[X_{n+1}]$, which implies that

$$k[\{X_1,\dots,X_{n+1}\}]/(P) = R_n[X_{n+1}]/(P).$$

(3) Let $x := X_{n+1} + (P)$ be the image of X_{n+1}. Then $R_n[X_{n+1}]/(P)$ is a free R_n-module having $\{1, x, \dots, x^{m-1}\}$ as a basis.

(3.15) REMARK: (1) Let k be an infinite field. Let $F \in k[\{X_1,\dots,X_n\}]$ be a non-unit power series different from zero, of order m, and let F_m be its initial form. Since k is infinite, we may find $(\gamma_1,\dots,\gamma_{n-1}) \in k^{n-1}$ such that $F_m(\gamma_1,\dots,\gamma_{n-1},1) \neq 0$. Consider the linear change of variables $\varphi: R_n \to R_n$ defined by $\varphi(X_i) = X_i + \gamma_i X_n$ for every $i \in \{1,\dots,n-1\}$ and $\varphi(X_n) = X_n$ [cf. (1.4) and (2.4)(8)]. Now $\varphi(F_m)(0,\dots,0,X_n) = F_m(\gamma_1,\dots,\gamma_{n-1},1)X_n^m$, hence $\varphi(F)$ is X_n-regular of order m. This shows that every non-zero power series can be made regular in X_n of order equal to its own order, by applying a linear change of variables.

Since the affine space k^n is not a union of finitely many hypersurfaces [since k is infinite], the same is true simultaneously for a finite number of power series.

(2) It can be shown that, even in the case of a finite field, every non-zero formal power series in $k[\![X_1,\dots,X_n]\!]$ can be made regular in X_n of order equal to its own order, by applying a suitable k-algebra automorphism of $k[\![X_1,\dots,X_n]\!]$ [cf. [204], vol. II, Ch. VII, Lemma 3]. We tacitly assume this result.

(3.16) Proposition: The power series ring $k[\{X\}]$ is a discrete valuation ring, and its maximal ideal is generated by X. In particular, $k[\{X\}]$ is noetherian and factorial.

Proof: The order function $o: k[\{X\}] \to \mathbb{N}_0 \cup \{\infty\}$ can be extended in a unique way as a discrete valuation $v: k(\{X\}) \to \mathbb{Z}_\infty$ and $k[\{X\}]$ is the ring of v [cf. chapter I, (3.20) and (3.28)].

(3.17) Proposition: *The ring R_n is noetherian.*

We prove it by induction on n. The ring R_1 is noetherian [cf. (3.16)]. Let $n > 1$, and assume that $k[\{X_1, \ldots, X_{n-1}\}]$ is noetherian. Let $\mathfrak{a} \neq R_n$ be a non-zero ideal of R_n and choose $F \in \mathfrak{a}$, $F \neq 0$. By applying a k-automorphism [cf. (3.15)] and the Weierstraß preparation theorem [cf. (3.9)], we may assume that F is a Weierstraß polynomial with respect to X_n of degree $m \geq 1$. By (3.14)(3) we see that $R_n/(F)$ is a finitely generated $k[\{X_1, \ldots, X_{n-1}\}]$-module, hence it is a noetherian ring [cf. [63], Cor. 1.3], and therefore the ideal $\mathfrak{a}/(F)$ is a finitely generated ideal of $R_n/(F)$. It follows that $\mathfrak{a}$ is a finitely generated ideal of R_n, hence R_n itself is a noetherian ring.

(3.18) Proposition: *The ring R_n is factorial.*

We prove it by induction on n. The ring R_1 is factorial [cf. (3.16)]. Let $n > 1$, and assume that $k[\{X_1, \ldots, X_{n-1}\}]$ is factorial; then $k[\{X_1, \ldots, X_{n-1}\}][X_n]$ is also factorial by Gauß's lemma.
Let $F \in R_n$ be a non-unit different from zero. By (3.15) we may assume that F is X_n-regular of positive order, and, by (3.9), that $F = UP$ where $U \in R_n$ is a unit and $P \in k[\{X_1, \ldots, X_{n-1}\}][X_n]$ is a Weierstraß polynomial. Since $k[\{X_1, \ldots, X_{n-1}\}][X_n]$ is factorial, we may write, uniquely up to units, $P = P_1 \cdots P_h$ where $P_1, \ldots, P_h \in k[\{X_1, \ldots, X_{n-1}\}][X_n]$ are irreducible. By (3.13) this is a factorization of P in R_n into a product of irreducible elements. Let $P = F_1 \cdots F_m$ be another factorization of P into irreducible elements in R_n. Then the elements $F_1, \ldots, F_m$ are X_n-regular. For every $i \in \{1, \ldots, m\}$ we can write $F_i = V_i P_i'$ where V_i is a unit of R_n and $P_i' \in k[\{X_1, \ldots, X_{n-1}\}][X_n]$ is an irreducible Weierstraß polynomial [cf. (3.9)]. By (3.12) we have $P = P_1' \cdots P_m'$, hence a factorization of P into a product of irreducible elements of $k[\{X_1, \ldots, X_{n-1}\}][X_n]$. Since the latter ring is factorial, we have $m = h$ and, after relabelling, $U_i P_i = P_i'$ for every $i \in \{1, \ldots, h\}$ where $U_1, \ldots, U_h$ are units of $k[\{X_1, \ldots, X_{n-1}\}]$. Therefore we have $F_i = V_i U_i P_i$ for every $i \in \{1, \ldots, h\}$ which means that $P = P_1 \cdots P_h$ is a factorization of P in R_n into irreducible factors which is unique up to relabelling and units.

(3.19) Corollary: *Let $F \in k[\{X_1, \ldots, X_{n+1}\}]$ be a power series which is X_{n+1}-regular of positive order m. If F is an irreducible element of $k[\{X_1, \ldots, X_{n+1}\}]$, then the quotient field of the domain $k[\{X_1, \ldots, X_{n+1}\}]/(F)$ is a simple algebraic extension of degree m of the quotient field of $k[\{X_1, \ldots, X_n\}]$.*

Proof: Since $k[\{X_1, \ldots, X_{n+1}\}]$ is factorial by (3.18), the ideal generated by F is a prime ideal. By Weierstraß preparation theorem we have $F = UP$ where

$U \in k[\{X_1, \ldots, X_{n+1}\}]$ is a unit and $P \in R_n[X_{n+1}]$ is the Weierstraß polynomial associated with F. Now the result follows from (3.14).

(3.20) Theorem: *The ring R_n is an n-dimensional regular local ring.*

Proof: R_n is a noetherian ring by (3.17). The ideals generated by $X_1, \ldots, X_j$, $j \in \{0, \ldots, n\}$, form a strictly ascending chain of prime ideals, and the maximal ideal of R_n is minimally generated by n elements, hence $n \leq \dim(R_n) \leq \operatorname{emdim}(R_n) = n$, hence we have $\operatorname{emdim}(R_n) = \dim(R_n)$, and R_n is regular [cf. II(4.19)].

(3.21) REMARK: From the result in (3.20) we now can deduce the factoriality of R_n [cf. (3.18)] also by applying Th. 19.19 in Eisenbud's book [63].

(3.22) Theorem: *Assume that k is algebraically closed. Let $F \in R_n[Z]$ be a monic polynomial of positive degree, and assume that $\overline{F}(Z) \in k[Z]$ has at least two different zeroes; then we have $\overline{F}(Z) = \prod_{l=1}^{m}(Z - \gamma_l)^{s_l}$, where $m \geq 2$, $\gamma_1, \ldots, \gamma_m$ are pairwise different elements in k and $s_1, \ldots, s_m \in \mathbb{N}$. Then there exist monic polynomials $F_1, \ldots, F_m \in R_n[Z]$ such that $F = F_1 \cdots F_m$ and $\overline{F_l}(Z) = (Z - \gamma_l)^{s_l}$ for every $l \in \{1, \ldots, m\}$.*

Proof [by recursion]: Set $G(Z) := F(\gamma_1 + Z) \in R_n[Z]$; G is monic of the same degree as F and

$$\overline{G}(Z) = \overline{F}(\gamma_1 + Z) = Z^{s_1} \prod_{l=2}^{m}(Z + \gamma_1 - \gamma_l)^{s_l}.$$

Therefore G is Z-regular of order s_1. Let $H \in R_n[Z]$ be the Weierstraß polynomial associated with G; then we have $G = UH$, where $U \in R_n[Z]$ is monic [cf. (3.11)] and $\deg_Z(U) = \deg_Z(G) - \deg_Z(H)$. Set $F_1(Z) := H(Z - \gamma_1)$ and $F_2(Z) := U(Z - \gamma_1)$. Then F_1 and F_2 are monic polynomials in $R_n[Z]$, and we have

$$F = F_1 F_2, \quad \overline{F_1}(Z) = (Z - \gamma_1)^{s_1}, \quad \overline{F_2}(Z) = \prod_{l=2}^{m}(X - \gamma_j)^{s_j}.$$

Repeating the above argument with F_2 yields the result.

(3.23) DEFINITION: Let A be a quasilocal ring with residue field k. Then A is called henselian if for every monic polynomial $F \in A[T]$ such that $\omega(F) = \overline{G}\,\overline{H}$, where $\overline{G}, \overline{H} \in k[T]$ are monic and relatively prime, there exist monic polynomials $G, H \in A[T]$ with $F = GH$ and with $\omega(F) = \overline{F}$, $\omega(G) = \overline{G}$.

(3.24) REMARK: Let A be a henselian ring, and let $\mathfrak{a}$ be an ideal of A contained in the maximal ideal of A. Then $A/\mathfrak{a}$ is a henselian ring.

(3.25) Corollary: *Let k be algebraically closed. Then R_n is a Henselian ring.*

4   The Category of Formal and Analytic Algebras

(4.0) For the sake of simplicity, we assume in this section that k *is algebraically closed.*

4.1   Local k-algebras

(4.1) REMARK: Let B be a local k-algebra with maximal ideal $\mathfrak{n}$ and residue field k. Then we have $B = k \, 1_B \oplus \mathfrak{n}$ as a direct sum of k-vector spaces.

Let $\{y_1, \ldots, y_h\}$ be a system of generators of the ideal $\mathfrak{n}$, and let $r \in \mathbb{N}_0$. Any $y \in B$ admits a representation

$$y = \sum_{\substack{(i_1, \ldots, i_h) \in \mathbb{N}_0^h \\ i_1 + \cdots + i_h < r}} \gamma_{i_1, \ldots, i_h} y_1^{i_1} \cdots y_h^{i_h} + z_r \tag{*}$$

where $z_r \in \mathfrak{n}^r$ and $\gamma_{i_1, \ldots, i_h} \in k$ for all $(i_1, \ldots, i_h) \in \mathbb{N}_0^h$ such that $i_1 + \cdots + i_h < r$; from this it follows easily that *the k-vector space $B / \mathfrak{n}^r$ is generated by the images of the elements* $y_1^{i_1} \cdots y_h^{i_h}$ where $(i_1, \ldots, i_h) \in \mathbb{N}_0^h$ and $0 \le i_1 + \cdots + i_h < r$.

(4.2) **Proposition:** *Let A, B be local k-algebras having k as residue field, let $\mathfrak{m}$ be the maximal ideal of A and $\mathfrak{n}$ be the maximal ideal of B. Let $\varphi \colon A \to B$ be a k-algebra homomorphism. Then φ is a local homomorphism, i.e., we have $\varphi(\mathfrak{m})B = \mathfrak{m}B \subset \mathfrak{n}$. Moreover, let $\{x_1, \ldots, x_n\}$ be a system of generators of $\mathfrak{m}$; then φ is determined uniquely by the elements $\varphi(x_1), \ldots, \varphi(x_n)$.*

Proof (1) We have $A = k \, 1_A \oplus \mathfrak{m}$ as a direct sum of k-vector spaces. Let $x \in \mathfrak{m}$; since we also have $B = k \, 1_B \oplus \mathfrak{n}$, we have $\varphi(x) = \gamma + y$ where $\gamma \in k$ and $y \in \mathfrak{n}$, and $\varphi(x - \gamma) = y$. Suppose that $\gamma \ne 0$; then $x - \gamma$ is a unit of A, hence y is a unit of B contradicting the choice of y. Therefore we have $\gamma = 0$.
(2) It is enough to consider the case $\varphi(x_1) = \cdots = \varphi(x_n) = 0$. Let $x \in A$; for every $r \in \mathbb{N}$ we have a representation

$$x = \sum_{\substack{(i_1, \ldots, i_n) \in \mathbb{N}_0^n \\ i_1 + \cdots + i_n < r}} \gamma_{i_1, \ldots, i_n} x_1^{i_1} \cdots x_n^{i_n} + z_r$$

where $z_r \in \mathfrak{m}^r$ and $\gamma_{i_1, \ldots, i_n} \in k$ for all $(i_1, \ldots, i_n) \in \mathbb{N}_0^n$ such that $i_1 + \cdots + i_n < r$, hence $\varphi(x) = \varphi(z_r) \in \mathfrak{n}^r$ [since φ is local]. This implies that $\varphi(x)$ lies in the intersection of all powers of $\mathfrak{n}$, hence we have $\varphi(x) = 0$ by Krull's intersection theorem [cf. [63], Cor. 5.4].

4.2   Morphisms of Formal and Analytic Algebras

(4.3) DEFINITION: A k-algebra is called a *formal k-algebra* if it is a homomorphic image of a ring of formal power series over k.

(4.4) REMARK: (1) Let A be a formal k-algebra; we have $A = k[\![\,X_1, \ldots, X_n\,]\!]/\mathfrak{a}$ for some $n \in \mathbb{N}$ and an ideal $\mathfrak{a}$ of $k[\![\,X_1, \ldots, X_n\,]\!]$. Then A is local, its maximal ideal $\mathfrak{m}$ is the ideal generated by $x_1 := X_1 + \mathfrak{a}, \ldots, x_n := X_n + \mathfrak{a}$, and k is its residue field. It is clear that A is complete in its $\mathfrak{m}$-adic topology, and it is Hausdorff. Moreover, being a homomorphic image of a henselian ring [cf. (3.25)], it is also a henselian ring [cf. (3.24)].

(2) Let A' be a formal k-algebra, and let $x_1, \ldots, x_n$ be elements in the maximal ideal of A'. Then there exists a unique k-algebra homomorphism $\varphi \colon k[\![\,X_1, \ldots, X_n\,]\!] \to A'$ such that $\varphi(X_i) = x_i$ for every $i \in \{1, \ldots, n\}$ [cf. (1.2)]. The image of φ will be denoted by $k[\![\,x_1, \ldots, x_n\,]\!]$. In the situation of (1) we therefore have $A = k[\![\,x_1, \ldots, x_n\,]\!]$.

(3) Let A be a formal k-algebra. Then elements $x_1, \ldots, x_n$ lying in the maximal ideal of A are said to be formally independent over k if the unique continuous k-algebra homomorphism $\varphi \colon k[\![\,X_1, \ldots, X_n\,]\!] \to A$ such that $\varphi(X_i) = x_i$ for every $i \in \{1, \ldots, n\}$ is injective.

(4.5) EXAMPLE: Let $R = k[\,X_1, \ldots, X_n\,]$ be a polynomial ring over k, and let $\mathfrak{m}$ be the maximal ideal of R generated by $X_1, \ldots, X_n$. Then $\widehat{R} := k[\![\,X_1, \ldots, X_n\,]\!]$ is the completion of R with respect to the $\mathfrak{m}$-adic topology, and also the completion of $R_{\mathfrak{m}}$ with respect to its $\mathfrak{m}R_{\mathfrak{m}}$-adic topology.

Let $F_1, \ldots, F_m$ be elements in $\mathfrak{m}$, set $S := k[\,F_1, \ldots, F_m\,]$, and let $\mathfrak{n}$ be the ideal of S generated by $F_1, \ldots, F_m$. We endow S with the $\mathfrak{n}$-adic topology. Let $\varphi \colon k[\,T_1, \ldots, T_m\,] \to S$ be the surjective k-algebra homomorphism with $\varphi(T_i) = F_i$ for every $i \in \{1, \ldots, m\}$. Then φ admits a unique continuous extension $\widehat{\varphi} \colon k[\![\,T_1, \ldots, T_m\,]\!] \to \widehat{S}$ which is surjective, hence $\widehat{S} = k[\![\,F_1, \ldots, F_m\,]\!]$. On the other hand, the inclusion map $S \hookrightarrow R$ is a uniformly continuous map between metric spaces since $\mathfrak{n}^s \subset \mathfrak{m}^s$ for every $s \in \mathbb{N}_0$, hence has a unique continuous extension $\widehat{S} \to \widehat{R}$ which is injective, and therefore we can consider $k[\![\,F_1, \ldots, F_m\,]\!]$ as a subring of $k[\![\,X_1, \ldots, X_n\,]\!]$. Clearly $k[\![\,F_1, \ldots, F_m\,]\!]$ can also be considered as the closure of the subring $k[\,F_1, \ldots, F_m\,]$ of R in $\widehat{R}$.

We endow $S_{\mathfrak{n}}$ with the $\mathfrak{n}S_{\mathfrak{n}}$-adic topology. The completion of $S_{\mathfrak{n}}$ can be identified with $\widehat{S}$. Since $\mathfrak{m} \cap S = \mathfrak{n}$, we have an injective k-algebra homomorphism $S_{\mathfrak{n}} \hookrightarrow R_{\mathfrak{m}}$ which induces the inclusion $k[\![\,F_1, \ldots, F_m\,]\!] \subset k[\![\,X_1, \ldots, X_n\,]\!]$.

(4.6) DEFINITION: A k-algebra is called an analytic k-algebra if it is a homomorphic image of a ring of convergent power series over k [where now $k = \mathbb{C}$].

(4.7) REMARK: Let A be an analytic k-algebra; we have $A = k[\langle X_1, \ldots, X_n \rangle]/\mathfrak{a}$ for some $n \in \mathbb{N}$ and an ideal $\mathfrak{a}$ of $k[\langle X_1, \ldots, X_n \rangle]$. Then A is local, its maximal ideal $\mathfrak{m}$ is the ideal generated by $X_1 + \mathfrak{a}, \ldots, X_n + \mathfrak{a}$, k is its residue field, and the canonical map $k[\langle X_1, \ldots, X_n \rangle] \to A$ is continuous with respect to the natural topologies of these local rings. It is clear that A is Hausdorff in its $\mathfrak{m}$-adic topology. Moreover, A is a henselian ring [cf. (3.24) and (3.25)].

(4.8) Proposition: *Let A be an analytic k-algebra, and let $x_1, \ldots, x_n$ be elements in the maximal ideal of A. Then there exists a unique continuous k-algebra homomorphism $\varphi: k[\langle X_1, \ldots, X_n \rangle] \to A$ such that $\varphi(X_i) = x_i$ for every $i \in \{1, \ldots, n\}$.*

Proof [Existence]: Let $\psi: k[\langle U_1, \ldots, U_m \rangle] \to A$ be a continuous surjective k-algebra homomorphism, and choose non-units $G_1, \ldots, G_n \in k[\langle U_1, \ldots, U_m \rangle]$ such that $\psi(G_j) = x_j$ for every $j \in \{1, \ldots, n\}$. Let $\omega: k[\langle T_1, \ldots, T_n \rangle] \to k[\langle U_1, \ldots, U_m \rangle]$ be defined by $\omega(T_j) = G_j$ for every $j \in \{1, \ldots, n\}$ [cf. (2.4)(6)]; clearly ω is continuous. Then $\varphi := \psi \circ \omega$ has the asserted property.
[Uniqueness] Note that $k[X_1, \ldots, X_n]$ lies dense in $k[\langle X_1, \ldots, X_n \rangle]$ [cf. (2.4)(5)].

(4.9) Notation: In the situation of (4.8) we write $\operatorname{im}(\varphi) = k[\langle x_1, \ldots, x_n \rangle]$.

(4.10) Notation: To avoid repetitions, if A is a formal or analytic k-algebra, and if $x_1, \ldots, x_n$ are elements of the maximal ideal of A, then the image of the k-algebra homomorphism $\varphi: k[\{X_1, \ldots, X_n\}] \to A$ with $\varphi(X_1) = x_1, \ldots, \varphi(X_n) = x_n$ [cf. (4.4)(2) and (4.8)] will be denoted by $k[\{x_1, \ldots, x_n\}]$.

(4.11) Notation: Let A be an analytic k-algebra. Then elements $x_1, \ldots, x_n$ lying in the maximal ideal of A are said to be analytically independent over k if the unique continuous k-algebra homomorphism $\varphi: k[\langle X_1, \ldots, X_n \rangle] \to A$ such that $\varphi(X_i) = x_i$ for every $i \in \{1, \ldots, n\}$ [cf. (4.8)] is injective.

(4.12) Notation: We denote by C_f the category of formal and by C_a the category of analytic k-algebras, the morphisms being k-algebra homomorphisms [they are local by (4.2)]. If there is no need to distinguish between these two categories, we just write C.

(4.13) Proposition: *Let $n \in \mathbb{N}_0$, $h \in \mathbb{N}$ and $R_{n+h} = k[\{X_1, \ldots, X_n, Y_1, \ldots, Y_h\}]$. Let M be an R_{n+h}-module which, when considered as an $k[\{X_1, \ldots, X_n\}]$-module, is quasifinite. If M is a finitely generated R_{n+h}-module, then M even is a finitely generated $k[\{X_1, \ldots, X_n\}]$-module.*

Proof [by recursion on h]: (1) First, we consider the case $h = 1$. Let $\mathfrak{n}$ be the maximal ideal of $R_n := k[\{X_1, \ldots, X_n\}]$. Now $M/\mathfrak{n}M$ is, by assumption, a finite-dimensional k-vector space; by B(6.1) with $A := k$, $\mathfrak{a} := k$, $B := R_{n+1}$, $\mathfrak{b} := Y_1$ and $M = N := M/\mathfrak{n}M$, there exist $p \in \mathbb{N}$ and $\gamma_1, \ldots, \gamma_p \in k$ such that $F := Y_1^p + \gamma_1 Y_1^{p-1} + \cdots + \gamma_p \in \operatorname{Ann}_{R_{n+1}}(M/\mathfrak{n}M)$, i.e., that $FM \subset \mathfrak{n}M$. Now M is a finitely generated R_{n+1}-module; again, by B(6.1) with $A = B := R_{n+1}$, $\mathfrak{a} := \mathfrak{n}R_{n+1}$, $\mathfrak{b} := F$ and $M = N := M$, there exist $q \in \mathbb{N}$ and $A_1, \ldots, A_q \in \mathfrak{n}$ such that $G := F^q + A_1 F^{q-1} + \cdots + A_q \in \operatorname{Ann}_{R_{n+1}}(M)$. The power series G is Y_1-regular since $G(0, \ldots, 0, Y_1) = F^q(0, \ldots, 0, Y_1) \neq 0$. Again M is a finitely generated $R_{n+1}/(G)$-module, and the ring $R_{n+1}/(G)$ is a finitely generated R_n-module [cf. (3.14)], hence M is a finitely generated R_n-module [cf. [63], Ex. 4.1].

(2) Let $h > 1$ and $l \in \{2, \ldots, h\}$; we assume that the assertion of the proposition holds true for finitely generated R_{n+l-1}-modules which are quasifinite when considered as R_n-modules. Let M be a finitely generated R_{n+l}-module which is quasifinite when considered as R_n-module. Then M is, a fortiori, a quasifinite R_{n+l-1}-module, hence, by (1), a finitely generated R_{n+l-1}-module, and therefore a finitely generated R_n-module, by our assumption.

Now we can prove the assertion by recursion.

(4.14) Theorem: *A quasifinite morphism in C is finite.*

Proof: Let $A, B \in C$, and let $\varphi \colon A \to B$ be a quasifinite k-algebra homomorphism. We choose a surjective k-algebra homomorphism $\alpha \colon k[\{X_1, \ldots, X_m\}] \to A$. Since φ is finite iff $\alpha \circ \varphi$ is finite, we may assume that $A = k[\{X_1, \ldots, X_m\}]$. We choose a surjective k-algebra homomorphism $\beta \colon k[\{Y_1, \ldots, Y_n\}] \to B$, and we define a k-algebra homomorphism $\psi \colon k[\{X_1, \ldots, X_m, Y_1, \ldots, Y_n\}] \to B$ by $\psi(X_i) := \varphi(X_i)$ for every $i \in \{1, \ldots, m\}$ and $\psi(Y_j) := \beta(Y_j)$ for every $j \in \{1, \ldots, n\}$ [cf. (4.4)(2) and (4.8), and note that φ is a local homomorphism]. Then ψ is surjective [since β is surjective], hence it is finite, and since φ is quasifinite, φ is finite by (4.13).

(4.15) Corollary: *Let $\varphi \colon A \to B$ be a morphism in C, let $\mathfrak{m}$ be the maximal ideal of A, and let $y_1, \ldots, y_h \in B$. Then we have $B = Ay_1 + \cdots + Ay_h$ iff the images $\overline{y}_1, \ldots, \overline{y}_h$ in $B/\mathfrak{m}B$ of the elements $y_1, \ldots, y_h$ generate the k-vector space $B/\mathfrak{m}B$.*

Proof: (1) We assume that $\overline{y}_1, \ldots, \overline{y}_h$ generate the k-vector space $B/\mathfrak{m}B$. Then φ is quasifinite, hence it is finite by (4.14). Since we have $B = Ay_1 + \cdots + Ay_h + \mathfrak{m}B$, we get $B = Ay_1 + \cdots + Ay_h$ by Nakayama's lemma [cf. [63], Cor. 4.8].

(2) The other assertion is trivially true.

(4.16) Corollary: *Let $\varphi \colon A \to B$ be a morphism in C, let $\mathfrak{m}$ be the maximal ideal of A, let $\mathfrak{n}$ be the maximal ideal of B, and assume that we have $\mathfrak{n}^r \subset \mathfrak{m}B$ for some $r \in \mathbb{N}$. Let $\{y_1, \ldots, y_h\}$ be a system of generators of the ideal $\mathfrak{n}$; then the set $\{y_1^{i_1} \cdots y_h^{i_h} \mid (i_1, \ldots, i_h) \in \mathbb{N}_0^h$ and $0 \leq i_1 + \cdots + i_h < r\}$ is a system of generators of the A-module B. In particular, B is a finitely generated A-module, and we have $B = A[y_1, \ldots, y_h]$.*

Proof: The images in $B/\mathfrak{n}^r$ of the elements $y_1^{i_1} \cdots y_h^{i_h}$ where $(i_1, \ldots, i_h) \in \mathbb{N}_0^h$ and $0 \leq i_1 + \cdots + i_h < r$, generate the k-vector space $B/\mathfrak{n}^r$ [cf. (4.1)], hence their images in $B/\mathfrak{m}B$ generate the k-vector space $B/\mathfrak{m}B$. The assertion follows from (4.15).

(4.17) Proposition: *Let $\varphi \colon A \to B$ be a morphism in C, let $\mathfrak{m}$ be the maximal ideal of A and let $\mathfrak{n}$ be the maximal ideal of B. If we have $\mathfrak{m}B = \mathfrak{n}$, then φ is surjective.*

Proof: This follows from (4.16) with $r = 1$.

(4.18) REMARK: Let $A \in C$, set $n := \mathrm{emdim}(A)$, and let $\{x_1, \ldots, x_n\}$ be a system of generators of the maximal ideal of A. We set $R_n := k[\{X_1, \ldots, X_n\}]$; let $\varphi: R_n \to A$ be the k-algebra homomorphism defined by $\varphi(X_i) = x_i$ for every $i \in \{1, \ldots, n\}$ [cf. (4.4)(2) and (4.8)]. Then φ is surjective by (4.17), hence we have $A = k[\{x_1, \ldots, x_n\}]$.

Now we assume, in addition, that A is regular, hence that $\dim(A) = n$. Then φ is injective, since we have $\dim(R_n) = n$ [cf. (3.20)]. Thus, any minimal system of generators of the maximal ideal of a regular $A \in C$ is formally resp. analytically independent over k. This implies the following result:

(4.19) Corollary: Let $A \in C$ and set $n := \dim(A)$. Then A is a regular local ring iff $A \cong k[\{X_1, \ldots, X_n\}]$.

(4.20) REMARK: (1) Let $\varphi: A \to B$ be a morphism in C, let $\mathfrak{m}$ be the maximal ideal of A and let $\mathfrak{n}$ be the maximal ideal of B. Then φ induces a linear map of k-vector spaces $\tau_\varphi: \mathfrak{m}/\mathfrak{m}^2 \to \mathfrak{n}/\mathfrak{n}^2$. If $A = B$, and if φ is the identity map of A, then τ_φ is the identity map of $\mathfrak{m}/\mathfrak{m}^2$. Let $\psi: B \to C$ be a morphism in C; then we have $\tau_{\psi \circ \varphi} = \tau_\psi \circ \tau_\varphi$. Thus, the correspondence $A \mapsto \mathfrak{m}/\mathfrak{m}^2$ where $A \in C$ and $\mathfrak{m}$ is the maximal ideal of A, is a covariant functor from the category C to the category of finite-dimensional k-vector spaces.

(2) Let, in particular, $A = k[\{X_1, \ldots, X_m\}]$, $B = k[\{Y_1, \ldots, Y_n\}]$; then $AX_1 + \cdots + AX_m$ is the maximal ideal $\mathfrak{m}$ of A, and $BY_1 + \cdots + BY_n$ is the maximal ideal $\mathfrak{n}$ of B. Now $\mathfrak{m}/\mathfrak{m}^2$ is an m-dimensional k-vector space having $\{x_1 := X_1 + \mathfrak{m}^2, \ldots, x_m := X_m + \mathfrak{m}^2\}$ as a basis, and $\mathfrak{n}/\mathfrak{n}^2$ is an n-dimensional k-vector space having $\{y_1 := Y_1 + \mathfrak{n}^2, \ldots, y_n := Y_n + \mathfrak{n}^2\}$ as a basis. Let $\varphi: A \to B$ be a k-algebra homomorphism; the elements $F^{(1)} := \varphi(X_1), \ldots, F^{(m)} := \varphi(X_m)$ lie in $\mathfrak{n}$, and if we write $F_1^{(i)} = \sum_{j=1}^{n} \gamma_{ij} Y_j$ for every $i \in \{1, \ldots, m\}$, i.e., we have $J_\varphi = (\gamma_{ij})$, then we have

$$\tau_\varphi(x_i) = \sum_{j=1}^{n} \gamma_{ij} y_j \quad \text{for every } i \in \{1, \ldots, m\},$$

i.e., the matrix of τ_φ with respect to the bases of $\mathfrak{m}/\mathfrak{m}^2$ and $\mathfrak{n}/\mathfrak{n}^2$ given above is the Jacobian matrix $J_\varphi \in M(m, n; k)$ of φ. For any $x = \xi_1 x_1 + \cdots + \xi_m x_m \in \mathfrak{m}/\mathfrak{m}^2$ where $\xi_1, \ldots, \xi_m \in k$ we have $\tau_\varphi(x) = \eta_1 y_1 + \cdots + \eta_n y_n$ with $(\eta_1, \ldots, \eta_n) = (\xi_1, \ldots, \xi_m) J_\varphi$.

(4.21) Corollary: Let $\varphi: A \to B$ be a morphism in C. If τ_φ is surjective, then φ is surjective.

Proof: Let $\mathfrak{m}$ be the maximal ideal of A, and let $\mathfrak{n}$ be the maximal ideal of B. Since τ_φ is surjective, we have $\mathfrak{n} = \mathfrak{m}B + \mathfrak{n}^2$, hence we have $\mathfrak{n} = \mathfrak{m}B$ by Nakayama's lemma [cf. [63], Cor. 4.8], and φ is surjective by (4.17).

(4.22) Theorem: Let $A \in C$, and let $\varphi: A \to k[\{Y_1, \ldots, Y_n\}]$ be a morphism in C. If τ_φ is bijective, then φ is an isomorphism of k-algebras.

Proof: Let $\mathfrak{m}$ be the maximal ideal of A, and let $\mathfrak{n}$ be the maximal ideal of $k[\{Y_1,\ldots,Y_n\}]$. The homomorphism φ is surjective by (4.21). Therefore there exist elements $x_1,\ldots,x_n \in \mathfrak{m}$ with $\varphi(x_i) = Y_i$ for every $i \in \{1,\ldots,n\}$. Now we have $\tau_\varphi(x_i + \mathfrak{m}^2) = Y_i + \mathfrak{n}^2$ for every $i \in \{1,\ldots,n\}$; since τ_φ is an isomorphism of k-vector spaces and since $\{Y_1 + \mathfrak{n}^2,\ldots,Y_n + \mathfrak{n}^2\}$ is a basis of the k-vector space $\mathfrak{n}/\mathfrak{n}^2$, the set $\{x_1 + \mathfrak{m}^2,\ldots,x_n + \mathfrak{m}^2\}$ is a basis of the k-vector space $\mathfrak{m}/\mathfrak{m}^2$, and therefore $\{x_1,\ldots,x_n\}$ is a minimal system of generators of $\mathfrak{m}$. Let $\psi\colon k[\{Y_1,\ldots,Y_n\}] \to A$ be the morphism defined by $\psi(Y_i) = x_i$ for every $i \in \{1,\ldots,n\}$ [cf. (4.4) and (4.8)]. Then $\psi \circ \varphi \colon A \to A$ is a morphism with $\psi(\varphi(x_i)) = x_i$ for every $i \in \{1,\ldots,n\}$, hence $\psi \circ \varphi$ is the identity on A [cf. (4.2)], and therefore φ is injective.

(4.23) REMARK: Assume, in addition, that under the assumptions of (4.22) the ring A is also a ring of power series. Then the result of (4.22) is the Jacobi inversion theorem: *A morphism $\varphi\colon k[\{X_1,\ldots,X_m\}] \to k[\{Y_1,\ldots,Y_n\}]$ is an isomorphism iff we have $m = n$ and the Jacobian matrix J_φ has rank equal to n.*

4.3 Integral Extensions

(4.24) Theorem: *Let $A \in C$, and let B be an overring of A which is integral over A. If B is a domain, then B is quasilocal.*

Proof: Let $\mathfrak{m}$ be the maximal ideal of A. Since B is integral over A, there exist maximal ideals $\mathfrak{n}$ of B such that $\mathfrak{n} \cap A = \mathfrak{m}$ [cf. [63], Prop. 4.15 and Cor. 4.17]. Let $\mathfrak{n} \subset B$ be one of them; we shall show that every non-unit of B is contained in $\mathfrak{n}$, hence that $\mathfrak{n}$ is the only maximal ideal of B.
Thus, let $b \in B$ be a non-unit and let $F = Z^m + a_1 Z^{m-1} + \cdots + a_m \in A[Z]$ be an equation of integral dependence of b over A of minimal degree. From

$$a_m = -b(a_{m-1} + a_{m-2}b + \cdots + a_1 b^{m-2} + b^{m-1})$$

we conclude that a_m is a non-unit of B, hence that a_m is a non-unit of A, and that therefore $a_m \in \mathfrak{m}$. If $m = 1$, then we are done. Now we consider the case $m \geq 2$, and we suppose that not all the elements $a_{m-1},\ldots,a_1$ lie in $\mathfrak{m}$; we choose $l \in \{0,\ldots,m-2\}$ such that $a_m,\ldots,a_{m-l} \in \mathfrak{m}$, $a_{m-l-1} \notin \mathfrak{m}$. Let $\omega\colon A \to k$ be the canonical k-algebra homomorphism. Then we get

$$\omega(F)(Z) = Z^m + \omega(a_1)Z^{m-1} + \cdots + \omega(a_{m-l-1})Z^{l+1} = Z^{l+1}\widetilde{G}$$

where $\widetilde{G} \in k[Z]$ and $Z \nmid \widetilde{G}$ [because $\omega(a_{m-l-1}) \neq 0$]. Since A is henselian [cf. (4.4) and (4.7)], there exist monic polynomials $G_1, G_2 \in A[Z]$ with $\deg_Z(G_1) = l + 1$, $\deg_Z(G_2) = m - l - 1$, and such that $F = G_1 G_2$. Now we have $0 = F(b) = G_1(b)G_2(b)$, and this implies that either $G_1(b) = 0$ or $G_2(b) = 0$ [because B is a domain]. Thus, there exist equations of integral dependence of b over A of degree less than m which is in contradiction with the choice of m. Therefore we have $a_1,\ldots,a_m \in \mathfrak{m}$, hence $b^m \in \mathfrak{m}B \subset \mathfrak{n}$, and therefore $b \in \mathfrak{n}$ [because $\mathfrak{n}$ is a prime ideal].

(4.25) Proposition: *Let $A \in C$, let $\mathfrak{m}$ be the maximal ideal of A, let B be a quasilocal ring with maximal ideal $\mathfrak{n}$, and let $A \to B$ be a homomorphism. If $b \in \mathfrak{n}$ is integral over A, and if $F = Z^m + a_1 Z^{m-1} + \cdots + a_m \in A[Z]$ is a polynomial of minimal degree such that $F(b) = 0$, then the elements $a_1, \ldots, a_m$ lie in $\mathfrak{m}$, i.e., F is a Weierstraß polynomial.*

Proof: Suppose that not all of $a_1, \ldots, a_m$ lie in $\mathfrak{m}$. Just as in the proof of (4.24) we find a factorization $F = G_1 G_2$ where $G_1, G_2 \in A[Z]$ are monic of smaller degree than F and where the constant term a of G_2 is a unit of A [in the notation of the proof of (4.24) we have $\omega(G_2) = \widetilde{G}$]. Then we have $G_2(b) - a 1_B \in \mathfrak{n}$ [since $b \in \mathfrak{n}$], and therefore $G_2(b)$ is a unit of B [since B is quasilocal], hence we obtain $G_1(b) = 0$. Thus, there exist equations of integral dependence of b over A of degree less than m; this is in contradiction with the choice of m.

(4.26) Corollary: *Let $A \to B$ be a morphism in C, let $\mathfrak{m}$ be the maximal ideal of A and let $\mathfrak{n}$ be the maximal ideal of B. If the ideal $\mathfrak{n}$ can be generated by elements $y_1, \ldots, y_h$ which are integral over A, then we have $B = A[y_1, \ldots, y_h]$, hence B is a finitely generated A-module.*

Proof: By (4.25) there exists $l \in \mathbb{N}$ such that $y_i^l \in \mathfrak{m}B$ for every $i \in \{1, \ldots, h\}$. Then we have $\mathfrak{n}^{lh} \subset \mathfrak{m}B$, whence $B = A[y_1, \ldots, y_h]$ by (4.16), and therefore B is a finitely generated A-module [cf. [63], Cor. 4.5].

4.4 Noether Normalization

(4.27) Proposition: *Let $A \in C$ and set $m := \dim(A)$. Then there exists an injective finite k-algebra homomorphism $k[\{Y_1, \ldots, Y_m\}] \to A$.*

Proof: (1) The set L of non-negative integers l such that there exists a finite k-algebra homomorphism $k[\{Y_1, \ldots, Y_l\}] \to A$ is not empty. Let l be the smallest integer in L, and let $\varphi \colon k[\{Y_1, \ldots, Y_l\}] \to A$ be a finite k-algebra homomorphism. Suppose that φ is not injective; then we have $l \geq 1$. We choose a non-zero element $F \in k[\{Y_1, \ldots, Y_l\}]$ in the kernel of φ. By (3.15) we may assume that F is Y_l-regular. By (3.14) the canonical k-algebra homomorphism $\psi \colon k[\{Y_1, \ldots, Y_{l-1}\}] \to k[\{Y_1, \ldots, Y_l\}]/(F)$ is finite. Now the induced k-algebra homomorphism $\overline{\varphi} \colon k[\{Y_1, \ldots, Y_l\}]/(F) \to A$ is finite, and therefore the composition $\overline{\varphi} \circ \psi \colon k[\{Y_1, \ldots, Y_{l-1}\}] \to A$ is a finite k-algebra homomorphism. But this is in contradiction with the choice of l.
(2) By (1) the k-algebra $R := k[\{Y_1, \ldots, Y_l\}]$ can be considered as a subalgebra of A, and A is integral over R [cf. [63], Cor. 4.5]. Since we have $\dim(R) = l$ by (3.14), we have $\dim(A) = l$ [cf. [63], Prop. 9.2], hence we obtain $m = l$.

(4.28) NOETHER NORMALIZATION: In this remark we give a more constructive version of proposition (4.27).
(1) Let $R_n = k[\{X_1, \ldots, X_n\}]$, let $\mathfrak{a}$ be a proper non-zero ideal of R_n, and set $m := \dim(R_n/\mathfrak{a})$.

(a) We set $Y_i^{(0)} := X_i$ for every $i \in \{1, \ldots, n\}$ and $\mathfrak{b}^{(0)} := \mathfrak{a}$. Let $F_n \in \mathfrak{a}$ be a non-zero element. We choose elements $\gamma_1^{(0)}, \ldots, \gamma_{n-1}^{(0)} \in k$ such that, setting

$$Y_i^{(1)} := Y_i^{(0)} + \gamma_i^{(0)} Y_n^{(0)} \quad \text{for every } i \in \{1, \ldots, n-1\}, \quad Y_n^{(1)} := Y_n^{(0)},$$

the linear change of variables $\varphi^{(1)} : R_n \to R_n$ defined by $\varphi^{(1)}(Y_i^{(0)}) = Y_i^{(1)}$ for every $i \in \{1, \ldots, n\}$ has the property that $\varphi^{(1)}(F_n)$ is $Y_n^{(1)}$-regular of order equal to its own order [cf. (3.15)]. Let $G_n^{(1)} \in k[\{Y_1^{(1)}, \ldots, Y_{n-1}^{(1)}\}][Y_n^{(1)}]$ be the Weierstraß polynomial associated with $\varphi^{(1)}(F_n)$; its degree is equal to its order. We set $\mathfrak{b}^{(1)} := \varphi^{(1)}(\mathfrak{b}^{(0)})$; note that $G_n^{(1)} \in \mathfrak{b}^{(1)}$.

(b) We assume that, for some $p \in \{1, \ldots, n-1\}$ and for every $q \in \{1, \ldots, p\}$, we have constructed

(i) elements $Y_1^{(q)}, \ldots, Y_n^{(q)} \in k[X_1, \ldots, X_n]$ which are linearly independent homogeneous polynomials of degree 1; let $\varphi^{(q)} : R_n \to R_n$ be the linear change of variables defined by $\varphi^{(q)}(Y_i^{(q-1)}) = Y_i^{(q)}$ for every $i \in \{1, \ldots, n\}$ [cf. (4.23)],

(ii) ideals $\mathfrak{b}^{(q)} \in k[\{Y_1^{(q)}, \ldots, Y_n^{(q)}\}]$ such that $\varphi^{(q)}(\mathfrak{b}^{(q-1)}) = \mathfrak{b}^{(q)}$,

(iii) Weierstraß polynomials $G_i^{(q)} \in k[\{Y_1^{(q)}, \ldots, Y_{i-1}^{(q)}\}][Y_i^{(q)}]$ such that the degree of $G_i^{(q)}$ is equal to its order and that $G_i^{(q)} \in \mathfrak{b}^{(q)}$ for $i \in \{n-q+1, \ldots, n\}$.

If $\mathfrak{b}^{(p)} \cap k[\{Y_1^{(p)}, \ldots, Y_{n-p}^{(p)}\}] = \{0\}$, then we have $\dim(k[\{Y_1^{(p)}, \ldots, Y_n^{(p)}\}]/\mathfrak{b}^{(p)}) = n - p$.

In fact, we set $Y_i := Y_i^{(p)}$ for every $i \in \{1, \ldots, n\}$ and $\mathfrak{b} := \mathfrak{b}^{(p)}$, and we let $y_1, \ldots, y_n$ be the images of $Y_1, \ldots, Y_n$ in $k[\{Y_1, \ldots, Y_n\}]/\mathfrak{b}$. Obviously, the elements $y_1, \ldots, y_{n-p}$ are formally resp. analytically independent over k, y_{n-p+1} is integral over the power series ring $k[\{y_1, \ldots, y_{n-p}\}]$, and we have $k[\{y_1, \ldots, y_{n-p+1}\}] = k[\{y_1, \ldots, y_{n-p}\}][y_{n-p+1}]$ [cf. (4.26)]. By recursion, it is clear that, for $i \in \{n-p+1, \ldots, n\}$, y_i is integral over $k[\{y_1, \ldots, y_{i-1}\}] = k[\{y_1, \ldots, y_{n-p}\}][y_{n-p+1}, \ldots, y_{i-1}]$, hence the elements $y_{n-p+1}, \ldots, y_n$ are integral over $k[\{y_1, \ldots, y_{n-p}\}]$ [cf. [63], Cor. 4.5], and therefore we have $k[\{y_1, \ldots, y_n\}] = k[\{y_1, \ldots, y_{n-p}\}][y_{n-p+1}, \ldots, y_n]$ [cf. (4.26)].

If $\mathfrak{b}^{(p)} \cap k[\{Y_1^{(p)}, \ldots, Y_{n-p}^{(p)}\}] \neq \{0\}$, then we choose a non-zero F in this intersection. If $n - p \geq 2$, then we choose elements $\gamma_1^{(p)}, \ldots, \gamma_{n-p-1}^{(p)} \in k$ such that, setting

$$Y_i^{(p+1)} = \begin{cases} Y_i^{(p)} + \gamma_i^{(p)} Y_{n-p}^{(p)} & \text{for } i \in \{1, \ldots, n-p-1\}, \\ Y_i^{(p)} & \text{for } i \in \{n-p, \ldots, n\}, \end{cases}$$

the linear change of variables $\varphi^{(p+1)} : R_n \to R_n$ defined by $\varphi^{(p+1)}(Y_i^{(p)}) = Y_i^{(p+1)}$ for every $i \in \{1, \ldots, n\}$, has the property that $\varphi^{(p+1)}(F) \in k[\{Y_1^{(p+1)}, \ldots, Y_{n-p}^{(p+1)}\}]$ is $Y_{n-p}^{(p+1)}$-regular of order equal to its own order.

Let $G_{n-p}^{(p+1)} \in k[\{Y_1^{(p+1)}, \ldots, Y_{n-p-1}^{(p+1)}\}][Y_{n-p}^{(p+1)}]$ be the Weierstraß polynomial associated with $\varphi^{(p+1)}(F)$; the degree of $G_{n-p}^{(p+1)}$ is equal to its order. We set $G_i^{(p+1)} := \varphi^{(p+1)}(G_i^{(p)})$ for every $i \in \{n-p+1, \ldots, n\}$ and $\mathfrak{b}^{(p+1)} := \varphi^{(p+1)}(\mathfrak{b}^{(p)})$.

Then, for every $i \in \{n-p,\ldots,n\}$, $G_i^{(p+1)} \in k[\{Y_1^{(p+1)},\ldots,Y_{i-1}^{(p+1)}\}][Y_i^{(p+1)}]$ is a Weierstraß polynomial of degree equal to its order, and we have $G_i^{(p+1)} \in \mathfrak{b}^{(p+1)}$. If $n-p=1$, then $F \in k[\{Y_1^{(p)}\}]$ is a power series which is $Y_1^{(p)}$-regular, and there is no need to make a change of variables.

Since $k[\{Y_1^{(p+1)},\ldots,Y_n^{(p+1)}\}]/\mathfrak{b}^{(p+1)}$ and $k[\{X_1,\ldots,X_n\}]/\mathfrak{a}$ are isomorphic k-algebras, we find after $n-m$ steps that $k[\{Y_1^{(n-m)},\ldots,Y_m^{(n-m)}\}] \cap \mathfrak{b}^{(n-m)} = \{0\}$.

(2) Let A be a formal resp. analytic k-algebra, and set $m := \dim(A)$, $n := \operatorname{emdim}(A)$. Then, using (1), we may write $A = k[\{X_1,\ldots,X_n\}]/\mathfrak{a}$ with $\mathfrak{a} \cap k[\{X_1,\ldots,X_m\}] = \{0\}$ and, for $i \in \{m+1,\ldots,n\}$, there exists a Weierstraß polynomial $F_i \in k[\{X_1,\ldots,X_{i-1}\}][X_i] \cap \mathfrak{a}$ which has degree equal to its order. Let $x_1,\ldots,x_n$ be the images of $X_1,\ldots,X_n$ in A. Then the elements $x_1,\ldots,x_m$ are formally resp. analytically independent over k, and we have $k[\{x_1,\ldots,x_i\}] = k[\{x_1,\ldots,x_m\}][x_{m+1},\ldots,x_i]$ for every $i \in \{m+1,\ldots,n\}$. Note that the polynomial $F_i(x_1,\ldots,x_{i-1},Z)$ is an equation of integral dependence for x_i over $k[\{x_1,\ldots,x_m\}][x_{m+1},\ldots,x_{i-1}]$. Moreover, we have $A = k[\{x_1,\ldots,x_m\}][x_{m+1},\ldots,x_n]$.

(4.29) Proposition: *Let $A \in C$, and set $m := \dim(A)$, $n := \operatorname{emdim}(A)$. Then there exists a system $\{x_1,\ldots,x_n\}$ of generators of the maximal ideal $\mathfrak{m}$ of A such that, setting $\overline{x}_i := x_i + \mathfrak{m}^2 \in \operatorname{gr}_{\mathfrak{m}}(A)$ for every $i \in \{1,\ldots,n\}$, we have:*
(1) the elements $x_1,\ldots,x_m$ are formally resp. analytically independent over k, and the elements $x_{m+1},\ldots,x_n$ are integral over $k[\{x_1,\ldots,x_m\}]$,
(2) the elements $\overline{x}_1,\ldots,\overline{x}_m$ are algebraically independent over k, and the elements $\overline{x}_{m+1},\ldots,\overline{x}_n$ are integral over $k[\overline{x}_1,\ldots,\overline{x}_m]$.
In particular, we have

$$A = k[\{x_1,\ldots,x_m\}][x_{m+1},\ldots,x_n], \quad \operatorname{gr}_{\mathfrak{m}}(A) = k[\overline{x}_1,\ldots,\overline{x}_m][\overline{x}_{m+1},\ldots,\overline{x}_n].$$

Proof: By (4.28) we may assume that $A = k[\{X_1,\ldots,X_n\}]/\mathfrak{a} = k[\{x_1,\ldots,x_n\}]$ for some ideal $\mathfrak{a} \in k[\{X_1,\ldots,X_n\}]$; here $x_i := X_i + \mathfrak{a}$ for every $i \in \{1,\ldots,n\}$, the elements $x_1,\ldots,x_m$ are formally resp. analytically independent over k and for $j \in \{m+1,\ldots,n\}$ there exists a Weierstraß polynomial $F_j \in k[\{X_1,\ldots,X_{j-1}\}][X_j]$ of degree equal to its order such that $F_j \in \mathfrak{a}$.

Let $\omega \colon k[X_1,\ldots,X_n] \to \operatorname{gr}_{\mathfrak{m}}(A)$ be the homomorphism of graded k-algebras induced by the canonical homomorphism $k[\{X_1,\ldots,X_n\}] \to A$ [here we identify the polynomial ring $k[X_1,\ldots,X_n]$ with the associated graded ring of the local ring $k[\{X_1,\ldots,X_n\}]$; under this identification, the leading form of a power series is its initial form, cf. B(5.3)(3)]. The kernel $\mathfrak{s}$ of ω is generated by all the initial forms of elements in $\mathfrak{a}$ [cf. B(5.3)(3)]; in particular, the leading forms $\operatorname{In}(F_{m+1}),\ldots,\operatorname{In}(F_n)$ lie in $\mathfrak{s}$. Since, for $j \in \{m+1,\ldots,n\}$, we have

$$\operatorname{In}(F_j) = X_j^{q_j} + a_{1j}(X_1,\ldots,X_{j-1})X_j^{q_j-1} + \cdots + a_{q_j j}(X_1,\ldots,X_{j-1})$$

where $q_j \in \mathbb{N}$ and $a_{lj}(X_1,\ldots,X_{j-1}) \in k[X_1,\ldots,X_{j-1}]$ is homogeneous of degree l, $l \in \{1,\ldots,q_j\}$, it is clear that $\operatorname{gr}_{\mathfrak{m}}(A)$ is integral over $k[\overline{x}_1,\ldots,\overline{x}_m]$. We have $m = \dim(A) = \dim(\mathrm{gr}_\mathrm{m}(A))$ [cf. [63], Ex. 13.8] and $\dim(k[\overline{x}_1,\ldots,\overline{x}_m]) = \dim(\mathrm{gr}_\mathrm{m}(A))$ [cf. [63], Prop. 9.2], therefore we have $\dim(k[\overline{x}_1,\ldots,\overline{x}_m]) = m$, whence the elements $\overline{x}_1,\ldots,\overline{x}_m$ are algebraically independent over k.

5   Extensions of Formal and Analytic Algebras

(5.0) For the sake of simplicity, in this section we assume that k is *algebraically closed of characteristic* 0.

(5.1) Proposition: *Let* $R_n = k[\{X_1,\ldots,X_n\}]$, *let* $m \in \mathbb{N}$, *and, for every* $i \in \{1,\ldots,m\}$, *let* $F_i \in R_n[Y_i]$ *be a Weierstraß polynomial. Let* $\mathfrak{a} \subset R_n[Y_1,\ldots,Y_m]$ *be the ideal generated by* $F_1,\ldots,F_m$. *Then* $R_n[Y_1,\ldots,Y_m]/\mathfrak{a}$ *is a formal resp. analytic* k-*algebra, and its maximal ideal is generated by the images of the elements* $X_1,\ldots,X_n, Y_1,\ldots,Y_m$.

Proof: We set $A := R_n[Y_1,\ldots,Y_m]/\mathfrak{a}$; it is clear that $\mathfrak{a} \cap R_n = \{0\}$, hence that R_n can be considered as a subring of A. Let $\alpha\colon R_n[Y_1,\ldots,Y_m] \to A$ be the canonical homomorphism. It will be shown by recursion with respect to m: There exists a surjective k-algebra homomorphism $\varphi\colon k[\{X_1,\ldots,X_n,Y_1,\ldots,Y_m\}] \to A$ such that we have $\varphi(X_i) = X_i$ for every $i \in \{1,\ldots,n\}$ and $\varphi(Y_j) = \alpha(Y_j)$ for every $j \in \{1,\ldots,m\}$.

The case $m = 1$ has been dealt with in (3.14)(2). Let $p \in \{2,\ldots,m\}$, let $\mathfrak{a}' \subset R_n[Y_1,\ldots,Y_{p-1}]$ be the ideal generated by the elements $F_1,\ldots,F_{p-1}$, and set $A' := R_n[Y_1,\ldots,Y_{p-1}]/\mathfrak{a}'$. Furthermore, let $\alpha'\colon R_n[Y_1,\ldots,Y_{p-1}] \to A'$ be the canonical homomorphism; we assume by recursion that there exists a surjective k-algebra homomorphism $\varphi'\colon k[\{X_1,\ldots,X_n,Y_1,\ldots,Y_{p-1}\}] \to A'$ satisfying $\varphi'(X_i) = X_i$ for every $i \in \{1,\ldots,n\}$ and $\varphi'(Y_j) = \alpha'(Y_j)$ for every $j \in \{1,\ldots,p-1\}$. Now let $\beta\colon R_n[Y_1,\ldots,Y_{p-1}][Y_p] \to A'[Y_p]$ be the surjective k-algebra homomorphism which extends α' [i.e., we have $\beta(Y_p) = Y_p$]. We have $\ker(\beta) \subset \mathfrak{a}$; therefore there exists a unique k-algebra homomorphism $\gamma\colon A'[Y_p] \to A$ satisfying $\gamma \circ \beta = \alpha$. Note that γ is surjective. Since we have $\gamma(\beta(F_p)) = 0$, there exists a unique k-algebra homomorphism $\delta\colon A'[Y_p]/(\beta(F_p)) \to A$ satisfying $\delta \circ \pi = \gamma$ where $\pi\colon A'[Y_p] \to A'[Y_p]/(\beta(F_p))$ is the canonical homomorphism. Let $\psi\colon k[\{X_1,\ldots,X_n,Y_1,\ldots,Y_{p-1}\}][Y_p] \to A'[Y_p]$ be the extension of φ' [i.e., we have $\psi(Y_p) = Y_p$]. Clearly we have $(\pi \circ \psi)(F_p) = 0$, hence there exists a unique k-algebra homomorphism $\chi\colon k[\{X_1,\ldots,X_n,Y_1,\ldots,Y_{p-1}\}][Y_p]/(F_p) \to A$ satisfying $\chi \circ \sigma = \delta \circ \pi \circ \psi = \gamma \circ \psi$ where

$$\sigma\colon k[\{X_1,\ldots X_n,Y_1,\ldots,Y_{p-1}\}][Y_p] \to k[\{X_1,\ldots,X_n,Y_1,\ldots,Y_{p-1}\}][Y_p]/(F_p)$$

is the canonical homomorphism; χ is surjective, and we have $\chi \circ \sigma(X_i) = \alpha(X_i) = X_i$ for every $i \in \{1,\ldots,n\}$ and $\chi \circ \sigma(Y_j) = \alpha(Y_j)$ for every $j \in \{1,\ldots,m\}$. By (3.14)(2) there exists a surjective k-algebra homomorphism

$$\omega\colon k[\{X_1,\ldots,X_n,Y_1,\ldots,Y_p\}] \to k[\{X_1,\ldots,X_n,Y_1,\ldots,Y_{p-1}\}][Y_p]/(F_p),$$

and $\varphi := \chi \circ \omega$ does the job. (The reader who wants to follow the above proof is advised to produce some commutative diagrams.)

(5.2) Proposition: *Let* $R_n = k[\{X_1, \ldots, X_n\}]$, *and let* A *be a* k-*algebra which contains* R_n *as a subalgebra such that it is a finitely generated* R_n-*module. Assume that* A *is either quasilocal or a domain. Then* A *is a formal resp. analytic* k-*algebra.*

Proof: If A is a domain, then A is quasilocal [cf. (4.24)]. Hence we may assume, to begin with, that A is a local ring; let $\mathfrak{m}$ be its maximal ideal. Note that $\mathfrak{m} \cap R_n = \mathfrak{n}$, where $\mathfrak{n}$ is the maximal ideal of R_n [cf. [63], Cor. 4.17]. As the field $A/\mathfrak{m}$ is a finite extension of k [since A is a finitely generated R_n-module and $R_n/\mathfrak{n} = k$], we have $A/\mathfrak{m} = k$ [since k is algebraically closed] and therefore we have $A = k 1_A \oplus \mathfrak{m}$ [direct sum of k-vector spaces]. Since A is a finitely generated R_n-module, we can therefore assume that $A = R_n[y_1, \ldots, y_h]$ where $y_1, \ldots, y_h$ are elements of $\mathfrak{m}$. For every $i \in \{1, \ldots, m\}$ there exists a Weierstraß polynomial $F_i \in R_n[Y]$ such that $F_i(y_i) = 0$ [cf. (4.25)]. Now the assertion follows from (5.1).

(5.3) Theorem: *Let* $A \in C$ *be a domain, let* K *be the field of quotients of* A, *let* L *be a finite extension of* K, *and let* B *be the integral closure of* A *in* L. *Then* B *is a finitely generated* A-*module and, moreover,* B *lies in* C.

Proof: We set $n := \dim(A)$; then we can consider A as an overring of a power series ring $R_n = k[\{X_1, \ldots, X_n\}]$ such that A is a finitely generated R_n-module [cf. (4.27)]. Now K is a finite extension of the quotient field $Q(R_n)$ of R_n, hence L is a finite extension of $Q(R_n)$, and B is the integral closure of R_n in L. Note that R_n is integrally closed in $Q(R_n)$ [since R_n is factorial, cf. (3.18)]. Using III(4.5)(2) we see that B is a finitely generated R_n-module, hence B is a formal resp. analytic k-algebra by (5.2) and a priori B is a finitely generated A-module.

(5.4) Proposition: *Let* $A \in C$ *be a one-dimensional domain, let* K *be its field of quotients, let* L *be a finite extension of* K, *and let* B *be the integral closure of* A *in* L. *Then* $B \in C$, B *is a discrete valuation ring which is a finitely generated* A-*module, and we have* $B = k[\{t\}]$, *a ring of power series, where* t *is a generator of the maximal ideal of* B.

Proof: Now B is a formal resp. analytic k-algebra which is a finitely generated A-module [cf. (5.3)], and we have $\dim(B) = 1$ [cf. [63], Prop. 9.2]. Furthermore, B is an integrally closed local domain, hence B is a discrete valuation ring [cf. I(3.29)]. Therefore B is a regular formal resp. analytic k-algebra of dimension 1, and if t is a generator of the maximal ideal of B, it follows from (4.18) that $B = k[\{t\}]$, a ring of power series.

(5.5) Proposition: *Let* A *be an analytic* $\mathbb{C}$-*algebra which is a domain. Then:*
(1) *The completion of* A *is a domain and a formal* $\mathbb{C}$-*algebra; in particular,* A *is analytically irreducible.*
(2) A *is analytically normal.*

Proof: We prove (1) and (2) by induction on $\dim(A)$.

(a) Let B be the integral closure of A in its field of quotients. Then B is a finitely generated A-module and an analytic $\mathbb{C}$-algebra [cf. (5.3)]. Let $\mathfrak{m}$ be the maximal ideal of A and $\mathfrak{n}$ be the maximal ideal of B. The $\mathfrak{m}B$-adic and the $\mathfrak{n}$-adic topology of B are the same [cf. B(3.8)]. The completion $\widehat{A}$ of A and the completion $\widehat{B}$ of B are local rings, $\widehat{A}$ can be considered as a subring of $\widehat{B}$ [cf. [63], Lemma 7.15], and $\widehat{B}$ is a finitely generated $\widehat{A}$-module [cf. [63], Th. 7.2]. It is enough to show that $\widehat{A}$ is a formal $\mathbb{C}$-algebra, and that $\widehat{B}$ is a formal $\mathbb{C}$-algebra which is a domain and which is integrally closed.

(b) We assume that $\dim(A) = 1$. Then A contains a ring $R = \mathbb{C}[\langle x \rangle]$ of convergent power series such that A is a finitely generated R-module [cf. (4.27)]. The completion $\widehat{R}$ of R is a ring $\mathbb{C}[\![x]\!]$ of formal power series in one indeterminate. The quotient field $Q(A)$ of A is a finite extension of the quotient field of R, and B is the integral closure of R in $Q(A)$; by (5.4) we have $B = \mathbb{C}[\langle t \rangle]$, a ring of convergent power series which is a discrete valuation ring, hence we get $\widehat{B} = \mathbb{C}[\![t]\!]$, a ring of formal power series which is an integrally closed domain, and therefore $\widehat{A}$ is a domain. Moreover, $\widehat{A}$ is a finitely generated $\widehat{R}$-module [cf. [63], Th. 7.2], and it is therefore a formal $\mathbb{C}$-algebra by (5.2).

(c) Let $\dim(A) = n > 1$, and assume that the assertions have been proved for every analytic $\mathbb{C}$-algebra which is a domain and has dimension less than n. Let $\mathfrak{p}$ be a prime ideal of B of height one. Then $B/\mathfrak{p}$ is an analytic $\mathbb{C}$-algebra which is a domain and we have $\dim(B/\mathfrak{p}) < n$. By our induction assumption, the completion of $B/\mathfrak{p}$ is a domain. The n-dimensional analytic $\mathbb{C}$-algebra A contains a ring of convergent power series $R_n = \mathbb{C}[\langle x_1, \ldots, x_n \rangle]$ such that A is a finitely generated R_n-module [cf. (4.27)]; then B is a finitely generated R_n-module. Since the completion $\widehat{R}_n$ of R_n is a formal power series ring over $\mathbb{C}$ in n indeterminates [cf. (2.4)(5)], and since such a power series ring is a domain and integrally closed [cf. (3.18) and [63], Prop. 4.10], by III(6.9) $\widehat{B}$ is reduced and integrally closed in its ring of quotients. Since $\widehat{B}$ is local, it is a domain by B(3.5), and therefore $\widehat{A}$ as a subring of $\widehat{B}$ is a domain. Moreover, $\widehat{A}$ is a finitely generated $\widehat{R}_n$-module [cf. [63], Th. 7.2], and it is therefore a formal $\mathbb{C}$-algebra by (5.2).

Chapter V

Quasiordinary Singularities

INTRODUCTION: In this chapter we work over an algebraically closed field k of characteristic zero. Let R_n be the ring of formal resp. convergent power series over k in n indeterminates $X_1, \ldots, X_n$, let Q_n be the field of quotients of R_n, let $d > 1$ be a natural integer, and let $Q_{n,d}$ be the splitting field over Q_n of the polynomial $(T^d - X_1) \cdots (T^d - X_n) \in Q_n[T]$. The elements of the integral closure $R_{n,d} = k[\![X_1^{1/d}, \ldots, X_n^{1/d}]\!]$ of R_n in $Q_{n,d}$ are called fractionary power series (or Puiseux series). In section 1 we study, using Galois theory, the class of intermediate fields between Q_n and $Q_{n,d}$; this is a generalization of Kummer's theory of cyclic extensions. The theorem of Jung-Abhyankar in section 2 can be considered as a generalization to several variables of the well-known Newton-Puiseux theorem. It states: Let $F = Y^d + A_1 Y^{d-1} + \cdots + A_d$ with $A_1, \ldots, A_d \in R_n$ be an irreducible Weierstraß polynomial; then there exists an extension field $Q_{n,e}$ of Q_n such that F splits in $Q_{n,e}$ provided the discriminant of F has the form $X_1^{h_1} \cdots X_n^{h_n} U$ where $h_1, \ldots, h_n$ are non-negative integers and U is a unit of R_n—this condition holds automatically if $n = 1$. Such a Weierstraß polynomial will be called a quasiordinary Weierstraß polynomial. In section 3 we handle the convergent case. In section 4 we characterize those fractionary power series which are zeroes of a quasiordinary Weierstraß polynomial. In section 5 we describe an algorithm to find the Puiseux expansion of a zero of a quasiordinary Weierstraß polynomial. In the case $n = 1$ this algorithm becomes the classical Newton algorithm. In section 6 we prove some results on strictly generated semigroups which are needed in section 4.

1 Fractionary Power Series

1.1 Generalities

(1.0) In this chapter k always is an *algebraically closed field of characteristic zero*. The ring of formal resp. convergent power series $k[\{X_1, \ldots, X_n\}]$ over k in n indeterminates $X_1, \ldots, X_n$ will be denoted by R_n, and $Q_n := Q(R_n)$ is the field of quotients of R_n; let $\overline{Q}_n$ be an algebraic closure of Q_n. By C we denote the category of formal resp. analytic k-algebras (where we have $k = \mathbb{C}$ in the latter case).

The notations introduced in the last section of this chapter and the results of that section shall be used without any warning.

(1.1) CHOICE OF ROOTS: Let $i \in \{1, \ldots, n\}$. For every $d \in \mathbb{N}$ we denote by $X_i^{1/d} \in \overline{Q}_n$ a zero of the polynomial $T^d - X_i \in Q_n[T]$ chosen in such a way that $(X_i^{1/de})^e = X_i^{1/d}$ for every $e \in \mathbb{N}$.

(1.2) MONOMIALS: For every $d \in \mathbb{N}$ and $r = (\rho_1, \ldots, \rho_n) \in \mathbb{Z}^n$ we write

$$X^{r/d} = X_1^{\rho_1/d} \cdots X_n^{\rho_n/d} \in \overline{Q}_n;$$

such an expression shall be called a (fractionary) monomial if $r \geq 0$.

(1.3) Proposition: *Let $d_1, \ldots, d_n \in \mathbb{N}$. Then we have*

$$[Q_n(X_1^{1/d_1}, \ldots, X_n^{1/d_n}) : Q_n] = d_1 \cdots d_n,$$

the integral closure S of R_n in $Q_n(X_1^{1/d_1}, \ldots, X_n^{1/d_n})$ is a regular local ring in C, and admits $\{X_1^{1/d_1}, \ldots, X_n^{1/d_n}\}$ as a regular system of parameters, hence we have $S = k[\{X_1^{1/d_1}, \ldots, X_n^{1/d_n}\}]$.

This follows easily from IV(3.20) and IV(5.1).

(1.4) REMARK: Let $d \in \mathbb{N}$; we define $R_{n,d} := k[\{X_1^{1/d}, \ldots, X_n^{1/d}\}] \subset \overline{Q}_n$, and $Q_{n,d} := Q(R_{n,d})$. The ring $R_{n,d}$ is a ring of formal resp. convergent power series in "indeterminates" $X_1^{1/d}, \ldots, X_n^{1/d}$ [cf. IV(4.18)], and it is called the ring of fractionary power series (or Puiseux series) in n indeterminates with denominator d. In particular, we have $R_{n,d} = R_n[X_1^{1/d}, \ldots, X_n^{1/d}]$ [cf. IV(4.26)], and the field of quotients of $R_{n,d}$ is $Q_{n,d} = Q_n[X_1^{1/d}, \ldots, X_n^{1/d}]$. Moreover, $Q_{n,d}$ is a splitting field over Q_n of the polynomial $(T^d - X_1) \cdots (T^d - X_n) \in Q_n[T]$, hence $Q_{n,d}$ is a Galois extension of Q_n, and we have $[Q_{n,d} : Q_n] = d^n$ by (1.3). The set of monomials

$$\{X^{r/d} \mid r = (\rho_1, \ldots, \rho_n) \in \mathbb{N}_0^n \quad \text{and } 0 \leq \rho_i < d \text{ for every } i \in \{1, \ldots, n\}\}$$

is a basis of the free R_n-module $R_{n,d}$ and a Q_n-basis of the extension $Q_{n,d}$ of Q_n.

(1.5) THE GALOIS GROUP $\mathrm{Gal}(Q_{n,d}/Q_n)$: Let $\mathrm{Gal}(Q_{n,d}/Q_n)$ be the Galois group of $Q_{n,d}$ over Q_n, and let ε be a primitive d-th root of unity in k. For every $s = (\sigma_1, \ldots, \sigma_n) \in \mathbb{Z}^n$ let $\varphi_s \in \mathrm{Gal}(Q_{n,d}/Q_n)$ be defined by $\varphi_s(X_i^{1/d}) = \varepsilon^{\sigma_i} X_i^{1/d}$ for every $i \in \{1, \ldots, n\}$ [there exists a well defined k-algebra automorphism $R_{n,d} \to R_{n,d}$ defined by mapping $X_i^{1/d}$ to $\varepsilon^{\sigma_i} X_i^{1/d}$ for every $i \in \{1, \ldots, n\}$, cf. IV(1.2) and IV(4.8), and φ_s is its extension to $Q_{n,d}$]. Then the map $s \mapsto \varphi_s : \mathbb{Z}^n \to \mathrm{Gal}(Q_{n,d}/Q_n)$ is a homomorphism φ of groups having $d\mathbb{Z}^n$ as kernel, and as $\mathrm{Card}(\mathrm{Gal}(Q_{n,d}/Q_n)) = d^n = \mathrm{Card}(\mathbb{Z}^n/d\mathbb{Z}^n)$, the induced homomorphism $\mathbb{Z}^n/d\mathbb{Z}^n \to \mathrm{Gal}(Q_{n,d}/Q_n)$ is an isomorphism. For every $r,\ s \in \mathbb{Z}^n$ we have $\varphi_s(X^{r/d}) = \varepsilon^{\langle r,s \rangle} X^{r/d}$, and therefore we have $\varphi_s(X^{r/d}) = X^{r/d}$ iff $\langle r,s \rangle \in \mathbb{Z}d$, i.e., iff $\langle r,s \rangle_d = 0$.

Thus, we have shown the following proposition.

(1.6) Proposition: *The extension $Q_{n,d}/Q_n$ is an abelian Galois extension, and the homomorphism $s \to \varphi_s : \mathbb{Z}^n \to \mathrm{Gal}(Q_{n,d}/Q_n)$ is surjective and has kernel $d\mathbb{Z}^n$, hence $\mathrm{Gal}(Q_{n,d}/Q_n)$ is canonically isomorphic to $\mathbb{Z}^n/d\mathbb{Z}^n$.*

1.2   Intermediate Fields

(1.7) Theorem: *Any intermediate field L with $Q_n \subset L \subset Q_{n,d}$ is generated over Q_n by finitely many monomials, i.e., there exist $r_1, \ldots, r_h \in \mathbb{N}_0^n$ such that $L = Q_n[X^{r_1/d}, \ldots, X^{r_h/d}]$.*

Proof: Let $y \in Q_{n,d} = Q_n[X_1^{1/d}, \ldots, X_n^{1/d}]$ be a primitive element for L/Q_n. Then we have $y = \sum_{j=1}^h A_j X^{r_j/d}$ where $A_j \in Q_n$ and $r_j \in \mathbb{N}_0^n$ for every $j \in \{1, \ldots, h\}$. We may assume that the monomials $X^{r_1/d}, \ldots, X^{r_h/d}$ are linearly independent over Q_n. We have $L \subset Q_n[X^{r_1/d}, \ldots, X^{r_h/d}]$. An element of $\mathrm{Gal}(Q_{n,d}/Q_n)$ leaves y fixed iff it leaves fixed all the monomials $X^{r_1/d}, \ldots, X^{r_h/d}$. By Galois theory we now get $L = Q_n[X^{r_1/d}, \ldots, X^{r_h/d}]$.

(1.8) Remark: We set up a bijective correspondence between the set of intermediate fields L (with $Q_n \subset L \subset Q_{n,d}$) and the set of $\mathbb{Z}$-submodules N of $\mathbb{Z}^n$ containing $d\mathbb{Z}^n$.

(1) Let L be a field lying between Q_n and $Q_{n,d}$. We associate with L a $\mathbb{Z}$-submodule $N \subset \mathbb{Z}^n$ containing $d\mathbb{Z}^n$ in the following way: N is the $\mathbb{Z}$-module generated by the submodule $d\mathbb{Z}^n$ and all $r \in \mathbb{N}_0^d$ such that the monomials $X^{r/d}$ belong to L.

(2) Let N be a $\mathbb{Z}$-submodule of $\mathbb{Z}^n$ containing $d\mathbb{Z}^n$. We associate with N a field L lying between Q_n and $Q_{n,d}$ in the following way: L is the field extension of Q_n generated by all monomials $X^{r/d}$ with $r \in N \cap \mathbb{N}_0^n$.

(3) Let L be as in (1); then we have $L = Q_n[X^{r_1/d}, \ldots, X^{r_h/d}]$ with $r_1, \ldots, r_h \in \mathbb{N}_0^n$ [cf. (1.7)]. It is clear that the submodule N associated with L [cf. (1)] is the $\mathbb{Z}$-module $d\mathbb{Z}^n + \mathbb{Z}r_1 + \cdots + \mathbb{Z}r_h$.

(4) Let N be a $\mathbb{Z}$-submodule of $\mathbb{Z}^n$ containing $d\mathbb{Z}^n$ and let $\{r_1, \ldots, r_h\}$ with $r_i \in N \cap \mathbb{N}_0^n$ for $i \in \{1, \ldots, h\}$ be a system of generators of N [it is clear that N has a system of generators contained in $\mathbb{N}_0^n$]. The intermediate field L associated with N [cf. (2)] is the field $L = Q_n[X^{r_1/d}, \ldots, X^{r_h/d}]$.

The following result now does not need a proof:

(1.9) Proposition: *The correspondence between intermediate fields $Q_n \subset L \subset Q_{n,d}$ and $\mathbb{Z}$-submodules of $\mathbb{Z}^n$ containing $d\mathbb{Z}^n$ described in (1.8) is an inclusion-preserving bijective map.*

(1.10) Proposition: *Let $r_1, \ldots, r_h \in \mathbb{N}_0^n$, let $(\nu_i)_{1 \le i \le h}$ be the ν-sequence associated with d and $(r_i)_{1 \le i \le h}$, and let θ_{h+1} be the degree of this sequence. We set $L := Q_n[X^{r_1/d}, \ldots, X^{r_h/d}]$. Then we have*

$$[L : Q_n] = \nu_1 \cdots \nu_h = \frac{d^n}{\theta_{h+1}},$$

and $\{X^{r/d} \mid r = \alpha_1 r_1 + \cdots + \alpha_h r_h, \ (\alpha_1, \ldots, \alpha_h) \in Z(\nu)\}$ is a Q_n-basis of L.

Proof: Set $N := d\mathbb{Z}^n + \mathbb{Z}r_1 + \cdots + \mathbb{Z}r_h$, write $r_i = (\rho_{i1}, \ldots, \rho_{in})$ for $i \in \{1, \ldots, h\}$, and set

$$R := (dI_n, {}^t((\rho_{lj})_{1 \le l \le h, 1 \le j \le n})) \in M(n, n+h; \mathbb{Z}).$$

There exist $P \in \mathrm{GL}(n; \mathbb{Z})$, $Q \in \mathrm{GL}(n+h; \mathbb{Z})$ such that

$$PRQ = (\mathrm{diag}(\beta_1, \ldots, \beta_n), 0_{n,n+h}) \in M(n, n+h; \mathbb{Z})$$

[Smith canonical form of R, cf. (6.3)], and we have $\theta_{h+1} = \beta_1 \cdots \beta_n$. We write $P = (p_{ij})$, and we let $p_1 = {}^t(p_{11}, \ldots, p_{n1}), \ldots, p_n = {}^t(p_{1n}, \ldots, p_{nn})$ be the columns of P. Then the map

$$X_i \mapsto X_1^{p_{1i}} \cdots X_n^{p_{ni}} =: Y_i \text{ for } i \in \{1, \ldots, n\}$$

induces a k-automorphism of Q_n, and we have

$$L = Q_n[Y_1^{\beta_1/d}, \ldots, Y_n^{\beta_n/d}],$$

hence we have

$$[L : Q_n] = \frac{d^n}{\beta_1 \cdots \beta_n} = \frac{d^n}{\theta_{h+1}} = \nu_1 \cdots \nu_h.$$

[Note that, for $i \in \{1, \ldots, d\}$, we have $\beta_i \mid d$, hence that $Z^{d/\beta_i} - Y_i$ is the minimal polynomial of $Y_i^{\beta_i/d}$ over Q_n.]

As a next step, we want to calculate the Galois group $\mathrm{Gal}(Q_{n,d}/L)$ of an intermediate field $Q_n \subset L \subset Q_{n,d}$.

(1.11) Proposition: *(1) Let $Q_n \subset L \subset Q_{n,d}$, and let N be the $\mathbb{Z}$-submodule of $\mathbb{Z}^n$ associated to L. Then we have*

$$\mathrm{Gal}(Q_{n,d}/L) = \{\varphi_s \mid \langle r, s \rangle_d = 0 \text{ for every } r \in N\}.$$

(2) Let G be a subgroup of $\mathrm{Gal}(Q_{n,d}/Q_n)$, and set

$$N := \{r \in \mathbb{Z}^n \mid \langle r, s \rangle_d = 0 \text{ for every } s \in \mathbb{Z}^n \text{ with } \varphi_s \in G\}.$$

Then the fixed field $Q_{n,d}^G$ of G is the intermediate field associated with N.

Proof: Note that we have

$$\varphi_{s+s'} = \varphi_s \varphi_{s'} \text{ for all } s, s' \in \mathbb{Z}^n, \quad \varphi_s(X^{r/d}) = \varepsilon^{\langle r,s \rangle_d} X^{r/d} \text{ for all } r, s \in \mathbb{Z}^n. \quad (*)$$

(1) Let G be a subgroup of $\mathrm{Gal}(Q_{n,d}/Q_n)$, and set $\Gamma := \{s \in \mathbb{Z}^n \mid \varphi_s \in G\}$. Using the first part of $(*)$ it is clear that Γ is a $\mathbb{Z}$-submodule of $\mathbb{Z}^n$ which contains $d\mathbb{Z}^n$. Set $\Gamma^\perp := \{r \in \mathbb{Z} \mid \langle r, s \rangle_d = 0 \text{ for every } s \in \Gamma\}$; then $\Gamma^\perp$ is a $\mathbb{Z}$-submodule of $\mathbb{Z}^n$ which contains $d\mathbb{Z}^n$. Let L be the intermediate field associated with $\Gamma^\perp$. Using the second part of $(*)$ it is clear that $L = Q_{n,d}^G$ is the fixed field of G.
(2) This can be proved in a similar way.

(1.12) Now we have collected all the general results on fields between Q_n and $Q_{n,d}$ which shall be used in this chapter. The particular result in (1.14) below shall be needed in section 5. The starting point of the construction in (1.14) is the following result in (1.13) below.

(1.13) REMARK: Let us consider the case $h = 1$ in (1.10), i.e., $r = r_1 = (\rho_1, \ldots, \rho_n) \in \mathbb{Z}^n$ is non-negative and $L = Q_n(X^{r/d})$. Let $\theta = \theta_2 := \gcd(d, \rho_1, \ldots, \rho_n)$. For $i \in \{1, \ldots, n\}$ delete the i-th column in the matrix $(dI_n, {}^t r)$; the determinant of this matrix is $(-1)^{n+i} d^{n-1} \cdot \rho_i$, and the greatest common divisor of all these minors is $\gcd(d^n, d^{n-1} \cdot \rho_1, \ldots, d^{n-1} \cdot \rho_n) = d^{n-1}\theta$. Then we have $\nu_1 = d^n/d^{n-1}\theta = d/\theta$, and therefore the ν-sequence associated with d and (r_1) is (d/θ), and ν_1 is its degree. Then L/Q_n is a cyclic extension of degree ν_1, and $Y^{\nu_1} - X^{r/\theta} \in Q_n[Y]$ is the minimal polynomial of $X^{r/d}$ over Q_n.
Choose, e.g. by the Euclidean algorithm, integers $\sigma_1, \ldots, \sigma_n$ such that

$$\theta \equiv \sigma_1 \cdot \rho_1 + \cdots + \sigma_n \cdot \rho_n \pmod{d}.$$

Set $s := (\sigma_1, \ldots, \sigma_n)$, and consider $\varphi_s \in \mathrm{Gal}(Q_{n,d}/Q_n)$. We have $\varphi_s(X^{r/d}) = \varepsilon^\theta X^{r/d}$, and clearly the d/θ elements $\varphi_s(X^{r/d}), \ldots, \varphi_{(d/\theta)s}(X^{r/d})$ are pairwise different. Therefore the Galois group $\mathrm{Gal}(Q_n(X^{r/d})/Q_n)$ is generated by the element $\varphi_s | Q_n(X^{r/d})$; moreover, the set $\{\varphi_s^j \mid j \in \{1, \ldots, d/\theta\}\}$ is a system of representatives for the cosets of the subgroup $\mathrm{Gal}(Q_{n,d}/Q_n(X^{r/d}))$ of $\mathrm{Gal}(Q_{n,d}/Q_n)$.

(1.14) A SYSTEM OF REPRESENTATIVES: Let $d \in \mathbb{N}$, let $(r_i)_{i \geq 1}$ be an infinite sequence in $\mathbb{N}_0^n$, and let $r_i = (\rho_{i1}, \ldots, \rho_{in})$ for every $i \in \mathbb{N}$. Furthermore, we define $M_i := X^{r_i/d} \in Q_{n,d}$ and $Q_n(i) := Q_n(M_1, \ldots, M_i)$ for every $i \in \mathbb{N}$. Then we have an ascending chain of subfields $Q_n =: Q_n(0) \subset Q_n(1) \subset Q_n(2) \subset \cdots$ of $Q_{n,d}$, hence there exists $h \in \mathbb{N}$ with $Q_n(h) = Q_n(h+1) = \cdots$.
With respect to d and the sequence $(r_i)_{i \geq 1}$ we may define the sequences $(\theta_i)_{i \geq 1}$ and $(\nu_i)_{i \geq 1}$ in $\mathbb{N}$ as in (6.4) [since these sequences are defined recursively, we may start with an infinite sequence $(r_i)_{i \geq 1}$]. Note that $[Q_n(i) : Q_n] = \nu_1 \cdots \nu_i$ for every $i \in \mathbb{N}$ [cf. (1.10)]; in particular, we have $\nu_{h+1} = \nu_{h+2} = \cdots = 1$.
Likewise, if $(r_i)_{1 \leq i \leq h}$ is a finite sequence in $\mathbb{N}_0^n$, we have the ascending chain of subfields $Q_n = Q_n(0) \subset Q_n(1) \subset \cdots \subset Q_n(h)$, and we have the sequences

$(\theta_i)_{1 \le i \le h+1}$ and $(\nu_i)_{1 \le i \le h}$ in $\mathbb{N}$. In this case we have $[Q_n(i) : Q_n] = \nu_1 \cdots \nu_i$ for every $i \in \{1, \ldots, h\}$.

The Galois group $\mathrm{Gal}(Q_n(h)/Q_n)$ of $Q_n(h)$ over Q_n is isomorphic to the factor group $\mathrm{Gal}(Q_{n,d}/Q_n)/\mathrm{Gal}(Q_{n,d}/Q_n(h))$; we construct a system of representatives in $\mathrm{Gal}(Q_{n,d}/Q_n)$ for the cosets of $\mathrm{Gal}(Q_{n,d}/Q_n(h))$.

(1) Let $i = 1$. By (1.13) we can find a system of representatives for the cosets of the group $\mathrm{Gal}(Q_{n,d}/Q_n(X^{r_1/d}))$ in $\mathrm{Gal}(Q_{n,d}/Q_n)$.

(2) Let $i \in \{1, \ldots, h-1\}$, and define $\nu := \nu_1 \cdots \nu_i$; then we have $[Q_n(i) : Q_n] = \nu$ by (1.10). We assume that we already have constructed a system of representatives $\{\varphi^{(j)} \mid j \in \{1, \ldots, \nu\}\} \subset \mathrm{Gal}(Q_{n,d}/Q_n)$ for the cosets of the subgroup $\mathrm{Gal}(Q_{n,d}/Q_n(i))$ of $\mathrm{Gal}(Q_{n,d}/Q_n)$.

Set $N_i := d\mathbb{Z}^n + \mathbb{Z}r_1 + \cdots + \mathbb{Z}r_i$; we know that $\nu_{i+1}r_{i+1} \in N_i$ [cf. (6.5)], hence we have $M_{i+1}^{\nu_{i+1}} \in Q_n(i)$, and since $[Q_n(i+1) : Q_n(i)] = \nu_{i+1}$, we see that $Y^{\nu_{i+1}} - M_{i+1}^{\nu_{i+1}}$ is the minimal polynomial of M_{i+1} over $Q_n(i)$. This implies that $Q_n(i+1)$ is a cyclic extension of $Q_n(i)$ [cf. [117], Ch. VIII, §6, Th. 10]. Let us choose a representative $\psi \in \mathrm{Gal}(Q_{n,d}/Q_n(i))$ of a generator of the cyclic group $\mathrm{Gal}(Q_n(i+1)/Q_n(i))$ of order ν_{i+1}. Then $\{\varphi^{(j)}\psi^l \mid j \in \{1, \ldots, \nu\}, l \in \{1, \ldots, \nu_{i+1}\}\} \subset \mathrm{Gal}(Q_{n,d}/Q_n)$ is a system of representatives for the cosets of the subgroup $\mathrm{Gal}(Q_{n,d}/Q_n(i+1))$ of $\mathrm{Gal}(Q_{n,d}/Q_n)$.

(3) How does one find such an element ψ? Let us write $\psi = \varphi_s$ for some $s \in \mathbb{Z}^n$. Then φ_s lies in $\mathrm{Gal}(Q_{n,d}/Q_n(i))$, and it is a representative for a generator of the cyclic group $\mathrm{Gal}(Q_n(i+1)/Q_n(i))$ iff s satisfies the following system of congruences

$$\langle r_l, s \rangle \equiv 0 \pmod{d} \quad \text{for every } l \in \{1, \ldots, i\}, \qquad \langle r_{i+1}, s \rangle \equiv \frac{d}{\nu_{i+1}} \pmod{d}$$

[and there exist solutions by (2)]. We have to find integers $\sigma_1, \ldots, \sigma_{n+2}$ such that

$$\sum_{j=1}^{n} \rho_{lj}\sigma_j + d\sigma_{n+1} = 0 \quad \text{for every } l \in \{1, \ldots, i\},$$

$$\sum_{j=1}^{n} \rho_{i+1,j}\sigma_j + d\sigma_{n+2} = \frac{d}{\nu_{i+1}}.$$

This can be done, e.g., by the algorithm given in [172], Ch. 4, Cor. 5.3b.

(4) From (1)-(3) above we see: *a system of representatives for the cosets of the subgroup* $\mathrm{Gal}(Q_{n,d}/Q_n(h))$ *of* $\mathrm{Gal}(Q_{n,d}/Q_n)$ *may be constructed by recursion.*

(1.15) REMARK: Let $d \in \mathbb{N}$ and $r_1, \ldots, r_h \in \mathbb{N}_0^n$, and let N be the $\mathbb{Z}$-submodule of $\mathbb{Z}^n$ generated by $d\mathbb{Z}^n$ and $r_1, \ldots, r_h$. The results in (1.6)-(1.7) and (1.13)-(1.14) remain true for the case $R_n := k[X_1, \ldots, X_n]$, $R_{n,d} := k[X_1^{1/d}, \ldots, X_n^{1/d}]$; note that in this case we have $Q_n = k(X_1, \ldots, X_n)$, the field of rational functions in n indeterminates $X_1, \ldots, X_n$ over k, and that we have $Q_{n,d} = Q_n(X_1^{1/d}, \ldots, X_n^{1/d})$.

1.3   Intermediate Fields Generated by a Fractionary Power Series

(1.16) Until now we studied fields of the form $Q_n[X^{r_1/d}, \ldots, X^{r_h/d}]$; now we shall study fields of the form $Q_n(y)$ where $y \in R_{n,d}$ is a fractionary power series. We could also study fields $Q_n(y)$ with $y \in Q_{n,d}$; since $R_{n,d}$ is the integral closure of R_n in $Q_{n,d}$, there exists $F \in R_n$, $F \neq 0$, such that $Fy \in R_{n,d}$. Thus, it is enough to restrict our considerations to elements $y \in R_{n,d}$.

(1.17) DEGREE ORDERING: (1) Let $r = (\rho_1, \ldots, \rho_n) \in \mathbb{N}_0^n$; then $\deg(r) := \rho_1 + \cdots + \rho_n$ is called the degree of r. Let $r' = (\rho_1', \ldots, \rho_n') \in \mathbb{N}_0^n$; then we will write $r \prec r'$ if $\deg(r) < \deg(r')$ or if $\deg(r) = \deg(r')$ and r is smaller than r' in the lexicographic ordering of $\mathbb{N}_0^n$. Note that $\prec$ defines a total ordering of $\mathbb{N}_0^n$ which is compatible with its semigroup structure; it is called the grade-lex ordering of $\mathbb{N}_0^n$.
(2) Let $M = X^{r/d}$, $N = X^{s/d} \in R_{n,d}$ where r, $s \in \mathbb{N}_0^n$. We will write $M \prec N$ if $r \prec s$; note that $\prec$ is a total ordering on the set of monomials $\{X^{r/d} \mid r \in \mathbb{N}_0^n\}$.
(3) Let M_1, M_2 and N be monomials in $R_{n,d}$. If $M_1 \prec M_2$, then we have $M_1 N \prec M_2 N$ [hence $\prec$ is a monomial ordering, cf. [53], p. 54]. Therefore, if N_1, N_2 are monomials in $R_{n,d}$ such that $N_1 \prec N_2$, then we have $M_1 N_1 \prec M_2 N_2$.
(4) Let r, $r' \in \mathbb{N}_0^n$, and assume that $r < r'$ with respect to the partial ordering of $\mathbb{Z}_0^n$ defined in (6.2). Then we have $r \prec r'$.

(1.18) REPRESENTATION OF FRACTIONARY POWER SERIES: (1) Let $y \in R_{n,d}$ be a non-unit different from 0. It has a unique representation

$$y = \alpha_1 M_1 + \alpha_2 M_2 + \cdots = \sum_{i \in S(y)} \alpha_i M_i \qquad (*)$$

with $\alpha_i \in k^\times$ for every $i \in S(y) \subset \mathbb{N}$ and with monomials $M_1, M_2, \ldots$ in $R_{n,d}$ of positive degree and ordered in such a way that $M_1 \prec M_2 \prec \cdots$. The representation in $(*)$ is said to be the normal representation of y. We write $M_i = X^{r_i/d}$ for every $i \in S(y)$, and we define

$$\text{Supp}(y) = \{r_i \mid i \in S(y)\} \subset \mathbb{N}_0^n.$$

We call the term $M_1 = X^{r_1/d}$ the leading monomial, α_1 the leading coefficient, and $\alpha_1 M_1$ the leading term of y, respectively, and we write $\text{lt}(y) := \alpha_1 M_1$.
(2) Let $y_1, \ldots, y_h \in R_{n,d}$ be non-zero fractionary power series which are non-units, and let $0 \neq (l_1, \ldots, l_h) \in \mathbb{N}_0^h$. Then we have $\text{lt}(y_1^{l_1} \cdots y_h^{l_h}) = \text{lt}(y_1)^{l_1} \cdots \text{lt}(y_h)^{l_h}$, and a corresponding result holds for the leading monomial and the leading coefficient of the power product $y_1^{l_1} \cdots y_h^{l_h}$ of the fractionary power series $y_1, \ldots, y_h$.

(1.19) REMARK: Let $y = \sum_{i \in S(y)} \alpha_i X^{r_i/d} \in R_{n,d}$ be a non-unit different from 0.
(1) We have

$$\varphi_s(y) = \sum_{i \in S(y)} \alpha_i \epsilon^{\langle r_i, s \rangle} X^{r_i/d} \quad \text{for every } s \in \mathbb{Z}^n.$$

(2) Let L be a subfield of $Q_{n,d}$ containing Q_n. Then we have $y \in L$ iff we have $X^{r_i/d} \in L$ for every $i \in S(y)$.

In fact, let $s \in \mathbb{Z}^n$; then, by (1), every Q_n-automorphism of $Q_{n,d}$ which leaves y invariant must leave invariant all the monomials $X^{r_i/d}$, $i \in S(y)$. Since we have $y \in L$ iff we have $\varphi(y) = y$ for every $\varphi \in \mathrm{Gal}(Q_{n,d}/L)$, the assertion follows from (1.8).

(3) There exists $h \in S(y)$ such that y lies in $R_n[X^{r_1/d}, \ldots, X^{r_h/d}] \subset Q_n[y]$. In this case we have $Q_n[y] = Q_n[X^{r_1/d}, \ldots, X^{r_h/d}]$.

In fact, the ascending chain $(R_n[X^{r_1/d}, \ldots, X^{r_i/d}])_{i \in S(y)}$ of R_n-submodules of $R_{n,d}$ becomes stationary; we choose $h \in S(y)$ with $X^{r_i/d} \in R_n[X^{r_1/d}, \ldots, X^{r_h/d}]$ for every $i \in S(y)$. Then we have $y \in R_n[X^{r_1/d}, \ldots, X^{r_h/d}] \subset Q_n(y)$ [cf. (2)].

(4) Let h be as in (3), and let θ_{h+1} be the greatest common divisor of all n-rowed minors of the matrix $(dI_n, {}^t r_1, \ldots, {}^t r_h)$. Then θ_{h+1} is the greatest common divisor of all n-rowed minors of the matrix $(dI_n, {}^t r_1, \ldots, {}^t r_i)$ for every $i \in S(y)$, $i \geq h$. In particular, we have $[Q_n(y) : Q_n] = d^n/\theta_{h+1}$.

In fact, $d\mathbb{Z}^n$ and $r_1, \ldots, r_h$ generate the $\mathbb{Z}$-module N associated with the field $Q_n(y)$ by (3) [cf. (1.10)], and this implies that we have $r_i \in N$ for every $i \in S(y)$. Now the assertion is obvious.

The following result is an immediate consequence.

(1.20) Proposition: *Let* $y = \sum_{i \in S(y)} \alpha_i X^{r_i/d} \in R_{n,d}$ *be a non-unit different from* 0*, and let* $(\nu_i)_{i \in S(y)}$ *be the* ν*-sequence associated with* d *and* $(r_i)_{i \in S(y)}$*. Then there exists* $h \in S(y)$ *with* $Q_n(y) = Q_n[X^{r_1/d}, \ldots, X^{r_h/d}]$*, we have*

$$[Q_n(X^{r_1/d}, \ldots, X^{r_i/d}) : Q_n] = \nu_1 \cdots \nu_i \quad \text{for every } i \in S(y),$$

and we have $\nu_i = 1$ *for every* $i \in S(y)$ *with* $i > h$*. Furthermore, we have* $X^{r_i/d} \in Q_n(y)$ *for every* $i \in S(y)$*.*

The following characterization of the integral closure will be used in section 4.

(1.21) Proposition: *Let* $r_1, \ldots, r_h \in \mathbb{N}_0^n$*, and let* N *be the* $\mathbb{Z}$*-submodule of* $\mathbb{Z}^n$ *generated by* $d\mathbb{Z}^n$ *and the elements* $r_1, \ldots, r_h$*. An element* $z \in R_{n,d}$ *lies in the integral closure of* $R_n[X^{r_1/d}, \ldots, X^{r_h/d}]$ *iff* $\mathrm{Supp}(z)$ *is contained in the set of non-negative elements of* N*.*

Proof: Set $S := R_n[X^{r_1/d}, \ldots, X^{r_h/d}]$, and let L be the field of quotients of S; note that we have $L = Q_n[X^{r_1/d}, \ldots, X^{r_h/d}]$. Since $R_{n,d}$ is integrally closed, it is clear that $L \cap R_{n,d}$ is the integral closure of S. Let $z \in R_{n,d}$. Now we have $z \in L$ iff $r \in N$ for every $r \in \mathrm{Supp}(z)$ [cf. (1.19)(2)], hence iff $r \in N \cap \mathbb{N}_0^n$ for every $r \in \mathrm{Supp}(z)$.

The last result of this section is useful when one does explicit calculations.

(1.22) Proposition: *Let* $y \in R_{n,d} \setminus \{0\}$*, and set* $e := [Q_n(y) : Q_n]$*. Then the power series* y *lies already in* $R_{n,e}$*.*

Proof: We may assume that y is a non-unit. Let $y = \sum_{i \in S(y)} \alpha_i X^{r_i/d}$ be the normal representation of y, and let $r_i = (\rho_{i1}, \ldots, \rho_{in}) \in \mathbb{N}_0^n$ for every $i \in S(y)$. Let N be the submodule of $\mathbb{Z}^n$ generated by $d\mathbb{Z}^n$ and the elements r_i, $i \in S(y)$. Let $j \in \{1, \ldots, n\}$. We define $f_j := \gcd(\{\rho_{ij} \mid i \in S(y)\})$, and $\rho'_{ij} = \rho_{ij}/f_j$ for every $i \in S(y)$. Then we have

$$\frac{\rho_{ij}}{d} = \frac{f_j}{d}\rho'_{ij} \quad \text{for every } i \in S(y). \tag{$*$}$$

Let us choose $h \in S(y)$ such that $y \in R_n[X^{r_1/d}, \ldots, X^{r_h/d}]$ [cf. (1.19)(3)] and that f_j is a linear combination of $\rho_{1j}, \ldots, \rho_{hj}$ with integer coefficients [note that we may assume that this h works for all $j \in \{1, \ldots, n\}$]. Thus, we have

$$f_j = \alpha_{1j}\rho_{1j} + \cdots + \alpha_{hj}\rho_{hj}$$

where $(\alpha_{ij}) \in M(h, n; \mathbb{Z})$.
Let $i \in S(y)$ and $j \in \{1, \ldots, n\}$; we set

$$B_i := \begin{pmatrix} d & 0 & \cdots & 0 & \rho_{i1} \\ 0 & d & \cdots & 0 & \rho_{i2} \\ & & \ddots & & \\ 0 & 0 & & d & \rho_{in} \end{pmatrix} = (dI_n, {}^t r_i) \in M(n, n+1; \mathbb{Z}),$$

and we let $B_{ij} \in M(n; \mathbb{Z})$ be the matrix obtained from B_i by deleting the j-th column. Note that $\det(B_{ij}) = (-1)^{n-j}d^{n-1} \cdot \rho_{ij}$, hence we obtain

$$f_j d^{n-1} = (-1)^{n-j}\big(\alpha_{1j}\det(B_{1j}) + \cdots + \alpha_{hj}\det(B_{hj})\big).$$

Let θ_{h+1} be the greatest common divisor of all n-rowed minors of $(dI_n, {}^t(r_i)_{1 \le i \le h})$. Then we have $d^n/\theta_{h+1} = [Q_n(y) : Q_n] = e$ [cf. (1.10)], and θ_{h+1} is a divisor of $f_j d^{n-1}$. Now we get

$$\frac{\rho_{ij}}{d} = \frac{f_j d^{n-1}}{\theta_{h+1}}\frac{\rho'_{ij}}{e} \quad \text{for every } i \in S(y) \text{ and } j \in \{1, \ldots, n\},$$

hence we have shown that y lies in $R_{n,e}$.

2 The Jung-Abhyankar Theorem: Formal Case

(2.0) In the sequel, we use some results from valuation theory; they can be found in section 7 of chapter I. For any unexplained notation we also refer to loc. cit..

(2.1) In this section we work in the category C_f of formal k-algebras.

(2.2) Lemma: *Let L be a field of characteristic zero, and assume that L contains a primitive n-th root of unity for every $n \in \mathbb{N}$. Let $\overline{L}$ be an algebraic closure of L, and let M be a subfield of $\overline{L}$ containing L such that M is a finite Galois extension of L. Let R be a noetherian subring of L which contains the prime field of L, has L as field of quotients and is integrally closed in L. Let $\mathfrak{n}_1 := Rx_1, \ldots, \mathfrak{n}_h := Rx_h$ be principal prime ideals of R. Let $i \in \{1, \ldots, h\}$. Then $V_i := R_{\mathfrak{n}_i}$ is a discrete valuation ring; let W_i be an extension of V_i to M, set $e_i := e(W_i/V_i)$ and let $y_i \in \overline{L}$ satisfy $y_i^{e_i} = x_i$. Set $L' := L(y_1, \ldots, y_h)$, $M' := M(y_1, \ldots, y_h)$. For every $i \in \{1, \ldots, h\}$ let W_i' be an extension of W_i to M', and set $V_i' := W_i' \cap L'$; thus, V_i' is an extension of V_i to L'. Let $\mathfrak{n}$ be a prime ideal of R of height one which is different from $\mathfrak{n}_1, \ldots, \mathfrak{n}_h$. Then $V := R_{\mathfrak{n}}$ is a discrete valuation ring; let W' be an extension of V to M', and set $V' := W' \cap L'$, $W := W' \cap M$. Then:*

(1) The extensions L'/L, M'/M, M'/L and M'/L' are Galois extensions.
(2) We have $e(W_i'/W_i) = 1$ and $e(V_i'/V_i) = e_i$ for every $i \in \{1, \ldots, h\}$.
(3) We have $e(W'/W) = e(V'/V) = 1$ and $e(W'/V') = e(W/V)$.
(4) For all valuations in (2) and (3), the reduced ramification index and the ramification index are equal.

Proof: We may assume that $h = 1$, since the general case follows from this by induction [and by [117], Ch. VIII, Th. 5]. Furthermore, as R contains the prime field of L, all residue fields of the valuations considered have characteristic zero, hence ramification index and reduced ramification index of all these valuations are equal. Note, furthermore, that e_1 is independent of the choice of the extension W_1 of V_1 [cf. I(6.17)].

In proving the lemma, the reader should bear in mind the following commutative diagrams (with self-evident notations):

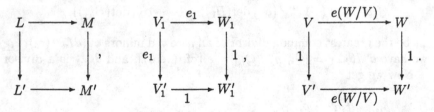

(a) Since $\mathfrak{n}$ is a prime ideal of height one, $R_{\mathfrak{n}}$ is a discrete valuation ring [cf. B(10.5)]. Likewise $R_{\mathfrak{n}_1}$ is a discrete valuation ring [since Rx_1 is a prime ideal of height 1 by [63], Th. 10.1].

(b) L contains a primitive e_1-th root of unity, $H := T^{e_1} - x_1 \in V_1[T]$ is an Eisenstein polynomial [cf. I(7.13)], and $L' = L(y_1)$ is a cyclic Galois extension of L of degree e_1 [cf. [117], Ch. VIII, §6, Th. 10]. Set $n := [M' : M]$ and $z_1 := y_1^n$; then n divides e_1, $M' = M(y_1)$ is a cyclic Galois extension of M of degree n and $F := T^n - z_1$ is the minimal polynomial of z_1 over M [cf. loc. cit.]. Now we have $M' = ML'$, the smallest subfield of $\overline{L}$ containing M and L', hence M' is a Galois extension of L [cf. [117], Ch. VIII, §1, Th. 5], and therefore M' is a Galois extension of L', and (1) has been proved.

(c) Let v, w, v', w' be the valuations defined by, respectively, the valuation rings V, W, V', W'. By I(7.6)(2) we have $\mathrm{dis}_Z(F) = (-1)^{n(n-1)/2}n^n z_1^{n-1}$. Clearly n^n is a unit of V [because R contains the prime field of L], and $x_1 = z_1^{e_1/n}$ is a unit of V, hence $z_1 \in W$ is a unit of W, and therefore $\mathrm{dis}_Z(F)$ is a unit of W. The polynomial $\overline{F} = T^n - \overline{z}_1 \in W/\mathfrak{m}(W)[T]$ is separable, and therefore we have $e(W'/W) = 1$ [cf. I(7.11)].
In the same way we can prove that $e(V'/V) = 1$. Now we get the equality $e(W/V) = e(W'/V')$ by I(5.6).
Let v_1, w_1, v_1', w_1' be the valuations defined by, respectively, the valuation rings V_1, W_1, V_1', W_1'. We choose $u_1 \in W_1$ such that $w_1(u_1) = 1$. Then y_1/u_1 is a primitive element for M'/M, and $G := Z^n - z_1/u_1^n \in M[Z]$ is the minimal polynomial of y_1/u_1 over M. Now we have $w_1(y_1^{e_1}) = w_1(x_1) = e_1 v_1(x_1) = e_1$ and $(e_1/n)w_1(z_1) = w_1(y_1^{e_1}) = e_1$, hence we have $w_1(z_1) = n$ and $w_1(z_1/u_1^n) = 0$. Therefore $\mathrm{dis}_Z(G)$ is a unit of W_1, and, as before, we obtain $e(W_1'/W_1) = 1$.
Lastly, we consider $L' = L(y_1)$. The polynomial $H = Z^{e_1} - x_1 \in V[Z]$ is an Eisenstein polynomial, and therefore V' is the only extension of V to L', and we have $e(V'/V) = e_1$ [cf. I(7.12)].

Now the proof of (2.2) is complete.

(2.3) Proposition: *Let N be a finite extension of Q_n, and let S be the integral closure of R_n in N. Assume that every prime ideal of R_n of height one is unramified in S. Then we have $S = R_n$ and $N = Q_n$.*

Proof: (1) Note that $S \in \mathcal{C}_f$, and that S is a finitely generated R_n-module [cf. IV(5.3)]. Let $\mathfrak{p}$ be a prime ideal of R_n of height one, set $A := (R_n)_\mathfrak{p}$ and $B := S_\mathfrak{p}$. Note that B is the integral closure of A in N [cf. [63], Prop. 4.13], and that it is a finitely generated A-module [cf. III(4.5)(2)]. Now B is unramified over A by assumption. The residue field of A is infinite, therefore, by III(3.16), there exists a monic separable polynomial $G \in A[T]$ and an element $b \in B$ such that $B = A[b]$, $G(b) = 0$ and $G'(b)$ is a unit of B.
(2) Let $j \in \{1, \ldots, n\}$. The partial derivative $D_j \colon R_n \to R_n$ has a unique extension to a k-derivation of the field of quotients Q_n of R_n [cf. III(1.4)(3)], and this derivation has a unique extension as a k-derivation of N [cf. III(1.5)] which will be also denoted by D_j. Since A is a localization of R_n, we have $D_j(A) \subset A$ [cf. III(1.4)(2)], and $0 = D_j(G(b)) = D_j(G)(b) + G'(b)D_j(b)$, hence $D_j(b) \in B$ and therefore $D_j(B) \subset B$.
(3) Let $\mathfrak{q}$ be a prime ideal of height 1 of S, and set $\mathfrak{p} := R_n \cap \mathfrak{q}$. Then $\mathfrak{p}$ is a prime ideal of height 1 of R_n [cf. [63], Th. 13.9], hence we have $D_j(S_\mathfrak{p}) \subset S_\mathfrak{p}$ by (2). The ring $S_\mathfrak{q}$ is a localization of $S_\mathfrak{p}$, hence we have $D_j(S_\mathfrak{q}) \subset S_\mathfrak{q}$ [cf. III(1.4)(2)]. The ring S is the intersection of the rings $S_\mathfrak{q}$ where $\mathfrak{q}$ runs through the set of prime ideals of S of height one [cf. [63], Cor. 11.4]. Therefore we have $D_j(S) \subset S$.
(4) Let $\mathfrak{n}$ be the maximal ideal of S, and let $\psi \colon S \to S/\mathfrak{n} = k$ be the canonical homomorphism. We define a surjective k-algebra homomorphism $\tau \colon S \to R_n$ which is an extension of the Taylor expansion [cf. IV(1.3)] $\mathrm{id}_{R_n} \colon R_n \to R_n$ in the

following way: For $f \in S$ we set

$$\tau(f) := \sum_{(i_1,\ldots,i_n)\in \mathbb{N}_0^n} \frac{1}{i_1!\cdots i_n!}\psi(D_1^{i_1}\cdots D_n^{i_n}(f))X_1^{i_1}\cdots X_n^{i_n}.$$

Clearly τ is k-linear, and we have $\mathrm{id}_{R_n} = \tau \circ \iota$ where $\iota\colon R_n \hookrightarrow S$ is the inclusion map, hence τ is surjective. Let $f, g \in S$. We have for all $(i_1,\ldots,i_n) \in \mathbb{N}_0^n$ [cf. III(1.2)(2)]

$$\frac{1}{i_1!\cdots i_n!}D_1^{i_1}\cdots D_n^{i_n}(fg)$$

$$= \sum_{l_1=0}^{i_1}\cdots\sum_{l_n=0}^{i_n} \frac{D_1^{l_1}(f)}{l_1!}\cdots\frac{D_n^{l_n}(f)}{l_n!}\cdot\frac{D_1^{i_1-l_1}(g)}{(i_1-l_1)!}\cdots\frac{D_n^{i_n-l_n}(g)}{(i_n-l_n)!}.$$

From this one immediately deduces that $\tau(fg) = \tau(f)\tau(g)$, i.e., that τ is a k-algebra homomorphism. Let $\mathfrak{a}$ be the kernel of τ. We have $S/\mathfrak{a} \cong R_n$. Suppose that $\mathfrak{a} \neq \{0\}$; then we get $\dim(S/\mathfrak{a}) < \dim(S)$, contradicting $\dim(R_n) = \dim(S)$. Therefore we have $\mathfrak{a} = \{0\}$, hence τ is injective,. Thus, τ is an isomorphism, and therefore the inclusion map ι is surjective, and this yields $S = R_n$ and $N = Q_n$.

(2.4) Theorem: *Let E be a finite extension of Q_n, and let C be the integral closure of R_n in E. We set $\mathfrak{n}_i := R_n X_i$ for every $i \in \{1,\ldots,n\}$. If there exists $h \in \{0,\ldots,n\}$ such that the set of prime ideals of height one of R_n which are ramified in C is contained in the set $\{\mathfrak{n}_1,\ldots,\mathfrak{n}_h\}$, then there exist natural integers $e_1,\ldots,e_h$ such that E is contained in the field $Q_n(X_1^{1/e_1},\ldots,X_h^{1/e_h})$, and that C is contained in the ring of formal power series $k[\![X_1^{1/e_1},\ldots,X_h^{1/e_h},X_{h+1},\ldots,X_n]\!]$.*

Proof: We may assume that $E \subset \overline{Q}_n$; let $M \subset \overline{Q}_n$ be the smallest Galois extension of Q_n containing E. There nothing to show if $h = 0$ [cf. (2.3)]. We now assume that $h \geq 1$.
(1) Let $i \in \{1,\ldots,h\}$. Now $V_i := (R_n)_{\mathfrak{n}_i}$ is a discrete valuation ring of Q_n [cf. I(3.29)]; let W_i be an extension of V_i to M and set $e_i := e(W_i/V_i)$ [e_i is independent of the choice of W_i extending V_i since M is a Galois extension of Q_n, cf. I(6.17)].
(2) Set $L := Q_n$, $L' := L(X_1^{1/e_1},\ldots,X_h^{1/e_h})$, and $M' := M(X_1^{1/e_1},\ldots,X_h^{1/e_h})$; then M' is a Galois extension of L' [cf. (2.2)]. Set $A := R_n$, and let A' (resp. B resp. B') be the integral closure of A in L' (resp. M resp. M'). Note that $A' = k[\![X_1^{1/e_1},\ldots,X_h^{1/e_h},X_{h+1},\ldots,X_n]\!]$ is a ring of formal power series [cf. (1.3)], and that $A \subset C \subset B$. We assert that $L' = M'$; to show this, it is enough to prove that every prime ideal of height 1 of A' is unramified in B' [cf. (2.3)].
(3) Let $\mathfrak{n}$ be a prime ideal of A which is unramified in C; then $\mathfrak{n}$ is also unramified in B. In fact, $C_\mathfrak{n}$ is unramified over $A_\mathfrak{n}$. Since $B_\mathfrak{n}$ is the integral closure of $C_\mathfrak{n}$ in M [cf. [63], Prop. 4.13], $B_\mathfrak{n}$ is unramified over $A_\mathfrak{n}$ by III(4.7).

(4) Let $\mathfrak{q}$ be a prime ideal of A' of height 1; then $V' := A'_{\mathfrak{q}}$ is a discrete valuation ring, and $\mathfrak{q}$ is unramified in B' iff $e(W'/V') = 1$ for every extension W' of V' to M' [cf. III(3.21), and note that $r(W'/V') = e(W'/V')$]. Set $\mathfrak{n} := \mathfrak{q} \cap A$. By [63], Th. 13.9, $\mathfrak{n}$ is a prime ideal of height one of A, $V := A_{\mathfrak{n}}$ is a discrete valuation ring, and V' is an extension of V to L'. Let W' be an extension of V' to M'.
(a) First, we consider the case $\mathfrak{n} \in \text{branch}(C/A)$. Then we have $\mathfrak{n} = \mathfrak{n}_i$ for some $i \in \{1, \ldots, h\}$ by hypothesis, and $e(W'/V') = 1$ by (2.2).
(b) Second, we consider the case $\mathfrak{n} \notin \text{branch}(C/A)$. Then $\mathfrak{n}$ is unramified in B by (3), i.e., $e((W' \cap M)/V) = 1$ [cf. III(3.21)], and by (2.2) we have $e(W'/V') = 1$.
(5) From (3) we get $E \subset L'$; hence we have proved the theorem.

(2.5) Corollary: [Jung-Abhyankar theorem, formal case] *Let F be an irreducible Weierstraß polynomial in $k[\![X_1, \ldots, X_n]\!][Z]$, and let L be the splitting field for F over Q_n in $\overline{Q}_n$. If $\text{dis}_Z(F) = X_1^{\mu_1} \cdots X_h^{\mu_h} U$ where $h \in \{0, \ldots, n\}$, $\mu_1, \ldots, \mu_h \in \mathbb{N}$ and $U \in k[\![X_1, \ldots, X_n]\!]$ is a unit, then there exist natural integers $e_1, \ldots, e_h$ such that L is contained in $Q_n[X_1^{1/e_1}, \ldots, X_h^{1/e_h}]$; in particular, every zero of F in $\overline{Q}_n$ lies in the ring of formal power series $k[\![X_1^{1/e_1}, \ldots, X_h^{1/e_h}, X_{h+1}, \ldots, X_n]\!]$, and we have $L = Q_n(y)$ where y is a zero of F.*

Proof: Let y be a zero of F in L, and set $E := Q_n(y)$, $d := [E : Q_n] = \deg_Z(F)$. Let C be the integral closure of R_n in E. We have $\text{dis}_Z(F) = D_{R_n[y]/R_n}(1, y, \ldots, y^{d-1}) = \det((\text{Tr}_{E/Q_n}(y^{i+j-2}))_{1 \leq i, j \leq d})$ [cf. B(10.22)(1)(*)], hence $\text{dis}_Z(F) \in \mathfrak{n}_{C/R_n}$; since a prime ideal of R_n is ramified in C iff it contains the Noether discriminant [cf. III(4.12)], we find that $\{R_n X_1, \ldots, R_n X_h\}$ contains the set of prime ideals of height one of R_n which are ramified in C. Therefore there exist natural integers $e_1, \ldots, e_h$ such that $E \subset Q_n[X_1^{1/e_1}, \ldots, X_h^{1/e_h}]$ [cf. (2.4)]. Now $Q_n[X_1^{1/e_1}, \ldots, X_h^{1/e_h}]$ is a Galois extension of Q_n [cf. (1.6)], hence $L \subset Q_n[X_1^{1/e_1}, \ldots, X_h^{1/e_h}]$, and therefore our first assertion follows. Since $R_n[X_1^{1/e_1}, \ldots, X_h^{1/e_h}]$ is the integral closure of R_n in $Q_n[X_1^{1/e_1}, \ldots, X_h^{1/e_h}]$, and since every zero of F is integral over R_n, the last assertion has also been proved.

3 The Jung-Abhyankar Theorem: Analytic Case

(3.0) Throughout this section we take $k = \mathbb{C}$, i.e., we work in the category C_a of analytic $\mathbb{C}$-algebras. For every $n \in \mathbb{N}$ we denote by $R_n = \mathbb{C}[\langle X_1, \ldots, X_n \rangle]$ the ring of convergent power series over $\mathbb{C}$ [cf. IV(2.4)(5)], and by Q_n its field of quotients.

(3.1) Proposition: *Let N be a finite extension of Q_n, and let S be the integral closure of R_n in N. Assume that every prime ideal of R_n of height one is unramified in S. Then we have $S = R_n$ and $N = Q_n$.*

Proof: Note that S is a finitely generated R_n-module and an analytic $\mathbb{C}$-algebra [cf. IV(5.3)]. The completion $\hat{S}$ of S is a formal $\mathbb{C}$-algebra and a domain, and

$\widehat{S}$ is the integral closure of $\widehat{R}_n = \mathbb{C}[\![X_1, \ldots, X_n]\!]$ in $Q(\widehat{S})$ [cf. IV(5.5)]. More-over, we have $\mathfrak{n}_{S/R_n} \cdot \widehat{R}_n = \mathfrak{n}_{\widehat{S}/\widehat{R}_n}$ by III(6.10). Now N is unramified over Q_n by III(3.10), i.e., the zero ideal of R_n is unramified in S, and therefore the ideal $\mathfrak{n}_{S/R_n}$ is not the zero ideal by III(4.12) and III(4.10). The rings R_n and $\widehat{R}_n$ are catenary [they are regular by IV(3.20), and therefore Cohen-Macaulay by [63], Cor. 19.15, hence catenary by [63], Cor. 18.10]. Assume that a prime ideal $\mathfrak{p}$ of $\widehat{R}_n$ of height one is ramified in $\widehat{S}$. Then we have $\mathfrak{n}_{\widehat{S}/\widehat{R}_n} \subset \mathfrak{p}$ [cf. III(4.12), and note that $\widehat{R}_n$ is integrally closed], hence we have $\mathfrak{p} \in \mathrm{Ass}(\widehat{R}_n/\mathfrak{n}_{\widehat{S}/\widehat{R}_n})$. This implies that we have $\dim(\widehat{R}_n/(\mathfrak{n}_{S/R_n} \cdot \widehat{R}_n)) = n - 1$, and since we have $\dim(R_n/\mathfrak{n}_{S/R_n}) = \dim(\widehat{R}_n/(\mathfrak{n}_{S/R_n} \cdot \widehat{R}_n))$ [cf. [63], Cor. 10.12], it follows that $\mathrm{Ass}(R_n/\mathfrak{n}_{S/R_n})$ contains prime ideals of height 1. But every such prime ideal is ramified in S [cf. III(4.12) and note that R_n is integrally closed] which is in con-tradiction with our hypothesis. Therefore every prime ideal of $\widehat{R}_n$ of height one is unramified in $\widehat{S}$. From this it follows, by (2.4), that $\widehat{R}_n = \widehat{S}$, and therefore we have $R_n = S$ [cf. [63], Lemma 7.15].

Now we can prove the following two assertions just as in section 2.

(3.2) Theorem: *Let E be a finite extension of Q_n, and let C be the integral closure of R_n in E. We set $\mathfrak{n}_i := R_n X_i$ for every $i \in \{1, \ldots, n\}$. If there exists $h \in \{0, \ldots, n\}$ such that the set of prime ideals of height one of R_n which are ramified in C is contained in the set $\{\mathfrak{n}_1, \ldots, \mathfrak{n}_h\}$, then there exist natural integers $e_1, \ldots, e_h$ such that E is contained in the field $Q_n(X_1^{1/e_1}, \ldots, X_h^{1/e_h})$ and that C is contained in the ring of convergent power series $\mathbb{C}[\langle X_1^{1/e_1}, \ldots, X_h^{1/e_h}, X_{h+1}, \ldots, X_n \rangle]$.*

(3.3) Corollary: [Jung-Abhyankar theorem, analytic case] *Let F be an irre-ducible Weierstraß polynomial in $\mathbb{C}[\langle X_1, \ldots, X_n \rangle][Z]$, and let L be the splitting field for F over Q_n in $\overline{Q}_n$. If $\mathrm{dis}_Z(F) = X_1^{\mu_1} \cdots X_h^{\mu_h} U$ where $h \in \{0, \ldots, n\}$, $\mu_1, \ldots, \mu_h \in \mathbb{N}$ and $U \in \mathbb{C}[\langle X_1, \ldots, X_n \rangle]$ is a unit, then there exist natural integers $e_1, \ldots, e_h$ such that $L \subset Q_n[X_1^{1/e_1}, \ldots, X_h^{1/e_h}]$; in particular, every zero of F in $\overline{Q}_n$ lies in the convergent ring of power series $\mathbb{C}[\langle X_1^{1/e_1}, \ldots, X_h^{1/e_h}, X_{h+1}, \ldots, X_n \rangle]$, and we have $L = Q_n(y)$ where y is a zero of F.*

4 Quasiordinary Power Series

(4.0) We keep the notations introduced in section 1.

(4.1) Let $y \in R_{n,d}$; the Galois group $\mathrm{Gal}(Q_n(y)/Q_n)$ of the abelian extension $Q_n(y)$ of Q_n [cf. (1.4)] will be denoted by $G(y)$, and ι is the unit element of the group $G(y)$.

(4.2) REMARK: Let

$$F = Y^d + A_1(X_1,\ldots,X_n)Y^{d-1} + \cdots + A_d(X_1,\ldots,X_n) \in R_n[Y]$$

be an irreducible Weierstraß polynomial, and assume that there exists $e \in \mathbb{N}$ and a fractionary power series $y \in Q_{n,e}$ such that $F(X_1,\ldots,X_n,y) = 0$. We may assume that $e = d$ [cf. (1.22)]. We show that y is a non-unit.
We write $y = H(X_1^{1/d},\ldots,X_n^{1/d}) \in R_{n,d}$, and let $y^{(1)} = y, y^{(2)},\ldots,y^{(d)}$ be the conjugates of y over Q_n. Then we have $y^{(j)} = H_j(X_1^{1/d},\ldots,X_n^{1/d}) \in R_{n,d}$ for $j \in \{1,\ldots,d\}$ [cf. (1.19)(1)], hence we have

$$F(X_1,\ldots,X_n,Y) = \prod_{j=1}^{d}(Y - H_j(X_1^{1/d},\ldots,X_n^{1/d})).$$

Since F is a Weierstraß polynomial, we have $F(0,\ldots,0,Y) = Y^d$, hence we have $Y^d = \prod_{j=1}^{d}(Y - H_j(0,\ldots,0))$, and this yields $H_j(0,\ldots 0) = 0$ for $j \in \{1,\ldots,d\}$.

(4.3) DEFINITION: An irreducible Weierstraß polynomial

$$F(X_1,\ldots,X_n,Y) = Y^d + A_1(X_1,\ldots,X_n)Y^{d-1} + \cdots + A_d(X_1,\ldots,X_n)$$

in $R_n[Y]$ will be called quasiordinary if its discriminant $\mathrm{dis}_Y(F) \in R_n$ has the form $\mathrm{dis}_Y(F) = X_1^{\mu_1}\cdots X_n^{\mu_n}U$ where $(\mu_1,\ldots,\mu_n) \in \mathbb{N}_0^n$ and $U \in R_n$ is a unit. Note that, if $d > 1$, then $\mathrm{dis}_Y(F)$ is not a unit of R_n by IV(3.4).

(4.4) REMARK: Let $F \in R_n[Y]$ be an irreducible quasiordinary Weierstraß polynomial of degree d, and let $y = y^{(1)},\ldots,y^{(d)}$ be its zeroes in $\overline{Q}_n$.
(1) By (2.5) resp. (3.3) and (1.22), we know that there exist fractionary power series $H_1,\ldots,H_d$ in $R_{n,d}$ such that $y^{(j)} = H_j(X_1^{1/d},\ldots,X_n^{1/d})$ for every $j \in \{1,\ldots,d\}$, and that these power series are non-units [cf. (4.2)].
(2) Now we have

$$F(X_1,\ldots,X_n,Y) = \prod_{j=1}^{d}(Y - H_j(X_1^{1/d},\ldots,X_n^{1/d})) = \prod_{\sigma \in G(y)}(Y - \sigma(y));$$

since

$$\mathrm{dis}_Y(F) = \prod_{1 \le i < j \le d}(H_j - H_i)^2 = (-1)^{d(d-1)/2}\prod_{\substack{\sigma,\tau \in G(y) \\ \sigma \ne \tau}}(\sigma(y) - \tau(y)) = NU$$

for some monomial $N \in R_n$ and some unit U of R_n, and since $R_{n,d}$ is factorial [cf. (1.3)], the following holds true. Let σ, $\tau \in G(y)$, $\sigma \ne \tau$; then there exist a monomial $M_{\sigma,\tau} \in R_{n,d}$ and a unit $U_{\sigma,\tau} \in R_{n,d}$ such that

$$\sigma(y) - \tau(y) = M_{\sigma,\tau}U_{\sigma,\tau}.$$

Moreover, since y and all its conjugates over Q_n are non-units in $R_{n,d}$, the monomial $M_{\sigma,\tau}$ has positive degree.

(4.5) REMARK: Let $y \in R_{n,d}$ be a non-unit, and assume that for every $\sigma \in G(y)$, $\sigma \neq \iota$, there exist a monomial $M_\sigma \in R_{n,d}$ and a unit $U_\sigma \in R_{n,d}$ such that

$$y - \sigma(y) = M_\sigma U_\sigma.$$

Then, for all $\sigma, \tau \in G(y)$, $\sigma \neq \tau$, there exist a monomial $M_{\sigma,\tau} \in R_{n,d}$ and a unit $U_{\sigma,\tau} \in R_{n,d}$ such that

$$\sigma(y) - \tau(y) = M_{\sigma,\tau} U_{\sigma,\tau} \tag{$*$}$$

[since $\sigma^{-1}\big(\sigma(y) - \tau(y)\big) = y - \sigma^{-1}\tau(y)$]. All the monomials M_σ and $M_{\sigma,\tau}$ have positive degree. This implies: *The minimal polynomial $F := \prod_{\sigma \in G(y)}(Y - \sigma(y)) \in R_n[Y]$ of y over Q_n is a quasiordinary irreducible Weierstraß polynomial.*
Proof: Just as above, we have

$$\mathrm{dis}_Y(F) = \pm \prod_{\substack{\sigma,\tau \in G(y) \\ \sigma \neq \tau}} (\sigma(y) - \tau(y)),$$

and from $(*)$ we get the claim.

(4.6) DEFINITION: A fractionary power series $y \in R_{n,d}$ which is a non-unit is called a quasiordinary branch if the minimal polynomial of y over Q_n is a quasiordinary Weierstraß polynomial [note that this minimal polynomial is at any rate an irreducible Weierstraß polynomial by IV(4.25)]. If y is a quasiordinary branch and if $\mathrm{Supp}(y)$ is finite, then y is said to be a quasiordinary polynomial.

Note that, in case $n = 1$, every fractionary power series which is a non-unit, is a quasiordinary branch.

(4.7) REMARK: Let $y \in R_{n,d}$ be a fractionary power series. The results in (4.4) and (4.5) show that the following statements are equivalent:
(1) The fractionary power series y is a quasiordinary branch.
(2) The fractionary power series y is a non-unit of $R_{n,d}$, and for every $\sigma \in G(y)$, $\sigma \neq \iota$, there exist a monomial $M_\sigma \in R_{n,d}$ and a unit $U_\sigma \in R_{n,d}$ such that we have $y - \sigma(y) = M_\sigma U_\sigma$.
If these conditions are satisfied, then the monomials M_σ, $\sigma \in G(y) \setminus \{\iota\}$, have positive degree.

(4.8) DEFINITION: Let $y \in R_{n,d}$ be a quasiordinary branch. Then the set

$$\{M_\sigma \mid \sigma \in G(y) \setminus \{\iota\}\}$$

of (4.7) is called the set of characteristic monomials of y.

(4.9) REMARK: Let $y \in R_{n,d}$ be a quasiordinary branch. There is a bunch of interesting properties which we are going to list.

(1) The set of characteristic monomials of y is empty iff $y \in R_n$ [in particular, this set is empty if $y = 0$].

(2) Any conjugate of y over Q_n is quasiordinary, and has the same set of characteristic monomials as y itself.

(3) Let $M = X^{m/d}$ [where $m \in \mathbb{N}_0^n$] be a characteristic monomial of y. Then we have $m \in \text{Supp}(y)$ [for every $\sigma \in G(y)$ there exists $s \in \mathbb{Z}^n$ such that $\sigma = \varphi_s$, cf. (1.6), and use (1.19)(1)].

(4) Let $M = X^{m/d}$, $M' = X^{m'/d}$ be two different characteristic monomials of y. Then we have $m < m'$ or $m' < m$.
Proof: We choose σ, $\sigma' \in G(y)$ such that $y - \sigma(y) = MU$, $y - \sigma'(y) = M'U'$ where U and U' are units of $R_{n,d}$. Then we have $\sigma \neq \sigma'$, hence we have $MU - M'U' = \sigma(y) - \sigma'(y) = M''U''$ for some monomial $M'' \in R_{n,d}$ and some unit $U'' \in R_{n,d}$ [cf. (4.4)(2)], whence $m < m'$ or $m' < m$ [since $R_{n,d}$ is factorial].

(5) Let $\{M_i = X^{m_i/d} \mid i \in \{1,\dots,h\}\}$ [where $h = 0$ if y has no characteristic monomials] be the set of pairwise different characteristic monomials of y. We will always assume that [cf. (4)]

$$0 =: m_0 < m_1 < \cdots < m_h;$$

the sequence $(m_1,\dots,m_h) \in \mathbb{N}_0^n$ will be called the sequence of characteristic exponents of y. The monomials $M_1,\dots,M_h$ lie in $Q_n(y)$ [cf. (3) and (1.19)]. For every $i \subset \{1,\dots,h\}$ we set $Q_n(i) := Q_n(M_1,\dots,M_i)$.

(6) Let $\sigma \in G(y) \setminus \{\iota\}$; then we have $y - \sigma(y) = MU$ where $M \in R_{n,d}$ is a monomial and $U \in R_{n,d}$ is a unit. We show that $\sigma(M) \neq M$. In fact, suppose that $\sigma(M) = M$; for every $l \in \mathbb{N}$ we have $y - \sigma^l(y) = M(U + \sigma(U) + \cdots + \sigma^{l-1}(U))$. Since $U(0) + \cdots + \sigma^{l-1}(U)(0) = lU(0) \neq 0$, we have $\sigma^l(y) \neq y$ for every $l \in \mathbb{N}$, and σ does not have finite order which is absurd.

(7) We show that $Q_n(h) = Q_n(y)$.
Proof: By (5) we know that $Q_n(h) \subset Q_n(y)$. We show that $y \in Q_n(h)$. Let $\sigma \in \text{Gal}(Q_{n,d}/Q_n(h))$; then we have $\sigma(M_i) = M_i$ for every $i \in \{1,\dots,h\}$. Suppose that $\sigma(y) \neq y$; then σ lies in $G(y)$, hence we have $y - \sigma(y) = M_i U$ for some $i \in \{1,\dots,h\}$ and some unit $U \in R_{n,d}$, and we have $\sigma(M_i) \neq M_i$ by (6) which is absurd. Therefore we have $\sigma(y) = y$, hence y lies in $Q_n(h)$.

(8) Let $i \in \{1,\dots,h\}$. Then we have $M_i \notin Q_n(i-1)$, hence $Q_n(i-1) \neq Q_n(i)$.
Proof: Let $\sigma \in G(y)$ be such that $y - \sigma(y) = M_i U$ where $U \in R_{n,d}$ is a unit. Then we have $\sigma(M_i) \neq M_i$ [cf. (6)]. Let $r \in \text{Supp}(y)$; if $\sigma(X^{r/d}) \neq X^{r/d}$, then we have $m_i \leq r$. Hence we have $\sigma(M_j) = M_j$ for every $j \in \{1,\dots,i-1\}$ [cf. (5)], and therefore we have $M_i \notin Q_n(i-1)$.

(9) Let $r \in \text{Supp}(y)$, and choose $i \in \{1,\dots,h\}$ such that $X^{r/d} \in Q_n(i)$, but that $X^{r/d} \notin Q_n(i-1)$ [cf. (8) and (1.19)(2)]. Then we have $m_i \leq r$.
Proof: We choose $\sigma \in \text{Gal}(Q_{n,d}/Q_n(i-1))$ such that $\sigma(X^{r/d}) \neq X^{r/d}$. Then we have $\sigma(M_i) \neq M_i$ by (8), and we have r, $m_i \in \text{Supp}(y - \sigma(y))$. Now we have $y - \sigma(y) = M_j U$ where $j \in \{1,\dots,h\}$ and $U \in R_{n,d}$ is a unit, whence $m_j \leq m_i$

and $m_j \le r$. The first inequality shows that $j \in \{1, \ldots, i\}$, and from $\sigma(M_j) \ne M_j$ we see that $j = i$ [cf. (8)], hence we have $m_i \le r$.

(4.10) Proposition: *Let $y \in R_{n,d} \setminus R_n$ be a fractionary power series, and let N be the $\mathbb{Z}$-submodule of $\mathbb{Z}^n$ associated with the subfield $Q_n(y)$ of $Q_{n,d}$. Then y is a quasiordinary branch iff there exists a system $\{m_1, \ldots, m_h\}$ of generators of N such that—N_i being the $\mathbb{Z}$-submodule of N generated by $d\mathbb{Z}^n$ and $m_1, \ldots, m_i$ for $i \in \{1, \ldots, h\}$ and $(\nu_i)_{1 \le i \le h}$ being the ν-sequence associated with d and $(m_i)_{1 \le i \le h}$— we have*

(1) $0 < m_1 < \cdots < m_h$,
(2) $m_i \in \mathrm{Supp}(y)$ for every $i \in \{1, \ldots, h\}$,
(3) N is generated by $d\mathbb{Z}^n$ and $m_1, \ldots, m_h$,
(4) if $r \in \mathrm{Supp}(y)$ and $i \in \{1, \ldots, h\}$ such that $r \in N_i$, but $r \notin N_{i-1}$, then $m_i \le r$,
(5) $\nu_i > 1$ for every $i \in \{1, \ldots, h\}$.

In this case $(m_1, \ldots, m_h)$ is the sequence of characteristic exponents of y. Furthermore, we have, setting $M_i := X^{m_i/d}$ for $i \in \{1, \ldots, h\}$,

$$[Q_n(M_1, \ldots, M_i) : Q_n] = \nu_1 \cdots \nu_i \quad \text{for every } i \in \{1, \ldots, h\};$$

in particular, we have $Q_n(y) = Q_n[M_1, \ldots, M_h]$.

Proof: (a) Assume that y is a quasiordinary branch, and let $(m_1, \ldots, m_h)$ be the sequence of characteristic exponents of y. Now (1) follows from (4.9)(5), (2) follows from (4.9)(3), (3) follows from (4.9)(7) and (1.10), (4) follows from (4.9)(9), and (5) follows from (4.9)(8) [cf. (1.10)(2)]. The last two claims follow from (1.10).
(b) Conversely, assume that $y \in R_{n,d}$ is a non-unit and that $m_1, \ldots, m_h$ satisfy (1)-(5). Let $y = \sum_{r \in \mathrm{Supp}(y)} \beta_r X^{r/d}$ be the normal representation of y. We define

$$H_i := \sum_{\substack{r \in N_i \setminus N_{i-1} \\ r \in \mathrm{Supp}(y)}} \beta_r X^{r/d} \quad \text{for every } i \in \{0, \ldots, h\}$$

$[N_{-1} = \{0\}]$. Then we have $H_0 \in R_n$ and $H_i = X^{m_i/d} U_i$ where $U_i \in R_{n,d}$ is a unit for $i \in \{1, \ldots, h\}$ [cf. (4) and (5)]. By (3) we find that $y = H_0 + H_1 + \cdots + H_h$. Let $\sigma \in G(y)$, $\sigma \ne \iota$. Then we have $\sigma(y) \ne y$, and therefore we have $\sigma(H_i) \ne H_i$ for at least one $i \in \{1, \ldots, h\}$. Let i be the smallest index with that property; by (1) we have $y - \sigma(y) = X^{m_i/d} U_\sigma$ where $U_\sigma \in R_{n,d}$ is a unit. Therefore y is a quasiordinary branch and m_i is a characteristic exponent of y [cf. (4.7)].
Now we consider the monomials $M_1, \ldots, M_h$. Let $i \in \{1, \ldots, h\}$. From (5) and (1.10) we get that $M_i \notin Q_n(M_1, \ldots, M_{i-1})$; then there exists $\sigma \in G(y)$ with $\sigma(M_i) \ne M_i$ and $\sigma(M_j) = M_j$ for $j \in \{1, \ldots, i-1\}$. Therefore we have $y - \sigma(y) = M_i U_\sigma$ where $U_\sigma \in R_{n,d}$ is a unit. Applying (4.7) yields that y is a quasiordinary branch and that $(m_1, \ldots, m_h)$ is the sequence of characteristic exponents of the quasiordinary branch y.

(4.11) Proposition: *Let L be a field between Q_n and $Q_{n,d}$. Then there exists a quasiordinary polynomial $y \in R_{n,d}$ such that $L = Q_n(y)$.*

Proof: There exist $r_1, \ldots, r_h \in \mathbb{N}_0^n$ such that $L = Q_n[X^{r_1/d}, \ldots, X^{r_h/d}]$ [cf. (1.7)].
We define $N_i := d\mathbb{Z}^n + \mathbb{Z}r_1 + \cdots + \mathbb{Z}r_i$ for $i \in \{0, \ldots, h\}$. If $N_h = N_0$, then $L = Q_n$,
and any non-unit $y \in R_n$ is a quasiordinary branch with $L = Q_n(y)$. In the other
case we may, by adding suitable multiples of $(d, \ldots, d)$ to $r_1, \ldots, r_h$, assume that
$r_i > 0$ for $i \in \{1, \ldots, h\}$. Let $i_1 \in \{1, \ldots, h\}$ be the smallest integer with $N_{i_1} \neq N_0$.
Then we define $r_1' := r_{i_1}$, and let (ν_1') be the ν-sequence associated with d and
(r_1'); note that $\nu_1' > 1$ [cf. (6.5)]. Again we may assume, by adding suitable
multiples of $(d, \ldots, d)$ to $r_{i_1+1}, \ldots, r_h$, that $r_{i_1+j} > \nu_1'r_1'$ for $j \in \{1, \ldots, h - i_1\}$.
Let $i_2 \in \{i_1 + 1, \ldots, h\}$ be the smallest integer with $N_{i_2} \neq N_{i_1}$. Then we define
$r_2' := r_{i_2}$, and let (ν_1', ν_2') be the ν-sequence associated with d and (r_1', r_2'); note that
$\nu_2' > 1$. Continuing and relabelling, we may assume: The ν-sequence $(\nu_i)_{1 \leq i \leq h}$
associated with d and the sequence $(r_i)_{1 \leq i \leq h}$ satisfies $\nu_i > 1$ for $i \in \{1, \ldots, h\}$ and
we have $r_i > \nu_{i-1}r_{i-1}$ for $i \in \{1, \ldots, h\}$ where $r_0 := 0$, $\nu_0 := 1$. Then the sequence
$(m_i)_{1 \leq i \leq h}$ defined by $m_0 := 0$ and $m_i := r_i - \nu_{i-1}r_{i-1} + m_{i-1}$ for $i \in \{1, \ldots, h\}$
satisfies $m_0 < m_1 < \cdots < m_h$ and has the same ν-sequence [cf. (6.7)], and if
we define $y := X^{m_1/d} + \cdots + X^{m_h/d}$, then y is a quasiordinary polynomial with
$Q_n(y) = Q_n[X^{m_1/d}, \ldots, X^{m_h/d}] = Q_n[X^{r_1/d}, \ldots, X^{r_h/d}]$ [cf. (4.10)].

(4.12) REMARK: The proof of (4.11) shows that, given elements $r_1, \ldots, r_h \in \mathbb{N}_0^n$
with $L = Q_n(X^{r_1/d}, \ldots, X^{r_h/d})$, we can construct explicitly a quasiordinary polynomial y such that $Q_n(y) = L$.

(4.13) NOTATION: Let $y = \sum_{r \in \mathrm{Supp}(y)} \beta_r X^{r/d} \in R_{n,d}$ be a quasiordinary branch;
we may assume that $[Q_n(y) : Q_n] = d$ [cf. (1.22)], hence that the minimal
polynomial of y over Q_n is a quasiordinary Weierstraß polynomial $F \in R_n[Y]$
of degree d. We are only interested in the case $d > 1$. Let $(m_1, \ldots, m_h)$ be the
sequence of characteristic exponents and $M_1 = X^{m_1/d}, \ldots, M_h = X^{m_h/d}$ be the
characteristic monomials of y.
(1) Let $(\theta_i)_{1 \leq i \leq h+1}$ be the divisor sequence and $(\nu_i)_{1 \leq i \leq h}$ be the ν-sequence associated with d and the sequence $(m_i)_{1 \leq i \leq h}$. Since $Q_n(y) = Q_n(h)$, we have
$d = \nu_1 \cdots \nu_h = d^n/\theta_{h+1}$ [cf. (1.10)], and therefore we have $\theta_{h+1} = d^{n-1}$.
(2) As in (6.7) we define the elements $r_1, \ldots, r_h$ by

$$r_0 := 0, \quad r_i := \nu_{i-1}r_{i-1} + m_i - m_{i-1} \quad \text{for every } i \in \{1, \ldots, h\}$$

[where $m_0 := 0$, $\nu_0 = 1$]. We call $(r_1, \ldots, r_h)$ the semigroup sequence of the
quasiordinary branch y. By (6.10) and (4.10) we know that the set of strict linear
combinations of d and $r_1, \ldots, r_h$ is the semigroup $d\mathbb{N}_0^n + \mathbb{N}_0 r_1 + \cdots + \mathbb{N}_0 r_h$ which
will be denoted by $\Gamma(y)$ and will be called the semigroup of the quasiordinary
branch y. Note that $Q_n(y) = Q_n[X^{r_1/d}, \ldots, X^{r_h/d}]$.

(4.14) REMARK: When proving (4.11) we constructed a sequence $(r_i)_{1 \leq i \leq h}$ and
a quasiordinary branch y; the sequence $(r_i)_{1 \leq i \leq h}$ is the semigroup sequence of the
quasiordinary branch y.

We study the integral closure of the ring $R_n[y]$.

(4.15) Proposition: *With notations as in* (4.13) *let S be the integral closure of $R_n[y]$, and let $\overline{\Gamma}$ be the saturation of $\Gamma(\dot{y})$. Then:*
(1) *S is the integral closure of $R_n[X^{r_1/d}, \ldots, X^{r_h/d}]$.*
(2) *An element $z \in R_{n,d}$ lies in S iff $\mathrm{Supp}(z) \subset \overline{\Gamma}$.*
(3) *Let Σ be the set of strict linear combinations of $r_1, \ldots, r_h$, and set $\Sigma' := \{s \bmod d \mid s \in \Sigma\}$. Then we have $\mathrm{Card}(\Sigma') = \nu_1 \cdots \nu_h = d$, and the set $\{X^{s/d} \mid s \in \Sigma'\}$ is a system of generators of the R_n-module S, hence S is a free R_n-module.*

Proof: (1) Since S is the integral closure of R_n in $Q_n(y)$, we have $S = Q_n(y) \cap R_{n,d}$; since the elements $X^{r_1/d}, \ldots, X^{r_h/d}$ are integral over R_n and lie in $Q_n(y)$, the assertion follows immediately from $Q_n(y) = Q_n[X^{r_1/d}, \ldots, X^{r_h/d}]$ [cf. (4.13)(2)].
(2) This is a consequence of (1.21) and (6.13).
(3) Let $z \in S$ be a non-zero element and let $r \in \mathrm{Supp}(z)$. Then we have $r \in \overline{\Gamma}$ by (2), and therefore $r = da_0 + s$ where $a_0 \in \mathbb{N}_0^n$ and $s \in \Sigma'$ [cf. (6.14)]. Thus, we can write $z = \sum_{s \in \Sigma'} f_s X^{s/d}$ where $f_s \in R_n$ for every $s \in \Sigma'$, i.e., $\{X^{s/d} \mid s \in \Sigma'\}$ is a system of generators of the R_n-module S, and therefore we have $\mathrm{Card}(\Sigma') \geq d$. On the other hand, we have $\mathrm{Card}(\Sigma') \leq d$ by (6.14), and therefore we have $\mathrm{Card}(\Sigma') = d$, hence S is a free R_n-module having $\{X^{s/d} \mid s \in \Sigma'\}$ as a basis.

(4.16) Remark: Let $L = Q_n[X^{r_1/d}, \ldots, X^{r_h/d}]$ be a field between Q_n and $Q_{n,d}$. The proof of (4.11) shows how to construct a quasiordinary branch $y \in R_{n,d}$ with $L = Q_n(y)$, and the proof of (4.15) shows how we can construct an R_n-basis for the integral closure of $R_n[y]$ in $Q_n(y)$.

(4.17) Corollary: *The integral closure of $R_n[y]$ is a Cohen-Macaulay ring.*

Proof: This follows from (4.15) and [63], Cor. 18.17.

(4.18) Corollary: *Let L be a field between Q_n and $Q_{n,d}$. The integral closure of R_n in L is a free R_n-module and a Cohen-Macaulay ring.*

Proof: By (4.11) there exists a quasiordinary branch $y \in R_{n,d}$ such that $L = Q_n(y)$, and since y is integral over R_n, the result follows from (4.15) and [63], Cor. 18.17.

(4.19) The case $n = 1$: We now specialize to the case $n = 1$. We shall write X instead of X_1, we set $R := k[\{X\}]$, and we denote the field of quotients of R by Q. For $d \in \mathbb{N}$ we write $R_d = R[X^{1/d}]$ and $Q_d = Q[X^{1/d}]$. We have $Q_d \subset Q_e$ if e divides d, hence

$$Q^* := \bigcup_{d \in \mathbb{N}} Q_d$$

is an algebraic extension of Q.
(1) Let $d \in \mathbb{N}$, $d > 1$, and let

$$F := Y^d + A_1(X)Y^{d-1} + \cdots + A_d \in R[Y] \quad \text{with } A_1, \ldots, A_d \in R \qquad (*)$$

be an irreducible polynomial. Set $L := Q[Y]/(F)$. We chose $\gamma \in k$ such that, replacing Y by $Y - \gamma$, we may assume that F is Y-regular of order $e > 0$. By Weierstraß preparation theorem [cf. IV(3.9)] we can write $F = UP$ where $U \in R$ is a unit and $P \in R[Y]$ is an irreducible Weierstraß polynomial. Therefore, we may assume, to begin with, that $L = Q[Y]/(F)$ where $F \in R[Y]$ is an irreducible quasiordinary Weierstrass polynomial.

(2) Conversely, let L be a finite extension of Q of degree d. Then there exists an irreducible quasiordinary Weierstraß polynomial F as in $(*)$ such that $L = Q[Y]/(F)$.

(3) Let F be as in $(*)$, assume that F is an irreducible Weierstraß polynomial, and let $y \in L$ be a zero of F. Then $y \in R_d$ is a fractionary power series, and since $[Q_d : Q] = d$, we have $L = Q(y) = Q_d$, hence y is a primitive element for Q_d over Q.

We can state

(4.20) Theorem: [Puiseux's Theorem] (1) *The field Q^* is an algebraic closure of the field $Q = k(\{X\})$.*
(2) *For every $d \in \mathbb{N}$ the integral closure of R in Q_d is the discrete valuation ring R_d.*

Proof: (1) We have shown in (4.19) that every extension of Q of finite degree d is contained in Q^*.
(2) Clearly R_d is a discrete valuation ring; and it is integral over R.

(4.21) Characteristic pairs: (1) Let $d > 1$, and let $y \in R_d$ with $[Q(y) : Q] = d$. Let $m_1 < \cdots < m_h$ be the characteristic sequence of the quasiordinary branch y. We change the notation to connect these numbers with what is classically called the characteristic sequence. We write $g := h$, $\beta_0 := d$, $\beta_i := m_i$ for $i \in \{1, \ldots, g\}$, and $t := X^{1/d}$. Then $(\beta_i)_{0 \leq i \leq g}$ is called the *characteristic sequence* of the pair $(t^d, y(t))$.
(2) Set

$$e_0 := d, \quad e_i := \frac{d}{\nu_1 \cdots \nu_i} \quad \text{for } i \in \{1, \ldots, g\}.$$

Then we have

$$e_i = \gcd(\beta_0, \ldots, \beta_i) \quad \text{for every } i \in \{0, \ldots, g\},$$

and $e_0 > e_1 > \cdots > e_g = 1$. Define

$$m_i := \frac{\beta_i}{e_i}, \quad n_i := \frac{e_{i-1}}{e_i} \quad \text{for every } i \in \{1, \ldots, g\}.$$

Then we have

$$\frac{\beta_i}{\beta_0} = \frac{m_i}{n_1 \cdots n_i} \quad \text{for every } i \in \{1, \ldots, g\},$$

and we have $\beta_0 = e_0 = n_1 e_1 = n_1 n_2 e_2 = \cdots = n_1 \cdots n_g$. Furthermore, we have

$$m_i n_{i+1} < m_{i+1} \quad \text{for every } i \in \{1, \ldots, g-1\},$$

$$\gcd(m_i, n_i) = 1, \quad n_i > 1 \quad \text{for every } i \in \{1, \ldots, g\}.$$

The g pairs $(m_1, n_1), \ldots, (m_g, n_g)$ of coprime integers are called the *characteristic pairs* of $(t^d, y(t))$. Define

$$h_0 := \left\lfloor \frac{m_1}{n_1} \right\rfloor, \quad h_i := \left\lfloor \frac{m_{i+1}}{n_{i+1}} \right\rfloor - m_i \quad \text{for every } i \in \{1, \ldots, g-1\}.$$

We can write

$$y(t) = \sum_{\mu=0}^{h_0} \gamma_\mu t^{n\mu} + \sum_{i=1}^{g-1} \sum_{\mu=0}^{h_i} \gamma_{i,\mu} t^{(m_i+\mu)n_{i+1}\cdots n_g} + \sum_{\mu=0}^{\infty} \gamma_{g,\mu} t^{m_g+\mu} \tag{$*$}$$

where the coefficients γ_μ, $\gamma_{i,\mu}$ lie in k and $\gamma_{i,0} \neq 0$ for every $i \in \{1, \ldots, g\}$ [note that $\beta_0 = d$, that, for every $i \in \{1, \ldots, g-1\}$, we have $\beta_i + \mu e_i = (m_i+\mu)n_{i+1}\cdots n_g$ for every $\mu \in \{0, \ldots, h_i\}$, and that $m_g = \beta_g$].
(3) The integral closure of $R[y]$ is the discrete valuation ring R_d [cf. (4.20)]. Let ν be the valuation of Q_d defined by R_d. It can be shown that $\Gamma(y) = \nu(R[y] \setminus \{0\})$ [cf., e.g., [40], Ch. IV], hence that $\Gamma(y)$ *is an invariant of the ring* $R[y]$.

5   A Generalized Newton Algorithm

(5.0) We keep the notations introduced in the earlier sections of this chapter.

5.1   The Algorithm

(5.1) REMARK: Let F be as in (4.2), and assume that there exists $y \in Q_{n,d}$ with $F(X_1, \ldots, X_n, y) = 0$. (This is the case if F is a quasiordinary Weierstraß polynomial, cf. (4.3) and (4.4).) Let $y = \sum_{i \in S(y)} \alpha_i M_i$ where $M_i = X^{r_i/d}$ for every $i \in S(y)$, be the normal representation of y. In the following, we will describe an algorithm to find M_i and α_i for every $i \in S(y)$ from the coefficients $A_1, \ldots, A_d$ of F. More precisely: we will see that the monomials $M_1, M_2, \ldots$ are determined uniquely while the coefficients $\alpha_1, \alpha_2, \ldots$ are determined only up to factors in k which are roots of unity.

(5.2) NOTATION: (1) With respect to d and the sequence $(r_i)_{i \in S(y)}$ we may define the sequence $(\nu_i)_{i \in S(y)}$ as in (6.4) [cf. the remark in (1.14)].
(2) We set $Q_n(0) := Q_n$ and $\delta_1 := [Q_n(y) : Q_n]$; furthermore, we set $Q_n(i) := Q_n(M_1, \ldots, M_i)$ and $\delta_{i+1} := [Q_n(y) : Q_n(i)]$ for $i \in S(y)$.
(3) By (1.10) we have $[Q_n(i) : Q_n] = \nu_1 \cdots \nu_i$ for every $i \in S(y)$; note that $\delta_i = \nu_i \delta_{i+1}$ for every $i \in S(y)$, and, in particular, $\delta_1 = d = \nu_1 \delta_2$. There exists $h \in S(y)$ with $Q_n(y) = Q_n(h)$ [cf. (1.20)]. Then we have $\nu_i = 1$ for every $i \in S(y)$ with $i > h$, and we have $\delta_i = \nu_i \cdots \nu_h$ for every $i \in \{1, \ldots, h\}$.

(5.3) DETERMINATION OF M_1 AND α_1: Note that

$$(-1)^d A_d(X_1, \ldots, X_n) = y^{(1)} \cdots y^{(d)}. \tag{*}$$

(1) Clearly, by $(*)$ and (1.18)(2)(a), the leading monomial of $A_d(X_1, \ldots, X_n)$ is M_1^d; thus, r_1 can be found by inspecting $\mathrm{Supp}(A_d)$, namely, r_1 is the smallest element in $\mathrm{Supp}(A_d)$. Furthermore, we can calculate ν_1 [cf. (1.13)] and δ_2 [since $d = \nu_1 \delta_2$].

(2) Now $Q_n(M_1)$ is a cyclic extension of Q_n of degree ν_1 and $[Q_n(y) : Q_n(M_1)] = \delta_2$. Let $\Gamma_1 \subset \mathrm{Gal}(Q_{n,d}/Q_n)$ be a system of representatives for the cosets of the subgroup $\mathrm{Gal}(Q_{n,d}/Q_n(M_1))$ [cf. (1.13)]; then we have $\nu_1 = \mathrm{Card}(\Gamma_1)$. Let $\varepsilon_1 \in k$ be a primitive ν_1-th root of unity. There are ν_1 pairwise different conjugates of M_1 over Q_n, namely $\varepsilon_1^l M_1$ for $l \in \{1, \ldots, \nu_1\}$.

(3) From (2) we see that the conjugates of y can be ordered in the following way:

$$y^{(j)} = \begin{cases} \alpha_1 \, M_1 + y_1^{(j)} & \text{for } j \in \{1, \ldots, \delta_2\}, \\ \alpha_1^{(j)} M_1 + y_1^{(j)} & \text{for } j \in \{\delta_2 + 1, \ldots, d\}; \end{cases}$$

here we have

$$\alpha_1^{(l\delta_2 + \lambda)} = \varepsilon_1^l \alpha_1 \quad \text{for } l \in \{1, \ldots, \nu_1 - 1\} \text{ and } \lambda \in \{1, \ldots, \delta_2\}.$$

Thus, there are ν_1 different choices for the leading coefficient of y, and any two such choices differ by a factor which is a ν_1 th root of unity. Clearly, if $y_1^{(1)} \neq 0$, then M_2 is the leading monomial of $y_1^{(j)}$ for every $j \in \{1, \ldots, d\}$.

(4) Let $S_1, \ldots, S_d \in \mathbb{Z}[T_1, \ldots, T_d]$ be the elementary symmetric polynomials in d indeterminates $T_1, \ldots, T_d$. For $l \in \{0, \ldots, d-1\}$ we are interested in the leading term of $A_{d-l}(X_1, \ldots, X_n) = (-1)^{d-l} S_{d-l}(y^{(1)}, \ldots, y^{(d)})$.

The minimal polynomial of $\alpha_1 M_1$ over Q_n is $Y^{\nu_1} - \alpha_1^{\nu_1} M_1^{\nu_1}$ [cf. (1.13)], hence

$$(Y^{\nu_1} - \alpha_1^{\nu_1} M_1^{\nu_1})^{\delta_2} = \sum_{l=0}^{\delta_2} \binom{\delta_2}{l} Y^{l\nu_1} (-\alpha_1^{\nu_1} M_1^{\nu_1})^{\delta_2 - l}$$

is the field polynomial of $\alpha_1 M_1 \in Q_n(y)$ over Q_n. Therefore M_1^{d-l} is the leading monomial of A_{d-l} iff $A_{d-l} \neq 0$ and $\nu_1 \mid (d-l)$. Let $\lambda \in \{1, \ldots, \delta_2\}$, and let β_λ be the leading coefficient of $A_{\lambda\nu_1}$ if $A_{\lambda\nu_1} \neq 0$ and define $\beta_\lambda := 0$ if $A_{\lambda\nu_1} = 0$. Comparing the coefficients of the leading monomial M_1^d in $0 = F(y)$ shows that α_1 is a zero of the polynomial

$$\Omega(T) = T^{\nu_1 \delta_2} + \beta_1 T^{\nu_1(\delta_2 - 1)} + \cdots + \beta_{\delta_2} = (T^{\nu_1} - \alpha_1^{\nu_1})^{\delta_2} \in k[T];$$

therefore the polynomial Ω may be used to calculate the leading coefficient of y.

(5) Now we make the substitution

$$Y_1 := Y - \alpha_1 M_1$$

to get $F_1(X_1^{1/d}, \ldots, X_n^{1/d}, Y_1) := F(X_1, \ldots, X_n, Y_1 + \alpha_1 M_1)$ where

$$F_1 = Y_1^d + A_{1,1}(X_1^{1/d}, \ldots, X_n^{1/d})Y_1^{d-1} + \cdots + A_{1,d}(X_1^{1/d}, \ldots, X_n^{1/d}) \in R_{n,d}[Y_1].$$

The d zeroes of F_1 are [cf. (3)]

$$y_1^{(j)} \text{ for } j \in \{1, \ldots, \delta_2\},$$
$$\beta_{1,1}^{(j)} M_1 + y_1^{(j)} \text{ for } j \in \{\delta_2 + 1, \ldots, d\}$$

[note that $\beta_{1,1}^{(j)} := \alpha_1^{(j)} - \alpha_1 \neq 0$ for every $j \in \{\delta_2 + 1, \ldots, d\}$].

(5.4) THE INDUCTION ASSUMPTION: Let $i \in \mathbb{N}$, and assume that the terms $M_1, \ldots, M_i$, $\alpha_1, \ldots, \alpha_i$ have been calculated. Let $\nu_1, \ldots, \nu_i$, $\delta_1, \ldots, \delta_{i+1}$ be the natural integers which are defined by $[Q_n(j) : Q_n] = \nu_1 \cdots \nu_j$ and $\delta_j := \delta_{j+1}\nu_j$ for $j \in \{1, \ldots, i\}$. Assume, furthermore, that in the course of calculating $M_1, \ldots, \alpha_i$ we have got the following data.
(1) A system of representatives $\Gamma_i \subset \mathrm{Gal}(Q_{n,d}/Q_n)$ for the cosets of the Galois group $\mathrm{Gal}(Q_{n,d}/Q_n(i))$. The set Γ_i has $\nu_1 \cdots \nu_i$ elements.
(2) Order the conjugates of y in such a way that $y^{(1)}, \ldots, y^{(\delta_{i+1})}$ are those conjugates of y the first i terms of which are left invariant by the elements of Γ_i, i.e.,

$$y^{(j)} = \begin{cases} \displaystyle\sum_{l=1}^{i} \alpha_l M_l + y_i^{(j)} & \text{for } j \in \{1, \ldots, \delta_{i+1}\}, \\ \displaystyle\sum_{l=1}^{i} \alpha_{i,l}^{(j)} M_l + y_i^{(j)} & \text{for } j \in \{\delta_{i+1} + 1, \ldots, d\} \end{cases}$$

where $\alpha_{i,l}^{(j)}$, $l \in \{1, \ldots, i\}$ and $j \in \{\delta_{i+1} + 1, \ldots, d\}$, are non-zero elements of k. If $y_i^{(1)} \neq 0$, then the leading monomial of $y_i^{(j)}$ is M_{i+1} for every $j = \{1, \ldots, d\}$.
(3) A polynomial

$$F_i = Y_i^d + A_{i,1}(X_1^{1/d}, \ldots, X_n^{1/d})Y_i^{d-1} + \cdots + A_{i,d}(X_1^{1/d}, \ldots, X_n^{1/d}) \in R_{n,d}[Y_i].$$

(4) For every $l \in \{1, \ldots, i\}$ and every $j \in \{\delta_{i+1} + 1, \ldots, \delta_l\}$ a sum of monomials

$$G_i^{(j)} := \sum_{\lambda=l}^{i} \beta_{i,\lambda}^{(j)} M_\lambda, \quad \beta_{i,\lambda}^{(j)} \in k \text{ for } \lambda \in \{l, \ldots, i\}, \beta_{i,l}^{(j)} \neq 0,$$

such that

$$y_i^{(j)} \text{ for } j \in \{1, \ldots, \delta_{i+1}\}, \quad G_i^{(j)} + y_i^{(j)} \text{ for } j \in \{\delta_{i+1} + 1, \ldots, d\}$$

are the d zeroes of F_i. Note that, for $l \in \{1, \ldots, i\}$ and $j \in \{\delta_{i+1}, \ldots, \delta_l\}$, M_l is the leading monomial of $G_i^{(j)} + y_i^{(j)}$.

(5.5) THE INDUCTION STEP: Starting with the data in (5.4), we will describe the next step. If $A_{i,d} = 0$, then $y_i^{(1)} = 0$, and we are done: $y = \sum_{l=1}^{i} \alpha_l M_l$ is a zero of F. Now we assume that $A_{i,d} \neq 0$.

(1) It is easy to calculate M_{i+1}. Namely, by (1.18)(2), and since $\delta_l - \delta_{l+1} = \delta_{l+1}(\nu_l - 1)$ for every $l \in \{1, \ldots, i\}$, the monomial

$$M_{i+1}^{\delta_{i+1}} M_i^{\delta_i - \delta_{i+1}} M_{i-1}^{\delta_{i-1} - \delta_i} \cdots M_1^{\delta_1 - \delta_2} = M_{i+1}^{\delta_{i+1}} \prod_{l=1}^{i} M_l^{\delta_{i+1}(\nu_l - 1)}$$

is the leading monomial of $A_{i,d}(X_1^{1/d}, \ldots, X_n^{1/d})$. Thus, we know M_{i+1}, and therefore we know ν_{i+1} and $\delta_{i+2} = \delta_{i+1}/\nu_{i+1}$.

(2) We know that $\mathrm{Gal}\big(Q_n(i+1)/Q_n(i)\big)$ is a cyclic group of order ν_{i+1}, and using the construction described in (1.14), we may find a representative $\psi \in \mathrm{Gal}\big(Q_{n,d}/Q_n(i)\big)$ for a generator of the Galois group $\mathrm{Gal}\big(Q_n(i+1)/Q_n(i)\big)$. Then the set $\Gamma_{i+1} := \{\Gamma_i \psi^l \mid l \in \{1, \ldots, \nu_{i+1}\}\}$ is a system of representatives for the cosets of the subgroup $\mathrm{Gal}(Q_{n,d}/Q_n(i+1))$ of $\mathrm{Gal}(Q_{n,d}/Q_n)$.

(3) Let $\varepsilon_{i+1} \in k$ be a primitive ν_{i+1}-th root of unity. The conjugates $y^{(1)}, \ldots, y^{(\delta_{i+1})}$ can be ordered in the following way:

$$y^{(j)} = \begin{cases} \sum_{l=1}^{i} \alpha_l M_l + \alpha_{i+1} M_{i+1} + y_{i+1}^{(j)} & \text{for } j \in \{1, \ldots, \delta_{i+2}\}, \\ \sum_{l=1}^{i} \alpha_l M_l + \alpha_{i+1}^{(j)} M_{i+1} + y_{i+1}^{(j)} & \text{for } j \in \{\delta_{i+2} + 1, \ldots, \delta_{i+1}\}; \end{cases}$$

here we have

$$\alpha_{i+1}^{(l\delta_{i+2} + \lambda)} = \varepsilon_{i+1}^{l} \alpha_{i+1} \quad \text{for } l \in \{1, \ldots, \nu_{i+1} - 1\} \text{ and } \lambda \in \{1, \ldots, \delta_{i+2}\}.$$

Thus, there are ν_{i+1} different choices for the leading coefficient α_{i+1} of $y_i^{(1)}$, and any two such choices differ by a factor which is a ν_{i+1}-th root of unity. Clearly, if $y_{i+1}^{(1)} \neq 0$, then M_{i+2} is the leading monomial of $y_{i+1}^{(1)}$ for every $j \in \{1, \ldots, \delta_{i+1}\}$. The zeroes of F_i can now be written in the following form:

$$y_i^{(j)} = \begin{cases} \alpha_{i+1} M_{i+1} + y_{i+1}^{(j)} & \text{for } j \in \{1, \ldots, \delta_{i+2}\}, \\ \alpha_{i+1}^{(j)} M_{i+1} + y_{i+1}^{(j)} & \text{for } j \in \{\delta_{i+2} + 1, \ldots, \delta_{i+1}\}, \\ G_i^{(j)} + \alpha_{i+1}^{(j)} M_{i+1} + y_{i+1}^{(j)} & \text{for } j \in \{\delta_{i+1} + 1, \ldots, d\}. \end{cases}$$

Let $S_1, \ldots, S_{\delta_{i+1}} \in \mathbb{Z}[T_1, \ldots, T_{\delta_{i+1}}]$ be the elementary symmetric polynomials in δ_{i+1} indeterminates $T_1, \ldots, T_{\delta_{i+1}}$. The coefficient of $Z^{\delta_{i+1} - \nu_{i+1}}$ in the polynomial

$$\left(Z^{\nu_{i+1}} - 1\right)^{\delta_{i+2}} = \prod_{l=1}^{\nu_{i+1}} \left(Z - \varepsilon_{i+1}^{l-1}\right)^{\delta_{i+2}}$$

is, by binomial expansion,

$$- \delta_{i+2} = (-1)^{\nu_{i+1}} S_{\nu_{i+1}}(1,\ldots,1,\varepsilon_{i+1},\ldots,\varepsilon_{i+1},\ldots,\varepsilon_{i+1}^{\nu_{i+1}-1},\ldots,\varepsilon_{i+1}^{\nu_{i+1}-1})$$

[where, on the right hand side, there are δ_{i+2} elements 1, δ_{i+2} elements ε_{i+1}, ..., δ_{i+2} elements $\varepsilon_{i+1}^{\nu_{i+1}-1}$].

The coefficient $A_{i,d-\delta_{i+1}+\nu_{i+1}}(X_1^{1/d},\ldots,X_n^{1/d})$ of $Y_i^{\delta_{i+1}-\nu_{i+1}}$ in F_i is, up to a sign, the $(d - \delta_{i+1} + \nu_{i+1})$-th symmetric function of the zeroes of F_i. The leading monomial of $A_{i,d-\delta_{i+1}+\nu_{i+1}}$ is, therefore, the product of the leading monomial of $S_{\nu_{i+1}}(y_i^{(1)},\ldots,y_i^{(\delta_{i+1})})$ and the leading monomial of $G_i^{(\delta_1)}\cdots G_i^{(\delta_{i+1}+1)}$, hence is equal to

$$M_{i+1}^{\nu_{i+1}} M_i^{\delta_i-\delta_{i+1}} M_{i-1}^{\delta_{i-1}-\delta_i} \cdots M_1^{\delta_1-\delta_2} = M_{i+1}^{\nu_{i+1}} \prod_{l=1}^{i} M_l^{\delta_{l+1}(\nu_l-1)},$$

and the leading coefficient β of $A_{i,d-\delta_{i+1}+\nu_{i+1}}$ is equal to

$$(-1)^{d-\delta_{i+1}+1} \delta_{i+2} \alpha_{i+1}^{\nu_{i+1}} \prod_{l=1}^{i} \prod_{\lambda=\delta_l}^{\delta_{l+1}+1} \beta_{i,l}^{(\lambda)},$$

hence we find that

$$\alpha_{i+1}^{\nu_{i+1}} = (-1)^{d-\delta_{i+1}+1} \frac{\beta}{\delta_{i+2}} \left(\prod_{l=1}^{i} \prod_{\lambda=\delta_l}^{\delta_{l+1}+1} \beta_{i,l}^{(\lambda)} \right)^{-1}. \qquad (*)$$

As α_{i+1} is determined up to multiplication by a ν_{i+1}-th root of unity, we may calculate α_{i+1} by using the last displayed equation; let α_{i+1} be an element of k satisfying $(*)$.

(4) Using the system of representatives Γ_{i+1}, we may calculate the coefficients $\alpha_{i+1,l}^{(j)}$ for every $l \in \{1,\ldots,i+1\}$, $j \in \{\delta_{i+1}+1,\ldots,d\}$, appearing in the following set of displayed equations

$$y^{(j)} = \begin{cases} \sum_{l=1}^{i+1} \alpha_l M_l + y_{i+2}^{(j)} & \text{for } j \in \{1,\ldots,\delta_{i+2}\}, \\ \sum_{l=1}^{i+1} \alpha_{i+1,l}^{(j)} M_l + y_{i+2}^{(j)} & \text{for } j \in \{\delta_{i+2}+1,\ldots,d\}. \end{cases}$$

(5) Now we make the substitution

$$Y_{i+1} := Y_i - \alpha_{i+1} M_{i+1},$$

and we get a monic polynomial $F_{i+1}(X_1^{1/d},\ldots,X_n^{1/d},Y_{i+1}) \in R_{n,d}[Y_{i+1}]$ of degree d, having as zeroes the elements

$$y_{i+1}^{(j)} \quad \text{for } j \in \{1,\ldots,\delta_{i+2}\}, \qquad G_{i+1}^{(j)} + y_{i+1}^{(j)} \quad \text{for } j \in \{\delta_{i+2}+1,\ldots,d\}$$

where for every $j \in \{\delta_{i+2} + 1, \ldots, \delta_{i+1}\}$

$$G_{i+1}^{(j)} := \beta_{i+1,i+1}^{(j)} M_{i+1}, \quad \beta_{i+1,i+1}^{(j)} := \alpha_{i+1}^{(j)} - \alpha_{i+1} \neq 0,$$

and

$$G_{i+1}^{(j)} := G_i^{(j)} - \alpha_{i+1} M_{i+1} \quad \text{for every } j \in \{\delta_{i+1} + 1, \ldots, \delta_1\}.$$

At this point of the algorithm we have the same situation as at the beginning of (5.4), with i replaced by $i + 1$.

(5.6) REMARK: We keep the notations used in (5.2)–(5.5).
(1) There exists $h \in \mathbb{N}$ such that $Q_n(y) = Q_n(M_1, \ldots, M_h)$ [cf. (1.20)]. Thus, Γ_h is a system of representatives for $\mathrm{Gal}(Q_{n,d}/Q_n(y))$, and we have constructed the Galois group $\mathrm{Gal}(Q_n(y)/Q_n)$.
(2) For every $i > h$ we have $\nu_i = 1$, and the coefficient α_i is determined uniquely.

(5.7) Corollary: Let $r_1 = (\rho_{11}, \ldots, \rho_{1n})$. The local ring $R_n[y]$ has multiplicity

$$e(R_n[y]) = \min(d, \rho_{11} + \cdots + \rho_{1n}).$$

Proof: Set $\rho := \rho_{11} + \cdots + \rho_{1n}$. For every $j \in \{0, \ldots, d-1\}$ the order of $A_{d-j} Y^j$ is at least $\rho(d-j)/d + j$ [by the reasoning in (5.3)(4)], hence $\geq d$ if $\rho \geq d$, and $\rho(d-j)/d + j = \rho + j(d-\rho)/d \geq \rho$ if $\rho < d$. Therefore we have $o(F) = \min(d, \rho)$, and the result follows from B(10.8).

(5.8) EXAMPLE: Let $0 < p < q$ be coprime integers, and set $F := Z^q - XY^p \in k[\![X, Y]\!][Z]$. Now F is irreducible by Eisenstein's criterion, and since $\mathrm{dis}_Z(F) = (-1)^{q(q-1)/2} q^q (XY^p)^{q-1}$ [cf. I(7.6)(2)], F is a quasiordinary Weierstraß polynomial. Let z be a quasiordinary branch defined by F; clearly z has only one characteristic monomial, namely $X^{1/q} Y^{p/q}$. Therefore the only characteristic exponent of z is $m_1 = (1, p)$, and the semigroup sequence of z has the only member $r_1 = m_1$. Then the integral closure S of $R_2 := k[\![X, Y]\!]$ in the field of quotients of $R_2[z]$ is, by (4.15) and (6.15)(2), a free R_2-module generated by the q elements $1, X^{1/q} Y^{p[1]/q}, \ldots, X^{(q-1)/q} Y^{p[q-1]/q}$.
We have $e(R_2[z]) = p + 1$ by (5.7) and $\mathrm{emdim}(R_2[z]) = 3$; in particular, $R_2[z]$ is not a regular local ring. We shall calculate the embedding dimension and the multiplicity of S in VI(4.23); clearly we have $\mathrm{emdim}(S) > 2$, hence S is not a regular local ring.

5.2   An Example

(5.9) EXAMPLE: Let $k = \mathbb{C}$, $n = 2$, and, for the sake of simplicity, set $X := X_1$, $Y := X_2$; consider $F \in \mathbb{C}[\![X, Y]\!][Z]$ where

$$F = Z^4 + (-2XY^3 - 4XY^4 - 2XY^5)Z^2 + (-4X^2Y^6 - 4X^2Y^7)Z$$
$$+ (X^2Y^6 + 4X^2Y^7 + 6X^2Y^8 + 4X^2Y^9 + X^2Y^{10} - X^3Y^9).$$

The discriminant of F is

$$\mathrm{dis}_Z(F) = -256X^7Y^{21}((1+Y)^4 + XY^3).$$

We will show below that F is irreducible in $\mathbb{C}[\![X,Y]\!][Z]$; hence F is an irreducible
Weierstraß polynomial, and $d = 4$. By (2.5) and (1.22) we know that there exists
a fractional power series $z \in Q_{2,4}$ satisfying $F(z) = 0$. We construct its normal
representation using the algorithm described above.
(1) The leading monomial of A_4 is X^2Y^6, hence we have $r_1 = (2,6)$ and $M_1 = X^{2/4}Y^{6/4}$. Now we have $\theta_1 = 16$, $\theta_2 = 8$, and therefore we get $\nu_1 = 2$, $\delta_2 = 2$.
The set $\Gamma_1 = \{\varphi_{(0,0)}, \varphi_{(1,0)}\} \subset \mathrm{Gal}(Q_{2,4}/Q_2)$ is a system of representatives for the
cosets of the Galois group $\mathrm{Gal}(Q_{2,4}/Q_2(M_1))$. Furthermore, $\nu_1 = 2$ is a divisor of
2 and 0; the leading term of A_2 is $-2XY^3$ and the leading term of A_0 is X^2Y^6.
Therefore we have

$$\Omega(T) = T^4 - 2T^2 + 1 = (T^2 - 1)^2,$$

and we may choose $\alpha_1 = 1$.
The four conjugates of z over Q_2 are

$$z^{(1)} = M_1 + z_1^{(1)}, \qquad\qquad z^{(2)} = M_1 + z_1^{(2)},$$
$$z^{(3)} = -M_1 + z_1^{(3)}, \qquad\qquad z^{(4)} = -M_1 + z_1^{(4)}.$$

(2) Now we make the substitution $Z_1 := Z - M_1$; we get

$$F_1 = Z_1^4 + 4X^{2/4}Y^{6/4}Z_1^3 + (4XY^{12/4} - 4XY^{16/4} - 2XY^{20/4})Z_1^2$$
$$+ (-8X^{6/4}Y^{22/4} - 4X^{6/4}Y^{26/4} - 4X^{8/4}Y^{24/4} - 4X^{8/4}Y^{28/4})Z_1$$
$$+ (4X^{8/4}Y^{32/4} + 4X^{8/4}Y^{36/4} + X^{8/4}Y^{40/4}$$
$$- 4X^{10/4}Y^{30/4} - 4X^{10/4}Y^{34/4} - X^{12/4}Y^{36/4}).$$

The zeroes of F_1 are $z_1^{(1)}$, $z_1^{(2)}$, $-2M_1 + z_1^{(3)}$, $-2M_1 + z_1^{(4)}$. The leading monomial
of $A_{1,4}$ is $X^{8/4}Y^{32/4}$, hence we have

$$M_2^2 \cdot M_1^2 = X^{8/4}Y^{32/4},$$

and therefore $r_2 = (2,10)$, $M_2 = X^{2/4}Y^{10/4}$. Since $\theta_3 = 8$, we have $\nu_2 = \theta_2/\theta_3 = 1$
and $\delta_3 = \delta_2/\nu_2 = 2$. The leading coefficient of $A_{1,4-2+1}$ is -8, hence we have

$$\alpha_2 = (-1)^3 \frac{(-8)}{2} \frac{1}{(-2)\cdot(-2)} = 1.$$

Furthermore, we have $\Gamma_2 = \Gamma_1$. Therefore we get

$$z^{(1)} = M_1 + M_2 + z_2^{(1)}, \qquad\qquad z^{(2)} = M_1 + M_2 + z_2^{(2)},$$
$$z^{(3)} = -M_1 - M_2 + z_2^{(3)}, \qquad\qquad z^{(4)} = -M_1 - M_2 + z_2^{(4)}$$

[note that $z^{(3)} = \varphi_{(1,0)}(z^{(1)})$, $z^{(4)} = \varphi_{(1,0)}(z^{(2)})$].

(3) Now we make the substitution $Z_2 := Z_1 - M_1$; we get

$$F_2 = Z_2^4 + (4X^{2/4}Y^{6/4} + 4X^{2/4}Y^{10/4})Z_2^3 + (4XY^{12/4}$$
$$+ 8XY^{16/4} + 4XY^{20/4})Z_2^2 + (-4X^{8/4}Y^{24/4} - 4X^{8/4}Y^{28/4})Z_2$$
$$- 4X^{10/4}Y^{30/4} - 8X^{10/4}Y^{34/4} - 4X^{10/4}Y^{38/4} - X^{12/4}Y^{36/4}.$$

The zeroes of F_2 are $z_2^{(1)}$, $z_2^{(2)}$, $-2M_1 - 2M_2 + z_2^{(3)}$, $-2M_1 - 2M_2 + z_2^{(4)}$. The leading monomial of $A_{2,4}$ is $X^{10/4}Y^{30/4}$, hence we have

$$M_3^2 M_1^2 = X^{10/4}Y^{30/4},$$

and therefore $r_3 = (3,9)$, $M_3 = X^{3/4}Y^{9/4}$. Since $\theta_4 = 4$, we have $\nu_3 = \theta_3/\theta_4 = 2$ and $\delta_4 = \delta_3/\nu_3 = 1$. This implies already that $Q_2(z) = Q_2(M_1, M_2, M_3)$. The leading coefficient of $A_{2,4-2+2}$ is -4, hence we have

$$\alpha_3^2 = (-1)^{4-2+1} \frac{(-4)}{1} \frac{1}{(-2) \cdot (-2)} = 1.$$

We choose $\alpha_3 = 1$. The element $\psi = \varphi_{(1,3)}$ is a representative for a generator of the Galois group of $Q_2(M_1, M_2, M_3)$ over $Q_2(M_1, M_2)$, hence we obtain $\Gamma_3 = \{\varphi_{(0,0)}, \varphi_{(1,0)}, \varphi_{(1,3)}, \varphi_{(2,3)}\}$. Two of the four conjugates of z over Q_2 are

$$z^{(1)} = M_1 + M_2 + M_3 + z_3^{(1)}, \quad z^{(2)} = M_1 + M_2 - M_3 + z_3^{(2)}.$$

Using the elements in Γ_3 we find the other two conjugates

$$z^{(3)} = -M_1 - M_2 - iM_3 + z_3^{(3)}, \quad z^{(4)} = -M_1 - M_2 + iM_3 + z_3^{(4)}.$$

(4) Now we make the substitution $Z_3 := Z_2 - M_3$; we get

$$F_3 = Z_3^4 + (4X^{2/4}Y^{6/4} + 4X^{2/4}Y^{10/4} + 4X^{3/4}Y^{9/4})Z_3^3$$
$$+ (4XY^{12/4} + 8XY^{16/4} + 4XY^{20/4} + 12X^{5/4}Y^{15/4}$$
$$+ 12X^{5/4}Y^{19/4} + 6X^{6/4}Y^{18/4})Z_3^2$$
$$+ (8X^{8/4}Y^{24/4} + 8X^{8/4}Y^{28/4} + 8X^{7/4}Y^{21/4} + 16X^{7/4}Y^{25/4}$$
$$+ 8X^{7/4}Y^{29/4} + 4X^{9/4}Y^{27/4})Z_3.$$

Since $A_{3,4} = 0$, we have $z_3^{(1)} = 0$, and this yields that

$$z = z^{(1)} = M_1 + M_2 + M_3 = X^{2/4}Y^{6/4} + X^{2/4}Y^{10/4} + X^{3/4}Y^{9/4}$$

is a zero of F. We have $[Q_2(z) : Q_2] = \nu_1 \nu_2 \nu_3 = 4$, and therefore F is irreducible in $Q_2[Z]$.

(5) We collect our results: With $M_1 := X^{2/4}Y^{6/4}$, $M_2 := X^{2/4}Y^{10/4}$, $M_3 := X^{3/4}Y^{9/4}$ we have the following: The conjugates $z = z^{(1)}$, $z^{(2)}$, $z^{(3)}$, $z^{(4)}$ of z over Q_2 are

$$z^{(1)} = M_1 + M_2 + M_3, \qquad\qquad z^{(2)} = M_1 + M_2 - M_3,$$
$$z^{(3)} = -M_1 - M_2 - iM_3, \qquad\qquad z^{(4)} = -M_1 - M_2 + iM_3,$$

hence $\{M_1, M_3\}$ is the set of characteristic monomials of z. Therefore the set of characteristic exponents is $\{m_1 = (2,6), m_2 = (3,9)\}$, we have $\nu_1 = \nu_2 = 2$, $r_1 = m_1$, $r_2 = \nu_1 r_1 + m_2 - m_2 = (5,15)$,

$$\Sigma = \{(0,0), r_1, r_2, r_1 + r_2\} = \{(0,0), (2,6), (5,15), (7,21)\},$$
$$\Sigma' = \{(0,0), (2,2), (1,3), (3,1)\},$$

and the integral closure of $\mathbb{C}[\![X,Y]\!][z]$ in its field of quotients is the $\mathbb{C}[\![X,Y]\!]$-module having $\{1, X^2Y^2, XY^3, X^3Y\}$ as a basis.

(5.10) REMARK: We consider the particular case $n = 1$. Then every irreducible Weierstraß polynomial $F \in k[\{X_1\}][Y]$ has a discriminant of the form $X_1^a U$ for some $a \in \mathbb{N}$ and a unit $U \in k[\{X_1\}]$, hence there exists a fractionary power series $y = \sum_{i>1} \alpha_i X_1^{r_i/d}$ with $F(X_1, y) = 0$. In this case our algorithm reduces to the classical Newton algorithm [cf. e.g., [35], p. 494].

6   Strictly Generated Semigroups

6.1   Generalities

(6.1) In this section n and h are positive integers. Elements of $\mathbb{Z}^n$ usually are denoted by lower-case italic letters, and the components of an element $r \in \mathbb{Z}^n$ are denoted by lower-case greek letters, thus we write $r = (\rho_1, \ldots, \rho_n) \in \mathbb{Z}^n$.
Let $d \in \mathbb{N}$ and $r = (\rho_1, \ldots, \rho_n) \in \mathbb{Z}^n$, and for every $j \in \{1, \ldots, n\}$ we write $\rho_j = \rho_j' d + \rho_j''$ where $\rho_j' \in \mathbb{Z}$, $\rho_j'' \in \mathbb{N}_0$ and $0 \le \rho_j'' < d$. Then we set $r \operatorname{div} d := (\rho_1', \ldots, \rho_n')$ and $r \bmod d := (\rho_1'', \ldots, \rho_n'')$.

(6.2) NOTATION: We endow the set $\mathbb{Z}^n$ with the partial ordering $\le$ defined by

$$r = (\rho_1, \ldots, \rho_n) \le (\sigma_1, \ldots, \sigma_n) = s \quad \text{if } \rho_i \le \sigma_i \quad \text{for every } i \in \{1, \ldots, n\};$$

if $r \le s$, but $r \ne s$, then we write $r < s$.
We say that $r \in \mathbb{Z}^n$ is non-negative if $r \ge 0$, and that it is positive if $r > 0$.

(6.3) REMARK: (1) Let $\{e_1, \ldots, e_n\}$ be the canonical $\mathbb{Z}$-basis of $\mathbb{Z}^n$, and let $N \ne \{0\}$ be a $\mathbb{Z}$-submodule of $\mathbb{Z}^n$ of rank p. Then there exist a $\mathbb{Z}$-basis $\{e_1', \ldots, e_n'\}$ of $\mathbb{Z}^n$ and positive integers $\beta_1, \ldots, \beta_p$ satisfying $\beta_1 \mid \beta_2 \mid \cdots \mid \beta_p$ such that $\{\beta_1 e_1', \ldots, \beta_p e_p'\}$ is a $\mathbb{Z}$-basis of N. The elements $\beta_1, \ldots, \beta_p$ are determined uniquely, and they are the invariant factors of N with respect to $\mathbb{Z}^n$.

More specifically: Let $r_1, \ldots, r_h \in \mathbb{Z}^n$ be a system of generators of N. Let $r_i = (\rho_{i1}, \ldots, \rho_{in})$ for $i \in \{1, \ldots, h\}$, and set $R := {}^t(\rho_{ij}) \in M(n, h; \mathbb{Z})$. Then there exist $P = (p_{ij}) \in \mathrm{GL}(n; \mathbb{Z})$ and $Q = (q_{ij}) \in \mathrm{GL}(h; \mathbb{Z})$ such that

$$PRQ = \begin{pmatrix} \mathrm{diag}(\beta_1, \ldots, \beta_p) & 0_{p,h-p} \\ 0_{n-p,p} & 0_{n-p,h-p} \end{pmatrix}. \qquad (*)$$

The matrix PRQ is often called the Smith canonical form of the matrix R. For every $q \in \{1, \ldots, p\}$ the product $\beta_1 \cdots \beta_q$ is the greatest common divisor of the minors of R of order q; in particular, if N has rank n, then $\mathrm{Card}(\mathbb{Z}^n/N) = \beta_1 \cdots \beta_n$, the greatest common divisor of all n-rowed minors of R. Let us define the basis $\{e_1', \ldots, e_n'\}$ of $\mathbb{Z}^n$ by $e_j = \sum_{i=1}^{n} p_{ij} e_i'$ for every $j \in \{1, \ldots, n\}$. The elements $r_1', \ldots, r_h'$, defined by $r_j' = \sum_{i=1}^{h} q_{ij} r_i$ for $j \in \{1, \ldots, h\}$, generate the $\mathbb{Z}$-module N, and we have

$$r_j' = \beta_j e_j' \quad \text{for } j \in \{1, \ldots, p\}, \ r_j' = 0 \quad \text{for } j \in \{p+1, \ldots, h\}.$$

For a proof of this result one may consult any textbook containing a chapter on modules over principal ideal domains, e.g., [32], Ch. VII, § 4, no. 5, Prop. 4.

(2) Let $d \in \mathbb{N}$. The perfect pairing of free $\mathbb{Z}$-modules of rank n

$$(r, s) \mapsto \langle r, s \rangle := \rho_1 \sigma_1 + \cdots + \rho_n \sigma_n : \mathbb{Z}^n \times \mathbb{Z}^n \to \mathbb{Z}$$

where $r = (\rho_1, \ldots, \rho_n)$, $s = (\sigma_1, \ldots, \sigma_n) \in \mathbb{Z}^n$, induces by composing it with the canonical map $\mathbb{Z} \to \mathbb{Z}_d \ (= \mathbb{Z}/\mathbb{Z}d)$, a $\mathbb{Z}$-bilinear pairing

$$(r, s) \mapsto \langle r, s \rangle_d : \mathbb{Z}^n \times \mathbb{Z}^n \to \mathbb{Z}_d$$

where $\langle r, s \rangle_d = \langle r, s \rangle \bmod d$ for every $r, s \in \mathbb{Z}^n$. The left and the right kernel of this pairing is the $\mathbb{Z}$-module $d\mathbb{Z}^n$; the group $\mathbb{Z}^n/d\mathbb{Z}^n$ is finite of order d^n.

For every submodule N of $\mathbb{Z}^n$ we define

$$N^\perp := \{s \in \mathbb{Z}^n \mid \langle r, s \rangle_d = 0 \text{ for every } r \in N\};$$

if N contains $d\mathbb{Z}^n$, then $N^\perp$ contains $d\mathbb{Z}^n$.

(3) Let N be a submodule of $\mathbb{Z}^n$ which contains $d\mathbb{Z}^n$; then N has rank n. Let $\beta_1, \ldots, \beta_n$ be the invariant factors of N with respect to $\mathbb{Z}^n$. Let $\{e_1', \ldots, e_n'\}$ be a $\mathbb{Z}$-basis of $\mathbb{Z}^n$ such that $\{\beta_1 e_1', \ldots, \beta_n e_n'\}$ is a $\mathbb{Z}$-basis of N. The condition $d\mathbb{Z}^n \subset N$ is equivalent to $\beta_i \mid d$ for $i \in \{1, \ldots, n\}$. We define $\gamma_i := d/\beta_i$ for $i \in \{1, \ldots, n\}$. Let $\{f_1', \ldots, f_n'\}$ be the basis of $\mathbb{Z}^n$ dual to this basis [cf. III(2.5)]. Then $N^\perp$ has $\{\gamma_1 f_1', \ldots, \gamma_n f_n'\}$ as a $\mathbb{Z}$-basis, and we have

$$\mathrm{Card}(N/d\mathbb{Z}^n) = \frac{d^n}{\beta_1 \cdots \beta_n}, \ \mathrm{Card}(N^\perp/d\mathbb{Z}^n) = \frac{d^n}{\gamma_1 \cdots \gamma_n};$$

in particular, we have $\mathrm{Card}(N/d\mathbb{Z}^n) \, \mathrm{Card}(N^\perp/d\mathbb{Z}^n) = d^n$.

(6.4) NOTATION: Let $d \in \mathbb{N}$ and $(r_i)_{1 \leq i \leq h}$ be a sequence of elements in $\mathbb{Z}^n$; let $r_i = (\rho_{i1}, \ldots, \rho_{in})$ for every $i \in \{1, \ldots, h\}$.

(1) For every $i \in \{1, \ldots, h+1\}$ let θ_i be the greatest common divisor of all minors of order n of the $(n, n+i-1)$-matrix $\left(dI_n, {}^t((\rho_{lj})_{1 \leq l \leq i-1, 1 \leq j \leq n})\right)$; in particular, we have $\theta_1 = d^n$ and $\theta_{h+1} \mid \theta_h \mid \cdots \mid \theta_1$. The sequence $(\theta_i)_{1 \leq i \leq h+1}$ is called the divisor sequence associated with d and the sequence $(r_i)_{1 \leq i \leq h}$.

(2) For every $i \in \{1, \ldots, h\}$ define $\nu_i := \theta_i/\theta_{i+1}$. The sequence $(\nu_i)_{1 \leq i \leq h}$ is called the ν-sequence associated with d and the sequence $(r_i)_{1 \leq i \leq h}$, and $\theta_{h+1} = d^n/(\nu_1 \cdots \nu_h) \in \mathbb{N}$ is called its degree. Note that $\operatorname{Card}(\mathbb{Z}^n/N) = \theta_{h+1}$, and that therefore $\operatorname{Card}(N/d\mathbb{Z}^n) = d^n/\theta_{h+1} = \nu_1 \cdots \nu_h$.

Moreover, we define

$$Z(\nu) := \{(\alpha_1, \ldots, \alpha_h) \in \mathbb{N}_0^h \mid 0 \leq \alpha_i < \nu_i \ \text{ for every } i \in \{1, \ldots, h\}\}.$$

Note that, for every $j \in \{1, \ldots, h\}$, the sequence $(\nu_i)_{1 \leq i \leq j}$ depends only on d and the sequence $(r_i)_{1 \leq i \leq j}$.

(6.5) Proposition: Let $d \in \mathbb{N}$ and $(r_i)_{1 \leq i \leq h}$ be a sequence in $\mathbb{Z}^n$; let $(\nu_i)_{1 \leq i \leq h}$ be the ν-sequence associated with d and $(r_i)_{1 \leq i \leq h}$. For every $i \in \{0, \ldots, h\}$ let N_i be the $\mathbb{Z}$-submodule of $\mathbb{Z}^n$ generated by $d\mathbb{Z}^n$ and the elements $r_1, \ldots, r_i$. The $\mathbb{Z}$-modules $N_0, \ldots, N_h$ are free of rank n. Furthermore, we have:

(1) For $i \in \{1, \ldots, h\}$ and $\gamma \in \mathbb{Z}$ we have $\gamma r_i \in N_{i-1}$ iff $\nu_i \mid \gamma$.

(2) Every $a \in N_h$ has a unique representation

$$a = da_0 + \sum_{i=1}^{h} \alpha_i r_i, \quad a_0 \in \mathbb{Z}^n, \ (\alpha_1, \ldots, \alpha_h) \in Z(\nu).$$

Proof: (1) Let $(\theta_j)_{1 \leq j \leq h+1}$ be the divisor sequence associated with d and $(r_j)_{1 \leq j \leq h}$. By (6.4) we see that $\operatorname{Card}(\mathbb{Z}^n/N_{j-1}) = \theta_j$ for $j \in \{1, \ldots, h+1\}$. Let N_i' be the $\mathbb{Z}$-submodule of $\mathbb{Z}^n$ generated by N_{i-1} and γr_i. Then $\operatorname{Card}(\mathbb{Z}^n/N_i') = \gcd(\theta_i, \gamma\theta_{i+1})$; thus, we have $N_i' = N_{i-1}$ iff $\gcd(\nu_i, \gamma) = \nu_i$, i.e., iff $\nu_i \mid \gamma$.

(2) [Uniqueness] Let $da_0 + \sum_{i=1}^{h} \alpha_i r_i = db_0 + \sum_{i=1}^{h} \beta_i r_i$ where $a_0, b_0 \in \mathbb{Z}^n$ and $(\alpha_1, \ldots, \alpha_h), (\beta_1, \ldots, \beta_h) \in Z(\nu)$. Assume that there exists $j \in \{1, \ldots, h\}$ such that $\alpha_j \neq \beta_j$, and choose j maximal having this property; furthermore, we may assume that $\alpha_j > \beta_j$. Then $0 = dc + \sum_{i=1}^{j} \gamma_i r_i$ for some $c \in \mathbb{Z}^n$ and some $\gamma_1, \ldots, \gamma_j \in \mathbb{Z}$ such that $0 < \gamma_j < \nu_j$. Now we have $\gamma_j r_j \in N_{j-1}$, contradicting the assertion in (1).

[Existence] From (1) we get that $\nu_i r_i \in N_{i-1}$ for every $i \in \{1, \ldots, h\}$. Let $a = da_0 + \sum_{i=1}^{h} \alpha_i r_i$ where $a_0 \in \mathbb{Z}^n$ and $\alpha_1, \ldots, \alpha_h$ are integers. We may write $\alpha_h = \beta_h \nu_h + \gamma_h$ where $\beta_h, \gamma_h \in \mathbb{Z}$ and $0 \leq \gamma_h < \nu_h$, hence we find that $a = da_0' + \sum_{i=1}^{h-1} \alpha_i' r_i + \gamma_h r_h$ for some $a_0' \in \mathbb{Z}^n$ and integers $\alpha_1', \ldots, \alpha_{h-1}'$. Continuing in this way we get the representation we were looking for.

This ends the proof of the proposition.

(6.6) REMARK: Let $(r_i)_{1 \leq i \leq h}$ be a sequence in $\mathbb{Z}^n$, and let $(\nu_i)_{1 \leq i \leq h}$ be the ν-sequence associated with d and $(r_i)_{1 \leq i \leq h}$. In the following, we are mainly interested in the $\mathbb{Z}$-module

$$N = d\mathbb{Z}^n + \mathbb{Z}r_1 + \cdots + \mathbb{Z}r_h$$

and the sets

$$\Sigma = \{\alpha_1 r_1 + \cdots + \alpha_h r_h \mid (\alpha_1, \ldots, \alpha_h) \in Z(\nu)\}, \quad \Gamma = d\mathbb{N}_0^n + \Sigma.$$

We construct a new sequence $(r_i')_{1 \leq i \leq h'}$ in the following way: If $\nu_1 = \cdots = \nu_h = 1$, then we define $h' = 0$. In the other case let i_1 be the smallest integer in $\{1, \ldots, h\}$ with $\nu_{i_1} > 1$; we define $r_1' := r_{i_1}$. Now let i_2 be the smallest integer in $\{i_1+1, \ldots, h\}$ with $\nu_{i_2} > 1$; we define $r_2' := r_{i_2}$. Continuing we get a sequence $(r_i')_{1 \leq i \leq h'}$; let $(\nu_i')_{1 \leq i \leq h'}$ be the ν-sequence associated with d and $(r_i')_{1 \leq i \leq h'}$. Then we have $\nu_i > 1$ for $i \in \{1, \ldots, h\}$, and from (6.5) we see that $N = d\mathbb{Z}^n + \mathbb{Z}r_1' + \cdots + \mathbb{Z}r_{h'}'$; furthermore, we get

$$\Sigma = \{\alpha_1 r_1' + \cdots + \alpha_{h'} r_{h'}' \mid (\alpha_1, \ldots, \alpha_{h'}) \in Z(\nu')\}.$$

Therefore we may and will for the rest of this section consider only sequences $(r_i)_{1 \leq i \leq h}$ in $\mathbb{Z}^n$ such that the ν-sequence associated with d and $(r_i)_{1 \leq i \leq h}$ satisfies the condition $\nu_i > 1$ for $i \in \{1, \ldots, h\}$.

(6.7) Proposition: Let $d \in \mathbb{N}$ and $(r_i)_{1 \leq i \leq h}$ be a sequence in $\mathbb{Z}^n$; let $(\nu_i)_{1 \leq i \leq h}$ be the ν-sequence associated with d and the sequence $(r_i)_{1 \leq i \leq h}$. Define $\nu_0 := 1$, $r_0 := m_0 := (0, \ldots, 0) \in \mathbb{Z}^n$, and by recursion

$$m_i := m_{i-1} + r_i - \nu_{i-1} r_{i-1} \quad \text{for every } i \in \{1, \ldots, h\}. \qquad (*)$$

Then:
(1) Each of the two sequences $(m_i)_{1 \leq i \leq h}$ and $(r_i)_{1 \leq i \leq h}$ determines the other sequence, and the ν-sequences associated with d and any of these two sequences are the same. In particular, we have

$$r_i = \nu_{i-1} r_{i-1} + m_i - m_{i-1} \quad \text{for every } i \in \{1, \ldots, h\}.$$

(2) The following statements are equivalent:
(i) The sequence $(m_i)_{0 \leq i \leq h}$ is linearly ordered, i.e., we have $m_0 < m_1 < \cdots < m_h$.
(ii) For every $i \in \{1, \ldots, h\}$ we have $r_i > \nu_{i-1} r_{i-1}$.
If these conditions are satisfied, then we have

$$r_i > \sum_{j=1}^{i-1} (\nu_j - 1) r_j \quad \text{for } i \in \{1, \ldots, h\}. \qquad (**)$$

Proof: (1) It is easy to see that the divisor sequence associated with d and $(m_i)_{1 \leq i \leq h}$ is the divisor sequence associated with d and $(r_i)_{1 \leq i \leq h}$, hence both sequences have the same ν-sequence. Now we have

$$r_i = \nu_{i-1} r_{i-1} + m_i - m_{i-1} \quad \text{for every } i \in \{1, \ldots, h\},$$

hence each of these two sequences determines the other sequence.

(2) Clearly the conditions in (i) and (ii) are equivalent by (1). Assume that these conditions are satisfied; since we have

$$r_i = (\nu_{i-1}-1)r_{i-1}+(\nu_{i-2}-1)r_{i-2}+\cdots+(\nu_1-1)r_1+m_i \quad \text{for every } i \in \{1,\ldots,h\},$$

we have proved also the last assertion.

6.2 Strictly Generated Semigroups

(6.8) REMARK: We keep the notations introduced in (6.5) and (6.7).

(1) Let $a = da_0 + \sum_{i=1}^{h}\alpha_i r_i$ where $a_0 \in \mathbb{Z}^n$ and $(\alpha_1,\ldots,\alpha_h) \in Z(\nu)$. Let $j \in \{1,\ldots,h\}$. Then, by uniqueness, $a \in N_{j-1}$ iff $\alpha_j = \cdots = \alpha_h = 0$.

(2) Note that, in the statements in (6.7), the index h may be replaced by any index $j \in \{1,\ldots,h\}$.

(6.9) DEFINITION: Let $d \in \mathbb{N}$ and $(r_i)_{1\leq i\leq h}$ be a sequence in $\mathbb{Z}^n$; let $(\nu_i)_{1\leq i\leq h}$ be the ν-sequence associated with d and the sequence $(r_i)_{1\leq i\leq h}$.

(1) The set

$$\Gamma := \{da_0 + \alpha_1 r_1 + \cdots + \alpha_h r_h \mid a_0 \in \mathbb{N}_0^n \text{ and } (\alpha_1,\ldots,\alpha_h) \in Z(\nu)\}$$

is called the set of strict linear combinations of d and $r_1,\ldots,r_h$. If Γ is a subsemigroup of $\mathbb{Z}^n$, then we say that Γ is strictly generated by $d, r_1,\ldots,r_h$.

(2) The set

$$\Sigma := \{\alpha_1 r_1 + \cdots + \alpha_h r_h \mid (\alpha_1,\ldots,\alpha_h) \in Z(\nu)\}$$

is called the set of strict linear combinations of $r_1,\ldots,r_h$.

(6.10) Proposition: *Let $d \in \mathbb{N}$ and $(r_i)_{1\leq i\leq h}$ be a sequence in $\mathbb{Z}^n$, and let $(\nu_i)_{1\leq i\leq h}$ be the ν-sequence associated with d and $(r_i)_{1\leq i\leq h}$. Let Σ be the set of strict linear combinations of $r_1,\ldots,r_h$, and let Γ be the set of strict linear combinations of d and $r_1,\ldots,r_h$. We assume that one of the following conditions holds:*

(i) $\nu_i r_i \in d\mathbb{N}_0^n + \mathbb{N}_0 r_1 + \cdots + \mathbb{N}_0 r_{i-1}$ for every $i \in \{1,\ldots,h\}$,

(ii) $r_1 > 0$ and $r_{i+1} > \nu_i r_i$ for every $i \in \{1,\ldots,h-1\}$.

Then we have $\Gamma = d\mathbb{N}_0^n + \mathbb{N}_0 r_1 + \cdots + \mathbb{N}_0 r_h$, hence Γ is a finitely generated subsemigroup of $\mathbb{N}_0^n$ having $e_1,\ldots,e_n,r_1,\ldots,r_h$ as a system of generators.

In particular, if (ii) holds, then Σ is totally ordered by the relation $<$.

Proof: (1) We assume that (i) holds. We define $\Gamma' := d\mathbb{N}_0^n + \mathbb{N}_0 r_1 + \cdots + \mathbb{N}_0 r_h$. It is clear that Γ is contained in Γ'. Conversely, let $a \in \Gamma'$; then we have $a = da_0 + \sum_{i=1}^{h}\alpha_i r_i$ where $a_0 \in \mathbb{N}_0^n$ and $\alpha_1,\ldots,\alpha_h \in \mathbb{N}_0$. We write $\alpha_h = \mu\nu_h + \gamma_h$ where $\mu, \gamma_h \in \mathbb{N}_0$ and $0 \leq \gamma_h < \nu_h$; now we have $\mu\nu_h r_h \in d\mathbb{N}_0^n + \mathbb{N}_0 r_1 + \cdots + \mathbb{N}_0 r_{h-1}$ by assumption. Continuing in this way, we find that a is a strict linear combination of d and $r_1,\ldots,r_h$.

(2) Now we assume that (ii) holds. We show that (ii) implies (i). In fact, we have $\nu_1 r_1 \in d\mathbb{Z}^n$ by (6.5), hence we have $\nu_1 r_1 \in d\mathbb{N}_0^n$ since $r_1 > 0$. For $i \in \{2, \ldots, h\}$ we have, again by (6.5),

$$\nu_i r_i = da_0 + \sum_{j=1}^{i-1} \alpha_j r_j \quad \text{with } a_0 \in \mathbb{Z}^n, \, 0 \leq \alpha_j < \nu_j \text{ for } j \in \{1, \ldots, i-1\},$$

hence we have by (6.7)

$$da_0 = \nu_i r_i - \sum_{j=1}^{i-1} \alpha_j r_j \geq r_i - \sum_{j=1}^{i-1} (\nu_j - 1) r_j > 0,$$

hence we have $a_0 \in \mathbb{N}_0^n$, and therefore we have $\nu_i r_i \in d\mathbb{N}_0^n + \mathbb{N}_0 r_1 + \cdots + \mathbb{N}_0 r_{i-1}$. Thus, we have shown that (ii) implies (i).

Now we show that Σ is totally ordered. Let $a = \sum_{i=1}^{h} \alpha_i r_i \neq \sum_{i=1}^{h} \beta_i r_i = b$ be elements of Σ. Let us choose $j \in \{1, \ldots, h\}$ such that $\alpha_i = \beta_i$ for $i = h, h - 1, \ldots, j+1$, and $\alpha_j \neq \beta_j$. We may assume that $\alpha_j > \beta_j$. Now we have

$$\alpha_j r_j \geq \beta_j r_j + r_j > \beta_j r_j + \sum_{i=1}^{j-1} (\nu_i - 1) r_i \geq \beta_j r_j + \sum_{i=1}^{j-1} \beta_i r_i$$

by (6.7)(∗∗).

(6.11) DEFINITION: Let M be a free $\mathbb{Z}$-module of finite rank, let Δ be a subsemigroup of M, and let $M(\Delta)$ be the $\mathbb{Z}$-submodule of M generated by Δ. We consider M as a $\mathbb{Z}$-submodule of $M \otimes_{\mathbb{Z}} \mathbb{Q}$. The subsemigroup $\bigcup_{t \in \mathbb{N}} (\frac{1}{t}\Delta) \cap M(\Delta)$ of M is called the saturation $\overline{\Delta}$ of Δ. The subsemigroup Δ is called saturated if $\Delta = \overline{\Delta}$; clearly, $\overline{\Delta}$ is saturated.

(6.12) REMARK: Let Δ be a finitely generated subsemigroup of M. Then the saturation $\overline{\Delta}$ of Δ is also a finitely generated semigroup [cf. VI(2.12)].

(6.13) Corollary: *Assume that* (i) *or* (ii) *of* (6.10) *holds. Let N be the $\mathbb{Z}$-submodule of $\mathbb{Z}^n$ generated by Γ. Then the set $\{r \in N \mid r \geq 0\}$ is the saturation $\overline{\Gamma}$ of the semigroup Γ.*

Proof: For $i \in \{0, \ldots, h\}$ let Γ_i be the subsemigroup of $\mathbb{N}_0^n$ generated by $d\mathbb{N}_0^n$ and $r_1, \ldots, r_i$, and let N_i be the $\mathbb{Z}$-submodule of $\mathbb{Z}^n$ generated by Γ_i. We show by recursion that the set of non-negative elements of N_i is the saturation $\overline{\Gamma}_i$ of Γ_i. Clearly, for every $i \in \{0, \ldots, h\}$, the saturation of Γ_i is contained in the set of non-negative elements of N_i.

The saturation of $\Gamma_0 = d\mathbb{N}_0^n$ is Γ_0, and it is the set of non-negative elements of N_0. Assume that for some $i \in \{1, \ldots, h\}$ we have shown that the set of non-negative elements of N_{i-1} is the saturation $\overline{\Gamma}_{i-1}$ of Γ_{i-1}. Let $r \in N_i$ be non-negative;

then we have $r = da_0 + \alpha_1 r_1 + \cdots + \alpha_i r_i$ where $a_0 \in \mathbb{Z}^n$ and $\alpha_1, \ldots, \alpha_i$ are non-negative integers satisfying $0 \leq \alpha_j < \nu_j$ for every $j \in \{1, \ldots, i\}$ [cf. (6.5)]. Since $\nu_i \alpha_i r_i \in N_{i-1}$, it follows that $\nu_i r \in N_{i-1}$, and since this element is non-negative, it lies in $\overline{\Gamma}_{i-1}$, hence there exists $\kappa \in \mathbb{N}$ such that $\kappa r \in \Gamma_{i-1} \subset \Gamma_i$, and therefore we have $r \in \overline{\Gamma}_i$.

(6.14) Corollary: *Assume that* (i) *or* (ii) *of* (6.10) *holds, and set*

$$\Sigma' := \{s \bmod d \mid s \in \Sigma\}.$$

Then $\mathrm{Card}(\Sigma') \leq \nu_1 \cdots \nu_h$, $\Sigma' \subset \overline{\Gamma}$, *and every* $r \in \overline{\Gamma}$ *has a unique representation* $r = da_0 + s'$ *where* $a_0 \in \mathbb{N}_0^n$ *and* $s' \in \Sigma'$. *In particular, the semigroup* $\overline{\Gamma}$ *is generated by the finite set* $\{e_1, \ldots, e_n\} \cup \Sigma'$.

Proof: The first statement is clear since $\mathrm{Card}(\Sigma) = \nu_1 \cdots \nu_h$, and we have $\Sigma' \subset \overline{\Gamma}$ by (6.13), since Σ' lies in the $\mathbb{Z}$-submodule of $\mathbb{Z}^n$ generated by $d\mathbb{Z}^n$ and $r_1, \ldots, r_h$, and for $s' \in \Sigma'$ we have $s' \geq 0$.
Let $r \in \overline{\Gamma}$; then we have $r \geq 0$, and there exist $a_0 \in \mathbb{Z}^n$ and $s \in \Sigma$ such that $r = da_0 + s$. Writing $s = ds'' + s'$ where $s'' := s \operatorname{div} d$ and $s' := s \bmod d \in \Sigma'$, we find that $r = d(a_0 + s'') + s'$ and that $a_0 + s'' \geq 0$ [since $r \geq 0$].
Assume that $da_0 + s' = db_0 + t' \geq 0$ where a_0, $b_0 \in \mathbb{N}_0^n$ and s', $t' \in \Sigma'$. Since $s' - t' = d(b_0 - a_0)$, it follows that $s' = t'$ and that $a_0 = b_0$.

(6.15) EXAMPLE: (1) In the following example, we have $\mathrm{Card}(\Sigma') < \nu_1 \cdots \nu_h$. Let $d > 1$ be a natural integer, and let us choose $r_1 = r = (\rho_1, \ldots, \rho_n) \in \mathbb{N}_0^n$ such that for $\theta := \gcd(\rho_1, \ldots, \rho_n, d)$ we have $1 < \theta < d$. Now (d/θ) is the ν-sequence associated with d and (r_1), and we have $1 < d/\theta = \mathrm{Card}(\Sigma') < d$ [cf. (1.13)].
(2) The following example shall play an important role in chapter VI. Let $0 < p < q$ be coprime natural integers; the ν-sequence associated with q and $(1, p)$ is (q) [cf. (1)]. Then we have $\Sigma = \{(i, ip) \mid 0 \leq i \leq q - 1\}$. For every $i \in \mathbb{Z}$ let $p[i] \in \{0, \ldots, q - 1\}$ be that integer with $ip \equiv p[i] \pmod{q}$. Since $\gcd(p, q) = 1$, the integers $p[0] = 0, p[1], \ldots, p[q - 1]$ are pairwise different. We have

$$\Sigma' = \{(0, 0), (1, p[1]), \ldots, (q - 1, p[q - 1])\},$$

and $\mathrm{Card}(\Sigma') = q$.

Chapter VI

The Singularity $Z^q = XY^p$

INTRODUCTION: In this chapter we describe a resolution process for the singularity of the normalization of the surface in $\mathbb{A}^3$ defined by the equation $Z^p = XY^q$ over an algebraically closed field k of characteristic zero; here $0 < p < q$ are integers and $\gcd(p, q) = 1$. These singularities arise in a natural way: In section 1 we show in (1.6) that if L is a finite extension of $Q = k((U, V))$, S is the integral closure of $R = k[U, V]$ in L, and the only prime ideals of R which are ramified in S are at most RU or RV, then $L = k((X, Y))[Z]/(Z^p - XY^q)$, and that S is the integral closure of $S_0 = k[X, Y][Z]/(Z^p - XY^q)$. Also, they arise as the only singular point of the normalization of a toric variety associated with a not nonsingular strongly convex rational polyhedral cone of dimension 2 [cf. section 4].

After some generalities on semigroup rings in section 2 we treat the continued fraction expansion and the Hirzebruch-Jung continued fraction expansion of rational numbers in section 3. Two-dimensional strongly convex rational polyhedral cones are dealt with in section 4; they give rise to a semigroup and an associated affine variety which is called a toric variety. Every two-dimensional strongly convex rational polyhedral cone can be described by integers p, q as above. In the last section we show that we can resolve the singularities of a two-dimensional toric variety by repeatedly blowing up points.

1 Hirzebruch-Jung Singularities

(1.0) Unless otherwise stated, in this chapter k is an algebraically closed field of characteristic 0, and p, q are integers such that $0 \leq p < q$ and $\gcd(q, p) = 1$; $\varepsilon \in k$ is a primitive q-th root of unity. We use the notation $p[0], \ldots, p[q-1]$ introduced in V(6.15)(2).

(1.1) NOTATION: (1) A complete local k-algebra which is k-isomorphic to a k-algebra $k[X, Y, Z]/(Z^q - XY^p)$ is called a Hirzebruch-Jung singularity of type $A_{q,p}$ or an $A_{q,p}$-singularity; note that $Z^q - XY^p$ is irreducible in $k[X, Y][Z]$ by

Eisenstein's criterion, hence it remains irreducible in $k[\![X,Y,Z]\!]$ [cf. IV(3.13)], and therefore $k[\![X,Y,Z]\!]/(Z^q - XY^p) = k[\![X,Y]\!][\![Z]\!]/(Z^q - XY^p)$ [cf. IV(3.14)(1)] is a domain. Remember that in V(5.8) we considered such rings. Every local k-algebra whose completion is a Hirzebruch-Jung singularity of type $A_{q,p}$ is also called a Hirzebruch-Jung singularity of type $A_{q,p}$; such a k-algebra is a domain [for any local ring A, the canonical homomorphism $A \to \widehat{A}$ from A into its completion $\widehat{A}$ is injective].

(2) Let $s \in \mathbb{N}_0$. A complete local k-algebra which is k-isomorphic to a k-algebra of the form $k[\![X,Y,Z]\!]/(Z^{s+1} - XY)$ is called a singularity of type A_s or an A_s-singularity. A local k-algebra is called an A_s-singularity if its completion is an A_s-singularity. An A_0-singularity is a regular local ring, and if $s \geq 1$, then an A_s-singularity is a Hirzebruch-Jung singularity of type $A_{s+1,1}$.

(3) Let X be a variety over k, and let $x \in X$. The variety X is said to have a Hirzebruch-Jung singularity of type $A_{q,p}$ (resp. a singularity of type A_s) in x, or x is said to be an $A_{q,p}$-singularity (resp. an A_s-singularity), if the local ring of x in X is a Hirzebruch-Jung singularity of type $A_{q,p}$ (resp. a singularity of type A_s).

(1.2) Proposition: *Let k be a field, and let R_n be either a polynomial ring or a ring of power series over k in n indeterminates, and let G be a group of k-automorphisms of R_n. Then the ring of invariants $R_n^G := \{F \in R_n \mid \varphi(F) = F$ for every $\varphi \in G\}$ is integrally closed.*

Proof: The ring R_n is factorial [for the case of a power series ring cf. IV(3.18)], hence integrally closed [cf. [63], Prop. 4.10]. Let $F_1, F_2 \in R_n^G$, $F_2 \neq 0$, and assume that $F_1 F_2^{-1} \in Q(R_n^G)$ is integral over R_n^G. Then $F_1 F_2^{-1}$ is integral over R_n, and since R_n is integrally closed, we see, in particular, that $F_1 F_2^{-1} \in R_n$. Now, for every $\varphi \in G$ we have

$$(F_1 F_2^{-1})F_2 = F_1 = \varphi(F_1) = \varphi((F_1 F_2^{-1})F_2) = \varphi(F_1 F_2^{-1})\varphi(F_2) = \varphi(F_1 F_2^{-1})F_2$$

[since $F_1, F_2 \in R_n^G$], and therefore we get $F_1 F_2^{-1} = \varphi(F_1 F_2^{-1})$. Thus, the element $F_1 F_2^{-1}$ lies in R_n^G.

(1.3) EXAMPLE: Let $T := k[U,V]$, the polynomial ring over k in two indeterminates U, V, and set $L := Q(T) = k(U,V)$, $R_0 := k[U^q, V^q]$ and $K_0 := Q(R_0) = k(U^q, V^q)$. Then L/K_0 is an abelian Galois extension with Galois group $\mathrm{Gal}(L/K_0) = \{\sigma_{ij} \mid 0 \leq i,j < q\} \cong \mathbb{Z}_q \times \mathbb{Z}_q$ where $\sigma_{ij}(U) = \varepsilon^i U$ and $\sigma_{ij}(V) = \varepsilon^j V$, $i, j \in \{0, \ldots, q-1\}$ [cf. V(1.15)]. Let $G := G(p,q)$ be the cyclic subgroup of $\mathrm{Gal}(L/K_0)$ generated by $\varphi := \sigma_{q-p,1}$. Note that G is a cyclic group of k-algebra automorphisms of T of order q.

(1) The fixed field of G is $K := k(U^q, V^q, UV^p)$, and we have $L = K[U]$. K is a cyclic extension of K_0 of degree q, and we have $K = K_0[UV^p]$.

In fact, we have $K \subset L^G$. The polynomial $H := Z^q - U^q \in K[Z]$ is irreducible and has zero U in L. There exist $a, b \in \mathbb{Z}$ with $ap + bq = 1$; then we get $V = (UV^p)^a (V^q)^b U^{-a}$, hence V lies in $K(U)$, and therefore L is the splitting field of H over K, and we have $[L:K] = q$, hence we have $K = L^G$. The minimal polynomial of $UV^p \in K$ over K_0 is $Z^q - U^q V^{pq}$; this proves the second assertion.

(2) We construct a system of generators of $S := T^G$. Let $F = \sum_{i,j} \gamma_{ij} U^i V^j$ be a polynomial in T: F remains invariant under φ iff each of the monomials $U^i V^j$ appearing in F remains invariant under φ. Let $i, j \in \mathbb{N}_0$, and write $i = hq + i'$, $j = lq + j'$ with non-negative integers h, l, and with i', $j' \in \{0, \ldots, q-1\}$. The monomial $U^i V^j$ remains invariant under φ iff $j' = -p[i']$. Therefore we have

$$S = T^G = k[U^q, V^q, UV^{p[1]}, U^2 V^{p[2]}, \ldots, U^{q-1} V^{p[q-1]}].$$

(3) Let $i, j \in \{1, \ldots, q-1\}$; the elements $U^i V^{p[i]}$ and $U^j V^{p[j]}$ are integral over R_0, and since $p[i] + p[j] \equiv pi + pj \pmod{q} \equiv p[(i+j) \pmod{q}]$, we see that we have $U^i V^{p[i]} \cdot U^j V^{p[j]} = (U^q)^s (V^q)^t \cdot U^{i'} V^{p[i']}$ for some s, $t \in \{0,1\}$ and $i' \in \{0, \ldots, q-1\}$. In particular, T^G is integral over R_0, and the R_0-module T^G is generated by the elements $U^i V^{p[i]}$, $i \in \{0, \ldots, q-1\}$.

(4) We set $R := k[U^q, V^q, UV^p]$; note that $Q(R) = K$. The elements U^q, V^q, UV^p belong to the system of generators of the k-algebra $S = T^G$ given in (2), and the rings R and T^G have the same field of quotients K; furthermore, T^G is integral over R by (3), hence it is the integral closure of R in its field of quotients K [since T^G is integrally closed, cf. (1.2)]. Note that R is integrally closed iff $R = S$, hence iff $p = 1$. Since we have $[K:K_0] = q$, we see that S is a free R_0-module which has the set $\{U^i V^{p[i]} \mid i \in \{0, \ldots, q-1\}\}$ as a basis.

(5) The ideal $\mathfrak{m} = (U^q, V^q, UV^p)$ of R is a maximal ideal, and the maximal ideal $\mathfrak{n} = (U^q, V^q, UV^{p[1]}, \ldots, U^{q-1} V^{p[q-1]})$ of S lies over $\mathfrak{m}$; both of these ideals lie over the maximal ideal $\mathfrak{m}_0$ of R_0 which is generated by U^q and V^q. Since every maximal ideal of T which lies over $\mathfrak{m}_0$ contains U and V, the ideal (U,V) of T is the only maximal ideal of T which lies over $\mathfrak{m}$, hence $\mathfrak{n}$ is the only maximal ideal of S which lies over $\mathfrak{m}$. In particular, the integral closure of $R_\mathfrak{m}$ is the local ring $S_\mathfrak{n}$, and it is a finitely generated $R_\mathfrak{m}$-module.

(6) Applying IV(4.5) we get: $\widehat{R} := k[\![U^q, V^q, UV^p]\!]$ is the completion of R with respect to its $\mathfrak{m}$-adic topology, and it is also the completion of $R_\mathfrak{m}$ in its natural topology. Note that $\widehat{R} \cong k[\![X, X, Z]\!]/(Z^q - XY^p)$, hence $R_\mathfrak{m}$ is a Hirzebruch-Jung singularity of type $A_{q,p}$.

(7) The completion of T with respect to the ideal (U,V) is the ring $\widehat{T} := k[\![U,V]\!]$, and $\widehat{R}_0 := k[\![U^q, V^q]\!]$ is the completion of R_0 with respect to the $\mathfrak{m}_0$-adic topology. We set $\widehat{L} := k((U,V))$, $\widehat{K}_0 := k((U^q, V^q))$. Just as above we get: $\widehat{L}/\widehat{K}_0$ is an abelian Galois extension with Galois group isomorphic to $\mathbb{Z}_q \times \mathbb{Z}_q$, $\widehat{K} := k((U^q, V^q))[UV^p]$ is the fixed field of G, G acts as a group of k-automorphisms of $\widehat{T}$, $\widehat{T}^G = k[\![U^q, V^q, UV^{p[1]}, U^2 V^{p[2]}, \ldots, U^{q-1} V^{p[q-1]}]\!]$ is the integral closure of $\widehat{R}$, and it is a free $\widehat{R}_0$-module which has $\{U^i V^{p[i]} \mid i \in \{0, \ldots, q-1\}\}$ as a basis. Thus, we have another proof of V(5.8). Again by IV(4.5) we see that $\widehat{T}^G$ is the completion of $S = T^G$ with respect to its $\mathfrak{n}$-adic topology and of $S_\mathfrak{n}$ in its natural topology.

(1.4) A PARTICULAR CLASS OF FIELD EXTENSIONS: Let $R := k[\![X,Y]\!]$ be the ring of formal power series over k in two indeterminates X, Y, let K be its field of quotients, let $d \in \mathbb{N}$ and let L be a subfield of $K[X^{1/d}, Y^{1/d}] = k((X^{1/d}, Y^{1/d}))$ containing K. We show that there exist positive integers s, t such that $X^{1/s}$ and $Y^{1/t}$ lie in L and integers $p \in \mathbb{N}_0$, $q \in \mathbb{N}$, $0 \le p < q$ and $\gcd(p,q) = 1$, such that with $U := X^{1/s}$, $V := Y^{1/t}$ we have

$$L = k((X^{1/s}, Y^{1/t}))[X^{1/sq}Y^{p/tq}] = k((U,V))[U^{1/q}V^{p/q}].$$

(1) The $\mathbb{Z}$-submodule N of $\mathbb{Z}^2$ associated with L [cf. V(1.8)] contains $d\mathbb{Z}^2$ and has rank 2, and can therefore be generated by two elements (α, β), $(\gamma, \delta) \in \mathbb{Z}^2$; then we have $L = K[X^{\alpha/d}Y^{\beta/d}, X^{\gamma/d}Y^{\delta/d}]$. We consider the matrix

$$S := \begin{pmatrix} \alpha & \beta \\ \gamma & \delta \end{pmatrix} \in M(2, \mathbb{Z});$$

note that $\det(S) \neq 0$. For any $T \in GL(2, \mathbb{Z})$ the rows of

$$TS =: \begin{pmatrix} \alpha' & \beta' \\ \gamma' & \delta' \end{pmatrix}$$

generate N, and we have $L = K[X^{\alpha'/d}Y^{\beta'/d}, X^{\gamma'/d}Y^{\delta'/d}]$.

By interchanging the rows of S if necessary, we may assume that $\det(S) > 0$. Let $\theta \in \mathbb{N}$ be the greatest common divisor of α and γ. Then there exist $\lambda, \kappa \in \mathbb{Z}$ such that $\lambda\alpha + \kappa\gamma = \theta$, hence

$$T := \begin{pmatrix} \lambda & \kappa \\ -\gamma/\theta & \alpha/\theta \end{pmatrix} \in GL(2; \mathbb{Z}) \text{ and } \det(T) = 1.$$

Then we have

$$TS = \begin{pmatrix} \alpha' & \beta' \\ 0 & \delta' \end{pmatrix} \text{ with } \alpha' = \theta \text{ and } \alpha'\delta' = \det(S) > 0.$$

In particular, we have $\delta' > 0$. If $\beta' \ge \delta'$, then we can write $\beta' = \mu\delta' + \beta''$ where $\mu \in \mathbb{N}$, $\beta' \in \mathbb{N}_0$ and $0 \le \beta'' < \delta'$. We have

$$\begin{pmatrix} 1 & -\mu \\ 0 & 1 \end{pmatrix} \begin{pmatrix} \alpha' & \beta' \\ 0 & \delta' \end{pmatrix} = \begin{pmatrix} \alpha' & \beta'' \\ 0 & \delta' \end{pmatrix}.$$

Therefore we may assume, to begin with, that

$$S = \begin{pmatrix} s' & pr \\ 0 & qr \end{pmatrix}, \quad q, r, s' \in \mathbb{N}, \ p \in \mathbb{N}_0, \ 0 \le p < q, \ \gcd(p,q) = 1.$$

Since the rows $(d, 0)$ and $(0, d)$ are linear combinations of the rows (s', pr) and $(0, qr)$ with integer coefficients, we see that s' divides d and that qr divides d. Therefore we may write $d = s's''$, $d = qrt$ with natural integers s'' and t. Moreover, there exists $b \in \mathbb{Z}$ with $s''p + bq = 0$. If $p = 0$, then we set $s := s''$, and if $p \neq 0$, then p divides b since p and q are coprime, hence $s'' = qs$ where $s = -b/p > 0$. Therefore we have

$$S = \begin{pmatrix} d/qs & pd/qt \\ 0 & d/t \end{pmatrix}.$$

(2) By (1) we now have

$$L = K[X^{1/qs}Y^{p/qt}, Y^{1/t}].$$

We have $Y^{1/t} \in L$ and $X^{1/s} = (X^{1/qs}Y^{p/qt})^q/(Y^{1/t})^p \in L$, and with $X' := X^{1/s}$, $Y' := Y^{1/t}$ we find that

$$L = k((X',Y'))[X'^{1/q}Y'^{p/q}].$$

(3) Since $W^q - X'Y'^p \in k[\![X',Y']\!][W]$ is irreducible by Eisenstein's criterion, we have $L = k((X',Y'))[W]/(W^q - X'Y'^p)$; in particular, L is a finite extension of $k((X',Y'))$ of degree q.
(4) We arrive at the same result if we replace $k((X,Y))$ by the field $k(X,Y)$ of rational functions in two indeterminates over k, and let L be a subfield of $k(X,Y)[X^{1/d}, Y^{1/d}]$ containing $k(X,Y)$ [cf. V(1.15)].

(1.5) REMARK: Let $R = k[\![U,V]\!]$ be the ring of formal power series in two indeterminates U, V over k, let $K = k((U,V))$ be the field of quotients of R, and let L be a finite extension of K. Let S be the integral closure of R in L; S is a finitely generated R-module and a formal k-algebra [cf. IV(5.3)]. *Assume that among the prime ideals of height one of R at most the ideals (U) and (V) are ramified in S.* By V(2.4) there exists $d \in \mathbb{N}$ such that $L \subset k((U^{1/d}, V^{1/d}))$; by V(1.22) we can even choose $d = [L:K]$.
(1) By (1.4) there exist natural integers s and t such that the elements $U' := U^{1/s}$, $V' := V^{1/t}$ lie in L and such that for some integers p, q as in (1.0), we can write $L = k((U',V'))[W]/(W^q - U'V'^p)$, a finite extension of $k((U',V'))$ of degree q, and therefore we have

$$L = k((U',V'))[U'^{1/q}V'^{p/q}] \subset k((U',V'))[U'^{1/q}, V'^{1/q}] = k((U'^{1/q}, V'^{1/q})).$$

Since U' and V' are integral over R, the elements U', V' lie in S; hence S is the integral closure in L of the ring

$$S_0 := k[\![U',V']\!][U'^{1/q}V'^{p/q}] \subset k[\![U',V']\!][U'^{1/q}, V'^{1/q}] = k[\![U'^{1/q}, V'^{1/q}]\!].$$

We see that S_0 contains $k[\![U',V']\!]$ as a subring and it is integral over it, and that $S_0 = k[\![U',V']\!][W]/(W^q - U'V'^p)$; moreover, L is the field of quotients of S_0.
(2) Comparing the result in (1.3)(4) with (1) we see: The k-algebras $k[\![U,V]\!]^G$ and S are isomorphic. In particular, S is a free submodule of the $k[\![U',V']\!]$-module $k[\![U'^{1/q}, V'^{1/q}]\!]$ with basis $\{U'^{i/q}V'^{p[i]/q} \mid i \in \{0,\ldots,q-1\}\}$.
(3) The complete local k-algebra S_0 considered in (1) is an $A_{q,p}$-singularity. Note that S_0 is integrally closed iff $p = 1$ [cf. (1.3)(4)].

In summary, we have proved

(1.6) Proposition: *Let L be a finite extension of $k((U,V))$ and S the integral closure of $k[\![U,V]\!]$ in L, and assume that the only prime ideals of $k[\![U,V]\!]$ of height 1 which are ramified in S are at most the prime ideals (U) or (V). Then there exist integers p and q with $0 \le p < q$ and $\gcd(p,q) = 1$ and elements $X, Y \in k((U,V))$ such that $L = k((X,Y))[Z]/(Z^q - XY^p)$, and that S is the integral closure of the ring $S_0 = k[\![X,Y]\!][Z]/(Z^q - XY^p)$ which is a free $k[\![X,Y]\!]$-module. Moreover, S is also a free $k[\![X,Y]\!]$-module, and the set $\{X^{i/q}Y^{p[i]/q} \mid i \in \{0, \dots, q-1\}\}$ is a basis of S.*

2 Semigroups and Semigroup Rings

2.1 Generalities

(2.0) In this section M is a free $\mathbb{Z}$-module of rank r.

(2.1) SEMIGROUP ALGEBRA: (1) Let $k[M]$ be the group algebra of M over k; it may be defined in the following way. First of all, we set $k[M] := k^{(M)}$, the set of maps $f \colon M \to k$ having finite support [i.e., for every $f \in k[M]$ we have $f(m) \ne 0$ for only finitely many $m \in M$]. Note that $k[M]$ is a k-vector space with basis $\{e(m) \mid m \in M\}$, where, for each $m \in M$, $e(m) \colon M \to k$ is the map defined by

$$e(m)(m') = \begin{cases} 1 & \text{if } m' = m, \\ 0 & \text{if } m' \ne m, \end{cases} \quad \text{for all } m' \in M;$$

thus, we have $k[M] = \bigoplus_{m \in M} ke(m)$. The k-vector space $k[M]$ is made into a [commutative and associative] k-algebra by the multiplication $e(m)e(m') := e(m + m')$ for all $m, m' \in M$, and $e(0)$ is the unit element of $k[M]$. It is clear that $k[M]$ is an M-graded commutative k-algebra.
(2) Let Γ be a subsemigroup of M; the subspace $k[\Gamma]$ of the k-vector space $k[M]$ having the set $\{e(m) \mid m \in \Gamma\}$ as a basis is a homogeneous subalgebra of the M-graded k-algebra $k[M]$ [because it is generated by homogeneous elements], and it is called the semigroup algebra of Γ over k.
(3) Let Γ be a subsemigroup of M, let A be a k-algebra, and let S be a multiplicatively closed set in A; then S is a semigroup. Let $\varphi \colon \Gamma \to S$ be a homomorphism of semigroups; then φ admits a unique extension $\tilde{\varphi} \colon k[\Gamma] \to A$ as a homomorphism of k-algebras.
The proof is easy and is left to the reader.

(2.2) Proposition: *The k-algebra $k[M]$ is isomorphic to the ring of Laurent polynomials $k[X_1, X_1^{-1}, \dots, X_r, X_r^{-1}]$; in particular, $k[M]$ is an integrally closed integral domain, and it is a factorial k-algebra of finite type of Krull dimension r. The irreducible normal affine variety associated with $k[M]$ can be considered as an open affine subset of $\mathbb{A}^r$.*

Proof: Let $\{m_1, \ldots, m_r\}$ be a $\mathbb{Z}$-basis of M, and set $\Gamma := \mathbb{N}_0 m_1 + \cdots + \mathbb{N}_0 m_r$. The set $\{e(m_1)^{a_1} \cdots e(m_r)^{a_r} \mid (a_1, \ldots, a_r) \in \mathbb{N}_0^r\}$ is multiplicatively closed, and $k[M]$ is the localization of $k[\Gamma]$ with respect to this set. Let $k[X_1, \ldots, X_r]$ be the polynomial ring over k in indeterminates $X_1, \ldots, X_r$, and let $\varphi \colon k[X_1, \ldots, X_r] \to k[\Gamma]$ be the k-algebra homomorphism defined by $\varphi(X_i) = e(m_i)$ for every $i \in \{1, \ldots, r\}$. We set $S := \{X_1^{i_1} \cdots X_r^{i_r} \mid (i_1, \ldots, i_r) \in \mathbb{N}_0^r\}$, and we let $\tilde{\psi} \colon k[\Gamma] \to k[X_1, \ldots, X_r]$ be the k-algebra homomorphism associated with the semigroup homomorphism $\psi \colon \Gamma \to S$ defined by $\psi(m_i) = X_i$ for every $i \in \{1, \ldots, r\}$ [cf. (2.1)(3)]. Then $\tilde{\psi}$ and φ are inverse to each other, hence φ is a k-algebra isomorphism. We consider the localization $k[X_1, X_1^{-1}, \ldots, X_r, X_r^{-1}]$ of $k[X_1, \ldots, X_r]$ with respect to S. Clearly φ can be extended to an isomorphism of k-algebras $k[X_1, X_1^{-1}, \ldots, X_r, X_r^{-1}] \to k[M]$.

The ring $k[X_1, \ldots, X_r]$ is integrally closed, because it is factorial; therefore also the localization $k[X_1, X_1^{-1}, \ldots, X_r, X_r^{-1}]$ of $k[X_1, \ldots, X_r]$ is integrally closed [cf. [63], Prop. 4.13]. The transcendence degree of the quotient field of $k[X_1, \ldots, X_r]$ over k is r; since $k[X_1, X_1^{-1}, \ldots, X_r, X_r^{-1}]$ has the same field of quotients, its transcendence degree over k is also r, and therefore the Krull dimension of the k-algebra $k[X_1, X_1^{-1}, \ldots, X_r, X_r^{-1}]$, which is a k-algebra of finite type, is r [cf. [63], Th. A, p. 286].

The affine variety associated with $k[M]$ [cf. A(1.12)] can be realized as the intersection of open affine subsets of $\mathbb{A}^r$, namely as $\mathbb{A}_{X_1}^r \cap \ldots \cap \mathbb{A}_{X_r}^r \subset \mathbb{A}^r$ [cf. A(1.19)], hence the last assertion follows from A(3.11).

(2.3) REMARK: Let Γ be a subsemigroup of M.

(1) Let $M(\Gamma)$ be the $\mathbb{Z}$-submodule of M generated by Γ; $M(\Gamma)$ is a free $\mathbb{Z}$-module of finite rank, and $k[\Gamma]$ and $k[M(\Gamma)]$ have the same field of quotients.

In fact, the first assertion follows immediately since submodules of a free module over a principal ideal domain are free [cf. [117], Ch. XV, § 2, Th. 1]. For every $m \in M(\Gamma)$, there exist $m', m'' \in \Gamma$ such that $m = m' - m''$, hence we have $e(m) = e(m')/e(m'')$. This holds, in particular, for a $\mathbb{Z}$-basis of $M(\Gamma)$.

(2) Γ *is a finitely generated semigroup iff* $k[\Gamma]$ *is a k-algebra of finite type. In particular, let* $\{m_1, \ldots, m_h\}$ *be a finite system of generators for* Γ. *Then* $\{m_1, \ldots, m_h\}$ *is an irredundant system of generators for the semigroup* Γ *iff* $\{e(m_1), \ldots, e(m_h)\}$ *is an irredundant system of generators for the k-algebra* $k[\Gamma]$.

Proof: Assume that the semigroup Γ is generated by the elements $m_1, \ldots, m_h$. Then we have $k[\Gamma] = k[e(m_1), \ldots, e(m_h)]$. Conversely, assume that $k[\Gamma]$ is a k-algebra of finite type. Then $k[\Gamma]$ is generated by homogeneous elements, and we may choose $m_1, \ldots, m_h \in \Gamma$ such that $k[\Gamma] = k[e(m_1), \ldots, e(m_h)]$. By using the M-grading of $k[M]$, we see: For each $m \in \Gamma$ there exists $(a_1, \ldots, a_h) \in \mathbb{N}_0^h$ such that $e(m) = e(m_1)^{a_1} \cdots e(m_h)^{a_h}$, hence we have $m = a_1 m_1 + \cdots + a_h m_h$, and therefore $\{m_1, \ldots, m_h\}$ is a system of generators for Γ. The last claim now follows easily.

(3) Assume that Γ is finitely generated, and let $M(\Gamma)$ be the subgroup of M generated by Γ. Then the Krull dimension $\dim(k[\Gamma])$ of $k[\Gamma]$ is equal to the rank

of the $\mathbb{Z}$-module $M(\Gamma)$ [since $k[\Gamma]$ and $k[M(\Gamma)]$ have the same field of quotients, cf. (1), which has transcendence degree over k equal to the rank of $M(\Gamma)$ by (2.2)].
(4) Let Γ be finitely generated, and let $\{m_1, \ldots, m_h\}$ be a system of generators of Γ. Let $k[T_1, \ldots, T_h]$ be the polynomial ring over k in h indeterminates $T_1, \ldots, T_h$, and let $\varphi \colon k[T_1, \ldots, T_h] \to k[\Gamma]$ be the surjective k-algebra homomorphism defined by $\varphi(T_i) := e(m_i)$ for every $i \in \{1, \ldots, h\}$. Let $\mathfrak{a}$ be the kernel of φ. For every pair of h-tuples $(a_1, \ldots, a_h)$, $(b_1, \ldots, b_h) \in \mathbb{N}_0^h$ such that $a_1 m_1 + \cdots + a_h m_h = b_1 m_1 + \cdots + b_h m_h$ we have

$$T_1^{a_1} \cdots T_h^{a_h} - T_1^{b_1} \cdots T_h^{b_h} \in \mathfrak{a}, \qquad\qquad (*)$$

and the ideal $\mathfrak{a}$ is generated by a finite number of binomials, i.e., of elements having the form given in $(*)$.
Proof: For every $(a_1, \ldots, a_h)$, $(b_1, \ldots, b_h) \in \mathbb{N}_0^h$ we have

$$\varphi(T_1^{a_1} \cdots T_h^{a_h} - T_1^{b_1} \cdots T_h^{b_h}) = e(a_1 m_1 + \cdots + a_h m_h) - e(b_1 m_1 + \cdots + b_h m_h),$$

hence those elements on the left-hand side of $(*)$ for which $a_1 m_1 + \cdots + a_h m_h = b_1 m_1 + \cdots + b_h m_h$, lie in $\mathfrak{a}$.
Now, let $F \in k[T_1, \ldots, T_h]$ be an element of $\mathfrak{a}$. Let us write

$$F = \sum_{(b_1, \ldots, b_h) \in \mathbb{N}_0^h} \alpha_{b_1, \ldots, b_h} T_1^{b_1} \cdots T_h^{b_h}$$

where $\alpha_{b_1, \ldots, b_h} \in k^\times$ for all $(b_1, \ldots, b_h) \in \mathbb{N}_0^h$. Then we get

$$0 = \varphi(F) = \sum_{(b_1, \ldots, b_h) \in \mathbb{N}_0^h} \alpha_{b_1, \ldots, b_h} e(b_1 m_1 + \cdots + b_r m_r).$$

Let us consider a particular term $\alpha_{b_1, \ldots, b_h} e(b_1 m_1 + \cdots + b_h m_h)$ of the right hand side of this equation. Since this term must be cancelled, there exists $(b_1', \ldots, b_h') \in \mathbb{N}_0^h$ with $b_1 m_1 + \cdots + b_h m_h = b_1' m_1 + \cdots + b_h' m_h$ and with $\alpha_{b_1, \ldots, b_h} = -\alpha_{b_1', \ldots, b_h'}$. By repeating this procedure we easily see that F is an element of the ideal generated by the elements on the left-hand side of $(*)$. Since $k[T_1, \ldots, T_h]$ is a noetherian ring, a finite number of elements as in $(*)$ generate the ideal $\mathfrak{a}$.

(2.4) REMARK: Let Γ be a subsemigroup of M. Assume that $\Gamma \cap (-\Gamma) = \{0\}$, and set

$$\mathfrak{m}_\Gamma := \sum_{m \in \Gamma \setminus \{0\}} k e(m);$$

the sum is direct, hence $\mathfrak{m}_\Gamma$ is a subspace of $k[\Gamma]$. Furthermore, $\mathfrak{m}_\Gamma$ is a homogeneous ideal of the M-graded ring $k[\Gamma]$, and it is even a proper ideal of $k\Gamma$ [for every m, $m' \in \Gamma \setminus \{0\}$ we have $m + m' \neq 0$, since otherwise we have $m = -m' \in \Gamma \cap (-\Gamma)$, hence we get $m = m' = 0$]. Therefore, since $k[\Gamma] = k e(0) \oplus \mathfrak{m}_\Gamma$, $\mathfrak{m}_\Gamma$ is a maximal ideal of $k[\Gamma]$, $k[\Gamma]/\mathfrak{m}_\Gamma = k$, and every proper homogeneous ideal of $k[\Gamma]$ is contained in $\mathfrak{m}_\Gamma$. This implies, in particular, that R is a local M-graded ring [cf. B(4.9) for this notion].

(2.5) DEFINITION: Let Γ be a subsemigroup of M. An element $m \in \Gamma \setminus \{0\}$ is called indecomposable if $m = m' + m''$ for some $m', m'' \in \Gamma$ implies that $m' = 0$ or that $m'' = 0$.

(2.6) **Proposition:** *Let Γ be a finitely generated subsemigroup of M with $\Gamma \cap (-\Gamma) = \{0\}$, and let $\{m_1, \ldots, m_h\}$ be a system of generators of Γ consisting of non-zero elements. The following statements are equivalent:*
(1) $\{m_1, \ldots, m_h\}$ *is an irredundant system of generators of Γ.*
(2) $\{m_1, \ldots, m_h\}$ *is the set of indecomposable elements of Γ.*
(3) $\{e(m_1), \ldots, e(m_h)\}$ *is a minimal system of generators of the ideal $\mathfrak{m}_\Gamma$.*
(4) $h = \operatorname{emdim}(k[\Gamma]_{\mathfrak{m}_\Gamma})$.
In particular, if these conditions are satisfied, then $\{m_1, \ldots, m_h\}$ is a minimal set of generators of Γ

Proof (1) $\Rightarrow$ (2): Let $\{m_1, \ldots, m_h\}$ be an irredundant system of generators of Γ. Obviously, every indecomposable $m \in \Gamma$ is one of the elements $m_1, \ldots, m_h$. We have to show that the elements $m_1, \ldots, m_h$ are indecomposable. Consider m_1, and suppose that $m_1 = m' + m''$ for some $m', m'' \in \Gamma$. Let us write

$$m' = a_1' m_1 + \cdots + a_h' m_h, \quad m'' = a_1'' m_1 + \cdots + a_h'' m_h \quad \text{with } a_1', \ldots, a_h'' \in \mathbb{N}_0.$$

Then we have $m_1 = (a_1' + a_1'') m_1 + \cdots + (a_h' + a_h'') m_h$. If both elements a_1', a_1'' are zero, then we get $m_1 = (a_2' + a_2'') m_2 + \cdots + (a_h' + a_h'') m_h$, contradicting the fact that $\{m_1, \ldots, m_h\}$ is an irredundant set of generators for Γ. Hence we have $a_1' + a_1'' \geq 1$, and we get

$$0 = (a_1' + a_1'' - 1) m_1 + (a_2' + a_2'') m_2 + \cdots + (a_h' + a_h'') m_h;$$

since we have $\Gamma \cap (-\Gamma) = \{0\}$ by assumption, we see that $a_1' + a_1'' = 1$ and $a_i' = a_i'' = 0$ for $i \in \{2, \ldots, h\}$, hence that $m' = 0$ or that $m'' = 0$. In the same way we may show that the elements $m_2, \ldots, m_h$ are indecomposable.
(2) $\Rightarrow$ (3): This is clear since $\mathfrak{m}_\Gamma$ is a homogeneous ideal.
(3) $\Rightarrow$ (1): This follows from (2.3)(2).
(3) $\Leftrightarrow$ (4): This follows from B(4.11) [and the fact that the k-vector spaces $\mathfrak{m}_\Gamma / \mathfrak{m}_\Gamma^2$ and $\mathfrak{n}/\mathfrak{n}^2$, $\mathfrak{n}$ being the maximal ideal of $k[\Gamma]_{\mathfrak{m}_\Gamma}$, are isomorphic].

2.2 Integral Closure of Semigroup Rings

(2.7) In this subsection let N be a free $\mathbb{Z}$-module of rank r, and set $M := \operatorname{Hom}_\mathbb{Z}(N, \mathbb{Z})$, a free $\mathbb{Z}$-module of rank r.

(2.8) THE GROUP T_N: We set $T_N := \operatorname{Hom}_\mathbb{Z}(M, k^\times)$, an (algebraic) torus group.
(1) Let $m \neq m'$ be elements of M. Then there exists $\tau \in T_N$ such that we have $\tau(m) \neq \tau(m')$.
Indeed, let $\{m_1, \ldots, m_r\}$ be a basis of the free $\mathbb{Z}$-module M, let $m = \sum_{i=1}^r a_i m_i$ and $m' = \sum_{i=1}^r a_i' m_i$ where $a_1, \ldots, a_r' \in \mathbb{Z}$, and choose $j \in \{1, \ldots, r\}$ such that

$a_j \neq a'_j$. Let $\gamma \in k^\times$ be not a root of unity, and define $\tau \in T_N$ by $\tau(m_i) = 1$ for every $i \in \{1,\ldots,r\}$, $i \neq j$, and $\tau(m_j) = \gamma$. Then we get $\tau(m) = \gamma^{a_j} \neq \gamma^{a'_j} = \tau(m')$.

(2) For every $m \in M \setminus \{0\}$ the orbit $T_N(m) = \{\tau(m) \mid \tau \in T_N\} \subset k^\times$ of m is an infinite set.

Indeed, let $\{m_1,\ldots,m_r\}$ be a basis of M, write $m = \sum_{i=1}^r a_i m_i$ and choose $j \in \{1,\ldots,r\}$ with $a_j \neq 0$. We define $\tau \in T_N$ as in (1). Then $\{\tau^l(m) \mid l \in \mathbb{Z}\}$ is an infinite set.

(2.9) THE ACTION OF T_N ON $k[M]$: Let $\tau \in T_N$. We define

$$\tau(e(m)) := \tau(m)e(m) \quad \text{for every } m \in M,$$

and we define the action of τ on the k-vector space $k[M]$ by linearity. For all m, $m' \in M$, we have

$$\tau(e(m)e(m')) = \tau(e(m+m')) = \tau(m+m')e(m+m')$$
$$= (\tau(m)e(m))(\tau(m')e(m')) = \tau(e(m))\tau(e(m'));$$

we see, therefore, that τ acts on the k-algebra $k[M]$ as a homogeneous k-algebra automorphism. It is easy to check that T_N acts on $k[M]$ as a group of homogeneous k-algebra automorphisms.

A k-subspace V of $k[M]$ is called T_N-stable if $\tau(v) \in V$ for every $\tau \in T_N$ and $v \in V$, i.e., if $T_N(V) \subset V$.

(2.10) Proposition: (1) *For every $m \in M$ we have*

$$\{w \in k[M] \mid \tau(w) = \tau(m)w \text{ for all } \tau \in T_N\} = ke(m).$$

(2) *Let V be a T_N-stable k-subspace of $k[M]$. Then we have*

$$V = \sum_{m \in M} (V \cap ke(m)) = \bigoplus_{m \in M, e(m) \in V} ke(m),$$

i.e., V is a homogeneous k-subspace of $k[M]$.

Proof: (1) Let $m \in M$, and let $w \in k[M]$ be a non-zero element satisfying $\tau(w) = \tau(m)w$ for all $\tau \in T_N$. We consider the representation $w = \sum_{m' \in M} \alpha_{m'} e(m')$ where $\alpha_{m'} \in k$ for every $m' \in M$, and only finitely many coefficients $\alpha_{m'}$ are different from 0. We have $\tau(w) = \sum_{m' \in M} \tau(m')\alpha_{m'} e(m')$ for every $\tau \in T_N$. Let $m' \in M$, $m' \neq m$, and choose $\tau \in T_N$ such that $\tau(m') \neq \tau(m)$ [cf. (2.8)(1)]; then we see that $\alpha_{m'} = 0$, and therefore we have $w \in ke(m)$.

(2) Let $h > 1$, and assume that the following has already been shown. For every $v \in V$ having a representation $v = \sum_{i=1}^l \alpha_i e(m_i)$ where $l < h$, $\alpha_1,\ldots,\alpha_l \in k^\times$, and $m_1,\ldots,m_l$ are pairwise different elements in M, the elements $e(m_1),\ldots,$

$e(m_l)$ lie in V. Now let $\alpha_1 e(m_1) + \cdots + \alpha_h e(m_h) =: v \in V$ where $\alpha_1, \ldots, \alpha_h \in k^\times$ and where $m_1, \ldots, m_h$ are pairwise different elements of M. We choose $\tau \in T_N$ such that $\tau(m_1) \neq \tau(m_2)$ [cf. (2.8)(1)]. We have

$$-\tau(m_2)v + \tau(v) = \alpha_1(\tau(m_1) - \tau(m_2))e(m_1) + \sum_{i=3}^{h} \alpha_i(\tau(m_i) - \tau(m_2))e(m_i),$$

hence we get $e(m_1) \in V$ by induction. Then, again by induction, we find that $e(m_2), \ldots, e(m_h) \in V$.

(2.11) Proposition: *Let Γ be a subsemigroup of M, and let $\overline{\Gamma}$ be its saturation. Then $k[\overline{\Gamma}]$ is the integral closure of $k[\Gamma]$; in particular, $k[\Gamma]$ is integrally closed iff $\Gamma = \overline{\Gamma}$.*

Proof: We set $M' := M(\Gamma)$; $k[\Gamma]$ and $k[M']$ have the same field of quotients [cf. (2.3)(1)]. Let R be the integral closure of $k[\Gamma]$. Since $k[M']$ is integrally closed [cf. (2.2)], we have $k[\Gamma] \subset R \subset k[M']$. Since $k[\Gamma]$ is T_N-stable, we see that R is T_N-stable [let $z \in R$, and let $X^h + y_1 X^{h-1} + \cdots + y_h \in k[\Gamma][X]$ be an equation of integral dependence of z over $k[\Gamma]$, then, for every $\tau \in T_N$, $X^h + \tau(y_1)X^{h-1} + \cdots + \tau(y_h)$ is an equation of integral dependence of $\tau(z)$ over $k[\Gamma]$], hence R is a homogeneous subring of $k[M]$ [cf. (2.10)(2)] and therefore also of $k[M']$. Let $m \in \overline{\Gamma}$; then there exists $d \in \mathbb{N}$ with $m \in (\frac{1}{d}\Gamma) \cap M'$. Now $e(m)$ lies in the quotient field of $k[\Gamma]$, and we have $e(m)^d = e(dm)$; since dm lies in Γ, the element $e(m)$ is integral over $k[\overline{\Gamma}]$. This shows that we have $k[\overline{\Gamma}] \subset R$. We show that $R \subset k[\overline{\Gamma}]$. Since both rings are homogeneous subrings of $k[M']$, it is enough to show the following: If $m \in M'$ and if $e(m)$ is integral over $k[\Gamma]$, then m lies in $\overline{\Gamma}$. We choose elements $y_1, \ldots, y_h \in k[\Gamma]$ with $e(m)^h + y_1 e(m)^{h-1} + \cdots + y_h = 0$; we may assume that, for every $j \in \{1, \ldots, h\}$, y_j is homogeneous of degree jm [cf. B(2.5)]. At least one of the elements $y_1, \ldots, y_h$ is not zero since $k[M]$ is a domain; hence there exists $j \in \{1, \ldots, h\}$ such that $y_j \neq 0$, and then we have $jm \in \Gamma$, i.e., we have $m \in \overline{\Gamma}$.

(2.12) Proposition: *Let Γ be a subsemigroup of M, and let $\overline{\Gamma}$ be its saturation. If Γ is finitely generated, so is $\overline{\Gamma}$.*

Proof: Since Γ is finitely generated, $k[\Gamma]$ is a k-algebra of finite type [cf. (2.3)(1)], hence the integral closure of $k[\Gamma]$ is a k-algebra of finite type [cf. B(3.6)]. Now $k[\overline{\Gamma}]$ is the integral closure of $k[\Gamma]$ by (2.11), and therefore it is a k-algebra of finite type, hence $\overline{\Gamma}$ is finitely generated by (2.3)(1).

3 Continued Fractions

(3.0) In the first part of this section we present, for the convenience of the reader, the results on continued fractions which are needed in the sequel. In the second part we deal with the Hirzebruch-Jung continued fraction expansion of a positive rational number.

3.1 Continued Fractions

(3.1) NOTATION: (1) Let $n \in \mathbb{N}_0$, a_0 be a real and $a_1, \ldots, a_n$ be real positive numbers. Define $[a_0] := a_0$ and

$$[a_0, a_1, \ldots, a_{j-1}, a_j] := [a_0, a_1, \ldots, a_{j-2}, a_{j-1} + \frac{1}{a_j}] \quad \text{for every } j \in \{1, \ldots, n\}.$$

Hence we have for $n \geq 1$

$$[a_0, \ldots, a_n] = a_0 + \cfrac{1}{a_1 + \cfrac{1}{a_2 + \cfrac{1}{\ddots + \cfrac{1}{a_n}}}},$$

a continued fraction.

We assume that $n \geq 1$; clearly we have $[a_1, \ldots, a_n] > 0$ and

$$[a_0, a_1, \ldots, a_n] = a_0 + \frac{1}{[a_1, a_2, \ldots, a_n]}. \tag{*}$$

(2) Let $n \in \mathbb{N}$, $a_0 \in \mathbb{Z}$ and $a_1, \ldots, a_n \in \mathbb{N}$, and assume that $a_n \geq 2$. Then we have

$$a_0 < [a_0, a_1, \ldots, a_n] < a_0 + 1, \tag{**}$$

and we have $a_i = \lfloor [a_i, a_{i+1}, \ldots, a_n] \rfloor$ for every $i \in \{0, 1, \ldots, n\}$.
Proof: If $n = 1$, then we have $a_0 < [a_0, a_1] = a_0 + 1/a_1 \leq a_0 + 1/2 < a_0 + 1$. If $n \geq 2$, then we assume that for all $a_0' \in \mathbb{Z}$ and all positive integers $a_1', \ldots, a_{n-1}'$ such that $a_{n-1}' \geq 2$ it has been proved that $a_0' < [a_0', a_1', \ldots, a_{n-1}'] < a_0' + 1$. We have $a_1 < [a_1, a_2, \ldots, a_n] < a_1 + 1$ by assumption, hence we have

$$a_0 < a_0 + \frac{1}{a_1 + 1} < a_0 + \frac{1}{[a_1, a_2, \ldots, a_n]} < a_0 + \frac{1}{a_1} \leq a_0 + 1,$$

and the assertion in (**) follows from (*).
Now we have $[a_n] = a_n$, hence we get $\lfloor [a_n] \rfloor = a_n$. Let $i \in \{0, \ldots, n-1\}$; by (**) we have $a_i < [a_i, a_{i+1}, \ldots, a_n] < a_i + 1$, hence we get $\lfloor [a_i, a_{i+1}, \ldots, a_n] \rfloor = a_i$.

(3.2) Proposition: *Let $a \in \mathbb{Z}$ and $b \in \mathbb{N}$. Then there exist uniquely determined $n \in \mathbb{N}_0$, $a_0 \in \mathbb{Z}$ and $a_1, \ldots, a_n \in \mathbb{N}$ with $a_n \geq 2$ if $n \geq 1$, such that*

$$\frac{a}{b} = [a_0, a_1, \ldots, a_n].$$

Proof: (1)(a) By the Euclidean algorithm there exist $n \in \mathbb{N}_0$ and $a_0 \in \mathbb{Z}$, $a_1, \ldots, a_n, b_1, \ldots, b_n \in \mathbb{N}$ such that

$$
\begin{aligned}
a &= a_0 b + b_1 \qquad \text{and} \quad b_1 < b_0 := b, \\
b &= a_1 b_1 + b_2 \qquad \text{and} \quad b_2 < b_1, \\
&\cdots\cdots\cdots\cdots\cdots \\
b_{n-2} &= a_{n-1} b_{n-1} + b_n \quad \text{and} \quad b_n < b_{n-1}, \\
b_{n-1} &= a_n b_n.
\end{aligned}
$$

Here we have $b_n = \gcd(a, b)$, and we have $a_n = b_{n-1}/b_n \geq 2$ if $n \geq 1$.
(b) If $n = 0$, then we have $a/b = a_0 = [a_0]$, and if $n = 1$, then we have $a/b = a_0 + 1/(b_0/b_1) = [a_0, b_0/b_1]$. Now let $n \geq 2$. Then we have $a/b = a_0 + 1/(b_0/b_1) = [a_0, b_0/b_1]$. We assume that we have shown already that

$$
a/b = [a_0, a_1, \ldots, a_j, b_j/b_{j+1}] \tag{$*$}
$$

for some $j \in \{0, \ldots, n-2\}$. Since $b_j/b_{j+1} = a_{j+1} + 1/(b_{j+1}/b_{j+2})$, we have

$$
\frac{a}{b} = \left[a_0, a_1, \ldots, a_j, a_{j+1} + \frac{1}{b_{j+1}/b_{j+2}}\right] = \left[a_0, a_1, \ldots, a_j, a_{j+1}, \frac{b_{j+1}}{b_{j+2}}\right].
$$

Therefore $(*)$ holds for every $j \in \{0,, \ldots, n-1\}$. Take $j = n-1$; then we have

$$
\frac{a}{b} = \left[a_0, a_1, \ldots, a_{n-1}, \frac{b_{n-1}}{b_n}\right] = [a_0, a_1, \ldots, a_{n-1}, a_n].
$$

(2) To show uniqueness of the continued fraction expansion, we proceed as follows. Let $n, n' \in \mathbb{N}_0$, $a_0, a_0' \in \mathbb{Z}$, $a_1, \ldots, a_n, a_1', \ldots, a_{n'}'$ be positive integers such that $a_n \geq 2$ if $n \geq 1$, $a_{n'}' > 2$ if $n' \geq 1$, and assume that $[a_0, a_1, \ldots, a_n] = [a_0', a_1', \ldots, a_{n'}']$. We may assume that $n \leq n'$. If $n = 0$, then we have also $n' = 0$ [because otherwise $a_0 = [a_0] = [a_0', a_1', \ldots, a_{n'}']$ is not an integer, cf. (3.1)(2)], and therefore we have $a_0 = a_0'$. If $n \geq 1$, then we have, by (3.1)(2), $a_0 = \lfloor [a_0, a_1, \ldots, a_n] \rfloor = \lfloor [a_0', a_1', \ldots, a_{n'}'] \rfloor = a_0'$. Since we have

$$
a_0 + \frac{1}{[a_1, a_2, \ldots, a_n]} = [a_0, a_1, \ldots, a_n] = [a_0', a_1', \ldots, a_{n'}'] = a_0' + \frac{1}{[a_1', a_2', \ldots, a_{n'}']},
$$

it follows that $[a_1, a_2, \ldots, a_n] = [a_1', a_2', \ldots, a_{n'}']$; hence we see, by induction, that $n - 1 = n' - 1$ and that $a_j = a_j'$ for every $j \in \{1, \ldots, n\}$.

(3.3) NOTATION: Let $a \in \mathbb{Z}$ and $b \in \mathbb{N}$.
(1) The continued fraction in (3.2) is called the continued fraction expansion of the rational number a/b.
(2) Now let $a, b \in \mathbb{N}$. We also have a unique expansion

$$
\frac{a}{b} = [a_0, a_1, \ldots, a_{2m}]
$$

where $a_{2m} \in \mathbb{N}$: if, in the notation of (3.2), n is even, let m be defined by $2m = n$, and if n is odd, then we have $a_n \geq 2$ and $a/b = [a_0, a_1, \ldots, a_{n-1}, a_n - 1, 1]$.

3.2 Hirzebruch-Jung Continued Fractions

(3.4) ANOTHER CONTINUED FRACTION EXPANSION: Let $n \in \mathbb{N}$ and $a_1, \ldots, a_n$ be real numbers such that $a_i \geq 2$ for every $i \in \{1, \ldots, n\}$.
(1) Define $[[\, a_1 \,]] := a_1$ and, for every $j \in \{2, \ldots, n\}$, define

$$[[\, a_1, \ldots, a_{j-1}, a_j \,]] := [[\, a_1, \ldots, a_{j-2}, a_{j-1} - \frac{1}{a_j} \,]].$$

Hence we have

$$[[\, a_1, \ldots, a_n \,]] = a_1 - \cfrac{1}{a_2 - \cfrac{1}{a_3 - \cfrac{1}{\ddots - \cfrac{1}{a_n}}}},$$

a continued fraction.
Clearly we have $[[\, a_1, \ldots, a_n \,]] > 0$, and if $n \geq 2$, then we have

$$[[\, a_1, \ldots, a_n \,]] = a_1 - \frac{1}{[[\, a_2, \ldots, a_n \,]]}.$$

(2) Let $n \geq 2$ and $a_1, \ldots, a_n$ be integers greater than 1. Then we have

$$a_1 - 1 < [[\, a_1, \ldots, a_n \,]] < a_1, \qquad\qquad (*)$$

and we have $a_i = \lceil [[\, a_i, \ldots, a_n \,]] \rceil$ for every $i \in \{1, \ldots, n\}$.
Proof: If $n = 2$, then we have $a_1 - 1 < [[\, a_1, a_2 \,]] = a_1 - 1/a_2 < a_1$. If $n \geq 3$, then we assume that for all integers $a_1', \ldots, a_{n-1}'$ greater than 1 it has been proved already that $a_1' - 1 < [[\, a_1', \ldots, a_{n-1}' \,]] < a_1'$. Then we have $a_2 - 1 < [[\, a_2, \ldots, a_n \,]] < a_2$ by assumption, hence we have

$$a_1 - 1 \leq a_1 - \frac{1}{a_2 - 1} < a_1 - \frac{1}{[[\, a_2, \ldots, a_n \,]]} < a_1 - \frac{1}{a_2} < a_1,$$

and the proof of the first assertion is complete.
Now we have $[[\, a_n \,]] = a_n$, hence we have $\lceil [[\, a_n \,]] \rceil = a_n$. Since $a_i - 1 < [[\, a_i, \ldots, a_n \,]] < a_i$ for every $i \in \{1, \ldots, n\}$, we find that $\lceil [[\, a_i, \ldots, a_n \,]] \rceil = a_i$.

(3.5) Let $a, b \in \mathbb{N}$ and assume that $a > b$. Then we can write

$$a = rb - s \quad \text{where } r \in \mathbb{N} \text{ and } r \geq 2, \ s \in \mathbb{N}_0 \text{ and } 0 \leq s < b,$$

and such a representation is unique.

(3.6) Proposition: Let $a, b \in \mathbb{N}$ and assume that $a > b$. Then there exist uniquely determined $n \in \mathbb{N}$ and positive integers $a_1, \ldots, a_n$ greater than 1 such that

$$\frac{a}{b} = [[\, a_1, \ldots, a_n \,]]. \qquad\qquad (*)$$

Proof: (1)(a) Define $q_1 := a$, $q_2 := b$. By (3.5) there exist $n \in \mathbb{N}$, positive integers $a_1, \ldots, a_n$ greater than 1 and positive integers $q_3, \ldots, q_{n+1}$ such that

$$q_{j-1} = a_{j-1}q_j - q_{j+1} \quad \text{for } j = 2, \ldots, n+1,$$

where we have $0 < q_{j+1} < q_j$ for every $j \in \{1, \ldots, n\}$ and $q_{n+2} = 0$. Clearly we have $q_{n+1} = \gcd(a, b)$.

(b) If $n = 1$, then we have $a/b = q_1/q_2 = a_1 = [[\, a_1 \,]]$. If $n = 2$, then we have

$$\frac{a}{b} = a_1 - \frac{q_3}{q_2} = a_1 - 1/(q_2/q_3) = [[\, a_1, q_2/q_3 \,]].$$

If $n \geq 3$, then we have $a/b = a_1 - q_3/q_2 = [[\, a_1, q_2/q_3 \,]]$. Assume that we have already shown that

$$a/b = [[\, a_1, \ldots, a_{j-1}, q_j/q_{j+1} \,]] \tag{$**$}$$

for some $j \in \{2, \ldots, n-1\}$. Since $q_j/q_{j+1} = a_j - q_{j+2}/q_{j+1}$, we find that

$$\frac{a}{b} = \left[\left[\, a_1, \ldots, a_{j-1}, a_j - \frac{1}{q_{j+1}/q_{j+2}} \,\right]\right] = \left[\left[\, a_1, \ldots, a_{j-1}, a_j, \frac{q_{j+1}}{q_{j+2}} \,\right]\right].$$

Hence $(**)$ holds for every $j \in \{1, \ldots, n\}$. Take $j = n$; then we get

$$\frac{a}{b} = [[\, a_1, \ldots, a_{n-1}, q_n/q_{n+1} \,]] = [[\, a_1, \ldots, a_n \,]].$$

(2) To show uniqueness of the continued fraction expansion we proceed as follows. Let $n, n' \in \mathbb{N}$, let $a_1, \ldots, a_n$ and $a_1', \ldots, a_{n'}'$ be integers greater than 1, and assume that $[[\, a_1, \ldots, a_n \,]] = [[\, a_1', \ldots, a_{n'}' \,]]$. We may assume that $n \leq n'$. If $n = 1$, then we get $n' = 1$ [because otherwise $[[\, a_1', \ldots, a_{n'}' \,]]$ is not an integer]. If $n > 1$, then we have $a_1 = \lceil [[\, a_1, \ldots, a_n \,]] \rceil = \lceil [[\, a_1', \ldots, a_{n'}' \,]] \rceil = a_1'$ [cf. (3.4)(2)], hence we get $a_1 = a_1'$, and $[[\, a_2, \ldots, a_n \,]] = [[\, a_2', \ldots, a_{n'}' \,]]$ by (3.4)(1). By recursion we get uniqueness of the representation $(*)$.

(3.7) NOTATION: From now on, we use the notations

$$[a_1, \ldots, a_s] \quad =: \quad a_1 + \overline{|a_2|} + \cdots + \overline{|a_s|},$$
$$[[a_1, \ldots, a_s]] \quad =: \quad a_1 - \overline{|a_2|} - \cdots - \overline{|a_s|}$$

where, in the first case, $a_1, \ldots, a_s$ are natural integers [the case that there is an a_0 which is an integer, cf. (3.2), will not occur], and we speak of a continued fraction expansion, and, in the second case, $a_1, \ldots, a_s$ are integers larger than 1, and we speak of a Hirzebruch-Jung continued fraction expansion.

(3.8) REMARK: Let $p, q \in \mathbb{N}$, $p < q$ and $\gcd(p, q) = 1$, and let

$$\frac{q}{q-p} = b_1 - \overline{|b_2|} - \cdots - \overline{|b_s|} \quad \text{with } b_i \geq 2 \text{ for every } i \in \{1, \ldots, s\}$$

be the Hirzebruch-Jung continued fraction expansion of $q/(q-p)$. Then we have $p = 1$ iff $b_1 = \cdots = b_s = 2$; in this case we have $s = q - 1$.

(3.9) REMARK: Let $p, q \in \mathbb{N}$, $p < q$ and $\gcd(p, q) = 1$, and let

$$\frac{q}{q-p} = b_1 - \overline{b_2} - \cdots - \overline{b_s} \quad \text{with } b_i \geq 2 \quad \text{for every } i \in \{1, \ldots, s\},$$

$$\frac{q}{p} = c_1 - \overline{c_2} - \cdots - \overline{c_t} \quad \text{with } c_j \geq 2 \quad \text{for every } j \in \{1, \ldots, t\}$$

be the Hirzebruch-Jung continued fraction expansions of $q/(q-p)$ and q/p, respectively.

(1) Given the expansion for $q/(q-p)$, i.e., the length s and the integers $b_1, \ldots, b_s$, the following program calculates the length t and the integers $c_1, \ldots, c_t$ of the Hirzebruch-Jung continued fraction expansion of q/p [in the program we write $b[i]$ instead of b_i and $c[j]$ instead of c_j].

```
1.   l := 1;  c[1] := 2;
2.   for i from 1 to s do
3.       for j from 2 to b[i] − 1 do
4.           l := l + 1;  c[l] := 2
5.       enddo;
6.       if i < s then c[l] := c[l] + 1 endif;
7.   enddo;
8.   t := l.
```

(2) We show the correctness of the program by induction with respect to s and b_1.
(a) Let $s = 1$. Then $q/(q-p) = b_1 \in \mathbb{N}$, hence $q - p = 1$ [since $\gcd(q, p) = \gcd(q, q - p) = 1$], and $q/p = q/(q-1)$. For every integer n such that $n \geq 2$, we have $n = 2(n-1) - (n-2)$ and $0 \leq n - 2 < n - 1$, and therefore we have

$$\frac{q}{q-1} = c_1 - \overline{c_2} - \cdots - \overline{c_t}$$

where $t = q - 1$ and $c_j = 2$ for every $j \in \{1, \ldots, t\}$. This is in accordance with lines 1, 2–5 and 8 of the program.
(b) Let $s > 1$, and assume that the program is correct for all $p', q' \in \mathbb{N}$ satisfying $0 < p' < q'$, $\gcd(q', p') = 1$, and being such that the Hirzebruch-Jung continued fraction expansion of $q'/(q' - p')$ has length $s - 1$. Now let $q/(q-p)$ have the Hirzebruch-Jung continued fraction expansion

$$\frac{q}{q-p} = b_1 - \overline{b_2} - \cdots - \overline{b_s}. \tag{*}$$

Assume that $b_1 = 2$. We may define uniquely integers p', q' satisfying $0 < p' < q'$ and $\gcd(q', p') = 1$ by

$$\frac{q'}{q'-p'} := b_2 - \overline{b_3} - \cdots - \overline{b_s};$$

now we have

$$\frac{q}{q-p} = 2 - \frac{1}{\dfrac{q'}{q'-p'}} = \frac{q'+p'}{q'},$$

hence $p' = p$, $q' = q - p$ [because $\gcd(q, q - p) = \gcd(q' + p', q') = 1$]. By our induction assumption, we have

$$\frac{q'}{p'} = c'_1 - \lceil c'_2 \rceil - \cdots - \lceil c'_{t'} \rceil$$

where t' and $c'_1, \ldots, c'_{t'}$ are calculated by the program, using as input $s - 1$ and $b_2, \ldots, b_s$, and therefore we have

$$\frac{q}{p} = \frac{q' + p'}{p'} = 1 + \frac{q'}{p'} = c'_1 + 1 - \lceil c'_2 \rceil - \cdots - \lceil c'_{t'} \rceil$$

in accordance with the program, since for $i = 1$ the action beginning in line 3 will not be executed, whereas in line 6 we get $c_1 = c'_1 + 1$.

Assume that $b_1 > 2$, and that the program is correct for all $p', q' \in \mathbb{N}$ satisfying $0 < p' < q'$, $\gcd(q', p') = 1$, and such that the Hirzebruch-Jung continued fraction expansion of $q'/(q' - p')$ has the form

$$\frac{q'}{q' - p'} = b'_1 - \lceil b'_2 \rceil - \cdots - \lceil b'_s \rceil$$

where $b'_1 < b_1$. We have [note that $q < 2p$]

$$\frac{q}{q - p} - 1 = \frac{p}{p - (2p - q)} = (b_1 - 1) - \lceil b_2 \rceil - \cdots - \lceil b_s \rceil,$$

hence we get $q' = p$, $p' = 2p - q$, and by our induction assumption we have

$$\frac{p}{2p - q} = c'_1 - \lceil c'_2 \rceil - \cdots - \lceil c'_{t'} \rceil$$

where t' and the integers $c'_1, \ldots, c'_{t'}$ are calculated by the program, using as input $s, b_1 - 1, b_2, \ldots, b_s$. From

$$\frac{q}{p} = 2 - \cfrac{1}{\cfrac{p}{2p - q}}$$

we get

$$\frac{q}{p} = c_1 - \lceil c_2 \rceil - \cdots - \lceil c_t \rceil$$

where $t = t' + 1$, $c_1 = 2$, and $c_j = c'_{j-1}$ for every $j \in \{2, \ldots, t\}$, again in accordance with the program [by lines 3, 4, and 8, we see that $t = t' + 1$].

(3.10) By the program in (3.9) we get

$$t = 1 + \sum_{i=1}^{s}(b_i - 2) = 1 - 2s + \sum_{i=1}^{s} b_i. \tag{$*$}$$

From this it follows that

$$(b_1 + \cdots + b_s) - s = (c_1 + \cdots + c_t) - t = s + t - 1. \tag{$**$}$$

Indeed, by $(*)$, the first and the third term in $(**)$ are equal. By symmetry, the second and the last term in $(**)$ are also equal.

4   Two-Dimensional Cones

4.1   Two-dimensional Cones and Semigroups

(4.1) In this section N is a free $\mathbb{Z}$-module of rank 2, and $M := \mathrm{Hom}_{\mathbb{Z}}(N, \mathbb{Z})$ is the $\mathbb{Z}$-module dual to N. Let

$$\langle\ ,\ \rangle \colon M \times N \to \mathbb{Z} \qquad\qquad (*)$$

be the canonical $\mathbb{Z}$-bilinear pairing; it is non-degenerate, i.e., the induced mappings $M \to \mathrm{Hom}_{\mathbb{Z}}(N, \mathbb{Z})$ and $N \to \mathrm{Hom}_{\mathbb{Z}}(M, \mathbb{Z})$ are bijective.

We set $N_{\mathbb{R}} := N \otimes_{\mathbb{Z}} \mathbb{R}$, $M_{\mathbb{R}} := M \otimes_{\mathbb{Z}} \mathbb{R}$; $M_{\mathbb{R}}$ and $N_{\mathbb{R}}$ are two-dimensional $\mathbb{R}$-vector spaces, $M_{\mathbb{R}}$ is the $\mathbb{R}$-vector space dual to $N_{\mathbb{R}}$, and $(*)$ gives rise to the canonical non-degenerate $\mathbb{R}$-bilinear pairing $\langle\ ,\ \rangle \colon M_{\mathbb{R}} \times N_{\mathbb{R}} \to \mathbb{R}$. We consider N as a subgroup of the (additive group underlying the $\mathbb{R}$-vector space) $N_{\mathbb{R}}$ and M as a subgroup of $M_{\mathbb{R}}$. Let $\{n_1, n_2\}$ be a $\mathbb{Z}$-basis of N, and let $\{m_1, m_2\}$ be the $\mathbb{Z}$-basis of M dual to the basis $\{n_1, n_2\}$, i.e., we have $\langle m_i, n_j \rangle = \delta_{ij}$ for $i, j \in \{1, 2\}$. Then $\{n_1, n_2\}$ is an $\mathbb{R}$-basis of $N_{\mathbb{R}}$, and $\{m_1, m_2\}$ is the $\mathbb{R}$-basis of $M_{\mathbb{R}}$ dual to this basis.

(4.2) NOTATION: Remember that a subset $\sigma \subset N_{\mathbb{R}}$ is called *convex* if we have $av + (1-a)v' \in \sigma$ for all $v, v' \in \sigma$ and $a \in \mathbb{R}$ with $0 \leq a \leq 1$, and that it is called a *cone* if $av \in \sigma$ and $v + v' \in \sigma$ for all $v, v' \in \sigma$ and $a \in \mathbb{R}_{\geq 0}$. A cone always is a convex set and contains the origin 0. A convex cone σ in $N_{\mathbb{R}}$ is called *strongly convex* if $\sigma \cap (-\sigma) = \{0\}$. If σ is a convex cone in $N_{\mathbb{R}}$, then $\sigma + (-\sigma)$ is a subspace of $N_{\mathbb{R}}$; its dimension is called the dimension of σ.

A convex cone $\sigma \subset N_{\mathbb{R}}$ is called a *polyhedral cone* if there exist $v_1, \ldots, v_r \in N_{\mathbb{R}}$ such that

$$\sigma = \mathbb{R}_{\geq 0}\, v_1 + \cdots + \mathbb{R}_{\geq 0}\, v_r; \qquad\qquad (*)$$

a convex polyhedral cone is called *rational* (with respect to N) if the elements in $(*)$ can be taken to lie in N.

(4.3) DEFINITION: A strongly convex rational polyhedral cone σ of dimension 2 is called *nonsingular* if there exists a $\mathbb{Z}$-basis $\{n_1, n_2\}$ of N such that we have $\sigma = \mathbb{R}_{\geq 0}\, n_1 + \mathbb{R}_{\geq 0}\, n_2$.

(4.4) DEFINITION: Let σ be a strongly convex rational polyhedral cone in $N_{\mathbb{R}}$. Then

$$\sigma^{\vee} := \{u \in M_{\mathbb{R}} \mid \langle u, v \rangle \geq 0 \text{ for all } v \in \sigma\}$$

is a convex cone which is called the cone dual to σ, and

$$\Gamma_{\sigma} := M \cap \sigma^{\vee} = \{m \in M \mid \langle m, v \rangle \geq 0 \text{ for all } v \in \sigma\}$$

is a subsemigroup of M, the semigroup associated with σ. Clearly Γ_{σ} is a saturated subgroup of M.

(4.5) PRIMITIVE ELEMENTS: (1) An element $n \in N \setminus \{0\}$ is called primitive if $n = an'$ for some $a \in \mathbb{N}$ and $n' \in N$ implies $a = 1$. It is easily seen that every $n \in N \setminus \{0\}$ has a unique representation $n = an'$ where $a \in \mathbb{N}$ and $n' \in N$ is primitive.

In particular, let $\{n_1, n_2\}$ be a $\mathbb{Z}$-basis of N; then n_1 and n_2 are primitive elements of N, and an element $a_1 n_1 + a_2 n_2$ where $a_1, a_2 \in \mathbb{Z}$ are not both are equal to 0, is primitive iff $\gcd(a_1, a_2) = 1$.

(2) Let $n \in N$ be primitive. Then there exists a primitive $n' \in N$ such that $\{n, n'\}$ is a $\mathbb{Z}$-basis of N.

Proof: Let $\{n_1, n_2\}$ be a $\mathbb{Z}$-basis of N. Then we have $n = a_1 n_1 + a_2 n_2$ for some $a_1, a_2 \in \mathbb{Z}$ with $\gcd(a_1, a_2) = 1$. We choose $b_1, b_2 \in \mathbb{Z}$ with $a_1 b_1 + a_2 b_2 = 1$; then the set $\{n, n'\}$ where $n' = b_2 n_1 - b_1 n_2$, is a $\mathbb{Z}$-basis of N, and n' is primitive.

(4.6) PRIMITIVE ELEMENTS AND CONES: Let σ be a strongly convex rational polyhedral cone in $N_{\mathbb{R}}$ of dimension 2.

(1) There exist primitive elements $n, \tilde{n} \in N$ such that $\sigma = \mathbb{R}_{\geq 0} \, n + \mathbb{R}_{\geq 0} \, \tilde{n}$.

Proof: We may choose primitive elements $n_1, \ldots, n_s \in N$ such that $\sigma = \mathbb{R}_{\geq 0} \, n_1 + \cdots + \mathbb{R}_{\geq 0} \, n_s$. The claim shall be proved by induction on s. If $s = 2$, then we are done. So let $s > 2$; we may assume that $\sigma' := \mathbb{R}_{\geq 0} \, n_1 + \cdots + \mathbb{R}_{\geq 0} \, n_{s-1}$ is a strongly convex rational polyhedral cone of dimension 2 [since otherwise, as it is easily checked, we have $\sigma' = \mathbb{R}_{\geq 0} \, n$ for some primitive element $n \in N$]. By the induction assumption, we have $\sigma' = \mathbb{R}_{\geq 0} \, n' + \mathbb{R}_{\geq 0} \, n''$ for some primitive elements $n', n'' \in N$ which are linearly independent over $\mathbb{Q}$. Now, there exist rational integers a, b and c [where $a, b \in \mathbb{Z}$, $c \in \mathbb{N}$, and not both of a and b are zero] such that $n_s = (a/c)n' + (b/c)n''$. If $a < 0$, then we have $b > 0$ [because otherwise cn_s and $-cn_s \in \sigma$] and $n'' \in \mathbb{R}_{\geq 0} \, n_s + \mathbb{R}_{\geq 0} \, n'$. Likewise, if $b < 0$, then we have $a > 0$ and $n' \in \mathbb{R}_{\geq 0} \, n_s + \mathbb{R}_{\geq 0} \, n''$. Thus, we have proved the claim.

(2) If σ is not nonsingular, then there exist a $\mathbb{Z}$-basis $\{n_1, n_2\}$ of N and integers p, q with $1 \leq p < q$ and $\gcd(p, q) = 1$ such that we have $\sigma = \mathbb{R}_{\geq 0} n_1 + \mathbb{R}_{\geq 0} n^*$ where $n^* = p n_1 + q n_2$.

Proof: By (1) there exist primitive elements $n_1, \tilde{n}_1 \in N$ such that $\sigma = \mathbb{R}_{\geq 0} \, n_1 + \mathbb{R}_{\geq 0} \, \tilde{n}_1$. By (4.5)(2) there exists a primitive element $n_1' \in N$ such that $\{n_1, n_1'\}$ is a $\mathbb{Z}$-basis of N. Then we have $\tilde{n}_1 = a n_1 + q n_1'$ where $a, q \in \mathbb{Z}$ and $q \neq 0$. If $q < 0$, then we replace n_1' by $-n_1'$; therefore we may assume that $q > 0$. Now we have $q > 1$ [since σ is not nonsingular] and a is not a multiple of q since $\tilde{n}_1$ is primitive. We write $a = hq + p$ where $h \in \mathbb{Z}$, $p \in \mathbb{N}$ and $p < q$. Then we have $\tilde{n}_1 = p n_1 + q n_2$ where $n_2 = n_1' + h n_1$. Now $\{n_1, n_2\}$ is a $\mathbb{Z}$-basis of N, and $\gcd(p, q) = 1$ because $\tilde{n}_1$ is primitive.

(4.7) EXAMPLE: Let $\{n_1, n_2\}$ be a $\mathbb{Z}$-basis of N, and let $\{m_1, m_2\}$ be the $\mathbb{Z}$-basis of M dual to this basis. Let $1 \leq p < q$ be natural integers with $\gcd(p, q) = 1$.

(1) We set $n^* := p n_1 + q n_2$ and $\sigma := \mathbb{R}_{\geq 0} \, n_1 + \mathbb{R}_{\geq 0} \, n^*$. Note that $\{n_1, n^*\}$ is an $\mathbb{R}$-basis of $N_{\mathbb{R}}$, and that σ is a strongly convex rational polyhedral cone of dimension 2 in $N_{\mathbb{R}}$. We set $\tilde{m}_1 := m_2$, $\tilde{m}_2 := m_1 - m_2$; then $\{\tilde{m}_1, \tilde{m}_2\}$ is a $\mathbb{Z}$-basis of M. We

set $\tilde{m}^* := (q-p)\tilde{m}_1 + q\tilde{m}_2 = qm_1 - pm_2$. The basis of the $\mathbb{R}$-vector space $M_{\mathbb{R}}$ dual to the basis $\{n_1, n^*\}$ of $N_{\mathbb{R}}$ is $\{\frac{1}{q}\tilde{m}^*, \frac{1}{q}\tilde{m}_1\}$, hence we have $\sigma^\vee = \mathbb{R}_{\geq 0}\,\tilde{m}^* + \mathbb{R}_{\geq 0}\,\tilde{m}_1$. In particular, $\sigma^\vee$ is a strongly convex rational polyhedral cone (with respect to M) of dimension 2 in $M_{\mathbb{R}}$.

(2) We set $\Gamma_\sigma := \sigma^\vee \cap M$. Then, using the notation introduced in (1.5)(2), the set of $q + 1$ elements

$$\left\{ \tilde{m}^*, \tilde{m}_1, \frac{1}{q}(\tilde{m}^* + p[1]\tilde{m}_1), \frac{1}{q}(2\tilde{m}^* + p[2]\tilde{m}_1), \ldots, \frac{1}{q}((q-1)\tilde{m}^* + p[q-1]\tilde{m}_1) \right\}$$

is a system of generators of the semigroup Γ_σ.

Proof: We have $\tilde{m}^*, \tilde{m}_1 \in \Gamma_\sigma$, and since we have

$$q^{-1}(i\tilde{m}^* + p[i]\tilde{m}_1) = im_1 + q^{-1}(p[i] - pi)m_2 \in M \quad \text{for every } i \in \{1, \ldots, q-1\},$$

these elements lie in Γ_σ. Conversely, let i, j be non-negative real numbers with $iq^{-1}\tilde{m}^* + jq^{-1}\tilde{m}_1 = im_1 + (j - pi)q^{-1}m_2 \in \mathbb{Z}m_1 + \mathbb{Z}m_2 = M$; then the real numbers i, j lie in $\mathbb{N}_0$, and we have $j \equiv p[i] \pmod q$. Hence there exists $l \in \mathbb{Z}$ such that $j = p[i] + lq$, and we have $l \geq 0$ because $j \geq 0$ and $0 \leq p[i] < q$. Now we have $iq^{-1}\tilde{m}^* + jq^{-1}\tilde{m}_1 = l\tilde{m}_1 + q^{-1}(i\tilde{m}^* + p[i]\tilde{m}_1)$.

(3) We determine the semigroup ring $k[\Gamma_\sigma]$. We have

$$\Gamma_1 := \mathbb{N}_0\,\tilde{m}^* + \mathbb{N}_0\,\tilde{m}_1 \subset \Gamma_\sigma \subset \mathbb{N}_0\,\frac{1}{q}\tilde{m}^* + \mathbb{N}_0\,\frac{1}{q}\tilde{m}_1 =: \Gamma_q;$$

note that $(1/q)\tilde{m}^*$, $(1/q)\tilde{m}_1$ are free over $\mathbb{Z}$. The k-algebra homomorphism

$$k[U, V] \to k[\Gamma_q], \text{ defined by } U \mapsto (1/q)e(\tilde{m}^*), V \mapsto (1/q)e(\tilde{m}_1),$$

is an isomorphism [cf. proof of (2.3)], and under this isomorphism the k-subalgebra of $k[U, V]$ generated by $U^q, V^q, UV^{p[1]}, \ldots, U^{q-1}V^{p[q-1]}$ is mapped onto the semigroup algebra $k[\Gamma_\sigma]$. This implies that the k-algebra $k[\Gamma_\sigma]$ is isomorphic to the k-algebra $S = T^G$ of invariants defined in (1.3), hence $k[\Gamma_\sigma]$ is the integral closure of $k[e(\tilde{m}^*), e(\tilde{m}_1), e((1/q)\tilde{m}^*)e((1/q)\tilde{m}_1)^p]$ ($\cong k[U^q, V^q][UV^p]$), the localization of the latter ring with respect to the ideal $\mathfrak{m}$ generated by $e(\tilde{m}^*)$, $e(\tilde{m}_1)$ and $e((1/q)\tilde{m}^*)e((1/q)\tilde{m}_1)^p$ is a Hirzebruch-Jung singularity of type $A_{q,p}$, there lies only one maximal ideal $\mathfrak{n}$ of $k[\Gamma_\sigma]$ over $\mathfrak{m}$, and $k[\Gamma_\sigma]_{\mathfrak{n}}$ is the integral closure of $k[e(\tilde{m}^*), e(\tilde{m}_1), e((1/q)\tilde{m}^*)e((1/q)\tilde{m}_1)^p]_{\mathfrak{m}}$.

(4.8) Proposition: Let σ be a strongly convex rational polyhedral cone of dimension 2 in $N_{\mathbb{R}}$. Then $\sigma^\vee$ is a strongly convex rational polyhedral cone of dimension 2 in $M_{\mathbb{R}}$, Γ_σ is a finitely generated saturated semigroup of M, and $k[\Gamma_\sigma]$ is an integral integrally closed k-algebra of finite type.

Proof: The claim of the proposition is clear if σ is nonsingular. If σ is not nonsingular, this follows immediately from (4.7) and (4.6)(2). From (2.11) we get that $k[\Gamma_\sigma]$ is integrally closed.

(4.9) DEFINITION: *Let σ be a strongly convex rational polyhedral cone of dimension 2 in $N_\mathbb{R}$. The irreducible normal affine variety associated with $k[\Gamma_\sigma]$ [cf. A(1.12)] is denoted by X_σ. Such a variety is also called a toric variety.*

(4.10) Proposition: *Let σ be a strongly convex rational polyhedral cone of dimension 2 in $N_\mathbb{R}$. The affine variety X_σ is nonsingular iff σ is a nonsingular cone.*

Proof: If σ is a nonsingular cone, then there exists a $\mathbb{Z}$-basis $\{n_1, n_2\}$ of N such that $\sigma = \mathbb{R}_{\geq 0}\, n_1 + \mathbb{R}_{\geq 0}\, n_2$. Let $\{m_1, m_2\}$ be the $\mathbb{Z}$-basis of M dual to the basis $\{n_1, n_2\}$. Then we have $\Gamma_\sigma = \mathbb{N}_0\, m_1 + \mathbb{N}_0\, m_2$, hence X_σ is $\mathbb{A}^2$.
If σ is not nonsingular, then there exist coprime integers p and q with $1 \leq p < q$ such that X_σ has a singular point whose local ring is the integral closure of a local domain which is a Hirzebruch-Jung singularity of type $A_{q,p}$ [cf. (4.7)].

4.2   The Boundary Polygon of σ and the Ideal of X_σ

(4.11) Lemma: *Let L be a free $\mathbb{Z}$-module of rank 2, and let $\{l_1, l_2\} \subset L$ be an $\mathbb{R}$-basis of $L_\mathbb{R} := L \otimes_\mathbb{Z} \mathbb{R}$. Let $T := \{al_1 + bl_2 \mid a, b \in \mathbb{R}_{\geq 0}, a + b \leq 1\}$ be the triangle in $L_\mathbb{R}$ with vertices $0, l_1, l_2$. Then we have $T \cap L = \{0, l_1, l_2\}$ iff $\{l_1, l_2\}$ is a $\mathbb{Z}$-basis of L.*

Proof: Assume that $\{l_1, l_2\}$ is a $\mathbb{Z}$-basis of L; then it is clear that $T \cap L$ consists only of 0, l_1 and l_2. Conversely, assume that $\{l_1, l_2\}$ is not a $\mathbb{Z}$-basis of L. Then there exist rational numbers a_1, a_2 such that $l := a_1 l_1 + a_2 l_2 \in L$, but $(a_1, a_2) \notin \mathbb{Z}^2$. Set $a_1' := a_1 - \lfloor a_1 \rfloor$, $a_2' := a_2 - \lfloor a_2 \rfloor$, and $l' := l - \lfloor a_1 \rfloor l_1 - \lfloor a_2 \rfloor l_2$. If $a_1' + a_2' \leq 1$, then $l' = a_1' l_1 + a_2' l_2$ lies in $T \cap L$, and if $a_1' + a_2' > 1$, then $l'' := l_1 + l_2 - l' = (1 - a_1')l_1 + (1 - a_2')l_2$ lies in $T \cap L$, and these points l' and l'' are different from the vertices of T.

(4.12) Proposition: *Let L be a free $\mathbb{Z}$-module of rank 2, and let $\{e_1, e_2\}$ be a $\mathbb{Z}$-basis of L. Let $r \in \mathbb{N}$, and let $a_1, \ldots, a_r \in \mathbb{N}$ be such that $a_i \geq 2$ for every $i \in \{1, \ldots, r\}$.*
Define subsets $\{f_0, \ldots, f_{r+1}\}$, $\{e_3, \ldots, e_{r+2}\}$ of L by

$$f_0 := e_1,$$
$$f_i := f_{i-1} + e_{i+1}, \ e_{i+2} := (a_i - 2)f_{i-1} + (a_i - 1)e_{i+1} \quad \text{for } i \in \{1, \ldots, r\},$$
$$f_{r+1} := f_r + e_{r+2}.$$

Then, for every $i \in \{1, \ldots, r+1\}$, the sets $\{f_{i-1}, e_{i+1}\}$ and $\{f_{i-1}, f_i\}$ are $\mathbb{Z}$-bases of L. Furthermore, we have

$$f_{i-1} + f_{i+1} = a_i f_i \quad \text{for every } i \in \{1, \ldots, r\}, \tag{*}$$

and, for every $i \in \{0, \ldots, r-1\}$ and $j \in \{0, \ldots, r+1\}$ satisfying $i + 2 < j$, we have

$$f_i + f_j = (a_{i+1} - 1)f_{i+1} + \sum_{\kappa=i+2}^{j-2} (a_\kappa - 2)f_\kappa + (a_{j-1} - 1)f_{j-1}. \tag{**}$$

Proof: Since $f_0 = e_1$, we see that $\{f_0, e_2\}$ is a $\mathbb{Z}$-basis of L. Let $i \in \{1, \ldots, r\}$, and assume that $\{f_{i-1}, e_{i+1}\}$ is a $\mathbb{Z}$-basis of L. Since $f_i = f_{i-1} + e_{i+1}$, $e_{i+2} = (a_i - 2)f_{i-1} + (a_i - 1)e_{i+1}$ and $(a_i - 1) - (a_i - 2) = 1$, it follows that also $\{f_i, e_{i+2}\}$ is a $\mathbb{Z}$-basis of L.

For every $i \in \{1, \ldots, r\}$ we have

$$f_{i-1} + f_{i+1} = f_{i-1} + f_i + e_{i+2} = f_{i-1} + f_i + (a_i - 2)f_{i-1} + (a_i - 1)e_{i+1}$$
$$= (a_i - 1)(f_{i-1} + e_{i+1}) + f_i = (a_i - 1)f_i + f_i = a_i f_i.$$

It is easy to see that $\{f_{i-1}, f_i\}$ is a $\mathbb{Z}$-basis of L for every $i \in \{1, \ldots, r+1\}$. The formula $(**)$ can easily be shown by induction on j.

(4.13) We keep the notation introduced in (4.12).
(1) We define uniquely integers p, q as in (1.0) by

$$\frac{q}{q - p} = a_1 - \lceil a_2 - \cdots - \lceil a_r \rceil;$$

note that $p \geq 1$. Let $\lambda_0 := q$, $\lambda_1 := q - p$. By the definition of the Hirzebruch-Jung continued fraction expansion for $q/(q - p)$, there exist integers $\lambda_2, \ldots, \lambda_{r+1}$ such that we have

$$0 = \lambda_{r+1} < \lambda_r = 1 < \lambda_{r-1} < \cdots < \lambda_2 < \lambda_1 < \lambda_0,$$

and that we have

$$\lambda_{i-1} = a_i \lambda_i - \lambda_{i+1} \quad \text{for every } i \in \{1, \ldots, r\}.$$

(2) We show that $f_{r+1} = pe_1 + qe_2$; in particular, f_0 and f_{r+1} are linearly independent over $\mathbb{R}$.
Proof: Set $e^* := pe_1 + qe_2$. Then we have $e^* = (\lambda_0 - \lambda_1)f_0 + \lambda_0 e_2$. Let $i \in \{1, \ldots, r\}$, and assume that we have already shown that $e^* = (\lambda_{i-1} - \lambda_i)f_{i-1} + \lambda_{i-1}e_{i+1}$. We see that

$$(\lambda_i - \lambda_{i+1})f_i + \lambda_i e_{i+2} = (\lambda_i - \lambda_{i+1})(f_{i-1} + e_{i+1}) + \lambda_i((a_i - 2)f_{i-1} + (a_i - 1)e_{i+1})$$
$$= (\lambda_{i-1} - \lambda_i)f_{i-1} + \lambda_{i-1}e_{i+1}.$$

Therefore we have $e^* = (\lambda_r - \lambda_{r+1})f_r + \lambda_r e_{r+2} = f_r + e_{r+2} = f_{r+1}$.
(3) Set $\mu_0 := 0$, $\mu_1 := 1$, and define $\mu_2, \ldots, \mu_{r+1}$ by $\mu_{i+1} = a_i \mu_i - \mu_{i-1}$ for every $i \in \{1, \ldots, r\}$. Note that $\mu_0 < \mu_1 < \cdots < \mu_{r+1}$ [we have $\mu_1 > \mu_0$, and if we have already shown that $\mu_i > \mu_{i-1}$ for some $i \in \{1, \ldots, r\}$, then it follows that $\mu_{i+1} - \mu_i = (a_i - 1)\mu_i - \mu_{i-1} \geq \mu_i - \mu_{i-1} > 0$].
(4) We show that we have

$$f_i = \frac{\lambda_i}{q}f_0 + \frac{\mu_i}{q}f_{r+1} \quad \text{for every } i \in \{0, \ldots, r+1\}.$$

Since $f_0 = e_1$ and $f_1 = f_0 + e_2 = (1/q)((q-p)f_0 + f_{r+1}) = (1/q)(\lambda_1 f_0 + \mu_1 f_{r+1})$, the assertion holds for $i = 0$ and $i = 1$. Let $j \in \{1, \ldots, r\}$, and assume that the assertion is true for every $i \in \{0, \ldots, j\}$. Then, by (4.12)(*), we have

$$f_{j+1} = a_j f_j - f_{j-1} = \frac{a_j \lambda_j - \lambda_{j-1}}{q} f_0 + \frac{a_j \mu_j - \mu_{j-1}}{q} f_{r+1} = \frac{\lambda_{j+1}}{q} f_0 + \frac{\mu_{j+1}}{q} f_{r+1}.$$

(5) From (4) we get $\mu_{r+1} = q$.
(6) Let $i \in \{0, \ldots, r+1\}$. A linear combination $\alpha_0 f_0 + \cdots + \alpha_{r+1} f_{r+1}$ where $\alpha_0, \ldots, \alpha_{r+1}$ are non-negative integers, can represent f_i only in the trivial way: we have $\alpha_i = 1$ and $\alpha_j = 0$ for all $j \in \{0, \ldots, r+1\}$, $j \neq i$.
In fact, since the sequence $(\lambda_i)_{0 \leq i \leq r+1}$ is strictly decreasing, we have $\alpha_0 = \cdots = \alpha_{i-1} = 0$, and since the sequence $(\mu_i)_{0 \leq i \leq r+1}$ is strictly increasing, we have $\alpha_{i+1} = \cdots = \alpha_{r+1} = 0$.

(4.14) We keep the notations of (4.12) and (4.13). We set $L_\mathbb{R} := L \otimes_\mathbb{Z} \mathbb{R}$ and $\sigma := \mathbb{R}_{\geq 0} f_0 + \mathbb{R}_{\geq 0} f_{r+1}$; σ is a strongly convex rational polyhedral cone (with respect to L) in $L_\mathbb{R}$ of dimension 2.
Let $\Gamma := \sigma \cap L$ be the semigroup associated with σ, let θ be the convex hull in $L_\mathbb{R}$ of the set $(\sigma \cap L) \setminus \{0\}$, and let $\partial\theta$ be the boundary polygon of θ; by abuse of language, we call $\partial\theta$ the boundary polygon of σ.

(4.15) Proposition: With notations as in (4.14) we have:
(1) The elements $f_0, f_1, \ldots, f_{r+1}$ in this order are the points of L lying on the compact edges of the boundary polygon $\partial\theta$.
(2) The set $\{f_0, \ldots, f_{r+1}\}$ is a minimal set of generators of Γ.
(3) Let $m \in \Gamma$ and $m \neq 0$; if m does not lie on one of the halflines $\mathbb{R}_{\geq 0} f_i$, $i \in \{0, \ldots, r+1\}$, then there exists exactly one index $j \in \{1, \ldots, r+1\}$ such that $m \in \mathbb{N}_0 f_{j-1} + \mathbb{N}_0 f_j$.
(4) We have $f_{r+1} = e^*$. For every $i \in \{1, \ldots, r\}$ the point f_i lies on the line through f_{i-1} and f_{i+1} iff $a_i = 2$, and it is a vertex of θ iff $a_i > 2$.
(5) For $i \in \{0, \ldots, r\}$ we set $\sigma_i := \mathbb{R}_{\geq 0} f_i + \mathbb{R}_{\geq 0} f_{i+1}$. Then σ_i is a nonsingular strongly convex rational polyhedral cone of dimension 2. The intersection of any two such different cones is either $\{0\}$ or a halfline $\mathbb{R}_{\geq 0} f_j$ for some $j \in \{1, \ldots, r\}$. Furthermore, we have $\sigma = \sigma_0 \cup \cdots \cup \sigma_r$.

Proof: (a) By (4.13)(4) the elements $f_0, \ldots, f_{r+1}$ lie in σ. Let $i \in \{0, \ldots, r\}$; since $\{f_i, f_{i+1}\}$ is a $\mathbb{Z}$-basis of L by (4.12), σ_i is nonsingular. Since the sequence $(\lambda_l)_{0 \leq l \leq r+1}$ is strictly decreasing and the sequence $(\mu_l)_{0 \leq l \leq r+1}$ is strictly increasing, for all $i, j \in \{0, \ldots, r\}$ such that $i \neq j$, we have that either $\sigma_i \cap \sigma_j = \{0\}$ or that $|i - j| = 1$; in the latter case, the intersection $\sigma_i \cap \sigma_j$ is equal to the halfline $\mathbb{R}_{\geq 0} f_j$ if $j = i + 1$, and to the halfline $\mathbb{R}_{\geq 0} f_i$ if $i = j + 1$. Furthermore, σ is the union of the cones $\sigma_0, \ldots, \sigma_r$.
(b) Let $i \in \{0, \ldots, r\}$. Since $\{f_i, f_{i+1}\}$ is a $\mathbb{Z}$-basis of L [cf. (4.12)], the only points of L lying in the triangle with vertices 0, f_i, f_{i+1} are its vertices [cf. (4.11)]. This

implies that $f_0, \ldots, f_{r+1}$ are *all* points of L lying on the compact edges of the boundary $\partial \theta$ of θ. Since we have $f_{i-1} + f_{i+1} = a_i f_i$, we get the assertion in (4).
(c) The semigroup Γ is the union of the semigroups $L \cap \sigma_i = \mathbb{N}_0 f_i + \mathbb{N}_0 f_{i+1}$, $i \in \{0, \ldots, r\}$. Thus, $\{f_0, \ldots, f_{r+1}\}$ is a set of generators for Γ. That all these element are, indeed, necessary to generate Γ, follows from (4.13)(6), hence this system is an irredundant system of generators of Γ. By (b) this system is also a minimal system of generators of Γ.

(4.16) We keep the notations introduced above. The following will be used in the proof of (4.17). Set $e'_1 := f_1$, $e'_2 := e_3$; then the set $\{e'_1, e'_2\}$ is a $\mathbb{Z}$-basis of L by (4.12). Let $a'_1, \ldots, a'_{r-1}$ be integers ≥ 2, and define subsets $\{f'_0, \ldots, f'_r\}$, $\{e'_3, \ldots, e'_{r+1}\}$ of L by

$$f'_0 := e'_1,$$
$$f'_i := f'_{i-1} + e'_{i+1}, \ e'_{i+2} := (a'_i - 2)f'_{i-1} + (a'_i - 1)e'_{i+1} \text{ for } i \in \{1, \ldots, r-1\},$$
$$f'_r := f'_{r-1} + e'_{r+1}.$$

Then, if we have $a'_1 = a_2, \ldots, a'_{r-1} = a_r$, then we have

$$f'_{i-1} = f_i \quad \text{for } i \in \{1, \ldots, r+1\}, \ e'_{i+1} = e_{i+2} \quad \text{for } i \in \{1, \ldots, r\},$$

as is easily seen by recursion.

(4.17) We keep the notations introduced above. For $i, j \in \{0, \ldots, r+1\}$, $i < j-1$, we define monomials $P_{ij} \in k[T_0, \ldots, T_{r+1}]$ by

$$P_{ij} := \begin{cases} T_{i+1}^{a_{i+1}} & \text{if } i+2 = j, \\ T_{i+1}^{a_{i+1}-1} \left(\prod_{\kappa=i+2}^{j-2} T_\kappa^{a_\kappa - 2} \right) T_{j-1}^{a_{j-1}-1} & \text{if } i+2 < j, \end{cases}$$

and set

$$Q_{ij} := T_i T_j - P_{ij}, \quad i, j \in \{0, \ldots, r+1\}, \ i < j-1. \tag{$*$}$$

Let $\varphi \colon k[T_0, \ldots, T_{r+1}] \to k[\Gamma]$ be the k-algebra homomorphism defined by $\varphi(T_i) = e(f_i)$ for every $i \in \{0, \ldots, r+1\}$ [cf. (4.15)(2)]. Then φ is surjective, and the ideal $\mathfrak{a} := \ker(\varphi)$ is generated by the monomials Q_{ij} in $(*)$.
Proof: Let $\mathfrak{a}'$ be the ideal generated by the monomials Q_{ij} in $(*)$.
(1) From (4.12)$(**)$ we immediately get that $Q_{ij} \in \mathfrak{a}$ for all $i, j \in \{0, \ldots, r+1\}$ satisfying $i < j-1$, and this yields $\mathfrak{a}' \subset \mathfrak{a}$.
(2) Claim: *For every* $0 \neq (\gamma_0, \ldots, \gamma_{r+1}) \in \mathbb{N}_0^{r+2}$, *there exist a uniquely determined* $i \in \{0, \ldots, r+1\}$ *and uniquely determined integers* $\alpha \in \mathbb{N}$, $\beta \in \mathbb{N}_0$ *with* $\gamma_0 f_0 + \cdots + \gamma_{r+1} f_{r+1} = \alpha f_i + \beta f_{i+1}$ *[where* $\beta = 0$ *if* $i = r+1$*], with* $\alpha + \beta \geq \gamma_0 + \cdots + \gamma_{r+1}$, *and with*

$$T_0^{\gamma_0} \cdots T_{r+1}^{\gamma_{r+1}} \equiv T_i^\alpha T_{i+1}^\beta \pmod{\mathfrak{a}'}.$$

Proof of claim: We consider the case $r = 1$. Let $(\gamma_0, \gamma_1, \gamma_2) \in \mathbb{N}_0^3$. By (4.15)(4) the element $m := \gamma_0 f_0 + \gamma_1 f_1 + \gamma_2 f_2$ lies in $\mathbb{N}_0 f_0 + \mathbb{N}_0 f_1$ or in $\mathbb{N}_0 f_1 + \mathbb{N}_0 f_2$.

We consider the first case. Since $f_2 = a_1 f_1 - f_0$, we see that $\gamma_0 \geq \gamma_2$ and $m = \alpha f_0 + \beta f_1$ where $\alpha = \gamma_0 - \gamma_2$, $\beta = \gamma_1 + \gamma_2 a_1$, hence that $\alpha + \beta \geq \gamma_0 + \gamma_1 + \gamma_2$. Since

$$T_0^{\gamma_0} T_1^{\gamma_1} T_2^{\gamma_2} - T_0^{\alpha} T_1^{\beta} = (T_0 T_2 - T_1^{a_1}) \sum_{\kappa=1}^{\gamma_2} T_0^{\alpha+\kappa-1} T_1^{\beta-\kappa a_1} T_2^{\kappa-1},$$

and since $P_{02} = T_1^{a_1}$, hence $Q_{02} = T_0 T_2 - T_1^{a_1}$, it follows that $T_0^{\gamma_0} T_1^{\gamma_1} T_2^{\gamma_2} \equiv T_0^{\alpha} T_1^{\beta}$ (mod $\mathfrak{a}'$). In the same way we can handle the second case.

Let $r \geq 2$, and assume that for $r - 1$ integers $a_1', \ldots, a_{r-1}'$ such that $a_i' \geq 2$ for every $i \in \{1, \ldots, r-1\}$, and for every $\mathbb{Z}$-basis $\{e_1', e_2'\}$ of L the following has been proved: Define elements $f_0', \ldots, f_r'$ with respect to $a_1', \ldots, a_{r-1}'$ and the $\mathbb{Z}$-basis $\{e_1', e_2'\}$ as in (4.16), and define the elements $Q_{ij}' \in k[T_0', \ldots, T_r']$ with respect to $a_1', \ldots, a_{r-1}'$ for all $i, j \in \{0, \ldots, r\}$ such that $i < j - 1$ as above; then, for every $(\gamma_0, \ldots, \gamma_r) \in \mathbb{N}_0^{r+1}$, there exist $\alpha, \beta \in \mathbb{N}_0$ and $i \in \{0, \ldots, r-1\}$ with $\gamma_0 f_0' + \cdots + \gamma_r f_r' = \alpha f_i' + \beta f_{i+1}'$, with $\alpha + \beta \geq \gamma_0 + \cdots + \gamma_r$, and with $T_0'^{\gamma_0} \cdots T_r'^{\gamma_r} \equiv T_i'^{\alpha} T_{i+1}'^{\beta}$ (mod $\mathfrak{a}''$) where $\mathfrak{a}'' \subset k[T_0', \ldots, T_r']$ is the ideal generated by the elements Q_{ij}' for all $i, j \in \{0, \ldots, r\}$ such that $i < j - 1$.

Let $(\gamma_0, \ldots, \gamma_{r+1}) \in \mathbb{N}_0^{r+2}$. First case: we assume that $\gamma_{r+1} = 0$. By the induction assumption, applied to the elements $a_1' := a_1, \ldots, a_{r-1}' := a_{r-1}$, the $\mathbb{Z}$-basis $\{e_1, e_2\}$ of L and the ideal $\mathfrak{a}_1' \subset k[T_0, \ldots, T_r]$ [where $T_j' := T_j$ for all $j \in \{0, \ldots, r\}$] generated by the elements $Q_{jl}' = Q_{jl}$ for all $j, l \in \{0, \ldots, r\}$ such that $j < l - 1$, there exist $i \in \{0, \ldots, r-1\}$ and $\alpha, \beta \in \mathbb{N}_0$ with $\gamma_0 f_0 + \cdots + \gamma_r f_r = \alpha f_i + \beta f_{i+1}$, with $\alpha + \beta \geq \gamma_0 + \cdots + \gamma_r$, and with $T_0^{\gamma_0} \cdots T_r^{\gamma_r} - T_i^{\alpha} T_{i+1}^{\beta} \in \mathfrak{a}_1'$.

Second case: we assume that $\gamma_0 = 0$; by the induction assumption, applied to the elements $a_1' := a_2, \ldots, a_{r-1}' := a_r$, the $\mathbb{Z}$-basis $\{e_1', e_2'\}$ of L defined in (4.16), and the ideal $\mathfrak{a}_2' \subset k[T_1, \ldots, T_{r+1}]$ [where $T_j' := T_{j+1}$ for all $j \in \{0, \ldots, r\}$] generated by the elements $Q_{jl}' = Q_{j+1,l+1}$ for all $j, l \in \{0, \ldots, r\}$ such that $j < l - 1$, and by (4.16), we find $i \in \{1, \ldots, r\}$ and $\alpha, \beta \in \mathbb{N}_0$ with $\gamma_1 f_1 + \cdots + \gamma_{r+1} f_{r+1} = \alpha f_i + \beta f_{i+1}$, with $\alpha + \beta \geq \gamma_1 + \cdots + \gamma_{r+1}$, and with $T_1^{\gamma_1} \cdots T_{r+1}^{\gamma_{r+1}} - T_i^{\alpha} T_{i+1}^{\beta} \in \mathfrak{a}_2'$. Hence, in both cases, there exist $i \in \{0, \ldots, r\}$ and $\alpha, \beta \in \mathbb{N}_0$ with $\gamma_0 f_0 + \cdots + \gamma_{r+1} f_{r+1} = \alpha f_i + \beta f_{i+1}$, with $\alpha + \beta \geq \gamma_0 + \cdots + \gamma_{r+1}$, and with $T_0^{\gamma_0} \cdots T_{r+1}^{\gamma_{r+1}} - T_i^{\alpha} T_{i+1}^{\beta} \in \mathfrak{a}'$.

Now let $\gamma_0 > 0$, $\gamma_{r+1} > 0$, and assume that $\gamma_0 \geq \gamma_{r+1}$. We apply the first case to the monomial $T_0^{\gamma_0 - \gamma_{r+1}} T_1^{\gamma_1} \cdots T_r^{\gamma_r} P_{0,r+1}^{\gamma_{r+1}} \in k[T_0, \ldots, T_r]$; we see that there exist $i \in \{0, \ldots, r-1\}$ and $\alpha, \beta \in \mathbb{N}_0$ with $\alpha + \beta \geq \gamma_0 + \cdots + \gamma_{r+1}$, with

$$\alpha f_i + \beta f_{i+1} = (\gamma_0 - \gamma_{r+1}) f_0 + ((a_1 - 1)\gamma_{r+1} + \gamma_1) f_1 +$$

$$+ \sum_{j=2}^{r-1} ((a_j - 2)\gamma_{r+1} + \gamma_j) f_j + ((a_r - 1)\gamma_{r+1} + \gamma_r) f_r$$

$$= \gamma_0 f_0 + \gamma_1 f_1 + \cdots + \gamma_{r+1} f_{r+1}$$

[cf. (4.12)(**) with $i := 0$ and $j := r + 1$] and with

$$T_0^{\gamma_0 - \gamma_{r+1}} T_1^{\gamma_1} \cdots T_r^{\gamma_r} P_{0,r+1}^{\gamma_{r+1}} \equiv T_i^\alpha T_{i+1}^\beta \quad (\text{mod } \mathfrak{a}'). \tag{$*$}$$

Clearly we have $P_{0,r+1} \equiv T_0 T_{r+1}$ (mod $\mathfrak{a}'$), hence $T_0^{\gamma_0 - \gamma_{r+1}} P_{0,r+1}^{\gamma_{r+1}} \equiv T_0^{\gamma_0} T_{r+1}^{\gamma_{r+1}}$ (mod $\mathfrak{a}'$). Using (*) we now see that $T_0^{\gamma_0} T_1^{\gamma_1} \cdots T_{r+1}^{\gamma_{r+1}} \equiv T_i^\alpha T_{i+1}^\beta$ (mod $\mathfrak{a}'$). Likewise, when $\gamma_0 \leq \gamma_{r+1}$, by the second case above, we find $i \in \{1, \ldots, r\}$ and α, $\beta \in \mathbb{N}_0$ with $\gamma_0 f_0 + \cdots + \gamma_{r+1} f_{r+1} = \alpha f_i + \beta f_{i+1}$, with $\alpha + \beta \geq \gamma_0 + \cdots + \gamma_{r+1}$, and with $T_0^{\gamma_0} T_1^{\gamma_1} \cdots T_{r+1}^{\gamma_{r+1}} \equiv T_i^\alpha T_{i+1}^\beta$ (mod $\mathfrak{a}'$). Thus, we have shown the existence part of (4.17).

Now let $(\gamma_0, \ldots, \gamma_{r+1})$ and $(\delta_0, \ldots, \delta_{r+1}) \in \mathbb{N}_0^{r+2}$ be such that $\gamma_0 f_0 + \cdots + \gamma_{r+1} f_{r+1} = \delta_0 f_0 + \cdots + \delta_{r+1} f_{r+1}$. By what we have just shown, there exist i, $j \in \{0, \ldots, r\}$ and $\alpha, \beta, \gamma, \delta \in \mathbb{N}_0$ with

$$\gamma_0 f_0 + \cdots + \gamma_{r+1} f_{r+1} = \alpha f_i + \beta f_{i+1} = \gamma f_j + \delta f_{j+1} = \delta_0 f_0 + \cdots + \delta_{r+1} f_{r+1},$$

and with

$$T_0^{\gamma_0} \cdots T_{r+1}^{\gamma_{r+1}} \equiv T_i^\alpha T_{i+1}^\beta \quad (\text{mod } \mathfrak{a}'), \quad T_0^{\delta_0} \cdots T_{r+1}^{\delta_{r+1}} \equiv T_j^\gamma T_{j+1}^\delta \quad (\text{mod } \mathfrak{a}').$$

The two convex cones $\mathbb{R}_{\geq 0} f_i + \mathbb{R}_{\geq 0} f_{i+1}$ and $\mathbb{R}_{\geq 0} f_j + \mathbb{R}_{\geq 0} f_{j+1}$ are either equal, or have only the point $\{0\}$ or a halfline in common [cf. (4.15)(4)]. Hence, from $\alpha f_i + \beta f_{i+1} = \gamma f_j + \delta f_{j+1}$, we see that $T_i^\alpha T_{i+1}^\beta = T_j^\gamma T_{j+1}^\delta$, and we have shown that

$$T_0^{\gamma_0} \cdots T_{r+1}^{\gamma_{r+1}} - T_0^{\delta_0} \cdots T_{r+1}^{\delta_{r+1}} \in \mathfrak{a}'. \tag{$*$}$$

By changing i and j if necessary, we may assume that $\alpha > 0$ and $\gamma > 0$; then $T_i^\alpha T_{i+1}^\beta = T_j^\gamma T_{j+1}^\delta$ implies $i = j$, $\alpha = \gamma$ and $\beta = \delta$. Thus, we have proved the uniqueness part of (2).

(3) By (2.3)(4) the ideal $\mathfrak{a}$ is generated by binomials

$$T_0^{\gamma_0} \cdots T_{r+1}^{\gamma_{r+1}} - T_0^{\delta_0} \cdots T_{r+1}^{\delta_{r+1}} \text{ with } \gamma_0 f_0 + \cdots + \gamma_{r+1} f_{r+1} = \delta_0 f_0 + \cdots + \delta_{r+1} f_{r+1}. \tag{$*$}$$

From (*) in (2) we see that $\mathfrak{a} \subset \mathfrak{a}'$.

(4) From (1) and (3) we now get $\mathfrak{a} = \mathfrak{a}'$.

(4.18) **Theorem:** Let N be a free $\mathbb{Z}$-module of rank 2, and let $\{n_1, n_2\}$ be a $\mathbb{Z}$-basis of N. Let M be the $\mathbb{Z}$-module dual to N, and let $\{m_1, m_2\}$ be the $\mathbb{Z}$-basis of M dual to the basis $\{n_1, n_2\}$. Set $\tilde{m}_1 := m_2$, $\tilde{m}_2 := m_1 - m_2$, so that $\{\tilde{m}_1, \tilde{m}_2\}$ is another $\mathbb{Z}$-basis of M. Let $p, q \in \mathbb{N}$ be as in (1.0) with $p > 0$. Set $n^* := p n_1 + q n_2$ and $\sigma := \mathbb{R}_{\geq 0} n_1 + \mathbb{R}_{\geq 0} n^*$. Set $\tilde{m}^* := (q - p)\tilde{m}_1 + q\tilde{m}_2$. The cone dual to σ is the cone $\sigma^\vee = \mathbb{R}_{\geq 0} \tilde{m}_1 + \mathbb{R}_{\geq 0} \tilde{m}^*$ [cf. (4.7)]. Let θ be the convex hull in $N_\mathbb{R}$ of the set $(\sigma \cap N) \setminus \{0\}$, and let $\partial \theta$ be the boundary polygon of θ. Let $\theta^\vee$ be the convex hull in $M_\mathbb{R}$ of the set $(\sigma^\vee \cap M) \setminus \{0\}$, and let $\partial \theta^\vee$ be the boundary polygon of $\theta^\vee$. Then:

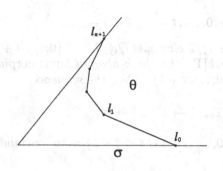
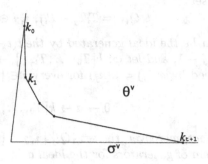

(1) *Let*

$$\frac{q}{q-p} = b_1 - \overline{\lceil b_2 \rceil} - \cdots - \overline{\lceil b_s \rceil} \quad \text{with } b_i \geq 2 \quad \text{for } i \in \{1, \ldots, s\}$$

be the Hirzebruch-Jung continued fraction expansion of $q/(q-p)$. *Define the following subsets* $\{l_0, \ldots, l_{s+1}\}$ *and* $\{n_3, \ldots, n_{s+2}\}$ *of* N *by*

$$l_0 := n_1,$$
$$l_i := l_{i-1} + n_{i+1}, \ n_{i+2} := (b_i - 2)l_{i-1} + (b_i - 1)n_{i+1} \quad \text{for } i \in \{1, \ldots, s\},$$
$$l_{s+1} := l_s + n_{s+2}.$$

Then we have $l_{s+1} = n^*$; *the set* $\{l_0, \ldots, l_{s+1}\}$ *has the properties listed in* (4.15) *[with* $L := N$, $r := s$ *and* $f_i := l_i$ *for* $i \in \{0, \ldots, s+1\}$*].*
(2) *Let*

$$\frac{q}{p} = c_1 - \overline{\lceil c_2 \rceil} - \cdots - \overline{\lceil c_t \rceil} \quad \text{with } c_j \geq 2 \quad \text{for } j \in \{1, \ldots, t\}$$

be the Hirzebruch-Jung continued fraction expansion of q/p. *Define the following subsets* $\{k_0, \ldots, k_{t+1}\}$ *and* $\{\tilde{m}_3, \ldots, \tilde{m}_{t+2}\}$ *of* M *by*

$$k_0 := \tilde{m}_1,$$
$$k_j := k_{j-1} + \tilde{m}_{j+1}, \ \tilde{m}_{j+2} := (c_j - 2)k_{j-1} + (c_j - 1)\tilde{m}_{j+1} \quad \text{for } j \in \{1, \ldots, t\},$$
$$k_{t+1} := k_t + \tilde{m}_{t+2}.$$

Then we have $k_{t+1} = \tilde{m}^*$; *the set* $\{k_0, \ldots, k_{t+1}\}$ *has the properties listed in* (4.15)
[with $L := M$, $r := t$, $f_i := k_i$ *for* $i \in \{0, \ldots, t+1\}$ *and* $\Gamma := \sigma^\vee \cap M = \Gamma_\sigma$*].*
(3) *For all* $i, j \in \{0, \ldots, t+1\}$, $i < j-1$, *define monomials* $P_{ij} \in k[T_0, \ldots, T_{t+1}]$
by

$$P_{ij} = \begin{cases} T_{i+1}^{c_{i+1}} & \text{if } i+2 = j, \\[2ex] T_{i+1}^{c_{i+1}-1} \left(\displaystyle\prod_{\kappa=i+2}^{j-2} T_\kappa^{c_\kappa-2} \right) T_{j-1}^{c_{j-1}-1} & \text{if } i+2 < j, \end{cases}$$

and set

$$Q_{ij} := T_i T_j - P_{ij}, \quad i, j \in \{0, \ldots, t+1\}, \quad i < j - 1.$$

Let $\mathfrak{a}$ be the ideal generated by the $t(t+1)/2$ elements Q_{ij}, $i, j \in \{0, \ldots, t+1\}$, $i < j - 1$, and let $\varphi \colon k[T_0, \ldots, T_{t+1}] \to k[\Gamma_\sigma]$ be the k-algebra homomorphism defined by $\varphi(T_j) = e(k_j)$ for every $j \in \{0, \ldots, t+1\}$. Then the sequence

$$0 \to \mathfrak{a} \to k[T_0, \ldots, T_{t+1}] \xrightarrow{\varphi} k[\Gamma_\sigma] \to 0 \qquad (*)$$

is exact; moreover, the set $\{Q_{ij} \mid i, j \in \{0, \ldots, t+1\}, \ i < j - 1\}$ is an irredundant system of generators for the ideal $\mathfrak{a}$.

Proof: All these statements, except the last one, follow from (4.12)–(4.17), applied to N and M, respectively. The last statement follows from (4.23) below.

(4.19) REMARK: Let L be a free $\mathbb{Z}$-module of rank 2, and let $\sigma \subset L \otimes_{\mathbb{Z}} \mathbb{R}$ be a strongly convex rational polyhedral cone (with respect to L) of dimension 2. If σ is not nonsingular, then there exist a $\mathbb{Z}$-basis $\{l_1, l_2\}$ of L and integers p, q as in (1.0) with $p > 0$ such that $\sigma = \mathbb{R}_{\geq 0}\, l_1 + \mathbb{R}_{\geq 0}\, l^*$ where $l^* = p l_1 + q l_2$ [cf. (4.5)(4)]. We show below [cf. (4.21)] that the normal irreducible affine variety X_σ associated with σ has exactly one singular point. If σ is nonsingular, then X_σ is just affine 2-space; in this case X_σ is a nonsingular variety.
Therefore we have covered in (4.18) the general case of a strongly convex rational polyhedral cone of dimension 2 in $\mathbb{R}^2$.

(4.20) We keep the notations introduced in (4.18).
(1) We know that

$$\Gamma_1 := \mathbb{N}_0 \tilde{m}_1 + \mathbb{N}_0 \tilde{m}^* \subset \Gamma_\sigma \subset \frac{1}{q} \mathbb{N}_0 \tilde{m}_1 + \frac{1}{q} \mathbb{N}_0 \tilde{m}^* =: \Gamma_q$$

[cf. (4.7)(3)]. Let us write $k[\Gamma_1] = k[U, V]$ [polynomial ring over k in two indeterminates U and V]; then we have $k[\Gamma_q] = k[U^{1/q}, V^{1/q}]$. We define non-negative integers $\alpha_0, \ldots, \alpha_{t+1}$, $\beta_0, \ldots, \beta_{t+1}$ by $\alpha_0 := 0$, $\alpha_1 := 1$, $\beta_0 := q$, $\beta_1 := p$, and recursively by

$$\alpha_{j+1} = c_j \alpha_j - \alpha_{j-1}, \quad \beta_{j+1} = c_j \beta_j - \beta_{j-1} \quad \text{for every } j \in \{1, \ldots, t\}.$$

Then we have [cf. (4.13)] $\alpha_{t+1} = q$, $\beta_{t+1} = 0$ and

$$k_j = \frac{\beta_j}{q} \tilde{m}_1 + \frac{\alpha_j}{q} \tilde{m}^* \quad \text{for every } j \in \{0, \ldots, t+1\}.$$

Therefore we get

$$k[U, V] \subset k[\Gamma_\sigma] = k[\{U^{\alpha_j/q} V^{\beta_j/q} \mid j \in \{0, \ldots, t+1\}\}] \subset k[U^{1/q}, V^{1/q}],$$

the set $\{U^{\alpha_j/q}V^{\beta_j/q} \mid j \in \{0,\ldots,t+1\}\}$ is a minimal system of generators of the k-algebra $k[\Gamma_\sigma]$ [cf. (2.6), and note that $\{k_0,\ldots,k_{t+1}\}$ is a minimal system of generators of Γ_σ], and the k-algebra homomorphism $\varphi\colon k[T_0,\ldots,T_{t+1}] \to k[\Gamma_\sigma]$ is defined by $\varphi(T_j) = U^{\alpha_j/q}V^{\beta_j/q}$ for every $j \in \{0,\ldots,t+1\}$. Remember that $k[\Gamma_\sigma]$ is the integral closure of $k[U,V][W]/(W^q - UV^p)$ [cf. (4.7)(3)]; in particular, $k[U,V][W]/(W^q - UV^p)$ is integrally closed iff $p = 1$.

(2) Consider the "matrix"

$$
\begin{pmatrix}
T_0 & T_1 & T_2 & \cdots & T_{t-1} & T_t \\
T_1 & T_2 & T_3 & \cdots & T_t & T_{t+1} \\
T_1^{c_1-2} & T_2^{c_2-2} & \multicolumn{4}{c}{\cdots\cdots\cdots\cdots} & T_t^{c_t-2}
\end{pmatrix} ; \quad (*)
$$

the kernel $\mathfrak{a}$ of φ, i.e., the ideal of relations of the elements V, $U^{\alpha_1/q}V^{\beta_1/q} = U^{1/q}V^{p/q},\ldots,U^{\alpha_{t+1}/q}V^{\beta_{t+1}/q} = U$, is irredundantly generated by the "generalized" minors of $(*)$, namely by the elements

$$
T_iT_j - T_{i+1}\big(T_{i+1}^{c_{i+1}-2}\cdots T_{j-1}^{c_{j-1}-2}\big)T_{j-1} \quad i,j \in \{0,\ldots,t+1\},\ i < j-1.
$$

(4.21) Corollary: *The affine variety X_σ associated with σ has one singular point, namely the point corresponding to the maximal ideal $\mathfrak{m}_{\Gamma_\sigma}$ of $k[\Gamma_\sigma]$.*

Proof: (1) The ring $k[\Gamma_\sigma]$ is integrally closed and two-dimensional [cf. (4.7)(3) and (2.3)(1) and note that $M(\Gamma_\sigma) = M$], hence the singular locus of the affine variety X_σ is a zero-dimensional closed subset of X_σ [cf. A(7.17)], and is therefore a finite set; the elements of this set correspond to maximal ideals of $k[\Gamma_\sigma]$. We consider X_σ via φ [cf. (4.17)] as a closed affine subvariety of the affine space $\mathbb{A}^{t+2}$. The maximal ideals of $k[\Gamma_\sigma] = k[e(k_0),\ldots,e(k_{t+1})]$ correspond uniquely to those points of $\mathbb{A}^{t+2}$ which lie in X_σ.

(2) T_N can be considered as a group of k-algebra automorphisms of $k[\Gamma_\sigma]$. If $\mathfrak{n}$ and $\mathfrak{n}'$ are maximal ideals of $k[\Gamma_\sigma]$ such that $\tau(\mathfrak{n}) = \mathfrak{n}'$ for some $\tau \in T_N$, then the local rings $k[\Gamma_\sigma]_\mathfrak{n}$ and $k[\Gamma_\sigma]_{\mathfrak{n}'}$ are isomorphic, hence both local rings are either regular or not regular.

(3) Let $(\xi_0,\ldots,\xi_{t+1}) \in \mathbb{A}^{t+2}$ be a point of X_σ, and consider its ideal in $k[\Gamma_\sigma]$, which is the ideal generated by $e(k_0) - \xi_0,\ldots,e(k_{t+1}) - \xi_{t+1}$; its image under the action of $\tau \in T_N$ on $k[\Gamma_\sigma]$ is the ideal of the point $(\xi_0/\tau(k_0),\ldots,\xi_{t+1}/\tau(k_{t+1})) \in \mathbb{A}^{t+2}$. Therefore, the maximal ideal $\mathfrak{m}_{\Gamma_\sigma}$ of $k[\Gamma_\sigma]$, corresponding to the origin in $\mathbb{A}^{t+2}$, is the only maximal ideal of $k[\Gamma_\sigma]$ which is kept fixed under the action of T_N [note that Γ_σ generates M, hence for $\tau \in T_N$ we have $\tau|\Gamma_\sigma = \mathrm{id}_{\Gamma_\sigma}$ iff $\tau|M = \mathrm{id}_M$] and for every other maximal $\mathfrak{n}$ of $k[\Gamma_\sigma]$, the orbit $T_N(\mathfrak{n}) = \{\tau(\mathfrak{n}) \mid \tau \in T_N\}$ of $\mathfrak{n}$ is an infinite set [cf. (2.9)(2)], hence the origin is the only singular point of X_σ [X_σ is not regular since σ is not nonsingular, cf. (4.10)].

(4.22) Corollary: *The integral domain $k[U,V][W]/(W^q - UV^p)$ is integrally closed in its field of quotients iff $p = 1$; in this case X_σ has a singularity of type*

A_{q-1} *in its singular point, and the completion of the local ring in this singular point is a domain and integrally closed.*

Proof: The first claim follows from (4.20)(1). Assume that $p = 1$; the completion of $k[\Gamma_\sigma]$ with respect to $\mathfrak{m}_\sigma$ is $k[U, V, W]/(W^q - UV)$, and from this and (1.5)(3) we get the last assertion.

(4.23) Corollary: *We set $S := k[\{U^{\alpha_j/q}V^{\beta_j/q} \mid j \in \{0, \ldots, t+1\}\}]$, considered as a subring of $k[U^{1/q}, V^{1/q}]$; S is the completion of the local ring $\mathcal{O}_{X_\sigma, x_\sigma}$ of X_σ at its singular point x_σ. Then*
(1) The completion $\widehat{\mathfrak{a}}$ of $\mathfrak{a}$ with respect to the $(T_0, \ldots, T_{t+1})$-adic topology of $k[T_0, \ldots, T_{t+1}]$ is the ideal of $k[T_0, \ldots, T_{t+1}]$ generated by the set $\{Q_{ij} \mid i, j \in \{0, \ldots, t+1\}, i < j-1\}$, and the sequence

$$0 \to \widehat{\mathfrak{a}} \to k[T_0, \ldots, T_{t+1}] \xrightarrow{\widehat{\varphi}} S \to 0 \qquad (**)$$

which we get from (4.18)($*$) is exact.
(2) The two-dimensional complete local ring S has embedding dimension $t + 2$, and its Hilbert function $H(S)$ is given by

$$H(S)(n) = (t+1)\binom{n}{2} + n - 1 \quad \text{for every } n \in \mathbb{N}_0;$$

the multiplicity $e(S)$ of S is $t + 1$.
(3) The set $\{Q_{ij} \mid i, j \in \{0, \ldots, t+1\}, i < j-1\}$ is a minimal set of generators for the ideal $\widehat{\mathfrak{a}}$.

Remember that S is the integral closure of $k[U, V][W]/(W^q - UV^p)$, i.e., it is the integral closure of a Hirzebruch-Jung singularity of type $A_{q,p}$.

Proof: The exactness of the sequence and the first claim made for $\widehat{\mathfrak{a}}$ follow from the fact that $k[T_0, \ldots, T_{t+1}]$ is a noetherian ring [cf. [63], Lemma 7.15]. Let $\mathfrak{m}$ be the maximal ideal of $k[T_0, \ldots, T_{t+1}]$. We have $Q_{ij} \in \mathfrak{m}^2$ for all $i, j \in \{0, \ldots, t+1\}$, $i < j-1$, hence $\widehat{\mathfrak{a}} \subset \mathfrak{m}^2$. Let $\mathfrak{n}$ be the maximal ideal of S, and let $F_0, \ldots, F_{t+1} \in k[T_0, \ldots, T_{t+1}]$ be such that $\widehat{\varphi}(F_0)\widehat{\varphi}(T_0) + \cdots + \widehat{\varphi}(F_{t+1})\widehat{\varphi}(T_{t+1}) \in \mathfrak{n}^2$; then $F_0 T_0 + \cdots + F_{t+1}T_{t+1} \in \mathfrak{m}^2 + \widehat{\mathfrak{a}} = \mathfrak{m}^2$, hence $F_j \in \mathfrak{m}$ for every $j \in \{0, \ldots, t+1\}$ and $\{\widehat{\varphi}(T_j) \mid j \in \{0, \ldots, t+1\}\}$ is a minimal system of generators for the maximal ideal $\mathfrak{n}$ of S by Nakayama's lemma [cf. [63], Cor. 4.8]. Therefore we see that $\text{emdim}(S) = t + 2$.
Let $\nu \in \mathbb{N}$ and $(\gamma_0, \ldots, \gamma_{t+1}) \in \mathbb{N}_0^{t+2}$ be such that $\gamma_0 + \cdots + \gamma_{t+1} = \nu$. By (4.17) there exist a unique $i \in \{0, \ldots, t+1\}$ and $\alpha, \beta \in \mathbb{N}_0$, $\alpha > 0$, with

$$\varphi(T_0^{\gamma_0}) \cdots \varphi(T_{t+1}^{\gamma_{t+1}}) = \varphi(T_i^\alpha)\varphi(T_{i+1}^\beta)$$

[if $i = t+1$, then we have $\beta = 0$], and with $\alpha + \beta \geq \gamma_0 + \cdots + \gamma_{t+1}$. This implies that a k-basis of $\mathfrak{n}^\nu/\mathfrak{n}^{\nu+1}$ is given by all elements $\varphi(T_i^\alpha T_{i+1}^\beta)$, $i \in \{0, \ldots, t+1\}$, $\alpha > 0$, $\alpha + \beta = \nu$ [$\beta = 0$ if $i = t+1$]. Counting these elements shows that

$$\dim_k(\mathfrak{n}^\nu/\mathfrak{n}^{\nu+1}) = (t+2) + (t+1)(\nu-1) = (t+1)\nu + 1;$$

this implies the remaining assertion in (2).

We show: the images in $\widehat{a}/m\widehat{a}$ of the $t(t + 1)/2$ elements Q_{ij}, $i, j \in \{0, \ldots, t+1\}$, $i < j - 1$, are linearly independent over k. Then the assertion in (3) follows from Nakayama's lemma [cf. [63], Cor. 4.8]. Indeed, let γ_{ij}, $i, j \in \{0, \ldots, t+1\}$, $i < j - 1$, be elements in k such that

$$\sum_{\substack{i,j=0 \\ i<j-1}}^{t+1} \gamma_{ij}Q_{ij} \in m\widehat{a}.$$

Consider the elements $Q_{0j} = T_0T_j - P_{0j}$ for every $j \in \{2, \ldots, t+1\}$. There exists a power series $A \in k[\![T_1, \ldots, T_{t+1}]\!]$ [note that T_0 does not appear in P_{0j} for $j \in \{2, \ldots, t+1\}$ nor in Q_{ij} for $i \in \{1, \ldots, t-1\}$, $j \in \{3, \ldots, t+1\}$, $i < j-1$] such that $\sum_{j=2}^{t+1} \gamma_{0j}T_0T_j + A \in m\widehat{a} \subset m^3$ [since $\widehat{a} \subset m^2$], and therefore $\gamma_{0j} = 0$ for every $j \in \{2, \ldots, t+1\}$. In a similar way, we can show that, for each $i \in \{0, \ldots, t-1\}$, the elements γ_{ij} for every $j \in \{i+2, \ldots, t+1\}$ are zero.

(4.24) REMARK: Let $\theta^\vee$ be the convex hull of $(\sigma^\vee \cap M) \setminus \{0\}$ in $M_\mathbb{R}$, and let vol_2 be Lebesgue measure of $M_\mathbb{R}$ normalized in such a way that $1/2$ is the area of a triangle with vertices at 0, n_1, n_2 where $\{n_1, n_2\}$ is any $\mathbb{Z}$-basis of M. Then we have

$$\frac{t+1}{2} = \mathrm{vol}_2(\sigma^\vee \setminus \theta^\vee),$$

since the right hand side is the sum of the areas of the triangles with vertices at 0, k_j and k_{j+1}, $j \in \{0, \ldots, t\}$. Thus, the multiplicity of S can be interpreted in geometric terms: It is twice as large as the Lebesgue measure of $\sigma^\vee \setminus \theta^\vee$.

5   Resolution of Singularities

(5.0) Let X_σ be the (affine) toric variety associated with a two-dimensional strongly convex rational polyhedral cone which is not nonsingular, and let x_σ be its only singular point [cf. (4.21)]. We show that the blow-up $\mathrm{Bl}_{x_\sigma}(X_\sigma)$ of X_σ with center x_σ [cf. A(13.6) for this notion] is a finite union of affine toric varieties having "better" singularities than X_σ; furthermore, we show that X_σ can be desingularized by repeatedly blowing up points.

5.1   Some Useful Formulae

(5.1) NOTATION: Let N be a free $\mathbb{Z}$-module of rank 2, and let $\{n_1, n_2\}$ be a $\mathbb{Z}$-basis of N. Let M be the $\mathbb{Z}$-module dual to N, and let $\{m_1, m_2\}$ be the $\mathbb{Z}$-basis of M dual to the basis $\{n_1, n_2\}$. We use the Hirzebruch-Jung continued fraction expansion of $q/(q - p)$ and q/p as in (4.18), and we also define $l_0, \ldots, l_{s+1}$ and $k_0, \ldots, k_{t+1}$ as in (4.18). Furthermore, we set $\sigma := \mathbb{R}_{\geq 0} l_0 + \mathbb{R}_{\geq 0} l_{s+1}$ and, for every $i \in \{0, \ldots, s\}$, we set $\sigma_i := \mathbb{R}_{\geq 0} l_i + \mathbb{R}_{\geq 0} l_{i+1}$; let θ be the convex hull of $(\sigma \cap N) \setminus \{0\}$ and let $\partial\theta$ be its boundary [cf. (4.18)].

(5.2) For every $i \in \{0, \dots, s+1\}$ we define non-negative integers μ_i, ν_i by

$$l_i = \nu_i n_1 + \mu_i n_2.$$

Then we have $\mu_0 = 0$, $\mu_1 = 1$, $\nu_0 = 1$, $\nu_1 = 1$, and we have [cf. (4.12)(*)]

$$\mu_{i+1} = b_i \mu_i - \mu_{i-1}, \quad \nu_{i+1} = b_i \nu_i - \nu_{i-1} \quad \text{for every } i \in \{1, \dots, s\}.$$

By (4.13)(3) and (4.13)(5) we get

$$0 = \mu_0 < 1 = \mu_1 < \mu_2 < \cdots < \mu_{s+1} = q;$$

it is easy to see that [for the last equality sign cf. (4.13)(2)]

$$1 = \nu_0 = \nu_1 \leq \nu_2 \leq \cdots \leq \nu_{s+1} = p.$$

(1) We have

$$\mu_{i+1}\nu_i - \mu_i \nu_{i+1} = 1 \quad \text{for every } i \in \{0, \dots, s\}.$$

Proof: We have $\mu_1 \nu_0 - \mu_0 \nu_1 = 1$. Let $i \in \{1, \dots, s\}$, and assume that we have shown already that $\mu_i \nu_{i-1} - \mu_{i-1} \nu_i = 1$; then we have

$$\mu_{i+1}\nu_i - \mu_i \nu_{i+1} = (b_i \mu_i - \mu_{i-1})\nu_i - (b_i \nu_i - \nu_{i-1})\mu_i$$
$$= \mu_i \nu_{i-1} - \mu_{i-1}\nu_i = 1.$$

(2) We have

$$\mu_i \nu_j - \mu_j \nu_i > 0 \quad \text{for all } i, j \in \{0, \dots, s+1\} \text{ such that } i > j.$$

Proof: We have, by (1),

$$\frac{\mu_i}{\nu_i} - \frac{\mu_{i-1}}{\nu_{i-1}} = \frac{1}{\nu_i \nu_{i-1}} \quad \text{for every } i \in \{1, \dots, s+1\},$$

hence we get

$$\frac{\mu_{i+h}}{\nu_{i+h}} - \frac{\mu_{i-1}}{\nu_{i-1}} = \sum_{j=0}^{h} \frac{1}{\nu_{i+j}\nu_{i+j-1}} \quad \text{for } i \in \{1, \dots, s+1\} \text{ and } h \in \{0, \dots, s+1-i\},$$

and the assertion is proved.

(3) In particular, since $\mu_1 = \nu_1 = 1$ and $\mu_{s+1} = q$, $\nu_{s+1} = p$, we have

$$q\nu_j \geq p\mu_j \text{ for every } j \in \{0, \dots, s+1\}, \quad \mu_j \geq \nu_j \text{ for every } j \in \{1, \dots, s+1\}.$$

(5.3) Note that $k_0 = m_2$, $k_1 = m_1$, hence that $\{k_1, k_0\}$ is the basis of M dual to $\{n_1, n_2\}$. Let $i \in \{0, \dots, s\}$.

(1) Let $\{m_0^{(i)}, m_1^{(i)}\}$ be the $\mathbb{Z}$-basis of M dual to the $\mathbb{Z}$-basis $\{l_i, l_{i+1}\}$ of N. By (5.2) we get

$$m_0^{(i)} = -\nu_{i+1}k_0 + \mu_{i+1}k_1, \quad m_1^{(i)} = \nu_i k_0 - \mu_i k_1,$$

hence, by (5.2)(1), we find that

$$k_0 = \mu_i m_0^{(i)} + \mu_{i+1}m_1^{(i)}, \quad k_1 = \nu_i m_0^{(i)} + \nu_{i+1}m_1^{(i)}.$$

(2) Note that $\Gamma_{\sigma_i} = \mathbb{N}_0\, m_0^{(i)} + \mathbb{N}_0\, m_1^{(i)}$.

5.2   The Case $p = 1$

(5.4) In this subsection we assume that $p = 1$.

(5.5) REMARK: (1) Then we have $\sigma = \mathbb{R}_{\geq 0}\, n_1 + \mathbb{R}_{\geq 0}\, (n_1 + qn_2)$ and

$$\frac{q}{q-1} = b_1 - \overline{|b_2} - \cdots - \overline{|b_s} \quad \text{where } s = q - 1 \text{ and } b_1 = \cdots = b_s = 2$$

[cf. (3.9)(2)]. The semigroup Γ_σ has the set $\{k_0, k_1, k_2\}$ as a minimal system of generators where

$$k_0 = \tilde{m}_1, \quad k_1 = \tilde{m}_1 + \tilde{m}_2, \quad k_2 = (q-1)\tilde{m}_1 + q\tilde{m}_2$$

[cf. (4.18) for $p = 1$], and the affine variety X_σ has one singular point x_σ which is an A_s-singularity [cf. (4.22)]. Note that k_0, k_2 are linearly independent over $\mathbb{Q}$, and that $qk_1 = k_0 + k_2$, which shows again that $k[\Gamma_\sigma] = k[U, V][W]/(W^q - UV)$.
(2) The projective variety $\mathrm{Bl}_{x_\sigma}(X_\sigma)$ is covered by three open affine subsets U_0, U_1, U_2 having, respectively, the coordinate rings R_0, R_1, R_2 where

$$R_j := k[\Gamma_\sigma][e(k_0)/e(k_j), e(k_1)/e(k_j), e(k_2)/e(k_j)] \quad \text{for } j = 0, 1, 2$$

[cf. A(14.1)(3)]. We show that U_0 and U_2 are nonsingular, and that U_1 is nonsingular if $q \leq 3$, and that it is (isomorphic to) a toric variety which has a singularity of type A_{q-3} if $q \geq 4$.
Indeed, the subsemigroup of M generated by $k_0, k_1, k_2, k_1 - k_0, k_2 - k_0$ is also generated by k_0 and $k_1 - k_0$ since $k_2 - k_0 = q(k_1 - k_0) + (q-2)k_0$, and the subsemigroup of M generated by $k_0, k_1, k_2, k_0 - k_2, k_1 - k_2$ is also generated by k_2 and $k_1 - k_2$ since $k_0 - k_2 = q(k_1 - k_2) + (q-2)k_2$; note that $\{k_0, k_1 - k_0\}$ and $\{k_2, k_1 - k_2\}$ are $\mathbb{Z}$-bases of M. The subsemigroup Γ_1 of M generated by $k_0, k_1, k_2, k_0 - k_1, k_2 - k_1$ is also generated by $k_0 - k_1, k_2 - k_1, k_1$, and we have $(q-2)k_1 = (k_0 - k_1) + (k_2 - k_1)$. If $q = 2$, then we have $\Gamma_1 = \mathbb{Z}(k_0 - k_1)$, hence U_1 is an open affine subset of $\mathbb{A}^1$ [cf. (2.2)]. If $q = 3$, then we have $\Gamma_1 = \mathbb{N}_0(k_0 - k_1) + \mathbb{N}_0(k_2 - k_1)$; since $k_0 - k_1, k_2 - k_1$ are linearly independent over $\mathbb{Q}$, the affine variety U_1 is nonsingular [cf. (4.10)]. If $q \geq 4$, then U_1 is (isomorphic to) a toric variety and has one singular point of type A_{q-3}.

Thus, we have shown:

(5.6) **Proposition:** *Assume that $p = 1$. Then X_σ has one singular point, namely a singularity of type A_{q-1}. By blowing up the singular point, one gets a nonsingular variety if $q \leq 3$, and a variety with exactly one singular point if $q \geq 4$; this singular point is a singularity of type A_{q-3}, and it has an open affine neighborhood which is (isomorphic to) a toric variety.*

(5.7) **Corollary:** *Assume that $p = 1$. Then X_σ can be desingularized by repeatedly blowing up points.*

Proof: This follows from (5.6) and A(14.4).

5.3   The General Case

(5.8) NOTATION: If $s = 1$, then we set $v := 1$ and $i(0) := 0$, $i(1) := 1$, $i(2) := 2$. If $s > 1$, then let v' be the number of those vertices of θ which are different from l_0, l_1, l_s, l_{s+1}—note that $0 \le v' \le s - 2$—, set $v := v' + 2$, and label these vertices as $l_{i(2)}, \ldots, l_{i(v-1)}$ where $i(2) < \cdots < i(v-1)$. We set $i(0) := 0$, $i(1) := 1$, $i(v) := s$, $i(v+1) = s + 1$. Note that l_1, l_s are, in general, not vertices of θ. By (4.18)(1) we have

$$b_{i(\alpha)+1} = \cdots = b_{i(\alpha+1)-1} = 2 \quad \text{for every } \alpha \in \{1, \ldots, v - 1\}.$$

From the recursion formula in (5.2) we find for every $\alpha \in \{1, \ldots, v-1\}$

$$\mu_{i(\alpha+1)} - \mu_{i(\alpha+1)-1} = \mu_{i(\alpha+1)-1} - \mu_{i(\alpha+1)-2} = \cdots = \mu_{i(\alpha)+1} - \mu_{i(\alpha)},$$
$$\nu_{i(\alpha+1)} - \nu_{i(\alpha+1)-1} = \nu_{i(\alpha+1)-1} - \nu_{i(\alpha+1)-2} = \cdots = \nu_{i(\alpha)+1} - \nu_{i(\alpha)}.$$

If $v > 2$, then we have $b_{i(2)} > 2, \ldots, b_{i(v-1)} > 2$ since $l_{i(2)}, \ldots, l_{i(v-1)}$ are vertices of θ.

(5.9) NOTATION: We define elements $r_0, \ldots, r_v \in M$ by requiring that

$$\langle r_0, l_0 \rangle = 0, \quad \langle r_0, l_1 \rangle = 1, \quad \langle r_v, l_s \rangle = 1, \quad \langle r_v, l_{s+1} \rangle = 0,$$

$$\langle r_\alpha, l_{i(\alpha)} \rangle = \langle r_\alpha, l_{i(\alpha)+1} \rangle = 1 \quad \text{for every } \alpha \in \{1, \ldots, v - 1\}$$

[note that $\{l_i, l_{i+1}\}$ is a $\mathbb{Z}$-basis of N for every $i \in \{0, \ldots, s\}$, cf. (4.18)(1)]. Let $\alpha \in \{1, \ldots, v - 1\}$; the points $l_{i(\alpha)}, l_{i(\alpha)+1}, \ldots, l_{i(\alpha+1)}$ lie on a line, hence we have

$$\langle r_\alpha, l_\beta \rangle = 1 \quad \text{for every } \beta \in \{i(\alpha), i(\alpha) + 1, \ldots, i(\alpha + 1)\}.$$

(1) Using (5.2)(1) and (5.3)(1), we see that

$$r_0 = k_0,$$
$$r_\alpha = -(\nu_{i(\alpha)+1} - \nu_{i(\alpha)})k_0 + (\mu_{i(\alpha)+1} - \mu_{i(\alpha)})k_1 \quad \text{for every } \alpha \in \{1, \ldots, v - 1\},$$
$$r_v = -\nu_{i(v)+1}k_0 + \mu_{i(v)+1}k_1.$$

(2) Let $\alpha \in \{1, \ldots, v - 1\}$. Using (5.3)(1) and the definition of $r_0, \ldots, r_v$, we get

$$r_0 = k_0 = m_0^{(1)} + b_1 m_1^{(1)}, \quad r_\alpha = m_0^{(i(\alpha))} + m_1^{(i(\alpha))}, \quad r_v = k_{t+1} = m_0^{(s)}. \qquad (*)$$

By (5.2) and (5.8) we get

$$r_\alpha = m_0^{(i(\alpha+1))} + (b_{i(\alpha+1)} - 1)m_1^{(i(\alpha+1))}. \qquad (**)$$

(5.10) Lemma: The $3 + (b_1 - 2) + \cdots + (b_s - 2) = t + 2$ elements

$$\frac{1}{b_1}((b_1 - \beta)r_0 + \beta r_1), \quad \beta \in \{0, \ldots, b_1\},$$

in case $s = 1$, and

$$\frac{1}{b_1 - 1}\left((b_1 - 1 - \beta)r_0 + \beta r_1\right), \qquad\qquad \beta \in \{0, \ldots, b_1 - 1\},$$

$$\frac{1}{b_{i(\alpha)} - 2}\left((b_{i(\alpha)} - 2 - \beta)r_{\alpha-1} + \beta r_\alpha\right), \alpha \in \{2, \ldots, v - 1\}, \beta \in \{1, \ldots, b_{i(\alpha)} - 2\},$$

$$\frac{1}{b_s - 1}\left((b_s - 1 - \beta)r_{v-1} + \beta r_v\right), \qquad\qquad \beta \in \{1, \ldots, b_s - 1\},$$

in case $s \geq 2$, in this order, are the elements $k_0, k_1, \ldots, k_{t+1}$ on the compact edges of the boundary polygon $\partial\theta$.

Proof: Note that $3 + (b_1 - 2) + \cdots + (b_s - 2) = t + 2$ by (3.10)(∗).
First, we consider the case $s = 1$. Here we have $b_1 = q$, $\mu_2 = q$, $\nu_2 = q - 1$, $i(0) = 0$, $i(1) = 1$, $i(2) = 2$, $t + 1 = q$, $r_0 = k_0$, $r_1 = k_{t+1}$, $c_1 = \cdots = c_t = 2$ [cf. (4.18)(2)], hence we have

$$k_j = \frac{(q - j)k_0 + jk_{t+1}}{q} \quad \text{for every } j \in \{0, \ldots, q\}$$

[cf. (4.13)(4)]. This is the formula given above.
Now assume that $s \geq 2$. We use induction with respect to s and b_1. We define q', p' as in (3.9)(2)(b); let s' be the length of the Hirzebruch-Jung continued fraction expansion of $q'/(q' - p')$, and let $l'_0, \ldots, l'_{s'+1}$ be the data corresponding to q', $q' - p'$. We define v' and $r'_0, \ldots, r'_{v'}$ with respect to $l'_0, \ldots, l'_{s'+1}$ in the same way as we defined v and $r_0, \ldots, r_v$ with respect to $l_0, \ldots, l_{s+1}$. Furthermore, let t' be the length of the Hirzebruch-Jung continued fraction expansion of q'/p', and let $k'_0, \ldots, k'_{t'+1}$ be the data corresponding to q', p'.
(1) $b_1 = 2$: then we have $c_1 \geq 3$, $s' = s - 1$, $t' = t$, $q' = q - p$, $p' = p$, $b'_1 = b_2$, $b'_2 = b_3, \ldots, b'_{s'} = b_s$, $c'_1 = c_1 - 1$, $c'_2 = c_2, \ldots, c'_{t'} = c_t$. We set $n'_1 := l_1 = n_1 + n_2$, $n'_2 := n_2$. Then $\{n'_1, n'_2\}$ is a $\mathbb{Z}$-basis of N. We find [cf. (4.13) and (4.18) for notations which are not explained here] that $l'_0 = n'_1 = l_1$, $l'_1 = n'_1 + n'_2 = l_2$ [as $b_1 = 2$, we see that $n_2 = n_3$], and that $l'_i = l_{i+1}$ for $i \in \{2, \ldots, s\}$. We set $m'_1 := m_1$, $m'_2 := -m_1 + m_2$. The $\mathbb{Z}$-basis $\{m'_1, m'_2\}$ of M is dual to $\{n'_1, n'_2\}$. We set $\tilde{m}'_1 := m'_2 = -\tilde{m}_2$, $\tilde{m}'_2 := m'_1 - m'_2 = 2\tilde{m}_2 + \tilde{m}_1$; then we have $k'_0 = \tilde{m}'_1 = -\tilde{m}_2$, $k'_1 = k'_0 + \tilde{m}'_2 = k_1$, $k'_2 = k'_1 + \tilde{m}'_3 = k_1 + \tilde{m}_3 = k_2$ [because $\tilde{m}'_3 = (c_1 - 3)k'_0 + (c_1 - 2)\tilde{m}'_2 = \tilde{m}_3$] and $k'_j = k_j$ for every $j \in \{3, \ldots, t+1\}$.
(2) $b_1 \geq 3$: then we have $c_1 = 2$, $s' = s$, $t' = t - 1$, $q' = p$, $p' = 2p - q$, $b'_1 = b_1 - 1$, $b'_2 = b_2, \ldots, b'_{s'} = b_s$, $c_2 = c'_1, \ldots, c_t = c'_{t'}$. We set $n'_1 := -n_2$, $n'_2 := n_1 + 2n_2$. Then we have $l'_0 = n'_1 = -n_1$, $l'_1 = n'_1 + n'_2 = l_1$, $l'_2 = l'_1 + n'_3 = l_2$, and $l'_i = l_i$ for every $i \in \{3, \ldots, s+1\}$. We set $m'_1 := -m_2 + 2m_1$, $m'_2 := m_1$. The $\mathbb{Z}$-basis $\{m'_1, m'_2\}$ of M is dual to $\{n'_1, n'_2\}$. We set $\tilde{m}'_1 := m'_2 = \tilde{m}_1 + \tilde{m}_2$, $\tilde{m}'_2 := m'_1 - m'_2 = \tilde{m}_2$. Then we have $k'_0 = \tilde{m}'_1 = \tilde{m}_1 + \tilde{m}_2 = k_1$, $k'_1 = \tilde{m}'_1 + \tilde{m}'_2 = k_2$, and $k'_j = k_{j+1}$ for $j \in \{2, \ldots, t\}$.
(3) $s = 2$, $b_1 = 2$: then we have $v = 2$. Now we have $k_0 = r_0$, $k_1 = r_1$ by (5.9)(1),

hence we get

$$k_\beta = \frac{1}{b_1 - 1}((b_1 - 1 - \beta)r_0 + \beta r_1), \quad \beta \in \{0, 1\}.$$

Since $r_1' = r_2$ and $l_1 + l_3 = b_2 l_2$, we get $r_0' = (b_2 r_1 - r_2)/(b_2 - 1)$, hence, by the case $s = 1$, we get

$$k_\beta = k_\beta' = \frac{1}{b_2}((b_2 - \beta)r_0' + \beta r_1') = \frac{1}{b_2 - 1}((b_2 - \beta)r_1 + (\beta - 1)r_2), \quad \beta \in \{2, \ldots, b_2\}.$$

(4) $s = 2$, $b_1 > 2$: from $l_0' + l_2' = (b_1 - 1)l_1'$ and $l_0 + l_2 = b_1 l_1$ we get $r_0' = (1/(b_2 - 1))((b_1 - 2)r_0 + r_1)$. By (5.9)(1) we find that

$$r_1' = -(b_1 - 3)k_0' + (b_1 - 2)k_1' = -(b_1 - 2)k_0 + (b_1 - 1)k_1 = r_1,$$
$$r_2' = -p'k_0' + q'k_1' = -(2p - q)k_1 + p(2k_1 - k_0) = -pk_0 + qk_1 = r_2.$$

We assume that the formulae are true for $b_1' = b_1 - 1$. Then we have

$$k_{\beta+1} = k_\beta' = \frac{1}{b_1 - 2}((b_1 - 2 - \beta)r_0' + \beta r_1')$$
$$= \frac{1}{b_1 - 1}((b_1 - 2 - \beta)r_0 + (\beta + 1)r_1), \quad \beta \in \{1, \ldots, b_1 - 2\},$$

and by (5.9)(1) the displayed equation holds for $\beta = 0$ and $\beta = -1$; furthermore, we have

$$k_{b_1 - 1 + \beta} = k_{b_1 - 2 + \beta}' = \frac{1}{b_2 - 1}((b_2 - 1 - \beta)r_1 + \beta r_2), \quad \beta \in \{1, \ldots, b_2 - 1\}.$$

(5) We assume that $s \geq 3$, and that the assertion is true for $s - 1$.
(a) $b_1 = 2$: then we have $k_0 = r_0$, $k_1 = r_1$ by (5.9)(1).
If $b_2 = 2$, then we have $v' = v$ and $r_\alpha' = r_\alpha$ for every $\alpha \in \{1, \ldots, v\}$, and this case is finished.
If $b_2 > 2$, then we have $v' = v - 1$, and $r_1' = r_2, \ldots, r_{v-1}' = r_v$. Since we have $\langle r_0', l_0' \rangle = 0$, $\langle r_0', l_1' \rangle = 1$ by definition, we get $r_0' = (1/(b_2 - 2))((b_2 - 1)r_1 - r_2)$, hence, again by induction, we get

$$k_{\beta+1} = k_{\beta+1}' = \frac{1}{b_2 - 1}((b_2 - 2 - \beta)r_0' + \beta r_1')$$
$$= \frac{1}{b_2 - 2}((b_2 - 2 - \beta)r_1 + \beta r_2), \quad \beta \in \{1, \ldots, b_2 - 2\}.$$

Now it is clear by induction that the assertion is also true in this case.
(b) $b_1 \geq 3$: then we have $v' = v$, and $r_\alpha' = r_\alpha$ for every $\alpha \in \{1, \ldots, v\}$. We assume that the formulae are true for $b_1 - 1$. We can use the same argument as in the case $s = 2$, $b_1 > 2$.

(5.11) NOTATION: We set

$$\tau_\alpha := \mathbb{R}_{\geq 0}\, l_{i(\alpha)} + \mathbb{R}_{\geq 0}\, l_{i(\alpha+1)} \quad \text{for every } \alpha \in \{0, \ldots, v\};$$

note that $\tau_0, \ldots, \tau_v$ are strongly convex rational polyhedral cones of dimension 2 in $N_\mathbb{R}$.

(5.12) Let $\alpha \in \{0, \ldots, v\}$. We set

$$\lambda_\alpha := i(\alpha + 1) - i(\alpha);$$

then we have $\lambda_0 = \lambda_v = 1$.
(1) Let $z_0, z_1 \in \mathbb{Z}$, and define recursively z_j for every $j \in \mathbb{N}$ by $z_{j+1} = 2z_j - z_{j-1}$; then we have $z_j = jz_1 + (1-j)z_0$ for every $j \in \mathbb{N}_0$.
(2) By (1) we have

$$l_j = (j - i(\alpha))l_{i(\alpha)+1} + (1 - (j - i(\alpha)))l_{i(\alpha)} \quad \text{for every } j \in \{i(\alpha), i(\alpha)+1, \ldots, i(\alpha+1)\}.$$

(3) By (2) we have

$$\tau_\alpha = \mathbb{R}_{\geq 0}\, l_{i(\alpha)} + \mathbb{R}_{\geq 0}\left(l_{i(\alpha)} + \lambda_\alpha(l_{i(\alpha)+1} - l_{i(\alpha)})\right).$$

Since $\{l_{i(\alpha)}, l_{i(\alpha)+1} - l_{i(\alpha)}\}$ is a $\mathbb{Z}$-basis of N, the only point of the toric variety X_{τ_α} which can be singular is a singularity of type $A_{\lambda_\alpha - 1}$ [cf. (5.5)(1)].

(5.13) We set

$$b_1^* := b_1 - 1, \quad b_\alpha^* := b_{i(\alpha)} - 2 \text{ for every } \alpha \in \{2, \ldots, v-1\}, \quad b_v^* := b_s - 1.$$

(5.14) Let $\alpha \in \{1, \ldots, v-1\}$. We set

$$s_\alpha := \frac{1}{b_\alpha^*}(r_{\alpha-1} - r_\alpha).$$

(1) We have by (5.12)(2), and since $l_{i(\alpha+1)-1} + l_{i(\alpha+1)+1} = b_{i(\alpha+1)}l_{i(\alpha+1)}$,

$$l_{i(\alpha)} = \left(\lambda_\alpha(b_{i(\alpha+1)} - 1) + 1\right)l_{i(\alpha+1)} - \lambda_\alpha l_{i(\alpha+1)+1},$$
$$l_{i(\alpha)+1} = \left((\lambda_\alpha - 1)(b_{i(\alpha+1)} - 1) + 1\right)l_{i(\alpha+1)} + (1 - \lambda_\alpha)l_{i(\alpha+1)+1}.$$

Now we have $\langle r_0, l_1 \rangle = 1$, $\langle r_0, l_2 \rangle = b_1 - 1$, and for $\alpha \in \{2, \ldots, v-2\}$ we get

$$\langle r_{\alpha-1}, l_{i(\alpha)} \rangle = 1, \quad \langle r_{\alpha-1}, l_{i(\alpha)+1} \rangle = b_{i(\alpha)-1}.$$

(2) By (1) we find that $\{r_\alpha, s_\alpha\}$ is dual to the $\mathbb{Z}$-basis $\{l_{i(\alpha)}, l_{i(\alpha)+1} - l_{i(\alpha)}\}$ of N.
(3) We set

$$m_{1,\alpha} := s_\alpha, \quad m_{2,\alpha} := r_\alpha - m_{1,\alpha}.$$

Then we have [cf. (4.18)]

$$\tau_\alpha^\vee = \mathbb{R}_{\geq 0}\, m_{1,\alpha} + \mathbb{R}_{\geq 0}\, ((\lambda_\alpha - 1)m_{1,\alpha} + \lambda_\alpha m_{2,\alpha});$$

a minimal system of generators for the subsemigroup Γ_{τ_α} of M is $\{m_{1,\alpha}, m_{2,\alpha}\}$ if $\lambda_\alpha = 1$, and is $\{m_{1,\alpha}, m_{1,\alpha} + m_{2,\alpha}, (\lambda_\alpha - 1)m_{1,\alpha} + \lambda_\alpha m_{2,\alpha}\}$ if $\lambda_\alpha > 1$.
(4) Note that $\Gamma_{\tau_0} = \mathbb{N}_0\, m_0^{(0)} + \mathbb{N}_0\, m_1^{(0)}$, and that $\Gamma_{\tau_v} = \mathbb{N}_0\, m_0^{(s)} + \mathbb{N}_0\, m_1^{(s)}$.

(5.15) Let $\alpha \in \{1, \ldots, v - 1\}$.
(1) We have

$$\lambda_\alpha r_\alpha = \frac{1}{b_{\alpha+1}^*}(r_{\alpha+1} - r_\alpha) + \frac{1}{b_\alpha^*}(r_{\alpha-1} - r_\alpha)$$

[evaluate both sides at $l_{i(\alpha)}$ and $l_{i(\alpha+1)}$].
(2) From (1) we get

$$m_{1,1} = \frac{1}{b_1^*}(r_0 - r_1), \quad (\lambda_\alpha - 1)m_{1,\alpha} + \lambda_\alpha m_{2,\alpha} = \frac{1}{b_{\alpha+1}^*}(r_{\alpha+1} - r_\alpha).$$

(5.16) For every $j \in \{0, \ldots, t+1\}$ let Γ_j be the subsemigroup of M generated by the elements $k_0, \ldots, k_{t+1}, k_0 - k_j, \ldots, k_{j-1} - k_j, k_{j+1} - k_j, \ldots, k_{t+1} - k_j$. For every $\alpha \in \{0, \ldots, v\}$ let $j(\alpha) \in \{0, \ldots, t+1\}$ be defined by $k_{j(\alpha)} = r_\alpha$ [cf. (5.10)]; we have $j(0) = 0$, $j(1) = b_1$ if $s = 1$, $j(1) = b_1 - 1$ if $s \geq 2$, and $j(v) = t+1$.
(1) $s = 1$: then we have $b_1 = q = t+1$. By (5.3)(1) we get $m_0^{(0)} = k_1 - k_0$, $m_1^{(0)} = k_0$, and by (5.10) we get $k_1 - k_0 = (1/q)(r_1 - r_0)$, hence $k_\beta - k_0 = \beta m_0^{(0)}$ for every $\beta \in \{1, \ldots, b_1\}$. Since $\Gamma_{\tau_0} = \mathbb{N}_0\, m_0^{(0)} + \mathbb{N}_0\, m_1^{(0)}$, we have shown that $\Gamma_0 = \Gamma_{\tau_0}$.
Let $j \in \{1, \ldots, t+1\}$. By (5.9)(2) we have $m_0^{(1)} = k_{t+1} = r_1$ and $r_0 = k_0 = m_0^{(1)} + b_1 m_1^{(1)}$, hence $k_\beta - k_j = (j - \beta)m_1^{(1)}$ for every $\beta \in \{0, \ldots, b_1\}$ and $k_j = m_0^{(1)} - jm_1^{(1)}$. Thus, we have $\Gamma_j = \mathbb{Z}\, m_0^{(1)} + \mathbb{N}_0\, m_1^{(1)}$ for every $j \in \{1, \ldots, t\}$, and $\Gamma_{t+1} = \Gamma_{\tau_1}$.
(2) $s \geq 2$: then we have $v \geq 2$. Just as above we see that $\Gamma_0 = \Gamma_{\tau_0}$.
Since $r_v = k_{t+1} = m_0^{(s)}$ and $m_1^{(s)} = (1/(b_1 - 1))(r_{v-1} - r_v) = k_t - k_{t+1}$, we see that $\Gamma_{\tau_v} \subset \Gamma_{t+1}$. Just as above we can show that $\Gamma_{t+1} \subset \Gamma_{\tau_v}$, and therefore we have $\Gamma_{t+1} = \Gamma_{\tau_v}$.
Let $\alpha \in \{1, \ldots, v - 1\}$; we show that $\Gamma_{j(\alpha)} = \Gamma_{\tau_\alpha}$.
(a) First, we show that $r_0, \ldots, r_v, r_0 - r_\alpha, \ldots, r_v - r_\alpha$ lie in Γ_{τ_α}. Since

$$r_{\alpha-1} - r_\alpha = b_\alpha^* m_{1,\alpha}, \quad r_\alpha = m_{1,\alpha} + m_{2,\alpha}, \quad r_{\alpha+1} - r_\alpha = b_{\alpha+1}^*((\lambda_\alpha - 1)m_{1,\alpha} + \lambda_\alpha m_{2,\alpha}),$$

we see that $r_{\alpha-1} - r_\alpha$, r_α, and $r_{\alpha+1} - r_\alpha \in \Gamma_{\tau_\alpha}$, and therefore $r_{\alpha-1}$ and $r_{\alpha+1}$ lie in Γ_{τ_α}.
Assume that $\alpha < v - 1$. Then we have [cf. (5.15)(1)]

$$\lambda_{\alpha+1} r_{\alpha+1} + \frac{1}{b_{\alpha+1}^*}(r_{\alpha+1} - r_\alpha) = \frac{1}{b_{\alpha+2}^*}(r_{\alpha+2} - r_{\alpha+1}),$$

and therefore $r_{\alpha+2}-r_{\alpha+1}$ lies in Γ_{τ_α}, hence $r_{\alpha+2}-r_\alpha = (r_{\alpha+2}-r_{\alpha+1})+(r_{\alpha+1}-r_\alpha)$ lies in Γ_{τ_α}. Continuing in this way, we find that the elements $r_{\alpha+2},\ldots,r_v$ and $r_{\alpha+2}-r_\alpha,\ldots,r_v-r_\alpha$ lie in Γ_{τ_α}.
Likewise, if $\alpha \geq 2$, then we have

$$\lambda_{\alpha-1}r_{\alpha-1} + \frac{1}{b_\alpha^*}(r_{\alpha-1}-r_\alpha) = \frac{1}{b_{\alpha-1}^*}(r_{\alpha-2}-r_{\alpha-1}),$$

and therefore $r_{\alpha-2}-r_{\alpha-1}$ lies in Γ_{τ_α}, hence $r_{\alpha-2}-r_\alpha = (r_{\alpha-2}-r_{\alpha-1})+(r_{\alpha-1}-r_\alpha)$ lies in Γ_{τ_α}. Continuing in this way, we find that the elements $r_0,\ldots,r_{\alpha-1}$ and $r_0-r_\alpha,\ldots,r_{\alpha-1}-r_\alpha$ lie in Γ_{τ_α}.
(b) We show that $k_\beta - r_\alpha$ lies in Γ_{τ_α} for every $\beta \in \{0,\ldots,t+1\}$. Since $k_0 = r_0$, we know already that $k_0 - r_\alpha \in \Gamma_{\tau_\alpha}$. Let $\gamma \in \{1,\ldots,v\}$ and $\beta \in \{1,\ldots,b_\gamma^*\}$; then we have by (5.10)

$$k_\beta - r_\alpha = \frac{1}{b_\gamma^*}((b_\gamma^* - \beta)(r_{\gamma-1}-r_\alpha) + \beta(r_\gamma - r_\alpha)),$$

hence we have $k_\beta - r_\alpha \in \Gamma_{\tau_\alpha}$ since $r_{\gamma-1} - r_\alpha, r_\gamma - r_\alpha \in \Gamma_{\tau_\alpha}$ by (a).
Thus, we have shown that $\Gamma_{j(\alpha)} \subset \Gamma_{\tau_\alpha}$ for every $\alpha \in \{1,\ldots,v-1\}$.
(c) By (5.10) and (5.15)(2) we have

$$k_{j(\alpha)+1} - k_{j(\alpha)} = \frac{1}{b_{\alpha+1}^*}(r_{\alpha+1}-r_\alpha) = (\lambda_\alpha - 1)m_{1,\alpha} + \lambda_\alpha m_{2,\alpha},$$

$$m_{1,\alpha} + m_{2,\alpha} = r_\alpha = k_{j(\alpha)},$$

$$k_{j(\alpha)-1} - k_{j(\alpha)} + \lambda_\alpha(m_{1,\alpha} + m_{2,\alpha}) = \frac{1}{b_{\alpha+1}^*}(r_\alpha - r_{\alpha+1}) + \lambda_\alpha(m_{1,\alpha} + m_{2,\alpha})$$

$$= m_{1,\alpha}.$$

Thus, we have shown that $\Gamma_{\tau_\alpha} \subset \Gamma_{j(\alpha)}$, hence we have $\Gamma_{j(\alpha)} = \Gamma_{\tau_\alpha}$.
(3) We set

$$m_\alpha := \frac{1}{b_{\alpha+1}^*}(r_\alpha - r_{\alpha+1}) \quad \text{for } \alpha \in \{1,\ldots,v-1\}.$$

We show: For $\alpha \in \{1,\ldots,v-1\}$ and $j \in \{j(\alpha)+1,\ldots,j(\alpha+1)-1\}$ we have

$$\Gamma_j = \mathbb{Z}\, m_\alpha + \mathbb{N}_0\, k_j.$$

In fact, we set $\Gamma_j' := \mathbb{Z}\, m_\alpha + \mathbb{N}_0\, k_j$; we have

$$k_\beta - k_j = \frac{\beta - j}{b_{\alpha+1}^*}(r_{\alpha+1} - r_\alpha) = (j - \beta)m_\alpha \quad \text{for } \beta \in \{j(\alpha),\ldots,j(\alpha+1)\}.$$

Therefore we have $\Gamma_j' \subset \Gamma_j$, and the elements $r_\alpha = k_{j(\alpha)},\ldots,k_{j(\alpha+1)} = r_{\alpha+1}$ lie in Γ_j'.

We assume that $\alpha < v - 2$, and let $\beta \in \{j(\alpha + 1), \ldots, j(\alpha + 2)\}$. Then we have

$$k_\beta - r_{\alpha+1} = \frac{(\beta - j(\alpha + 1))}{b^*_{\alpha+2}}(r_{\alpha+2} - r_{\alpha+1}) = (\beta - j(\alpha + 1))(\lambda_{\alpha+1} r_{\alpha+1} - m_\alpha),$$

hence $k_\beta - k_j = (k_\beta - r_{\alpha+1}) + (r_{\alpha+1} - k_j) \in \Gamma'_j$, and therefore the elements $k_{j(\alpha+1)+1}$, $\ldots, k_{j(\alpha+2)} = r_{\alpha+2}$ lie in Γ'_j. Continuing in this way, we find that $k_\beta - k_j$ lies in Γ'_j for every $\beta \in \{j(\alpha), \ldots, t+1\}$.

We assume that $\alpha \geq 2$, and let $\beta \in \{j(\alpha - 1), \ldots, j(\alpha)\}$. Then we have

$$k_\beta - r_\alpha = \frac{\beta - j(\alpha)}{b^*_{\alpha+1}}(r_{\alpha+1} - r_\alpha) = (j(\alpha) - \beta)m_\alpha,$$

hence $k_\beta - k_j = (k_\beta - r_\alpha) + (r_\alpha - k_j)$ lies in Γ'_j. Continuing in this way, we find that the elements $k_\beta - k_j$ lie in Γ'_j for every $\beta \in \{1, \ldots, j(\alpha)\}$.

(4) We have shown: Let $\alpha \in \{0, \ldots, v\}$. Then the toric variety associated with $k[\Gamma_{j(\alpha)}]$ has at most one singular point, and it is a singularity of type $A_{\lambda_\alpha - 1}$. The varieties associated with $k[\Gamma_j]$, $j \in \{0, \ldots, t+1\} \setminus \{j(0), \ldots, j(v)\}$, are nonsingular.

(5.17) **Proposition:** $\mathrm{Bl}_{x_\sigma}(X_\sigma)$ has exactly $\mathrm{Card}(\{\alpha \mid 1 \leq \alpha \leq v, \lambda_\alpha > 1\})$ singular points, each of them is a singularity of type A_λ for some $\lambda \in \mathbb{N}$, and admits an open affine neighborhood which is isomorphic to a toric variety.

Proof: $\mathrm{Bl}_{x_\sigma}(X_\sigma)$ is covered by $t+2$ open affine sets $U_0, \ldots, U_{t+1}$, where, for every $j \in \{1, \ldots, t+1\}$, U_j has coordinate ring $k[\Gamma_\sigma][e(k_0)/e(k_j), \ldots, e(k_{t+1})/e(k_j)] = k[\Gamma_j]$ [cf. A(14.1)(3)]. Of these varieties, only the varieties $U_{j(1)}, \ldots, U_{j(v)}$ can have a singular point and, for every $\alpha \in \{1, \ldots, v\}$, $U_{j(\alpha)}$ has a singular point iff $\lambda_\alpha > 1$, and in this case it is a singularity of type $A_{\lambda_\alpha - 1}$.

(5.18) **Corollary:** X_σ can be desingularized be repeatedly blowing up points.

Proof: This follows from (5.17) and A(14.4).

5.4 Counting Singularities of the Blow-up

(5.19) **Proposition:** The continued fraction expansion of $q/(q - p)$ having an odd number of terms has the form

$$\frac{q}{q - p} = a_1 + \lceil a_2 + \cdots + \lceil a_{2v-1}, \qquad (*)$$

and, moreover, we have

$$\lambda_\alpha = a_{2\alpha} \quad \text{for every } \alpha \in \{1, \ldots, v - 1\}.$$

Proof: (1) Note that, by (3.3)(2), the expansion (∗) is uniquely determined.
(2) The Hirzebruch-Jung continued fraction expansion of $q/(q-p)$,

$$\frac{q}{q-p} = b_1 - \overline{\lceil b_2} - \cdots - \overline{\lceil b_s},$$

can be found in the following way: we define $\pi_0 = q$, $\pi_1 = q - p$, $\pi_2, \ldots, \pi_s$ and $\pi_{s+1} = 0$ by

$$\pi_{j-1} = b_j \pi_j - \pi_{j+1} \quad \text{for } j \in \{1, \ldots, s\}, \quad 0 < \pi_{j+1} < \pi_j \quad \text{for } j \in \{0, \ldots, s-1\}.$$

The continued fraction expansion of $q/(q-p)$ having an odd number of terms,

$$\frac{q}{q-p} = a_1 + \overline{\lceil a_2} + \cdots + \overline{\lceil a_{2r-1}},$$

can be found in the following way: We define $q_0 = q$, $q_1 = q - p, q_2, \ldots, q_{2r-1}$ and $q_{2r} = 0$ by

$$q_{j-1} = a_j q_j + q_{j+1} \text{ for } j \in \{1, \ldots, 2r-1\} \text{ and } 0 < q_{j+1} < q_j \text{ for } j \in \{0, \ldots, 2r-3\},$$

and we have either $a_{2r-1} > 1$ or $a_{2r-1} = 1$ in which case we have $q_{2r-1} = q_{2r-2}$.
(3) Assume that $b_1 = 2$ and that $s \geq 2$.
(a) First, we consider the case where $b_1 = \cdots = b_s = 2$. Then we have $s = q - 1$ and $p = 1$ [cf. (3.8)], and we have $q \geq 3$ and

$$\frac{q}{q-1} = 1 + \overline{\lceil q-2} + \overline{\lceil 1},$$

hence we have $r = 2$, $a_1 = 1$, $a_2 = q - 2$, $a_3 = 1$.
(b) Now we consider the case where at least one of $b_2, \ldots, b_s$ is greater than 2, and we choose $u \in \{2, \ldots, s\}$ such that $b_1 = \cdots = b_{u-1} = 2$, $b_u > 2$. Then we have

$$\pi_{j-1} = 2\pi_j - \pi_{j+1} \quad \text{for every } j \in \{1, \ldots, u-1\},$$

hence $\pi_j = q - jp$ for every $j \in \{0, \ldots, u\}$, and

$$\pi_{u-1} = b_u \pi_u - \pi_{u+1}$$

where $0 \leq \pi_{u+1} = b_u \pi_u - \pi_{u-1} < \pi_u$. It is clear that

$$\frac{\pi_{u-1}}{\pi_u} = b_u - \overline{\lceil b_{u+1}} - \cdots - \overline{\lceil b_s}. \tag{∗}$$

Furthermore, since $p < q - p$ and $0 < q - up < p$, we have

$$q = (q - p) + p, \quad q - p = (u - 1)p + (q - up).$$

If $u = s$, then we have $\pi_{s-1} = b_s \pi_s$, hence we have $p = (b_s - 1)(q - sp)$, and therefore we obtain $r = 2$, $a_1 = 1$, $a_2 = s - 1$ and $a_3 = b_s - 1$. If $u < s$, then we have $\pi_{u-1} = b_u \pi_u - \pi_{u+1}$, hence we have $p = (b_u - 2)(q - up) + (\pi_u - \pi_{u+1})$, and therefore we obtain $r > 2$, $a_1 = 1$, $a_2 = u - 1$, $a_3 = b_u - 2$, and

$$\frac{\pi_{u-1}}{\pi_u} = a_3 + 1 + \overline{\left|a_4\right.} + \cdots + \overline{\left|a_{2r-1}\right.}. \qquad (**)$$

(4) Now we prove the assertion by induction with respect to s and b_1. If $s = 1$, then we have $r = v = 1$. Let $s \geq 2$, and assume that the assertion is true for all q, p such that $q/(q - p)$ has a Hirzebruch-Jung continued fraction expansion of length smaller than s. Let $q/(q - p)$ have a Hirzebruch-Jung continued fraction expansion of length s.

If $b_1 = \cdots = b_s = 2$, then we have $v = 2$, $i(1) = 1$, $i(2) = s = q - 1$, hence we get $\lambda_1 = q - 2$, and by (3)(b) we find that $r = 2$, $a_2 = q - 2$. Now we assume that $b_1 = 2$, and that at least one of $b_2, \ldots, b_s$ is larger than 2. We use the notation of (3)(b). Set $q' := \pi_{u-1}$ and $p' := \pi_u$, and let v' (resp. r') be defined with respect to q', p' in the same way as v (resp. r) was defined with respect to q, p. If $u = s$, then we have $v = 2$, $\lambda_1 = s - 1$, and we have $r = 2$, $a_2 = s - 1$ by (3)(b). If $u < s$, then we have $v' > 2$, $v' = v - 1$, and [cf. $(**)$] $r' = r - 1$. In this case $l_0, \ldots, l_u$ lie on a line and l_{u+1} does not lie on this line. Let $l'_0, \ldots, l'_{s-u+2}$ (resp. $\lambda'_1, \ldots, \lambda'_{v'}$) be defined as in (4.18) (resp. (5.12)) with respect to q', p'. Then we have $l'_0 = l_u$, $l'_1 = l_{u+1}, \ldots, l'_{s+u-2} = l_{s+1}$, and therefore we get $\lambda'_1 = \lambda_2, \ldots, \lambda'_{v'} = \lambda_v$. By our induction assumption, applied to q' and p', we get from $(*)$ and $(**)$ that $\lambda_2 = a_4, \ldots, \lambda_{v-1} = a_{2r}$. Furthermore, we have $\lambda_1 = i(2) - i(1) = u - 1$.
If $b_1 > 2$, then we have $q < 2p$, hence we get $a_1 > 1$, and we have

$$\frac{q}{q-p} - 1 = b_1 - 1 - \overline{\left|b_2\right.} - \cdots - \overline{\left|b_s\right.},$$
$$\frac{q}{q-p} - 1 = a_1 - 1 + \overline{\left|a_2\right.} + \cdots + \overline{\left|a_{2r-1}\right.}.$$

Using the result in (5.10)(2), and by induction, we see that the assertion is true in this case also.

(5.20) REMARK: The result in (5.19) can be interpreted in the following way: the continued fraction expansion in (5.19) allows us to determine the number $v - 1$ of singular points of $\mathrm{Bl}_{x_\sigma}(X_\sigma)$ and, moreover, to say that every singular point y_α of $\mathrm{Bl}_{x_\sigma}(X_\sigma)$ is a singularity of type $A_{a_{2\alpha}-1}$, $\alpha \in \{1, \ldots, v - 1\}$. We know that $\lambda_0 = \lambda_v = 1$.

Chapter VII

Two-Dimensional Regular Local Rings

INTRODUCTION: The theory of complete (= integrally closed) ideals of a two-dimensional regular local ring R was introduced by Zariski [cf. [191] and the appendix of [204]]. He used it in his paper [192] to resolve surface singularities. Similarly we shall use facts from this theory in section 7 of chapter VIII.

We usually do not assume that the residue field of R is infinite (under this additional assumption we could prove the fact that the product of complete ideals is complete in a much simpler way, cf., e.g., [98], Th. 3.7). We introduce the notion of ideal transform in section 1. In section 2 we define quadratic transforms of a two-dimensional regular local ring and study their properties. Complete ideals are introduced in section 3; one of the main results of the theory, the unique factorization of complete ideals, is proved in section 4. The predecessors of a simple ideal shall play a decisive role in the resolution process in section 7 of chapter VIII; they are studied in section 5. The connection between simple complete ideals of R and valuations of the second kind which dominate R is described in section 6; here we also prove the length formula of Hoskin-Deligne. Some results concerning proximity are proved in section 7. The result of (8.13) in section 8 is a local version of the theorem of resolution of singularities of curves embedded in a regular surface—and shall play a decisive role when proving this theorem in section 1 of chapter VIII.

1 Ideal Transform

1.1 Generalities

(1.0) Let $R \subset S$ be two factorial domains having the same field of quotients. We associate with an ideal $\mathfrak{a}$ of R an ideal $\mathfrak{a}^S$ of S, the transform of $\mathfrak{a}$ in S, and we study properties of this transform operation.

(1.1) Lemma: *Let R be an integral domain with field of quotients K, and let $\mathfrak{p}$ be a prime ideal of R such that the localization $R_\mathfrak{p}$ is a discrete valuation ring. Let S be a subring of K containing R, and set $\mathfrak{p}^S := \mathfrak{p}S_\mathfrak{p} \cap S$. The following statements are equivalent:*

(1) $\mathfrak{p}^S \neq S$.

(2) $\mathfrak{q} := \mathfrak{p}^S$ *is the unique prime ideal of S whose intersection with R is $\mathfrak{p}$, and we have $S_\mathfrak{q} = R_\mathfrak{p}$.*

(3) *There exists an ideal $\mathfrak{b}$ of S such that $\mathfrak{b} \cap R = \mathfrak{p}$.*

(3′) $\mathfrak{p}S \cap R = \mathfrak{p}$.

(4) $S_\mathfrak{p} \neq K$.

(5) $S_\mathfrak{p} = R_\mathfrak{p}$.

(5′) $S \subset R_\mathfrak{p}$.

Proof: Since $R_\mathfrak{p}$ is a discrete valuation ring, we have $\mathfrak{p} \neq \{0\}$; moreover, any subring of K properly containing $R_\mathfrak{p}$ is equal to K [cf. I(3.11)]. Set $M := R \setminus \mathfrak{p}$. Then we have $S_\mathfrak{p} = M^{-1}S$. From $R \subset S$ we get $R_\mathfrak{p} \subset S_\mathfrak{p}$, and this implies that either $S_\mathfrak{p} = K$ or $S_\mathfrak{p} = R_\mathfrak{p}$; for any ideal $\mathfrak{b}$ of S with $\mathfrak{b} \cap R = \mathfrak{p}$ we have $M \cap \mathfrak{b} = \emptyset$. The implications (2) $\Rightarrow$ (1), (2) $\Rightarrow$ (3), (2) $\Rightarrow$ (5′) and (4) $\Rightarrow$ (5) are clear. We have (1) $\Rightarrow$ (4) since (1) implies that $\mathfrak{p}S_\mathfrak{p} \neq S_\mathfrak{p}$, and we have (4) $\Rightarrow$ (3) since from $S_\mathfrak{p} \neq K$ we get $S_\mathfrak{p} = R_\mathfrak{p}$, hence $\mathfrak{p} = \mathfrak{p}S_\mathfrak{p} \cap R = (\mathfrak{p}S_\mathfrak{p} \cap S) \cap R = \mathfrak{b} \cap R$ with $\mathfrak{b} := \mathfrak{p}S_\mathfrak{p} \cap S$. Assume that (3) holds. Then $\mathfrak{p}$ is contracted from S, hence it is the contraction of its extension $\mathfrak{p}S$, and therefore we get (3) $\Rightarrow$ (3′). Since (3′) $\Rightarrow$ (3), we have (3) $\Leftrightarrow$ (3′). Since $R_\mathfrak{p} \subset S_\mathfrak{p}$, we have (5) $\Leftrightarrow$ (5′).
It remains to show that (5) $\Rightarrow$ (2). Assume that (5) holds; then we have $S \subset S_\mathfrak{p} = R_\mathfrak{p}$. Now $\mathfrak{q} := \mathfrak{p}S_\mathfrak{p} \cap S = \mathfrak{p}R_\mathfrak{p} \cap S$ is a prime ideal of S, and $\mathfrak{q} \cap R = \mathfrak{p}R_\mathfrak{p} \cap R = \mathfrak{p}$. Furthermore, since $M \subset S \setminus \mathfrak{q}$, we have $S_\mathfrak{p} \subset S_\mathfrak{q}$, and we have $S_\mathfrak{q} \neq K$ since $\mathfrak{q}$ is a prime ideal of S different from $\{0\}$, hence $R_\mathfrak{p} = S_\mathfrak{q}$. Let $\mathfrak{q}'$ be any prime ideal of S with $\mathfrak{q}' \cap R = \mathfrak{p}$; by the same argument we get $R_\mathfrak{p} = S_{\mathfrak{q}'}$, hence $S_\mathfrak{q} = S_{\mathfrak{q}'}$. This equality implies, as is easily checked, that $\mathfrak{q} = \mathfrak{q}'$.

The following result shall be needed later.

(1.2) Proposition: *Let R be a semilocal domain, and let $\mathfrak{m}_1, \ldots, \mathfrak{m}_h$ be the maximal ideals of R. If the rings $R_{\mathfrak{m}_i}$ are factorial for $i \in \{1, \ldots, h\}$, then R is also factorial.*

Proof: It is enough to show that every prime ideal of R of height one is principal [cf. [63], Prop. 3.11 and Th. 10.1]. Thus, let $\mathfrak{p}$ be a prime ideal of R of height 1. For $i \in \{1, \ldots, h\}$ we set $R_i := R_{\mathfrak{m}_i}$. Since R_i is factorial and $\mathfrak{p}R_i$ is either the unit ideal or a prime ideal of R_i of height 1, there exists an element $p_i \in \mathfrak{p}$ with $p_iR_i = \mathfrak{p}R_i$. For $i \in \{1, \ldots, h\}$ there exists an element $a_i \notin \mathfrak{m}_i$ with $a_i \in \mathfrak{m}_j$ for $j \in \{1, \ldots, h\}$ with $j \neq i$ [cf. [63], Lemma 3.3]; note that p_ia_i also is a generator of $\mathfrak{p}R_i$. We set $p := p_1a_1 + \cdots + p_ha_h$; then we have $p \in \mathfrak{p}$. Let $i \in \{1, \ldots, h\}$. Then we have $p_ja_j \in \mathfrak{p}\mathfrak{m}_iR_i$ for $j \in \{1, \ldots, h\}$, $j \neq i$, hence $\mathfrak{p}R_i = p_ia_iR_i \subset pR_i + \mathfrak{p}\mathfrak{m}_iR_i \subset \mathfrak{p}R_i$, and therefore $\mathfrak{p}R_i = pR_i + \mathfrak{p}\mathfrak{m}_iR_i$; by Nakayama's lemma [cf. [63], Cor. 4.8] we get $pR_i = \mathfrak{p}R_i$, and this implies that $\mathfrak{p} = Rp$ by B(2.5).

1.2   Ideal Transforms

(1.3) GREATEST COMMON DIVISOR: Let R be a factorial domain with field of quotients K.

(1) Let $\mathfrak{a}$ be a non-zero ideal of R. An element $x \in R$ is a greatest common divisor for the elements of $\mathfrak{a}$ iff the following two conditions are satisfied:

(i) We have $\mathfrak{a} \subset Rx$;

(ii) For every $y \in R$ with $\mathfrak{a} \subset Ry$ we have $Rx \subset Ry$.

Clearly every non-zero ideal of R admits a greatest common divisor of its elements. Let x be a greatest common divisor for the elements of $\mathfrak{a}$, and let $z \in R$ be a non-zero element. Then zx is a greatest common divisor for the elements of $z\mathfrak{a}$.

(2) Let $M \subset R$ be multiplicatively closed with $0 \notin M$, and set $S := M^{-1}R$; the ring S is factorial by I(3.30)(3). Let $\mathfrak{a}$ be a non-zero ideal of R, and let x be a greatest common divisor for the elements of $\mathfrak{a}$. Then x is also a greatest common divisor for the elements of the ideal $\mathfrak{a}S$ of S.

Proof: We have $\mathfrak{a}S \subset Sx$. Let $y \in S$ be an element with $\mathfrak{a}S \subset Sy$. We may assume that $y = p_1 \cdots p_h$ with irreducible elements $p_1, \ldots, p_h \in R$ which satisfy $p_i S \neq S$ for every $i \in \{1, \ldots, h\}$. Then we have [cf. I(3.30)(3)(b)]

$$\mathfrak{a} \subset (\mathfrak{a}S) \cap R \subset (Sy) \cap R = Ry,$$

hence we have $Rx \subset Ry$, and therefore we have $Sx \subset Sy$.

(3) Let $\mathfrak{a}$ be a non-zero ideal of R, and let $x \in R$ be a non-zero element. Then x is a greatest common divisor for the elements of $\mathfrak{a}$ iff

$$(R : \mathfrak{a})_K = Rx^{-1}.$$

In particular, 1 is a greatest divisor for the elements of $\mathfrak{a}$ iff $(R : \mathfrak{a})_K = R$.

Proof: Assume that x is a greatest common divisor for the elements of $\mathfrak{a}$. From $\mathfrak{a} \subset Rx$ we get $Rx^{-1} = (R : Rx)_K \subset (R : \mathfrak{a})_K$. Let $z \in (R : \mathfrak{a})_K$ be a non-zero element. Then we can write $z = z_1/z_2$ where $z_1, z_2 \in R$ are non-zero elements, and we have $\mathfrak{a}z_1 \subset Rz_2$. Now $z_1 x$ is a greatest common divisor for the elements of $\mathfrak{a}z_1$ [cf. (1)], hence $Rz_1 x \subset Rz_2$, and therefore $z = z_1/z_2 \in Rx^{-1}$, hence $(R : \mathfrak{a})_K \subset Rx^{-1}$. Thus, we have shown that $Rx^{-1} = (R : \mathfrak{a})_K$.

Conversely, assume that $(R : \mathfrak{a})_K = Rx^{-1}$. Then we have $x^{-1} \in (R : \mathfrak{a})_K$, hence $\mathfrak{a}x^{-1} \subset R$ and therefore we have $\mathfrak{a} \subset Rx$. Let $y \in R$ with $\mathfrak{a} \subset Ry$. Then we have $Ry^{-1} \subset (R : \mathfrak{a})_K = Rx^{-1}$, hence $Rx \subset Ry$, and therefore x is a greatest common divisor for the elements of $\mathfrak{a}$.

(4) Let $\mathfrak{a}$ be a non-zero ideal of R, and set

$$\mathfrak{b} := \mathfrak{a}(R : \mathfrak{a})_K.$$

Then $\mathfrak{b}$ is an ideal of R, and it is the unique ideal of R with the following properties:

(i) $(R : \mathfrak{b})_K = R$, i.e., 1 is a greatest common divisor for the elements of $\mathfrak{b}$;

(ii) $\mathfrak{a} = \mathfrak{b}z$ for some $z \in R$;

any element z as in (ii) is a greatest common divisor for the elements of $\mathfrak{a}$.

Proof: Let x be a greatest common divisor for the elements of $\mathfrak{a}$. Then we have $\mathfrak{a}(R:\mathfrak{a})_K = \mathfrak{a}x^{-1}$ [cf. (3)], hence we have

$$(R:\mathfrak{b})_K = (R:\mathfrak{a}x^{-1})_K = (R:\mathfrak{a})_K x = Rxx^{-1} = R,$$

and we have $\mathfrak{a} = \mathfrak{b}x$, hence $\mathfrak{b} = \mathfrak{a}x^{-1} \subset R$ [since $\mathfrak{a} \subset Rx$].

Let $\mathfrak{c}$ be a non-zero ideal of R with $(R:\mathfrak{c})_K = R$ and $\mathfrak{a} = \mathfrak{c}z$ for some $z \in R$. Then we have $(R:\mathfrak{a})_K = (R:\mathfrak{c})_K z^{-1} = Rz^{-1}$, hence we have $\mathfrak{a}(R:\mathfrak{a})_K = \mathfrak{a}z^{-1} = \mathfrak{c}$. Now we have proved that $\mathfrak{b}$ satisfies the assertions in (i) and (ii).

Let $z \in R$ be an element with $\mathfrak{a} = \mathfrak{b}z$. Then we have $(R:\mathfrak{a})_K = (R:\mathfrak{b})z^{-1} = Rz^{-1}$, and therefore z is a greatest common divisor for the elements of $\mathfrak{a}$ [cf. (3)].

(5)(a) Let $\mathfrak{a}$, $\mathfrak{b}$ be non-zero ideals of R such that 1 is a greatest common divisor for the elements of $\mathfrak{a}$ and of $\mathfrak{b}$. Then 1 is a greatest common divisor for the elements of $\mathfrak{ab}$.

In fact, we have $(R:\mathfrak{ab})_K = ((R:\mathfrak{a})_K : \mathfrak{b})_K = (R:\mathfrak{b})_K = R$; hence the assertion follows from (3).

(b) Let $\mathfrak{a}_1$, $\mathfrak{a}_2$ be non-zero ideals of R, and let, for $i = 1, 2$, x_i be a greatest common divisor for the elements of $\mathfrak{a}_i$. Then $x_1 x_2$ is a greatest common divisor for the elements of $\mathfrak{a}_1 \mathfrak{a}_2$.

In fact, for $i = 1, 2$ we have $\mathfrak{a}_i = \mathfrak{b}_i x_i$ where $\mathfrak{b}_i$ is an ideal having 1 as a greatest common divisor for its elements. Then 1 is a greatest common divisor for the elements of $\mathfrak{b}_1 \mathfrak{b}_2$ by (a), and from $\mathfrak{a}_1 \mathfrak{a}_2 = \mathfrak{b}_1 \mathfrak{b}_2 (x_1 x_2)$ we get the result by (4).

(6) Let $\mathfrak{a}$ be a non-zero ideal of R. By (4) we have a unique representation

$$\mathfrak{a} = \mathfrak{p}_1^{e_1} \cdots \mathfrak{p}_n^{e_n} \mathfrak{b} \tag{$*$}$$

where $n \in \mathbb{N}_0$, $\mathfrak{p}_1, \ldots, \mathfrak{p}_n$ are pairwise different principal prime ideals of R, $e_1, \ldots, e_n$ are natural integers and $\mathfrak{b}$ is a non-zero ideal of R with $(R:\mathfrak{b})_K = R$.

(1.4) DEFINITION: Let R be a factorial domain with field of quotients K, and let S be a factorial subring of K containing R. Let $\mathfrak{a}$ be a non-zero ideal of R, and write $\mathfrak{a} = \mathfrak{p}_1^{e_1} \cdots \mathfrak{p}_n^{e_n} \mathfrak{b}$ as in (1.3)(6)($*$). We define the transform of $\mathfrak{a}$ in S to be that ideal $\mathfrak{a}^S$ of S which is defined by

$$\mathfrak{a}^S := \mathfrak{q}_1^{e_1} \cdots \mathfrak{q}_n^{e_n} (\mathfrak{b}S)(S:\mathfrak{b}S)_K$$

where $\mathfrak{q}_i := \mathfrak{p}_i^S$ for every $i \in \{1, \ldots, n\}$ [cf. (1.1), and note that $R_{\mathfrak{p}_i}$ is a discrete valuation ring by I(3.30)(4)]. In particular, if $\mathfrak{a}$ is a non-zero principal prime ideal, then the ideal $\mathfrak{a}^S$ as defined here is the same as the ideal $\mathfrak{a}^S$ defined in (1.1). (Note that $\mathfrak{a}^R = \mathfrak{a}$.)

Some basic properties of the "transform" operation are given in the next proposition [for the notion of integral closure of an ideal cf. section 6 of appendix B].

(1.5) Proposition: Let R be a factorial domain with field of quotients K, and let $S \subset T$ be factorial subrings of K containing R. Let $\mathfrak{a}$, $\mathfrak{a}_1$, $\mathfrak{a}_2$ be non-zero ideals of R. Then:

(1) If $\mathfrak{a}$ is a principal prime ideal, then either $\mathfrak{a}S \cap R = \mathfrak{a}$, in which case $\mathfrak{a}^S$ is a principal prime ideal of S with $\mathfrak{a}^S \cap R = \mathfrak{a}$, or $\mathfrak{a}S \cap R \neq \mathfrak{a}$, in which case we have $\mathfrak{a}^S = S$.

(2) If $(R : \mathfrak{a})_K = R$, then $\mathfrak{a}^S = (\mathfrak{a}S)(S : \mathfrak{a}S)_K$, and therefore $(S : \mathfrak{a}^S)_K = S$ and $\mathfrak{a}S = z\mathfrak{a}^S$ where z is a greatest common divisor for the elements of $\mathfrak{a}S$.

(3) [Compatibility with products] $(\mathfrak{a}_1 \mathfrak{a}_2)^S = \mathfrak{a}_1^S \mathfrak{a}_2^S$.

(4) [Transitivity] $(\mathfrak{a}^S)^T = \mathfrak{a}^T$.

(5) If, in particular, S is a ring of fractions of R, then $\mathfrak{a}^S = \mathfrak{a}S$.

(6) If $\mathfrak{a}_2 \subset \mathfrak{a}_1 \subset \overline{\mathfrak{a}_2}$ (the integral closure of $\mathfrak{a}_2$ in R), then $\mathfrak{a}_2^S \subset \mathfrak{a}_1^S \subset \overline{(\mathfrak{a}_2^S)}$ (the integral closure of $\mathfrak{a}_2^S$ in S).

(7) We have $\overline{\mathfrak{a}}^S = \overline{\mathfrak{a}^S}$ [integral closure of $\mathfrak{a}^S$ in S].

Proof: (1) Let $\mathfrak{a} =: \mathfrak{p}$ be a non-zero principal prime ideal of R with $\mathfrak{p}S \cap R = \mathfrak{p}$. Then $\mathfrak{q} := \mathfrak{p}^S$ is a prime ideal of S with $\mathfrak{q} \cap R = \mathfrak{p}$, and we have $S_\mathfrak{q} = R_\mathfrak{p}$ [cf. (1.1)], hence $\mathfrak{q}$ is a prime ideal of height 1 of S, and therefore is a principal prime ideal [since S is factorial]. If $\mathfrak{p}S \cap R \neq \mathfrak{p}$, then we have $\mathfrak{p}^S = S$ by (1.1).

(2) If $(R : \mathfrak{a})_K = R$, then we have $\mathfrak{a}^S = (\mathfrak{a}S)(S : \mathfrak{a}S)_K$ by definition; the last assertions follow from (1.3)(2), applied to $\mathfrak{a}S$.

(3) Let $i \in \{1, 2\}$, and let $\mathfrak{a}_i = \mathfrak{b}_i x_i$ where x_i is a greatest common divisor for the elements of $\mathfrak{a}_i$ and $\mathfrak{b}_i := \mathfrak{a}_i(R : \mathfrak{a}_i)_K$; let $y_i \in S$ be a greatest common divisor for the elements of $\mathfrak{b}_i S$. Then we have $\mathfrak{a}_1 \mathfrak{a}_2 = \mathfrak{b} x_1 x_2$ with $\mathfrak{b} := \mathfrak{b}_1 \mathfrak{b}_2$. Since 1 is a greatest common divisor for the elements of $\mathfrak{b}_1$ and $\mathfrak{b}_2$, 1 is also a greatest common divisor for the elements of $\mathfrak{b}$ [cf. (1.3)(3)], hence $(R : \mathfrak{b})_K = 1$. Furthermore, we have [cf. (1.3)(1)] $(S : \mathfrak{b}_1 S)_K = S y_1^{-1}$ and $(S : \mathfrak{b}_2 S)_K = S y_2^{-1}$, and since $y_1 y_2$ is a greatest common divisor for the elements of $(\mathfrak{b}_1 S)(\mathfrak{b}_2 S) = \mathfrak{b}S$ [cf. (1.3)(2)], we have $(S : \mathfrak{b}S)_K = S y_1^{-1} y_2^{-1}$, and this yields $(S : \mathfrak{b}S)_K = (S : \mathfrak{b}_1 S)_K (S : \mathfrak{b}_2 S)_K$. Now it is clear that $\mathfrak{a}_1^S \mathfrak{a}_2^S = (\mathfrak{a}_1 \mathfrak{a}_2)^S$.

(4) Let $\mathfrak{p}$ be a non-zero principal prime ideal of R with $\mathfrak{p}S \cap R = \mathfrak{p}$. Then $\mathfrak{q} := \mathfrak{p}^S$ is the unique prime ideal of S with $\mathfrak{q} \cap R = \mathfrak{p}$ and with $S_\mathfrak{q} = R_\mathfrak{p}$. Assume that $\mathfrak{q}T \cap S = \mathfrak{q}$. Then $\mathfrak{q}^T$ is the unique prime ideal of T with $\mathfrak{q}^T \cap S = \mathfrak{q}$ and with $T_{\mathfrak{q}^T} = S_\mathfrak{q}$. Since $\mathfrak{q}^T \cap R = \mathfrak{q} \cap R = \mathfrak{p}$, it follows that $\mathfrak{p}^T = (\mathfrak{p}^S)^T$. It is easy to check that $\mathfrak{p}^T = (\mathfrak{p}^S)^T$ in the other two cases.

Let $\mathfrak{a}$ be a non-zero ideal of R with $(R : \mathfrak{a})_K = R$. We have $\mathfrak{a}^S = \mathfrak{a}S(S : \mathfrak{a}S)_K$ [cf. (2)], hence we have $(\mathfrak{a}^S)^T = (\mathfrak{a}^S T)(T : \mathfrak{a}^S T)_K$, again by (2). We write $\mathfrak{a}S = \mathfrak{a}^S z$ with some $z \in S$ [cf. (2)]. Then we have

$$(\mathfrak{a}^S)^T = (\mathfrak{a}^S T)(T : \mathfrak{a}^S T)_K = (\mathfrak{a}T)(T : z^{-1}\mathfrak{a}T)_K z^{-1} = (\mathfrak{a}T)(T : \mathfrak{a}T) = \mathfrak{a}^T.$$

Let $\mathfrak{a}$ be a non-zero ideal of R. We write $\mathfrak{a} = \mathfrak{p}_1^{e_1} \cdots \mathfrak{p}_n^{e_n} \mathfrak{b}$ as in (1.3)(6)(*). Then we have $\mathfrak{a}^S = \mathfrak{b}^S y$ where $Sy = \mathfrak{q}_1^{e_1} \cdots \mathfrak{q}_n^{e_n}$ with $\mathfrak{q}_i := \mathfrak{p}_i^S$. By what we have just shown, and by (2), we get $(\mathfrak{a}^S)^T = \mathfrak{a}^T$.

(5) Let $\mathfrak{p}$ be a non-zero principal prime ideal of R. Then $\mathfrak{p}S$ is either a prime ideal of S, in which case we have $\mathfrak{p}S \cap R = \mathfrak{p}$, or $\mathfrak{p}S = S$; in both cases we get $\mathfrak{p}^S = \mathfrak{p}S$. Let $\mathfrak{a}$ be an ideal of R with $(R : \mathfrak{a})_K = R$, i.e., 1 is a greatest common divisor for the elements of $\mathfrak{a}$. Then 1 is a greatest common divisor for the elements of $\mathfrak{a}S$ [cf.

I(3.30)(6)]. Therefore we have $(S : \mathfrak{a}S)_K = S$, hence $\mathfrak{a}^S = (\mathfrak{a}S)(S : \mathfrak{a}S)_K = \mathfrak{a}S$. Now the assertion follows from (3).

(6) Let $i \in \{1, 2\}$; we use the notation introduced at the beginning of the proof of (3). Let p be an irreducible element in R; then $R_{(p)}$ is a discrete valuation ring [cf. I(3.30)(2)]. We write $\mathfrak{a}_i = x_i \mathfrak{b}_i$ where x_i is a greatest common divisor for the elements of $\mathfrak{a}_i$ and $\mathfrak{b}_i := \mathfrak{a}_i(R : \mathfrak{a}_i)_K$. Then $x_i R_{(p)}$ is a greatest common divisor for the elements of $\mathfrak{a}_i R_{(p)}$ [cf. I(3.30)(6)], hence $x_i R_{(p)} = \mathfrak{a}_i R_{(p)}$. Since $R_{(p)}$ is a principal ideal ring, every ideal of $R_{(p)}$ is integrally closed. Every element of $\bar{\mathfrak{a}}_2$ is integral over $\mathfrak{a}_2$, hence $\bar{\mathfrak{a}}_2 R_{(p)}$ is integral over $\mathfrak{a}_2 R_{(p)}$, hence $\mathfrak{a}_2 R_{(p)} = \bar{\mathfrak{a}}_2 R_{(p)}$, and therefore we have $\mathfrak{a}_1 R_{(p)} = \mathfrak{a}_2 R_{(p)}$. This implies that $x_1 R_{(p)} = x_2 R_{(p)}$. Since this equality holds for every $p \in \mathbb{P}$, we get $Rx_1 = Rx_2 =: Rx$ for some $x \in R$ [cf. I(3.30)(4)(b)]. In particular, we have $(R : \mathfrak{a}_1)_K = Rx^{-1} = (R : \mathfrak{a}_2)_K$, and from $\mathfrak{a}_2 \subset \mathfrak{a}_1 \subset \bar{\mathfrak{a}}_2$ we get $\mathfrak{b}_2 \subset \mathfrak{b}_1 \subset \bar{\mathfrak{b}}_2$ with $\bar{\mathfrak{b}}_2 = x^{-1}\bar{\mathfrak{a}}_2$ [note that $\bar{\mathfrak{b}}_2$ is the integral closure of the ideal $x^{-1}\mathfrak{a}_2$]. The inclusion $\mathfrak{b}_2 \subset \mathfrak{b}_1 \subset \bar{\mathfrak{b}}_2$ implies that

$$\mathfrak{b}_2 S \subset \mathfrak{b}_1 S \subset \overline{\mathfrak{b}_2 S} \tag{*}$$

[since every element of $\mathfrak{b}_1$ is integral over $\mathfrak{b}_2$ and $\overline{\mathfrak{b}_2 S}$ is an ideal of S]. From (*) we get, applying the argument above to the ring S and the ideals $\mathfrak{b}_1 S$, $\mathfrak{b}_2 S$ of S, that $(S : \mathfrak{b}_1 S)_K = (S : \mathfrak{b}_2 S)_K$. Then we have, by (2),

$$\mathfrak{b}_2^S = (\mathfrak{b}_2 S)(S : \mathfrak{b}_2 S)_K \subset (\mathfrak{b}_1 S)(S : \mathfrak{b}_1 S)_K = \mathfrak{b}_1^S \subset \overline{\mathfrak{b}_2^S}$$

[with regard to the last inclusion, note that from $\mathfrak{b}_1 S \subset \overline{\mathfrak{b}_2 S}$ we see that every element of $(\mathfrak{b}_1 S)(S : \mathfrak{b}_1 S)_K$ is integral over $(\mathfrak{b}_2 S)(S : \mathfrak{b}_2 S)_K = \mathfrak{b}_2^S$]. By (3) we now have $\mathfrak{a}_2^S = (Rx)^S \mathfrak{b}_2^S \subset (Rx)^S \mathfrak{b}_1^S = \mathfrak{a}_1^S$ and $(Rx)^S \mathfrak{b}_1^S \subset (Rx)^S \overline{\mathfrak{b}_2^S} = \overline{(Rx)^S \mathfrak{b}_2^S} = \overline{(\mathfrak{b}_2)^S x} = \overline{\mathfrak{a}_2^S}$.

(7) This follows immediately from (6).

2 Quadratic Transforms and Ideal Transforms

2.1 Generalities

(2.0) In this subsection R is an n-dimensional regular local ring with $n \geq 2$, $\mathfrak{m}_R = \mathfrak{m}$ is its maximal ideal, and $\kappa_R = \kappa = R/\mathfrak{m}$ is its residue field. Note that R is a factorial domain [cf. [63], Th. 19.19]; let K be the field of quotients of R. Let $\{x_1, \ldots, x_n\}$ be a system of generators of $\mathfrak{m}$; then $(x_1, \ldots, x_n)$ is a regular sequence on R, and if we denote, as usual, by $\text{In}(z) \in \text{gr}_\mathfrak{m}(R)$ the leading form of $z \in R$, then we have $\text{gr}_\mathfrak{m}(R) = \kappa[\text{In}(x_1), \ldots, \text{In}(x_n)]$, and $\text{In}(x_1), \ldots, \text{In}(x_n)$ are algebraically independent over κ [cf. [21], Th. 11.22 or [63], Ex. 17.16]. Note that the ring R/Rx_1 is a regular local ring of dimension $n - 1$ [cf. [63], Cor. 10.15]. For a polynomial $f \in R[Z]$ (resp. a homogeneous polynomial $h \in R[X_1, \ldots, X_n]$) we denote by $\overline{f} \in \kappa[Z]$ (resp. $\overline{h} \in \kappa[X_1, \ldots, X_n]$) the polynomial (resp. the homogeneous polynomial) which we obtain by reducing the coefficients of f modulo $\mathfrak{m}$ (resp. the coefficients of h modulo $\mathfrak{m}$).

We denote by $\mathcal{M}(R) = \mathcal{M}$ the multiplicative monoid of non-zero ideals of R. For $\mathfrak{a}, \mathfrak{b} \in \mathcal{M}$ we say that $\mathfrak{a}$ divides $\mathfrak{b}$ if there exists $\mathfrak{c} \in \mathcal{M}$ with $\mathfrak{b} = \mathfrak{a}\mathfrak{c}$; this implies that $\mathfrak{b} \subset \mathfrak{a}$. Clearly $\mathfrak{a}$ divides $\mathfrak{b}$ iff $\mathfrak{b} = \mathfrak{a}(\mathfrak{b} : \mathfrak{a})$ [if we have $\mathfrak{b} = \mathfrak{a}\mathfrak{c}$ for some ideal $\mathfrak{c}$ of R, then $\mathfrak{c} \subset \mathfrak{b} : \mathfrak{a}$, hence $\mathfrak{b} = \mathfrak{a}\mathfrak{c} \subset \mathfrak{a}(\mathfrak{b} : \mathfrak{a}) \subset \mathfrak{b}$, and therefore $\mathfrak{b} = \mathfrak{a}(\mathfrak{b} : \mathfrak{a})$, and if $\mathfrak{b} = \mathfrak{a}(\mathfrak{b} : \mathfrak{a})$, then $\mathfrak{a}$ divides $\mathfrak{b}$].

(2.1) REMARK: In particular, if $n = 2$, then we have the following: Let $\mathfrak{a}$ be a non-zero ideal of R; we have $\mathfrak{a} = z\mathfrak{b}$ where $z \in R$ is a greatest common divisor for the elements of $\mathfrak{a}$ and $\mathfrak{b}$ is an ideal of R having 1 as a greatest common divisor for its elements [cf. (1.3)]. Since every prime ideal of R of height 1 is a principal ideal [since R is factorial], it follows that *a non-zero proper ideal of R is an $\mathfrak{m}$-primary ideal iff 1 is a greatest common divisor for its elements*.

(2.2) THE ORDER FUNCTION: For every nonzero $f \in R$ we define $\mathrm{ord}_R(f) = \mathrm{ord}(f) = s$ if $f \in \mathfrak{m}^s$, $f \notin \mathfrak{m}^{s+1}$ [note that the intersection of the powers of the maximal ideal $\mathfrak{m}$ is the zero ideal by Krull's intersection theorem, cf. [63], Cor. 5.4]. Since $\mathrm{gr}_\mathfrak{m}(R)$ is a domain, we have $\mathrm{ord}(fg) = \mathrm{ord}(f) + \mathrm{ord}(g)$ for all non-zero f, $g \in R$, and if $f + g \neq 0$, then we have $\mathrm{ord}(f + g) \geq \min(\{\mathrm{ord}(f), \mathrm{ord}(g)\})$. The canonical extension of ord to $K^\times$ [cf. I(3.20)] shall be denoted by $\nu_R = \nu$; ν is a discrete valuation of rank 1 of K, ν is non-negative on R and has center $\mathfrak{m}$ in R, hence the residue field κ of R can be considered as a subfield of the residue field of ν. The valuation ring of ν_R will be denoted by V_R.
Let $\mathfrak{a}$ be a non-zero ideal of R; then $\mathrm{ord}_R(\mathfrak{a}) = \mathrm{ord}(\mathfrak{a}) := \min\{\mathrm{ord}(a) \mid a \in \mathfrak{a}\}$ is called the order of $\mathfrak{a}$. Set $r := \mathrm{ord}(\mathfrak{a})$; note that $\mathfrak{a} \subset \mathfrak{m}^r$, but $\mathfrak{a} \not\subset \mathfrak{m}^{r+1}$. For non-zero ideals $\mathfrak{a}$, $\mathfrak{b}$ of R we have $\mathrm{ord}(\mathfrak{a}\mathfrak{b}) = \mathrm{ord}(\mathfrak{a}) + \mathrm{ord}(\mathfrak{b})$.
Clearly the powers $\mathfrak{m}^r$, $r \in \mathbb{N}$, are ν-ideals, and are therefore integrally closed in R [cf. B(6.16)(2)].

(2.3) **Proposition:** We set $A := R[x_2/x_1, \ldots, x_n/x_1] = R[\mathfrak{m}/x_1]$. Let, for $i \in \{2, \ldots, n\}$, τ_i be the ν-residue of x_i/x_1. Then:
(1) $\tau_2, \ldots, \tau_n$ are algebraically independent over κ, and the rational function field $\kappa(\tau_2, \ldots, \tau_n)$ is the residue field of ν. In particular, ν is of the second kind with respect to R, and we have $\mathrm{rank}(\nu) + \mathrm{tr.d}_\kappa(\nu) = \dim(R)$.
(2) ν is non-negative on A, has center $\mathfrak{m}A = Ax_1$ in A, Ax_1 is a prime ideal of height 1 of A, and A_{Ax_1} is the ring of the valuation ν.
(3) We have $A/\mathfrak{m}A = \kappa[\tau_2, \ldots, \tau_n]$, hence the prime ideals of A containing Ax_1 correspond uniquely to the prime ideals of the polynomial ring $\kappa[\tau_2, \ldots, \tau_n]$. For every maximal ideal $\mathfrak{q}$ of A with $x_1 \in \mathfrak{q}$ the local ring $A_\mathfrak{q}$ is regular and we have $\dim(A_\mathfrak{q}) = n$.
(4) The ring A is an n-dimensional integrally closed noetherian domain, and we have $\mathfrak{m}^m A \cap R = \mathfrak{m}^m$ for every $m \in \mathbb{N}$.

Proof: (a) Suppose that $Ax_1 = A$; then there exists $y \in A$ with $x_1 y = 1$. We choose a homogeneous polynomial $f \in R[X_2, \ldots, X_n]$ with $y = f(x_2/x_1, \ldots, x_n/x_1)$; let d be the degree of f. Then we have $x_1^d = x_1^{d+1}y = x_1 f(x_2, \ldots, x_n)$. This implies that $x_1^d \in \mathfrak{m}^{d+1}$, hence that $d = \nu(x_1^d) \geq \nu(\mathfrak{m}^{d+1}) = d + 1$ which is absurd. Therefore we have $\mathfrak{m}A = Ax_1$, $Ax_1 \cap R = \mathfrak{m}$, and ν is non-negative on A since $\nu(x_2/x_1) = \cdots = \nu(x_n/x_1) = 0$.

We show that $\tau_2, \ldots, \tau_n$ are algebraically independent over κ. Let $\overline{f} \in \kappa[Z_2, \ldots, Z_n]$ be a non-zero polynomial of degree d, and choose a polynomial $f \in R[Z_2, \ldots, Z_n]$ of degree d whose image in $\kappa[Z_2, \ldots, Z_n]$ is $\overline{f}$. We define the homogeneous polynomial $g \in R[X_1, \ldots, X_n]$ by $g(X_1, \ldots, X_n) = X_1^d f(X_2/X_1, \ldots, X_n/X_1)$; then we have $\nu(g(x_1, \ldots, x_n)) = d$ since not all coefficients of g lie in $\mathfrak{m}$ [cf. [21], Prop. 11.20 or [63], Ex. 17.16], hence the ν-residue of $g(x_1, \ldots, x_n)/x_1^d$ is not zero, which means that $\overline{f}(\tau_2, \ldots, \tau_n) \neq 0$.

We show that $\kappa(\tau_2, \ldots, \tau_n)$ is the residue field κ_ν of ν, hence that $\operatorname{tr.d}_\kappa(\nu) = n - 1$. In fact, every non-zero element in κ_ν is the image of a quotient u/v where u, $v \in R$ have the same ν-value r; there exist homogeneous polynomials g, $h \in R[X_1, \ldots, X_n]$ of degree r having not all their coefficients in $\mathfrak{m}$ such that $u = g(x_1, \ldots, x_n)$, $v = h(x_1, \ldots, x_n)$. We have $u/v = g(x_1, \ldots, x_n)/h(x_1, \ldots, x_n) = g(1, x_2/x_1, \ldots, x_n/x_1)/h(1, x_2/x_1, \ldots, x_n/x_1)$, and therefore the ν-image of u/v is $g(1, \tau_2, \ldots, \tau_n)/h(1, \tau_2, \ldots, \tau_n) \in \kappa(\tau_2, \ldots, \tau_n)$.

(b) By (a) we have $\mathfrak{m}A \cap R = \mathfrak{m}$. Let $\alpha: A \to \overline{A} := A/\mathfrak{m}A$ be the canonical homomorphism; then we have $\tau_i = \alpha(x_i/x_1)$ for $i \in \{2, \ldots, n\}$, $\alpha(R) = \kappa$ and $\overline{A} = \kappa[\tau_2, \ldots, \tau_n]$. In particular, $\overline{A}$ is a polynomial ring over a field in $n - 1$ indeterminates, hence we have $\dim(\overline{A}) = n - 1$ [cf. [63], Cor. 10.13].

(c) Let $f \in R[X_1, \ldots, X_n]$ be homogeneous of degree r; if $f(x_1, \ldots, x_n)/x_1^r \in A$ lies in the center of ν in A, then all the coefficients of f lie in $\mathfrak{m}$ since $\tau_2, \ldots, \tau_n$ are algebraically independent over κ, and therefore we have $f(x_1, \ldots, x_n)/x_1^r \in Ax_1$. Since $Ax_1 = \mathfrak{m}A$, it follows that Ax_1 is the center of ν in A, hence Ax_1 is a prime ideal of height 1 by Krull's principal ideal theorem [cf. [63], Th. 10.1], and A_{Ax_1} is contained in the ring of ν. Since A_{Ax_1} is a one-dimensional local domain whose maximal ideal is generated by one element, it is a discrete valuation ring [cf. I(3.29)], hence it is the ring of the discrete valuation ν. Moreover, since A is an R-algebra of finite type, and since regular local rings are universally catenary [cf. [63], Cor. 18.10], we have $\dim(A) = \operatorname{ht}(Ax_1) + \dim(A/Ax_1) = 1 + (n - 1) = n$.

(d) Let $\mathfrak{q}$ be a maximal ideal of A with $x_1 \in \mathfrak{q}$. Then $\overline{\mathfrak{q}} := \mathfrak{q}/Ax_1$ is a maximal ideal in $\overline{A}$, and $\overline{\mathfrak{q}}$ can be generated by $n - 1$ elements [cf. [63], Ex. 4.27 or [204], vol. II, Ch. VII, § 7, Th. 24], hence $\mathfrak{q}$ can be generated by n elements. Since A is an R-algebra of finite type, and since regular local rings are universally catenary, we have $n = \operatorname{ht}(\mathfrak{q}) + \dim(A/\mathfrak{q})$, hence $\dim(A_\mathfrak{q}) = n$, and therefore $A_\mathfrak{q}$ is an n-dimensional regular local ring.

(e) Since R is integrally closed and all the powers $\mathfrak{m}^m$ are integrally closed in R, the ring A is integrally closed [cf. B(6.9)].

(f) We have $\mathfrak{m}^m A = Ax_1^m$ for every $m \in \mathbb{N}$. Let $z \in Ax_1^m \cap R$. Then we have $\nu(z) \geq m$, hence we get $z \in \mathfrak{m}^m$ [since $\mathfrak{m}^m$ is a ν-ideal].

(2.4) REMARK: Note that now we have equality in (3) of Abhyankar's theorem, cf. I(11.9).

2.2   Quadratic Transforms and the First Neighborhood

(2.5) For the rest of this chapter R is a *two-dimensional regular local ring*.

(2.6) QUADRATIC TRANSFORMS: (1) We consider the canonical homomorphism of graded rings

$$\varphi \colon \mathcal{R}(\mathfrak{m}, R) = \bigoplus_{n \geq 0} \mathfrak{m}^n T^n \to \bigoplus_{n \geq 0} \mathfrak{m}^n / \mathfrak{m}^{n+1} = \mathrm{gr}_{\mathfrak{m}}(R)$$

from the Rees ring $\mathcal{R}(\mathfrak{m}, R) \subset R[T]$ to the associated graded ring $\mathrm{gr}_{\mathfrak{m}}(R)$. Note that $\mathrm{gr}_{\mathfrak{m}}(R) = \kappa[\overline{x}, \overline{y}]$ [here $\mathfrak{m} = Rx + Ry$, $\overline{x} := \mathrm{In}(x) = x \bmod \mathfrak{m}^2$, $\overline{y} := \mathrm{In}(y) = y \bmod \mathfrak{m}^2$]. The kernel of φ is the ideal

$$J := \bigoplus_{n \geq 0} \mathfrak{m}^{n+1} T^n;$$

it is a homogeneous prime ideal of $\mathcal{R}(\mathfrak{m}, R)$. Let $\mathbb{P}_R = \mathbb{P}$ be the set of homogeneous prime ideals of $\mathrm{gr}_{\mathfrak{m}}(R)$ of height 1, i.e., $\mathbb{P}$ is the set of closed points of $\mathrm{Proj}(\mathrm{gr}_{\mathfrak{m}}(R))$. Every $p \in \mathbb{P}$ is a principal ideal, generated by an irreducible homogeneous element $\overline{f} \in \kappa[\overline{x}, \overline{y}]$. We set $\deg(p) := \deg(\overline{f})$. Let $p = (\overline{f}) \in \mathbb{P}$ where $\overline{f} \in \mathrm{gr}_{\mathfrak{m}}(R)$ is homogeneous of degree m, and choose $f \in \mathfrak{m}^m$ with $\mathrm{In}(f) = \overline{f}$; then we have $\mathrm{ord}_R(f) = m$ and $\mathfrak{n}'_p := \varphi^{-1}(p) = J + f T^m \mathcal{R}(\mathfrak{m}, R)$. Now $\mathfrak{n}'_p$ is a closed point of $\mathrm{Proj}(\mathcal{R}(\mathfrak{m}, R))$, and the local ring of this point is $S_p := \mathcal{R}(\mathfrak{m}, R)_{(\mathfrak{n}'_p)}$. Every such ring is called a *quadratic transform* of R.

(2) Let $p = (\overline{f})$ and f be as in (1). Either $\overline{x}$ or $\overline{y}$ do not lie in p. We consider the case that $\overline{x}$ does not lie in p. Then we have $xT \notin \mathfrak{n}'_p$, and in $\mathcal{R}(\mathfrak{m}, R)_{(xT)} = R[\mathfrak{m}/x] =: A$ we have the following [cf. B(5.7)]: The prime ideal of A determined by J is the ideal Ax, the maximal ideal $\mathfrak{n}_p$ of A determined by $\mathfrak{n}'_p$ is the ideal generated by x and f/x^m [since, for $n \in \mathbb{N}$, the homogeneous elements of $\mathfrak{n}'_p$ of degree n are linear combinations of terms $x^i y^j T^n$, $i + j = n + 1$, and $x^i y^j f^l T^n$, $i + j + lm = n$, with coefficients in R], and we have $\mathcal{R}(\mathfrak{m}, R)_{(\mathfrak{n}'_p)} = A_{\mathfrak{n}_p}$ [cf. B(5.6)], and $S_p := A_{\mathfrak{n}_p}$ is a two-dimensional regular local ring [cf. (2.3)(3)] with field of quotients K, and its maximal ideal is generated by x and f/x^m. Moreover, $\mathfrak{m} S_p = S_p x$ is a prime ideal of S_p, and S_p dominate R. The residue field of S_p is $\kappa[T]/(\overline{f}(1, \overline{y}))$ [note that $\mathfrak{n}_p \cap R = \mathfrak{m}$], hence *the residue field of S_p is a finite extension of κ of degree* $\deg(p)$. We set $[S_p : R] := \deg(p)$. Furthermore, the valuation ring V_R of ν_R contains S_p and has center $S_p x$ in S_p [since V_R has center Ax in A by (2.3)(2)].

(3) Let, as in (2), $p = (\overline{f}) \in \mathbb{P}$, let $\overline{g} \in \mathrm{gr}_{\mathfrak{m}}(R)$ be homogeneous of degree $d > 0$ and assume that $\overline{g} \notin p$, or, equivalently, that $\overline{f}$ does not divide $\overline{g}$. We choose $g \in \mathfrak{m}^d$ with $\mathrm{In}(g) = \overline{g}$. Then $\mathfrak{n}'_p$ determines a prime ideal $\mathfrak{n}''_p$ in $\mathcal{R}(\mathfrak{m}, R)_{(gT^d)} = R[\mathfrak{m}^d/g]$ [cf. B(5.5) and B(5.6)], and we have $S_p = R[\mathfrak{m}^d/g]_{\mathfrak{n}''_p}$. We show: S_p consists of all elements z/w with z, $w \in R$, $\mathrm{ord}(z) \geq \mathrm{ord}(w)$ and $\mathrm{In}(w) \notin p$.
Proof: In fact, let z, w be as stated. Set $s := \mathrm{ord}(w)$; the condition $\overline{f} \nmid \mathrm{In}(w)$ is equivalent to $wT^s \notin \mathfrak{n}'_p$. Upon replacing, if necessary, w by w^n and z by

$w^{n-1}z$ for some $n \in \mathbb{N}$, we may assume that $\mathrm{ord}(w)$ is a multiple of d, say td. By this replacement the condition $wT^{\mathrm{ord}(w)} \notin \mathfrak{n}'_p$ is not affected. Now we have $z/w = (z/g^t)/(w/g^t)$, we have $z/g^t, w/g^t \in R[\mathfrak{m}^d/g]$, and, moreover, $w/g^t \notin \mathfrak{n}''_p$. This means that $z/w \in S_p$. On the other hand, every element of S_p can be written as $(z/g^t)/(w/g^t)$ where $z, w \in R$, $\mathrm{ord}(z) \geq td$, $\mathrm{ord}(w) = td$ for some $t \in \mathbb{N}$ and $wT^{td} \notin \mathfrak{n}'_p$, hence $\overline{f}$ does not divide the element $\mathrm{In}(w)$.

(2.7) DEFINITION: The set $\{S_p \mid p \in \mathbb{P}\}$ of quadratic transforms of R is called the first neighborhood of R, and shall be denoted by $N_1(R)$.

(2.8) REMARK: Let $p \in \mathbb{P}$ and $n \in \mathbb{N}$. Then we have $\mathfrak{m}^n S_p \cap R = \mathfrak{m}^n$.
In fact, let us choose $x \in \mathfrak{m} \setminus \mathfrak{m}^2$ with $(\mathrm{In}(x)) \neq p$. Then S_p is a localization of $A := R[\mathfrak{m}/x]$ with respect to a maximal ideal of A containing Ax, and therefore we have $Ax^n = S_p x^n \cap A$ [cf. [63], Th. 3.10, and note that $\mathrm{Ass}(A/Ax^n) = \{Ax\}$]. Now the result follows from (2.3)(4).

(2.9) INTERSECTION OF QUADRATIC TRANSFORMS: Let $p_1, \ldots, p_t \in \mathbb{P}$ be pairwise different, and set

$$S := S_{p_1} \cap \cdots \cap S_{p_t}.$$

By prime avoidance [cf. [63], Lemma 3.3] we can choose a homogeneous element $\overline{g} \in \mathrm{gr}_{\mathfrak{m}}(R)$ of positive degree with $\overline{g} \notin p_1 \cup \cdots \cup p_t$; we set $d := \deg(\overline{g})$, and we choose $g \in \mathfrak{m}^d$ with $\mathrm{In}(g) = \overline{g}$. We set

$$A := R[\mathfrak{m}^d/g] = \mathcal{R}(\mathfrak{m}, R)_{(gT^d)};$$

A is integrally closed by B(6.9). In A we have the maximal ideals $\mathfrak{n}''_{p_1}, \ldots, \mathfrak{n}''_{p_t}$ of height two determined by the prime ideals $\mathfrak{n}'_{p_1}, \ldots, \mathfrak{n}'_{p_t}$ of $\mathcal{R}(\mathfrak{m}, R)$ [note that, for $i \in \{1, \ldots, t\}$, we have $gT^d \notin \mathfrak{n}'_{p_i}$ and $S_{p_i} = A_{\mathfrak{n}''_{p_i}}$] and the prime ideal $\mathfrak{p}$ determined by J; clearly we have $\mathfrak{p} = \mathfrak{m}A$. Now we have

$$S = A_{\mathfrak{n}''_{p_1}} \cap \cdots \cap A_{\mathfrak{n}''_{p_t}}.$$

We set $\Sigma := A \setminus (\mathfrak{n}''_{p_1} \cup \cdots \cup \mathfrak{n}''_{p_t})$; then $\Sigma^{-1}A$ is a two-dimensional integrally closed semilocal ring with maximal ideals $\mathfrak{n}_1 := \Sigma^{-1}\mathfrak{n}''_{p_1}, \ldots, \mathfrak{n}_t := \Sigma^{-1}\mathfrak{n}''_{p_t}$. Since $\Sigma^{-1}A$ is the intersection of its localizations $(\Sigma^{-1}A)_{\mathfrak{n}_{p_i}} = A_{\mathfrak{n}''_{p_i}}$, $i \in \{1, \ldots, t\}$ [cf. B(2.5)], we have $S = \Sigma^{-1}A$ and

$$S_{\mathfrak{n}_i} = S_{p_i} \text{ for } i \in \{1, \ldots, t\}.$$

(There should be no confusion with regard to the different meanings of the subscripts $\mathfrak{n}_i$ resp. p_i of S.) By (2.6)(3) it is easy to check that S is the set of all elements u/v where $u, v \in R$, $\mathrm{ord}(u) \geq \mathrm{ord}(v)$, and $\mathrm{In}(v) \notin p_1 \cup \cdots \cup p_t$. The localizations $S_{\mathfrak{n}_1}, \ldots, S_{\mathfrak{n}_t}$ are factorial, hence S is also factorial [cf. (1.2)]. Moreover, since $\mathfrak{m}S = \mathfrak{p}S$, $\mathfrak{m}S$ is a prime ideal of S, and since $(\mathfrak{m}S)S_{\mathfrak{n}_i} = \mathfrak{m}S_{\mathfrak{n}_i}$ is a principal prime ideal of $S_{\mathfrak{n}_i} = S_{p_i}$ for $i \in \{1, \ldots, t\}$, the ideal $\mathfrak{m}S$ is a prime ideal of S of height 1, hence it is a principal ideal $\mathfrak{m}S = Sz$ for some $z \in S$.

Thus, we have proved:

(2.10) Proposition: *A finite intersection S of quadratic transforms of R is a two-dimensional factorial semilocal domain; $\mathfrak{m}S$ is a principal prime ideal of S.*

2.3   Ideal Transforms

(2.11) IDEAL TRANSFORM: (1) Let $p \in \mathbb{P}$ and set $S := S_p$; let $\mathfrak{n}$ be the maximal ideal of S. Since R and S are factorial rings, for every ideal $\mathfrak{a}$ of R the ideal transform $\mathfrak{a}^S$ [cf. (1.4)] of $\mathfrak{a}$ in S is defined. In the following, we give an explicit description of $\mathfrak{a}^S$.

(2) Let $\{x, y\}$ be a regular system of parameters of R with $\mathrm{In}(x) \notin p$. Set $A := R[\mathfrak{m}/x]$; then we have $S = A_{\mathfrak{n}_p}$ and $\mathfrak{m}S = Sx$ [cf. (2.6)(2)]. Let $\mathfrak{q}$ be a prime ideal of height 1 of A different from Ax. Then $x \notin \mathfrak{q}$, hence we have $A_{\mathfrak{q}} = (A_x)_{\mathfrak{q}A_x}$. Furthermore, we have $A_x = R_x$ [since $A \subset R_x$], and $(R_x)_{\mathfrak{q}R_x} = R_{\mathfrak{q}\cap R}$, hence we have $A_{\mathfrak{q}} = R_{\mathfrak{q}\cap R}$, and therefore $\mathfrak{q} \cap R$ is a prime ideal of height 1 of R.

Let $\mathfrak{a}$ be an $\mathfrak{m}$-primary ideal of R of order n; note that 1 is a greatest common divisor for the elements of $\mathfrak{a}$ [cf. (2.1)]. The only prime ideal of A of height 1 which contains $\mathfrak{a}A$ is Ax since $\mathfrak{m}^h \subset \mathfrak{a}$ for some $h \in \mathbb{N}$, hence $x^h \in \mathfrak{a}A$, and we have $\mathfrak{a}A \subset Ax^n$, $\mathfrak{a}A \not\subset Ax^{n+1}$ [since $Ax^m \cap R = \mathfrak{m}^m$ for every $m \in \mathbb{N}$ by (2.3)(4)]. In particular, no height one prime ideal of A contains the ideal $x^{-n}\mathfrak{a}A$ of A. Furthermore, this implies that Sx is the only prime ideal of S of height 1 which contains $\mathfrak{a}S$, $\mathfrak{a}S \subset Sx^n$, $\mathfrak{a}S \not\subset Sx^{n+1}$ and x^n is a greatest common divisor for the elements of $\mathfrak{a}S$. Therefore we have $(S : \mathfrak{a}S)_K = Sx^{-n}$, hence $\mathfrak{a}^S = S \cdot \mathfrak{a}/x^n$. Moreover, $\mathfrak{a}^S$ is either the unit ideal or it is an $\mathfrak{n}$-primary ideal of S [cf. (2.1)].

Now let $h \in R$ be irreducible; note that R_{Rh} is a discrete valuation ring. Then we have either $Sh \cap R = Rh$ in which case $(Rh)^S$ is a principal prime ideal of S, or we have $Rh \cap S \neq Rh$ in which case $(Rh)^S = S$ [cf. (1.5)]. Set $n := \mathrm{ord}(h)$. If $h = ux$ with a unit u of R, then we have $n = 1$ and $Sh \cap R = Sx \cap R = \mathfrak{m} \neq Rh$, hence $S = (Rh)^S = S \cdot h/x$. Now we consider the case that x and h are not associated in R. If $\mathrm{In}(h) \notin p$, then $x^n/h \in S$, hence h/x^n is a unit of S [cf. (2.6)(3)]. Therefore we have $Sh = Sx^n = \mathfrak{m}^n S$, hence $Sh \cap R = Ax^n \cap R = \mathfrak{m}^n \neq Rh$, and therefore we have $S = (Rh)^S = S \cdot h/x^n$. If $\mathrm{In}(h) \in p$, then we have $(Sh) \cap R = Rh$. In fact, let $u, v \in R$ with $\mathrm{ord}(u) \geq \mathrm{ord}(v)$ and $\mathrm{In}(v) \notin p$, and $(u/v)h =: z \in R$. From $hu = vz$ and $h \nmid v$ [since $\mathrm{In}(v) \notin p$ and $\mathrm{In}(h) \in p$] we get that $h \mid z$ and therefore that $u/v \in R$. In this case we have $S_{Rh} = R_{Rh}$ [cf. I(3.30)]. Furthermore, we have $R_{Rh}h = R_{Rh} \cdot h/x^n = S_{Rh} \cdot h/x^n$. Now h/x^n is irreducible in S_{Rh} and $h/x^n \in S$. We show that h/x^n is also irreducible in S. Suppose, on the contrary, that h/x^n is not irreducible in S. Then we have $h/x^n = (u/v)(u'/v')$ where u, v, u' and $v' \in R$, $\deg(u) = \deg(v)$, $\deg(u') = \deg(v')$, $\mathrm{In}(v)$ and $\mathrm{In}(v')$ lie not in p, and u/v, u'/v' are non-units of S. Since $h \in \mathrm{In}(p)$, we see that h divides neither v nor v'. From $hvv' = x^n uu'$ it follows therefore that h divides exactly one of the elements u, u', say h divides u, hence $u = hu''$ with $u'' \in R$. Then we have $vv' = x^n u'u''$, and therefore $\mathrm{In}(u') \notin p$, hence u'/v' is a unit of S, in contradiction with our

assumption. Hence h/x^n is irreducible in S, and therefore $S \cdot h/x^n$ is a prime ideal of S. Thus, we have $S \cdot h/x^n = S_{Rh} \cdot h/x^n \cap S$, i.e., we have $(Rh)^S = S \cdot h/x^n$, and we have shown: If $h \in R$ is irreducible and $\operatorname{ord}(h) = n$, then $(Rh)^S = S \cdot h/x^n$, and h/x^n is either irreducible in S—and this is the case if $\operatorname{In}(h) \in p$—or it is a unit of S—and this is the case if $\operatorname{In}(h) \notin p$.
(3) Let $f, g \in R$ be irreducible and not associated, and assume that $(Rf)^S$ and $(Rg)^S$ are prime ideals of S. Then we have $(Rf)^S \neq (Rg)^S$ and $Sx \neq (Rf)^S$.
Proof: Suppose that $(Rf)^S = (Rg)^S$. Set $m := \operatorname{ord}(f)$, $n := \operatorname{ord}(g)$, and assume, as we may, that $m \geq n$. There exist $u, v \in R$ with $\operatorname{In}(u) \notin p$, $\operatorname{In}(v) \notin p$, $\deg(u) = \deg(v)$, and $f/x^m = (g/x^n)(u/v)$. Then we have $fv = gux^{m-n}$. Since f and g are irreducible and not associated, and x divides neither f nor g, we see that f and u, as well as g and v are associated, contradicting $\operatorname{In}(f) \in p$ and $\operatorname{In}(g) \in p$ [cf. (2)]. Similarly, one can prove the second assertion in (3).

(2.12) RESULTS ON IDEAL TRANSFORMS: We apply the results of (2.11) and get:
(1) Since ideal transforms are compatible with products [cf. (1.5)], we have the following:
(a) Let $f \in m$, $f \neq 0$, and set $n := \operatorname{ord}(f)$. Let $p \in \mathbb{P}$ and choose $x \in m \setminus m^2$ with $\operatorname{In}(x) \notin p$; then we have $(Rf)^{S_p} = S_p \cdot f/x^n$. Moreover, we have shown: Let

$$\operatorname{gr}_m(R)\operatorname{In}(f) = \prod_{q \in \mathbb{P}} q^{n_q(f)}$$

be the factorization of the homogeneous principal ideal $\operatorname{gr}_m(R)\operatorname{In}(f)$ of $\operatorname{gr}_m(R)$ into a product of homogeneous principal prime ideals of $\operatorname{gr}_m(R)$. Then we have $(Rf)^{S_p} \neq S_p$ iff $n_p(f) > 0$.
(b) Let a be an ideal of R of order n. Let $p \in \mathbb{P}$ and choose $x \in m \setminus m^2$ with $\operatorname{In}(x) \notin p$. Then we have $a^{S_p} = S_p \cdot a/x^n$ [since every non-principal proper ideal of R has the form gb with $g \in R$ and an m-primary ideal b of R, cf. (1.3) and (2.1)].
(2) Let $p_1, \ldots, p_t \in \mathbb{P}$, and let S be the intersection of the rings $S_{p_1}, \ldots, S_{p_t}$. Then S is factorial, and there exists $z \in S$ with $mS = Sz$ [cf. (2.10)]. Let a be an ideal of R of order n. Then we have

$$a^S = S \cdot a/z^n. \tag{$*$}$$

In fact, let $\{x, y\}$ be a system of generators of m and let $i \in \{1, \ldots, t\}$. Then we have either $\operatorname{In}(x) \notin p_i$—in which case we have $(mS)S_{p_i} = S_{p_i}x$, hence $S_{p_i}z = S_ix$—or $\operatorname{In}(y) \notin p_i$—in which case we have $(mS)S_{p_i} = S_{p_i}y$, hence $S_{p_i}z = S_{p_i}y$. By (1) we have $a^{S_{p_i}} = S_{p_i} \cdot a/z^n$ for every $i \in \{1, \ldots, t\}$, and since $a^{S_{p_i}} = a^S S_{p_i}$ for every $i \in \{1, \ldots, t\}$ [cf. (1.5)] and $a/z^n \subset a^S$, we get $(*)$ [cf. B(2.5)].

(2.13) CHARACTERISTIC IDEAL OF AN IDEAL: Let a be an non-zero ideal of R, and set $r := \operatorname{ord}(a)$. The homogeneous ideal $c(a)$ of $\operatorname{gr}_m(R)$ which is generated by all the elements $\operatorname{In}(f) = f \bmod m^{r+1}$ where $f \in a$ is of order r, is a principal ideal; it is called the characteristic ideal of a. We have $c(a) = (\bar{g})$ for some homogeneous

$\overline{g} \in \mathrm{gr}_{\mathfrak{m}}(R)$, and we set $\deg(c(\mathfrak{a})) =: \deg(\overline{g})$. We have $0 \leq \deg(c(\mathfrak{a})) \leq r$. Every generator of $c(\mathfrak{a})$ is called a characteristic form of $\mathfrak{a}$. Note that $c(\mathfrak{a}) = (1)$ if $\mathfrak{a}$ is a power of $\mathfrak{m}$.

If $\mathfrak{b}$ is another non-zero ideal of R, then we have $c(\mathfrak{ab}) = c(\mathfrak{a})c(\mathfrak{b})$, and if $\mathfrak{a} \subset \mathfrak{b}$ and $\mathrm{ord}(\mathfrak{a}) = \mathrm{ord}(\mathfrak{b})$, then $c(\mathfrak{b})$ divides $c(\mathfrak{a})$, and therefore $\deg(c(\mathfrak{b})) \leq \deg(c(\mathfrak{a}))$. In particular, if $\mathfrak{a} = Rf$ is a non-zero principal ideal of R, then we have $c(\mathfrak{a}) = (\mathrm{In}(f))$.

(2.14) Proposition: *Let $p \in \mathbb{P}$, and let $S := S_p$ be the quadratic transform of R defined by p. Let $\mathfrak{a}$ be a non-zero ideal of R, and let p^h be the highest power of p which divides $c(\mathfrak{a})$. Then we have $s := \mathrm{ord}_S(\mathfrak{a}^S) \leq h$, and if $h > 0$, then we have $s > 0$. In particular, we have $\mathfrak{a}^S = S$ iff $p \nmid c(\mathfrak{a})$.*

Proof: We set $\mathfrak{b} := \mathfrak{a}^S$ and $r := \mathrm{ord}(\mathfrak{a})$. Let $x \in \mathfrak{m} \setminus \mathfrak{m}^2$ be such that $\mathrm{In}(x) \notin p$. Then we have $\mathfrak{b} = S \cdot \mathfrak{a} x^{-r}$ [cf. (2.12)]. Let $p = (\overline{g})$, set $n := \deg(\overline{g})$, and choose $g \in \mathfrak{m}^n$ with $\mathrm{In}(g) = \overline{g}$. There exists an element $z \in \mathfrak{a}$ of order r such that $\mathrm{In}(z) \in c(\mathfrak{a}) \subset p^h$ but that $\mathrm{In}(z) \notin p^{h+1}$. Then we have $\mathrm{In}(z) = \overline{g}^h \overline{w}$ where $\overline{w} \in \mathrm{gr}_{\mathfrak{m}}(R)$ is homogeneous of degree $r - hn$ and $\overline{g} \nmid \overline{w}$. We choose $w \in \mathfrak{m}^{r-hn}$ with $\mathrm{In}(w) = \overline{w}$; then we have $z - g^h w \in \mathfrak{m}^{r+1}$, hence in S we have $z/x^r - (g/x^n)^h \cdot (w/x^{r-hn}) \in Sx$. From $\overline{g} \nmid \overline{w}$ we get that w/x^{r-hn} is a unit of S [cf. (2.6)(3)]. Since the maximal ideal $\mathfrak{n}$ of S is generated by x and g/x^n, we see that $z/x^r \notin \mathfrak{n}^{h+1}$, and since $z/x^r \in \mathfrak{b}$, we have $\mathfrak{b} \not\subset \mathfrak{n}^{h+1}$, hence $s \leq h$, and therefore $\mathfrak{b} = S$ if $h = 0$. Now we assume that $h > 0$. If we apply the argument just given to any element $z \in \mathfrak{a}$ of order $\geq r$, then we get $z/x^r \in Sx$, and this means that $s > 0$. Therefore we have $\mathfrak{b} \neq S$ iff p divides $c(\mathfrak{a})$.

(2.15) Corollary: *Let $\mathfrak{a}$ be a non-zero ideal of R, and let*

$$c(\mathfrak{a}) = \prod_{q \in \mathbb{P}} q^{n_q(\mathfrak{a})}$$

be the factorization of the ideal $c(\mathfrak{a})$. For $p \in \mathbb{P}$ we have $\mathfrak{a}^{S_p} \neq S_p$ iff $n_p(\mathfrak{a}) > 0$.

(2.16) Lemma: *Let $x \in \mathfrak{m} \setminus \mathfrak{m}^2$, and let $\mathfrak{n}_1, \ldots, \mathfrak{n}_h$ be pairwise different maximal ideals of $A := R[\mathfrak{m}/x]$ containing x. For $i \in \{1, \ldots, h\}$ let $\mathfrak{q}_i$ be an $\mathfrak{n}_i$-primary ideal of A. For every $s \in \mathbb{N}$ we have $\mathrm{Ass}(A/x^s \mathfrak{q}_1 \cdots \mathfrak{q}_h) = \{Ax, \mathfrak{n}_1, \ldots, \mathfrak{n}_h\}$.*

Proof: We set $\mathfrak{q} := \mathfrak{q}_1 \cdots \mathfrak{q}_h = \mathfrak{q}_1 \cap \cdots \cap \mathfrak{q}_h$ [note that the ideals $\mathfrak{q}_1, \ldots, \mathfrak{q}_h$ are pairwise comaximal by B(10.1)]; then we have $\{\mathfrak{n}_1, \ldots, \mathfrak{n}_h\} = \mathrm{Ass}(A/\mathfrak{q})$. Now we have $Ax \in \mathrm{Ass}(A/x^s \mathfrak{q})$ since Ax is a minimal prime overideal of $x^s \mathfrak{q}$ [cf. [63], Th. 3.1]. Let $i \in \{1, \ldots, h\}$; since $\mathfrak{n}_i \in \mathrm{Ass}(A/\mathfrak{q})$, there exists $z_i \in A$ with $\mathfrak{q} : A z_i = \mathfrak{n}_i$, hence with $x^s \mathfrak{q} : Ax^s z_i = \mathfrak{n}_i$, and therefore we have $\mathfrak{n}_i \in \mathrm{Ass}(A/x^s \mathfrak{q})$. No height one prime ideal of A different from Ax contains $x^s \mathfrak{q}$. Since A is two-dimensional [cf. (2.3)(4)], it is enough to show that no maximal ideal of A different from the ideals $\mathfrak{n}_1, \ldots, \mathfrak{n}_h$ lies in $\mathrm{Ass}(A/x^s \mathfrak{q})$. Let $\mathfrak{n} \notin \{\mathfrak{n}_1, \ldots, \mathfrak{n}_h\}$ be a maximal ideal of A which contains $x^s \mathfrak{q}$. Since $\mathfrak{q} \not\subset \mathfrak{n}$, we have $x \in \mathfrak{n}$, and therefore we have

$n = Ax + A(g/x^n)$ where $g \in \mathfrak{m}^n$ for some $n \in \mathbb{N}$ and $g/x^n \notin \mathfrak{n}_i$ for $i \in \{1, \ldots, h\}$ [cf. (2.6)(2)]. We show that $x^s \mathfrak{q} : \mathfrak{n} = x^s \mathfrak{q}$ which implies that $\mathfrak{n} \notin \mathrm{Ass}(A/x^s \mathfrak{q})$. Let $z \in x^s \mathfrak{q} : \mathfrak{n}$. Then we have $z \cdot (g/x^n) \in x^s \mathfrak{q} \subset Ax^s$. Since $g/x^n \notin Ax$, we have $z \in Ax^s$, hence $z = x^s w$ with $w \in A$, and $w \cdot (g/x^n) \in \mathfrak{q}$. Since $g/x^n \notin \mathfrak{n}_i$ for $i \in \{1, \ldots, h\}$, we have $w \in \mathfrak{q}_i$ for $i \in \{1, \ldots, h\}$, hence $w \in \mathfrak{q}$, and therefore we have $z \in x^s \mathfrak{q}$.

(2.17) TRANSFORM OF IDEALS: Let $x \in \mathfrak{m} \setminus \mathfrak{m}^2$, and set $A := R[\mathfrak{m}/x]$. Let $\mathfrak{a}$ be a non-zero ideal of R, and set $r := \mathrm{ord}(\mathfrak{a})$. We have $\mathfrak{a}A \subset \mathfrak{m}^r A = Ax^r$; we set $\mathfrak{a}' := x^{-r} \mathfrak{a}A$, and we call $\mathfrak{a}'$ the *transform of $\mathfrak{a}$ in A*. In the following, we study properties of $\mathfrak{a}'$.
(1) Let $p \in \mathbb{P}$, and assume that $\mathrm{In}(x) \notin p$; then we have $S_p = A_{\mathfrak{n}_p}$ and $\mathfrak{a}^{S_p} = \mathfrak{a}' S_p$ [cf. (2.12)(1)(b)].
(2) In particular, assume that $\mathfrak{a}$ is an $\mathfrak{m}$-primary ideal. We have $\mathfrak{a}A = x^r \mathfrak{a}'$. There exists $h \in \mathbb{N}$ with $h > r$ and with $\mathfrak{m}^h \subset \mathfrak{a}$, hence every prime ideal of A which contains $\mathfrak{a}'$, also contains the principal prime ideal Ax. No height 1 prime ideal of A contains $\mathfrak{a}'$ [cf. (2.11)(2)]. Let us consider the case that $\mathfrak{a}'$ is a proper ideal of A; then $\mathrm{Ass}(A/\mathfrak{a}')$ consists of those maximal ideals of A which contain $\mathfrak{a}'$, each of them contains x and therefore $\mathrm{Ass}(A/\mathfrak{a}') = \{\mathfrak{n}_{p_1}, \ldots, \mathfrak{n}_{p_h}\}$ where $p_1, \ldots, p_h \in \mathbb{P}$ are the different prime ideal factors of $c(\mathfrak{a})$ different from $(\mathrm{In}(x))$ [cf. (2.15) and (2.6)(2)]. Let $\mathfrak{a}' = \mathfrak{q}_1 \cap \cdots \cap \mathfrak{q}_h = \mathfrak{q}_1 \cdots \mathfrak{q}_h$ where $\mathfrak{q}_i$ is an $\mathfrak{n}_{p_i}$-primary ideal for $i \in \{1, \ldots, h\}$, be the primary decomposition of $\mathfrak{a}'$. Then [cf. B(10.1) and [63], Cor. 9.1] $A/\mathfrak{a}'$ is artinian, and

$$A/\mathfrak{a}' \cong A/\mathfrak{q}_1 \times \cdots \times A/\mathfrak{q}_h.$$

We set $\Sigma := A \setminus (\mathfrak{n}_{p_1} \cup \cdots \cup \mathfrak{n}_{p_h})$, $S := \Sigma^{-1} A$ and $S_i := S_{p_i}$ for $i \in \{1, \ldots, h\}$; then we have $S = S_1 \cap \cdots \cap S_h$ [cf. (2.9)], $\mathfrak{a}'S \cap A = \mathfrak{a}'$ and $\mathfrak{a}S \cap A = \mathfrak{a}A$ [cf. (2.16) and [63], Th. 3.10]. Since $\mathfrak{m}S = Sx$, we find that $\mathfrak{a}'S = \mathfrak{a}^S$ and that $\mathfrak{a}'S_i = \mathfrak{a}^{S_i}$ for $i \in \{1, \ldots, h\}$ [cf. (1.5)]. Since S is semilocal, we have $\mathfrak{a}'S_1 \cap \cdots \cap \mathfrak{a}'S_h = \mathfrak{a}'S$ [cf. B(2.5)]. Therefore we have $(\mathfrak{a}'S_1 \cap \cdots \cap \mathfrak{a}'S_h) \cap A = \mathfrak{a}'$, and a similar result holds with $\mathfrak{a}'$ replaced by $x^r \mathfrak{a}' = \mathfrak{a}A$.
The image of Σ in $A/\mathfrak{a}'$ is the set of units of $A/\mathfrak{a}'$; therefore we have $A/\mathfrak{a}' \cong S/\mathfrak{a}^S$. Let $i \in \{1, \ldots, h\}$. Since $x^r \mathfrak{a}'S_i = \mathfrak{a}S_i = x^r \mathfrak{a}^{S_i}$, we have $\mathfrak{a}^{S_i} = \mathfrak{a}'S_i = \mathfrak{q}_i S_i$, and this implies that $A/\mathfrak{q}_i \cong S_i/\mathfrak{a}^{S_i}$. In particular, we have

$$S/\mathfrak{a}^S \cong S_1/\mathfrak{a}^{S_1} \times \cdots \times S_h/\mathfrak{a}^{S_h}.$$

(3) Now let $\mathfrak{a} = Rf$ be a non-zero ideal in R, set $r := \mathrm{ord}(f)$, and let $p_1, \ldots, p_h$ be the pairwise different factors of $c(\mathfrak{a}) = (\mathrm{In}(f))$ which are different from $(\mathrm{In}(x))$. The only maximal ideals of A of the form $\mathfrak{n}_p$ for some $p \in \mathbb{P}$, $p \neq (\mathrm{In}(x))$, with $f/x^r \in \mathfrak{n}_p$ are the ideals $\mathfrak{n}_{p_1}, \ldots, \mathfrak{n}_{p_h}$.

2.4 Valuations Dominating R

(2.18) DIRECTIONAL IDEAL OF A VALUATION: Let v be a valuation of K dominating R with valuation ring V; note that $V \neq K$. Then κ can be considered as a subfield of the residue field κ_v of V, and $\mathfrak{m}$ is a v-ideal with $v(\mathfrak{m}) > 0$. Note that, by Abhyankar's theorem [cf. I(11.9)], we have

$$\text{rat.}\,\text{rank}(v) + \text{tr.}\,\text{d}_\kappa(\kappa_v) \leq \dim(R) = 2,$$

hence we have either $\text{tr.}\,\text{d}_\kappa(\kappa_v) = 1$—in this case V is of the second kind with respect to R, κ_v is a finitely generated extension of κ, $\text{rank}(v) = \text{rat.}\,\text{rank}(v) = 1$, v is discrete of rank 1 [cf. I(11.9)(3)], or we have $\text{tr.}\,\text{d}_\kappa(\kappa_v) = 0$—in this case V is of the first kind with respect to R. If, in particular, the value group of v is archimedean, then every non-zero proper v-ideal of R is an $\mathfrak{m}$-primary ideal [cf. I(11.7)].

For the rest of this subsection we assume that $V \neq V_R$.

We choose x, $y \in \mathfrak{m}$ with $\mathfrak{m} = Rx + Ry$, and we set $\overline{x} := \text{In}(x)$, $\overline{y} := \text{In}(y)$. We may assume that $v(x) \leq v(y)$, hence that $v(\mathfrak{m}) = v(x)$. Clearly we have $v(z) \geq \nu_R(z)v(\mathfrak{m})$ for every $z \in R$, $z \neq 0$. Since $V \neq V_R$, there exists $z \in R \setminus \{0\}$ with $v(z) > \nu_R(z)v(\mathfrak{m})$. We choose $f \in R[X,Y]$ with $z = f(x,y)$. Then we have $v(f(1,y/x)) > 0$. This means that the v-residue ζ of y/x is algebraic over κ. Let $g \in \kappa[Z]$ be the minimal polynomial of ζ over κ. Set $d := \deg(g)$, and define the homogeneous polynomial $\overline{g} \in \kappa[\overline{x},\overline{y}] = \text{gr}_\mathfrak{m}(R)$ by $\overline{g}(\overline{x},\overline{y}) := \overline{x}^d g(\overline{y}/\overline{x})$; then $\overline{g}$ is irreducible and $\overline{g}(1,\zeta) = 0$. Every irreducible homogeneous polynomial $\overline{h} \in \text{gr}_\mathfrak{m}(R)$ with $\overline{h}(1,\zeta) = 0$ has the form $\overline{h} = \gamma \overline{g}$ with $\gamma \in \kappa^\times$, and it is called a *directional form* of v or of V. We set $p := (\overline{g})$. We have $p \neq (\overline{x})$, and we set $p(V) = p(v) := p \in \mathbb{P}$ and call it the *directional ideal* of v or of V.

Set $A := R[\mathfrak{m}/x]$; we have $V \supset A$ and $v(\mathfrak{n}_p) > 0$. Therefore $\mathfrak{n}_p$ is the center of v in A, hence V dominates S_p. If $v(y) = v(x)$, then we have $D := R[\mathfrak{m}/y] \subset V$, $p \neq (\overline{y})$, and the center of V in B is the maximal ideal of B corresponding to p, hence we have $S_p = B_{\mathfrak{m}(V) \cap B}$, and therefore S_p is the only quadratic transform of R which is contained in V.

Thus, we have shown:

(2.19) Proposition: *Let v be a valuation of K dominating R, let V be the valuation ring of v and assume that $V \neq V_R$. Let $p = p(v)$ be the directional ideal of V. Then V dominates S_p and S_p is the only quadratic transform of R which is contained in V. Moreover, if $\text{rat.}\,\text{rank}(v) = 2$, then V is of the first kind with respect to R.*

3 Complete Ideals

3.1 Generalities

(3.0) In this section we keep the notations and hypotheses introduced in section 2. For any finite non-empty subset $P \subset \mathbb{P}$ we set

$$S_P := \bigcap_{p \in P} S_p.$$

Remember that S_P is a semilocal factorial domain, that $\{\mathfrak{m}(S_p) \cap S_P \mid p \in P\}$ is the set of maximal ideals of S_P, and that we have $\mathfrak{c} = \bigcap_{p \in P} \mathfrak{c}S_p$ for every ideal $\mathfrak{c}$ of S_P [cf. (2.10) and B(2.5)].

(3.1) COMPLETE IDEALS: Let S be a two-dimensional regular local ring with maximal ideal $\mathfrak{n}$. Let $\mathfrak{a}$ be an ideal of S. Since S is integrally closed, the integral closure of $\mathfrak{a}$ in S is also the integral closure of $\mathfrak{a}$ in the field of quotients of S [cf. B(6.9)]. An ideal $\mathfrak{a}$ of S which is an integrally closed ideal of S will be called complete following the time-honored tradition set up by Zariski. We use the notion of complete ideals also for the rings S_P. Every principal ideal of S is complete since S is integrally closed. Let $\mathfrak{a}$ be a non-zero non-principal proper ideal of S. Then we have $\mathfrak{a} = a\mathfrak{b}$ where a is a greatest common divisor for the elements of $\mathfrak{a}$, and $\mathfrak{b}$ is an $\mathfrak{n}$-primary ideal [cf. (2.1)]. Since S is integrally closed, we know [cf. B(6.16)(6)] that $\mathfrak{a}$ is complete iff $\mathfrak{b}$ is complete. Thus, in studying complete ideals of S, it is often enough to consider only complete $\mathfrak{n}$-primary ideals of S.

The following general result shall be very useful.

(3.2) Proposition: *Let $\mathfrak{a}$, $\mathfrak{b}$ be integrally closed ideals of an integrally closed noetherian domain A, and let $\mathfrak{c}$ be a non-zero ideal of A. If we have $\mathfrak{ac} = \mathfrak{bc}$, then we have $\mathfrak{a} = \mathfrak{b}$.*

Proof: Let $a \in \mathfrak{a}$; then we have $a\mathfrak{c} \subset \mathfrak{ac} = \mathfrak{bc}$, hence a is integral over $\mathfrak{b}$ by B(6.2)(3), hence a lies in $\mathfrak{b}$ and therefore we have $\mathfrak{a} \subset \mathfrak{b}$. By symmetry, we also obtain $\mathfrak{b} \subset \mathfrak{a}$, hence we have $\mathfrak{a} = \mathfrak{b}$.

(3.3) REMARK: (1) Let $T \supset R$ be a subring of K, and let $\mathfrak{b}$ be an ideal of T which is integrally closed in T. Then $\mathfrak{a} := R \cap \mathfrak{b}$ is complete.
In fact, if $z \in R$ is integral over $\mathfrak{a}$, then, considering an equation of integral dependence for z over $\mathfrak{a}$, one sees that z is integral over $\mathfrak{b}$ [since $\mathfrak{a}^i \subset \mathfrak{b}^i$ for $i \in \mathbb{N}_0$], hence that $z \in \mathfrak{b} \cap R = \mathfrak{a}$. (Note that this argument works nor only for R but also for any integrally closed subring of T having field of quotients K.)
(2) Let $\mathfrak{a}$ be a non-zero ideal of R, and set $r := \mathrm{ord}(\mathfrak{a})$. Then we have $\mathfrak{a} \subset \mathfrak{m}^r$, $\mathfrak{a} \not\subset \mathfrak{a}^{r+1}$, and since $\mathfrak{m}^r$ is integrally closed [cf. (2.2)], the integral closure of $\mathfrak{a}$ has the same order as $\mathfrak{a}$.

(3.4) Proposition: *Let $\mathfrak{a}$ be a complete ideal of R, and let $P \subset \mathbb{P}$ be a finite non-empty subset. Then $\mathfrak{a}S_P$ and $\mathfrak{a}^{S_P}$ are complete ideals of S_P.*

Proof: We set $T := S_P$. We have $\mathfrak{a}T = z^r\mathfrak{a}^T$ where $r := \mathrm{ord}(\mathfrak{a})$ and $z \in T$ with $\mathfrak{m}T = Tz$. Since T is integrally closed [cf. (2.10)], we know that $\mathfrak{a}T$ is complete iff $\mathfrak{a}^T$ is complete [cf. B(6.16)(6)]. We show that $\mathfrak{a}T$ is complete. Let $u \in T$ be integral over $\mathfrak{a}T$. Then we have an equation $u^n + a_1 u^{n-1} + \cdots + a_n = 0$ where $a_i \in (\mathfrak{a}T)^i = \mathfrak{a}^iT$ for $i \in \{1, \ldots, n\}$. This means that there exist $b_i \in \mathfrak{a}^i$ and $w \in R$ with $\mathrm{In}(w) \notin p$ for every $p \in P$ and with $\mathrm{ord}(b_i) \geq \mathrm{ord}(w)$ and

$a_i = b_i/w$ for $i \in \{1, \ldots, n\}$ [cf. (2.9)]. Set $u' := uw$. Then we get an equation $u'^n + b_1 u'^{n-1} + b_2 w u'^{n-2} + \cdots + b_n w^{n-1} = 0$, hence u' is integral over $\mathfrak{a}$, and therefore we have $uw = u' \in \mathfrak{a}$, hence we have $u = u'/w \in \mathfrak{a}T$.

(3.5) Corollary: *Let* $x \in \mathfrak{m} \setminus \mathfrak{m}^2$, *and set* $A := R[\mathfrak{m}/x]$. *Let* $\mathfrak{a}$ *be a complete ideal of* R. *Then* $\mathfrak{a}A$ *and the transform* $\mathfrak{a}'$ *of* $\mathfrak{a}$ *in* A *are integrally closed ideals of* A.

Proof: Since A is integrally closed [cf. (2.3)(4)], the proof given above works in this case also [one uses a power of x instead of w].

3.2 Complete Ideals as Intersections

(3.6) Lemma: *Let* B *be a noetherian domain with field of quotients* L, *and let* $\mathfrak{p} \neq \{0\}$ *be a prime ideal of* B. *Let* $\{x_1, \ldots, x_n\}$ *be a system of non-zero generators of* $\mathfrak{p}$, *and set* $B_i := B[\mathfrak{p}/x_i]$ *for* $i \in \{1, \ldots, n\}$. *Then there exist* $i \in \{1, \ldots, n\}$ *and a prime ideal* $\mathfrak{q}$ *of height 1 of* B_i *with* $\mathfrak{q} \cap B = \mathfrak{p}$.

Proof: The rings $B_1, \ldots, B_n$ are B-algebras of finite type, and therefore noetherian rings. Let V be a valuation ring of L having center $\mathfrak{p}$ in B [cf. I(3.5)], and choose $i \in \{1, \ldots, n\}$ with $Vx_i = \mathfrak{p}V \subset \mathfrak{m}(V)$ [cf. I(2.4)]. We have $B_i \subset V$; we set $\mathfrak{p}_i := \mathfrak{m}(V) \cap B_i$. Now we have $\mathfrak{p}B_i = B_i x_i \subset \mathfrak{p}_i$, and clearly $\mathfrak{p}_i \cap B = \mathfrak{p}$. Therefore we have $B_i x_i \cap B = \mathfrak{p}$. Now let $\mathfrak{q} \subset \mathfrak{p}_i$ be a minimal prime ideal of $B_i x_i$. By Krull's principal ideal theorem [cf. [63], Th. 10.1] we have $\mathrm{ht}(\mathfrak{q}) = 1$, and clearly we have $\mathfrak{q} \cap B = \mathfrak{p}$.

(3.7) Proposition: *Let* $\mathfrak{a}$ *be an ideal of* R, *and let* $\bar{\mathfrak{a}}$ *be the integral closure of* $\mathfrak{a}$ *in* R. *Then:*
(1) *For every valuation ring* V *of* K *with* $V \supset R$ *we have* $\mathfrak{a}V = \bar{\mathfrak{a}}V$.
(2) *For every* $z \in K \setminus \bar{\mathfrak{a}}$ *there exists a valuation ring* V *of* K *being of the second kind with respect to* R *and such that* $z \notin \mathfrak{a}V$.

Proof: (1) Since R is integrally closed, the integral closure of $\mathfrak{a}$ in K is also the integral closure of $\mathfrak{a}$ in R; therefore (1) is just a restatement of part of B(6.14).
(2) We consider the ring

$$B := R[z^{-1}\mathfrak{a}] = \bigcup_{n \geq 0} (\mathfrak{a} + Rz)^n/z^n;$$

B is an R-subalgebra of K of finite type, hence it is noetherian. We set $\mathfrak{p} := \mathfrak{m} + z^{-1}\mathfrak{a}B$. We have

$$\mathfrak{m}(\mathfrak{a} + Rz)^n/z^n \subset \mathfrak{m} + z^{-1}\mathfrak{a}(\mathfrak{a} + Rz)^{n-1}/z^{n-1} \subset \mathfrak{p} \quad \text{for } n \in \mathbb{N};$$

therefore $\mathfrak{p}$ is an ideal of B. We have $1 \notin \mathfrak{p}$, since otherwise there would exist $w \in \mathfrak{m}$ and $n \in \mathbb{N}$ with $1 + w \in z^{-1}\mathfrak{a}(\mathfrak{a} + Rz)^n/z^n$, hence $(1 + w)z^{n+1}$ would lie in $\mathfrak{a}^{n+1} + z\mathfrak{a}^n + \cdots + z^n\mathfrak{a}$, and z would be integral over $\mathfrak{a}$ [note that $1 + w$ is a unit of R]. Therefore we have $\mathfrak{p} \cap R = \mathfrak{m}$. We have $B = R + \mathfrak{p}$ since $(\mathfrak{a} + Rz)^n/z^n = R + z^{-1}\mathfrak{a}(\mathfrak{a} + Rz)^{n-1}/z^{n-1}$ for $n \in \mathbb{N}$. Therefore we have $B/\mathfrak{p} = (R + \mathfrak{p})/\mathfrak{p} \cong R/\mathfrak{m}$, hence $\mathfrak{p}$ is a maximal ideal of B. There exists an R-algebra C of finite type with $B \subset C \subset K$ and a prime ideal $\mathfrak{q}$ of C of height 1 with $\mathfrak{q} \cap B = \mathfrak{p}$ [cf. (3.6)]. Since regular local rings are universally catenary [cf. [63], Cor. 18.10], we have $\operatorname{ht}(\mathfrak{q}) + \dim(C/\mathfrak{q}) = \dim(C)$. Now $C/\mathfrak{q}$ is an integral $R/\mathfrak{m}$-algebra of finite type, and therefore we have $\dim(C/\mathfrak{q}) = \operatorname{tr.}\operatorname{d}_{R/\mathfrak{m}} Q(C/\mathfrak{q})$ [cf. [63], Th. A on p. 286]. Let $\mathfrak{n}$ be a maximal ideal of C with $\mathfrak{q} \subset \mathfrak{n}$. Then we have $\mathfrak{n} \cap B = \mathfrak{p}$ and $\dim(C_\mathfrak{n}) = \dim(R)$ [cf. [63], Th. 13.8, and note that R and C have K as field of quotients, hence $K \otimes_R C = K$ and $\dim(K) = 0$]. We have $\dim(C_\mathfrak{n}) = \dim(C)$ [since C is catenary], and therefore we get at last

$$\operatorname{tr.}\operatorname{d}_{R/\mathfrak{m}}(C_\mathfrak{q}/\mathfrak{q}C_\mathfrak{q}) = \dim(R) - 1 = 1$$

[note that $Q(C/\mathfrak{q}) = C_\mathfrak{q}/\mathfrak{q}C_\mathfrak{q}$]. Let V be a valuation ring of K having center $\mathfrak{q}C_\mathfrak{q}$ in $C_\mathfrak{q}$ [cf. I(3.5)]; then V is of the second kind with respect to R. Furthermore, we have $z^{-1}\mathfrak{a} \subset \mathfrak{m}(V)$, hence z does not lie in $\mathfrak{a}V$.

(3.8) Proposition: *Let $\mathfrak{a}$ be an ideal of R. If $\mathfrak{a}$ is complete, then we have*

$$\mathfrak{a} = \bigcap_{p \in \mathbb{P}} \mathfrak{a}S_p. \tag{$*$}$$

Proof: The left hand side of $(*)$ is contained in the right hand side. Let $z \in K \setminus \mathfrak{a}$; it is enough to show that there exists $p \in \mathbb{P}$ with $z \notin \mathfrak{a}S_p$. By (3.7) there exists a valuation ring V of K dominating R with $z \notin \mathfrak{a}V$. If $V = V_R$, then every quadratic transform of R is contained in V_R, and if $V \neq V_R$, then there exists a quadratic transform S_p of R which is dominated by V by (2.19), and we have $z \notin \mathfrak{a}S_p$.

(3.9) Remark: Let $\{x, y\}$ be a system of generators of $\mathfrak{m}$. For every complete ideal $\mathfrak{a}$ of R we have

$$\mathfrak{a} = \mathfrak{a}R[\mathfrak{m}/x] \cap \mathfrak{a}R[\mathfrak{m}/y] = \bigcap_{p \in \mathbb{P}} \mathfrak{a}S_p.$$

This follows immediately from (3.8).

(3.10) Corollary: *Let $\mathfrak{a}$ be a complete non-zero ideal of R, and let $P \subset \mathbb{P}$ be a finite non-empty subset. Set $r := \operatorname{ord}(\mathfrak{a})$. If P contains all $p \in \mathbb{P}$ which divide $c(\mathfrak{a})$, then we have*

$$\mathfrak{a} = \mathfrak{a}S_P \cap R = \mathfrak{a}S_P \cap \mathfrak{m}^r.$$

Proof: We have

$$\mathfrak{m}^r S_P \cap R = \left(\bigcap_{p \in P} \mathfrak{m}^r S_p \right) \cap R = \mathfrak{m}^r$$

[for the first equality sign cf. (3.0), and for the second equality sign cf. (2.8)]. On the other hand, we have $aS_P \subset \mathfrak{m}^r S_P$, hence $aS_P = aS_P \cap \mathfrak{m}^r S_P$, and therefore we have $aS_P \cap R = aS_P \cap \mathfrak{m}^r$. For every $p \notin P$ we have $aS_p = \mathfrak{m}^r S_p$ [cf. (2.15)], hence we obtain from (3.8) [note that $\mathfrak{m}^r$ is complete]

$$aS_P \cap \mathfrak{m}^r = \bigcap_{p \in P} aS_p \cap \bigcap_{p \in \mathbb{P}} \mathfrak{m}^r S_p = \bigcap_{p \in \mathbb{P}} aS_p = a.$$

(3.11) Corollary: *Let a be a complete non-zero ideal of R. We have $c(a) = (1)$ iff a is a power of $\mathfrak{m}$.*

Proof: We set $r := \mathrm{ord}(a)$. If $c(a) = (1)$, then we have $aS_p = \mathfrak{m}^r S_p$ for every $p \in \mathbb{P}$ [cf. (2.15)], hence $a = \mathfrak{m}^r$ by (3.8). Conversely, if $a = \mathfrak{m}^r$, then we have $c(a) = (1)$.

(3.12) Proposition: *Let a be a complete $\mathfrak{m}$-primary ideal of R which is not a power of $\mathfrak{m}$. Set $r := \mathrm{ord}(a)$ and $P := \{p \in \mathbb{P} \mid p \mid c(a)\}$. Then $P \neq \emptyset$, and the canonical map*

$$\mathfrak{m}^r/a \to \mathfrak{m}^r S_P/aS_P$$

is an isomorphism of R-modules.

Proof: Since we have $aS_P \cap \mathfrak{m}^r = a$ [cf. (3.10)], this map is injective.
(1) Let $\{x, y\}$ be a system of generators of $\mathfrak{m}$, set $A := R[\mathfrak{m}/x]$, $B := R[\mathfrak{m}/y]$ and $C := R[\mathfrak{m}^2/xy] = R[y/x, x/y]$. Note that C is a localization of A with respect to the multiplicatively closed system $\{(y/x)^n \mid n \in \mathbb{N}_0\}$; likewise, C is a localization of B with respect to the multiplicatively closed system generated by x/y.
(a) Let a' be the transform of a in A. First, we assume that a' is not the unit ideal. Then the set $\{p_1, \ldots, p_h\}$ of prime ideals of $\mathbb{P}$ which divide $c(a)$ and are different from $(\mathrm{In}(x))$ is not empty and $\mathrm{Ass}(A/a') = \{n_1, \ldots, n_h\}$ where $n_i = n_{p_i}$ for $i \in \{1, \ldots, h\}$ [cf. (2.15) and (2.17)(2)]. We consider the commutative diagram

$$
\begin{array}{ccc}
\mathfrak{m}^r A/aA & \xrightarrow{\varphi} & \mathfrak{m}^r C/aC \\
\downarrow & & \downarrow \\
A/a' & \xrightarrow{\widetilde{\varphi}} & C/a'C
\end{array}
$$

where the vertical maps are isomorphisms of R-modules. Set $n := (x, y/x) \subset A$. If $n \notin \{n_1, \ldots, n_h\}$, then y/x is a unit of A/a', and $\widetilde{\varphi}$ is an isomorphism of A-modules. Now we consider the case that n is one the ideals $n_1, \ldots, n_h$, say $n = n_1$. Let $a' = q_1 \cap \cdots \cap q_h$ be the primary decomposition of a' where q_i is n_i-primary, $i \in \{1, \ldots, h\}$ [cf. (2.17)(2)]. Then $a'C = q_2 C \cap \cdots \cap q_h C$ is the primary decomposition of $a'C$ [cf. [63], Th. 3.10], we have $A/q_i \cong C/q_i C$ for $i \in \{2, \ldots, h\}$ [since the image of y/x in A/q_i is a unit] and the map $\widetilde{\varphi}: A/a' \to C/a'C$ is the projection

$$A/a' \cong A/q_1 \times \cdots \times A/q_h \to C/q_2 C \times \cdots \times C/q_h C = C/a'C$$

onto the last $h - 1$ factors [if $h = 1$, then $C = \mathfrak{a}'C$], hence $\widetilde{\varphi}$ is surjective also.
Second, if $\mathfrak{a}'$ is the unit ideal, then $\widetilde{\varphi}$ is a map between null rings.
Thus, the map $\widetilde{\varphi}$ is surjective, and therefore φ is surjective.
(b) Likewise, the canonical map $\psi\colon \mathfrak{m}^r B/\mathfrak{a}B \to \mathfrak{m}^r C/\mathfrak{a}C$ is surjective.
(2) Let $\alpha\colon \mathfrak{m}^r/\mathfrak{a} \to \mathfrak{m}^r A/\mathfrak{a}A$ and $\beta\colon \mathfrak{m}^r/\mathfrak{a} \to \mathfrak{m}^r B/\mathfrak{a}B$ be the canonical R-linear
maps. We consider the commutative diagram

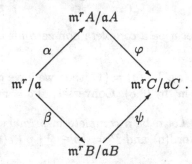

We show: For every pair $(f, g) \in \mathfrak{m}^r A \times \mathfrak{m}^r B$ with $\varphi(f \bmod \mathfrak{a}A) = \psi(g \bmod \mathfrak{a}B)$
there exists $h \in \mathfrak{m}^r$ with $\alpha(h \bmod \mathfrak{a}) = f \bmod \mathfrak{a}A$ and $\beta(h \bmod \mathfrak{a}) = g \bmod \mathfrak{a}B$.
In fact, $\varphi(f \bmod \mathfrak{a}A) = \psi(g \bmod \mathfrak{a}B)$ means that $f - g \in \mathfrak{a}C$. We write

$$f - g = \sum_{i,j} a_{ij}(y/x)^i(x/y)^j \text{ with } a_{ij} \in \mathfrak{a},$$

hence we have $f - g = f_1 - g_1$ where

$$f_1 = \sum_{i \geq j} a_{ij}(y/x)^{i-j} \in \mathfrak{a}A, \quad g_1 = \sum_{i < j}(-a_{ij})(x/y)^{j-i} \in \mathfrak{a}B.$$

We set $f' := f - f_1 \in \mathfrak{m}^r A$, $g' := g - g_1 \in \mathfrak{m}^r B$; then we have $f \bmod \mathfrak{a}A = f' \bmod \mathfrak{a}A$, $g \bmod \mathfrak{a}B = g' \bmod \mathfrak{a}B$, and $f' = g' \in \mathfrak{m}^r A \cap \mathfrak{m}^r B = \mathfrak{m}^r$ [cf. (3.9) and note that $\mathfrak{m}^r$ is complete by (2.2)]. Now we set $h := f'$.
(3) Since $c(\mathfrak{a}) \neq (1)$, at least one of the transforms of $\mathfrak{a}$ in A and in B is not the unit ideal. Let us assume, as we may, that $\mathfrak{a}'$ is not the unit ideal. Then we have

$$S_P = A_{\mathfrak{n}_1} \cap \cdots \cap A_{\mathfrak{n}_h} \quad \text{if } (\operatorname{In}(x)) \nmid c(\mathfrak{a}),$$

and if $(\operatorname{In}(x)) \mid c(\mathfrak{a})$, and we denote by $\mathfrak{n}'$ the maximal ideal of B generated by y
and x/y, then we have

$$S_P = A_{\mathfrak{n}_1} \cap \cdots \cap A_{\mathfrak{n}_h} \cap B_{\mathfrak{n}'}.$$

We set

$$S_i := A_{\mathfrak{n}_i} \quad \text{for } i \in \{1, \ldots, h\}, \ S' := B_{\mathfrak{n}'}.$$

Then we have

$$\mathfrak{m}^r S_P/\mathfrak{a}S_P \cong \begin{cases} \mathfrak{m}^r S_1/\mathfrak{a}S_1 \times \cdots \times \mathfrak{m}^r S_h/\mathfrak{a}S_h & \text{if } (\operatorname{In}(x)) \nmid c(\mathfrak{a}), \\ \mathfrak{m}^r S_1/\mathfrak{a}S_1 \times \cdots \times \mathfrak{m}^r S_h/\mathfrak{a}S_h \times \mathfrak{m}^r S'/\mathfrak{a}S' & \text{if } (\operatorname{In}(x)) \mid c(\mathfrak{a}). \end{cases}$$

If $(\text{In}(x)) \nmid c(\mathfrak{a})$, then we have $\mathfrak{m}^r S_P/\mathfrak{a}S_P \cong \mathfrak{m}^r A/\mathfrak{a}A$, and the map $\mathfrak{m}^r/\mathfrak{a} \to \mathfrak{m}^r S_P/\mathfrak{a}S_P$ is surjective by (2). Likewise, interchanging the roles of A and B, if $(\text{In}(y)) \nmid c(\mathfrak{a})$, then we have $\mathfrak{m}^r S_P/\mathfrak{a}S_P \cong \mathfrak{m}^r B/\mathfrak{a}B$, and the map $\mathfrak{m}^r/\mathfrak{a} \to \mathfrak{m}^r S_P/\mathfrak{a}S_P$ is surjective. Now we consider the case that both $(\text{In}(x))$ and $(\text{In}(y))$ divide $c(\mathfrak{a})$. The map $\psi\colon \mathfrak{m}^r B/\mathfrak{a}B \to \mathfrak{m}^r C/\mathfrak{a}C$ maps the factor $\mathfrak{m}^r B_{\mathfrak{n}'}/\mathfrak{a}B_{\mathfrak{n}'}$ of $\mathfrak{m}^r B/\mathfrak{a}B$ to 0. Therefore, given $f \in \mathfrak{m}^r A$ and $g' \in \mathfrak{m}^r B_{\mathfrak{n}'}$, there exists $g \in \mathfrak{m}^r B$ such that we have $\varphi(f \bmod \mathfrak{a}A) = \psi(g \bmod \mathfrak{a}B)$ and such that the image of $g \bmod \mathfrak{a}B$ in $\mathfrak{m}^r B_{\mathfrak{n}'}/\mathfrak{a}B_{\mathfrak{n}'}$ equals $g' \bmod \mathfrak{a}B_{\mathfrak{n}'}$. By (2) we see that also in this case the map $\mathfrak{m}^r/\mathfrak{a} \to \mathfrak{m}^r S_P/\mathfrak{a}S_P$ is surjective.

(3.13) Corollary: *Let $\mathfrak{a}$ be a complete $\mathfrak{m}$-primary ideal of R, and set $r := \text{ord}(\mathfrak{a})$, $P := \{p \in \mathbb{P} \mid p \mid c(\mathfrak{a})\}$. Then we have*

$$\ell_R(\mathfrak{m}^r/\mathfrak{a}) = \sum_{p \in P} [S_p : R]\ell_{S_p}(S_p/\mathfrak{a}^{S_p}) = \sum_{p \in \mathbb{P}} [S_p : R]\ell_{S_p}(S_p/\mathfrak{a}^{S_p}).$$

Proof: The assertion is trivially true if $\mathfrak{a} = \mathfrak{m}^r$, since in this case both sides are equal to 0; therefore we may assume that $\mathfrak{a} \neq \mathfrak{m}^r$. We have $\mathfrak{m}^r S_P/\mathfrak{a}S_P = \prod_{p \in P} \mathfrak{m}^r S_p/\mathfrak{a}S_p$, and for $p \in P$ we have $\mathfrak{m}^r S_p/\mathfrak{a}S_p \cong S_p/\mathfrak{a}^{S_p}$. Therefore we have by (3.12) [since $S_p/\mathfrak{m}_p$ is the only simple S_p-module, and $\ell_R(S_p/\mathfrak{m}_p) = [S_p : R]$]

$$\ell_R(\mathfrak{m}^r/\mathfrak{a}) = \sum_{p \in P} \ell_R(S_p/\mathfrak{a}^{S_p}) = \sum_{p \in \mathbb{P}} [S_p : R]\ell_{S_p}(S_p/\mathfrak{a}^{S_p}).$$

(3.14) Corollary: *Let $\mathfrak{a}$ be a complete $\mathfrak{m}$-primary ideal of R, and set $r := \text{ord}(\mathfrak{a})$. Then we have*

$$\ell_R(R/\mathfrak{a}) = \frac{r(r+1)}{2} + \sum_{p \mid c(\mathfrak{a})} [S_p : R]\ell_{S_p}(S_p/\mathfrak{a}^{S_p})$$

$$= \frac{r(r+1)}{2} + \sum_{p \in \mathbb{P}} [S_p : R]\ell_{S_p}(S_p/\mathfrak{a}^{S_p}).$$

Proof: We have $\ell_R(R/\mathfrak{a}) = \ell_R(R/\mathfrak{m}^r) + \ell_R(\mathfrak{m}^r/\mathfrak{a})$, and therefore the result follows from (3.13) since $\ell_R(R/\mathfrak{m}^r) = r(r+1)/2$.

(3.15) Corollary: *Let $\mathfrak{a}$ be a complete $\mathfrak{m}$-primary ideal of R. Then we have $\ell_R(R/\mathfrak{a}) > \ell_S(S/\mathfrak{a}^S)$ for every quadratic transform S of R.*

(3.16) REMARK: In (3.8) we showed that a complete ideal $\mathfrak{a}$ of R is the intersection of all its extensions $\mathfrak{a}S_p$ where $p \in \mathbb{P}$. In theorem (3.27) below we prove the existence of complete ideals having a prescribed behavior in finitely many of the rings S_p.

3.3   When Does $\mathfrak{m}$ Divide a Complete Ideal?

The next two results are useful when studying contracted ideals. We denote, as usual, by $\mu(\mathfrak{a})$ the minimal number of generators of an ideal $\mathfrak{a}$ of a local ring.

(3.17) Lemma: *Let $\mathfrak{a}$ be an $\mathfrak{m}$-primary ideal of R. Then we have*

$$\ell_R((\mathfrak{a}:\mathfrak{m})/\mathfrak{a}) = \dim_\kappa((\mathfrak{a}:\mathfrak{m})/\mathfrak{a}) = \mu(\mathfrak{a}) - 1.$$

Proof: $(\mathfrak{a}:\mathfrak{m})/\mathfrak{a}$ is an ideal of the local artinian ring $R/\mathfrak{a}$ and it is annihilated by its maximal ideal, hence its dimension as κ-vector space is equal to its length as $R/\mathfrak{a}$-module. Let $\mathfrak{m} = Rx + Ry$. The sequence

$$0 \to R \xrightarrow{\delta_2} R^2 \xrightarrow{\delta_1} \mathfrak{m} \to 0$$

where $\delta_2(1_R) = (-y, x)$, $\delta_1((1,0)) = x$, $\delta_1((0,1)) = y$, is easily seen to be exact [it is the Koszul complex on (x, y)], and it gives rise, as is easy to check, to an exact sequence

$$0 \to (\mathfrak{a}:\mathfrak{m})/\mathfrak{a} \to R/\mathfrak{a} \to (R/\mathfrak{a})^2 \to \mathfrak{m}/\mathfrak{a}\mathfrak{m} \to 0.$$

Therefore we have

$$\ell_{R/\mathfrak{a}}((\mathfrak{a}:\mathfrak{m})/\mathfrak{a}) = \ell_{R/\mathfrak{a}}(\mathfrak{m}/\mathfrak{m}\mathfrak{a}) - \ell_{R/\mathfrak{a}}(R/\mathfrak{a}) = \ell_{R/\mathfrak{a}}(\mathfrak{a}/\mathfrak{m}\mathfrak{a}) - \ell_{R/\mathfrak{a}}(R/\mathfrak{m}) = \mu(\mathfrak{a}) - 1$$

[note that $R \supset \mathfrak{m} \supset \mathfrak{a} \supset \mathfrak{m}\mathfrak{a}$].

(3.18) Proposition: *Let $\mathfrak{a}$ be a $\mathfrak{m}$-primary ideal of R. Then we have $\mu(\mathfrak{a}) \leq \operatorname{ord}(\mathfrak{a}) + 1$.*

Proof: We set $n := \mu(\mathfrak{a})$; we have $n \geq 2$. By Hilbert's syzygy theorem [cf. [63], Cor. 19.6] there exists a minimal resolution

$$0 \to F_2 \xrightarrow{\varphi_2} F_1 \xrightarrow{\varphi_1} \mathfrak{a}$$

of $\mathfrak{a}$ where F_1 is a free R-module of rank n; note that $\varphi_2(F_2) \subset \mathfrak{m}F_1$. The sequence

$$0 \to F_2 \xrightarrow{\varphi_2} F_1 \xrightarrow{\varphi_1} R \to R/\mathfrak{a} \to 0$$

is exact. The theorem of Hilbert-Burch [cf. [63], Th. 20.15] says in this situation: F_2 is free of rank $n - 1$, and if $\mathfrak{a}_{n-1}(\varphi_2)$ is the ideal of R generated by the $(n-1)$-minors of the matrix P describing φ_2 with respect to bases of F_2 and F_1, then we have $\mathfrak{a} = a\mathfrak{a}_{n-1}(\varphi_2)$ where $a \in R$. In our case we have $\mathfrak{a}_{n-1}(\varphi_2) \subset \mathfrak{m}^{n-1}$, hence we have $\mu(\mathfrak{a}) = n \leq \operatorname{ord}(\mathfrak{a}) + 1$.

(3.19) Proposition: *Let $\mathfrak{a}$ be a complete $\mathfrak{m}$-primary ideal of R. Then:*
(1) $\mathfrak{m}\mathfrak{a}$ is complete.
(2) We have $\mu(\mathfrak{a}) = \operatorname{ord}(\mathfrak{a}) + 1$.

Proof: We set $r := \operatorname{ord}(\mathfrak{a})$. Let $\mathfrak{b}$ be the integral closure of $\mathfrak{m}\mathfrak{a}$; we have $\operatorname{ord}(\mathfrak{b}) = r + 1$ by (3.3)(2) and $\mathfrak{b} \subset \mathfrak{a}$ since $\mathfrak{a}$ is complete. Let S be a quadratic transform of R. By (1.5)(7) we have $\overline{(\mathfrak{m}\mathfrak{a})^S} = \overline{\mathfrak{b}^S}$, and since $(\mathfrak{m}\mathfrak{a})^S = \mathfrak{a}^S$ and $\mathfrak{b}^S$ are complete ideals of S by (3.4), we have $\mathfrak{a}^S = \mathfrak{b}^S$. Now we obtain by (3.14)

$$\ell_R(\mathfrak{a}/\mathfrak{b}) = \ell_R(R/\mathfrak{b}) - \ell_R(R/\mathfrak{a}) = \frac{(r+1)(r+2)}{2} - \frac{r(r+1)}{2} = r + 1.$$

On the other hand, we have $\ell_R(\mathfrak{a}/\mathfrak{m}\mathfrak{a}) \leq r + 1$ by (3.18), hence we have

$$r + 1 = \ell_R(\mathfrak{a}/\mathfrak{b}) \leq \ell_R(\mathfrak{a}/\mathfrak{m}\mathfrak{a}) \leq r + 1,$$

and from this we get $\mathfrak{b} = \mathfrak{m}\mathfrak{a}$ and $\ell_R(\mathfrak{a}/\mathfrak{m}\mathfrak{a}) = r + 1$.

(3.20) Corollary: *Let $\mathfrak{a}$ be a complete $\mathfrak{m}$-primary ideal of R. For every $s \in \mathbb{N}$ the ideal $\mathfrak{m}^s\mathfrak{a}$ is complete.*

Proof: By induction using (3.19).

(3.21) Corollary: *Let $\mathfrak{a}$ be a complete $\mathfrak{m}$-primary ideal of R. Then we have*

$$\ell_R((\mathfrak{a} : \mathfrak{m})/\mathfrak{a}) = \operatorname{ord}(\mathfrak{a}).$$

Proof: This follows from (3.17) and (3.19).

(3.22) Proposition: (1) *Let $\mathfrak{a}$ be a complete $\mathfrak{m}$-primary ideal of R. Then $\mathfrak{m}$ divides $\mathfrak{a}$ iff $\operatorname{ord}(\mathfrak{a}) > \deg(c(\mathfrak{a}))$.*
(2) *Let $\mathfrak{a}$, $\mathfrak{b}$ be complete $\mathfrak{m}$-primary ideals of R. If there exists $s \in \mathbb{N}$ with $\operatorname{ord}(\mathfrak{b}) = s + \operatorname{ord}(\mathfrak{a})$ and with $\mathfrak{m}^s\mathfrak{a} \subset \mathfrak{b}$, then $\mathfrak{m}$ divides $\mathfrak{b}$.*

Proof: (1) The assertion is true if $\mathfrak{a} = \mathfrak{m}$; we now consider the case $\mathfrak{a} \neq \mathfrak{m}$. We have $\mathfrak{a} = \mathfrak{m}(\mathfrak{a} : \mathfrak{m})$ iff $\mathfrak{m}$ divides $\mathfrak{a}$ [cf. (2.0)].
We set $\mathfrak{b} := \mathfrak{a} : \mathfrak{m}$; then $\mathfrak{b}$ is a complete $\mathfrak{m}$-primary ideal of R [cf. B(6.16)(5)], and we have $\mathfrak{m}\mathfrak{b} \subset \mathfrak{a}$. We consider the exact sequence of R-modules

$$0 \to \mathfrak{a}/\mathfrak{m}\mathfrak{b} \to \mathfrak{b}/\mathfrak{m}\mathfrak{b} \to \mathfrak{b}/\mathfrak{a} \to 0.$$

We have $\dim_\kappa(\mathfrak{b}/\mathfrak{a}) = \mu(\mathfrak{a}) - 1$ by (3.17), and we have $\mu(\mathfrak{a}) - 1 = \operatorname{ord}(\mathfrak{a})$ by (3.19). We have $\dim_\kappa(\mathfrak{b}/\mathfrak{m}\mathfrak{b}) = \mu(\mathfrak{b}) = \operatorname{ord}(\mathfrak{b}) + 1$ [cf. (3.19)], and since $\mathfrak{a} \subset \mathfrak{b}$, we have $\operatorname{ord}(\mathfrak{a}) \geq \operatorname{ord}(\mathfrak{b})$. Therefore we have

$$\dim_\kappa(\mathfrak{a}/\mathfrak{m}\mathfrak{b}) = \dim_\kappa(\mathfrak{b}/\mathfrak{m}\mathfrak{b}) - \dim_\kappa(\mathfrak{b}/\mathfrak{a}) = \operatorname{ord}(\mathfrak{b}) + 1 - \operatorname{ord}(\mathfrak{a}) \leq 1,$$

hence $\dim_\kappa(\mathfrak{a}/\mathfrak{m}\mathfrak{b}) \leq 1$ with equality iff $\operatorname{ord}(\mathfrak{b}) = \operatorname{ord}(\mathfrak{a})$. This means that $\mathfrak{a} \neq \mathfrak{m}\mathfrak{b}$ iff $\operatorname{ord}(\mathfrak{b}) = \operatorname{ord}(\mathfrak{a})$, and in this case we have $\dim_\kappa(\mathfrak{a}/\mathfrak{m}\mathfrak{b}) = 1$.
(a) We assume that $\mathfrak{a} \neq \mathfrak{m}\mathfrak{b}$. We set $r := \operatorname{ord}(\mathfrak{a}) = \operatorname{ord}(\mathfrak{b})$; then we have $\mathfrak{m}\mathfrak{b} \subset \mathfrak{m}^{r+1}$, hence we get

$$1 = \dim_\kappa(\mathfrak{a}/\mathfrak{m}\mathfrak{b}) \geq \dim_\kappa(\mathfrak{a}/(\mathfrak{a} \cap \mathfrak{m}^{r+1})) = \dim_\kappa((\mathfrak{a} + \mathfrak{m}^{r+1})/\mathfrak{m}^{r+1}),$$

and therefore we have $\dim_\kappa((\mathfrak{a} + \mathfrak{m}^{r+1})/\mathfrak{m}^{r+1}) = 1$, which means that the ideal $(\mathfrak{a} + \mathfrak{m}^{r+1})/\mathfrak{m}^{r+1}$ in $\mathrm{gr}_\mathfrak{m}(R)$ is generated by a homogeneous polynomial of degree r, and therefore we have $\deg(c(\mathfrak{a})) = r = \mathrm{ord}(\mathfrak{a})$.

(b) We assume that $\mathfrak{a} = \mathfrak{mb}$. Then we have $c(\mathfrak{a}) = c(\mathfrak{m})c(\mathfrak{b}) = c(\mathfrak{b})$, and therefore we have $\deg(c(\mathfrak{a})) = \deg(c(\mathfrak{b})) \leq \mathrm{ord}(\mathfrak{b}) < \mathrm{ord}(\mathfrak{a})$.

(2) We have $\mathrm{ord}(\mathfrak{m}^s\mathfrak{a}) = \mathrm{ord}(\mathfrak{b})$; since $\mathfrak{a}$ is complete, the ideal $\mathfrak{m}^s\mathfrak{a}$ is also complete [cf. (3.20)]. From $\mathfrak{m}^s\mathfrak{a} \subset \mathfrak{b}$ we get $c(\mathfrak{b}) \supset c(\mathfrak{m}^s\mathfrak{a})$ [cf. (2.13)], hence $\deg(c(\mathfrak{m}^s\mathfrak{a})) \geq \deg(c(\mathfrak{b}))$. We have $\mathrm{ord}(\mathfrak{m}^s\mathfrak{a}) > \deg(c(\mathfrak{m}^s\mathfrak{a}))$ by (1), hence $\mathrm{ord}(\mathfrak{b}) > \deg(c(\mathfrak{b}))$, and therefore $\mathfrak{m}$ divides $\mathfrak{b}$, again by (1).

(3.23) REMARK: Let $\mathfrak{a}$ be a complete $\mathfrak{m}$-primary ideal of R of order r.
(1) The proof of (3.22)(1) shows that $\mathrm{ord}(\mathfrak{a} : \mathfrak{m})$ is equal to $r - 1$ or r, namely $\mathrm{ord}(\mathfrak{a} : \mathfrak{m}) = r - 1$ if $\mathfrak{m}(\mathfrak{a} : \mathfrak{m}) = \mathfrak{a}$, and $\mathrm{ord}(\mathfrak{a} : \mathfrak{m}) = r$ if $\mathfrak{m}(\mathfrak{a} : \mathfrak{m}) \neq \mathfrak{a}$. In the latter case we have

$$\ell_R(\mathfrak{a}/(\mathfrak{m}^{r+1} \cap \mathfrak{a})) = 1, \quad \mathrm{ord}(\mathfrak{m}^{r+1} \cap \mathfrak{a}) = r + 1 \tag{$*$}$$

[the last equation in $(*)$ follows from the fact that $\mathfrak{a}$ contains elements of order $r + 1$].
(2) Assume that there exists an ideal $\mathfrak{b} \subset \mathfrak{a}$ such that $\ell_R(\mathfrak{a}/\mathfrak{b}) = 1$ and $\mathrm{ord}(\mathfrak{b}) = r + 1$. Then we have $\mathfrak{b} = \mathfrak{a} \cap \mathfrak{m}^{r+1}$. If, in addition, $\mathfrak{b}$ is a complete $\mathfrak{m}$-primary ideal, then we have $\mathfrak{m}(\mathfrak{b} : \mathfrak{m}) = \mathfrak{b}$ and $\mathrm{ord}(\mathfrak{b} : \mathfrak{m}) = r$.
Proof: We have $\mathfrak{b} \subset \mathfrak{a}$, $\mathfrak{b} \subset \mathfrak{m}^{r+1}$, and since $\mathfrak{a} \not\subset \mathfrak{m}^{r+1}$, we have $\mathfrak{b} = \mathfrak{a} \cap \mathfrak{m}^{r+1}$ [since $\ell_R(\mathfrak{a}/\mathfrak{b}) = 1$]. Now we assume that $\mathfrak{b}$ is a complete $\mathfrak{m}$-primary ideal. We have $\mathfrak{ma} \subset \mathfrak{b}$ since $\ell_R(\mathfrak{a}/\mathfrak{b}) = 1$, and we have $\mathrm{ord}(\mathfrak{b}) = r + 1$ by assumption. From (3.22)(2) we get that $\mathfrak{m}$ divides $\mathfrak{b}$, hence $\mathfrak{m}(\mathfrak{b} : \mathfrak{m}) = \mathfrak{b}$ [cf. (2.0)] and $\mathrm{ord}(\mathfrak{b} : \mathfrak{m}) = r$.
(3) Let $\mathfrak{a}$ be not a power of $\mathfrak{m}$. We assume that there exists a complete $\mathfrak{m}$-primary ideal $\mathfrak{b} \subset \mathfrak{a}$ with $\ell_R(\mathfrak{a}/\mathfrak{b}) = 1$ and $\mathrm{ord}(\mathfrak{b}) = r + 1$. If, in addition, κ is an infinite field, then we have $\mathrm{ord}(\mathfrak{a} : \mathfrak{m}) = \mathrm{ord}(\mathfrak{a})$ and $\mathfrak{b} : \mathfrak{m} = \mathfrak{a} : \mathfrak{m}$.
Proof: Let $c(\mathfrak{a}) = p_1^{e_1} \cdots p_h^{e_h}$ be the factorization of $c(\mathfrak{a})$ [note that $h \geq 1$]. The κ-subspaces $\mathrm{gr}_\mathfrak{m}(R)_1 \cap p_i$, $i \in \{1, \ldots, h\}$, are different from $\mathrm{gr}_\mathfrak{m}(R)_1$, hence their union is different from $\mathrm{gr}_\mathfrak{m}(R)_1$ [since κ is an infinite field], and therefore there exists $x \in \mathfrak{m} \setminus \mathfrak{m}^2$ with $(\mathrm{In}(x)) \neq p_i$ for $i \in \{1, \ldots, h\}$. We set $\mathfrak{c} := Rx + \mathfrak{m}^2$. Then $\mathfrak{c}$ is an $\mathfrak{m}$-primary ideal of R with $\mathfrak{m}^2 \subset \mathfrak{c}$, $\mathrm{ord}(\mathfrak{c}) = 1$ and $c(\mathfrak{c}) = (\mathrm{In}(x))$. From $\mathrm{ord}(\mathfrak{a}) = \mathrm{ord}(\mathfrak{c}\mathfrak{m}^{r-1})$ we get $\mathfrak{a} \not\subset \mathfrak{c}\mathfrak{m}^{r-1}$ [cf. (2.13) and note that $c(\mathfrak{c})$ does not divide $c(\mathfrak{a})$]. We have $\mathfrak{b} \subset \mathfrak{m}^{r+1} = \mathfrak{m}^2\mathfrak{m}^{r-1} \subset \mathfrak{c}\mathfrak{m}^{r-1}$, and since $\mathfrak{b} \subset \mathfrak{a}$, but $\mathfrak{a} \not\subset \mathfrak{c}\mathfrak{m}^{r-1}$, we have $\mathfrak{b} = \mathfrak{a} \cap \mathfrak{c}\mathfrak{m}^{r-1}$ [since $\ell_R(\mathfrak{a}/\mathfrak{b}) = 1$].
We know by (1) that $\mathrm{ord}(\mathfrak{a} : \mathfrak{m})$ is either equal to r or $r - 1$. Suppose that $\mathrm{ord}(\mathfrak{a} : \mathfrak{m}) = r - 1$. Then we have $\mathfrak{c}(\mathfrak{a} : \mathfrak{m}) \subset \mathfrak{c}\mathfrak{m}^{r-1}$ and $\mathfrak{c}(\mathfrak{a} : \mathfrak{m}) \subset \mathfrak{m}(\mathfrak{a} : \mathfrak{m}) = \mathfrak{a}$. This implies that we have $\mathfrak{c}(\mathfrak{a} : \mathfrak{m}) \subset \mathfrak{a} \cap \mathfrak{c}\mathfrak{m}^{r-1} = \mathfrak{b}$, hence $r = \mathrm{ord}(\mathfrak{c}(\mathfrak{a} : \mathfrak{m})) \geq \mathrm{ord}(\mathfrak{b}) = r + 1$ which is absurd. Therefore we have $\mathrm{ord}(\mathfrak{a} : \mathfrak{m}) = r$.
We show that $\mathfrak{b} : \mathfrak{m} = \mathfrak{a} : \mathfrak{m}$. We have $\mathfrak{b} : \mathfrak{m} \subset \mathfrak{a} : \mathfrak{m} \subset \mathfrak{m}^r$, hence $\mathfrak{b} = \mathfrak{m}(\mathfrak{b} : \mathfrak{m}) \subset \mathfrak{m}(\mathfrak{a} : \mathfrak{m}) \subset \mathfrak{a} \cap \mathfrak{m}^{r+1} = \mathfrak{b}$ by (2). Thus, we have shown that $\mathfrak{m}(\mathfrak{a} : \mathfrak{m}) = \mathfrak{m}(\mathfrak{b} : \mathfrak{m})$. The ideals $\mathfrak{a} : \mathfrak{m}$ and $\mathfrak{b} : \mathfrak{m}$ are complete [by B(6.16)(5) since $\mathfrak{a}$ and $\mathfrak{b}$ are complete], hence we have $\mathfrak{a} : \mathfrak{m} = \mathfrak{b} : \mathfrak{m}$ [by (3.2)].

(3.24) Corollary: *Let $\mathfrak{a}$ be a complete $\mathfrak{m}$-primary ideal of R, set $h := \mathrm{ord}(\mathfrak{a}) - \deg(c(\mathfrak{a}))$, and assume that $h > 0$. For every $j \in \{1, \ldots, h\}$ set $\mathfrak{b}_j := \mathfrak{a} : \mathfrak{m}^j$. Then the ideals $\mathfrak{b}_1, \ldots, \mathfrak{b}_{h-1}$ are complete $\mathfrak{m}$-primary ideals, $\mathfrak{b}_h$ is a complete $\mathfrak{m}$-primary ideal which is not divisible by $\mathfrak{m}$ if $\mathfrak{a}$ is not a power of $\mathfrak{m}$ and it is equal to R otherwise, and we have $\mathfrak{a} = \mathfrak{m}^j \mathfrak{b}_j$ for every $j \in \{0, \ldots, h\}$.*

Proof: (1) Let l be a positive integer; if $\mathfrak{m}^l$ divides $\mathfrak{a}$, then we have $\mathfrak{a} = \mathfrak{m}^l \mathfrak{b}$ with $\mathfrak{b} := \mathfrak{a} : \mathfrak{m}^l$ [cf. (2.0)].
(2) The first assertion follows from B(6.16)(5). If $h = 1$, then $\mathfrak{m}$ divides $\mathfrak{a}$ [cf. (3.22)(1)], and therefore we have $\mathfrak{a} = \mathfrak{m}\mathfrak{b}_1$; we have $\mathfrak{a} = \mathfrak{m}$ iff $\mathfrak{b}_1 = R$, and if $\mathfrak{a} \neq \mathfrak{m}$, then $\mathfrak{b}_1$ is $\mathfrak{m}$-primary, and since $\mathrm{ord}(\mathfrak{b}_1) - \deg(c(\mathfrak{b}_1)) = \mathrm{ord}(\mathfrak{a}) - 1 - \deg(c(\mathfrak{a})) = 0$, $\mathfrak{m}$ does not divide $\mathfrak{b}_1$. We get the assertion of the corollary by repeated application of (3.22)(1).

(3.25) Corollary: *Let $\mathfrak{a}$ and $\mathfrak{b}$ be complete $\mathfrak{m}$-primary ideals of R. If $\mathfrak{m}$ divides $\mathfrak{a}\mathfrak{b}$, then $\mathfrak{m}$ divides one of the ideals $\mathfrak{a}$, $\mathfrak{b}$.*

Proof: We have $\mathrm{ord}(\mathfrak{a}\mathfrak{b}) = \mathrm{ord}(\mathfrak{a}) + \mathrm{ord}(\mathfrak{b})$ and $\deg(c(\mathfrak{a}\mathfrak{b})) = \deg(c(\mathfrak{a})) + \deg(c(\mathfrak{b}))$. Since $\mathfrak{m}$ divides $\mathfrak{a}\mathfrak{b}$, we have $\deg(c(\mathfrak{a}\mathfrak{b})) < \mathrm{ord}(\mathfrak{a}\mathfrak{b})$, hence $\deg(c(\mathfrak{a})) < \mathrm{ord}(\mathfrak{a})$ in which case $\mathfrak{m}$ divides $\mathfrak{a}$ [cf. (3.22)(1)], or $\deg(c(\mathfrak{b})) < \mathrm{ord}(\mathfrak{b})$ in which case $\mathfrak{m}$ divides $\mathfrak{b}$.

(3.26) Corollary: *Let $\mathfrak{a}$ be a complete $\mathfrak{m}$-primary ideal of R, and let $p \in \mathbb{P}$. If $\mathfrak{a} = \mathfrak{a}S_p \cap R$, then $c(\mathfrak{a})$ is a power of p. Furthermore, setting $r := \mathrm{ord}(\mathfrak{a})$, $s := \deg(c(\mathfrak{a}))$, $\mathfrak{b} := \mathfrak{a} : \mathfrak{m}^{r-s}$, we have the following: the ideal $\mathfrak{b}$ is a complete $\mathfrak{m}$-primary ideal of R or $\mathfrak{b} = R$, and we have*

$$\mathfrak{a} = \mathfrak{b}\mathfrak{m}^{r-s}, \quad \mathfrak{b}S_p \cap R = \mathfrak{b}.$$

Proof: If p does not divide $c(\mathfrak{a})$, then we have $\mathfrak{a}S_p = \mathfrak{m}^r S_p$ by (2.15), hence $\mathfrak{a} = \mathfrak{m}^r$ [cf. (2.8)], and therefore $c(\mathfrak{a}) = (1)$, $\mathfrak{b} = R$. Now we assume that p divides $c(\mathfrak{a})$; let l be the highest power of p which divides $c(\mathfrak{a})$, and set $t := \deg(p^l)$. Then we have $t \leq s \leq r$. We set $\mathfrak{c} := \mathfrak{a} : \mathfrak{m}^{r-t}$. Then $\mathfrak{c}$ is complete and $\mathfrak{m}$-primary and we have $\mathfrak{a} = \mathfrak{c}\mathfrak{m}^{r-t}$ [cf. (3.24)], and therefore we have $\mathrm{ord}(\mathfrak{c}) = t$ and $c(\mathfrak{a}) = c(\mathfrak{c})$, hence $s \leq t$, hence $s = t$, $\mathfrak{b} = \mathfrak{c}$, and therefore we have $c(\mathfrak{a}) = c(\mathfrak{b}) = p^l$; this implies that $\mathfrak{b} = \mathfrak{b}S_p \cap R$ [cf. (3.10)].

3.4 An Existence Theorem

Now we prove the promised existence theorem [cf. (3.16)].

(3.27) Theorem: *Let $P \subset \mathbb{P}$ be finite and non-empty, for $p \in P$ let $\mathfrak{c}_p$ be a complete $\mathfrak{m}(S_p)$-primary ideal of S_p, and set $\mathfrak{c}_p = S_p$ for every $p \in \mathbb{P} \setminus P$. Then there exists a unique complete $\mathfrak{m}$-primary ideal $\mathfrak{a}$ of R which is not divisible by $\mathfrak{m}$ with the following two properties:*
(i) $\mathfrak{a}^{S_p} = \mathfrak{c}_p$ for every $p \in \mathbb{P}$,

(ii) *if $\tilde{a}$ is a complete $\mathfrak{m}$-primary ideal of R with $\tilde{a}^{S_p} = c_p$ for every $p \in \mathbb{P}$, then we have $\tilde{a} = \mathfrak{m}^t a$ for some $t \in \mathbb{N}_0$.*

Proof: Let $\{x, y\}$ be a system of generators of $\mathfrak{m}$, and set $B_1 := R[y/x]$, $B_2 := R[x/y]$. For $i \in \{1, 2\}$ we set $P_i := \{p \in P \mid S_p \supset B_i\}$. First, we assume that P_1 and P_2 are not empty, and we set

$$S_i := \bigcap_{p \in P_i} S_p, \quad c_i := \bigcap_{p \in P_i} (S_i \cap c_p), \quad \mathfrak{b}_i := c_i \cap B_i \text{ for } i \in \{1, 2\}.$$

Let $i \in \{1, 2\}$. For every $p \in P_i$ let $\mathfrak{n}_{ip}$ be the maximal ideal of B_i defined by p, and set $\Sigma_i := B_i \setminus \bigcup_{p \in P_i} \mathfrak{n}_{ip}$; then we have $\Sigma_i^{-1} B_i = S_i$, and we have $\mathfrak{b}_i = \bigcap_{p \in P_i} \mathfrak{q}_{ip}$ where $\mathfrak{q}_{ip}$ is $\mathfrak{n}_{ip}$-primary and $\mathfrak{b}_i S_p = c_p$ for every $p \in P_i$ [cf. (2.17)]. Let $\{f_{i1}, \ldots, f_{ih}\}$ be a system of generators of $\mathfrak{b}_i$. We choose $n \in \mathbb{N}$ with $x^n f_{1j}$, $y^n f_{2j} \in R$ for $j \in \{1, \ldots, h\}$ and $x^n \in \mathfrak{q}_{1p}$ for $p \in P_1$, $y^n \in \mathfrak{q}_{2p}$ for $p \in P_2$. Then we have $y^n = (y/x)^n x^n \in \mathfrak{q}_{1p}$ for $p \in P_1$, $x^n = (x/y)^n y^n \in \mathfrak{q}_{2p}$ for $p \in P_2$, and therefore the elements x^n, y^n, $x^n f_{ij}$ for $j \in \{1, \ldots, h\}$ and $i = 1, 2$ are a system of generators of $\mathfrak{m}^n c_p$ for every $p \in \mathbb{P}$. Second, if $P_1 = \emptyset$ or $P_2 = \emptyset$, it is clear that there exist $n \in \mathbb{N}$ and finitely may elements in R which are a system of generators of $\mathfrak{m}^n c_p$ for every $p \in \mathbb{P}$.

For every $l \in \mathbb{N}$ we set

$$a_l := \bigcap_{p \in \mathbb{P}} \mathfrak{m}^l c_p \subset \bigcap_{p \in \mathbb{P}} S_p = R;$$

then we have $a_l S_p \subset \mathfrak{m}^l c_p$ for every $p \in \mathbb{P}$, hence, in particular, $a_l \neq R$, $\mathfrak{m}^l c_p$ is a complete ideal of S_p [since $\mathfrak{m}^l S_p$ is a principal ideal of S_p], and we have

$$\mathfrak{m}^n c_p = a_n S_p \quad \text{for every } p \in \mathbb{P}.$$

We choose $r \in \mathbb{N}$ minimal along all integers $l \in \mathbb{N}$ such that, for every $p \in \mathbb{P}$, $\mathfrak{m}^l c_p$ is the integral closure of $a_l S_p$ in S_p, and we set $a := a_r$. We show that a satisfies the assertions (i) and (ii) of (3.27).

Let $l \in \mathbb{N}$. If $z \in R$ is integral over a_l, then z is integral over $a_l S_p \subset \mathfrak{m}^l c_p$ for every $p \in \mathbb{P}$, hence $z \in \bigcap_{p \in \mathbb{P}} \mathfrak{m}^l c_p = a_l$, and therefore a_l is complete. In particular, since a_r is complete, for every $p \in P$ the ideal $a S_p$ is a complete ideal of S_p [cf. (3.4)], and therefore we have $a S_p = \mathfrak{m}^r c_p$ for every $p \in \mathbb{P}$. By our choice of n we have, for every $p \in \mathbb{P}$, $\mathfrak{m}^n S_p \subset c_p$, hence $\mathfrak{m}^{n+l} S_p \subset \mathfrak{m}^l c_p$, and therefore, by (3.8), we have

$$\mathfrak{m}^{n+l} = \bigcap_{p \in \mathbb{P}} \mathfrak{m}^{n+l} S_p \subset \bigcap_{p \in \mathbb{P}} \mathfrak{m}^l c_p = a_l.$$

Since $a = a_r \neq R$, we see that a is a complete $\mathfrak{m}$-primary ideal of R. We have

$$a = \bigcap_{p \in \mathbb{P}} a S_p = \bigcap_{p \in \mathbb{P}} \mathfrak{m}^r c_p \subset \bigcap_{p \in \mathbb{P}} \mathfrak{m}^r S_p = \mathfrak{m}^r.$$

Suppose that $\mathfrak{a} \subset \mathfrak{m}^{r+1}$. Then, for every $p \in \mathbb{P}$, we have $\mathfrak{m}^r \mathfrak{c}_p = \mathfrak{a} S_p \subset \mathfrak{m}^{r+1} S_p$, hence $\mathfrak{c}_p \subset \mathfrak{m} S_p$ [note that $\mathfrak{m} S_p$ is a principal ideal of S_p], contrary to our assumption [remember that $\mathbb{P}$ is infinite]. Therefore we have $\mathrm{ord}(\mathfrak{a}) = r$. For every $p \in \mathbb{P}$ we have

$$\mathfrak{a}^{S_p} = (\mathfrak{m} S_p)^{-r} \mathfrak{a} S_p = (\mathfrak{m} S_p)^{-r} \mathfrak{m}^r \mathfrak{c}_p = \mathfrak{c}_p;$$

hence (i) is satisfied. Now we prove (ii) [from which, in particular, the uniqueness of $\mathfrak{a}$ follows]. Set $s := \mathrm{ord}(\widetilde{\mathfrak{a}})$. We have, for $p \in \mathbb{P}$, $(\mathfrak{m} S_p)^{-s} \widetilde{\mathfrak{a}} S_p = \widetilde{\mathfrak{a}}^{S_p} = \mathfrak{c}_p$, hence $\mathfrak{m}^s \mathfrak{c}_p = \widetilde{\mathfrak{a}} S_p$, and therefore, again by (3.8), $\widetilde{\mathfrak{a}} = \bigcap_{p \in \mathbb{P}} \widetilde{\mathfrak{a}} S_p = \bigcap_{p \in \mathbb{P}} \mathfrak{m}^s \mathfrak{c}_p = \mathfrak{a}_s$, hence $s \geq r$ by definition of r, and therefore $\mathfrak{m}^s \mathfrak{c}_p = \mathfrak{m}^{s-r} \mathfrak{a} S_p$ for every $p \in \mathbb{P}$, hence $\widetilde{\mathfrak{a}} = \bigcap_{p \in \mathbb{P}} \mathfrak{m}^s \mathfrak{c}_p = \bigcap_{p \in \mathbb{P}} \mathfrak{m}^{s-r} \mathfrak{a} S_p = \mathfrak{m}^{s-r} \mathfrak{a}$ [note that $\mathfrak{m}^{s-r} \mathfrak{a}$ is complete by (3.20)]. Lastly, we show that $\mathfrak{m}$ does not divide $\mathfrak{a}$. Suppose that $\mathfrak{m}$ divides $\mathfrak{a}$. Then we have $\mathfrak{a} = \mathfrak{m}\mathfrak{b}$ where $\mathfrak{b}$ is complete and $\mathfrak{m}$-primary, and since $\mathfrak{b}^{S_p} = \mathfrak{a}^{S_p}$ for every $p \in \mathbb{P}$, we get $\mathfrak{b} = \mathfrak{m}^t \mathfrak{a}$ for some $t \in \mathbb{N}_0$ by (ii), hence $\mathfrak{a} = \mathfrak{m}^{t+1} \mathfrak{a}$ which is absurd.

4 Factorization of Complete Ideals

4.1 Preliminary Results

(4.0) We keep the notations of the earlier sections of this chapter.

(4.1) SIMPLE IDEALS: Let A be an integrally closed noetherian domain. A non-zero proper ideal $\mathfrak{a}$ of A is called *simple* if it is not the product of two proper ideals of A.
(1) We show: *Every non-zero proper ideal of A is a product of simple ideals of A.* In fact, suppose not; then the set of non-zero proper ideals of A which are not a product of simple ideals is not empty, hence it has a maximal element $\mathfrak{a}$. Then we have $\mathfrak{a} = \mathfrak{b}\mathfrak{c}$ where $\mathfrak{b}$, $\mathfrak{c}$ are proper ideals of A. We have $\mathfrak{a} \subset \mathfrak{b}$ and $\mathfrak{a} \subset \mathfrak{c}$; if both ideals $\mathfrak{b}$, $\mathfrak{c}$ properly contain $\mathfrak{a}$, then $\mathfrak{b}$ and $\mathfrak{c}$ are products of simple ideals, hence $\mathfrak{a}$ is also a product of simple ideals, contradicting the choice of $\mathfrak{a}$. Thus, one of these ideals is equal to $\mathfrak{a}$, say $\mathfrak{b} = \mathfrak{a}$. From $\mathfrak{a} = \mathfrak{a}\mathfrak{c}$ we get $\mathfrak{a} = \mathfrak{a}\mathfrak{c}^n$ for every $n \in \mathbb{N}$, hence $\mathfrak{a} \subset \mathfrak{c}^n$ for every $n \in \mathbb{N}$, and therefore we have $\mathfrak{a} = \{0\}$ [by Krull's intersection theorem, cf. [63], Cor. 5.4], contradicting the choice of $\mathfrak{a}$.
(2) We show: *Every non-zero proper integrally closed ideal of A is a product of simple integrally closed ideals of A.*
Let $\mathfrak{a}$ be a non-zero proper integrally closed ideal of A. If $\mathfrak{a}$ is not simple, then $\mathfrak{a}$ can be factored into a product of simple integrally closed ideals. In fact, if $\mathfrak{a} = \mathfrak{a}_1 \cdots \mathfrak{a}_h$ where $\mathfrak{a}_1, \ldots, \mathfrak{a}_h$ are proper ideals of A, and if, for any ideal $\mathfrak{b}$ of A, we denote by $\overline{\mathfrak{b}}$ its integral closure in A, then we have $\mathfrak{a} = \overline{\mathfrak{a}} \supset \overline{\mathfrak{a}}_1 \cdots \overline{\mathfrak{a}}_h \supset \mathfrak{a}_1 \cdots \mathfrak{a}_h = \mathfrak{a}$ [cf. B(6.16)(4)], whence $\mathfrak{a} = \overline{\mathfrak{a}}_1 \cdots \overline{\mathfrak{a}}_h$, and $\overline{\mathfrak{a}}_1, \ldots, \overline{\mathfrak{a}}_h$ are proper integrally closed ideals. Now we can argue as in (1), and we have proved the assertion.

(4.2) REMARK: A non-zero proper complete ideal of R is a product of simple complete ideals of R as we just showed. The main results of this section are the following: This factorization is unique [cf. (4.17)]; moreover, a product of complete ideals is complete [cf. (4.13)].

For the proof of the next proposition the following two lemmas shall be useful.

(4.3) Lemma: Let $S = \bigoplus_{n \geq 0} S_n$ be the graded polynomial ring over a field k in two indeterminates. Let $h \geq 2$, and let $f_1, \ldots, f_h$ be non-zero homogeneous polynomials in S which are pairwise relatively prime. We set $s_i := \deg(f_i)$, $g_i := (f_1 \cdots f_h)/f_i$ for $i \in \{1, \ldots, h\}$, and $s := s_1 + \cdots + s_h$. Then the homogeneous ideal $Sg_1 + \cdots + Sg_h$ contains S_t for every $t \geq s - 1$.

Proof: For every $n \in \mathbb{N}_0$ we have $S_1 S_n = S_{n+1}$, hence it is enough to show that $S_0 \subset Sg_1 + \cdots + Sg_h$ if $s = 0$, and that $S_{s-1} \subset Sg_1 + \cdots + Sg_h$ if $s \geq 1$.
(a) We consider the case $h = 2$. If one of the integers s_1, s_2 is equal to zero, then $S_0 \subset Sg_1 + Sg_2$. Now let $s_1 \geq 1$, $s_2 \geq 1$. For $i \in \{1, 2\}$ let $a_i \in S_{s_i - 1}$. If $a_1 g_1 + a_2 g_2 = 0$, then g_2 divides a_1, hence $a_1 = 0$ [since $\deg(a_1) = s_1 - 1 < \deg(g_2)$], and also $a_2 = 0$. Therefore the k-linear map $(a_1, a_2) \mapsto a_1 g_1 + a_2 g_2 : S_{s_1 - 1} \oplus S_{s_2 - 1} \to S_{s_1 + s_2 - 1}$ is injective, and since $\dim_k(S_{s_1 - 1}) + \dim_k(S_{s_2 - 1}) = s_1 + s_2 = \dim_k(S_{s_1 + s_2 - 1})$, it is also surjective.
(b) Let $h \geq 2$, and assume that the assertion has been proved for h pairwise relatively prime non-zero homogeneous polynomials $f_1, \ldots, f_h \in S$. Let $f_{h+1} \in S$ be a non-zero homogeneous polynomial such that f_i and f_{h+1} are relatively prime for $i \in \{1, \ldots, h\}$. We set $s_i := \deg(f_i)$ and $g_i := (f_1 \cdots f_{h+1})/f_i$ for $i \in \{1, \ldots, h+1\}$.
The cases $s_1 + \cdots + s_h = 0$ or $s_{h+1} = 0$ are easy to settle; hence we may assume that $s_1 + \cdots + s_h \geq 1$ and that $s_{h+1} \geq 1$. We apply (a) to the relatively prime polynomials g_{h+1} and f_{h+1}, and we find that

$$S_{s_1 + \cdots + s_{h+1} - 1} = \{bf_{h+1} + a_{h+1}g_{h+1} \mid a \in S_{s_1 + \cdots + s_h - 1}, \, b \in S_{s_{h+1} - 1}\}.$$

By induction, applied to $f_1, \ldots, f_h$, we see that every $b \in S_{s_1 + \cdots + s_h - 1}$ can be written as

$$b = a_1(f_1 \cdots f_h)/f_1 + \cdots + a_h(f_1 \cdots f_h)/f_h, \; a_i \in S_{s_1 + \cdots + s_h - 1 - s_i} \text{ for } i \in \{1, \ldots, h\}.$$

Therefore we have

$$S_{s_1 + \cdots + s_{h+1} - 1} = S_{s_1 + \cdots + s_{h+1} - 1 - s_1} g_1 + \cdots + S_{s_1 + \cdots + s_{h+1} - 1 - s_{h+1}} g_{h+1}.$$

(4.4) Lemma: Let $q_1, \ldots, q_h$ be pairwise relatively prime homogenous principal ideals of $\mathrm{gr}_{\mathfrak{m}}(R)$, and set $r := \deg(q_1) + \cdots + \deg(q_h)$. Let $f \in R$ be of order r and such that $(\mathrm{In}(f)) = q_1 \cdots q_h$. For every $j \in \mathbb{N}$ there exist elements $f_{1j}, \ldots, f_{hj}$ in R such that $(\mathrm{In}(f_{ij})) = q_i$ for $i \in \{1, \ldots, h\}$, and that $f - f_{1j} \cdots f_{hj} \in \mathfrak{m}^{r+j}$.

Proof: The assertion is trivially true if $h = 1$. Now we assume that $h \geq 2$. Let $i \in \{1, \ldots, h\}$. We set $s_i := \deg(q_i)$, and we choose $\overline{f}_i \in \mathrm{gr}_{\mathfrak{m}}(R)$ with $q_i = (\overline{f}_i)$ and

$f_i \in R$ with $\text{In}(f_i) = \overline{f}_i$. For $j = 1$ we set $f_{ij} := f_i$. Let $j \geq 1$, and assume that there exist elements $f_{ij} \in R$ with $\text{In}(f_{ij}) = \overline{f}_i$ and with $f - f_{1j} \cdots f_{hj} \in \mathfrak{m}^{r+j}$. If $f - f_{1j} \cdots f_{hj} \in \mathfrak{m}^{r+j+1}$, then we set $f_{i,j+1} := f_{ij}$ for $i \in \{1,\ldots,h\}$. In the other case $\text{In}(f - f_{1j} \cdots f_{hj})$ is homogeneous of degree $r + j$; by the preceding lemma there exist homogeneous polynomials $\overline{a}_1,\ldots,\overline{a}_h \in \text{gr}_\mathfrak{m}(R)$ with $\deg(\overline{a}_i) = r+j-s_i$ for $i \in \{1,\ldots,h\}$ such that

$$\text{In}(f - f_{1j} \cdots f_{hj}) = \sum_{i=1}^{h} \overline{a}_i (\overline{f}_1 \cdots \overline{f}_h)/\overline{f}_i.$$

For $i \in \{1,\ldots,h\}$ we choose an element $a_i \in R$ with $\text{In}(a_i) = \overline{a}_i$, hence with $\text{ord}(a_i) = \deg(\overline{a}_i)$, and we set $f_{i,j+1} := f_{ij} + a_i$. Then we have

$$f_{1,j+1} \cdots f_{h,j+1} = f_{1j} \cdots f_{hj} + \sum_{i=1}^{h} a_i(f_1 \cdots f_h)/f_i + b,$$

and b is a sum of terms of the form $f_{i_1j} \cdots f_{i_kj} a_{i_{k+1}} \cdots a_{i_h}$ where $k < h - 1$ and where $(i_1,\ldots,i_k,i_{k+1}\ldots,i_h)$ is a permutation of $(1,\ldots,h)$. We have

$$\text{ord}(b) \geq s_{i_1} + \cdots + s_{i_k} + (s+j-s_{i_{k+1}}) + \cdots + (s+j-s_{i_h}) > s+j+1,$$

and therefore the elements $f_{1,j+1},\ldots,f_{h,j+1}$ satisfy the assertion of the lemma.

(4.5) Proposition: *Let $\mathfrak{a}$ be a complete $\mathfrak{m}$-primary ideal of R, and let $p_1,\ldots,p_h$ be pairwise different elements of $\mathbb{P}$. We set $P := \{p_1,\ldots,p_h\}$, $r := \text{ord}(\mathfrak{a})$ and $s := \deg(c(\mathfrak{a}))$. We assume that $\mathfrak{a} = \mathfrak{a}S_P \cap R$. Then:*
(1) Let $i \in \{1,\ldots,h\}$. We set

$$\mathfrak{a}_i := \mathfrak{a}S_{p_i} \cap R.$$

Then $\mathfrak{a}_i$ is a complete $\mathfrak{m}$-primary ideal of R, $\text{ord}(\mathfrak{a}_i) = r$, $c(\mathfrak{a}_i) = p_i^{e_i}$ is a power of p_i [with $e_i \geq 0$], and we have

$$\mathfrak{a} = \mathfrak{a}_1 \cap \cdots \cap \mathfrak{a}_h.$$

(2) Let $i \in \{1,\ldots,h\}$. We set

$$s_i := \deg(c(\mathfrak{a}_i)), \quad \mathfrak{b}_i := \mathfrak{a}_i : \mathfrak{m}^{r-s_i}.$$

Then $\mathfrak{b}_i$ is a complete $\mathfrak{m}$-primary ideal of R or $\mathfrak{b}_i = R$, and we have $\mathfrak{a}_i = \mathfrak{m}^{r-s_i}\mathfrak{b}_i$ and $\text{ord}(\mathfrak{b}_i) = s_i$.
(3) We have $s_1 + \cdots + s_h = s \leq r$ and

$$\mathfrak{m}^{r-s}\mathfrak{b}_1 \cdots \mathfrak{b}_h = \mathfrak{a}.$$

(4) Every $p \in \mathbb{P}$ which divides $c(\mathfrak{a})$ is contained in P.

(5) *Relabelling the ideals $p_1, \ldots, p_h$ in such a way that $e_i > 0$ for $i \in \{1, \ldots, l\}$, and $e_i = 0$ for $i \in \{l+1, \ldots, h\}$ [where $0 \le l \le h$], we have*

$$\mathfrak{a} = \mathfrak{m}^{r-s} \mathfrak{b}_1 \cdots \mathfrak{b}_l, \quad c(\mathfrak{a}) = p_1^{e_1} \cdots p_l^{e_l},$$

and for $i \in \{1, \ldots, l\}$ we have $c(\mathfrak{b}_i) = p_i^{e_i}$.

Proof: We set $S_i := S_{p_i}$ for $i \in \{1, \ldots, h\}$ and $S := S_1 \cap \cdots \cap S_h$.
(1) and (2): We have $\mathfrak{a}S = \mathfrak{a}S_1 \cap \cdots \cap \mathfrak{a}S_h$ [cf. (3.0)], and $\mathfrak{a}S \cap R = \mathfrak{a}$ by assumption, hence we have

$$\mathfrak{a} = \mathfrak{a}_1 \cap \cdots \cap \mathfrak{a}_h.$$

Let $i \in \{1, \ldots, h\}$. Since $\mathfrak{a}S_i$ is a complete ideal of S_i [cf. (3.4)], $\mathfrak{a}_i$ is also a complete ideal of R [cf. (3.3)]. From $\mathfrak{a} \subset \mathfrak{a}_i$ we see that $\mathfrak{a}_i$ contains a power of $\mathfrak{m}$, we have $\mathrm{ord}(\mathfrak{a}_i) \le r$, and since $\mathfrak{a}_i \subset \mathfrak{a}S_i$, we have $\mathrm{ord}(\mathfrak{a}_i) = \nu_R(\mathfrak{a}_i) \ge \nu_R(\mathfrak{a}) = r$, and therefore we obtain $\mathrm{ord}(\mathfrak{a}_i) = r$, hence $\mathfrak{a}_i$ is $\mathfrak{m}$-primary. We have $\mathfrak{a}_i = \mathfrak{a}_i S_i \cap R$ [since $\mathfrak{a}_i$ is contracted from S_i, it is the contraction of its extension], and therefore $c(\mathfrak{a}_i) =: p_i^{e_i}$ is a power of p_i, $\mathfrak{b}_i$ is complete and $\mathfrak{m}$-primary or $\mathfrak{b}_i = R$, and we have $\mathfrak{b}_i \mathfrak{m}^{r-s_i} = \mathfrak{a}_i$ [cf. (3.26)], hence $\mathrm{ord}(\mathfrak{b}_i) = s_i$ and $c(\mathfrak{b}_i) = c(\mathfrak{a}_i)$.
(3) We set $t := s_1 + \cdots + s_h$ and $\mathfrak{b} := \mathfrak{b}_1 \cdots \mathfrak{b}_h$. Since $c(\mathfrak{a}_i) = p_i^{e_i}$ divides $c(\mathfrak{a})$, also $c(\mathfrak{a}_1 \cdots \mathfrak{a}_h)$ divides $c(\mathfrak{a})$, hence $t = s_1 + \cdots + s_h \le \deg(c(\mathfrak{a})) \le r$. Since $\mathfrak{m}^{r-t} \mathfrak{b} \subset \mathfrak{b}_i \mathfrak{m}^{r-s_i} = \mathfrak{a}_i$ for $i \in \{1, \ldots, h\}$, we have $\mathfrak{m}^{r-t} \mathfrak{b} \subset \mathfrak{a}_1 \cap \cdots \cap \mathfrak{a}_h$, hence

$$\mathfrak{m}^{r-t} \mathfrak{b} \subset \mathfrak{a}. \tag{$*$}$$

For $i \in \{1, \ldots, h\}$ the ideal $\mathfrak{b}_i$ is $\mathfrak{m}$-primary or $\mathfrak{b}_i = R$; therefore $\mathfrak{m}^{r-t} \mathfrak{b} \subset \mathfrak{a}$ is $\mathfrak{m}$-primary, hence there exists $j \in \mathbb{N}$ with $\mathfrak{m}^{r+j} \subset \mathfrak{m}^{r-t} \mathfrak{b}$.
We have $c(\mathfrak{b}) = p_1^{e_1} \cdots p_h^{e_h}$. Clearly we have $\mathrm{ord}(\mathfrak{m}^{r-t} \mathfrak{b}) = \mathrm{ord}(\mathfrak{a})$, hence $c(\mathfrak{a})$ divides $c(\mathfrak{b})$ [cf. (2.13)]; on the other hand, $c(\mathfrak{b}) = c(\mathfrak{a}_1 \cdots \mathfrak{a}_h)$ divides $c(\mathfrak{b})$. Therefore we have

$$c(\mathfrak{a}_1 \cdots \mathfrak{a}_h) = c(\mathfrak{a}) = p_1^{e_1} \cdots p_h^{e_h},$$

and, in particular, $s = t = s_1 + \cdots + s_h$.
Our aim is to show equality in $(*)$. We relabel the ideals $p_1, \ldots, p_h$ as in (5). If $l = 0$, then we have $s = 0$ and $\mathfrak{a} = \mathfrak{m}^r$ by (3.11), hence $\mathfrak{a}_1 = \cdots = \mathfrak{a}_h = \mathfrak{m}^r$ by (2.8), and therefore $\mathfrak{b}_1 = \cdots = \mathfrak{b}_h = R$. In particular, we have equality in $(*)$.
Now we consider the case $l > 0$. Then $c(\mathfrak{a}) = p_1^{e_1} \cdots p_l^{e_l}$. Let $f \in \mathfrak{a}$ be of order r. The ideal $(\mathrm{In}(f))$ therefore has the form $p_1^{e_1'} \cdots p_l^{e_l'} q$ where $e_1' \ge e_1, \ldots, e_l' \ge e_l$, q is generated by a homogeneous element of $\mathrm{gr}_{\mathfrak{m}}(R)$ [$q = (1)$ is allowed] and none of the ideals $p_1, \ldots, p_l$ divides q; moreover, there exists an element $f \in \mathfrak{a}$ of order r with $(\mathrm{In}(f)) = p_1^{e_1} \cdots p_l^{e_l}$.
(a) Let $f \in \mathfrak{a}$ be of order r and with $(\mathrm{In}(f)) = p_1^{e_1} \cdots p_l^{e_l} q$. By the lemma above, there exist elements $f_1, \ldots, f_l$ and g in R such that $f - f_1 \cdots f_l g \in \mathfrak{m}^{r+j} \subset \mathfrak{a}$ and with $(\mathrm{In}(f_i)) = p_i^{e_i}$ for $i \in \{1, \ldots, l\}$ and $(\mathrm{In}(g)) = q$. Now we have $f_1 \cdots f_h g \in \mathfrak{a}$. Let $i \in \{1, \ldots, l\}$. We choose $x \in \mathfrak{m} \setminus \mathfrak{m}^2$ such that $(\mathrm{In}(x)) \ne p_i$. Then the element $f_1 \cdots f_{i-1} f_{i+1} \cdots f_l g / x^{r-s_i}$ is a unit of S_i [cf. (2.6)(3)], hence $f_i / x^{s_i} \in \mathfrak{a}S_i =$

$x^{-r} \mathfrak{a} S_i$, and therefore we have $f_i \mathfrak{m}^{r-s_i} \in \mathfrak{a} S_i \cap R = \mathfrak{a}_i$, hence $f_i \in \mathfrak{a}_i : \mathfrak{m}^{r-s_i} = \mathfrak{b}_i$. Therefore we have $f_1 \cdots f_l \in \mathfrak{b}$. Since $\mathrm{ord}(g) = \deg(\mathrm{In}(g)) = r - s$, we have $g \in \mathfrak{m}^{r-s}$, hence $f_1 \cdots f_l g$ lies in $\mathfrak{m}^{r-s} \mathfrak{b}$.

(b) Let f' be an element of $\mathfrak{a}$ of order r with $(\mathrm{In}(f')) = p_1^{e_1'} \cdots p_h^{e_h'} q$ where $e_i' > e_i$ for at least one $i \in \{1, \ldots, l\}$. Then $f + f'$ satisfies the assumptions of (a) for f, hence $f + f' \in \mathfrak{m}^{r-s} \mathfrak{b}$, and therefore f' lies in $\mathfrak{m}^{r-s} \mathfrak{b}$.

(c) Lastly, let $f' \in \mathfrak{a}$ and $\mathrm{ord}(f') > r$. Then $f + f'$ satisfies the assumptions of (a) for f, and again we find that f' lies in $\mathfrak{m}^{r-s} \mathfrak{b}$. Thus, we have shown that $\mathfrak{a}$ is contained in $\mathfrak{m}^{r-s} \mathfrak{b}$, hence we have $\mathfrak{a} = \mathfrak{m}^{r-s} \mathfrak{b}$.

(4) and (5) follow immediately.

(4.6) Corollary: *If $\mathfrak{a}$ is a simple complete $\mathfrak{m}$-primary ideal different from $\mathfrak{m}$, then $c(\mathfrak{a})$ is a positive power of $p \in \mathbb{P}$, and $\mathfrak{a}^{S_p}$ is a simple complete $\mathfrak{m}(S_p)$-primary ideal of S_p.*

Proof: The first assertion follows from (4.5)(5). (Another proof follows from (3.27).) We set $S := S_p$, and we have to show that $\mathfrak{a}^S$ is a simple ideal of S. Suppose that $\mathfrak{a}^S$ is not simple, and write $\mathfrak{a}^S = \mathfrak{b}_1 \cdots \mathfrak{b}_h$ with $h > 1$ where $\mathfrak{b}_1, \ldots, \mathfrak{b}_h$ are complete $\mathfrak{m}(S)$-primary ideals of S [cf. (4.1)(2)]. By (3.27)(i) there exist complete $\mathfrak{m}$-primary ideals $\mathfrak{a}_1, \ldots, \mathfrak{a}_h$ of R, not divisible by $\mathfrak{m}$, with $\mathfrak{a}_i^S = \mathfrak{b}_i$ for $i \in \{1, \ldots, h\}$. By (3.27)(ii) we find that $\mathfrak{a}_1 \cdots \mathfrak{a}_h = \mathfrak{m}^t \mathfrak{a}$ for some $t \in \mathbb{N}_0$ which is absurd since $\mathfrak{a}$ is simple.

We want to show that a product of complete $\mathfrak{m}$-primary ideals of R is complete, again. We can do this now in a special case.

(4.7) Lemma: *Let $\mathfrak{a}_1, \ldots, \mathfrak{a}_n$ be complete $\mathfrak{m}$-primary ideals of R. If the ideals $c(\mathfrak{a}_1), \ldots, c(\mathfrak{a}_n)$ are pairwise coprime, then $\mathfrak{a}_1 \cdots \mathfrak{a}_n$ is a complete $\mathfrak{m}$-primary ideal.*

Proof: Since the product of a complete ideal with a power of the maximal ideal is complete, again [cf. (3.20)], and since, if $\mathfrak{a}_i$ is not a power of $\mathfrak{m}$, we can write $\mathfrak{a}_i = \mathfrak{m}^{t_i} \mathfrak{a}_i'$ with $t_i \in \mathbb{N}_0$ and a complete $\mathfrak{m}$-primary ideal $\mathfrak{a}_i'$ which satisfies $\mathrm{ord}(\mathfrak{a}_i') = \deg(c(\mathfrak{a}_i'))$ for $i \in \{1, \ldots, n\}$ [cf. (3.22)], we may assume, to begin with, that $\mathrm{ord}(\mathfrak{a}_i) = \deg(c(\mathfrak{a}_i))$ for $i \in \{1, \ldots, n\}$. By (4.5)(5) and the assumption that, for $i, j \in \{1, \ldots, n\}$, $i \neq j$, the ideals $c(\mathfrak{a}_i)$ and $c(\mathfrak{a}_j)$ have no non-trivial common divisor, it is easily seen that it is enough to show the following: Let $\mathfrak{b}_1, \ldots, \mathfrak{b}_h$ be complete $\mathfrak{m}$-primary ideals where, for $i \in \{1, \ldots, h\}$, we have $c(\mathfrak{b}_i) = p_i^{e_i}$ with $p_i \in \mathbb{P}$ and $e_i > 0$, with $\mathrm{ord}(\mathfrak{b}_i) = \deg(c(\mathfrak{b}_i)) =: s_i$, and with pairwise different $p_1, \ldots, p_h$. Then $\mathfrak{a}' := \mathfrak{b}_1 \cdots \mathfrak{b}_h$ is complete.

We set $s := s_1 + \cdots + s_h$; then we have $\mathrm{ord}(\mathfrak{a}') = s$. For $i \in \{1, \ldots, h\}$ we set $S_i := S_{p_i}$, and we set $S := S_1 \cap \cdots \cap S_h$ and $\mathfrak{a} := \mathfrak{a}' S \cap R$. Then we have $\mathfrak{a} S = \mathfrak{a}' S$ [since an extended ideal is the extension of its contraction] and $\mathfrak{a}' \subset \mathfrak{a}$, hence $r := \mathrm{ord}(\mathfrak{a}) \leq s$, and $\mathfrak{a}$ is $\mathfrak{m}$-primary or $\mathfrak{a} = R$.

Let $i \in \{1, \ldots, h\}$; we have $\mathfrak{b}_j S_i = \mathfrak{m}^{s_j} S_i$ for every $j \in \{1, \ldots, h\}$ with $j \neq i$ [cf. (2.14)], hence $\mathfrak{a}' S_i = \mathfrak{m}^{s-s_i} \mathfrak{b}_i S_i$, and therefore $\mathfrak{a}' S_i$ is a complete ideal of S_i [cf.

(3.4) and (3.20)], hence also $\mathfrak{a}'S = \mathfrak{a}'S_1 \cap \cdots \cap \mathfrak{a}'S_h$ [cf. (3.0)] is a complete ideal of S [cf. proof of (3.3)(1)], hence $\mathfrak{a} = \mathfrak{a}'S \cap R$ is a complete ideal of R [cf. (3.3)(1)]. We have $\mathfrak{m}^r \mathfrak{a}^{S_i} = (\mathfrak{m}^r \mathfrak{a}^S) = (\mathfrak{a}S)S_i = (\mathfrak{a}'S)S_i = \mathfrak{a}'S_i = \mathfrak{m}^{s-s_i} \mathfrak{b}_i S_i = \mathfrak{m}^s \mathfrak{b}_i^{S_i}$, hence $\mathfrak{a}^{S_i} = \mathfrak{m}^{s-r} \mathfrak{b}_i^{S_i}$ [since $\mathfrak{m}S_i$ is a principal ideal of S_i]. Now $\mathfrak{b}_i^{S_i}$ is an $\mathfrak{m}(S_i)$-primary ideal of S_i [cf. (2.11) and (2.14)], the ideal $\mathfrak{a}^{S_i}$ is either equal to S_i or it is also an $\mathfrak{m}(S_i)$-primary ideal of S_i and $\mathfrak{m}^{r-s}S_i$ is a principal ideal of S_i; therefore we have $r = s$, and $\mathfrak{a}$ is m-primary. We set $\mathfrak{a}_i := \mathfrak{a}S_i \cap R$; then $\mathfrak{a}_i$ is a complete ideal of R, we have $\mathfrak{a}_i S_i = \mathfrak{a}S_i = \mathfrak{a}'S_i = \mathfrak{m}^{s-s_i} \mathfrak{b}_i S_i$, and since $\mathfrak{m}^{s-s_i} \mathfrak{b}_i$ is contracted from S_i by (3.10), and $\mathfrak{a}_i$ is contracted from S_i by definition, we have $\mathfrak{a}_i = \mathfrak{m}^{s-s_i} \mathfrak{b}_i$. This implies that $\mathfrak{a}_i = \mathfrak{m}^{s-s_i}(\mathfrak{a} : \mathfrak{m}^{s-s_i})$ [cf. (2.0)], hence that $\mathfrak{a}_i : \mathfrak{m}^{s-s_i} = \mathfrak{b}_i$ [cf. (3.2)], hence we get $\mathfrak{a} = \mathfrak{m}^{r-t} \mathfrak{b}_1 \cdots \mathfrak{b}_h$ where $t := \deg(c(\mathfrak{a}))$ by (4.5), hence $t = s_1 + \cdots + s_h = s = r$, and therefore we get $\mathfrak{a} = \mathfrak{a}'$; this means that $\mathfrak{a}'$ is complete.

4.2   Contracted Ideals

Now we are going to show that a product of complete ideals of R is complete; we need the following four lemmas (with the exception of lemma (4.11) which fits into this context but is needed only later).

(4.8) CONTRACTED IDEALS: Let $x \in \mathfrak{m} \setminus \mathfrak{m}^2$, and set $A := R[\mathfrak{m}/x]$. Let $\mathfrak{a}$ be an ideal of R which is contracted from A; then $\mathfrak{a}$ is the contraction of its extension, hence we have $\mathfrak{a} = \mathfrak{a}A \cap R$. We have $\mathfrak{m}^n = Ax^n \cap R$ for every $n \in \mathbb{N}$ [cf. (2.3)(4)]. A proper ideal of R which is contracted from A is not a power of Rx.

(4.9) Lemma: Let $x \in \mathfrak{m} \setminus \mathfrak{m}^2$, and set $A := R[\mathfrak{m}/x]$. An ideal $\mathfrak{a}$ of R is contracted from A iff $\mathfrak{a} : Rx = \mathfrak{a} : \mathfrak{m}$.

Proof: (1) We assume that $\mathfrak{a}A \cap R = \mathfrak{a}$. Let $a \in \mathfrak{a} : Rx$. For every $z \in \mathfrak{m}$ we have $az = (ax) \cdot (z/x) \in \mathfrak{a}A \cap R = \mathfrak{a}$, hence $a\mathfrak{m} \subset \mathfrak{a}$, showing that $\mathfrak{a} : Rx \subset \mathfrak{a} : \mathfrak{m}$. Since $x \in \mathfrak{m}$, we have $\mathfrak{a} : \mathfrak{m} \subset \mathfrak{a} : Rx$, hence we have shown that $\mathfrak{a} : Rx = \mathfrak{a} : \mathfrak{m}$.
(2) We assume that $\mathfrak{a} : Rx = \mathfrak{a} : \mathfrak{m}$. There exists $y \in R$ such that $\mathfrak{m} = Rx + Ry$. Let $a \in \mathfrak{a}A \cap R$; then we can write $a = r/x^n$ with $n \in \mathbb{N}_0$ and $r \in \mathfrak{a}\mathfrak{m}^n$. If $n > 0$, then we can write $r = by^n + cx$ with $b \in \mathfrak{a}$ and $c \in \mathfrak{a}\mathfrak{m}^{n-1}$; since $r = ax^n$, we see that $b = xd$ for some $d \in R$ [since (x, y) is a regular sequence on R]. This implies that $r = x((yd)y^{n-1} + c)$. We have $b \in \mathfrak{a}$, hence $d = b/x \in \mathfrak{a} : Rx = \mathfrak{a} : \mathfrak{m}$, and therefore we have $yd \in \mathfrak{a}$, hence $r' := (yd)y^{n-1} + c \in \mathfrak{a}\mathfrak{m}^{n-1}$, and we have $a = r'/x^{n-1}$. By recursion we find that $a \in \mathfrak{a}$.

(4.10) Lemma: Let $x \in \mathfrak{m} \setminus \mathfrak{m}^2$, and set $A := R[\mathfrak{m}/x]$. The product of a finite number of ideals of R contracted from A is also contracted from A.

Proof: By induction, it is enough to prove the lemma for two ideals $\mathfrak{a}, \mathfrak{b}$. Let $\mathfrak{a}$ and $\mathfrak{b}$ be proper ideals of R which are contracted from A. Then $\mathfrak{a}$ and $\mathfrak{b}$ are not equal to powers of Rx [cf. (4.8)]. By (4.9) it is enough to show that $\mathfrak{a}\mathfrak{b} : Rx = \mathfrak{a}\mathfrak{b} : \mathfrak{m}$.

Let $z \in R$ with $zx \in \mathfrak{a}\mathfrak{b}$. Now R/Rx is a discrete valuation ring, hence $\mathfrak{a}(R/Rx)$ is a principal non-zero ideal of R/Rx. We choose $a \in \mathfrak{a}$ with $\mathfrak{a}(R/Rx) = a(R/Rx)$. Let $a' \in \mathfrak{a}$; then there exist u, $v \in R$ with $a' = ua + vx$, hence $v \in \mathfrak{a} : Rx$, and since $x(\mathfrak{a} : Rx) \subset \mathfrak{a}$, it follows that $\mathfrak{a} = Ra + x(\mathfrak{a} : Rx)$. Likewise, there exists $b \in \mathfrak{b}$ with $\mathfrak{b} = Rb + x(\mathfrak{b} : Rx)$. Therefore we have $\mathfrak{a}\mathfrak{b} = Rab + xa(\mathfrak{b} : Rx) + xb(\mathfrak{a} : Rx)$ [note that $xa(\mathfrak{b} : Rx) \subset \mathfrak{a}\mathfrak{b}$ and $xb(\mathfrak{a} : Rx) \subset \mathfrak{a}\mathfrak{b}$]. We have $ab \notin Rx$; for every $c \in R$ with $cx \in \mathfrak{a}\mathfrak{b}$ we therefore have $c \in Rab + \mathfrak{a}(\mathfrak{b} : Rx) + \mathfrak{b}(\mathfrak{a} : Rx)$, hence $c\mathfrak{m} \subset \mathfrak{a}\mathfrak{b}$ [since $\mathfrak{b} : Rx = \mathfrak{b} : \mathfrak{m}$ and $\mathfrak{a} : Rx = \mathfrak{a} : \mathfrak{m}$ by (4.9), we have $\mathfrak{m}(\mathfrak{b} : Rx) \subset \mathfrak{b}$ and $\mathfrak{m}(\mathfrak{a} : Rx) \subset \mathfrak{a}$]. This shows that $\mathfrak{a}\mathfrak{b} : Rx \subset \mathfrak{a}\mathfrak{b} : \mathfrak{m}$, and therefore we have $\mathfrak{a}\mathfrak{b} : Rx = \mathfrak{a}\mathfrak{b} : \mathfrak{m}$.

The following result shall play an important role in section 7 of chapter VIII.

(4.11) Lemma: *Assume that R has infinite residue field κ. Then:*
(1) *Let $\mathfrak{a}$ be an $\mathfrak{m}$-primary ideal of R. Then there exists $x \in \mathfrak{m} \setminus \mathfrak{m}^2$ such that $\mathfrak{a}$ is contracted from $A = R[\mathfrak{m}/x]$ iff $\mu(\mathfrak{a}) = \mathrm{ord}(\mathfrak{a}) + 1$.*
(2) *Let $\mathfrak{a}_1, \ldots, \mathfrak{a}_n$ be complete $\mathfrak{m}$-primary ideals of R. Then there exists $x \in \mathfrak{m} \setminus \mathfrak{m}^2$ such that $\mathfrak{a}_1, \ldots, \mathfrak{a}_n$ are contracted from $A = R[\mathfrak{m}/x]$.*

Proof: (1) We set $r := \mathrm{ord}(\mathfrak{a})$, and let $\{x_1, y_1\}$ be a system of generators of $\mathfrak{m}$. Then we have $\mathrm{gr}_{\mathfrak{m}}(R) = \kappa[\mathrm{In}(x_1), \mathrm{In}(y_1)]$. Let $f \in \mathfrak{a}$ be of order r; since κ is an infinite field, there exists a linear combination of $\mathrm{In}(x_1)$ and $\mathrm{In}(y_1)$ with coefficients in κ which does not divide $\mathrm{In}(f)$ in $\mathrm{gr}_{\mathfrak{m}}(R)$, hence there exists $x \in \mathfrak{m} \setminus \mathfrak{m}^2$ such that $\mathrm{In}(x)$ does not divide $\mathrm{In}(f)$ in $\mathrm{gr}_{\mathfrak{m}}(R)$. We choose $y \in R$ with $\mathfrak{m} = Rx + Ry$. This means that $f = \sum_{i+j=r} a_{ij}x^iy^j$ with $a_{ij} \in R$ for all i, $j \in \mathbb{N}_0$ with $i + j = r$, and a_{0r} is a unit of R. Then we have $\mathfrak{a} + Rx = Rf + Rx$. The ring $\overline{R} = R/Rx$ is a discrete valuation ring [cf. (2.0)], and the image $\overline{f}$ of f in $\overline{R}$ has order r in $\overline{R}$. We have $R/(\mathfrak{a} + Rx) = \overline{R}/\overline{R}\overline{f}$, and therefore we have $\ell_R(R/(\mathfrak{a} + Rx)) = r$. Since $(\mathfrak{a} + Rx)/\mathfrak{a} \cong R/(\mathfrak{a} : Rx)$, and since $R/\mathfrak{a}$ is an R-module of finite length, we have $\ell_R((\mathfrak{a} : Rx)/\mathfrak{a}) = r$. Since $\mathfrak{a} : \mathfrak{m} \subset \mathfrak{a} : Rx$, we have $\ell_R((\mathfrak{a} : \mathfrak{m})/\mathfrak{a}) \leq r$ with equality iff $\mathfrak{a}$ is contracted from $A = R[\mathfrak{m}/x]$ [cf. (4.9)]. Since $\ell_R((\mathfrak{a} : \mathfrak{m})/\mathfrak{a}) = \mu(\mathfrak{a}) - 1$ by (3.17), we get the assertion.
(2) We have $\mu(\mathfrak{a}_i) = \mathrm{ord}(\mathfrak{a}_i) + 1$ for $i \in \{1, \ldots, n\}$ by (3.19). For $i \in \{1, \ldots, n\}$ we choose $f_i \in \mathfrak{a}_i$ of order $\mathrm{ord}(\mathfrak{a}_i)$. Since κ is an infinite field, there exists $x \in \mathfrak{m} \setminus \mathfrak{m}^2$ such that $\mathrm{In}(x)$ does not divide $\mathrm{In}(f_i)$ in $\mathrm{gr}_{\mathfrak{m}}(R)$ for every $i \in \{1, \ldots, n\}$. By the proof of (1) this implies that the ideals $\mathfrak{a}_1, \ldots, \mathfrak{a}_n$ are contracted from $R[\mathfrak{m}/x]$.

(4.12) Lemma: *Let $p \in \mathbb{P}$, and let $\mathfrak{a}_1, \ldots, \mathfrak{a}_n$ be complete $\mathfrak{m}$-primary ideals of R. If $c(\mathfrak{a}_i)$ is a power of p for every $i \in \{1, \ldots, n\}$, then $(\mathfrak{a}_1 \cdots \mathfrak{a}_n)S_p \cap R = \mathfrak{a}_1 \cdots \mathfrak{a}_n$.*

Proof: We set $\mathfrak{a} := \mathfrak{a}_1 \cdots \mathfrak{a}_n$. We choose $x \in \mathfrak{m} \setminus \mathfrak{m}^2$ with $(\mathrm{In}(x)) \neq p$; then $S := S_p$ is a localization of $A := R[\mathfrak{m}/x]$ with respect to a maximal ideal $\mathfrak{n}$ of A which contains x. Let $i \in \{1, \ldots, n\}$; we have $\mathfrak{a}_i S \cap R = \mathfrak{a}_i$ [cf. (3.10)], hence $\mathfrak{a}_i A \cap R = \mathfrak{a}_i$, hence $\mathfrak{a}_i : Rx = \mathfrak{a}_i : \mathfrak{m}$ [cf. (4.9)], and therefore we have $\mathfrak{a}A \cap R = \mathfrak{a}$ [cf. (4.10)]. Since $\mathfrak{a}$ is an $\mathfrak{m}$-primary ideal, and since $c(\mathfrak{a})$ is a power of p, we have $\mathfrak{a}S \cap A = \mathfrak{a}A$ [cf. (2.17)(2)], and therefore we obtain $\mathfrak{a}S \cap R = \mathfrak{a}$.

Now we can put our results together and prove:

(4.13) Theorem: *The product of complete ideals of R is complete.*

Proof: Clearly it is enough to prove the theorem for a product $\mathfrak{a} = \mathfrak{a}_1 \cdots \mathfrak{a}_n$ of complete $\mathfrak{m}$-primary ideals $\mathfrak{a}_1, \ldots \mathfrak{a}_n$. We know that every complete $\mathfrak{m}$-primary ideal of R is a product of simple complete $\mathfrak{m}$-primary ideals of R [cf. (4.1)(2)], even if we do not know yet that the factors in such a product are determined uniquely. Nevertheless, it makes sense to prove by induction on l—and this will prove our assertion—: If $\mathfrak{a}$ is a product of complete $\mathfrak{m}$-primary ideals $\mathfrak{a}_1, \ldots, \mathfrak{a}_n$ and each of the factors $\mathfrak{a}_1, \ldots, \mathfrak{a}_n$ is a product of simple complete $\mathfrak{m}$-primary ideals having colength $\leq l$, then $\mathfrak{a}$ is complete. If the ideals $\mathfrak{a}_1, \ldots, \mathfrak{a}_n$ are products of simple complete $\mathfrak{m}$-primary ideals of colength 1, then the ideals $\mathfrak{a}_1, \ldots, \mathfrak{a}_n$ are powers of $\mathfrak{m}$, hence also $\mathfrak{a}$ is a power of $\mathfrak{m}$, and therefore $\mathfrak{a}$ is complete [cf. (2.2)]. Let $l > 1$, and assume that $\mathfrak{a} = \mathfrak{a}_1 \cdots \mathfrak{a}_n$ is complete if the ideals $\mathfrak{a}_1, \ldots, \mathfrak{a}_n$ are products of simple complete $\mathfrak{m}$-primary ideals having colength $< l$. Now we consider a product $\mathfrak{a} = \mathfrak{a}_1 \cdots \mathfrak{a}_n$ of complete $\mathfrak{m}$-primary ideals where all the ideals $\mathfrak{a}_1, \ldots, \mathfrak{a}_n$ can be written as a product of simple complete $\mathfrak{m}$-primary ideals having colength $\leq l$. If all the factors $\mathfrak{a}_1, \ldots, \mathfrak{a}_n$ of $\mathfrak{a}$ are powers of $\mathfrak{m}$, then $\mathfrak{a}$ is complete. In the other case, for $i \in \{1, \ldots, n\}$ we write every ideal $\mathfrak{a}_i$ which is not a power of $\mathfrak{m}$ as a product $\mathfrak{a}_i = \mathfrak{m}^{t_i} \mathfrak{b}_{i1} \cdots \mathfrak{b}_{in_i}$ where $t_i \in \mathbb{N}_0$, where the ideals $\mathfrak{b}_{i1}, \ldots, \mathfrak{b}_{in_i}$ are complete $\mathfrak{m}$-primary ideals which are not divisible by $\mathfrak{m}$, and where $c(\mathfrak{b}_{ij})$ is a positive power of a prime ideal $p_{ij} \in \mathbb{P}$, $j \in \{1, \ldots, n_i\}$ [cf. (3.10) and (4.5)]; note that we can write any of the ideals $\mathfrak{b}_{ij}$ as a product of simple complete $\mathfrak{m}$-primary ideals having colength $\leq l$. In $\mathfrak{a}_1 \cdots \mathfrak{a}_n$ we group together all those factors $\mathfrak{b}_{ij}$ for which the characteristic ideal is a power of the same prime ideal of $\mathbb{P}$, hence we choose pairwise different $q_1, \ldots, q_k \in \mathbb{P}$ and write $\mathfrak{a} = \mathfrak{m}^t \mathfrak{c}_1 \cdots \mathfrak{c}_k$ where, for $i \in \{1, \ldots, k\}$, $\mathfrak{c}_i = \mathfrak{c}_{i1} \cdots \mathfrak{c}_{ik_i}$ is a product of complete $\mathfrak{m}$-primary ideals none of which is divisible by $\mathfrak{m}$, and for which the characteristic ideal is a positive power of q_i, and where we can write the ideals $\mathfrak{c}_{ij}$ as a product of simple complete $\mathfrak{m}$-primary ideals having colength $\leq l$. It is enough to show that the ideals $\mathfrak{c}_1, \ldots, \mathfrak{c}_k$ are complete [cf. (4.7) and (3.20)].

Let $i \in \{1, \ldots, k\}$ and set $S_i := S_{q_i}$; let $\mathfrak{n}_i$ be the maximal ideal of S_i. Since $\mathfrak{c}_i = \mathfrak{c}_i S_i \cap R$ [cf. (4.12)], it is enough to show that $\mathfrak{c}_i S_i$ is a complete ideal of S_i [cf. (3.3)(1)]. We consider the transform $\mathfrak{c}_i^{S_i}$ of $\mathfrak{c}_i$ which is the product of the complete $\mathfrak{n}_i$-primary ideals $\mathfrak{c}_{ij}^{S_i}$. We write $\mathfrak{c}_i$ as a product of simple complete $\mathfrak{m}$-primary ideals of R having colength $\leq l$; then the transform in S_i of each of these simple $\mathfrak{m}$-primary ideals is simple [cf. (4.6)] and has colength $< l$ [cf. (3.14)], hence, by our induction assumption, applied to S_i and $\mathfrak{c}_i^{S_i}$, the ideal $\mathfrak{c}_i^{S_i}$ of S_i is complete, and therefore $\mathfrak{c}_i S_i$ is also a complete ideal of S_i [cf. B(6.16)(6)].

(4.14) Corollary: *The set of non-zero complete ideals of R (resp. the set of complete $\mathfrak{m}$-primary ideals of R) is under multiplication a monoid with cancellation law. Every non-zero complete ideal (resp. every complete $\mathfrak{m}$-primary ideal) of R*

has a factorization into a product of simple complete (resp. simple complete m-primary) ideals of R.

Proof: The first assertions follow from (4.13) and (3.2), while the last assertion was proved in (4.1)(2).

(4.15) NOTATION: Let T be a two-dimensional regular local subring of K with field of quotients K; let $\mathfrak{n}$ be its maximal ideal, and let $\mathbb{P}_T$ be the set of non-zero principal homogeneous prime ideals of $\mathrm{gr}_{\mathfrak{n}}(T)$. In the following, we use these notations: $\mathcal{MC}(T)$ is the monoid of complete non-zero ideals of T and $\mathcal{MCP}(T)$ is the submonoid of complete ideals of T of finite colength. $\mathcal{SCP}(T)$ is the set of simple complete $\mathfrak{n}$-primary ideals of T, and, for $p \in \mathbb{P}_T$, $\mathcal{SCP}(T,p)$ is the subset of $\mathcal{SCP}(T)$ of those ideals whose characteristic ideal is a positive power of p, and $\mathcal{MCP}(T,p)$ is the submonoid of $\mathcal{MCP}(T)$ which is generated by the elements of the set $\mathcal{SCP}(T,p)$.

(4.16) Corollary: *Let $p \in \mathbb{P}$ and set $S := S_p$. The map $\mathfrak{a} \mapsto \mathfrak{a}^S : \mathcal{SCP}(R,p) \to \mathcal{MCP}(S)$ is injective and $\mathcal{SCP}(S)$ is its image.*

Proof: We only have to show [cf. (4.6)]: If $\mathfrak{c}_p$ is a simple complete $\mathfrak{m}(S)$-primary ideal of S, then the unique complete $\mathfrak{m}$-primary ideal $\mathfrak{a}$ of R with $\mathfrak{a}^S = \mathfrak{c}_p$ and $\mathfrak{a}^{S_q} = S_q$ for every $q \in \mathbb{P}$ different from p and which is not divisible by $\mathfrak{m}$ [cf. (3.27)] is simple. Suppose that $\mathfrak{a}$ is not simple. Then we have $\mathfrak{a} = \mathfrak{a}_1 \cdots \mathfrak{a}_h$ with $h > 1$ where $\mathfrak{a}_1, \ldots, \mathfrak{a}_h$ are simple complete and $\mathfrak{m}$-primary ideals [cf. (4.1)(2)]. We label these ideals in such a way that $\mathfrak{a}_1^S = \mathfrak{c}_p$ and $\mathfrak{a}_2^S = \cdots = \mathfrak{a}_h^S = S$. Then we have $\mathfrak{a}_1 = \mathfrak{m}^t \mathfrak{a}$ for some $t \in \mathbb{N}_0$ by (3.27)(ii), hence $\mathfrak{m}^t \mathfrak{a}_2 \cdots \mathfrak{a}_h = R$ which is absurd.

4.3 Unique Factorization

We quote Zariski [cf. [204], vol. 2, p. 385]: "The culminating point of our theory of complete ideals is a theorem of unique factorization of complete ideals into simple complete ideals."
These results do not hold in regular local rings of higher dimension. For a certain generalization one may consult Lipman's paper [131].

(4.17) Theorem: *Every complete ideal of R has a unique factorization into a product of simple complete ideals of R, i.e., $\mathcal{MC}(R)$ is a free monoid which is generated by the simple complete $\mathfrak{m}$-primary ideals of R and the prime ideals of height 1 of R.*

Proof: It is enough to prove the assertion for complete $\mathfrak{m}$-primary ideals of R [cf. (3.1)]. Since $\mathcal{MCP}(R)$ has cancellation law [cf. (4.14)], it is enough to show the following: If a simple complete $\mathfrak{m}$-primary ideal $\mathfrak{a}$ of R divides a product $\mathfrak{b}_1 \cdots \mathfrak{b}_h$ of simple complete $\mathfrak{m}$-primary ideals $\mathfrak{b}_1, \ldots, \mathfrak{b}_h$ of R, i.e., if $\mathfrak{a}\mathfrak{c} = \mathfrak{b}_1 \cdots \mathfrak{b}_h$ for some ideal $\mathfrak{c}$ of R, then we have $\mathfrak{a} = \mathfrak{b}_i$ for some $i \in \{1, \ldots, h\}$. We prove this by induction on $\ell_R(R/\mathfrak{a})$. If $\ell_R(R/\mathfrak{a}) = 1$, then $\mathfrak{a} = \mathfrak{m}$, and the result follows from (3.25). Assume that $l > 1$, and that the result is true for simple complete $\mathfrak{m}$-primary ideals $\mathfrak{a}$ of R with $\ell_R(R/\mathfrak{a}) < l$. Now let $\mathfrak{a}$ be a simple complete $\mathfrak{m}$-primary ideal of R with $\ell_R(R/\mathfrak{a}) = l$. Then $c(\mathfrak{a})$ is a positive power of p for some $p \in \mathbb{P}$ [cf. (4.16)], and, for $i \in \{1, \ldots, h\}$, $c(\mathfrak{b}_i)$ is either a positive power of p or is not divisible by p [cf. (4.16)]. We set $S := S_p$. We have $\mathfrak{a}^S \mathfrak{c}^S = \mathfrak{b}_1^S \cdots \mathfrak{b}_h^S$ [cf. (1.5)], and, for $i \in \{1, \ldots, h\}$, $\mathfrak{b}_i^S$ is a simple complete $\mathfrak{m}(S)$-primary ideal of S or $\mathfrak{b}_i^S = S$ [cf. (2.14) and (4.16)]. Now $\mathfrak{a}^S$ is a simple complete $\mathfrak{m}(S)$-primary ideal of S [cf. (4.16)] with $\ell_S(S/\mathfrak{a}^S) < \ell_R(R/\mathfrak{a})$ [cf. (3.15)]. By our induction assumption we have $\mathfrak{a}^S = \mathfrak{b}_i^S$ for some $i \in \{1, \ldots, h\}$; in this case we have $\mathfrak{a} = \mathfrak{b}_i$ by (3.27).

5   The Predecessors of a Simple Ideal

(5.0) In this section we keep the notations and hypotheses introduced in section 2. In particular, let ν_R be the valuation of K defined by the order function ord_R of R, and let V_R be the valuation ring of ν_R.

(5.1) VALUATION ASSOCIATED WITH A SIMPLE COMPLETE IDEAL: (1) Let $\wp \neq \mathfrak{m}$ be a simple complete $\mathfrak{m}$-primary ideal of R. We have $c(\wp) = p^l$ where $p \in \mathbb{P}$ and $l \in \mathbb{N}$, the ring $S := S_p$ is the only quadratic transform of R with $\wp^S \neq S$, $\wp^S$ is a simple complete $\mathfrak{m}(S)$-primary ideal of S, and $\wp S \cap R = \wp$ [cf. (3.10) and (4.16)]. Furthermore, we have $\ell_R(R/\wp) > \ell_S(S/\wp^S)$ [cf. (3.15)].
We set $R_0 := R$, $\wp^{(0)} := \wp$, and $R_1 := S$, $\wp^{(1)} := \wp^S$. If $\wp^S \neq \mathfrak{m}(S)$—and this is the case iff $\ell_S(S/\wp^S) > 1$—then we can repeat the procedure described above, and we get a sequence $R = R_0 \subset R_1 \subset \cdots \subset R_h$ of two-dimensional regular local subrings of K where R_{i+1} is a quadratic transform of R_i for $i \in \{0, \ldots, h-1\}$, and a sequence $(\wp^{(i)})_{0 \leq i \leq h}$ where $\wp^{(i)}$ is a simple complete $\mathfrak{m}(R_i)$-primary ideal of R_i, $\wp^{(h)}$ is the maximal ideal of R_h and $\wp^{(i+1)} = \wp^{(i)R_{i+1}}$, $i \in \{0, \ldots, h-1\}$. In particular, we have [cf. (1.5)(4)] $\wp^{(i)} = \wp^{R_i}$ for every $i \in \{0, \ldots, h\}$. We say that $\wp$ is a simple ideal of rank h, and we set $v_\wp := \nu_{R_h}$, the valuation of K defined by the order function of R_h; it is a discrete valuation of K of rank 1 having center $\mathfrak{m}$ in R, and κ can be considered as a subfield of $\kappa_{v_\wp}$, the residue field of the valuation $v_\wp$.
If $\wp = \mathfrak{m}$, then we set $v_\wp := \nu_R$, and the rank of $\wp$ is defined to be 0.
Thus, we have shown: *Let $\wp$ be a simple complete $\mathfrak{m}$-primary ideal of R. Then there exists a uniquely determined sequence*

$$R = R_0 \subset R_1 \subset \cdots \subset R_h$$

[where $h \geq 0$] such that R_{i+1} is a quadratic transform of R_i for $i \in \{1, \ldots, h-1\}$, that $\wp^{R_i}$ is a simple complete $\mathfrak{m}(R_i)$-primary ideal of R_i for $i \in \{0, \ldots, h\}$ and that $\wp^{R_h} = \mathfrak{m}(R_h)$.
(2) Let $\wp$ be a simple complete $\mathfrak{m}$-primary ideal of R of rank h. Then $\wp$ is a $v_\wp$-ideal of R.

The proof is by induction on the rank h of $\wp$. If $h = 0$, then we have $\wp = \mathfrak{m}$. Now we consider the case $h > 0$; then we have $\wp \neq \mathfrak{m}$. Let R_1 be the unique quadratic transform of R such that the ideal transform $\wp^{(1)}$ of $\wp$ in R_1 is a proper ideal; then $\wp^{(1)}$ is a simple complete $\mathfrak{m}(R_1)$-primary ideal of R_1 of rank $h - 1$, $\wp R_1 \cap R = \wp$ [cf. (3.10)], and it is clear that $v_\wp = v_{\wp^{(1)}}$. By our induction assumption the ideal $\wp^{(1)}$ is a $v_\wp$-ideal of R_1.

Let $p \in \mathbb{P}$ be that prime ideal with $R_1 = S_p$. We choose $x \in \mathfrak{m} \setminus \mathfrak{m}^2$ such that $(\mathrm{In}(x)) \neq p$; then we have $\wp R_1 = x^r \wp^{(1)} R_1$ with $r := \mathrm{ord}(\wp)$. We set $\mathfrak{q} := \{ z \in R \mid v_\wp(z) \geq v_\wp(\wp) \}$; $\mathfrak{q}$ is a $v_\wp$-ideal of R, and clearly $\mathfrak{q} \supset \wp$. We have to show that $\mathfrak{q} = \wp$. We first show that $\wp = \mathfrak{m}^r \cap \mathfrak{q}$; to prove this it is enough to show that $\mathfrak{m}^r \cap \mathfrak{q} \subset \wp$. Let $z \in \mathfrak{m}^r \cap \mathfrak{q}$. We have $z/x^r =: z_1 \in R_1$. Since $z \in \mathfrak{q}$, we have $v_\wp(x^r z_1) \geq v_\wp(\wp) = v_\wp(x^r \wp^{(1)})$, hence $v_\wp(z_1) \geq v_\wp(\wp^{(1)})$ and therefore $z_1 \in \wp^{(1)}$ [since $\wp^{(1)}$ is a $v_\wp$-ideal], and thus $z \in x^r \wp^{(1)} \cap R = \wp$, i.e., $\wp = \mathfrak{m}^r \cap \mathfrak{q}$. Now we show that $\mathfrak{q} \subset \mathfrak{m}^r$. Suppose not; then we have $s := \mathrm{ord}(\mathfrak{q}) < r$. We have $\mathfrak{m}^{r-s}\mathfrak{q} \subset \mathfrak{m}^r$ [since $\mathfrak{q} \subset \mathfrak{m}^s$] and $\mathfrak{m}^{r-s}\mathfrak{q} \subset \mathfrak{q}$, hence $\mathfrak{m}^{r-s}\mathfrak{q} \subset \mathfrak{m}^r \cap \mathfrak{q} = \wp$; from $\mathrm{ord}(\mathfrak{m}^{r-s}\mathfrak{q}) = \mathrm{ord}(\wp)$ we get $\deg(c(\wp)) \leq \deg(c(\mathfrak{q})) \leq \mathrm{ord}(\mathfrak{q}) = s < r$ [cf. (2.13)], and therefore $\mathfrak{m}$ divides $\wp$ by (3.22)(1), contradicting the assumption on $\wp$.

(5.2) v-IDEALS OF R: Let v be a valuation of K dominating R with valuation ring V, and assume that $V \neq V_R$. We use the notations of (2.18). In particular, set $p := p(v)$ and $S := S_p$. The following results shall play a role when we study the predecessors of an ideal [cf. (5.3) below].

(1) Let $z \in R$, $z \neq 0$. Then we have $v(z) > \nu_R(z)v(\mathfrak{m})$ iff $\mathrm{In}(z) \in p$.

Proof: We set $r := \nu_R(z)$. We write $z = f(x, y)$ where $f \in R[X, Y]$. Then we have $v(z) = rv(\mathfrak{m}) + v(f(1, y/x))$. Therefore we have $v(z) > rv(\mathfrak{m})$ iff $\overline{f}(1, \zeta) = 0$, hence iff $\overline{g}(\mathrm{In}(x), \mathrm{In}(y))$ divides $\overline{f}(\mathrm{In}(x), \mathrm{In}(y))$ in $\kappa[\mathrm{In}(x), \mathrm{In}(y)]$.

(2) Let $\mathfrak{a}$ be an $\mathfrak{m}$-primary v-ideal of R. Then $c(\mathfrak{a})$ is a power of p.

Proof: Since $\mathfrak{a}$ is a v-ideal, it is a complete ideal of R [cf. B(6.16)(2)] and we have $\mathfrak{a} = \mathfrak{a}V \cap R$, hence is contracted from S and we have $\mathfrak{a} = \mathfrak{a}S \cap R$, and therefore $c(\mathfrak{a})$ is a power of p [cf. (3.26)].

(3) Let $\mathfrak{a}$ be an $\mathfrak{m}$-primary v-ideal of R and set $r := \mathrm{ord}(\mathfrak{a})$. Let $h \in \mathbb{N}$, and set $\mathfrak{q} := \mathfrak{m}^h \mathfrak{a}V \cap R$. Then we have $\mathfrak{m}^h \mathfrak{a} = \mathfrak{m}^{r+h} \cap \mathfrak{q}$.

Proof: We set $\mathfrak{d} := \mathfrak{m}^{r+h} \cap \mathfrak{q}$; $\mathfrak{d}$ is an $\mathfrak{m}$-primary ideal of R which is contracted from S, and we have $\mathfrak{m}^h \mathfrak{a} \subset \mathfrak{d}$. Now $\mathfrak{m}^h \mathfrak{a} \subset \mathfrak{d} \subset \mathfrak{m}^{r+h}$ implies that $\mathrm{ord}(\mathfrak{d}) = r + h$, hence that $\deg(c(\mathfrak{d})) \leq \deg(c(\mathfrak{a})) \leq \mathrm{ord}(\mathfrak{a}) = r$, and therefore we have $\mathrm{ord}(\mathfrak{d}) - \deg(c(\mathfrak{d})) \geq h$. This implies that [cf. (3.24)] $\mathfrak{d} = \mathfrak{m}^h \mathfrak{e}$ with $\mathfrak{e} := \mathfrak{d} : \mathfrak{m}^h$. We have $\mathfrak{m}^h \mathfrak{a} \subset \mathfrak{d} \subset \mathfrak{q}$ and $v(\mathfrak{m}^h \mathfrak{a}) = v(\mathfrak{q})$, and therefore we have $v(\mathfrak{d}) = v(\mathfrak{q})$. Now we get

$$v(\mathfrak{m}^h) + v(\mathfrak{e}) = v(\mathfrak{d}) = v(\mathfrak{q}) = v(\mathfrak{m}^h \mathfrak{a}) = v(\mathfrak{m}^h) + v(\mathfrak{a}),$$

hence $v(\mathfrak{e}) = v(\mathfrak{a})$, and therefore $\mathfrak{e} \subset \mathfrak{a}$ [since $\mathfrak{a}$ is a v-ideal]. Therefore we have $\mathfrak{d} = \mathfrak{m}^h \mathfrak{e} \subset \mathfrak{m}^h \mathfrak{a}$, and therefore $\mathfrak{d} = \mathfrak{m}^h \mathfrak{a}$.

(4) Let $\mathfrak{a}$ be an $\mathfrak{m}$-primary v-ideal of R. Then $\mathfrak{a}^S$ is a v-ideal of S.

Proof: Set $r := \mathrm{ord}(\mathfrak{a})$. Let $z \in S$, $z \neq 0$, with $v(z) \geq v(\mathfrak{a}^S)$. We have to show that $z \in \mathfrak{a}^S$. We write $x^r z = u_1/u_2$ with $u_1, u_2 \in R$, $\mathrm{In}(u_2) \notin p$, and $\mathrm{ord}(u_1) \geq$

$\operatorname{ord}(u_2) =: h$ [cf. (2.6)(3)]. Now we have to show that $u_1/u_2 \in \mathfrak{a}S = x^r \mathfrak{a}^S$. We have $v(u_1/u_2) = v(x^r z) \geq v(x^r \mathfrak{a}^S) = v(\mathfrak{a}S) = v(\mathfrak{a})$. Since $u_2 \in \mathfrak{m}^h$, we have $v(u_1) \geq v(\mathfrak{m}^h \mathfrak{a})$. Now we have $\nu_R(z) \geq 0$, hence $\nu_R(u_1/u_2) \geq r$, and therefore $\nu_R(u_1) \geq r + \nu_R(u_2) = r + h$, hence $u_1 \in \mathfrak{m}^{r+h}$. From this and from $v(u_1) \geq v(\mathfrak{m}^h \mathfrak{a})$ we get by (3) that $u_1 \in \mathfrak{m}^h \mathfrak{a}$, hence that $u_1/u_2 = (x^h/u_2)\mathfrak{a}'$ with $\mathfrak{a}' \in \mathfrak{a}S$, hence that u_1/u_2 lies in $\mathfrak{a}S$.

(5) Let $\mathfrak{b}$ be an $\mathfrak{m}(S)$-primary v-ideal of S, and let $\mathfrak{a} \in \mathcal{MCP}(R,p)$ be the ideal with $\mathfrak{a}^S = \mathfrak{b}$ [cf. (4.16)(2)]. *Then $\mathfrak{a}$ is a v-ideal of R.*
Proof: We set $r := \operatorname{ord}(\mathfrak{a})$; we have $\mathfrak{a}S = x^r \mathfrak{b}$ and $\mathfrak{a}S \cap R = \mathfrak{a}$. We set $\mathfrak{q} := \mathfrak{a}V \cap R$. Just as in the first part of the proof of (5.1)(2) we get $\mathfrak{a} = \mathfrak{m}^r \cap \mathfrak{q}$. Since $\mathfrak{m}$ does not divide $\mathfrak{a}$, we have $\deg(c(\mathfrak{a})) = \operatorname{ord}(\mathfrak{a})$ [cf. (3.22)]. We use this to show that $\mathfrak{q} \subset \mathfrak{m}^r$. Suppose not; then we have $s := \operatorname{ord}(\mathfrak{q}) < r$ and $\mathfrak{m}^{r-s}\mathfrak{q} \subset \mathfrak{m}^r \cap \mathfrak{q} = \mathfrak{a}$, and from $\operatorname{ord}(\mathfrak{m}^{r-s}\mathfrak{q}) = \operatorname{ord}(\mathfrak{a})$ we get $\deg(c(\mathfrak{a})) \leq \deg(c(\mathfrak{q})) \leq \operatorname{ord}(\mathfrak{q}) = s < r = \operatorname{ord}(\mathfrak{a})$, contradicting the equality above.

(6) Let $\mathfrak{b}_1$, $\mathfrak{b}_2$ be $\mathfrak{m}(S)$-primary v-ideals of S, and let $\mathfrak{a}_1, \mathfrak{a}_2 \in \mathcal{MCP}(R,p)$ be the $\mathfrak{m}$-primary v-ideals with $\mathfrak{a}_i^S = \mathfrak{b}_i$ for $i = 1,2$ [cf. (5)]. *If $\mathfrak{a}_1 \subsetneqq \mathfrak{a}_2$, then we have $\mathfrak{b}_1 \subsetneqq \mathfrak{b}_2$.*
Proof: Let $i \in \{1,2\}$ and set $a_i := \operatorname{ord}(\mathfrak{a}_i)$; then we have $\mathfrak{a}_iS \cap R = \mathfrak{a}_i$ and $\mathfrak{a}_iS = x^{a_i}\mathfrak{a}_i^S = x^{a_i}\mathfrak{b}_i$. Note that $a_1 \geq a_2$. Suppose that the assertion is not true. Then we have $\mathfrak{b}_1 \supset \mathfrak{b}_2$ [cf. B(6.13)(1) and note that $\mathfrak{b}_1, \mathfrak{b}_2$ are v-ideals]. Since $\mathfrak{m}^{a_1-a_2}\mathfrak{a}_2S = x^{a_1}\mathfrak{b}_2 \subset x^{a_1}\mathfrak{b}_1 = \mathfrak{a}_1S$, and since $\mathfrak{m}^{a_1-a_2}\mathfrak{a}_2$ and $\mathfrak{a}_1$ are contracted from S [cf. (3.10)], we get $\mathfrak{m}^{a_1-a_2}\mathfrak{a}_2 \subset \mathfrak{a}_1$. This implies that $a_1 = \deg(c(\mathfrak{a}_1)) \leq \deg(c(\mathfrak{a}_2)) = a_2$ [cf. (3.22)], hence that $a_1 = a_2$, and therefore we have $\mathfrak{a}_2 \subset \mathfrak{a}_1$, contradicting $\mathfrak{a}_1 \subsetneqq \mathfrak{a}_2$.

(7) Let $\mathfrak{a}$ be an $\mathfrak{m}$-primary v-ideal of R, and set $r := \operatorname{ord}(\mathfrak{a})$. *If $\mathfrak{a}^S = S$, then we have $\mathfrak{a} = \mathfrak{m}^r$, and if $\mathfrak{a}^S \neq S$ and $\mathfrak{a}_1$ is the unique complete $\mathfrak{m}$-primary ideal with $\mathfrak{a}_1^S = \mathfrak{a}^S$ [cf. (3.27)], then we have $\mathfrak{a} = \mathfrak{m}^t \mathfrak{a}_1$ for some $t \in \mathbb{N}_0$.*
Proof: If $\mathfrak{a}^S = S$, then we have $\mathfrak{a}S = x^r S$, hence $\mathfrak{m}^r \subset \mathfrak{a}$, and therefore $\mathfrak{a} = \mathfrak{m}^r$, and if $\mathfrak{a}^S \neq S$, then we have $\mathfrak{m}^t \mathfrak{a}_1 = \mathfrak{a}$ for some $t \in \mathbb{N}_0$ by (3.27).

(5.3) THE PREDECESSORS OF A SIMPLE IDEAL: Let $\wp$ be a simple complete $\mathfrak{m}$-primary ideal of R of rank h. If $h > 0$, then we have $c(\wp) = p^n$ with $p \in \mathbb{P}$ and $n \in \mathbb{N}$. Let $(R_i)_{0 \leq i \leq h}$ with $R_0 := R$ be the sequence defined by $\wp$ [cf. (5.1)]. Let $v_\wp$ be the valuation of K defined by the order function of R_h. Since $v_\wp$ is an archimedean valuation, we know that every non-zero proper $v_\wp$-ideal of R is an $\mathfrak{m}$-primary ideal [cf. (2.18)]. Let $i \in \{1, \ldots, h\}$; R_i is a quadratic transform of R_{i-1}, and we set $\mathfrak{n}_i := \mathfrak{m}(R_i)$. By recursion, using (4.16)(2), for $i \in \{1, \ldots, h\}$ there exists a unique simple complete $\mathfrak{m}$-primary ideal $\wp_i \in SCP(R,p)$ with $\wp_i^{R_i} = \mathfrak{n}_i$; note that $\wp_h = \wp$, and that $c(\wp_i)$ is a positive power of p for $i \in \{1, \ldots, h\}$. We set $\wp_0 := \mathfrak{m}$. Then we have:
(1) $\wp_0 \supsetneqq \wp_1 \supsetneqq \cdots \supsetneqq \wp_h = \wp$;
(2) *The ideals $\wp_0, \ldots, \wp_h$ are simple $\mathfrak{m}$-primary $v_\wp$-ideals of R, and every $\mathfrak{m}$-primary $v_\wp$-ideal of R is a product of these $h + 1$ simple $v_\wp$-ideals.*
(3) *Every power of $\wp_h = \wp$ is a $v_\wp$-ideal.*

Proof [by induction on h]: If $h = 0$, then we have $\wp = \mathfrak{m}$, hence (1) is satisfied trivially, and (2), (3) hold since the only $\mathfrak{m}$-primary ν_R-ideals in R are the powers of the maximal ideal $\mathfrak{m}$. Assume that $h > 0$. In $S := R_1$ we have the simple complete $\mathfrak{m}(S)$-primary ideals $\mathfrak{b}_1 := \wp_1^S, \ldots, \mathfrak{b}_h := \wp_h^S$ [cf. (4.16)(2)], and we have $(\wp_i^S)^{R_i} = \mathfrak{n}_i$ for $i \in \{1, \ldots, h\}$. Now $\mathfrak{b}_h$ is a simple complete $\mathfrak{m}(S)$-primary ideal of S of rank $h - 1$, and we have $v_{\mathfrak{b}_h} = v_\wp$. Therefore, by induction, we have

$$\mathfrak{m}(S) = \mathfrak{b}_1 \supsetneq \mathfrak{b}_2 \supsetneq \cdots \supsetneq \mathfrak{b}_h,$$

the ideals $\mathfrak{b}_1, \ldots, \mathfrak{b}_h$ are simple $\mathfrak{m}(S)$-primary $v_\wp$-ideals of S, and every $\mathfrak{m}(S)$-primary $v_\wp$-ideal of S is a product of these ideals. From (5.2)(6) we get that $\wp_1 \supsetneq \wp_2 \supsetneq \cdots \supsetneq \wp_h$, and since $\wp_0 \supsetneq \wp_1$, we have proved (1). Now $\wp_0$ is a $v_\wp$-ideal, and, by (5.2)(5), each of the ideals $\wp_1, \ldots, \wp_h$ is a $v_\wp$-ideal of R. Let $\mathfrak{c}$ be an $\mathfrak{m}$-primary $v_\wp$-ideal of R. If $\mathfrak{c}^S = S$, then we have $\mathfrak{c} = \mathfrak{m}^r$ [with $r := \mathrm{ord}(\mathfrak{c})$, cf. (5.2)(7)]. Now we consider the case that $\mathfrak{c}^S \neq S$. Then $\mathfrak{c}^S$ is an $\mathfrak{m}(S)$-primary $v_\wp$-ideal of S [cf. (5.2)(4)], and we have, by induction,

$$\mathfrak{c}^S = \mathfrak{b}_1^{l_1} \cdots \mathfrak{b}_h^{l_h} \quad \text{with } l_1, \ldots, l_h \in \mathbb{N}_0.$$

Let $\mathfrak{c}_1$ be the unique ideal in $\mathcal{MCP}(R, p)$ with $\mathfrak{c}_1^S = \mathfrak{c}^S$ [cf. (4.16)(2)]; it is clear that $\mathfrak{c}_1 = \wp_1^{l_1} \cdots \wp_h^{l_h}$. Then we have [cf. (5.2)(7)]

$$\mathfrak{c} = \wp_0^{l_0} \wp_1^{l_1} \cdots \wp_h^{l_h} \quad \text{with } l_0 \in \mathbb{N}_0,$$

and this is the assertion in (2). With respect to (3), by induction we get that every power $\mathfrak{b}_h^l$ of $\mathfrak{b}_h$ is a $v_\wp$-ideal, hence also that $\wp_h^l$ is a $v_\wp$-ideal [cf. (5.2)(5)].

(5.4) DEFINITION: Let $\wp$ be a simple $\mathfrak{m}$-primary of R of rank $h > 0$. The ideals $\wp_0, \wp_1, \ldots, \wp_{h-1}$ defined in (5.3) are called the predecessors of $\wp$. We have $c(\wp) = p^n$ with $p \in \mathbb{P}$ and $n \in \mathbb{N}$, and $c(\wp_i)$ is a positive power of p for $i \in \{1, \ldots, h-1\}$.

(5.5) Lemma: Let $\mathfrak{a}$ be a complete $\mathfrak{m}$-primary ideal of R, let $P \subset \mathbb{P}$ be a finite non-empty subset which contains every $p \in \mathbb{P}$ which divides $c(\mathfrak{a})$, and choose $z \in S_P$ with $\mathfrak{m}S_P = S_P z$. Let $\mathfrak{b}$ be any non-zero ideal of R, and set $\mathfrak{c} := \mathfrak{a} : \mathfrak{b}$. Set $r := \mathrm{ord}(\mathfrak{a})$, $s := \mathrm{ord}(\mathfrak{b})$, and $t := \mathrm{ord}(\mathfrak{c})$. Then we have $s + t \geq r$, $\mathfrak{c}$ is a complete ideal of R, every $p \in \mathbb{P}$ which divides $c(\mathfrak{c})$ also divides $c(\mathfrak{a})$, and for the ideals $\mathfrak{a}^{S_P}$, $\mathfrak{b}^{S_P}$, $\mathfrak{c}^{S_P}$ of S_P we have

$$\mathfrak{c}^{S_P} = \mathfrak{a}^{S_P} : z^{s+t-r} \mathfrak{b}^{S_P}.$$

Proof: We write $P = \{p_1, \ldots, p_h\}$, and set $S := S_P$ and $S_i := S_{p_i}$ for $i \in \{1, \ldots, h\}$. The ideal $\mathfrak{c}$ is complete by B(6.16)(5). Since $\mathfrak{b}\mathfrak{c} \subset \mathfrak{a}$, we have $s + t \geq r$ and $\mathfrak{c}^S \subset \mathfrak{a}^S : z^{s+t-r}\mathfrak{b}^S =: \widetilde{\mathfrak{c}}$; note that $\widetilde{\mathfrak{c}}$ is a complete ideal of S [cf. loc. cit.]. Let $i \in \{1, \ldots, h\}$; now S_i is a localization of S, and therefore we have, on the one side, $\mathfrak{a}^S S_i : z^{s+t-r}\mathfrak{b}^S S_i = \widetilde{\mathfrak{c}}S_i$, and, on the other side, $\mathfrak{a}^S S_i = \mathfrak{a}^{S_i}$ and $\mathfrak{b}^S S_i = \mathfrak{b}^{S_i}$ [cf. (1.5)]. Therefore the ideal $\widetilde{\mathfrak{c}}S_i$ is either a complete $\mathfrak{m}(S_i)$-primary ideal of S_i or it is equal to S_i. This implies that $\ell_S(S/\widetilde{\mathfrak{c}})$ is finite. By (3.27) there exists a complete ideal $\mathfrak{d}$ of R having finite colength and with $\mathfrak{d}^S = \widetilde{\mathfrak{c}}$ which is not divisible by $\mathfrak{m}$, and with $\mathfrak{d}S \cap R = \mathfrak{d}$. We set $q := \mathrm{ord}(\mathfrak{d})$.

(1) We have $q \leq t$.

In fact, suppose that $q > t$. Since $\mathfrak{c}^S \subset \widetilde{\mathfrak{c}}$, we have $\mathfrak{m}^{q-t}\mathfrak{c} \subset z^{q-t}\mathfrak{c}S = z^q\mathfrak{c}^S \subset z^q\widetilde{\mathfrak{c}} = z^q\mathfrak{d}^S = \mathfrak{d}S$, hence $\mathfrak{m}^{q-t}\mathfrak{c} \subset \mathfrak{d}S \cap R = \mathfrak{d}$. This implies that $\mathfrak{m}$ divides $\mathfrak{d}$ [cf. (3.22)(2)], contradicting the choice of $\mathfrak{d}$.

(2) We have $\mathfrak{c} \subset \mathfrak{c}S = z^t\mathfrak{c}^S \subset z^t\mathfrak{d}^S = \mathfrak{m}^{t-q}\mathfrak{d}S$, hence $\mathfrak{c} \subset \mathfrak{m}^{t-q}\mathfrak{d} \cap R = \mathfrak{m}^{t-q}\mathfrak{d}$ since also $\mathfrak{m}^{t-q}\mathfrak{d}$ is contracted from S [cf. (3.10)]. On the other hand, by definition of $\widetilde{\mathfrak{c}}$, we have $z^{s+t-r}\mathfrak{b}^S\mathfrak{d}^S \subset \mathfrak{a}^S$, hence $\mathfrak{m}^{t-q}\mathfrak{b}\mathfrak{d} \subset \mathfrak{a}S \cap R = \mathfrak{a}$, hence $\mathfrak{m}^{t-q}\mathfrak{d} \subset \mathfrak{a} : \mathfrak{b} = \mathfrak{c}$. Now we have shown that $\mathfrak{m}^{t-q}\mathfrak{d} = \mathfrak{c}$, hence that $\mathfrak{c}^S = \mathfrak{d}^S = \widetilde{\mathfrak{c}}$.

(3) Let $p \in \mathbb{P}$ divide the ideal $c(\mathfrak{c})$. Then we have $\mathfrak{c}^{S_p} \neq S_p$ [cf. (2.15)], and since $\mathfrak{c}^{S_p} = \mathfrak{a}^{S_p} : z^{s+t-r}\mathfrak{b}^{S_p}$, we find that $\mathfrak{a}^{S_p} \neq S_p$, and therefore p is a divisor of $c(\mathfrak{a})$, again by (2.15).

(5.6) REMARK: Let $r \in \mathbb{N}$, let $\mathfrak{b}$ be a non-zero ideal of R, and set $s := \mathrm{ord}(\mathfrak{b})$. Then we have $\mathfrak{m}^r : \mathfrak{b}\mathfrak{m}^{r-s}$ if $s < r$ and $\mathfrak{m}^r : \mathfrak{b} = R$ otherwise.

Proof: We have $\mathfrak{m}^r : \mathfrak{b} \supset \mathfrak{m}^r : \mathfrak{m}^s$. If $s \geq r$, then we have $\mathfrak{m}^r : \mathfrak{b} = R$. If $s < r$, then we have $\mathfrak{m}^r : \mathfrak{b} \supset \mathfrak{m}^{r-s}$, hence $\mathfrak{b}\mathfrak{m}^{r-s} \subset \mathfrak{b}(\mathfrak{m}^r : \mathfrak{b}) \subset \mathfrak{m}^r$, which implies that $\mathrm{ord}(\mathfrak{m}^r : \mathfrak{b}) = r - s$, hence that $\mathfrak{m}^r : \mathfrak{b} = \mathfrak{m}^{r-s}$.

(5.7) Lemma: *Let $\mathfrak{a}$ be a complete $\mathfrak{m}$-primary ideal of R, and let $\mathfrak{b}$ be any ideal of R. Then $\mathfrak{a} : \mathfrak{b}$ is complete, and every simple complete $\mathfrak{m}$-primary factor of $\mathfrak{a} : \mathfrak{b}$ is a predecessor of, or coincides with a simple complete $\mathfrak{m}$-primary factor of $\mathfrak{a}$.*

Proof: (1) For any complete $\mathfrak{m}$-primary ideal $\mathfrak{d}$ which is not a power of $\mathfrak{m}$ we define its rank $h(\mathfrak{d})$ to be the minimum of the ranks of those of its simple complete factors which are different from $\mathfrak{m}$; if $\mathfrak{d}$ is a power of $\mathfrak{m}$, then we set $h(\mathfrak{d}) = 0$. If $\mathfrak{d}$ is not a power of $\mathfrak{m}$ and if S is a quadratic transform of R with $\mathfrak{d}^S \neq S$, then $\mathfrak{d}^S$ is a complete $\mathfrak{m}(S)$-primary ideal of S, and we have $h(\mathfrak{d}) > h(\mathfrak{d}^S)$.

(2) We prove the assertion by induction on $h(\mathfrak{a})$. The assertion is true if $h(\mathfrak{a}) = 0$ [cf. (5.6)]. Now let $h(\mathfrak{a}) > 0$, i.e., $\mathfrak{a}$ is not a power of $\mathfrak{m}$. Let $c(\mathfrak{a}) = p_1^{e_1} \cdots p_l^{e_l}$ be the factorization of $c(\mathfrak{a})$; note that $l > 0$. We set $S_i := S_{p_i}$ for $i \in \{1, \ldots, l\}$ and $S := S_1 \cap \cdots \cap S_l$, and we choose $z \in S$ with $\mathfrak{m}S = Sz$. Furthermore, we set $\mathfrak{n}_i := \mathfrak{m}(S_i) \cap S$; note that $\mathfrak{n}_1, \ldots, \mathfrak{n}_l$ are the maximal ideals of S [cf. (2.9)]. We set $\mathfrak{c} := \mathfrak{a} : \mathfrak{b}$. Since $\mathfrak{m}$ is either a factor of $\mathfrak{a}$ or equal to a predecessor of a simple factor of $\mathfrak{a}$, it is enough to consider only the case that $\mathfrak{c}$ is not a power of $\mathfrak{m}$ [this includes the case that $\mathfrak{c} = R$]. We write

$$\mathfrak{a} = \mathfrak{m}^{a_0}\mathfrak{a}_1 \cdots \mathfrak{a}_l, \quad \mathfrak{c} = \mathfrak{m}^{c_0}\mathfrak{c}_1 \cdots \mathfrak{c}_l;$$

here we have $a_0, c_0 \in \mathbb{N}_0$, and, for $i \in \{1, \ldots, l\}$, $\mathfrak{a}_i$ is a product of simple complete $\mathfrak{m}$-primary ideals of $\mathcal{SCP}(R, p_i)$ and $\mathfrak{c}_i$ is a product of simple complete $\mathfrak{m}$-primary ideals of $\mathcal{SCP}(R, p_i)$ or $\mathfrak{c}_i = R$ [cf. (5.5)]. With $r := \mathrm{ord}(\mathfrak{a})$, $s := \mathrm{ord}(\mathfrak{b})$ and $t := \mathrm{ord}(\mathfrak{c})$ we have for the ideals $\mathfrak{a}^S$, $\mathfrak{b}^S$ and $\mathfrak{c}^S$ of S [cf. (5.5)]

$$\mathfrak{c}^S = \mathfrak{a}^S : z^{s+t-r}\mathfrak{b}^S.$$

For $i \in \{1, \ldots, l\}$ the ideal $\mathfrak{a}_i^S$ is an $\mathfrak{n}_i$-primary ideal of S, and therefore we have [cf. B(10.1)] $\mathfrak{a}_1^S \cdots \mathfrak{a}_l^S = \mathfrak{a}_1^S \cap \cdots \cap \mathfrak{a}_l^S$. This implies that

$$\mathfrak{c}^S = \mathfrak{c}_1^S \cdots \mathfrak{c}_l^S = \mathfrak{a}^S : z^{s+t-r}\mathfrak{b}^S = \left(\bigcap_{i=1}^l \mathfrak{a}_i^S\right) : z^{s+t-r}\mathfrak{b}^S$$

$$= \bigcap_{i=1}^l (\mathfrak{a}_i^S : z^{s+t-r}\mathfrak{b}^S) = \prod_{i=1}^l (\mathfrak{a}_i^S : z^{s+t-r}\mathfrak{b}^S)$$

[since, for $i \in \{1, \ldots, l\}$, $\mathfrak{a}_i^S : z^{s+t-r}\mathfrak{b}^S$ is either equal to S or it is an $\mathfrak{n}_i$-primary ideal of S]. For $i, j \in \{1, \ldots, l\}$ we have $(\mathfrak{a}_i^S : z^{s+t-r}\mathfrak{b}^S)S_j = \mathfrak{a}_i^{S_j} : z^{s+t-r}\mathfrak{b}^{S_j}$ since S_j is a localization of S, and we have $\mathfrak{a}_i^{S_j} : z^{s+t-r}\mathfrak{b}^{S_j} = S_j$ if $i \neq j$. Therefore we have $\mathfrak{c}^{S_i} = \mathfrak{c}_i^{S_i} = \mathfrak{a}_i^{S_i} : z^{s+t-r}\mathfrak{b}^{S_i}$, and by induction, applied to the $\mathfrak{m}(S_i)$-primary ideal $\mathfrak{a}_i^{S_i}$ of S_i of rank $< h(\mathfrak{a})$ and the ideal $z^{s+t-r}\mathfrak{b}^{S_i}$ of S_i, we obtain that every simple complete $\mathfrak{m}(S_i)$-primary factor of $\mathfrak{c}^{S_i}$ is a predecessor of, or coincides with a simple complete $\mathfrak{m}(S_i)$-primary factor of $\mathfrak{a}_i^{S_i}$. This proves the lemma.

(5.8) Proposition: *We assume that κ is an infinite field. For any two complete $\mathfrak{m}$-primary ideals $\mathfrak{a}$, $\mathfrak{b}$ of R with*

$$\mathfrak{b} \subset \mathfrak{a}, \quad \ell_R(\mathfrak{a}/\mathfrak{b}) = 1 \text{ and } \mathrm{ord}(\mathfrak{b}) = \mathrm{ord}(\mathfrak{a}) + 1$$

we have: Every simple complete $\mathfrak{m}$-primary factor of $\mathfrak{b}$ is a predecessor of, or coincides with a simple complete $\mathfrak{m}$-primary factor of $\mathfrak{a}$.

Proof: If $\mathfrak{a}$ would be a power of $\mathfrak{m}$, say $\mathfrak{a} = \mathfrak{m}^r$, then $\mathfrak{b} \subset \mathfrak{m}^{r+1} \subset \mathfrak{m}^r$, and $1 = \ell_R(\mathfrak{m}^r/\mathfrak{b}) \geq \ell_R(\mathfrak{m}^r/\mathfrak{m}^{r+1}) = r + 1$ which is absurd. Therefore $\mathfrak{a}$ is not a power of $\mathfrak{m}$. We have $\mathfrak{b} : \mathfrak{m} = \mathfrak{a} : \mathfrak{m}$ by (3.23)(3). It is enough to show the following: Every simple complete $\mathfrak{m}$-primary factor of $\mathfrak{a} : \mathfrak{m}$ is a predecessor of, or coincides with a simple complete $\mathfrak{m}$-primary factor of $\mathfrak{a}$. But this is an immediate consequence of (5.7).

6 The Quadratic Sequence

(6.1) In this section we change our point of view. Until now we studied a fixed two-dimensional regular local ring with field of quotients K. Now we start with a field K, and we consider all two-dimensional regular local subrings of K having K as field of quotients.

(6.2) NOTATION: Let K be a fixed field. We denote by Ω the set of all two-dimensional regular local subrings of K having K as field of quotients. For $S \in \Omega$ let $\mathfrak{m}_S$ be its maximal ideal, $\kappa_S = S/\mathfrak{m}_S$ its residue field, ord_S its order function which gives rise to the discrete valuation ν_S of K with valuation ring V_S, and let $\mathbb{P}_S$ be the set of homogeneous prime ideals of height 1 of $\mathrm{gr}_{\mathfrak{m}_S}(S)$.

(6.3) Lemma: *Let S, $T \in \Omega$. Then:*
(1) *If $S \subset T$, then T dominates S. In particular, κ_S is a subfield of κ_T.*
(2) *If $S \subset T$ and $\mathfrak{m}_S T = \mathfrak{m}_T$, then we have $S = T$.*
(3) *If $S \subsetneq T$, then we have $V_S \neq V_T$.*

Proof: (1) Suppose that T does not dominate S. Then $\mathfrak{p} := \mathfrak{m}_T \cap S$ is a prime ideal of S of height 1 [we have $\mathfrak{m}_T \cap S \neq \{0\}$ since S and T have the same field of quotients], and therefore $\mathfrak{p} = Sx$ for some $x \in S$. V_T is a discrete valuation ring of K which dominates T, hence $\mathfrak{m}(V_T) \cap S = \mathfrak{p}$. Now $S_\mathfrak{p}$ is a discrete valuation ring of K, and we have $S_\mathfrak{p} \subset V_T$, hence $S_\mathfrak{p} = V_T$. In particular, we have $\nu_T(x) = 1$, hence x is a regular parameter of T. We choose $y \in T$ such that $\mathfrak{m}_T = Tx + Ty$. Since $\nu_T(y) = 1$, we have $y = x(z/w)$ with z, $w \in S$, and we may assume that neither w nor z is divisible by x in S. Therefore we have $\nu_T(w) = \nu_T(z) = 0$. The relation $wy = xz$ implies that x divides w in T, hence $\nu_T(w) > 0$, contradicting $\nu_T(w) = 0$. Therefore we have proved that T dominates S.
(2) This follows from B(9.6) [note that a regular local ring is analytically normal, hence, in particular, analytically irreducible, cf. III(6.3)].
(3) Suppose that $V_S = V_T =: V$, hence that $\nu_S = \nu_T$. Let $\{x, y\}$ be a regular system of parameters of S, and let ζ be the image of y/x in $\kappa_V := V/\mathfrak{m}(V)$. Then ζ is transcendental over κ_S and we have $\kappa_V = \kappa_S(\zeta)$ [cf. (2.3)(1)]. By the same argument we get that κ_V is purely transcendental over κ_T of transcendence degree 1. Therefore ζ is transcendental over κ_T. We have $\nu_T(x) = \nu_T(y) = 1$. Suppose that $y \bmod \mathfrak{m}_T^2$ is contained in the subspace of $\mathfrak{m}_T/\mathfrak{m}_T^2$ generated by $x \bmod \mathfrak{m}_T^2$, hence that $y = ax + z$ with a unit a of T and $z \in \mathfrak{m}_T^2$. Since $\nu_T(z/x) \geq 1$, this yields that $\zeta \in \kappa_T$, in contradiction with the fact that ζ is transcendental over κ_T. Therefore $y \bmod \mathfrak{m}_T^2$ does not lie in the subspace of $\mathfrak{m}_T/\mathfrak{m}_T^2$ generated by $x \bmod \mathfrak{m}_T^2$. Considering x/y instead of y/x, we see that $x \bmod \mathfrak{m}_T^2$ does not lie in the subspace of $\mathfrak{m}_T/\mathfrak{m}_T^2$ generated by $y \bmod \mathfrak{m}_T^2$. Therefore $\{x, y\}$ is a regular system of parameters of T, and therefore we have $\mathfrak{m}_S T = \mathfrak{m}_T$. From (2) we now get $S = T$, contrary to our assumption.

(6.4) Proposition: *Let S, $T \in \Omega$ with $S \subset T$. Then there exists a uniquely determined sequence*
$$S =: S_0 \subset S_1 \subset \cdots \subset S_n := T$$
[with $n \geq 0$] where, for $i \in \{1, \ldots, n\}$, S_i is a quadratic transform of S_{i-1}.

Proof: There is nothing to show if $S = T$; hence we assume that $S \neq T$. Then T dominates S and $V_S \neq V_T$ by (6.3); note that V_T dominates T.
(1) We show that there exists a quadratic transform S_1 of S with $S_1 \subset T$. Let $\{x, y\}$ be a regular system of parameters of S, and set $u := y/x$.
(a) If $u \in T$, then we have $A := S[u] \subset T$, and since T dominates S, the ideal $\mathfrak{m}_T \cap A = \mathfrak{m}(V_T) \cap A$ is a prime ideal of A which contains $\mathfrak{m}_S A = Ax$. Suppose that $\mathfrak{m}(V_T) \cap A = Ax$. Then we have $A_{Ax} \subset V_T$, and since $V_S = A_{Ax}$ [cf. (2.3)(2)], we get $V_S = V_T$, contradicting our assumption. Therefore we have $\mathfrak{m}_T \cap A = \mathfrak{n}_p$ where $p \in \mathbb{P}_S$, and the quadratic transform S_1 of S defined by p is contained in T.

(b) Likewise, if $1/u \in T$, then there exists a quadratic transform S_1 of S with $S_1 \subset T$.

(c) We show that either u or $1/u$ lie in T. Suppose that this is not the case. We set $B := T[u]$ and $\mathfrak{q} := \mathfrak{m}_T B$. The ideal $\mathfrak{q}$ is a prime ideal of B, the $\mathfrak{q}$-residue $\bar{u}$ of u is transcendental over κ_T and $B/\mathfrak{q} = \kappa_T[\bar{u}]$ [cf. I(3.34)]. Let V be a valuation ring of K having center $\mathfrak{q}$ in B [cf. I(3.5)]. Then $\kappa_T[\bar{u}]$ is a subring of $V/\mathfrak{m}(V)$. Moreover, V dominates S, and since $\bar{u}$ is also transcendental over κ_S, we have $V = V_S$ by (2.19). We have $\nu_S(x) = 1$, and since $\nu_S(\mathfrak{m}_T) > 0$, we have $\nu_S(\mathfrak{m}_T) = 1$, hence $x \notin \mathfrak{m}_T^2$, and since $\nu_S(y) = 1$, we also see that $y \notin \mathfrak{m}_T^2$. Just as in (3) of the proof of (6.3) we obtain that $y \bmod \mathfrak{m}_T^2$ does not lie in the subspace of $\mathfrak{m}_T/\mathfrak{m}_T^2$ generated by $x \bmod \mathfrak{m}_T^2$.
Applying this reasoning to $B' := T[1/u]$, we see that also $x \bmod \mathfrak{m}_T^2$ does not lie in the subspace of $\mathfrak{m}_T/\mathfrak{m}_T^2$ generated by $y \bmod \mathfrak{m}_T^2$.
Now we have shown that $\{x, y\}$ is a regular system of parameters of T, i.e., that $\mathfrak{m}_S T = \mathfrak{m}_T$. This implies that $S = T$ by (6.3)(2), contradicting our assumption. Therefore we have $u \in T$ or $1/u \in T$

(2) If $S_1 \neq T$, we can continue this process. We have to show that after a finite number of steps we obtain a ring S_n such that $S_n = T$.
Suppose that there is an infinite sequence $S =: S_0 \subset S_1 \subset S_2 \subset \cdots \subset T \subset V_T$ such that, for $i \geq 1$, S_i is a quadratic transform of S_{i-1}, and that T contains all the rings S_i. Let $u \in V_T$ be such that the ν_T-residue of u is transcendental over κ_T. We write $u = x_0/y_0$ with $x_0, y_0 \in S_0$; it is clear that $x_0, y_0 \in \mathfrak{m}_{S_0}$. Now $\mathfrak{m}_{S_0} S_1$ is a proper principal ideal $S_1 z_1$ of S_1; hence we have $x_0 = z_1 x_1$, $y_0 = z_1 y_1$ with $x_1, y_1 \in S_1$, hence $u = x_1/y_1$, and again we have $x_1, y_1 \in \mathfrak{m}_{S_1}$. Note that $\nu_T(x_0) > \nu_T(x_1)$. Continuing, we get a strictly decreasing sequence $(\nu_T(x_i))_{i\in\mathbb{N}}$ of positive integers; this is absurd.

(3) With regard to uniqueness, let S' be any quadratic transform of S which is contained in T. Then V_T dominates S' [cf. (6.3)], and therefore $S' = S_1$ by (2.19). By repeating this argument, we see that, for $i \in \{1, \ldots, n-1\}$, S_i is that quadratic transform of S_{i-1} which is contained in T.

(6.5) DEFINITION: Let $S, T \in \Omega$ with $S \subset T$. The sequence

$$S =: S_0 \subset S_1 \subset \cdots \subset S_n := T$$

of (6.4) is called the quadratic sequence between S and T; n is called the length of the sequence.

(6.6) NOTATION: Let $R \in \Omega$ and $n \in \mathbb{N}_0$. We denote by $N_n(R)$ the set of $S \in \Omega$ with $S \supset R$ and such that the quadratic sequence between R and S has length n, and set $N(R) := \bigcup_{n\in\mathbb{N}_0} N_n(R)$. Note that $N_0(R) = \{R\}$, that $N_1(R)$ is the first neighborhood of R, and that $N(R)$ is the set of all two-dimensional regular local subrings of K which contain R.

(6.7) Proposition: Let $S, T \in \Omega$ with $S \subset T$. Then:

(1) κ_T is a finite extension of κ_S.
(2) Let $\mathfrak{a}$ be an $\mathfrak{m}_S$-primary ideal of S. Then $\mathfrak{a}^T$ is an $\mathfrak{m}_T$-primary ideal of T or $\mathfrak{a}^T = T$.

Proof: Both assertions are true if T is a quadratic transform of S [cf. (2.6)(2) and (2.11)(2)]. In the general case, we consider the quadratic sequence between S and T and use induction [we use the multiplicative behavior of field degrees and also (1.5)(4)].

(6.8) NOTATION: Let S, $T \in \Omega$ with $S \subset T$. The degree $[\kappa_T : \kappa_S]$ shall be denoted by $[T : S]$. (This agrees with the notation introduced in (2.6)(3).)

(6.9) Corollary: [Hoskin-Deligne] Let $\mathfrak{a}$ be a complete $\mathfrak{m}$-primary ideal of R. Then we have

$$\ell_R(R/\mathfrak{a}) = \sum_{S \in N(R)} [S : R] \frac{\mathrm{ord}_S(\mathfrak{a}^S)(\mathrm{ord}_S(\mathfrak{a}^S) + 1)}{2}, \qquad (*)$$

and there exists $n \in \mathbb{N}$ such that $\mathfrak{a}^S = S$ for every $S \in \bigcup_{i \geq n} N_i(R)$.

Proof: The formula in $(*)$ follows by repeated application of (3.14); since $\ell_R(R/\mathfrak{a})$ is finite, the second assertion follows immediately.

(6.10) SIMPLE COMPLETE IDEALS, QUADRATIC SEQUENCES, VALUATIONS OF THE SECOND KIND: Let $R \in \Omega$; we associate with R the following three sets:
(1) The set $N(R)$ of rings $S \in \Omega$ with $R \subset S$;
(2) The set $\mathcal{SCP}(R)$ of simple complete $\mathfrak{m}_R$-primary ideals of R;
(3) The set $\mathcal{V}(R)$ of valuation rings V of K which dominate R and are of the second kind with respect to R [such a V is discrete of rank 1 by (2.18)].

We show that there are bijective maps between these three sets.

THE SETS $\mathcal{SCP}(R)$ AND $N(R)$: To each simple complete $\mathfrak{m}_R$-primary ideal $\wp$ of R there corresponds a quadratic sequence $R = R_0 \subset \cdots \subset R_h = S_\wp$ such that $\wp^{S_\wp} = \mathfrak{m}_{S_\wp}$ [cf. (5.1)]; we associate with $\wp$ the ring $S_\wp$. Conversely, let $S \in N(R)$, and let $R =: R_0 \subset R_1 \subset \cdots \subset R_h := S$ be the quadratic sequence between R and S. Set $\wp^{(h)} := \mathfrak{m}_S$ and, for $i \in \{1, \ldots, h\}$, let $\wp^{(i-1)} \subset R_{i-1}$ be that simple complete $\mathfrak{m}_{R_{i-1}}$-primary ideal with $(\wp^{(i-1)})^{R_i} = \wp^{(i)}$ [cf. (3.27)]; we associate with S the simple complete $\mathfrak{m}$-primary ideal $\wp := \wp^{(0)}$ of R. This sets up a bijective map between the set $\mathcal{SCP}(R)$ of simple complete $\mathfrak{m}_R$-primary ideals of R and the set $N(R)$ of rings $S \in \Omega$ with $S \supset R$; in particular, for $\wp \in \mathcal{SCP}(R)$ the length of the quadratic sequence between R and $S_\wp$ is equal to the rank of $\wp$.
Let $\wp \in \mathcal{SCP}(R)$ be of rank h, let $\wp_0 \supset \wp_1 \supset \cdots \supset \wp_h := \wp$ be the set of simple $v_\wp$-ideals defined by $\wp$ [cf. (5.3)], and let $R =: R_0 \subset R_1 \subset \cdots \subset R_h := S_\wp$ be the quadratic sequence between R and $S_\wp$; then we have $S_{\wp_i} = R_i$ for $i \in \{0, \ldots, h\}$.
THE SETS $N(R)$ AND $\mathcal{V}(R)$: Let $S \in N(R)$; we associate with S the discrete valuation ring V_S. Then V_S dominates R and it is of the second kind with respect to

R [since V_S is of the second kind with respect to S, cf. (2.3)(1)], hence $V_S \in \mathcal{V}(R)$. Conversely, let $V \in \mathcal{V}(R)$. If $V = V_R$, then we associate R with V. If $V \neq V_R$, then V has a directional ideal $p(V) \in \mathbb{P}$ [cf. (2.18)], and setting $R_1 := S_{p(V)}$, then V is still of the second kind with respect to R_1. If $V = V_{R_1}$, then we associate R_1 with V. Otherwise we repeat this process. Suppose that there exists an infinite sequence $R =: R_0 \subset R_1 \subset R_2 \subset \cdots$ in Ω such that for $i \geq 1$ the ring R_i is a quadratic transform of R_{i-1}, and V dominates all the rings R_i. Just as in (2) of the proof of (6.4) this leads to an absurd result. Therefore there exists $h \in \mathbb{N}_0$ such that $V = V_{R_h}$. Now we associate with V the ring R_h. Thus, we have a surjective map $N(R) \to \mathcal{V}(R)$. To show that this map is injective, we have to assert: Let S, $T \in N(R)$ and $S \neq T$; then we have $V_S \neq V_T$. We consider the quadratic sequences

$$R =: R'_0 \subset R'_1 \subset \cdots \subset R'_m := S, \quad R =: R''_0 \subset R''_1 \subset \cdots \subset R''_n := T.$$

We prove the assertion by induction on $\min(m, n)$. We may assume that $m \geq n$. If $n = 0$, then we have $V_S \neq V_R$ by (6.3). Now we assume that $n > 0$, and we choose $i \in \{0, \ldots, n\}$ minimal with $R'_i = R''_i$. If $i = 0$, then we have $R'_1 \neq R''_1$, hence $V_S \neq V_T$ since V_S and V_T contain only one quadratic transform of R by (2.19). If $i \geq 1$, then we replace R by $R'_i = R''_i$, and, by induction, we again get that $V_S \neq V_T$.

(6.11) Corollary: (1) Let $\mathfrak{a}$ be an ideal of R and let $\bar{\mathfrak{a}}$ be the integral closure of $\mathfrak{a}$. Then we have

$$\bar{\mathfrak{a}} = \bigcap_{S \in N(R)} \mathfrak{a} V_S.$$

(2) Let $\mathfrak{a}$, $\mathfrak{b}$ be complete ideals of R. Then we have $\mathfrak{a} \subset \mathfrak{b}$ iff $\mathfrak{a} V_S \subset \mathfrak{b} V_S$ for every $S \in N(R)$.

Proof: (1) follows from B(6.15), (3.7) and (6.10), and (2) is a consequence of (1).

(6.12) Corollary: Let $R_0 \subset R_1 \subset \cdots$ be a sequence in Ω such that R_{i+1} is a quadratic transform of R_i for every $i \in \mathbb{N}_0$. Then $V := \bigcup R_i$ is a valuation ring of K which is of the first kind with respect to R_0. Moreover, V is the only valuation ring of K which dominates all the rings R_i, $i \geq 0$.

Proof: The ring V is integrally closed and quasilocal with maximal ideal $\mathfrak{n} := \bigcup \mathfrak{m}_{R_i}$. Clearly V dominates all the rings R_i; in particular, $V/\mathfrak{n}$ is algebraic over $R_0/\mathfrak{m}_{R_0}$ [since the residue fields of all the rings R_i, $i \geq 0$, are finite extensions of the residue field of R]. Suppose that V is not a valuation ring of K. By I(3.34)(2) there exists a valuation ring W of K which dominates V and which is of the second kind with respect to V. Then W dominates R_0 and it is also of the second kind with respect to R_0, and therefore we have $W = V_{R_i}$ for some $i \in \mathbb{N}_0$ [cf. (6.10)], hence W does not dominate R_{i+1} [cf. (2.6)(2)]. This contradiction shows that V is a valuation ring of K, and it is of the first kind with respect to R_0.

If W' is any valuation ring of K which dominates all the rings R_i, $i \geq 0$, then W' dominates V, hence we have $W' = V$ by I(3.8).

7 Proximity

(7.0) We keep the notations of the last section.

(7.1) DEFINITION: Let $R, S \in \Omega$ with $R \subsetneqq S$; we say that S is proximate to R and we write $S \succ R$ or $R \prec S$ if the valuation ring V_R contains S.

(7.2) REMARK: (1) Let $R \in \Omega$; then we have $R \prec S$ for every $S \in N_1(R)$.
(2) Let $R \in \Omega$ and $S_1 \in N_1(R)$. There exists a unique quadratic transform S_2 of S_1 which is contained in V_R.
In fact, we have $\mathfrak{m}_R S_1 = S_1 x$ with $x \in R$ and $\mathfrak{m}(V_R) \cap S_1 = S_1 x$, and x is part of a regular system of parameters of S_1 [cf. (2.6)(2)]. We choose $u \in S_1$ with $(x, u) = \mathfrak{m}_{S_1}$. Then we have $\nu_R(u) = 0$, and the only quadratic transforms of S_1 which are contained in V_R are localizations of $B := S_1[x/u]$ with respect to maximal ideals of B containing u. It is easy to check that $\mathfrak{m}(V_R) \cap B = B \cdot x/u$, hence $B \cdot x/u$ is a prime ideal of B, and therefore $B_{B \cdot x/u} \subset V_R$ is a discrete valuation ring [cf. B(10.5) and note that B is integrally closed by (2.3)(4)], hence we have $B_{B \cdot x/u} = V_R$. The only maximal ideal of B containing u and x/u is the ideal $\mathfrak{n}$ generated by u and x/u. Therefore $S_2 := B_\mathfrak{n}$ is the only quadratic transform of S_1 contained in V_R. Note that $\mathfrak{m}_{S_2} = (u, x/u)$, $[S_2 : S_1] = 1$ and $\mathfrak{m}(V_R) \cap S_2 = S_2 \cdot x/u$.
(3) The argument in (2) can be applied to $S_2[x/u^2]$ etc., and it leads to the following result. Let $R \in \Omega$ and $p \in \mathbb{P}_R$, and set $R_1 := S_p$; then there exists a unique infinite sequence

$$R =: R_0 \subset R_1 \subset R_2 \subset R_3 \subset \cdots \subset V_R$$

where R_{i+1} is a quadratic transform of R_i for $i \geq 0$. For $i \geq 1$ we have $\mathfrak{m}_{R_i} = (u, x/u^{i-1})$, $\mathfrak{m}(V_R) \cap R_i = R_i \cdot x/u^{i-1}$ and $[R_{i+1} : R_i] = 1$ for $i \geq 1$. The union $\bigcup R_i$ is a valuation ring of K which dominates all the rings R_i, $i \geq 0$, it is contained in V_R, and it is of the first kind with respect to R [cf. (6.12)]. In (7.6) below we shall describe the valuation defined by this valuation ring. Note that $R \prec R_i$ for every $i \in \mathbb{N}$.
(4) Let $S \subsetneqq T$ be rings in Ω, and let $S =: S_0 \subset S_1 \subset \cdots \subset S_n := T$ be the quadratic sequence between S and T. Then T is proximate to S_{n-1}, and if $n \geq 2$, then T is proximate to at most one other ring S_i, $i \in \{0, \ldots, n-2\}$.
Proof: Assume that $n \geq 2$ and that T is proximate to one of the rings $S_0, \ldots, S_{n-2}$. We choose $i \in \{0, \ldots, n-2\}$ minimal with $S_i \prec T$. There is nothing to show if $i = n - 2$; hence we assume that $i \leq n - 3$. We apply to S_i, S_{i+1} and V_{S_i} the construction in (3). Then we obtain $x \in S_i$, $u \in S_{i+1}$ with $\nu_{S_i}(u) = 0$ and $\mathfrak{m}_{S_{i+j}} = (u, x/u^{j-1})$ for $j \in \{1, \ldots, n-i\}$; in particular, we get $\mathfrak{m}_T =$

$(u, x/u^{n-i-1})$. For $j \in \{1, \ldots, n-i-2\}$ we have $\nu_{S_{i+j}}(u) = \nu_{S_{i+j}}(x/u^{j-1}) = 1$, hence $\nu_{S_{i+j}}(x/u^{n-i-1}) = 1 - (n - j - i) \leq -1$, and therefore T is not proximate to any of the rings $S_{i+1}, \ldots, S_{n-2}$.

(7.3) Proposition: *Let $S \subsetneq T$ in Ω, and let $S =: S_0 \subset S_1 \subset \cdots \subset S_n := T$ be the quadratic sequence between S and T. We have $\mathfrak{m}_T = (x, y)$ with $\mathfrak{m}_{S_{n-1}} T = Tx$. For every non-zero ideal $\mathfrak{a}$ of S we have $\mathfrak{a}T = x^c y^d \mathfrak{a}^T$ where $c := \mathrm{ord}_{S_{n-1}}(\mathfrak{a}S_{n-1})$, and if there exists $i \in \{0, \ldots, n-2\}$ with $S_i \prec T$, then we have $d := \mathrm{ord}_{S_i}(\mathfrak{a}S_i)$, and $d := 0$ otherwise.*

Proof: The assertion is clear if $n = 1$. Now let $n = 2$; we have $\mathfrak{m}_{S_0} S_1 = S_1 x_1$, and we choose $y_1 \in S_1$ with $\mathfrak{m}_{S_1} = (x_1, y_1)$. If $S_0 \prec S_2$, then $\mathfrak{m}_{S_1} S_2 = S_2 y_1$, and $\{x := y_1, y := x_1/y_1\}$ is a system of generators of $\mathfrak{m}_{S_2}$ [cf. (7.2)(2)]. With $c := \mathrm{ord}_{S_1}(\mathfrak{a}S_1)$ and $d := \mathrm{ord}_{S_0}(\mathfrak{a})$ we get $\mathfrak{a}S_2 = x_1^d \mathfrak{a}^{S_1} S_2 = x^c y^d \mathfrak{a}^{S_2}$. If S_2 is not proximate to S_0, then S_2 is not the ring S_p with $p = \mathrm{In}(x_1) \in \mathrm{gr}_{\mathfrak{m}_1}(S_1)$, hence S_2 is a localization of $S_1[y_1/x_1]$. Therefore we have $\mathfrak{m}_{S_1} S_2 = S_2 x_1$, and with $x := x_1$ we obtain $\mathfrak{a}S_2 = x^c \mathfrak{a}^{S_2}$ with $c := \mathrm{ord}_{S_1}(\mathfrak{a}S_1)$.

Now let $n \geq 3$, and assume that the assertion holds for all $S \subsetneq T$ such that the quadratic sequence between S and T has length $n - 1$. Let $S := S_0 \subset S_1 \subset \cdots \subset S_n := T$ be a quadratic sequence of length n. First, we consider the case that there exists $i \in \{0, \ldots, n-2\}$ with $S_i \prec T$. If $i = n-2$, then we can argue as above. If $i < n-2$, then we choose $t, u \in S_{i+1}$ with $\mathfrak{m}_{S_i} S_{i+1} = S_{i+1} t$ and $\mathfrak{m}_{S_{i+1}} = (t, u)$. Then we have $\mathfrak{m}_{S_{n-1}} = (u, t/u^{n-2-i})$, $\mathfrak{m}_{S_{n-1}} S_n = S_n u$ and $\mathfrak{m}_{S_n} = (u, t/u^{n-1-i})$ by (7.2)(3). Set $d := \mathrm{ord}_{S_i}(\mathfrak{a}S_i)$, $c' := \mathrm{ord}_{S_{n-2}}(\mathfrak{a}S_{n-2})$. Then we have, by induction, $\mathfrak{a}S_{n-1} = u^{c'}(t/u^{n-2-i})^d \mathfrak{a}^{S_{n-1}}$, and therefore $\mathfrak{a}S_n = u^c(t/u^{n-1-i})^d \mathfrak{a}^{S_n}$ with $c := c' + d + \mathrm{ord}_{S_{n-1}}(\mathfrak{a}^{S_{n-1}}) = \mathrm{ord}(\mathfrak{a}S_{n-1})$.

Now we consider the case that T is proximate only to S_{n-1}. If S_{n-1} is proximate only to S_{n-2}, then we choose $x_{n-1} \in S_{n-1}$ with $\mathfrak{m}_{S_{n-2}} S_{n-1} = S_{n-1} x_{n-1}$; we have, by induction, $\mathfrak{a}S_{n-1} = x_{n-1}^{c'} \mathfrak{a}^{S_{n-1}}$ with $c' := \mathrm{ord}_{S_{n-2}}(\mathfrak{a}S_{n-2})$. Since S_n is not proximate to S_{n-2}, we have $\mathfrak{m}_{S_{n-1}} S_n = S_n x_{n-1}$, and with $x := x_{n-1}$ we get $\mathfrak{a}S_n = x^c \mathfrak{a}^{S_n}$ with $c := c' + \mathrm{ord}_{S_{n-1}}(\mathfrak{a}^{S_{n-1}}) = \mathrm{ord}_{S_{n-1}}(\mathfrak{a}S_{n-1})$. Now we assume that there exists $i \in \{0, \ldots, n-3\}$ with $S_i \prec S_{n-1}$. In S_{i+1} we choose t and u as above, and we have $\mathfrak{m}_{S_{n-2}} S_{n-1} = S_{n-1} u$, $\mathfrak{m}_{S_{n-1}} = (u, t/u^{n-2-i})$. By induction, we have $\mathfrak{a}S_{n-1} = u^{c'}(t/u^{n-2-i})^d \mathfrak{a}^{S_{n-1}}$ with $c' := \mathrm{ord}_{S_{n-2}}(\mathfrak{a}S_{n-2})$ and $d := \mathrm{ord}_{S_i}(\mathfrak{a}S_i)$. Since S_n is not proximate to S_{n-2}, we have $\mathfrak{m}_{S_{n-1}} S_n = S_n \cdot u/t^{n-2-i}$, and $u/(t/u^{n-2-i})$ is a unit of S_n. Therefore we get with $x := t/u^{n-2-i}$ that $\mathfrak{a}S_n = x^c \mathfrak{a}^{S_n}$ with $c := c' + d + \mathrm{ord}_{S_{n-1}}(\mathfrak{a}^{S_{n-1}}) = \mathrm{ord}_{S_{n-1}}(\mathfrak{a}S_{n-1})$.

(7.4) Corollary: *Let $S \subsetneq T$ in Ω, let $\wp$ be a simple complete $\mathfrak{m}_S$-primary ideal of S, and let $\mathfrak{m}_S =: \wp_0 \supset \wp_1 \supset \cdots \supset \wp_h := \wp$ be the sequence of simple $v_\wp$-ideals defined by $\wp$. If $\wp T$ is a principal ideal, then $\wp_i T$ is a principal ideal for every $i \in \{0, \ldots, h-1\}$.*

Proof: Let $i \in \{1, \ldots, h-1\}$. If $\wp T$ is a principal ideal of T, then $\wp^T = T$ by (7.3), hence $T \not\subset S_\wp$, the ring in Ω associated with $\wp$, and therefore $T \not\subset S_{\wp_i}$ [cf. (6.10)], hence we have $\wp_i^T = T$ and therefore $\wp_i T$ is a principal ideal, again by (7.3).

Now we describe the valuation mentioned in (7.2).

(7.5) A PARTICULAR CLASS OF VALUATIONS: Let $p \in \mathbb{P}_R$. We define a valuation

$$\nu_p : K \to \mathbb{Z} \times \mathbb{Z}$$

with value group $\mathbb{Z} \times \mathbb{Z}$ where $\mathbb{Z} \times \mathbb{Z}$ is ordered lexicographically, in the following way. Let $f \in R$ be a non-zero element, and let $(\text{In}(f)) = \prod_{q \in \mathbb{P}_R} q^{n_q(f)}$ be the factorization of the homogeneous principal ideal $(\text{In}(f))$ in $\text{gr}_{\mathfrak{m}_R}(R)$. We define

$$\nu_p(f) := (\text{ord}_R(f), n_p(f)).$$

Let $f, g \in R$ be non-zero elements. Since $(\text{In}(fg)) = (\text{In}(f))(\text{In}(g))$, we have

$$\nu_p(fg) = \nu_p(f) + \nu_p(g).$$

Assume, furthermore, that $f + g \neq 0$. If $\text{ord}_R(f + g) > \min(\{\text{ord}_R(f), \text{ord}_R(g)\})$, then we have $\nu_p(f + g) > \min(\{\nu_p(f), \nu_p(g)\})$. Now we consider the case that $\text{ord}_R(f + g) = \min(\{\text{ord}_R(f), \text{ord}_R(g)\})$. If $\text{ord}_R(f) \neq \text{ord}_R(g)$, say $\text{ord}_R(f) > \text{ord}_R(g)$, then we have $\text{In}(f + g) = \text{In}(g)$, hence $n_p(f + g) = n_p(g)$, which implies that $n_p(f + g) \geq \min(\{n_p(f), n_p(g)\})$. If, on the other hand, $\text{ord}_R(f) = \text{ord}_R(g)$, then we have $\text{In}(f + g) = \text{In}(f) + \text{In}(g)$, and again we find that $n_p(f + g) \geq \min(\{n_p(f), n_p(g)\})$. Therefore we have in both cases

$$\nu_p(f + g) \geq \min(\{\nu_p(f), \nu_p(g)\}).$$

The canonical extension of ν_p to $K^\times$ [cf. I(3.20)] shall also be denoted by ν_p, and it is a valuation of K.

Let V_p be the valuation ring of ν_p. We choose $f \in R$ with $(\text{In}(f)) = p$, and set $h := \text{ord}_R(f) = \deg(p)$. Then we have $\nu_p(f) = (h, 1)$. Let $\mathfrak{m}_R = (x, y)$; we may assume that $(\text{In}(x)) \neq p$. Then $(1, 0) = \nu_p(x) \leq \nu_p(y)$, hence $\nu_p(\mathfrak{m}_R) = (1, 0)$. Let $(m, n) \in \mathbb{Z} \times \mathbb{Z}$; then we have $\nu_p(f^n x^{m-hn}) = (m, n)$. Therefore ν_p is surjective, hence ν_p is a discrete valuation of rank 2, and $V_p \subsetneq V_R$. Since $(0, 1)$ is the smallest positive element in $\mathbb{Z} \times \mathbb{Z}$, we have $\nu_p(\mathfrak{m}(V_p)) = (0, 1) < (1, 0) = \nu_p(\mathfrak{m})$, hence V_p dominates R. Then $[\kappa_{\nu_p} : \kappa_R]$ is finite by Abhyankar's theorem [cf. I(11.9)], hence V_p is of the first kind with respect to R. It is easy to see that p is the directional ideal of V_p, hence that the quadratic transform $S_p = R[y/x]_{(x, f/x^h)}$ of R is the first quadratic dilatation of R with respect to ν_p [cf. VIII(3.1) for the notion of quadratic dilatation].

(7.6) Proposition: *With notations as in (7.5) we have:*
(1) *Let $T \in \Omega$ with $R \subsetneq T \subset V_p$. Then we have $S_p \subset T$.*
(2) *We set $\overline{f} := \text{In}(\overline{f})$, $\zeta := \text{In}(y)/\text{In}(x) \in Q_+(\text{gr}_{\mathfrak{m}_R}(R))_0$. Then we have $\kappa_{\nu_R} = \kappa_R(\zeta)$, $\overline{f}(1, \zeta) \in \kappa_R[\zeta]$ is irreducible and denoting by $\overline{v}_p$ the valuation*

of the rational function field $\kappa_R(\zeta)$ in one variable over κ_R defined by the discrete valuation ring $\kappa_R[\operatorname{In}(x), \operatorname{In}(y)]_{(p)} = \kappa_R[\zeta]_{(\bar{f}(1,\zeta))}$, we have $\nu_p = \nu_R \circ \bar{\nu}_p$, and $\kappa_R[\zeta]/(\bar{f}(1,\zeta))$ is the residue field of ν_p; in particular, ν_p and S_p have the same residue field.

(3) Let $T \in \Omega$ with $S_p \subset T \subset V_R$. Then we have $T \subset V_p$. Moreover, V_p is the only valuation ring of K dominating T and contained in V_R, the inclusion map $T \hookrightarrow V_p$ induces an isomorphism $T/\mathfrak{m}_T \to V_p/\mathfrak{m}(V_p)$ of residue fields, and we have $\nu_p(\mathfrak{m}_T) = (0,1)$ and $[T : S_p] = 1$.

(4) Let $R =: R_0 \subsetneqq R_1 \subsetneqq R_2 \cdots \subset V_R$ be the sequence of $(7.2)(3)$. Then we have $V_p = \bigcup_{i \geq 0} R_i$ and $\{R_i \mid i \geq 1\} = \{T \in \Omega \mid S_p \subset T \subset V_R\}$.

Proof: (1) Since S_p is the only quadratic transform of R contained in V_p [cf. (2.19)], we have $S_p \subset T$ by (6.3).

(2) From (2.3) we obtain $\kappa_{\nu_R} = \kappa_R(\zeta)$; for the definition of $\bar{\nu}_p$ cf. I(10.3). Now $V_p \subset V_R$, hence $\mathfrak{m}(V_R)$ is a prime ideal of V_p, and since V_p has rank 2, we have $(V_p)_{\mathfrak{m}(V_R)} = V_R$ [cf. I(3.8)]. Let $\varphi \colon V_R \to V_R/\mathfrak{m}(V_R) = \kappa_R(\zeta)$ be the canonical homomorphism. Then $\varphi(V_p) = V_p/\mathfrak{m}(V_R)$ is a valuation ring of $\kappa_R(\zeta)$ which contains κ_R [cf. I(3.12)]. Since $\bar{f}(1,\zeta) \in \mathfrak{m}(V_p)/\mathfrak{m}(V_R)$, it is clear that $\varphi(V_p)$ is the valuation ring of $\bar{\nu}_p$, and therefore we obtain $\nu_p = \nu_R \circ \bar{\nu}_p$.

(3) We have $\mathfrak{m}_R S_p = S_p x = \mathfrak{m}(V_R) \cap S_p =: \mathfrak{p}$; $S_p/\mathfrak{p}$ is a discrete valuation ring [cf. VII(2.0)], and $\mathfrak{p}$ is a prime ideal of height 1 of S_p. Therefore $\mathfrak{q} := \mathfrak{m}(V_R) \cap T$ is also a prime ideal of T of height 1 [since T dominates S_p by (6.3) and $\mathfrak{q} \cap S_p = \mathfrak{p}$]. We have $\varphi(S_p) \subset \varphi(T) \subset \kappa_{\nu_R}$, and $\varphi(S_p) \cong S_p/\mathfrak{p}$, $\varphi(T) \cong T/\mathfrak{q}$. Since $(S_p)_{\mathfrak{p}} = V_R$, the field of quotients of $\varphi(S_p)$ is κ_{ν_R}, and since $T/\mathfrak{q}$ is not a field, it follows that $\varphi(S_p) = \varphi(T)$. Since $S_p \subset V_p$, we get $\varphi(S_p) = \varphi(V_p)$, and therefore we have $\varphi(S_p) = \varphi(T) = \varphi(V_p)$.

The valuation rings W of K which dominate T and are contained in V_R correspond uniquely to the valuation rings of κ_{ν_R} which dominate $\varphi(T)$ [cf. I(3.12)]; the correspondence is given by $W \mapsto \varphi(W)$. Since $\varphi(T)$ is a discrete valuation ring of κ_{ν_R}, there exists exactly one valuation ring W of K contained in V_R and dominating T, and since $V_p = \varphi^{-1}(\varphi(V_p))$ is a valuation ring which contains T, and since V_p dominates S_p and T dominates S_p, the valuation ring V_p dominates T, and we must have $W = V_p$.

The ideals of the valuation ring $W = V_p$ are totally ordered by inclusion [cf. I(2.2)]. Since V_R does not dominate T, we cannot have $\mathfrak{m}_T W \subset \mathfrak{m}(V_R)$, hence we have $\mathfrak{m}(V_R) \subsetneqq \mathfrak{m}_T W$. Since $\varphi(T) = \varphi(W)$, the ideal $\varphi(\mathfrak{m}_T W)$ is the maximal ideal of $\varphi(W)$, hence $\mathfrak{m}_T W$ is the maximal ideal $\mathfrak{m}(V_p)$ of $W = V_p$, and $\nu_p(\mathfrak{m}_T) = \nu_p(\mathfrak{m}(V_p)) = (0,1)$. Moreover, since $\varphi(S_p) = \varphi(T)$, the inclusion $S_p \hookrightarrow T$ of local rings induces an isomorphism $S_p/\mathfrak{p} \cong T/\mathfrak{q}$, hence an isomorphism of residue fields, and since $\varphi(T) = \varphi(V_p)$, the inclusion $T \hookrightarrow V_p$ of quasilocal rings induces an isomorphism $T/\mathfrak{q} \cong V_p/\mathfrak{m}(V_R)$, hence an isomorphism of residue fields, and therefore we have $[T : S_p] = 1$ and $[V_p/\mathfrak{m}(V_p) : T/\mathfrak{m}_T] = 1$.

(4) Since $\bigcup_{i \geq 0} R_i$ is a valuation ring of K contained in V_R by $(7.2)(3)$, and since it dominates R, it is equal to V_p by (2). Let $T \in \Omega$ with $S_p \subset T \subset V_R$. We consider

the quadratic sequence between S_p and T; by the uniqueness part of (7.2)(3), we see that $T = R_i$ for some $i \geq 1$.

8   Resolution of Embedded Curves

(8.1) NOTATION: Let $R \in \Omega$ and $f \in R$ be a non-unit different from 0. We say that f is reduced (resp. analytically reduced resp. analytically irreducible) if the local ring R/Rf is reduced (resp. analytically unramified resp. analytically irreducible). Note that f is reduced iff f has no multiple factors, that f is analytically reduced iff f has, considered as an element of the completion of $\widehat{R}$, no multiple factors—in this case f is reduced—, and that f is analytically irreducible iff f is irreducible in $\widehat{R}$—in this case f is irreducible.

(8.2) THE RESIDUE CLASS RING R/Rf: Let $R \in \Omega$, set $\mathfrak{m} := \mathfrak{m}_R$, let $f \in R$ be a non-unit different from 0, set $n := \operatorname{ord}_R(f)$ and $\overline{R} := R/Rf$; $\overline{R}$ is a one-dimensional local CM-ring with maximal ideal $\overline{\mathfrak{m}} = \mathfrak{m}/Rf$ and multiplicity $e(\overline{R}) = n$ [cf. B(10.8)]. The canonical homomorphism $\varphi \colon R \to \overline{R}$ gives rise to a commutative diagram with exact rows [cf. B(5.3)]

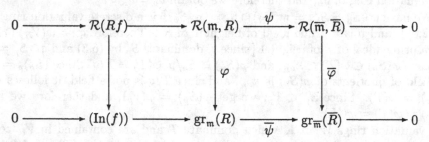

Let $(\operatorname{In}(f)) = p_1^{l_1} \cdots p_h^{l_h}$ be the factorization of the ideal $(\operatorname{In}(f))$ in $\operatorname{gr}_{\mathfrak{m}}(R)$ where $p_1, \ldots, p_h \in \mathbb{P}_R$ are pairwise different and $l_1, \ldots, l_h$ are natural integers. For $i \in \{1, \ldots, h\}$ let $S_i := S_{p_i}$ be the quadratic transform of R determined by p_i [cf. (2.6)]. By prime avoidance [cf. [63], Lemma 3.3] there exist $t \in \mathbb{N}$ and an element $g \in R$ with $\underline{\operatorname{ord}(g)} = t$ such that $\operatorname{In}(g) \notin p_1 \cup \cdots \cup p_h$. Let $\overline{g}$ be the image of g in $\overline{R}$ and $\overline{\operatorname{In}(g)}$ be the image of $\operatorname{In}(g) \in \operatorname{gr}_{\mathfrak{m}}(R)$ in $\operatorname{gr}_{\overline{\mathfrak{m}}}(\overline{R})$. The surjective homomorphism of graded rings $\psi \colon \mathcal{R}(\mathfrak{m}, R) \longrightarrow \mathcal{R}(\overline{\mathfrak{m}}, \overline{R})$ gives rise to a commutative diagram of surjective homomorphisms of rings

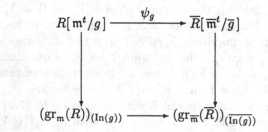

[note that $\overline{\text{In}(g)} = \overline{g} \bmod \overline{\mathfrak{m}}^{t+1}$]. For every $i \in \{1,\ldots,h\}$ let $\overline{p}_i$ be the image of p_i in $\text{gr}_{\overline{\mathfrak{m}}}(\overline{R})$; it is clear that

$$\text{Proj}(\text{gr}_{\overline{\mathfrak{m}}}(\overline{R})) = \text{Ass}(\text{gr}_{\overline{\mathfrak{m}}}(\overline{R})) = \{\overline{p}_1,\ldots,\overline{p}_h\}, \quad \overline{\text{In}(g)} \notin \bigcup_{i=1}^{s} \overline{p}_i.$$

Therefore $\overline{g} \in \overline{\mathfrak{m}}^t$ is transversal for $\overline{\mathfrak{m}}$ of order t [cf. II(1.15)]. Replacing g by a power of g, we may assume that $\overline{R}[\overline{\mathfrak{m}}^t/\overline{g}] = \overline{R}^{\overline{\mathfrak{m}}}$ [cf. II(4.3)]. For every $i \in \{1,\ldots,h\}$ let $\mathfrak{n}_i''$ be the maximal ideal in $R[\mathfrak{m}^t/g]$ determined by p_i; remember that $S_i = R[\mathfrak{m}^t/g]_{\mathfrak{n}_i''}$ [cf. (1.5)(3)]. We consider the canonical homomorphism $\overline{\varphi}: \mathcal{R}(\overline{\mathfrak{m}},\overline{R}) \to \text{gr}_{\overline{\mathfrak{m}}}(\overline{R})$. For every $i \in \{1,\ldots,h\}$ we set $\overline{\mathfrak{n}}_i' := \overline{\varphi}^{-1}(\overline{p}_i)$, and let $\overline{\mathfrak{n}}_i$ be the maximal ideal of $\overline{R}[\overline{\mathfrak{m}}^t/\overline{g}]$ determined by $\overline{\mathfrak{n}}_i'$; it is clear that $\psi_g(\mathfrak{n}_i'') = \overline{\mathfrak{n}}_i$. Now $\overline{R}^{\overline{\mathfrak{m}}}$ is semilocal with maximal ideals $\overline{\mathfrak{n}}_1,\ldots,\overline{\mathfrak{n}}_h$, hence the elements of $\Sigma := R[\mathfrak{m}^t/g] \setminus \bigcup_{i=1}^{h} \mathfrak{n}_i$ are mapped to units of $\overline{R}^{\overline{\mathfrak{m}}}$. Therefore the surjective homomorphism $R[\mathfrak{m}^t/g] \longrightarrow \overline{R}[\overline{\mathfrak{m}}^t/\overline{g}]$ can be extended to a surjective homomorphism

$$\omega: S := \bigcap_{i=1}^{h} S_i \longrightarrow \overline{R}^{\overline{\mathfrak{m}}}$$

[remember that S is a two-dimensional semilocal domain with maximal ideals $\mathfrak{m}(S_1) \cap S, \ldots, \mathfrak{m}(S_h) \cap S$, and that $\mathfrak{m}S = Sz$ is a principal prime ideal of S by (2.9)]. We show: The kernel of ω is the ideal $(Rf)^S$.
The kernel of $\psi: \mathcal{R}(\mathfrak{m},R) \to \mathcal{R}(\overline{\mathfrak{m}},\overline{R})$ is the homogeneous ideal which has $\mathcal{R}(f)_h = (\mathfrak{m}^h \cap Rf)T^h$ as its h-th component; note that $\mathfrak{m}^h \cap Rf = \mathfrak{m}^{h-n}f$ for every $h \in \mathbb{N}_0$ [where $\mathfrak{m}^{h-n} = R$ for $h \leq n$]. This implies that $\psi_g(f) = 0$. We consider an element $u/g^s \in R[\mathfrak{m}^t/g]$ where $s \in \mathbb{N}$ and $u \in \mathfrak{m}^{st}$; we may assume that $st \geq n$. From $\psi_g(u/g^s) = 0$ we get $u \in \mathfrak{m}^{st-n}f$, hence $Su \subset Sz^{st} \cdot f/z^n$, and therefore $u/g^s \in (Rf)^S \cdot f/z^n$. Therefore we have proved our assertion.

We collect the results which we have proved in the preceding paragraph:

(8.3) **Proposition:** Let $R \in \Omega$, let $f \in \mathfrak{m}_R$, $f \neq 0$, and let $S_1,\ldots,S_h$ be the finitely many quadratic transforms T of R with $(Rf)^T \neq T$. Set $\overline{R} := R/Rf$, let $\overline{\mathfrak{m}}$ be the maximal ideal of $\overline{R}$, and let $\overline{R}^{\overline{\mathfrak{m}}}$ be the blow-up of $\overline{R}$ with respect to $\overline{\mathfrak{m}}$. Set $S := S_1 \cap \cdots \cap S_h$. Then we have a surjective homomorphism $S \to \overline{R}^{\overline{\mathfrak{m}}}$ with kernel $(Rf)^S$; under this homomorphism, the maximal ideals of S correspond bijectively to the maximal ideals of $\overline{R}^{\overline{\mathfrak{m}}}$. Every point of $\text{Proj}(\text{gr}_{\overline{\mathfrak{m}}}(\overline{R}))$ is closed, and the maximal ideals of $\overline{R}^{\overline{\mathfrak{m}}}$ correspond bijectively to the points of $\text{Proj}(\text{gr}_{\overline{\mathfrak{m}}}(\overline{R}))$.

(8.4) REMARK: With the assumptions of (8.3), we have also the following: Let $i \in \{1,\ldots,h\}$, and let $\overline{\mathfrak{m}}_i$ be the image of $\mathfrak{m}(S_i) \cap S$ in $\overline{R}^{\overline{\mathfrak{m}}}$; then ω induces a surjective homomorphism $\omega_i: S_i = S_{\mathfrak{m}(S_i) \cap S} \to \overline{R}^{\overline{\mathfrak{m}}}_{\overline{\mathfrak{m}}_i} =: A_i$ with kernel $(Rf)^{S_i}$ [cf. (1.5)(5)].

Note that $A_1, \ldots, A_h \in \Delta_1(\overline{R})$ are *all* local rings in the first neighborhood of $\overline{R}$ [cf. II(5.2)], and that $\mathrm{ord}_R(f) = e(R/Rf) \geq e(A_i) = e(S_i/(Rf)^{S_i}) = \mathrm{ord}_{S_i}((Rf)^{S_i})$ by II(5.8) and B(10.8).

(8.5) Now we assume, in addition, that $f \in R$ *is analytically reduced*; then the integral closure W of $\overline{R}$ in its ring of quotients is a finitely generated $\overline{R}$-module [cf. II(3.22)]. Let S be a quadratic transform of R. If $(Rf)^S \neq S$, then any generator of the principal ideal $(Rf)^S$ is also analytically reduced. In fact, we have $S/(Rf)^S = A$ for some local ring A in the first neighborhood of $\overline{R} := R/Rf$ [cf. (8.4)]. Since A is a localization of $\overline{R}^m$, the integral closure of A is a finitely generated A-module, hence A is an analytically reduced ring [cf. II(3.22)] which implies the assertion. Moreover, let $\{\mathfrak{p}_1, \ldots, \mathfrak{p}_l\}$ be the set of maximal ideals of W; then $V_1 := W_{\mathfrak{p}_1}, \ldots, V_l := W_{\mathfrak{p}_l}$ are discrete valuation rings [cf. II(2.6)].
(1) Let $i \in \{1, \ldots, l\}$; we consider the branch sequence of $\overline{R}$ along V_i [cf. II(5.5)]. We have a commutative diagram—with $R_0 := R$, $\overline{R}_0 := \overline{R}$ and $m \in \mathbb{N}$ minimal with $\overline{R}_m := V_i$

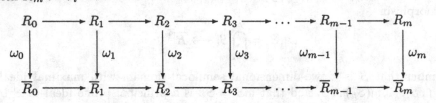

For $j \in \{1, \ldots, m\}$ the ring R_j is a quadratic transform of R_{j-1}, the ring $\overline{R}_j$ is a localization with respect to a maximal ideal of the blow-up of $\overline{R}_{j-1}$, and ω_j is surjective and has kernel $(Rf)^{R_j}$. In particular, we see that $1 = e(V_j) = \mathrm{ord}(Rf)^{R_m}$ [cf. B(10.8)].
(2) Let $(R_i)_{i \geq 0}$ with $R_0 := R$ be an increasing sequence in Ω such that R_i is a quadratic transform of R_{i-1} for every $i \in \mathbb{N}$. Then there exists $m \in \mathbb{N}$ such that $(Rf)^{R_m} = R_m$ or $(Rf)^{R_m}$ has order 1.

(8.6) NOTATION: Let $R \in \Omega$, and let $f, g \in R$. We call

$$\iota_R(Rf, Rg) := \ell_R(R/(Rf + Rg))$$

the intersection multiplicity of the ideals Rf, Rg. Note that $\iota_R(Rf, Rg)$ is finite exactly in the following two cases: one of the elements f, g is a unit or $Rf + Rg$ is an $\mathfrak{m}_R$-primary ideal.

(8.7) Proposition: *Let $R \in \Omega$, and let $f, g, h \in R$. Then:*
(1) $\iota_R(Rf, Rg) = \iota_R(Rg, Rf)$.
(2) $\iota_R(Rgh, Rf) = \iota_R(Rg, Rf) + \iota_R(Rh, Rf)$.
(3) *Let $f_1, f_2 \in R$ be of order 1 and not associated. If $\iota_R(Rf_1, Rf_2) \geq 2$, then there exists only one quadratic transform S of R with $(Rf_1 f_2)^S \neq S$; for this S we have $[S : R] = 1$ and $(Rf_1)^S \neq S$ and $(Rf_2)^S \neq S$.*

Proof: (1) This follows from the definition.

(2) We may assume that f and gh have no common non-trivial factor in R, that f is not a unit and that $f \neq 0$. Then $\overline{R} := R/Rf$ is a one-dimensional local CM-ring, and the image $\overline{g}$ of g and the image $\overline{h}$ of h in $\overline{R}$ are regular elements. By II(3.1) we have $\ell_{\overline{R}}(\overline{R}/(\overline{g})) = \ell_{\overline{R}}(\overline{R}\,\overline{h}/\overline{R}\,\overline{g}\overline{h})$, and since $\overline{R}/(\overline{h}) \cong (\overline{R}/\overline{R}\,\overline{g}\overline{h})/(\overline{R}\,\overline{h}/\overline{R}\,\overline{g}\overline{h})$, we have $\ell_{\overline{R}}(\overline{R}\,\overline{h}/\overline{R}\,\overline{g}\overline{h}) = \ell_{\overline{R}}(\overline{R}/(\overline{g}\overline{h})) - \ell_{\overline{R}}(\overline{R}/(\overline{h}))$, hence the assertion follows.

(3) The assumption implies that the elements $\text{In}(f_1)$, $\text{In}(f_2) \in \text{gr}_{\mathfrak{m}_R}(R)$ are homogeneous of degree 1, and that $(\text{In}(f_1)) = (\text{In}(f_2))$. The result follows from (2.15).

(8.8) Proposition: [M. Noether] Let $R \in \Omega$, and let $f, g \in R$. Then:
(1) $\iota_R(Rf, Rg)$ is finite iff $\iota_S((Rf)^S, (Rg)^S)$ is finite for every quadratic transform S of R, and in this case we have

$$\iota_R(Rf, Rg) = \text{ord}_R(f)\,\text{ord}_R(g) + \sum_{S \in N_1(R)} [S:R]\iota_S((Rf)^S, (Rg)^S)).$$

(2) We have

$$\sum_{S \in N_1(R)} [S:R]\iota_S((Rf)^S, \mathfrak{m}_R S) = \text{ord}_R(f).$$

Proof: First, we assume that $\{f, g\}$ is a regular sequence in R. We set $m := \text{ord}(f)$, $n := \text{ord}(g)$. We use the notations introduced in (8.3) and (8.4). The image $\overline{g}$ of g in $\overline{R}$ is a regular element of $\overline{R}$ which lies in the maximal ideal of $\overline{R}$; by II(5.7) we have

$$\ell_R(R/(Rf + Rg)) = \ell_{\overline{R}}(\overline{R}/\overline{R}\overline{g}) = \sum_{i=1}^{h} f(A_i)\ell_{A_i}(A_i/A_i\overline{g}).$$

For every $i \in \{1, \ldots, h\}$ we have by II(3.2)

$$\ell_{A_i}(A_i/A_i\overline{g}) = \ell_{A_i}(A_i/A_i z^n (Rg)^{S_i})$$
$$= n\ell_{A_i}(A_i/A_i z) + \ell_{A_i}(A_i/A_i(Rg)^{S_i})$$
$$= n\ell_{A_i}(A_i/A_i z) + \ell_{S_i}(S_i/((Rf)^{S_i} + (Rg)^{S_i})),$$

and since $f(A_i) = [A_i/\mathfrak{n}(A_i) : R/\mathfrak{m}] = [S_i : R]$, we get

$$\ell_R(R/(Rf+Rg)) = n\sum_{i=1}^{h} f(A_i)\ell_{A_i}(A_i/A_i z) + \sum_{i=1}^{h}[S_i:R]\ell_{S_i}(S_i/((Rf)^{S_i}+(Rg)^{S_i})).$$

We have $m = e(\overline{R})$ by B(10.8) and $e(\overline{R}) = \ell_{\overline{R}}(\overline{S}/\mathfrak{m}\overline{S})$ by II(4.12), and $\mathfrak{m}\overline{S} = \overline{S}z$. Since we have

$$\ell_{\overline{R}}(\overline{S}/\overline{S}z) = \sum_{i=1}^{h} f(A_i)\ell_{A_i}(A_i/A_i z)$$

[cf. B(10.10)], we get (2), and (1) follows immediately.

Second, if $\{f, g\}$ is not a regular sequence, then $\{(Rf)^S, (Rg)^S\}$ is not a regular sequence for any $S \in N_1(R)$, hence both sides are not finite or 0.

(8.9) Corollary: *Let $R \in \Omega$, and let $\{f, g\}$ be a regular sequence in R. Then we have*

$$\iota_R(Rf, Rg) = \sum_{S \in N(R)} [S : R] \operatorname{ord}_S((Rf)^S) \operatorname{ord}_S((Rg)^S).$$

Proof: By repeated application of (8.8)(1) we find for $n \in \mathbb{N}$

$\iota_R(Rf, Rg)$

$$= \sum_{i=0}^{n-1} \sum_{S \in N_i(R)} [S : R] \operatorname{ord}_S((Rf)^S) \operatorname{ord}_S((Rg)^S) + \sum_{S \in N_n(R)} [S : R] \iota_S((Rf)^S, (Rg)^S).$$

Since $\iota_R(Rf, Rg)$ is finite, there exists $n \in \mathbb{N}$ with

$$\sum_{S \in N_n(R)} [S : R] \iota_S((Rf)^S, (Rg)^S) = 0,$$

hence for every $S \in N_n(R)$ the ideal $(Rf)^S + (Rg)^S$ of S is the unit ideal, and therefore one of the integers $\operatorname{ord}_S((Rf)^S)$, $\operatorname{ord}_S((Rg)^S)$ is zero, hence for every $T \in N_m(R)$ with $m \geq n$ one of the integers $\operatorname{ord}_T((Rf)^T)$, $\operatorname{ord}_T((Rg)^T)$ is zero.

(8.10) NOTATION: Let $R \in \Omega$, and let $\mathfrak{a}$ be an ideal of R.
(1) Let S be a quadratic transform of R. Then the ideal $\mathfrak{m}_R \cdot \mathfrak{a}^S$ of S is called the total transform of $\mathfrak{a}$ in S.
(2) Let $n > 1$, and assume that we have already defined the total transform of $\mathfrak{a}$ in S for every $S \in N_{n-1}(R)$. Let $S \in N_n(R)$, and let $R =: R_0 \subset R_1 \subset \cdots \subset R_{n-1} \subset R_n := S$ be the quadratic sequence between R and S. Let $\tilde{\mathfrak{a}}$ be the total transform of $\mathfrak{a}$ in R_{n-1}. Then $\mathfrak{m}_{S_{n-1}} \cdot \tilde{\mathfrak{a}}^S$ is called the total transform of $\mathfrak{a}$ in S. Note that, if $\mathfrak{a}$ is a principal ideal, then the total transform of $\mathfrak{a}$ in S is a principal ideal of S.

(8.11) PROPERTIES OF THE TOTAL TRANSFORM: (1) Let $T \supsetneq R$ with $T, R \in \Omega$. Then there exists a regular system of parameters $\{x, y\}$ of T such that the total transform of any ideal $\mathfrak{a}$ of R in T has the form $xy^e \mathfrak{a}^T$ with $e \in \{0, 1\}$.
Proof: We prove the assertion by induction on the length n of the quadratic sequence between R and T. The assertion is clear in the case $n = 1$. Assume that $n > 2$, and that the assertion holds for $n - 1$. Let $S \in N_{n-1}(R)$ and let T be a quadratic transform of S. By induction, there exists a regular system of parameters $\{u, v\}$ of S such that the total transform $\tilde{\mathfrak{a}}$ of $\mathfrak{a}$ in S has the form $uv^e \mathfrak{a}^S$ where $e \in \{0, 1\}$. Let $\mathfrak{n}$ be the maximal ideal of S.
If $\mathfrak{n}T = Tu$, then we set $x := u$; if v/u is not a unit of T, then $\{x, y := v/u\}$ is a regular system of parameters of T, and we have $\mathfrak{n} \cdot (uv^e \mathfrak{a}^S)^T = xy^e \mathfrak{a}^T$, and if v/u is a unit of T, then we choose $y \in T$ such that $\{x, y\}$ is a regular system of parameters of T, and we get $\mathfrak{n} \cdot (uv^e \mathfrak{a}^S)^T = x \mathfrak{a}^T$. If $\mathfrak{n}T = Tv$, then we set $x := v$. If u/v is not a unit of T, then $\{x, y := u/v\}$ is a regular system of parameters of T, and we

have $n \cdot (uv^e a^S)^T = xya^T$, and if u/v is a unit of T, then we choose $y \in T$ such that $\{x, y\}$ is a regular system of parameters of T, and we get $n \cdot (uv^e a^S)^T = xa^T$.
(2) Now we assume that $f \in R$ is reduced (resp. analytically reduced). We show: Any generator of the total transform of Rf in T is reduced (resp. analytically reduced).
By induction, it is enough to show: Let $S \in N_1(R)$; then any generator of the total transform of Rf is S is reduced (resp. analytically reduced). If f is reduced, then this follows from the multiplicative behavior of forming the ideal transform, and from (2.11)(3). Now let f be analytically reduced, and choose $x \in R$ with $m_R S = Sx$. Let f_S be a generator of $(Rf)^S$. Then xf_S is a generator of the total transform of Rf in S; if f_S is not a unit of S, then x does not divide f_S in S [cf. (2.11)(3)], hence $Sx + Sf_S$ is an m_S-primary ideal of S. Then $\widehat{S}x + \widehat{S}f_S$ is an $m(\widehat{S})$-primary ideal of $\widehat{S}$, and therefore xf_S is a reduced element of $\widehat{S}$, hence xf_S is analytically reduced. If f_S is a unit of S, then xf_S is analytically reduced.

(8.12) NOTATION: Let $R \in \Omega$. A reduced non-unit f of R is said to define a curve C in R; C is called regular if $\mathrm{ord}(f) = 1$. The irreducible factors of f are called the irreducible components of C. C is called a normal crossing curve if either C is regular or C has two irreducible components C_1, C_2, C_1 and C_2 are regular, and if C_1 is defined by f_1 and C_2 is defined by f_2, then we have $\iota_R(Rf_1, Rf_2) = 1$. Note that in the latter case the condition to be a normal crossing curve is equivalent to $\{f_1, f_2\}$ being a regular system of parameters of R.

(8.13) Proposition: Let $(R_i)_{i \geq 0}$ be an increasing sequence in Ω such that R_i is a quadratic transform of R_{i-1} for every $i \in \mathbb{N}$. Let $f \in R$ be analytically reduced. Then there exists $m \in \mathbb{N}$ such that the total transform of Rf in R_m is a normal crossing curve.

Proof: (1) Let $S \in \Omega$ and let $g \in S$ be reduced; we set $\mathrm{ord}_S(g) =: n$, and we assume that g is a product of irreducible factors $g_1, \ldots, g_n$ of order 1. We define the integer

$$\theta_S(Sg) = \frac{(\mathrm{ord}(g) - 1)(\mathrm{ord}(g) - 2)}{2} + \sum_{1 \leq j < l \leq \mathrm{ord}_S(g)} (\iota_S(Sg_j, Sg_l) - 1).$$

Then we have $\theta_S(Sg) \geq 0$, and we have $\theta_S(Sg) = 0$ iff either $\mathrm{ord}_S(g) = 1$—in this case the curve defined by g is regular—or $\mathrm{ord}_S(g) = 2$ and $\iota_S(Sg_1, Sg_2) = 1$. Therefore we have $\theta_S(Sg) = 0$ iff the curve C defined by g is a normal crossing curve.
It is clear that

$$\theta_S(Sg) = \sum_{1 \leq j < l \leq \mathrm{ord}_S(g)} \iota_S(Sg_j, Sg_l) - (\mathrm{ord}_S(g) - 1). \tag{$*$}$$

(2) Let S and g be as in (1), and for every $T \in N_1(S)$ let g_T be a generator of the total transform of Rf in T. We show:

$$\sum_{T \in N_1(S)} \theta_T(Tg_T) = \theta_S(Sg) - \frac{(\mathrm{ord}_S(g) - 1)(\mathrm{ord}_S(g) - 2)}{2}.$$

In fact, for any quadratic transform T of S we choose a regular parameter x_T with $m_S T = T x_T$. Let n'_T be the number of proper ideals in the set $\{(Sg_1)^T, \ldots, (Sg_n)^T\}$; we have $n'_T \in \{0, \ldots, n\}$. Then we have

$$\theta_T(Tg_T) = \sum_{1 \le j < l \le \mathrm{ord}_S(g)} \iota_T((Sg_j)^T, (Sg_l)^T) + \sum_{j=1}^{n} \iota_T((Sg_j)^T, x_T T) - n'_T$$

$$= \sum_{1 \le j < l \le \mathrm{ord}_S(g)} \iota_T((Sg_j)^T, (Sg_l)^T)$$

[since, for $j \in \{1, \ldots, n\}$, there is only one $T \in N_1(S)$ with $(Sg_j)^T \ne T$, and for this T we have $[T : S] = 1$, we have $\iota_T((Sg_j)^T, Tx_T) = 0$ if $(Sg_j)^T = T$ and $\iota_T((Sg_j)^T, Tx_T) = 1$ if $(Sg_j)^T \ne T$ by (8.8)(2)]. Now we get from (8.8)(1)

$$\sum_{T \in N_1(S)} \theta(Tg_T) = \sum_{1 \le j < l \le \mathrm{ord}_S(g)} \left(\sum_{T \in N_1(S)} \iota_T((Sg_j)^T, (Sg_l)^T) \right)$$

$$= \sum_{1 \le j < l \le \mathrm{ord}_S(g)} (-1 + \iota_S(Sg_j, Sg_l))$$

$$= -\frac{(\mathrm{ord}(g) - 1)\,\mathrm{ord}(g)}{2} + \sum_{1 \le j < l \le \mathrm{ord}_S(g)} \iota_S(Sg_j, Sg_l)$$

$$= \theta_S(gS) - \frac{(\mathrm{ord}_S(g) - 1)(\mathrm{ord}_S(g) - 2)}{2},$$

which is our assertion.

(3) We now assume that $\theta_S(Sg) > 0$. If $\mathrm{ord}_S(g) \ge 3$, then we have $\theta_T(Tg_T) < \theta_S(Sg)$ for every quadratic transform T of S by (2). Let us consider the case $\mathrm{ord}_S(g) = 2$. Then we have $g = g_1 g_2$. Since by assumption we have $0 < \theta_S(Sg) = \iota_S(Sg_1, Sg_2) - 1$, there is only one quadratic transform T of S with $(Sg)^T \ne T$, and for this T we have $[T : S] = 1$ and $(Sg_1)^T \ne T$ and $(Sg_2)^T \ne T$ [cf. (8.7)(3)], hence we have $\mathrm{ord}_T((Sg)^T) = 2$. Then we have $\mathrm{ord}(g_T) = 3$. Let U be any quadratic transform of T, and let g_U be a generator of the total transform of Sg in U; by (2) we have $\theta_U(Ug_U) < \theta_T(Tg_T) = \theta_S(Sg)$. For all other quadratic transforms T' of S we have $\theta_{T'}(T'g_{T'}) = 0$.

(4) From (2), (3), (8.5) and (8.11) we get our assertion.

Chapter VIII

Resolution of Singularities

INTRODUCTION: In the first section we resolve the singularities of a curve by a finite number of blowing up points. Furthermore, we consider a curve C embedded in a regular surface X, and we show: by a finite sequence of blowing up points we get a regular surface X' such that the total transform of C in X' is a divisor with normal crossings [embedded resolution of curves in a regular surface].

In the rest of this chapter we give two different proofs of resolution of singularities for a surface over an algebraically closed field of characteristic 0. First, using the result mentioned in the last paragraph and the possibility of resolving the singular point of a toric variety by repeatedly blowing up points, we show in section 2 how to resolve the singularities of an irreducible surface.

For the second proof we need more algebraic background, namely the uniformization theorem. To prepare the ground for the proof of the uniformization theorem, we treat in sections 3 and 4 quadratic dilatations with respect to a valuation; in the case of two-dimensional regular local rings, a quadratic dilatation also is, in general, a quadratic transform as defined in chapter VII. Valuations of fields of algebraic functions in two variables are classified in section 5. We prove in section 6 a local version of the uniformization theorem and the uniformization theorem.

Second, in the last section we use the uniformization theorem to show: An irreducible surface can be desingularized by repeatedly blowing up points and normalizing.

1 Blowing up Curve Singularities

(1.0) In this section we shall prove the theorem of resolution of singularities for curves and for curves embedded in a regular surface [embedded resolution]. Here, a *curve* is a pure one-dimensional variety, a *surface* is a pure two-dimensional variety, a *regular surface* is an irreducible regular surface, and a *curve on a regular surface* is a closed subset which is a curve. Note that a curve on a regular surface

303

is a divisor [since X is regular, the notions of Cartier and Weil divisor coincide].
We work over an algebraically closed ground field k of arbitrary characteristic.

(1.1) Theorem: [Resolution of singularities of curves] *Given a curve C, then
there exists a finite sequence of blowing ups*

$$\pi \colon D := C_n \longrightarrow C_{n-1} \longrightarrow \cdots \longrightarrow C_0 := C$$

of C [with suitable centers] such that
(1) D is a regular curve,
*(2) if $C^{(1)}, \ldots, C^{(h)}$ are the irreducible components of C, then D also has h irreducible components, and they can be labelled $D^{(1)}, \ldots, D^{(h)}$ in such a way that
the restriction of π induces a birational morphism $D^{(i)} \to C^{(i)}$ for $i \in \{1, \ldots, h\}$,*
(3) π induces an isomorphism $D \setminus \pi^{-1}(\mathrm{Sing}(C)) \to C \setminus \mathrm{Sing}(C)$.
Therefore π is a desingularization of C.

(1.2) REMARK: Before we give the proof of (1.1), we will point out the connection
with the concepts introduced in chapter II, sections 4 and 5. We use without
further warning the notations introduced in appendix A, sections 13 and 14.
Thus, let C be a curve, let $p \in C$, set $A := \mathcal{O}_{C,p}$, and let $\mathfrak{m}$ be the maximal ideal of
A. The ring A is a one-dimensional reduced local ring, hence $\mathfrak{m} \notin \mathrm{Ass}(A)$ [since,
on the one hand, the nilradical is the zero ideal, and, on the other hand, it is the
intersection of the minimal prime ideals of A] and therefore $\mathfrak{m}$ contains regular
elements, hence A is a CM-ring [note that C has no zero-dimensional irreducible
components by definition because of our usage of the word curve]. Since the
residue field $A/\mathfrak{m} = k$ is infinite, the maximal ideal $\mathfrak{m}$ admits transversal elements
[cf. II(1.18)]. The integral closure $S(A)$ of A in its full ring of quotients is a finitely
generated A-module [cf. B(3.7) and note that A is the localization with respect
to a prime ideal of a reduced k-algebra of finite type].
Let $(A_i)_{i \geq 0}$ with $A_0 := A$ be the blow-up sequence of A [cf. II(4.7)], and let
$m(A) := m$ be the smallest non-negative integer with $A_m = S(A)$ [the existence
of $m(A)$ follows from II(4.8)]. Note that $A_0 = S(A)$ iff $A = A_0$ is a discrete
valuation ring [cf. II(2.5)], hence $m(A) = 0$ iff A is a discrete valuation ring.
Let $\pi \colon \mathrm{Bl}_p(C) \to C$ be the blowing up of C with center $p \in C$, and set $C' :=
\mathrm{Bl}_p(C)$; let E' be the exceptional fibre of π [we use the language of A(13.6)].
First, let us remark that C' has as many irreducible components $C'^{(1)}, \ldots, C'^{(h)}$
as C, and that they can be labelled in such a way that the restriction of π induces,
for $i \in \{1, \ldots, h\}$, a morphism $C'^{(i)} \to C^{(i)}$ which is birational [cf. A(13.9)].
Furthermore, E' is a finite set, and the set of points of E' corresponds to the set
of maximal non-irrelevant homogeneous ideals of $\mathrm{gr}_\mathfrak{m}(A)$, i.e., to the set of closed
points of $\mathrm{Proj}(\mathrm{gr}_\mathfrak{m}(A))$ [note that $\dim(\mathrm{gr}_\mathfrak{m}(A)) = 1$ by [63], Ex. 13.8].
Let $A^\mathfrak{m} = A_1$ be the quadratic transform of A, i.e., the blow-up of A with respect
to $\mathfrak{m}$. By II(4.4)(3) and A(14.2)(6) there exists a bijective map between the set
E' and the set of maximal ideals of $A^\mathfrak{m}$, and if $q \in E'$ corresponds to the maximal
ideal $\mathfrak{n}$ of $A^\mathfrak{m}$, then $\mathcal{O}_{C',q}$ and $(A^\mathfrak{m})_\mathfrak{n}$ are isomorphic rings. From II(5.1) we see
that $m(\mathcal{O}_{C',q}) < m(\mathcal{O}_{C,p})$.

(1.3) Now we consider the case that C is affine. Here we can give a more concrete description of the bijective map $E' \to \operatorname{Max}(A^m)$ defined in (1.2).

(1) First of all, we show that C' is affine. In fact, set $R := \mathcal{O}(C)$ and $\mathfrak{m} := \mathcal{I}_C(\{p\})$. Let $\mathfrak{m}$ be generated by n elements; then the closed subset $\operatorname{Bl}_\mathfrak{m}(C)$ of $\mathbb{P}^{n-1} \times C$ can be identified with C' [cf. A(14.1)]. The morphism $\pi\colon C' \to C$ has finite fibres [since $\dim(C') = 1$ and no irreducible component of C' is a fibre of π], and is therefore finite [cf. A(12.5)]; this implies that C' is affine [cf. A(2.6)(1)], and therefore $R' := \mathcal{O}(C')$ is a finite R-algebra.

(2) Let C be irreducible; then C' is irreducible. We choose n non-zero elements $x_1, \ldots, x_n$ which generate $\mathfrak{m}$. We show that

$$\mathcal{O}(C') = \bigcap_{i=1}^n R[\,\mathfrak{m}/x_i\,] \subset K := Q(R).$$

In fact, we have $C' = \bigcup_{i=1}^n U_i$ where $U_i \subset C'$ is open and affine and $\mathcal{O}(U_i) = R[\,\mathfrak{m}/x_i\,]$ for $i \in \{1, \ldots, n\}$ [cf. A(14.1)(4)]. We have $\mathcal{O}(U_i) = \bigcap_{q \in U_i} \mathcal{O}_{C', q}$ for $i \in \{1, \ldots, n\}$ [cf. A(1.7)]; this implies the assertion since $\mathcal{O}(C') = \bigcap_{q \in C'} \mathcal{O}_{C', q}$.

(3) Now we consider the case that C is not irreducible. Let $C^{(1)}, \ldots, C^{(h)}$ be the irreducible components of C, and set $\mathfrak{c}_j := \mathcal{I}_C(C^{(j)})$ for $j \in \{1, \ldots, h\}$; we have $\mathcal{O}(C^{(j)}) = R/\mathfrak{c}_j$. Note that $\{\mathfrak{c}_1, \ldots, \mathfrak{c}_h\}$ is the set of minimal prime ideals of the reduced ring R. Then $\Sigma := R \setminus (\mathfrak{c}_1 \cup \cdots \cup \mathfrak{c}_h)$ is the set of regular elements of R, $K := Q(R) = \Sigma^{-1}R$ is the ring of quotients of R, and with $\mathfrak{d}_j := \mathfrak{c}_j K$ and $K_j := K/\mathfrak{d}_j = Q(R/\mathfrak{c}_j)$ for $j \in \{1, \ldots, h\}$ we have an isomorphism

$$\varphi\colon K \xrightarrow{\sim} K_1 \times \cdots \times K_h;$$

φ induces an injective homomorphism $R \to R/\mathfrak{c}_1 \times \cdots \times R/\mathfrak{c}_h$. For $j \in \{1, \ldots, h\}$ let $\varphi_j\colon K \to K_j$ be the projection of K to the j-th factor.

Since C is a curve, $\mathfrak{m}$ is an ideal of R containing regular elements, hence $\mathfrak{m}$ can be generated by regular elements [cf. I(1.12)]. We assume that $x_1, \ldots, x_n$ is a system of generators of $\mathfrak{m}$ consisting of regular elements.

Let $C'^{(1)}, \ldots, C'^{(h)}$ be the irreducible components of C'; we label them in such a way that $C'^{(j)} = \operatorname{Bl}_p(C^{(j)})$ for $j \in \{1, \ldots, h\}$ [cf. A(14.1)]. Note that $C'^{(j)} = C^{(j)}$ if $p \notin C^{(j)}$. By (2) we have

$$\mathcal{O}(C'^{(j)}) = \bigcap_{i=1}^n R/\mathfrak{c}_j[\varphi_j(\mathfrak{m})/\varphi_j(x_i)] \quad \text{for } j \in \{1, \ldots, h\}.$$

Therefore we have

$$R' = \mathcal{O}(C') = \bigcap_{i=1}^n R[\,\mathfrak{m}/x_i\,] \subset K.$$

(4) We set $A := R_\mathfrak{m}$, $B := R'_\mathfrak{m}$; let $\mathfrak{n} := \mathfrak{m}R_\mathfrak{m}$ be the maximal ideal of A. We may assume that the image of x_1 in A is transversal; then we have $\mathfrak{n}^{l+1} = x_1\mathfrak{n}^l$ for all large integers l. Now we have $1/x_1 = z/x_i^{l+1}$ with $z \in \mathfrak{n}^l$, hence we have $A[\mathfrak{n}/x_1] \subset A[\mathfrak{n}/x_i]$ for $i \in \{2, \ldots, n\}$. Therefore we have $B = A^n$ [cf. II(4.4)(2)].

(1.4) PROOF OF (1.1): For every $p \in C$ we set $m(p) := m(\mathcal{O}_{C,p})$, and we set $M(C) := \max(\{m(p) \mid p \in C\})$. Since C has only finitely many singular points [note that $\mathrm{Sing}(C)$ is zero-dimensional or empty by A(6.11)], it is clear that $M(C)$ is a non-negative integer. If $M(C) = 0$, we are done. Otherwise, we set $N(C) := \mathrm{Card}(\{p \in C \mid m(p) = M(C)\})$. We choose $p \in C$ with $m(p) = M(C)$, and we consider the blowing up $\pi \colon C' \to C$ of C with center p. Since π is an isomorphism outside p, we have $m(q) = m(\pi(q))$ for every $q \in C'$ which does not lie over p. For every $q \in C'$ which lies over p we have $m(q) < m(p)$ [cf. (1.2)]. Thus, we have either $M(C') = 0$, or $M(C) \geq M(C') > 0$ and $N(C') < N(C)$, and if $N(C) = 1$, then we have $M(C) > M(C')$. Now it is clear that after a finite number of blowing up points we arrive at a curve D with $M(D) = 0$, hence at a nonsingular curve. The assertion in (3) is clear from this construction.

(1.5) REMARK: Let C be a curve on a regular surface X, and let $p \in C$. Let $\pi \colon X_1 \to X$ be the blowing up of X with center p. Then X_1 is a regular surface [cf. A(14.5)], and the strict transform C_1 of C is a curve on X_1; it is irreducible if C is irreducible.
Let

$$X_n \xrightarrow{\pi_n} X_{n-1} \xrightarrow{\pi_{n-1}} \cdots \xrightarrow{\pi_2} X_1 \xrightarrow{\pi_1} X_0 := X \qquad (*)$$

be a sequence of point blowing ups, and assume that, for $i \in \{1, \ldots, n-1\}$, the center of π_{i+1} lies on C_i, the strict transform of C_{i-1}, and let C_{i+1} be the strict transform of C_i, $i \in \{1, \ldots, n-1\}$. Then, by abuse of language, C_n is called the strict transform of C in X_n. Note that C_n is irreducible if C is irreducible.

(1.6) Corollary: *Let C be a curve on regular surface X. Then there exists a finite sequence of blowing ups $X_n \to X_{n-1} \to \cdots \to X_0 := X$ of X [with suitable centers] such that the strict transform C_n of C in X_n is nonsingular. The restriction $\pi|C_n \colon C_n \to C$ is a desingularization of C.*

Proof: The proof of (1.1) given in (1.4) works also in this case.

(1.7) Now we want to prove another theorem on resolution for curves embedded in a regular surface. This time we will not only consider the strict transform of our curve—in this case the theorem is just (1.6)—, but we will consider the total transform of our curve [cf. (1.8) below for a definition].

(1.8) TOTAL TRANSFORM: Let C be a curve on a regular surface X, let $p \in C$, and let $\pi \colon X_1 \to X$ be the blowing up of X with center p. Let C_1 be the strict transform of C, and let E be the exceptional divisor of π. Then E is not an irreducible component of C_1 [cf. A(13.8)(4)], and $\tilde{C}_1 := C_1 \cup E$ is a curve on X_1, hence a divisor. Note that E is a nonsingular curve since $E \cong \mathbb{P}^1$ [cf. A(13.5)]. The curve $\tilde{C}_1$ is called the total transform of C in X_1. Note that $\pi^{-1}(C) = \tilde{C}$, and that the irreducible components of $\tilde{C}$ are E and the irreducible components of C_1.

Let $i \in \{1, \ldots, n-1\}$. Let us assume, in the situation of (1.5) above, that the total transform $\tilde{C}_i$ of C in X_i has been defined. Let the center of π_{i+1} lie on $\tilde{C}_i$, and let $\tilde{C}_{i+1}$ be the total transform of $\tilde{C}_i$ in X_{i+1}; then $\tilde{C}_{i+1}$ is called the total transform of C in X_{i+1}.

Then $\tilde{C}_n$ is a curve on the regular surface X_n which is called, by abuse of language, the total transform of C. Set $\pi := \pi_1 \circ \cdots \circ \pi_n$; note that $\pi^{-1}(C) = \tilde{C}_n$.

(1.9) Theorem: [Embedded resolution of curves on surfaces] *Let C be a curve on a regular surface X. Then there exists a finite sequence of blowing ups*

$$\pi \colon X_n \longrightarrow X_{n-1} \longrightarrow \cdots \longrightarrow X_0 := X$$

of X [with suitable centers] such that the total transform $\tilde{C}$ of C in X_n is a divisor with normal crossings. Moreover, there exists a finite set $F \subset C$ such that π induces an isomorphism $\tilde{C} \setminus \pi^{-1}(F) \to C \setminus F$.

Before proving the theorem, we need some preparations: We show that locally we are in the situation which we studied in section 8 of chapter VII.

(1.10) BLOWING UP CURVES ON A REGULAR SURFACE: Let X be a regular surface, let $p \in X$, and let $\pi \colon \mathrm{Bl}_p(X) \to X$ be the blowing up of X with center p. We want to study the local behavior in p; hence we may assume that $X \subset \mathrm{A}^n$ is affine, and that p is the origin.

(1) Let $R := k[y_1, \ldots, y_n]$ be the affine ring of A^n, and let $\mathfrak{p}$ be the prime ideal of R defining X; let $\{f^{(1)}, \ldots, f^{(m)}\}$ be a system of generators of $\mathfrak{p}$. We may assume that the tangent plane $T_{X,p}$ to X at p is defined by $y_3 = \cdots = y_n = 0$. Then, relabelling the polynomials $f^{(1)}, \ldots, f^{(m)}$, we may assume, to begin with, that, for every $i \in \{1, \ldots, n-2\}$, we have $f^{(i)} = y_{i+2} - g^{(i)}$ where $g^{(i)} \in R$ and, when writing $g^{(i)}$ as a sum of homogeneous polynomials, the homogeneous polynomial of lowest degree of $g^{(i)}$ has degree ≥ 2.

We set $\overline{R} := R/\mathfrak{p}$ and $\overline{y}_j := y_j + \mathfrak{p}$ for $j \in \{1, \ldots, n\}$. We set $S := \mathcal{O}_{\mathrm{A}^n, p} = R_{(y_1, \ldots, y_n)}$; let $\mathfrak{n}$ be the maximal ideal of S, set $\overline{S} := \mathcal{O}_{X,p} = S/\mathfrak{p}S$ and let $\overline{\mathfrak{n}} = \mathfrak{n}/\mathfrak{p}S$ be the maximal ideal of $\overline{S}$. Then $\{\overline{y}_1, \overline{y}_2\}$ is a regular system of parameters of $\overline{S}$ since $\overline{y}_3, \ldots, \overline{y}_n$ lie in the square of the maximal ideal $\overline{\mathfrak{n}}$ of $\overline{S}$.

(2) Note that $\{y_1, \ldots, y_n\}$ is a regular system of parameters of S, and that therefore $\{y_1, y_2, f^{(1)}, \ldots, f^{(n-2)}\}$ is a regular system of parameters of S, too. Then $S/(f^{(1)}, \ldots, f^{(n-2)})$ is a two-dimensional regular local ring. The kernel of the canonical homomorphism $S \to \overline{S}$ contains the elements $f^{(1)}, \ldots, f^{(n-2)}$. Since the induced homomorphism $S/(f^{(1)}, \ldots, f^{(n-2)}) \to \overline{S}$ is surjective, it must be an isomorphism [for dimension reasons], and therefore $\{f^{(1)}, \ldots, f^{(n-2)}\}$ is a system of generators of the kernel of the homomorphism $S \to \overline{S}$.

Now we consider the homomorphism $\mathrm{gr}_{\mathfrak{n}}(S) \to \mathrm{gr}_{\overline{\mathfrak{n}}}(\overline{S})$ induced by the homomorphism $S \to \overline{S}$. Let $j \in \{1, \ldots, n\}$. Since y_j is mapped to $\overline{y}_j \in \overline{S}$, the element

$\widetilde{y}_j := y_j + \mathfrak{n}^2$ will be mapped to $\overline{y}_j + \overline{\mathfrak{n}}^2$. Since $\overline{y}_3 + \overline{\mathfrak{n}}^2 = \cdots = \overline{y}_n + \overline{\mathfrak{n}}^2 = 0$, we have a surjective k-algebra homomorphism

$$k[\widetilde{y}_1, \ldots, \widetilde{y}_n] = \mathrm{gr}_{\mathfrak{n}}(S) \to \mathrm{gr}_{\overline{\mathfrak{n}}}(\overline{S}) = k[\overline{y}_1 + \overline{\mathfrak{n}}^2, \overline{y}_2 + \overline{\mathfrak{n}}^2]; \qquad (*)$$

the elements $w_1 := \overline{y}_1 + \overline{\mathfrak{n}}^2$, $w_2 := \overline{y}_2 + \overline{\mathfrak{n}}^2$ are algebraically independent over k, and the homomorphism in $(*)$ is just the k-algebra homomorphism defined by $\widetilde{y}_1 \mapsto w_1$, $\widetilde{y}_2 \mapsto w_2$, and $\widetilde{y}_j \mapsto 0$ for $j \in \{3, \ldots, n\}$.

(3) Let $C \subset X$ be a curve passing through p. We are going to work out C in a neighborhood of p. We set $e := e_p(C)$. The ideal $\mathcal{I}_{X,p}(C)$ of C in the two-dimensional regular local ring $\overline{S}$ is a radical ideal of height 1, hence it is generated by a reduced element φ, and we have $\mathrm{ord}_{\overline{S}}(\varphi) = e$ [cf. B(10.8)]; note that φ is even analytically reduced by B(3.7). We choose $f \in R$ such that $\overline{S}f = \overline{S}\varphi$. Since $f^{(1)}, \ldots, f^{(n-2)}$ lie in $\mathfrak{p}$, f and the polynomial $f(y_1, y_2, g^{(1)}, \ldots, g^{(n-2)})$ have the same image in $\overline{S}$. Thus, we may assume that, as a sum of homogeneous polynomials, we can write $f = f_e + f_{e+1} + \cdots$ where $f_e \in k[y_1, y_2]$ is of order e. Furthermore, by making a linear change of variables of the form $y_1' := y_1 + \beta y_2$, $y_2' := y_2$, and after multiplying f by a non-zero element of k, we may assume: The coordinate ring of $\mathbb{A}^n$ is $k[y_1, \ldots, y_n]$, the images of y_1, y_2 in $\overline{S}$ are a regular system of parameters for $\overline{S}$, and $f_e = \prod_{i=1}^{e}(y_2 - \beta_i y_1)$, $\beta_1, \ldots, \beta_e \in k$. Let $\alpha_1, \ldots, \alpha_r$ be the pairwise different elements among the elements $\beta_1, \ldots, \beta_e$; then we can write

$$f_e = (y_2 - \alpha_1 y_1)^{e_1} \cdots (y_r - \alpha_r y_1)^{e_r}$$

where $e_1, \ldots, e_r$ are positive integers and $e_1 + \cdots + e_r = e$.

Note the following: We have $\mathrm{gr}_{\overline{\mathfrak{n}}}(\overline{S}) = k[w_1, w_2]$, the initial form in $\mathrm{gr}_{\overline{\mathfrak{n}}}(\overline{S})$ of the image of f in $\overline{S}$ is $f_e(w_1, w_2)$, hence the associated graded ring of $\mathcal{O}_{C,p}$ with respect to its maximal ideal is $k[w_1, w_2]/(f_e(w_1, w_2))$ [cf. B(5.3)(3)].

(4) We consider an open covering of $\mathrm{Bl}_p(\mathbb{A}^n)$ as in A(14.1), and study the behavior on A_1 with affine coordinate ring $k[y_1, z_2, \ldots, z_n]$; π induces the k-algebra homomorphism $\omega \colon k[y_1, \ldots, y_n] \to k[y_1, z_2, \ldots, z_n]$ where $\omega(y_1) = y_1$ and $\omega(y_j) = y_1 z_j$ for $j \in \{2, \ldots, n\}$.

Let X' (resp. C') be the strict transform of X (resp. C) in $\mathrm{Bl}_p(\mathbb{A}^n)$, and let E_1 be the exceptional fibre in $\mathrm{Bl}_p(\mathbb{A}^n)$. Now X' is the closure of $\pi^{-1}(X \setminus \{p\})$ in $\mathrm{Bl}_p(\mathbb{A}^n)$, and C' is the closure of $\pi^{-1}(X \setminus \{p\})$ in X' [cf. A(13.7)].

According to A(14.2), the points of E_1 correspond to the points $(\gamma_1 : \gamma_2 : \cdots : \gamma_n)$ of $\mathbb{P}^{n-1}$, the points of $E_1 \cap A_1$ correspond to the points $(\gamma_1 : \gamma_2 : \cdots : \gamma_n)$ of $\mathbb{P}^{n-1}$ with $\gamma_1 \neq 0$, and the points of $E := E_1 \cap X'$ correspond to the points $(\gamma_1 : \gamma_2 : 0 : \cdots : 0)$ of $\mathbb{P}^{n-1}$ with $\gamma_1 \neq 0$. Note that E is a curve on X' isomorphic to $\mathbb{P}^1$; $E' := C' \cap E$ is a set consisting of r points $q_1, \ldots, q_r$ which all lie in A_1; they correspond to the points $(1 : \alpha_1 : 0 : \cdots : 0), \ldots, (1 : \alpha_r : 0 : \cdots : 0)$ of $\mathbb{P}^{n-1}$.

Let $\rho \in \{1, \ldots, r\}$. The point q_ρ of A_1 is defined by the maximal ideal generated by $y_1, z_2 - \alpha_\rho, z_3, \ldots, z_n$, the images of these elements in the local ring of q_ρ on $\mathrm{Bl}_p(\mathbb{A}^n)$ are a regular system of parameters, and therefore the images of the elements $y_1, z_2 - \alpha_\rho$ are a regular system of parameters of $\mathcal{O}_{X', q_\rho}$.

We choose an open affine neighborhood U of p in $\mathbb{A}^n$ such that $C \cap (U \cap X)$ is the set of points of $U \cap X$ where $f|U \cap X$ vanishes. Let $\rho \in \{1, \ldots, r\}$, and let V be an open affine neighborhood of q_ρ in A_1 which among the points $q_1, \ldots, q_r$ contains only q_ρ, and such that $\pi(V) \subset U$. We have $\omega(f) = y_1^e f'$ where f' is the strict transform of f in $k[y_1, z_2, \ldots, z_n]$. Now $C' \cap (V \cap X')$ is the set of points of $V \cap X'$ where $f|U \cap X \circ \pi|V \cap X'$ vanishes, hence the set of points of $V \cap X'$ where $\omega(f)|V \cap X = y_1^e f'|V \cap X'$ vanishes, and since $y_1|V \cap X'$ vanishes only in q_ρ, we see that $f'|V \cap X'$ is a local equation for the divisor C' on $V \cap X'$. Therefore the image φ' of f' in $\mathcal{O}_{X',q_\rho}$ is an equation for C' in $\mathcal{O}_{X',q_\rho}$.

The ring $\overline{S}_\rho := \mathcal{O}_{X',q_\rho}$ is the quadratic transform of $\overline{S}$ with respect to the prime ideal $(\overline{y}_1, \overline{y}_2 - \alpha_\rho \overline{y}_1)$ of $\mathrm{gr}_{\overline{n}}(\overline{S})$ [cf. VII(2.6)]. We have $\mathrm{ord}_{\overline{S}}(\varphi) = e$, and therefore we have $(\overline{S}\varphi)^{\overline{S}_\rho} = \overline{S}_\rho \overline{y}_1^{-e} \varphi$, i.e., $\overline{S}_\rho \varphi' = (\overline{S}\varphi)^{\overline{S}_\rho}$. We express this fact loosely by saying: the transform in $\overline{S}_\rho$ of an equation for C in $\overline{S}$ is an equation in $\overline{S}_\rho$ for the strict transform C' of C.

Let $\widetilde{C}$ be the total transform of C in X'. Since y_1 is a local equation for E on A_1, $y_1 f'|V \cap X'$ is a local equation for $\widetilde{C}$ on $V \cap X'$, hence the image of $y_1 f'$ in $\overline{S}_\rho$ is a generator of the total transform of the ideal $\overline{S}\varphi$ in $\overline{S}_\rho$. We express this fact loosely by saying: the total transform in $\overline{S}_\rho$ of an equation for C in $\overline{S}$ is an equation in $\overline{S}_\rho$ for the total transform $\widetilde{C}$ of C.

The following should also be clear: C has a normal-crossing point in p iff the "curve" in $\overline{S}$ defined by an equation for C in $\overline{S}$ is a normal crossing curve in the sense of VII(8.12).

(1.11) PROOF OF THEOREM (1.9): There exists a finite sequence $\widetilde{X} := X_h \xrightarrow{\pi_h} X_{h-1} \to \cdots \to X_1 \xrightarrow{\pi_1} X_0 := X$ of blowing ups [with suitable centers] such that the total transform $\widetilde{C}$ of C in $\widetilde{X}$ has nonsingular irreducible components [cf. (1.6), A(14.5) and A(13.9)]. Therefore, to begin with, we may assume that C has nonsingular irreducible components. Now C has only a finite number of singular points; therefore, to prove the theorem, it is enough to show the following:

CLAIM: Set $X_0 := X$, $C_0 := C$, and let

$$\cdots \to X_2 \xrightarrow{\pi_2} X_1 \xrightarrow{\pi_1} X_0$$

be a sequence of point blowing ups such that, if we denote by p_{i-1} the center of π_i, $i > 0$, and write $C_i = (\pi_1 \circ \cdots \circ \pi_i)^{-1}(C_0)$, one has:
(1) For each $i > 0$, $\pi_i(p_i) = p_{i-1}$.
(2) For each $i \geq 0$, p_i is a singular point of C_i which is not a normal-crossing point of the divisor C_i.
Then the sequence is necessarily finite.

The ring $R := \mathcal{O}_{X_0, p_0}$ is a regular local ring, and the equation of C in R is an analytically reduced element φ. The claim follows immediately from the considerations in (1.10) and from VII(8.13). The last assertion is also clear.

(1.12) REMARK: One can define the notion of a quadratic transformation of $\mathbb{P}^2$; this is a birational map. Then one can show: Applying a finite number of quadratic transformations to an irreducible projective curve in $\mathbb{P}^2$ leads to an irreducible projective curve in $\mathbb{P}^2$ which as singularities has at most ordinary multiple points [cf. [69], Ch. 7, Th. 2]. Thus, if we want to work only in the projective plane, we can, in general, not resolve singularities of plane projective curves.

2   Resolution of Surface Singularities I: Jung's Method

(2.0) In this section we assume that *the ground field k is algebraically closed of characteristic zero.* Let X be an irreducible surface. We shall resolve the singularities of X by using the fact that we can resolve the singularity of a two-dimensional toric variety by repeatedly blowing up points [cf. chapter VI, (5.7) and (5.18)]. This method of resolving singularities should bear the name of H. E. W. Jung, as we pointed out in the preface.

(2.1) Theorem: *Let X be an irreducible surface. Then there exists a desingularization $\pi\colon X' \to X$ of X.*

Proof: We can consider X as an open subset of an irreducible projective surface. Therefore, by appendix A, (8.2) and (8.3), we may assume, to begin with, that X is an irreducible normal projective surface. There exists a finite surjective morphism $\varphi\colon X \to Z := \mathbb{P}^2$ [cf. A(12.4)]. The branch locus $D := \mathrm{branch}(X/Z)$ is a pure one-dimensional closed subset of Z [cf. A(10.10) and note that X is normal by assumption, and that Z is regular], i.e., it is a curve. We apply embedded resolution to $D \subset Z$, as follows. By (1.9) there exists a proper morphism $\pi\colon Z' \to Z$, obtained as a finite sequence of blowing ups of points, and a finite set $F \subset D$ such that π induces an isomorphism $Z' \setminus \pi^{-1}(F) \to Z \setminus F$, and that $\pi^{-1}(D)$, the total transform of D in Z', is a divisor with normal crossings. We set $S := \varphi^{-1}(F)$; note that S is a finite sets of points [by A(2.7)].
We consider the projective variety $X \times_Z Z'$. Let $p\colon X \times_Z Z' \to X$, $q\colon X \times_Z Z' \to Z'$ be the projections; p is proper by A(4.23), q is finite by A(3.14)(8), and both maps are surjective and closed. We set $W := Z \setminus F$, $W' := \pi^{-1}(W)$ and $U := \varphi^{-1}(W)$; note that $U = X \setminus S$ is irreducible, and that the restriction of π induces an isomorphism $W' \to W$. We consider $U \times_W W'$ as an open subset of $X \times_Z Z'$ [cf. A(3.14)(4)]; we have $p^{-1}(U) = q^{-1}(W') = U \times_W W'$, $p(U \times_W W') = U$ and $q(U \times_W W') = W'$. Let X' be the closure of $U \times_W W'$. Since p and q are closed maps, we have $\overline{p(U \times_W W')} \subset p(X')$, $\overline{q(U \times_W W')} \subset q(X')$, hence the restrictions $p' := p|X'\colon X' \to X$, $q' := q|X'\colon X' \to Z'$ are surjective. Now p induces an isomorphism $U \times_W W' \to U$ [cf. A(3.14)(7)], hence X' is an irreducible projective surface; $p'\colon X' \to X$ is proper by A(4.19), and $X' \setminus p'^{-1}(S) \to X \setminus S$ is an isomorphism. The morphism $q'\colon X' \to Z'$ is finite by A(2.6). Clearly we have $\deg(q') = \deg(\varphi) =: m$ [since p' and π are birational]. Let $w' \in W'$ and set $w := \pi(w') \in W$; there lie exactly m points of X over w [cf. A(10.8)], and these points lie in U; therefore there lie exactly m points of X' over w', hence q' is

unramified over w' [cf. A(10.8)]. Thus, we have branch$(q') \subset \pi^{-1}(D)$; moreover, branch(q') is a proper closed subset of Z' by A(10.9). Thus, replacing X' by X and Z' by Z, it is enough to show the following:

Let X be an irreducible projective surface, Z a regular projective surface, $\varphi \colon X \to Z$ a finite surjective morphism such that branch(φ) is contained in a curve $D \subset Z$ with nonsingular irreducible components and having only normal crossings as singularities. Then there exists a desingularization of X.

We will show: *Every singular point of $\overline{X}$, the normalization of X, can be resolved by repeatedly blowing up points.* (Cf. below for a more precise formulation.) Using A(8.3) yields the claim since $\mathrm{Sing}(\overline{X})$ is finite [cf. A(7.17)].

Let $\pi \colon \overline{X} \to X$ be the normalization of X. Let $z \in Z \setminus D$; then, for every $x \in X$ lying over z, $\mathcal{O}_{Z,z} \to \mathcal{O}_{X,x}$ is unramified, hence $\mathcal{O}_{X,x}$ is a regular local ring [note that $\dim(\mathcal{O}_{Z,z}) = \dim(\mathcal{O}_{X,x}) = 2$], hence the only singular points of $\overline{X}$ can be those lying over D.

Let K be the field of rational functions on Z, and let L be the field of rational functions on X which is also the field of rational functions on $\overline{X}$; L is a finite extension of K [cf. A(10.2)]. Let $z \in D$ be the image of a singular point of $\overline{X}$; we choose an open affine neighborhood W of z in Z such that, $u, v \in A := \mathcal{O}(W)$ being representatives for a system of generators of the maximal ideal of $\mathcal{O}_{W,z} = \mathcal{O}_{Z,z}$, we have $\mathcal{I}_W(D) = Au$ or $\mathcal{I}_W(D) = Auv$.

Set $U := \varphi^{-1}(W)$, $B := \mathcal{O}(U)$, $V := \pi^{-1}(U)$, $C := \mathcal{O}(V)$; U and V are affine, B is integral over A, and C is the integral closure of B and A in L [cf. appendix A, (2.6) and (7.11)]; in particular, C is finite over A [cf. B(3.6)]. We have branch$(\varphi) \cap W \subset D \cap W$, hence $\mathcal{I}_W(D) \subset \mathrm{branch}(B/A) = \mathfrak{n}_{B/A}$ [cf. III(4.12) for the equality sign]. In particular, A_{uv} is unramified in B_{uv} [cf. III(3.2)(5)]. If a maximal ideal $\mathfrak{n}$ of B neither contains u nor v, then $\mathfrak{n}$ is unramified over A, hence $B_{\mathfrak{n}}$ is regular, and therefore $\mathfrak{n}$ is unramified in C [cf. A(7.11)]. This means that B_{uv} is unramified in C_{uv}, hence that A_{uv} is unramified in C_{uv} [cf. III(3.5)(3)], and therefore we have $(\mathfrak{n}_{C/A})_{uv} = A_{uv}$.

Set $\mathfrak{q} := \mathcal{I}_W(\{z\}) \subset A$, $R := A_{\mathfrak{q}} = \mathcal{O}_{Z,z}$ and $S := C_{\mathfrak{q}}$; then S is the integral closure of R in L [cf. [63], Prop. 4.13], and it is finite over R [cf. B(3.6)]. Let $\{x_1, \ldots, x_h\} \subset V$ be the set of points of $\overline{X}$ lying over z, and, for $i \in \{1, \ldots, h\}$, set $\mathfrak{n}_i := \mathcal{I}_V(\{x_i\})$. Then $\{\mathfrak{n}_1, \ldots, \mathfrak{n}_h\}$ is the set of maximal ideals of S; they lie over the maximal ideal $\mathfrak{m}$ of R, and we have $S_{\mathfrak{n}_i} = \mathcal{O}_{\overline{X}, x_i}$ for $i \in \{1, \ldots, h\}$ [cf. A(7.12)]; these rings are integrally closed. Moreover, we have $\mathfrak{n}_{S/R} = \mathfrak{n}_{C/A} R$ and $(\mathfrak{n}_{S/R})_{uv} = R_{uv}$ [cf. III(4.12)].

Now we pass to completions. The completion $\widehat{R}$ of R is a ring of formal power series $k[\![u, v]\!]$ [cf. [63], Prop. 10.16]. Since S is a finitely generated R-module, the $\mathfrak{m}S$-adic completion $\widehat{S}$ of S is the completion of S with respect to its Jacobson radical, hence $\widehat{S} = \widehat{S}_1 \times \cdots \times \widehat{S}_h$ where, for $i \in \{1, \ldots, h\}$, $\widehat{S}_i$ is the completion of the local ring $S_i := S_{\mathfrak{n}_i}$ with respect to its maximal ideal [cf. B(8.4)]. Moreover, by the results of chapter III, (6.9), (6.10) and (6.14), the rings $\widehat{S}_1, \ldots, \widehat{S}_h$ are integrally closed local domains, $\widehat{S}$ is the integral closure of $\widehat{R}$ in $Q(\widehat{S})$, and

$n_{S/R}\widehat{S} = n_{\widehat{S}/\widehat{R}}$, hence $\widehat{S}_{uv}$ is unramified over $\widehat{R}_{uv}$, and therefore, by III(3.8), the rings $(\widehat{S}_1)_{uv}, \ldots, (\widehat{S}_h)_{uv}$ are unramified over $\widehat{R}_{uv}$. This implies, in particular, the following: Let $i \in \{1, \ldots, h\}$, and assume that x_i is a singular point of $\overline{X}$. The ring $\widehat{S}_i$ is the integral closure of $\widehat{R}$ in the field of quotients $\widehat{L}_i$ of $\widehat{S}_i$ which is a finite extension of $k((u,v))$. Now we are in the situation of VI(1.5). By VI(4.23), $\widehat{S}_i$ is the completion of the local ring of an affine variety T_{σ_i} at its singular point p_i, σ_i being a strongly convex rational polyhedral cone of dimension 2. By chapter VI, (5.7) and (5.18), we can desingularize T_{σ_i} in the following way: $\mathrm{Bl}_{p_i}(T_{\sigma_i})$ has a finite number of singular points, they lie over p_i, and each of them admits an open affine neighborhood which is isomorphic to a toric variety [cf. VI(5.17)]. This set of singular points of $\mathrm{Bl}_{p_i}(T_{\sigma_i})$ corresponds uniquely to the set of singular points of $\mathrm{Bl}_{x_i}(\overline{X})$ lying over x_i, and the completion of the local ring on $\mathrm{Bl}_{p_i}(T_{\sigma_i})$ of such a point is isomorphic to the completion of the local ring on $\mathrm{Bl}_{x_i}(\overline{X})$ of the point of $\mathrm{Bl}_{x_i}(\overline{X})$ corresponding to it [cf. A(14.7)], hence the singular points of $\mathrm{Bl}_{x_i}(\overline{X})$ lying over x_i are normal [cf. III(6.15)]. There exists a finite sequence of point blowing ups

$$T_{il_i} \xrightarrow{\pi'_{il_i}} T_{il_i-1} \xrightarrow{\pi'_{il_i-1}} \cdots \xrightarrow{\pi'_{i2}} T_{i1} \xrightarrow{\pi'_{i1}} T_{i0} = T_{\sigma_i},$$

in which, for $i \in \{1, \ldots, l_i\}$, T_{ij} is a two-dimensional irreducible normal projective variety having only a finite number of singular points, each of them lies over p_i and it admits an open affine neighborhood which is isomorphic to a toric variety, T_{il_i} is regular, and we have $T_{ij} = \mathrm{Bl}_{p_{i,j-1}}(T_{i,j-1})$ for $j \in \{1, \ldots, l_i\}$ where $p_{i,j-1}$ is a singular point of $T_{i,j-1}$ with $p_{i0} := p_i$. By using A(14.7) we get a sequence

$$Y_{il_i} \xrightarrow{\pi_{il_i}} Y_{il_i-1} \xrightarrow{\pi_{il_i-1}} \cdots \xrightarrow{\pi_{i2}} Y_{i1} \xrightarrow{\pi_{i1}} Y_{i0} = \overline{X},$$

where $Y_{ij} = \mathrm{Bl}_{x_{i,j-1}}(Y_{i,j-1})$, $x_{i,j-1}$ is a singular point of $Y_{i,j-1}$ lying over x_i, $x_{i0} = x_i$, and $Y_i := Y_{il_i}$ is an irreducible normal projective surface with the following property: Every point of Y_i lying over x_i is regular (we say loosely that we have resolved the point x_i of $\overline{X}$) and

$$\pi_i := \pi_{i1} \circ \cdots \circ \pi_{il_i} : Y_i \to \overline{X}$$

is proper and induces an isomorphism $Y_i \setminus \pi_i^{-1}(\{x_i\}) \to \overline{X} \setminus \{x_i\}$. If $i' \in \{1, \ldots, h\}$ with $i' \neq i$ and if $x_{i'}$ is also a singular point of $\overline{X}$, then we can apply the above procedure to the point y_i of Y_i corresponding to $x_{i'}$ [note that π_i is an isomorphism outside of x_i], getting a sequence of point blowing ups $Y_{i,i'} \to \cdots \to Y_i$ and no point of $Y_{i,i'}$ lying over y_i is singular. After at most h such procedures we have a proper morphism $\widetilde{Y}_z \to \overline{X}$ with Y_z being an irreducible normal projective surface which is a composition of point blowing ups, and if we consider the morphism $\widetilde{Y}_z \to \overline{X} \to Z$, then every point of $\widetilde{Y}_z$ lying over z is a regular point of $\widetilde{Y}_z$.

Now let $z' \neq z$ be another point in D such that z' is the image of $x' \in \mathrm{Sing}(\overline{X})$. Let $y' \in \widetilde{Y}_z$ correspond to x'. Clearly we can apply the above procedure to y' and z'. After a finite number of steps we have proved the above assertion.

This ends the proof of theorem (2.1).

3   Quadratic Dilatations

3.1   Quadratic Dilatations

(3.1) QUADRATIC DILATATION: Let R be a local integral domain with maximal ideal $\mathfrak{m}$ and field of quotients K, and let v be a valuation of K with valuation ring V dominating R. We assume that $R \neq K$.

(1) Let $\{x_1, \ldots, x_m\}$ be a system of generators of $\mathfrak{m}$; the finitely generated ideal $\mathfrak{m}V$ is a principal ideal [cf. I(2.4)(2)], hence we have $\mathfrak{m}V = Vx_i$ for some $i \in \{1, \ldots, m\}$. We set $A := R[x_1/x_i, \ldots, x_m/x_i]$ and $\mathfrak{p} := \mathfrak{m}(V) \cap A$. Let $\{y_1, \ldots, y_n\}$ be another system of generators of $\mathfrak{m}$; then we have $\mathfrak{m}V = Vy_j$ for some $j \in \{1, \ldots, n\}$. We set $B := R[y_1/y_j, \ldots, y_n/y_j]$ and $\mathfrak{q} := \mathfrak{m}(V) \cap B$. Then we have $A_\mathfrak{p} = B_\mathfrak{q}$, $Ax_i = \mathfrak{m}A$, $By_j = \mathfrak{m}B$ and $\mathfrak{m}A_\mathfrak{p} = A_\mathfrak{p}x_i = B_\mathfrak{q}y_j = \mathfrak{m}B_\mathfrak{q}$.

Proof: First, we assume that $m \leq n$ and that $x_k = y_k$ for $k \in \{1, \ldots, m\}$; then we have $x_k/y_j \in B$ for $k \in \{1, \ldots, m\}$. We set $C := R[y_1/x_i, \ldots, y_n/x_i]$. Since $Vx_i = Vy_j$, we see that $x_i/y_j \notin \mathfrak{q}$, hence that x_i/y_j is a unit of $B_\mathfrak{q}$, and since $y_l/x_i = (y_l/y_j) \cdot (y_j/x_i) \in B_\mathfrak{q}$ for $l \in \{1, \ldots, n\}$, we have $C \subset B_\mathfrak{q}$. We set $\mathfrak{r} := \mathfrak{q}B_\mathfrak{q} \cap C = \mathfrak{m}(V) \cap C$; then we have $C_\mathfrak{r} \subset B_\mathfrak{q}$. Similarly, since y_j/x_i is a unit of $C_\mathfrak{r}$, we have $y_l/y_j = (y_l/x_i) \cdot (x_i/y_j) \subset C_\mathfrak{r}$ for every $l \in \{1, \ldots, n\}$, hence $B \subset C_\mathfrak{r}$ and therefore $B_\mathfrak{q} \subset C_\mathfrak{r}$. Thus, we have shown that $B_\mathfrak{q} = C_\mathfrak{r}$. On the other hand, since the elements $y_1, \ldots, y_n$ are linear combinations of the elements $x_1, \ldots, x_m$ with coefficients in R, we have $C \subset A$, and clearly $A \subset C$, whence $A = C$ and $A_\mathfrak{p} = C_\mathfrak{r}$. Therefore we have shown that $A_\mathfrak{p} = B_\mathfrak{q}$ in this case. Since $\mathfrak{m}A = Ax_i$ and $\mathfrak{m}B = By_j$, the last assertion follows also.

The general case now can be obtained by considering the following particular system $\{x_1, \ldots, x_m, y_1, \ldots, y_n\}$ of generators of $\mathfrak{m}$.

(2) The local ring $S := A_\mathfrak{p} \subset V$ of (1) is called the first quadratic dilatation of R with respect to v or along V. The center of v in S is the maximal ideal $\mathfrak{n}$ of S, hence $\mathfrak{n} \cap R = \mathfrak{m}$, and $\mathfrak{m}S$ is a principal ideal of S with $\mathfrak{m}S \cap R = \mathfrak{m}$. The field of quotients of S is K, and S is essentially of finite type over R.

(3) Note that, if $\mathfrak{m}$ is a principal ideal, then we have $S = R$.

(4) From (1) we get: Since $\mathfrak{m}S = Sx$ for some $x \in \mathfrak{m}$, then with $A := R[\mathfrak{m}/x]$ and $\mathfrak{p} := A \cap \mathfrak{n}$ we have $S = A_\mathfrak{p}$. This implies, in particular, the following: Let w be any valuation of K which has center $\mathfrak{n}$ in S; then S is also the first quadratic dilatation of R with respect to w.

(3.2) NOTATION: (1) Let R, v and V be as in (3.1). Set $R_0 := R$, and let $(R_i)_{i \geq 0}$ be the sequence of local subrings of K such that, for every $i \in \mathbb{N}$, R_i is the first quadratic dilatation of R_{i-1} with respect to v. Then, for $h \in \mathbb{N}$, R_h is called the h-th quadratic dilatation of R with respect to v, and every ring R_i, $i \in \mathbb{N}$, is called a quadratic dilatation of R with respect to v. The sequence $(R_i)_{i \geq 0}$ is called the sequence of quadratic dilatations of R with respect to v or along V.

(2) Let R be as in (3.1). If a subring S of K is a quadratic dilatation of R with respect to some valuation v of K having center $\mathfrak{m}$ in R, then S is called a quadratic dilatation of R.

(3) Let R and v be as in (3.1). If R is essentially of finite type over a subfield k of K, then any quadratic dilatation of R is also essentially of finite type over k.

(3.3) NORMAL QUADRATIC DILATATION: Let k be a field, let R be a local integral domain which is a k-algebra essentially of finite type with maximal ideal $\mathfrak{m}$ and field of quotients K, and let v be a valuation of K having center $\mathfrak{m}$ in R; let V be the valuation ring of the valuation v.

(1) We keep the notations of (3.1)(1). Let $\overline{A}$ (resp. $\overline{B}$) be the integral closure of A (resp. B) in K [note that $\overline{A}$ and $\overline{B}$ are k-algebras essentially of finite type by B(3.6)]. Then we have $\overline{A} \subset V$ (resp. $\overline{B} \subset V$) [since V is integrally closed by I(2.4)]; we set $\overline{\mathfrak{p}} := \overline{A} \cap \mathfrak{m}(V)$ (resp. $\overline{\mathfrak{q}} := \overline{B} \cap \mathfrak{m}(V)$). We show that $\overline{A}_{\overline{\mathfrak{p}}} = \overline{B}_{\overline{\mathfrak{q}}}$.

Proof: We use the notations of the first part of the proof of (3.1)(1); let $\overline{C}$ be the integral closure of C. Clearly we have $\overline{C} \subset \overline{B}_{\overline{\mathfrak{q}}}$, hence $\overline{C}_{\overline{\mathfrak{r}}} \subset \overline{B}_{\overline{\mathfrak{q}}}$ [where $\overline{\mathfrak{r}} := \overline{C} \cap \mathfrak{p}_v$]. Similarly, we find that $\overline{B} \subset \overline{C}_{\overline{\mathfrak{r}}}$, hence that $\overline{B}_{\overline{\mathfrak{q}}} \subset \overline{C}_{\overline{\mathfrak{r}}}$. Thus, we have shown that $\overline{B}_{\overline{\mathfrak{q}}} = \overline{C}_{\overline{\mathfrak{r}}}$. Since $A = C$, we have $\overline{A}_{\overline{\mathfrak{p}}} = \overline{C}_{\overline{\mathfrak{r}}}$. Therefore we have shown that $\overline{A}_{\overline{\mathfrak{p}}} = \overline{B}_{\overline{\mathfrak{q}}}$ in this case. The general case now follows easily.

(2) The local ring $\overline{A}_{\overline{\mathfrak{p}}}$ is called the *first normal quadratic dilatation* of R with respect to v or along V. It is an integrally closed local k-algebra essentially of finite type, and $\mathfrak{m}\overline{A}_{\overline{\mathfrak{p}}}$ is a principal ideal.

(3) Note the following: Let S be the first quadratic dilatation of R with respect to v, let $\mathfrak{n}$ be the maximal ideal of S, and let $\overline{S}$ be the integral closure of S. Then the first normal quadratic dilatation of R with respect to v is a localization of $\overline{S}$ with respect to a maximal ideal $\overline{\mathfrak{n}}$ of $\overline{S}$ lying over $\mathfrak{n}$.

(4) Set $R_0 := R$. The sequence $(R_i)_{i \geq 0}$ where, for every $i \in \mathbb{N}$, R_i is the first normal quadratic dilatation of R_{i-1} with respect to v, is called the sequence of normal quadratic dilatations of R with respect to v [cf. (3.2)(1)].

(5) If a subring S of K is a first normal quadratic dilatation of R with respect to some valuation of K having center $\mathfrak{m}$ in R, then S is called a first normal quadratic dilatation of R.

3.2   Quadratic Dilatations and Algebraic Varieties

(3.4) REMARK: In this subsection we assume that k *is an algebraically closed field*. Let X be an irreducible projective variety defined over k with field of functions K. Then K is a field of algebraic functions in $\dim(X)$ variables over k.

(1) Let v be a valuation of the algebraic function field K. Then there exists a unique irreducible closed subset $Y \subset X$ such that v dominates the local ring $\mathcal{O}_{X/Y}$ of Y on X. If v is zero-dimensional, then $Y = \{p\}$ is a point; by abuse of language we often say in this case: v dominates p.

[Existence] We choose $n \in \mathbb{N}$ such that $X \subset \mathbb{P}^n$ is closed and, $k[x_0, \ldots, x_n]$ being the homogeneous coordinate ring of X, such that the elements $x_0, \ldots, x_n$ are not 0. Let $i \in \{0, \ldots, n\}$; the ring $A_i := k[x_0/x_i, \ldots, x_n/x_i]$ can be considered as a subring of K having K as field of quotients. There exists $j \in \{0, \ldots, n\}$ such that A_j is contained in the ring of v. In fact, if $v(x_i/x_0) \geq 0$ for every

$i \in \{1, \ldots, n\}$, then the ring of v contains A_0. Otherwise, we choose $j \in \{1, \ldots, n\}$ with $v(x_j/x_0) \leq v(x_i/x_0)$ for every $i \in \{1, \ldots, n\}$. Then we have $v(x_i/x_j) = v(x_i/x_0) + v(x_0/x_j) \geq 0$ for every $i \in \{0, \ldots, n\}$, hence the ring of v contains A_j. The center of v in A_j is a prime ideal of A_j, corresponding to an irreducible closed subset Y of X, and, if v is zero-dimensional, it is a maximal ideal by I(10.9)(2), corresponding to a point p of X. Therefore there exists $Y \subset X$, as asserted.

[Uniqueness] Let $Y \subset X$ be closed and irreducible, and let v dominate the local ring $\mathcal{O}_{X/Y}$ of Y on X. There exists $j \in \{0, \ldots, n\}$ with $v(x_i/x_j) \geq 0$ for every $i \in \{0, \ldots, n\}$; it is enough to show that for every such j we have $\mathcal{O}_{X/Y} = B_{\mathfrak{p}}$ where $B := k[x_0/x_j, \ldots, x_n/x_j]$ and $\mathfrak{p}$ is the center of v in B.

We set $R := \mathcal{O}_{X/Y}$; let V be the valuation ring of v, and let $\mathfrak{m}$ be its maximal ideal. We choose $l \in \{0, \ldots, n\}$ such that $R = C_{\mathfrak{q}}$ with $C := k[x_0/x_l, \ldots, x_n/x_l]$ and $\mathfrak{q} := \mathfrak{m} \cap C$. Now we choose $j \in \{0, \ldots, n\}$ with $x_i/x_j \in V$ for every $i \in \{0, \ldots, n\}$ [cf. above], and set $B := k[x_0/x_j, \ldots, x_n/x_j]$, $\mathfrak{p} := \mathfrak{m} \cap B$. Then x_j/x_l and x_l/x_j lie in V, hence x_j/x_l is a unit of V, and since $x_j/x_l \in R$ and V dominates R, we see that x_j/x_l is a unit of R, hence that $x_l/x_j \in R$. This means that $x_i/x_j = (x_i/x_l) \cdot (x_l/x_j) \in R$ for $i \in \{0, \ldots, n\}$, hence that $B \subset R$; since V dominates R and $\mathfrak{p} = \mathfrak{m} \cap B$, we find that $\mathfrak{p} = \mathfrak{m}(R) \cap B$, hence that $B_{\mathfrak{p}} \subset R$. The element x_l/x_j lies in $B_{\mathfrak{p}}$, V dominates $B_{\mathfrak{p}}$, and x_l/x_j is a unit of V; therefore x_l/x_j is a unit of $B_{\mathfrak{p}}$, hence $x_j/x_l \in B_{\mathfrak{p}}$, and therefore $x_i/x_l = (x_i/x_j) \cdot (x_j/x_l) \in B_{\mathfrak{p}}$ for every $i \in \{0, \ldots, n\}$ which means that $C \subset B_{\mathfrak{p}}$. Since V dominates $B_{\mathfrak{p}}$ and $\mathfrak{q} = \mathfrak{m} \cap C$, we get $\mathfrak{q} = \mathfrak{m}(B_{\mathfrak{p}}) \cap C$, and therefore $C_{\mathfrak{q}} \subset B_{\mathfrak{p}}$. Now $R = C_{\mathfrak{q}}$ yields $R = B_{\mathfrak{p}}$.

(2) Conversely, let $Y \subset X$ be closed and irreducible (resp. $Y = \{p\}$, a point); then there exists a valuation (resp. a zero-dimensional valuation) of K which dominates $\mathcal{O}_{X/Y}$ (resp. p).

In fact, there exists $j \in \{0, \ldots, n\}$ such that with $B := k[x_0/x_j, \ldots, x_n/x_j]$ [cf. above], we have $\mathcal{O}_{X/Y} = B_{\mathfrak{p}}$ where $\mathfrak{p}$ is a prime ideal of B (resp. $\mathcal{O}_{X,p} = B_{\mathfrak{m}}$ where $\mathfrak{m}$ is a maximal ideal of B). There exists a valuation of K having center $\mathfrak{p}$ in B [cf. I(3.5)] (resp. a zero-dimensional valuation v of K having center $\mathfrak{m}$ in B [cf. I(10.9)]); now v dominates $\mathcal{O}_{X/Y}$ (resp. $\mathcal{O}_{X,p}$).

(3.5) QUADRATIC DILATATION AND BLOWING UP: Let $X \subset \mathbb{P}^n$ be an irreducible projective variety passing through $p \in \mathbb{P}^n$; set $R := \mathcal{O}_{X,p}$, and let K be the field of rational functions on X. Let $\pi\colon \mathrm{Bl}_p(X) \to X$ be the blowing up of X with center p; $\mathrm{Bl}_p(X)$ is an irreducible projective variety with field of functions K [cf. A(13.8)]. Let v be a zero-dimensional valuation of the algebraic function field K dominating p [cf. (3.4)(2)].

Without loss of generality, we may assume that $p = (1:0:\cdots:0)$. In the following, we use without further warning the notations introduced in appendix A, (13.11)–(14.2), cf. also A(14.4), and we set $\overline{y}_i := y_i + \mathfrak{b}$ for $j \in \{1, \ldots, n\}$, $\overline{z}_j := z_j + \mathfrak{b}_1$ for $j \in \{2, \ldots, n\}$. We assume that $\overline{y}_1 \neq 0$; then we have $\overline{z}_j = \overline{y}_j/\overline{y}_1$ for $j \in \{2, \ldots, n\}$.

(1) Let $q \in \mathrm{Bl}_p(X)$ be the point which is dominated by v [cf. (3.4)(1)]; then v dom-

inates $\pi(q)$, hence by uniqueness we have $\pi(q) = p$, hence $q = (p, (\beta_1 : \cdots : \beta_n))$.
We show that $S := \mathcal{O}_{\mathrm{Bl}_p(X),q}$ is the first quadratic dilatation of R with respect
to v.

We may assume that $\beta_1 \neq 0$. We know that R is the localization of $k[\bar{y}_1, \ldots, \bar{y}_n]$
with respect to the maximal ideal generated by $\bar{y}_1, \ldots, \bar{y}_n$, and that S is the
localization of $k[\bar{y}_1, \bar{z}_2, \ldots, \bar{z}_n]$ with respect to the maximal ideal generated by
$\bar{y}_1, \bar{z}_2 - \beta_2/\beta_1, \ldots, \bar{z}_n - \beta_n/\beta_1$. Therefore the ring $A := R[\bar{y}_2/\bar{y}_1, \ldots, \bar{y}_n/\bar{y}_1]$ is
contained in the ring of v, and the center of v in A is the ideal generated by
$\bar{y}_1, \bar{y}_2/\bar{y}_1 - \beta_2/\beta_1, \ldots, \bar{y}_n/\bar{y}_1 - \beta_n/\beta_1$. Since $A = k[\bar{y}_1, \bar{y}_2/\bar{y}_1, \ldots, \bar{y}_n/\bar{y}_1]$, the
assertion follows immediately.

(2) Let Y be the normalization of $\mathrm{Bl}_p(X)$; Y is an irreducible projective variety
having K as field of rational functions [cf. A(7.10)]. Let $y \in Y$ be the point which
is dominated by v [cf. (3.4)(1)]; y lies over q. Let $\bar{S}$ be the integral closure of S; it
is a finitely generated S-module [cf. B(3.6)], hence it is, in particular, a semilocal
ring [cf. B(3.8)]. The localizations of $\bar{S}$ with respect to maximal ideals of $\bar{S}$ are, on
the one side, integrally closed [cf. [63], Prop. 4.13], and are, on the other side, the
local rings of the points of Y lying over q [cf. A(7.12)]. Then we have $\mathcal{O}_{Y,y} = \bar{S}_{\mathfrak{m}}$
where $\mathfrak{m}$ is the center of v in $\bar{S}$, hence $\mathcal{O}_{Y,y}$ is the first normal quadratic dilatation
of R with respect to v.

4 Quadratic Dilatations of Two-Dimensional Regular Local Rings

(4.0) In this section K is a field. Sometimes we specify a subfield k of K; in
this case valuations of K are tacitly assumed to be trivial on k. We denote by
$\mathrm{lneft}_k(K)$ the family of local integrally closed (= normal) k-subalgebras of K which
are essentially of finite type over k and have K as field of quotients.

(4.1) REMARK: Let R be a two-dimensional regular local ring with field of quotients K, and let V be a valuation ring of K. Let $\{x, y\}$ be a regular system of
parameters of R. If $V = V_R$ where V_R is the valuation ring of K defined by the
order function of R, then the first quadratic dilatation of R along V is the localization of $A := R[\mathfrak{m}/x] = R[y/x]$ with respect to the center Ax of V_R in A, hence
V_R is the first quadratic dilatation of R along V [cf. VII(2.3)]. Now assume that
$V \neq V_R$, and let p be the directional ideal of V. *The first quadratic dilatation of R
along V is the quadratic transform S_p of R, and it is the only quadratic transform
of R which is contained in V;* this follows immediately from the considerations in
VII(2.18). We use this observations to prove the following proposition.

(4.2) Proposition: *Let R be a two-dimensional regular local ring with field of
quotients K, and let V be a valuation ring of K which dominates R. Then:*
*(1) If V is of the second kind with respect to R, then the sequence of quadratic
dilatations of R along V is finite; it is of the form $R =: R_0 \subsetneq R_1 \subsetneq \cdots \subsetneq R_h \subsetneq V$*

where $h \geq 0$ and, for $i \in \{1, \ldots, h\}$, R_i is a quadratic transform of R_{i-1}. Moreover, in this case V is the discrete valuation ring associated with R_h.
(2) If V is of the first kind with respect to R, then the sequence of quadratic dilatations of R along V is not finite; it is of the form $R =: R_0 \subsetneqq R_1 \subsetneqq \cdots V$ where, for $i \in \mathbb{N}$, R_i is a quadratic transform of R_{i-1}, and $V = \bigcup_{i \geq 0} R_i$.

Proof: (1) is an immediate consequence of chapter VII, (2.19) and (6.10).
(2) The first part of (2) is a consequence of VII(2.19) since V is of the first kind with respect to R. The second part of (2) follows from VII(6.12).

(4.3) REMARK: With the same notations as in (4.1), let $(R_i)_{i \geq 0}$ with $R_0 := R$ be the sequence of quadratic dilatations of R_0 along V. Let $\wp$ be a simple complete m-primary ideal of R. If $\wp R_l$ is a principal ideal of R_l for some $l \in \mathbb{N}$, then $\wp' R_l$ is a principal ideal of R_l for every predecessor $\wp'$ of $\wp$.
The assertion is trivially true if $V = V_R$. Now we consider the case $V \neq V_R$. We prove the assertion by induction on the rank h of $\wp$. Let p be the directional form of V. The assertion is trivial if $h = 0$. Now let $h > 0$. Let $\mathfrak{m} =: \wp_0 \supset \wp_1 \supset \cdots \supset \wp_{h-1}$ be the predecessors of $\wp$. First, we consider the case that $c(\wp)$ is a power of p; then the transform $\wp^{R_1}$ of $\wp$ in R_1 has rank $h - 1$, and its predecessors are $\mathfrak{m}(R_1) = \wp_1^{R_1} \supset \wp_2^{R_1} \supset \cdots \supset \wp_{h-1}^{R_1}$. There exist $x \in \mathfrak{m} \setminus \mathfrak{m}^2$ and natural integers $s_1, \ldots, s_h$ such that $\wp_i R_1 = x^{s_i} \wp_i^{R_1}$ for $i \in \{1, \ldots, h\}$. The ideal $\wp_h R_l = x^{s_h} \wp_h^{R_1} R_l$ is a principal ideal by assumption, hence the ideals $\wp_i^{R_1} R_l$, $i \in \{1, \ldots, h - 1\}$, are principal ideals of R_l by induction. Therefore $\wp_1 R_l = x^{s_1} \wp_1^{R_1} R_l, \ldots, \wp_{h-1} R_l = x^{s_{h-1}} \wp_{h-1}^{R_1} R_l$ are principal ideals of R_l. In the other case, all the ideals $\wp_1 R_1, \ldots, \wp_h R_1$ are principal ideals by VII(2.15) and VII(5.4).

(4.4) Proposition: Let $(R_i)_{i \geq 0}$ be a strictly increasing sequence of two-dimensional regular local rings having K as common field of quotients and, for $i \in \mathbb{N}_0$, let $\mathfrak{m}_i$ be the maximal ideal of R_i. If, for every $i \in \mathbb{N}_0$, R_{i+1} is a quadratic dilatation of R_i, then $V := \bigcup_{i \geq 0} R_i$ is the ring of a valuation v of K such that, for $i \in \mathbb{N}_0$, v dominates R_i and $\mathrm{tr.\,d}_{R_i/\mathfrak{m}_i}(\kappa_v) = 0$, and, for $i \in \mathbb{N}_0$, V is the only valuation ring of K dominating R_i.

Proof: We may assume that, for $i \in \mathbb{N}_0$, R_{i+1} is a first quadratic dilatation of R_i [by considering also the rings in the sequence of quadratic dilatations from R_i to R_{i+1}]. Let $i \in \mathbb{N}$. Since $\dim(R_i) = 2$, we see that R_i is a quadratic transform of R_{i-1} [cf. (4.1)]; now the assertion follows from VII(6.12).

(4.5) Lemma: Let $(R_i)_{i \geq 0}$ be an increasing sequence of integrally closed quasilocal proper subrings of K having K as common field of quotients, and, for $i \in \mathbb{N}_0$, let $\mathfrak{m}_i$ be the maximal ideal of R_i and assume that $\mathfrak{m}_i = R_i \cap \mathfrak{m}_{i+1}$. We set $S := \bigcup_{i \geq 0} R_i$. If S is not a valuation ring of K, then there exists a valuation v of K dominating R_i and such that v is of the second kind with respect to R_i for every $i \in \mathbb{N}_0$.

Proof: Clearly S is integrally closed, and $\mathfrak{n} := \bigcup_{i \geq 0} \mathfrak{m}_i$ is a proper ideal of S. It is easy to see that S is quasilocal, and that $\mathfrak{n}$ is its maximal ideal. For every $i \in \mathbb{N}_0$ we can consider $R_i/\mathfrak{m}_i$ as a subfield of $S/\mathfrak{n}$ and of $R_{i+1}/\mathfrak{m}_{i+1}$. Then we have $S/\mathfrak{n} = \bigcup_{i \geq 0} R_i/\mathfrak{m}_i$. Since S is not a valuation ring, there exists a valuation v of K having center $\mathfrak{n}$ in S and such that κ_v is not algebraic over $S/\mathfrak{n}$ [cf. I(3.34)]. We have $\mathfrak{p}_v \cap R_i = \mathfrak{m}_i$, and κ_v is not algebraic over $R_i/\mathfrak{m}_i$ for every $i \in \mathbb{N}_0$.

(4.6) Proposition: *Let K be a field of algebraic functions in two variables over k. Let $R_0 \in \mathrm{lneft}_k(K)$ with $\dim(R_0) = 2$. Let $(R_i)_{i \geq 0}$ be a strictly increasing sequence of two-dimensional local subrings of K such that, for every $i \in \mathbb{N}$, R_i is a first normal quadratic dilatation of R_{i-1}. Then:*
(1) $V := \bigcup_{i \geq 0} R_i$ is the ring of a zero-dimensional valuation v of K over k, the center of v in R_i is the maximal ideal $\mathfrak{m}_i$ of R_i for every $i \in \mathbb{N}_0$ and, moreover, V is the only valuation of K with these properties.
(2) For every $i \in \mathbb{N}$, R_i is the first normal quadratic dilatation of R_{i-1} with respect to v.

Proof: (a) Let $i \in \mathbb{N}_0$. R_i is a k-algebra essentially of finite type with $\dim(R_i) = 2$, hence $R_i/\mathfrak{m}_i$ is a finite extension of k [cf. B(7.5)].
(b) As before, V is integrally closed and quasilocal with maximal ideal $\mathfrak{n} := \bigcup_{i \geq 0} \mathfrak{m}_i$. V dominates all the rings R_i, and has residue field $V/\mathfrak{n} = \bigcup_{i \geq 0} R_i/\mathfrak{m}_i$. In particular, $V/\mathfrak{n}$ is algebraic over k.
(c) We show that V is the ring of a valuation of K. In fact, suppose not; then there exists a valuation w of K having center $\mathfrak{m}_i$ in R_i and with $\mathrm{tr.\,d}_{R_i/\mathfrak{m}_i}(\kappa_w) > 0$ for every $i \in \mathbb{N}_0$ [cf. (4.5)]; note that w is a discrete valuation of rank 1 [cf. I(11.9)]. Since V is not a valuation ring, there exists $x \in K$ with $x \notin V$ and $1/x \notin V$. We write $x = y_0/z_0$ with $y_0, z_0 \in R_0$; note that y_0 and z_0 lie in $\mathfrak{m}_0$ [since x and $1/x$ do not belong to R_0], hence we have $w(y_0) > 0$ and $w(z_0) > 0$. Since $R_1 \supset R_0$, there exists $x_1 \in \mathfrak{m}_0$ with $R_0[\mathfrak{m}_0/x_1] \subset R_1$, hence $y_1 := y_0/x_1$ and $z_1 := z_0/x_1$ lie in R_1. We have $w(y_0) > w(y_1)$, $w(z_0) > w(z_1)$ and $x = y_1/z_1$. Since x and $1/x$ lie not in R_1, the elements y_1 and z_1 lie in $\mathfrak{m}_1$. Since $(R_i)_{i \geq 0}$ is strictly increasing, we can repeat the construction, and we obtain sequences $(y_i)_{i \geq 0}$, $(z_i)_{i \geq 0}$ in V with $w(y_0) > w(y_1) > \cdots > 0$, $w(z_0) > w(z_1) > \cdots > 0$ which is absurd since w is a discrete valuation of rank 1. Therefore V is the ring of a valuation v of K, and v dominates R_i for $i \in \mathbb{N}_0$.
From (a)-(c) we get all the assertions of (1) except the last one; this assertion can be proved in the same way as the corresponding result in (4.4). Assertion (2) is clear.

(4.7) Lemma: *Let R be a two-dimensional integrally closed local domain with maximal ideal $\mathfrak{m}$ and field of quotients K, and let $\mathfrak{p}$ be a prime ideal of R of height 1. Then $R_\mathfrak{p}$ is a discrete valuation ring; let w be the valuation of K defined by $R_\mathfrak{p}$. The set of valuations of K which have center $\mathfrak{m}$ in R and are composite with w is finite and not empty, each such valuation v is discrete of rank 2, and κ_v is a finite extension of $R/\mathfrak{m}$.*

Proof: Using B(10.5) we see that $R_{\mathfrak{p}}$ is a discrete valuation ring. Now $R/\mathfrak{p}$ is a one-dimensional local domain. Let $\kappa_w = Q(R/\mathfrak{p})$ be the residue field of w, and let $\{\bar{v}_1, \ldots, \bar{v}_h\}$ be the set of valuations of κ_w belonging to $R/\mathfrak{p}$. These valuations are the only valuations of κ_w dominating $R/\mathfrak{p}$, and they are discrete of rank 1 [cf. II(2.10)]. For $i \in \{1, \ldots, h\}$ we set $v_i := w \circ \bar{v}_i$. Each of the valuations $v_1, \ldots, v_h$ of K is discrete of rank 2 and has center $\mathfrak{m}$ in R; these valuations are the only valuations of K which are composite with w, have rank 2 and have center $\mathfrak{m}$ in R [cf. I(3.12)]. The last statement follows from I(11.9).

(4.8) Proposition: Let $R \in \mathrm{Ineft}_k(K)$ be two-dimensional and regular with maximal ideal $\mathfrak{m}$. Let $\mathfrak{p}$ be a prime ideal of R of height 1, and let w be the valuation of K with valuation ring $R_{\mathfrak{p}}$. Let v be a valuation of K which is composite with w and has center $\mathfrak{m}$ in R; then we have $A_v \subset A_w$. For every $i \in \mathbb{N}$ let R_i be the i-th quadratic dilatation of R with respect to v, and let $\mathfrak{p}_i$ be the center of w in R_i. Then:

(1) v is discrete of rank 2.

(2) R_i is a two-dimensional regular local ring, $\mathfrak{p}_i$ is a prime ideal of height 1 of R_i, and there exists $i_0 \in \mathbb{N}$ such that for every $i \geq i_0$ the ring R_i admits a regular system of parameters $\{x_i, y_i\}$ with $R_i x_i = \mathfrak{p}_i$, and such that $v(x_i) = (1, a_i)$ with $a_i \in \mathbb{Z}$ and $v(y_i) = (0, 1)$.

Proof: (a) For every $i \in \mathbb{N}$ let $\mathfrak{m}_i$ be the maximal ideal of R_i. From (4.7) we see: v is discrete of rank 2 and κ_v is a finite extension of $R/\mathfrak{m}$. Therefore R_i, being a quadratic transform of R_{i-1} [cf. (4.2)], is a two-dimensional regular local ring. Since $\mathfrak{p} = \mathfrak{p}_i \cap R$ is the center of w in R, and since R_i dominates R [cf. VII(6.3)], we have $\mathrm{ht}(\mathfrak{p}_i) = 1$.

(b) Let $\{x, y\}$ be a system of generators of $\mathfrak{m}$. We may assume that $v(x) \leq v(y)$. Then we have $w(x) \leq w(y)$ [since $A_v \subset A_w$]. This implies that $x \notin \mathfrak{p}$, since otherwise we would have $0 < w(x) \leq w(y)$, hence x and y would lie in $\mathfrak{p}$. We set $A := R[y/x]$; let $\mathfrak{q}_v$ be the center of v in A and $\mathfrak{q}_w$ be the center of w in A. Note that $\mathfrak{q}_w \subset \mathfrak{q}_v$. We set $\overline{R} := R/\mathfrak{p}$, $\overline{\mathfrak{m}} := \mathfrak{m}/\mathfrak{p}$, the maximal ideal of $\overline{R}$, $\overline{A} := A/\mathfrak{q}_w$, $\overline{\mathfrak{q}} := \mathfrak{q}_v/\mathfrak{q}_w$ and, for $i \in \mathbb{N}$, $\overline{R}_i := R_i/\mathfrak{p}_i$, $\overline{\mathfrak{m}}_i := \mathfrak{m}_i/\mathfrak{p}_i$. Let $\overline{v}$ be the valuation of κ_w induced by v, and let $\overline{x}$ resp. $\overline{y}$ be the image of x resp. y in κ_w. Then $\overline{R}_i$ is a one-dimensional local domain with field of quotients κ_w and $\overline{v}$ has center $\overline{\mathfrak{m}}_i$ in $\overline{R}_i$, hence $\overline{v}$ is discrete of rank 1 [cf. II(2.10)]. We have an increasing chain $\overline{R} =: \overline{R}_0 \subset \overline{R}_1 \subset \cdots$ of subrings of κ_w having κ_w as common field of quotients; note that $\overline{R}_i$ dominates $\overline{R}_{i-1}$ for every $i \in \mathbb{N}$. We have $R_1 = A_{\mathfrak{q}_v}$, $\mathfrak{m}_1 = \mathfrak{q}_v R_1$, and $\overline{x} \in \overline{R}$ is not zero. Therefore we have $\overline{\mathfrak{m}} = \overline{R}\overline{x} + \overline{R}\overline{y}$, $\overline{A} = \overline{R}[\overline{y}/\overline{x}]$, $\overline{\mathfrak{q}}$ is the center of $\overline{v}$ in $\overline{A}$, and $\overline{R}_1 = \overline{A}_{\overline{\mathfrak{q}}}$, hence $\overline{\mathfrak{m}}\overline{A} = \overline{A}\overline{x}$ and $\overline{\mathfrak{m}}\overline{R}_1 = \overline{R}_1\overline{z}_1$ for some $\overline{z}_1 \in \overline{R}_1$. Similarly we get, for $i \in \mathbb{N}$, that $\overline{\mathfrak{m}}_i\overline{R}_{i+1} = \overline{R}_{i+1}\overline{z}_{i+1}$ with $\overline{z}_{i+1} \in \overline{R}_{i+1}$.

(c) We show that $\bigcup_{i \geq 0} \overline{R}_i = A_{\overline{v}}$, the valuation ring of $\overline{v}$. Let $c \in A_{\overline{v}}$, and write $c = a/b$ with $a, b \in \overline{R}$, $b \neq 0$. If $b \notin \overline{\mathfrak{m}}$, then we have $c \in \overline{R}$. Now we consider the case that $b \in \overline{\mathfrak{m}}$. Since $0 \leq \overline{v}(c) = \overline{v}(a/b)$, we have $\overline{v}(a) \geq \overline{v}(b)$, hence $a \in \overline{\mathfrak{m}}$. We can write $a = a_1\overline{z}_1$, $b = b_1\overline{z}_1$ with $a_1, b_1 \in \overline{R}_1$. We have $c = a_1/b_1$. If $b_1 \notin \overline{\mathfrak{m}}_1$, then we have $c \in \overline{R}_1$. If $b_1 \in \overline{\mathfrak{m}}_1$, then we have $\overline{v}(a_1) \geq \overline{v}(b_1)$, hence $a_1 \in \overline{\mathfrak{m}}$, and we can write $a_1 = a_2 \overline{z}_2$, $b_1 = b_2 \overline{z}_2$ with a_2, $b_2 \in \overline{R}_2$, and we have $a = a_2 \overline{z}_1 \overline{z}_2$, $b = b_2 \overline{z}_1 \overline{z}_2$ and $c = a_2/b_2$. Continuing, we see: If $i \in \mathbb{N}$ and $c \notin \overline{R}_{i-1}$, then we have $a = a_i \overline{z}_1 \cdots \overline{z}_i$, $b = b_i \overline{z}_1 \cdots \overline{z}_i$ with a_i, $b_i \in \overline{R}_i$. For $n \in \{1, \ldots, i\}$ the center of $\overline{v}$ in $\overline{R}_n$ is $\overline{\mathfrak{m}}_n$, hence we have $\overline{v}(\overline{z}_n) > 0$, and therefore we obtain $\overline{v}(b) \geq i$. Thus, this construction cannot have infinitely many steps. Therefore, there exists $i \in \mathbb{N}$ with $c \in \overline{R}_i$, and we have proved the assertion.

(d) Let $i \in \mathbb{N}_0$. Now $\overline{R}_i$ is a one-dimensional local domain. Let $\overline{\mathcal{V}}_i$ be the set of valuations of κ_w belonging to $\overline{R}_i$; $\overline{\mathcal{V}}_i$ is a finite non-empty set of discrete valuations of rank 1, and $\overline{\mathcal{V}}_i \supset \overline{\mathcal{V}}_{i+1}$. Note that $\overline{v} \in \overline{\mathcal{V}}_i$ for every $i \in \mathbb{N}_0$, and that $S_i :=$ $\bigcap_{\nu \in \overline{\mathcal{V}}_i} A_\nu$ is the integral closure of $\overline{R}_i$. We set $\{\nu_1, \ldots, \nu_h\} = \overline{\mathcal{V}}_0 \setminus \{\overline{v}\}$. We consider the case that $h \geq 1$; let $j \in \{1, \ldots, h\}$. There exists $a_j \in A_{\overline{v}}$ with $a_j \notin A_{\nu_j}$ [since $A_{\overline{v}} \not\subset A_{\nu_j}$], and by (c) there exists $i_j \in \mathbb{N}$ with $a_j \in \overline{R}_{i_j}$. We choose $i_0 \in \mathbb{N}$ with $i_0 \geq i_j$ for $j \in \{1, \ldots, h\}$. Then we have $a_j \in \overline{R}_{i_0}$ for $j \in \{1, \ldots, h\}$; this implies that $\overline{\mathcal{V}}_{i_0} = \{\overline{v}\}$. In particular, $A_{\overline{v}}$ is the integral closure of $\overline{R}_{i_0}$. Since $\overline{R}_{i_0}$ is essentially of finite type over the field k, $A_{\overline{v}}$ is a finitely generated $\overline{R}_{i_0}$-module by B(3.6), hence the increasing chain $\overline{R}_0 \subset \overline{R}_1 \subset \cdots \subset A_{\overline{v}}$ becomes stationary, and there exists $i_1 \in \mathbb{N}$ with $\overline{R}_{i_1} = \overline{R}_{i_1+1} = \cdots = A_{\overline{v}}$ [cf. (c)].

(e) Let $i \geq i_1$, and choose $\overline{y}_i \in \overline{R}_i$ with $\overline{v}(\overline{y}_i) = 1$. Let $y_i \in R_i$ be an element with image $\overline{y}_i$. Then we have $v(y_i) = (0, 1)$ [cf. I(3.33)]. Since $\mathfrak{p}_i$ is a prime ideal of height 1 of the factorial domain R_i [note that R_i is a regular local ring], $\mathfrak{p}_i = R_i x_i$ is a principal ideal. Since $(R_i)_{\mathfrak{p}_i} = A_w$ [since $(R_i)_{\mathfrak{p}_i}$ is a discrete valuation ring which is contained in A_w], we have $v(x_i) = (1, a_i)$ with $a_i \in \mathbb{Z}$. We show that $\mathfrak{m}_i = R_i x_i + R_i y_i$. In fact, let $z \in \mathfrak{m}_i$, and let $\overline{z} \in \overline{\mathfrak{m}}_i$ be the image of z_i in $\overline{R}_i$; we can write $\overline{z} = \overline{t}\overline{y}_i$ with $\overline{t} \in \overline{R}_i$. We choose $t \in R_i$ having image $\overline{t}$. Then we get $z - ty_i \in \mathfrak{p}_i$, hence $z \in R_i x_i + R_i y_i$.

5   Valuations of Algebraic Function Fields in Two Variables

(5.0) In this section k is a field of arbitrary characteristic, and K is an algebraic function field over k. Every valuation v of K is tacitly assumed to be trivial on k; remember that, by definition, $\dim(v) = \operatorname{tr.} \mathrm{d}_k(\kappa_v)$.

(5.1) REMARK: Let K be an algebraic function field in one variable over k. There exist non-trivial valuations of K, and every non-trivial valuation of K is discrete of rank 1 [cf. I(10.4)].

(5.2) CLASSIFICATION OF VALUATIONS OF FUNCTION FIELDS IN TWO VARIABLES: Let K be an algebraic function field in two variables over k, and let v be a valuation of K with value group Γ. We have $\dim(v) = 1$ or $\dim(v) = 0$ [cf. I(10.6)].
(1) If $\dim(v) = 1$, then v is a discrete valuation of rank 1 [cf. I(10.7)], hence Γ is isomorphic as an ordered group to the ordered group $\mathbb{Z}$ of integers.

(2) If $\dim(v) = 0$, then we have the two possibilities $\operatorname{rank}(v) = 2$ or $\operatorname{rank}(v) = 1$ [cf. I(10.6)].

(a) If $\operatorname{rank}(v) = 2$, then v is discrete of rank 2 [cf. I(10.6)], i.e., Γ is isomorphic to $\mathbb{Z} \times \mathbb{Z}$, ordered lexicographically.

(b) Now we consider the case $\operatorname{rank}(v) = 1$. In this case Γ is a subgroup of the reals [cf. B(1.21)]. The rational rank of Γ is either 2 or 1 [cf. I(10.6)].

(b1) If $\operatorname{rat.rank}(\Gamma) = 2$, then there exist rationally independent real numbers ρ_1, $\rho_2 \in \Gamma$ such that $\Gamma = \mathbb{Z}\rho_1 \oplus \mathbb{Z}\rho_2$ [cf. I(10.6)].

(b2) It remains the case that $\operatorname{rat.rank}(\Gamma) = 1$. If Γ is discrete, then Γ is isomorphic to $\mathbb{Z}$. If Γ is not discrete, then we may assume that Γ is a subgroup of the rationals [if Γ contains an irrational number ρ, then we have $\mathbb{Z}\rho \neq \Gamma$, but $\Gamma \otimes_{\mathbb{Z}} \mathbb{Q} = \mathbb{Q}\rho$; for every $\sigma \in \Gamma \setminus \mathbb{Z}\rho$ there exist integers m, n with $m \neq 0$ and $\sigma = \frac{n}{m}\rho$, and $\Gamma \cdot 1/\rho \subset \mathbb{Q}$ is order-isomorphic to Γ].

In the rest of this section we show that all the possibilities mentioned above do really occur.

(5.3) REMARK: Let K be an algebraic function field in $r \geq 2$ variables over k. Then K admits valuations of dimension $r - 1$ [cf. I(10.8)]. Such valuations are discrete of rank 1 [cf. I(10.7)], hence this result takes care of case 1 in (5.2).

(5.4) Let $r \in \mathbb{N}$, and let K be the field of rational functions in r variables over k. Let Γ be a totally ordered abelian group. For any r elements $\gamma_1, \ldots, \gamma_r \in \Gamma$ there exists a valuation $v: K \to \Gamma_\infty$ with $v(K^\times) = \mathbb{Z}\gamma_1 + \cdots + \mathbb{Z}\gamma_r$ [cf. I(9.1)].

(1) Let $\Gamma := \mathbb{Z}^r$, endowed with lexicographic ordering; note that $\operatorname{rank}(\Gamma) = \operatorname{rat.rank}(\Gamma) = r$. Let $\{e_1, \ldots, e_r\}$ be the canonical $\mathbb{Z}$-basis of Γ, and let $v: K \to \Gamma_\infty$ be a valuation with $v(K^\times) = \mathbb{Z}e_1 + \cdots + \mathbb{Z}e_r$. Then we have $v(K^\times) = \Gamma$, hence $\operatorname{rank}(v) = \operatorname{rat.rank}(v) = r$ and $\dim(v) = 0$. This takes care of case (2)(a) in (5.2).

(2) Let $\rho_1, \ldots, \rho_r \in \mathbb{R}$ be rationally independent, and let $v: K \to \mathbb{R}_\infty$ be a valuation with $v(K^\times) = \mathbb{Z}\rho_1 + \cdots + \mathbb{Z}\rho_r$; then we have $\operatorname{rank}(v) = 1$, $\operatorname{rat.rank}(v) = r$ and $\dim(v) = 0$. This takes care of case (2)(b1) in (5.2).

(5.5) NON-DISCRETE VALUATIONS OF RANK 1 AND RATIONAL RANK 1: We have to consider the case mentioned in (5.2)(2)(b2).

(1) Let $\Gamma \subset \mathbb{Q}$ be a non-discrete subgroup with $1 \in \Gamma$. We write the elements of Γ always in reduced form m/n with $m \in \mathbb{Z}$, $n \in \mathbb{N}$, $\gcd(m, n) = 1$. We divide the prime numbers which occur in the denominators n of elements of Γ into two classes P and Q: The first class P contains those prime numbers p which occur in the denominators n of elements of Γ only to a bounded power: For $p \in P$ there exists $\mu_p \in \mathbb{N}$ such that p^{μ_p} occurs in the denominator n of some element of Γ, but $p^{\mu_p + 1}$ does not occur in the denominator of any element of Γ. In the second class Q we put the remaining primes. If Q is empty, then P is infinite [since Γ is non-discrete]. We show: Γ *consists of all rational numbers whose denominators are of the form* $p_1^{\mu_1} p_2^{\mu_2} \cdots q_1^{\nu_1} q_2^{\nu_2} \cdots$ *where* $p_1, p_2, \ldots \in P$, $q_1, q_2, \ldots \in Q$, $\mu_1, \mu_2, \ldots$

are non-negative integers with $\mu_1 \leq \mu_{p_1}$, $\mu_2 \leq \mu_{p_2}, \ldots$ and $\nu_1, \nu_2, \ldots$ are arbitrary non-negative integers.

Proof: (a) If $a/b \in \Gamma$ where $a \in \mathbb{Z}$, $b \in \mathbb{N}$ and $\gcd(a,b) = 1$, then there exist integers a', b' with $aa' + bb' = 1$, hence $1/b = b' + a'(a/b) \in \Gamma$ [since $1 \in \Gamma$ by assumption], and therefore all integral multiples of $1/b$ belong to Γ.

(b) If Γ contains a_1/b_1, a_2/b_2 with a_1, $a_2 \in \mathbb{Z}$, b_1, $b_2 \in \mathbb{N}$ and $\gcd(a_1,b_1) = \gcd(a_2,b_2) = 1$, and if $\gcd(b_1,b_2) = 1$, then $(a_1 b_2 + a_2 b_1)/b_1 b_2 \in \Gamma$, and since $a_1 b_2 + a_2 b_1$ and $b_1 b_2$ are relatively prime, we have $1/(b_1 b_2) \in \Gamma$ by (a).

Now the assertion follows from (a) and (b).

(2) Given a non-discrete subgroup Γ of $\mathbb{Q}$ as in (1), we construct a valuation v with value group Γ of $K := k(x,y)$, the field of rational functions in two variables over k, as follows: we set

$$P' := \{p_1^{\mu_{p_1}}, p_2^{\mu_{p_2}}, \ldots \mid p_1, p_2, \ldots \in P\},$$
$$Q' := \{q_1^{\nu_1}, q_2^{\nu_2}, \ldots \mid q_1, q_2, \ldots \in Q, \nu_1, \nu_2, \ldots \in \mathbb{N}\}.$$

The set $P' \cup Q'$ is countable; we write $P' \cup Q' = \{m_1, m_2, \ldots\}$. Then, by the assertion in (1), it is clear that

$$\Gamma = \left\{ \frac{n_s}{m_1 \cdots m_s} \mid n_s \in \mathbb{Z}, s \in \mathbb{N} \right\}$$

where now the elements $n_s/m_1 \cdots m_s$ in general are not in reduced form.

(a) Let $(c_i)_{i \in \mathbb{N}}$ be a sequence in $k^{\times}$. We define a sequence $(x_i)_{i \in \mathbb{N}}$ in K by recursion, as follows. We set $x_1 := x$, $x_2 := y$, and $x_{i+2} := (x_i - c_i x_{i+1}^{m_i})/x_{i+1}^{m_i}$ for every $i \in \mathbb{N}$. We set $R_i := k[x_i, x_{i+1}]$, $\mathfrak{q}_i := R_i x_i + R_i x_{i+1} \subset R_i$ for every $i \in \mathbb{N}$. Let $i \in \mathbb{N}$; since $x_i = x_{i+1}^{m_i}(x_{i+2} + c_i) \in \mathfrak{q}_{i+1} \subset R_{i+1}$, we have $R_1 \subset R_2 \subset \cdots$, and we see that $\{x_i, x_{i+1}\}$ is a transcendence basis of K over k, hence $\mathfrak{q}_i$ is a maximal ideal of R_i, $\mathrm{ht}(\mathfrak{q}_i) = 2$, $\mathfrak{q}_{i+1} \cap R_i = \mathfrak{q}_i$, $S_i = (R_i)_{\mathfrak{q}_i}$ is a two-dimensional regular local ring and $\{x_i, x_{i+1}\}$ is a regular system of parameters of S_i, and if $\mathfrak{n}_i$ denotes the maximal ideal of S_i, then we have $S_i \subset S_{i+1}$, $\mathfrak{n}_{i+1} \cap S_i = \mathfrak{n}_i$ and $S_i/\mathfrak{n}_i = k$. We set $S := \bigcup_{i \geq 1} S_i$, $\mathfrak{n} := \bigcup_{i \geq 1} \mathfrak{n}_i$; then S is a proper quasilocal subring of K, $\mathfrak{n}$ is its maximal ideal, and we have $S/\mathfrak{n} = k$. There exists a valuation v of K which has center $\mathfrak{n}$ in S [cf. I(3.5)]; set $\Delta := v(K^{\times})$. Then we have $v(x_i - c_i x_{i+1}^{m_i}) > v(x_{i+1}^{m_i})$ for every $i \in \mathbb{N}$ [since $v(x_{i+2}) > 0$], whence $v(x_i) = m_i v(x_{i+1})$ and therefore

$$v(x_1) = m_1 m_2 \cdots m_i v(x_{i+1}) \quad \text{for every } i \in \mathbb{N}. \tag{$*$}$$

We have $\mathrm{rank}(\Delta) = 2$ or $\mathrm{rank}(\Delta) = 1$ [cf. I(10.6)]. If $\mathrm{rank}(\Delta) = 2$, then Δ is isomorphic to $\mathbb{Z} \times \mathbb{Z}$, ordered lexicographically, and if $\mathrm{rank}(\Delta) = 1$ and $\mathrm{rat.\,rank}(\Delta) = 2$, then Δ is isomorphic as a $\mathbb{Z}$-module to $\mathbb{Z}^2$ [cf. I(10.6)]. Since $v(x_{i+1}) = v(x_1)/m_1 m_2 \cdots m_i$ for every $i \in \mathbb{N}$ and $v(x_1) > 0$, and since the sequence $(m_1 \cdots m_i)_{i \geq 1}$ is not bounded above, the two possibilities for Δ just mentioned are excluded. Therefore we have $\mathrm{rat.\,rank}(\Delta) = 1$. Thus, replacing Δ by $(1/v(x_1))\Delta$, we can consider Δ as a non-discrete subgroup of $\mathbb{Q}$ with $v(x_1) = 1 \in \Delta$.

(b) *We show that $\Delta = \Gamma$ and that $A_v = S$.*

(i) Let $i \in \mathbb{N}$; we show that S_{i+1} is a quadratic dilatation of S_i with respect to v. In fact, set $\xi^{(1)} := x_{i+1}$, $\eta^{(1)} := x_i$, and define recursively

$$\xi^{(j+1)} := \xi^{(j)}, \quad \eta^{(j+1)} := \eta^{(j)}/\xi^{(j)} \quad \text{for } j \in \{1, \dots, m_i - 2\},$$

$$\xi^{(m_i)} := \xi^{(m_i-1)}, \quad \eta^{(m_i)} := (\eta^{(m_i-1)} - c_i \xi^{(m_i-1)})/\xi^{(m_i)}.$$

Then we have $\xi^{(j)} = x_{i+1}$ for $j \in \{1, \dots, m_i\}$ and $\eta^{(j)} = x_i/x_{i+1}^j$ for $j \in \{1, \dots, m_i - 1\}$, whereas $\eta^{(m_i)} = x_{i+2}$. Note that, for $j \in \{1, \dots, m_i\}$, $\{\xi^{(j)}, \eta^{(j)}\}$ is a transcendence basis of K over k. We have

$$v(\eta^{(j)}) = (m_i - j)v(\xi^{(j)}) > v(\xi^{(j)}) \quad \text{for } j \in \{1, \dots, m_i - 1\}.$$

Set $A^{(j)} := k[\xi^{(j)}, \eta^{(j)}]$, $\mathfrak{p}^{(j)} := \xi^{(j)} A^{(j)} + \eta^{(j)} A^{(j)}$, $T^{(j)} := (A^{(j)})_{\mathfrak{p}^{(j)}}$ for $j \in \{1, \dots, m_i\}$. Then we have $A^{(j+1)} = A^{(j)}[\eta^{(j)}/\xi^{(j)}] \subset T^{(j)}[\eta^{(j)}/\xi^{(j)}] \subset T^{(j+1)}$, hence $T^{(j+1)}$ is the first quadratic dilatation of $T^{(j)}$ with respect to v for $j \in \{1, \dots, m_i - 1\}$, and therefore $S_{i+1} = T^{(m_i)}$ is a quadratic dilatation of $S_i = T^{(1)}$ with respect to v. Now S is the valuation ring A_v by (4.4).

(ii) Let $f(x_1, x_2) \in k[x_1, x_2]$ be a non-zero polynomial with $f(0,0) = 0$. We show that there exist $h \in \mathbb{N}$ and $n \in \mathbb{N}$ with

$$f(x_1, x_2) = x_h^n f_h(x_h, x_{h+1}) \quad \text{with } f_h(x_h, x_{h+1}) \in R_h, \ f_h(0,0) \neq 0. \quad (**)$$

We choose $\mu \in \mathbb{N}$ with $v(x_1^\mu) \geq v(f(x_1, x_2))$, and we set $z := x_1^\mu/f(x_1, x_2)$. Then we have $z \in A_v$, hence by (i) there exists $h \in \mathbb{N}$ with $z \in S_h$, i.e., there exist polynomials $p(x_h, x_{h+1})$, $q(x_h, x_{h+1}) \in R_h$ with $q(0,0) \neq 0$ and with $z = p(x_h, x_{h+1})/q(x_h, x_{h+1})$. Since we have $x_1 = x_h^\nu g(x_h, x_{h+1})$ with $\nu \in \mathbb{N}$ and $g(x_h, x_{h+1}) \in R_h$, $g(0,0) \neq 0$ [as is easily seen by induction], we have

$$f(x_1, x_2) p(x_h, x_{h+1}) = x_h^{\mu\nu} g(x_h, x_{h+1})^\mu q(x_h, x_{h+1})$$

$$= x_h^{\mu\nu} r(x_h, x_{h+1}) \quad \text{where } r(x_h, x_{h+1}) \in R_h, \ r(0,0) \neq 0.$$

Let us consider the factorization of $f(x_1, x_2)$ in R_h into a product of irreducible factors. The last displayed equation shows that the only irreducible factor $g(x_h, x_{h+1})$ of $f(x_1, x_2)$ in R_h with $g(0,0) = 0$ is a power of x_h; this implies the assertion.

(iii) Let $f(x_1, x_2) \in k[x_1, x_2]$ with $f(0,0) \neq 0$. It is easily seen by induction: For every $h \in \mathbb{N}$ we can write

$$f(x_1, x_2) = x_h^n f_h(x_h, x_{h+1}) \quad \text{with } f_h(x_h, x_{h+1}) \in R_h, \ f_h(0,0) \neq 0, \ n \in \mathbb{N}.$$

(iv) We have to show that $\Delta = \Gamma$. By (*) in (a) we have $\Delta \supset \Gamma$. A non-zero element $z \in K$ can be written as a quotient $z = f(x_1, x_2)/g(x_1, x_2)$ with polynomials $f(x_1, x_2)$, $g(x_1, x_2) \in k[x_1, x_2]$. By what we have just shown in (ii) and (iii), there exist $h \in \mathbb{N}$, polynomials $f_h(x_h, x_{h+1})$, $g_h(x_h, x_{h+1}) \in R_h$ with $f_h(0,0) \neq 0$, $g_h(0,0) \neq 0$ and $n \in \mathbb{Z}$ with

$$z = \frac{f(x_1, x_2)}{g(x_1, x_2)} = x_h^n \frac{f_h(x_h, x_{h+1})}{g_h(x_h, x_{h+1})}.$$

We have $v(x_h) = 1/(m_1 \cdots m_{h-1})$ for $h \in \mathbb{N}$ [cf. (*) in (a) above], hence $v(z)$ is an integral multiple of $1/(m_1 \cdots m_{h-1})$, i.e., $v(z) \in \Gamma$, hence $\Gamma \supset \Delta$, and therefore we have $\Delta = \Gamma$.

6 Uniformization

(6.0) In this section K is a field; sometimes we specify a subfield k of K. We keep the notation $\mathrm{lneft}_k(K)$ introduced in (4.0).

6.1 Classification of Valuations and Local Uniformization

(6.1) CLASSIFICATION OF VALUATIONS: Let R be a two-dimensional regular local ring with maximal ideal $\mathfrak{m}$, field of quotients K and residue field κ which we assume to be *algebraically closed*. We classify the valuation of K which dominate R.

Let v be a valuation of K dominating R; let V be the valuation ring of v. We have $\mathrm{tr.\,d}_\kappa(\kappa_v) = 1$ or $\mathrm{tr.\,d}_\kappa(\kappa_v) = 0$ [cf. I(11.9)].

(1) If $\mathrm{tr.\,d}_\kappa(\kappa_v) = 1$, then v is a discrete valuation of rank 1 [cf. I(11.9)], hence Γ_v is isomorphic as an ordered group to the ordered group $\mathbb{Z}$ of integers.

(2) If $\mathrm{tr.\,d}_\kappa(\kappa_v) = 0$, then we have the two possibilities $\mathrm{rank}(v) = 2$ or $\mathrm{rank}(v) = 1$ [cf. I(11.9)].

(a) If $\mathrm{rank}(v) = 2$, then v is discrete of rank 2 [cf. I(11.9)], i.e., Γ_v is isomorphic to $\mathbb{Z} \times \mathbb{Z}$, ordered lexicographically. The isolated subgroup of Γ_v of rank 1 is $\{0\} \times \mathbb{Z}$ [cf. B(1.25)]; then we can write $v = w \circ \overline{v}$ where w is a discrete valuation of K of rank 1 and $\overline{v}$ is a discrete valuation of κ_w [cf. chapter I, (3.23) and (3.26)]. Let W be the valuation ring of w; then V is contained in W, hence W contains R. If W does not dominate R, then $\mathfrak{p} := \mathfrak{m}(W) \cap R$ is a prime ideal of R of height 1, hence $R_\mathfrak{p}$ is a discrete valuation ring, we have $R_\mathfrak{p} = W$, and v is a valuation of the type described in (4.7).

Now we consider the case that W dominates R. Let $\{x, y\}$ be a system of generators of $\mathfrak{m}$, and assume that $v(x) \leq v(y)$. Let g be a directional form and let $p = (g)$ be the directional ideal of v; note that g is linear since κ is algebraically closed. If $W = V_R$ where V_R is the valuation ring of K defined by the order function of R, then we have $\kappa_w = \kappa(\zeta)$ where ζ is the v-image of y/x [cf. VII(2.3)], and $\overline{v}(g(1, \zeta)) = 0$, and therefore we have $v = \nu_p$ where ν_p is the valuation defined in VII(7.5). In this case we can choose x and y in such a way that $v(x) = (1, 0)$ and $\mathrm{In}(y) = p$, hence that $v(y) = (1, 1)$. Now let R_1 be the quadratic dilatation of R with respect to v; note that $R_1 = S_p$ is a quadratic transform of R [cf. (4.1)], and that $\{y, y/x\}$ is a regular system of parameters of R_1; in particular, we have $v(y/x) = (0, 1)$ and $R_1 \subset W$.

Again we have the two possibilities: Either W does not dominate R_1 in which case W is the localization of R_1 with respect to a prime ideal of height 1, hence v is a valuation of the type discussed in (4.7), or W dominates R_1. Then we have either $W = V_{R_1}$, hence v is of the type discussed in VII(7.5) with respect to R_1, or we have $W \neq V_{R_1}$. In the latter case we repeat the construction. Taking into account

VII(6.10) and (4.8), we can state: *There exists a quadratic dilatation S of R with respect to v which is a two-dimensional regular local ring having a regular system of parameters $\{x, y\}$ with $v(x) = (1, a)$ with $a \in \mathbb{Z}$ and $v(y) = (0, 1)$.*

(b) Now we consider the case $\text{rank}(v) = 1$. In this case Γ_v is a subgroup of the additive group $\mathbb{R}$ [cf. B(1.21)]. The rational rank of Γ_v is either 2 or 1 [cf. I(11.9)].

(b1) If $\text{rat.rank}(\Gamma_v) = 2$, then there exist rationally independent real numbers ρ_1, $\rho_2 \in \Gamma_v$ such that $\Gamma_v = \mathbb{Z}\rho_1 \oplus \mathbb{Z}\rho_2$ [cf. I(11.9)].

(b2) It remains the case that $\text{rat.rank}(\Gamma_v) = 1$. If Γ_v is discrete, then Γ_v is isomorphic to $\mathbb{Z}$. If Γ_v is not discrete, then we may assume that Γ_v is a subgroup of the rationals [cf. (5.2)(2)(b2)].

(6.2) Theorem: [Local uniformization] *Let $R \in \text{lneft}_\kappa(K)$ be two-dimensional and regular with maximal ideal $\mathfrak{m}$ and algebraically closed residue field κ. Let v be a valuation of K dominating R and with $\text{tr.d}_\kappa(\kappa_v) = 0$. Let $f \in R$ be non-zero. Then there exist a quadratic dilatation S of R with respect to v which is a two-dimensional regular local ring and a regular system of parameters $\{x, y\}$ of S such that $f = x^a y^b u$ where a, $b \in \mathbb{N}_0$, u is a unit of S, and where the following conditions are satisfied:*

(a) *If $\text{rank}(v) = 2$, then $\Gamma_v = \mathbb{Z} \times \mathbb{Z}$, ordered lexicographically, and we have $v(x) = (1, c)$ with $c \in \mathbb{Z}$, $v(y) = (0, 1)$; in particular $\{v(x), v(y)\}$ is a $\mathbb{Z}$-basis of Γ_v.*

(b) *If $\text{rank}(v) = 1$, then*

(b1) *if $\text{rat.rank}(v) = 2$, then either $v(x)$ and $v(y)$ are rationally independent — and in this case $\{v(x), v(y)\}$ is a $\mathbb{Z}$-basis of Γ_v — or $v(x)$ and $v(y)$ are rationally dependent—and in this case we have $b = 0$,*

(b2) *if $\text{rat.rank}(v) = 1$, then we have $b = 0$.*

Proof: (a) We assume that $\text{rank}(v) = 2$. Then v is discrete of rank 2 by I(11.9); we may assume that $\Gamma_v = \mathbb{Z} \times \mathbb{Z}$, ordered lexicographically. By (6.1)(2)(a) there exists a quadratic dilatation S_0 of R with respect to v which is a two-dimensional regular local ring and a system of generators $\{x_0, y_0\}$ of the maximal ideal $\mathfrak{n}_0$ of S_0 such that $v(x_0) = (1, c)$ with $c \in \mathbb{Z}$ and $v(y_0) = (0, 1)$. For every $i \in \mathbb{N}$ let S_i be the i-th quadratic dilatation of S_0 with respect to v. We have $v(x_0) > iv(y_0)$ for every $i \in \mathbb{N}$, hence, for $i \in \mathbb{N}$, $\{x_i := x_0/y_0^i, y_i := y_0\}$ is a system of generators for the maximal ideal $\mathfrak{n}_i$ of S_i. For $\mu \neq \mu' \in \mathbb{N}_0$ and ν, $\nu' \in \mathbb{N}_0$ the v-values of $x_0^\mu y_0^\nu$ and $x_0^{\mu'} y_0^{\nu'}$ are not equal, and for $\mu \in \mathbb{N}_0$ and $\nu \neq \nu' \in \mathbb{N}_0$ the v-values of $x_0^\mu y_0^\nu$ and $x_0^\mu y_0^{\nu'}$ are not equal. Since every non-zero element $h \in S_0$ is a linear combination of monomials $x_0^\mu y_0^\nu$ with coefficients which are units of S_0, we see that $v(h) = (m, n)$ with $m > 0$ iff x_0 divides h [remember that S_0 is factorial since it is a regular local ring]. In S_0 we write $f = x_0^{e_0} g_0$ with $e_0 \in \mathbb{N}_0$, $g_0 \in S_0$ and x_0 does not divide g_0. Then we have $v(g_0) = (0, n)$ with $n \in \mathbb{Z}$. By induction, we define elements $g_i \in S_i$, $i \in \mathbb{N}$, by writing $g_{i-1} = y_i^{e_i} g_i$ with $e_i \in \mathbb{N}_0$, $g_i \in S_i$ and y_i does not divide g_i. We have $\mathfrak{n}_{i-1} S_i = S_i y_i$ for $i \in \mathbb{N}$. Let $i \in \mathbb{N}$; if $g_{i-1} \in \mathfrak{n}_{i-1}$, then we have $e_i > 0$. In particular, if $l \in \mathbb{N}$ and $g_{i-1} \in \mathfrak{n}_{i-1}$ for $i \in \{1, \ldots, l\}$, then we have $v(g_0) \geq (0, l)$. Therefore there exists $l_0 \in \mathbb{N}$ such that g_{l_0} is a unit of S_{l_0}. Since $x_0 = x_{l_0} y_{l_0}^{l_0}$ and $y_i = y_{l_0}$ for $i \in \{0, \ldots, l_0\}$, we see that we have $f = x_{l_0}^a y_{l_0}^b u$ with $a, b \in \mathbb{N}_0$ and a unit $u \in S_{l_0}$.

(b) Now we have $\operatorname{rank}(v) = 1$. We set $R_0 := R$, $\mathfrak{m}_0 := \mathfrak{m}$. Let $(R_i)_{i \geq 0}$ be the sequence of quadratic dilatations of R with respect to v. By (4.4) we have $\bigcup_{i \geq 0} R_i = A_v$. Let $i \in \mathbb{N}_0$; R_{i+1} is a quadratic transform of R_i [cf. (4.1)], and we can choose a system of generators $\{x_i, y_i\}$ of the maximal ideal $\mathfrak{m}_i$ of R_i such that $\mathfrak{m}_i R_{i+1} = R_{i+1} x_{i+1}$ and $y_{i+1} = (y_i - u_i x_i)/x_i$, i.e., we have $x_i = x_{i+1}$ and $y_i = x_i(y_{i+1} + u_i)$, or that $x_{i+1} = y_i$ and $y_{i+1} = (x_i - u_i y_i)/y_i$, i.e., we have $y_i = x_{i+1}$, $x_i = y_i(x_{i+1} + u_i)$; here u_i is a unit of R_{i-1} [cf. proof of VII(2.19) and note that κ is algebraically closed, hence that the irreducible polynomials in $\operatorname{gr}_{\mathfrak{m}_i}(R_i)$ are linear].

We set $f_0 := f$. We define $f_i \in R_i$ inductively in the following way: If $i \in \mathbb{N}$ and $f_{i-1} \in R_{i-1}$ is defined, then we define $f_i \in R_i$ by $f_{i-1} = x_i^{m_i} f_i$ where x_i does not divide f_i [remember that R_i is factorial since it is a regular local domain]. We consider the case that f_i is not a unit of R_i; let $f_i = f_{i1} \cdots f_{ih_i}$ be a factorization of f_i in R_i into a product of irreducible elements. For $j \in \{1, \ldots, h_i\}$ the ideal $R_i f_{ij}$ is a prime ideal of height 1 of R_i; let w_{ij} be the valuation of K defined by the discrete valuation ring $(R_i)_{(f_{ij})}$. Let $\mathcal{V}_i$ be the set of valuations of K which have center $\mathfrak{m}_i$ in R_i and are composite with w_{ij} for some $j \in \{1, \ldots, h_i\}$. Now $\mathcal{V}_i$ is a finite non-empty set by (4.7). Let $\nu \in \mathcal{V}_i$; then we have $\nu = w_{ij} \circ \overline{v}$ for some w_{ij} with $j \in \{1, \ldots, h_i\}$, and ν is a valuation of rank 2. We have $w_{ij}(x_i) = 0$. Let $\mathfrak{p}_{ij}$ be the center of w_{ij} in R_{i-1}; note that $\mathfrak{p}_{ij}$ is not the zero ideal [since w_{ij} is not the trivial valuation of K], and that $x_i \notin \mathfrak{p}_{ij}$. Therefore we see that $\mathfrak{p}_{ij} \neq \mathfrak{m}_{i-1}$, hence we get $\operatorname{ht}(\mathfrak{p}_{ij}) = 1$. We have $f_{i-1} \in \mathfrak{p}_{ij}$, hence f_{i-1} is not a unit of R_{i-1}. Let $f_{i-1} = f_{i-1,1} \cdots f_{i-1,h_{i-1}}$ be the factorization of f_{i-1} in R_{i-1} into a product of irreducible elements. Since $\mathfrak{p}_{ij}$ is a principal prime ideal [note that R_{i-1} is a factorial domain], there exists $j' \in \{1, \ldots, h_{i-1}\}$ such that $\mathfrak{p}_{ij} = R_{i-1} f_{i-1,j'}$, hence that the valuation $w_{i-1,j'}$ of K defined by the discrete valuation ring $(R_{i-1})_{(f_{i-1,j'})}$ is the valuation w_{ij}, i.e., we have $\nu \in \mathcal{V}_{i-1}$. Thus, we have shown that $\mathcal{V}_i \subset \mathcal{V}_{i-1}$. Suppose that $\bigcap_{i \geq 0} \mathcal{V}_i \neq \emptyset$, and let ω be a valuation in this intersection. Now ω dominates R_i for every $i \in \mathbb{N}$, hence we get $A_v = A_\omega$ by (4.4). We have $\operatorname{rank}(v) = 1$ and $\operatorname{rank}(\omega) = 2$. This contradiction shows that the intersection is empty.

For every $i \in \mathbb{N}$ the set $\mathcal{V}_i$ is finite; therefore there exists $i_0 \in \mathbb{N}$ with $\mathcal{V}_i = \emptyset$ for every $i \geq i_0$. This means that f_{i_0} is a unit of R_{i_0}. Therefore we have $f = x_{i_0}^{a_1} y_{i_0}^{b_1} u'$ where $a_1, b_1 \in \mathbb{N}_0$ and u' is a unit of R_{i_0}.

We consider separately the following two cases.

Case (i): $\operatorname{rat.rank}(v) = 1$, or $\operatorname{rat.rank}(v) = 2$ and $v(x_{i_0})$, $v(y_{i_0}) \in \Gamma_v \subset \Gamma_v \otimes_{\mathbb{Z}} \mathbb{Q}$ are linearly dependent over $\mathbb{Q}$ [we identify Γ_v with its image in $\Gamma_v \otimes_{\mathbb{Z}} \mathbb{Q}$]. If $\operatorname{rat.rank}(v) = 1$, then we may assume, in addition, that $\Gamma_v \subset \mathbb{Q}$ [cf. (6.1)(2)(b2)]. Therefore, in both cases we can write $v(x_{i_0})/v(y_{i_0}) = n_0/n_1$ with coprime positive integers n_0, n_1. If $n_0 = n_1 = 1$, then there exists a system of parameters in R_{i_0+1} of the form $\{x_{i_0+1} = x_{i_0}, y_{i_0+1} = (y_{i_0} - u^* x_{i_0})/x_{i_0}\}$ where u^* is a unit of R_{i_0}, hence we get $f = x_{i_0+1}^a u$ where $a = a_1 + b_1$ and $u \in R_{i_0+1}$ is a unit. In the general case we have $v(x_{i_0}) \neq v(y_{i_0})$. Let

$$n_0 = s_1 n_1 + n_2, \ n_1 = s_2 n_2 + n_3, \ldots, n_{k-1} = s_k n_k$$

with $n_1 > n_2 > \cdots > n_k = 1$ be the Euclidean algorithm for the integers n_0, n_1. We set

$$z_0 := x_{i_0}, \ z_1 := y_{i_0}, \ z_{i+1} := \frac{z_{i-1}}{z_i^{s_i}} \quad \text{for } i \in \{1, \ldots, k-1\};$$

then we have

$$\frac{v(z_i)}{v(z_{i+1})} = \frac{n_i}{n_{i+1}} \quad \text{for } i \in \{0, \ldots, k-1\},$$

hence we have $v(z_{k-1}) = v(z_k^{s_k})$. Furthermore, we set $\xi_0 := x_{i_0}, \eta_0 := y_{i_0}, s_0 := 0$ and

$$\xi_{s_1 + \cdots + s_{i-1} + j} := z_i, \ \eta_{s_1 + \cdots + s_{i-1} + j} := \frac{z_{i-1}}{z_i^j} \quad \text{for } i \in \{1, \ldots, k\}, j \in \{1, \ldots, s_i\}.$$

We set $t := s_1 + \cdots + s_k$. The sequence of quadratic dilatations of $S_0 := R_{i_0}$ with respect to v takes the form

$$S_0 \subset S_1 \subset \cdots \subset S_{s_1} \subset S_{s_1 + 1} \subset \cdots \subset S_{s_1 + s_2} \subset \cdots \subset S_{t-1}$$

where, for $i \in \{1, \ldots, k-1\}$, $j \in \{1, \ldots, s_i\}$, and for $i = k$, $j \in \{1, \ldots, s_k - 1\}$, $\{\xi_i, \eta_i\}$ is a regular system of parameters in S_i. We have $\xi_{t-1} = z_k$, $\eta_{t-1} = z_{k-1}/z_k^{s_k - 1}$. We choose a unit $u_{t-1} \in S_{t-1}$ with $v(z_{k-1}/z_k^{s_k} - u_{t-1}) > 0$, and we set $\xi_t := \xi_{t-1}$, $\eta_t := (\eta_{t-1} - u_{t-1}\xi_t)/\xi_t$; then $\{\xi_t, \eta_t\}$ is a regular system of parameters for the quadratic dilatation S_t of S_{t-1} with respect to v. Note that $\eta_{t-1} = \xi_t(\eta_t + u_{t-1})$. Now it is easy to check that we have $\xi_0 = \xi_{t-1}^{a'} \eta_{t-1}^{b'}$, $\eta_0 = \xi_{t-1}^{a''} \eta_{t-1}^{b''}$ with $a', a'', b', b'' \in \mathbb{N}_0$. Therefore we have $x_{i_0} = x_{i_0 + t}^{a_2} u_1$, $y_{i_0} = x_{i_0 + t}^{b_2} u_2$ where $a_2, b_2 \in \mathbb{N}$ and u_1, u_2 are units of $R_{i_0 + t}$, hence we have $f = x_{i_0 + t}^a u$ with $a := a_1 a_2 + b_1 b_2 \geq 0$, $u := u' u_1 u_2$, and u is a unit of $R_{i_0 + t}$.

Case (ii): $\text{rat. rank}(v) = 2$ and the elements $v(x_{i_0})$, $v(y_{i_0})$ are linearly independent over $\mathbb{Q}$, i.e., we have $\Gamma_v \otimes_{\mathbb{Z}} \mathbb{Q} = \mathbb{Q}v(x_{i_0}) + \mathbb{Q}v(y_{i_0})$. It is enough to show that $\{p := v(x_{i_0}), q := v(y_{i_0})\}$ is a $\mathbb{Z}$-basis of Γ_v. Since $p \neq q$, we may assume that $p < q$. Let Σ be a system of representatives of $\kappa = R_{i_0}/\mathfrak{m}_{i_0}$ in R_{i_0}. Let $z \in R_{i_0}$ and set $r := v(z)$. Since $p > 0$, there exist $l \in \mathbb{N}$ with $r < lp$. There exists a polynomial $F = \sum_{i+j \leq l} \sigma_{ij} X^i Y^j \in R_{i_0}[X, Y]$ of degree $\leq l$ with coefficients $\sigma_{ij} \in \Sigma$ such that $z^* := z - F(x_{i_0}, y_{i_0}) \in \mathfrak{m}_{i_0}^l$. We have $v(\mathfrak{m}_{i_0}^l) = lp > r$, hence $v(z^*) > r$, and therefore we have $v(z) = v(F(x_{i_0}, y_{i_0}))$ [cf. I(2.10)]. Since $v(\sigma_{ij}) = 0$ if $\sigma_{ij} \neq 0$, and since p and q are linearly independent over $\mathbb{Q}$, the elements in the set $\{\sigma_{ij} x_{i_0}^i y_{i_0}^j \mid i, j \in \{0, \ldots, l\}, \sigma_{ij} \neq 0\}$ have pairwise different v-values, and therefore there exist $s, t \in \{0, \ldots, l\}$ with $v(\sigma_{st} x_{i_0}^s y_{i_0}^t) < v(\sigma_{ij} x_{i_0}^i y_{i_0}^j)$ for all $i, j \in \{0, \ldots, l\}$ with $\sigma_{ij} \neq 0$ and $(i, j) \neq (s, t)$. This means that $r = v(z) = v(F(x_{i_0}, y_{i_0})) = v(\sigma_{st} x_{i_0}^s y_{i_0}^t) = ps + qt$ [cf. I(2.10)]. Therefore we have shown that $v(R_{i_0} \setminus \{0\}) \subset \mathbb{N}_0 p + \mathbb{N}_0 q$, hence that $v(K^{\times}) = \mathbb{Z}p + \mathbb{Z}q$.

6.2   Existence of Subrings Lying Under a Local Ring

(6.3) Proposition: Let K be an algebraic function field in r variables over a perfect field k. Let L be a finite extension of K, let $S \in \mathrm{lneft}_k(L)$ with $\dim(S) = r$, and let $\mathfrak{n}$ be the maximal ideal of S. The following statements are equivalent:
(1) There exists $R \in \mathrm{lneft}_k(K)$ such that S lies over R.
(2) $(K \cap \mathfrak{n})S$ is an $\mathfrak{n}$-primary ideal of S.
(3) There exist elements $x_1, \ldots, x_n \in K \cap \mathfrak{n}$ such that $Sx_1 + \cdots + Sx_n$ is an $\mathfrak{n}$-primary ideal of S.
If these conditions are satisfied, then R is determined uniquely, namely $R = S \cap K$, and we have $\dim(R) = r$.

Proof: Clearly (2) and (3) are equivalent since S is noetherian.
(1) $\Rightarrow$ (3): Let $\mathfrak{m}$ be the maximal ideal of R. Let B be the integral closure of R in L which is a k-algebra essentially of finite type [cf. B(3.6)]. Let $\mathfrak{n}_1, \ldots, \mathfrak{n}_l$ be the maximal ideals of B labelled in such a way that $S = B_{\mathfrak{n}_1}$. Then we have $\mathfrak{m}B = \mathfrak{q}_1 \cap \cdots \cap \mathfrak{q}_l$ where $\mathfrak{q}_i$ is $\mathfrak{n}_i$-primary for $i \in \{1, \ldots, l\}$. Now we have $\mathfrak{m}S = \mathfrak{q}_1 S$, hence $\mathfrak{m}S$ is $\mathfrak{n}$-primary; since $\mathfrak{n} \cap R = \mathfrak{m}$ [cf. B(7.3)(2)], we can take for $x_1, \ldots, x_n$ a system of generators of $\mathfrak{m}$.
(3) $\Rightarrow$ (1): We choose elements $x_1, \ldots, x_n \in K \cap \mathfrak{n}$ such that $\mathfrak{q} := Sx_1 + \cdots + Sx_n$ is an $\mathfrak{n}$-primary ideal of S. We set $B' := k[x_1, \ldots, x_n]$; then $K' := k(x_1, \ldots, x_n)$ is the field of quotients of B', and $\mathfrak{q}' := \mathfrak{q} \cap B' = B'x_1 + \cdots + B'x_n$ is a maximal ideal of B' [since $B'/\mathfrak{q}' = k$]. Let A' be the integral closure of B' in K'; A' is a k-algebra of finite type by B(3.6). Note that $A' \subset S$. We set $\mathfrak{p}' := A' \cap \mathfrak{n}$; since $\mathfrak{p}' \cap B' = \mathfrak{n} \cap B' = \mathfrak{q}$, we see that $\mathfrak{p}'$ is a maximal ideal of A' [cf. [63], Cor. 4.17]. We set $R' := A'_{\mathfrak{p}'}$; then $\mathfrak{m}' := \mathfrak{p}'R'$ is the maximal ideal of R'. Moreover, we have $\dim(R') = \mathrm{tr.\,d}_k(K') =: s$ [cf. [63], Th. A, p. 286]. Note that $R' \in \mathrm{lneft}_k(K')$, and that $R' \subset S$. We have $\mathfrak{q}'S \subset \mathfrak{m}'S \subset \mathfrak{n}$, and $\mathfrak{q}'S = \mathfrak{q}$ is $\mathfrak{n}$-primary; hence $\mathfrak{m}'S$ is $\mathfrak{n}$-primary also. Since $\dim(R') = s$, there exists an $\mathfrak{m}'$-primary ideal $\mathfrak{a}$ in R' which can be generated by s elements [cf. [63], Cor. 10.7], and there exists $m \in \mathbb{N}$ with $\mathfrak{m}'^m \subset \mathfrak{a} \subset \mathfrak{m}'$. Also, there exists $n \in \mathbb{N}$ with $\mathfrak{n}^n \subset \mathfrak{m}'S$, hence $\mathfrak{n}^{nm} \subset \mathfrak{m}'^m S \subset \mathfrak{a}S \subset \mathfrak{n}$, and therefore $\mathfrak{a}S$ is $\mathfrak{n}$-primary. This implies, by [63], Cor. 10.7, that $r = \dim(S) \leq s$. On the other hand, we have $s = \mathrm{tr.\,d}_k(K') \leq \mathrm{tr.\,d}_k(K) = r$, hence $r = s$, and therefore L is a finite extension of K', and also K is a finite extension of K'. Let B be the integral closure of A' in L; $B \subset S$ is a k-algebra of finite type by B(3.6), and since $(\mathfrak{n} \cap B) \cap A' = \mathfrak{p}'$, the ideal $\mathfrak{n} \cap B$ is a maximal ideal of B, hence $\widetilde{S} := B_{\mathfrak{n} \cap B} \in \mathrm{lneft}_k(L)$ lies over R', $\widetilde{\mathfrak{n}} := (\mathfrak{n} \cap B)\widetilde{S}$ is the maximal ideal of $\widetilde{S}$, and $\dim(\widetilde{S}) = \dim(R') = r$ [cf. B(7.5)]. Now $A := B \cap K$ is the integral closure of A' in K, it is a k-algebra of finite type, and $A \cap (\mathfrak{n} \cap B)$ is a maximal ideal of A, hence $R := A_{A \cap (\mathfrak{n} \cap B)} \in \mathrm{lneft}_k(K)$. Note that $\widetilde{S}$ lies over R, and that R lies over R', hence that $\widetilde{S}$ lies over R' [cf. B(7.3)]. Furthermore, S dominates $\widetilde{S}$, both rings have the same field of quotients, and we have $\dim(\widetilde{S}) = \dim(S) = r$. Note that $\widetilde{\mathfrak{n}} \cap R' = \mathfrak{m}'$, hence that $\mathfrak{m}'S \subset \widetilde{\mathfrak{n}}S$, and that therefore $\widetilde{\mathfrak{n}}S$ is an $\mathfrak{n}$-primary ideal of S. Since k is perfect, $\widetilde{S}$ is analytically irreducible by III(6.14). From B(9.6) we

now get $\widetilde{S} = S$. Since S lies over R, we have $R = S \cap K$ by B(7.3), hence R is determined uniquely. The last assertion follows from B(7.5).

(6.4) Theorem: *Let K be an algebraic function field in two variables over an algebraically closed field k. Let L be a finite extension of K and let w be a zero-dimensional valuation of the algebraic function field L over k. Let $S \in \mathrm{lneft}_k(L)$ be two-dimensional and regular. If w dominates S, then there exist a quadratic dilatation S^* of S with respect to w and $R^* \in \mathrm{lneft}_k(K)$ such that S^* lies over R^*.*

Proof: Let $\{\xi, \eta\}$ be a transcendence basis of K over k. By replacing ξ by $1/\xi$ or η by $1/\eta$ if necessary, we may assume that $w(\xi) \geq 0$, $w(\eta) \geq 0$. Replacing S by a quadratic dilatation with respect to w, we may assume that ξ and η lie in S [cf. (4.4)]. Set $R := S \cap K$; clearly R is integrally closed in K [since S is integrally closed in L]. Since ξ and η lie in K and S, these elements lie in R, and therefore K is algebraic over the field of quotients of R. Let $z \in K$; then there exists $c \in R \setminus \{0\}$ such that $cz \in K$ is integral over R, hence $cz \in R$, and therefore K is the field of quotients of R.

Set $v := w|K$. Note that $\dim(v) = 0$ [since L is a finite extension of K, hence κ_w is a finite extension of κ_v by I(6.10)]; moreover, we have $\mathrm{rank}(v) = \mathrm{rank}(w)$, $\mathrm{rat.\,rank}(v) = \mathrm{rat.\,rank}(w)$ and Γ_w/Γ_v is a torsion group [cf. I(6.11)]. We choose $z_1 \in K$ with $v(z_1) > 0$. If $\mathrm{rat.\,rank}(v) = 1$, then we set $z_2 := 1$, and if $\mathrm{rat.\,rank}(v) = 2$, then we choose $z_2 \in K$ such that $v(z_2) > 0$ and that $v(z_1)$, $v(z_2)$ are linearly independent over $\mathbb{Q}$ in $\Gamma_v \otimes_{\mathbb{Z}} \mathbb{Q}$; then $v(z_1)$, $v(z_2)$ are linearly independent in $\Gamma_w \otimes_{\mathbb{Z}} \mathbb{Q}$. We set $z := z_1 z_2 \in K$; note that $v(z) > 0$. By (4.4) we may replace S by a quadratic dilatation of S with respect to w which contains z_1 and z_2, hence we may assume that z_1 and z_2 lie in S; note that z is not a unit of S. By (6.2) there exists a quadratic dilatation S^* of S with respect to w such that $z_1, z_2 \in S^*$ and that $z = x^a y^b u$ where $\{x, y\}$ is a regular system of parameters of S^*, $a, b \in \mathbb{N}_0$, u is a unit of S^*, and with the following additional property: either we have $a > 0$ and $b = 0$, or the elements $w(x)$ and $w(y)$ are rationally independent. We set $R^* := S^* \cap K$ and $\mathfrak{p}^* := \mathfrak{n}^* \cap R^*$ where $\mathfrak{n}^*$ is the maximal ideal of S^*; note that R^* is integrally closed, that K is the field of quotients of R^* [proof as above], and that $z \in \mathfrak{p}^*$ since $z \in \mathfrak{n}^*$. We set $\mathfrak{q}^* := \mathfrak{p}^* S^*$. We show that in both cases $\mathfrak{q}^*$ is an $\mathfrak{n}^*$-primary ideal of S^*; then we have $R^* \in \mathrm{lneft}_k(K)$ and S^* lies over R^* by (6.3).

Case 1: $a > 0$ and $b = 0$. Since K is the field of quotients of R^* and since y is algebraic over K, there exists a non-zero $c \in R^*$ such that cy is integral over R^*. Let $X^n + f_1 X^{n-1} + \cdots + f_n \in K[X]$ be the minimal polynomial of cy over K; then the coefficients $f_1, \ldots f_n$ lie in R^* [cf. III(2.12)(2)], and we have $f_n \neq 0$. We set $f_0 := 1$ and $g_i := f_{n-i} c^i$ for $i \in \{0, \ldots, n\}$. The elements $g_0, \ldots, g_n$ lie in R^*, and from $g_0 = -y(g_n y^{n-1} + \cdots + g_1)$ we see that the non-zero element $g_0 \in S^*$ is divisible by y. We write $g_0 = x^l h^*$ with $l \in \mathbb{N}_0$, $h^* \in S^*$ and x does not divide h^*; note that y divides h^*. We set $h := g_0^a/z^l$. Then we have $h \in K$ [since $z \in K$] and $h = h^{*a}/u^l \in S^*$, hence $h \in S^* \cap K = R^*$, and h is not divisible by

x in S^*. From $h \in yS^* \subset \mathfrak{n}^*$ we get $h \in \mathfrak{n}^* \cap R^* = \mathfrak{p}^*$. Suppose that $\mathfrak{q}^*$ is not an $\mathfrak{n}^*$-primary ideal of S^*. Then there exists a non-unit $t \in S^*$ with $\mathfrak{q}^* \subset S^*t$ [since S^* is a two-dimensional regular local ring and $\mathfrak{q}^*$ is not $\mathfrak{n}^*$-primary, every prime ideal in $\mathrm{Ass}(S^*/\mathfrak{q}^*)$ has height 1, and every such prime ideal is a principal ideal]. We have $z = x^a u \in \mathfrak{q}^* \subset S^*t$, and therefore we have $S^*t = S^*x^{a'}$ with $1 \leq a' \leq a$. We have $h \in \mathfrak{p}^* \subset \mathfrak{p}^*S^* = \mathfrak{q}^* \subset S^*t = S^*x^{a'}$, hence h is divisible by x in S^*. This contradiction shows that $\mathfrak{q}^*$ is $\mathfrak{n}^*$-primary.

Case 2: $w(x)$ and $w(y)$ are rationally independent. From z_1, $z_2 \in S \subset S^*$ and $z_1 z_2 = z = x^a y^b u$ in S^* we see that $z_1 = x^{a_1} y^{b_1} u_1$, $z_2 = x^{a_2} y^{b_2} u_2$ with $a_1, a_2, b_1, b_2 \in \mathbb{N}_0$, $a = a_1 + a_2$, $b = b_1 + b_2$, and units u_1, $u_2 \in S^*$. We set $d := \left| \begin{smallmatrix} a_1 & b_1 \\ a_2 & b_2 \end{smallmatrix} \right|$, $z_3 := z_1^{a_2} z_2^{-a_1}$ and $z_4 := z_1^{b_2} z_2^{-b_1}$; the elements z_3, z_4 lie in K, and we have $z_3 = y^{-d} u_3$ and $z_4 = x^d u_4$ with units $u_3 := u_1^{a_2} u_2^{-a_1}$, $u_4 := u_1^{b_2} u_2^{-b_1}$ of S^*. Suppose that $d = 0$. Then z_3 and z_4 are units of S^*, hence $a_2 w(z_1) - a_1 w(z_2) = 0$, $b_2 w(z_1) - b_1 w(z_2) = 0$. Since $w(z_1) = v(z_1)$, $w(z_2) = v(z_2)$ are rationally independent by the choice made above, we obtain $a_1 = a_2 = b_1 = b_2 = 0$, hence $a = a_1 + a_2 = 0$, $b = b_1 + b_2 = 0$, and therefore $z = z_1 z_2$ is a unit of S^*. Since S^* dominates S and $z \in S$, we see that z is a unit of S, and this contradicts the choice of z. Therefore we have $d \neq 0$. From $z_3 = y^{-d} u_3$ and $z_4 = x^d u_4$, and the fact that $w(x)$ and $w(y)$ are rationally independent, we now see that the elements $w(z_3) = v(z_3)$ and $w(z_4) = v(z_4)$ are rationally independent. We set

$$z_1^* := \begin{cases} z_3 & \text{if } d < 0, \\ z_3^{-1} & \text{if } d > 0, \end{cases} \qquad z_2^* := \begin{cases} z_4 & \text{if } d > 0, \\ z_4^{-1} & \text{if } d < 0. \end{cases}$$

We have $z_1^* = y^{|d|} u_1'$, $z_2^* = x^{|d|} u_2'$ with units u_1', $u_2' \in S^*$, hence z_1^* and z_2^* lie in S^*. Since z_3, $z_4 \in K$, we have z_1^*, $z_2^* \in K$, hence $z_i^* \in \mathfrak{n}^* \cap K = \mathfrak{p}^*$ for $i \in \{1,2\}$. Therefore we have $\mathfrak{n}^{*2|d|} \subset \mathfrak{q}^*$, hence $\mathfrak{q}^*$ is $\mathfrak{n}^*$-primary.

6.3   Uniformization

(6.5) NOTATION: Let K be an algebraic function field in two variables over an algebraically closed field k. Let v be a valuation of the algebraic function field K. We say that v *can be uniformized* if there exists a regular $R \in \mathrm{lneft}_k(K)$ which is dominated by v and with $\mathrm{tr.\,d}_k(\kappa) = \mathrm{tr.\,d}_k(\kappa_v)$ (κ is the residue field of R).

(6.6) REMARK: Let K be an algebraic function field in two variables over an algebraically closed field k.

Let A be a subalgebra of K of finite type which has K as field of quotients, let $\mathfrak{p}$ be a non-zero prime ideal of A, set $R := A_\mathfrak{p}$ and let $\kappa = Q(A/\mathfrak{p})$ be the residue field of R. We have $\dim(R) = 1$—and in this case we have $\mathrm{tr.\,d}_k(\kappa) = 1$—or $\dim(R) = 2$—and in this case we have $\kappa = k$ since $\mathfrak{p}$ is a maximal ideal of A, hence $A/\mathfrak{p}$ is a finitely generated field extension of k with $\mathrm{tr.\,d}_k(A/\mathfrak{p}) = 0$, hence an algebraic extension of k [cf. [63], Th. A on p. 286 and Cor. 13.4], hence $\kappa = k$ since k is algebraically closed.

Let v be a valuation of K. We have $\dim(v) = 1$ or $\dim(v) = 0$. If $\dim(v) = 1$, then v can be uniformized iff there exists a one-dimensional regular $R \in \mathrm{lneft}_k(K)$ which is dominated by v. The existence of such an R follows from I(10.7). If $\dim(v) = 0$, then v can be uniformized iff there exists a two-dimensional regular $R \in \mathrm{lneft}_k(K)$ which is dominated by v. If k has characteristic 0, we shall prove the existence of such an R in theorem (6.9) below.

(6.7) Lemma: *Let K be an algebraic function field in two variables over an algebraically closed field k of characteristic 0. Let L be a cyclic Galois extension of K of prime degree p. Let v be a valuation of K with $\mathrm{tr.}\,\mathrm{d}_k(v) = 0$ and $\mathrm{rat.}\,\mathrm{rank}(v) = 1$ which has only one extension w to L. If v can be uniformized, then w can be uniformized.*

Proof: By hypothesis, there exists a two-dimensional regular $R \in \mathrm{lneft}_k(K)$ with residue field k such v dominates R. Let $\mathfrak{m}$ be the maximal ideal of R. The local rings in $\mathrm{qln}(L)$ lying over R are localizations of the integral closure S of R in L with respect to maximal ideals of S. Since v admits only one extension w to L, the valuation ring A_w is the integral closure of A_v in L [cf. I(6.1)], A_w dominates A_v [cf. I(5.2)], we have $S \subset A_w$ and $\mathfrak{m}S \subset \mathfrak{n} := S \cap \mathfrak{m}(A_w)$; hence $\mathfrak{n}$ is a maximal ideal of S [cf. [63], Prop. 9.2]. We have $\sigma(\mathfrak{m}(A_w)) = \mathfrak{m}(A_w)$ for every $\sigma \in \mathrm{Gal}(L/K)$, hence $\sigma(\mathfrak{n}) = \mathfrak{n}$ for every $\sigma \in \mathrm{Gal}(L/K)$. Since the maximal ideals of S are conjugate under $\mathrm{Gal}(L/K)$ [cf. [63], Prop. 13.10], S has only one maximal ideal $\mathfrak{n}$, hence S is the only ring in $\mathrm{qln}(L)$ lying over R, and it is therefore the integral closure of R in L. Since $S \in \mathrm{lneft}_k(L)$ and $\dim(S) = 2$ [cf. B(7.5)], the local ring S has residue field k by (6.6).

Since k is algebraically closed of characteristic 0, k contains a primitive p-th root of unity; then there exists a primitive element z for L/K such that the minimal polynomial for z over K has the form $F = Z^p - r$ for some $r \in K^\times$. Clearly we may assume that $r \in R$. By (6.2), replacing R by a quadratic dilatation of R with respect to v, we may assume that $r = x^a u$ where $\{x, y\}$ is a regular system of parameters of R, $a \in \mathbb{N}_0$ and u is a unit of R; by the argument above, the integral closure S of R in L is local. Suppose that $a \equiv 0 \pmod{p}$. Then $z' := zx^{-a/p}$ is a primitive element for L/K, and $Z^p - u$ is the minimal polynomial for z' over K. This implies that $D_{L/K}(1, z', \ldots, z'^{p-1}) = \pm p^p u^{p-1}$ [cf. I(7.6)] is a unit of R, hence $[S/\mathfrak{n} : R/\mathfrak{m}]_{\mathrm{sep}} = [L : K] = p$ [cf. III(5.4)]. On the other hand, we have $R/\mathfrak{m} = S/\mathfrak{n} = k$. This contradiction shows that p does not divide a.

We choose a positive integer b with $ab \equiv 1 \pmod{p}$. Since b is not divisible by p, also $z' := z^b x^{(1-ab)/p}$ is a primitive element for L/K, and $Z^p - xu^b$ is the minimal polynomial of z' over K. Since u^b is a unit of R, also $\{xu^b, y\}$ is a regular system of parameters of R. Therefore we may assume: $\{x, y\}$ is a regular system of parameters for R, and $Z^p - x$ is the minimal polynomial of a primitive element z for L/K.

We set $B' := R[z]$; note that $B' \subset S$ since z is integral over R, and that $Q(B') = K(z) = L$. Set $\mathfrak{q}' := B' \cap \mathfrak{n}$, $B := B'_{\mathfrak{q}'}$ and $\mathfrak{q} := \mathfrak{q}' B_{\mathfrak{q}'}$. Then $Q(B) = L$, and since $R \subset B \subset S$, we see that B is integral over R, hence $\dim(B) = 2$ [cf. [63],

Prop. 9.2]. We have $x = z^p \in \mathfrak{q}'$, hence z lies in $\mathfrak{q}'$. Let $\mathfrak{b}'$ be the ideal of B' generated by the elements y, z; then we have $\mathfrak{m} \subset \mathfrak{b}' \subset \mathfrak{q}'$. Now we have $B'/\mathfrak{b}' = k$, and therefore we have $\mathfrak{q}' = \mathfrak{b}'$, hence $\mathfrak{q}$ is generated by y and z. This implies that B is a regular local ring, hence B is integrally closed, and therefore we have $B = S$, hence w can be uniformized.

(6.8) Lemma: *Let K be an algebraic function field in r variables over a perfect field k. Let L be a finite extension of K, let $R \in \mathrm{lneft}_k(K)$, and let $S \in \mathrm{lneft}_k(L)$ lie over R. Let $\mathfrak{m}$ (resp. $\mathfrak{n}$) be the maximal ideal of R (resp. of S). We assume that $\dim(R) = r$, and that $R/\mathfrak{m} = S/\mathfrak{n} = k$. If $\mathfrak{m}S = \mathfrak{n}$ and if S is regular, then also R is regular.*

Proof: The local ring R is analytically irreducible [cf. III(6.9)]. Since $\dim(R) = \dim(S)$ [cf. B(7.5)], and since S is a quasifinite R-module by assumption, one can show, just as in the proof of B(9.6)(1), that the canonical homomorphism $\varphi \colon \widehat{R} \to \widehat{S}$, induced by the local homomorphism $R \hookrightarrow S$, is injective. We have $\mathfrak{m}\widehat{S} = \widehat{\mathfrak{n}}$. Since S is regular, we have $\widehat{S} = k[\![x_1, \ldots, x_r]\!]$ where $x_1, \ldots, x_r \in \mathfrak{m}$ [cf. [63], Prop. 10.16]. Therefore we have $k[\![x_1, \ldots, x_r]\!] \subset \widehat{R} \subset \widehat{S} = k[\![x_1, \ldots, x_r]\!]$, hence $\widehat{R} = \widehat{S}$, and therefore $\widehat{R}$ is regular, hence also R is regular.

After these preparations, we prove the main theorem (6.9) of this subsection. It shall play the decisive role when proving theorem (7.9) from which we derive the existence of a desingularization of an irreducible surface. For a more geometric version of the main theorem we refer to [192], Fundamental Lemma on p. 650 (and its generalization to varieties of higher dimension in [193]).

(6.9) Theorem: [Uniformization] *Let K be an algebraic function field in two variables over an algebraically closed field k of characteristic 0. Let v be a valuation of the algebraic function field K with $\dim(v) = 0$. Then v can be uniformized.*

Proof: (1) By (5.2) we have rat. $\mathrm{rank}(v) = 2$ or rat. $\mathrm{rank}(v) = 1$.
(a) If rat. $\mathrm{rank}(v) = 2$, then Γ_v is a free $\mathbb{Z}$-module which is minimally generated by two elements $v(x_1)$, $v(x_2)$ where $x_1, x_2 \in K$ [cf. B(1.39)]. By I(9.3) the set $\{x_1, x_2\}$ is a transcendence basis of K over k.
(b) If rat. $\mathrm{rank}(v) = 1$, let $\{x_1, x_2\}$ be an arbitrary transcendence basis of K over k. By replacing, if necessary, x_1 resp. x_2 by its inverse, we may assume that $v(x_1) \geq 0$ and $v(x_2) \geq 0$. Since k is algebraically closed and $\dim(v) = 0$, we have $\kappa_v = k$. Therefore, replacing x_i by $x_i - \gamma_i$ for some $\gamma_i \in k$, we may even assume that $v(x_i) > 0$ for $i \in \{1, 2\}$.
In both cases we set $K_1 := k(x_1, x_2)$, $R_1 := k[x_1, x_2]_{(x_1, x_2)}$ and denote by $\mathfrak{m}_1$ the maximal ideal of R_1; then v has center $\mathfrak{m}_1$ in R_1. Note that R_1 is a two-dimensional regular local ring with residue field k.
(2) Let L be the smallest Galois extension of K_1 containing K, and let w be an extension of v to L. Set $v_1 := v|K_1$. If rat. $\mathrm{rank}(v) = 2$, then we have $\Gamma_v = \Gamma_{v_1}$ by (1)(a). Let $w_1 = w, w_2, \ldots, w_n$ be a complete set of extensions of v_1 to L.

Then $D := A_{w_1} \cap \cdots \cap A_{w_n}$ is the integral closure of A_{v_1} in L, and $\mathfrak{n} := \mathfrak{n}_1 :=$ $D \cap \mathfrak{m}(A_{w_1}), \ldots, \mathfrak{n}_n := D \cap \mathfrak{m}(A_{w_n})$ are the pairwise different maximal ideals of D [cf. I(6.1)]. By B(10.1) there exists $z \in D$ such that $z \in \mathfrak{n}_1, z \notin \mathfrak{n}_j$ for $j \in \{2, \ldots, n\}$. Let $X^m + a_1 X^{m-1} + \cdots + a_m \in K_1[X]$ be the minimal polynomial of z over K_1; note that $a_1, \ldots, a_m$ lie in A_{v_1} [cf. III(2.12)(2)]. Replacing R_1 by a quadratic dilatation of R_1 with respect to v_1, we may assume that the elements $a_1, \ldots, a_m$ lie in R_1 [cf. (4.4)]. Let B be the integral closure of R_1 in L; note that $z \in B \subset D$. We set $\mathfrak{q} := B \cap \mathfrak{m}(A_w) = B \cap \mathfrak{n}$, $S := B_\mathfrak{q}$. Then S is the local subring of L lying over R_1 such that w dominates S; note that $\dim(S) = 2$ by B(7.5) and that S has residue field k by (6.6). From $z \in B \cap \mathfrak{n} = \mathfrak{q}, z \notin B \cap \mathfrak{n}_j$ for $j \in \{2, \ldots, n\}$, we get $\mathfrak{q} \neq B \cap \mathfrak{n}_j$ for $j \in \{2, \ldots, n\}$, hence $\mathfrak{n}$ is the only maximal ideal of D lying over $\mathfrak{q}$. Note that $R_1 \subset A_{v_1}$, $S \subset A_w$, and that A_{v_1} dominates R_1.

(i) Let $G_1 := \mathrm{Gal}(L/K_1)$ be the Galois group of L over K_1, and let $\sigma \in G_1$. If $\sigma(\mathfrak{q}) = \mathfrak{q}$, then we have $\sigma(\mathfrak{n}) = \mathfrak{n}$ [since $\mathfrak{n}$ is the only maximal ideal of D lying over $\mathfrak{q}$]. Therefore we have $G_Z(w/v_1) \supset G_Z(S/R_1)$. On the other hand, we have $G_Z(w/v_1) \subset G_Z(S/R_1)$ by B(7.10)(2). Therefore we have $G_Z(w/v_1) = G_Z(S/R_1)$.

(ii) We set $R := S \cap K$, $\mathfrak{m} := \mathfrak{n} \cap R$. Then S lies over R and R lies over R_1 [cf. B(7.3)], hence we have $R \in \mathrm{lneft}_k(K)$ by B(3.6); note that $\dim(R) = 2$ and that R has residue field k. We set $G := \mathrm{Gal}(L/K)$.

(iii) Let $(K_1)_Z$ be the fixed field of $G_Z(S/R_1)$, and K_Z be the fixed field of $G_Z(S/R)$. By B(7.8) K_Z is the compositum of $(K_1)_Z$ and K and $G_Z(S/R) = G_Z(S/R_1) \cap G$, $G_Z(w/v) = G_Z(w/v_1) \cap G$. Therefore we have $G_Z(w/v) = G_Z(S/R)$ by (i).

(iv) The residue field $\kappa_{v_1} - k$ of v_1 is algebraically closed; by I(8.3)(4) we have $G_T(w/v_1) = G_Z(w/v_1)$, and $G_T(w/v_1)$ is abelian by I(8.16). Therefore $G_Z(w/v_1) = G_Z(S/R_1) = \mathrm{Gal}(L/(K_1)_Z)$ is abelian, hence K_Z is an abelian Galois extension of $(K_1)_Z$. By Galois theory there exists a sequence of fields $(K_1)_Z =: L_0 \subset L_1 \subset \cdots \subset L_h := K_Z$ such that L_i/L_{i-1} is a cyclic extension of prime degree for $i \in \{1, \ldots h\}$.

Let $i \in \{0, \ldots, h\}$. We set $S_i := S \cap L_i$ and $\omega_i := w|L_i$. Note that S_i lies over R_1, hence $S_i \in \mathrm{lneft}_k(L_i)$ [cf. B(3.6)], $\dim(S_i) = 2$ [cf. B(7.5)] and S_i has residue field k. Now $\mathfrak{m}_1 S_0$ is the maximal ideal of S_0 by B(7.7); since R_1 is a two-dimensional regular local ring, we see that S_0 is also a two-dimensional regular local ring. Therefore ω_0 can be uniformized. Let $i \in \{1, \ldots, h\}$; since w is the only extension of ω_0 to L [cf. I(8.3)(1)], ω_i is the only extension of ω_{i-1} to L_i. If $\mathrm{rat.\,rank}(v) = 1$, then we can apply (6.7) successively to the extensions L_1/L_0, $L_2/L_1, \ldots, L_h/L_{h-1}$, and get: ω_h can be uniformized. If $\mathrm{rat.\,rank}(v) = 2$, then we have $\Gamma_v = \Gamma_{v_1}$ [cf. above], $\Gamma_v = \Gamma_{\omega_h}$ and $\Gamma_{v_1} = \Gamma_{\omega_0}$ by I(8.7), hence $\Gamma_{\omega_h} = \Gamma_{\omega_0}$, and since ω_h is the only extension of ω_0 to L_h, we see that L_0 is the fixed field of $G_Z(\omega_h/\omega_0)$ by B(7.7)(2). Since $\dim(v) = 0$, we have $\dim(w) = 0$, and since k is algebraically closed, the residue field of w is k. Therefore the residue fields of ω_0 and ω_h are equal to k, hence we have $G_Z(\omega_h/\omega_0) = G_T(\omega_h/\omega_0)$ by B(7.11)(1). We have $\Gamma_{\omega_h}/\Gamma_{\omega_0} = G_T(\omega_h/\omega_0)$ [cf. I(8.16)], hence $G_T(\omega_h/\omega_0)$ is trivial, and this yields $L_0 = L_h$. Therefore $\omega_h = \omega_0$ can be uniformized.

(v) Hence in both cases ω_h can be uniformized, i.e., there exists a two-dimensional regular local ring $\widetilde{S} \in \mathrm{lneft}_k(K_Z)$ with residue field k which is dominated by ω_h. Since $S_h \subset A_{\omega_h}$, and S_h is essentially of finite type over k, we can, replacing $\widetilde{S}$ by a quadratic dilatation of $\widetilde{S}$ with respect to ω_h, assume that $S_h \subset \widetilde{S}$ [cf. (4.4)]; note that a first quadratic dilatation of $\widetilde{S}$ is two-dimensional and regular by (4.1) and has residue field k by VII(6.7). Let $S_L \in \mathrm{lneft}_k(L)$ be the ring lying over $\widetilde{S}$ which is dominated by w; S_L is two-dimensional and has residue field k. Clearly we have $S \subset S_L \subset A_w$, and S_L dominates S. From B(7.10) we get $G_Z(w/\omega_h) \subset G_Z(S_L/\widetilde{S}) \subset G_Z(S/S_h)$. From B(7.9) we get $G_Z(w/\omega_h) = G_Z(w/v) \cap G$ and $G_Z(S/S_h) = G_Z(S/R) \cap G$. From (iii) we therefore get $G_Z(w/\omega_h) = G_Z(S_L/\widetilde{S}) = G_Z(S/S_h)$. Since w is the only extension of ω_h to L, the field L_h is the fixed field of $G_Z(w/\omega_h)$ by B(7.7)(2), hence L_h is also the fixed field of $G_Z(S_L/\widetilde{S})$; since $\mathfrak{m}(\widetilde{S})S_L = \mathfrak{m}(S_L)$ by B(7.7)(3), we see that S_L is a two-dimensional regular local ring with residue field k. Applying (4.1) yields: Every quadratic dilatation of S_L with respect to w is two-dimensional and regular and has residue field k.

Replacing S_L by a quadratic dilatation with respect to w we may assume by (6.4): There exists $R_K \in \mathrm{lneft}_k(K)$ such that S_L lies over R_K; note that $R_K = S_L \cap K$ [cf. B(7.3)]. We have $R \subset R_K \subset A_v$, $\dim(R_K) = 2$, and R_K has residue field k. Since R_K dominates R and A_w dominates R_K, from B(7.10) we get $G_Z(w/v) \subset G_Z(S_L/R_K) \subset G_Z(S/R)$, hence $G_Z(w/v) = G_Z(S_L/R_K) = G_Z(S/R)$ [cf. (iii)]. Since K_Z is the fixed field of $G_Z(S/R)$, we see that K is also the fixed field of $G_Z(S_L/R_K)$, and therefore we have $\mathfrak{m}(R_K)S_L = \mathfrak{m}(S_L)$ by B(7.11)(2). Since S_L is regular, also R_K is regular by (6.8).

7 Resolution of Surface Singularities II: Blowing up and Normalizing

(7.0) In this section k is an *algebraically closed field of characteristic zero*, and K is a field of algebraic functions in two variables over k. All varieties are varieties over k. All valuations of K will tacitly be assumed to be valuations over k.

7.1 Principalization

(7.1) REES RING AND INTEGRAL CLOSURE: Let R be an integrally closed domain, and let $\mathfrak{a}$ be an ideal of R. Let $\mathfrak{b}$ be the integral closure of $\mathfrak{a}$ in R. We assume that $\mathfrak{b}^n$ *is integrally closed in R for every* $n \in \mathbb{N}$; then $\mathfrak{b}^n$ is the integral closure of $\mathfrak{a}^n$ for every $n \in \mathbb{N}$ [cf. B(6.16)(4)], hence $\mathcal{R}(\mathfrak{b}, R)$ is the integral closure of $\mathcal{R}(\mathfrak{a}, R)$ [cf. B(6.8)]. Under this assumption, we have the following: Let $\mathfrak{q}$ be a homogeneous prime ideal of $\mathcal{R}(\mathfrak{a}, R)$ which is not irrelevant such that $\mathcal{R}(\mathfrak{a}, R)_{(\mathfrak{q})}$ is integrally closed. We choose $x \in \mathfrak{a}$ with $xT \notin \mathfrak{q}$. Then $\mathcal{R}(\mathfrak{a}, R)_{(\mathfrak{q})}$ is a localization of $\mathcal{R}(\mathfrak{a}, R)_{(xT)} = R[\mathfrak{a}/x]$ with respect to a prime ideal $\mathfrak{p}$ [cf. B(5.6)(2)]. Now $\mathcal{R}(\mathfrak{b}, R)_{xT}$ is integral over $\mathcal{R}(\mathfrak{a}, R)_{xT}$ [cf. [63], Prop. 4.13], hence also $(\mathcal{R}(\mathfrak{b}, R)_{xT})_0 = \mathcal{R}(\mathfrak{b}, R)_{(xT)} = R[\mathfrak{b}/x]$ is integral over $(\mathcal{R}(\mathfrak{a}, R)_{xT})_0 =$

$\mathcal{R}(\mathfrak{a}, R)_{(xT)} = R[\mathfrak{a}/x]$ [cf. B(4.24)(1)]. Furthermore, $\mathcal{R}(\mathfrak{b}, R)_{(xT)} = R[\mathfrak{b}/x]$ is integrally closed [cf. B(4.30)(2)], hence $R[\mathfrak{b}/x]_{\mathfrak{p}} = R[\mathfrak{a}/x]_{\mathfrak{p}}$ [since every localization of $R[\mathfrak{b}/x]$ with respect to a multiplicatively closed system is integrally closed by [63], Prop. 4.13]. Now $R[\mathfrak{b}/x]_{\mathfrak{p}}$ is the localization of $R[\mathfrak{b}/x]$ with respect to a prime ideal $\mathfrak{p}'$ [cf. B(2.3)(3)], hence $R[\mathfrak{b}/x]_{\mathfrak{p}} = R[\mathfrak{b}/x]_{\mathfrak{p}'} = \mathcal{R}(\mathfrak{b}, R)_{(\mathfrak{q}')}$ with a homogeneous prime ideal $\mathfrak{q}'$ of $\mathcal{R}(\mathfrak{b}, R)$ [cf. B(5.6)(2)], hence we have $\mathcal{R}(\mathfrak{a}, R)_{(\mathfrak{q})} = \mathcal{R}(\mathfrak{b}, R)_{(\mathfrak{q}')}$.

(7.2) Lemma: *Let $S \in \mathrm{lneft}_k(K)$ with $\dim(S) = 2$, and let $\mathfrak{n}$ be the maximal ideal of S. Let Σ be the set of zero-dimensional valuations v of K with $A_v \supset S$. Then Σ is not empty, and we have:*
(1) Every $v \in \Sigma$ dominates S, and we have $S = \bigcap_{v \in \Sigma} A_v$, $\mathfrak{n} = \bigcap_{v \in \Sigma} \mathfrak{p}_v$.
(2) Let w be a one-dimensional valuation of K with $A_w \supset S$; then there exists $v \in \Sigma$ with $A_v \subset A_w$.

Proof: We can write $S = A_{\mathfrak{p}}$ where A is a k-subalgebra of K of finite type and $\mathfrak{p}$ is a prime ideal of A [cf. B(2.3)(3)]; since $\dim(S) = 2$, we see that $\mathfrak{p}$ is a maximal ideal of A [cf. B(7.5)], hence $\Sigma \neq \emptyset$ [cf. I(10.9)]. Since $k \subset S$, we know that S is the intersection of all valuation rings of K which contain the ring S [cf. I(3.4)]. Let $v \in \Sigma$; then $\mathfrak{p}_v \cap S$ is the maximal ideal of S since $\kappa_v = k$, hence v dominates S. Let w be a one-dimensional valuation of K with $S \subset A_w$; κ_w is a field of algebraic functions in one variable over k [cf. I(10.7)]. Let $\varphi \colon A_w \to \kappa_w$ be the canonical homomorphism. Since S is a localization of a k-algebra of finite type, $\varphi(S)$ is local and essentially of finite type over k, hence $\varphi(S)$ is the localization of a k-algebra of finite type with respect to a prime ideal. By I(10.9) there exists a valuation $\overline{v}$ of the function field κ_w with $A_{\overline{v}} \supset \varphi(S)$; $\overline{v}$ is a zero-dimensional valuation of κ_w. Therefore $v := w \circ \overline{v}$ is a zero-dimensional valuation of K, and we have $S \subset A_v \subset A_w$. For any non-trivial valuation v of the function field K we have $\dim(v) = 0$ or $\dim(v) = 1$ [cf. I(10.6)], hence we have proved the assertions of the lemma.

(7.3) Proposition: *Let R and S in $\mathrm{lneft}_k(K)$ with $\dim(R) = \dim(S) = 2$, R regular with maximal ideal $\mathfrak{m}$, and assume that S dominates R but $S \neq R$. Then there exist complete $\mathfrak{m}$-primary ideals $\mathfrak{a}$, $\mathfrak{b}$ of R with $\mathfrak{b} \subset \mathfrak{a}$, $\mathfrak{a}/\mathfrak{b}$ is a simple R-module, and $S = \mathcal{R}(\mathfrak{a}, R)_{(\mathfrak{q})}$ where $\mathfrak{q} = \bigoplus_{n > 0} \mathfrak{q}_n$, with $\mathfrak{q}_0 := \mathfrak{m}$ and $\mathfrak{q}_n := \mathfrak{a}^{n-1}\mathfrak{b}T^n$ for every $n \in \mathbb{N}$, is a homogeneous non-irrelevant prime ideal of $\mathcal{R}(\mathfrak{a}, R)$. Moreover, writing $\mathfrak{a} = \mathfrak{b} + k\,a_0$ with $a_0 \in R$, the ideal $\mathfrak{b}/a_0 \cdot S$ is the maximal ideal of S.*

Proof: (1) Let $\mathfrak{n}$ be the maximal ideal of S; note that $S/\mathfrak{n} = k$ by B(7.5). There exist $y_1, \ldots, y_h \in K$ such that, setting $B := R[y_1, \ldots, y_h]$, we have $S = B_{\mathfrak{n} \cap B}$ and $\mathfrak{p} := \mathfrak{n} \cap B$ is a maximal ideal of B [cf. B(7.5)]; therefore the elements $\gamma_1 := y_1 \bmod \mathfrak{p}, \ldots, \gamma_h := y_h \bmod \mathfrak{p}$ lie in k. Replacing y_j by $y_j - \gamma_j$ for $j \in \{1, \ldots, h\}$, we may assume that $y_j \in \mathfrak{p}$ for $j \in \{1, \ldots, h\}$. Since $Q(R) = K$, there exist elements $a_0, a_1, \ldots, a_h \in R$, $a_0 \neq 0$, with $y_j = a_j/a_0$ for $j \in \{1, \ldots, h\}$. Clearly we may assume that $a_0, a_1, \ldots, a_h$ have greatest common divisor 1 in R. We set

$\mathfrak{a} := Ra_0 + \cdots + Ra_h$; then 1 is a greatest common divisor for the elements of $\mathfrak{a}$, and we have $B = R[\mathfrak{a}/a_0]$. Since $B \neq R$ [as $S \neq R$], we have $a_0 \in \mathfrak{m}$. Suppose that $\mathfrak{a} = R$; then we have $B = R[\mathfrak{a}/a_0] = R[1/a_0]$, hence we get $\mathfrak{m}S = S$, contradicting the assumption that S dominates R. Therefore $\mathfrak{a}$ is an $\mathfrak{m}$-primary ideal of R [cf. VII(2.1)]. Since $S = R[\mathfrak{a}/a_0]_\mathfrak{p}$, we have $S = \mathcal{R}(\mathfrak{a}, R)_{(\mathfrak{q})}$ where $\mathfrak{q}$ is a homogeneous prime ideal of $\mathcal{R}(\mathfrak{a}, R)$ with $a_0 T \notin \mathfrak{q}$ [cf. B(5.6)]; since powers of complete ideals of R are complete [cf. VII(4.13)], we may replace $\mathfrak{a}$ by its integral closure [cf. (7.1)]; hence we may assume, to begin with, that $\mathfrak{a}$ is a complete $\mathfrak{m}$-primary ideal of R, generated by $a_0, \ldots, a_h$, and that $B = R[\mathfrak{a}/a_0]$.

(2) Let Σ be the set of zero-dimensional valuations v of K with $A_v \supset S$. By (7.2) we know that v dominates S for every $v \in \Sigma$, and that $S = \bigcap_{v \in \Sigma} A_v$. Since S dominates R, every $v \in \Sigma$ dominates R also. For every $v \in \Sigma$ we have $\mathfrak{a}/a_0 \subset A_v$, hence $\mathfrak{a}$ is contained in $\mathfrak{a}'_v := \{z \in R \mid v(z) \geq v(a_0)\}$. We set $\mathfrak{a}' := \bigcap_{v \in \Sigma} \mathfrak{a}'_v$. Now $\mathfrak{a}'_v$ is a v'-ideal, hence it is integrally closed, and therefore $\mathfrak{a}'$ is a complete ideal of R [cf. B(6.16)] which contains $\mathfrak{a}$. For every $a' \in \mathfrak{a}'$ we have $v(a'/a_0) \geq 0$ for every $v \in \Sigma$, hence a'/a_0 lies in S, and therefore we have $B' := R[\mathfrak{a}'/a_0] \subset S$, hence $S = B'_{\mathfrak{n} \cap B'}$. Replacing $\mathfrak{a}$ by $\mathfrak{a}'$ and B by B', we see that we may assume that $B = R[\mathfrak{a}/a_0]$ where $\mathfrak{a}$ is complete and that $B_{\mathfrak{n} \cap B} = S$. Likewise, for $v \in \Sigma$ we define $\mathfrak{b}_v := \{z \in R \mid v(z) > v(a_0)\}$; then $\mathfrak{b} := \bigcap_{v \in \Sigma} \mathfrak{b}_v$ is a complete ideal of R. Note that $\mathfrak{b} \subset \mathfrak{a}$, and that $v(\mathfrak{a}) = v(a_0)$ and that $v(\mathfrak{b}) > v(\mathfrak{a})$ for every $v \in \Sigma$. We show that $\mathfrak{a}/\mathfrak{b}$ is a simple R-module which implies that $\mathfrak{b}$ is an $\mathfrak{m}$-primary ideal. In fact, let $a \in \mathfrak{a}$. We have $a/a_0 \in S$, hence there exists $\gamma \in k$ with $a/a_0 - \gamma \in \mathfrak{n}$, whence $v(a - \gamma a_0) > v(a_0)$ for every $v \in \Sigma$ [since the valuations in Σ dominate S], and therefore we have $a \in \mathfrak{b} + k a_0$, hence $\mathfrak{a} = \mathfrak{b} + k a_0$; since $a_0 \notin \mathfrak{b}$, the assertion has been proved.

(3) We have

$$\mathfrak{a}^n = \mathfrak{a}^{n-1}\mathfrak{b} + k a_0^n, \quad a_0^n \notin \mathfrak{a}^{n-1}\mathfrak{b} \quad \text{for every } n \in \mathbb{N} \tag{$*$}$$

[note that $v(\mathfrak{a}^{n-1}\mathfrak{b}) = (n-1)v(a_0) + v(\mathfrak{b}) > v(a_0^n)$ for every $v \in \Sigma$]; in particular, we have $\mathfrak{a}^n \neq \mathfrak{a}^{n-1}\mathfrak{b}$ for every $n \in \mathbb{N}$.

We set

$$\mathfrak{q}_0 := \mathfrak{m}, \quad \mathfrak{q}_n := \mathfrak{a}^{n-1}\mathfrak{b}T^n \subset \mathcal{R}(\mathfrak{a}, R)_n \quad \text{for every } n \in \mathbb{N}.$$

Then $\mathfrak{q} := \bigoplus_{n \geq 0} \mathfrak{q}_n$ is a homogeneous prime ideal in $\mathcal{R}(\mathfrak{a}, R)$ which is not irrelevant, and we obtain $S = \mathcal{R}(\mathfrak{a}, R)_{(\mathfrak{q})}$.

In fact, it is easy to check that $\mathfrak{q}$ is a prime ideal of $\mathcal{R}(\mathfrak{a}, R)$ [cf. ($*$) above and use B(4.1)]. Since $\mathfrak{a}^n \neq \mathfrak{a}^{n-1}\mathfrak{b}$ for every $n \in \mathbb{N}$, the ideal $\mathfrak{q}$ is not irrelevant. We have $a_0 T \notin \mathfrak{q}$; then $\mathfrak{p} := (\mathfrak{b}/a_0) \cdot R[\mathfrak{a}/a_0]$ is the prime ideal in $R[\mathfrak{a}/a_0]$ corresponding to $\mathfrak{q}$. We have $\mathfrak{b}/a_0 \subset \mathfrak{n}$ since $\mathfrak{n} = \{z \in S \mid v(z) > 0 \text{ for every } v \in \Sigma\}$ [cf. (7.2)]. Conversely, let $z \in R[\mathfrak{a}/a_0] \cap \mathfrak{n}$, hence $v(z) > 0$ for every $v \in \Sigma$. We can write $z = a/a_0^n$ for some $n \in \mathbb{N}$ and $a \in \mathfrak{a}^n$; then we have $v(a) > v(a_0^n) = v(\mathfrak{a}^n)$, hence $a \in \mathfrak{a}^{n-1}\mathfrak{b}$ by ($*$), hence $z \in (\mathfrak{b}/a_0) \cdot R[\mathfrak{a}/a_0]$. Thus, we have shown that $\mathfrak{p} = R[\mathfrak{a}/a_0] \cap \mathfrak{n}$, hence that $S = R[\mathfrak{a}/a_0]_\mathfrak{p}$ and $\mathfrak{n} = (\mathfrak{b}/a_0) \cdot S$, and, in particular, that $S = \mathcal{R}(\mathfrak{a}, R)_{(\mathfrak{q})}$.

(7.4) Corollary: *Let ν_R be the valuation of K determined by the order function of R and let v be a zero-dimensional valuation of K which dominates S. If the first quadratic dilatation R_1 of R with respect to v is not contained in S, then we have $\nu_R(\mathfrak{b}) = \nu_R(\mathfrak{a}) + 1$.*

Proof: Let $\mathfrak{n}$ be the maximal ideal of S. For typographical reasons, if $x \in \mathfrak{m} \setminus \mathfrak{m}^2$ and $\overline{g} \in \mathrm{gr}_{\mathfrak{m}}(R)$ is a form of degree 1 not associated with $\mathrm{In}(x)$, then the maximal ideal of $A := R[\mathfrak{m}/x]$ determined by $\overline{g}$ [cf. VII(2.6)] shall be denoted by $\mathfrak{n}(\overline{g})$. Since $\mathfrak{ma} \subset \mathfrak{b} \subset \mathfrak{a}$ [cf. (7.3)], we have $1 + \nu_R(\mathfrak{a}) \geq \nu_R(\mathfrak{b}) \geq \nu_R(\mathfrak{a})$, hence we have $\nu_R(\mathfrak{b}) = \nu_R(\mathfrak{a}) + 1$ or $\nu_R(\mathfrak{b}) = \nu_R(\mathfrak{a})$. We show that $\nu_R(\mathfrak{b}) = \nu_R(\mathfrak{a})$ implies that R_1 is contained in S, contradicting our assumption.

Thus, suppose that $\nu_R(\mathfrak{b}) = \nu_R(\mathfrak{a}) =: r$. Let x_1, y_1 be a regular system of parameters of R with $v(x_1) < v(y_1)$, i.e., we have $v(\mathfrak{m}) = v(x_1)$. The construction in the proof of VII(4.11) shows that there exists a linear combination $x := a_1 x_1 + b_1 y_1$ of x_1, y_1 with units a_1, b_1 of R such that every simple complete $\mathfrak{m}$-primary factor of $\mathfrak{a}$ and of $\mathfrak{b}$ different from $\mathfrak{m}$ is contracted from $A = R[\mathfrak{m}/x]$. Since v dominates R, we still have $v(\mathfrak{m}) = v(x)$; hence with $y := y_1$ we have $\mathfrak{m} = Rx + Ry$ and $v(y) > v(x)$. Let $\alpha'_j \mathrm{In}(x) + \beta'_j \mathrm{In}(y)$, $j \in \{1, \ldots, l\}$, be the pairwise different non-associated characteristic forms associated with the various simple complete $\mathfrak{m}$-primary factors of $\mathfrak{a}$ and of $\mathfrak{b}$ different from $\mathfrak{m}$. Then we have $\beta'_j \neq 0$ for $j \in \{1, \ldots, l\}$. We include among these forms the homogeneous polynomial $\overline{g} := \mathrm{In}(y)$, also there may be no simple complete $\mathfrak{m}$-primary factor of $\mathfrak{a}$ or of $\mathfrak{b}$ having characteristic form $\mathrm{In}(y)$. We label these forms as $\overline{g}_1 := \mathrm{In}(y), \overline{g}_2 := \alpha_2 \mathrm{In}(x) + \beta_2 \mathrm{In}(y), \ldots, \overline{g}_h := \alpha_h \mathrm{In}(x) + \beta_h \mathrm{In}(y)$; note that $\alpha_2, \ldots, \alpha_h$ are different from 0. We now write

$$\mathfrak{a} = \mathfrak{m}^{r_0} \mathfrak{a}_1 \cdots \mathfrak{a}_h, \quad \mathfrak{b} = \mathfrak{m}^{s_0} \mathfrak{b}_1 \cdots \mathfrak{b}_h$$

where $r_0, s_0 \in \mathbb{N}_0$, and where, for $i \in \{1, \ldots, h\}$, $\mathfrak{a}_i$ is the product of those simple complete $\mathfrak{m}$-primary factors of $\mathfrak{a}$ which have characteristic form $\overline{g}_i$ [$\mathfrak{a}_i = R$ if $\mathfrak{a}$ does not have a simple complete $\mathfrak{m}$-primary factor with characteristic form $\overline{g}_i$]; similarly for $\mathfrak{b}$ and $\mathfrak{b}_1, \ldots, \mathfrak{b}_h$. Note that the ideals $\mathfrak{a}_1, \ldots, \mathfrak{a}_h$ and $\mathfrak{b}_1, \ldots, \mathfrak{b}_h$ are contracted from A by VII(4.10). We set $r_i := \nu_R(\mathfrak{a}_i)$, $s_i := \nu_R(\mathfrak{b}_i)$ for $i \in \{1, \ldots, h\}$. Then we have $\nu_R(\mathfrak{a}) = r_0 + r_1 + \cdots + r_h$, $\nu_R(\mathfrak{b}) = s_0 + s_1 + \cdots + s_h$.

(1) We show that $r_i \leq s_i$ and $\mathfrak{b}_i \subset \mathfrak{m}^{s_i - r_i} \mathfrak{a}_i$ for every $i \in \{1, \ldots, h\}$, and that $s_0 \leq r_0$.

In fact, let $i \in \{1, \ldots, h\}$. There is nothing to show if $\mathfrak{a}_i = R$. Thus, assume that $\mathfrak{a}_i \neq R$, and let $\mathfrak{a}'_i$ be the transform of $\mathfrak{a}_i$ in A. Since $\mathfrak{a}_i$ is contracted from A, we have $\mathfrak{a}_i = \mathfrak{a}_i A \cap R = x^{r_i} \mathfrak{a}'_i \cap R$, $\mathfrak{a}'_i$ is $\mathfrak{n}(\overline{g}_i)$-primary and we have $\mathfrak{a}'_i A_{\mathfrak{n}(\overline{g}_i)} \cap A = \mathfrak{a}'_i$ [cf. VII(2.17)]. Let $\mathfrak{a}'$ be the transform of $\mathfrak{a}$ in A; then we have $\mathfrak{a}' A_{\mathfrak{n}(\overline{g}_i)} = \mathfrak{a}'_i A_{\mathfrak{n}(\overline{g}_i)}$ [cf. VII(2.15)]. Let $\mathfrak{b}'$ be the transform of $\mathfrak{b}$ in A. Since $\nu_R(\mathfrak{a}) = \nu_R(\mathfrak{b})$ and $\mathfrak{b} \subset \mathfrak{a}$, we have $\mathfrak{b}' A_{\mathfrak{n}(\overline{g}_i)} \subset \mathfrak{a}' A_{\mathfrak{n}(\overline{g}_i)}$. Let $\mathfrak{b}'_i$ be the transform of $\mathfrak{b}_i$ in A. We have $\mathfrak{b}' A_{\mathfrak{n}(\overline{g}_i)} = \mathfrak{b}'_i A_{\mathfrak{n}(\overline{g}_i)}$, hence $\mathfrak{b}'_i \neq A$, and therefore $\mathfrak{b}'_i$ is $\mathfrak{n}(\overline{g}_i)$-primary, $\mathfrak{b}'_i A_{\mathfrak{n}(\overline{g}_i)} \cap A = \mathfrak{b}'_i$, $x^{s_i} \mathfrak{b}'_i \cap R = \mathfrak{b}_i$ and $\mathfrak{b}'_i \subset \mathfrak{a}'_i$. Suppose that $r_i > s_i$. Then we have $x^{r_i - s_i} x^{s_i} \mathfrak{b}'_i \subset x^{r_i} \mathfrak{a}'_i$, and since $\mathfrak{m}^{r_i - s_i} \mathfrak{b}_i$ is contracted from A [cf.

VII(4.10)$\,$], we have $\mathfrak{m}^{r_i - s_i}\mathfrak{b}_i \subset \mathfrak{a}_i$, hence $\mathfrak{m}$ divides $\mathfrak{a}_i$ [$\,$cf. VII(3.22)(2)$\,$], contrary
to the definition of $\mathfrak{a}_i$. Therefore we have $s_i \geq r_i$, hence $x^{s_i}\mathfrak{b}_i' \subset x^{s_i - r_i}x^{r_i}\mathfrak{a}_i'$, and,
as above, $\mathfrak{b}_i \subset \mathfrak{m}^{s_i - r_i}\mathfrak{a}_i$. Since $\nu_R(\mathfrak{b}) = \nu_R(\mathfrak{a})$, hence $r_0 + \cdots + r_h = s_0 + \cdots + s_h$,
we get $s_0 \leq r_0$.

(2) We have $\mathfrak{a} = \mathfrak{b} + k\,\mathfrak{a}_0$, hence $\dim_k((\mathfrak{a} + \mathfrak{m}^{r+1})/\mathfrak{m}^{r+1})$ and $\dim_k((\mathfrak{b} + \mathfrak{m}^{r+1})/\mathfrak{m}^{r+1})$
differ at most by 1. On the other hand, for $i \in \{1, \ldots, h\}$ the k-vector space
$(\mathfrak{a}_i + \mathfrak{m}^{r_i+1})/\mathfrak{m}^{r_i+1}$ is generated by $\overline{g}_i^{r_i}$ [$\,$since $\deg(c(\mathfrak{a}_i)) = \nu_R(\mathfrak{a}_i)$ by VII(3.22)(1)$\,$];
similarly for $\mathfrak{b}_i$. Therefore we have $\dim_k((\mathfrak{a} + \mathfrak{m}^{r+1})/\mathfrak{m}^{r+1}) = \dim_k(\mathfrak{m}^{r_0}/\mathfrak{m}^{r_0+1}) = $
r_0, $\dim_k((\mathfrak{b} + \mathfrak{m}^{r+1})/\mathfrak{m}^{r+1}) = s_0$, and since $\mathfrak{b} \subset \mathfrak{a}$, we get $r_0 = s_0$ or $r_0 = s_0 + 1$.

(a) If $r_0 = s_0$, then we have $r_j = s_j$ for every $j \in \{1, \ldots, h\}$, and $\mathfrak{b}_i \neq \mathfrak{a}_i$ for
exactly one $i \in \{1, \ldots, h\}$ [$\,$note that, if $\mathfrak{c} \subsetneq \mathfrak{d}$ and $\mathfrak{c}' \subsetneq \mathfrak{d}'$ are complete $\mathfrak{m}$-primary
ideals of R, then we have $\mathfrak{c}\mathfrak{c}' \subsetneq \mathfrak{d}\mathfrak{c}' \subsetneq \mathfrak{d}\mathfrak{d}'$ by unique factorization of complete ideals,
cf. VII(4.17)$\,$].

(b) If $r_0 = s_0 + 1$, then we have $s_i = r_i + 1$ for one $i \in \{1, \ldots, h\}$, while $r_j = s_j$
for every $j \in \{1, \ldots, h\}$, $j \neq i$, hence $\mathfrak{b}_i \subset \mathfrak{m}\mathfrak{a}_i$ and, as above, $\mathfrak{b}_j = \mathfrak{a}_j$ for
$j \in \{1, \ldots, h\}$, $j \neq i$.

Therefore, in both cases, there exists exactly one $i \in \{1, \ldots, h\}$ with $\mathfrak{b}_i \neq \mathfrak{a}_i$.

(c) We show that $\mathfrak{b}_1 \neq \mathfrak{a}_1$. Suppose that $\mathfrak{b}_1 = \mathfrak{a}_1$, hence that $s_1 = r_1$. Let
$j \in \{2, \ldots, h\}$. Since $v(y) > v(x)$ and $\alpha_j \neq 0$, we have $v(\alpha_j x + \beta_j y)^{r_j} = v(x^{r_j}) = $
$v(\mathfrak{m}^{r_j})$. Since $\mathfrak{a}_j + \mathfrak{m}^{r_j+1} = R(\alpha_j x + \beta_j y)^{r_j} + \mathfrak{m}^{r_j+1}$, we get $v(\mathfrak{a}_j) = v(\mathfrak{m}^{r_j})$, and
similarly $v(\mathfrak{b}_j) = v(\mathfrak{m}^{s_j})$. Since $s_1 = r_1$, we have $s_0 + s_2 + \cdots + s_h = r_0 + r_2 + \cdots + r_h$.
We have $v(\mathfrak{a}) - v(\mathfrak{a}_1) = (r_0 + r_2 + \cdots + r_h)v(\mathfrak{m}) = (s_0 + s_2 + \cdots + s_h)v(\mathfrak{m}) = $
$v(\mathfrak{b}) - v(\mathfrak{b}_1)$, hence $v(\mathfrak{a}) = v(\mathfrak{b})$, contradicting the definition of $\mathfrak{b}$ [$\,$cf. (2) in the
proof of (7.2)$\,$].

(3) Let v' be a zero-dimensional valuation of K having center $\mathfrak{n}$ in S. Since S
dominates R, we have $v'(\mathfrak{m}) > 0$. Furthermore, we have $v'(\mathfrak{a}) < v'(\mathfrak{b})$ by definition
of $\mathfrak{b}$.

Suppose that $v'(y) < v'(x)$; then we have $v'(\mathfrak{m}) = v'(y)$. We have $\mathfrak{a}_1 + \mathfrak{m}^{r_1+1} = $
$y^{r_1}R + \mathfrak{m}^{r_1+1}$, and therefore $v'(\mathfrak{a}_1) = r_1 v'(\mathfrak{m})$, and, similarly, $v'(\mathfrak{b}_1) = s_1 v'(\mathfrak{m})$.
If $r_0 = s_0$, then we have $r_i = s_i$ for $i \in \{1, \ldots, h\}$ and $\mathfrak{a}_i = \mathfrak{b}_i$ for $i \in \{2, \ldots, h\}$ [$\,$cf.
(2)(a) and (2)(c)$\,$]. From $v'(\mathfrak{a}) < v'(\mathfrak{b})$ we get $v'(\mathfrak{a}_1) < v'(\mathfrak{b}_1)$. From $r_1 = s_1$ we get
$v'(\mathfrak{a}_1) = v'(\mathfrak{b}_1)$. This contradiction shows that in this case we have $v'(y) \geq v'(x)$.
If $r_0 = s_0 + 1$, then we have $s_1 = r_1 + 1$ and $\mathfrak{a} = \mathfrak{m}^{s_0}(\mathfrak{m}\mathfrak{a}_1)\mathfrak{b}_2 \cdots \mathfrak{b}_h$ [$\,$cf. (2)(b)
and (2)(c)$\,$]. From $v'(\mathfrak{a}) < v'(\mathfrak{b})$ we get $v'(\mathfrak{m}\mathfrak{a}_1) < v'(\mathfrak{b}_1)$. This inequality yields
$(r_1 + 1)v'(\mathfrak{m}) < s_1 v'(\mathfrak{m}) = (r_1 + 1)v'(\mathfrak{m})$, hence $r_1 + 1 < r_1 + 1$ since $v'(\mathfrak{m}) > 0$
which is absurd. Therefore also in this case we have $v'(y) \geq v'(x)$.
Since S is the intersection of all valuation rings of zero-dimensional valuations of
K having center $\mathfrak{n}$ in S [$\,$cf. (7.2)$\,$], we have $y/x \in S$, hence $R[\mathfrak{m}/x] \subset S$; since
S dominates R, also v dominates R, and the first quadratic dilatation of R with
respect to v is $R_1 = R[\mathfrak{m}/x]_{R[\mathfrak{m}/x]\cap\mathfrak{n}} \subset S$, as we wanted to show.
This ends the proof of (7.4).

(7.5) Corollary: *Let $\mathfrak{n}$ be the maximal ideal of S. Let v be a zero-dimensional
valuation of K which dominates S. If the first quadratic dilatation R_1 of R with*

respect to v is not contained in S, and if there exists an l-th quadratic dilatation R_l of R with respect to v with $S \subset R_l$, then $\mathfrak{n} R_l$ is a principal ideal of R_l.

Proof: By (7.4) we have $\nu_R(\mathfrak{b}) = \nu_R(\mathfrak{a}) + 1$. Using the notation in the proof of (7.3), we have $\mathfrak{a} R[\mathfrak{a}/a_0] = a_0 R[\mathfrak{a}/a_0]$, and since $R[\mathfrak{a}/a_0] \subset S$, the ideal $\mathfrak{a} S$ is a principal ideal of S, hence $\mathfrak{a} R_l$ is a principal ideal of R_l. From (4.3) and VII(5.8) we get that the ideal $\mathfrak{b} R_l$ is a principal ideal of R_l; since $\mathfrak{n} = \mathfrak{b}/a_0 \cdot S$, it follows that $\mathfrak{n} R_l$ is a principal ideal of R_l.

7.2 Tangential Ideals

(7.6) TANGENT LINES AND MAXIMAL PRIMARY IDEALS: Let $X \subset \mathbb{A}^n$ be an affine algebraic variety passing through the origin, let $k[T_1, \ldots, T_n]$ be the affine coordinate ring of $\mathbb{A}^n$, set $\mathfrak{a} := \mathcal{I}(X)$ and let $A = k[T_1, \ldots, T_n]/\mathfrak{a} = k[t_1, \ldots, t_n]$ be the affine coordinate ring of X. Let R be the local ring of the origin on X, and, for $i \in \{1, \ldots, n\}$, let x_i be the image of t_i in R. The maximal ideal $\mathfrak{n}$ of R is generated by the n elements $x_1, \ldots, x_n$, and we have $R = A_{\mathfrak{m}}$ where $\mathfrak{m}$ is the maximal ideal of A generated by $t_1, \ldots, t_n$.

Since $R/\mathfrak{n} = k$, we know that every element $x \in \mathfrak{n}$ admits a representation $x = \beta_1 x_1 + \cdots + \beta_n x_n + x'$ with $\beta_1, \ldots, \beta_n \in k$ and $x' \in \mathfrak{n}^2$.

(1) Let $L \subset \mathbb{A}^n$ be a line through the origin, and let $(\gamma_1, \ldots, \gamma_n) \in k^n$ be a non-zero vector determining L, i.e., the line L is the intersection of all hyperplanes $\mathcal{Z}((\alpha_1 T_1 + \cdots + \alpha_n T_n))$ in $\mathbb{A}^n$ where $(\alpha_1, \ldots, \alpha_n) \in k^n$ is a non-zero vector with $\gamma_1 \alpha_1 + \cdots + \gamma_n \alpha_n = 0$. The elements $\alpha_1 x_1 + \cdots + \alpha_n x_n + x'$ in $\mathfrak{n}$ [where $\alpha_1, \ldots, \alpha_n \in k$ and $x' \in \mathfrak{n}^2$] with $\gamma_1 \alpha_1 + \cdots + \gamma_n \alpha_n = 0$ form a maximal $\mathfrak{n}$-primary ideal of R [cf. B(10.16) for the definition of a maximal $\mathfrak{n}$-primary ideal]. In fact, it is easy to check that the set $\mathfrak{q}$ of these elements is an ideal containing $\mathfrak{n}^2$ and properly contained in $\mathfrak{n}$, hence it is an $\mathfrak{n}$-primary ideal [since $\mathfrak{n}$ is a maximal ideal]. We have $\gamma_i \neq 0$ for some $i \in \{1, \ldots, n\}$, hence $x_i \notin \mathfrak{q}$. For every $j \in \{1, \ldots, n\}$, $j \neq i$, we have $x'_j := x_j - (\gamma_j/\gamma_i)x_i \in \mathfrak{q}$, hence $\mathfrak{q}$ is a maximal $\mathfrak{n}$-primary ideal, generated by $x'_1, \ldots, x'_{i-1}, x'_{i+1}, \ldots, x'_n$ and $\mathfrak{n}^2$.

(2) Let $\mathfrak{q}$ be a maximal $\mathfrak{n}$-primary ideal of R; then $\mathfrak{n}/\mathfrak{q}$ is a one-dimensional k-vector space, and we have $\mathfrak{n}^2 \subset \mathfrak{q}$ [cf. B(10.17)(2)]. Let $z \in \mathfrak{n} \setminus \mathfrak{q}$; then every element $x \in \mathfrak{n}$ has a unique representation $x = \gamma z + y$ with $\gamma \in k$ and $y \in \mathfrak{q}$. In particular, we have $x_i = \gamma_i z + y_i$ with $\gamma_i \in k$ and $y_i \in \mathfrak{q}$ for $i \in \{1, \ldots, n\}$. Since $\mathfrak{q} \neq \mathfrak{n}$, we have $(\gamma_1, \ldots, \gamma_n) \neq 0$. Let $x \in \mathfrak{n}$; we write $x = \alpha_1 x_1 + \cdots + \alpha_n x_n + x'$ with $\alpha_1, \ldots, \alpha_n \in k$ and $x' \in \mathfrak{n}^2$. Then we have

$$x = (\alpha_1 \gamma_1 + \cdots + \alpha_n \gamma_n)z + \alpha_1 y_1 + \cdots + \alpha_n y_n + x';$$

we see that $x \in \mathfrak{q}$ iff $\alpha_1 \gamma_1 + \cdots + \alpha_n \gamma_n = 0$. The one-dimensional subspace of k^n with basis $\{(\gamma_1, \ldots, \gamma_n)\}$ is uniquely determined by $\mathfrak{q}$.

(3) From (1) and (2) it follows that there exists a bijective map between the set of lines in $\mathbb{A}^n$ passing through the origin and the set of maximal $\mathfrak{n}$-primary ideals of R.

(4) Let $\mathfrak{q}$ be a maximal $\mathfrak{n}$-primary ideal of R, and let $L \subset \mathbb{A}^n$ be the line through the origin corresponding to $\mathfrak{q}$. We show: L *is a tangent line to X at the origin iff*

$$\mathfrak{n}^l \neq \mathfrak{n}^{l-1}\mathfrak{q} \quad \text{for every } l \in \mathbb{N}. \tag{$*$}$$

In fact, let $\mathfrak{q}$ correspond to $(\gamma_1, \ldots, \gamma_n) \in k^n \setminus \{0\}$, and assume, as we may, that $\gamma_1 \neq 0$. The elements $x_1' := x_1, x_2' := x_2 - (\gamma_2/\gamma_1)x_1, \ldots, x_n' := x_n - (\gamma_n/\gamma_1)x_1$ generate $\mathfrak{n}$, and $\mathfrak{q}$ is generated by $x_2', \ldots, x_n'$ and $\mathfrak{n}^2$. After having made a linear change of coordinates, we may assume, therefore, that $\gamma_1 = 1$, $\gamma_2 = \cdots = \gamma_n = 0$, and that $\mathfrak{q}$ *is generated by* $x_2, \ldots, x_n$ *and* $\mathfrak{n}^2$. Then the line corresponding to $\mathfrak{q}$ is the line $L = \mathcal{Z}((T_2, \ldots, T_n))$.

We know that L is a tangent line to X at the origin iff $\text{In}(\mathfrak{a}) \subset (T_2, \ldots, T_n)$ [cf. A(6.13)(4)]. We assume that L is not a tangent line to X at the origin. Then there exists $F \in \mathfrak{a}$ with $\text{In}(F) \notin (T_2, \ldots, T_n)$. We set $r := \deg(F)$, and we write $F = F_r + \cdots + F_{r+s}$ where $F_i \in k[T_1, \ldots, T_n]$ is homogeneous of degree i for $i \in \{r, \ldots, r+s\}$, and $\text{In}(F) = F_r$. We have $F_r = \alpha_0 T_1^r + G_2 T_2 + \cdots + G_n T_n$ with $\alpha_0 \in k^\times$ and $G_2, \ldots, G_n \in k[T_1, \ldots, T_n]$ homogeneous of degree $r - 1$. Now we have $0 = F_r(x_1, \ldots, x_n) + \cdots + F_{r+s}(x_1, \ldots, x_n)$ and $F_{r+i}(x_1, \ldots, x_n) = \alpha_i x_1^{r+i} + y_i$ with $\alpha_i \in k$ and $y_i \in \mathfrak{n}^{r-1}\mathfrak{q}$ for $i \in \{1, \ldots, s\}$, we have $x_1^r(\alpha_0 + \alpha_1 x_1 + \cdots + \alpha_s x_1^s) \in \mathfrak{n}^{r-1}\mathfrak{q}$, hence $x_1^r \in \mathfrak{n}^{r-1}\mathfrak{q}$, and therefore we have $\mathfrak{n}^r = \mathfrak{n}^{r-1}\mathfrak{q}$. Conversely, assume that $\mathfrak{n}^r = \mathfrak{n}^{r-1}\mathfrak{q}$ for some $r \in \mathbb{N}$. Then there exist $y_2, \ldots, y_n \in \mathfrak{n}^{r-1}$ such that $x_1^r = y_2 x_2 + \cdots + y_n x_n$. This implies that there exist homogeneous polynomials $G_2, \ldots, G_n \in k[T_1, \ldots, T_n]$ of degree $r - 1$ and a polynomial $H \in k[T_1, \ldots, T_n]$ with $H(0) = 0$ such that $F := (T_1^r + G_2 T_2 + \cdots + G_n T_n)(1 + H) \in \mathfrak{a}$, and we have $\text{In}(F) = T_1^r + G_2 T_2 + \cdots + G_n T_n \notin (T_2, \ldots, T_n)$, hence L is not a tangent line to X at the origin.

(5) A maximal $\mathfrak{n}$-primary ideal $\mathfrak{q}$ of R is called a *tangential ideal* if the line in $\mathbb{A}^n$ through the origin corresponding to it is a tangent line to X at the origin; note that $\mathfrak{q}$ is a tangential ideal iff $(*)$ in (4) holds.

(7.7) ANOTHER CHARACTERIZATION OF TANGENTIAL IDEALS: We keep the notations of (7.6) and we assume, in addition, that X is irreducible. Let F be the field of rational functions on X.

(1) Let v be a zero-dimensional valuation of F which dominates R; then we have $v(\mathfrak{n}) = \min(v(x_1), \ldots, v(x_n))$. The v-ideal $\mathfrak{q} := \{x \in R \mid v(x) > v(\mathfrak{n})\}$ of R is an $\mathfrak{n}$-primary ideal [since $\mathfrak{n}^2 \subset \mathfrak{q}$]; it is properly contained in $\mathfrak{n}$. Let $z \in \mathfrak{n} \setminus \mathfrak{q}$; for $x \in \mathfrak{n}$ we have $v(x/z) \geq 0$, hence there exists $\gamma \in k$ with $v(x/z - \gamma) > 0$ [since v is zero-dimensional], i.e., we have $x - \gamma z \in \mathfrak{q}$. Therefore $\mathfrak{q}$ is a maximal $\mathfrak{n}$-primary ideal [cf. B(10.16)].

(2) Let $\mathfrak{q}$ be a maximal $\mathfrak{n}$-primary ideal of R; we assume that there exists a zero-dimensional valuation v of F dominating R such that $\mathfrak{q}$ is a v-ideal [i.e., we have $\mathfrak{q}A_v \cap R = \mathfrak{q}$]. Then we have $\mathfrak{q} = \{x \in R \mid v(x) > v(\mathfrak{n})\}$. In fact, since $\mathfrak{q}$ is a v-ideal different from $\mathfrak{n}$, we have $v(\mathfrak{q}) > v(\mathfrak{n})$ [since $\mathfrak{n}$ is a v-ideal]. We choose $z \in \mathfrak{n}$ with $v(\mathfrak{n}) = v(z)$. Let $x \in \mathfrak{n}$ and $v(x) > v(\mathfrak{n})$. We have $x = \gamma z + y$ with

$\gamma \in k$ and $y \in \mathfrak{q}$, and since $v(x) > v(z)$, we have $\gamma = 0$, hence $x \in \mathfrak{q}$. This proves our assertion.

(3) Let $\mathfrak{q}$ be a maximal $\mathfrak{n}$-primary ideal of R. Then $\mathfrak{q}$ is a tangential ideal iff there exists a zero-dimensional valuation v of F dominating R such that $\mathfrak{q}$ is a v-ideal. In fact, to prove this assertion, we may assume, as above, that we have $x_1 \notin \mathfrak{q}$, and that $\mathfrak{q}$ is generated by $x_2, \ldots, x_n$ and $\mathfrak{n}^2$.

(a) Assume that $\mathfrak{q}$ is a v-ideal where v is a zero-dimensional valuation v of F which dominates R. We have $v(\mathfrak{n}) > v(\mathfrak{q})$, and since $v(\mathfrak{n}) = v(x_1)$, we have $v(x_j) > v(x_1)$ for $j \in \{2, \ldots, n\}$, and therefore $v(\mathfrak{n}^{l-1}\mathfrak{q}) \geq (l-1)v(\mathfrak{n}) + v(\mathfrak{q}) > lv(x_1) = v(x_1^l)$, hence $x_1^l \notin \mathfrak{n}^{l-1}\mathfrak{q}$ for every $l \in \mathbb{N}$; this means that $(*)$ in (7.6)(4) holds.

(b) Assume that $(*)$ in (7.6)(4) holds. We have $x_1 \neq 0$ since $x_1 \notin \mathfrak{q}$. We set $A' := A[\mathfrak{m}/x_1] = k[x_1, x_2/x_1, \ldots, x_n/x_1]$; let $\mathfrak{m}'$ be the ideal in A' which is generated by $x_1, x_2/x_1, \ldots, x_n/x_1$.
Suppose that $\mathfrak{m}' = A'$. Then there exist $m \in \mathbb{N}$ and elements $w_1, \ldots, w_n \in \mathfrak{m}^m$ with $1 = (w_1/x_1^m)x_1 + (w_2/x_1^m)\cdot(x_2/x_1) + \cdots + (w_n/x_1^m)\cdot(x_n/x_1)$, hence $x_1^{m+1} = w_1 x_1^2 + w_2 x_2 + \cdots + w_n x_n \in \mathfrak{n}^m \mathfrak{q}$, hence we have $\mathfrak{n}^m = \mathfrak{n}^{m-1}\mathfrak{q}$, contradicting $(*)$. Now $\mathfrak{m}'$ is a maximal ideal of A'; let v be a zero-dimensional valuation of F which dominates $A'_{\mathfrak{m}'}$ [cf. I(10.9)]. Since $\mathfrak{m}A' = x_1 A' \subset \mathfrak{m}'$, we have $\mathfrak{m}' \cap A = \mathfrak{m}$, hence we see that $A'_{\mathfrak{m}'}$ dominates $A_{\mathfrak{m}} = R$. We have $v(x_1) > 0$, $v(x_j/x_1) > 0$ for $j \in \{2, \ldots, n\}$, and therefore $v(\mathfrak{q}) > v(x_1) = v(\mathfrak{n})$. Let $z \in \mathfrak{n}$ and $v(z) > v(\mathfrak{n})$; then we have $z = \alpha_2 x_2 + \cdots + \alpha_n x_n + z'$ with $\alpha_2, \ldots, \alpha_n \in k$ and $z' \in \mathfrak{n}^2$, hence z lies in $\mathfrak{q}$, and therefore $\mathfrak{q}$ is a v-ideal.

(7.8) TANGENTIAL IDEALS AND BLOWING UP: Let $X \subset \mathbb{P}^n$ be an irreducible projective variety with field of rational functions F, and assume that X passes through $p = (1 : 0 : \cdots : 0)$. Then $\mathcal{O}_{X,p}$ is also the ring of p, considered as a point of the affine variety $X \cap \{(\xi_0 : \xi_1 : \cdots : \xi_n) \mid \xi_0 \neq 0\}$ contained in $\mathbb{A}^n$. Let $\bar{\mathfrak{n}}$ be the maximal ideal of $\mathcal{O}_{X,p}$, and let Σ be the non-empty set of zero-dimensional valuations of F which dominate $\mathcal{O}_{X,p}$ [cf. I(10.9)].
Let $\pi: X' \to X$ be the blowing up of X with center p, and let $E' = \pi^{-1}(\{p\})$ be the exceptional fibre. Now $X' \subset \mathbb{P}^n \times \mathbb{P}^{n-1}$ is an irreducible projective variety with field of rational functions F [cf. A(13.8)].
In the following, we use the notations introduced in A(13.11). We set $\bar{y}_i := y_i + \mathfrak{b}$ for $j \in \{1, \ldots, n\}$, $\bar{z}_j := z_j + \mathfrak{b}_1$ for $j \in \{2, \ldots, n\}$. We assume that $\bar{y}_1 \neq 0$; then we have $\bar{z}_j = \bar{y}_j/\bar{y}_1$ for $j \in \{2, \ldots, n\}$.
(1) Let $q = (p, (\beta_1 : \cdots : \beta_n))$ be a point in E'. Let Σ'_q be the non-empty set of zero-dimensional valuations of F which dominate $\mathcal{O}_{X',q}$; note that $\Sigma'_q \subset \Sigma$. Let $\mathfrak{q}$ be the maximal $\bar{\mathfrak{n}}$-primary ideal of $\mathcal{O}_{X,p}$ corresponding to $(\beta_1, \ldots, \beta_n)$ [cf. (7.6)(3)]. We show: $\mathfrak{q}$ is a v-ideal for every $v \in \Sigma'_q$, and therefore $\mathfrak{q}$ is, in particular, a tangential ideal of $\mathcal{O}_{X,p}$ [cf. (7.7)(3)].
For the proof we assume, as we may, that $\beta_1 \neq 0$. The ideal $\mathfrak{q}$ is generated by $\bar{y}_2 - (\beta_2/\beta_1)\bar{y}_1, \ldots, \bar{y}_n - (\beta_n/\beta_1)\bar{y}_1$ and $\bar{\mathfrak{n}}^2$. The ring $\mathcal{O}_{X',q}$ is the localization of $k[\bar{y}_1, \bar{z}_2 - \beta_2/\beta_1, \ldots, \bar{z}_n - \beta_n/\beta_1]$ with respect to the maximal ideal generated by $\bar{y}_1, \bar{z}_2 - \beta_2/\beta_1, \ldots, \bar{z}_n - \beta_n/\beta_1$. Let $v \in \Sigma'_q$. Then we have $v(\bar{y}_1) > 0$, $v(\bar{z}_j - \beta_j/\beta_1) =$

$v(\overline{y}_j/\overline{y}_1 - \beta_j/\beta_1) > 0$ for $j \in \{2,\ldots,n\}$ which means that $v(\overline{y}_j - (\beta_j/\beta_1)\overline{y}_1) > v(\overline{y}_1)$ for $j \in \{2,\ldots,n\}$, hence $v(\mathfrak{q}) > v(\overline{y}_1) = v(\overline{n})$. Let $\overline{y} \in \overline{n}$ and $v(\overline{y}) > v(\overline{n})$. We have $\overline{y} = \alpha_2\overline{y}_2 + \cdots + \alpha_n\overline{y}_n + \overline{y}'$ with $\alpha_2,\ldots,\alpha_n \in k$ and $\overline{y}' \in \overline{n}^2$, hence $\overline{y} \in \mathfrak{q}$. Therefore $\mathfrak{q}$ is a v-ideal.

(2) Let $(\beta_1,\ldots,\beta_n) \in k^n \setminus \{0\}$, and let $\mathfrak{q}$ be the maximal $\overline{n}$-primary ideal of $\mathcal{O}_{X,p}$ corresponding to $(\beta_1,\ldots,\beta_n)$. We show: If $\mathfrak{q}$ is a tangential ideal, then the point $q := (p,(\beta_1 : \cdots : \beta_n)) \in \mathbb{P}^n \times \mathbb{P}^{n-1}$ lies in E', and $\mathfrak{q}$ is a v-ideal for every $v \in \Sigma'_q$. For the proof we assume, as we may, that $\beta_1 \neq 0$. Then $\mathfrak{q}$ is generated by the elements $\overline{y}_2 - (\beta_2/\beta_1)\overline{y}_1,\ldots,\overline{y}_n - (\beta_n/\beta_1)\overline{y}_1$ and $\overline{n}^2$. There exists $v \in \Sigma$ such that $\mathfrak{q}$ is a v-ideal [cf. (7.7)(3)]; in particular, we have $v(\overline{y}_j - (\beta_j/\beta_1)\overline{y}_1) > v(\overline{y}_1)$ for $j \in \{2,\ldots,n\}$. Now $p \in X$ is the point of X which is dominated by v, v dominates exactly one point $q \in X'$ [cf. (3.4)], and clearly we have $q \in E'$. We have $v(\overline{y}_1) > 0$, $v(\overline{z}_j - (\beta_j/\beta_1)) > 0$ for $j \in \{2,\ldots,n\}$, hence $(\overline{y}_1, \overline{z}_2 - \beta_2/\beta_1, \ldots, \overline{z}_n - \beta_n/\beta_1)$ is the center of v in $k[\overline{y}_1, \overline{z}_2, \ldots, \overline{z}_n]$, hence $q = (p,(\beta_1 : \cdots : \beta_n))$.

(3) We consider the bijective map between the set of lines in $\mathbb{A}^n$ passing through the origin and the set of maximal $\overline{n}$-primary ideals of $\mathcal{O}_{X,p}$ [cf. (7.6)(3)]. From (1) and (2) we get: If $(\beta_1,\ldots,\beta_n) \in k^n \setminus \{0\}$ corresponds to $\mathfrak{q}$, then $q := (p,(\beta_1 : \cdots : \beta_n)) \in \mathbb{P}^n \times \mathbb{P}^{n-1}$ lies in E' iff $\mathfrak{q}$ is a tangential ideal of $\mathcal{O}_{X,p}$; in this case $\mathfrak{q}$ is a v-ideal for every zero-dimensional valuation v of F which dominates $\mathcal{O}_{X',q}$.

(4) Let $q \in E'$ correspond to the tangential ideal $\mathfrak{q}$ of $\mathcal{O}_{X,p}$. We consider the Rees ring $\mathcal{R}(\overline{n}, \mathcal{O}_{X,p}) = \bigoplus_{m \geq 0} \overline{n}^m T^m \subset \mathcal{O}_{X,p}[T]$, and we show: The homogeneous ideal $\mathfrak{p} = \bigoplus_{m \geq 0} \mathfrak{p}_m$ of $\mathcal{R}(\overline{n}, \mathcal{O}_{X,p})$ with $\mathfrak{p}_0 := \overline{n}$ and $\mathfrak{p}_m := \overline{n}^{m-1}\mathfrak{q}T^m$ for every $m \in \mathbb{N}$ is a prime ideal which is not irrelevant, and we have

$$\mathcal{O}_{X',q} = \mathcal{R}(\overline{n}, \mathcal{O}_{X,p})_{(\mathfrak{p})}.$$

Proof: Note that $\mathfrak{q}$ and $\overline{n}$ are v-ideals for every zero-dimensional valuation of F which dominates $\mathcal{O}_{X',q}$. There exists $z \in \overline{n}$ with $\overline{n} = kz + \mathfrak{q}$. Just as in (3) of the proof of (7.3) it follows that $\mathfrak{p}$ is a homogeneous prime ideal in $\mathcal{R}(\overline{n}, \mathcal{O}_{X,p})$, and from (7.7)(3) it follows that $\mathfrak{p}$ is not irrelevant.

We may assume that $q = (p,(1: 0: \cdots : 0))$; then $\mathfrak{q}$ is the ideal generated by $\overline{y}_2,\ldots,\overline{y}_n, \overline{n}^2$, and $\mathcal{O}_{X',q}$ is the localization of $k[\overline{y}_1, \overline{z}_2, \ldots, \overline{z}_n]$ with respect to the prime ideal generated by $\overline{y}_1, \overline{z}_2, \ldots, \overline{z}_n$. An element z' in $\mathcal{O}_{X',q}$ has the form

$$z' = \frac{F(\overline{y}_1, \overline{z}_2, \ldots, \overline{z}_n)}{G(\overline{y}_1, \overline{z}_2, \ldots, \overline{z}_n)} \text{ with } F(T_1,\ldots,T_n), G(T_1,\ldots,T_n) \in k[T_1,\ldots,T_n]$$

and $G(0,\ldots,0) \neq 0$. Making the substitution $\overline{z}_j = \overline{y}_j/\overline{y}_1$ for $j \in \{2,\ldots,n\}$ shows that we can write

$$z' = \frac{f(\overline{y}_1,\ldots,\overline{y}_n)}{g(\overline{y}_1,\ldots,\overline{y}_n)} = \frac{f_r(\overline{y}_1,\ldots,\overline{y}_n) + f_{r+1}(\overline{y}_1,\ldots,\overline{y}_n) + \cdots}{g_r(\overline{y}_1,\ldots,\overline{y}_n) + g_{r+1}(\overline{y}_1,\ldots,\overline{y}_n) + \cdots} \tag{*}$$

where $f_i(T_1,\ldots,T_n)$, $g_i(T_1,\ldots,T_n) \in k[T_1,\ldots,T_n]$ are homogeneous of degree i for $i = r, r+1, \ldots$, and where

$$g_r(T_1,\ldots,T_n) = \alpha T_1^r + \cdots \quad \text{with } \alpha \in k^\times. \tag{**}$$

Conversely, every quotient (*) for which (**) holds is an element of $\mathcal{O}_{X',q}$. It is clear that the elements $f(\overline{y}_1, \ldots, \overline{y}_n)$, $g(\overline{y}_1, \ldots, \overline{y}_n)$ in (*) lie in $\overline{\mathfrak{n}}^r$.

Let $r \in \mathbb{N}$, $g = g_r + g_{r+1} + \cdots \in k[T_1, \ldots, T_n]$ where g_i is homogeneous of degree i for $i = r, r+1, \ldots$ and $g_r \neq 0$. We write $g_r = \alpha T_1^r + \cdots$ with $\alpha \in k$. Set $\overline{w} := g(\overline{y}_1, \ldots, \overline{y}_n)$; then $\overline{w} \in \overline{\mathfrak{n}}^r$. We show that $\alpha \neq 0$ iff $\overline{w} \notin \overline{\mathfrak{n}}^{r-1}\mathfrak{q}$. In fact, note that $\overline{\mathfrak{n}}^{r+1} \subset \overline{\mathfrak{n}}^{r-1}\mathfrak{q}$ [since $\mathfrak{n}^2 \subset \mathfrak{q}$]. If $\alpha \neq 0$, then $\overline{w} \in \overline{\mathfrak{n}}^{r-1}\mathfrak{q}$ would imply that $\overline{y}_1^r \in \overline{\mathfrak{n}}^{r-1}\mathfrak{q}$, hence $\overline{\mathfrak{n}}^r = \overline{\mathfrak{n}}^{r-1}\mathfrak{q}$, contradicting (7.7)(3). Conversely, if $\overline{w} \notin \overline{\mathfrak{n}}^{r-1}\mathfrak{q}$, then $\alpha \neq 0$. Thus, we have proved the assertion.

7.3 The Main Result

(7.9) Theorem: *Let X be an irreducible normal projective surface and let*

$$\cdots \longrightarrow X_2 \longrightarrow X_1 \longrightarrow X_0 := X \qquad (*)$$

be a sequence such that, for $i \geq 1$, X_i is the normalization of a surface obtained by blowing up a singular point of X_{i-1}. Then (*) is a finite sequence.

Proof: Let K be the field of rational functions on X.

We set $\mathcal{S}_0 = \{X_0\}$, and let, for $i \in \mathbb{N}_0$, $\mathcal{S}_{i+1}$ be the set of all normalizations of all those surfaces which can be obtained by blowing up a singular point of a surface in $\mathcal{S}_i$. Since the set of singular points of an irreducible normal surface is finite [cf. A(7.17)(2)], it is easily seen by induction that $\mathcal{S}_n$ is a finite set for every $n \in \mathbb{N}_0$. Let $n \in \mathbb{N}_0$, and let $\mathcal{T}_n$ be the set of all local subrings Q of K such that there exists a sequence

$$\mathcal{S}_0 \subsetneqq \mathcal{S}_1 \subsetneqq \cdots \subsetneqq \mathcal{S}_n = Q \qquad (**)$$

where, for every $i \in \{0, \ldots, n\}$, $S_i = \mathcal{O}_{X_i, x_i}$ with $X_i \in \mathcal{S}_i$ and $x_i \in \mathrm{Sing}(X_i)$, and, for $i \in \{1, \ldots, n\}$, X_i is the normalization of $X_i' := \mathrm{Bl}_{x_{i-1}}(X_{i-1})$ where $x_{i-1} \in \mathrm{Sing}(X_{i-1})$, and x_i lies over a point x_i' of the exceptional fibre of X_i'. Every element of $\mathcal{T}_n$ is the local ring of a singular point on one of the finitely many members of $\mathcal{S}_n$; therefore $\mathcal{T}_n$ is a finite set. Note that $\mathcal{T}_0 = \{\mathcal{O}_{X_0, x} \mid x \in \mathrm{Sing}(X_0)\}$. Suppose that there were infinitely many sequences of the form (**). Since $\mathcal{T}_0$ is finite, there would be infinitely many such sequences beginning with a specific $S_0 \in \mathcal{T}_0$; then $S_0 = \mathcal{O}_{X_0, x_0}$ with $x_0 \in \mathrm{Sing}(X_0)$. Since $\mathcal{T}_1$ is finite, there would be some $S_1 \in \mathcal{T}_1$ such that among those sequences beginning with S_0, there would be infinitely many which begin with $S_0 \subsetneqq S_1 \subsetneqq \cdots$. Since $\mathcal{T}_2$ is finite, there would be some $S_2 \in \mathcal{T}_2$ such that among the sequences which begin with $S_0 \subsetneqq S_1 \subsetneqq \cdots$ there would be infinitely many beginning with $S_0 \subsetneqq S_1 \subsetneqq S_2 \subsetneqq \cdots$. Continuing in this manner, we define $S_3, S_4, \ldots$, and so obtain a strictly increasing infinite sequence $(S_i)_{i \geq 0}$ where, for $i \geq 1$, $S_i = \mathcal{O}_{X_i, x_i}$ with $x_i \in \mathrm{Sing}(X_i)$, X_i is the normalization of $X_i' := \mathrm{Bl}_{x_{i-1}}(X_{i-1})$ where $x_{i-1} \in \mathrm{Sing}(X_{i-1})$, and x_i lies over a point x_i' of the exceptional fibre of X_i'. In particular, S_i is a normal quadratic dilatation of S_{i-1} for every zero-dimensional valuation of K which dominates S_{i-1} [cf. (3.5)]. For every $i \in \mathbb{N}_0$ let $\mathfrak{n}_i$ be the maximal ideal of S_i.

Now $V := \bigcup_{i \geq 0} S_i$ is the ring of a zero-dimensional valuation v of K which dominates S_i for every $i \geq 0$ [cf. (4.6)]. By the theorem of uniformization (6.9) there exists a two-dimensional regular local k-subalgebra R of K having K as field of quotients and being essentially of finite type over k which is dominated by v. Let $(R_i)_{i \geq 0}$ be the sequence of quadratic dilatations of $R_0 := R$ with respect to v [cf. (3.2)]. Since v is zero-dimensional, for every $i \in \mathbb{N}$ the ring R_i is a two-dimensional regular local ring and $(R_i)_{i \geq 0}$ is a strictly increasing sequence [cf. (4.2)]. Therefore the ring $\bigcup_{i \geq 0} R_i$ is the ring of a zero-dimensional valuation of K which dominates R_i for every $i \geq 0$ [cf. (4.4)], hence it must be equal to V by (4.4). Since, for every $i \in \mathbb{N}_0$, the k-algebras R_i are essentially of finite type [cf. (3.1)], and also the k-algebras S_i are essentially of finite type, for every $i \in \mathbb{N}_0$ there exist $a(i)$, $b(i) \in \mathbb{N}$ with $R_i \subset S_{a(i)}$, $S_i \subset R_{b(i)}$.

We choose $i \in \mathbb{N}$ with $R_0 \subset S_i$; since $V \not\subset S_i$, not every quadratic dilatation of R_0 with respect to v is contained in S_i. By replacing R_0 by an appropriate quadratic dilatation R_j of R_0 with respect to v and relabelling, we may assume that $R_0 \subsetneq S_i$, but that the first quadratic dilatation R_1 of R_0 with respect to v is not contained in S_i. We choose $l \geq 1$ with $S_i \subset R_l$. Then the extended ideal $\mathfrak{n}_i R_l$ is a principal ideal [cf. (7.5)], i.e., $\mathfrak{n}_i R_l = R_l w$ with $w \in R_l$. Let x'_{i+1} be the point of $X'_{i+1} = \mathrm{Bl}_{x_i}(X_i)$ lying under x_{i+1} and set $S'_{i+1} := \mathcal{O}_{X'_{i+1}, x'_{i+1}}$; then we have $S_{i+1} = \overline{S}_{\mathfrak{m}(V) \cap \overline{S}}$ where $\overline{S}$ is the integral closure of S'_{i+1} [cf. (3.5)(2)]. We show that $S'_{i+1} \subset R_l$. Let $\mathfrak{q} \subset S_i$ be the maximal $\mathfrak{n}_i$-primary ideal of S_i corresponding to the point $x'_{i+1} \in \mathrm{Bl}_{x_i}(X_i)$; it is a tangential ideal and a v-ideal [cf. (7.8)(3)], and therefore we have $v(\mathfrak{q}) > v(\mathfrak{n}_i)$. The non-zero elements of S'_{i+1} have the form z_1/z_2 with $z_1 \in \mathfrak{n}_i^m$, $z_2 \in \mathfrak{n}_i^m \setminus \mathfrak{n}_i^{m-1}\mathfrak{q}$ for some $m \in \mathbb{N}$ [cf. (7.8)(4)]; we have $z_1 = w^m z'_1$, $z_2 = w^m z'_2$ with $z'_1, z'_2 \in R_l$. Since $z_2 \notin \mathfrak{n}_i^{m-1}\mathfrak{q}$, we have $v(z_2) = mv(\mathfrak{n}_i) = mv(w)$ [cf. (7.7)(2)], hence $v(z'_2) = 0$, and therefore z'_2 is a unit of R_l. This means that $z = z'_1/z'_2 \in R_l$, hence that $S'_{i+1} \subset R_l$, hence that $\overline{S} \subset R_l$ since R_l is integrally closed. Now, since v dominates R_l, we have $\mathfrak{m}(V) \cap \overline{S} = \mathfrak{m}(R_l) \cap \overline{S}$; therefore we see that $S_{i+1} = \overline{S}_{\mathfrak{m}(R_l) \cap \overline{S}}$ is contained in R_l. Repeating this argument we get $S_j \subset R_l$ for every $j \geq i$, hence $V \subset R_l$, contrary to the fact that $(R_i)_{i \geq 0}$ is strictly increasing. Therefore there exist only finitely many sequences of the form $(**)$, hence $(*)$ is a finite sequence.

(7.10) **Corollary:** [Resolution of singularities] *Let X be an irreducible projective surface. There exists a desingularization $\pi\colon Y \to X$ where Y is an irreducible regular projective surface and π is the composition of a finite number of morphisms which are either of the form $\overline{Z} \to Z$ where Z is an irreducible projective surface and $\overline{Z}$ is its normalization, or of the form $\mathrm{Bl}_z(Z) \to Z$ where Z is an irreducible normal projective surface and $z \in \mathrm{Sing}(Z)$.*

Appendix A

Results from Classical Algebraic Geometry

In this first appendix we treat only those aspects of classical algebraic geometry which are needed in this book. After introducing the notions of (affine and projective) varieties and stating, mostly without proof, some results on varieties and morphisms of varieties in section 1, we study affine and finite morphisms in section 2. Products and fibre products in the category of varieties is the contents of section 3. When defining the notion of resolution of singularities in section 8, we need the concept of proper morphisms; these are introduced in section 4. Regular and singular points of a variety as well as the classical Jacobian criterion for regular points are treated in section 6. One process of resolution of singularities uses repeatedly blowing up and normalization. Normalizing is the contents of section 7, while blowing up shall be dealt with in sections 13 and 14; section 5 describes the connection between affine algebraic cones and projective varieties which is used when constructing the normalization of a projective variety. Dimension of fibres and ramification for finite maps is treated in sections 9 and 10. In section 11 we introduce the group of Cartier divisors on an irreducible variety X and the group of Weil divisors on an irreducible normal variety X, and we show that these groups coincide if X is an irreducible nonsingular variety.

1 Generalities

1.1 Ideals and Varieties

(1.0) NOTATION: In this chapter k is an algebraically closed field of arbitrary characteristic, and m, $n \in \mathbb{N}_0$. We denote by $\mathbb{A}^n_k = \mathbb{A}^n$ (resp. $\mathbb{P}^n_k = \mathbb{P}^n$) the n-dimensional affine (resp. projective) space over k. $\mathbb{A}^n$ and $\mathbb{P}^n$ are endowed with the Zariski topology. The coordinates of a point in $\mathbb{A}^n$ are denoted by $(\xi_1, \xi_2, \ldots, \xi_n)$, and the homogeneous coordinates of a point in $\mathbb{P}^n$ are denoted by $(\xi_0 : \xi_1 : \cdots : \xi_n)$.

345

The elements F of the polynomial ring $k[T_1, \ldots, T_n]$ will be considered as functions $F \colon \mathbb{A}^n \to k$; for any subset M of $k[T_1, \ldots, T_n]$ we denote by $\mathcal{Z}(M) \subset \mathbb{A}^n$ the zero set of M, and for any $X \subset \mathbb{A}^n$ we denote by $\mathcal{I}(X) \subset k[T_1, \ldots, T_n]$ the ideal of polynomials vanishing on X.

For any subset M of the graded polynomial ring $k[T_0, \ldots, T_n]$ we denote by $\mathcal{Z}_+(M)$ the zero set in $\mathbb{P}^n$ of the homogeneous elements of M, and for any $X \subset \mathbb{P}^n$ we denote by $\mathcal{I}_+(X)$ the ideal in $k[T_0, \ldots, T_n]$ generated by all homogeneous polynomials vanishing on X; $\mathcal{I}_+(X)$ is a homogeneous ideal.

A good source for the following results from algebraic geometry which are given without proof are the first three sections of the first chapter of Hartshorne's "Algebraic Geometry" [82] or the first two chapters of the first volume of Shafarevich's "Basic Algebraic Geometry" [174].

(1.1) NOTATION: (1) A closed subset X of $\mathbb{A}^n$ is called an affine variety. An open subset of an affine variety is called a quasi-affine variety; equivalently, a quasi-affine variety is a locally closed subset of some $\mathbb{A}^n$.

(2) A closed subset X of $\mathbb{P}^n$ is called a projective variety. An open subset of a projective variety is called a quasi-projective variety.

(3) A variety is an affine or quasi-affine or projective or quasi-projective variety.

(4) Note that every locally closed subset of a variety is a variety.

(1.2) REMARK: These definitions differ from those used in [82] where a variety is always assumed to be irreducible.

(1.3) IDEALS AND VARIETIES: (1) Let $X \subset \mathbb{A}^n$ be closed. Then we have $X = \mathcal{Z}(\mathcal{I}(X))$, and the ideal $\mathcal{I}(X)$ is a radical ideal. The reduced k-algebra

$$A(X) := k[T_1, \ldots, T_n]/\mathcal{I}(X)$$

is called the affine coordinate ring of X. The variety X is irreducible iff $\mathcal{I}(X)$ is a prime ideal. Note that an element $f \in A(X)$ can be considered as a function $f \colon X \to k$; moreover, it is a regular function on X [cf. below], and if $F \in k[T_1, \ldots, T_n]$ has f as its image, then we have $F|X = f$.

(2) Let $X \subset \mathbb{P}^n$ be closed. Then we have $X = \mathcal{Z}_+(\mathcal{I}_+(X))$, and the ideal $\mathcal{I}_+(X)$ is a homogeneous non-irrelevant radical ideal. The reduced graded k-algebra

$$S(X) := k[T_0, T_1, \ldots, T_n]/\mathcal{I}_+(X)$$

is called the homogeneous coordinate ring of X. The variety X is irreducible iff $\mathcal{I}_+(X)$ is a prime ideal.

(3) Let $\mathfrak{a}$ be an ideal in $R := k[T_1, \ldots, T_n]$. Then we have $\mathcal{I}(\mathcal{Z}(\mathfrak{a})) = \mathrm{rad}(\mathfrak{a})$ [Hilbert's Nullstellensatz]. Moreover, the map $X \mapsto \mathcal{I}(X)$ is an inclusion-reversing bijective map from the family of closed subsets of $\mathbb{A}^n$ to the family of radical ideals of R; the irreducible closed subsets of $\mathbb{A}^n$ correspond to prime ideals of R.

(4) Let $\mathfrak{a}$ be a homogeneous ideal in $S := k[T_0, \ldots, T_n]$ and assume that $\mathcal{Z}_+(\mathfrak{a}) \neq \emptyset$. Then we have $\mathcal{I}_+(\mathcal{Z}_+(\mathfrak{a})) = \mathrm{rad}(\mathfrak{a})$ [homogeneous version of Hilbert's Nullstellensatz]. Moreover, the map $X \mapsto \mathcal{I}_+(X)$ is an inclusion-reversing bijective map from the family of non-empty closed subsets of $\mathbb{P}^n$ to the family of non-irrelevant homogeneous radical ideals of S; the irreducible closed subsets of $\mathbb{P}^n$ correspond to non-irrelevant homogeneous prime ideals of S. The empty set of $\mathbb{P}^n$ is the zero set of every irrelevant homogeneous ideal of S.

(1.4) REGULAR FUNCTIONS AND MORPHISMS: (1) Let X be a variety. The definition of a regular function on X is the same as in [82], p. 15 [where it is defined only for irreducible varieties]. The ring of global regular functions on X is denoted by $\mathcal{O}(X)$; it is a reduced k-algebra of finite type. In particular, if X is affine, then we have a canonical injective k-algebra homomorphism $A(X) \to \mathcal{O}(X)$.
(2) Let X and Y be varieties. A morphism $\varphi \colon X \to Y$ is defined as in [82], p. 15 [where it is defined only for irreducible varieties]; φ induces a k-algebra homomorphism $\varphi^* \colon \mathcal{O}(Y) \to \mathcal{O}(X)$.

(1.5) LOCAL RINGS: (1) Let X be a variety, and let $x \in X$. The local ring of x in X shall be denoted by $\mathcal{O}_{X,x}$, and $e_x(X)$ is the multiplicity of the local ring $\mathcal{O}_{X,x}$.
(2) Let $\varphi \colon X \to Y$ be a morphism of varieties. For each $x \in X$ the morphism φ induces a local homomorphism of k-algebras $\varphi_x^* \colon \mathcal{O}_{Y,\varphi(x)} \to \mathcal{O}_{X,x}$.

(1.6) THE CATEGORY OF VARIETIES: The varieties form a category $\textbf{\textit{Var}}$; the set of morphisms between varieties X and Y shall be denoted by $\mathrm{Mor}(X,Y)$. As usual, a morphism $\varphi \colon X \to Y$ between varieties X and Y is called an isomorphism if φ is bijective and if $\varphi^{-1} \colon Y \to X$ is a morphism.
Note that the map $X \mapsto \mathcal{O}(X)$ defines a contravariant functor from the category of varieties to the category of reduced k-algebras of finite type.
A variety isomorphic to an affine (resp. quasi-affine) variety is said to be an affine (resp. quasi-affine) variety, and a variety isomorphic to a projective (resp. quasi-projective) variety is said to be a projective (resp. quasi-projective) variety.
Let U be an open subset of a variety X; if U is an affine variety, then U is called an open affine subset of X.

1.2 Rational Functions and Maps

(1.7) RATIONAL FUNCTIONS: Let X be an irreducible variety, and let $\mathcal{U}$ be the family of non-empty open subsets of X. On the set of pairs $\{(U, f) \mid U \in \mathcal{U}, f \in \mathcal{O}(U)\}$ one defines the obvious equivalence relation; the equivalence class of a pair (U, f) is called the rational function defined by f, and we say that f is defined on U. The set of equivalence classes can be made into a field, the field of rational functions $K(X)$ on X. Note that $K(X)$ is a finitely generated extension field of k. For every rational function $f \in K(X)$ there exists a maximal $U \in \mathcal{U}$ such that f is defined on U; this U is called the domain of definition $\delta(f)$ of f. Note that there are injective homomorphisms of k-algebras $\mathcal{O}(X) \to \mathcal{O}_{X,x} \to K(X)$ for $x \in X$ which will be used to identify the k-algebras $\mathcal{O}(X)$ and $\mathcal{O}_{X,x}$ with subalgebras of the k-algebra $K(X)$. After this identification we have $\mathcal{O}(X) = \bigcap_{x \in X} \mathcal{O}_{X,x}$, and $K(X)$ is the field of quotients of $\mathcal{O}_{X,x}$ for every $x \in X$. Furthermore, for every $f \in K(X)$ we have $\delta(f) = \{x \in X \mid f \in \mathcal{O}_{X,x}\}$.

(1.8) RATIONAL MAPS: Let X and Y be irreducible varieties. A rational map $\varphi\colon X \to Y$ is an equivalence class of pairs $\langle U, \varphi_U \rangle$ where U is a non-empty open subset of X and $\varphi_U\colon U \to Y$ is a morphism, the equivalence relation being the obvious one; we say that φ is defined on U. There exists a largest open subset $\delta(\varphi)$ of X on which φ is defined, the domain of definition of φ. The rational map is dominant if for some (and therefore for every) pair $\langle U, \varphi_U \rangle$ the morphism φ_U is dominant [cf. (1.14) below].
A birational map $\varphi\colon X \to Y$ is a rational map $\varphi\colon X \to Y$ which admits an inverse; in this case X and Y are called birationally equivalent. X and Y are birationally equivalent iff the fields of rational functions $K(X)$ and $K(Y)$ are k-isomorphic fields.

1.3 Coordinate Ring and Local Rings

(1.9) Theorem: Let $X \subset \mathbb{A}^n$ be closed with affine coordinate ring $A(X)$. Then:
(1) The canonical map $A(X) \to \mathcal{O}(X)$ is an isomorphism of k-algebras.
(2) For each point $x \in X$ let $\mathfrak{m}_x \subset A(X)$ be the ideal consisting of all $f \in A(X)$ with $f(x) = 0$. Then $x \mapsto \mathfrak{m}_x$ is a bijective map between the set of points of X and the set of maximal ideals of $A(X)$.
(3) For each $x \in X$, the canonical map $A(X)_{\mathfrak{m}_x} \to \mathcal{O}_{X,x}$ is an isomorphism of k-algebras.
(4) If X is irreducible, then the k-algebras $Q(A(X))$ and $K(X)$ are isomorphic.

(1.10) Theorem: Let $X \subset \mathbb{P}^n$ be closed with homogeneous coordinate ring $S(X)$. Then:
(1) $\mathcal{O}(X) = k$.
(2) For each point $x \in X$ let $\mathfrak{m}_x \subset S(X)$ be the homogeneous ideal consisting of all homogeneous $f \in S(X)$ such that $f(x) = 0$. The local k-algebras $\mathcal{O}_{X,x}$ and $S(X)_{(\mathfrak{m}_x)}$ are isomorphic.
(3) If X is irreducible, then the k-algebras $K(X)$ and $S(X)_{((0))}$ are isomorphic.

(1.11) Proposition: Let X be a variety, and let Y be an affine variety. Then there is a natural bijective map

$$\alpha\colon \operatorname{Mor}(X,Y) \longrightarrow \operatorname{Hom}_{k\text{-alg}}(A(Y), \mathcal{O}(X)).$$

Note, in particular, that affine varieties X and Y are isomorphic iff the k-algebras $\mathcal{O}(X)$ and $\mathcal{O}(Y)$ are isomorphic.

(1.12) Corollary: *The functor* $X \mapsto \mathcal{O}(X)$ *induces an arrow-inverting equivalence of categories between the category of affine varieties and the category of reduced k-algebras of finite type, and the irreducible affine varieties correspond to integral k-algebras of finite type.*

(1.13) RELATIVE VERSION: (1) Let $X \subset \mathbb{A}^n$ be an affine variety; then we have $\mathcal{O}(X) = \mathcal{O}(\mathbb{A}^n)/\mathcal{I}(X)$. Let Y be a closed subset of X. Then we have $\mathcal{I}(X) \subset \mathcal{I}(Y)$ in $\mathcal{O}(\mathbb{A}^n)$, Y is affine, and the ideal $\mathcal{I}_X(Y) := \mathcal{I}(Y)\mathcal{O}(X)$ of $\mathcal{O}(X)$ is the ideal of all regular functions on X which vanish on Y. Clearly $\mathcal{I}_X(Y)$ is prime iff Y is irreducible.

Let $\mathfrak{a} \subset \mathcal{O}(X)$ be an ideal; then $\mathcal{Z}_X(\mathfrak{a}) := \{x \in X \mid f(x) = 0 \text{ for all } f \in \mathfrak{a}\}$ is a closed subset of X, and $\mathrm{rad}(\mathfrak{a}) = \mathcal{I}_X(\mathcal{Z}_X(\mathfrak{a}))$. Moreover, the map $Y \mapsto \mathcal{I}_X(Y)$ is an inclusion-reversing bijective map between the family of closed subsets of X and the family of radical ideals of $\mathcal{O}(X)$; the irreducible closed subsets of X correspond to prime ideals of $\mathcal{O}(X)$. For $x \in X$ we set $\mathfrak{m}_x := \mathcal{I}_X(\{x\})$; then $\mathfrak{m}_x$ is a prime ideal of $\mathcal{O}(X)$, and we have $\mathcal{O}_{X,x} = \mathcal{O}(X)_{\mathfrak{m}_x}$.

(2) Let $\varphi\colon X \to Y$ be a morphism of affine varieties, and let $\varphi^*\colon \mathcal{O}(Y) \to \mathcal{O}(X)$ be the induced k-algebra homomorphism. Let $Z \subset X$ and $W \subset Y$ be closed. Then we have

$$\overline{\varphi(Z)} = \mathcal{Z}_Y(\varphi^{*-1}(\mathcal{I}_X(Z))), \quad \mathcal{I}_Y(\overline{\varphi(Z)}) = \varphi^{*-1}(\mathcal{I}_X(Z)),$$

$$\varphi^{-1}(W) = \mathcal{Z}_X(\mathcal{I}_Y(W)\mathcal{O}(X)), \quad \mathcal{I}_X(\varphi^{-1}(W)) = \mathrm{rad}(\mathcal{I}_Y(W)\mathcal{O}(X)).$$

For $x \in X$ we set $\mathfrak{m}_x := \mathcal{I}_X(\{x\}) \subset \mathcal{O}(X)$, $y := \varphi(x)$ and $\mathfrak{n}_y := \mathcal{I}_Y(\{y\}) \subset \mathcal{O}(Y)$. Then $\mathfrak{n}_y = \varphi^{*-1}(\mathfrak{m}_x)$, and the induced homomorphism $\mathcal{O}(Y)_{\mathfrak{n}_y} = \mathcal{O}_{Y,y} \to \mathcal{O}_{X,x} = \mathcal{O}(X)_{\mathfrak{m}_x}$ is the homomorphism in (1.5)(2).

1.4 Dominant Morphisms and Closed Embeddings

(1.14) DEFINITION: Let $\varphi\colon X \to Y$ be a morphism of varieties. Then φ is called dominant (resp. a closed embedding resp. an open embedding) if $\varphi(X)$ is dense in Y (resp. $\varphi(X)$ is a closed resp. open subset of Y and $X \to \varphi(X)$ is an isomorphism of varieties).

(1.15) Proposition: *Let* $\varphi\colon X \to Y$ *be a morphism of affine varieties, and let* $\varphi^*\colon \mathcal{O}(Y) \to \mathcal{O}(X)$ *be the induced homomorphism. Then:*
(1) φ *is dominant iff* φ^* *is injective.*
(2) φ *is a closed embedding iff* φ^* *is surjective.*

(1.16) Corollary: *Let* $\varphi\colon X \to Y$ *be a morphism of varieties with Y affine, and let* $\varphi^*\colon \mathcal{O}(Y) \to \mathcal{O}(X)$ *be the induced homomorphism. Then we have* $\mathcal{I}_Y(\overline{\varphi(X)}) = \ker(\varphi^*)$. *In particular, if φ is dominant, then φ^* is injective.*

(1.17) Proposition: *Let* $\varphi\colon X \to Y$ *be a dominant morphism of irreducible varieties. For every $x \in X$ the induced homomorphism* $\varphi_x^*\colon \mathcal{O}_{Y,\varphi(x)} \to \mathcal{O}_{X,x}$ *is* *injective. In particular, φ induces a k-algebra homomorphism $\varphi_K^*\colon K(Y) \to K(X)$ of the fields of rational functions.*

1.5   Elementary Open Sets

(1.18) COORDINATE COVERING OF $\mathbb{P}^n$: Let $i \in \{0,\ldots,n\}$ and set $U_i := \mathbb{P}^n \setminus \mathcal{Z}_+(T_i)$; note that U_i is open in $\mathbb{P}^n$, and that $U_0 \cup \cdots \cup U_n = \mathbb{P}^n$. The map $(\alpha_0\colon \cdots \colon \alpha_n) \mapsto (\alpha_0/\alpha_i,\ldots,\alpha_{i-1}/\alpha_i,\alpha_{i+1}/\alpha_i,\ldots,\alpha_n/\alpha_i)\colon U_i \to \mathbb{A}^n$ is an isomorphism of varieties, hence U_i is an affine variety, and we have a covering of $\mathbb{P}^n$ by open affine subsets.
Let X be a subset of $\mathbb{P}^n$; then $X = (X \cap U_0) \cup \cdots \cup (X \cap U_n)$, and X is open (resp. closed) in $\mathbb{P}^n$ iff $X \cap U_0,\ldots,X \cap U_n$ are open (resp. closed) subsets of $\mathbb{P}^n$. If X is a closed subset of $\mathbb{P}^n$, then $(U_i \cap X)_{0 \le i \le n}$ is an open affine covering of X.

(1.19) ELEMENTARY OPEN SETS: Let X be a variety, and let $f \ne 0$ be a regular function on X. The non-empty open subset $X_f := \{x \in X \mid f(x) \ne 0\}$ of X is called an elementary open subset of X; the inclusion $X_f \hookrightarrow X$ induces a k-algebra homomorphism $\mathcal{O}(X) \to \mathcal{O}(X_f)$. Since $f|X_f$ is a unit of $\mathcal{O}(X_f)$, we have an induced homomorphism $\mathcal{O}(X)_f \to \mathcal{O}(X_f)$ mapping g/f^n to $(g|X_f)/(f|X_f)^n$. We show: *This homomorphism is injective.* In fact, if $(g|X_f)/(f|X_f)^n$ is zero on X_f, then we have $g|X_f = 0$, hence fg is zero on X, and therefore $g/f^n = fg/f^{n+1} = 0$ in $\mathcal{O}(X)_f$.
(1) Now we assume that X is affine.
(a) Any open subset U of X can be covered by finitely many sets of the form X_f with $f \in \mathcal{O}(X)$. In fact, $X' := X \setminus U$ is closed, hence $\mathcal{I}_X(X') = (f_1,\ldots,f_h)$ with regular functions $f_1,\ldots,f_h \in \mathcal{O}(X)$, and $U = X_{f_1} \cup \cdots \cup X_{f_h}$.
(b) Hilbert's Nullstellensatz takes the following form: Let $f_1,\ldots,f_h$ be regular functions on X; then we have $X = X_{f_1} \cup \cdots \cup X_{f_h}$ iff the elements $f_1,\ldots,f_h$ generate the unit ideal of $\mathcal{O}(X)$.
(c) Let f be a regular function on X. Then X_f is affine, and $\mathcal{O}(X)_f \to \mathcal{O}(X_f)$ is an isomorphism.
In fact, let $X \subset \mathbb{A}^n$ be closed; we choose $F \in k[T_1,\ldots,T_n] \subset k[T_1,\ldots,T_{n+1}]$ with $F|X = f$. Let $p\colon \mathbb{A}^{n+1} \to \mathbb{A}^n$ be the projection. Note that $Y := p^{-1}(X) \cap \mathcal{Z}_{\mathbb{A}^{n+1}}((1 - T_{n+1}F)) \subset \mathbb{A}^{n+1}$ is affine. Now p induces a morphism $\pi\colon Y \to X_f$. It is easily checked that the map $x = (\xi_1,\ldots,\xi_n) \mapsto (\xi_1,\ldots,\xi_n,1/f(x))\colon X_f \to Y$ is inverse to π, hence X_f is affine. We have to show that $\mathcal{O}(X)_f \to \mathcal{O}(X_f)$ is surjective. Let $k[t_1,\ldots,t_{n+1}]$ be the affine coordinate ring of Y; note that we have $F(t_1,\ldots,t_n)t_{n+1} = 1$. Let w be any regular function on X_f. Then $\pi^*(w) \in k[t_1,\ldots,t_{n+1}]$, and there exists $s \in \mathbb{N}$ with $F(t_1,\ldots,t_n)^s\pi^*(w) \in k[t_1,\ldots,t_n]$; therefore there exists $G \in k[T_1,\ldots,T_n]$ with $G(t_1,\ldots,t_n) = F(t_1,\ldots,t_n)^s\pi^*(w)$. Now $(G|X)/(F|X)^s = g/f^s$ with $g := G|X$ is mapped to w.
(d) The open affine subsets of X are a basis of the topology of X [cf. (a) and (c)].
(2) Let X be quasi-affine. Then any open subset U of X is a finite union of open affine subsets of X [since X is an open subset of some affine variety]. Therefore X has a basis of its topology consisting of open affine sets.

(3) Let X be projective. Then X has a finite covering by open affine sets [cf. (1.18)], hence X has a basis of its topology consisting of open affine sets.

(4) A quasi-projective variety X has a basis of its topology consisting of open affine sets.

(1.20) **Lemma:** Let X be a variety, and let $f \in \mathcal{O}(X)$, $f \neq 0$. The canonical homomorphism $\mathcal{O}(X)_f \to \mathcal{O}(X_f)$ is an isomorphism.

Proof: Since this homomorphism is injective [cf. (1.19)], we have only to show that it is surjective. Thus, let $g \in \mathcal{O}(X_f)$, and choose open affine subsets $U_1, \ldots, U_r$ of X which cover X. Let $i \in \{1, \ldots, r\}$. We set $f_i := f|U_i$; the k-algebra homomorphism $\mathcal{O}(U_i)_{f_i} \to \mathcal{O}(X_f \cap U_i)$ is an isomorphism [cf. (1.19)(1)(c)], hence we can write $(f_i^{m_i}|X_f \cap U_i)(g|X_f \cap U_i) = h_i|X_f \cap U_i$ with an integer $m_i \in \mathbb{N}$ and a regular function $h_i \in \mathcal{O}(U_i)$. Setting $m := \max(\{m_1, \ldots, m_r\})$ and replacing h_i by $f_i^{m-m_i}h_i$, we may assume that $m_1 = \cdots = m_r = m$. Since $f|X_f$ is a unit of $\mathcal{O}(X_f)$, we have $h_i|X_f \cap U_i \cap U_j = h_j|X_f \cap U_i \cap U_j$ for $i, j \in \{1, \ldots, r\}$.
Let $i, j \in \{1, \ldots, r\}$. We set $f_{ij} := f|U_i \cap U_j$. The canonical k-algebra homomorphism $\mathcal{O}(U_i \cap U_j)_{f_{ij}} \to \mathcal{O}(X_f \cap U_i \cap U_j)$ is injective [cf. (1.19)]; the elements $(h_i|U_i \cap U_j)/1 \in \mathcal{O}(U_i \cap U_j)_{f_{ij}}$ and $(h_j|U_i \cap U_j)/1 \in \mathcal{O}(U_i \cap U_j)_{f_{ij}}$ have the same image in $\mathcal{O}(X_f \cap U_i \cap U_j)$. Therefore we can write $f_{ij}^{n_{ij}}(h_i|U_i \cap U_j) = f_{ij}^{n_{ij}}(h_j|U_i \cap U_j)$ with $n_{ij} \in \mathbb{N}$; clearly we may assume that all the exponents n_{ij} are equal, i.e., $n_{ij} = n$. Now we have $(f_i^n h_i)|U_i \cap U_j = (f_j^n h_j)|U_i \cap U_j$, hence there exists a regular function h on X with $h|U_i = f_i^n h_i$ for $i \in \{1, \ldots, r\}$. Now we use the first part of the proof, and we find that $(f_i^{m+n}|X_f \cap U_i)(g|X_f \cap U_i) = (f_i^n|X_f \cap U_i)(h_i|X_f \cap U_i) = h|X_f \cap U_i$ for $i \in \{1, \ldots, r\}$, hence that $(f^{m+n}|X_f)g = h|X_f$, and therefore g is the image of $h/f^{m+n} \in \mathcal{O}(X)_f$.

(1.21) **Proposition:** Let X be a variety, and let $f_1, \ldots, f_n$ be regular functions on X which generate the unit ideal of $\mathcal{O}(X)$. If $X_{f_1}, \ldots, X_{f_n}$ are affine, then so is X.

Proof: We have $\mathcal{O}(X_{f_i}) = \mathcal{O}(X)_{f_i}$ for every $i \in \{1, \ldots, n\}$ by (1.20), hence, in particular, $\mathcal{O}(X)_{f_1}, \ldots, \mathcal{O}(X)_{f_n}$ are k-algebras of finite type. Using the fact that the elements $f_1, \ldots, f_n$ generate the unit ideal of $\mathcal{O}(X)$, it is easy to see that also $\mathcal{O}(X)$ is a k-algebra of finite type; clearly $\mathcal{O}(X)$ is reduced. Let X' be the affine variety associated with $\mathcal{O}(X)$, and let $\varphi: X \to X'$ be the morphism associated with the identity map $\mathcal{O}(X) \to \mathcal{O}(X)$ [cf. (1.11)]. Now, for every $i \in \{1, \ldots, n\}$, there is an induced morphism $\varphi_i: X_{f_i} \to X'_{f'_i}$; we leave it to the reader to prove that φ_i is an isomorphism. Since $X' = X'_{f'_1} \cup \cdots \cup X'_{f'_n}$ [note that $f_1, \ldots, f_n$ generate the unit ideal of $\mathcal{O}(X')$], we see that φ is an isomorphism.

1.6 Varieties as Topological Spaces

(1.22) IRREDUCIBLE COMPONENTS: Let X be a variety. Then X is a noetherian space, i.e., the family $\mathfrak{F}(X)$ of closed subsets of X satisfies the descending chain condition, X is, in particular, quasicompact, and X has only finitely many irreducible components (an irreducible component of X is a maximal element in the family of irreducible subsets of X).

The variety is called irreducible in $x \in X$ if the point x lies in only one irreducible component of X.

(1.23) Proposition: *Let X be a variety, and let $x \in X$. Then the local ring $\mathcal{O}_{X,x}$ is a domain iff X is irreducible in x. If this is the case, and if Z is the irreducible component of X with $x \in Z$, then we have $\mathcal{O}_{X,x} = \mathcal{O}_{Z,x}$.*

Proof: Let $X_1, \ldots, X_h$ be the irreducible components of X passing through x, and let $U \subset X$ be an open affine neighborhood of x. Then $U \cap X_1, \ldots, U \cap X_h$ are the irreducible components of U passing through x, hence to prove the assertion we may assume that $X \subset \mathbf{A}^n$ is closed. In $\mathcal{O}(\mathbf{A}^n)$ we set $\mathfrak{m} := \mathcal{I}(\{x\})$, $\mathfrak{a} := \mathcal{I}(X)$ and $\mathfrak{p}_i := \mathcal{I}(X_i)$ for every $i \in \{1, \ldots, h\}$. Then we have $\mathfrak{a} = \mathfrak{p}_1 \cap \cdots \cap \mathfrak{p}_h$ (irredundant primary decomposition). In $\mathcal{O}_{X,x} = A(X)_{\mathfrak{m}/\mathfrak{a}}$ we have $\{0\} = \mathfrak{p}_1 \mathcal{O}_{X,x} \cap \cdots \cap \mathfrak{p}_h \mathcal{O}_{X,x}$ as an irredundant primary decomposition; therefore $\mathcal{O}_{X,x}$ is a domain iff $h = 1$.

(1.24) Remark: Let X be a variety, and let $X_1, \ldots, X_h$ be the irreducible components of X, set $Z_i := \bigcup_{j \neq i} X_j$ and $U_i := X \setminus Z_i$ for $i \in \{1, \ldots, h\}$. Then $U_i \subset X_i$ is open in X, not empty [since $X_i \not\subset Z_i$] and irreducible, and $\bigcup_i U_i$ is dense in X. Let $x \in X$; then X is irreducible in x iff there exists a unique $i \in \{1, \ldots, h\}$ with $x \in U_i$; this shows again that in this case we have $\mathcal{O}_{X,x} = \mathcal{O}_{X_i,x}$.

(1.25) Dimension and codimension: Let X be a variety.

(1) The dimension of X is the supremum of the set of lengths of chains of irreducible closed subsets of X, and it is denoted by $\dim(X)$. Let $x \in X$; we define $\dim_x(X) = \inf(\{\dim(U) \mid U$ open neighbourhood of $x\})$. The following elementary facts shall be used.

(a) The dimension of X is the supremum of the dimension of its irreducible components, hence is finite [cf. (1.26) below]; X is said to be pure-dimensional of dimension d, if $\dim(Z) = d$ for every irreducible component Z of X.

(b) We have $\dim(X) = \sup(\{\dim_x(X) \mid x \in X\})$.

(c) Let $x \in X$, and let $X_1, \ldots, X_h$ be those irreducible components of X which contains x. Then we have $\dim_x(X) = \max(\{\dim_x(X_i) \mid i \in \{1, \ldots, h\}\})$.

(2)(a) Let $Y \subset X$ be irreducible and closed. Then we define the codimension of Y in X to be the supremum of the set of lengths of chains of irreducible closed subsets of X having Y as smallest element.

(b) Let $Y \subset X$ be closed. The codimension of Y in X is defined as the infimum of the set of codimensions in X of the irreducible components of Y; it is denoted by $\text{codim}(Y, X)$. Y is said to be pure d-codimensional if $\text{codim}(Z, X) = d$ for every irreducible component Z of Y.

(3) X is called catenary if for any irreducible closed subsets $Y \subsetneq Z$ of X any strictly ascending chain $Y =: Y_0 \subsetneq Y_1 \subsetneq \cdots \subsetneq Y_n := Z$ of closed irreducible subsets has the same length n.

(1.26) Proposition: *Let X be a variety. Then:*
(1) *Let X be irreducible. Then X is catenary, $\dim(X)$ is equal to the transcendence degree of the field of rational functions $K(X)$ over k, and we have $\dim(X) = \dim(\mathcal{O}_{X,x})$ for every $x \in X$. Moreover, we have $\operatorname{codim}(Y,X) = \dim(X) - \dim(Y)$ for every closed $Y \subset X$. In particular, if $Y \neq X$, then we have $\dim(Y) < \dim(X)$.*
(2) *We have $\dim(X) = \max(\{\dim(\mathcal{O}_{X,x}) \mid x \in X\})$ and $\dim(\mathcal{O}_{X,x}) = \dim_x(X)$ for every $x \in X$.*
(3) *If X is irreducible in $x \in X$, then $\mathcal{O}_{X,x}$ is catenary.*

1.7 Local Ring on a Subvariety

(1.27) REMARK: (1) Let X be a variety, and let Y be a closed subset of X. Let $y \in Y$; we denote by $\mathcal{I}_{X,y}(Y)$ the ideal of those elements of $\mathcal{O}_{X,y}$ which have a representative g on an open neighborhood U of y with $g(U \cap Y) = \{0\}$. If, in particular, X is affine and $Y \subset X$ is closed, then $\mathcal{I}_{X,y}(Y) = \mathcal{I}_X(Y)\mathcal{O}_{X,y}$. In the general case, let U be an open affine neighborhood of y in X; then we have $\mathcal{I}_U(U \cap Y)\mathcal{O}_{X,y} = \mathcal{I}_{X,y}(Y)$. Note that $\mathcal{I}_{X,y}(Y)$ is prime iff Y is irreducible in y. The canonical k-algebra homomorphism $\mathcal{O}_{X,y} \to \mathcal{O}_{Y,y}$ induced by the inclusion $Y \hookrightarrow X$ is surjective with kernel $\mathcal{I}_{X,y}(Y)$.
(2) Let X be a variety and let $x \in X$. Then $Y \mapsto \mathcal{I}_{X,x}(Y)$ is an inclusion-reversing bijective map from the set of irreducible closed subsets of X which contain x to the set of prime ideals of $\mathcal{O}_{X,x}$.
(3) Let $\varphi \colon X \to Y$ be a morphism of varieties, let $y \in Y$, $x \in \varphi^{-1}(\{y\})$, let $Z \subset X$ be closed with $x \in Z$, and let $W \subset Y$ be closed with $y \in W$. Then we have

$$\mathcal{I}_{Y,y}(\overline{\varphi(Z)}) = \varphi_x^{*-1}(\mathcal{I}_{X,x}(Z)), \quad \mathcal{I}_{X,x}(\varphi^{-1}(W)) = \operatorname{rad}(\mathcal{I}_{Y,y}(W)\mathcal{O}_{X,x}).$$

This follows easily from (1.13)(2).

(1.28) LOCAL RING ON A SUBVARIETY: Let X be an irreducible variety, let Y be an irreducible closed subset of X, and let $Z \subset X$ be closed with $Y \subset Z$. Remember that the field of rational functions $K(X)$ consists of equivalence classes of pairs (U, f) where $U \subset X$ is open and f is a regular function on U [cf. (1.7)]. Now we define $\mathcal{O}_{X/Y}$ to be the set of classes of pairs (U, f) where $U \subset X$ is open, $U \cap Y \neq \emptyset$ and f is a regular function on U, and we define $\mathcal{I}_{X/Y}(Z)$ to be the subset of those classes (U, f) where $f(z) = 0$ for every $z \in Z \cap U$. It is immediate that $\mathcal{O}_{X/Y}$ is a subring of $K(X)$, that $\mathcal{I}_{X/Y}(Z)$ is a radical ideal in $\mathcal{O}_{X/Y}$ and that $\mathfrak{m}_{X/Y} := \mathcal{I}_{X/Y}(Y)$ is the only maximal ideal in $\mathcal{O}_{X/Y}$ [if for a pair (U, f) with $U \subset X$ open, $U \cap Y \neq \emptyset$ and f regular on U, we have $f(x) \neq 0$ for some $x \in U \cap Y$, then there exists an open neighborhood $V \subset U$ of x with $f(z) \neq 0$ for every $z \in V$, and if we denote by g the restriction of f to V, then the class of $(V, 1/g)$ lies in $\mathcal{O}_{X/Y}$ and the product of the classes of (U, f) and $(V, 1/g)$ is 1], i.e., $\mathcal{O}_{X/Y}$ is a quasilocal domain.
(1) There is a well-defined k-algebra homomorphism $\mathcal{O}_{X/Y} \to K(Y)$ which maps the class of (U, f) in $\mathcal{O}_{X/Y}$ to the class of $(U \cap Y, f|U \cap Y)$ in $K(Y)$; it is easy to check that this homomorphism is surjective with kernel $\mathfrak{m}_{X/Y}$.

(2) Let y be any element of Y. The ideal $\mathcal{I}_{X,y}(Y)$ is a prime ideal [cf. (1.27)]. We show that

$$\mathcal{O}_{X/Y} = (\mathcal{O}_{X,y})_{\mathcal{I}_{X,y}(Y)}, \quad \mathcal{I}_{X/Y}(Z) = \mathcal{I}_{X,y}(Z)\mathcal{O}_{X/Y} = \mathcal{I}_{X,y}(Z)_{\mathcal{I}_{X,y}(Y)}, \quad (*)$$

hence, in particular, that $\mathcal{O}_{X/Y}$ is a local domain and that $\dim(\mathcal{O}_{X/Y}) = \operatorname{codim}(Y, X)$. Moreover, the map $Z \mapsto \mathcal{I}_{X/Y}(Z)$ is an inclusion-reversing bijective map from the set of closed subsets of X which contain Y to the set of radical ideals of $\mathcal{O}_{X/Y}$.

Proof: Let Z be a closed subset of X with $Z \supset Y$. We set $\Sigma := \mathcal{O}_{X,y} \setminus \mathcal{I}_{X,y}(Y)$. Clearly $\mathcal{O}_{X,y}$ is a subring of $\mathcal{O}_{X/Y}$, and the elements of Σ are invertible in $\mathcal{O}_{X/Y}$, hence $\Sigma^{-1}\mathcal{O}_{X,y}$ is a subring of $\mathcal{O}_{X/Y}$, and $\mathcal{I}_{X,y}(Z)_{\mathcal{I}_{X,y}(Y)} = \Sigma^{-1}\mathcal{I}_{X,y}(Z) \subset \mathcal{I}_{X/Y}(Z)$. Conversely, take any element of $\mathcal{O}_{X/Y}$, i.e., take the class of a pair (U, f) where $U \subset X$ is open, $U \cap Y \neq \emptyset$, and f is a regular function on U. Let $V \subset X$ be an open affine neighborhood of y; then $V \cap Y \neq \emptyset$, and therefore $(U \cap Y) \cap (V \cap Y) = U \cap V \cap Y \neq \emptyset$ [since Y is irreducible]. Let $z \in U \cap V \cap Y$, and let V_g [g regular on V] be an open neighborhood of z. We have, in particular, $g(z) \neq 0$, hence the class of (V, g) lies in Σ. Since $z \in V_g \cap U$, we can find a regular function h on V with $z \in V_h \subset V_g \cap U$; hence we may assume that already $V_g \subset U$. The restriction of f to V_g is regular, hence has the form h/g for some regular h on V [note that $\mathcal{O}(V_g) = \mathcal{O}(V)_g$], and therefore the class of $(V_g, f|V_g \cdot g|V_g)$ lies in $\mathcal{O}_{X,y}$. In particular, if the element defined by (U, f) already lies in $\mathcal{I}_{X/Y}(Z)$, then the class of $(V_g, f|V_g \cdot g|V_g)$ lies in $\mathcal{I}_{X,y}(Z)$. Thus, we have shown $(*)$, and from the proof there easily follows the last assertion. The local ring $\mathcal{O}_{X,y}$ is catenary [cf. (1.26)(3)], hence we have $\operatorname{ht}(\mathcal{I}_{X,y}(Y)) + \dim(\mathcal{O}_{X,y}/\mathcal{I}_{X,y}(Y)) = \dim(\mathcal{O}_{X,y})$, and from $(*)$ and (1.26)(1) we get $\dim(\mathcal{O}_{X/Y}) = \operatorname{ht}(\mathcal{I}_{X,y}(Y)) = \dim(X) - \dim(Y) = \operatorname{codim}(Y, X)$.

(3) The local ring $\mathcal{O}_{X/Y}$ is called the local ring of X along Y or on Y. If $Y = \{y\}$ consists of only one point y, then we have $\mathcal{O}_{X/Y} = \mathcal{O}_{X,y}$.

(4) Now we assume, in particular, that X is affine, and we set $\mathfrak{p} := \mathcal{I}_X(Y)$. Let $y \in Y$ and set $\mathfrak{m} := \mathcal{I}_X(\{y\})$. Then we have $\mathcal{O}_{X,y} = \mathcal{O}(X)_{\mathfrak{m}}$ and $\mathcal{I}_{X,y}(Y) = \mathfrak{p}\mathcal{O}_{X,y}$, and therefore we obtain $\mathcal{O}_{X/Y} = \mathcal{O}(X)_{\mathfrak{p}}$.

(5) From (4) we easily get the following: Let $U \subset X$ be open and affine with $U \cap Y \neq \emptyset$. Then we have $\mathcal{O}_{X/Y} = \mathcal{O}(U)_{\mathcal{I}_U(U \cap Y)}$.

(1.29) LOCAL RING ON A SUBVARIETY IN THE PROJECTIVE CASE: Let $X \subset \mathbb{P}^n$ be irreducible and closed, let $Y \subset X$ be an irreducible closed subset, and let Z be a closed subset of X containing Y. Let $S = S(X)$ be the homogeneous coordinate ring of X, and set $\mathfrak{p} := \mathcal{I}_+(Y)S$, a homogeneous prime ideal in the graded integral domain S, and $\mathfrak{a} := \mathcal{I}_+(Z)S$, a homogeneous radical ideal in S contained in $\mathfrak{p}$. Then we have

$$\mathcal{O}_{X/Y} = S_{(\mathfrak{p})}, \quad \mathcal{I}_{X/Y}(Z) = (\mathfrak{a}S_{(\mathfrak{p})})_0.$$

In fact, let $y \in Y$ and set $\mathfrak{q} := \mathcal{I}_+(\{y\})S$. We have $\mathfrak{p} \subset \mathfrak{q}$. Now, on the one hand, we have $\mathcal{O}_{X/Y} = (\mathcal{O}_{X,y})_{\mathcal{I}_{X,y}(Y)}$ and $\mathcal{I}_{X/Y}(Z) = \mathcal{I}_{X,y}(Z)_{\mathcal{I}_{X,y}(Y)}$ [cf. (1.28)(2)], and, on the other hand, we have $\mathcal{O}_{X,y} = S_{(\mathfrak{q})} = (S_{\langle \mathfrak{q} \rangle})_0$, $\mathcal{I}_{X,y}(Y) = (\mathfrak{p}S_{\langle \mathfrak{q} \rangle})_0$ and $\mathcal{I}_{X,y}(Z) = (\mathfrak{a}S_{\langle \mathfrak{q} \rangle})_0$; therefore we have

$$\mathcal{O}_{X/Y} = (S_{(\mathfrak{q})})_{(\mathfrak{p}S_{\langle \mathfrak{q} \rangle})_0} = ((S_{\langle \mathfrak{q} \rangle})_{\langle \mathfrak{p}S_{\langle \mathfrak{q} \rangle} \rangle})_0 = (S_{\langle \mathfrak{p} \rangle})_0 = S_{(\mathfrak{p})},$$
$$\mathcal{I}_{X/Y}(Z) = ((\mathfrak{a}S_{\langle \mathfrak{q} \rangle})_0)_{(\mathfrak{p}S_{\langle \mathfrak{q} \rangle})_0} = (\mathfrak{a}S_{\langle \mathfrak{p} \rangle})_0, \quad \mathfrak{m}_{X/Y} = (\mathfrak{p}S_{\langle \mathfrak{p} \rangle})_0.$$

(1.30) LOCAL RINGS ON SUBVARIETIES AND MORPHISMS: Let $\varphi \colon X \to Y$ be a morphism of irreducible varieties, let $X_1 \subset X$ be irreducible and closed, and let Y_1 be the closure of $\varphi(X_1)$ in Y; note that Y_1 is irreducible. Then we have an induced homomorphism $\mathcal{O}_{Y/Y_1} \to \mathcal{O}_{X/X_1}$ which is local; it is injective if φ is dominant.
(1) First, we show that we may assume that X and Y are affine. In fact, let $V \subset Y$ be open and affine with $V \cap Y_1 \neq \emptyset$, and choose an open affine $U \subset X$ with $U \cap X_1 \neq \emptyset$ and $\varphi(U) \subset V$. We set $\mathfrak{q} := \mathcal{I}_V(V \cap Y_1)$ and $\mathfrak{p} := \mathcal{I}_U(U \cap X_1)$; then we have $\mathcal{O}_{X/X_1} = \mathcal{O}(U)_{\mathfrak{p}}$ and $\mathcal{O}_{Y/Y_1} = \mathcal{O}(V)_{\mathfrak{q}}$.
(2) Let X and Y be affine, let $\mathfrak{p} \subset \mathcal{O}(X)$ be the ideal defining X_1, and let $\mathfrak{q} \subset \mathcal{O}(Y)$ be the ideal defining Y_1. Now $\mathfrak{q}$ is the preimage of $\mathfrak{p}$ since Y_1 is the closure of $\varphi(X_1)$ [cf. (1.13)(2)], and we have an induced local homomorphism $\mathcal{O}_{Y/Y_1} \to \mathcal{O}_{X/X_1}$ of local rings. If φ is dominant, then this homomorphism is injective [cf. (1.15)].

(1.31) FUNCTORIALITY: Let $\varphi \colon X \to Y$, $\psi \colon Y \to Z$ be morphisms of irreducible varieties, let $X_1 \subset X$ be irreducible and closed, let Y_1 be the closure of $\varphi(X_1)$ in Y, and let Z_1 be the closure of $\psi(Y_1)$ in Z. Then the homomorphism $\mathcal{O}_{Z/Z_1} \to \mathcal{O}_{X/X_1}$ induced by $\psi \circ \varphi$ is the composition of the homomorphism $\mathcal{O}_{Y/Y_1} \to \mathcal{O}_{X/X_1}$ induced by φ and the homomorphism $\mathcal{O}_{Z/Z_1} \to \mathcal{O}_{Y/Y_1}$ induced by ψ.

2   Affine and Finite Morphisms

(2.1) DEFINITION: A morphism $\varphi \colon X \to Y$ of varieties is called affine if for every open affine $V \subset Y$ the open set $\varphi^{-1}(V) \subset X$ is affine.

It is immediate that the composition of affine morphisms is affine, and that the identity map of a variety is affine.

(2.2) Proposition: A morphism $\varphi \colon X \to Y$ of varieties is affine iff every $y \in Y$ admits an open affine neighborhood W in Y such that $\varphi^{-1}(W)$ is affine.

Proof: Clearly the condition is necessary. We show that the condition is also sufficient. Let $V \subset Y$ be an open affine subset. Let $y \in V$, and choose an open affine neighborhood W of y in Y such that $\varphi^{-1}(W)$ is affine. There exists $g \in \mathcal{O}(W)$ such that $y \in W_g \subset V$ and that $\varphi^{-1}(W_g) = \varphi^{-1}(W)_{\varphi^*(g)}$ is an open affine subset of X [cf. (1.19)(1)]. There exists a regular $h \in \mathcal{O}(V)$ with $y \in V_h \subset W_g$, hence $V_h = W_{g'}$ with $g' := gh'$ where $h' \in \mathcal{O}(W)$ with $h|W_g = h'/g^n$, and therefore $\varphi^{-1}(V_h) = \varphi^{-1}(V)_{\varphi^*(h)} = \varphi^{-1}(W)_{\varphi^*(gh')}$ is affine. We may assume, therefore, that for every $y \in V$ there exists a regular $g_y \in \mathcal{O}(V)$ such that $y \in V_{g_y}$ and that

$\varphi^{-1}(V_{g_\nu}) = \varphi^{-1}(V)_{\varphi^*(g_\nu)}$ is affine. Since V is quasicompact, there exist regular functions $g_1, \ldots, g_h \in \mathcal{O}(V)$ such that $\varphi^{-1}(V_{g_i})$ is affine for every $i \in \{1, \ldots, h\}$ and that $V = V_{g_1} \cup \cdots \cup V_{g_h}$. Now $g_1, \ldots, g_h$ generate the unit ideal of $\mathcal{O}(V)$ by Hilbert's Nullstellensatz [cf. (1.19)(1)]. We set $f_i := \varphi^*(g_i) \in \mathcal{O}(\varphi^{-1}(V))$ for every $i \in \{1, \ldots, h\}$. Then $f_1, \ldots, f_h$ generate the unit ideal of $\mathcal{O}(\varphi^{-1}(V))$, and $\varphi^{-1}(V_{g_i}) = \varphi^{-1}(V)_{f_i}$ is affine for every $i \in \{1, \ldots, h\}$, hence $\varphi^{-1}(V)$ is affine by (1.21).

(2.3) NOTATION: Let $\varphi: X \to Y$ be a morphism of varieties, and let $V \subset Y$ be open and affine. Then φ is said to be finite over V if $\varphi^{-1}(V)$ is affine and if $\varphi^*: \mathcal{O}(V) \to \mathcal{O}(\varphi^{-1}(V))$ is a finite homomorphism.

(2.4) DEFINITION: A morphism $\varphi: X \to Y$ of varieties is called finite if it is finite over every open affine $V \subset Y$.

(2.5) Proposition: (1) *Let $\varphi: X \to Y$ be a morphism of varieties. Then φ is finite iff every $y \in Y$ admits an open affine neighborhood W in Y such that φ is finite over W.*
(2) *In particular, a morphism $\varphi: X \to Y$ of affine varieties is finite iff the induced homomorphism $\varphi^*: \mathcal{O}(Y) \to \mathcal{O}(X)$ is finite.*

Proof: (1) Clearly the condition is necessary. We show that the condition is also sufficient.
(a) Let $W \subset Y$ be open and affine such that φ is finite over W, and let $g \in \mathcal{O}(W)$. Then $\mathcal{O}(W) \to \mathcal{O}(\varphi^{-1}(W))$ is a finite homomorphism, hence the induced homomorphism $\mathcal{O}(W_g) = \mathcal{O}(W)_g \to \mathcal{O}(\varphi^{-1}(W))_{\varphi^*(g)} = \mathcal{O}(\varphi^{-1}(W_g))$ also is finite.
(b) Let $V \subset Y$ be open and affine, and let $y \in V$. There exists an open affine neighborhood W of y such that φ is finite over W. Just as in the proof of (2.2), there exists an open and affine $W' \subset V$ containing y such that $W' = W_g = V_h$ where $g \in \mathcal{O}(W)$ and $h \in \mathcal{O}(V)$. Therefore, to begin with, we may assume that for every $y \in V$ there exists $g_y \in \mathcal{O}(V)$ such that φ is finite over V_{g_y}.
(c) Since V is quasicompact, there exist $g_1, \ldots, g_h \in \mathcal{O}(V)$ such that $g_1, \ldots, g_h$ generate the unit ideal of $\mathcal{O}(V)$ and that $\mathcal{O}(V)_{g_i} \to \mathcal{O}(\varphi^{-1}(V))_{\varphi^*(g_i)}$ is a finite homomorphism for every $i \in \{1, \ldots, h\}$. Now $\mathcal{O}(\varphi^{-1}(V))$ is a finitely generated $\mathcal{O}(V)$-module by B(2.6).
(2) This follows at once from the local characterization in (1) and from B(2.6).

(2.6) Corollary: (1) *Let $\varphi: X \to Y$ be a finite morphism of varieties. Then φ is affine.*
(2) *Let X be a variety, and let $Z \subset X$ be closed. The inclusion morphism $Z \hookrightarrow X$ is finite.*
(3) *Let $\varphi: X \to Y$, $\psi: Y \to Z$ be finite morphisms of varieties. Then $\psi \circ \varphi: X \to Z$ is finite.*

Proof: (1) This follows immediately from the definitions [cf. (2.1) and (2.4)].
(2) We may assume that X is affine. Then Z is affine, $\mathcal{O}(Z) = \mathcal{O}(X)/\mathcal{I}_X(Z)$ is a finitely generated $\mathcal{O}(X)$-module, and for every $f \in \mathcal{O}(X)$ the homomorphism $\mathcal{O}(X)_f \to \mathcal{O}(Z)_f$ is finite. The result follows from (2.5).
(3) This is immediate by (1) and (2.5).

(2.7) Proposition: *Let $\varphi\colon X \to Y$ be a finite morphism. Then:*
(1) *For every $y \in Y$ the set $\varphi^{-1}(\{y\})$ is finite, and non-empty if φ is dominant.*
(2) *The map φ is closed.*

Proof: (1) It is enough to consider the case where X and Y are affine [cf. (2.5)]. We consider the homomorphism $\varphi^*\colon \mathcal{O}(Y) \to \mathcal{O}(X)$. Let $y \in Y$; the points of X in $\varphi^{-1}(\{y\})$ correspond uniquely to the prime ideals of $\mathcal{O}(X)$ lying over the maximal ideal $\mathfrak{m}_y$ of $\mathcal{O}(Y)$. Since φ^* is a finite homomorphism, the result follows from B(3.8).
Now we assume that φ is dominant. Then $\mathcal{O}(Y)$ can be considered as a subring of $\mathcal{O}(X)$ [cf. (1.15)(1)], and the result follows from [63], Prop. 4.15 and Cor. 4.17.
(2) Let $Z \subset X$ be closed. Since the inclusion $Z \hookrightarrow X$ is a finite morphism [cf. (2.6)(2)], the composition $Z \to X \to Y$ is a finite morphism [cf. (2.6)(3)], hence it is enough to show that $\varphi(X)$ is closed in Y. It is enough to consider the case where Y is affine [since Y is quasicompact]. Since $X \to Y$ is finite, it is clear that the induced morphism $X \to \overline{\varphi(X)}$ also is finite; thus, we may assume that $\varphi(X)$ is dense in Y. Since Y is affine, X also is affine [cf. (2.6)(1)], and since $\varphi(X)$ is dense in Y, the induced homomorphism $\varphi^*\colon \mathcal{O}(Y) \to \mathcal{O}(X)$ is injective [cf. (1.15)(1)], hence we get $\varphi(X) = Y$ by [63], Prop. 4.15.

(2.8) Proposition: *Let $\varphi\colon X \to Y$ be a finite morphism. Then $\dim(X) = \dim(\varphi(X))$; in particular, if φ is surjective, then $\dim(X) = \dim(Y)$.*

Proof: Since $\varphi(X)$ is a closed subset of Y [cf. (2.7)(2)], it is enough to assume that φ is surjective. Furthermore, it is enough to consider the case that X and Y are affine [cf. (2.6)]. In this case $\varphi^*\colon \mathcal{O}(Y) \to \mathcal{O}(X)$ is a finite injective homomorphism [cf. (2.5)(2) and (1.15)(1)], and the result follows from [63], Prop. 9.2.

3 Products

(3.1) PRODUCTS: Let C be a category, and let X, Y be objects of C. A triple (P, p, q) where P is an object of C and $p\colon P \to X$, $q\colon P \to Y$ are morphisms, is called a product of X and Y, if the following holds true: For every W in C and morphisms $\varphi\colon W \to X$, $\psi\colon W \to Y$, there exists a unique morphism $(\varphi, \psi)\colon W \to P$ such that $p \circ (\varphi, \psi) = \varphi$ and $q \circ (\varphi, \psi) = \psi$.
It is clear that a product, if it exists, is unique up to isomorphism. More precisely: If (P, p, q) and (P', p', q') are products of X and Y, then there exists a unique morphism $u\colon P \to P'$ such that diagram 1 on the next page is commutative, and

u is an isomorphism. A product of X and Y usually is denoted by $(X \times Y, p, q)$ or by $(X \times Y, \mathrm{pr}_1, \mathrm{pr}_2)$ or just by $X \times Y$.

(3.2) FUNCTORIAL PROPERTIES: Let C be a category in which for any two objects there exists a product. Then one has the following functorial properties:
(1) Let $\varphi \colon X \to X'$, $\psi \colon Y \to Y'$ be morphisms in C, and let $(X \times Y, p, q)$ (resp. $(X' \times Y', p', q')$) be the product of X and Y (resp. of X' and Y'). Then there exists a unique morphism $\varphi \times \psi \colon X \times Y \to X' \times Y'$ with $\varphi \circ p = p' \circ \varphi \times \psi$ and with $\psi \circ q = q' \circ \varphi \times \psi$.
(2) Let $\varphi \colon Z \to X$, $\psi \colon Z \to Y$ be morphisms in C, and let $(X \times Y, p, q)$ be a product of X and Y. Then there exists a unique morphism $(\varphi, \psi) \colon Z \to X \times Y$ with $p \circ (\varphi, \psi) = \varphi$ and with $q \circ (\varphi, \psi) = \psi$.
(3) The morphism in (1) and (2) have the usual functorial properties: Let $\varphi' \colon X' \to X''$, $\psi' \colon Y' \to Y''$ be morphisms; then we have $(\varphi' \times \psi') \circ (\varphi \times \psi) = (\varphi' \circ \varphi) \times (\psi' \circ \psi)$ and $\varphi' \times \psi' \circ (\varphi, \psi) = (\varphi' \circ \varphi, \psi' \circ \psi)$.

(3.3) FIBRE PRODUCTS: Let C be a category, and let S be an object of C. We define a new category $C_{/S}$ of S-objects of C: The objects of $C_{/S}$ are pairs (X, φ) where X is an object of C and $\varphi \colon X \to S$ is a morphism in C; an S-morphism $(X, \varphi) \to (Y, \psi)$ is a morphism $\alpha \colon X \to Y$ in C with $\psi \circ \alpha = \varphi$. By abuse of language we often say: Let $\varphi \colon X \to S$ be an object of $C_{/S}$.
Let $\varphi \colon X \to S$, $\psi \colon Y \to S$ be objects of $C_{/S}$. We define a fibre product of X and Y over S to be a product of (X, φ) and (Y, ψ) in the category $C_{/S}$. Thus, it is a pair (Z, χ) together with S-morphisms $p \colon Z \to X$, $q \colon Z \to Y$ such that for any object (W, ω) of $C_{/S}$ and S-morphisms $\alpha \colon W \to X$, $\beta \colon W \to Y$ there exists a unique S-morphism $\gamma \colon W \to Z$ with $p \circ \gamma = \alpha$ and with $q \circ \gamma = \beta$. Note that in the commutative diagram 2 below we have $\omega = \chi \circ \gamma$.

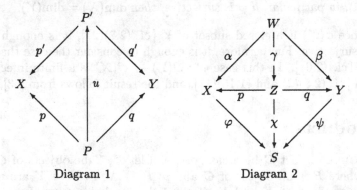

Diagram 1   Diagram 2

A fibre product, if it exists, is unique up to isomorphism. It shall be denoted by $X \times_S Y$.

(3.4) NOTATION: We now consider the category C and for any variety S the category of S-varieties $C_{/S}$. If $S = \{s\}$ consists of only one point, then we identify $C_{/S}$ with C. In this case the fibre product over S is just the product in the category C.

We assume that the reader is acquainted with the following results.

(3.5) Proposition: *In the category* **Var** *there exist products. More specifically:*
(1) *If X and Y are quasi-affine (resp. affine resp. projective), then $X \times Y$ is quasi-affine (resp. affine resp. projective). Moreover, if $X' \subset X$ and $Y' \subset Y$ are open (resp. closed resp. locally closed) then $X' \times Y'$ is open (resp. closed resp. locally closed) in $X \times Y$.*
(2) *If X and Y are affine, then the k-algebras $\mathcal{O}(X \times Y)$ and $\mathcal{O}(X) \otimes_k \mathcal{O}(Y)$ are isomorphic. More precisely: If $X \subset \mathbb{A}^m$, and $\mathcal{I}(X)$ is generated by polynomials $F_1, \ldots, F_r \in k[T_1, \ldots, T_m]$, and $Y \subset \mathbb{A}^n$, and $\mathcal{I}(Y)$ is generated by polynomials $G_1, \ldots, G_s \in k[U_1, \ldots, U_n]$, then $\mathcal{I}(X \times Y)$ is the ideal in $k[T_1, \ldots, T_m, U_1, \ldots, U_n]$ generated by the polynomials $F_1, \ldots, F_r, G_1, \ldots, G_s$.*

(3.6) REMARK: In particular, when proving this proposition, one starts with the affine case, and one shows: Let $X \subset \mathbb{A}^m$ and $Y \subset \mathbb{A}^n$ be closed subsets. Then $X \times Y$ is the cartesian product of the sets X and Y, considered as a subset of $\mathbb{A}^{m+n}$, and carrying the induced topology. The affine coordinate ring $\mathcal{O}(X \times Y)$ is the k-algebra $\mathcal{O}(X) \otimes_k \mathcal{O}(Y)$.
The projective case is settled using the Segre embedding $\mathbb{P}^m \times \mathbb{P}^n \to \mathbb{P}^{mn+m+n}$.

(3.7) Proposition: *Let X, Y be varieties. Then:*
(1) *If X and Y are irreducible, then $X \times Y$ is irreducible.*
(2) *We have $\dim(X \times Y) = \dim(X) + \dim(Y)$.*

(3.8) ASSOCIATIVITY OF PRODUCTS: Let X, Y, Z be varieties.
(1) Then $X \times (Y \times Z)$ and $(X \times Y) \times Z$ are canonically isomorphic, and these products will be denoted by $X \times Y \times Z$.
(2) Furthermore, the products $X \times Y \times Z$ and $X \times Z \times Y$ are isomorphic.

(3.9) NOTATION: Let X be a variety. We define the diagonal morphism

$$\delta_X := (\mathrm{id}_X, \mathrm{id}_X) \colon X \to X \times X;$$

the image $\Delta_X := \delta_X(X) \subset X \times X$ is called the diagonal of $X \times X$.

(3.10) Proposition: *Let X be a variety. Then $\delta_X \colon X \to X \times X$ is a closed embedding.*

Taking these results as starting point, we have the following additional results.

(3.11) Corollary: *Let X be a variety, and let $U \subset X$ and $V \subset X$ be open and affine. Then $U \cap V$ is open and affine.*

Proof: δ_X induces an isomorphism $\delta: U \cap V \to \delta_X(U \cap V) = \Delta_X \cap (U \times V)$ [cf. (3.10)]. $U \times V$ is an open affine subset of $X \times X$ by (3.5). Δ_X is closed in $X \times X$, hence $\Delta_X \cap (U \times V)$ is closed in $U \times V$, and is therefore an affine variety, hence $U \cap V$ is affine.

(3.12) Corollary: *Let* $\varphi: X \to Y$ *be a morphism of varieties. The graph*

$$\Gamma_\varphi := \{(x, \varphi(x)) \mid x \in X\} \subset X \times Y$$

of φ *is a closed subset of* $X \times Y$, *and the restriction of* $\mathrm{pr}_1: X \times Y \to X$ *to* Γ_φ *induces an isomorphism* $\Gamma_\varphi \to X$.

Proof: The first assertion follows from $\Gamma_\varphi = (\varphi \times \mathrm{id}_Y)^{-1}(\Delta_Y)$. Now the morphism $(\mathrm{id}_X, \varphi): X \to X \times Y$ maps X on Γ_φ, and it is inverse to the projection, hence the second assertion also is proved.

(3.13) EXISTENCE OF FIBRE PRODUCTS: (1) Let $\boldsymbol{Ens}$ be the category of sets, let S be a set, and let (X, φ), (Y, ψ) be in $\boldsymbol{Ens}_{/S}$. Fibre products exist in $\boldsymbol{Ens}$. In fact, set $Z := \{(x, y) \in X \times Y \mid \varphi(x) = \psi(y)\} \subset X \times Y$, and let $p: Z \to X$ (resp. $q: Z \to Y$) be the restriction of the first (resp. second) projection. Then (Z, p, q) is a product of X and Y in $\boldsymbol{Ens}_{/S}$.
(2) Now let S be a variety and (X, φ), (Y, ψ) be objects in $\boldsymbol{Var}_{/S}$. Then $\Gamma_\varphi \times Y$ and $\Gamma_\psi \times X$ are closed in $X \times Y \times S$ [cf. (3.5) and (3.12)]; we identify the set $X \times_S Y$ with the closed subset $\Gamma_\varphi \times Y \cap \Gamma_\psi \times X$ of $X \times Y \times S$. Let $p: X \times_S Y \to X$ (resp. $q: X \times_S Y \to Y$) be the restriction of the projection $X \times Y \times S \to X$ (resp. $X \times Y \times S \to Y$). It is not hard to check that $(X \times_S Y, p, q)$ is a product of X and Y in the category $\boldsymbol{Var}_{/S}$.
It is clear that in case $S = \{s\}$ the fibre product $X \times_S Y$ is just the product $X \times Y$ of X and Y in $\boldsymbol{Var}$.

(3.14) PROPERTIES OF THE FIBRE PRODUCT: Let S be a variety, and let (X, φ), (Y, ψ) be in $\boldsymbol{Var}_{/S}$. Let $(X \times_S Y, p, q)$ be the fibre product.
(1) We set $\widetilde{\varphi} := \varphi \circ p: X \times Y \to S$ and $\widetilde{\psi} := \psi \circ q: X \times Y \to S$. Then the closed subset $Z = \widetilde{\varphi}^{-1}(S) \cap \widetilde{\psi}^{-1}(S)$ of $X \times Y$ is an S-variety which can be identified with the S-variety $X \times_S Y$.
In fact, the restriction to $X \times_S Y$ of the projection $X \times Y \times S \to X \times Y$ induces an S-morphism $\alpha: X \times_S Y \to Z$, and the S-morphisms $p': Z \to X$, $q': Z \to Y$, induced by the projections $X \times Y \to X$ and $X \times Y \to Y$, give rise to an S-morphism $\beta: Z \to X \times_S Y$. It is easy to check that these morphisms are inverse to each other, and that $p \circ \beta = p'$, $q \circ \beta = q'$, and $p' \circ \alpha = p$, $q' \circ \alpha = q$.
(2) From (1) we get [cf. (3.5)]: If X and Y are quasi-affine (resp. affine resp. projective), then $X \times_S Y$ is quasi-affine (resp. affine resp. projective).
(3) Let $X' \subset X$ and $Y' \subset Y$ be locally closed (resp. closed resp. open), and let $\varphi': X' \to S$ be the composition $X' \hookrightarrow X \overset{\varphi}{\to} S$ which is a morphism of varieties.

Define $\psi': Y' \to S$ in a similar way. Then $X' \times_S Y'$ is a closed subset of $X' \times Y' \times S$; the latter set is locally closed (resp. closed resp. open) in $X \times Y \times S$ [cf. (3.5)], hence $X' \times_S Y'$ is locally closed (resp. closed resp. open) in $X \times_S Y$.

(4) Let W be an open set in S, and let U be open in X with $U \subset \varphi^{-1}(W)$, V be open in Y with $V \subset \psi^{-1}(W)$. Then the fibre product $U \times_W V$ can be identified with the open subset $p'^{-1}(U) \cap q'^{-1}(V)$ of the variety Z defined in (1). In particular, this implies that $U \times_W V$ can be identified with $U \times_S V$.

(5) Let $(W_i)_{1 \leq i \leq r}$ be an open covering of S, let $(U_i)_{1 \leq i \leq r}$ be an open covering of X with $U_i \subset \varphi^{-1}(W_i)$ (resp. $(V_i)_{1 \leq i \leq r}$ be an open covering of Y with $V_i \subset \psi^{-1}(W_i)$) for $i \in \{1, \ldots, r\}$. Then $(U_i \times_{W_i} V_i)_{1 \leq i \leq r}$ is an open covering of $X \times_S Y$.

(6) Let X, Y, S be affine, and set $B := \mathcal{O}(X)$, $C := \mathcal{O}(Y)$, $A := \mathcal{O}(S)$; then $X \times_S Y$ is affine and we have $\mathcal{O}(X \times_S Y) = (B \otimes_A C)_{\mathrm{red}}$.

In fact, let $\pi: B \otimes_A C \to (B \otimes_A C)_{\mathrm{red}}$ be the canonical homomorphism of A-algebras, and let D be a reduced A-algebra. It is easy to see that π induces a bijective map $\mathrm{Hom}_{A\text{-alg}}((B \otimes_A C)_{\mathrm{red}}, D) \to \mathrm{Hom}_{A\text{-alg}}(B \otimes_A C, D)$. Let $u: B \to B \otimes_A C$ and $v: C \to B \otimes_A C$ be the canonical A-algebra homomorphisms defined by $u(b) = b \otimes 1$ for every $b \in B$ and $v(c) = 1 \otimes c$ for every $c \in C$. Then $f \mapsto (f \circ \pi \circ u, f \circ \pi \circ v): \mathrm{Hom}_{A\text{-alg}}((B \otimes_A C)_{\mathrm{red}}, D) \to \mathrm{Hom}_{A\text{-alg}}(B, D) \times \mathrm{Hom}_{A\text{-alg}}(C, D)$ is a bijective map. Since $(B \otimes_A C)_{\mathrm{red}}$ is a k-algebra of finite type, it is clear that the affine variety Z associated with $(B \otimes_A C)_{\mathrm{red}}$ [cf. (1.12)] together with the S-morphisms $Z \to X$, $Z \to Y$ induced by u and v is a product of X and Y in the category $\mathbf{Var}_{/S}$.

(7) Assume that ψ is an isomorphism; then the projection $p: X \times_S Y \to X$ is an isomorphism.

In fact, we set $\chi := \psi^{-1} \circ \varphi: X \to Y$, and let $\rho := (\mathrm{id}_X, \varphi): X \to X \times_S Y$ be the induced S-morphism. It is easy to check that ρ and p are inverse to each other.

(8) Assume that φ is finite; then the projection $q: X \times_S Y \to Y$ is finite.

In fact, there exist open affine coverings $(U_i)_{1 \leq i \leq r}$ of X, $(V_i)_{1 \leq i \leq r}$ of Y and $(W_i)_{1 \leq i \leq r}$ of S with $U_i \subset \varphi^{-1}(W_i)$, $V_i \subset \psi^{-1}(W_i)$ for $i \in \{1, \ldots, r\}$. Then $(U_i \times_{W_i} V_i)_{1 \leq i \leq r}$ is an open covering of $X \times_S Y$ by (5). Let $i \in \{1, \ldots, r\}$; since $\mathcal{O}(W_i) \to \mathcal{O}(U_i)$ is finite, it follows that $\mathcal{O}(V_i) \to (\mathcal{O}(U_i) \otimes_{\mathcal{O}(W_i)} \mathcal{O}(V_i))_{\mathrm{red}}$ is finite, and since $\mathcal{O}(U_i \times_{W_i} V_i) = (\mathcal{O}(U_i) \otimes_{\mathcal{O}(W_i)} \mathcal{O}(V_i))_{\mathrm{red}}$ [cf. (6)], the claim follows from (2.5).

(3.15) Let $\varphi: X \to Y$ be a morphism of varieties, let $Y' \subset Y$ be locally closed, and define $X' := \varphi^{-1}(Y')$. Then X' is a locally closed subset of X, and it is easy to check that X' and $X \times_Y Y'$ are isomorphic varieties.

4 Proper Morphisms

(4.0) In the sequel, we use some notions and results from topology; the reader should consult, e.g., [34], Ch. II, § 4, no. 1 and no. 2.

4.1   Space of Irreducible Closed Subsets

(4.1) NOTATION: Let X be a topological space. We denote by $\mathfrak{O}(X)$ (resp. $\mathfrak{F}(X)$)) the family of open (resp. closed) subsets of X. For any $Z \subset X$ we denote by $\overline{Z}$ the closure of Z in X. Remember that X is a T_0-space (resp. a T_1-space), if for every pair of distinct points of X there exists an open subset of X which contains exactly one of these points (resp. if every point of X is closed).

(4.2) Proposition: *Let X be a topological space, and let X_0 be a subset of X. The following statements are equivalent:*
(1) *The map $F \mapsto F \cap X_0 : \mathfrak{F}(X) \to \mathfrak{F}(X_0)$ is injective (hence bijective).*
(2) *The map $U \mapsto U \cap X_0 : \mathfrak{O}(X) \to \mathfrak{O}(X_0)$ is injective (hence bijective).*
(3) *For every closed subset Z of X we have $Z = \overline{Z \cap X_0}$.*

Proof: The assertions in (1) and (2) that the maps are surjective are trivial. (1) $\Leftrightarrow$ (2) and (3) $\Rightarrow$ (1) are evident.
(1) $\Rightarrow$ (3): Since $Z \cap X_0$ is closed in X_0, we have $\overline{(Z \cap X_0)} \cap X_0 = Z \cap X_0$, hence we see that $Z = \overline{Z \cap X_0}$.

(4.3) DEFINITION: If a subset X_0 of a topological space X satisfies the equivalent conditions of (4.2), then X_0 is said to be *very dense* in X.

Note that in this case X_0 is dense in X by (4.2)(3).

(4.4) Proposition: *Let $\varphi\colon X_0 \to X$ be a continuous map of topological spaces. The following statements are equivalent:*
(1) *The map $U \mapsto \varphi^{-1}(U) : \mathfrak{O}(X) \to \mathfrak{O}(X_0)$ is bijective.*
(2) *The map $F \mapsto \varphi^{-1}(F) : \mathfrak{F}(X) \to \mathfrak{F}(X_0)$ is bijective.*
(3) *The topology of X_0 is the inverse image of the topology of X, and $\varphi(X_0)$ is very dense in X.*

Proof: Clearly (1) and (2) are equivalent, and (1) implies that the topology of X_0 is the inverse image of the topology of X. If $\varphi(X_0)$ is not very dense in X, then there exist open subsets $U \neq U'$ of X with $U \cap \varphi(X_0) = U' \cap \varphi(X_0)$, hence with $\varphi^{-1}(U) = \varphi^{-1}(U')$, which shows that (1) implies (3). Conversely, (3) implies that the maps $U \mapsto U \cap \varphi(X_0) : \mathfrak{O}(X) \to \mathfrak{O}(\varphi(X_0))$ and $V \mapsto \varphi^{-1}(V) : \mathfrak{O}(\varphi(X_0)) \to \mathfrak{O}(X_0)$ are bijective; then also their composition $U \mapsto \varphi^{-1}(U) : \mathfrak{O}(X) \to \mathfrak{O}(X_0)$ is bijective.

(4.5) DEFINITION: A continuous map $\varphi\colon X_0 \to X$ of topological spaces is called a *quasi-homeomorphism* if it satisfies the equivalent conditions of (4.4).

(4.6) REMARK: (1) Let $\varphi\colon X_0 \to X$ be a quasi-homeomorphism. The inverse map of the map $\mathfrak{F}(X) \to \mathfrak{F}(X_0)$ is the map $F \mapsto \overline{\varphi(F)} : \mathfrak{F}(X_0) \to \mathfrak{F}(X)$. In particular, if φ is bijective, then φ is a homeomorphism.
(2) The composition of quasi-homeomorphisms also is a quasi-homeomorphism.

(4.7) DEFINITION: (1) Let X be a topological space and Z an irreducible closed subset of X; a generic point for Z is a point $z \in Z$ with $Z = \overline{\{z\}}$.
(2) A topological space is a Zariski space if it is noetherian and if every irreducible closed subset has a unique generic point.
(3) Let X be a topological space.
(a) Let $x, y \in X$. If $x \in \overline{\{y\}}$, then we say that y specializes to x, written $y \rightsquigarrow x$; we also say that x is a specialization of y.
(b) A subset $Y \subset X$ is said to be stable under specialization if it contains every specialization of any of its points.

(4.8) PROPERTIES OF ZARISKI SPACES: Let X be a Zariski space.
(1) Every minimal non-empty closed subset Y of X consists of one point [since irreducible components of Y are closed in Y, we see that Y is irreducible, hence $Y = \overline{\{y\}}$ for a unique $y \in Y$, and if $z \in \overline{\{y\}}$, then $\overline{\{z\}} \subset \overline{\{y\}}$, hence $\overline{\{z\}} = \overline{\{y\}}$ by minimality of Y, and therefore $z = y$ by uniqueness of generic points].
(2) X is a T_0-space [for points $x \neq y$ of X we have $\overline{\{x\}} \neq \overline{\{y\}}$, hence $\overline{\{x\}} \not\subset \overline{\{y\}}$ or $\overline{\{y\}} \not\subset \overline{\{x\}}$, and therefore we have $x \notin \overline{\{y\}}$ or $y \notin \overline{\{x\}}$].
(3) If X is irreducible and x is its generic point, then x lies in every non-empty open subset U of X [otherwise, we would have $x \in X \setminus U$ for some non-empty open $U \subset X$, hence $X = \overline{\{x\}} \subset X \setminus U$].
(4) The relation $y \rightsquigarrow x$ is a (partial) ordering on X, the minimal elements of X are the closed points of X, and the maximal elements of X are the generic points of irreducible components of X [if $Z = \overline{\{z\}}$ is an irreducible component of X, and if $y \rightsquigarrow z$, then we have $Z = \overline{\{z\}} \subset \overline{\{y\}}$ and therefore $Z = \overline{\{y\}}$ and $z = y$, and if z is a maximal element of X and $Z := \overline{\{z\}}$, then Z is an irreducible component of X since otherwise we could choose an irreducible component $Z' = \overline{\{z'\}}$ of X with $Z \subset Z'$, hence $z' \rightsquigarrow z$, and therefore $z' = z$].
(5) A closed subset Y of X is stable under specialization [let $y \in Y$ and $y \rightsquigarrow z$, then we have $z \in \overline{\{y\}} \subset Y$].

(4.9) THE SPACE $t(X)$: Let X be a topological space, and let $t(X)$ be the set of irreducible closed subsets of X. If F is a closed subset of X, then $t(F)$ is a subset of $t(X)$. Furthermore, if F, F' are closed subsets of X, then we have $t(F \cup F') = t(F) \cup t(F')$ [if Z is an irreducible closed subset of X with $Z \subset F \cup F'$, then $Z = (Z \cap F) \cup (Z \cap F')$, hence $Z \subset F$ or $Z \subset F'$], and for any family $(F_i)_{i \in I}$ of closed subsets of X we have $t(\bigcap_{i \in I} F_i) = \bigcap_{i \in I} t(F_i)$. So we can define a topology on $t(X)$ by taking as closed sets the subsets of $t(X)$ of the form $t(F)$ where F is a closed subset of X. In particular, if Y is a closed subset of X, then $t(Y)$ is a closed subset of $t(X)$, and the topology induced on $t(Y)$ is the topology of the space $t(Y)$.
If $V \subset X$ is open, then the set $\tilde{V} \subset t(X)$ consisting of all irreducible closed subsets of X which meet V is open, and every open set in $t(X)$ has this form. Furthermore, we have the following: If U, V are open in X, then $(U \cap V)\tilde{} = \tilde{U} \cap \tilde{V}$, and if $(V_i)_{i \in I}$ is a family of open subsets of X, then $(\bigcup_{i \in I} V_i)\tilde{} = \bigcup_{i \in I} \tilde{V}_i$. Therefore we have

$\mathfrak{O}(t(X)) = \{\widetilde{V} \mid V \in \mathfrak{O}(X)\}$, and if $(V_i)_{i \in I}$ is a basis for the open sets of X, then $(\widetilde{V}_i)_{i \in I}$ is a basis for the open sets of $t(X)$.

Let U be an open subset of X; the subspace $\widetilde{U}$ of $t(X)$ and the space $t(U)$ are homeomorphic.

If X is a noetherian space, then $t(X)$ is a Zariski space. In fact, the irreducible closed subsets of $t(X)$ are the sets $t(F)$ where $F \subset X$ is irreducible and closed, and it is immediate that $t(F)$ is the closure of the point $F \in t(F)$, and that this is the only point of $t(F)$ having this property. Clearly $t(X)$ is noetherian, and therefore $t(X)$ is a Zariski space.

(4.10) THE FUNCTOR t: Let $\varphi \colon X \to Y$ be a continuous map of topological spaces, and let $t(\varphi) \colon t(X) \to t(Y)$ be defined by $t(\varphi)(F) = \overline{\varphi(F)}$ for every irreducible closed subset F of X [note that $\varphi(F)$ is irreducible since φ is continuous, hence $\overline{\varphi(F)}$ is irreducible and closed]. For every closed subset $G \subset Y$ we have $t(\varphi)^{-1}(t(G)) = t(\varphi^{-1}(G))$. In fact, let $F \subset X$ be irreducible and closed. If $F \in t(\varphi^{-1}(G))$, then $\overline{\varphi(F)} \subset G$, hence $t(\varphi)(F) \in t(G)$, and if $\overline{\varphi(F)} \subset G$, then $F \subset \varphi^{-1}(G)$. Therefore $t(\varphi) \colon t(X) \to t(Y)$ is continuous. Note also that for open $V \subset Y$ we have $(\varphi^{-1}(V))\widetilde{\ } = t(\varphi)^{-1}(\widetilde{V})$.

Let $\psi \colon Y \to Z$ be a continuous map of topological spaces; then we have $t(\psi \circ \varphi) = t(\psi) \circ t(\varphi)$ since by continuity of ψ we have $\psi(\overline{\varphi(F)}) \subset \overline{\psi(\varphi(F))}$ for every irreducible closed $F \subset X$.

Clearly, for every topological space X we have $t(\mathrm{id}_X) = \mathrm{id}_{t(X)}$.

Thus, we have defined a covariant functor $t \colon \boldsymbol{Top} \to \boldsymbol{Top}$ of topological spaces.

(4.11) THE MAP α_X: We now study the map $\alpha_X \colon X \to t(X)$ defined by $\alpha_X(x) = \overline{\{x\}}$ for every $x \in X$. It is continuous, since we have $\alpha_X^{-1}(t(F)) = F$ for every closed $F \subset X$ [note that a closed subset F of X contains $\overline{\{x\}}$ iff $x \in F$].

The map α_X has the following properties:

(1) The map $\alpha_X \colon X \to t(X)$ is a quasi-homeomorphism. In fact, $t(F) \mapsto \alpha_X^{-1}(t(F)) \colon \mathfrak{F}(t(X)) \to \mathfrak{F}(X)$ is a bijective map, since we have $\alpha_X^{-1}(t(F)) = F$ and $\overline{\alpha_X(F)} = t(F)$ for every closed $F \subset X$.

(2) If X is a T_1-space, then $X \to \alpha_X(X)$ is a homeomorphism [note that $X \to \alpha_X(X)$ is bijective, and use (4.6)].

(3) If X is a Zariski space, then $\alpha_X \colon X \to t(X)$ is a homeomorphism [note that $x \mapsto \overline{\{x\}} \colon X \to t(X)$ is bijective].

(4) If $\varphi \colon X \to Y$ is continuous, then we have $\alpha_Y \circ \varphi = t(\varphi) \circ \alpha_X$.

(4.12) REMARK: Let $\varphi \colon X \to Y$ be a continuous map of T_1-spaces.

(1) Let $F \subset X$ be closed; if $t(\varphi)(t(F))$ is closed in $t(Y)$, then $\overline{\varphi(F)}$ is closed in Y. In fact, for every irreducible closed subset $F' \subset F$ we have $\overline{\varphi(F')} \subset \overline{\varphi(F)}$, hence $t(\varphi)(t(F)) \subset t(\overline{\varphi(F)})$. By assumption, there exists a closed subset Z of Y with $t(\varphi)(t(F)) = t(Z)$, and we have $Z \subset \overline{\varphi(F)}$. On the other hand, for every $x \in F$ we have $\overline{\{x\}} \in t(F)$, and since $t(\varphi)(\overline{\{x\}}) = \overline{\{\varphi(x)\}}$, we have $\varphi(x) \in Z$, hence $\varphi(F) \subset Z$, and therefore $\overline{\varphi(F)} = Z$. Let $z \in Z$; then there exists an irreducible

closed $F' \subset F$ with $\overline{\varphi(F')} = \{z\}$, hence we have $\varphi(x) = z$ for any $x \in F'$, and therefore $\varphi(F) = Z$ is a closed subset of Y.

(2) If $\varphi(X)$ is dense in Y, then $t(\varphi)(t(X))$ is dense in $t(Y)$.

In fact, let V be a non-empty open subset of Y; then there exists $x \in X$ with $\varphi(x) \in V$, hence $\{\varphi(x)\}$ lies in $t(\varphi)(t(X)) \cap \widetilde{V}$, and therefore $t(\varphi)(t(X))$ is dense in $t(Y)$.

4.2 Varieties and the Functor t

(4.13) VARIETIES AND THE FUNCTOR t: In the sequel, we use results on spectra of rings [cf., e.g., [21], Ch. I, Ex. 15 and 21 or [34], Ch. II, §4, no. 3].

(1) Let X be an affine variety and $\mathcal{O}(X)$ its ring of regular functions. The map $F \mapsto \mathcal{I}_X(F)$ between the family of irreducible closed subsets of X and the set of prime ideals of $\mathcal{O}(X)$ [cf. (1.13)] induces a homeomorphism $t(X) \to \operatorname{Spec}(\mathcal{O}(X))$ where the spectrum of $\mathcal{O}(X)$ is endowed with the Zariski topology. Usually we identify $t(X)$ and $\operatorname{Spec}(\mathcal{O}(X))$. Note that $\alpha_X \colon X \to t(X)$ induces a homeomorphism $X \to \alpha_X(X)$.

(2) Let X be a variety. Then X is a noetherian T_1-space, hence $t(X)$ is a Zariski-space [cf. (4.9)]. Moreover, the family $\{\widetilde{V} \mid V \subset X \text{ open affine}\}$ is a basis for the topology of $t(X)$ [cf. (1.19) and (4.9)].

(3) Let X be an irreducible variety, and let $\zeta \in t(X)$; then $\zeta = Z$ is an irreducible closed subset Z of X, and the local ring $\mathcal{O}_{X/Z}$ of X along Z is called the local ring of ζ in X; it will be denoted by $\mathcal{O}_{X,\zeta}$.

(4) Let $\varphi \colon X \to Y$ be a morphism of affine varieties, and let $\varphi^* \colon \mathcal{O}(Y) \to \mathcal{O}(X)$ be the induced homomorphism of k-algebras. The continuous map $\operatorname{Spec}(\mathcal{O}(X)) \to \operatorname{Spec}(\mathcal{O}(Y))$ induced by φ^* is nothing else than the map $t(\varphi) \colon t(X) \to t(Y)$ of (4.10), as the reader may check without any difficulties. This implies, in particular, that φ^* is injective iff $t(\varphi)(t(X))$ is dense in $t(Y)$.

(5) Let $\varphi \colon X \to Y$ be a morphism of irreducible varieties, let $\xi \in t(X)$, and set $\eta := t(\varphi)(\xi) \in t(Y)$. Then φ induces a local homomorphism of local rings $\mathcal{O}_{Y,\eta} \to \mathcal{O}_{X,\xi}$ [cf. (1.30)] which is injective if φ is dominant.

(4.14) **Lemma:** *Let $\varphi \colon X \to Y$ be a morphism of varieties. Then:*
(1) $t(\varphi)(t(X))$ *is closed in $t(Y)$ iff $t(\varphi)(t(X))$ is stable under specialization.*
(2) *If $t(\varphi)(t(X))$ is closed in $t(Y)$, then $\varphi(X)$ is closed in Y.*

Proof: (1) If $t(\varphi)(t(X))$ is closed, then it is stable under specialization [cf. (4.8)(5)]. Now we assume that $t(\varphi)(t(X))$ is stable under specialization. Since $t(\varphi)(t(X)) \subset t(\overline{\varphi(X)})$ and $t(\overline{\varphi(X)})$ is a closed subset of $t(Y)$ [cf. (4.9)], we may replace Y by $\overline{\varphi(X)}$. Then $\varphi(X)$ is dense in Y, hence $t(\varphi)(t(X))$ is dense in $t(Y)$ [cf. (4.12)(2)]. Let $\eta \in t(Y)$. We choose an open affine subset V of Y with $\eta \in t(V)$ [cf. (4.13)(2)], and set $U := \varphi^{-1}(V)$; then $\varphi(U)$ is dense in V. We may replace Y by V and X by U; hence we may assume that Y is affine and that $\varphi(X)$ is dense in Y. Now we can write $X = U_1 \cup \cdots \cup U_h$, a finite union of open affine sets. There exists $i \in \{1, \ldots, h\}$ such that $\eta \in \overline{t(\varphi)(t(U_i))}$. We set $V_i := \overline{\varphi(U_i)}$; note

that V_i is affine. Furthermore, we set $B := \mathcal{O}(V_i)$ and $A := \mathcal{O}(U_i)$. The induced homomorphism $B \to A$ is injective [cf. (1.15)(1)]. Let $\mathfrak{q} \in \operatorname{Spec}(B)$ be the prime ideal corresponding to η, and let $\mathfrak{q}'$ be a minimal prime ideal of B contained in $\mathfrak{q}$. Then $\mathfrak{q}'$ corresponds to a point η' of $t(V_i)$ which specializes to η. $B_{\mathfrak{q}'}$ is a field; we consider the induced homomorphism $B_{\mathfrak{q}'} \to A_{\mathfrak{q}'}$. Let $\mathfrak{p}'_0$ be a prime ideal of $A_{\mathfrak{q}'}$; then $\mathfrak{p}'_0 \cap B_{\mathfrak{q}'} = \{0\}$. Let $\mathfrak{p}'$ be the prime ideal of A corresponding to $\mathfrak{p}'_0$; then we have $\mathfrak{p}' \cap B = \mathfrak{q}'$, hence $\mathfrak{p}'$ corresponds to a point $\xi' \in t(U_i)$ with $t(\varphi)(\xi') = \eta'$, hence $\eta \in t(\varphi)(t(X))$ since $t(\varphi)(t(X))$ is stable under specialization.
(2) This follows immediately from (4.12)(1).

(4.15) Notation: Let X be a variety.
(1) Let $\xi_1 \in t(X)$, i.e., $\xi_1 = X_1$ where $X_1 \subset X$ is irreducible and closed. In this case the field $K(X_1)$ of rational functions on X_1 shall be denoted by $\kappa(\xi_1)$.
(2) Let $\xi_0, \xi_1 \in t(X)$ and $\xi_1 \rightsquigarrow \xi_0$. Now $\xi_0 = X_0$, $\xi_1 = X_1$ where X_0, X_1 are irreducible closed subsets of X with $X_0 \subset X_1$. The local ring $\mathcal{O}_{X_1/X_0}$ of X_1 along X_0 is called the local ring of ξ_0 in ξ_1, and we denote it by $\mathcal{O}_{\xi_1/\xi_0}$; it is a subring of $\kappa(\xi_1)$ which contains k and has $\kappa(\xi_1)$ as field of quotients.
(3) Let $\xi_1 \in t(X)$. We say that a pair (K, α) where K is a field containing k as a subfield and $\alpha \colon \kappa(\xi_1) \to K$ is a k-algebra homomorphism, determines ξ_1 or that ξ_1 is determined by (K, α).
(4) Let $\xi_1 \in t(X)$ be determined by (K, α), and let ξ_0 be a specialization of ξ_1. In this situation we have the following: Let $\xi_0 = X_0$, $\xi_1 = X_1$ where $X_0 \subset X_1$; X_0 and X_1 are irreducible closed subsets of X. Let U be an open affine subset of X with $U \cap X_0 \neq \emptyset$ and set $\mathfrak{p}_1 := \mathcal{I}_U(X_1 \cap U)$, $\mathfrak{p}_0 := \mathcal{I}_U(X_0 \cap U)$. Then we have $\mathfrak{p}_1 \subset \mathfrak{p}_0$, $\kappa(\xi_1) = K(X_1) = Q(\mathcal{O}(U)/\mathfrak{p}_1)$, and $\mathcal{O}_{\xi_1/\xi_0} = \mathcal{O}_{X_1/X_0} = (\mathcal{O}(U)/\mathfrak{p}_1)_{\mathfrak{p}_0/\mathfrak{p}_1}$. Via $\alpha \colon \kappa(\xi_1) \to K$ we consider $\mathcal{O}_{\xi_1/\xi_0}$ as a subring of K. Let R be a valuation ring of K containing k; it makes sense to say that R dominates $\mathcal{O}_{\xi_1/\xi_0}$. Assume this to be the case; since $\mathcal{O}_{\xi_1/\xi_0} = (\mathcal{O}(U)/\mathfrak{p}_1)_{\mathfrak{p}_0/\mathfrak{p}_1} \subset R$, we have $\mathcal{O}(U)/\mathfrak{p}_1 \subset R$, and $\mathfrak{p}_0/\mathfrak{p}_1$ is the center of R in the ring $\mathcal{O}(U)/\mathfrak{p}_1$.

(4.16) Remark: Let $\varphi \colon X \to Y$ be a morphism of varieties, let $\xi_1 \in t(X)$, and set $\eta_1 := t(\varphi)(\xi_1) \in t(Y)$. Then $\xi_1 = X_1$ is an irreducible closed subset of X and $\eta_1 = Y_1$ is an irreducible closed subset of Y. Let $\xi_0 \in t(X)$ be a specialization of ξ_1, and set $\eta_0 := t(\varphi)(\xi_0)$. Then η_0 is a specialization of η_1.
The induced morphism $\varphi_1 \colon X_1 \to Y_1$ is dominant. Therefore φ_1 induces a k-algebra homomorphism $\kappa(\varphi_1) \colon \kappa(\eta_1) = K(Y_1) \hookrightarrow K(X_1) = \kappa(\xi_1)$ [cf. (1.17)]. In particular, if (K, α) determines ξ_1, then η_1 is determined by $(K, \alpha \circ \kappa(\varphi_1))$.

(4.17) **Proposition:** *Let X be a variety, let K be a field, and let R be a valuation ring of K containing k as a subfield. Let $\xi_1 \in t(X)$ be a point determined by (K, α). There exists at most one specialization ξ_0 of ξ_1 such that R dominates the local ring $\mathcal{O}_{\xi_1/\xi_0}$.*

Proof: Let $\xi_1 = X_1$, and let $\xi_0 = X_0$ be a specialization of ξ_1. Note that X_1 is an irreducible closed subset of X, and that we have the k-algebra homomorphism

$\alpha \colon \kappa(\xi_1) \hookrightarrow K$.

(1) First, we assume that X is affine. Then X_1 is an irreducible affine variety, and $\kappa(\xi_1)$ is the quotient field of $\mathcal{O}(X_1)$. Let $\mathfrak{p}_0 \in \operatorname{Spec}(\mathcal{O}(X_1))$ be the prime ideal of $\mathcal{O}(X_1)$ corresponding to ξ_0. We have $\mathcal{O}_{X_1/X_0} = \mathcal{O}(X_1)_{\mathfrak{p}_0}$ [cf. (1.28)(4)]; if R dominates $\mathcal{O}_{X_1/X_0}$, then R contains $\mathcal{O}_{X_1/X_0}$ and $\mathfrak{p}_0 \mathcal{O}_{X_1/X_0}$ is the center of R in $\mathcal{O}_{X_1/X_0}$, hence R contains also $\mathcal{O}(X_1)$ and $\mathfrak{p}_0$ is the center of R in $\mathcal{O}(X_1)$. Therefore ξ_0 is the only specialization of ξ_1 such that R dominates the local ring of ξ_0 in ξ_1.

(2) Now we assume that $X \subset \mathbb{A}^n$ is quasi-affine; let $\overline{X}$ be the closure of X in $\mathbb{A}^n$ and let $\overline{X}_1$ (resp. $\overline{X}_0$) be the closure of X_1 (resp. X_0) in $\overline{X}$. Then X_1 is open in $\overline{X}_1$, and $\kappa(\xi_1)$ is the field of quotients of $\mathcal{O}(\overline{X}_1)$. There exists $f \in \mathcal{O}(\overline{X}_1)$ such that $U := (\overline{X}_1)_f$ is open and affine with $U \cap X_0 \neq \emptyset$. Then we have $\mathcal{O}_{X_1/X_0} = (\mathcal{O}(U))_{\mathcal{I}_U(X_0 \cap U)}$ [cf. (1.28)(5)], hence we have $R \supset \mathcal{O}(U) = (\overline{X}_1)_f \supset \mathcal{O}(\overline{X}_1)$, and the center of R in $\mathcal{O}(\overline{X}_1)$ is the ideal defining $\overline{X}_0$. Since we have $X_0 = X_1 \cap \overline{X}_0$, we have settled this case also.

(3) Now let $X \subset \mathbb{P}^n$ be closed; then X_1 is an irreducible closed subset of $\mathbb{P}^n$. Let $\xi_0' = X_0'$ also be a specialization of ξ_1. We assume that R dominates $\mathcal{O}_{X_1/X_0}$ and $\mathcal{O}_{X_1/X_0'}$; then we get $\xi_0 = \xi_0'$ by VIII(3.4).

Similarly, if $X \subset \mathbb{P}^n$ is quasi-projective, we may argue as in (2).

4.3 Proper Morphisms

(4.18) DEFINITION: Let $\varphi \colon X \to Y$ be a morphism of varieties. Then φ is called proper if for any field K, any valuation ring R of K which contains k as a subfield, and any point $\xi_1 \in t(X)$ determined by a pair (K, α), we have the following: set $\eta_1 := t(\varphi)(\xi_1)$—note that η_1 is determined by $(K, \alpha \circ \kappa(\varphi_1))$—; if R dominates the local ring in η_1 of a specialization η_0 of η_1, then there exists a specialization ξ_0 of ξ_1 with $t(\varphi)(\xi_0) = \eta_0$ and such that R dominates the local ring of ξ_0 in ξ_1 [note that ξ_0 and η_0 are determined uniquely by (4.17)].

A variety X is called complete if $X \to *$, where $*$ is the one-point variety corresponding to k, is a proper morphism.

(4.19) Proposition: (1) *The composition of proper morphisms is proper.*
(2) *A closed embedding is proper.*
(3) *Let* $\varphi \colon X \to Y$, $\psi \colon Y \to Z$ *be morphisms of varieties. If* $\psi \circ \varphi$ *is proper, then* φ *is proper.*

Proof: (1) follows immediately from the definitions and from (1.31).
(2) Let $\varphi \colon X \to Y$ be a closed embedding. Let $\xi_1 = X_1 \in t(X)$ be determined by (K, α), set $\eta_1 := t(\varphi)(\xi_1)$ and $\eta_1 = Y_1$, i.e., $X_1 = Y_1$ [since $X \to Y$ is a closed embedding], and let R be a valuation ring of K containing k which dominates the local ring in η_1 of a specialization $\eta_0 = Y_0 \in t(Y)$ of η_1. We choose an open affine subset V of Y with $V \cap Y_0 \neq \emptyset$, and set $U := X \cap V$; U is an open affine subset of X with $U \cap X_1 \neq \emptyset$. The induced map $\mathcal{O}(V) \to \mathcal{O}(U)$ is surjective; let $\mathfrak{b}$ be

its kernel. We set $\mathfrak{q}_i := \mathcal{I}_V(V \cap Y_i)$ for $i \in \{0, 1\}$ and $\mathfrak{p}_1 := \mathcal{I}_U(U \cap X_1)$. Since $X_1 = Y_1$, the ideal $\mathfrak{q}_1$ contains $\mathfrak{b}$, hence $\mathfrak{q}_0$ contains $\mathfrak{b}$. We define $\mathfrak{p}_0 := \mathfrak{q}_0/\mathfrak{b}$, and set $\xi_0 = X_0$ where X_0 is the closure in X of the irreducible closed subset X_0' of $U \cap X$ defined by $\mathfrak{p}_0$. The local ring of η_0 in η_1 is the local ring of ξ_0 in ξ_1.

(3) Let K be a field, and let R be a valuation ring of K containing k as a subfield. Let (K, α) determine a point $\xi_1 \in t(X)$, set $\eta_1 := t(\varphi)(\xi_1)$, and assume that R dominates the local ring in η_1 of a specialization η_0 of η_1. Set $\zeta_1 := t(\psi)(\eta_1)$, $\zeta_0 := t(\psi)(\eta_0)$. Then ζ_0 is a specialization of ζ_1, and R dominates the local ring of ζ_0 in ζ_1 [cf. (1.30)]. Since $\psi \circ \varphi$ is proper, there exists a unique specialization ξ_0 of ξ_1 such that $t(\psi \circ \varphi)(\xi_0) = \zeta_0$ and such that R dominates the local ring of ξ_0 in ξ_1. Then $t(\varphi)(\xi_0)$ is a specialization of $t(\varphi)(\xi_1) = \eta_1$, and R dominates the local ring of $t(\varphi)(\xi_0)$ in η_1. Therefore we have $t(\varphi)(\xi_0) = \eta_0$ [cf. (4.17)].

(4.20) Proposition: *A finite morphism is proper.*

Proof: Let $\varphi \colon X \to Y$ be a finite morphism of varieties. Let K be a field, and let R be a valuation ring of K containing k as a subfield. Let $\xi_1 \in t(X)$ be determined by (K, α), and set $\eta_1 := t(\varphi)(\xi_1)$. Let η_0 be a specialization of η_1 such that R dominates the local ring of η_0 in η_1. Let $\eta_1 = Y_1$, $\eta_0 = Y_0$ and $\xi_1 = X_1$. Let $V \subset Y$ be open and affine with $V \cap Y_0 \neq \emptyset$, and such that $U := \varphi^{-1}(V)$ is affine; then $\mathcal{O}(V) \to \mathcal{O}(U)$ is finite. Set $\mathfrak{q}_0 := \mathcal{I}_V(V \cap Y_0)$ and $\mathfrak{q}_1 := \mathcal{I}_V(V \cap Y_1)$; then we have $\mathcal{O}_{Y_1/Y_0} = (\mathcal{O}(V)/\mathfrak{q}_1)_{\mathfrak{q}_0/\mathfrak{q}_1}$, hence R contains $\mathcal{O}(V)/\mathfrak{q}_1$ and $\mathfrak{q}_0/\mathfrak{q}_1$ is the center of R in $\mathcal{O}(V)/\mathfrak{q}_1$. We set $\mathfrak{p}_1 := \mathcal{I}_U(U \cap X_1)$; since $U \cap X_1 \to V \cap Y_1$ is dominant, $\mathfrak{q}_1$ is the preimage of $\mathfrak{p}_1$ in $\mathcal{O}(U)$, and the induced homomorphism $\mathcal{O}(V)/\mathfrak{q}_1 \to \mathcal{O}(U)/\mathfrak{p}_1$ is injective and finite. The quotient field of $\mathcal{O}(U)/\mathfrak{p}_1$ is $K(X_1) = \kappa(\xi_1)$ which can be considered via α as a subfield of K. Since $R \supset \mathcal{O}(V)/\mathfrak{q}_1$ and $\mathcal{O}(U)/\mathfrak{p}_1$ is integral over $\mathcal{O}(V)/\mathfrak{q}_1$, we have $\mathcal{O}(U)/\mathfrak{p}_1 \subset R$ [cf. I(3.4)]. The preimage $\mathfrak{p}_0 \subset \mathcal{O}(U)$ of the center of R in $\mathcal{O}(U)/\mathfrak{p}_1$ defines an irreducible closed subset X_0' of U with $X_0' \subset X_1 \cap U$, and we have an injective homomorphism $\mathcal{O}(V)/\mathfrak{q}_0 = \mathcal{O}(Y_0 \cap V) \to \mathcal{O}(X_0') = \mathcal{O}(U)/\mathfrak{p}_0$, hence a dominant morphism $X_0' \to Y_0 \cap V$. Now let $\xi_0 = X_0$ be the closure of X_0' in X. Then $X_0 \cap U = X_0'$, ξ_0 is a specialization of ξ_1, $t(\varphi)(\xi_0) = \eta_0$, and R dominates the local ring $(\mathcal{O}(U)/\mathfrak{p}_1)_{\mathfrak{p}_0/\mathfrak{p}_1}$ of ξ_0 in ξ_1.

(4.21) Proposition: *Let $\varphi \colon X \to Y$ be proper. Then $t(\varphi)\colon t(X) \to t(Y)$ is a closed map; in particular, $\varphi \colon X \to Y$ is a closed map.*

Proof: Let $Z \subset X$ be closed; we have to show that $t(\varphi)(t(Z))$ is stable under specialization [cf. (4.14)]. Let $\eta_1 \in t(\varphi)(t(Z))$, choose $\zeta_1 \in t(Z)$ with $\eta_1 = t(\varphi)(\zeta_1)$, and let η_0 be a specialization of η_1. Let $\mathcal{O}$ be the local ring of η_0 in η_1. Now $\varphi \colon X \to Y$ induces a k-algebra homomorphism $\kappa(\varphi_1)\colon \kappa(\eta_1) \to \kappa(\zeta_1) =: K$, i.e., $(K, \kappa(\varphi_1))$ determines η_1. There exists a valuation ring R of K which dominates $\mathcal{O}$ [cf. I(3.5)], and R contains k as a subfield. Since φ is proper, there exists a specialization ζ_0 of ζ_1 with $t(\varphi)(\zeta_0) = \eta_0$, and such that R dominates the local ring of ζ_0 in ζ_1. Since $\zeta_0 \in t(Z)$ by (4.8)(5), we have proved the first assertion; the second assertion follows from (4.14)(2).

(4.22) REMARK: Let $\varphi\colon X \to Y$ be a finite morphism. Then φ is proper by (4.20), hence closed by (4.21); thus, we have another proof of (2.7)(2).

(4.23) Proposition: *Let $\varphi\colon X \to Y$ be a proper morphism of varieties. For every morphism $\psi\colon Y' \to Y$ the induced morphism $\varphi'\colon X \times_Y Y' \to Y'$ is proper.*

Proof: We set $X' := X \times_Y Y'$. Let K be a field, and let R be a valuation ring of K containing k as a subfield. Let (K, α) determine the point $\xi_1' \in t(X')$, set $\eta_1' := t(\varphi')(\xi_1')$, and assume that R dominates the local ring in η_1' of a specialization η_0' of η_1'. We consider $\kappa(\xi_1')$ via α as a subfield of K. Set $\xi_1 := t(\mathrm{pr}_1)(\xi_1') \in t(X)$, $\eta_0 := t(\psi)(\eta_0') \in t(Y)$ and $\eta_1 := t(\psi)(\eta_1') \in t(Y)$. Then η_0 is a specialization of η_1, we have $t(\varphi)(\xi_1) = \eta_1$, $(K, \alpha \circ \kappa(\varphi_1))$ determines the point $\xi_1 \in t(X)$, and R dominates the local ring of η_0 in η_1 [cf. (1.30)]. Let $\xi_0 \in t(X)$ be the specialization of ξ_1 with $t(\varphi)(\xi_0) = \eta_0$ and such that R dominates the local ring of ξ_0 in ξ_1.
Let $\xi_0 = X_0$, $\eta_0' = Y_0'$, $\eta_0 = Y_0$ and $\xi_1' = X_1'$, $\xi_1 = X_1$, $\eta_1' = Y_1'$, $\eta_1 = Y_1$. We choose non-empty open affine subsets $V' \subset Y'$, $U \subset X$ and $V \subset Y$ such that $\varphi(U) \subset V$ and $\psi(V') \subset V$, and with $X_1' \cap (U \times_V V') \neq \emptyset$. Note that $U \times_V V'$ is affine and that $\mathcal{O}(U \times_V V') = (\mathcal{O}(U) \otimes_{\mathcal{O}(V)} \mathcal{O}(V'))_{\mathrm{red}}$ [cf. (3.14)(6)].
Let $\widetilde{\mathfrak{q}}_1'$ be the prime ideal in $\mathcal{O}(U \times_V V')$ of $X_1' \cap (U \times_V V')$, let $\mathfrak{q}_0'$ (resp. $\mathfrak{q}_1'$) be the prime ideal in $\mathcal{O}(V')$ of $Y_0' \cap V'$ (resp. $Y_1' \cap V'$), let $\mathfrak{p}_0$ (resp. $\mathfrak{p}_1$) be the prime ideal in $\mathcal{O}(U)$ of $X_0 \cap U$ (resp. $X_1 \cap U$) and let $\mathfrak{q}_0$ (resp. $\mathfrak{q}_1$) be the prime ideal in $\mathcal{O}(V)$ of $Y_0 \cap V$ (resp. $Y_1 \cap V$). Clearly $\kappa(\xi_1') \subset K$ is the quotient field of $\mathcal{O}(U \times_V V')/\widetilde{\mathfrak{q}}_1'$. We have injective homomorphisms of $\mathcal{O}(V)/\mathfrak{q}_1$-algebras $\mathcal{O}(U)/\mathfrak{p}_1 \to \mathcal{O}(U \times_V V')/\widetilde{\mathfrak{q}}_1'$, $\mathcal{O}(V')/\mathfrak{q}_1' \to \mathcal{O}(U \times_V V')/\widetilde{\mathfrak{q}}_1'$, and the $\mathcal{O}(V)/\mathfrak{q}_1$-algebra $\mathcal{O}(U \times_V V')/\widetilde{\mathfrak{q}}_1'$ is generated by the images of $\mathcal{O}(U)/\mathfrak{p}_1$ and $\mathcal{O}(V')/\mathfrak{q}_1'$. Since both of these rings are contained in R, we see that also $\mathcal{O}(U \times_V V')/\widetilde{\mathfrak{q}}_1'$ is contained in R; let $\widetilde{\mathfrak{q}}_0' \supset \widetilde{\mathfrak{q}}_1'$ be the prime ideal of $\mathcal{O}(U \times_V V')$ such that $\widetilde{\mathfrak{q}}_0'/\widetilde{\mathfrak{q}}_1'$ is the center of R in $\mathcal{O}(U \times_V V')/\widetilde{\mathfrak{q}}_1'$. Let X_0^* be the irreducible closed subset of $U \times_V V'$ defined by the ideal $\widetilde{\mathfrak{q}}_0'$ and let $\xi_0' = X_0'$ be the closure of X_0^* in X'. Then ξ_0' is a specialization of ξ_1', and R dominates the local ring $(\mathcal{O}(U \times_V V')/\mathfrak{p}_1')_{\mathfrak{p}_0'/\mathfrak{p}_1'}$ of ξ_0' in ξ_1'. It is easy to check that $t(\mathrm{pr}_1)(\xi_0') = \xi_0$ and that $t(\mathrm{pr}_2)(\xi_0') = \eta_0'$.

(4.24) Corollary: (1) *Let $\varphi\colon X \to Y$ be a morphism of varieties. If X is complete, then $\varphi(X)$ is closed in Y.*
(2) *If X and Y are complete varieties, then $X \times Y$ is complete.*

Proof: (1) The morphism $X \times Y \to Y$ is proper, and $\varphi(X)$ is the image of the graph Γ_φ of φ which is closed [cf. (3.12)].
(2) The morphisms $X \times Y \to Y$ and $Y \to *$ are proper, hence $X \times Y$ is complete [cf. (4.19)(1)].

(4.25) Proposition: $\mathbb{P}^n$ *is complete.*

Proof: Let K be a field and let R be a valuation ring of K containing k as a subfield; let v be the valuation of K defined by R [cf. I(2.13)]. Let $X_1 \subset \mathbb{P}^n$ be irreducible and closed, and let $\xi_1 = X_1 \in t(X)$ be determined by (K, α). Let $\mathfrak{p}_1 \subset k[T_0, \ldots, T_n]$ be the homogeneous prime ideal defining X_1; we may assume that $k[T_0, \ldots, T_n]/\mathfrak{p}_1 = k[t_0, \ldots, t_h]$ where $t_i := T_i + \mathfrak{p}_1$, $i \in \{0, \ldots, h\}$, are different from 0. Now we have $K(X_1) = k(t_1/t_0, \ldots, t_h/t_0)$. We choose $i \in \{0, \ldots, h\}$ with $v(t_i/t_0) \leq v(t_j/t_0)$ for $j \in \{0, \ldots, h\}$; then $k[t_0/t_i, \ldots, t_h/t_i]$ is contained in R. Let $\mathfrak{p}_0''$ be the center of R in $k[t_0/t_i, \ldots, t_h/t_i]$, and let $\mathfrak{p}_0'$ be the homogeneous prime ideal of $k[t_0, \ldots, t_h]$ corresponding to $\mathfrak{p}_0''$ [cf. B(4.20)]. Now $\mathfrak{p}_0'$ defines an irreducible closed subset $X_0 \subset X_1$, and by (1.29) the local ring of X_1 along X_0 is the ring $k[t_0, \ldots, t_h]_{(\mathfrak{p}')} = k[t_0/t_i, \ldots, t_h/t_i]_{\mathfrak{p}_0''}$, hence R dominates $\mathcal{O}_{X_1/X_0}$.

(4.26) Corollary: (1) *A projective variety is complete.*
(2) *Let $\varphi: X \to Y$ be a morphism of varieties. If X is projective, then φ is a closed map.*

Proof: (1) follows from (4.25) and (4.19), and (2) follows from (4.24).

5   Algebraic Cones and Projective Varieties

(5.1) NOTATION: Let $C \subset \mathbb{A}^n$ be closed . We say that C is an affine algebraic cone with vertex in the origin 0 if for every point $x \in C \setminus \{0\}$ the line through 0 and x is contained in C.

(5.2) Proposition: *Let $C \subset \mathbb{A}^n$ be closed. The following statements are equivalent:*
(1) $C = \mathcal{Z}(\mathfrak{a})$ *where $\mathfrak{a} \subset k[T_1, \ldots, T_n]$ is a homogeneous ideal.*
(2) $\mathcal{I}(C)$ *is a homogeneous ideal in $k[T_1, \ldots, T_n]$.*
(3) C *is an affine algebraic cone with vertex 0.*

Proof: (1) implies (2) since $\mathcal{I}(C) = \mathrm{rad}(\mathfrak{a})$, and $\mathrm{rad}(\mathfrak{a})$ is homogeneous since $\mathfrak{a}$ is homogeneous [cf. [63], Ex. 3.5], and (2) implies (1) since $C = \mathcal{Z}(\mathcal{I}(C))$.
The equivalence of (2) and (3) follows from the trivial observation that a polynomial $F \in k[T_1, \ldots, T_n]$ vanishes for all points of a line through the origin iff all the homogeneous parts of F have this property.

(5.3) NOTATION: Let $C \subset \mathbb{A}^n$ be an affine algebraic cone with vertex 0. The lines $L \subset C$ passing through 0 are called the generating lines of C.
In particular, if $\dim(C) = 1$, then C is a finite union of lines through 0 [since every generating line of C is irreducible, hence it is an irreducible component of C].

(5.4) DEFINITION: Let $C \subset \mathbb{A}^m$ and $D \subset \mathbb{A}^n$ be affine algebraic cones with vertex 0.
Let $\varphi: C \to D$ be a morphism; it is called a morphism of cones if

$$\varphi(\lambda x) = \lambda(\varphi(x)) \quad \text{for every } x \in C, \lambda \in k.$$

Thus, φ is a morphism of cones if $\varphi(0) = 0$ and if, for every generating line L of C, $\varphi(L)$ is a generating line of D or $\varphi(L) = \{0\}$.
Now it is easy to see: A morphism $\varphi\colon C \to D$ is a morphism of cones iff the induced homomorphism $\varphi^*\colon \mathcal{O}(D) \to \mathcal{O}(C)$ is a homomorphism of graded k-algebras.

(5.5) REMARK: The canonical map

$$\pi\colon \mathbb{A}^{n+1} \setminus \{0\} \to \mathbb{P}^n \quad \text{with } \pi(\xi_0,\ldots,\xi_n) = (\xi_0\colon \cdots \colon \xi_n)$$

is a surjective morphism. *We show that π is open.*
Let $F \in k[T_0,\ldots,T_n]$ be a non-zero polynomial and let $W := \mathbb{A}_F^{n+1}$ be the open affine set of points where F does not vanish. It is enough to show that $\pi(W \setminus \{0\})$ is open in $\mathbb{P}^n$. We write $F = F_0 + \cdots + F_h$ where $F_l \in R[T_0,\ldots,T_n]$ is homogeneous of degree l, $l \in \{0,\ldots,h\}$. For $l \in \{0,\ldots,h\}$ with $F_l \neq 0$ and $i \in \{0,\ldots,n\}$ we may consider $g_{il} = F_l(T_0/T_i,\ldots,T_n/T_i)$ as a regular function on the open set $U_i := \mathbb{P}^n \setminus \mathcal{Z}_+(T_i)$ of $\mathbb{P}^n$. We show that $\pi(W \setminus \{0\})$ is the union of all the open sets $(U_i)_{g_{il}}$, $i \in \{0,\ldots,n\}$, $l \in \{0,\ldots,h\}$ and $F_l \neq 0$. Namely, if $w = (\xi_0,\ldots,\xi_n) \in W$ and $\xi_i \neq 0$ for some $i \in \{0,\ldots,n\}$, then we have $F_j(\xi_0,\ldots,\xi_n) \neq 0$ for some $j \in \{0,\ldots,h\}$, hence $g_{ij}(\xi_0/\xi_i,\ldots,\xi_n/\xi_i) \neq 0$ and therefore $\pi(w) \in (U_i)_{g_{ij}}$. Conversely, let $x = (\xi_0\colon \cdots \colon \xi_n) \in \mathbb{P}^n$ lie in $(U_i)_{g_{ij}}$ where $i \in \{0,\ldots,n\}$ and $j \in \{0,\ldots,h\}$ with $F_j \neq 0$. Then we have $F_j(\xi_0,\ldots,\xi_n) \neq 0$. This implies that there exists $\lambda \in k^\times$ with $\sum_{l=0}^h \lambda^l F_l(\xi_0,\ldots,\xi_n) \neq 0$, hence $w := (\lambda\xi_0,\ldots,\lambda\xi_n) \in W$, $w \neq 0$, and we have $\pi(w) = (\xi_0\colon \cdots \colon \xi_n)$.

(5.6) ALGEBRAIC CONES AND PROJECTIVE VARIETIES: (1) Let $X \subset \mathbb{P}^n$ be closed. Then

$$\mathbb{C}(X) := \pi^{-1}(X) \cup \{0\} \subset \mathbb{A}^{n+1}$$

is an algebraic cone with vertex in the origin which is called the cone associated with X. Let $\pi_X\colon \mathbb{C}(X) \setminus \{0\} \to X$ be the morphism induced by the restriction of π.
(2) Let $C \subset \mathbb{A}^{n+1}$ be an algebraic cone. Then

$$\mathbb{P}(C) := \pi(C \setminus \{0\}) \subset \mathbb{P}^n$$

is a closed subset of $\mathbb{P}^n$ by (5.5); it is called the projectivization of C.
(3) Let X be as in (1), and let $\mathcal{I}_+(X) = \mathfrak{a} \subset k[T_0,\ldots,T_n]$ be the ideal of X; we have $\mathcal{I}(\mathbb{C}(X)) = \mathfrak{a}$ [cf. (5.2)]. Note that the ring $\mathcal{O}(\mathbb{C}(X)) = k[T_0,\ldots,T_n]/\mathfrak{a}$ of regular functions on $\mathbb{C}(X)$ is the homogeneous coordinate ring of X, and that $X = \mathbb{P}(\mathbb{C}(X))$.
(4) Let C be as in (2). Then $\mathbb{P}(C)$ is closed in $\mathbb{P}^n$, and we have $C = \mathbb{C}(\mathbb{P}(C))$. In particular, $\mathcal{I}_{\mathbb{A}^{n+1}}(C)$ is a homogeneous ideal in $k[T_0,\ldots,T_n]$, and we have $\mathcal{O}(C) = k[T_0,\ldots,T_n]/\mathcal{I}_{\mathbb{A}^{n+1}}(C)$, a graded ring. Set $\pi_C := \pi|C \setminus \{0\}$; $\pi_C\colon C \setminus \{0\} \to \mathbb{P}(C)$ is a morphism of varieties.
(5) It is obvious that $X \mapsto \mathbb{C}(X)$ is an inclusion-preserving bijective map from the family of closed subsets X of $\mathbb{P}^n$ to the family of algebraic cones C in $\mathbb{A}^{n+1}$, the inverse map being $C \mapsto \mathbb{P}(C)$. In particular, we have $\mathbb{C}(\emptyset) = \{0\}$ and $\mathbb{P}(\{0\}) = \emptyset$.

(6) The following is easily checked. Let $C \neq \{0\}$ be an algebraic cone. If C is irreducible, then $\mathbb{P}(C)$ is irreducible. Conversely, if $X \subset \mathbb{P}^n$ is irreducible and closed, then $\mathbb{C}(X)$ is irreducible and $\mathbb{C}(X) \neq \{0\}$.

(5.7) REMARK: In (1.19) we studied for an affine variety X the affine set X_f. In this remark we study the projective case; it shall be needed for the proof of (5.8). Let $X \subset \mathbb{P}^n$ be closed and non-empty, let S be the homogeneous coordinate ring of X, and let $f \in S_1$ be non-zero. Note that $S = \mathcal{O}(\mathbb{C}(X))$ is the affine coordinate ring of $\mathbb{C}(X)$. Now $U := \{x \in X \mid f(x) \neq 0\}$ is an open subset of X. We show: U is affine, and $\mathcal{O}(U)$ is canonically isomorphic to $S_{(f)} \cong S/((1-f))$ [cf. B(4.16) for the last isomorphism].
The map $(\xi_0 : \cdots : \xi_n) \mapsto (\xi_0/f(x), \ldots, \xi_n/f(x)) : U \to \mathcal{Z}_{\mathbb{C}(X)}((1-f))$ is a morphism, and $(\xi_0, \ldots, \xi_n) \mapsto (\xi_0 : \cdots : \xi_n) : \mathcal{Z}_{\mathbb{C}(X)}((1-f)) \to U$ is the inverse morphism, hence U is affine. The canonical surjective map $\pi : C(X)_f \to U$, defined by $(\xi_0, \ldots, \xi_n) \mapsto (\xi_0 : \cdots : \xi_n)$, defines an injective homomorphism $\pi^* : \mathcal{O}(U) \to \mathcal{O}(\mathbb{C}(X)_f) = S_f$ [cf. (1.19)]. Let $g \in S_p$; then g/f^p defines a regular function on U, and its image under π^* is g/f^p. Conversely, let $w \in \mathcal{O}(U)$; we can write $\pi^*(w) = \sum_{i \in \mathbb{Z}} v_i(f/1)^i$ with $v_i \in S_{(f)}$. Let $x \in U$ and $\lambda \in k^\times$. Then we have $\sum v_i(x)\lambda^i f^i(x) = \sum v_i(\lambda x)f^i(\lambda x) = \pi^*(w)(\lambda x) = w(\pi(\lambda x)) = w(\pi(x)) = \pi^*(w)(x) = \sum v_i(x)f^i(x)$, hence $v_i = 0$ for every $i \in \mathbb{Z}$, $i \neq 0$, hence $\pi^*(\mathcal{O}(U)) = S_{(f)}$.

(5.8) Proposition: Let $X \subset \mathbb{P}^m$ and $X' \subset \mathbb{P}^n$ be non-empty closed subsets. Assume that there exists a morphism $\varphi : \mathbb{C}(X) \to \mathbb{C}(X')$ of algebraic cones, and set $G(\varphi) := \pi_X(\mathbb{C}(X) \setminus \varphi^{-1}(\{0\}))$, an open subset of X. Then φ induces by restriction a morphism $\varphi' : \mathbb{C}(X) \setminus \varphi^{-1}(\{0\}) \to \mathbb{C}(X') \setminus \{0\}$, and there exists a unique morphism $\widetilde{\varphi} : G(\varphi) \to X'$ such that

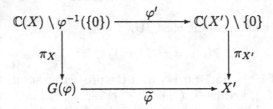

is commutative. If φ is finite, then we have $G(\varphi) = X$ and $\widetilde{\varphi}$ is finite.

[By abuse of notation, let π_X be the restriction of π to $\mathbb{C}(X) \setminus \varphi^{-1}(\{0\})$.]
Proof: Let $S := k[T_0, \ldots, T_m]/\mathcal{I}(\mathbb{C}(X))$ (resp. $S' := k[T_0, \ldots, T_n]/\mathcal{I}(\mathbb{C}(X'))$) be the affine coordinate ring of $\mathbb{C}(X)$ (resp. of $\mathbb{C}(X')$). Note that $\varphi^* : S' \to S$ is finite iff φ is finite [cf. (2.5)(2)]. Let $x = (\xi_0, \ldots, \xi_n) \in \mathbb{C}(X)$ with $\varphi(x) \neq 0$, i.e., $x = (\xi_0 : \cdots : \xi_n) \in G(\varphi)$, and let L be the line through the origin in $\mathbb{A}^{n+1}$ and the point $(\xi_0, \ldots, \xi_n) \in \mathbb{C}(X)$. Then $\varphi(L)$ is not the origin in $\mathbb{A}^{m+1}$, hence taking any point $(\xi_0', \ldots, \xi_m') \in \varphi(L)$ different from the origin gives us a well-defined point

$x' := \tilde{\varphi}(x) \in X'$. The map $\tilde{\varphi} \colon G(\varphi) \to X'$ is a well-defined morphism, and it is the only morphism making the above diagram commutative.

Now we assume that φ is a finite map. Then we have $G(\varphi) = X$ [since no generating line of $\mathbb{C}(X)$ can be mapped to the origin of $\mathbb{A}^{m+1}$ [cf. (2.7)(1)]. We have to show that $\tilde{\varphi}$ is finite. Let $f' \in S_1$, $f := \varphi^*(f')$, and set $U' := \{x' \in X' \mid f'(x') \neq 0\}$, $U := \{x \in X \mid f(x) \neq 0\}$. We have $\mathcal{O}(U') = S'_{(f')}$, $\mathcal{O}(U) = S_{(f)}$ [cf. (5.7)], and $U = \tilde{\varphi}^{-1}(U')$ [as one easily checks]. Since $S' \to S$ is a finite and graded homomorphism [cf. (2.5)(2)], we see that S_f is an $S'_{f'}$-module which is generated by finitely many homogeneous elements of degree 0 since the element f of S_f of degree 1 is a unit of S_f, hence also the homomorphism $S'_{(f')} \to S_{(f)}$ is finite. Since X' can be covered by finitely many sets of the form U, we have proved the assertion.

(5.9) REMARK: We use the result of (5.8) in the following way: Let $S' = \bigoplus_{p \geq 0} S'_p$ and $S = \bigoplus_{p \geq 0} S_p$ be reduced graded k-algebras with $S'_0 = S_0 = k$ which as rings are generated by finitely many elements of degree 1. Let $S = k[f_0, \ldots, f_n]$ with elements $f_0, \ldots, f_n \in S_1$ and let $C \subset \mathbb{A}^{n+1}$ be the algebraic cone with vertex 0 and affine coordinate ring S; let $S' = k[f'_0, \ldots, f'_m]$ with $f'_0, \ldots, f'_m \in S'_1$ and let $C' \subset \mathbb{A}^{m+1}$ be the algebraic cone with vertex 0 and affine coordinate ring S'. Assume that there exists a finite homomorphism of graded rings $\alpha \colon S' \to S$, and let $\varphi \colon C \to C'$ be the morphism with $\varphi^* = \alpha$. Then φ is finite [cf. (2.5)(2)]. Setting $X := \mathbb{P}(C)$ and $X' := \mathbb{P}(C')$, we see that we have a finite morphism $\tilde{\varphi} \colon X \to X'$ fitting into a commutative diagram as above.

6 Regular and Singular Points

(6.1) NOTATION: Let X be a variety.

(1) A point x of X is said to be nonsingular or regular if the local ring $\mathcal{O}_{X,x}$ is a regular local ring [note that a regular local ring is a domain, cf. [63], Cor. 10.14]. A point of X which is not regular is said to be singular or non-regular.

(2) The set of singular points of X is denoted by $\mathrm{Sing}(X)$, and the complement $X \setminus \mathrm{Sing}(X)$ of regular points of X is denoted by $\mathrm{Reg}(X)$.

(3) The variety X is called nonsingular or regular if $\mathrm{Sing}(X) = \emptyset$.

(4) By [63], Cor. 19.14, $\mathbb{A}^n$ and $\mathbb{P}^n$ are nonsingular varieties.

(6.2) REMARK: Let X be a variety, and let $x \in X$. Then $x \in \mathrm{Reg}(X)$ iff X is irreducible in x and x is a regular point of the component containing it [this follows from (1.23) and (1.24)].

(6.3) ZARISKI TANGENT SPACE: Let X be a variety, and let $x \in X$; let $\mathfrak{m}_{X,x}$ be the maximal ideal of the local ring $\mathcal{O}_{X,x}$.

(1) We have $\mathcal{O}_{X,x}/\mathfrak{m}_{X,x} = k$, hence we can consider k as an $\mathcal{O}_{X,x}$-module: For every $f \in \mathcal{O}_{X,x}$ and $\alpha \in k$ we have $f \cdot \alpha = f(x)\alpha$.

The k-vector space $\mathrm{Der}_k(\mathcal{O}_{X,x}, k)$ of k-derivations $\mathcal{O}_{X,x} \to k$ is called the Zariski tangent space to X at x.

(2) The finite-dimensional k-vector space $\mathfrak{m}_{X,x}/\mathfrak{m}_{X,x}^2$ is called the Zariski cotangent space to X at x.

(3) For every $\mathcal{O}_{X,x}$-module M we can identify the k-vector spaces $\mathrm{Hom}_{\mathcal{O}_{X,x}}(M, k)$ and $\mathrm{Hom}_k(M/\mathfrak{m}_{X,x}M, k)$, hence we have an isomorphism of k-vector spaces [cf. [63], section 16.1]

$$\mathrm{Hom}_k(\Omega_{\mathcal{O}_{X,x}/k}/\mathfrak{m}_{X,x}\Omega_{\mathcal{O}_{X,x}/k}, k) \to \mathrm{Der}_k(\mathcal{O}_{X,x}, k).$$

(4) Every k-derivation $d\colon \mathcal{O}_{X,x} \to k$ vanishes on $\mathfrak{m}_{X,x}^2$, since for $f, g \in \mathfrak{m}_{X,x}$ we have $d(f \cdot g) = g(x)d(f) + f(x)d(g) = 0$. Therefore the restriction $d|\mathfrak{m}_{X,x}$ of a k-derivation $d\colon \mathcal{O}_{X,x} \to k$ induces a k-linear map $\mathfrak{m}_{X,x}/\mathfrak{m}_{X,x}^2 \to k$, and we have a k-linear map of k-vector spaces

$$\delta_{X,x}\colon \mathrm{Der}_k(\mathcal{O}_{X,x}, k) \to \mathrm{Hom}_k(\mathfrak{m}_{X,x}/\mathfrak{m}_{X,x}^2, k).$$

(6.4) Proposition: *Let X be a variety, and let $x \in X$. The k-linear map $\delta_{X,x}$ is an isomorphism of k-vector spaces; in particular, $\mathrm{Der}_k(\mathcal{O}_{X,x}, k)$ is a finite-dimensional k-vector space, and $\dim_k(\mathrm{Der}_k(\mathcal{O}_{X,x}, k)) = \mathrm{emdim}(\mathcal{O}_{X,x})$.*

Proof: (1) Let $d \in \mathrm{Der}_k(\mathcal{O}_{X,x}, k)$ and assume that $\delta_{X,x}(d) = 0$. For every $f \in \mathcal{O}_{X,x}$ we have $f - f(x) \in \mathfrak{m}_{X,x}$, hence $0 = \delta_{X,x}(d)(f - f(x)) = d(f)$, and therefore $d = 0$ and $\delta_{X,x}$ is injective.

(2) Let $\sigma\colon \mathfrak{m}_{X,x}/\mathfrak{m}_{X,x}^2 \to k$ be k-linear. Since $fg - f(x)g(x) = (f - f(x))(g - g(x)) + (f - f(x))g(x) + f(x)(g - g(x))$, the map $f \mapsto \sigma\left((f - f(x)) \bmod \mathfrak{m}_{X,x}^2\right)\colon \mathcal{O}_{X,x} \to k$ is a k-derivation of $\mathcal{O}_{X,x}$ in k which is mapped by $\delta_{X,x}$ to σ.

(3) With regard to the last assertion, note that the k-vector space dual to the k-vector space $\mathfrak{m}_{X,x}/\mathfrak{m}_{X,x}^2$ has dimension equal to $\dim_k(\mathfrak{m}_{X,x}/\mathfrak{m}_{X,x}^2)$.

(6.5) Corollary: *Let X be a variety and $x \in X$. Then x is a regular point of X iff $\dim_k(\mathrm{Der}_k(\mathcal{O}_{X,x}, k)) = \dim(\mathcal{O}_{X,x})$.*

Proof: Note that x is a regular point of X iff $\mathrm{emdim}(\mathcal{O}_{X,x}) = \dim(\mathcal{O}_{X,x})$.

(6.6) FUNCTORIALITY: (1) Let $\varphi\colon X \to Y$ be a morphism, let $x \in X$ and set $y := \varphi(x)$, and let $\varphi_x^*\colon \mathcal{O}_{Y,y} \to \mathcal{O}_{X,x}$ be the induced homomorphism of local rings.

(a) φ induces a k-linear map

$$d\varphi_x\colon \mathrm{Der}_k(\mathcal{O}_{X,x}, k) \to \mathrm{Der}_k(\mathcal{O}_{Y,y}, k)$$

by defining $d\varphi_x(d) = d \circ \varphi_x^*$ for every $d \in \mathrm{Der}_k(\mathcal{O}_{X,x}, k)$.

(b) φ induces a k-linear map

$$\mathfrak{m}_{Y,y}/\mathfrak{m}_{Y,y}^2 \to \mathfrak{m}_{X,x}/\mathfrak{m}_{X,x}^2.$$

(c) The following diagram

$$\text{Der}_k(\mathcal{O}_{X,x}, k) \xrightarrow{\quad d\varphi_x \quad} \text{Der}_k(\mathcal{O}_{Y,y}, k)$$

$$\delta_{X,x} \downarrow \qquad\qquad\qquad\qquad \downarrow \delta_{Y,y}$$

$$\text{Hom}_k(\mathfrak{m}_{X,x}/\mathfrak{m}_{X,x}^2, k) \longrightarrow \text{Hom}_k(\mathfrak{m}_{Y,y}/\mathfrak{m}_{Y,y}^2, k)$$

is commutative, as one checks easily.

(2) Let $\psi\colon Y \to Z$ be a morphism of varieties; then we have $d(\psi \circ \varphi)_x = d\psi_y \circ d\varphi_x$. If $Y = X$ and $\varphi = \text{id}_X$, then $d\varphi_x$ is the identity map.

(6.7) AFFINE TANGENT SPACE: Let $X \subset \mathbb{A}^n$ be an affine variety, let $k[T_1, \ldots, T_n]$ be the affine coordinate ring of $\mathbb{A}^n$, and set $\mathfrak{a} := \mathcal{I}(X)$; let $F_1, \ldots, F_m$ be a system of generators of $\mathfrak{a}$. Let $x = (\xi_1, \ldots, \xi_n) \in X$.

(1) The canonical injection $\iota\colon X \hookrightarrow \mathbb{A}^n$ induces the local surjective homomorphism $\iota_x^*\colon \mathcal{O}_{\mathbb{A}^n, x} \to \mathcal{O}_{X,x}$ having kernel $\mathfrak{a}_x = \mathfrak{a}\mathcal{O}_{\mathbb{A}^n, x}$ and the injective k-linear map

$$d\iota_x\colon \text{Der}_k(\mathcal{O}_{X,x}, k) \to \text{Der}_k(\mathcal{O}_{\mathbb{A}^n, x}, k).$$

Clearly we have

$$\text{im}(d\iota_x) = \{ D \in \text{Der}_k(\mathcal{O}_{\mathbb{A}^n, x}, k) \mid D(f) = 0 \text{ for every } f \in \mathfrak{a}_x \}.$$

(2) The $\mathcal{O}_{\mathbb{A}^n, x}$-module $\Omega_{\mathcal{O}_{\mathbb{A}^n, x}/k}$ of k-differentials of $\mathcal{O}_{\mathbb{A}^n, x}$ is a free module of rank n [cf. [63], Prop. 16.1 and Prop. 16.9], hence $\text{Der}_k(\mathcal{O}_{\mathbb{A}^n, x}, k)$ is an n-dimensional k-vector space [cf. (6.3)(3)] which admits a basis $(D_j)_{1 \leq j \leq n}$ where, for $j \in \{1, \ldots, n\}$, D_j is induced by the partial derivative $\partial/\partial T_j\colon k[T_1, \ldots, T_n] \to k[T_1, \ldots, T_n]$.

(3) The affine subspace

$$T_{X,x} = \{ (\zeta_1, \ldots, \zeta_n) \mid \frac{\partial F}{\partial T_1}(x)(\zeta_1 - \xi_1) + \cdots + \frac{\partial F}{\partial T_n}(x)(\zeta_n - \xi_n) = 0 \text{ for each } F \in \mathfrak{a} \}$$

of $\mathbb{A}^n$ passes through x; it is called the *affine tangent space* to X at x. The associated linear subspace of k^n is denoted by $\Theta_{X,x}$, hence

$$\Theta_{X,x} = \{ (\zeta_1, \ldots, \zeta_n) \in k^n \mid \frac{\partial F}{\partial T_1}(x)\zeta_1 + \cdots + \frac{\partial F}{\partial T_n}(x)\zeta_n = 0 \text{ for each } F \in \mathfrak{a} \}.$$

In particular, we have $T_{\mathbb{A}^n, x} = \mathbb{A}^n$ and $\Theta_{\mathbb{A}^n, x} = k^n$.

(4) The k-linear isomorphism $(\zeta_1, \ldots, \zeta_n) \mapsto \sum_{j=1}^{n} \zeta_j D_j\colon k^n \to \text{Der}_k(\mathcal{O}_{\mathbb{A}^n, x}, k)$ induces an isomorphism of k-vector spaces

$$\Theta_{X,x} \to \text{Der}_k(\mathcal{O}_{X,x}, k).$$

In fact, we have to show the following: Let $D = \sum_{j=1}^{n} \zeta_j D_j$ with $z = (\zeta_1, \ldots, \zeta_n) \in k^n$; then $D \in \text{im}(d\iota_x)$ iff $z \in \Theta_{X,x}$. If $z \in \Theta_{X,x}$, then we have

$$D(F_i) = \sum_{j=1}^{n} \zeta_j \frac{\partial F_i}{\partial T_j}(x) = 0 \quad \text{for all } i \in \{1, \ldots, m\};$$

since for $f \in \mathfrak{a}_x$ there exist $g_1, \ldots, g_m \in \mathcal{O}_{\mathbb{A}^n, x}$ with $f = g_1 F_1 + \cdots + g_m F_m$, we have $D(F) = \sum_{i=1}^{m} \sum_{j=1}^{n} g_i(x) \zeta_j \partial F_i / \partial T_j(x) = 0$. Conversely, if $D \in \operatorname{im}(d\iota_x)$, then we have $z \in \Theta_{X,x}$ [cf. (1)].

(5) Note, in particular, that the proof in (4) shows that $\Theta_{X,x}$ consists of all $z = (\zeta_1, \ldots, \zeta_n) \in k^n$ which satisfy

$$\sum_{j=1}^{n} \frac{\partial F_i}{\partial T_j}(x) \zeta_j = 0 \quad \text{for } i \in \{1, \ldots, m\},$$

hence $T_{X,x}$ is the intersection of the hyperplanes which are the zero-sets of the m linear polynomials

$$L_i := \sum_{j=1}^{n} \frac{\partial F_i}{\partial T_j}(x)(T_j - \xi_j) \quad \text{for } i \in \{1, \ldots, m\}.$$

(6.8) Proposition: *Let $X \subset \mathbb{A}^n$ be closed, let $\mathfrak{a} = \mathcal{I}(X) \subset k[T_1, \ldots, T_n]$ be the ideal of X, and let $F_1, \ldots, F_m$ be a system of generators of $\mathfrak{a}$. For every $x \in X$ the matrix*

$$\left(\frac{\partial F_i}{\partial T_j}(x) \right)_{1 \le i \le m, 1 \le j \le n} \in M(m, n; k)$$

has rank equal to $n - \operatorname{emdim}(\mathcal{O}_{X,x})$, hence its rank is $\le n - \dim(\mathcal{O}_{X,x})$.

Proof: With notations as in (6.7), we have $\operatorname{rank}(\partial F_i / \partial T_j(x)) = n - \dim(\Theta_{X,x}) = n - \operatorname{emdim}(\mathcal{O}_{X,x})$ [cf. (6.7)(5) and (6.4)].

(6.9) Corollary: [Jacobian criterion] *x is a regular point of X iff*

$$\operatorname{rank}(\partial F_i / \partial T_j(x)) + \dim(\mathcal{O}_{X,x}) = n,$$

hence iff $\dim(T_{X,x}) = \dim_x(X)$.

(6.10) Corollary: *For any variety X the set $\operatorname{Reg}(X)$ of regular points of X is an open subset of X.*

Proof: Let $x \in \operatorname{Reg}(X)$; then X is irreducible in x [cf. (1.23)], hence it is enough to prove the assertion for an affine irreducible variety [cf. (1.24)]. Thus, we may use the notation introduced in (6.7), and assume that X is irreducible; note that in this case we have $\dim(X) = \dim(\mathcal{O}_{X,y})$ for every $y \in X$ [cf. (1.26)]. Since $\operatorname{rank}(\partial F_i / \partial T_j(x)) = n - \dim(X)$ [cf. (6.9)], the matrix $(\partial F_i / \partial T_j)$ has a quadratic submatrix of order $n - \dim(X)$ whose determinant in x does not vanish; by continuity, this holds true in an open neighborhood U of x, hence every point of U is a regular point of X by (6.9).

(6.11) Proposition: *Let X be a non-empty variety. The set of regular points of X is an open dense subset of X, hence we have $\dim(\operatorname{Sing}(X)) < \dim(X)$.*

Proof: (1) It is enough to consider the case that X is irreducible and affine [cf. (1.24)].

(2) Let $X \subset \mathbb{A}^n$ be irreducible and closed, set $\dim(X) := p-1$ where $p \in \{1, \ldots, n\}$, let $\mathfrak{p} = \mathcal{I}(X) \subset \mathcal{O}(\mathbb{A}^n) = k[T_1, \ldots, T_n]$, and let $\mathcal{O}(X) = k[t_1, \ldots, t_n]$. After a linear change of coordinates we may assume that $t_1, \ldots, t_{p-1}$ are algebraically independent over k, and that $t_p, \ldots, t_n$ are integral over $k[t_1, \ldots, t_{p-1}]$ and separable algebraic over the quotient field $k(t_1, \ldots, t_{p-1})$ of $k[t_1, \ldots, t_{p-1}]$ [cf. B(10.2)]. For every $i \in \{p, \ldots, n\}$ there exists $F_i \in k[T_1, \ldots, T_{p-1}][T_i]$ such that $F_i(t_1, \ldots, t_{p-1})(T_i)$ is the minimal polynomial of t_i over $k(t_1, \ldots, t_{p-1})$. Then we have $F_p, \ldots, F_n \in \mathfrak{p}$ and, for every $i \in \{p, \ldots, n\}$, $\partial F_i/\partial T_i(t_1, \ldots, t_{p-1}, t_i) \neq 0$, i.e., $\partial F_i/\partial T_i \notin \mathfrak{p}$.

Let $x \in \mathbb{A}^n$; the matrix $(\partial F_i/\partial T_j(x))_{p \leq i \leq n, 1 \leq j \leq n}$ has rank equal to $n - p + 1$ if x is not a zero of $H := \prod_{i=p}^n \partial F_i/\partial T_i$, hence, by the Jacobian criterion [cf. (6.9)], the non-regular points of X lie on the hypersurface $Y := \mathcal{Z}(H)$. Since $X \not\subset Y$, $X \cap Y$ is a proper closed subset of X which implies the assertion.

(6.12) **Proposition:** *Let X and Y be varieties. Then we have*

$$\mathrm{Reg}(X \times Y) = \mathrm{Reg}(X) \times \mathrm{Reg}(Y).$$

Proof: Clearly, it is enough to consider the case that $X \subset \mathbb{A}^m$ and $Y \subset \mathbb{A}^n$ are closed subsets. Let $\mathfrak{a} = \mathcal{I}(X)$ and $F_1, \ldots, F_r \in \mathcal{O}(\mathbb{A}^m) = k[T_1, \ldots, T_m]$ be a system of generators of $\mathfrak{a}$, and let $\mathfrak{b} = \mathcal{I}(Y)$ and $G_1, \ldots, G_s \in \mathcal{O}(\mathbb{A}^n) = k[U_1, \ldots, U_n]$ be a system of generators of $\mathfrak{b}$. Now $\mathcal{I}(X \times Y) = \mathfrak{c}$ where $\mathfrak{c} \subset k[T_1, \ldots, T_m, U_1, \ldots, U_n]$ is the ideal generated by $F_1, \ldots, F_r$, $G_1, \ldots, G_s$ [cf. (3.5)(2)], and we have

$$\mathrm{rank}\begin{pmatrix} (\partial F_i/\partial T_j(x)) & 0 \\ 0 & (\partial G_h/\partial U_l(y)) \end{pmatrix} = \mathrm{rank}(\partial F_i/\partial T_j(x)) + \mathrm{rank}(\partial G_h/\partial U_l(y));$$

the assertion follows immediately from (6.9).

(6.13) TANGENT CONE TO AN AFFINE VARIETY: Let $X \subset \mathbb{A}^n$ be closed, let $\mathcal{O}(\mathbb{A}^n) = k[T_1, \ldots, T_n]$, and let $x = (\xi_1, \ldots, \xi_n) \in X$. For the sake of simplicity we assume that x is the origin. Let $\mathcal{I}(X) = \mathfrak{a} \subset k[T_1, \ldots, T_n]$ be the ideal of X; then $A = k[T_1, \ldots, T_n]/\mathfrak{a} = k[t_1, \ldots, t_n]$ is the affine coordinate ring of X. The ideal $\mathfrak{m} = (t_1, \ldots, t_n)$ defines the origin in X, and $\mathcal{O}_{X,x} = A_{\mathfrak{m}}$ is its local ring in X; let $\mathfrak{m}_{X,x}$ be the maximal ideal of $\mathcal{O}_{X,x}$. Let $F \in k[T_1, \ldots, T_n]$ be a non-zero polynomial; we write it $F = F_r + F_{r+1} + \cdots$ as a sum of homogenous terms with $F_r \neq 0$, and call $F_r := \mathrm{In}(F)$ the initial term of F and $o(F) := r$ its order [with respect to the maximal ideal $(T_1, \ldots, T_n)$]. Let $\mathrm{In}(\mathfrak{a})$ be the ideal of $k[T_1, \ldots, T_n]$ which is generated by all initial forms of elements of $\mathfrak{a}$.

(1) The canonical surjective homomorphism of graded k-algebras

$$k[T_1, \ldots, T_n] \to \mathrm{gr}_{\mathfrak{m}_{X,x}}(\mathcal{O}_{X,x}),$$

defined by mapping T_i to $t_i/1 \mod \mathfrak{m}_{X,x}^2$ for $i \in \{1,\dots,n\}$, has kernel $\mathrm{In}(\mathfrak{a})$.

In fact, if $F \in k[T_1,\dots,T_n]$ is homogeneous of degree r, then $F(t_1/1,\dots,t_n/1) \in \mathfrak{m}_{X,x}^{r+1}$ iff $F \in (T_1,\dots,T_n)^{r+1} + \mathfrak{a}$ [note that $(i_A^m)^{-1}(\mathfrak{m}_{X,x})^d = \mathfrak{m}^d$ for every $d \in \mathbb{N}$ since $\mathfrak{m}$ is a maximal ideal], and clearly the homogeneous elements of degree r in this ideal are polynomials $\mathrm{In}(F)$ where $F \in \mathfrak{a}$ and $o(F) = r$.

(2) We set $C_{X,x} := \mathcal{Z}(\mathrm{In}(\mathfrak{a})) \subset \mathbb{A}^n$ and call it the *affine tangent cone* to X at 0; it is an affine algebraic cone in $\mathbb{A}^n$ with vertex at the origin. The generating lines of $C_{X,x}$ are called *tangent lines* to X at x. Note that $\mathcal{O}(C_{X,x}) = (k[T_1,\dots,T_n]/\mathrm{In}(\mathfrak{a}))_{\mathrm{red}} = \mathrm{gr}_{\mathfrak{m}_{X,x}}(\mathcal{O}_{X,x})_{\mathrm{red}}$ [cf. (1)].

(3) We show that $C_{X,x} \subset T_{X,x}$, hence the generating lines of $C_{X,x}$ are lines in the affine space $T_{X,x}$, the tangent space to X at x, passing through x.

In fact, every non-zero $F \in \mathfrak{a}$ can be written as $F = F_1 + \cdots + F_r$ where F_j is homogeneous of degree j for $j \in \{1,\dots,r\}$ and $F_1 = \sum_{i=1}^n \partial F/\partial T_i(x)T_i$. Let $\tilde{\mathfrak{a}}$ be the ideal in $\mathcal{O}(\mathbb{A}^n)$ generated by the linear polynomials $\sum_{i=1}^n \partial F/\partial T_i(x)T_i$ with $F \in \mathfrak{a}$. Then $T_{X,x} = \mathcal{Z}(\tilde{\mathfrak{a}})$, and since $\tilde{\mathfrak{a}} \subset \mathrm{In}(\mathfrak{a})$, we have $C_{X,x} \subset T_{X,x}$.

(4) Let $L \subset \mathbb{A}^n$ be a line passing through the origin; then $L = \mathcal{Z}((F_1,\dots,F_{n-1}))$ where $F_1,\dots,F_{n-1} \in k[T_1,\dots,T_n]$ are linear homogenous polynomials. Now we have $L \subset C_{X,x}$ iff $\mathrm{In}(\mathfrak{a}) \subset (F_1,\dots,F_{n-1})$, as one easily checks.

(6.14) Proposition: *Let X be an affine variety, and let $x \in X$. Then we have*

$$\dim(C_{X,x}) = \dim_x(X).$$

Proof: We have $\dim_x(X) = \dim(\mathcal{O}_{X,x})$, $\dim(\mathcal{O}_{X,x}) = \dim(\mathrm{gr}_{\mathfrak{m}_{X,x}}(\mathcal{O}_{X,x}))$ [cf. [63], Ex. 13.8, 3] and $\dim(\mathrm{gr}_{\mathfrak{m}_{X,x}}(\mathcal{O}_{X,x})) = \dim((\mathrm{gr}_{\mathfrak{m}_{X,x}}(\mathcal{O}_{X,x}))_{\mathrm{red}})$. The assertion follows from (6.13)(2).

(6.15) Proposition: *Let $X \subset \mathbb{A}^n$ be closed, and let $x \in X$. Then x is a regular point of X iff $T_{X,x} = C_{X,x}$. Let this be the case; then, in particular, a line $L \subset \mathbb{A}^n$ passing through x is a tangent line to X at x iff $L \subset T_{X,x}$.*

Proof: We have $C_{X,x} \subset T_{X,x}$ [cf. (6.13)], hence $\dim(C_{X,x}) = \dim(T_{X,x})$ iff $C_{X,x} = T_{X,x}$ since $T_{X,x}$ is irreducible. Since $\dim(C_{X,x}) = \dim_x(X)$ [cf. (6.14)], and x is regular iff $\dim(T_{X,x}) = \dim_x(X)$ [cf. (6.9)], we have proved the assertion.

7   Normalization of a Variety

(7.1) Veronese Embedding: Let $n \in \mathbb{N}_0$ and $d \in \mathbb{N}$, and set $\nu_{n,d} := \binom{n+d}{n} - 1$.

(1) The set $I := \{(j_0,\dots,j_n) \in \mathbb{N}_0^{n+1} \mid j_0 + \cdots + j_n = d\}$ has $\nu_{n,d} + 1$ elements. We endow I with the inverse lexicographic ordering, and label the homogeneous coordinates of $\mathbb{P}^{\nu_{n,d}}$ accordingly; thus, a point of $\mathbb{P}^{\nu_{n,d}}$ shall be written as $(\xi_{(d,0,\dots,0)} : \cdots : \xi_{(j_0,\dots,j_n)} : \cdots : \xi_{(0,\dots,0,d)})$. The homogeneous coordinate ring of $\mathbb{P}^{\nu_{n,d}}$ is the polynomial ring $k[\{U_{(j_0,\dots,j_n)} \mid (j_0,\dots,j_n) \in I\}]$.

(2) The map

$$v_d: \mathbb{P}^n \to \mathbb{P}^{\nu_{n,d}} \text{ with } v_d((\xi_0: \cdots : \xi_n)) = (\xi_{(d,0,\ldots,0)}: \cdots : \xi_{(j_0,\ldots,j_n)}: \cdots : \xi_{(0,\ldots,0,d)})$$

where $\xi_{(j_0,\ldots,j_n)} := \xi_0^{j_0} \cdots \xi_n^{j_n}$ for all $(j_0,\ldots,j_n) \in I$, is well-defined. In fact, if we multiply all $\xi_0,\ldots,\xi_n$ by $\lambda \in k^\times$, then $\xi_{(j_0,\ldots,j_n)}$ is multiplied by λ^d, and at least one of the elements $\xi_{(d,0,\ldots,0)}, \xi_{(0,d,0,\ldots,0)}, \ldots, \xi_{(0,\ldots,0,d)}$ is not 0, hence $v_d((\xi_0: \cdots : \xi_n))$ is indeed a point of $\mathbb{P}^{\nu_{n,d}}$. The map v_d is called the d-th Veronese map; it is obvious that v_d is a morphism.

(3) Let $S := k[T_0,\ldots,T_n]$ be the homogeneous coordinate ring of $\mathbb{P}^n$; we set

$$M_{(j_0,\ldots,j_n)} = T_0^{j_0} \cdots T_n^{j_n} \quad \text{for every } (j_0,\ldots,j_n) \in I.$$

We define a k-algebra homomorphism

$$\theta_d: k[\{U_{(j_0,\ldots,j_n)} \mid (j_0,\ldots,j_n) \in I\}] \to k[T_0,\ldots,T_n]$$

by $\theta_d(U_{(j_0,\ldots,j_n)}) = M_{(j_0,\ldots,j_n)}$ for every $(j_0,\ldots,j_n) \in I$; θ_d is homogeneous of degree d, and its kernel $\mathfrak{a}$ is a homogeneous prime ideal.
Let $(j_0,\ldots,j_n), (l_0,\ldots,l_n), (p_0,\ldots,p_n), (q_0,\ldots,q_n) \in I$ with

$$j_0 + l_0 = p_0 + q_0, \ldots, j_n + l_n = p_n + q_n. \tag{$*$}$$

Then we have

$$U_{(j_0,\ldots,j_n)} U_{(l_0,\ldots,l_n)} - U_{(p_0,\ldots,p_n)} U_{(q_0,\ldots,q_n)} \in \mathfrak{a}. \tag{$**$}$$

It can be shown that $\mathfrak{a}$ is generated by all the quadratic polynomials in $(**)$ [cf. [79], Kapitel III, Abschn. III, Satz].

(4) It is clear that $v_d(\mathbb{P}^n) \subset \mathcal{Z}_+(\mathfrak{a})$. Conversely, let $(\xi_{(d,\ldots,0)}: \cdots : \xi_{(0,\ldots,d)}) \in \mathcal{Z}_+(\mathfrak{a})$. Now at least one coordinate of the form $\xi_{(0,\ldots,d,\ldots,0)}$ is different from zero [cf. $(**)$]. Assume that $\xi_{(d,0,\ldots,0)} \neq 0$, and define $(\xi_0: \cdots : \xi_n) \in \mathbb{P}^n$ by

$$\xi_0 := \xi_{(d,0,\ldots,0)}, \ \xi_1 := \xi_{(d-1,1,0,\ldots,0)}, \ \xi_2 := \xi_{(d-1,0,1,0,\ldots,0)}, \ldots, \xi_n := \xi_{(d-1,0,\ldots,0,1)}.$$

Then we have $v_d((\xi_0: \cdots : \xi_n)) = (\xi_{(d,0,\ldots,0)}: \cdots : \xi_{(0,\ldots,d)})$ as may be easily shown by using $(**)$. It is not difficult to check that this gives a well-defined map $\mathcal{Z}_+(\mathfrak{a}) \cap (\mathbb{P}^{\nu_{n,d}} \setminus \mathcal{Z}_+(U_{(d,0,\ldots,0)})) \to \mathbb{P}^n$, and that these maps define a morphism $\omega: \mathcal{Z}_+(\mathfrak{a}) \to \mathbb{P}^n$ such that $v_d \circ \omega$ is the identity map on $\mathcal{Z}_+(\mathfrak{a})$ and $\omega \circ v_d$ is the identity map on $\mathbb{P}^n$. This implies that $\mathcal{Z}_+(\mathfrak{a}) = v_d(\mathbb{P}^n)$, hence $v_d(\mathbb{P}^n)$ is closed in $\mathbb{P}^{\nu_{n,d}}$, and $v_d: \mathbb{P}^n \to \mathbb{P}^{\nu_{n,d}}$ is a closed embedding; it is called the d-uple embedding of $\mathbb{P}^n$.

(5) Let $X \subset \mathbb{P}^n$ be closed. The closed set

$$X^{(d)} := v_d(X) \subset \mathbb{P}^{\nu_{n,d}}$$

is said to be the d-th Veronese transform of X. The last assertion in (4) implies that X and $X^{(d)}$ are isomorphic.

(6) From the definition of v_d we immediately see that

$$\theta_d\colon k[\,U_{(d,0,\ldots,0)},\ldots,U_{(j_0,\ldots,j_n)},\ldots,U_{(0,\ldots,d)}\,] \to (k[\,T_0,\ldots,T_n\,])^{(d)} = S^{(d)}$$

is a surjective homomorphism of graded k-algebras.
Let $X \subset \mathbb{P}^n$ be closed, and let $\mathfrak{b} = \mathcal{I}_+(X) \subset S$ be the ideal of X; then $S(X) = S/\mathfrak{b}$ is the homogeneous coordinate ring of X. Set $\mathfrak{b}^{(d)} := \sum_{m\geq 0} \mathfrak{b}_{md} \subset S^{(d)}$. By (4) we see that $\theta_d^{-1}(\mathfrak{b}^{(d)}) = \mathcal{I}_+(X^{(d)})$, hence *the homogeneous coordinate ring of the closed subset $X^{(d)}$ of $\mathbb{P}^{\nu_{n,d}}$ is $S(X)^{(d)}$.*

(7.2) DEFINITION: Let X be a variety.
(1) Let $x \in X$; X is called normal in x if $\mathcal{O}_{X,x}$ is a domain which is integrally closed in its field of quotients. $\mathrm{Nor}(X)$ is the set of points of X in which X is normal.
(2) X is said to be normal if X is normal in every of its points.

(7.3) FACTORIALITY: It is well known that a regular local ring A is factorial [cf. [63], Th. 19.19]; in the case that $A = \mathcal{O}_{X,x}$ where x is a point of a variety X, this fact can be proved very simple [cf. [174], vol. I, Ch. II, Th. 2, and Appendix, section 7]. Since a factorial domain is integrally closed [cf. [63], Prop. 4.10], we have $\mathrm{Reg}(X) \subset \mathrm{Nor}(X)$, hence $\mathrm{Nor}(X)$ is dense in X [cf. (6.11)]. In particular, a regular variety is normal.

(7.4) Proposition: *Let X be an irreducible affine variety. Then X is normal iff $\mathcal{O}(X)$ is integrally closed in $K(X)$, the field of rational functions on X.*

Proof: A domain A is integrally closed in its field of quotients iff $A_\mathfrak{m}$ is integrally closed in its field of quotients for every maximal ideal $\mathfrak{m}$ of A [cf. B(2.5) and [63], Prop. 4.13]; now we can use (1.9).

(7.5) Proposition: *Let $X \subset \mathbb{P}^n$ be irreducible and closed. If the homogeneous coordinate ring $S(X)$ of X is integrally closed, then X is normal.*

Proof: X is covered by $U_0 \cap X, \ldots, U_n \cap X$ [cf. (1.18)]. Let $i \in \{0,\ldots,n\}$; the open subset $U_i \cap X$ of X is affine, and we have $\mathcal{O}(U_i \cap X) = S(X)_{(t_i)}$ where t_i is the image of T_i in $S(X)$ [cf. (5.7)]. Now $S(X)_{(t_i)}$ is integrally closed [cf. B(4.30)(2)], hence $U_i \cap X$ is normal by (7.4), and therefore X is normal.

(7.6) NOTATION: An irreducible projective variety $X \subset \mathbb{P}^n$ is called arithmetically normal if its homogeneous coordinate ring $S(X)$ is integrally closed. Thus, the result of (7.5) can be stated as follows: An irreducible projective arithmetically normal variety is normal.

(7.7) Proposition: *Let $\varphi\colon Y \to X$, $\psi\colon Z \to X$ be morphisms of irreducible varieties. Assume that φ is dominant, that Y is normal, and that ψ is finite and birational. Then there exists a unique morphism $\chi\colon Y \to Z$ such that $\varphi = \psi \circ \chi$.*

Proof: (1) First, we consider the case where X is affine. Since ψ is finite, also Z is affine [cf. (2.6)(1)]. We have the commutative diagrams [cf. (1.17)]

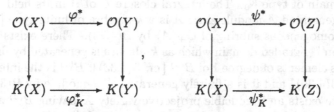

where φ^* is injective [since φ is dominant, cf. (1.15)] and ψ_K^* is an isomorphism. Set $\omega := \varphi_K^* \circ \psi_K^{*-1} \colon K(Z) \to K(Y)$. Since ψ is finite, $\mathcal{O}(Z)$ is a finitely generated $\mathcal{O}(X)$-module, hence $\mathcal{O}(X) \subset \psi_K^{-1}(\mathcal{O}(Z))$ is an integral extension, and therefore $\varphi^*(\mathcal{O}(X)) \subset \omega(\mathcal{O}(Z))$ is an integral extension. Since $\varphi^*(\mathcal{O}(X)) \subset \mathcal{O}(Y)$ and $\mathcal{O}(Y)$ is integrally closed by hypothesis, we have $\omega(\mathcal{O}(Z)) \subset \mathcal{O}(Y)$. Thus, the restriction of ω to $\mathcal{O}(Z)$ induces a k-algebra homomorphism $\omega \colon \mathcal{O}(Z) \to \mathcal{O}(Y)$, and obviously $\varphi^* = \omega \circ \psi^*$. Clearly ω is the only homomorphism with this property. Let $\chi \colon Y \to Z$ be the morphism defined by ω [cf. (1.11)]; then $\psi \circ \chi = \varphi$ [this is immediate if Y is affine; in the other case one can use an open affine covering of Y to get the assertion]. Moreover, $\chi \colon Y \to Z$ is the only morphism satisfying $\psi \circ \chi = \varphi$ [this is immediate if Y is affine by (1.11), since ω is uniquely determined; in the other case one can use an open affine covering of Y to get the assertion].
(2) In the general case, we write $X = X_1 \cup \cdots \cup X_h$ where $X_1, \ldots, X_h$ are open affine subsets of X. Let $i \in \{1, \ldots, h\}$. The induced morphism $\varphi^{-1}(X_i) \to X_i$ is dominant, and the induced morphism $\psi^{-1}(X_i) \to X_i$ is finite by definition, hence by (1) there exists a unique morphism $\chi_i \colon \varphi^{-1}(X_i) \to \psi^{-1}(X_i)$ such that $\psi(\chi_i(y)) = \varphi(y)$ for every $y \in \varphi^{-1}(X_i)$. Since $X_i \cap X_j$ is affine for all $i, j \in \{1, \ldots, h\}$ [cf. (3.11)], the restrictions of χ_i and χ_j to $\varphi^{-1}(X_i) \cap \varphi^{-1}(X_j) = \varphi^{-1}(X_i \cap X_j)$ are equal by (1). Hence there exists a unique map $\chi \colon Y \to Z$ with $\varphi = \psi \circ \chi$, and χ is a morphism.

(7.8) DEFINITION: Let X be an irreducible variety. A pair $(\overline{X}, \pi)$, where $\overline{X}$ is an irreducible normal variety and $\pi \colon \overline{X} \to X$ is a finite birational morphism, is called a normalization of X.
By abuse of language we usually say: Let $\pi \colon \overline{X} \to X$ be a normalization of X.

(7.9) REMARK: Let X be an irreducible variety.
(1) Let $\pi \colon \overline{X} \to X$ be a normalization of X. Let $\varphi \colon Y \to X$ be a dominant morphism where Y is irreducible and normal. Then there exists a unique morphism $\chi \colon Y \to \overline{X}$ such that $\pi \circ \chi = \varphi$ [cf. (7.7)].
(2) In particular, a normalization of X—if it exists—is unique up to isomorphism.

(7.10) Theorem: *Let X be an irreducible variety. Then there exists a normalization $\pi \colon \overline{X} \to X$. In particular, if X is affine (resp. quasi-affine resp. projective), then $\overline{X}$ is affine (resp. quasi-affine resp. projective).*

Proof: (1) First, we consider the case where X is projective, i.e., $X \subset \mathbb{P}^n$ is irreducible and closed. The homogeneous coordinate ring $A := S(X)$ of X is a $\mathbb{Z}$-graded domain of type $\mathbb{N}_0$. The integral closure B of A in its field of quotients is a finitely generated A-module since A is a k-algebra of finite type [cf. B(3.6)], and it is a homogeneous subring of $Q_+(A)$ by B(4.28). There exists $d \in \mathbb{N}$ such that $B^{(d)}$ is an $\mathbb{N}_0$-graded domain which as k-algebra is generated by finitely many homogeneous elements of degree 1 of $B^{(d)}$ [cf. B(4.33)]; $B^{(d)}$ is the integral closure of $A^{(d)}$ [cf. B(4.30)] and it is a finitely generated $A^{(d)}$-module by B(3.6).

By (5.9) there exists an irreducible projective variety $\overline{X}$ having $B^{(d)}$ as its homogeneous coordinate ring; in particular, $\overline{X}$ is normal [cf. (7.5)].

The graded k-algebra $A^{(d)}$ is the homogeneous coordinate ring of the d-th Veronese transform $X^{(d)} \subset \mathbb{P}^{\nu_n,d}$ of X [cf. (7.1)(6)]. The inclusion map $A^{(d)} \hookrightarrow B^{(d)}$ induces a finite birational morphism $\overline{\varphi} \colon \overline{X} \to X^{(d)}$ [cf. (5.9)]. Let $v_{d,X} \colon X \to X^{(d)}$ be defined by restricting v_d. Then $v_{d,X}$ is an isomorphism [cf. (7.1)(5)], and $\pi := v_{d,X}^{-1} \circ \overline{\varphi} \colon \overline{X} \to X$ is a normalization of X.

(2) Thus, we have shown that an irreducible projective variety admits a normalization which is projective. Now assume that $X \subset \mathbb{P}^n$ is irreducible and quasi-projective. Then the closure Y of X in $\mathbb{P}^n$ is irreducible, hence admits a normalization $\pi \colon \overline{Y} \to Y$ where $\overline{Y}$ is projective [cf. (1)]. It is clear that $\pi^{-1}(X)$ is a normalization of X.

(3) In (2) we have shown that every irreducible variety admits a normalization.
(a) Now let X be an irreducible affine variety, and let $\pi \colon \overline{X} \to X$ be a normalization of X. Then $\overline{X} = \pi^{-1}(X)$ is affine since a finite morphism is affine [cf. (2.6)(1)].
(b) Assume that $X \subset \mathbb{A}^n$ is an irreducible quasi-affine variety. Then the closure Y of X in $\mathbb{A}^n$ is irreducible, hence admits a normalization $\pi \colon \overline{Y} \to Y$ by (a) where $\overline{Y}$ is affine. It is clear that $\pi^{-1}(X)$ is a normalization of X.

(7.11) Corollary: *Let X be an irreducible affine variety, and let $\pi \colon \overline{X} \to X$ be its normalization. Let $A = \mathcal{O}(X)$, and let $K(X)$ be the field of rational functions on X. Then $\mathcal{O}(\overline{X})$ is the integral closure of A in $K(X)$. Moreover, if $x \in \mathrm{Nor}(X)$, then there lies only one point $\overline{x} \in \overline{X}$ over x, and the local homomorphism $\mathcal{O}_{X,x} \to \mathcal{O}_{\overline{X},\overline{x}}$ is an isomorphism.*

Proof: Since π is finite and birational, we can identify $B := \mathcal{O}(\overline{X})$ with a subring of $K(X)$, and the induced homomorphism $\pi^* \colon A \to B$ is injective and finite. Since B is integrally closed, we see that B is the integral closure of A in $K(X)$. Let $x \in \mathrm{Nor}(X)$, and set $\mathfrak{p} := \mathcal{I}_X(x)$. Then $B_{\mathfrak{p}}$ is the integral closure of $A_{\mathfrak{p}}$ [cf. [63], Prop. 4.13]; since $A_{\mathfrak{p}}$ is integrally closed, we have $A_{\mathfrak{p}} = B_{\mathfrak{p}}$, and this implies the last assertion.

(7.12) Corollary: *Let X be an irreducible variety, let $\pi \colon \overline{X} \to X$ be its normalization, and let $x \in X$. The set of points of $\overline{X}$ lying over x is finite and not empty. Let $y_1, \ldots, y_h$ be the points of $\overline{X}$ lying over x; then $\mathcal{O}_{\overline{X}, y_1} \cap \cdots \cap \mathcal{O}_{\overline{X}, y_h}$ is the integral closure of $\mathcal{O}_{X,x}$.*

Proof: Since π is finite, there exists an open affine neighborhood U of x such that setting $V := \pi^{-1}(U)$, V is affine and $A := \mathcal{O}(U) \to \mathcal{O}(V) =: B$ is finite [cf. (2.5)]; moreover, B is the integral closure of A [cf. (7.11)], and B is semilocal [cf. B(3.8)]. Setting $\mathfrak{m} := \mathcal{I}_U(\{x\})$, we see: The finitely many maximal ideals $\mathfrak{n}_1, \ldots, \mathfrak{n}_h$ of B lying over $\mathfrak{m}$ correspond uniquely to the points $y_1, \ldots, y_h$ of $\overline{X}$ lying over x, and we may assume that $\mathcal{O}_{\overline{X}, y_i} = B_{\mathfrak{n}_i}$ for $i \in \{1, \ldots, h\}$. $B_{\mathfrak{m}}$ is the integral closure of $A_{\mathfrak{m}} = \mathcal{O}_{X,x}$, and $B_{\mathfrak{m}} = \mathcal{O}_{\overline{X}, y_1} \cap \cdots \cap \mathcal{O}_{\overline{X}, y_h}$ [cf. B(2.5)].

(7.13) REMARK: Let A be a domain, let B be its integral closure, and let $\mathfrak{p}$ be a prime ideal of A. Assume that $A_{\mathfrak{p}}$ is integrally closed; then we have $A_{\mathfrak{p}} = B_{\mathfrak{p}}$ [cf. [63], Prop. 4.13]. If, furthermore, B is a finitely generated A-module, then $C := B/A$ is a finitely generated A-module with $C_{\mathfrak{p}} = \{0\}$, hence there exists $s \in A \setminus \mathfrak{p}$ with $C_s = \{0\}$ [cf. [63], Prop. 2.1], and therefore we have $B_s = A_s$.

(7.14) Proposition: *Let X be an irreducible variety, let $\pi \colon \overline{X} \to X$ be its normalization, and let $x \in X$. The following statements are equivalent:*
(1) X is normal in x;
(2) there exists an open neighborhood U of x such that π induces an isomorphism $\pi^{-1}(U) \to U$.
In particular, $\mathrm{Nor}(X)$ is open and dense in X, and the induced morphism $\pi|\pi^{-1}(\mathrm{Nor}(X)) \colon \pi^{-1}(\mathrm{Nor}(X)) \to \mathrm{Nor}(X)$ is an isomorphism.

Proof (1) $\Rightarrow$ (2): Since π is finite, there exists an open affine neighborhood U of x such that $\pi^{-1}(U)$ is affine and that $\pi^{-1}(U) \to U$ is finite [cf. (2.5)(1)]. Hence we may assume, to begin with, that X and $\overline{X}$ are affine. Then $\pi \colon \overline{X} \to X$ is surjective, and $\mathcal{O}(\overline{X})$ is the integral closure of $\mathcal{O}(X)$ [cf. (7.11)]. Since $\mathcal{O}_{X,x}$ is integrally closed, there exists $f \in \mathcal{O}(X)$ with $f(x) \neq 0$ and $\mathcal{O}(X_f) = \mathcal{O}(X)_f \overset{\sim}{\to} \mathcal{O}(\overline{X})_{\pi^*(f)} = \mathcal{O}(\pi^{-1}(X_f))$ [cf. (7.13) and (1.20)], hence $\pi^{-1}(X_f) \to X_f$ is an isomorphism.
(2) $\Rightarrow$ (1): This is clear.

(7.15) REMARK: Let X be a variety, and let $x \in X$. Then x lies in $\mathrm{Nor}(X)$ iff X is irreducible in x and x is a normal point of the irreducible component containing it [this follows from (1.23) and (1.24)].

(7.16) Proposition: *The set $\mathrm{Nor}(X)$ of normal points of a variety X is open and dense in X.*

Proof: Using the notation of (1.24) we have $\mathrm{Nor}(X) = \mathrm{Nor}(U_1) \cup \cdots \cup \mathrm{Nor}(U_h)$; now the result follows from (7.14).

(7.17) Proposition: *Let X be an irreducible variety. The following statements are equivalent:*
(1) X is a normal variety.
(2) $\mathrm{codim}(\mathrm{Sing}(X), X) > 1$ and $\mathrm{codim}(X \setminus \delta(f), X) = 1$ for every rational function $f \in K(X) \setminus \mathcal{O}(X)$.

Proof: We may assume that X is affine; let $A = \mathcal{O}(X)$ be the ring of regular functions on X.

(1) $\Rightarrow$ (2): In this case A is integrally closed [cf. (7.4)]. First, we show that every irreducible closed subset of X of codimension 1 contains regular points of X. Thus, let $Y \subset X$ be irreducible and closed with $\operatorname{codim}(Y, X) = 1$. Then $\mathfrak{p} := \mathcal{I}_X(Y)$ is a prime ideal of height 1 of A, hence $A_\mathfrak{p}$ is a discrete valuation ring [cf. B(10.5)]. Therefore there exists $g \in A$ with $gA_\mathfrak{p} = \mathfrak{p}A_\mathfrak{p}$, and since $\mathfrak{p}$ is finitely generated, there exists $h \in A \setminus \mathfrak{p}$ with $gA_h = \mathfrak{p}A_h$. Now $X_h \cap Y \neq \emptyset$, hence there exists $y \in \operatorname{Reg}(Y) \cap X_h$ [cf. (6.11)]. Since $h(y) \neq 0$, we have $g\mathcal{O}_{X,y} = \mathfrak{p}\mathcal{O}_{X,y} = \mathcal{I}_{X,y}(Y)$, and since $\mathcal{O}_{Y,y} = \mathcal{O}_{X,y}/\mathcal{I}_{X,y}(Y)$ [cf. (1.27)] and $\mathcal{O}_{Y,y}$ is a regular local ring, it follows easily that $\mathcal{O}_{X,y}$ is a regular local ring [note that $\dim(\mathcal{O}_{X,y}) = 1 + \dim(\mathcal{O}_{Y,y})$].

Now we show that the second condition in (2) holds. Suppose that there exists a rational function $f \in K(X) \setminus A$ with $\operatorname{codim}(X \setminus \delta(f), X) > 1$, and let Z be an irreducible component of $X \setminus \delta(f)$ with $\operatorname{codim}(Z, X) > 1$. There exists an open affine subset U of X which meets no other component of $X \setminus \delta(f)$ except Z [apply the construction in (1.24) to $X \setminus \delta(f)$]. Replacing X by U we may assume that $X \setminus \delta(f)$ is irreducible. Let $\mathfrak{p}$ be a prime ideal of A of height 1, and set $Y := \mathcal{Z}_X(\mathfrak{p}) \subset X$. Then Y is irreducible and closed, and we have $\operatorname{codim}(Y, X) = 1$. Therefore Y is not contained in $X \setminus \delta(f)$, hence $Y \cap \delta(f)$ is not empty. We choose $y \in Y \cap \delta(f)$; we have $f \in \mathcal{O}_{X,y} \subset A_\mathfrak{p}$ [cf. (1.28)(4)]. Thus, we have $f \in A$ [cf. [63], Cor. 11.4], which is in contradiction with the choice of f as an element of K, which does not lie in A.

(2) $\Rightarrow$ (1): First, we show that $A_\mathfrak{p}$ is integrally closed for every prime ideal $\mathfrak{p}$ of A of height 1. In fact, let $\mathfrak{p}$ be such a prime ideal, and set $Y := \mathcal{Z}_X(\mathfrak{p})$. Then we have $\operatorname{codim}(Y, X) = 1$, hence there exists $y \in \operatorname{Reg}(X) \cap Y$. Now $\mathcal{O}_{X,y}$ is a regular local ring, hence integrally closed [cf. (7.3)], and since $A_\mathfrak{p}$ is a localization of $\mathcal{O}_{X,y}$ [cf. (1.28)], we have proved the assertion.

Now let $f \in K(X)$, and assume that $f \in A_\mathfrak{p}$ for every prime ideal $\mathfrak{p}$ of A of height 1. Let Y be an irreducible closed subset of X of codimension 1. Then $\mathfrak{p} := \mathcal{I}_X(Y)$ is a prime ideal of height 1, and since $f \in A_\mathfrak{p}$, we have $Y \subset \delta(f)$. Therefore we have $\operatorname{codim}(X \setminus \delta(f), X) > 1$, hence $f \in A$. This implies that A is the intersection of all rings $A_\mathfrak{p}$ where $\mathfrak{p}$ runs through the set of prime ideals of A of height 1, hence A is integrally closed.

This ends the proof of the proposition.

8   Desingularization of a Variety

(8.1) DEFINITION: Let X be an irreducible variety. A pair $(\widetilde{X}, \varphi)$ with $\widetilde{X}$ an irreducible regular variety and $\varphi \colon \widetilde{X} \to X$ a proper morphism such that there exists a non-empty open set $U \subset X$ with $\varphi^{-1}(U) \to U$ being an isomorphism is called a desingularization of X.

By abuse of language we usually say: Let $\varphi \colon \widetilde{X} \to X$ be a desingularization of X.

If there exists a desingularization of X, then we say that X can be desingularized.

(8.2) Let X be an irreducible variety, and let $\varphi\colon \widetilde{X} \to X$ be a desingularization of X. Let $Y \subset X$ be open, and set $\widetilde{Y} := \varphi^{-1}(Y)$. Identifying $\widetilde{Y}$ with $\widetilde{X} \times_X Y$ [cf. (3.15)], we see that the restriction $\psi := \varphi | \widetilde{Y} \colon \widetilde{Y} \to Y$ is proper [cf. (4.23)], and it is easy to check that $\psi\colon \widetilde{Y} \to Y$ is a desingularization of Y.
In other words, this means: If X can be desingularized, then every open subset of X can be desingularized.

(8.3) REMARK: Let X be an irreducible variety, and let $\pi\colon \overline{X} \to X$ be the normalization of X.
(1) Let $\varphi\colon \widetilde{X} \to X$ be a desingularization of X. Let $\chi\colon \widetilde{X} \to \overline{X}$ be the morphism with $\pi \circ \chi = \varphi$ [cf. (7.9) and note that $\widetilde{X}$ is normal by (7.3) and φ is dominant]; then $\chi\colon \widetilde{X} \to \overline{X}$ is a desingularization of $\overline{X}$.
In fact, χ is proper by (4.19)(3), and the assertion follows from (7.11).
(2) Let $\psi\colon Y \to \overline{X}$ be a desingularization of $\overline{X}$. Then $\varphi = \pi \circ \psi\colon Y \to X$ is a desingularization of X since φ is proper [cf. (4.19)], and since there exists a non-empty open subset U of X such that $\pi^{-1}(U) \to U$ is an isomorphism [cf. (7.11)].
(3) From (1) and (2) we get: X can be desingularized iff its normalization $\overline{X}$ can be desingularized.

9   Dimension of Fibres

(9.1) Proposition: *Let $\varphi\colon X \to Y$ be a morphism. Let $y \in \varphi(X)$; for every irreducible component Z of $\varphi^{-1}(\{y\})$ we have $\operatorname{codim}(Z, X) \le \dim_y(\varphi(X))$.*

Proof: We may replace Y by $\overline{\varphi(X)}$. We set $d := \dim_y(Y)$. Let $z \in Z$; then we have $\varphi(z) = y$. Using (1.27)(2), we obtain $\operatorname{ht}(\mathcal{I}_{X,z}(Z)) = \operatorname{codim}(Z, X)$; therefore it is enough to show that $\operatorname{codim}(Z, X) \le d$. Now $\mathcal{I}_{X,z}(Z)$ is a minimal element in $\operatorname{Ass}(\mathcal{O}_{X,z}/\mathcal{I}_{X,z}(\varphi^{-1}(\{y\})))$; we have $\mathcal{I}_{X,z}(\varphi^{-1}(\{y\})) = \operatorname{rad}(\mathcal{I}_{Y,y}(\{y\}))$ by (1.27)(3) and $\mathfrak{m}_{Y,y} = \mathcal{I}_{Y,y}(\{y\})$, and therefore $\mathcal{I}_{X,z}(Z)$ is a minimal prime of $\mathfrak{m}_{Y,y}\mathcal{O}_{X,z}$. Let $g_1, \ldots, g_d$ be a system of parameters of $\mathcal{O}_{Y,y}$. Then we have $\operatorname{rad}((g_1, \ldots, g_d)) = \mathfrak{m}_{Y,y}$, hence $\operatorname{rad}(\varphi_z^*(g_1)\mathcal{O}_{X,z} + \cdots + \varphi_z^*(g_d)\mathcal{O}_{X,z}) = \mathfrak{m}_{Y,y}\mathcal{O}_{X,z}$; by Krull's principal theorem [cf. [63], Th. 10.2] we obtain $\operatorname{ht}(\mathcal{I}_{X,z}(Z)) \le d$

(9.2) Lemma: *Let B be an integral domain, and let A be a subring of B such that B is of finite type over A. Then there exist a non-zero $s \in A$ and elements $x_1, \ldots, x_r \in B$ which are algebraically independent over $Q(A)$ such that B_s is a finitely generated $A_s[x_1, \ldots, x_r]$-module.*

Proof: We have $B = A[t_1, \ldots, t_n]$; we set $K := Q(A)$, $L := Q(B)$, $\Sigma := A \setminus \{0\}$. Then we have $K = \Sigma^{-1}A$, L has finite transcendence degree r over K, and $\Sigma^{-1}B = K[t_1, \ldots, t_n]$, a K-algebra of finite type. By Noether normalization [cf. [63], Th. 13.3] there exist $x_1, \ldots, x_r \in \Sigma^{-1}B$ which are algebraically independent over K and such that $\Sigma^{-1}B$ is a finitely generated $K[x_1, \ldots, x_r]$-module. We can write $x_i = x_i'/s$ with $x_i' \in B$ for $i \in \{1, \ldots, r\}$ and $s \in \Sigma$, hence we may assume that

$x_1, \ldots, x_r \in B$. The elements $t_1, \ldots, t_n$ are integral over $K[x_1, \ldots, x_r]$. For every $j \in \{1, \ldots, n\}$ let $T^{p_j} + f_{j1}T^{p_j-1} + \cdots + f_{jp_j}$ with $f_{j1}, \ldots, f_{jp_j} \in K[x_1, \ldots, x_r]$ be an equation of integral dependence for t_j over $K[x_1, \ldots, x_r]$. We write $f_{jl} = g_{jl}/s$ with $g_{jl} \in A[x_1, \ldots, x_r]$ for $j \in \{1, \ldots, n\}$, $l \in \{1, \ldots, p_j\}$, and $s \in \Sigma$. We have $B_s = A_s[t_1, \ldots, t_n]$, and since $t_1, \ldots, t_n$ are integral over $A_s[x_1, \ldots, x_r]$, the assertion follows from [63], Cor. 4.5.

The geometric version of lemma (9.2) is the following

(9.3) Proposition: *Let $\varphi \colon X \to Y$ be a morphism of affine varieties with X irreducible. Then $r := \dim(X) - \dim(\overline{\varphi(X)})$ is a non-negative integer, and there exists a non-empty open affine subset V of Y such that $\varphi^{-1}(V)$ is affine, and that the restriction $\varphi|\varphi^{-1}(V)$ can be factored as $\varphi^{-1}(V) \xrightarrow{\varphi'} V \times \mathbb{A}^r \xrightarrow{\mathrm{pr}_1} V$; here $\varphi' \colon \varphi^{-1}(V) \to V \times \mathbb{A}^r$ is a finite dominant morphism and $\mathrm{pr}_1 \colon V \times \mathbb{A}^r \to V$ is the projection on the first factor.*

Proof: We may replace Y by $\overline{\varphi(X)}$; hence we may assume that Y is irreducible also, and that φ is dominant. Then $\varphi^* \colon A := \mathcal{O}(Y) \to \mathcal{O}(X) =: B$ is injective [cf. (1.15)], and we can consider $\varphi^*(A)$ as a subring of the integral domain B. Using (9.2) we see that there exists a non-zero regular function f on Y such $B_{\varphi^*(f)}$ is finite over $\varphi^*(A)_{\varphi^*(f)}[g_1, \ldots, g_r]$ where $g_1, \ldots, g_r \in B$ are algebraically independent over the field of quotients of $\varphi^*(A)$. We set $V := Y_f$; then $\varphi^{-1}(V) = X_{\varphi^*(f)}$ is affine. Since $\dim(X) = \dim(X_{\varphi^*(f)}) = \dim(B_{\varphi^*(f)})$ and $\dim(Y) = \dim(Y_f)$, we have $r = \mathrm{tr.d}_{Q(\varphi^*(A))}(Q(B)) = \mathrm{tr.d}_k(Q(B)) - \mathrm{tr.d}_k(Q(A)) = \dim(X) - \dim(Y)$. We define $\alpha \colon A_f[T_1, \ldots, T_r] \to \varphi^*(A)_{\varphi^*(f)}[g_1, \ldots, g_r] \to B_{\varphi^*(f)}$ in the obvious way, and let $\varphi' \colon X_{\varphi^*(f)} \to Y_f \times \mathbb{A}^r$ be the morphism with $\varphi'^* = \alpha$; φ' is finite and dominant. Now we clearly have $\varphi|X_{\varphi^*(f)} = \varphi' \circ \mathrm{pr}_1$.

(9.4) Proposition: *Let $\varphi \colon X \to Y$ be a morphism of varieties with X irreducible, and set $r := \dim(X) - \dim(\overline{\varphi(X)})$. Then r is a non-negative integer, and we have:*
(1) $\dim(\varphi^{-1}(\{y\})) \geq r$ for every $y \in \varphi(X)$.
(2) There exists $W \subset \varphi(X)$ which is an open dense subset of $\overline{\varphi(X)}$ such that

$$\varphi^{-1}(\{y\}) = r \quad \text{for every } y \in W.$$

Proof: Clearly we may replace Y by $\overline{\varphi(X)}$, hence we may assume that Y is irreducible also. Note that now $\dim_y(Y) = \dim(Y)$ for every $y \in Y$.
(1) For every $y \in \varphi(X)$ we have [cf. (1.26) and (9.1)]

$$\dim(\varphi^{-1}(\{y\})) = \dim(X) - \mathrm{codim}(\varphi^{-1}(\{y\}), X) \geq \dim(X) - \dim_y(Y) = r.$$

(2) Let V be a non-empty open affine subset of Y, and choose non-empty open affine subsets $U_1, \ldots, U_m$ of X with $\varphi^{-1}(V) = U_1 \cup \cdots \cup U_m$. Let $i \in \{1, \ldots, m\}$. Note that $\dim(V) = \dim(Y)$ and that $\dim(X) = \dim(U_i)$ for every $i \in \{1, \ldots, m\}$; therefore we have $r \geq 0$ by (9.3). We apply (9.3) to $\varphi_i := \varphi|U_i \colon U_i \to V$; then there exists an open affine subset $V_i \subset V$ such that $\varphi_i|\varphi_i^{-1}(V_i) \colon \varphi_i^{-1}(V_i) \to V_i$

can be factored as $\varphi_i^{-1}(V_i) \xrightarrow{\varphi_i'} V_i \times \mathbb{A}^r \xrightarrow{\mathrm{pr}_1} V_i$ where φ_i' is finite and dominant. For $y \in V_i$ we have $\mathrm{pr}_1^{-1}(\{y\}) = \{y\} \times \mathbb{A}^r$, hence $\dim(\mathrm{pr}_1^{-1}(\{y\})) = r$. On the other hand, the restriction of φ_i' induces a finite dominant morphism $\varphi_i^{-1}(\{y\}) = \varphi_i'^{-1} \circ \mathrm{pr}_1^{-1}(\{y\}) \to \mathrm{pr}_1^{-1}(\{y\})$ [cf. (2.6)]; therefore we have $\dim(\varphi_i^{-1}(\{y\})) = r$ [cf. (2.8)]. Now $W := V_1 \cap \cdots \cap V_m$ is open and dense in Y. Let $y \in W$, and let Z be an irreducible component of $\varphi^{-1}(\{y\})$ with $\dim(Z) = \dim(\varphi^{-1}(\{y\}))$. As $Z \subset \varphi^{-1}(\{y\}) \subset \varphi^{-1}(V) = U_1 \cup \cdots \cup U_m$, there exists $i \in \{1, \ldots, m\}$ with $Z \cap U_i \neq \emptyset$; note that $Z \cap U_i$ is open and dense in Z, hence $\dim(Z) = \dim(Z \cap U_i)$. Therefore we have $\dim(\varphi^{-1}(\{y\})) = \dim(Z) = \dim(Z \cap U_i) \leq \varphi_i^{-1}(\{y\}) = r$.

10 Quasifinite Morphisms and Ramification

10.1 Quasifinite Morphisms

(10.1) DEFINITION: Let X and Y be irreducible varieties. A morphism $\varphi \colon X \to Y$ is called quasifinite if there exist non-empty open subsets $U \subset X$, $V \subset Y$ with $\varphi(U) \subset V$ and such that the induced morphism $U \to V$ is dominant and finite. If φ has this property, then φ is dominant [since $\varphi(\overline{U}) \subset \overline{\varphi(U)}$ by continuity], and U, V can be assumed to be affine [if U and V are as above, then we may take U to be affine, and if we choose $V' \subset V$ to be non-empty open and affine, then $U' := U \cap \varphi^{-1}(V')$ is affine by (3.11)].

(10.2) Proposition: Let $\varphi \colon X \to Y$ be a morphism of irreducible varieties. The following statements are equivalent:
(1) φ is quasifinite.
(2) φ is dominant and $K(X)$ is a finite extension of $K(Y)$.
(3) φ is dominant and we have $\dim(X) = \dim(Y)$.
(4) There exists $y \in Y$ such that $\varphi^{-1}(\{y\})$ is a non-empty finite set and we have $\dim(X) = \dim(Y)$.

Proof (1) $\Rightarrow$ (2): We know that φ is dominant [cf. (10.1)]. We choose non-empty open affine subsets $U \subset X$, $V \subset Y$ with $\varphi(U) \subset V$ and such that $U \to V$ is finite. Then $\mathcal{O}(V)$ is a subring of $\mathcal{O}(U)$ [cf. (1.16)], we have $\mathcal{O}(U) = \mathcal{O}(V)[a_1, \ldots, a_h]$ where $a_1, \ldots, a_h$ are integral over $\mathcal{O}(V)$, hence $K(X) = Q(\mathcal{O}(U))$ is a finite extension of $K(Y) = Q(\mathcal{O}(V))$.
(2) $\Rightarrow$ (3): We know that $\dim(X)$ (resp. $\dim(Y)$) is equal to the transcendence degree of $K(X)$ over k (resp. of $K(Y)$ over k) [cf. (1.26)], hence $\dim(X) = \dim(Y)$.
(3) $\Rightarrow$ (4): We may assume that X and Y are affine, and that $B := \mathcal{O}(Y) \subset \mathcal{O}(X) =: A$ [cf. (1.15)]; A and B are integral domains. Moreover, A is a B-algebra of finite type and therefore $Q(A)$ is a finitely generated extension of $Q(B)$. Since $\dim(B) = \mathrm{tr.d}_k(B) = \dim(Y) = \dim(X) = \mathrm{tr.d}_k(A) = \dim(A)$, we see that $Q(A)$ is a finite extension of $Q(B)$, and therefore there exists $s \in B$ such that A_s is a finite B_s-algebra [cf. (9.2)]. Let $\mathfrak{q}$ be a maximal ideal of B_s; then $(A_s)_{\mathfrak{q}}$

is a finitely generated $(B_s)_q$-module, hence semilocal by B(3.8). This implies the assertion.

(4) $\Rightarrow$ (1): Let $y \in Y$ be such that $\varphi^{-1}(\{y\})$ is a non-empty finite set. Let $V \subset Y$ be an open affine neighborhood of y, and let $U \subset \varphi^{-1}(V)$ be an open affine subset of X which meets $\varphi^{-1}(\{y\})$. The restriction $U \to V$ satisfies the statements in (4), hence we may assume that X and Y are affine. We have $\dim(\varphi^{-1}(\{y\})) = 0$, hence $\dim(X) = \dim(\overline{\varphi(X)})$ [cf. (9.4)(1)], and since $\overline{\varphi(X)}$ is an irreducible closed subset of Y, we have $\overline{\varphi(X)} = Y$ [since $\dim(X) = \dim(Y)$ by hypothesis]. From (9.3) we obtain the existence of an open affine subset V of Y such that the restriction of φ induces a finite dominant morphism $\varphi^{-1}(V) \to V$, i.e., φ is quasifinite.

(10.3) DEFINITION: Let $\varphi \colon X \to Y$ be a quasifinite morphism of irreducible varieties. Then

$$\deg(\varphi) := [\,K(X) \colon K(Y)\,]$$

[cf. the equivalence of (1) and (2) in (10.2)] is called the degree $\deg(\varphi)$ of φ. The morphism φ is called separable (resp. inseparable) if the finite extension $K(X)$ of $K(Y)$ is a separable (resp. inseparable) extension of $K(Y)$.

(10.4) Corollary: Let $\varphi \colon X \to Y$, $\psi \colon Y \to Z$ be quasifinite morphisms of irreducible varieties. Then $\varphi \circ \psi$ is quasifinite, and $\deg(\varphi \circ \psi) = \deg(\varphi)\deg(\psi)$.

(10.5) REMARK: Let $B \subset A$ be integral k-algebras of finite type with B integrally closed and A integral over B. Let $\mathfrak{n}$ be a maximal ideal of B; then we have $B/\mathfrak{n} = k$.
(1) For every polynomial $F = T^m + b_1 T^{m-1} + \cdots + b_m \in B[T]$ let $\overline{F} \in k[T]$ be the polynomial which we get by reducing the coefficients of F modulo $\mathfrak{n}$.
(2) Let $a \in A$, and let $F = T^m + b_1 T^{m-1} + \cdots + b_m$ be the minimal polynomial of a over $Q(B)$; note that we have $F \in B[T]$ by B(2.12)(2).
(3) We keep the notations of (2), and we show: For every zero $\gamma \in k$ of $\overline{F} \in k[T]$ there exists a maximal ideal $\mathfrak{m}$ of A with $\mathfrak{m} \cap B = \mathfrak{n}$ and with $a \bmod \mathfrak{m} = \gamma$.
Proof: It is enough to show that $\mathfrak{n}A + (a - \gamma)A$ is a proper ideal of A. Since A is integral over $B' := B[a]$, it is even enough to show that $\mathfrak{n}B' + (a - \gamma)B'$ is a proper ideal of B' [by lying over, cf. [63], Prop. 4.15]. Suppose that this ideal is the unit ideal of B'. Then there exist elements $b_1', \ldots, b_h' \in \mathfrak{n}$ and polynomials $G_1, \ldots, G_h$, $G \in B[T]$ with $b_1' G_1(a) + \cdots + b_h' G_h(a) + (a - \gamma)G(a) = 1$. We consider the polynomial $H := b_1' G_1 + \cdots + b_h' G_h + (T - \gamma)G - 1 \in B[T]$. Since F is monic, we can write $H = H'F + H''$ with polynomials H', $H'' \in B[T]$ and $H'' = 0$ or $\deg(H'') < \deg(F)$. Since $H(a) = 0$, we have $H''(a) = 0$, and since F is the minimal polynomial of a over $Q(B)$, we have $H'' = 0$, and therefore $H = H'F$. Since $\overline{F}(\gamma) = 0$, it follows that $F(\gamma) \in \mathfrak{n}$, hence $b_1' G_1(\gamma) + \cdots + b_h' G_h(\gamma) - 1 = H(\gamma) = H'(\gamma)F(\gamma) \in \mathfrak{n}$, and therefore $1 \in \mathfrak{n}$, contradicting $\mathfrak{n} \neq B$.

(10.6) Proposition: Let $\varphi \colon X \to Y$ be a finite dominant morphism of irreducible varieties with Y normal. For every $y \in Y$ we have

$$1 \le \mathrm{Card}(\varphi^{-1}(\{y\})) \le \deg(\varphi).$$

Proof: Clearly we may assume that Y is affine; then $X = \varphi^{-1}(Y)$ is also affine. Set $A := \mathcal{O}(X)$, $B := \mathcal{O}(Y)$; then $B \subset A$, and A is a finitely generated B-module, hence, in particular, integral over B. Set $\mathfrak{n} := \mathcal{I}_Y(\{y\})$; there exist prime ideals of A lying over $\mathfrak{n}$, and each of them is maximal [cf. [63], Prop. 4.15 and Cor. 4.17], hence $\varphi^{-1}(\{y\})$ is not empty. Since $B_\mathfrak{n} \to A_\mathfrak{n}$ is finite, the set of maximal ideals of $A_\mathfrak{n}$ lying over $\mathfrak{n}B_\mathfrak{n}$ is finite [cf. B(3.8)], hence $\varphi^{-1}(\{y\})$ is a non-empty finite set. Set $m := \mathrm{Card}(\varphi^{-1}(\{y\}))$, and let $\varphi^{-1}(\{y\}) = \{x_1, \ldots, x_m\}$. We choose an element $a \in A$ with $a(x_i) \neq a(x_j)$ for $i, j \in \{1, \ldots, m\}$, $i \neq j$. [The existence of such an element is easy to prove: X is a closed subset of some $\mathbb{A}^p$, and there exist polynomials in $k[T_1, \ldots, T_p]$ which have different values at the points $x_1, \ldots, x_m \in \mathbb{A}^p$.] Let $F \in B[T]$ be the minimal polynomial of a over $Q(B)$ [cf. (10.5)(2)]; then we have $\deg(F) \leq \deg(\varphi)$. We have $\overline{F}(a(x_i)) = 0$ for $i \in \{1, \ldots, m\}$ [where $\overline{F}$ is defined as in (10.5)(1)], hence $m \leq \deg(\overline{F}) = \deg(F) \leq \deg(\varphi)$. This implies the assertion.

10.2 Ramification

(10.7) DEFINITION: Let $\varphi \colon X \to Y$ be a finite morphism of varieties. Let $y \in Y$, let V be an open affine neighborhood of y in Y, and set $\mathfrak{q} := \mathcal{I}_V(\{y\}) \subset \mathcal{O}(V)$; the morphism φ is called unramified over y if $\mathfrak{q}$ is unramified in $\mathcal{O}(\varphi^{-1}(V))$, otherwise φ is called ramified over y. The set of points $y \in Y$ over which φ is ramified is called the branch locus of φ or discriminant locus of φ, and shall be denoted by branch(φ) or branch(X/Y).
Note the following: $\varphi^{-1}(V)$ is affine since φ is finite, and $\mathcal{O}(V) \to \mathcal{O}(\varphi^{-1}(V))$ is finite. It is clear that the definition of φ being unramified (resp. ramified) in y does not depend on the choice of V as an open affine neighborhood of y in Y.

(10.8) Proposition: *Let* $\varphi \colon X \to Y$ *be a finite dominant morphism of irreducible varieties with* Y *normal, and let* $y \in Y$. *Then* φ *is unramified over* y *iff* $\mathrm{Card}(\varphi^{-1}(\{y\})) = \deg(\varphi)$.

Proof: We may assume that X and Y are affine; set $A := \mathcal{O}(X)$, $B := \mathcal{O}(Y)$, $\mathfrak{q} := \mathcal{I}_Y(\{y\})$, $n := \deg(\varphi)$ and $m := \mathrm{Card}(\varphi^{-1}(\{y\}))$. We use the notations in the proof of (10.6).
(1) Assume that $m = n$, and choose a and F as in the proof of (10.6). Then we have $\deg(F) = n$, hence $Q(A) = Q(B)(a)$. Since $D_{A/B}(1, a, \ldots, a^{n-1})$ is equal to the discriminant $\mathrm{dis}(F)$ of F [cf. B(10.22)], we see that $\mathrm{dis}(F) \in \mathfrak{n}_{A/B}$. We have to show that $\mathfrak{n}_{A_\mathfrak{q}/B_\mathfrak{q}} = B_\mathfrak{q}$. Since $\overline{F}$ has n different zeroes in k, we have $\mathrm{dis}(\overline{F}) = \mathrm{dis}(F)(y) \neq 0$ which means that $\mathrm{dis}(F)$ is a unit in $B_\mathfrak{q}$, and therefore we have $\mathfrak{n}_{A_\mathfrak{q}/B_\mathfrak{q}} = B_\mathfrak{q}$.
(2) Now we assume that $\mathfrak{q}$ is unramified in A. Then, on the one hand, $A_\mathfrak{q}$ is a free $B_\mathfrak{q}$-module of rank n, and we have $\dim_k(A_\mathfrak{q}/\mathfrak{q}A_\mathfrak{q}) = n$ [cf. III(3.18)]. On the other hand, $A_\mathfrak{q}/\mathfrak{q}A_\mathfrak{q}$ in unramified over k [cf. III(3.6)], and therefore isomorphic to k^n [cf. III(3.13)]. This implies that $A_\mathfrak{q}/\mathfrak{q}A_\mathfrak{q}$ has exactly n prime ideals. Thus, there lie exactly n prime ideals of A over $\mathfrak{q}$, and each of them is maximal [cf.

[63], Cor. 4.17]; they correspond to the points of X lying over y. This means that $m = n$.

(10.9) Proposition: *Let* $\varphi\colon X \to Y$ *be a finite dominant morphism of irreducible varieties with* Y *normal. Then:*
(1) branch(φ) *is a closed subset of* Y.
(2) *If* φ *is separable, then* branch(φ) *is a proper subset of* Y, *and for every* y *in the non-empty open set* $Y \setminus$ branch(φ) *we have* $\mathrm{Card}(\varphi^{-1}(\{y\})) = \deg(\varphi)$.

Proof: Again we may assume that X and Y are affine; set $n := \deg(\varphi)$.
(1) Let φ be unramified over $y \in Y$; then we have $n = \mathrm{Card}(\varphi^{-1}(\{y\}))$ by (10.8). We choose a and F as in the proof of (10.6); then we have $\mathrm{dis}(\overline{F}) \neq 0$. Since $\mathrm{dis}(\overline{F}) = \mathrm{dis}(F)(y)$, there exists an open neighborhood V of y in Y with $\mathrm{dis}(F)(y') \neq 0$ for every $y' \in V$. Let $y' \in V$; then the polynomial which we get from F by reducing it modulo $\mathfrak{q}'$, where $\mathfrak{q}' := \mathcal{I}_Y(\{y'\})$, has n pairwise different zeroes $\gamma_1, \ldots, \gamma_n$ in k, which means that there exist n points $x_1, \ldots, x_n \in \varphi^{-1}(\{y'\})$ with $a(x_j) = \gamma_j$ for $j \in \{1, \ldots, n\}$ [cf. (10.5)(3)]. Therefore V lies in the complement of branch(φ), hence branch(φ) is closed.
(2) In this case $K(X)$ is a finite separable extension of $K(Y)$. Let $a \in A$ be a primitive element for $K(X)$ over $K(Y)$, and let F be the minimal polynomial for a over $K(Y)$; its coefficients lie in B. Now F is a separable polynomial, hence $\mathrm{dis}(F) \neq 0$, and therefore there exist points $y \in Y$ with $\mathrm{dis}(F)(y) \neq 0$, and it is clear that φ is unramified over every such point [cf. the last part of (1) in the proof of (10.6)].

The next result shall play an important role when we desingularize irreducible surfaces by Jung's method in section 2 of chapter VIII.

(10.10) Theorem: [Purity of branch locus] *We assume that* k *has characteristic 0. Let* $\varphi\colon X \to Y$ *be a finite dominant morphism of irreducible varieties with* X *normal and* Y *regular. If* branch(X/Y) *is not empty, then it is pure 1-codimensional.*

Proof: Set $r := \dim(Y)$. Now branch(X/Y) is a non-empty closed subset of Y [cf. (10.9)], and we have $\mathrm{codim}(\mathrm{branch}(X/Y), Y) = r - \dim(\mathrm{branch}(X/Y))$ [cf. (1.26)]. Therefore we have to show that $\dim(Z) = r - 1$ for every irreducible component Z of branch(X/Y).
The theorem being local on Y, we may assume that X and Y are affine. Set $A := \mathcal{O}(X)$, $B := \mathcal{O}(Y)$; then A is a finite B-module, and $Q(A)$ is a finite separable extension of $Q(B)$ since k has characteristic 0. Therefore $Q(A)$ is unramified over $Q(B)$, hence $\mathfrak{n}_{A/B} \neq \{0\}$ [cf. chapter III, (3.10) and (4.10)]. Now A is integrally closed since X is normal [cf. (7.4)]; B is also integrally closed since Y is regular, hence normal [cf. (7.3)]. Set $\mathfrak{b} := \mathcal{I}_Y(\mathrm{branch}(X/Y)) \subset B$; then we have branch($A/B$) $= V(\mathfrak{b})$, and we know that branch(A/B) $= V(\mathfrak{n}_{A/B})$ [cf. III(4.12)]. Let $\mathrm{Ass}(B/\mathfrak{b}) = \{\mathfrak{q}_1, \ldots, \mathfrak{q}_h\}$; this is also the set of minimal elements in $\mathrm{Ass}(B/\mathfrak{n}_{A/B})$. Since $\dim(\mathrm{branch}(X/Y)) = \min(\{\dim(B/\mathfrak{q}_i) \mid i \in \{1, \ldots, h\}\})$,

and since $r = \dim(B/\mathfrak{q}_i) + \mathrm{ht}(\mathfrak{q}_i)$ for every $i \in \{1, \ldots, h\}$ [as B is catenary, cf. [63], Cor. 13.4], we have to show that $\mathrm{ht}(\mathfrak{q}_i) = 1$ for every $i \in \{1, \ldots, h\}$. All these prime ideals have positive height [since $\mathfrak{n}_{A/B} \neq \{0\}$]. Therefore the claim of the theorem follows from the following assertion:

Let A and B be as above. Let $\mathfrak{n}$ be a maximal ideal of B; if every prime ideal $\mathfrak{q}$ of B of height 1 with $\mathfrak{q} \subset \mathfrak{n}$ is unramified in A, then $\mathfrak{n}$ is unramified in A.

Thus, let $\mathfrak{n}$ be as stated, and let $\mathfrak{m}$ be any maximal ideal of A with $\mathfrak{m} \cap B = \mathfrak{n}$. We have to show that $B_\mathfrak{n} \to A_\mathfrak{m}$ is unramified [cf. III(3.2)]; since both local rings have k as residue field, we must show that $\mathfrak{n}A_\mathfrak{m}$ is the maximal ideal of $A_\mathfrak{m}$ [cf. III(3.14)].

Note that $B_\mathfrak{n}$ is an r-dimensional regular local ring; $A_\mathfrak{m}$ is an r-dimensional local domain, and it is integrally closed since A is integrally closed, hence $A_\mathfrak{m}$ is the intersection of all localizations at height 1 prime ideals $\mathfrak{p}$ of A with $\mathfrak{p} \subset \mathfrak{m}$ [cf. [63], Cor. 11.4]. We have a local homomorphism $f \colon B_\mathfrak{n} \to A_\mathfrak{m}$ which extends to a homomorphism $\widehat{f} \colon (B_\mathfrak{n})^\widehat{\ } \to (A_\mathfrak{m})^\widehat{\ }$ of completions.

Let $\{x_1, \ldots, x_r\}$ be a regular system of parameters of $B_\mathfrak{n}$. We have $k[x_1, \ldots, x_r] \subset B_\mathfrak{n} \subset (B_\mathfrak{n})^\widehat{\ } = k[\![x_1, \ldots, x_r]\!]$ [cf. [63], Prop. 10.16], and $x_1, \ldots, x_r$ are algebraically independent over k [cf. [cf. [21], Cor. 11.21 or [63], Ex. 13.16]. In particular, $\{x_1, \ldots, x_r\}$ is a transcendence basis of $Q(B)$ over k and $Q(B)$ is a finite separable extension of $k(x_1, \ldots, x_r)$, hence $Q(A)$ also is a finite separable extension of $k(x_1, \ldots, x_r)$. Let $D_j := \partial/\partial x_j \colon k[x_1, \ldots, x_r] \to k[x_1, \ldots, x_r]$ for $j \in \{1, \ldots, r\}$ be the partial derivatives; the unique extension of D_j to a derivation of $Q(A)$ into $Q(A)$ [cf. chapter III, (1.4) and (1.5)] shall be denoted also by D_j. Note that $D_j(Q(B)) \subset Q(B)$.

Let $j \in \{1, \ldots, r\}$. Now D_j can be extended in a unique way to a derivation of $k[\![x_1, \ldots, x_r]\!]$, and since $k[\![x_1, \ldots, x_r]\!] \cap Q(B) = B_\mathfrak{n}$ [cf. [34], Ch. I, §3, no. 5, Prop. 10], we see that $D_j(B_\mathfrak{n}) \subset B_\mathfrak{n}$. Let $\mathfrak{p}$ be a prime ideal of A of height 1 with $\mathfrak{p} \subset \mathfrak{m}$; then $\mathfrak{q} := \mathfrak{p} \cap B$ is a prime ideal of B of height 1 [by going-down, cf. [63], Th. 13.9]. Since $B_\mathfrak{q}$ is a localization of $B_\mathfrak{n}$, we get $D_j(B_\mathfrak{q}) \subset B_\mathfrak{q}$ [cf. III(1.4)]. Since $B_\mathfrak{q} \to A_\mathfrak{q}$ is finite and unramified by assumption, we have $A_\mathfrak{q} = B_\mathfrak{q}[a]$ where a is a zero of a monic separable polynomial $G \in B_\mathfrak{q}[T]$ and $G'(a)$ is a unit of $A_\mathfrak{q}$ [cf. III(3.16)]. From $0 = D_j(G)(a) + G'(a)D_j(a)$ we see that $D_j(a) \in A_\mathfrak{q}$, hence $D_j(A_\mathfrak{q}) \subset A_\mathfrak{q}$. Since $A_\mathfrak{p}$ is a localization of $A_\mathfrak{q}$, we finally get $D_j(A_\mathfrak{p}) \subset A_\mathfrak{p}$ [cf. III(1.4)]. This being true for every prime ideal of height 1 of A contained in $\mathfrak{m}$, we have $D_j(A_\mathfrak{m}) \subset A_\mathfrak{m}$.

We have the Taylor homomorphism $\tau \colon A_\mathfrak{m} \to k[\![x_1, \ldots, x_r]\!]$ [cf. proof of V(2.3)] which extends to a local homomorphism of completions $\widehat{\tau} \colon (A_\mathfrak{m})^\widehat{\ } \to k[\![x_1, \ldots, x_r]\!]$. The composition $\widehat{\tau} \circ \widehat{f} \colon k[\![x_1, \ldots, x_r]\!] = (B_\mathfrak{n})^\widehat{\ } \to (B_\mathfrak{n})^\widehat{\ }$ is the identity, hence $\widehat{\tau}$ is surjective. We know that $(A_\mathfrak{m})^\widehat{\ }$ is an r-dimensional local domain [cf. III(6.14)], hence $\widehat{\tau}$ being surjective implies that it is also injective. Therefore $\widehat{\tau}$ is an isomorphism, and $\widehat{f}$ is its inverse. Thus, we have shown that $(A_\mathfrak{m})^\widehat{\ }$ is a regular local ring, and $\{x_1, \ldots, x_r\}$ is a regular system of parameters of $(A_\mathfrak{m})^\widehat{\ }$. Now $A_\mathfrak{m} \to (A_\mathfrak{m})^\widehat{\ }$ is faithfully flat, hence $\{x_1, \ldots, x_r\}$ is a regular system of parameters of $A_\mathfrak{m}$, which means that $\mathfrak{n}A_\mathfrak{m}$ is the maximal ideal of $A_\mathfrak{m}$.

11 Divisors

(11.1) CARTIER DIVISORS: Let X be an irreducible variety, and let $K = K(X)$ be the field of rational functions on X. For every non-empty open subset $U \subset X$ we have $K(U) = K(X)$.

(1) Note that, if $U \subset V$ are non-empty open subsets of X, then there is a homomorphism of groups $\mathcal{O}(V)^{\times} \to \mathcal{O}(U)^{\times}$ defined by restriction.

(2) We consider systems $(U_i, f_i)_{i \in I}$ where $(U_i)_{i \in I}$ is an open covering of X, $U_i \neq \emptyset$ and $f_i \in K^{\times}$ for every $i \in I$, and, for all i, $j \in I$, the rational function f_i/f_j is defined on $U_i \cap U_j$ and it is a unit of the ring $\mathcal{O}(U_i \cap U_j)$; we express this by saying that $f_i/f_j \in \mathcal{O}(U_i \cap U_j)^{\times}$. Two such systems $(U_i, f_i)_{i \in I}$ and $(V_j, g_j)_{j \in J}$ are called equivalent if $f_i/g_j \in \mathcal{O}(U_i \cap V_j)^{\times}$ for all $(i,j) \in I \times J$. It is immediate that this really is an equivalence relation on the set $\{(U_i, f_i)_{i \in I}\}$ of such systems. The equivalence classes are called Cartier divisors on X.

Let $(U_i, f_i)_{i \in I}$ represent a Cartier divisor on X, and let $i \in I$; then we say that f_i is a local equation for this divisor on U_i.

(3) Let $(U_i, f_i)_{i \in I}$, $(V_j, g_j)_{j \in J}$ represent Cartier divisors on X; for every $i \in I$, $j \in J$ we set $W_{ij} := U_i \cap V_j$, $f_{ij} := f_i|W_{ij}$ and $g_{ji} := g_j|W_{ij}$. Then the system $(W_{ij}, f_{ij} \cdot g_{ji})_{(i,j) \in I \times J}$ represents a Cartier divisor on X which is called the sum of the given Cartier divisors [it is immediate that this sum is well-defined]. It is easy to check that the set of Cartier divisors on X under this operation becomes an abelian group, the group $C(X)$ of Cartier divisors on X; it will be considered as an additively written group.

(4) A Cartier divisor on X is called principal if it can be represented by (X, f) where $f \in K^{\times}$. Clearly the set of principal Cartier divisors on X is a subgroup of the group $C(X)$ of Cartier divisors on X.

(11.2) For the rest of this section we assume that X is an irreducible *normal* variety. Under this assumption we introduce the abelian group of Weil divisors $\mathrm{Div}(X)$ on X.

(11.3) DEFINITION: (1) An irreducible closed subset of X of codimension 1 is called a prime divisor on X.

(2) The free abelian group having the set of prime divisors on X as a $\mathbb{Z}$-basis is called the group of Weil divisors or just group of divisors on X. It will be denoted by $\mathrm{Div}(X)$. The elements of $\mathrm{Div}(X)$ are called divisors on X.

(11.4) NOTATION: (1) Let $D \in \mathrm{Div}(X)$; then $D = \sum n_Y Y$ is a finite sum where Y runs through the set of prime divisors on X, $n_Y \in \mathbb{Z}$ and only finitely many n_Y are different from zero. The prime divisors Y such that $n_Y \neq 0$ are called the components of D, and n_Y is called the multiplicity of Y in D; if $n_Y = 1$, then Y is called a simple component of D.

(2) A divisor $D = \sum n_Y Y$ on X is called positive if $n_Y \geq 0$ for every prime divisor Y. Note that the sum of two positive divisors is a positive divisor, hence $\mathrm{Div}(X)$ carries in a canonical way a structure as a (partially) ordered abelian group.

(11.5) NOTATION: Let Y be a prime divisor on X, and let $\mathcal{O}_{X/Y}$ be the local ring of X along Y [cf. (1.28)]. Now $\dim(\mathcal{O}_{X/Y}) = \mathrm{codim}(Y, X) = 1$ [cf. (1.28)(2)], and since X is normal, the ring $\mathcal{O}_{X/Y}$ is integrally closed [note that $\mathcal{O}_{X/Y}$ is a localization of $\mathcal{O}_{X,y}$ for $y \in Y$, cf. loc.cit., and use [63], Prop. 4.13], hence $\mathcal{O}_{X/Y}$ is a discrete valuation ring of $K(X)$, the field of rational functions on X [cf. I(3.28)]. Let v_Y be the valuation of $K(X)$ defined by $\mathcal{O}_{X/Y}$; v_Y is called the valuation of Y.

(11.6) REMARK: Let X be affine and A its coordinate ring; then A is an integrally closed noetherian domain [cf. (7.4)]. Let Y be a prime divisor on X, and set $\mathfrak{p} := \mathcal{I}_X(Y)$. Then $\mathfrak{p}$ is a prime ideal of height 1 of A, and $\mathcal{O}_{X/Y} = A_\mathfrak{p}$ [cf. (1.28)(4)]. Therefore we have $v_Y = v_\mathfrak{p}$ and the groups $\mathrm{Div}(X)$ and $\mathrm{Div}(A)$ [cf. B(10.5)] can be identified. Let $f \in A$ be non-zero; $\{\mathfrak{p} \in P(A) \mid v_\mathfrak{p}(f) > 0\}$ is finite [cf. loc.cit.].

(11.7) Proposition: *Let $f \in K(X)$ be non-zero; then $v_Y(f) \neq 0$ for at most finitely many prime divisors Y on X.*

Proof: We choose an open affine subset U of X contained in $\delta(f)$; clearly U is normal. For every prime divisor Y on X such that $U \cap Y \neq \emptyset$, this set is an irreducible closed subset of U, we have $\mathcal{O}_{X/Y} = \mathcal{O}_{U/(U\cap Y)}$ [cf. (1.28)(2)], and $U \cap Y$ is a prime divisor on U. There are only finitely many prime divisors Y on U for which $v_Y(f) \neq 0$ [cf. (11.6)]. If $U = X$, then this remark establishes the claim. Otherwise, the set $Z := X \setminus U$ is a proper non-empty closed subset of X, hence $\dim(Z) < \dim(X)$, and therefore every prime divisor on X contained in Z is an irreducible component of Z. Thus, Z contains only finitely many prime divisors on X. This establishes the claim.

(11.8) NOTATION: Let $f \in K(X)$, $f \neq 0$. Then $(f) := \sum v_Y(f)Y$ is a Weil divisor on X [cf. (11.7)]. Every divisor of this form is called a principal Weil divisor. Clearly the principal divisors are a subgroup of $\mathrm{Div}(X)$.

(11.9) Proposition: *Let X be an irreducible nonsingular variety. Then the group $\mathrm{Div}(X)$ of Weil divisors on X is isomorphic to the group $C(X)$ of Cartier divisors on X, and, furthermore, under this isomorphism the subgroup of $\mathrm{Div}(X)$ consisting of principal Weil divisors on X corresponds to the subgroup of $C(X)$ consisting of principal Cartier divisors on X.*

Proof: (1) Now X is normal since it is regular [cf. (7.3)], and the group of Weil divisors on X is defined.

(2) Let $(U_i, f_i)_{i\in I}$ represent a Cartier divisor on X. We define the associated Weil divisor $\sum n_Y Y$ as follows. For every prime divisor Y on X, take the coefficient n_Y of Y to be $v_Y(f_i)$ where $i \in I$ is any index for which $Y \cap U_i \neq \emptyset$. If $j \in I$ is another index for which $Y \cap U_j \neq \emptyset$, then $f_i/f_j \in \mathcal{O}(U_i \cap U_j)^\times$, and since $\mathcal{O}_{X/Y}$ is a localization of $\mathcal{O}(U_i \cap U_j)$, the element f_i/f_j is a unit of $\mathcal{O}_{X/Y}$, and therefore we have $v_Y(f_i) = v_Y(f_j)$. Thus, we obtain a well-defined Weil divisor $D = \sum n_Y Y$

on X [note that the sum is finite by (11.7) since we could have started with a finite affine covering $(U_i)_{i \in I}$ of X]. It is easy to check that this gives rise to a homomorphism $C(X) \to \mathrm{Div}(X)$.

(3) Let $D = \sum n_Y Y$ be a Weil divisor on X, and let $x \in X$. We consider the sum $D_x := \sum_{x \in Y} n_Y Y$. For every prime divisor Y on X with $x \in Y$, the ideal $\mathcal{I}_{X,x}(Y)$ of Y in $\mathcal{O}_{X,x}$ is a prime ideal of height 1, hence D_x can be considered as an element of $\mathrm{Div}(\mathcal{O}_{X,x})$. Since $\mathcal{O}_{X,x}$ is factorial, there exists $f_x \in K(X)$ such that $v_Y(f_x) = n_Y$ for every prime divisor Y of X with $x \in Y$ [cf. B(10.7)]. The set of prime divisors Y on X with $x \notin Y$ and either $n_Y \neq 0$ or $v_Y(f_x) \neq 0$ is finite [cf. (11.7)]; hence there exists an open affine neighborhood U_x of x with the following property: For every $z \in U_x$, if Y is a prime divisor on X with $z \in Y$, then we have $n_Y = v_Y(f_x)$. It is easy to see that $(U_x, f_x)_{x \in X}$ represents a Cartier divisor on X. Since for any open affine $U \subset X$, if $f, f' \in K(X)$ define the same [principal] Weil divisor on U, we have $f/f' \in \mathcal{O}(U)^\times$ [cf. [63], Th. 11.2], it is clear that we have constructed a map $\mathrm{Div}(X) \to C(X)$. It is easily seen that this map is a homomorphism of groups.

(4) The homomorphisms constructed in (2) and (3) are inverse to each other, hence the groups of Weil divisors and Cartier divisors on X are isomorphic. Moreover, it is clear that under this isomorphism the subgroup of principal Weil divisors of $\mathrm{Div}(X)$ is mapped isomorphically to the subgroup of principal Cartier divisors of $C(X)$.

(11.10) REMARK: Let X be an irreducible nonsingular variety. The result of (11.9) implies the following. Let D be a positive divisor on X, and let $x \in X$. Then there exists an open neighborhood U of x and a non-zero function $f \in \mathcal{O}(U)$ which is a local equation for the Cartier divisor D on U. The image of f in $\mathcal{O}_{U,x} = \mathcal{O}_{X,x}$ is called an equation for D at x or in $\mathcal{O}_{X,x}$. We say that D passes through x if the equation for D in $\mathcal{O}_{X,x}$ is not a unit of $\mathcal{O}_{X,x}$.

(11.11) REMARK: Let X be an irreducible nonsingular variety, and let Y be an irreducible *nonsingular* subvariety of X of codimension 1. Let $y \in Y$, and let φ be an equation for the divisor Y in $\mathcal{O}_{X,x}$. Then $\mathcal{O}_{Y,y} = \mathcal{O}_{X,x}/\varphi\mathcal{O}_{X,x}$ is a regular local ring, hence φ is part of a regular system of parameters of $\mathcal{O}_{X,x}$.

(11.12) NORMAL CROSSING: Assume that X is an irreducible nonsingular variety, and let D be a positive divisor on X with only simple components, each of which is a *nonsingular variety*.

(1) Let $x \in X$ be a point through which D passes. D is called a divisor with normal crossings in x, if h components $Y_1, \ldots, Y_h$ of D meet in x, then the elements $f_1, \ldots, f_h$, where, for $i \in \{1, \ldots, h\}$, f_i is an equation for Y_i at x, form part of a regular system of parameters of $\mathcal{O}_{X,x}$.

(2) D is called a divisor with normal crossings or a normal crossing divisor if D is a divisor with normal crossing at every point through which it passes.

If x is a point through which D passes, and D is not a divisor with normal crossings in x, then we say that x is a not normal-crossing point for D.

12   Some Results on Projections

(12.1) PROJECTION: Let $r \in \{0, \ldots, n-1\}$, and let $L_i \colon k^{n+1} \to k$ for $i \in \{0, \ldots, r\}$ be $r + 1$ linearly independent linear forms. Then $W_1 = \{z \in \mathbb{P}^n \mid L_0(z) = \cdots = L_r(z) = 0\}$ is an $(n - r - 1)$-dimensional projective subspace of $\mathbb{P}^n$, and the map $z \mapsto (L_0(z) \colon \cdots \colon L_r(z))$ is a rational map π from $\mathbb{P}^n$ to an r-dimensional projective space which has $\mathbb{P}^n \setminus W_1$ as domain of definition. It is called the projection with center W_1. Let W_2 be an r-dimensional projective subspace of $\mathbb{P}^n$ with $W_1 \cap W_2 = \emptyset$. Then we can realize the projection with center W_1 in the following way: By making a linear change of coordinates we may assume that for $(\zeta_0, \ldots, \zeta_n) \in k^{n+1}$ we have $L_i(\zeta_0, \ldots, \zeta_n) = \zeta_i$ for $i \in \{0, \ldots, r\}$ and that

$$W_1 = \{(\zeta_0 \colon \cdots \colon \zeta_n) \in \mathbb{P}^n \mid \zeta_0 = \cdots = \zeta_r = 0\},$$
$$W_2 = \{(\zeta_0 \colon \cdots \colon \zeta_n) \in \mathbb{P}^n \mid \zeta_{r+1} = \cdots = \zeta_n = 0\}.$$

The rational map π is defined on $\mathbb{P}^n \setminus W_1$, hence it induces the surjective morphism

$$\pi \colon \mathbb{P}^n \setminus W_1 \to W_2 = \mathbb{P}^r$$

with $\pi((\zeta_0 \colon \cdots \colon \zeta_n)) = (\zeta_0 \colon \cdots \colon \zeta_r)$ for $(\zeta_0 \colon \cdots \colon \zeta_n) \in \mathbb{P}^n \setminus W_1$. In this situation we say that π *is the projection with center* W_1 *onto* W_2. If X is any closed subset of $\mathbb{P}^n$ with $X \cap W_1 = \emptyset$, then the restriction of π defines a morphism $X \to W_2$. The geometric meaning of projection is the following: For every $z \in \mathbb{P}^n \setminus W_1$ let $W_1(z) := W_1 \vee \{z\}$ be the $(n - r)$-dimensional projective subspace of $\mathbb{P}^n$ spanned by W_1 and z. Then we have $\{\pi(z)\} = W_1(z) \cap W_2$ [note that $y = (\eta_0 \colon \cdots \colon \eta_n) \in W_1(z)$ iff $y = \lambda z + \mu w$ where $w \in W_1$ and $\lambda, \mu \in k$ are not both equal to 0]. If, in particular, $W_1 = \{p\}$, a point, then W_2 can be chosen as a hyperplane $H = \mathbb{P}^{n-1}$ of $\mathbb{P}^n$ with $p \notin H$, and the projection map $\pi \colon \mathbb{P}^n \setminus \{p\} \to H$ is given by $\pi(y) = py \cap H$ for every $y \in \mathbb{P}^n \setminus \{p\}$ [$\{p\} \vee \{y\} =: py$ is the line through p and y]. In this case we say that π is the projection with center p onto $\mathbb{P}^{n-1}$.

(12.2) Let $k[T_0, \ldots, T_n]$ be the homogeneous coordinate ring of $\mathbb{P}^n$, let Y be an affine variety and set $R := \mathcal{O}(Y)$. Furthermore, we set $S := R[T_0, \ldots, T_n]$; note that $S = \bigoplus_{n \geq 0} S_n$ carries a canonical grading with $S_0 = R$. Let $p \in \mathbb{N}_0$ and $F \in S_p$; for every point $(z, y) = ((\zeta_0 \colon \cdots \colon \zeta_n), y) \in \mathbb{P}^n \times Y$ it makes sense to say that $F(z, y) \neq 0$ (resp. that $F(z, y) = 0$). For $i \in \{0, \ldots, n\}$ we set $U_i := (\mathbb{P}^n \setminus \mathcal{Z}_+(T_i)) \times Y \subset \mathbb{P}^n \times Y$; note that U_i is affine and that its ring of regular functions $\mathcal{O}(U_i) = k[T_0/T_i, \ldots, T_n/T_i] \otimes_k \mathcal{O}(Y) = R[T_0/T_i, \ldots, T_n/T_i] = S_{(T_i)}$ [cf. (3.5)]; moreover, $(U_i)_{0 \leq i \leq n}$ is an open affine covering of $\mathbb{P}^n \times Y$. Let $X \subset \mathbb{P}^n \times Y$ be closed, and for $p \in \mathbb{N}_0$ let $\mathfrak{a}_p$ be the R-submodule of S_p consisting of those $F \in S_p$ with

$$F(T_0/T_i, \ldots, T_n/T_i) \in \mathcal{I}_{U_i}(X \cap U_i) \subset R[T_0/T_i, \ldots, T_n/T_i] \text{ for } i \in \{0, \ldots, n\}.$$

It is easy to see that $\mathfrak{a} := \sum_{p \geq 0} \mathfrak{a}_p$ is a homogeneous ideal of S. For $F \in S_p$, we have $F \in \mathfrak{a}_p$ iff $F(z, y) = 0$ for every $(z, y) \in X$. The ideal $\mathfrak{a}$ is called the ideal of X in S, and the graded R-algebra $R[T_0, \ldots, T_n]/\mathfrak{a}$ is called the homogeneous coordinate ring of X.

(1) Let $i \in \{0, \ldots, n\}$; we show that $\mathfrak{a}_{(T_i)} = \mathcal{I}_{U_i}(X \cap U_i)$.

Clearly we have $\mathfrak{a}_{(T_i)} \subset \mathcal{I}_{U_i}(X \cap U_i)$. Conversely, let $g \in \mathcal{I}_{U_i}(X \cap U_i)$; there exists $p \in \mathbb{N}$ with $T_i^p g =: F' \in S_p$. Let $j \in \{0, \ldots, n\}$, and set $g_j := F'/T_j^p \in R[T_0/T_j, \ldots, T_n/T_j]$. Then g_j vanishes on $X \cap U_i \cap U_j$, and if g_j does not vanish in a point of $X \cap U_j$, then $(T_i/T_j)g_j$ vanishes in this point. Therefore $F := T_i \cdot F' \in \mathfrak{a}_{p+1}$ and $F(T_0/T_i, \ldots, T_n/T_i) = g$, hence $g \in \mathfrak{a}_{(T_i)}$, and therefore $\mathcal{I}_{U_i}(X \cap U_i) \subset \mathfrak{a}_{(T_i)}$.

(2) The result in (1) implies, in particular, that for every $i \in \{0, \ldots, n\}$ we have $\mathcal{O}(X \cap U_i) = (R[T_0, \ldots, T_n]/\mathfrak{a})_{(T_i)}$.

(12.3) Proposition: *let W_1 be an $(n - r - 1)$-dimensional projective subspace of $\mathbb{P}^n$, let Y be an affine variety, and let $X \subset \mathbb{P}^n \times Y$ be a closed subset with $X \cap (W_1 \times Y) = \emptyset$. Let $\pi: (\mathbb{P}^n \times Y) \setminus (W_1 \times Y) \to \mathbb{P}^r \times Y$ be the morphism induced by the projection $\mathbb{P}^n \setminus W_1 \to \mathbb{P}^r$ with center W_1. Then π induces by restriction a finite morphism $X \to \mathbb{P}^r \times Y$. In particular, $\pi(X)$ is a closed subset of $\mathbb{P}^r \times Y$, and $X \to \pi(X)$ is finite.*

Proof: We set $R := \mathcal{O}(Y)$.

(1) We may assume that W_1 and $W_2 = \mathbb{P}^r$ are as in (12.1). For every $j \in \{0, \ldots, r\}$ we consider the affine subset $V_j := (\mathbb{P}^r \setminus \mathcal{Z}_+(T_j)) \times Y \subset \mathbb{P}^r \times Y$, and for every $i \in \{0, \ldots, n\}$ we consider the affine subset $U_i := (\mathbb{P}^n \setminus \mathcal{Z}_+(T_i)) \times Y \subset \mathbb{P}^n \times Y$. Note that $(V_j)_{0 \leq j \leq r}$ is an open affine covering of $\mathbb{P}^r \times Y$, and that $\pi^{-1}(V_j) = U_j$ for $j \in \{0, \ldots, r\}$. We set $S := R[T_0, \ldots, T_n]$ and define $\mathfrak{a}$ as in (12.2); we have $\mathcal{O}(U_i) = R[T_0/T_i, \ldots, T_n/T_i] = S_{(T_i)} =: R_i$ and $\mathcal{O}(X \cap U_i) = (S/\mathfrak{a})_{(T_i)}$ [cf. (12.2)(2)].

(2) Let $j \in \{0, \ldots, r\}$; then π induces a morphism $\pi_j: X \cap U_j \to V_j$. It is enough to show that π_j is finite [cf. (2.5)(1)].

(3) Let $\mathfrak{b}$ be the ideal of S generated by $T_0, \ldots, T_r$. Let $i \in \{0, \ldots, n\}$, and set $\mathfrak{a}_i := \mathfrak{a}_{(T_i)} = \mathcal{Z}(X \cap U_i)$, $\mathfrak{b}_i := \mathfrak{b}_{(T_i)}$; $\mathfrak{b}_i$ is the ideal in R_i generated by $T_0/T_i, \ldots, T_r/T_i$. We have $(X \cap U_i) \cap \mathcal{Z}_{U_i}(\mathfrak{b}_i) = \emptyset$ [since $X \cap (W_1 \times Y) = \emptyset$], hence we have $\mathfrak{a}_i + \mathfrak{b}_i = R_i$ [by Hilbert's Nullstellensatz for the affine variety U_i]. Then we can write

$$1 = f_i + \sum_{j=0}^{r} (T_j/T_i) \cdot g_{ij} \text{ in } R_i \tag{$*$}$$

where $f_i \in \mathfrak{a}_i$ and $g_{ij} \in R_i$ for $j \in \{0, \ldots, r\}$. There exist $p_i \in \mathbb{N}$ and a homogeneous $F_i \in R[T_0, \ldots, T_n]$ of degree p_i with $F_i/T_i^{p_i} = f_i$ [cf. (12.2)], hence we get

$$F_i = T_i^{p_i} + \sum_{j=0}^{r} T_j G_{ij} \tag{$**$}$$

where the elements $G_{i0} \ldots, G_{ir}$ lie in $R[T_0, \ldots, T_n]$ and are homogeneous of degree $p_i - 1$. By multiplying $(**)$ by a power of T_i we may assume that $F_0, \ldots, F_n$ have the same degree p, hence, in particular, that we have $p_1 = \cdots = p_n = p$ in $(**)$.

(4) Let $j \in \{0, \ldots, r\}$. The morphism $X \cap U_j \to V_j$ induces a homomorphism

$$R[T_0/T_j, \ldots, T_r/T_j] = \mathcal{O}(V_j) \to \mathcal{O}(X \cap U_j) = R_j/\mathfrak{a}_j.$$

The ideal $\mathfrak{a}_j$ contains the elements $F_0/T_j^p, \ldots, F_n/T_j^p$. Taking into account the particular form of the polynomials $F_0, \ldots, F_n$, it is clear that the elements

$$\prod_{i=0}^{n} (T_i/T_j)^{a_i} \quad \text{with } a_i \in \{0, \ldots, p-1\} \text{ for } i \in \{0, \ldots, n\}$$

are a system of generators of the $\mathcal{O}(V_j)$-module $\mathcal{O}(X \cap U_j)$, hence the homomorphism $\mathcal{O}(V_j) \to \mathcal{O}(X \cap U_j)$ is finite.
(6) From (2.7) we get that $\pi(X)$ is closed.

(12.4) Corollary: *Let X be an irreducible projective variety, and set $m :=$ $\dim(X)$. Then there exists a finite surjective morphism $\varphi\colon X \to \mathbb{P}^m$.*

Proof: There is nothing to prove if $X = \mathbb{P}^m$. Otherwise, there exists $n \in \mathbb{N}$ such that X is a proper closed subset of $\mathbb{P}^n$. We choose $x \in \mathbb{P}^n$, $x \notin X$; the restriction $\psi\colon X \to \mathbb{P}^{n-1}$ of the projection with center x induces a finite morphism $X \to \psi(X)$ [cf. (12.3)]. Now $\psi(X)$ is closed in $\mathbb{P}^{n-1}$; if $\psi(X) \neq \mathbb{P}^{n-1}$, we can apply the same argument, and after a finite number of steps we get a finite surjective morphism $X \to \mathbb{P}^h$ for some $h \in \mathbb{N}_0$ [since a composition of finite maps is finite, cf. (2.6)(3)]. Clearly h is the dimension of X [cf. (2.8)].

(12.5) Proposition: (1) *Let Y be a variety, and let $X \subset \mathbb{P}^n \times Y$ be closed; let $\pi\colon X \to Y$ be the restriction of the second projection. If the fibres of π are non-empty and finite, then π is a finite morphism.*
(2) *Let $\varphi\colon X \to Y$ be a morphism of varieties with X projective. If the fibres of φ are non-empty and finite, then φ is a finite morphism.*

Proof: (1) The statement is true if $n = 0$ [even without the condition on the fibres, cf. (2.6)]; therefore let us assume that $n > 0$.
(a) Let $y \in Y$; now $\mathbb{P}^n \times \{y\}$ is not contained in X since the fibres of π are finite, hence there exists $z \in \mathbb{P}^n$ with $(z, y) \notin X$. Since $\pi(X \cap (\{z\} \times Y))$ is closed in Y and does not contain y, we can replace Y be an open affine neighborhood Y' of y which does not meet $\pi(X \cap (\{z\} \times Y))$, and X by $X \cap (\mathbb{P}^n \times Y')$, and we may assume therefore that $X \cap (\{z\} \times Y) = \emptyset$ and Y is affine. The projection $\mathbb{P}^n \setminus \{z\} \to \mathbb{P}^{n-1}$ with center z induces a morphism $\psi\colon (\mathbb{P}^n \times Y) \setminus (\{z\} \times Y) \to \mathbb{P}^{n-1} \times Y$, the restriction $X \to \mathbb{P}^{n-1} \times Y$ is finite and $\psi(X)$ is closed in $\mathbb{P}^{n-1} \times Y$ [cf. (12.3)], hence the morphism $X \to Y$ has been factored in the form $X \to \psi(X) \to Y$ with $X \to \psi(X)$ being finite and surjective, and $\psi(X) \to Y$ having non-empty finite fibres.
(b) Now we apply the procedure described in (a) to the map $\psi(X) \to Y$. After at most n steps we have the following: There exists an open affine neighborhood V of y such that π is finite over V. This implies the assertion.

(2) Let $X \subset \mathbb{P}^n$ be closed. We identify X with the closed subset $\Gamma_\varphi \subset X \times Y$ [cf. (3.12)]; hence X can be considered as a closed subset of $\mathbb{P}^n \times Y$, and φ is the restriction of the second projection $\mathbb{P}^n \times Y \to Y$. Now the result follows from (1).

13   Blowing up

(13.0) We resolve singularities in chapters VI and VIII essentially by normalizing and by blowing up points. Normalizing was dealt with in section 7. In this and the next section we deal with blowing up.

(13.1) PROJECTION FROM A POINT: Let $n \in \mathbb{N}$, $p \in \mathbb{P}^n$, and H a hyperplane of $\mathbb{P}^n$ with $p \notin H$. Let $\mathrm{pr} \colon \mathbb{P}^n \setminus \{p\} \to H$ be the projection with center p onto H.
(1) By a projective frame associated with pr or p we mean any frame in $\mathbb{P}^n$ such that, if $(\eta_0 : \eta_1 : \cdots : \eta_n)$ are homogeneous coordinates in $\mathbb{P}^n$ with respect to this frame, then $p = (1 : 0 : \cdots : 0)$ and H has the equation $Y_0 = 0$ [this can always be achieved by making a linear change of coordinates]. Remark that, in this case,

$$\mathrm{pr}(\alpha_0 : \alpha_1 : \cdots : \alpha_n) = (\alpha_1 : \cdots : \alpha_n) \quad \text{for every } (\alpha_0 : \alpha_1 : \cdots : \alpha_n) \in \mathbb{P}^n \setminus \{p\},$$

provided we identify H with $\mathbb{P}^{n-1}$ by means of this frame. In the sequel, we always tacitly assume that we are in this situation.
(2) For every $y \in \mathbb{P}^{n-1}$ let $L_y = py$, the line through p and y; the fibre $\mathrm{pr}^{-1}(\{y\})$ is $L_y \setminus \{p\}$. The set

$$\Gamma := \left\{ ((\alpha_0 : \alpha_1 : \cdots : \alpha_n), (\alpha_1 : \cdots : \alpha_n)) \mid (\alpha_1, \ldots, \alpha_n) \neq (0, \ldots, 0) \right\} \subset \mathbb{P}^n \times \mathbb{P}^{n-1}$$

is the graph of pr.
(3) Using the results of (12.3) [with $r = n - 1$ and Y consisting of one point], we get: Let $X \subset \mathbb{P}^n$ be closed and $p \notin X$; then the restriction $\mathrm{pr}\,|X \colon X \to \mathbb{P}^{n-1}$ is finite, and $\mathrm{pr}(X)$ is closed.

(13.2) DEFINITION: The closure $\overline{\Gamma}$ of Γ in $\mathbb{P}^n \times \mathbb{P}^{n-1}$ is said to be the blow-up of $\mathbb{P}^n$ with center p and it is denoted by $\mathrm{Bl}_p(\mathbb{P}^n)$; the restriction $\pi \colon \mathrm{Bl}_p(\mathbb{P}^n) \to \mathbb{P}^n$ of the first projection $\mathrm{pr}_1 \colon \mathbb{P}^n \times \mathbb{P}^{n-1} \to \mathbb{P}^n$ is called the blowing up of $\mathbb{P}^n$ with center p.

(13.3) Theorem: $\mathrm{Bl}_p(\mathbb{P}^n)$ is an irreducible closed subset in $\mathbb{P}^n \times \mathbb{P}^{n-1}$, and π is a proper morphism.

Proof: Let $y = (\eta_0 : \cdots : \eta_n) \in \mathbb{P}^n$, $z = (\zeta_1 : \cdots : \zeta_n) \in \mathbb{P}^{n-1}$, and consider $((\eta_0 : \cdots : \eta_n), (\zeta_1 : \cdots : \zeta_n))$, the bi-homogeneous coordinates of the point $(y, z) \in \mathbb{P}^n \times \mathbb{P}^{n-1}$. Let $W \subset \mathbb{P}^n \times \mathbb{P}^{n-1}$; the ideal in $k[Y_0, \ldots, Y_n, Z_1, \ldots, Z_n] = k[Y, Z]$ generated by all bi-homogeneous polynomials vanishing on W is denoted, by abuse of notation, by $\mathcal{I}_+(W)$. Let $\mathfrak{a} \subset k[Y, Z]$ be the ideal generated by the bi-homogeneous polynomials $Y_i Z_j - Y_j Z_i$, $i, j \in \{1, \ldots, n\}$. Then $\mathfrak{a}$ is a prime ideal [cf. B(10.3)(2) with $A = k[Y_0]$] and, since all generators of $\mathfrak{a}$ vanish on Γ, one has $\mathfrak{a} \subset \mathcal{I}_+(\Gamma)$. Conversely, let us take a bi-homogeneous $f \in k[Y, Z]$ vanishing on Γ; by repeated application of the division algorithm, we may write for some $s, q \in \mathbb{N}_0$

$$Y_1^s f = f_2 \cdot (Y_1 Z_2 - Y_2 Z_1) + \cdots + f_n \cdot (Y_1 Z_n - Y_n Z_1) + g(Y_0, \ldots, Y_n) Z_1^q$$

with $f_2, \ldots, f_n \in k[Y, Z]$. Since $g(Y_0, \ldots, Y_n) Z_1^q$ must vanish on all points of Γ, one must have $g(Y_0, \ldots, Y_n) = 0$, hence $Y_1^s f \in \mathfrak{a}$, so $f \in \mathfrak{a}$. This proves that $\mathfrak{a} = \mathcal{I}_+(\Gamma) = \mathcal{I}_+(\overline{\Gamma})$, and the first part of the theorem.
Since $\mathbb{P}^{n-1}$ is complete, the morphism $\mathrm{pr}_1 : \mathbb{P}^n \times \mathbb{P}^{n-1} \to \mathbb{P}^n$ is proper [cf. (4.25) and (4.23)], and since $\overline{\Gamma}$ is closed in $\mathbb{P}^n \times \mathbb{P}^{n-1}$, the morphism $\pi : \overline{\Gamma} \to \mathbb{P}^n$, being a composition of proper morphisms, is proper [cf. (4.19)].

(13.4) PROPERTIES OF $\mathrm{Bl}_p(\mathbb{P}^n)$: We are going to describe some interesting properties of the irreducible variety $\overline{\Gamma} = \mathrm{Bl}_p(\mathbb{P}^n)$ and the morphism $\pi : \overline{\Gamma} \to \mathbb{P}^n$.
(1) $\Gamma = \pi^{-1}(\mathbb{P}^n \setminus \{p\})$, so Γ is open in $\overline{\Gamma}$, and π induces an isomorphism $\pi' : \Gamma \to \mathbb{P}^n \setminus \{p\}$. Furthermore, $\pi(\overline{\Gamma} \setminus \Gamma) = \{p\}$ and $E := \overline{\Gamma} \setminus \Gamma = \pi^{-1}(\{p\}) = \{p\} \times \mathbb{P}^{n-1}$, hence E is an $(n-1)$-dimensional irreducible nonsingular closed subset of $\overline{\Gamma}$ which is called the exceptional fibre of π.
In fact, to prove the first assertion, we only need to show that $\pi^{-1}(\mathbb{P}^n \setminus \{p\}) \subset \Gamma$. If $((\alpha_0 : \alpha_1 : \cdots : \alpha_n), (\beta_1 : \cdots : \beta_n)) \in \pi^{-1}(\mathbb{P}^n \setminus \{p\})$, and if, say, $\alpha_1 \neq 0$, then the system of homogeneous linear equations

$$-\alpha_j Z_1 + \alpha_1 Z_j = 0 \quad \text{for every } j \in \{2, \ldots, n\}$$

has a unique solution in $\mathbb{P}^{n-1}$, namely $(\beta_1 : \cdots : \beta_n) = (\alpha_1 : \cdots : \alpha_n)$. Furthermore, since Γ is the graph of $\mathrm{pr} : \mathbb{P}^n \setminus \{p\} \to \mathbb{P}^{n-1}$, we know by (3.12) that the projection to the first factor induces an isomorphism $\Gamma \to \mathbb{P}^n \setminus \{p\}$. The remaining assertions are now obvious [since $\mathbb{P}^{n-1}$ is irreducible and nonsingular].
(2) To deal properly with the next properties, one must take open affine coverings of $\mathbb{P}^n$ and $\overline{\Gamma}$, and express equations in these affine charts.
The open affine covering of $\mathbb{P}^n$ which we use is given by the charts $C_i : Y_i \neq 0$ for $i \in \{0, \ldots, n\}$; then $\mathcal{O}(C_i) = k[Y_0/Y_i, \ldots, Y_n/Y_i]$ [cf. (1.18)].
The open covering of $\overline{\Gamma}$ which we use consists of those parts of $\overline{\Gamma}$ which are contained in the following affine charts of $\mathbb{P}^n \times \mathbb{P}^{n-1}$:
(a) The charts $A_i : Y_0 \neq 0$, $Z_i \neq 0$ for $i \in \{1, \ldots, n\}$; then we have

$$\mathcal{O}(A_i) = k\left[\frac{Y_1}{Y_0}, \ldots, \frac{Y_n}{Y_0}, \frac{Z_1}{Z_i}, \ldots, \frac{Z_n}{Z_i}\right];$$

(b) the charts $B_i : Y_i \neq 0$, $Z_i \neq 0$ for $i \in \{1, \ldots, n\}$; then we have

$$\mathcal{O}(B_i) = k\left[\frac{Y_0}{Y_i}, \ldots, \frac{Y_n}{Y_i}, \frac{Z_1}{Z_i}, \ldots, \frac{Z_n}{Z_i}\right].$$

Note that

$$E \subset \bigcup_{i=1}^{n}(A_i \cap \overline{\Gamma}), \quad \Gamma \subset \bigcup_{i=1}^{n}(B_i \cap \overline{\Gamma}), \quad E \cap \bigcup_{i=1}^{n}(B_i \cap \overline{\Gamma}) = \emptyset.$$

Let $i \in \{1,\dots,n\}$. In the ring $\mathcal{O}(A_i)$, the ideal of $\overline{\Gamma}$ is the ideal

$$\mathfrak{a}_i' = \left(\left\{\frac{Y_j}{Y_0}\cdot\frac{Z_l}{Z_i} - \frac{Y_l}{Y_0}\cdot\frac{Z_j}{Z_i}\,\Big|\, j,l \in \{1,\dots,n\}\right\}\right) = \left(\left\{\frac{Y_j}{Y_0} - \frac{Y_i}{Y_0}\cdot\frac{Z_j}{Z_i}\,\Big|\, j \in \{1,\dots,n\}\right\}\right),$$

and in the ring $\mathcal{O}(B_i)$, the ideal of $\overline{\Gamma}$ is the ideal

$$\mathfrak{a}_i'' = \left(\left\{\frac{Y_j}{Y_i}\frac{Z_l}{Z_i} - \frac{Y_l}{Y_i}\cdot\frac{Z_j}{Z_i}\,\Big|\, j,l \in \{1,\dots,n\}\right\}\right) = \left(\left\{\frac{Y_j}{Y_i} - \frac{Z_j}{Z_i}\,\Big|\, j \in \{1,\dots,n\}\right\}\right).$$

(3) We describe the morphism π in these charts. Let $i \in \{1,\dots,n\}$.
(a) In the affine piece A_i, π coincides with the morphism of affine varieties corresponding to the ring homomorphism

$$\varphi_i\colon \mathcal{O}(C_0) \to \mathcal{O}(A_i)/\mathfrak{a}_i' = \mathcal{O}(\overline{\Gamma}\cap A_i) = k\left[\frac{Y_i}{Y_0},\frac{Z_1}{Z_i},\dots,\frac{Z_n}{Z_i}\right]$$

where $\varphi_i(Y_j/Y_0) = (Y_i/Y_0)\cdot(Z_j/Z_i)$ for every $j \in \{1,\dots,n\}$, which is injective.
(b) In the affine piece B_i, π coincides with the morphism of affine varieties corresponding to the ring homomorphism

$$\psi_i\colon \mathcal{O}(C_i) \to \mathcal{O}(B_i)/\mathfrak{a}_i'' = \mathcal{O}(\overline{\Gamma}\cap B_i) = k\left[\frac{Y_0}{Y_i},\dots,\frac{Y_n}{Y_i}\right],$$

which is just the identity.
(4) We have $\dim(\overline{\Gamma}) = n$, and $\overline{\Gamma}$ is nonsingular.
In fact, by (3)(a) and (3)(b), $\overline{\Gamma}$ is covered by open affine sets which are isomorphic to affine space $\mathbb{A}^n$, and $\mathbb{A}^n$ is nonsingular [cf. (6.1)].
(5) For $i \in \{1,\dots,n\}$ we have $E \cap (\overline{\Gamma}\cap A_i) = Z_{\overline{\Gamma}\cap A_i}((Y_i/Y_0))$.
(6) Let $\mathrm{pr}_2\colon \mathbb{P}^n \times \mathbb{P}^{n-1} \to \mathbb{P}^{n-1}$ be the second projection; the morphism

$$\mathrm{pr} = \mathrm{pr}_2 \circ \pi'^{-1}\colon \mathbb{P}^n \setminus \{p\} \to \mathbb{P}^{n-1}$$

is a proper morphism [cf. (1), (4.23) and (4.25)].
(7) π is birational [clear by (3)(a) and (3)(b)].
(8) As a final remark, note that $E = \{p\} \times \mathbb{P}^{n-1}$ is contained in the full inverse image, by π, of the affine piece C_0, so E can be interpreted as the hyperplane at infinity of $\mathbb{A}^n \subset \mathbb{P}^n$ [for which p is the origin] or, equivalently, as the set of [directions of] lines through p.

Thus, we have shown:

(13.5) Proposition: *The morphism* $\pi\colon \mathrm{Bl}_p(\mathbb{P}^n) \to \mathbb{P}^n$ *is a proper birational surjective morphism,* $\mathrm{Bl}_p(\mathbb{P}^n)$ *is an* n-*dimensional irreducible regular projective variety,* $E = \pi^{-1}(\{p\})$ *is a divisor on* $\mathrm{Bl}_p(\mathbb{P}^n)$ *which is isomorphic to* $\{p\} \times \mathbb{P}^{n-1}$, *and* π *induces an isomorphism* $\mathrm{Bl}_p(\mathbb{P}^n) \setminus E \to \mathbb{P}^n \setminus \{p\}$.

(13.6) DEFINITION: Let $X \in \mathbb{P}^n$ be locally closed, and let $p \in X$; let Γ_X be the graph of $\mathrm{pr}\,|(X \setminus \{p\})$. The closure of Γ_X in $X \times \mathbb{P}^{n-1}$ is called the blow-up of X with center p and shall be denoted by $\mathrm{Bl}_p(X)$. The restriction $\pi_X := \mathrm{pr}_1 \,|\, \mathrm{Bl}_p(X)$ is called the blowing up of X with center p, and $E_X := E \cap \mathrm{Bl}_p(X)$ is called the exceptional fibre.
We often write π instead of π_X.

(13.7) REMARK: Let $\pi\colon \mathrm{Bl}_p(\mathbb{P}^n) \to \mathbb{P}^n$ be the blowing up of $\mathbb{P}^n$ with center p, and let $X \subset \mathbb{P}^n$ be locally closed with $p \in X$.
(1) If $X = \{p\}$, then we have $\mathrm{Bl}_p(X) = E_X = \emptyset$.
(2) Assume that $X \neq \{p\}$. We have $\Gamma_X = \Gamma \cap (X \times \mathbb{P}^{n-1})$, and therefore $\Gamma_X = \pi^{-1}(X \setminus \{p\})$. Since $\pi^{-1}(X) = \overline{\Gamma} \cap (X \times \mathbb{P}^{n-1})$ is closed in $X \times \mathbb{P}^{n-1}$, we see that $\mathrm{Bl}_p(X)$ is the closure of Γ_X in $\pi^{-1}(X)$. Moreover, we have $E_X = \pi_X^{-1}(\{p\})$.
(3) Assume that X is closed and $X \neq \{p\}$. Then $\mathrm{Bl}_p(X)$ is the closure of Γ_X in $\mathrm{Bl}_p(\mathbb{P}^n)$.
(4) Let $Y \subset X$ be closed with $p \in Y$. Since $Y \times \mathbb{P}^{n-1}$ is closed in $X \times \mathbb{P}^{n-1}$, we see that $Y' := \mathrm{Bl}_p(Y)$ can be identified with a closed subset of $\mathrm{Bl}_p(X)$. We call Y' the strict transform of Y in X.

(13.8) Proposition: *Let* $X \subset \mathbb{P}^n$ *be locally closed and contain more than one point, let* $p \in X$, *and let* $\pi_X\colon \mathrm{Bl}_p(X) \to X$ *be the blowing up of* X *with center* p. *Then:*
(1) *The morphism* π_X *is proper and surjective.*
(2) *The morphism* π_X *induces an isomorphism* $\mathrm{Bl}_p(X) \setminus E_X \to X \setminus \{p\}$.
(3) *If* X *is irreducible, then* $\mathrm{Bl}_p(X)$ *is irreducible; in this case* π_X *is birational.*
(4) *We have* $\dim(\mathrm{Bl}_p(X)) = \dim(X)$.

Proof: (1) The projection $\mathrm{pr}_1\colon X \times \mathbb{P}^{n-1} \to X$ is proper since $\mathbb{P}^{n-1}$ is complete [cf. (4.25) and (4.23)], and $\mathrm{Bl}_p(X)$ is closed in $X \times \mathbb{P}^{n-1}$; now π_X as a composition of proper morphisms is proper [cf. (4.19)(1)].
(2) Since $\mathrm{Bl}_p(X) \setminus E_X$ is the graph of $\mathrm{pr}\colon X \setminus \{p\} \to \mathbb{P}^{n-1}$, we see that $\mathrm{Bl}_p(X) \setminus E_X$ and $X \setminus \{p\}$ are isomorphic [cf. (3.12)].
(3) Since $\pi_X^{-1}(X \setminus \{p\})$ is an open dense subset of $\mathrm{Bl}_p(X)$, we see that $\mathrm{Bl}_p(X)$ is irreducible if X is so. By (2) the morphism π_X is birational.
(4) Since $\mathrm{Bl}_p(X) \setminus E_X$ is an open dense subset of $\mathrm{Bl}_p(X)$, no irreducible component of $\mathrm{Bl}_p(X)$ can be contained in E_X. This implies the result by (2).

(13.9) Corollary: *Let* $X \subset \mathbb{P}^n$ *be closed, let* $Z_1, \ldots, Z_h$ *be the irreducible components of* X, *let* $p \in X$, *and assume that* $\{p\}$ *is not a component of* X. *Let* $\pi\colon \mathrm{Bl}_p(X) \to X$ *be the blowing up of* X *with center* p. *Let* $i \in \{1, \ldots, h\}$, *and let*

Z_i' be the strict transform of Z_i in $\mathrm{Bl}_p(X)$ if $p \in Z_i$, and $Z_i' := \pi^{-1}(Z_i)$ otherwise. Then $Z_1', \ldots, Z_h'$ are the irreducible components of $\mathrm{Bl}_p(X)$, and, for $i \in \{1, \ldots, h\}$, the restriction $Z_i' \to Z_i$ is birational.

Proof: $Z_1', \ldots, Z_h'$ are irreducible and closed subsets of $\mathrm{Bl}_p(X)$, and, for $i \in \{1, \ldots, h\}$, $Z_i' \to Z_i$ is birational, $\dim(Z_i) = \dim(Z_i')$ and $\pi(Z_i') = Z_i$ [cf. (13.8)]. The last equation shows that there are no inclusion relations between the sets $Z_1', \ldots, Z_h'$; the union of these sets is $\mathrm{Bl}_p(X)$. Therefore $Z_1', \ldots, Z_h'$ are the irreducible components of $\mathrm{Bl}_p(X)$.

(13.10) Proposition: Let $X \subset \mathbb{P}^n$ be locally closed, and let $U \subset X$ be an open subset of X with $p \in U$ and $\{p\} \neq U$. Then we have $\mathrm{Bl}_p(U) = \pi_X^{-1}(U)$, hence $\mathrm{Bl}_p(U)$ is the open subset $(U \times \mathbb{P}^{n-1}) \cap \mathrm{Bl}_p(X)$ of $\mathrm{Bl}_p(X)$.

Proof: We have to show that $\pi_X^{-1}(\{p\}) \subset \mathrm{Bl}_p(U)$. Thus, let $q \in \pi_X^{-1}(\{p\})$, and let $W \subset X \times \mathbb{P}^{n-1}$ be open and $q \in W$. Since $U \times \mathbb{P}^{n-1}$ is open in $X \times \mathbb{P}^{n-1}$ and $q \in U \times \mathbb{P}^{n-1}$, we may assume that $W \subset U \times \mathbb{P}^{n-1}$. Since $W \cap \Gamma_X$ is not empty, it is clear that $W \cap \Gamma_U$ is not empty, hence q lies in the closure of Γ_U in $U \times \mathbb{P}^{n-1}$.

(13.11) Equations: Let $X \subset \mathbb{P}^n$ be a closed subset passing through p, and assume that $\{p\}$ is not an irreducible component of X; let $X' = \mathrm{Bl}_p(X)$ be the blow-up of X with center p.

(1) We derive local equations for $E' := E \cap \mathrm{Bl}_p(X)$.

Let $i \in \{1, \ldots, n\}$, and assume that $X' \cap A_i \neq \emptyset$. Let $\mathfrak{b}_i$ be the ideal of X' in the ring $\mathcal{O}(\overline{\Gamma} \cap A_i) = k[Y_i/Y_0, Z_1/Z_i, \ldots, Z_n/Z_i]$; then $Y_i/Y_0 + \mathfrak{b}_i$ is an equation for E' on $X' \cap A_i$.

It only remains to show that $Y_i/Y_0 \notin \mathfrak{b}_i$. In fact, if $Y_i/Y_0 \in \mathfrak{b}_i$, then one would have $X' \cap A_i \subset E \cap A_i$ and, since $X' \cap A_i \neq \emptyset$, this would mean that $X' \cap A_i$ has an irreducible component contained in E', contrary to our assumptions.

(2) Now we are going to solve this problem: given equations for X, find equations for X'. This must be done in the affine pieces A_i, B_i, $i \in \{1, \ldots, n\}$ but, the latter being non-interesting because $\psi_i: \mathcal{O}(C_i) \to \mathcal{O}(B_i)/\mathfrak{a}_i''$ is an isomorphism for every $i \in \{1, \ldots, n\}$, we shall do it only in the former. We fix one of these pieces, say A_1, and assume that $X' \cap A_1 \neq \emptyset$. As a side remark, when do we have $X' \cap A_1 = \emptyset$? Exactly when X' is contained in the hyperplane $\mathcal{Z}_+(Z_1)$, equivalently, when X is contained in the hyperplane $\mathcal{Z}_+(Y_1)$.

We are given the ideal $\mathfrak{b}$ of X in $\mathcal{O}(C_0) = k[Y_1/Y_0, \ldots, Y_n/Y_0]$, and we want to compute the ideal $\mathfrak{b}_1$ of $X' \cap A_1$ in the ring $\mathcal{O}(\overline{\Gamma} \cap A_1)$ of $\overline{\Gamma} \cap A_1$, i.e., in the ring $k[Y_1/Y_0, Z_2/Z_1, \ldots, Z_n/Z_1]$. For the sake of brevity, we shall write $Y_j/Y_0 =: y_j$ for $j \in \{1, \ldots, n\}$, $Z_l/Z_1 =: z_l$ for $l \in \{2, \ldots, n\}$.

(a) Let us take $f \in k[y] = k[y_1, \ldots, y_n]$, $f \neq 0$, such that $f(0) = 0$; then f can be written as

$$f = f_r(y_1, y_2, \ldots, y_n) + \cdots + f_{r+s}(y_1, y_2, \ldots, y_n)$$

where f_l is homogeneous of degree l for $l \in \{r, \ldots, r+s\}$, and $f_r \neq 0$, $f_{r+s} \neq 0$. We call the strict transform of f the polynomial

$$f'(y_1, z_2, \ldots, z_n) = \frac{f(y_1, y_1 z_2, \ldots, y_1 z_n)}{y_1^r}$$

$$= f_r(1, z_2, \ldots, z_n) + \cdots + y_1^s f_{r+s}(1, z_2, \ldots, z_n) \in \mathcal{O}(\overline{\Gamma} \cap A_1).$$

[Remark that, by homogeneity, $f_r(1, z_2, \ldots, z_n) \neq 0$, $f_{r+s}(1, z_2, \ldots, z_n) \neq 0$ so, in particular, $f'(y_1, z_2, \ldots, z_n) \neq 0$).]

(b) Now, the assertion is that the ideal $\mathfrak{b}_1$ of $X' \cap A_1$ in $\mathcal{O}(\overline{\Gamma} \cap A_1)$ is the one generated by the strict transforms of the elements in $\mathfrak{b}$ [remark that $\mathfrak{b}$ contains no polynomial with a non-zero constant term since $p \in X \cap C_0$]. In fact, let us call $\mathfrak{b}'$ the ideal in $\mathcal{O}(\overline{\Gamma} \cap A_1)$ generated by these strict transforms. We recall that $y_1 \notin \mathfrak{b}'$, so $X' \cap A_1 \not\subset E \cap A_1$, and therefore the set $(X' \cap A_1) \setminus (E \cap A_1)$ is an open dense subset of $X' \cap A_1$. Thus, to find all the polynomials vanishing on $X' \cap A_1$, it is enough to find all the polynomials vanishing at the points $((1 : \alpha_1 : \alpha_2 : \cdots : \alpha_n), (1 : \beta_2 : \cdots : \beta_n)) \in X'$ such that $\alpha_1 \neq 0$ [remark that, in this case, $\alpha_j = \alpha_1 \beta_j$, $j \in \{2, \ldots, n\}$]. If $f \in \mathfrak{b}$, and f' is the strict transform of f, as before, we have $\alpha_1^r f'(\alpha_1, \beta_2, \ldots, \beta_n) = f(\alpha_1, \alpha_1 \beta_2, \ldots, \alpha_1 \beta_n) = f(\alpha_1, \alpha_2, \ldots, \alpha_n) = 0$, so $f'(\alpha_1, \beta_2, \ldots, \beta_n) = 0$, hence $f' \in \mathfrak{b}_1$, and therefore we have shown that $\mathfrak{b}' \subset \mathfrak{b}_1$. Conversely, let $g \in \mathcal{O}(\overline{\Gamma} \cap A_1)$ be a non-zero polynomial vanishing at all these points. Let d be the degree of g; then $y_1^d g$ can be written as $y_1^d g(y_1, z_2, \ldots, z_n) = g_1(y_1, y_1 z_2, \ldots, y_1 z_n) = g_1(y_1, y_2, \ldots, y_n)$ where $g_1 \in \mathcal{O}(C_0)$ is a homogeneous polynomial of degree d, and g is the strict transform of g_1. Now, we have $0 = \alpha_1^d g(\alpha_1, \beta_2, \ldots, \beta_n) = g_1(\alpha_1, \alpha_2, \ldots, \alpha_n)$, hence $g_1 \in \mathfrak{b}$, so $g \in \mathfrak{b}'$, and therefore we have $\mathfrak{b}_1 \subset \mathfrak{b}'$. This proves our assertion.

(c) The induced homomorphism $k[y_1, \ldots, y_n]/\mathfrak{b} \to k[y_1, z_2, \ldots, z_n]/\mathfrak{b}_1$ is injective since $X' \cap A_1 \to X \cap C_0$ is surjective.

14 Blowing up: The Local Rings

(14.0) Let X be a variety, let $x \in X$, and let $\pi_X : X' := \mathrm{Bl}_x(X) \to X$ be the blowing up of X with center x [cf. (14.4) below for this notion]. In this section we show, among other things, that there is a bijective map $\pi^{-1}(\{x\}) \to \mathrm{Max}(\mathrm{gr}_{\mathcal{O}_{X,x}}(\mathcal{O}_{X,x}))$, and that for $q \in \pi^{-1}(\{x\})$ the local ring $\mathcal{O}_{X',q}$ is isomorphic to the local ring of a closed point of $\mathrm{Proj}(\mathcal{R}(\mathfrak{m}_{\mathcal{O}_{X,x}}, \mathcal{O}_{X,x}))$. We shall see that for this it is enough to study the blow-up of affine varieties.

The reader is advised to get acquainted with (5.2)–(5.7) of appendix B.

(14.1) BLOWING UP AFFINE VARIETIES: Let $R = k[Y_1, \ldots, Y_n]$ be the coordinate ring of $\mathbb{A}^n$. We consider affine space $\mathbb{A}^n$ as an open subset of $\mathbb{P}^n$ via the embedding $(\eta_1, \ldots, \eta_n) \mapsto (1 : \eta_1 : \cdots : \eta_n)$; in particular, the origin of $\mathbb{A}^n$ corresponds to $p = (1 : 0 \cdots : 0) \in \mathbb{P}^n$. We set $\Delta := \mathrm{Bl}_p(\mathbb{A}^n)$ and $\theta := \pi_{\mathbb{A}^n} : \Delta \to \mathbb{A}^n$; we have $\Delta = (\mathbb{A}^n \times \mathbb{P}^{n-1}) \cap \mathrm{Bl}_p(\mathbb{P}^n)$ by (13.10). Note that Δ is the closed subset of

$\mathbb{A}^n \times \mathbb{P}^{n-1}$ consisting of those points whose coordinates $((\eta_1, \ldots, \eta_n), (\zeta_1 : \cdots : \zeta_n))$ satisfy $\eta_i \zeta_j - \eta_j \zeta_i = 0$, $i, j \in \{1, \ldots, n\}$, and it is an open subset of $\mathrm{Bl}_p(\mathbb{P}^n)$. Therefore Δ is an n-dimensional irreducible regular variety [cf. (13.5)], and $E = E_{\mathbb{A}^n} = \{p\} \times \mathbb{P}^{n-1}$. Moreover, θ induces an isomorphism $\Delta \setminus \theta^{-1}(\{p\}) \to \mathbb{A}^n \setminus \{p\}$.

(1) Let $Y \subset \mathbb{A}^n$ be a closed set passing through the origin, and assume that $\{p\}$ is not a component of Y; it is not hard to see that $Y' := \mathrm{Bl}_p(Y) \subset \mathbb{P}^n \times \mathbb{P}^{n-1}$ is the closure of $\theta^{-1}(Y \setminus \{p\})$ in $\mathrm{Bl}_p(\mathbb{A}^n)$. Set $E' := \pi_Y^{-1}(\{p\}) = E \cap Y'$.

(2) We can cover $\mathbb{A}^n \times \mathbb{P}^{n-1}$ by the affine charts $A_i : Z_i \neq 0$ for $i \in \{1, \ldots, n\}$; then we have

$$\mathcal{O}(A_i) = k\left[Y_1, \ldots, Y_n, \frac{Z_1}{Z_i}, \ldots, \frac{Z_n}{Z_i}\right].$$

Let $i \in \{1, \ldots, n\}$; it is easy to check that

$$\mathcal{O}(\Delta \cap A_i) = k\left[Y_i, \frac{Z_1}{Z_i}, \ldots, \frac{Z_n}{Z_i}\right] = k\left[Y_i, \frac{Y_1}{Y_i}, \ldots, \frac{Y_n}{Y_i}\right],$$

and that the ideal of $E \cap A_i$ in $\mathcal{O}(\Delta \cap A_i)$ is generated by Y_i.

We usually consider only the affine chart A_1; let us write, for the sake of brevity, $y_j := Y_j$ for $j \in \{1, \ldots, n\}$, $z_l := Z_l/Z_1$ for $l \in \{2, \ldots, n\}$. Then $\mathfrak{m} = (y_1, \ldots, y_n)$ is the ideal in $R = k[y_1, \ldots, y_n]$ which defines the origin in $\mathbb{A}^n$, $\mathcal{O}(A_1) = R[z_2, \ldots, z_n]$ and $\mathcal{O}(\Delta \cap A_1) = k[y_1, y_2/y_1, \ldots, y_n/y_1] = R[y_2/y_1, \ldots, y_n/y_1] = \mathcal{R}(\mathfrak{m}, R)_{(y_1 T)}$.

(3) Let $\mathfrak{b} = \mathcal{I}(Y)$ be the ideal of Y in $k[y_1, \ldots, y_n]$; just as in (13.11) we can show that the ideal of $Y' \cap A_1$ in the ring $\mathcal{O}(\Delta \cap A_1)$ is the ideal $\mathfrak{b}_1$ generated by the strict transforms of the elements in $\mathfrak{b}$.

(4) Let $\overline{R} := R/\mathfrak{b}$ be the affine coordinate ring of Y. Define $\overline{\mathfrak{m}} := \mathfrak{m}/\mathfrak{b}$ and $\overline{y}_i := y_i + \mathfrak{b}$ for $i \in \{1, \ldots, n\}$; then we have $\overline{\mathfrak{m}} = \overline{R}\overline{y}_1 + \cdots + \overline{R}\overline{y}_n$.
Set $\overline{A}_1 := (Y \times \mathbb{P}^{n-1}) \cap A_1$; then we have $\mathcal{O}(\overline{A}_1) = \mathcal{O}(Y) \otimes_k k[z_2, \ldots, z_n] = \overline{R}[z_2, \ldots, z_n]$ [cf. (3.5)(3)]. We assume that $\overline{A}_1 \cap Y' \neq \emptyset$; then we have $\overline{y}_1 \neq 0$ [cf. (13.11)(2)]. In $\mathcal{O}(\overline{A}_1 \cap Y')$ we have $\overline{y}_j = \overline{y}_1 z_j$ for $j \in \{2, \ldots, n\}$, hence

$$\mathcal{O}(\overline{A}_1 \cap Y') = \overline{R}[\overline{y}_2/\overline{y}_1, \ldots, \overline{y}_n/\overline{y}_1] = \mathcal{R}(\overline{\mathfrak{m}}, \overline{R})_{(\overline{y}_1 T)}.$$

Using this result, it is not difficult so show that $\mathcal{R}(\overline{\mathfrak{m}}, \overline{R})$ is the homogeneous coordinate ring of $\mathrm{Bl}_p(Y)$ in the sense of (12.2).

(14.2) We keep the notations introduced above. Set $S := \mathcal{O}_{\mathbb{A}^n, p}$ (resp. $\overline{S} := \mathcal{O}_{Y, p}$) and let $\mathfrak{n}$ (resp. $\overline{\mathfrak{n}}$) be the maximal ideal of S (resp. $\overline{S}$). We shall identify a point q of the exceptional fibre E (resp. E') with a homogeneous ideal of $\mathrm{gr}_{\mathfrak{n}}(S)$ (resp. $\mathrm{gr}_{\overline{\mathfrak{n}}}(\overline{S})$), and we will show that the local ring $\mathcal{O}_{\Delta, q}$ (resp. $\mathcal{O}_{Y', q}$) for $q \in E$ (resp. $q \in E'$) is isomorphic to a homogeneous localization of $\mathcal{R}(\mathfrak{n}, S)$ (resp. $\mathcal{R}(\overline{\mathfrak{n}}, \overline{S})$) [cf. (14.3) below].

(1) Set $\widetilde{y}_j := y_j \bmod \mathfrak{m}^2$ for every $j \in \{1, \ldots, n\}$; then $\mathrm{gr}_{\mathfrak{m}}(R) = k[\widetilde{y}_1, \ldots, \widetilde{y}_n]$, a polynomial ring. *The map*

$$\chi : E = \{p\} \times \mathbb{P}^{n-1} \to \mathrm{Max}(\mathrm{Proj}(\mathrm{gr}_{\mathfrak{m}}(R)))$$

from the exceptional fibre to the set of maximal homogeneous non-irrelevant ideals of the graded polynomial ring $\mathrm{gr}_{\mathfrak{m}}(R)$ *which maps a point* $q = (p, (\beta_1 : \cdots : \beta_n))$ *of* E *to the homogeneous ideal* $\mathfrak{q}_0$ *of* $\mathrm{gr}_{\mathfrak{m}}(R)$ *generated by* $\beta_i \widetilde{y}_j - \beta_j \widetilde{y}_i$, $i, j \in \{1, \ldots, n\}$, *is bijective.*

Let $q = (p, (\beta_1 : \cdots : \beta_n)) \in E$ correspond to the homogeneous prime ideal $\mathfrak{q}_0$ of $\mathrm{gr}_{\mathfrak{m}}(R)$ which is generated by $\beta_i \widetilde{y}_j - \beta_j \widetilde{y}_i$, $i, j \in \{1, \ldots, n\}$. We consider the canonical homomorphism

$$\mathcal{R}(\mathfrak{m}, R) = \bigoplus_{s \geq 0} \mathfrak{m}^s T^s \to \bigoplus_{s \geq 0} \mathfrak{m}^s / \mathfrak{m}^{s+1} = \mathrm{gr}_{\mathfrak{m}}(R);$$

its kernel is the ideal $J = \bigoplus_{s \geq 0} \mathfrak{m}^{s+1} T^s$. Let $\mathfrak{q} \subset \mathcal{R}(\mathfrak{m}, R)$ be the preimage of $\mathfrak{q}_0$ under this homomorphism; then $\mathfrak{q}$ is the prime ideal generated by the elements $y_i T^0$, $(\beta_i y_j - \beta_j y_i) T$ for $i, j \in \{1, \ldots, n\}$.

(2) Assume that $q = (p, (\beta_1 : \cdots : \beta_n)) \in E$ lies in A_1, hence that $\beta_1 \neq 0$. In this case we have $\mathcal{O}_{\Delta, q} = k[y_1, y_2/y_1, \ldots, y_n/y_1]_{(y_1, y_2/y_1 - \beta_2/\beta_1, \ldots, y_n/y_1 - \beta_n/\beta_1)}$. Now $\mathcal{R}(\mathfrak{m}, R)_{(y_1 T)} \cong R[y_2/y_1, \ldots, y_n/y_1]$ [cf. B(5.6)], and since $y_1 T \notin \mathfrak{q}$ [note that $\beta_1 \neq 0$], we see that the prime ideal $\mathfrak{p}$ of $R[y_2/y_1, \ldots, y_n/y_1] = k[y_1, y_2/y_1, \ldots, y_n/y_1]$ corresponding to $\mathfrak{q}$ is the ideal generated by y_1, $y_2/y_1 - \beta_2/\beta_1, \ldots, y_n/y_1 - \beta_n/\beta_1$, and therefore that

$$R[y_2/y_1, \ldots, y_n/y_1]_{\mathfrak{p}} = k[y_1, y_2/y_1, \ldots, y_n/y_1]_{(y_1, y_2/y_1 - \beta_2/\beta_1, \ldots, y_n/y_1 - \beta_n/\beta_1)},$$

hence we have k-algebra isomorphisms

$$\alpha_q \colon \mathcal{O}_{\Delta, q} \xrightarrow{\cong} R[y_2/y_1, \ldots, y_n/y_1]_{\mathfrak{p}} \xrightarrow{\cong} \mathcal{R}(\mathfrak{m}, R)_{(\mathfrak{q})}$$

[cf. B(5.6)(2) for the last isomorphism].

(3) Let $\mathfrak{a} := \mathrm{gr}(\mathfrak{b}) \subset \mathrm{gr}_{\mathfrak{m}}(R)$ be the homogeneous ideal generated by all leading forms of the elements of $\mathfrak{b}$ with respect to $\mathfrak{m}$ [cf. B(5.3)(2)]. We show: *The point* $q = (p, (\beta_1 : \cdots : \beta_n)) \in E$ *lies in* Y' *iff* $\mathfrak{q}_0 \supset \mathfrak{a}$.
For the proof we shall assume, as in (2), that $\beta_1 \neq 0$. Now, on the one hand, we have $q \in Y'$ iff $(0, \beta_2/\beta_1, \ldots, \beta_n/\beta_1)$ annihilates all polynomials in $\mathfrak{b}_1$, and it is enough to consider only strict transforms of elements of $\mathfrak{b}$ since $\mathfrak{b}_1$ is generated by such strict transforms [cf. (14.1)(3)]. On the other hand, we have $\mathfrak{q}_0 \supset \mathfrak{a}$ iff $\widetilde{f}(\beta_1, \ldots, \beta_n) = 0$ for every homogeneous $\widetilde{f} \in \mathfrak{a}$, and it is enough to consider only those homogeneous polynomials which are leading forms with respect to $\mathfrak{m}$ of polynomials in $\mathfrak{b}$ [since $\mathfrak{a}$ is generated by such polynomials].
Let $f \in \mathfrak{b}$ be written as in (13.11)(2) as a sum of forms

$$f = f_r(y_1, y_2, \ldots, y_n) + \cdots + f_{r+s}(y_1, y_2, \ldots, y_n), \tag{$*$}$$

and let f' be the strict transform of f. Now $f'(0, \beta_2/\beta_1, \ldots, \beta_n/\beta_1) = 0$ iff $f_r(\beta_1, \beta_2, \ldots, \beta_n) = 0$ which is our assertion.
(4) Note that $\mathrm{gr}_{\overline{\mathfrak{m}}}(\overline{R}) = \mathrm{gr}_{\mathfrak{m}}(R)/\mathfrak{a}$. Let $q \in E \cap Y'$; we assume that $q \in A_1$. Set $\overline{\mathfrak{q}}_0 := \mathfrak{q}_0/\mathfrak{a}$, and let $\overline{\mathfrak{q}} \subset \mathcal{R}(\overline{\mathfrak{m}}, \overline{R})$ be the preimage of $\overline{\mathfrak{q}}_0$ under the canonical

homomorphism $\mathcal{R}(\overline{\mathfrak{m}}, \overline{R}) \to \operatorname{gr}_{\overline{\mathfrak{m}}}(\overline{R})$. Note that now $\mathfrak{q}$ also is the preimage of $\overline{\mathfrak{q}}$ under the canonical surjective homomorphism $\mathcal{R}(\mathfrak{m}, R) \to \mathcal{R}(\overline{\mathfrak{m}}, \overline{R})$ with kernel $\mathcal{R}(\mathfrak{b})$, and that $\mathfrak{q} \supset \mathcal{R}(\mathfrak{b})$. Therefore, by (3) there exists a unique k-algebra homomorphism $\overline{\alpha}_q \colon \mathcal{O}_{Y',q} \to \mathcal{R}(\overline{\mathfrak{m}}, \overline{R})_{(\overline{\mathfrak{q}})}$ making the diagram

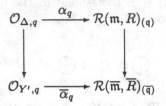

commutative. We show: $\overline{\alpha}_q$ *is an isomorphism.*
In fact, it is enough to show that $\alpha_q(\mathfrak{b}_1 \mathcal{O}_{\Delta,q}) = \mathcal{R}(\mathfrak{b})_{(\mathfrak{q})}$, the ideal of $\mathcal{R}(\mathfrak{m}, R)_{(\mathfrak{q})}$ generated by the elements of $\mathcal{R}(\mathfrak{b})\mathcal{R}(\mathfrak{m}, R)_{(\mathfrak{q})}$ of degree zero. The ideal $\mathfrak{b}_1$ is generated by the strict transforms of elements of $\mathfrak{b}$. Thus, let $f' \in \mathfrak{b}_1$ be the strict transform of an element $f \in \mathfrak{b}$ as in $(*)$. Then $\alpha_q(f'(y_1, y_2/y_1, \ldots, y_n/y_1)/1)$ is the image of $f(y_1, y_2, \ldots, y_n)T^r/(y_1 T)^r \in \mathcal{R}(\mathfrak{m}, R)_{(y_1 T)}$ in $\mathcal{R}(\mathfrak{m}, R)_{(\mathfrak{q})}$; since $f(y_1, y_2, \ldots, y_n) \in \mathfrak{b} \cap \mathfrak{m}^r$, it is clear that $\alpha_q(f'(y_1, y_2/y_1, \ldots, y_n/y_1)/1) \in \mathcal{R}(\mathfrak{b})_{(\mathfrak{q})}$. Conversely, the ideal $\mathcal{R}(\mathfrak{b})_{(\mathfrak{q})}$ of $R(\mathfrak{m}, R)_{(\mathfrak{q})}$ is generated by the images of elements $f(y_1, y_2, \ldots, y_n)T^r/(y_1 T)^r \in \mathcal{R}(\mathfrak{m}, R)_{(y_1 T)}$ where $r \in \mathbb{N}$ and $f(y_1, \ldots, y_n) \in \mathfrak{b} \cap \mathfrak{m}^r$.
(5) Set $\Sigma := R \setminus \mathfrak{m}$; then $S = \Sigma^{-1} R$, and the maximal ideal $\mathfrak{n}$ of S is generated by $y_1, \ldots, y_n$. We have $\overline{S} = S/\mathfrak{b}S$ and $\overline{\mathfrak{n}} = \mathfrak{n}/\mathfrak{b}S$. Furthermore, we have $\mathcal{R}(\mathfrak{n}, S) = \Sigma^{-1}\mathcal{R}(\mathfrak{m}, R)$ [cf. B(5.3)(4)]. Localization with respect to Σ induces commutative diagrams

$$
\begin{array}{ccc}
\mathcal{R}(\mathfrak{m}, R) & \longrightarrow & \mathcal{R}(\mathfrak{m}, R)_{(\mathfrak{q})} \\
\downarrow & & \downarrow \cong \\
\mathcal{R}(\mathfrak{n}, S) & \longrightarrow & \mathcal{R}(\mathfrak{n}, S)_{(\mathfrak{q}')}
\end{array}
\qquad
\begin{array}{ccc}
\mathcal{R}(\mathfrak{m}, R)_{(\mathfrak{q})} & \overset{\cong}{\longrightarrow} & \mathcal{R}(\mathfrak{n}, S)_{(\mathfrak{q}')} \\
\downarrow & & \downarrow \\
\mathcal{R}(\overline{\mathfrak{m}}, \overline{R})_{(\overline{\mathfrak{q}})} & \underset{\cong}{\longrightarrow} & \mathcal{R}(\overline{\mathfrak{n}}, \overline{S})_{(\overline{\mathfrak{q}}')}
\end{array}
$$

where $\mathfrak{q}'$ is the image of $\mathfrak{q}$ in $\mathcal{R}(\mathfrak{n}, S)$ and $\overline{\mathfrak{q}}'$ is the image of $\overline{\mathfrak{q}}$ in $\mathcal{R}(\overline{\mathfrak{n}}, \overline{S})$. Denoting now $\mathfrak{q}'$ by $\mathfrak{q}$ and $\overline{\mathfrak{q}}'$ by $\overline{\mathfrak{q}}$, we can state:

(14.3) Proposition: *Let* $p \in \mathbb{A}^n$, *let* $\pi_{\mathbb{A}^n} \colon \operatorname{Bl}_p(\mathbb{A}^n) \to \mathbb{A}^n$ *be the blowing up of* $\mathbb{A}^n$ *with center* p, *let* $Y \subset \mathbb{A}^n$ *be closed, and assume that* Y *passes through* p *and that* $\{p\}$ *is not a component of* Y; *let* $\pi_Y \colon Y' := \operatorname{Bl}_p(Y) \to Y$ *be blowing up of* Y *with center* p. *Set* $E := \pi_{\mathbb{A}^n}^{-1}(\{p\})$ *and* $E' := E \cap Y'$. *Set* $S := \mathcal{O}_{\mathbb{A}^n, p}$, $\overline{S} := \mathcal{O}_{Y, p}$, *let* $\mathfrak{n}$ *be the maximal ideal of* S *and* $\overline{\mathfrak{n}}$ *be the maximal ideal of* $\overline{S}$. *Then:*
(1) *The points of* E (*resp.* E') *correspond bijectively to the closed points of* $\operatorname{Proj}(\operatorname{gr}_{\mathfrak{n}}(S))$ (*resp.* $\operatorname{Proj}(\operatorname{gr}_{\overline{\mathfrak{n}}}(\overline{S}))$).
(2) *Let* $q \in E$ (*resp.* $q \in E'$) *correspond to the prime ideal* $\mathfrak{q}_0$ *of* $\operatorname{gr}_{\mathfrak{n}}(S)$ (*resp. to the prime ideal* $\overline{\mathfrak{q}}_0$ *of* $\operatorname{gr}_{\overline{\mathfrak{n}}}(\overline{S})$), *and let* $\mathfrak{q}$ (*resp.* $\overline{\mathfrak{q}}$) *be the preimage of* $\mathfrak{q}_0$ (*resp.* $\overline{\mathfrak{q}}_0$) *under the canonical homomorphism* $\mathcal{R}(\mathfrak{n}, S) \to \operatorname{gr}_{\mathfrak{n}}(S)$ (*resp. the canonical*

homomorphism $\mathcal{R}(\overline{n}, \overline{S}) \to \mathrm{gr}_{\overline{n}}(\overline{S})$; then the rings $\mathcal{O}_{\mathrm{Bl}_{A^n},q}$ and $\mathcal{R}(n, S)_{(q)}$ (resp. $\mathcal{O}_{Y',q}$ and $\mathcal{R}(\overline{n}, \overline{S})_{(\overline{q})}$) are isomorphic.

Proof: Using the results of (14.2), it is enough to remark that the graded rings $\mathrm{gr}_{\mathrm{m}}(R)$ and $\mathrm{gr}_{\mathrm{n}}(S)$ (resp. $\mathrm{gr}_{\overline{\mathrm{m}}}(\overline{R})$ and $\mathrm{gr}_{\overline{\mathrm{n}}}(\overline{S})$) are isomorphic [for any k-algebra A and any maximal ideal m of A, we have, setting $\mathrm{n} := \mathrm{m}A_{\mathrm{m}}$, an isomorphism of k-vector spaces $\mathrm{m}^i/\mathrm{m}^{i+1} \xrightarrow{\cong} \mathrm{n}^i/\mathrm{n}^{i+1}$ for $i \in \mathbb{N}_0$].

(14.4) Remark: Let X be a variety, and let $p \in X$. Then X is a locally closed subset of some $\mathbb{P}^n$, which is called the ambient space, and we may assume that $p = (1 : 0 : \cdots : 0)$. Let $X' := \mathrm{Bl}_p(X) \subset \mathbb{P}^n \times \mathbb{P}^{n-1}$ be the blow-up of X with center p. We are mostly interested in the behavior of the local rings $\mathcal{O}_{X',q}$ at points $q \in X'$ lying over p. For this, by (13.10), we may assume that we are in the situation of (14.1), hence these local rings do not depend on the embedding of X in the ambient space $\mathbb{P}^n$ [cf. (14.3)(2)]. By abuse of language we shall often say: Let $\mathrm{Bl}_p(X)$ be the blow-up of X without any reference to an ambient space.

(14.5) Corollary: Let X be a regular variety, and let $p \in X$. Then $\mathrm{Bl}_p(X)$ is regular.

Proof: Since $\pi \colon \mathrm{Bl}_p(X) \to X$ induces an isomorphism outside of the set $\mathrm{Bl}_p(X) \setminus \pi^{-1}(\{p\})$, we have only to consider the local rings of points of $\mathrm{Bl}_p(X)$ lying over p. The result follows from (14.3), VII(2.3) and appendix B, (5.5)–(5.7).

(14.6) Blowing up and completion: We keep the notations introduced in (14.2). In particular, let $q \in E$ correspond to the prime ideal q_0 of $\mathrm{gr}_{\mathrm{n}}(S)$ generated by the elements $\beta_j \widetilde{y}_i - \beta_i \widetilde{y}_j$, $i, j \in \{1, \ldots, n\}$, and let $q \subset \mathcal{R}(\mathrm{m}, R)$ be the preimage of q_0. We assume that $\beta_1 \neq 0$.
(1) The preimage q of q_0 is generated by the n elements $y_1 T^0, (\beta_1 y_2 - \beta_2 y_1)T, \ldots,$ $(\beta_1 y_n - \beta_n y_1)T$, and therefore the maximal ideal of $\mathcal{R}(n, S)_{(q)}$ is generated by $y_1, y_2/y_1 - \beta_2/\beta_1, \ldots, y_n/y_1 - \beta_n/\beta_1$. Now $\mathcal{O}_{\Delta,q}$ is an n-dimensional regular local ring [since Δ is regular], hence $\mathcal{R}(n, S)_{(q)}$ is a regular local ring, and the system of generators of its maximal ideal mentioned above is a regular system of parameters.
(2) The completion $\widehat{S}$ of S is the ring of formal power series $k[[y_1, \ldots, y_n]]$ [cf. IV(1.1)]; let $\widehat{n}$ be its maximal ideal. We identify the rings $\mathrm{gr}_{\widehat{n}}(\widehat{S})$ and $\mathrm{gr}_{\mathrm{n}}(S)$ [cf. [63], Cor. 7.13]; the preimage $\widetilde{q} \subset \mathcal{R}(\widehat{n}, \widehat{S})$ of q_0 is generated by the elements $y_1 T^0, (\beta_1 y_2 - \beta_2 y_1)T, \ldots, (\beta_1 y_n - \beta_n y_1)T$. Therefore the local ring $\mathcal{R}(\widehat{n}, \widehat{S})_{(\widetilde{q})}$ has dimension n, at most. On the other hand, the induced homomorphism $\mathcal{R}(\widehat{n}, \widehat{S})_{(\widetilde{q})} \to \mathrm{gr}_{\mathrm{n}}(S)_{(q_0)}$ is surjective with non-zero kernel, and we have $\dim(\mathrm{gr}_{\mathrm{n}}(S)_{(q_0)}) = n - 1$. Therefore $\mathcal{R}(\widehat{n}, \widehat{S})_{(\widetilde{q})}$ is a regular local ring of dimension n, and $y_1, y_2/y_1 - \beta_2/\beta_1, \ldots, y_n/y_1 - \beta_n/\beta_1$ is a regular system of parameters. The canonical homomorphism $S \to \widehat{S}$ is flat, hence $\mathcal{R}(n, S) \otimes_S \widehat{S} = \mathcal{R}(\widehat{n}, \widehat{S})$ [cf. B(5.3)(6)], and we have a local homomorphism $\mathcal{R}(n, S)_{(q)} \to \mathcal{R}(\widehat{n}, \widehat{S})_{(\widetilde{q})}$; under this homomorphism, a regular system of parameters of $\mathcal{R}(n, S)_{(q)}$ is mapped to a

regular system of parameters of $\mathcal{R}(\widehat{n}, \widehat{S})_{(\widehat{q})}$. Since k is the residue field of these local rings, we get an isomorphism of completions [cf. [63], Prop. 10.17]

$$(\mathcal{R}(n, S)_{(q)})\widehat{\ } \to (\mathcal{R}(\widehat{n}, \widehat{S})_{(\widehat{q})})\widehat{\ }.$$

(3) Assume, in particular, that $q \in Y'$. Then we have $q_0 \supset \mathfrak{a}$ [cf. (14.2)(3)]. Let $\widehat{\overline{S}}$ be the completion of $\overline{S}$, and $\widehat{\overline{n}}$ be the maximal ideal of $\widehat{\overline{S}}$. We identify $\mathrm{gr}_{\overline{n}}(\overline{S})$ and $\mathrm{gr}_{\widehat{\overline{n}}}(\widehat{\overline{S}})$ [cf. [63], Cor. 7.13]. Let $\widetilde{\overline{q}} \subset \mathcal{R}(\widehat{\overline{n}}, \widehat{\overline{S}})$ be the preimage of $\overline{q}_0$. Since $S \to \widehat{S}$ is a faithfully flat homomorphism [cf. cf. [34], Ch. III, §3, no. 5, Prop. 9], we have $\mathfrak{b}\widehat{S} \cap \widehat{n}^s = (\mathfrak{b}S \cap n^s)\widehat{S}$ for every $s \in \mathbb{N}_0$ [cf. [34], Ch. I, §3, no. 5, Prop. 10], hence the kernel of the canonical homomorphism $\mathcal{R}(\widehat{n}, \widehat{S}) \to \mathcal{R}(\widehat{\overline{n}}, \widehat{\overline{S}})$ is $\mathcal{R}(\mathfrak{b}\widehat{S})$ [note that $\mathfrak{b}\widehat{S}$ is the kernel of the homomorphism $\widehat{S} \to \widehat{\overline{S}}$]. The diagrams in (14.2), (4) and (5), give rise to the following commutative diagram of local rings and local homomorphisms

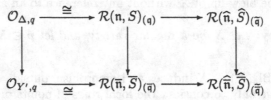

with injective horizontal and surjective vertical maps; the kernel of the right vertical map is the image of the kernel of the middle vertical map. Passing to completions induces therefore an isomorphism

$$(\mathcal{O}_{Y',q})\widehat{\ } \to (\mathcal{R}(\widehat{\overline{n}}, \widehat{\overline{S}})_{(\widetilde{\overline{q}})})\widehat{\ }$$

[note that if $A \to B$ is a local homomorphism of local rings which is surjective with kernel $\mathfrak{a}$, then $\widehat{A} \to \widehat{B}$ is surjective and $\mathfrak{a}\widehat{A}$ is its kernel].

The results of (14.6) can be formulated in the following proposition:

(14.7) Proposition: *Let X, Y be varieties, let $x \in X$ be such that $\{x\}$ is not an irreducible component of X, and let $y \in Y$ be such that $\{y\}$ is not an irreducible component of Y. We set $X' := \mathrm{Bl}_x(X)$ and $Y' := \mathrm{Bl}_y(Y)$. Set $A := \mathcal{O}_{X,x}$ and let $\mathfrak{m}$ be its maximal ideal, and set $B := \mathcal{O}_{Y,y}$ and let $\mathfrak{n}$ be its maximal ideal. Assume that the k-algebras $\widehat{A}$ and $\widehat{B}$ are isomorphic; then there is an induced isomorphism of graded k-algebras $\mathrm{gr}_{\mathfrak{m}}(A) \xrightarrow{\cong} \mathrm{gr}_{\mathfrak{n}}(B)$. Let $x' \in X'$ lie over x, $y' \in Y'$ lie over y and correspond to the same ideal of $\mathrm{gr}_{\mathfrak{m}}(A) = \mathrm{gr}_{\mathfrak{n}}(B)$. Then the rings $(\mathcal{O}_{X',x'})\widehat{\ }$ and $(\mathcal{O}_{Y',y'})\widehat{\ }$ are isomorphic as k-algebras.*

Appendix B

Miscellaneous Results

1 Ordered Abelian Groups

(1.1) REMARK: Let Γ be an additively written abelian group with neutral element 0.

(1) Let $\prec$ be a total ordering of Γ. Remember that Γ is called an ordered group if for γ_1, $\gamma_2 \in \Gamma$ with $\gamma_1 \prec \gamma_2$ we have $\gamma_1 + \delta \prec \gamma_2 + \delta$ for every $\delta \in \Gamma$. Then $\Pi := \{\gamma \in \Gamma \mid \gamma \succcurlyeq 0\}$ is called the set of positive elements of Γ; Π satisfies $\Pi \cup (-\Pi) = \Gamma$, $\Pi \cap (-\Pi) = \{0\}$ and $\Pi + \Pi \subset \Pi$.

(2) Let $\Pi \subset \Gamma$ be a subset with $\Pi \cup (-\Pi) = \Gamma$, $\Pi \cap (-\Pi) = \{0\}$ and $\Pi + \Pi \subset \Pi$. For elements $\gamma \neq \delta$ of Γ we define $\gamma \prec \delta$ iff $\delta - \gamma \in \Pi$. Then $\prec$ is a total ordering of Γ, and $(\Gamma, \prec)$ is an ordered group having Π as set of positive elements.

(3) Let $(\Gamma, \prec)$ be a totally ordered abelian group. For any subset Δ of Γ we denote by Δ_+ the set of elements $\delta \in \Delta$ with $\delta \succcurlyeq 0$.

(1.2) REMARK: Let $(\Gamma, \prec)$ and $(\Gamma', \prec')$ be totally ordered abelian groups. A homomorphism $\varphi \colon \Gamma \to \Gamma'$ is said to be a homomorphism of ordered groups, if for all γ, $\delta \in \Gamma$ with $\gamma \preceq \delta$ we have $\varphi(\gamma) \preceq' \varphi(\delta)$. This condition is equivalent to $\varphi(\Gamma_+) \subset \Gamma'_+$.

1.1 Isolated Subgroups

(1.3) NOTATION: In this subsection $(\Gamma, \prec)$ is a totally ordered abelian group.

(1.4) DEFINITION: (1) A non-empty subset Δ of Γ is called a segment if, for every $\alpha \in \Delta$, all the elements of Γ which lie between α and $-\alpha$, also belong to Δ.

(2) A proper subgroup of Γ is called isolated if it is a segment.

(1.5) REMARK: (1) The set of segments of Γ is totally ordered by inclusion.

409

In fact, if Δ, Δ' are segments of $\cdot\Gamma$ with $\Delta \not\subset \Delta'$, then we choose $\alpha \in \Delta$ with $\alpha \notin \Delta'$. Let $\beta \in \Delta'$; then α does not lie between β and $-\beta$, hence β lies between α and $-\alpha$, whence $\beta \in \Delta$, and therefore we have $\Delta' \subset \Delta$.

(2) A segment Δ of Γ is a subgroup of Γ iff the sum of positive elements of Δ lies again in Δ.

In fact, this condition is necessary. We show that it also is sufficient. Let α, $\beta \in \Delta$. First, we consider the case $\alpha \preccurlyeq \beta$. Let $\alpha \preccurlyeq 0$; if $\beta \succcurlyeq 0$, then $\beta - \alpha$ lies in Δ, and if $\beta \prec 0$, then we have $0 \preccurlyeq \beta - \alpha \preccurlyeq -\beta - \alpha \in \Delta$, hence $\beta - \alpha$ lies in Δ since Δ is a segment. Now let $\alpha \succ 0$; then we have $0 \preccurlyeq \beta - \alpha \prec \beta$, hence $\beta - \alpha$ lies in Δ since Δ is a segment. Second, we consider the case $\beta \preccurlyeq \alpha$. Then we have $\alpha - \beta \in \Delta$ by the first part, hence $-(\alpha - \beta) = \beta - \alpha$ lies in Δ.

(1.6) Proposition: *The set of isolated subgroups of a totally ordered abelian group is totally ordered by inclusion.*

Proof: This follows immediately from (1.5)(1).

(1.7) DEFINITION: If Γ has only finitely many isolated subgroups, then the cardinality of the set of isolated subgroups of Γ is called the rank of Γ and it is denoted by $\mathrm{rank}(\Gamma)$; in the other case we say that Γ has infinite rank, and we set $\mathrm{rank}(\Gamma) = \infty$.

(1.8) Proposition: *Let Δ be an isolated subgroup of Γ. Then the group Γ/Δ can be endowed with a total ordering such that the canonical homomorphism $\omega\colon \Gamma \to \Gamma/\Delta$ is a homomorphism of ordered groups, and this ordering is the only ordering of Γ/Δ for which ω is a homomorphism of ordered groups. Moreover, the map $\Delta' \mapsto \omega(\Delta')$ is an inclusion-preserving bijective map from the set of isolated subgroups of Γ containing Δ to the set of isolated subgroups of Γ/Δ.*

Proof: Set $\overline{\Gamma} := \Gamma/\Delta$. Let Π be the set of positive elements of Γ; then we have $\Pi + \Pi \subset \Pi$, $\Pi \cap (-\Pi) = \{0\}$ and $\Pi \cup (-\Pi) = \Gamma$ [cf. (1.1)].
We define $\overline{\Pi} := \omega(\Pi) \subset \overline{\Gamma}$. Then we have $\overline{\Pi} + \overline{\Pi} \subset \overline{\Pi}$ and $\overline{\Pi} \cup (-\overline{\Pi}) = \overline{\Gamma}$. Let γ_1, $\gamma_2 \in \Pi$ with $\omega(\gamma_1) = \omega(-\gamma_2)$. Then we have $\gamma_1 + \gamma_2 \in \Delta$, and since $0 \preccurlyeq \gamma_i \preccurlyeq \gamma_1 + \gamma_2$ for $i \in \{1, 2\}$, we have $\gamma_i \in \Delta$ for $i \in \{1, 2\}$ [since Δ is an isolated subgroup of Γ], and therefore $\overline{\Pi} \cap (-\overline{\Pi}) = \{0\}$. Defining $\overline{\Pi}$ as the set of positive elements of $\overline{\Gamma}$, we see that $\overline{\Gamma}$ is a totally ordered group [cf. (1.1)], and that $\omega\colon \Gamma \to \overline{\Gamma}$ is a homomorphism of ordered groups [cf. (1.2)]. It is clear that this ordering of $\overline{\Gamma}$ is the only ordering of $\overline{\Gamma}$ for which ω is a homomorphism of ordered groups.
The last assertion is easy to check.

(1.9) REMARK: In the sequel, when Δ is an isolated subgroup of Γ, the group Γ/Δ will always be considered as a totally ordered group with ordering as defined in (1.8).

(1.10) Corollary: *Let Δ be an isolated subgroup of Γ. Then we have $\mathrm{rank}(\Gamma) = \mathrm{rank}(\Delta) + \mathrm{rank}(\Gamma/\Delta)$.*

Proof: This follows immediately from (1.8) and (1.6).

(1.11) Proposition: *Let Γ' be a subgroup of Γ. Then:*
(1) Every isolated subgroup Δ' of Γ' has the form $\Delta' = \Delta \cap \Gamma'$ where Δ is an isolated subgroup of Γ, and therefore we have $\mathrm{rank}(\Gamma') \leq \mathrm{rank}(\Gamma)$.
(2) If Γ/Γ' is a torsion group, then $\Delta \mapsto \Delta \cap \Gamma'$ is a bijective map from the set of isolated subgroups of Γ to the set of isolated subgroups of Γ', and therefore we have $\mathrm{rank}(\Gamma) = \mathrm{rank}(\Gamma')$.

Proof: (1) Let Δ' be an isolated subgroup of Γ', and let Δ be the set of elements δ of Γ such that $m\delta \in \Delta'$ for some integer $m \neq 0$. It is immediate that Δ is an isolated subgroup of Γ with $\Delta \cap \Gamma' = \Delta'$.
(2) By (1) it is enough to show that $\Delta \mapsto \Delta \cap \Gamma'$ is an injective map from the set of isolated subgroups of Γ to the set of isolated subgroups of Γ'.
Thus, let Δ be an isolated subgroup of Γ; clearly, $\Delta' := \Delta \cap \Gamma'$ is a subgroup of Γ' and a segment in Γ'. Let $\beta \in \Gamma$ and $\beta \notin \Delta$. Since Γ/Γ' is a torsion group, there exists an integer $m \neq 0$ with $m\beta \in \Gamma'$. Since Δ is a segment of Γ and $\beta \notin \Delta$, we have $m\beta \notin \Delta$, hence $\Delta' \neq \Gamma'$, and therefore Δ' is an isolated subgroup of Γ'. Let $\beta \in \Delta$. Then $m\beta \in \Delta$ for every $m \in \mathbb{Z}$, and there exists an integer $m \neq 0$ with $m\beta \in \Gamma'$, hence $m\beta \in \Delta'$. On the other hand, if $\beta \in \Gamma$ is an element with $m\beta \in \Delta'$ for some integer $m \neq 0$, then $m\beta \in \Delta$ and therefore $\beta \in \Delta$ [since Δ is a segment]. Thus, we have shown that Δ is the set of all $\beta \in \Gamma$ such that $m\beta \in \Delta'$ for some integer $m \neq 0$. This implies that the map $\Delta \mapsto \Delta \cap \Gamma$ is injective.

1.2 Initial Index

(1.12) NOTATION: In this subsection $(\Gamma, \prec)$ is a totally ordered abelian group; we assume that $\Gamma \neq \{0\}$.

(1.13) NOTATION: Let $(S, <)$ be a totally ordered set. A subset M of S is called major if the following holds: For $x \in M$ and $y \in S$ with $x \leq y$ we have $y \in M$.

(1.14) DEFINITION: Let Δ be a subgroup of Γ of finite index. The number of major subsets of Γ which consist of strictly positive elements and contain all strictly positive elements of Δ, is called the initial index of Δ in Γ and it is denoted by $\varepsilon(\Gamma, \Delta)$.

(1.15) Proposition: *With notations as in (1.14) we have:*
(1) If the set of strictly positive elements of Γ does not have a smallest element, then $\varepsilon(\Gamma, \Delta) = 1$.
(2) If the set of strictly positive elements of Γ has a smallest element γ, then $\varepsilon(\Gamma, \Delta) = (\Gamma' : (\Gamma' \cap \Delta)) < \infty$ *where Γ' is the subgroup of Γ generated by γ.*

Proof: (1) Let $\gamma \in \Gamma$ and $\gamma \succ 0$. Since Γ has no smallest strictly positive element, the set of elements γ' of Γ with $0 \prec \gamma' \prec \gamma$ is not finite, and since the subgroup

Δ has finite index, there exist strictly positive elements $\gamma_1 \neq \gamma_2$ of Γ lying in the same coset of Δ and being strictly smaller than γ, and we may label them such that $0 \prec \gamma_1 \prec \gamma_2 \prec \gamma$, hence $\delta := \gamma_2 - \gamma_1 \prec \gamma$ is a strictly positive element of Δ. Now, if a major subset M of Γ contains all strictly positive elements of Δ, then it contains δ and therefore γ, hence it contains every strictly positive element of Γ, hence we have $\varepsilon(\Gamma, \Delta) = 1$.

(2) Let γ be the smallest strictly positive element of Γ. We choose $n \in \mathbb{N}$ minimal with $n\gamma \in \Delta$. Then n is also the smallest positive integer with $\delta \succcurlyeq n\gamma$ for every strictly positive $\delta \in \Delta$ [otherwise, there exists $i \in \{1, \ldots, n-1\}$ and $\delta \in \Delta$ with $i\gamma \preccurlyeq \delta \prec (i+1)\gamma$, hence $0 \preccurlyeq \delta - i\gamma \prec \gamma$, hence $\delta = i\gamma$, contrary to the choice of n]. For $i \in \{1, \ldots, n\}$ we set $M(i) := \{\gamma' \in \Gamma \mid \gamma' \succcurlyeq i\gamma\}$. Then $M(1), \ldots, M(n)$ are pairwise different major subsets in Γ which consist of strictly positive elements of Γ and contain all strictly positive elements of Δ. Let M be a major subset of Γ which consists of strictly positive elements of Γ and contains all strictly positive elements of Δ. Then $n\gamma \in M$. We choose $i \in \mathbb{N}$ minimal with $i\gamma \in M$. Let $\gamma' \in M$, and suppose that $\gamma' \prec i\gamma$. Then there exists $j \in \mathbb{N}$, $j < i$, with $j\gamma \preccurlyeq \gamma' \prec (j+1)\gamma$, hence with $0 \preccurlyeq \gamma' - j\gamma \prec \gamma$, hence we get $\gamma' = j\gamma \in M$, contrary to the choice of i. Therefore we have $\gamma' \succcurlyeq i\gamma$, i.e., we have $M = M(i)$. Since $n = (\Gamma' : (\Gamma' \cap \Delta))$, we have proved the assertion in (2).

(1.16) Corollary: *The initial index $\varepsilon(\Gamma, \Delta)$ divides $(\Gamma : \Delta)$; if Γ is isomorphic, as an ordered group, to the ordered group $\mathbb{Z}$, then we have $\varepsilon(\Gamma, \Delta) = (\Gamma : \Delta)$.*

1.3   Archimedean Ordered Groups

(1.17) NOTATION: In this subsection $(\Gamma, \prec)$ is a totally ordered abelian group.

(1.18) DEFINITION: Γ is called archimedean if for every pair of elements $\alpha, \beta \in \Gamma$ with $\beta \succ 0$ there exists an integer n with $n\beta \succ \alpha$.

(1.19) Proposition: *Γ has rank 1 iff it is archimedean.*

Proof: (1) Suppose that Γ is archimedean, but has not rank 1; then there exists an isolated subgroup $\Delta \neq \{0\}$ of Γ. Let $\delta \in \Delta$ be a strictly positive element; then we have $\delta \prec \gamma$ for every positive $\gamma \in \Gamma \setminus \Delta$. Let $\gamma \in \Gamma \setminus \Delta$ be positive, and choose $n \in \mathbb{Z}$ with $\gamma \prec n\delta$; this implies that $\gamma \in \Delta$, contradicting the choice of γ.
(2) Suppose that Γ has rank 1, but is not archimedean. Then there exist elements $\alpha, \beta \in \Gamma$ with $\beta \succ \alpha \succ 0$ and $n\alpha \preccurlyeq \beta$ for every positive integer n. Let Δ' be the set of all positive elements $\gamma \in \Gamma$ such that $\gamma \prec n\alpha$ for some integer n. Δ' is closed under addition. Furthermore, if $\gamma' \in \Gamma_+$ and $\gamma' \prec \gamma$ for some $\gamma \in \Delta'$, then $\gamma' \in \Delta'$. Thus, the set $\Delta := \{\gamma, -\gamma \mid \gamma \in \Delta'\}$ is a proper segment of Γ [since $\beta \notin \Delta$] and a subgroup of Γ [cf. (1.5)(2)] different from $\{0\}$ [since $\alpha \in \Delta$], hence an isolated subgroup of Γ with $\{0\} \subsetneqq \Delta \subsetneqq \Gamma$, and Γ has not rank 1, which is in contradiction with our assumption.

(1.20) Corollary: *Every subgroup of the additive group $\mathbb{R}$ has rank 1.*

(1.21) Proposition: *If Γ has rank 1, then there exists an injective order-preserving homomorphism $\varphi\colon \Gamma \to \mathbb{R}$, i.e., Γ is isomorphic, as an ordered group, to a subgroup of $\mathbb{R}$.*

Proof: We fix a positive element $\alpha \in \Gamma$. Let $\beta \in \Gamma$; we divide the set of rational numbers $\mathbb{Q} = \{m/n \mid m \in \mathbb{Z}, n \in \mathbb{N}\}$ into two classes C, C', as follows: $m/n \in C$ if $m\alpha \preccurlyeq n\beta$, and $m/n \in C'$ otherwise. The archimedean property of Γ [cf. (1.19)] implies that neither class is empty. We show that C and C' define a Dedekind cut. Let m'/n' be an element of the upper class C', and let m/n be an element of the lower class C. Then we have $m'\alpha \succ n'\beta$, $m\alpha \preccurlyeq n\beta$, hence $m'n\alpha \succ n'n\beta$, $mn'\alpha \preccurlyeq nn'\beta$, hence $mn' < m'n$, and we have $m'/n' > m/n$, as required. Let ρ be the real number defined by this cut: we set $\varphi(\beta) := \rho$. It is easy to check that $\varphi\colon \Gamma \to \mathbb{R}$ is an injective order-preserving homomorphism.

(1.22) Remark: The following remark shall be useful: A subgroup Γ of $\mathbb{R}$ is order-isomorphic to the subgroup $\mathbb{Z}$ of $\mathbb{R}$ iff Γ has a smallest strictly positive element γ, and if this is the case, then we have $\Gamma = \mathbb{Z}\gamma$ [since $\mathbb{Z}\gamma$ is a subgroup and a segment of Γ].

(1.23) Notation: Assume that Γ has finite rank h, and let $\{0\} = \Gamma_0 \subset \Gamma_1 \subset \cdots \subset \Gamma_{h-1}$ be the isolated subgroups of Γ. The factor groups $\Gamma_1/\Gamma_0, \ldots, \Gamma_h/\Gamma_{h-1}$, [where $\Gamma_h = \Gamma$] are ordered groups of rank 1, hence they are isomorphic as ordered groups to subgroups of $\mathbb{R}$ [cf. (1.21)]. Γ is called discrete of rank h if, for $i \in \{1, \ldots, h\}$, Γ_i/Γ_{i-1} is isomorphic, as an ordered group, to the group $\mathbb{Z}$ of integers. A discrete group of rank 1 usually is called a discrete group; it is isomorphic, as an ordered group, to the group $\mathbb{Z}$ of integers.

(1.24) Lexicographic ordering: Let $\Gamma_1, \ldots, \Gamma_h$ be finitely many totally ordered abelian groups. The direct sum $\Delta := \Gamma_1 \oplus \cdots \oplus \Gamma_h$ can be made into a totally ordered abelian group by ordering it lexicographically: For elements $(\gamma_1, \ldots, \gamma_h) \neq (\gamma_1', \ldots, \gamma_h')$ in Δ we define $(\gamma_1, \ldots, \gamma_h) \succ (\gamma_1', \ldots, \gamma_h')$ if in the difference $(\gamma_1 \ldots, \gamma_h) - (\gamma_1', \ldots, \gamma_h') = (\gamma_1 - \gamma_1', \ldots, \gamma_h - \gamma_h')$ the first non-zero entry is positive.
Let $i \in \{1, \ldots, h\}$, and let Δ_i be an isolated subgroup of Γ_i. Then the set

$$\{(0, \ldots, 0, \delta_i, \gamma_{i+1}, \ldots, \gamma_h) \mid \delta_i \in \Delta_i, \gamma_j \in \Gamma_j \text{ for } j \in \{i+1, \ldots, h\}\}$$

is an isolated subgroup of Δ. In particular, for $i \in \{1, \ldots, h\}$ we set $\Delta_0 := \{0\}$ and

$$\Delta_i := \{(0, \ldots, 0, \gamma_{h-i+1}, \ldots, \gamma_h) \mid \gamma_j \in \Gamma_j \text{ for } j \in \{h-i+1, \ldots, h\}\}.$$

The subgroups $\Delta_0 \subsetneqq \Delta_1 \subsetneqq \cdots \subsetneqq \Delta_{h-1}$ are isolated subgroups of Δ, and we have $\Delta_{i+1}/\Delta_i \cong \Gamma_{h-i+1}$ (as ordered groups) for $i \in \{0, \ldots, h-1\}$.

(1.25) REMARK: Let $h \in \mathbb{N}$, and set $\Delta := \mathbb{Z}^h$, ordered lexicographically. Then Δ is a discrete group of rank h.
In fact, for $i \in \{0, \ldots, h\}$ we set

$$\Delta_i := \{(0, \ldots, 0, \gamma_{h-i+1}, \ldots, \gamma_h) \mid \gamma_j \in \mathbb{Z} \text{ for } j \in \{h-i+1, \ldots, h\}\}.$$

Then $\{0\} = \Delta_0 \subsetneq \Delta_1 \subsetneq \cdots \subsetneq \Delta_{h-1}$ are the isolated subgroups of Δ, and we have $\Delta_{i+1}/\Delta_i \cong \mathbb{Z}$ for $i \in \{0, \ldots, h-1\}$ (as ordered groups).

(1.26) Proposition: *Let Γ be a discrete group of rank h. Then Γ is isomorphic, as an ordered group, to $\mathbb{Z}^h$, ordered lexicographically.*

Proof: Set $\Gamma_h := \Gamma$, and let $\{0\} =: \Gamma_0 \subset \Gamma_1 \subset \cdots \subset \Gamma_{h-1}$ be the isolated subgroups of Γ. There exists an isomorphism $\varphi_1 \colon \Gamma_1 \to \mathbb{Z}$ of ordered groups. Assume that for some $i \in \{1, \ldots, h-1\}$ we have φ_1 extended to an isomorphism of ordered groups $\varphi_i \colon \Gamma_i \to \mathbb{Z}^i$, where $\mathbb{Z}^i$ is ordered lexicographically. The group Γ_{i+1}/Γ_i is isomorphic to $\mathbb{Z}$ as an ordered group; we choose $\gamma_{i+1} \in \Gamma_{i+1}$ such that the image of γ_{i+1} in Γ_{i+1}/Γ_i is positive and a generator. Every $\gamma \in \Gamma_{i+1}$ has a unique representation as $\gamma = m\gamma_{i+1} + \gamma'$ with $m \in \mathbb{Z}$ and $\gamma' \in \Gamma_i$, and we have $\gamma \succcurlyeq 0$ iff either $m > 0$ or $m = 0$ and $\gamma' \succcurlyeq 0$. It is easy to check that

$$\varphi_{i+1} \colon \Gamma_{i+1} \to \mathbb{Z} \times \mathbb{Z}^i \text{ with } \varphi_{i+1}(m\gamma_{i+1} + \gamma') = (m, \varphi_i(\gamma'))$$

is an isomorphism of Γ_{i+1} with the lexicographically ordered group $\mathbb{Z} \oplus \mathbb{Z}^i$ [where $\mathbb{Z}^i$ is ordered lexicographically]. Thus, after h steps, we find an isomorphism $\varphi_h \colon \Gamma \to \mathbb{Z}^h$ of ordered groups.

(1.27) Proposition: *Let Γ' be a subgroup of Γ. If Γ/Γ' is finite, then Γ is discrete of rank h iff Γ' is discrete of rank h.*

Proof: From (1.11)(2) we get $\operatorname{rank}(\Gamma) = \operatorname{rank}(\Gamma')$; more precisely, the map $\Delta \mapsto \Delta \cap \Gamma'$ is a bijective map from the set of isolated subgroups of Γ to the set of isolated subgroups of Γ'. Set $h := \operatorname{rank}(\Gamma) = \operatorname{rank}(\Gamma')$, and let $\Delta_1 \subset \Delta_2$ be isolated subgroups of Γ with $\operatorname{rank}(\Delta_2/\Delta_1) = 1$. Set $\Delta_1' := \Delta \cap \Gamma'$ and $\Delta_2' := \Delta_2 \cap \Gamma'$; then Δ_2'/Δ_1' can be considered as a subgroup of Δ_2/Δ_1, and we have $\operatorname{rank}(\Delta_2'/\Delta_1') = 1$. Since Δ_2' is a subgroup of finite index of Δ_2, the subgroup Δ_2'/Δ_1' of Δ_2/Δ_1 has finite index. We have to show that Δ_2/Δ_1 is isomorphic, as an ordered group, to the ordered group $\mathbb{Z}$ iff Δ_2'/Δ_1' is isomorphic, as an ordered group, to the ordered group $\mathbb{Z}$. Δ_2/Δ_1 is isomorphic, as an ordered group, to a subgroup G of $\mathbb{R}$ [cf. (1.21)]; let $G' \subset G$ be the image of Δ_2'/Δ_1', and note that G/G' is a finite group. If the two ordered groups G and $\mathbb{Z}$ are isomorphic, then G' is also isomorphic to $\mathbb{Z}$. Conversely, assume that G' is order-isomorphic to $\mathbb{Z}$, hence has a smallest strictly positive element g'. Since G/G' is finite, we have $G = (g_1 + G') \cup \cdots \cup (g_p + G')$ (disjoint union) where we may assume that $g_i \geq 0$ for $i \in \{1 \ldots, p\}$. Then $g := \min(\{g_1 + g', \ldots, g_p + g'\})$ is the smallest strictly positive element in G, hence G is order-isomorphic to $\mathbb{Z}$.

1.4 The Rational Rank of an Abelian Group

(1.28) REMARK: (1) Let Γ be a torsion-free abelian group. Then the map $\gamma \mapsto \gamma \otimes 1 : \Gamma \to \Gamma \otimes_{\mathbb{Z}} \mathbb{Q} =: \Gamma^*$ is injective, and we can consider Γ as a subgroup of Γ^*.
(2) Let $(\Gamma, \prec)$ be a totally ordered abelian group. Then Γ is torsion-free [for any non-zero element $\gamma \in \Gamma$ and $n \in \mathbb{N}$ we have $n\gamma \succcurlyeq \gamma$ if $\gamma \succ 0$, and $n\gamma \preccurlyeq \gamma$ if $\gamma \prec 0$], hence Γ can be considered as a subgroup of $\Gamma^* := \Gamma \otimes_{\mathbb{Z}} \mathbb{Q}$.

(1.29) DEFINITION: Let Γ be an abelian group. The rational rank of Γ is defined as $\dim_{\mathbb{Q}}(\Gamma \otimes_{\mathbb{Z}} \mathbb{Q})$; it will be denoted by $\mathrm{rat.\,rank}(\Gamma)$.

(1.30) Proposition: *Let Γ be an abelian group, and let Γ' be a subgroup of Γ.*
(1) We have

$$\mathrm{rat.\,rank}(\Gamma) = \mathrm{rat.\,rank}(\Gamma') + \mathrm{rat.\,rank}(\Gamma/\Gamma').$$

(2) If, in addition, Γ is a totally ordered abelian group, then we have

$$\mathrm{rank}(\Gamma') \leq \mathrm{rank}(\Gamma) \leq \mathrm{rank}(\Gamma') + \mathrm{rat.\,rank}(\Gamma/\Gamma').$$

In particular, we always have $\mathrm{rank}(\Gamma) \leq \mathrm{rat.\,rank}(\Gamma)$.

Proof: (1) Since $\mathbb{Q}$ is a flat $\mathbb{Z}$-module, we have an exact sequence of $\mathbb{Q}$-vector spaces

$$0 \to \Gamma' \otimes_{\mathbb{Z}} \mathbb{Q} \to \Gamma \otimes_{\mathbb{Z}} \mathbb{Q} \to (\Gamma/\Gamma') \otimes_{\mathbb{Z}} \mathbb{Q} \to 0,$$

and the first assertion follows.
(2) The first inequality has already been proved [cf. (1.11)] To prove the second inequality, we show by induction on n: If $\{0\} = \Gamma_0 \subsetneqq \Gamma_1 \subsetneqq \cdots \subsetneqq \Gamma_n$ is a sequence of subgroups of Γ where $\Gamma_0, \ldots, \Gamma_{n-1}$ are isolated subgroups of Γ, and Γ_n is either an isolated subgroup of Γ or $\Gamma_n = \Gamma$, then we have

$$n \leq \mathrm{rank}(\Gamma') + \mathrm{rat.\,rank}(\Gamma/\Gamma'). \tag{$*$}$$

This is trivially true if $n = 0$. Let $n > 0$, and assume that $(*)$ holds with n replaced by $n - 1$. We apply the induction assumption to Γ_{n-1} and its subgroup $\Gamma' \cap \Gamma_{n-1}$, and we get

$$n - 1 \leq \mathrm{rank}(\Gamma' \cap \Gamma_{n-1}) + \mathrm{rat.\,rank}(\Gamma_{n-1}/(\Gamma' \cap \Gamma_{n-1})). \tag{$**$}$$

(a) If $\Gamma' \cap \Gamma_{n-1} = \Gamma'$, equivalently if $\Gamma' \subset \Gamma_{n-1}$, then we have

$$n - 1 \leq \mathrm{rank}(\Gamma') + \mathrm{rat.\,rank}(\Gamma_{n-1}/\Gamma').$$

The group Γ/Γ_{n-1} is totally ordered, hence torsion-free, and since $\Gamma_{n-1} \subsetneqq \Gamma$, we have $\mathrm{rat.\,rank}(\Gamma/\Gamma_{n-1}) \geq 1$. We get $\mathrm{rat.\,rank}(\Gamma_{n-1}/\Gamma') + 1 \leq \mathrm{rat.\,rank}(\Gamma/\Gamma')$ by (1), and this yields inequality $(*)$.
(b) If $\Gamma' \cap \Gamma_{n-1} \subsetneqq \Gamma'$, then $\Gamma' \cap \Gamma_{n-1}$ is an isolated subgroup of Γ', hence $\mathrm{rank}(\Gamma') \geq \mathrm{rank}(\Gamma' \cap \Gamma_{n-1}) + 1$, and since $\Gamma_{n-1}/(\Gamma' \cap \Gamma_{n-1}) \cong (\Gamma_{n-1} + \Gamma')/\Gamma' \subset \Gamma/\Gamma'$, we have $\mathrm{rat.\,rank}(\Gamma_{n-1}/(\Gamma' \cap \Gamma_{n-1})) \leq \mathrm{rat.\,rank}(\Gamma/\Gamma')$. Using these inequalities in $(**)$ yields inequality $(*)$ in this case also.
It is clear that $(*)$ yields the assertion.

(1.31) Proposition: *Let Γ be a totally ordered abelian group. The following statements are equivalent:*
(1) *Γ has finite rational rank.*
(2) *Γ has finite rank, and if $\{0\} = \Gamma_0 \subset \cdots \subset \Gamma_{n-1}$ are the isolated subgroups of Γ, then the groups $\Gamma_1/\Gamma_0, \ldots, \Gamma_n/\Gamma_{n-1}$ [where $\Gamma_n := \Gamma$] have finite rational rank.*
If Γ satisfies these conditions, then we have

$$\mathrm{rat.\,rank}(\Gamma) = \mathrm{rat.\,rank}(\Gamma_1/\Gamma_0) + \cdots + \mathrm{rat.\,rank}(\Gamma_n/\Gamma_{n-1}).$$

Proof: This follows immediately from (1.30).

(1.32) Definition: An abelian group Γ is called divisible if for $\gamma \in \Gamma$ and $n \in \mathbb{N}$ there exists $\delta \in \Gamma$ with $n\delta = \gamma$.

(1.33) Remark: Let Γ be a divisible abelian group, and let Δ be a subgroup of Γ.
(1) The factor group Γ/Δ is divisible.
(2) If Γ is, in addition, a totally ordered group with ordering $\prec$, and Δ is an isolated subgroup, then Δ is also divisible [let $\delta \in \Delta_+$ and $n \in \mathbb{N}$, and let $\gamma \in \Gamma$ with $n\gamma = \delta$, then we have $0 \preccurlyeq \gamma \preccurlyeq n\gamma = \delta$, hence γ lies in Δ].

(1.34) Remark: Let Γ be an abelian group which is torsion-free and divisible.
(1) For $\gamma \in \Gamma$ and $n \in \mathbb{N}$ there exists a unique $\delta \in \Gamma$ with $n\delta = \gamma$.
(2) The canonical map $\gamma \mapsto \gamma \otimes 1 : \Gamma \to \Gamma \otimes_{\mathbb{Z}} \mathbb{Q}$ is an isomorphism of groups, hence Γ carries in a canonical way a structure as a $\mathbb{Q}$-vector space.

(1.35) Proposition: *Let Γ be a torsion-free abelian group, and set $\Gamma^* := \Gamma \otimes_{\mathbb{Z}} \mathbb{Q}$.*
(1) *The $\mathbb{Q}$-vector space Γ^* is a torsion-free divisible abelian group, and Γ^*/Γ is a torsion group. Moreover, the map $\Delta \mapsto \Delta^* := \Delta \otimes_{\mathbb{Z}} \mathbb{Q}$ is an inclusion-preserving bijective map from the set of divisible subgroups of Γ to the set of subspaces of Γ^*, and the inverse map is given by $\Delta^* \mapsto \Delta^* \cap \Gamma$.*
(2) *Let Ω be a torsion-free divisible abelian group which contains Γ as a subgroup. There exists an unique $\mathbb{Q}$-linear map $\varphi \colon \Gamma^* \to \Omega$ with $\varphi(\gamma) = \gamma$ for every $\gamma \in \Gamma$; it is injective, and it is surjective iff Ω/Γ is a torsion group.*

Proof: (1) This is clear.
(2) Let $\gamma^* \in \Gamma^*$, choose $n \in \mathbb{N}$ with $n\gamma^* \in \Gamma$, and let $\omega \in \Omega$ be such that $n\omega = n\gamma^*$. It is immediate that ω is determined uniquely by γ^*; let $\varphi \colon \Gamma^* \to \Omega$ be defined by $\varphi(\gamma^*) = \omega$. Then φ is a $\mathbb{Q}$-linear map. We have $\varphi(\gamma) = \gamma$ for every $\gamma \in \Gamma$, and there is only one $\mathbb{Q}$-linear map $\Gamma^* \to \Omega$ with this property. The map φ is injective; it is surjective iff Ω/Γ is a torsion group.

(1.36) Remark: Let $(\Gamma, \prec)$ be a totally ordered abelian group. Then the ordering of Γ can be extended uniquely to an ordering of $\Gamma^* := \Gamma \otimes_{\mathbb{Z}} \mathbb{Q}$. Moreover, the map $\Delta \mapsto \Delta^* := \Delta \otimes_{\mathbb{Z}} \mathbb{Q}$ is an inclusion-preserving bijective map from the set of isolated subgroups of Γ to the set of isolated subgroups of Γ^*, and $\Delta^* \mapsto \Gamma \cap \Delta^*$ is the inverse map. In particular, we have $\operatorname{rank}(\Gamma) = \operatorname{rank}(\Gamma^*)$.

In fact, if $\gamma^* \neq \delta^* \in \Gamma^*$, then we choose $n \in \mathbb{N}$ with $n\gamma^*, n\delta^* \in \Gamma$, and define $\gamma^* \prec \delta^*$ if $n\gamma^* \prec n\delta^*$. It is clear that this gives rise to an ordering of Γ^* which extends the ordering of Γ, and this is the only extension of the ordering of Γ to an ordering of Γ^*. Let Δ be an isolated subgroup of Γ; then $\Delta^* := \Delta \otimes_{\mathbb{Z}} \mathbb{Q}$ is an isolated subgroup of Γ^*. In fact, if $\delta^* \in \Delta^*$ is positive and $\gamma^* \in \Gamma^*$ with $0 \prec \gamma^* \preccurlyeq \delta^*$, then we can choose $n \in \mathbb{N}$ with $\gamma := n\gamma^* \in \Gamma$ and $\delta := n\delta^* \in \Delta$, and since $0 \prec \gamma \preccurlyeq \delta$, we get $\gamma \in \Delta$, hence $n\gamma^* \in \Delta^*$, and therefore $\gamma^* \in \Delta^*$. Since Δ is divisible [cf. (1.33)(2)], we have $\Delta^* \cap \Gamma = \Delta$ by (1.35). Conversely, it is easy to check that if Δ^* is an isolated subgroup of Γ^*, then $\Delta := \Delta^* \cap \Gamma$ is an isolated subgroup of Γ.

(1.37) Proposition: *Let $(\Gamma, \prec)$ be a totally ordered abelian group. The following statements are equivalent:*

(1) Γ *is a finitely generated free $\mathbb{Z}$-module.*

(2) Γ *has finite rank, and if $\{0\} = \Gamma_0 \subset \Gamma_1 \subset \cdots \subset \Gamma_{n-1}$ are the isolated subgroups of Γ, then $\Gamma_1/\Gamma_0, \ldots, \Gamma_n/\Gamma_{n-1}$ [with $\Gamma_n := \Gamma$] are finitely generated free $\mathbb{Z}$-modules, and*

$$\Gamma \cong \Gamma_1/\Gamma_0 \oplus \Gamma_2/\Gamma_1 \oplus \cdots \oplus \Gamma_n/\Gamma_{n-1} \quad \text{as } \mathbb{Z}\text{-modules.}$$

If Γ satisfies these conditions, then we have

$$\operatorname{rat.\,rank}(\Gamma) = \sum_{i=1}^{n} \operatorname{rat.\,rank}(\Gamma_i/\Gamma_{i-1}).$$

Proof: $(2) \Rightarrow (1)$ is obvious.

$(1) \Rightarrow (2)$: Since Γ is a finitely generated free $\mathbb{Z}$-module, Γ has finite rational rank. We argue by induction on the rational rank of Γ. There is nothing to show if $\Gamma = \{0\}$. Let $\Gamma \neq \{0\}$ and $\operatorname{rat.\,rank}(\Gamma) = 1$. Then we have $\operatorname{rank}(\Gamma) = 1$ [cf. (1.30)(2)], and we are done. Let $\operatorname{rat.\,rank}(\Gamma) > 1$. If $\operatorname{rank}(\Gamma) = 1$, we are done. Otherwise, let Γ_1 be the smallest non-zero isolated subgroup of Γ. Then Γ_1 is a finitely generated free $\mathbb{Z}$-module, and there exist a basis $\{\gamma_1, \ldots, \gamma_s\}$ of Γ, $h \in \{1, \ldots, s - 1\}$ and positive integers $m_1 \mid m_2 \mid \cdots \mid m_h$ such that $\{m_1\gamma_1, \ldots, m_h\gamma_h\}$ is a basis of Γ_1 [cf. [32], Ch. VII, §4, no. 3, Th. 1]. We may assume that $\gamma_1, \ldots, \gamma_h \in \Gamma_+$. Then we have $0 \prec \gamma_i \preccurlyeq m_i\gamma_i$ for $i \in \{1, \ldots, h\}$, hence $\gamma_1, \ldots, \gamma_h \in \Gamma_1$ [since Γ_1 is a segment] and $m_1 = \cdots = m_h = 1$. We set $\overline{\Gamma} := \Gamma/\Gamma_1$. We have $\operatorname{rat.\,rank}(\overline{\Gamma}) = s - h < \operatorname{rat.\,rank}(\Gamma)$, and $\overline{\Gamma}$ is a finitely generated free $\mathbb{Z}$-module having $\{\overline{\gamma}_{h+1}, \ldots, \overline{\gamma}_s\}$ as a basis [where, for $i \in \{h + 1, \ldots, s\}$, $\overline{\gamma}_i$ is the image of γ_i in $\overline{\Gamma}$]. Using the induction hypothesis, we see that $\overline{\Gamma}$ has finite rank, say $n - 1$, hence Γ has rank n, and if $\{0\} = \overline{\Gamma}_1 \subset \overline{\Gamma}_2 \subset \cdots \subset \overline{\Gamma}_{n-1}$ are the isolated subgroups of $\overline{\Gamma}_n := \overline{\Gamma}$, then $\overline{\Gamma}_2/\overline{\Gamma}_1, \ldots, \overline{\Gamma}_n/\overline{\Gamma}_{n-1}$ are finitely generated free $\mathbb{Z}$-modules, and we have $\overline{\Gamma} \cong \overline{\Gamma}_2/\overline{\Gamma}_1 \oplus \cdots \oplus \overline{\Gamma}_n/\overline{\Gamma}_{n-1}$. Let $\Gamma_2, \ldots, \Gamma_{n-1}$ be the isolated subgroups of Γ with $\Gamma_i/\Gamma_1 = \overline{\Gamma}_i$ for $i \in \{2, \ldots, n-1\}$ [cf. (1.8)]. Then

we have $\Gamma_i/\Gamma_{i-1} \cong (\Gamma_i/\Gamma_1)/(\Gamma_{i-1}/\Gamma_1) \cong \overline{\Gamma}_i/\overline{\Gamma}_{i-1}$ for $i \in \{2,\ldots,n\}$, and since $\Gamma \cong \Gamma_1 \oplus \overline{\Gamma}$, we get $\Gamma \cong \Gamma_1/\Gamma_0 \oplus \Gamma_2/\Gamma_1 \oplus \cdots \oplus \Gamma_n/\Gamma_{n-1}$.
If these conditions are satified, then the last assertion is evident.

(1.38) Proposition: *Let Γ be a totally ordered abelian group of finite rank n. Then:*
(1) Γ is isomorphic, as an ordered group, to a subgroup of the lexicographically ordered group $\mathbb{R}^n$.
(2) If, in addition, Γ is divisible, then there exist non-zero subgroups $\Gamma_1,\ldots,\Gamma_n$ of $\mathbb{R}$ such that Γ is isomorphic, as an ordered group, to $\Gamma_1 \oplus \cdots \oplus \Gamma_n$, ordered lexicographically.

Proof: Using (1.35) and (1.36), we see that it is enough to prove (2). We argue by induction on the rank of Γ. If $\mathrm{rank}(\Gamma) = 1$, then Γ is order-isomorphic to a subgroup of $\mathbb{R}$ [cf. (1.21)]. Now assume that $\mathrm{rank}(\Gamma) =: n > 1$, and that the assertion is true for totally ordered divisible abelian groups of smaller rank. Let Δ be the largest isolated subgroup of Γ. Then Δ and Γ/Δ are totally ordered and divisible [cf. (1.33)]. Since $\mathrm{rank}(\Delta) = n - 1$, by induction there exist subgroups $\Gamma_2,\ldots\Gamma_n$ of $\mathbb{R}$ such that Δ is isomorphic, as an ordered group, to $\Gamma_2 \oplus \cdots \oplus \Gamma_n$, ordered lexicographically. Now $\mathrm{rank}(\Gamma/\Delta) = 1$, hence Γ/Δ is order-isomorphic to a subgroup Γ_1 of $\mathbb{R}$. Let $\pi\colon \Gamma \to \Gamma/\Delta$ be the canonical homomorphism. Let $\overline{B}$ be a basis of the $\mathbb{Q}$-vector space Γ/Δ; for every $\overline{b} \in \overline{B}$ choose an element $b \in \Gamma$ with $\pi(b) = \overline{b}$, and let $B \subset \Gamma$ be the set of all these elements b. Let Γ_1' be the subspace of Γ generated by B. The restriction $\pi|\Gamma_1'\colon \Gamma_1' \to \Gamma/\Delta$ induces an isomorphism of ordered groups; we have $\Gamma = \Gamma_1' \oplus \Delta$, and it is easy to check that the ordering of Γ induces the lexicographic ordering on $\Gamma_1' \oplus \Delta$. Therefore Γ is isomorphic, as an ordered group, to $\Gamma_1 \oplus \cdots \oplus \Gamma_n$, ordered lexicographically.

(1.39) Proposition: *Let Γ be a totally ordered abelian group. The following statements are equivalent:*
(1) Γ is a group of finite rank which is discrete.
(2) Γ is a finitely generated free $\mathbb{Z}$-module, and we have $\mathrm{rank}(\Gamma) = \mathrm{rat.\,rank}(\Gamma)$.
If Γ satisfies these conditions, then Γ is order-isomorphic to $\mathbb{Z}^n$, ordered lexicographically.

Proof: (1) $\Rightarrow$ (2) follows from (1.26).
(2) $\Rightarrow$ (1): We argue by induction on $\mathrm{rank}(\Gamma)$. If $\mathrm{rank}(\Gamma) = \mathrm{rat.\,rank}(\Gamma) = 1$, then, on the one hand, Γ is isomorphic, as an ordered group, to a subgroup of $\mathbb{R}$ [cf. (1.21)], and, and the other hand, the groups Γ and $\mathbb{Z}$ are isomorphic, hence Γ is a cyclic group, hence Γ and $\mathbb{Z}$ are isomorphic as ordered groups [cf. (1.22)]. Let $n > 1$, and assume that the assertion is true for all totally ordered abelian groups having rank smaller then n. Let Γ be a totally ordered abelian group which is a finitely generated free $\mathbb{Z}$-module with $\mathrm{rank}(\Gamma) = \mathrm{rat.\,rank}(\Gamma) = n$. Let Δ be the largest isolated subgroup of Γ; then $\mathrm{rank}(\Delta) = n - 1$. We have [cf. (1.10)

and (1.30)]

$$\text{rat.}\,\text{rank}(\Gamma) = \text{rank}(\Gamma) = \text{rank}(\Delta) + \text{rank}(\Gamma/\Delta)$$
$$\leq \text{rat.}\,\text{rank}(\Delta) + \text{rat.}\,\text{rank}(\Gamma/\Delta) = \text{rat.}\,\text{rank}(\Gamma),$$

hence $\text{rank}(\Delta) = \text{rat.}\,\text{rank}(\Delta) = n-1$ and $\text{rank}(\Gamma/\Delta) = \text{rat.}\,\text{rank}(\Gamma/\Delta) = 1$. Just as in the proof of (1.37) there exists a basis $\{\gamma_1, \ldots, \gamma_n\}$ of the free $\mathbb{Z}$-module Γ such that $\Delta = \mathbb{Z}\gamma_2 \oplus \cdots \oplus \mathbb{Z}\gamma_n$. In particular, Δ is a finitely generated free $\mathbb{Z}$-module and therefore, by induction, it is isomorphic, as an ordered group, to $\mathbb{Z}^{n-1}$, ordered lexicographically. The factor group Γ/Δ is isomorphic, as an ordered group, to $\mathbb{Z}$. We have $\Gamma = \mathbb{Z}\gamma_1 \oplus \Delta$, and the ordering of Γ induces the lexicographic ordering on $\mathbb{Z}\gamma_1 \oplus \Delta$. Therefore Γ is isomorphic, as an ordered group, to $\mathbb{Z}^n$, ordered lexicographically.

2   Localization

(2.0) In this section we lay down the notations used with respect to localization, and prove some results which are needed later on.

(2.1) (1) Let A be a ring, and let S be a multiplicatively closed subset of A [i.e., we have $1 \in S$, and if $s, t \in S$, then $st \in S$]. The ring of fractions of A with respect to S is denoted by $S^{-1}A$, and $i_A^S \colon A \to S^{-1}A$ with $i_A^S(a) = a/1$ for every $a \in A$ is the canonical homomorphism of rings. Let M be an A-module; the A-module of fractions of M with respect to S is denoted by $S^{-1}M$, and $i_M^S \colon M \to S^{-1}M$ with $i_M^S(m) = m/1$ for every $m \in M$ is the canonical map. $S^{-1}M$ has a structure as an $S^{-1}A$-module. Properties of the process of forming fractions are listed in [21], Ch. 3, [34], Ch. II, § 2 and [63], Ch. 2.
(2) Let $f \in A$ and $S := \{f^n \mid n \in \mathbb{N}_0\}$; we write A_f and M_f instead of $S^{-1}A$ and $S^{-1}M$. The ring homomorphism $A \to A[T] \to A[T]/(1-fT)$ induces an A-algebra homomorphism $A_f \to A[T]/(1-fT)$ [cf. [21], Prop. 3.1 or [63], Ex. 2.2]; it is immediate that this is an isomorphism. In particular, A_f is an A-algebra of finite type.
(3) Let $\mathfrak{p} \in \text{Spec}(A)$; $S := A \setminus \mathfrak{p}$ is multiplicatively closed, and we write $A_\mathfrak{p}$ and $M_\mathfrak{p}$ and $i_A^\mathfrak{p}$ and $i_M^\mathfrak{p}$ instead of $S^{-1}A$ and $S^{-1}M$ and i_A^S and i_M^S.
(4) The support $\text{Supp}_A(M) = \text{Supp}(M) \subset \text{Spec}(A)$ of M is the set of prime ideals $\mathfrak{p}$ of A such that $M_\mathfrak{p} \neq \{0\}$, and $\text{Ann}_A(M) = \text{Ann}(M)$ is the annihilator of M; if M is a finitely generated A-module, then we have $\text{Supp}(M) = V(\text{Ann}(M))$ [cf. [63], Cor. 2.2].

(2.2) Let A be a ring, let S be a multiplicatively closed subset of A, and let M be an A-module.
(1) A submodule N of M is called saturated (with respect to S) if $sx \in N$ for some $s \in S$ and $x \in M$ implies that $x \in N$. If N is a submodule of M, then the

submodule $(i_M^S)^{-1}(S^{-1}N)$ is the smallest saturated submodule of M containing N; it is called the saturation of N.

(2) The map $N \mapsto S^{-1}N$ is a bijective inclusion-preserving map from the set of saturated submodules of M to the set of submodules of the $S^{-1}A$-module $S^{-1}M$; the inverse map is the map $N' \mapsto (i_M^S)^{-1}(N')$ [cf. [34], Ch. II, §2, no. 4, Prop. 10].

(2.3) (1) Let $\varphi \colon A \to B$ be a homomorphism of rings, let $S \subset A$ and $T \subset B$ be multiplicatively closed, and assume that $\varphi(S) \subset T$. Then there exists a unique homomorphism of rings $\varphi_{A,B}^{S,T} \colon S^{-1}A \to T^{-1}B$ such that $i_B^T \circ \varphi = \varphi_{A,B}^{S,T} \circ i_A^S$.

(2) Let A be a ring, let S be a multiplicatively closed subset of A, let $\mathfrak{a}$ be an ideal of A, and let $\mathfrak{a}' := S^{-1}\mathfrak{a} = \mathfrak{a}(S^{-1}A)$ be the extended ideal. Let $\varphi \colon A \to A/\mathfrak{a}$ be the natural homomorphism. Then $\varphi_{A,\mathfrak{a}A}^{S,\varphi(S)} \colon S^{-1}A \to (\varphi(S))^{-1}(A/\mathfrak{a})$ is surjective and has kernel $\mathfrak{a}'$, hence the rings $(\varphi(S))^{-1}(A/\mathfrak{a})$ and $(S^{-1}A)/\mathfrak{a}'$ are canonically isomorphic. The proof of this result is straightforward.

(3) Let A be a ring and let S be a multiplicatively closed subset of A. Assume that $B := S^{-1}A$ is quasilocal with maximal ideal $\mathfrak{n}$, and set $\mathfrak{m} := (i_A^S)^{-1}(\mathfrak{n})$; then $\mathfrak{m}$ is a prime ideal of A, and we have $A_{\mathfrak{m}} = B$, as one easily checks.

(2.4) Let $\varphi \colon A \to B$ be a homomorphism of rings and let M be a finitely generated B-module. Assume that $M_{\mathfrak{n}}/\varphi^{-1}(\mathfrak{n})M_{\mathfrak{n}} = \{0\}$ for every maximal ideal $\mathfrak{n}$ of B. Then we have $M = \{0\}$.

Indeed, let $\mathfrak{n}$ be a maximal ideal of B. By assumption, we have $M_{\mathfrak{n}} = \varphi^{-1}(\mathfrak{n})M_{\mathfrak{n}}$; since $\varphi^{-1}(\mathfrak{n})B_{\mathfrak{n}} \subset \mathfrak{n}B_{\mathfrak{n}}$, we see that $M_{\mathfrak{n}} = \{0\}$ by Nakayama's lemma [cf. [63], Cor. 4.8], hence that $M = \{0\}$ by [63], Lemma 2.8.

(2.5) Proposition: *Let A be a domain, let K be its field of quotients, and let L be a field extension of K. Let $\mathrm{Max}(A)$ be the set of maximal ideals of A. Then:*
(1) *Let M be an A-submodule of L. The module of fractions $M_{\mathfrak{m}}$ is an $A_{\mathfrak{m}}$-submodule of L for every $\mathfrak{m} \in \mathrm{Max}(A)$, and we have $M = \bigcap_{\mathfrak{m} \in \mathrm{Max}(A)} M_{\mathfrak{m}}$.*
(2) *Let M and N be A-submodules of L with $M \subset N$. If we have $M_{\mathfrak{m}} = N_{\mathfrak{m}}$ for every $\mathfrak{m} \in \mathrm{Max}(A)$, then we have $M = N$.*

Proof: (1) Let M be an A-submodule of L. Clearly we have $M \subset M_{\mathfrak{m}}$ for every $\mathfrak{m} \in \mathrm{Max}(A)$. Let $x \in L$ be such that $x \in M_{\mathfrak{m}}$ for every $\mathfrak{m} \in \mathrm{Max}(A)$. For each $\mathfrak{m} \in \mathrm{Max}(A)$ there exists $s_{\mathfrak{m}} \in A \setminus \mathfrak{m}$ such that $s_{\mathfrak{m}}x \in M$. Let $\mathfrak{a}$ be the ideal generated by the elements $s_{\mathfrak{m}}$, $\mathfrak{m} \in \mathrm{Max}(A)$. Then $\mathfrak{a} = A$ [otherwise, there would exist a maximal ideal $\mathfrak{n}$ of A such that $\mathfrak{a} \subset \mathfrak{n}$; this contradicts $s_{\mathfrak{n}} \in \mathfrak{a}$, $s_{\mathfrak{n}} \notin \mathfrak{n}$]. Therefore, there exist $h \in \mathbb{N}$ and $a_1, \ldots, a_h \in A$ such that $1 = a_1 s_{\mathfrak{m}_1} + \cdots + a_h s_{\mathfrak{m}_h}$, hence we have $x = a_1(xs_{\mathfrak{m}_1}) + \cdots + a_h(xs_{\mathfrak{m}_h}) \in M$.
(2) This follows from (1).

(2.6) Proposition: *Let A be a ring, let M be an A-module, and let $f_1, \ldots, f_h \in A$ be elements which generate the unit ideal. Assume that, for $i \in \{1, \ldots, h\}$, M_{f_i} is a finitely generated A_{f_i}-module. Then M is a finitely generated A-module.*

Proof: We choose elements $m_1, \ldots, m_p$ such that $\{m_1/1, \ldots, m_p/1\}$ is a system of generators of the A_{f_i}-module M_{f_i} for $i \in \{1, \ldots, h\}$, and set $N := Am_1 + \cdots + Am_p$. By [63], Ex. 2.19, we get $M = N$.

3   Integral Extensions

(3.1) REMARK: The reader should be familiar with the notion of integral extension. We add some results which are used later on.
need the following two results for the proof of (3.4) below.

(3.2) Proposition: Let A be a ring, let $B_1, \ldots, B_n$ be A-algebras, and define $B := B_1 \times \cdots \times B_n$.
(1) An element $b = (b_1, \ldots, b_n) \in B$ is integral over A iff b_i is integral over A for every $i \in \{1, \ldots, n\}$.
(2) The ring B is integral over A iff the rings $B_1, \ldots, B_n$ are integral over A.

Proof: (1) Obviously the condition is necessary. Conversely, assume that $b_1, \ldots, b_n$ are integral over A. For every $i \in \{1, \ldots, n\}$ the A-subalgebra $A[b_i]$ of B_i is a finitely generated A-module, hence the A-subalgebra $\prod_{i=1}^{n} A[b_i]$ of B is a finitely generated A-module; since $A[b]$ is contained in this subalgebra, b is integral over A [cf. [63], Cor. 4.6].
(2) This follows immediately from (1).

(3.3) Corollary: Let A be a ring, let $B_1, \ldots, B_n$ be A-algebras, and define $B := B_1 \times \cdots \times B_n$. For every $i \in \{1, \ldots, n\}$ let C_i be the integral closure of A in B_i. Then $C_1 \times \cdots \times C_n$ is the integral closure of A in B.

Proof: This follows immediately from (3.2).

(3.4) Corollary: Let A be a reduced noetherian ring, and let $\mathfrak{p}_1, \ldots, \mathfrak{p}_n$ be the minimal prime ideals of A. Then, for every $i \in \{1, \ldots, n\}$, $Q(A)/\mathfrak{p}_i Q(A)$ is the field of quotients of $A/\mathfrak{p}_i$, and if B_i is the integral closure of $A/\mathfrak{p}_i$ in $Q(A)/\mathfrak{p}_i Q(A)$, then, under the canonical isomorphism $Q(A) \to Q(A)/\mathfrak{p}_1 Q(A) \times \cdots \times Q(A)/\mathfrak{p}_n Q(A)$, the integral closure of A in $Q(A)$ is mapped onto $B_1 \times \cdots \times B_n$.
In particular, if, for every $i \in \{1, \ldots, n\}$, B_i is a finitely generated $A/\mathfrak{p}_i$-module, then the integral closure of A in $Q(A)$ is a finitely generated A-module.

Proof: We have $\mathrm{Reg}(A) = A \setminus (\mathfrak{p}_1 \cup \cdots \cup \mathfrak{p}_n)$ [cf. [63], Th. 3.1 and Ex. 11.10]; set $\mathfrak{P}_i := \mathrm{Reg}(A)^{-1} \mathfrak{p}_i$ for $i \in \{1, \ldots, n\}$. Now $\{0\} = \mathfrak{p}_1 \cap \cdots \cap \mathfrak{p}_n$ is an irredundant primary decomposition in A, hence $\{0\} = \mathfrak{P}_1 \cap \cdots \cap \mathfrak{P}_n$ is an irredundant primary decomposition in $Q(A)$ [cf. [63], Th. 3.10], and therefore $\mathrm{Ass}(Q(A)) = \{\mathfrak{P}_1, \ldots, \mathfrak{P}_n\}$. Since every regular element of $Q(A)$ is invertible, it is clear that $\{\mathfrak{P}_1, \ldots, \mathfrak{P}_n\}$ is the set of maximal ideals of $Q(A)$, and $Q(A) \to Q(A)/\mathfrak{P}_1 \times \cdots \times Q(A)/\mathfrak{P}_n$ is an isomorphism [cf. (10.1)]. Let $i \in \{1, \ldots, n\}$; since $Q(A)/\mathfrak{P}_i = (\mathrm{Reg}(A))^{-1}(A/\mathfrak{p}_i)$

[cf. (2.3)(2)], we have $Q(A/\mathfrak{p}_i) = Q(A)/\mathfrak{P}_i$. The remaining assertions now follow from (3.3).

(3.5) Corollary: *Let A be a reduced local ring which is integrally closed in its ring of quotients. Then A is a domain.*

Proof: Using the notation of (3.4) we have $A \cong B_1 \times \cdots \times B_n$, hence we have $n = 1$ [since A is local] and $\mathfrak{p}_1 = \{0\}$.

(3.6) Proposition: *Let k be a field, and let A be an integral k-algebra of finite type (resp. essentially of finite type). Let K be the field of quotients of A, and let L be a finite extension of K. The integral closure of A in L is a finitely generated A-module, hence of finite type (resp. essentially of finite type) over k.*

Proof: We assume that A is of finite type over k; then the integral closure of A in L is a finitely generated A-module [cf. [63], Th. 4.14]. Now we assume that A is essentially of finite type over k. We have $A = \Sigma^{-1}R$ where R is a k-subalgebra of K of finite type and Σ is a multiplicative system in R. The integral closure S of R is a finitely generated R-module, and $\Sigma^{-1}S$ is the integral closure of A in L by [63], Prop. 4.13.

(3.7) Proposition: *Let k be a field, and let A be a reduced k-algebra of finite type. For every prime ideal $\mathfrak{q}$ of A the integral closure of $A_\mathfrak{q}$ in its ring of quotients is a finitely generated $A_\mathfrak{q}$-module*

Proof: Let $\{0\} = \mathfrak{p}_1 \cap \cdots \cap \mathfrak{p}_n$ be the irredundant primary decomposition of the ideal $\{0\}$ as an intersection of prime ideals; note that $\{\mathfrak{p}_1, \ldots, \mathfrak{p}_n\}$ is the set if minimal prime ideals of A. Let $i \in \{1, \ldots, n\}$; the integral closure of $A/\mathfrak{p}_i$ is a finitely generated $A/\mathfrak{p}_i$-module by (3.6), hence, for $\mathfrak{q} \in V(\mathfrak{p}_i)$, the integral closure of $(A/\mathfrak{p}_i)_\mathfrak{q}$ is a finitely generated $(A/\mathfrak{p}_i)_\mathfrak{q}$-module [cf. [63], Prop. 4.13].
Let $\mathfrak{q}$ be a prime ideal of A, and let the prime ideals $\mathfrak{p}_1, \ldots, \mathfrak{p}_n$ be labelled in such a way that $\mathfrak{q}$ contains exactly the prime ideals $\mathfrak{p}_1, \ldots, \mathfrak{p}_{n'}$. The minimal prime ideals of the reduced ring $A_\mathfrak{q}$ are those generated by the images of $\mathfrak{p}_1, \ldots, \mathfrak{p}_{n'}$. The integral closure of $A_\mathfrak{q}/\mathfrak{p}_i A_\mathfrak{q} = (A/\mathfrak{p}_i)_{\mathfrak{q}/\mathfrak{p}_i}$, $i \in \{1, \ldots, n'\}$, is a finitely generated module over this ring. The assertion now follows from (3.4).

(3.8) Lemma: *Let $\varphi \colon A \to B$ be a finite homomorphism, and let $\mathfrak{m}$ be a maximal ideal of A. Every ideal $\mathfrak{n}$ of B such that $\varphi^{-1}(\mathfrak{n}) = \mathfrak{m}$ is a maximal ideal, the set of such ideals is finite, and if it is not empty, then a power of their intersection lies in $\mathfrak{m}B$. In particular, if A is quasisemilocal, then B is quasisemilocal.*

Proof: We set $\mathfrak{a} := \ker(\varphi)$; we may assume that $\mathfrak{a}$ lies in $\mathfrak{m}$ [since otherwise there exist no ideals $\mathfrak{n}$ of B with $\varphi^{-1}(\mathfrak{n}) = \mathfrak{m}$]. We set $\overline{A} := A/\mathfrak{a}$ and $\overline{\mathfrak{m}} := \mathfrak{m}/\mathfrak{a}$. The ideal $\overline{\mathfrak{m}}$ is a maximal ideal of $\overline{A}$, the ring $\overline{A}$ is a subring of B, and B is integral over $\overline{A}$ [cf. [63], Cor. 4.5]. A proper ideal $\mathfrak{n}$ of B is a maximal ideal of B iff we have $\mathfrak{n} \supset \overline{\mathfrak{m}}B$ [cf. [63], Cor. 4.17]. Now the residue class ring $B/\overline{\mathfrak{m}}B$ is a finite-dimensional $A/\mathfrak{m}$-vector space, hence it is an artinian ring, and therefore $B/\overline{\mathfrak{m}}B$ has only finitely many prime ideals, each of them is a maximal ideal, and their intersection is a nilpotent ideal [cf. [63], Ex. 1.13, Cor. 2.12 and Th. 2.14]. The claim follows immediately.

4 Some Results on Graded Rings and Modules

4.1 Generalities

(4.0) Let Γ be a totally ordered abelian group, let $R = \bigoplus_{\gamma \in \Gamma} R_\gamma$ be a Γ-graded ring and let $M = \bigoplus_{\gamma \in \Gamma} M_\gamma$ be a Γ-graded R-module.
(1) If Δ is a subsemigroup of Γ, and if $R_\gamma = \{0\}$ for every $\gamma \in \Gamma \setminus \Delta$, then $R = \bigoplus_{\delta \in \Delta} R_\delta$ is called a Γ-graded ring of type Δ.
(2) If N is a homogeneous submodule of M, i.e., N is generated by homogeneous elements, then M/N can be considered as a graded module by defining $(M/N)_\gamma = M_\gamma/N_\gamma$, and the canonical homomorphism $\varphi \colon M \to N$ maps M_γ to $(M/N)_\gamma$. If, in particular, we have $M = R$ and $N = \mathfrak{a}$, then the ring $R/\mathfrak{a}$ is a graded ring.
(3) If R is noetherian, then we refer to [63], Ex. 3.5 for results on primary decomposition in this case. The following result is useful.

(4.1) Proposition: *Let $R = \bigoplus_{\gamma \in \Gamma}$ be a Γ-graded ring. Then:*
(1) Let $\mathfrak{p} \neq R$ be a homogeneous ideal of R having the following property: If x, $y \in R$ are homogeneous elements of R such that $x \notin \mathfrak{p}$ and $y \notin \mathfrak{p}$, then $xy \notin \mathfrak{p}$. Then $\mathfrak{p}$ is a prime ideal.
(2) Let $\mathfrak{q} \neq R$ be a homogeneous ideal of R having the following property: If x, $y \in R$ are homogeneous elements of R such that $xy \in \mathfrak{q}$ and $x \notin \mathfrak{q}$, then $y \in \mathrm{rad}(\mathfrak{q})$. Then $\mathfrak{q}$ is a primary ideal.

Proof: Let x, $y \in R$, and write $x = x_{\gamma_1} + \cdots + x_{\gamma_h}$, $y = y_{\delta_1} + \cdots + y_{\delta_k}$ where $\gamma_1 \prec \cdots \prec \gamma_h$, $\delta_1 \prec \cdots \prec \delta_k$ and $x_{\gamma_i} \in R_{\gamma_i}$, $y_{\delta_j} \in R_{\delta_j}$ for every $i \in \{1, \ldots, h\}$, $j \in \{1, \ldots, k\}$.
(1) Assume that x, $y \notin \mathfrak{p}$. Choose $p \in \{1, \ldots, h\}$, $q \in \{1, \ldots, k\}$ such that x_{γ_p} is the "first" component of x which does not lie in $\mathfrak{p}$ and y_{δ_q} is the "first" component of y which does not lie in $\mathfrak{p}$. Then $x_{\gamma_p} y_{\delta_q} \notin \mathfrak{p}$, and therefore

$$z := (x - (x_{\gamma_1} + \cdots + x_{\gamma_{p-1}}))((y - (y_{\delta_1} + \cdots + y_{\delta_{q-1}})) \notin \mathfrak{p}$$

[since the homogeneous component of degree $\gamma_p + \delta_q$ of z is $x_{\gamma_p} y_{\delta_q}$, and $\mathfrak{p}$ is homogeneous]. On the other hand, it is clear that $z - xy \in \mathfrak{p}$, hence that $xy \notin \mathfrak{p}$.
(2) Assume that $xy \in \mathfrak{q}$ and $x \notin \mathfrak{q}$. Choose $p \in \{1, \ldots, h\}$ such that x_{γ_p} is the first component of x which does not lie in $\mathfrak{q}$. Then $x_{\gamma_p} + \cdots + x_{\gamma_h}$ lies not in $\mathfrak{q}$ but $z := (x_{\gamma_p} + \cdots + x_{\gamma_h})y$ lies in $\mathfrak{q}$. The homogeneous component of z of degree $\gamma_p + \delta_1$ is $x_{\gamma_p} y_{\delta_1}$, hence $x_{\gamma_p} y_{\delta_1} \in \mathfrak{q}$, and therefore $y_{\delta_1} \in \mathrm{rad}(\mathfrak{q})$ by assumption. Assume that

it has already been shown that $y_{\delta_1}, \ldots, y_{\delta_j} \in \text{rad}(\mathfrak{q})$ for some $j \in \{1, \ldots, k-1\}$. Choose $m \in \mathbb{N}$ such that $(y_{\delta_1} + \cdots + y_{\delta_j})^m \in \mathfrak{q}$. Then we have

$$(x_{\gamma_p} + \cdots + x_{\gamma_h})(y - (y_{\delta_1} + \cdots + y_{\delta_j}))^m \in \mathfrak{q}$$

since $(x_{\gamma_p} + \cdots + x_{\gamma_h})y \in \mathfrak{q}$, hence we get $y_{\delta_{j+1}}^m \in \text{rad}(\mathfrak{q})$ since $x_{\gamma_p} \notin \mathfrak{q}$, hence we have $y_{\delta_{j+1}} \in \text{rad}(\mathfrak{q})$.

4.2 M-Graded Rings and M-Graded Modules

(4.2) NOTATION: Let M be a free $\mathbb{Z}$-module of rank r. Then M is isomorphic to the group $\mathbb{Z}^r$; this group can be considered as a totally ordered group using the lexicographic ordering, hence M can also be considered as a totally ordered group. (It is well known that every torsion-free abelian group Δ can be endowed with a total ordering $\prec$ such that $(\Delta, \prec)$ is a totally ordered group, cf. [156], Ch. 2, Th. 22.) In chapter VI we shall use M-graded rings; therefore we collect in this subsection some results on such rings. The notation M will be used throughout.

(4.3) DEFINITION: An M-graded ring $R = \bigoplus_{m \in M} R_m$ is called simple if every non-zero homogeneous element of R is a unit [equivalently, if R has no proper homogeneous ideals except $\{0\}$].

(4.4) EXAMPLE: Let k be a field and L a subgroup of M; let $k[L]$ be the group ring of L over k [cf. chapter VI, section 2, for unexplained notation; note, in particular, that loc. cit. there is a definition of $k[L]$; if s is the rank of L, then the k-algebra $k[L]$ is isomorphic to the ring of Laurent polynomials $k[T_1, T_1^{-1}, \ldots, T_s, T_s^{-1}]]$. Then $k[L]$ is an M-graded ring of type L, and for every $m \in M$ we have

$$k[L]_m = \begin{cases} ke(m) & \text{if } m \in L, \\ \{0\} & \text{if } m \notin L. \end{cases}$$

Clearly $k[L]$ is a simple M-graded ring: if $c \in k^\times$ and $m \in L$, then the inverse of $ce(m) \neq 0$ is $c^{-1}e(-m)$.

(4.5) NOTATION: Let R be an M-graded ring, and set

$$L(R) := \{m \in M \mid R_m \text{ contains a unit of } R\}.$$

It is clear that $L(R)$ is a subgroup of M. Note that, in case R is a simple M-graded ring, we have $L(R) = \{m \in M \mid R_m \neq \{0\}\}$.

(4.6) REMARK: Let R be an M-graded ring.
(1) Let $m \in L(R)$ and let $u \in R_m$ be a unit of R. Then $u^{-1} \in R_{-m}$ and $R_m = R_0 u$. In fact, let $x \in R_m$; then we have $xu^{-1} \in R_0$ and therefore $x \in R_0 u$. Thus, we have $R_m \subset R_0 u \subset R_m$, hence $R_m = R_0 u$.

(2) Assume that $R \neq \{0\}$ is a simple M-graded ring. Then R_0 is a field, and $\dim_{R_0}(R_m) = 1$ for every $m \in L(R)$.
In fact, since $R_0 \neq \{0\}$ and every element of R_0 is invertible in R, hence invertible in R_0, R_0 is a field. The last claim follows from (1).

(4.7) Proposition: *Let $R \neq \{0\}$ be an M-graded ring, and set $L := L(R)$. The following statements are equivalent:*
(1) *R is a simple M-graded ring.*
(2) *R_0 is a field and the M-graded R_0-algebras R and $R_0[L]$ are isomorphic.*
(3) *Every M-graded R-module is free.*

Proof (1) $\Rightarrow$ (2): R_0 is a field by (4.6)(2). Let $U(R)$ be the group of invertible elements of R. We construct a homomorphism $\varphi \colon L \to U(R)$ of groups such that $\varphi(m) \in R_m$ for every $m \in L$. Namely, let $\{m_1, \ldots, m_h\}$ be a $\mathbb{Z}$-basis of L, and define φ by $\varphi(m_i) = e(m_i)$ for every $i \in \{1, \ldots, h\}$. Since $e(m_i) \in R_{m_i}$ is invertible, φ has the desired properties. Using the universal property of the group ring $R_0[L]$ [cf. VI(2.1)(3)], φ can be extended to a homomorphism $\tilde{\varphi} \colon R_0[L] \to R$ of M-graded R_0-algebras. Since we have $\dim_{R_0}(R_m) = 1$ for every $m \in L$ [cf. (4.6)(2)], it is clear that $\tilde{\varphi}$ is an isomorphism.
(2) $\Rightarrow$ (3): Obviously R can be considered as an L-graded ring [$R_m = \{0\}$ for every $m \in M \setminus L$]. Let $\{m_i \mid i \in I\}$ be a system of representatives for the classes of M/L; then we have $M = \bigcup_{i \in I}(m_i + L)$, a disjoint union.
Let P be an M-graded R-module. Let $i \in I$, and define $P_i := \bigoplus_{m \in L} P_{m+m_i}$. Now P_i is an L-graded abelian group [by definition we have $(P_i)_m := P_{m+m_i}$ for every $m \in L$], and for m, $m' \in L$ we have $R_m \cdot (P_i)_{m'} = R_m P_{m'+m_i} \subset P_{m+m'+m_i} = (P_i)_{m+m'}$, hence P_i is an L-graded R-module. Since $P = \bigoplus_{i \in I} P_i$ as R-modules, in order to show that P is a free M-graded R-module, it is enough to consider only the case that $M = L$.
Thus, let $\{x_j \mid j \in J\}$ be an R_0-basis of P_0 [note that R_0 is a field], let $m \in M$ and $x \in P_m$. Choose a non-zero homogeneous element $u \in R_m$; then u is a unit of R, and $u^{-1}x \in P_0$. This implies that $\{x_j \mid j \in J\}$ is a system of generators of the R-module P, and it is obviously an R-free system.
(3) $\Rightarrow$ (1): Let $\mathfrak{a}$ be a proper homogeneous ideal of R. The $R/\mathfrak{a}$ is a non-zero free R-module, hence $\operatorname{Ann}_R(R/\mathfrak{a}) = \{0\}$, and therefore we have $\mathfrak{a} = \{0\}$.

(4.8) DEFINITION: Let R be an M-graded ring, and let $\mathfrak{m}$ be a proper homogeneous ideal of R; $\mathfrak{m}$ is called a maximal homogeneous ideal if $R/\mathfrak{m}$ is a simple M-graded ring, equivalently, if $\mathfrak{m}$ is a maximal element in the set of homogeneous ideals of R different from R. Clearly such an ideal is a prime ideal.

(4.9) NOTATION: An M-graded ring R is called a quasilocal M-graded ring if R has a unique maximal homogeneous ideal. If, in addition, R is noetherian, then R is called a local M-graded ring.

(4.10) Proposition: *Let R be a quasilocal M-graded ring, and let $\mathfrak{m}$ be its maximal homogeneous ideal. Then R_0 is a quasilocal ring and $\mathfrak{m} \cap R_0$ is the maximal ideal of R_0.*

Proof: $\mathfrak{m} \cap R_0$ is a maximal ideal of R_0 since $R_0/(\mathfrak{m} \cap R_0)$ is a field by (4.7). Let $x \in R_0 \setminus (\mathfrak{m} \cap R_0)$, and suppose that x is not a unit of R_0. Then x is not a unit of R, hence Rx is a proper homogeneous ideal of R, and therefore [since Rx is contained in a maximal homogeneous ideal of R] $Rx \subset \mathfrak{m}$—contradicting the choice of x.

(4.11) Proposition: [Nakayama's lemma] *Let R be a quasilocal M-graded ring, let $\mathfrak{m}$ be its maximal homogeneous ideal, and let P be a finitely generated M-graded R-module. Then:*
(1) $P/\mathfrak{m}P$ is a free $R/\mathfrak{m}$-module.
(2) If $P = \mathfrak{m}P$, then $P = \{0\}$.
(3) Let $\{x_1, \ldots, x_h\}$ be a homogeneous system of elements of P. Then $\{x_1, \ldots, x_h\}$ is an irredundant system of generators of P iff the images of these elements in $P/\mathfrak{m}P$ are a basis of the free $R/\mathfrak{m}$-module $P/\mathfrak{m}P$.

Proof: (1) follows immediately from (4.7).
(2) Let $\{x_1, \ldots, x_h\}$ be a homogeneous system of generators of P. Then we have $x_i = \sum_{j=1}^h z_{ij}x_j$ for every $i \in \{1, \ldots, h\}$ where the elements z_{ij}, $i, j \in \{1, \ldots, h\}$, are homogeneous elements of $\mathfrak{m}$. Set $d := \det(\delta_{ij} - z_{ij})$, and let $i \in \{1, \ldots, h\}$. The usual proof of Nakayama's lemma [cf. e.g., [21], Prop. 2.4 or look at the proof of (6.1) below] yields $dx_i = 0$. Now $d = 1 + z$ with $z \in \mathfrak{m}$. Write $z = \sum_{m \in M} z_m$ as a sum of homogeneous elements. Then we get $z_m x_i = 0$ for every $m \in M$, $m \neq 0$, and $(1 + z_0)x_i = 0$. Since $z_0 \in \mathfrak{m}$ and R_0 is quasilocal with maximal ideal $\mathfrak{m} \cap R_0$ [cf. (4.10)], the element $1 + z_0$ is a unit of R_0, hence we have $x_i = 0$.
(3) This follows immediately from (1) and (2).

(4.12) NOTATION: Under the assumptions of the proposition, it makes sense to call a finite irredundant system of homogeneous generators of P a minimal system of generators.

4.3 Homogeneous Localization

(4.13) HOMOGENEOUS LOCALIZATION: Let $R = \bigoplus_{\gamma \in \Gamma} R_\gamma$ be a Γ-graded ring, and let Σ be a multiplicatively closed set in R consisting of homogeneous elements.
(1) Consider the ring $T := \Sigma^{-1}R$. An element of T is called homogeneous if it can be written as x/f where $x \in R$ is homogeneous and $f \in \Sigma$; if $x/f \neq 0$, then we define $\deg(x/f) = \deg(x) - \deg(f)$. It is easily seen that the degree of a homogeneous non-zero element of T is well-defined; the zero element of T is, by definition, homogeneous of any degree.
For every $\gamma \in \Gamma$ let T_γ be the set of homogeneous elements of T of degree γ. Then we have $T = \sum_{\gamma \in \Gamma} T_\gamma$, and the sum is direct. In fact, let $z_1/f_1, \ldots, z_h/f_h$

be elements of T which are homogeneous of degree $\gamma_1, \ldots, \gamma_h$ respectively where $\gamma_1, \ldots, \gamma_h \in \Gamma$ are pairwise different, and assume that $z_1/f_1 + \cdots + z_h/f_h = 0$. Set $f := f_1 \cdots f_h$ and $y_i := f_1 \cdots f_{i-1} f_{i+1} \cdots f_h z_i$ for every $i \in \{1, \ldots, h\}$; then we have $(y_1 + \cdots + y_h)/f = 0$, hence there exists $g \in \Sigma$ such that $g(y_1 + \cdots + y_h) = 0$. Since $\deg(gy_i) = \gamma_i + \deg(f) + \deg(g)$ for every $i \in \{1, \ldots, h\}$, it follows that $gy_1 = \cdots = gy_h = 0$, hence that $z_1/f_1 = \cdots = z_h/f_h = 0$. Therefore the ring $\Sigma^{-1}R$ can be considered as a Γ-graded ring. The subring $(\Sigma^{-1}R)_0$ of $\Sigma^{-1}R$ will be denoted by $R_{(\Sigma)}$.

(2) Let $f \in R$ be a homogeneous element, and set $\Sigma := \{f^n \mid n \in \mathbb{N}_0\}$. Then the ring $(\Sigma^{-1}R)_0$ will be denoted by $R_{(f)}$.

(3) Let $\mathfrak{p}$ be a homogeneous prime ideal of R, and let Σ be the set of homogeneous elements of R which are not contained in $\mathfrak{p}$. Then $\Sigma^{-1}R$ will be denoted by $R_{(\mathfrak{p})}$ and $(\Sigma^{-1}R)_0$ will be denoted by $R_{(\mathfrak{p})}$.

Let $f \in \Sigma$; then $\mathfrak{p}R_f$ is a prime ideal in R_f, and its intersection with $R_{(f)}$ is a prime ideal $\mathfrak{q}$ of $R_{(f)}$. Let Σ' be the set of homogeneous elements of R_f which are not contained in $\mathfrak{p}R_f$; it is clear that the map $\Sigma'^{-1}R_f \to \Sigma^{-1}R$, defined by mapping $(x/f^h)/(y/f^k)$ to $xf^k/(yf^h)$ where h, $k \in \mathbb{N}_0$, $x \in R$, $y \in \Sigma$, defines an isomorphism of graded rings.

(4.14) REMARK: Let $S = \bigoplus_{n \in \mathbb{Z}} S_n$ be a $\mathbb{Z}$-graded ring, and let $\Sigma \subset S$ be a multiplicatively closed set consisting of homogeneous elements. Then $\Sigma^{-1}S$ is a $\mathbb{Z}$-graded ring [cf. (4.13)].

(1) Let $f \in \Sigma$ be homogeneous of degree m. Then, for every $n \in \mathbb{Z}$, $x \mapsto (f/1)x : (\Sigma^{-1}S)_n \to (\Sigma^{-1}S)_{m+n}$ is an isomorphism of $S_{(\Sigma)}$-modules [the inverse map is given by $x \mapsto (1/f)x$].

(2) Let $f \in \Sigma$ be homogeneous of degree 1. The $S_{(\Sigma)}$-algebra homomorphism $S_{(\Sigma)}[T] \to \Sigma^{-1}S$, defined by mapping T to f, extends to a homomorphism of $\mathbb{Z}$-graded rings $S_{(\Sigma)}[T, T^{-1}] = \bigoplus_{n \in \mathbb{Z}} S_{(\Sigma)} T^n \to \Sigma^{-1}S$ which maps T^{-1} to $1/f$; this map is an isomorphism by (1). (We use the convention that if $S_{(\Sigma)}$ is the null ring, then $S_{(\Sigma)}[T, T^{-1}]$ is also the null ring.)

(3) For every $d \in \mathbb{N}$ we define

$$S^{(d)} := \bigoplus_{n \in \mathbb{Z}} S_{dn}.$$

Then $S^{(d)}$ is a $\mathbb{Z}$-graded ring, $\Sigma \cap S^{(d)}$ is a multiplicatively closed set of homogeneous elements of $S^{(d)}$, and it is easy to see that $(\Sigma \cap S^{(d)})^{-1} S^{(d)} = (\Sigma^{-1}S)^{(d)}$.

(4.15) For the rest of this subsection let $S = \bigoplus_{n \geq 0} S_n$ be a $\mathbb{Z}$-graded ring of type $\mathbb{N}_0$.

(4.16) Proposition: Let $d \in \mathbb{N}$. Then, for $f \in S_d$, there exists a canonical isomorphism of rings $S_{(f)} \to S^{(d)}/(f - 1)S^{(d)}$.

Proof: Let $x \in S_{nd}$ and $h \in \mathbb{N}$; then $f^h x \equiv x \pmod{(f-1)S^{(d)}}$ [as follows from $f^h - 1 = (f-1)(f^{h-1} + \cdots + 1)$], hence we have $x \equiv 0 \pmod{(f-1)S^{(d)}}$ if $f^h x = 0$. Now the map $x/f^n \mapsto x \pmod{(f-1)S^{(d)}}$: $S_{(f)} \to S^{(d)}/(f-1)S^{(d)}$ [where $x \in S_{nd}$] is well-defined.

On the other hand, let $x \in S_{nd}$ and $x = (f-1)y$ where $y = y_{hd} + y_{(h+1)d} + \cdots + y_{kd}$ with $y_{jd} \in S_{jd}$ for $j \in \{h, h+1, \ldots, k\}$ and $y_{hd} \neq 0$. Then we have $h = n$, $y_{hd} = -x$, $y_{(j+1)d} = f y_{jd}$ for $j \in \{h, \ldots, k-1\}$ and $f y_{kd} = 0$, hence $f^{k-h+1} x = 0$. Now the map $x \pmod{(f-1)S^{(d)}} \mapsto x/f^n$: $S^{(d)}/(f-1)S^{(d)} \to S_{(f)}$ [where $x \in S_{nd}$] is well-defined.

It is easy to check that these maps are homomorphisms of rings, and that they are inverse to each other.

(4.17) Proposition: Let $\mathfrak{a} = \bigoplus_{n \geq 0} \mathfrak{a}_n$ be a homogeneous ideal of S, let $n_0 \in \mathbb{N}$, and assume that there exists an integer $d \geq n_0$ with $\mathfrak{a}_d \neq S_d$. The following statements are equivalent:

(1) There exists a homogeneous prime ideal $\mathfrak{p} = \bigoplus_{n \geq 0} \mathfrak{p}_n$ of S such that $\mathfrak{p}_n = \mathfrak{a}_n$ for every integer $n \geq n_0$.

(2) For all homogeneous elements x, y of S of degree at least n_0, if $xy \in \mathfrak{a}$, then we have $x \in \mathfrak{a}$ or $y \in \mathfrak{a}$.

Moreover, if these conditions are satisfied, then $\mathfrak{p}$ is determined uniquely.

Proof: It is clear that (1) implies (2). Now assume that (2) holds; we choose an element $a \in S_d \setminus \mathfrak{a}_d$, and set $\mathfrak{p} := \mathfrak{a}: aS$; $\mathfrak{p}$ is an ideal of S. Let $x \in S$ and write $x = x_p + x_{p+1} + \cdots + x_{p+q}$ as a sum of homogeneous elements; if $xa = x_p a + x_{p+1} a + \cdots + x_{p+q} a \in \mathfrak{a}$, then we have $x_p a, \ldots, x_{p+q} a \in \mathfrak{a}$ [since $\mathfrak{a}$ is homogeneous] and therefore $x_p, \ldots, x_{p+q} \in \mathfrak{p}$, hence $\mathfrak{p}$ is a homogeneous ideal. Since $a \notin \mathfrak{a}$, we see that $1 \notin \mathfrak{p}$, hence $\mathfrak{p} \neq S$. Let $x \in S_m$, $y \in S_n$ be such that $x \notin \mathfrak{p}$, $y \notin \mathfrak{p}$. Then $xa \notin \mathfrak{a}_{m+d}$, $ya \notin \mathfrak{a}_{n+d}$, and therefore, by assumption, $xya^2 \notin \mathfrak{a}_{m+n+2d}$. This implies that $xya \notin \mathfrak{a}_{m+n+d}$, hence that $xy \notin \mathfrak{p}$, and therefore $\mathfrak{p}$ is a prime ideal [cf. (4.1)(1)]. Let $n \geq n_0$ be an integer, and let $x \in S_n$; then $x \in \mathfrak{a}_n$ iff $xa \in \mathfrak{a}_{n+d}$ by assumption, and therefore $\mathfrak{p}_n = S_n \cap \mathfrak{p} = \mathfrak{a}_n$. Hence $\mathfrak{p}$ is a homogeneous prime ideal of S which satisfies (1).

Let $\mathfrak{p}'$ be a homogeneous prime ideal of S satisfying $\mathfrak{p}' \cap S_n = \mathfrak{a}_n$ for every integer $n \geq n_0$, and let $x \in \mathfrak{p}$ be homogeneous of degree $m \in \mathbb{N}_0$. Since $a \notin \mathfrak{p}'$ and $xa \in \mathfrak{p}_{m+d} = \mathfrak{a}_{m+d} = \mathfrak{p}' \cap S_{m+d}$, it follows that $x \in \mathfrak{p}'$ [since $\mathfrak{p}'$ is a prime ideal], and therefore $\mathfrak{p} \subset \mathfrak{p}'$. Conversely, let $x \in \mathfrak{p}'$ be homogeneous of degree $m \in \mathbb{N}_0$. Then $xa \in \mathfrak{p}' \cap S_{m+d} = \mathfrak{a}_{m+d}$, and therefore $x \in \mathfrak{p}$ [by definition of $\mathfrak{p}$], hence $\mathfrak{p}' \subset \mathfrak{p}$. Thus, we have shown that $\mathfrak{p} = \mathfrak{p}'$.

(4.18) Notation: (1) We define $S_+ := \bigoplus_{n > 0} S_n$; every homogeneous ideal $\mathfrak{a}$ of S with $\mathrm{rad}(\mathfrak{a}) \supset S_+$ is called irrelevant.

(2) We define $\mathrm{Proj}(S)$ to be the set of non-irrelevant homogeneous prime ideals $\mathfrak{p}$ of S, and $\mathrm{Max}_+(S)$ denotes the set of maximal elements of $\mathrm{Proj}(S)$. The elements of $\mathrm{Proj}(S)$ are called points, and the elements of $\mathrm{Max}_+(S)$ are called closed points; for $\mathfrak{p} \in \mathrm{Proj}(S)$ the ring $S_{(\mathfrak{p})}$ is called the ring of the point $\mathfrak{p}$ on $\mathrm{Proj}(S)$.

In particular, let k be an algebraically closed field and $S = k[T_0, \ldots, T_n]$; the set of closed points of $\mathrm{Proj}(S)$ corresponds bijectively to the set of points of projective n-space $\mathbb{P}_k^n$ over k.

(3) For every homogeneous ideal $\mathfrak{a}$ of S we set

$$V_+(\mathfrak{a}) := \{\mathfrak{p} \in \mathrm{Proj}(S) \mid \mathfrak{p} \supset \mathfrak{a}\}.$$

(4) For every homogeneous $f \in S$ of positive degree we set

$$D_+(f) := \{\mathfrak{p} \in \mathrm{Proj}(S) \mid f \notin \mathfrak{p}\}.$$

(4.19) REMARK: Let $F \subset S$ be a system of homogeneous elements of positive degree which generate S as S_0-algebra. Then we have $\mathrm{Proj}(S) = \bigcup_{f \in F} D_+(f)$.

(4.20) **Proposition:** *Let $f \in S$ be homogeneous of positive degree. Then the map*

$$\mathfrak{p} \mapsto \mathfrak{p}S_f \cap S_{(f)} : D_+(f) \to \mathrm{Spec}(S_{(f)})$$

is an inclusion-preserving bijective map, and the rings $(S_{(f)})_{\mathfrak{q}}$ and $S_{(\mathfrak{p})}$ where $\mathfrak{q} := \mathfrak{p}S_f \cap S_{(f)}$ are canonically isomorphic.

Proof: We denote by d the degree of f. Let $\mathfrak{q} \in \mathrm{Spec}(S_{(f)})$; we define a homogeneous prime ideal $\mathfrak{p} \in \mathrm{Proj}(S)$ which does not contain f in the following way. For every $n \in \mathbb{N}_0$ let $\mathfrak{p}_n$ be the set of elements $x \in S_n$ such that $x^d/f^n \in \mathfrak{q}$. Let x, $y \in S_n$ and $x^d/f^n, y^d/f^n \in \mathfrak{q}$. Then $(x+y)^{2d}/f^{2n} \in \mathfrak{q}$, hence $(x+y)^d/f^n \in \mathfrak{q}$ since $\mathfrak{q}$ is a prime ideal, and therefore $x + y \in \mathfrak{p}_n$. It follows that $\mathfrak{p}_n$ is a subgroup of S_n. Clearly $S_m\mathfrak{p}_n \subset \mathfrak{p}_{n+m}$; hence $\mathfrak{p} := \sum_{n \geq 0} \mathfrak{p}_n$ is a homogeneous ideal of S such that $f \notin \mathfrak{p}$. Using the fact that $\mathfrak{q}$ is a prime ideal it follows immediately that the condition in (4.17)(2) is satisfied, hence $\mathfrak{p}$ is a homogeneous prime ideal of S, and even $\mathfrak{p} \in D_+(f)$. Since $\mathfrak{p}S_f \cap S_{(f)} = \mathfrak{q}$, we have shown that the map is surjective. Let $\mathfrak{p}' \in D_+(f)$ be another prime ideal with $\mathfrak{p}'S_f \cap S_{(f)} = \mathfrak{q}$. For any $n \in \mathbb{N}$ and $x \in \mathfrak{p}'_n$ we have $x^d/f^n \in \mathfrak{q}$, hence $x \in \mathfrak{p}$, and therefore $\mathfrak{p}' \subset \mathfrak{p}$. Conversely, let $x \in \mathfrak{p}_n$; then $x^d/f^n \in \mathfrak{q}$, hence $x^d/f^n = y/f^m$ where $y \in \mathfrak{p}'_{md}$, hence $x^d \in \mathfrak{p}'$ and therefore $x \in \mathfrak{p}'$. Thus, we have shown that $\mathfrak{p} = \mathfrak{p}'$, and therefore the map is injective.
Let $\mathfrak{p} \in D_+(f)$ and $\mathfrak{q} = \mathfrak{p}S_f \cap S_{(f)}$. An element $(x/f^h)/(y/f^k)$ of $(S_{(f)})_{\mathfrak{q}}$ where $x \in S_{hd}$ and $y \in S_{kd} \setminus \mathfrak{p}_{kd}$ can be written as $xf^k/yf^h \in S_{(\mathfrak{p})}$, and an element $x/y \in S_{(\mathfrak{p})}$ where $x \in S_h$, $y \in S_h \setminus \mathfrak{p}_h$ can be written as $(xy^{d-1}/f^h)/(y^d/f^h) \in (S_{(f)})_{\mathfrak{q}}$.

(4.21) **Corollary:** *Let $\mathfrak{p} \in \mathrm{Proj}(S)$. Then $S_{(\mathfrak{p})}$ is quasilocal.*

(4.22) REMARK: S is a noetherian ring iff S_0 is noetherian and S_+ is a finitely generated ideal of S [cf. [63], Ex. 1.4]. Assume that S is a noetherian ring; then $S^{(d)}$ is noetherian for every $d \in \mathbb{N}$. In particular, for $d \in \mathbb{N}$ and $f \in S_d$ the ring $S_{(f)}$ is noetherian [cf. (4.16)], and therefore $S_{(\mathfrak{p})}$ for $\mathfrak{p} \in \mathrm{Proj}(S)$ is a local ring.

4.4   Integral Closure of Graded Rings

(4.23) NOTATION: In this subsection let $A = \bigoplus_{n \in \mathbb{Z}} A_n$ be a $\mathbb{Z}$-graded ring, and let $B = \bigoplus_{n \in \mathbb{Z}} B_n$ be a $\mathbb{Z}$-graded ring which contains A as a homogeneous subring.

(4.24) REMARK: (1) Let $x \in B$ be a homogeneous element which is integral over A, and let

$$x^h + a_1 x^{h-1} + \cdots + a_n = 0 \quad \text{with } a_1, \ldots, a_h \in A$$

be an equation of integral dependence for x over A. Set $m := \deg(x)$, and let a_i' be the homogeneous component of a_i of degree mi, $i \in \{1, \ldots, h\}$. Then we have $x^h + a_1' x^{h-1} + \cdots + a_h' = 0$, hence x satisfies also an equation of integral dependence with homogenous coefficients. (This argument works also if $\mathbb{Z}$ is replaced by any totally ordered abelian group.)

(2) Let $A[T]$ be the polynomial ring over A in an indeterminate T. We consider the multiplicative closed set $\Sigma := \{T^n \mid n \in \mathbb{N}_0\}$, and let $\Sigma^{-1} A[T] = A[T, T^{-1}]$ be the ring of Laurent polynomials over A. Then $A[T, T^{-1}]$ is a free A-module with basis $\{T^n \mid n \in \mathbb{Z}\}$. Let $x \in A$ be homogeneous, and for $n \in \mathbb{Z}$ let x_n be the homogeneous component of x of degree n. We define

$$\varphi_A(x) = \sum_{n \in \mathbb{Z}} a_n T^n;$$

the map $\varphi_A : A \to A[T, T^{-1}]$ clearly is an injective homomorphism of rings.

(4.25) **Proposition:** *The integral closure $\overline{A}$ of A in B is a homogeneous subring of B. If $A_n = \{0\}$ for $n < 0$ and if B is reduced, then $\overline{A}_n = \{0\}$ for $n < 0$.*

Proof: We consider the commutative diagram

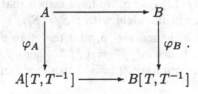

Let $x \in B$ be integral over A; clearly $\varphi_B(x)$ is integral over $A[T, T^{-1}]$. Then there exists $m \in \mathbb{N}$ such that $T^m \varphi_B(x) \in B[T]$ is integral over $A[T]$ [cf. [63], Prop. 4.13], hence the coefficients of the polynomial $T^m \varphi_B(x)$ are integral over A [cf. [63], Ex. 4.17]. These coefficients are, by definition, the homogeneous components of x; thus, we have shown that the homogeneous components of x are integral over A, and this means that $\overline{A}$ is a homogeneous subring of B.

Now we assume that $A_n = \{0\}$ for every $n < 0$. Let $n < 0$ and let $x \in \overline{A}_n$; then, by (4.24)(1), there exist $h \in \mathbb{N}$ and elements $a_1 \in A_n, a_2 \in A_{2n}, \ldots, a_h \in A_{hn}$ such that $x^h + a_1 x^{h-1} + \cdots + a_h = 0$. Then we have $x^h = 0$; if B is reduced, then we get $x = 0$, and this implies the last assertion of (4.25).

(4.26) Corollary: (1) Let $d \in \mathbb{N}$. Then $\overline{A}^{(d)}$ is the integral closure of $A^{(d)}$ in $B^{(d)}$.
(2) Let $\Sigma \subset A$ be a multiplicatively closed subset of A consisting of homogeneous elements. Then $\overline{A}_{(\Sigma)}$ is the integral closure of $A_{(\Sigma)}$ in $B_{(\Sigma)}$.

The proof is simple and is left to the reader.

(4.27) Remark: Let A be an integral domain, and let Σ be the set of non-zero homogeneous elements of A. Then we define

$$Q_+(A) := \Sigma^{-1}A;$$

$Q_+(A)$ is called the graded ring of fractions of A. Clearly every non-zero homogeneous element of $Q_+(A)$ is invertible, hence $Q_+(A)$ is a simple $\mathbb{Z}$-graded ring. From (4.14)(3) we see that for every $d \in \mathbb{N}$ we have $Q_+(A^{(d)}) = (Q_+(A))^{(d)}$. We consider the case that the subgroup $L(Q_+(A))$ of $\mathbb{Z}$ [cf. (4.5)] is not $\{0\}$; let $q \in \mathbb{N}$ be a generator of $L(Q_+(A))$, and let $f \in (Q_+(A))_q$ be a unit. Set $K_0 := (Q_+(A))_0$. K_0 is a field, and the map $K_0[T, T^{-1}] \to Q_+(A)$ defined by $T \mapsto f$ and $T^{-1} \mapsto f^{-1}$ is an isomorphism [cf. (4.7)]. Since $K_0[T, T^{-1}]$ is integrally closed [cf. [63], Prop. 4.13 and note that $K_0[T]$ is integrally closed by [63], Prop. 4.10 since it is factorial], we see that $Q_+(A)$ is integrally closed.

(4.28) Proposition: Assume that A is an integral domain. The integral closure $\overline{A}$ of A is a homogeneous subring of $Q_+(A)$; if $A_n = \{0\}$ for every $n < 0$, then $\overline{A}_n = \{0\}$ for every $n < 0$.

Proof: $Q_+(A)$ is integrally closed [cf. (4.27)], hence $\overline{A}$ is the integral closure of A in $Q_+(A)$, and it is therefore a homogeneous subring of $Q_+(A)$ by (4.25). The last assertion follows also from (4.25).

(4.29) Corollary: If every homogeneous element of $Q_+(A)$ which is integral over A already lies in A, then A is integrally closed

Proof: We have $\overline{A}_n \subset A$ for every $n \in \mathbb{Z}$, hence $\overline{A} = A$.

(4.30) Corollary: Let $f \in A_1$ be non-zero. Then:
(1) K_0 is the field of quotients of $A_{(f)}$.
(2) If A is integrally closed, then A_0 and $A_{(f)}$ are integrally closed.

Proof: (1) Clearly $A_{(f)}$ is a subring of K_0, hence $Q(A_{(f)}) \subset K_0$. Let $z \in K_0$ be a non-zero element; then we can write $z = y/x$ where y, x are non-zero homogeneous elements of A of the same degree p. We have $z = f^p y / f^p x = (y/f^p)(x/f^p)^{-1}$, hence $z \in Q(A_{(f)})$. Thus, we have shown that $K_0 = Q(A_{(f)})$.
(2) The quotient field $Q(A_0)$ of A_0 is a subfield of K_0, hence every element of $Q(A_0)$ which is integral over A_0 [hence also over A] lies in A_0. Let $y \in K_0$ be integral over $A_{(f)}$. Then we have $y^h + a_1 y^{h-1} + \cdots + a_h = 0$ with $h \in \mathbb{N}$ and $a_1, \ldots, a_h \in A_{(f)}$. We choose a positive integer q with $f^q a_1, \ldots, f^q a_h \in A$. Then

we have $yf^q \in Q_+(A)$ and $(yf^q)^h + (a_1 f^q)(yf^q)^{h-1} + \cdots + a_h f^{hq} = 0$, hence yf^q is integral over A, hence $yf^q \in A$, hence $y \in A_f$ and therefore y lies in $A_{(f)}$.

(4.31) Corollary: *For every $d \in \mathbb{N}$ the integral closure of $A^{(d)}$ is $\overline{A}^{(d)}$. In particular, if A is integrally closed, then $A^{(d)}$ is integrally closed.*

Proof: This follows from (4.26) and $Q_+(A^{(d)}) = (Q_+(A))^{(d)}$.

We conclude this section with the three following useful results.

(4.32) Proposition: *Let $A = \bigoplus_{n \geq 0} A_n$ be an A_0-algebra of finite type, and let $M = \bigoplus_{n \in \mathbb{Z}} M_n$ be a finitely generated graded A-module. For every $n \in \mathbb{Z}$ the A_0-module M_n is finitely generated, and $M_n = \{0\}$ for all sufficiently small integers n.*

Proof: We choose homogeneous elements $a_1, \ldots, a_r$ of positive degree in A with $A = A_0[a_1, \ldots, a_r]$; set $d_i := \deg(a_i)$ for $i \in \{1, \ldots, r\}$. We choose homogeneous elements $x_1, \ldots, x_s \in M$ with $M = Ax_1 + \cdots + Ax_s$; set $e_j := \deg(x_j)$ for $j \in \{1, \ldots, s\}$. Let $n \in \mathbb{Z}$; clearly M_n is generated by all $a_1^{m_1} \cdots a_r^{m_r} x_j$ where $m_1, \ldots, m_r$ are positive integers with $m_1 d_1 + \cdots m_r d_1 + e_j = n$ and $j \in \{1, \ldots, s\}$. Let $n \in \mathbb{Z}$; for $j \in \{1, \ldots, s\}$ there are only finitely many $(m_1, \ldots, m_r) \in \mathbb{N}^r$ with $m_1 d_1 + \cdots + m_r d_1 + e_j = n$. Furthermore, we have $M_n = \{0\}$ for all $n \in \mathbb{Z}$ with $n < d_i$ for $i \in \{1, \ldots, r\}$.

(4.33) Proposition: *Let $A = \bigoplus_{n \geq 0} A_n$ be a graded domain. We assume that $A_0 = k$ is an algebraically closed field and that $A = k[x_1, \ldots, x_h]$ with $x_1, \ldots, x_h \in A_1$. Let $B = \bigoplus_{n \in \mathbb{Z}} B_n$ be a homogeneous subring of $Q_+(A)$ which contains A and which is a finitely generated A-module. Then there exists $d \in \mathbb{N}$ such that $B^{(d)} = k[y_1, \ldots, y_l]$ where $y_1, \ldots, y_l \in B_d = (B^{(d)})_1$.*

Proof: By (4.28)(1) we have $B_n = \{0\}$ for all negative integers n. Since B is a finitely generated A-module, there exist homogenous elements $b_1, \ldots, b_p \in B$ of non-negative degree such that $B = Ab_1 + \cdots + Ab_p$. Let $j \in \{1, \ldots, p\}$; let $d_j \in \mathbb{N}_0$ be the degree of b_j, and choose $d \in \mathbb{N}$ such that $d \geq d_1, \ldots, d \geq d_p$.
(1) We show that $B_{n+1} = A_1 B_n$ for every $n \geq d$. In fact, let $x \in B_{n+1}$; then we have $x = a_1 b_1 + \cdots + a_p b_p$ with elements $a_1, \ldots, a_p \in A$ which we can assume to be homogenous, hence $\deg(a_i) = n + 1 - \deg(b_i) > 0$ for $i \in \{1, \ldots, p\}$. We have $a_i = a_i' a_i''$ with $a_i' \in A_1$ and $a_i'' \in A_{n-\deg(b_i)}$ for $i \in \{1, \ldots, p\}$. Therefore we have $x \in A_1 B_n \subset B_{n+1}$, hence $B_{n+1} = A_1 B_n$.
(2) From (1) we get $B_{n+k} = A_k B_n$ for every $n \geq d$ and $k \in \mathbb{N}$, hence $B_{n+k} = B_k B_n$, and this implies that $B_{md} = (B_d)^m$ for every $m \in \mathbb{N}$, hence that $B^{(d)} = B_0[B_d] = B_0[y_1, \ldots, y_l]$ with $y_1, \ldots, y_l \in B_d$ [since B_d is a finitely generated B_0-module by (4.32)].
(3) Let $b \in B_0$; then there exist $c_1, \ldots, c_q \in A_0$ such that $b^q + c_1 b^{q-1} + \cdots + c_q = 0$ [cf. (4.24)]. Since $A_0 = k$ and since k is algebraically closed, it follows that $b \in k$, hence $B_0 = k$ and $B^{(d)} = k[y_1, \ldots, y_l]$.

(4.34) Proposition: *Let* $S = \bigoplus_{n\geq 0} S_n$ *with* $S = S_0[y_1,\ldots,y_h]$ *and* $y_1,\ldots,y_h \in$ S_1. *Let* $x_1,\ldots,x_m \in S_+$ *be homogeneous, and set* $\mathfrak{A} := Sx_1 + \cdots + Sx_m$, $R :=$ $S_0[x_1,\ldots,x_m] = \bigoplus_{n\geq 0} R_n$. *The following statements are equivalent:*
(1) *S is a finitely generated R-module.*
(2) $\mathfrak{A}$ *is an irrelevant ideal of S.*

Proof: First, we show that it is enough to prove the equivalence of (1) and (2) under the additional assumption that the elements $x_1,\ldots,x_m$ all have the same degree. In fact, set $d_i := \deg(x_i)$ for $i \in \{1,\ldots,m\}$, $d := d_1 \cdots d_m$ and $z_i := x_i^{d/d_i}$ for $i \in$ $\{1,\ldots,m\}$; then we have $\deg(z_1) = \cdots = \deg(z_m) = d$. Set $\mathfrak{B} := Sz_1 + \cdots + Sz_m$; then we have $\mathfrak{B} \subset \mathfrak{A} \subset \mathrm{rad}(\mathfrak{B})$, hence $\mathfrak{A}$ is irrelevant iff $\mathfrak{B}$ is irrelevant. On the other hand, we have $R_0[z_1,\ldots,z_m] \subset R_0[x_1,\ldots,x_m]$, and the latter ring is a finitely generated $R_0[z_1,\ldots,z_m]$-module [cf. [63], Cor. 4.5], hence S is a finitely generated R-module iff S is a finitely generated $R_0[z_1,\ldots,z_m]$-module. Thus, we may assume that $\deg(x_1) = \cdots = \deg(x_m) =: d$, hence $R = \bigoplus_{n\geq 0} R_{nd}$.
(1) $\Rightarrow$ (2): y_1 is integral over R, hence we have an equation [cf. (4.24)]

$$y_1^s + a_1 y_1^{s-1} + \cdots + a_s = 0 \quad \text{with } a_i \in R_{si} \text{ for } i \in \{1,\ldots,s\}. \tag{$*$}$$

Since $R_n = \{0\}$ if n is not divisible by d, the relation $(*)$ takes the form $y_1^s + a_d y_1^{s-d} + a_{2d} y_1^{s-2d} + \cdots = 0$, hence y_1^s lies in $\mathfrak{A}$. Similarly, one may show that a power of each of the elements $y_2,\ldots,y_h$ lies in $\mathfrak{A}$, hence that $S_+ \subset \mathrm{rad}(\mathfrak{A})$.
(2) $\Rightarrow$ (1): There exists $p \in \mathbb{N}$ with $(S_+)^p \subset \mathfrak{A}$. We show: the set $B := \{y_1^{j_1}\cdots y_h^{j_h} \mid$ $j_1 + \cdots + j_h < p\}$ is a system of generators of the R-module S. Let $y_1^{i_1}\cdots y_h^{i_h}$ be a power product of $y_1,\ldots,y_h$. We show that $y_1^{i_1}\cdots y_h^{i_h} = \sum_{b\in B} r_b b$ with $r_b \in R$ for every $b \in B$ by induction on $i_1 + \cdots + i_h$. There is nothing to show if $i_1 + \cdots + i_h < p$. Now let $i_1 + \cdots + i_h \geq p$ and write $y_1^{i_1}\cdots y_h^{i_h} = y_1^{k_1}\cdots y_h^{k_h} y_1^{l_1}\cdots y_h^{l_h}$ with $k_1 + \cdots + k_h = p$. We have $y_1^{k_1}\cdots y_h^{k_h} \in S_p \subset (S_+)^p \subset \mathfrak{A}$, hence $y_1^{k_1}\cdots y_h^{k_h} =$ $s_1 x_1 + \cdots + s_m x_m$ with $s_1,\ldots,s_m \in S_{p-d}$. Now $y_1^{i_1}\cdots y_h^{i_h} = \sum_j s_j y_1^{l_1}\cdots y_h^{l_h} x_j$, and $s_j y_1^{l_1}\cdots y_h^{l_h}$ is homogeneous of degree $(p-d)+(i_1+\cdots+i_h-p) < i_1+\cdots+i_h$, hence, by induction, $s_j y_1^{l_1}\cdots y_h^{l_h} = \sum_{b\in B} r_{jb} b$ with $r_{jb} \in R$ for $j \in \{1,\ldots,m\}$, hence $y_1^{i_1}\cdots y_h^{i_h} = \sum_{b\in B} r_b b$ with $r_b := \sum_j r_{jb} x_j \in R$ for every $b \in B$.

5   Properties of the Rees Ring

(5.0) In this section A is a ring, B is an overring of A, and $A[T]$ (resp. $B[T]$) is the ring of polynomials over A (resp. B).

(5.1) DEFINITION: Let $\mathfrak{a}$ be an A-submodule of B. The subring

$$\mathcal{R}(\mathfrak{a}, B) = \bigoplus_{i\in\mathbb{N}_0} \mathfrak{a}^i T^i$$

of the polynomial ring $B[T]$ is a homogeneous subring of the graded ring $B[T]$. If $B = A$, then $\mathfrak{a}$ is an ideal of A, and $\mathcal{R}(\mathfrak{a}, A)$ is called the Rees ring of A with respect to $\mathfrak{a}$. (In [63] this ring is called the blow-up ring while $\bigoplus_{i \in \mathbb{Z}} \mathfrak{a}^i T^i$ where $\mathfrak{a}^i = A$ for $i < 0$, is called the Rees ring.) Note that $\mathcal{R}(\mathfrak{a}, A)$ carries in a canonical way a structure as a $\mathbb{Z}$-graded A-algebra of type $\mathbb{N}_0$.

(5.2) REMARK: Let $\mathfrak{a}$ be an ideal of A, and let

$$\mathrm{gr}_{\mathfrak{a}}(A) := \bigoplus_{i \in \mathbb{N}_0} \mathfrak{a}^i / \mathfrak{a}^{i+1}$$

be the associated graded ring [cf. [63], section 5.1]. The homomorphism of graded rings

$$\varphi_{\mathfrak{a}} : \mathcal{R}(\mathfrak{a}, A) = \bigoplus_{i \in \mathbb{N}_0} \mathfrak{a}^i T^i \to \bigoplus_{i \in \mathbb{N}_0} \mathfrak{a}^i / \mathfrak{a}^{i+1} = \mathrm{gr}_{\mathfrak{a}}(A),$$

defined by mapping an element $a T^i \in \mathfrak{a}^i T^i$ to $a \bmod \mathfrak{a}^{i+1} \in \mathfrak{a}^i / \mathfrak{a}^{i+1}$ is surjective, and its kernel is the homogeneous ideal $J = \bigoplus_{i \in \mathbb{N}_0} \mathfrak{a}^{i+1} T^i$ of $\mathcal{R}(\mathfrak{a}, A)$.

(5.3) REES RING AND ASSOCIATED GRADED RING: Let $\mathfrak{a}$ be an ideal of A. We collect in this remark some properties of the rings $\mathcal{R}(\mathfrak{a}, A)$ and $\mathrm{gr}_{\mathfrak{a}}(A)$.
(1) Assume that $\mathfrak{a} = A a_1 + \cdots + A a_h$ is a finitely generated ideal. Then the homogeneous elements $a_1 T, \ldots, a_h T \in \mathcal{R}(\mathfrak{a}, A)_1$ generate $\mathcal{R}(\mathfrak{a}, A)$ as A-algebra, and the homogeneous elements $a_1 \bmod \mathfrak{a}^2, \ldots, a_h \bmod \mathfrak{a}^2 \in \mathrm{gr}_{\mathfrak{a}}(A)_1$ generate $\mathrm{gr}_{\mathfrak{a}}(A)$ as $A/\mathfrak{a}$-algebra. In particular, if A is a noetherian ring, then $\mathcal{R}(\mathfrak{a}, A)$ and $\mathrm{gr}_{\mathfrak{a}}(A)$ are noetherian rings.
(2) Let $x \in A$. If $x \in \mathfrak{a}^i \setminus \mathfrak{a}^{i+1}$ for some $i \in \mathbb{N}_0$, then $\mathrm{In}(x) := x \bmod \mathfrak{a}^{i+1} \in \mathfrak{a}^i / \mathfrak{a}^{i+1}$ is called the leading form of x [with respect to $\mathfrak{a}$], and if $x \in \mathfrak{a}^i$ for every $i \in \mathbb{N}$, then we say that x has leading form 0 [with respect to $\mathfrak{a}$].
(3) Let $\mathfrak{b}$ be an ideal of A. Then we define

$$\mathcal{R}(\mathfrak{b}) := \bigoplus_{i \in \mathbb{N}_0} (\mathfrak{a}^i \cap \mathfrak{b}) T^i \subset \mathcal{R}(\mathfrak{a}, A), \quad \mathrm{gr}(\mathfrak{b}) := \bigoplus_{i \in \mathbb{N}_0} (\mathfrak{a}^i \cap \mathfrak{b} + \mathfrak{a}^{i+1}) / \mathfrak{a}^{i+1} \subset \mathrm{gr}_{\mathfrak{a}}(A).$$

Since $(\mathfrak{a}^i \cap \mathfrak{b}) \cdot (\mathfrak{a}^j \cap \mathfrak{b}) \subset \mathfrak{a}^{i+j} \cap \mathfrak{b}$ for all $i, j \in \mathbb{N}_0$, it is clear that $\mathcal{R}(\mathfrak{b})$ is a homogeneous ideal of $\mathcal{R}(\mathfrak{a}, A)$ and that $\mathrm{gr}(\mathfrak{b})$ is a homogeneous ideal of $\mathrm{gr}_{\mathfrak{a}}(A)$. Note, in particular, the following: If $\mathfrak{b} = A f$ is a principal ideal, and $\overline{f}$ is the leading form of f, then we have $\mathrm{gr}(\mathfrak{b}) = \mathrm{gr}_{\mathfrak{a}}(A) \overline{f}$.
Define $\overline{A} := A/\mathfrak{b}$, $\overline{\mathfrak{a}} := (\mathfrak{a} + \mathfrak{b})/\mathfrak{b}$. Since $\mathfrak{a}^i / (\mathfrak{a}^i \cap \mathfrak{b}) \cong (\mathfrak{a}^i + \mathfrak{b})/\mathfrak{b} \cong \overline{\mathfrak{a}}^i$ and $(\mathfrak{a}^i / \mathfrak{a}^{i+1}) / ((\mathfrak{a}^i \cap \mathfrak{b} + \mathfrak{a}^{i+1}) / \mathfrak{a}^{i+1}) \cong \mathfrak{a}^i / (\mathfrak{a}^i \cap \mathfrak{b} + \mathfrak{a}^{i+1}) \cong (\mathfrak{a}^i / (\mathfrak{a}^i \cap \mathfrak{b})) / ((\mathfrak{a}^i \cap \mathfrak{b} + \mathfrak{a}^{i+1}) / (\mathfrak{a}^i \cap \mathfrak{b})) \cong ((\mathfrak{a}^i + \mathfrak{b})/\mathfrak{b}) / ((\mathfrak{a}^{i+1} + \mathfrak{b})/\mathfrak{b}) \cong \overline{\mathfrak{a}}^i / \overline{\mathfrak{a}}^{i+1}$ for every $i \in \mathbb{N}_0$, we have a commutative diagram of surjective homomorphisms of graded rings

$$
\begin{array}{ccc}
\mathcal{R}(\mathfrak{a}, A) & \xrightarrow{\varphi_{\mathfrak{a}}} & \mathrm{gr}_{\mathfrak{a}}(A) \\
\downarrow & & \downarrow \\
\mathcal{R}(\overline{\mathfrak{a}}, \overline{A}) & \xrightarrow{\varphi_{\overline{\mathfrak{a}}}} & \mathrm{gr}_{\overline{\mathfrak{a}}}(\overline{A})
\end{array}
$$

where the kernel of the left vertical map is the ideal $\mathcal{R}(\mathfrak{b})$ of $\mathcal{R}(\mathfrak{a}, A)$ and the kernel of the right vertical map is the ideal $\mathrm{gr}(\mathfrak{b})$ of $\mathrm{gr}_{\mathfrak{a}}(A)$.

(4) Let $A \to A'$ be a flat homomorphism. Then the induced homomorphisms $\mathcal{R}(\mathfrak{a}, A) \otimes_A A' \to \mathcal{R}(\mathfrak{a}A', A')$ and $\mathrm{gr}_{\mathfrak{a}}(A) \otimes_A A' \to \mathrm{gr}_{\mathfrak{a}A'}(A')$ are isomorphisms.

(5.4) NOTATION: Let $\mathfrak{a}$ be an ideal of A, and let $x \in \mathfrak{a}$. Then $A[\mathfrak{a}/x] \subset A_x$ denotes the smallest A-subalgebra of A_x containing all the elements a/x, $a \in \mathfrak{a}$. Note that x is a regular element of A_x in case $A_x \neq \{0\}$, and that $\mathfrak{a}A[\mathfrak{a}/x] = xA[\mathfrak{a}/x]$.

(5.5) Proposition: *Let $\mathfrak{a}$ be an ideal of A, and let $x \in \mathfrak{a}$. Then there exists a unique A-algebra homomorphism*

$$\psi \colon \mathcal{R}(\mathfrak{a}, A)_{(xT)} \to A[\mathfrak{a}/x] \subset A_x,$$

and ψ is an isomorphism.

Proof: Let $m \in \mathbb{N}_0$, $a \in \mathfrak{a}^m$; if $\varphi \colon \mathcal{R}(\mathfrak{a}, A)_{(xT)} \to A_x$ is an A-algebra homomorphism, then

$$x^m \varphi(aT^m/x^m T^m) = \varphi(a \cdot x^m T^m/x^m T^m) = \varphi(a/1) = a\varphi(1/1) = a/1,$$

hence $\varphi(aT^m/x^m T^m) = a/x^m \in A[\mathfrak{a}/x]$. Define $\psi(aT^m/x^m T^m) := a/x^m$. Then ψ is well-defined: let $n \in \mathbb{N}$, $b \in \mathfrak{a}^n$, and assume that $aT^m/x^m T^m = bT^n/x^n T^n$; then there exists $l \in \mathbb{N}$ satisfying $x^l T^l(ax^n - bx^m)T^{m+n} = 0$, hence we have $a/x^m = b/x^n$ in A_x. It is immediate that ψ is an injective A-algebra homomorphism, and that $A[\mathfrak{a}/x]$ is the image of ψ.

(5.6) LOCALIZATION: We keep the notations of (5.5).

(1) Let $\mathfrak{b}$ be an ideal of A; then the image by ψ of the ideal $\mathcal{R}(\mathfrak{b})\mathcal{R}(\mathfrak{a}, A)_{xT} \cap \mathcal{R}(\mathfrak{a}, A)_{(xT)} \subset \mathcal{R}(\mathfrak{a}, A)_{(xT)}$ in $A[\mathfrak{a}/x]$ is the ideal generated by all the elements b/x^n where $n \in \mathbb{N}$ and $b \in \mathfrak{a}^n \cap \mathfrak{b}$.

(2) Let $\mathfrak{q}$ be a homogeneous prime ideal of $\mathcal{R}(\mathfrak{a}, A)$ which does not contain xT. Then $\mathfrak{q}' := \mathfrak{q}\mathcal{R}(\mathfrak{a}, A)_{xT} \cap \mathcal{R}(\mathfrak{a}, A)_{(xT)}$ is a prime ideal of $\mathcal{R}(\mathfrak{a}, A)_{(xT)}$, and we can identify $\mathcal{R}(\mathfrak{a}, A)_{(\mathfrak{q})}$ and $(\mathcal{R}(\mathfrak{a}, A)_{(xT)})_{\mathfrak{q}'}$ [cf. (4.20)]. Let $\mathfrak{p}$ be the image of $\mathfrak{q}'$ in $A[\mathfrak{a}/x]$. Then there exists a unique homomorphism of A-algebras $\mathcal{R}(\mathfrak{a}, A)_{(\mathfrak{q})} \to A[\mathfrak{a}/x]_{\mathfrak{p}}$ making the diagram

$$
\begin{array}{ccc}
\mathcal{R}(\mathfrak{a}, A)_{(xT)} & \xrightarrow{\ \psi\ } & A[\mathfrak{a}/x] \\
\downarrow & & \downarrow \\
\mathcal{R}(\mathfrak{a}, A)_{(\mathfrak{q})} & \xrightarrow[\cong]{\ } & A[\mathfrak{a}/x]_{\mathfrak{p}}
\end{array}
$$

commutative [the vertical maps are the canonical ones], and it is an isomorphism. Conversely, let $\mathfrak{p}$ be a prime ideal of $A[\mathfrak{a}/x]$. Then there exists a unique homogeneous prime ideal $\mathfrak{q}$ of $\mathcal{R}(\mathfrak{a}, R)$ not containing xT with $\mathcal{R}(\mathfrak{a}, R)_{(\mathfrak{q})} \cong R[\mathfrak{a}/x]_{\mathfrak{p}}$.

(5.7) THE EXCEPTIONAL SET E: Let $\mathfrak{a}$ be an ideal of A.

(1) The elements xT, $x \in \mathfrak{a}$, generate the A-algebra $\mathcal{R}(\mathfrak{a}, A)$, hence [cf. (4.19)]

$$\mathrm{Proj}(\mathcal{R}(\mathfrak{a}, A)) = \bigcup_{x \in \mathfrak{a}} D_+(xT).$$

For every $x \in \mathfrak{a}$ there exists a bijective map $D_+(xT) \to \mathrm{Spec}(A[\mathfrak{a}/x])$ [cf. (4.20) and (5.5)].

(2) Let J be the kernel of $\varphi_{\mathfrak{a}}$ [cf. (5.2)], and set $E := V_+(J) \subset \mathrm{Proj}(\mathcal{R}(\mathfrak{a}, A))$; the set E is called the exceptional set of $\mathrm{Proj}(\mathcal{R}(\mathfrak{a}, A))$.

(a) Let $x \in \mathfrak{a}$. The ideal $J\mathcal{R}(\mathfrak{a}, A)_{xT} \cap \mathcal{R}(\mathfrak{a}, A)_{(xT)} \subset \mathcal{R}(\mathfrak{a}, A)_{(xT)}$ is generated by the set $\{aT^n/(xT)^n \mid n \in \mathbb{N}, a \in \mathfrak{a}^{n+1}\}$, hence its image by ψ [cf. (5.5)] is the principal ideal $xA[\mathfrak{a}/x] = \mathfrak{a}A[\mathfrak{a}/x]$ [note that, for $a \in \mathfrak{a}^{n+1}$, we have $a/x^n = x \cdot (a/x^{n+1})$ and $a/x^{n+1} \in A[\mathfrak{a}/x]$].

(b) Let $\mathfrak{q} \in \mathrm{Proj}(\mathcal{R}(\mathfrak{a}, A))$, and choose $x \in \mathfrak{a}$ such that $xT \notin \mathfrak{q}$. Let $\mathfrak{p}$ be the prime ideal of $A[\mathfrak{a}/x]$ corresponding to $\mathfrak{q}$ [cf. (1)]. Then we have $\mathfrak{q} \in E$ iff $x/1 \in \mathfrak{p}$.

In fact, if $\mathfrak{q} \in E$, then $\mathfrak{p}$ contains the ideal $\mathfrak{a}A[\mathfrak{a}/x] = xA[\mathfrak{a}/x]$ [cf. (a)], hence we have $x/1 \in \mathfrak{p}$. Conversely, let $x/1 \in \mathfrak{p}$. Let $n \in \mathbb{N}_0$, and consider an arbitrary element $aT^n \in J_n$ where $a \in \mathfrak{a}^{n+1}$; since $a/x^n = x \cdot (a/x^{n+1}) \in \mathfrak{p}$, it follows that $aT^n \in \mathfrak{q}$ [cf. the proof of (4.20)], and therefore $\mathfrak{q} \supset J_n$, hence $\mathfrak{q}$ contains J.

(3) Let $\mathfrak{q} \in \mathrm{Proj}(\mathcal{R}(\mathfrak{a}, A))$, and choose $x \in \mathfrak{a}$ such that $xT \notin \mathfrak{q}$. Let $\mathfrak{p}$ be the prime ideal of $A[\mathfrak{a}/x]$ corresponding to $\mathfrak{q}$, and set $B := A[\mathfrak{a}/x]_{\mathfrak{p}}$.

(a) Assume that $\mathfrak{q} \notin E$, and let $\mathfrak{p}_0$ be the preimage of $\mathfrak{p}$ under the canonical homomorphism $A \to A[\mathfrak{a}/x]$. Then the canonical A-algebra homomorphism $A_{\mathfrak{p}_0} \to A[\mathfrak{a}/x]_{\mathfrak{p}} = B$ is an isomorphism.

In fact, since $\mathfrak{q} \notin E$, we have $x/1 \notin \mathfrak{p}$ [cf. (2)(b)]. Therefore an element $a/x^n \in A[\mathfrak{a}/x]$, $a \in \mathfrak{a}^n$, belongs to $\mathfrak{p}$ iff a belongs to $\mathfrak{p}_0$. Now the assertion follows immediately.

(b) Assume that $\mathfrak{q} \in E$, and let $\mathfrak{q}_0$ be the image of $\mathfrak{q}$ under $\varphi_{\mathfrak{a}}$ in $\mathrm{gr}_{\mathfrak{a}}(A)$. Then there exists a unique A-algebra homomorphism $\mathrm{gr}_{\mathfrak{a}}(A)_{(\mathfrak{q}_0)} \to B/xB$ making the diagram

$$
\begin{array}{ccc}
\mathcal{R}(\mathfrak{a}, A)_{(\mathfrak{q})} & \longrightarrow & A[\mathfrak{a}/x]_{\mathfrak{p}} = B \\
\downarrow & & \downarrow \\
\mathrm{gr}_{\mathfrak{a}}(A)_{(\mathfrak{q}_0)} & \overset{\cong}{\longrightarrow} & B/xB
\end{array}
$$

commutative, and it is an isomorphism.

6 Integral Closure of Ideals

6.1 Generalities

(6.0) In this section A is a ring, B is an overring of A, and $A[T]$ (resp. $B[T]$) is the ring of polynomials over A (resp. over B).

Let M, N be A-submodules of B; then MN is the A-submodule of B which is generated by all products xy where $x \in M$ and $y \in N$. If $M \subset N$ and L is an A-submodule of B, then we have $LM \subset LN$. If M and N are finitely generated A-modules, so is MN. For every $i \in \mathbb{N}$ the A-module M^i is defined; we set $M^0 := A$.

(6.1) Lemma: *Let $\mathfrak{a}$ be an A-submodule of B, let N be a B-module, let M be a finitely generated A-submodule of N, and let $b \in B$. Assume that $bM \subset \mathfrak{a}M$. Then there exist $h \in \mathbb{N}$, $a_1 \in \mathfrak{a}, \ldots, a_h \in \mathfrak{a}^h$ such that $(b^h + a_1 b^{h-1} + \cdots + a_h)M = \{0\}$.*

Proof: Write $M = Ax_1 + \cdots + Ax_h$; there exist $a_{ij} \in \mathfrak{a}$ for i, $j \in \{1, \ldots, h\}$ such that $bx_i = \sum_{j=1}^{h} a_{ij}x_j$ for every $i \in \{1, \ldots, h\}$, hence [δ_{ij} is Kronecker's symbol]

$$\sum_{j=1}^{h}(b\delta_{ij} - a_{ij})x_j = 0 \quad \text{for every } i \in \{1, \ldots, h\}. \tag{*}$$

Let (c_{ij}) be the adjoint matrix of $(b\delta_{ij} - a_{ij})$ and set $d := \det((b\delta_{ij} - a_{ij}))$. Then

$$0 = \sum_{i=1}^{h}\sum_{j=1}^{h} c_{ki}(b\delta_{ij} - a_{ij})x_j = \sum_{j=1}^{h} \delta_{kj} dx_j = dx_k \quad \text{for every } k \in \{1, \ldots, h\}$$

by $(*)$, hence $dM = \{0\}$. Expanding out the determinant shows that $d = b^h + a_1 b^{h-1} + \cdots + a_h$ where $a_i \in \mathfrak{a}^i$ for every $i \in \{1, \ldots, h\}$.

6.2 Integral Closure of Ideals

(6.2) Proposition: *Let $\mathfrak{a}$ be an A-submodule of B, and let $x \in B$. The following statements are equivalent:*
(1) There exist $n \in \mathbb{N}$ and elements $a_1 \in \mathfrak{a}$, $a_2 \in \mathfrak{a}^2, \ldots, a_n \in \mathfrak{a}^n$ such that

$$x^n + a_1 x^{n-1} + \cdots + a_n = 0. \tag{*}$$

(2) The element $xT \in B[T]$ is integral over $\mathcal{R}(\mathfrak{a}, B)$.
If, in addition, B is a domain, then the above conditions are equivalent to
(3) There exists a finitely generated non-zero A-submodule M of B with $xM \subset \mathfrak{a}M$.

Proof (1) $\Rightarrow$ (2): Multiply $(*)$ by T^n to get

$$(xT)^n + (a_1 T)(xT)^{n-1} + \cdots + (a_n T^n) = 0; \tag{**}$$

since $a_i T^i \in \mathcal{R}(\mathfrak{a}, B)$ for every $i \in \{1, \ldots, n\}$, $(**)$ is an equation of integral dependence for xT over $\mathcal{R}(\mathfrak{a}, B)$.

$(2) \Rightarrow (1)$: Assume that xT is integral over $\mathcal{R}(\mathfrak{a}, B)$. There exist $n \in \mathbb{N}$ and homogeneous $b_1, \ldots, b_n \in \mathcal{R}(\mathfrak{a}, B)$ with $\deg(b_i) = i$ for $i \in \{1, \ldots, n\}$ such that $(xT)^n + b_1(xT)^{n-1} + \cdots + b_n = 0$ [cf. (4.24)]; for $i \in \{1, \ldots, n\}$ we have $b_i = a_i T^i$ with elements $a_1 \in \mathfrak{a}, \ldots, a_n \in \mathfrak{a}^n$. Therefore there exists an equation of the form $(**)$, and cancelling T^n gives the result.

Assume that B is a domain. By (6.1) it is clear that (3) implies (1). Conversely, assume that (1) holds. It is easy to see that there exists a finitely generated A-submodule $\mathfrak{a}_1 \subset \mathfrak{a}$ such that $a_i \in \mathfrak{a}_1^i$ for every $i \in \{1, \ldots, n\}$. We set $M := \mathfrak{a}_1^{n-1} + \mathfrak{a}_1^{n-2} x + \cdots + \mathfrak{a}_1 x^{n-2} + A x^{n-1}$; then we have $xM \subset \mathfrak{a}M$ and $M \neq \{0\}$.

(6.3) DEFINITION: Let $\mathfrak{a}$ be an A-submodule of B. An element $x \in B$ is called integral over $\mathfrak{a}$ if it satisfies the equivalent conditions (1) and (2) of (6.2).

(6.4) Corollary: *Let $\mathfrak{a}$ be an A-submodule of B. Then the set of elements of B which are integral over $\mathfrak{a}$ is an A-submodule of B containing $\mathfrak{a}$.*

Proof: This follows immediately from (6.2) since the subset of elements of $B[T]$ which are integral over the subring $\mathcal{R}(\mathfrak{a}, B)$ is a subring of $B[T]$ [cf. [63], Th. 4.2].

(6.5) NOTATION: Let $\mathfrak{a}$ be an A-submodule of B. The A-submodule $\overline{\mathfrak{a}}_B$ of elements of B which are integral over $\mathfrak{a}$ is called the integral closure of $\mathfrak{a}$ in B, and $\mathfrak{a}$ is said to be integrally closed in B if it coincides with its integral closure in B. If $B = A$, then we write $\overline{\mathfrak{a}}$ instead of $\overline{\mathfrak{a}}_B$. Any A-submodule $\mathfrak{a}'$ of B such that $\mathfrak{a} \subset \mathfrak{a}' \subset \overline{\mathfrak{a}}_B$ is called integral over $\mathfrak{a}$.

If $\mathfrak{a}$ is an ideal of A which is integrally closed in A, then $\mathfrak{a}$ is said to be integrally closed.

(6.6) Corollary: (1) *Let $\mathfrak{a}$ be an A-submodule of B. Then $\overline{\mathfrak{a}}_B$ is integrally closed in B.*

(2) *Let $\mathfrak{a}, \mathfrak{b}, \mathfrak{c}$ be A-submodules of B. If $\mathfrak{b}$ is integral over $\mathfrak{a}$ and $\mathfrak{c}$ is integral over $\mathfrak{b}$, then $\mathfrak{c}$ is integral over $\mathfrak{a}$.*

(3) *Let $\mathfrak{a}, \mathfrak{b}$ be A-submodules of B. If $\mathfrak{a} \subset \mathfrak{b}$, and if $\mathfrak{b}$ admits a system of generators which are integral over $\mathfrak{a}$, then $\mathfrak{b}$ is integral over $\mathfrak{a}$.*

(4) *Let $\mathfrak{a}$ be an ideal of A, and assume that A is integrally closed in B. Then the integral closure of $\mathfrak{a}$ in A is equal to the integral closure of $\mathfrak{a}$ in B.*

(5) *Let $S \subset A$ be multiplicatively closed. If $\mathfrak{a}$ is an integrally closed ideal of A, then $S^{-1}\mathfrak{a}$ is an integrally closed ideal of $S^{-1}A$.*

Proof: (a) Let $\mathfrak{a}, \mathfrak{b}$ be A-submodules of B. If $\mathfrak{b}$ is integral over $\mathfrak{a}$, then the ring $\mathcal{R}(\mathfrak{b}, B)$ is integral over the ring $\mathcal{R}(\mathfrak{a}, B)$. In fact, consider any element $F = \sum_{j=0}^{h} b_j' T^j \in \mathcal{R}(\mathfrak{b}, B)$ where $b_j' \in \mathfrak{b}^j$ for every $j \in \{0, \ldots, h\}$. Now $b_0' \in A$, hence b_0' is integral over $\mathcal{R}(\mathfrak{a}, B)$ and, for every $j \in \{1, \ldots, h\}$, b_j' is a sum of elements of the form $b_1 \cdots b_j$ where $b_1, \ldots, b_j \in \mathfrak{b}$. Since $b_1 T, \ldots, b_j T$ are integral over $\mathcal{R}(\mathfrak{a}, B)$, it

is clear that F is integral over $\mathcal{R}(\mathfrak{a}, B)$. Conversely, assume that the ring $\mathcal{R}(\mathfrak{b}, B)$ is integral over the ring $\mathcal{R}(\mathfrak{a}, B)$. For every $b \in \mathfrak{b}$ the element $bT \in \mathcal{R}(\mathfrak{b}, B)$ is integral over $\mathcal{R}(\mathfrak{a}, B)$, hence b is integral over $\mathfrak{a}$ [cf. (6.2)].
(b) The assertions in (1), (2), and (3) follow immediately.
(4) Consider the chain of rings $\mathcal{R}(\mathfrak{a}, A) \subset A[T] = \bigoplus_{n \geq 0} AT^n \subset \bigoplus_{n \geq 0} BT^n = B[T]$. Since $A[T]$ is integrally closed in $B[T]$ [cf. [63], Ex. 4.17], and since every element of $B[T]$ which is integral over $\mathcal{R}(\mathfrak{a}, A)$ is integral over $A[T]$ also, the assertion in (4) follows immediately.
(5) This is easy to check.

(6.7) Proposition: (1) *Let x be a regular element of A. An element $z \in A$ is integral over the ideal Ax iff the element z/x of the ring of quotients $Q(A)$ of A is integral over A.*
(2) *Let $\mathfrak{a}$ be a finitely generated ideal of A which contains a regular element of A, let $\mathfrak{b}$ an ideal of A and $z \in A$. If $z\mathfrak{a} \subset \mathfrak{a}\mathfrak{b}$, then z is integral over $\mathfrak{b}$.*

Proof: (1) Assume that z is integral over Ax. Then there exist $n \in \mathbb{N}$ and $a_1, \ldots, a_n \in A$ such that $z^n + a_1 x z^{n-1} + \cdots + a_n x^n = 0$. Dividing by x^n gives an equation of integral dependence for the element $z/x \in Q(A)$ over the ring A. In the same way we can prove the other implication.
(2) Use (6.1) with $B := A$, $M = N := \mathfrak{a}$, $\mathfrak{a} := \mathfrak{b}$, $b := z$ and note that $\operatorname{Ann}_A(\mathfrak{a}) = \{0\}$ since $\mathfrak{a}$ contains regular elements [cf. also the last part of (6.2)].

(6.8) Proposition: *Let A be an integrally closed domain, and let $\mathfrak{a}$ be an ideal of A. For every $n \in \mathbb{N}$ let $\overline{\mathfrak{a}^n}$ be the integral closure of $\mathfrak{a}^n$ in A. Then $\bigoplus_{n \geq 0} \overline{\mathfrak{a}^n} T^n$ is the integral closure of $\mathcal{R}(\mathfrak{a}, A)$ in $A[T]$.*

Proof: Let $d \in \mathbb{N}_0$, and let $xT^d \in \bigoplus_{n \geq 0} \overline{\mathfrak{a}^n} T^n$ be homogeneous of degree d, i.e., we have $x \in \overline{\mathfrak{a}^d}$. Then x is integral over $\mathfrak{a}^d$, hence xT^d is integral over $\bigoplus_{n \geq 0} \mathfrak{a}^{dn} T^{dn} \subset \mathcal{R}(\mathfrak{a}, A)$, hence x is integral over $\mathcal{R}(\mathfrak{a}, A)$.
The integral closure of $\mathcal{R}(\mathfrak{a}, A)$ in $A[T]$ is a homogeneous subring of $A[T]$ [cf. (4.25)]. Let $d \in \mathbb{N}$, $x \in A$, and let xT^d be integral over $\mathcal{R}(\mathfrak{a}, A)$. Then $xT^d \in (A[T])^{(d)} = A[T^d]$ is integral over $(\mathcal{R}(\mathfrak{a}, A))^{(d)} = \bigoplus_{n \geq 0} \mathfrak{a}^{nd} T^{nd}$ [cf. (4.26)], hence x is integral over $\mathfrak{a}^d$ by (6.2), i.e., $xT^d \in \bigoplus \overline{\mathfrak{a}^{dn}} T^{dn} \subset \bigoplus \overline{\mathfrak{a}^n} T^n$.

(6.9) Proposition: *Let A be an integrally closed domain, and let K be its field of quotients. Let $\mathfrak{a}$ be an ideal of A. If all the powers $\mathfrak{a}^n$, $n \in \mathbb{N}$, are integrally closed in A, then for every non-zero $a \in \mathfrak{a}$ the ring $A[\mathfrak{a}/a]$ is integrally closed in K.*

Proof: The field of quotients of $\mathcal{R}(\mathfrak{a}, A)$ is $K(T)$. Now $\mathcal{R}(\mathfrak{a}, A)$ is integrally closed in $A[T]$ by (6.8), $A[T]$ is integrally closed in $K[T]$ by [63], Ex. 4.17, and $K[T]$ is integrally closed in $K(T)$ since it is factorial [cf. [63], Prop. 4.10]. Therefore $\mathcal{R}(\mathfrak{a}, A)$ is integrally closed, hence $\mathcal{R}(\mathfrak{a}, A)_{(aT)}$ is integrally closed [cf. (4.30)(2)], and therefore $A[\mathfrak{a}/a]$ is integrally closed [cf. (5.5)].

6.3   Integral Closure of Ideals and Valuation Theory

(6.10) NOTATION: For the rest of this subsection let A be an integral domain with field of quotients K, and let $\mathfrak{R}(A)$ be the set of valuations v of K with $A_v \supset A$. If $\mathfrak{a}$ is a finitely generated ideal of A, then $v(\mathfrak{a}) := \min(\{v(a) \mid a \in \mathfrak{a}\})$ is well defined.

(6.11) DEFINITION: An ideal $\mathfrak{a}$ of A is called a valuation ideal if there exists a valuation $v \in \mathfrak{R}(A)$ and an ideal $\mathfrak{a}_v$ of A_v with $\mathfrak{a}_v \cap A = \mathfrak{a}$. In this case we say, in particular, that $\mathfrak{a}$ is a v-ideal, or that $\mathfrak{a}$ belongs to the valuation v.

(6.12) REMARK: Let $v \in \mathfrak{R}(A)$, and let $\mathfrak{a}$ be an ideal of A. The following statements are equivalent:
(1) There exists an ideal $\mathfrak{a}_v$ of A_v with $\mathfrak{a}_v \cap A = \mathfrak{a}$.
(2) If $a, b \in A$, $a \in \mathfrak{a}$ and $v(b) \geq v(a)$, then we have $b \in \mathfrak{a}$.
(3) We have $\mathfrak{a}A_v \cap A = \mathfrak{a}$.
Proof: We have $b = (b/a) \cdot a \in \mathfrak{a}_v \cap A$, hence (1) implies (2). Now assume that (2) holds, and let $b \in \mathfrak{a}A_v$. Then there exist $a_1, \ldots, a_n \in \mathfrak{a}$, $c_1, \ldots, c_n \in A_v$ with $b = a_1c_1 + \cdots + a_nc_n$. We choose $i \in \{1, \ldots, n\}$ with $v(a_i) \leq v(a_j)$ for $j \in \{1, \ldots, n\}$. Then we have $v(b) \geq v(a_i)$; if $b \in A$, then we have $b \in \mathfrak{a}$, hence (2) implies (3). It is clear that (3) implies (1).

(6.13) REMARK: Let $v \in \mathfrak{R}(A)$.
(1) The set of ideals of A_v is totally ordered by inclusion [cf. I(2.2)], hence the set of v-ideals of A is also totally ordered by inclusion.
(2) If $\mathfrak{a}$, $\mathfrak{b}$ are v-ideals, and $\mathfrak{c}$ is an arbitrary ideal of A, then $\mathfrak{a} \cap \mathfrak{b}$ and $\mathfrak{a} : \mathfrak{c}$ are v-ideals of A.

(6.14) Proposition: *The integral closure $\bar{\mathfrak{a}}_K$ of $\mathfrak{a}$ in K is the intersection*

$$\bar{\mathfrak{a}}_K = \bigcap_{v \in \mathfrak{R}(A)} \mathfrak{a}A_v,$$

and $\bar{\mathfrak{a}}_K$ is an ideal of the integral closure of A.

Proof: Let $x \in \bar{\mathfrak{a}}_K$. Then there exists a finitely generated non-zero A-submodule M of K with $xM \subset \mathfrak{a}M$ [cf. (6.2)]. Let $z_1, \ldots, z_h$ be a system of generators of M; for every $i \in \{1, \ldots, h\}$ we can write

$$xz_i = \sum_{j=1}^{h} a_{ij}z_j \quad \text{with } a_{i1}, \ldots, a_{ih} \in \mathfrak{a}.$$

Let $v \in \mathfrak{R}(A)$ and choose $k, l \in \{1, \ldots, h\}$ such that $v(a_{kl}) \leq v(a_{ij})$ for all i, $j \in \{1, \ldots, h\}$. Then x/a_{kl} is integral over A_v [cf. [63], Cor. 4.6], hence $x/a_{kl} \in A_v$ [cf. I(3.2)], whence $x \in a_{kl}A_v \subset \mathfrak{a}A_v$.
Conversely, let $x \in K$ and assume that $x \in \mathfrak{a}A_v$ for every $v \in \mathfrak{R}(A)$. We consider the subring $B = A[\mathfrak{a}/x]$ of K which contains A. Let v be a valuation of K with

$A_v \supset B$. Then we have $v \in \mathfrak{R}(A)$, and there exist $a_1, \ldots, a_h \in \mathfrak{a}$, $c_1, \ldots, c_h \in A_v$ with $x = a_1 c_1 + \cdots + a_h c_h$. As above, we see that there exists $a \in \mathfrak{a}$ with $v(x/a) \geq 0$, i.e., $v(x) \geq v(a)$. This implies that the maximal ideal of A_v does not contain the ideal of B generated by the elements a/x, $a \in \mathfrak{a}$. Since this is true for every valuation v of K with $A_v \supset B$, the ideal of B generated by the elements a/x, $a \in \mathfrak{a}$, is the ring B itself [cf. I(3.6)]. Thus, there exist elements $a_1, \ldots, a_h \in \mathfrak{a}$ and polynomials $f_1, \ldots, f_h \in A[T_1, \ldots, T_h]$ with

$$1 = \frac{a_1}{x} f_1\left(\frac{a_1}{x}, \ldots, \frac{a_h}{x}\right) + \cdots + \frac{a_h}{x} f_h\left(\frac{a_1}{x}, \ldots, \frac{a_h}{x}\right). \tag{*}$$

There exists $p \in \mathbb{N}$ such that for $i \in \{1, \ldots, h\}$ we have $x^p f_i(a_1/x, \ldots, a_p/x) \in A$, hence

$$x^p f_i\left(\frac{a_1}{x}, \ldots, \frac{a_h}{x}\right) = a_{0i} x^p + \cdots + a_{pi} \quad \text{with } a_{ji} \in \mathfrak{a}^j \text{ for } j \in \{0, \ldots, p\}.$$

Multiplying (*) by x^{p+1} yields $x^{p+1} = a_0' x^p + \cdots + a_p'$ with $a_j' = \sum_{l=1}^h a_l a_{jl} \in \mathfrak{a}^{j+1}$ for $j \in \{0, \ldots, p\}$, hence x is integral over $\mathfrak{a}$. The last assertion of the proposition now follows from the fact that the integral closure of A is the intersection of all the rings A_v with $v \in \mathfrak{R}(A)$ [cf. I(3.4)].

(6.15) Corollary: *Assume that A is integrally closed. For every integrally closed ideal $\mathfrak{a}$ of A we have*

$$\mathfrak{a} = \bigcap_{v \in \mathfrak{R}(A)} \mathfrak{a} A_v = \bigcap_{v \in \mathfrak{R}(A)} (\mathfrak{a} A_v \cap A),$$

hence every integrally closed ideal of A is an intersection of valuation ideals of A.

(6.16) REMARK: Assume that A is integrally closed.
(1) Let $(\mathfrak{a}_i)_{i \in I}$ be a family of integrally closed ideals of A, and set $\mathfrak{b} := \bigcap_{i \in I} \mathfrak{a}_i$. From (6.15) we get

$$\mathfrak{b} \subset \bigcap_{v \in \mathfrak{R}(A)} \mathfrak{b} A_v \subset \bigcap_{i \in I}\left(\bigcap_{v \in \mathfrak{R}(A)} \mathfrak{a}_i A_v\right) = \bigcap_{i \in I}\left(\bigcap_{v \in \mathfrak{R}(A)} (\mathfrak{a}_i A_v \cap A)\right) = \bigcap_{i \in I} \mathfrak{a}_i = \mathfrak{b},$$

hence $\mathfrak{b} = \bigcap_{v \in \mathfrak{R}(A)} \mathfrak{b} A_v$, and therefore $\mathfrak{b}$ is integrally closed in A by (6.14).
(2) Let $v \in \mathfrak{R}(A)$, let $\mathfrak{a}$ be a v-ideal of A, and let $\mathfrak{b}$ be its integral closure in A. Then $\mathfrak{a} \subset \mathfrak{b} = \mathfrak{b} \cap A \subset \mathfrak{a} A_v \cap A = \mathfrak{a}$, hence $\mathfrak{b} = \mathfrak{a}$, and $\mathfrak{a}$ is integrally closed in A.
(3) Any intersection of valuation ideals of A is integrally closed in A [cf. (1) and (2)], hence *an ideal of A is integrally closed iff it is an intersection of valuation ideals of A* [cf. (6.15)].
(4) Let $\mathfrak{a}$, $\mathfrak{b}$ be ideals of A. Then we have $\overline{\mathfrak{a}\mathfrak{b}} = \overline{\overline{\mathfrak{a}}\,\overline{\mathfrak{b}}}$.
In fact, for every $v \in \mathfrak{R}(A)$ we have $\overline{\mathfrak{a}} \subset \mathfrak{a} A_v$, $\overline{\mathfrak{b}} \subset \mathfrak{b} A_v$, hence $\overline{\mathfrak{a}}\,\overline{\mathfrak{b}} \subset \bigcap_{v \in \mathfrak{R}(A)} (\mathfrak{a}\mathfrak{b}) A_v$, i.e., $\overline{\mathfrak{a}}\,\overline{\mathfrak{b}} \subset \overline{\mathfrak{a}\mathfrak{b}}$ [cf. (6.14)], and therefore $\overline{\overline{\mathfrak{a}}\,\overline{\mathfrak{b}}} \subset \overline{\mathfrak{a}\mathfrak{b}}$. Since $\mathfrak{a}\mathfrak{b} \subset \overline{\mathfrak{a}}\,\overline{\mathfrak{b}}$, we have $\overline{\mathfrak{a}\mathfrak{b}} \subset \overline{\overline{\mathfrak{a}}\,\overline{\mathfrak{b}}}$.
Thus, we have shown that $\overline{\mathfrak{a}\mathfrak{b}} = \overline{\overline{\mathfrak{a}}\,\overline{\mathfrak{b}}}$.

(5) Let $\mathfrak{a}$, $\mathfrak{b}$ be ideals of A. If $\mathfrak{a}$ is integrally closed, then $\mathfrak{a} : \mathfrak{b}$ is integrally closed. In fact, let $x \in A$ be integral over $\mathfrak{a} : \mathfrak{b}$. Then there exists a finitely generated non-zero A-submodule M of K with $xM \subset (\mathfrak{a} : \mathfrak{b})M$ [cf. (6.2)], hence $x\mathfrak{b}M \subset \mathfrak{a}M$, and therefore every element of $x\mathfrak{b}$ is integral over $\mathfrak{a}$, hence $x\mathfrak{b} \subset \mathfrak{a}$, and therefore we have $x \in \mathfrak{a} : \mathfrak{b}$.

(6) Let $\mathfrak{a}$ be an ideal of A, and let $x \in A$, $x \neq 0$. Then $\mathfrak{a}$ is integrally closed iff $x\mathfrak{a}$ is integrally closed.

(6.17) Proposition: *Let $\mathfrak{a}$ be a finitely generated ideal of A, and let $\{a_1, \dots, a_h\}$ be a system of non-zero generators of $\mathfrak{a}$. For every $i \in \{1, \dots, h\}$ set $A_i := A[\mathfrak{a}/a_i]$, and let B_i be the integral closure of A_i. Then we have $\bar{\mathfrak{a}}_K = \bigcap_{i=1}^{h} a_i B_i$.*

Proof: Let $x \in \bar{\mathfrak{a}}_K$. Let $i \in \{1, \dots, h\}$, and let V be a valuation ring of K with $B_i \subset V$. We have $\mathfrak{a}B_i = a_i B_i$, hence $x \in \mathfrak{a}V = a_i V$ [cf. (6.14)], and therefore we have $x/a_i \in V$. This implies that $x/a_i \in B_i$ [cf. I(3.4)], hence that $x \in a_i B_i$. Conversely, let $x \in \bigcap a_i B_i$, and let $v \in \mathfrak{R}(A)$. There exists $i \in \{1, \dots, h\}$ with $v(a_j/a_i) \geq 0$ for $j \in \{1, \dots, h\}$. Then we have $A[\mathfrak{a}/a_i] \subset A_v$, hence $B_i \subset A_v$ [cf. I(3.4)], hence $x \in a_i B_i \subset \mathfrak{a}V$, and therefore $x \in \bar{\mathfrak{a}}_K$ [cf. (6.14)].

7 Decomposition Group and Inertia Group

(7.1) NOTATION: Let K be a field. We denote by $\mathrm{qln}(K)$ the set of quasilocal integrally closed (= normal) subrings of K having K as field of quotients.

(7.2) NOTATION: Let K be a field, and let L be a finite extension of K. Let $A \in \mathrm{qln}(K)$, and let S be the integral closure of A in L. Let $\mathfrak{m}$ be the maximal ideal of A. For every maximal ideal $\mathfrak{q}$ of S we have $\mathfrak{q} \cap A = \mathfrak{m}$. Any ring $B = S_{\mathfrak{q}}$ where $\mathfrak{q}$ is a maximal ideal of S is said to lie over A. Note that such a ring B is integrally closed [cf. [63], Prop. 4.13], hence lies in $\mathrm{qln}(L)$, and dominates A.

(7.3) Proposition: *Let K be a field, and let L be a finite extension of K. Let $A \in \mathrm{qln}(K)$. Then:*

(1) The integral closure of A in L is quasisemilocal. The set of rings in $\mathrm{qln}(L)$ lying over A is not empty and has at most $[L : K]_{sep}$ elements.

(2) For any $B \in \mathrm{qln}(L)$ lying over A we have $B \cap K = A$ and $\mathfrak{n} \cap K = \mathfrak{m}$ where $\mathfrak{n}$ (resp. $\mathfrak{m}$) is the maximal ideal of B (resp. of A).

(3) Let L' be a subfield of L containing K, let $B \in \mathrm{qln}(L)$ lie over A, and set $B' := B \cap L'$. Then $B' \in \mathrm{qln}(L')$ lies over A, and B lies over B'. Conversely, let $C' \in \mathrm{qln}(L')$ lie over A. Then there exists $C \in \mathrm{qln}(L)$ lying over A with $C \cap L' = C'$.

Proof: Let M be a finite normal extension of K containing L, and let S (resp. T) be the integral closure of A in L (resp. in M). Note that T is the integral closure of S in M. Let $G := \mathrm{Gal}(M/K)$ be the group of K-automorphisms of M; we have

$\mathrm{ord}(G) \leq [M : K]$. Let $\mathfrak{Q}$ be a maximal ideal of T; then $\{\sigma(\mathfrak{Q}) \mid \sigma \in G\}$ is the set of all maximal ideals of T [cf. [63], Prop. 13.10], hence T has at most $[M : K]$ maximal ideals.

(1) Since every maximal ideal of T lies over a maximal ideal of S, the set of maximal ideals of S is finite, hence S is quasisemilocal. Since there exist at most $[L : K]_{\mathrm{sep}}$ pairwise different K-embeddings of L into M, it is clear that S has at most $[L : K]_{\mathrm{sep}}$ maximal ideals.

(2) Let q be the maximal ideal of S with $B = S_{\mathfrak{q}}$, and label the maximal ideals $\mathfrak{Q}_1 = \mathfrak{Q}, \mathfrak{Q}_2, \ldots, \mathfrak{Q}_t$ of T in such a way that $\mathfrak{Q} \cap S = \mathfrak{q}$. We set $D := T_{\mathfrak{Q}}$, and let $\mathfrak{N}$ be the maximal ideal of D; then we have $B \subset D$, and D dominates B. Therefore it is enough to show that $D \cap K = A$ and that $\mathfrak{N} \cap K = \mathfrak{m}$.

Let $x \in D \cap K$. For every $j \in \{1, \ldots, t\}$ there exists $\sigma_j \in G$ with $\sigma_j(\mathfrak{Q}) = \mathfrak{Q}_j$, hence with $\sigma(T_{\mathfrak{Q}}) = T_{\mathfrak{Q}_j}$ [since $\sigma_j(T) = T$]. Since $\sigma(x) = x$ for every $\sigma \in G$, we have $x \in T_{\mathfrak{Q}_1} \cap T_{\mathfrak{Q}_2} \cap \cdots \cap T_{\mathfrak{Q}_t} = T$ [cf. (2.5)]. Therefore we have $x \in T \cap K = A$, and we have shown that $D \cap K = A$. This implies that $\mathfrak{N} \cap K \subset A$; since $\mathfrak{m} \subset \mathfrak{N} \cap A$ and no element of $\mathfrak{N} \cap K$ is invertible in A, it follows that $\mathfrak{N} \cap K = \mathfrak{m}$.

(3) Let q be the maximal ideal of S with $B = S_{\mathfrak{q}}$, and set $\mathfrak{n} := \mathfrak{q} B$. Let S' be the integral closure of A in L', and set $\mathfrak{q}' := \mathfrak{q} \cap S'$. Then $\mathfrak{q}'$ is a maximal ideal of S', $S'_{\mathfrak{q}'} \in \mathrm{qln}(L')$ lies over A, $S_{\mathfrak{q}}$ is the integral closure of $S'_{\mathfrak{q}'}$ in L and $B = (S_{\mathfrak{q}'})_{\mathfrak{n} \cap S_{\mathfrak{q}'}}$, showing that B lies over $S'_{\mathfrak{q}'}$, hence that $S'_{\mathfrak{q}'} = B \cap L'$ by (2). Thus, we have shown that $B' = B \cap L' \in \mathrm{qln}(L')$ lies over A, and that B lies over B'.

Conversely, let $\mathfrak{q}'$ be the maximal ideal of S' with $C' = S'_{\mathfrak{q}'}$, and let q be a maximal ideal of S with $\mathfrak{q} \cap S' = \mathfrak{q}'$. Then $C := S_{\mathfrak{q}}$ is also a localization of the integral closure $S_{\mathfrak{q}'}$ of C' in L with respect to a maximal ideal of $S_{\mathfrak{q}'}$, hence C lies over C', and therefore $C' = C \cap L'$ by (2).

(7.4) Proposition: *Let L/K be a finite extension of fields, let $A \in \mathrm{qln}(K)$, let $\mathfrak{m}$ be the maximal ideal of A and $\kappa = A/\mathfrak{m}$ the residue field of A. Then:*

(1) *Let S be the integral closure of A in L. Let $a \in S$, and let $\bar{a}$ be the image of a in $S/\mathfrak{m}S$. Then we have $\dim_\kappa(\kappa[\bar{a}]) \leq [L : K]$.*

(2) *Let $B \in \mathrm{qln}(L)$ lie over A. The residue field λ of B is an algebraic extension of the residue field κ of A: We have $\dim_\kappa(\kappa[\zeta]) \leq [L : K]$ for every $\zeta \in \lambda$.*

Proof: (1) The minimal polynomial $F \in K[T]$ of a over K has its coefficients in A [cf. III(2.12)(2)]. Reducing this polynomial modulo the maximal ideal of A gives a monic polynomial $\bar{F} \in \kappa[T]$, and we have $\bar{F}(\bar{a}) = 0$, hence $\dim_\kappa(\kappa[\bar{a}]) \leq \deg(F) \leq [L : K]$.

(2) Let q be the maximal ideal of S with $S_{\mathfrak{q}} = B$. We have $S/\mathfrak{q} = B/\mathfrak{q}B = \lambda$; the result follows from (1), since the κ-algebra λ is a homomorphic image of $S/\mathfrak{m}S$.

(7.5) REMARK: Let K be an algebraic function field in r variables over a field k. Let R be a local r-dimensional k-subalgebra of K essentially of finite type, set $\mathfrak{m} := \mathfrak{m}(R)$, and let A be a k-subalgebra of K such that R is a localization of A with respect to a multiplicatively closed system. Then $\mathfrak{n} := \mathfrak{m} \cap A$ is a maximal ideal of A, $\dim(A) = r$, $A_\mathfrak{n} = R$, and the residue field $R/\mathfrak{m} = \kappa$ is a finite extension of k.

In fact, we have $A_\mathfrak{n} = R$ [cf. (2.3)(3)]. We have $\mathrm{ht}(\mathfrak{n}) = \dim(R) = r$ and $r = \mathrm{tr.d}_k(K) = \mathrm{ht}(\mathfrak{n}) + \dim(A/\mathfrak{n})$ [cf. [63], Cor. 13.4], hence $\mathfrak{n}$ is a maximal ideal of A, $A/\mathfrak{n} = R/\mathfrak{m} = \kappa$ is a finite extension of k [cf. [63], Cor. 13.12], and $\dim(A) = r$.

Let L be a finite extension of K, and let $S \subset L$ be a local ring lying over R. The integral closure B of A in L is a k-algebra of finite type [cf. (3.6)], and S is a localization of $B_\mathfrak{n}$ with respect to a maximal ideal $\mathfrak{n}$ of B, hence we have $\dim(S) = r$. We have $S/\mathfrak{m}(S) = B/\mathfrak{n}$, $B/\mathfrak{n}$ is an extension of k of finite type and $\mathrm{tr.d}_k(B/\mathfrak{n}) = 0$, hence $S/\mathfrak{m}(S)$ is a finite extension of k.

(7.6) NOTATION: For the rest of this section we consider the following situation. Let K be a field, let L be a finite Galois extension of K with Galois group G, let $A \in \mathrm{qln}(K)$, let S be the integral closure of A in L, and let $B \in \mathrm{qln}(L)$ lie over A; let $\mathfrak{q}$ be the maximal ideal of S with $B = S_\mathfrak{q}$. Let $\mathfrak{m}$ be the maximal ideal and $\kappa = A/\mathfrak{m}$ the residue field of A, and let $\mathfrak{n}$ be the maximal ideal and $\lambda = B/\mathfrak{n}$ the residue field of B; note that λ contains κ as a subfield.

We define
$$G_Z = G_Z(B/A) := \{\sigma \in G \mid \sigma(B) = B\}.$$

Clearly G_Z is a subgroup of G; it is called the decomposition group or splitting group of B over A ["Zerlegungsgruppe" in german]. We set $g := (G : G_Z)$. The fixed field of G_Z shall be denoted by K_Z; we set $A_Z := B \cap K_Z$ and let κ_Z be the residue field of A_Z. Note that $G_Z = \{\sigma \in G \mid \sigma(\mathfrak{q}) = \mathfrak{q}\}$; this yields, in particular, that S has g maximal ideals.

(7.7) Proposition: *Using the notation of* (7.6) *we have:*
(1) The number of rings in $\mathrm{qln}(L)$ *lying over* A *is equal to* $(G : G_Z)$. *For every* $\sigma \in G$ *we have* $G_Z(\sigma(B)/A) = \sigma^{-1} G_Z(B/A)\sigma$.
(2) K_Z *is the smallest field* K' *between* K *and* L *such that* B *is the only ring in* $\mathrm{qln}(L)$ *lying over* $B \cap K'$.
(3) A_Z *and* A *have the same residue field* $\kappa_Z = \kappa$, *and* $\mathfrak{m}A_Z$ *is the maximal ideal of* A_Z.

Proof: Let $\mathfrak{q}_1 = \mathfrak{q}$, $\mathfrak{q}_2, \ldots, \mathfrak{q}_g$ where $g = (G : G_Z)$ be the maximal ideals of S. Now $S_Z := S \cap K_Z$ is the integral closure of A in K_Z; setting $\mathfrak{p} := \mathfrak{q} \cap S_Z$ we have $(S_Z)_\mathfrak{p} = B \cap K_Z = A_Z$ [cf. (7.3)].

(1) The first assertion was already stated in (7.6), and the second assertion follows immediately from the definition of the decomposition group.

(2) By the Chinese remainder theorem [cf. (10.1)] there exists $a \in S$ with $a \equiv 0$ (mod $\mathfrak{q}_1$), $a \equiv 1$ (mod $\mathfrak{q}_j$) for $j \in \{2, \ldots, g\}$.

(a) We have $\sigma(\mathfrak{q}_1) = \mathfrak{q}_1$ for every $\sigma \in G_Z$, hence $\mathrm{N}_{L/K_Z}(a) = \prod_{\sigma \in G_Z} \sigma(a) \in \mathfrak{q}_1 \cap K_Z = \mathfrak{p}$ [cf. (7.3)].

(b) Let K' be as stated. We set $\mathfrak{a} := \mathfrak{q}_2 \cdots \mathfrak{q}_g = \mathfrak{q}_2 \cap \cdots \cap \mathfrak{q}_g$; then we have $a - 1 \in \mathfrak{q}_2 \cap \cdots \cap \mathfrak{q}_g = \mathfrak{a}$. Let $j \in \{2, \ldots, g\}$; for $\sigma \in G_Z$ we have $\sigma(\mathfrak{q}_j) \in \{\mathfrak{q}_2, \ldots, \mathfrak{q}_g\}$, hence

$\sigma(\mathfrak{a}) = \mathfrak{a}$, hence $\sigma(a) \equiv 1 \pmod{\mathfrak{a}}$. Now we get $N_{L/K_Z}(a) = \prod_{\sigma \in G_Z} \sigma(a) \equiv 1$ (mod $\mathfrak{a}$), and therefore $N_{L/K_Z}(a) \equiv 1 \pmod{\mathfrak{q}_j}$. This shows that none of the ideals $\mathfrak{q}_2, \ldots, \mathfrak{q}_g$ lies over the ideal $\mathfrak{p}$ of S_Z, and therefore we have shown that $K' \subset K_Z$, hence $G_Z \subset \mathrm{Gal}(L/K')$. To prove the other inclusion, let $\sigma \in \mathrm{Gal}(L/K')$. Suppose that $\sigma(\mathfrak{q}) = \mathfrak{q}_j$ for some $j \in \{2, \ldots, g\}$. We have $\mathfrak{q} \cap K' = \sigma(\mathfrak{q}) \cap K' = \mathfrak{q}_j \cap K'$, hence $\mathfrak{q}_j$ lies over the prime ideal $\mathfrak{q} \cap (K' \cap S)$, contrary to the definition of K'. Therefore we have $\sigma(\mathfrak{q}) = \mathfrak{q}$, hence $\sigma \in G_Z$, which implies that $G_Z = \mathrm{Gal}(L/K')$, hence that $K_Z = K'$.

(3) First, we show that $A/\mathfrak{m} = S_Z/\mathfrak{p}$, hence that A and A_Z have the same residue field. Let $a \in S_Z$. By the Chinese remainder theorem, and by (2), there exists $b \in S_Z$ with $b \equiv a \pmod{\mathfrak{p}}$, $b \equiv 1 \pmod{\mathfrak{q}_j \cap S_Z}$ for $j \in \{2, \ldots, g\}$, hence with $b \equiv a \pmod{\mathfrak{q}}$, $b \equiv 1 \pmod{\mathfrak{q}_j}$ for $j \in \{2, \ldots, g\}$. We choose $\tau_1 = \mathrm{id}_L$, $\tau_2, \ldots, \tau_g \in G$ such that $G = \tau_1 G_Z \cup \cdots \cup \tau_g G_Z$ (disjoint union of left cosets of G_Z in G). Then $\{\tau_1, \tau_2, \ldots, \tau_g\}$ is the set of pairwise different K-monomorphisms of K_Z in L, and therefore we have $N_{K_Z/K}(b) = \tau_1(b) \cdots \tau_g(b) =: c \in A$ [cf., e.g., [204], vol. I, Ch. II, § 10, (19), and use the fact that b is integral over A, hence that the characteristic polynomial of $b \in K_Z$ has its coefficients in A by III(2.12)(2)]. We show that $c \equiv a \pmod{\mathfrak{p}}$. We have $\tau_1(b) = b \equiv a \pmod{\mathfrak{q}}$. Let $i \in \{2, \ldots, g\}$; there exists $j \in \{2, \ldots, g\}$ with $\tau_j(\mathfrak{q}_i) = \mathfrak{q}$, hence we have $\tau_j(b) \equiv 1 \pmod{\mathfrak{q}}$. Thus, we have shown that $c \equiv a \pmod{\mathfrak{p}}$.

Now we show that $\mathfrak{m}A_Z$ is the maximal ideal of A_Z. Let $\mathfrak{p}_1 := \mathfrak{p}, \mathfrak{p}_2, \ldots, \mathfrak{p}_h$ be the maximal ideals of S_Z. Let $\mathfrak{m}S_Z = \tilde{\mathfrak{p}}_1 \cap \cdots \cap \tilde{\mathfrak{p}}_h = \tilde{\mathfrak{p}}_1 \cdots \tilde{\mathfrak{p}}_h$ be a primary decomposition of $\mathfrak{m}S_Z$ where $\tilde{\mathfrak{p}}_i$ is a $\mathfrak{p}_i$-primary ideal of S_Z for $i \in \{1, \ldots, h\}$ [cf. (10.21)].

(a) We show: If $z \in \mathfrak{p}_1$, but $z \notin \mathfrak{p}_i$ for $i \in \{2, \ldots, h\}$, then we have $z \in \tilde{\mathfrak{p}}_1$. In fact, set $w := \tau_2(z) \cdots \tau_g(z) \in S$. Let $\sigma \in G_Z$. The elements $\sigma\tau_2, \ldots, \sigma\tau_g$ lie in pairwise different left cosets of G_Z in G, hence $\sigma(w) = w$. This yields $w \in S \cap K_Z = S_Z$. Suppose that $w \in \mathfrak{p}_1 \subset \mathfrak{q}$. Then there exists $i \in \{2, \ldots, g\}$ with $\tau_i(z) \in \mathfrak{q}$, hence z lies in $\tau_i^{-1}(\mathfrak{q}) = \mathfrak{q}_j$ for some $j \in \{2, \ldots, g\}$; since $\mathfrak{q}_j \cap S_Z$ is one of the ideals $\mathfrak{p}_2, \ldots, \mathfrak{p}_h$, this contradicts the choice of z. Now we have $zw = N_{K_Z/K}(z) \in \mathfrak{m}$ [if zw would be a unit of A, then zw would be a unit of S_Z], hence $zw \in \tilde{\mathfrak{p}}_1$, and therefore $z \in \tilde{\mathfrak{p}}_1$.

(b) Set $\mathfrak{r} := \mathfrak{p}_1 \cap \cdots \cap \mathfrak{p}_h = \mathfrak{p}_1 \cdots \mathfrak{p}_h$. We choose $u \in S_Z$ with $u \equiv 0 \pmod{\mathfrak{p}_1}$ and $u \equiv 1 \pmod{\mathfrak{p}_i}$ for $i \in \{2, \ldots, h\}$. Then we have $u \in \tilde{\mathfrak{p}}_1$ by (a). Let $r \in \mathfrak{r}$. Then we have $r + u \in \mathfrak{p}_1$, $r + u \notin \mathfrak{p}_i$ for $i \in \{2, \ldots, h\}$, hence $r + u \in \tilde{\mathfrak{p}}_1$ by (a), and therefore $r \in \tilde{\mathfrak{p}}_1$. Thus, we have shown that $\mathfrak{r} \subset \tilde{\mathfrak{p}}_1$, hence that $\mathfrak{p}_1 \cdots \mathfrak{p}_h \subset \tilde{\mathfrak{p}}_1$, hence that $\mathfrak{p}(S_Z)_\mathfrak{p} \subset \tilde{\mathfrak{p}}_1(S_Z)_\mathfrak{p} = \mathfrak{m}(S_Z)_\mathfrak{p} \subset \mathfrak{p}(S_Z)_\mathfrak{p}$, showing that $\mathfrak{m}A_Z$ is the maximal ideal $\mathfrak{p}(S_Z))_\mathfrak{p}$ of $A_Z = (S_Z)_\mathfrak{p}$.

(7.8) THE INERTIA GROUP: For $z \in B$ let $\bar{z}$ be its image in λ. We define a homomorphism of groups $\varphi \colon G_Z \to \mathrm{Aut}_\kappa(\lambda)$ in the following way: Let $\sigma \in G_Z$, $\zeta \in \lambda$ and $z \in B$ with $\bar{z} = \zeta$. Then we set $\varphi(\sigma)(\zeta) = \overline{\sigma(z)}$. The map $\varphi(\sigma)$ is well-defined [if $w \in B$ and $\bar{w} = \zeta$, then we have $w - z \in \mathfrak{n}$, hence $\sigma(w) - \sigma(z) = \sigma(w - z) \in \sigma(\mathfrak{n}) = \mathfrak{n}$, and therefore $\overline{\sigma(w)} = \overline{\sigma(z)}$]. Clearly $\varphi(\sigma) \in \mathrm{Aut}_\kappa(\lambda)$, and φ

is a homomorphism of groups. We define

$$G_T := G_T(B/A) := \ker(\varphi) = \{\sigma \in G_Z \mid \sigma(z) - z \in \mathfrak{n} \text{ for every } z \in B\}.$$

G_T is a normal subgroup of G_Z which is called the inertia group of B/A ["Trägheitsgruppe" in german]. We denote by K_T the fixed field of G_T; note that $K_Z \subset K_T$. Furthermore, we set $A_T := B \cap K_T$, and let κ_T be the residue field of the quasilocal ring A_T.

(2) We set $H := \{\sigma \in G \mid \sigma(z) - z \in \mathfrak{q} \text{ for every } z \in S\}$. For $\sigma \in H$ we have $\sigma(\mathfrak{q}) = \mathfrak{q}$, hence $H \subset G_Z$, and it is easy to check that H is a subgroup of G_Z: If $\sigma, \tau \in H$ and $z \in S$, then we have $\sigma\tau(z) - z = \sigma(\tau(z) - z) + (\sigma(z) - z)$, and $\sigma^{-1}(z) - z = \sigma^{-1}(z - \sigma(z))$. Clearly we have $G_T \subset H$. We show that $G_T = H$. In fact, let $\sigma \in H$ and $z \in B$. We write $z = z_1/z_2$ with $z_1 \in S$, $z_2 \in S \setminus \mathfrak{q}$. Then we have

$$\sigma(z) - z = \frac{\sigma(z_1)}{\sigma(z_2)} - \frac{z_1}{z_2} = \frac{\sigma(z_1)z_2 - z_1\sigma(z_2)}{z_2\sigma(z_2)} = \frac{z_2(\sigma(z_1) - z_1) - z_1(\sigma(z_2) - z_2)}{z_2\sigma(z_2)}.$$

Since $\sigma \in G_Z$ and $z_2 \notin \mathfrak{q}$, we have $\sigma(z_2) \notin \mathfrak{q}$, and therefore $\sigma(z) - z \in \mathfrak{n}$, hence $\sigma \in G_T$.

(3) Since B is the only ring in $\mathrm{qln}(L)$ lying over A_Z [cf. (7.7)], B is the integral closure of A_Z in L, hence we have $G_Z(B/A_Z) = G_Z$, and $G_T(B/A_Z) = G_T$.

(7.9) Proposition: Let K' be a subfield of L containing K, and set $A' := B \cap K'$ and $G' := \mathrm{Gal}(L/K')$. Then we have $G_Z(B/A') = G_Z(B/A) \cap G'$, $G_T(B/A') = G_T(B/A) \cap G'$, and the fixed field of $G_Z(B/A')$ (resp. of $G_T(B/A')$) is the compositum of K_Z with K' (resp. of K_T with K').

Proof: This follows immediately from Galois theory.

(7.10) Proposition: Let $A_1 \subset A_2$ lie in $\mathrm{qln}(K)$, and let $B_1 \subset B_2$ lie in $\mathrm{qln}(L)$ with $B_i \cap K = A_i$ for $i \in \{1, 2\}$.
(1) We have $G_T(B_2/A_2) \subset G_T(B_1/A_1)$.
(2) If A_2 dominates A_1, then we have $G_Z(B_2/A_2) \subset G_Z(B_1/A_1)$.

Proof: Let S_i be the integral closure of A_i in L, $\mathfrak{m}_i$ the maximal ideal of A_i, and $\mathfrak{n}_i$ the maximal ideal of B_i, $i \in \{1, 2\}$. Then we have $S_1 \subset S_2$. We set $\mathfrak{p}_i := \mathfrak{n}_i \cap S_i$, $i \in \{1, 2\}$, $\mathfrak{q} := \mathfrak{p}_2 \cap S_1 = \mathfrak{n}_2 \cap S_1$. We have $\mathfrak{n}_2 \cap B_1 \subset \mathfrak{n}_1$, hence $\mathfrak{q} \subset \mathfrak{n}_1 \cap S_1 = \mathfrak{p}_1$. Let $\sigma \in G_T(B_2/A_2)$. We use (7.8)(2): For $x \in S_1$ we have $x - \sigma(x) \in \mathfrak{p}_2 \cap S_1 = \mathfrak{q} \subset \mathfrak{p}_1$, showing that $G_T(B_2/A_2) \subset G_T(B_1/A_1)$.
Now we assume that A_2 dominates A_1, i.e., that $\mathfrak{m}_2 \cap A_1 = \mathfrak{m}_1$. We have $\mathfrak{q} \cap A_1 = (\mathfrak{n}_2 \cap S_1) \cap A_1 = \mathfrak{n}_2 \cap A_1 = (\mathfrak{n}_2 \cap A_2) \cap A_1 = \mathfrak{m}_2 \cap A_1 = \mathfrak{m}_1$, i.e., we have $\mathfrak{q} = \mathfrak{p}_1$ [since $\mathfrak{q}$ and $\mathfrak{p}_1$ lie over $\mathfrak{m}_1$ and $\mathfrak{q} \subset \mathfrak{p}_1$], hence $\mathfrak{p}_1 = \mathfrak{p}_2 \cap S_1$. Let $\sigma \in G_Z(B_2/A_2)$. For $x \in \mathfrak{p}_1$ we have $x \in \mathfrak{p}_2$, hence $\sigma(x) \in \mathfrak{p}_2$, and therefore $\sigma(x) \in \mathfrak{p}_2 \cap S_1 = \mathfrak{p}_1$, showing that $\sigma \in G_Z(B_1/A_1)$.

(7.11) Theorem: *With notations as in* (7.8) *we have:*
(1) λ/κ *is a normal extension,* $\varphi\colon G_Z \to \mathrm{Gal}(\lambda/\kappa)$ *is surjective,* $\ker(\varphi) = G_T$,
and K_T/K_Z *is a Galois extension. Moreover,* κ_T/κ_Z *is a Galois extension, we
have* $[K_T : K_Z] = [\kappa_T : \kappa_Z] = [\kappa_T : \kappa]$, *and* φ *induces an isomorphism*
$\mathrm{Gal}(K_T/K_Z) \xrightarrow{\sim} \mathrm{Gal}(\lambda/\kappa) = \mathrm{Gal}(\kappa_T/\kappa)$;
(2) $\mathfrak{m}A_T$ *is the maximal ideal of* A_T *and* κ_T *is the separable closure of* κ *in* λ.

Proof: The residue field κ of A coincides with the residue field κ_Z of A_Z by (7.7).
All the subgroups of $\mathrm{Gal}(L/K)$ considered in the theorem are subgroups of G_Z, and
all the subfields of L considered in the theorem contain K_Z; therefore we may—in
proving the theorem—replace K by K_Z and G_Z by G. Hence we may assume that
B is the only ring in $\mathrm{qln}(L)$ lying over A; in particular, B is the integral closure
of A in L. For $z \in B$ we denote by $\bar{z}$ its image in λ.
(a) λ is a normal extension of κ. In fact, let $\zeta \in \lambda$ and $z \in B$ with $\bar{z} = \zeta$. Let
$f \in \kappa[X]$ be the minimal polynomial of ζ over κ [note that λ is an algebraic
extension of κ by (7.4)], and let $F \in K[X]$ be the minimal polynomial of z over
K. Since z is integral over A, we have $F \in A[X]$ [cf. III(2.12)(2)]. Let $\overline{F} \in \kappa[X]$
be the polynomial which we get from F by reducing its coefficients modulo $\mathfrak{m}$.
Since $0 = \overline{F}(\bar{z}) = \overline{F}(\zeta)$, f divides $\overline{F}$. Since L/K is a Galois extension, we have

$$F = (X - z_1) \cdots (X - z_n) \text{ with } z_1 = z, z_2, \ldots, z_n \in L.$$

Since $F \in A[X]$ and B is the integral closure of A in L, we have $z_2, \ldots, z_n \in B$.
Now this implies that $\overline{F} = (X - \bar{z}_1) \cdots (X - \bar{z}_n)$, hence f splits in λ, and therefore
λ is a normal extension of κ.
(b) Let κ_{sep} be the separable closure of κ in λ. We have $[\kappa_{\mathrm{sep}} : \kappa] = [\lambda : \kappa]_{\mathrm{sep}} \leq$
$[L : K]$ [cf. III(5.4)(1)]. Let $g \in \kappa[X]$ be irreducible and have a zero in κ_{sep}.
Then g splits in λ and all its zeroes are separable over κ, hence g splits in κ_{sep},
and therefore $\kappa_{\mathrm{sep}}/\kappa$ is a finite Galois extension. Every element of $\mathrm{Gal}(\kappa_{\mathrm{sep}}/\kappa)$
has a unique extension as a κ-automorphism of λ; let $\rho\colon \mathrm{Gal}(\lambda/\kappa) \to \mathrm{Gal}(\kappa_{\mathrm{sep}}/\kappa)$
be the isomorphism defined by restriction. Let $\zeta \in \kappa_{\mathrm{sep}}$ be a primitive element
for $\kappa_{\mathrm{sep}}/\kappa$, and let $\zeta = \zeta_1, \zeta_2, \ldots, \zeta_m$ be the zeroes of the minimal polynomial f of
ζ over κ. Define F as in (a), and label the zeroes $z_1, \ldots, z_n$ of F in L in such a
way that $\bar{z}_i = \zeta_i$ for $i \in \{1, \ldots, m\}$. For every $i \in \{1, \ldots, m\}$ there exists $\sigma_i \in G$
with $\sigma_i(z_1) = z_i$, hence $\varphi(\sigma_i)(\zeta_1) = \zeta_i$. This shows that $\rho \circ \varphi\colon G \to \mathrm{Gal}(\lambda/\kappa) \to$
$\mathrm{Gal}(\kappa_{\mathrm{sep}}/\kappa)$ is surjective; its kernel is by definition the inertia group G_T, and
therefore we have $(G : G_T) = [\kappa_{\mathrm{sep}} : \kappa]$.
(c) Since G_T is a normal subgroup of G, K_T/K is a Galois extension with Galois
group G/G_T, and we have $[K_T : K] = (G : G_T) = [\kappa_{\mathrm{sep}} : \kappa]$.
(d) We have $G_T(B/A_T) = G_Z(B/A_T) = G_T = \mathrm{Gal}(L/K_T)$. Applying (b) to the
Galois extension L/K_T, we see that we have a canonical surjective homomorphism
$\mathrm{Gal}(L/K_T) \to \mathrm{Gal}(\lambda/\kappa_T)$ with kernel $G_T = \mathrm{Gal}(L/K_T)$, hence $\mathrm{Gal}(\lambda/\kappa_T) =$
$\{\mathrm{id}_\lambda\}$, and therefore λ/κ_T is purely inseparable, whence $\kappa_{\mathrm{sep}} \subset \kappa_T$. On the other
hand, we have $[\kappa_T : \kappa]_{\mathrm{sep}} = [\kappa_{\mathrm{sep}} : \kappa] = [K_T : K] \geq [\kappa_T : \kappa]_{\mathrm{sep}}$ [cf. III(5.4)(1)],
hence $\kappa_T = \kappa_{\mathrm{sep}}$; moreover, since $[K_T : K] = [\kappa_T : \kappa]_{\mathrm{sep}}$, we see from III(5.5)
that $\mathfrak{m}A_T$ is the maximal ideal of A_T.

(e) We have the surjective homomorphism $\rho \circ \varphi \colon G \to \mathrm{Gal}(\kappa_T/\kappa)$ [cf. (b) and (d)], and, by restriction, a surjective homomorphism $G \to \mathrm{Gal}(K_T/K)$; both homomorphisms have the same kernel $\mathrm{Gal}(L/K_T)$. Therefore we have an induced isomorphism $\mathrm{Gal}(K_T/K) \to \mathrm{Gal}(\kappa_T/\kappa)$. It is easy to see that this isomorphism is the canonical map $\varphi_T \colon \mathrm{Gal}(K_T/K) \to \mathrm{Gal}(\kappa_T/\kappa)$ [defined in the same way as we defined $\varphi \colon G \to \mathrm{Gal}(\lambda/\kappa)$ in (7.8)].

8   Decomposable Rings

(8.1) DEFINITION: A quasisemilocal ring B is said to be decomposable if the canonical homomorphism $B \to \prod_{\mathfrak{n} \in \mathrm{Max}(B)} B_{\mathfrak{n}}$ is an isomorphism.

(8.2) REMARK: Let B be a quasisemilocal ring.
(1) Assume that $B = \prod_{i=1}^{n} B_i$ is a finite product of quasilocal rings. Then B is decomposable, we have $n = \mathrm{Card}(\mathrm{Max}(B))$, and for every $i \in \{1, \ldots, n\}$ there exists a unique maximal ideal $\mathfrak{n}_i$ of B with $B_i = B_{\mathfrak{n}_i}$.
In fact, if we denote, for every $i \in \{1, \ldots, n\}$, the maximal ideal of B_i by $\mathfrak{m}_i$, then $\mathfrak{n}_i := B_1 \times \cdots \times \mathfrak{m}_i \times \cdots \times B_n$ is a maximal ideal of B and $B_{\mathfrak{n}_i} = B_i$, and every maximal ideal of B is of this form, hence B is decomposable.
(2) Assume that B is decomposable. Corresponding to the isomorphism $B \to \prod_{\mathfrak{n} \in \mathrm{Max}(B)} B_{\mathfrak{n}}$ we have a family $(e_{\mathfrak{n}})_{\mathfrak{n} \in \mathrm{Max}(B)}$ of orthogonal idempotents of B, i.e., we have

$$1_B = \sum_{\mathfrak{n} \in \mathrm{Max}(B)} e_{\mathfrak{n}}, \ e_{\mathfrak{n}}^2 = e_{\mathfrak{n}} \text{ and } e_{\mathfrak{n}} e_{\mathfrak{n}'} = 0 \text{ for all } \mathfrak{n}, \mathfrak{n}' \in \mathrm{Max}(B) \text{ with } \mathfrak{n} \neq \mathfrak{n}'.$$

We call the family $(e_{\mathfrak{n}})_{\mathfrak{n} \in \mathrm{Max}(B)}$ the family of elementary idempotents of B. Let $\mathfrak{n} \in \mathrm{Max}(B)$; for every idempotent e of B we have $e e_{\mathfrak{n}} = e_{\mathfrak{n}}$ or $e e_{\mathfrak{n}} = 0$ [$1_{B_{\mathfrak{n}}}$ and 0 are the only idempotents of $B_{\mathfrak{n}}$ since $B_{\mathfrak{n}}$ is quasilocal]. In particular, for every idempotent $e \in B$ there exists a uniquely determined subset $L(e) \subset \mathrm{Max}(B)$ with $e = \sum_{\mathfrak{n} \in L(e)} e_{\mathfrak{n}}$.

(8.3) REMARK: Let $B_1, \ldots, B_n$ be quasilocal rings; for $i \in \{1, \ldots, n\}$ let $\mathfrak{n}_i$ be the maximal ideal of B_i. We set $B := B_1 \times \cdots \times B_n$. Then B is a quasisemilocal ring with maximal ideals $\mathfrak{m}_1, \ldots, \mathfrak{m}_n$ where $\mathfrak{m}_i = B_1 \times \cdots \times B_{i-1} \times \mathfrak{n}_i \times B_{i+1} \times \cdots \times B_n$ for $i \in \{1, \ldots, n\}$; in particular, B is decomposable.

(8.4) COMPLETE SEMILOCAL RINGS: Let A be a semilocal ring with maximal ideals $\mathfrak{m}_1, \ldots, \mathfrak{m}_r$, and let $\mathfrak{r} = \mathfrak{m}_1 \cap \cdots \cap \mathfrak{m}_r$ be its Jacobson radical.
(1) For every natural integer n we have a canonical isomorphism

$$A/\mathfrak{r}^n \xrightarrow{\ \cong\ } A/\mathfrak{m}_1^n \times \cdots \times A/\mathfrak{m}_r^n$$

[cf. (10.1)]. Therefore, for all natural integers $m \leq n$, we have a commutative diagram of homomorphisms of rings where the horizontal maps are the canonical isomorphisms and the vertical maps are the canonical ones, namely

$$A/\mathfrak{r}^n \xrightarrow{\ \cong\ } A/\mathfrak{m}_1^n \times \cdots \times A/\mathfrak{m}_r^n$$

$$A/\mathfrak{r}^m \xrightarrow[\cong]{\ \ } A/\mathfrak{m}_1^m \times \cdots \times A/\mathfrak{m}_r^m$$

Note that, for $\rho \in \{1, \ldots, r\}$ and every $n \in \mathbb{N}$, $A/\mathfrak{m}_\rho^n$ is a local ring, isomorphic to $A_{\mathfrak{m}_\rho}/\mathfrak{m}_\rho^n A_{\mathfrak{m}_\rho}$. Thus, we have an induced isomorphism of projective limits [note that projective limits and direct products commute]

$$\varprojlim_n A/\mathfrak{r}^n \xrightarrow{\ \cong\ } \varprojlim_n A_{\mathfrak{m}_1}/\mathfrak{m}_1^n A_{\mathfrak{m}_1} \times \cdots \times \varprojlim_n A_{\mathfrak{m}_r}/\mathfrak{m}_r^n A_{\mathfrak{m}_r},$$

and therefore we have proved [cf. (8.3)]: *The completion $\widehat{A}$ of A in its $\mathfrak{r}$-adic topology is the direct product of the completions of the local rings $A_{\mathfrak{m}_\rho}$, $\rho \in \{1, \ldots, r\}$, i.e., $\widehat{A}$ is decomposable.* In particular, $\widehat{A}$ is semilocal, and $\mathfrak{m}_1\widehat{A}, \ldots, \mathfrak{m}_r\widehat{A}$ are its maximal ideals.

(2) Now assume, in particular, that A is complete. Then A is the direct product of the complete rings $A_{\mathfrak{m}_1}, \ldots, A_{\mathfrak{m}_r}$. In particular, a complete semilocal ring which is a domain is a local ring.

9   The Dimension Formula

(9.1) DEFINITION: Let $R \subset S$ be integral domains, let $\mathfrak{q}$ be a prime ideal of S, and set $\mathfrak{p} := R \cap \mathfrak{q}$. If $\mathrm{ht}(\mathfrak{p})$ and $\mathrm{tr.}\,d_{Q(R)}(Q(S))$ are finite, then we call

$$\delta(\mathfrak{q}, R) := \mathrm{tr.}\,d_{Q(R)}(Q(S)) + (\mathrm{ht}(\mathfrak{p}) - \mathrm{ht}(\mathfrak{q})) - \mathrm{tr.}\,d_{Q(R/\mathfrak{p})}(Q(S/\mathfrak{q}))$$

the defect of $\mathfrak{q}$ over R. $\delta(\mathfrak{q}, R)$ is an integer or $-\infty$.

(9.2) Lemma: [Additivity] *In the situation of (9.1), let R_0 be a subring of R, and set $\mathfrak{p}_0 := \mathfrak{p} \cap R_0$. Then, if $\mathrm{ht}(\mathfrak{p}_0)$, $\mathrm{ht}(\mathfrak{p})$, and $\mathrm{tr.}\,d_{Q(R_0)}(Q(S))$ are finite, then we have*

$$\delta(\mathfrak{q}, R_0) = \delta(\mathfrak{q}, R) + \delta(\mathfrak{p}, R_0). \tag{*}$$

Proof: From the assumptions in (9.2) we see that $\delta(\mathfrak{q}, R_0)$, $\delta(\mathfrak{q}, R)$ and $\delta(\mathfrak{p}, R_0)$ are defined. Since all three terms of the sum defining $\delta(\mathfrak{q}, R)$ are additive, (*) follows immediately.

(9.3) Theorem: [Dimension formula of I. S. Cohen] *Let $R \subset S$ be integral domains, let $\mathfrak{q}$ be a prime ideal of S, and set $\mathfrak{p} := R \cap \mathfrak{q}$. If R is noetherian, then we have*

$$\mathrm{ht}(\mathfrak{q}) + \mathrm{tr.}\,d_{Q(R/\mathfrak{p})}(Q(S/\mathfrak{q})) \leq \mathrm{ht}(\mathfrak{p}) + \mathrm{tr.}\,d_{Q(R)}(Q(S)). \tag{*}$$

If $\mathrm{ht}(\mathfrak{p})$ and $\mathrm{tr.}\,d_{Q(R)}(Q(S))$ are finite, then the assertion of the theorem is equivalent to $\delta(\mathfrak{q}, R) \geq 0$.

Proof: There is nothing to show if either $\mathrm{ht}(\mathfrak{p})$ or $\mathrm{tr.}\,\mathrm{d}_{Q(R)}(Q(S))$ is not finite. Thus, we assume that $\mathrm{ht}(\mathfrak{p})$ and $\mathrm{tr.}\,\mathrm{d}_{Q(R)}(Q(S))$ are finite; then $\delta(\mathfrak{q}, R)$ is defined, and, by definition, the assertion of the theorem is equivalent to $\delta(\mathfrak{q}, R) \geq 0$.

(1) It is enough to prove the assertion under the additional hypothesis that S is an R-algebra of finite type.

In fact, let m and n be any non-negative integers with $m \leq \mathrm{ht}(\mathfrak{q})$ and with $n \leq \mathrm{tr.}\,\mathrm{d}_{Q(R/\mathfrak{p})}(Q(S/\mathfrak{q}))$. Then there exists a strictly increasing chain of prime ideals $\mathfrak{q}_0 \subset \mathfrak{q}_1 \subset \cdots \subset \mathfrak{q}_m := \mathfrak{q}$ of S. For every $i \in \{1, \ldots, m\}$ we choose $x_i \in \mathfrak{q}_i$, $x_i \notin \mathfrak{q}_{i-1}$. Furthermore, there exist elements $y_1, \ldots, y_n \in S$ such that the images $\overline{y}_1, \ldots, \overline{y}_n$ of these elements in $S/\mathfrak{q}$ are algebraically independent over $Q(R/\mathfrak{p})$. We set $R' := R[x_1, \ldots, x_m, y_1, \ldots, y_n] \subset S$, $\mathfrak{p}' := R' \cap \mathfrak{q}$, and, for $i \in \{0, \ldots, m\}$, $\mathfrak{p}'_i := \mathfrak{q}_i \cap R'$. Then we have $\mathfrak{p} = \mathfrak{p}' \cap R$, $\mathfrak{p}'_0 \subset \mathfrak{p}'_1 \subset \cdots \subset \mathfrak{p}'_m = \mathfrak{p}'$ is a strictly increasing chain of prime ideals of R' [since $x_i \in \mathfrak{p}_i$, but $x_i \notin \mathfrak{p}_{i-1}$], hence $m \leq \mathrm{ht}(\mathfrak{p}')$, and clearly $n \leq \mathrm{tr.}\,\mathrm{d}_{Q(R/\mathfrak{p})}(Q(R'/\mathfrak{p}'))$. Now $\delta(\mathfrak{p}', R)$ is defined. Assume that we already have shown that $\delta(\mathfrak{p}', R) \geq 0$. Then we have $\mathrm{ht}(\mathfrak{p}') + \mathrm{tr.}\,\mathrm{d}_{Q(R/\mathfrak{p})}(Q(R'/\mathfrak{p}')) \leq \mathrm{ht}(\mathfrak{p}) + \mathrm{tr.}\,\mathrm{d}_{Q(R)}(Q(R'))$, hence $m + n \leq \mathrm{ht}(\mathfrak{p}) + \mathrm{tr.}\,\mathrm{d}_{Q(R)}(Q(R'))$. This holds for all $m \leq \mathrm{ht}(\mathfrak{q})$ and all $n \leq \mathrm{tr.}\,\mathrm{d}_{Q(R/\mathfrak{p})}(Q(S/\mathfrak{q}))$; since we have $\mathrm{tr.}\,\mathrm{d}_{Q(R)}(Q(R')) \leq \mathrm{tr.}\,\mathrm{d}_{Q(R)}(Q(S))$, $(*)$ follows immediately.

(2) We may assume now that S is an R-algebra of finite type. Since the defect is additive by (9.2), it is enough, by induction on the number of generators of the R-algebra S, to consider the case that $S = R[x]$ is generated by one element x.

(3) Let $S = R[X]$, a polynomial ring over R. We show that we even have equality in $(*)$. We have $\mathrm{tr.}\,\mathrm{d}_{Q(R)}(Q(S)) = 1$. Let $\mathfrak{q}$ be a prime ideal of S, set $\mathfrak{p} := \mathfrak{q} \cap R$ and $\mathfrak{p}' = \mathfrak{p}S$. If $\mathfrak{q} = \mathfrak{p}'$, then $\mathrm{ht}(\mathfrak{p}') = \mathrm{ht}(\mathfrak{p})$ by [63], Ex. 10.2; in this case we have $S/\mathfrak{q} = (R/\mathfrak{p})[X]$, hence $\mathrm{tr.}\,\mathrm{d}_{Q(R/\mathfrak{p})}(Q(S/\mathfrak{q})) = 1$. If $\mathfrak{q} \neq \mathfrak{p}'$, then we have $\mathfrak{q} \supsetneqq \mathfrak{p}'$, $\mathrm{ht}(\mathfrak{q}) = \mathrm{ht}(\mathfrak{p}) + 1$ by [63], Ex. 10.2; the prime ideal $\mathfrak{q}/\mathfrak{p}'$ of $S/\mathfrak{p}' = (R/\mathfrak{p})[X]$ contains a polynomial $\overline{r}_0 + \overline{r}_1 X + \cdots + \overline{r}_h X^h$ of positive degree with $\overline{r}_0, \overline{r}_1, \ldots, \overline{r}_h \in R/\mathfrak{p}$, hence $\mathrm{tr.}\,\mathrm{d}_{Q(R/\mathfrak{p})}(Q(S/\mathfrak{q})) = 0$. This completes the proof in the case $S = R[X]$.

(4) Now we treat the case where $S = R[x]$ and x is not transcendental over R. We consider the canonical R-algebra homomorphism $R[X] \to R[x]$; let $\mathfrak{P}$ be its kernel and $\mathfrak{Q}$ the inverse image of $\mathfrak{q}$. We have $\mathfrak{P} \subset \mathfrak{Q}$, $\mathfrak{P} \cap R = \{0\}$, $\mathfrak{Q} \cap R = \mathfrak{p}$ and $\mathfrak{Q}/\mathfrak{P} = \mathfrak{q}$. We have $\mathrm{ht}(\mathfrak{Q}) + \mathrm{tr.}\,\mathrm{d}_{Q(R/\mathfrak{p})}(Q(R[X]/\mathfrak{Q})) = \mathrm{ht}(\mathfrak{p}) + 1$ by (3). Since $\mathrm{ht}(\mathfrak{Q}) \geq \mathrm{ht}(\mathfrak{q}) + 1$ [every strictly increasing chain of prime ideals between $\{0\}$ and $\mathfrak{q}$ in S gives rise to a strictly increasing chain of prime ideals between $\mathfrak{P}$ and $\mathfrak{Q}$ of the same length in $R[X]$], and $R[X]/\mathfrak{Q} = R[x]/\mathfrak{q}$, we get $(*)$.

(9.4) Corollary: *Let A be a local domain with maximal ideal* $\mathfrak{m}$. *For every quasilocal domain B with maximal ideal* $\mathfrak{n}$ *dominating A we have*

$$\dim(B) + \mathrm{tr.}\,\mathrm{d}_{A/\mathfrak{m}}(B/\mathfrak{n}) \leq \dim(A) + \mathrm{tr.}\,\mathrm{d}_{Q(A)}(Q(B)).$$

(9.5) Theorem: [I. S. Cohen] *Let R be a noetherian integral domain with field of quotients K, let L be a finitely generated extension field of K, let S be a subring of L having L as field of quotients with $R \subset S$, let $\mathfrak{q}$ be a prime ideal of S and set*

$\mathfrak{p} := \mathfrak{q} \cap R$. Then $\delta(\mathfrak{q}, R)$ is defined. If $\delta(\mathfrak{q}, R) = 0$, then the quotient field of $S/\mathfrak{q}$ is a finitely generated extension field of the quotient field of $R/\mathfrak{p}$.

Proof: $\mathrm{ht}(\mathfrak{p})$ is finite since R is noetherian, and $\mathrm{tr.d}_{Q(R)}(Q(S)) = \mathrm{tr.d}_K(L)$ is finite by assumption. Therefore $\delta(\mathfrak{q}, R)$ is defined. Moreover, by (9.3), $\mathfrak{q}$ has finite height and the transcendence degree of $Q(S/\mathfrak{q})$ over $Q(R/\mathfrak{p})$ is finite. Set $m := \mathrm{ht}(\mathfrak{q})$, and let $\{0\} = \mathfrak{q}_0 \subset \cdots \subset \mathfrak{q}_m = \mathfrak{q}$ be a strictly increasing chain of prime ideals of S. For every $i \in \{1, \ldots, m\}$ we choose $x_i \in \mathfrak{q}_i$, $x_i \notin \mathfrak{q}_{i-1}$. Set $n := \mathrm{tr.d}_{Q(R/\mathfrak{p})}(Q(S/\mathfrak{q}))$; we choose $y_1, \ldots, y_n \in S$ such that the images $\overline{y}_1, \ldots, \overline{y}_n$ of these elements in $S/\mathfrak{q}$ are algebraically independent over $Q(R/\mathfrak{p})$. Since L is a finitely generated extension field of K, we can write $L = K(z_1, \ldots, z_r)$ with elements $z_1, \ldots, z_r \in S$. Now we set $R' := R[x_1, \ldots, x_m, y_1, \ldots, y_n, z_1, \ldots, z_r] \subset S$, an R-algebra of finite type, and therefore R' is a noetherian ring. Furthermore, we set $\mathfrak{p}' := \mathfrak{q} \cap R'$, $\mathfrak{p}'_i := \mathfrak{q}_i \cap R'$ for $i \in \{0, \ldots, m\}$. Note that $R'/\mathfrak{p}'$ is an $R/\mathfrak{p}$-algebra of finite type, hence that $Q(R'/\mathfrak{p}')$ is a finitely generated extension field of $Q(R/\mathfrak{p})$; therefore, it is enough to prove that $Q(S/\mathfrak{q})$ is a finitely generated extension field of $Q(R'/\mathfrak{p}')$. By additivity of defect [cf. (9.2)] we have $0 = \delta(\mathfrak{q}, R) = \delta(\mathfrak{q}, R') + \delta(\mathfrak{p}', R)$, and since $\delta(\mathfrak{q}, R')$ and $\delta(\mathfrak{p}', R)$ are not negative by (9.3), we have $\delta(\mathfrak{q}, R') = 0$. Clearly we have $Q(R') = Q(S)$, and $\mathrm{tr.d}_{Q(R'/\mathfrak{p}')}(Q(S/\mathfrak{q})) = 0$ since $R'/\mathfrak{p}' \supset (R/\mathfrak{p})[\overline{y}_1, \ldots, \overline{y}_n]$. From $\delta(\mathfrak{q}, R') = 0$ we therefore get $\mathrm{ht}(\mathfrak{p}') = \mathrm{ht}(\mathfrak{q}) = m$, and $\{0\} = \mathfrak{p}'_0 \subset \mathfrak{p}'_1 \subset \cdots \subset \mathfrak{p}'_m = \mathfrak{p}'$ is a strictly increasing chain of prime ideals of R' since $x_i \in \mathfrak{p}'_i$, $x_i \notin \mathfrak{p}'_{i-1}$ for $i \in \{1, \ldots, m\}$.

We write R instead of R'; we have to prove, according to the last paragraph, the assertion of the theorem under the following additional hypotheses: $Q(R) = Q(S)$, $\mathrm{tr.d}_{Q(R/\mathfrak{p})}(Q(S/\mathfrak{q})) = 0$, $\mathrm{ht}(\mathfrak{p}) = \mathrm{ht}(\mathfrak{q}) = m$, there exists a strictly increasing chain of prime ideals $\{0\} = \mathfrak{q}_0 \subset \cdots \subset \mathfrak{q}_m$ in S, and, setting $\mathfrak{p}_i := \mathfrak{q}_i \cap R$ for $i \in \{0, \ldots, m\}$, $\{0\} = \mathfrak{p}_0 \subset \mathfrak{p}_1 \subset \cdots \subset \mathfrak{p}_m = \mathfrak{p}$ is a strictly increasing chain of prime ideals of R.

We prove now under these additional hypotheses the theorem by induction on $m := \mathrm{ht}(\mathfrak{q})$. If $m = 0$, then the assertion of the theorem holds trivially. Let $m > 0$, and assume that the assertion holds for prime ideals of S of height $m - 1$. Now let $\mathrm{ht}(\mathfrak{q}) = m$; we use the notations introduced in the preceding paragraph. We set $R' := R_{\mathfrak{p}_1}$, $S' := S_{\mathfrak{q}_1}$. Note that R' is a one-dimensional local domain, S' is an R'-submodule of $Q(R')$ and $\ell_{R'}(S'/\mathfrak{q}_1 S') < \infty$; using II(3.1)(2) we therefore get $[S'/\mathfrak{q}_1 S' : R'/\mathfrak{p}_1 R'] < \infty$. Now we set $\overline{S} := S/\mathfrak{q}_1$, $\overline{R} := R/\mathfrak{p}_1$, $\overline{\mathfrak{q}} := \mathfrak{q}/\mathfrak{q}_1$, $\overline{\mathfrak{p}} := \mathfrak{p}/\mathfrak{p}_1$. We have $\mathrm{ht}(\overline{\mathfrak{q}}) = \mathrm{ht}(\overline{\mathfrak{p}}) = m - 1$. Setting $\overline{\mathfrak{q}}_i := \mathfrak{q}_i/\mathfrak{q}_1$, $\overline{\mathfrak{p}}_i := \mathfrak{p}_i/\mathfrak{p}_1$ for $i \in \{1, \ldots, m\}$, we have strictly increasing chains of prime ideals $\{0\} = \overline{\mathfrak{q}}_1 \subset \cdots \subset \overline{\mathfrak{q}}_m$ in $\overline{S}$, $\{0\} = \overline{\mathfrak{p}}_1 \subset \cdots \subset \overline{\mathfrak{p}}_m$ in $\overline{R}$, and we have $\overline{\mathfrak{q}}_i \cap \overline{R} = \overline{\mathfrak{p}}_i$ for $i \in \{1, \ldots, m\}$. Since we have $Q(\overline{S}) = S'/\mathfrak{q}_1 S'$ and $Q(\overline{R}) = R'/\mathfrak{p}_1 R'$, we have $\mathrm{tr.d}_{Q(\overline{R})}(Q(\overline{S})) = 0$. Furthermore, since $\overline{S}/\overline{\mathfrak{q}} \cong S/\mathfrak{q}$, $\overline{R}/\overline{\mathfrak{p}} \cong R/\mathfrak{p}$, we have $\mathrm{tr.d}_{Q(\overline{R}/\overline{\mathfrak{p}})}(Q(\overline{S}/\overline{\mathfrak{q}})) = 0$. Therefore the rings $\overline{R}$, $\overline{S}$ and the ideals $\overline{\mathfrak{q}}$, $\overline{\mathfrak{p}}$ satisfy the assumptions of the preceding paragraph, hence the induction assumption yields that $Q(S/\mathfrak{q})$ is a finite extension of $Q(R/\mathfrak{p})$.

The following result shall play a decisive role at several places, e.g. in the proof of the important lemma VII(6.3), and in the proof of VIII(6.3) which is needed for the proof of the uniformization theorem VIII(6.9).

(9.6) Proposition: *Let $A \hookrightarrow B$ be a local homomorphism of local domains which have the same field of quotients; let $\mathfrak{m}$ be the maximal ideal of A and $\mathfrak{n}$ be the maximal ideal of B. Assume that A is integrally closed and analytically irreducible, and that $\dim(A) = \dim(B)$. If either B is a quasifinite A-module, or if $\mathfrak{m}A$ is $\mathfrak{n}$-primary, then we have $A = B$.*

Proof: (1) We assume that B is a quasifinite A-module. Let $\widehat{A}$ (resp. $\widehat{B}$) be the completion of A (resp. B). Then $A \to B$ induces a local homomorphism $\alpha \colon \widehat{A} \to \widehat{B}$ which is finite [cf. (10.13)], hence we have $\dim(\alpha(\widehat{A})) = \dim(\widehat{B})$ [cf. [63], Prop. 9.2]. Since we have $\dim(A) = \dim(\widehat{A})$, $\dim(B) = \dim(\widehat{B})$ [cf. [63], Cor. 10.12], we have $\dim(\widehat{A}) = \dim(\alpha(\widehat{A}))$. Since $\widehat{A}$ is a domain and $\alpha(\widehat{A}) = \widehat{A}/\ker(\alpha)$, it follows that α is injective, hence that $\widehat{A}$ can be considered as a subring of $\widehat{B}$, and $\widehat{B}$ is integral over $\widehat{A}$.

Let $z \in B$. Since $Q(A) = Q(B)$, we can write $z = x/y$ with x, $y \in A$. Since z is integral over $\widehat{A}$, there exist $n \in \mathbb{N}$ and elements $a_1, \ldots, a_n \in \widehat{A}$ with $z^n + a_1 z^{n-1} + \cdots + a_n = 0$. Therefore we have $x^n + a_1 y x^{n-1} + \cdots + a_n y^n = 0$. Let $\mathfrak{a} \subset A$ be the ideal generated by yx^{n-1}, $y^2 x^{n-2}, \ldots, y^n$. We have $x^n \in \mathfrak{a}\widehat{A} \cap A = \mathfrak{a}$ [cf. [34], Ch. I, §3, no. 5, Prop. 10], hence there exist $b_1, \ldots, b_n \in A$ with $x^n + b_1 y x^{n-1} + \cdots + b_n y^n = 0$ which means that $z = x/y$ is integral over A, hence that $z \in A$, and that therefore $B \subset A$, hence that we have $A = B$.

(2) Now we assume that $\mathfrak{m}B$ is an $\mathfrak{n}$-primary ideal. We show that then B is a quasifinite A-module, hence that $B = A$ by (1). By (9.4) we have $\mathrm{tr.}\,\mathrm{d}_{A/\mathfrak{m}}(B/\mathfrak{n}) = 0$, hence $\delta(\mathfrak{n}, A) = 0$, and therefore $B/\mathfrak{n}$ is a finite extension of $A/\mathfrak{m}$ [cf. (9.5)]. There exists $r \in \mathbb{N}$ with $\mathfrak{n}^r \subset \mathfrak{m}B$. Now $B/\mathfrak{n}^r$ is a B-module of finite length; the only simple B-module is—up to isomorphism—$B/\mathfrak{n}$ which is an A-module of finite length $[B/\mathfrak{n} : A/\mathfrak{m}]$. Therefore $B/\mathfrak{n}^r$ is an A-module of finite length, hence also $B/\mathfrak{m}B$; in particular, $B/\mathfrak{m}B$ is a finitely generated A-module, hence B is a quasifinite A-module.

10   Miscellaneous Results

10.1   The Chinese Remainder Theorem

(10.1) Let A be a ring.
(1) Ideals $\mathfrak{a}$ and $\mathfrak{b}$ of A are called comaximal or coprime if $\mathfrak{a} + \mathfrak{b} = A$. Ideals $\mathfrak{a}_1, \ldots, \mathfrak{a}_h$ of A are called pairwise comaximal if, for all i, $j \in \{1, \ldots, h\}$, $i \neq j$, the ideals $\mathfrak{a}_i$ and $\mathfrak{a}_j$ are comaximal.
(2) Let $\mathfrak{a}_1, \ldots, \mathfrak{a}_h$ be pairwise comaximal ideals of A. The Chinese remainder theorem [cf. [63], Ex. 2.6] states that the canonical homomorphism $A \to A/\mathfrak{a}_1 \times \cdots \times A/\mathfrak{a}_h$ is surjective and that it has kernel $\mathfrak{a}_1 \cdots \mathfrak{a}_h$.

(3) A useful supplement to (2) is the following result: If $\mathfrak{a}_1, \ldots, \mathfrak{a}_h$ are pairwise comaximal ideals of A, then we have [cf. [21], Prop. 1.10]

$$\mathfrak{a}_1 \cap \cdots \cap \mathfrak{a}_h = \mathfrak{a}_1 \cdots \mathfrak{a}_h.$$

(4) If $\mathfrak{a}_1, \ldots, \mathfrak{a}_h$ are pairwise comaximal, and $n_1, \ldots, n_h \in \mathbb{N}_0$, then $\mathfrak{a}_1^{n_1}, \ldots, \mathfrak{a}_h^{n_h}$ are pairwise comaximal.

By induction, it is enough to consider the case $h = 2$. Thus, let $\mathfrak{a}$, $\mathfrak{b}$ be comaximal ideals, and let m, n be positive integers. We chose $a \in \mathfrak{a}$, $b \in \mathfrak{b}$ with $a + b = 1$. Then we have $(a + b)^{m+n-1} = 1$, and since $a^i \in \mathfrak{a}^m$ for $i \geq m$ and $b^j \in \mathfrak{b}^n$ for $j \geq n$, we easily get the result by expanding $(a + b)^{m+n-1}$.

10.2 Separable Noether Normalization

In III(6.13) we proved for a perfect field k a result on the existence of a separating transcendence basis. If k is algebraically closed, we can do better.

(10.2) Proposition: *Let $A := k[t_1, \ldots, t_n]$ be an integral k-algebra over an algebraically closed field k. Then there exist elements $t_1', \ldots, t_n'$ in A which are linear combinations of $t_1, \ldots, t_n$ with coefficients in k and in integer $p \in \{1, \ldots, n+1\}$ such that $t_1', \ldots, t_{p-1}'$ are algebraically independent over k, $t_p', \ldots, t_n'$ are integral over $k[t_1', \ldots, t_{p-1}']$, $A = k[t_1', \ldots, t_n']$, and $Q(A)$ is a finite separable extension of the field of rational functions $k(t_1', \ldots, t_{p-1}')$.*

Proof: We consider the canonical k-algebra homomorphism $k[T_1, \ldots, T_n] \to k[t_1, \ldots, t_n]$ defined by mapping T_i to t_i for $i \in \{1, \ldots, n\}$. Let $\mathfrak{p}$ be its kernel. There is nothing to show if $\mathfrak{p} = \{0\}$; we consider therefore the case $\mathfrak{p} \neq \{0\}$.
(1) We choose an irreducible $F \in \mathfrak{p}$. We prove the existence of $(\alpha_1, \ldots, \alpha_{n-1}) \in k^{n-1}$ such that, if we consider the linear change of variables

$$T_1' = T_1 - \alpha_1 T_n, \ldots, T_{n-1}' = T_{n-1} - \alpha_{n-1} T_n, T_n' = T_n,$$

and, if for any $H \in k[T_1, \ldots, T_n]$ we write

$$H'(T_1', \ldots, T_n') = H(T_1' + \alpha_1 T_n', \ldots, T_{n-1}' + \alpha_{n-1} T_n', T_n'),$$

then we have

$$\frac{\partial F'}{\partial T_n'} \neq 0. \tag{$*$}$$

We have

$$\frac{\partial F'}{\partial T_n'} = \alpha_1 \frac{\partial F}{\partial T_1} + \cdots + \alpha_{n-1} \frac{\partial F}{\partial T_{n-1}} + \frac{\partial F}{\partial T_n}.$$

Since F is irreducible and k is algebraically closed, some partial derivative of F, say $\partial F/\partial T_j$ where $j \in \{1, \ldots, n\}$, must be different from zero. If $\partial F/\partial T_1 = \cdots = \partial F/\partial T_{n-1} = 0$, then $\partial F/\partial T_n \neq 0$. In the other case, there exist $j \in \{1, \ldots, n-1\}$

and $(\beta_1, \ldots, \beta_n) \in \mathbb{A}^n$ such that $\partial F/\partial T_j$ does not have a zero in this point. Consider the hyperplane $Z \subset \mathbb{A}^{n-1}$ defined as the zero set of the linear polynomial

$$\sum_{l=1}^{n-1} X_l \frac{\partial F}{\partial T_l}(\beta_1, \ldots, \beta_n) + \frac{\partial F}{\partial T_n}(\beta_1, \ldots, \beta_n);$$

Z is a proper closed subset of $\mathbb{A}^{n-1}$; therefore the set of points $(\alpha_1, \ldots, \alpha_{n-1}) \in \mathbb{A}^{n-1}$ not satisfying $(*)$ is a proper closed subset of $\mathbb{A}^{n-1}$, which is exactly what we wanted. Let $\mathfrak{p}'$ be the ideal of $k[T_1', \ldots, T_n']$ consisting of all polynomials H' where $H \in \mathfrak{p}$; it is a prime ideal.

(2) Now we repeat the basic construction in (1) with the ideals $\mathfrak{p}' \cap k[T_1', \ldots, T_{n-1}']$, $\mathfrak{p}^{(2)} \cap k[T_1^{(2)}, \ldots, T_{n-2}^{(2)}], \ldots$ as long as $\mathfrak{p}^{(j)} \cap k[T_1^{(j)}, \ldots, T_{n-j}^{(j)}] \neq \{0\}$; here $j \in \{1, \ldots, n\}$. Let $p \in \{1, \ldots, n\}$ be the first index with $\mathfrak{p}^{(p)} \cap k[T_1^{(p)}, \ldots, T_{n-p}^{(p)}] = \{0\}$.

(3) Assume that this process is already performed at the beginning. This amounts to the existence of $p \in \{1, \ldots, n\}$ and irreducible polynomials $F_p \in k[T_1, \ldots, T_p], F_{p+1} \in k[T_1, \ldots, T_{p+1}], \ldots, F_n \in k[T_1, \ldots, T_n]$ lying in $\mathfrak{p}$ such that
(a) $\mathfrak{p} \cap k[T_1, \ldots, T_{p-1}] = \{0\}$;
(b) let $j \in \{p, \ldots, n\}$; each F_j lies in $k[T_1, \ldots, T_j] \cap \mathfrak{p}$ and $\partial F_j/\partial T_j \neq 0$.

(4) In other words, the conditions in (3) mean the following. Set $t_j := T_j + \mathfrak{p}$ for every $j \in \{1, \ldots, n\}$. Then $k[T_1, \ldots, T_n]/\mathfrak{p} = k[t_1, \ldots, t_n]$ is integral over $k[t_1, \ldots, t_{p-1}]$ [by transitivity of integral extensions], and the elements $t_1, \ldots, t_{p-1}$ are algebraically independent over k. Furthermore, t_p is separable over $k(t_1, \ldots, t_{p-1})$, t_{p+1} is separable over $k(t_1, \ldots, t_p), \ldots$, and t_n is separable over $k(t_1, \ldots, t_{n-1})$. Therefore $k(t_1, \ldots, t_n)$ is a finite separable extension of $k(t_1, \ldots, t_{p-1})$ [by transitivity of finite separable extensions].

10.3   The Segre Ideal

When one studies Segre embedding $\mathbb{P}^m \times \mathbb{P}^n \to \mathbb{P}^{mn+m+n}$, it is useful to know the following result (10.3) below. We use it also when blowing up points in appendix A, section 13. The kernel of the homomorphism in (2) of the following lemma often is called the Segre ideal.

(10.3) Lemma: Let A be a ring, and let m, n be positive integers. Then:
(1) The kernel of the A-algebra homomorphism

$$A[\{T_{ij} \mid i \in \{1, \ldots, m\}, j \in \{1, \ldots, n\}\}] \to A[X_1, \ldots, X_m, Y_1, \ldots, Y_n],$$

defined by $T_{ij} \mapsto X_i Y_j$ for $i \in \{1, \ldots, m\}$, $j \in \{1, \ldots, n\}$, is the ideal $\mathfrak{a}$ generated by

$$T_{ij}T_{rs} - T_{is}T_{rj}, \quad i, r \in \{1, \ldots, m\}, j, s \in \{1, \ldots, n\}. \tag{$*$}$$

(2) Let R be one of the rings $A[X_1, \ldots, X_n]$ or $A[\![X_1, \ldots, X_n]\!]$. The kernel of the R-algebra homomorphism $R[Y_1, \ldots, Y_n] \to R[T]$, defined by $Y_j \mapsto X_j T$ for $j \in \{1, \ldots, n\}$, is the ideal $\mathfrak{a}$ generated by the elements $X_i Y_j - X_j Y_i$, $i, j \in \{1, \ldots, n\}$. In particular, if A is a domain, then the ideal $\mathfrak{a}$ in (1) and (2) is a prime ideal.

Proof: (1) We set $L := \{(i,j) \mid i \in \{1,\ldots,m\}, j \in \{1,\ldots,n\}\}$. We consider products of the form $T_{i_1 j_1} \cdots T_{i_r j_r}$ with $(i_1,j_1),\ldots,(i_r,j_r) \in L$; we always assume that $i_1 \leq \cdots \leq i_r$. Such a product shall be called standard if we have also $j_1 \leq \cdots \leq j_r$.

(a) Assume that $T_{i_1 j_1} \cdots T_{i_r j_r}$ is not standard, and choose $\rho \in \{1,\ldots,r-1\}$ with $j_1 \leq \cdots \leq j_{\rho-1}$ and $j_\rho > j_{\rho+1}$. Then the expression

$$T_{i_1 j_1} \cdots T_{i_r j_r} - T_{i_1 j_1} \cdots T_{i_{\rho-1} j_{\rho-1}} T_{i_{\rho+2} j_{\rho+2}} \cdots T_{i_r j_r} (T_{i_\rho j_\rho} T_{i_{\rho+1} j_{\rho+1}} - T_{i_\rho j_{\rho+1}} T_{i_{\rho+1} j_\rho})$$

has the form $T_{i_1 j_1} \cdots T_{i_\rho j_{\rho+1}} T_{i_{\rho+1} j_\rho} \cdots T_{i_r j_r}$. It is clear that after a finite number of steps we arrive at a standard product, i.e., for $T_{i_1 j_1} \cdots T_{i_r j_r}$ there exists an element $G \in \mathfrak{a}$, homogeneous of degree r, such that $T_{i_1 j_1} \cdots T_{i_r j_r} - G = T_{i_1 j'_1} \cdots T_{i_r j'_r}$ where $T_{i_1 j'_1} \cdots T_{i_r j'_r}$ is standard.

(b) Let $T_{i_1 j_1} \cdots T_{i_r j_r}$ and $T_{i'_1 j'_1} \cdots T_{i'_r j'_r}$ be different standard products. It is immediate that $X_{i_1} Y_{j_1} \cdots X_{i_r} Y_{j_r} \neq X_{i'_1} Y_{j'_1} \cdots X_{i'_r} Y_{j'_r}$.

(c) We have $\mathfrak{a} \subset \ker(\varphi)$. Let $F \in \ker(\varphi)$ be homogeneous of degree r. By (a) there exists a homogeneous $G \in \mathfrak{a}$ of degree r such that $F - G$ is a linear combination with coefficients in A of standard products of degree r. By (b) it follows that $F - G = 0$, hence that $F \in \mathfrak{a}$.

(2) Clearly we have $\mathfrak{a} \subset \ker(\varphi)$. We consider $R[Y_1,\ldots,Y_n]$ and $R[T]$ as $\mathbb{Z}$-graded rings of type $\mathbb{N}_0$; then φ is homogeneous of degree 0.

(a) Let $(i_1,\ldots,i_n), (j_1,\ldots,j_n), (i'_1,\ldots,i'_n), (j'_1,\ldots,j'_n) \in \mathbb{N}_0^n$, and define

$$H_{i_1,\ldots,i_n,j_1,\ldots,j_n,i'_1,\ldots,i'_n,j'_1,\ldots,j'_n} := X_1^{i_1} \cdots X_n^{i_n} Y_1^{j_1} \cdots Y_n^{j_n} - X_1^{i'_1} \cdots X_n^{i'_n} Y_1^{j'_1} \cdots Y_n^{j'_n}.$$

We assume that $i_1 + \cdots + i_n = i'_1 + \cdots + i'_n > 0$, $j_1 + \cdots + j_n = j'_1 + \cdots + j'_n > 0$, and that $i_\nu + j_\nu = i'_\nu + j'_\nu =: l_\nu$ for $\nu \in \{1,\ldots,n\}$; then the degree l of this homogeneous binomial is ≥ 2. We show that $H_{i_1,\ldots,i_n,j_1,\ldots,j_n,i'_1,\ldots,i'_n,j'_1,\ldots,j'_n} \in \mathfrak{a}$. In fact, this is trivial if $l = 2$. Let $l > 2$, and assume that the assertion is true for all binomials of degree $< l$ which are $\neq 0$. We consider a binomial $H = H_{i_1,\ldots,i_n,j_1,\ldots,j_n,i'_1,\ldots,i'_n,j'_1,\ldots,j'_n}$ of degree l. Then we have $(i_1,\ldots,i_n) \neq (i'_1,\ldots,i'_n)$, hence there exist $r, s \in \{1,\ldots,n\}$ with $i_r < i'_r$, $i_s > i'_s$. Then we have $j_r > j'_r$, $j_s < j'_s$, and with

$$M := X_r Y_s, \quad \widetilde{M} := X_1^{i_1} \cdots X_r^{i_r-1} X_{r+1}^{i_{r+1}} \cdots X_n^{i_n} Y_1^{j_1} \cdots Y_s^{j_s-1} Y_{s+1}^{j_{s+1}} \cdots Y_n^{j_n},$$

$$N := X_s Y_r, \quad \widetilde{N} := X_1^{i_1} \cdots X_s^{i_s-1} X_{s+1}^{i_{s+1}} \cdots X_n^{i_n} Y_1^{j_1} \cdots Y_r^{j_r-1} Y_{r+1}^{j_{r+1}} \cdots Y_n^{j_n}$$

we have $H = N\widetilde{N} - M\widetilde{M} = (N - M)\widetilde{N} + M(\widetilde{N} - \widetilde{M}) \in \mathfrak{a}$ since $N - M \in \mathfrak{a}$ and, by induction, $\widetilde{N} - \widetilde{M} \in \mathfrak{a}$.

(b) Let $q \in \mathbb{N}$, and let $F \in \ker(\varphi)$ be homogeneous of degree q. We have

$$F = \sum_{\substack{(j_1,\ldots,j_n)\in \mathbb{N}_0^n \\ j_1+\cdots+j_n=q}} F_{j_1,\ldots,j_n} Y_1^{j_1} \cdots Y_n^{j_n} \in R[Y_1,\ldots,Y_n],$$

and we write the elements $F_{j_1,\ldots,j_n} \in R$ as a sum of forms

$$\sum_{p=0}^{\infty} F_{j_1,\ldots,j_n}^{(p)} \quad \text{with } F_{j_1,\ldots,j_n}^{(p)} \in A[\,X_1,\ldots,X_n\,] \quad \text{homogeneous of degree } p.$$

The condition $F \in \ker(\varphi)$ is equivalent to

$$\sum_{\substack{(j_1,\ldots,j_n)\in\mathbb{N}_0^n \\ j_1+\cdots+j_n=q}} F_{j_1,\ldots,j_n} X_1^{j_1} \cdots X_n^{j_n} = 0. \tag{$*$}$$

Let $p \in \mathbb{N}$, and write, for every $(j_1,\ldots,j_n) \in \mathbb{N}_0^n$ with $j_1 + \cdots + j_n = q$,

$$F_{j_1,\ldots,j_n}^{(p)} = \sum_{\substack{(i_1,\ldots,i_n)\in\mathbb{N}_0^n \\ i_1+\cdots+i_n=p}} \alpha_{i_1,\ldots,i_n,j_1,\ldots,j_n}^{(p)} X_1^{i_1} \cdots X_n^{i_n}$$

with coefficients $\alpha_{i_1,\ldots,i_n,j_1,\ldots,j_n}^{(p)} \in A$. From $(*)$ we get, for every $(l_1,\ldots,l_n) \in \mathbb{N}_0^n$,

$$\sum_{\substack{i_1+\cdots+i_n=p,\,j_1+\cdots+j_n=q \\ i_1+j_1=l_1,\ldots,i_n+j_n=l_n}} \alpha_{i_1,\ldots,i_n,j_1,\ldots,j_n}^{(p)} = 0. \tag{$**$}$$

For every $p \in \mathbb{N}_0$ we choose fixed n-tuples $(i_1'^{(p)},\ldots,i_n'^{(p)})$, $(j_1'^{(p)},\ldots,j_n'^{(p)})$ with $i_1'^{(p)} + \cdots + i_n'^{(p)} = p$, $j_1'^{(p)} + \cdots + j_n'^{(p)} = q$. From $(**)$ we get, setting $i_1'^{(p)} + j_1'^{(p)} = l_1'^{(p)},\ldots,i_n'^{(p)} + j_n'^{(p)} = l_n'^{(p)}$,

$$F = \sum_{p} \sum_{\substack{i_1+\cdots+i_n=p,\,j_1+\cdots+j_n=q \\ i_1+j_1=l_1^{(p)},\ldots,i_n+j_n=l_n^{(p)}}} \alpha_{i_1,\ldots,i_n,j_1,\ldots,j_n}^{(p)} H_{i_1,\ldots,i_n,j_1,\ldots,j_n,i_1'^{(p)},\ldots,i_n'^{(p)},j_1'^{(p)},\ldots,j_n'^{(p)}},$$

hence $F \in \mathfrak{a}$ [cf. (a)].

10.4   Adjoining an Indeterminate

(10.4) Let A be a ring and $A[\,X\,]$ the polynomial ring over A.

(1) Let $\mathfrak{a}$ be an ideal of A; remember that $\mathfrak{a}A[\,X\,] \cap A = \mathfrak{a}$ and that $(A/\mathfrak{a})[\,X\,] \cong A[\,X\,]/\mathfrak{a}A[\,X\,]$. In particular, if $\mathfrak{p}$ is a prime ideal of A, then $\mathfrak{p}A[\,X\,]$ is a prime ideal of $A[\,X\,]$.

(2) Now we assume that A is a quasisemilocal ring with maximal ideals $\mathfrak{m}_1,\ldots,\mathfrak{m}_h$. Set $\Sigma := A[\,X\,] \setminus (\mathfrak{m}_1 A[\,X\,] \cup \cdots \cup \mathfrak{m}_h A[\,X\,])$. Then Σ is a multiplicatively closed system in $A[\,X\,]$; define $A(X) := \Sigma^{-1} A[\,X\,]$, and set $\mathfrak{n}_i := \mathfrak{m}_i(\Sigma^{-1} A[\,X\,])$ for every $i \in \{1,\ldots,h\}$. Note that $A(X)$ is quasisemilocal with maximal ideals $\mathfrak{n}_1,\ldots,\mathfrak{n}_h$. Clearly, for every $i \in \{1,\ldots,h\}$, the ring $A(X)/\mathfrak{n}_i$ is isomorphic to $(A/\mathfrak{m}_i)(X)$, the field of rational functions over the field $A/\mathfrak{m}_i$. Moreover, $A \to A(X)$ is faithfully flat. Indeed, let $\mathfrak{p}$ be a prime ideal in A; then $\mathfrak{p}A[\,X\,]$ is a prime ideal in $A[\,X\,]$ which does not meet Σ, hence $\mathfrak{p}A(X)$ is a prime ideal in $A(X)$, and $\mathfrak{p}A(X) \cap A = \mathfrak{p}$. Now the assertion follows from [34], Ch. II, §2, no. 5, Cor. 4 of Prop. 11.

10.5 Divisor Group and Class Group

(10.5) THE GROUP OF DIVISORS: Let A be an integrally closed noetherian domain with field of quotients K, and let $P(A)$ be the set of prime ideals of A of height 1. For any $\mathfrak{p} \in P(A)$ the ring $A_\mathfrak{p}$ is a discrete valuation ring of K [since it is local, one-dimensional and integrally closed, cf. I(3.28)]; let $v_\mathfrak{p}$ be the valuation of K defined by $A_\mathfrak{p}$.

(1) The free abelian group having the elements of $P(A)$ as a $\mathbb{Z}$-basis is called the group of divisors $\mathrm{Div}(A)$ of A; the elements of $\mathrm{Div}(A)$ are called divisors of A.

(2) Let $D \in \mathrm{Div}(A)$; then $D = \sum n_\mathfrak{p} \mathfrak{p}$ where $\mathfrak{p}$ runs through the elements of $P(A)$, $n_\mathfrak{p} \in \mathbb{Z}$ and only finitely many $n_\mathfrak{p}$ are not equal to zero.

(3) Let $f \in A \setminus \{0\}$; for any $\mathfrak{p} \in P(A)$ we have $v_\mathfrak{p}(f) > 0$ iff $f \in \mathfrak{p}$, i.e., iff $\mathfrak{p} \in \mathrm{Ass}(A/fA)$ [cf. [63], Th. 3.1 and Th. 10.1]; therefore $\sum_{\mathfrak{p} \in P(A)} v_\mathfrak{p}(f)\mathfrak{p} \in D(A)$. Clearly

$$f \mapsto \mathrm{div}(f) := \sum_{\mathfrak{p} \in P(A)} v_\mathfrak{p}(f)\mathfrak{p} \colon K^\times \to \mathrm{Div}(A)$$

is a homomorphism div of groups; the image of $K^\times$ is called the group of principal divisors of A, $\mathrm{div}(f)$ is called a principal divisor and $\mathrm{Cl}(A) := \mathrm{Div}(A)/\mathrm{div}(K^\times)$ is called the divisor class group of A.

(10.6) **Proposition:** *Let A be an integrally closed noetherian domain. Then every prime ideal of A of height 1 is a principal ideal iff $\mathrm{Cl}(A) = \{0\}$.*

Proof: (1) Assume that every prime ideal in $P(A)$ is a principal ideal, and let $\mathfrak{p} = Af$ be in $P(A)$. Then $\mathrm{div}(f) = \mathfrak{p}$, and therefore every divisor of A is a principal divisor.

(2) Assume that $\mathrm{Cl}(A) = \{0\}$, and let $\mathfrak{p} \in P(A)$. Choose $f \in Q(A)^\times$ with $\mathrm{div}(f) = \mathfrak{p}$. Then $v_\mathfrak{p}(f) = 1$ and $v_\mathfrak{q}(f) = 0$ for every $\mathfrak{q} \in P(A) \setminus \{\mathfrak{p}\}$, hence $f \in A$ [cf. [63], Th. 11.2] and $f \in \mathfrak{p}$. Let $g \in \mathfrak{p}$; then $v_\mathfrak{p}(g) \geq 1$ and $v_\mathfrak{q}(g) \geq 0$ for every $\mathfrak{q} \in P(A) \setminus \{\mathfrak{p}\}$. Thus, $g/f \in A$ [cf. loc. cit] and therefore $g \in Af$; this implies that $\mathfrak{p} = Af$.

(10.7) **Corollary:** *Let A be a noetherian domain. Then A is factorial iff A is integrally closed and $\mathrm{Cl}(A) = \{0\}$.*

Proof: A factorial domain is integrally closed. Therefore the result follows immediately from (10.5), (10.6) and [63], Cor. 10.6.

10.6 Calculating a Multiplicity

(10.8) **Proposition:** *Let R be a regular local ring with maximal ideal $\mathfrak{m}$, and let $v = v_\mathfrak{m}$ be the order function of R. For every non-zero $x \in \mathfrak{m}$ we have*

$$e(R/Rx) = v(x).$$

Proof: Since R is a regular local ring, v is a valuation of $Q(R)$ [cf. VII(2.2)], the field of quotients of R, and $\mathfrak{m}^n$ is a v-ideal for every $n \in \mathbb{N}_0$.
We set $\overline{R} := R/Rx$ and $d := \dim(R)$; we have $\dim(\overline{R}) = \dim(R) - 1$ [cf. [63], Cor. 10.9 and Th. 12.4]. Let $n > v(x)$; we consider the exact sequence

$$0 \to (Rx + \mathfrak{m}^n)/\mathfrak{m}^n \to R/\mathfrak{m}^n \to R/(Rx + \mathfrak{m}^n) \to 0.$$

We have $(Rx + \mathfrak{m}^n)/\mathfrak{m}^n \cong Rx/(\mathfrak{m}^n \cap Rx)$ and $R/(Rx + \mathfrak{m}^n) \cong \overline{R}/\mathfrak{m}^n\overline{R}$. We also have $\mathfrak{m}^n \cap Rx = x\mathfrak{m}^{n-v(x)}$ for every $n \geq v(x)$ [we have $x\mathfrak{m}^{n-v(x)} \subset \mathfrak{m}^n \cap Rx$, and for $y \in \mathfrak{m}^n \cap Rx$ we write $y = rx$, hence $v(r) = v(y) - v(x) \geq n - v(x)$, hence $y \in x\mathfrak{m}^{n-v(x)}$], hence $Rx/(\mathfrak{m}^n \cap Rx) = Rx/x\mathfrak{m}^{n-v(x)} \cong R/\mathfrak{m}^{n-v(x)}$. Therefore we have

$$\ell_R((R/Rx)/\mathfrak{m}^n(R/Rx)) + \ell_R(R/\mathfrak{m}^{n-v(x)}) = \ell_R(R/\mathfrak{m}^n).$$

We consider the Hilbert functions, and have for all large enough n

$$H(\overline{R})(n) + H(R)(n - v(x)) = H(R)(n).$$

Now we have

$$H(R)(n) - H(R)(n - v(x)) = \binom{n+d}{d} - \binom{n+d-v(x)}{d}$$

$$= \frac{v(x)}{(d-1)!}n^{d-1} + \text{lower terms},$$

$$H(\overline{R})(n) = \frac{e(\overline{R})}{(d-1)!}n^{d-1} + \text{lower terms},$$

and therefore we obtain $e(\overline{R}) = v(x)$.

10.7 A Length Formula

The following result is essentially contained in [63], Th. 2.13 and Cor. 2.17.

(10.9) Proposition: *Let A be a noetherian ring and M a finitely generated non-zero A-module. Then M is an artinian A-module iff every prime ideal in $\mathrm{Ass}(M)$ is a maximal ideal. If this is the case, then we have $\mathrm{Supp}(M) = \mathrm{Ass}(M)$, M has finite length $\ell_A(M)$ and*

$$\ell_A(M) = \sum_{\mathfrak{p} \in \mathrm{Supp}(M)} \ell_{A_\mathfrak{p}}(M_\mathfrak{p}) = \sum_{\mathfrak{p} \in \mathrm{Spec}(A)} \ell_{A_\mathfrak{p}}(M_\mathfrak{p}).$$

(10.10) REMARK: (1) Let A (resp. B) be a local ring with maximal ideal $\mathfrak{m}$ (resp. $\mathfrak{n}$), and let $A \to B$ be a local homomorphism [i.e., $\mathfrak{m}B \subset \mathfrak{n}$] which is quasifinite. Then, in particular, the field $B/\mathfrak{n}$ is a finite extension of the field $A/\mathfrak{m}$. Every simple B-module is isomorphic to $B/\mathfrak{n}$, and we have $\ell_A(B/\mathfrak{n}) = \dim_{A/\mathfrak{m}}(B/\mathfrak{n}) =$

$[B/\mathfrak{n}\colon A/\mathfrak{m}]$, hence $\ell_A(N) = [B/\mathfrak{n}\colon A/\mathfrak{m}]\ell_B(N)$ for every B-module N of finite length.

(2) Let A be local with maximal ideal $\mathfrak{m}$, and let $\varphi\colon A \to B$ be a finite homomorphism of noetherian rings with $\ker(\varphi) \subset \mathfrak{m}$. Then B is semilocal [cf. (3.8)]; let $\mathfrak{n}_1,\ldots,\mathfrak{n}_h$ be the maximal ideals of B. The induced homomorphism $A \to B_{\mathfrak{n}_i}$ is local for every $i \in \{1,\ldots,h\}$, and it is quasifinite by III(3.11). Let N be a B-module of finite length. Now we obtain from (1) and (10.9)

$$\ell_A(N) = \sum_{i=1}^{h}[B/\mathfrak{n}_i\colon A/\mathfrak{m}]\ell_{B_{\mathfrak{n}_i}}(N_{\mathfrak{n}_i}).$$

10.8 Quasifinite Modules

(10.11) DEFINITION: (1) Let A be a local ring with maximal ideal $\mathfrak{m}$. An A-module M is called quasifinite if $\dim_{A/\mathfrak{m}}(M/\mathfrak{m}M) < \infty$.
(2) A local homomorphism $A \to B$ of local rings is called quasifinite if B is a quasifinite A-module.

(10.12) Proposition: Let A be a complete local ring with maximal ideal $\mathfrak{m}$, and let M be a quasifinite A-module. If the $\mathfrak{m}$-adic topology of M is Hausdorff, then M is a finitely generated A-module.

Proof: Let $y_1,\ldots,y_h \in M$ be such that their images in $M/\mathfrak{m}M$ are a system of generators, and set $N := Ay_1 + \cdots + Ay_h$. Let $x \in M$; then there exist elements $a_1^{(1)},\ldots,a_h^{(1)} \in A$ with $x-(a_1^{(1)}y_1+\cdots+a_h^{(1)}y_h) \in \mathfrak{m}M$. We set $x^{(1)} := a_1^{(1)}y_1+\cdots+a_h^{(1)}y_h \in N$. Let $n \in \mathbb{N}$, and assume that we already have constructed elements $x^{(1)},\ldots,x^{(n)} \in N$ having the form $x^{(j)} = a_1^{(j)}y_1+\cdots+a_h^{(j)}y_h$ with $a_1^{(j)},\ldots,a_h^{(j)} \in \mathfrak{m}^{j-1}$ for $j \in \{1,\ldots,n\}$ such that $x - (x^{(1)} + \cdots + x^{(n)}) \in \mathfrak{m}^nM$. Then we can write $x - (x^{(1)}+\cdots+x^{(n)}) = b_1z_1+\cdots+b_lz_l$ with $b_1,\ldots,b_l \in \mathfrak{m}^n$, $z_1,\ldots,z_l \in M$. There exist elements $w_1,\ldots,w_l \in N$ with $z_1 - w_1,\ldots,z_l - w_l \in \mathfrak{m}M$; we set $x^{(n+1)} := b_1w_1 + \cdots + b_lw_l$. We can write $x^{(n+1)} = a_1^{(n+1)}y_1+\cdots+a_h^{(n+1)}y_h$ with $a_1^{(n+1)},\ldots,a_h^{(n+1)} \in \mathfrak{m}^n$, and we have $x-(x^{(1)}+\cdots+x^{(n+1)}) = b_1(z_1-w_1)+\cdots+ b_l(z_l-w_l) \in \mathfrak{m}^{n+1}M$. Thus, for every $n \in \mathbb{N}$ we have elements $a_1^{(n)},\ldots,a_h^{(n)} \in \mathfrak{m}^{n-1}$ such that, setting $x^{(n)} = a_1^{(n)}y_1+\cdots+a_h^{(n)}y_h$, we have $x-(x^{(1)}+\cdots+x^{(n)}) \in \mathfrak{m}^nM$. We set $a_i := \lim_{n\to\infty} a_i^{(n)}$ for $i \in \{1,\ldots,h\}$, $x' := a_1y_1+\cdots+a_hy_h \in N$. Then we have $x - x' \in \mathfrak{m}^nM$ for every $n \in \mathbb{N}$, hence $x = x'$, and therefore we have $M \subset N$, hence $M = N$.

(10.13) REMARK: Let A (resp. B) be a local ring with maximal ideal $\mathfrak{m}$ (resp. $\mathfrak{n}$) and completion $\widehat{A}$ (resp. $\widehat{B}$), and let $A \to B$ be a local homomorphism. We assume that B is a quasifinite A-module. Then $B/\mathfrak{m}B$ is an artinian ring, hence there exists $k \in \mathbb{N}$ with $\mathfrak{n}^k \subset \mathfrak{m}B \subset \mathfrak{n}$ [cf. [63], Cor. 2.17], and therefore the $\mathfrak{n}$-adic and $\mathfrak{m}B$-adic topology on B are identical. In particular, the local homomorphism $A \to$

B induces a homomorphism of completions $\widehat{A} \to \widehat{B}$, and since $\widehat{B}$ is a quasifinite $\widehat{A}$-module, it is therefore a finitely generated $\widehat{A}$-module by (10.12).

10.9 Maximal Primary Ideals

(10.14) NOTATION: In this subsection A is a noetherian ring.

(10.15) **Proposition:** Let $\mathfrak{a} \neq A$, $\mathfrak{b}$ be ideals of A. Then we have $\mathfrak{a} : \mathfrak{b} = \mathfrak{a}$ iff $\mathfrak{b}$ is not contained in a prime ideal of the set $\mathrm{Ass}(A/\mathfrak{a})$.

Proof: Let

$$\mathfrak{a} = \mathfrak{q}_1 \cap \cdots \cap \mathfrak{q}_h \quad \text{with } \mathfrak{p}_i = \mathrm{rad}(\mathfrak{q}_i) \text{ for } i \in \{1, \ldots, h\} \qquad (*)$$

be an irredundant primary decomposition of $\mathfrak{a}$.
(1) Assume that $\mathfrak{b} \not\subset \mathfrak{p}_i$ for $i \in \{1, \ldots, h\}$. We have $(\mathfrak{a} : \mathfrak{b})\mathfrak{b} \subset \mathfrak{a} \subset \mathfrak{q}_i$, hence we have $\mathfrak{a} : \mathfrak{b} \subset \mathfrak{q}_i$ for $i \in \{1, \ldots, h\}$, which means that $\mathfrak{a} : \mathfrak{b} \subset \mathfrak{a}$; since $\mathfrak{a} \subset \mathfrak{a} : \mathfrak{b}$, we have shown that $\mathfrak{a} : \mathfrak{b} = \mathfrak{a}$.
(2) Assume that $\mathfrak{a} : \mathfrak{b} = \mathfrak{a}$, and suppose that $\mathfrak{b}$ is contained in some prime ideal of the set $\mathrm{Ass}(A/\mathfrak{a})$, say $\mathfrak{b} \subset \mathfrak{p}_1$. We have $\mathfrak{a} : \mathfrak{b}^l = \mathfrak{a}$ for every $l \in \mathbb{N}$, and there exists $l \in \mathbb{N}$ with $\mathfrak{b}^l \subset \mathfrak{p}_1^l \subset \mathfrak{q}_1$ [cf. [63], Ex. 1.13] which implies that $\mathfrak{q}_1 : \mathfrak{b}^l = A$, whence

$$\mathfrak{a} = \mathfrak{a} : \mathfrak{b}^l = \bigcap_{i=1}^{h}(\mathfrak{q}_i : \mathfrak{b}^l) = \bigcap_{i=2}^{h}(\mathfrak{q}_i : \mathfrak{b}^l) \supset \bigcap_{i=2}^{h} \mathfrak{q}_i \supset \mathfrak{a};$$

this means that $(*)$ is not an irredundant primary decomposition of $\mathfrak{a}$.

(10.16) DEFINITION: Let $\mathfrak{p}$ be a prime ideal of A. A $\mathfrak{p}$-primary ideal $\mathfrak{q}$ of A is called a maximal $\mathfrak{p}$-primary ideal if it is maximal in the set of $\mathfrak{p}$-primary ideals of A which are different from $\mathfrak{p}$.

(10.17) **Proposition:** We have:
(1) Let $\mathfrak{p}$ be a prime ideal of A, and let $\mathfrak{q}$ be a maximal $\mathfrak{p}$-primary ideal of A. Then we have $\mathfrak{q} : \mathfrak{p} = \mathfrak{p}$.
(2) Let $\mathfrak{m}$ be a maximal ideal of A, and let $\mathfrak{q}$ be a maximal $\mathfrak{m}$-primary ideal of A. Then $\mathfrak{m}/\mathfrak{q}$ is a one-dimensional $A/\mathfrak{m}$-vector space.

Proof: (1) We have $\mathfrak{q} : \mathfrak{p} \supsetneq \mathfrak{q}$ by (10.15); it is easy to check that $\mathfrak{q} : \mathfrak{p}$ is a $\mathfrak{p}$-primary ideal, hence that $\mathfrak{q} : \mathfrak{p} = \mathfrak{p}$ by maximality of $\mathfrak{q}$.
(2) Since $\mathfrak{q} : \mathfrak{m} = \mathfrak{m}$ by (1), we have $\mathfrak{m}^2 \subset \mathfrak{q}$. Every ideal lying between $\mathfrak{m}^2$ and $\mathfrak{m}$ is an $\mathfrak{m}$-primary ideal [since $\mathfrak{m}$ is maximal], hence $\mathfrak{m}/\mathfrak{q}$ is a one-dimensional $R/\mathfrak{m}$-vector space.

10.10   Primary Decomposition in Non-Noetherian Rings

(10.18) In this subsection A is a ring.

(10.19) NOTATION: For any prime ideal $\mathfrak{p}$ of A we set $S^{(\mathfrak{p})} := A \setminus \mathfrak{p}$. For any ideal $\mathfrak{a}$ of A we set $\mathfrak{a}^{(\mathfrak{p})} := (i_A^{\mathfrak{p}})^{-1}(\mathfrak{a}A_{\mathfrak{p}})$; note that $\mathfrak{a}^{(\mathfrak{p})}$ is the saturation of $\mathfrak{a}$ with respect to $S^{(\mathfrak{p})}$.

(10.20) Lemma: Let $\mathfrak{a} \neq A$ be an ideal of A, and let $\mathfrak{p} \in V(\mathfrak{a})$ be a minimal element. Then $\mathfrak{a}^{(\mathfrak{p})}$ is a $\mathfrak{p}$-primary ideal, and every $x \in \mathfrak{p}$ is a zero-divisor for $A/\mathfrak{a}$.

Proof: We show that $\mathfrak{a}^{(\mathfrak{p})} \subset \mathfrak{p} \subset \mathrm{rad}(\mathfrak{a}^{(\mathfrak{p})})$ and that, if $x, y \in A$, $x \notin \mathfrak{p}$ and $xy \in \mathfrak{a}^{(\mathfrak{p})}$, then we have $y \in \mathfrak{a}^{(\mathfrak{p})}$; this implies that $\mathfrak{a}^{(\mathfrak{p})}$ is a $\mathfrak{p}$-primary ideal of A. Let $x \in \mathfrak{a}^{(\mathfrak{p})}$. Then there exists $s \in S^{(\mathfrak{p})}$ with $sx \in \mathfrak{a}$, hence $sx \in \mathfrak{p}$ and therefore $x \in \mathfrak{p}$, showing that $\mathfrak{a}^{(\mathfrak{p})} \subset \mathfrak{p}$. Let $x \in \mathfrak{p}$, and set $S =: \{x^n s \mid n \in \mathbb{N}_0, s \in S^{(\mathfrak{p})}\}$. Now S is multiplicatively closed. We show that $S \cap \mathfrak{a} \neq \emptyset$. In fact, suppose that $S \cap \mathfrak{a} = \emptyset$. Then there exists a prime ideal $\mathfrak{p}'$ of A with $\mathfrak{a} \subset \mathfrak{p}'$ and with $\mathfrak{p}' \cap S = \emptyset$ [cf. [63], Prop. 2.11]. Since $S^{(\mathfrak{p})} \subset S$, we have $\mathfrak{a} \subset \mathfrak{p}' \subset \mathfrak{p}$, hence $\mathfrak{p}' = \mathfrak{p}$ [since $\mathfrak{p}$ is a minimal element of $V(\mathfrak{a})$], and therefore $x \notin \mathfrak{p}$, contrary to the choice of x. Therefore there exists $n \in \mathbb{N}_0$ and $s \in S^{(\mathfrak{p})}$ with $x^n s \in \mathfrak{a}$, hence $x^n \in \mathfrak{a}^{(\mathfrak{p})}$, and therefore $x \in \mathrm{rad}(\mathfrak{a}^{(\mathfrak{p})})$, showing that $\mathfrak{p} \subset \mathrm{rad}(\mathfrak{a}^{(\mathfrak{p})})$.
Now let $x, y \in A$ with $x \in S^{(\mathfrak{p})}$ and $xy \in \mathfrak{a}^{(\mathfrak{p})}$. There exists $s \in S^{(\mathfrak{p})}$ with $sxy \in \mathfrak{a}$. Then $sx \in S^{(\mathfrak{p})}$ yields $y \in \mathfrak{a}^{(\mathfrak{p})}$.
Now let $x \in \mathfrak{p}$. From $\mathfrak{p} \subset \mathrm{rad}(\mathfrak{a}^{(\mathfrak{p})})$ we see that $x^n \in \mathfrak{a}^{(\mathfrak{p})}$ for some $n \in \mathbb{N}$, hence there exists $s \in S^{(\mathfrak{p})}$ with $sx^n \in \mathfrak{a}$, and therefore x^n is a zero-divisor for $A/\mathfrak{a}$. We choose $m \in \mathbb{N}$ minimal such that x^m is a zero-divisor for $A/\mathfrak{a}$. Suppose that $m > 1$. There exists $t \in A \setminus \mathfrak{a}$ with $tx^m \in \mathfrak{a}$. Since $tx \notin \mathfrak{a}$ [as x is not a zero divisor for $A/\mathfrak{a}$], we see that x^{m-1} is a zero-divisor for $A/\mathfrak{a}$, contrary to the choice of m. Therefore x is a zero-divisor for $\mathfrak{a}$.

(10.21) Proposition: Let $\mathfrak{a} \neq A$ be an ideal of A. If $V(\mathfrak{a})$ is finite and consists only of maximal ideals of A, then

$$\mathfrak{a} = \bigcap_{\mathfrak{p} \in V(\mathfrak{a})} \mathfrak{a}^{(\mathfrak{p})} \tag{$*$}$$

is an irredundant primary decomposition of $\mathfrak{a}$; $(*)$ is the only irredundant primary decomposition of $\mathfrak{a}$.

Proof: Let $\mathfrak{p}_1, \ldots, \mathfrak{p}_m$ be the pairwise different elements of $V(\mathfrak{a})$. Let $i \in \{1, \ldots, m\}$; from (10.20) we see that $\mathfrak{a}^{(\mathfrak{p}_i)}$ is a $\mathfrak{p}_i$-primary ideal. Let $x \in \mathfrak{a}^{(\mathfrak{p}_1)} \cap \cdots \cap \mathfrak{a}^{(\mathfrak{p}_m)}$, and set $\mathfrak{b} := \mathfrak{a} : A$. Let $i \in \{1, \ldots, m\}$; since $x \in \mathfrak{a}^{(\mathfrak{p}_i)}$, we see that there exists $s_i \in S^{(\mathfrak{p}_i)}$ with $s_i x \in \mathfrak{a}$, hence $s_i \in \mathfrak{b}$, and therefore $\mathfrak{b} \not\subset \mathfrak{p}_i$. This implies that $\mathfrak{b} \not\subset \mathfrak{p}_1 \cup \cdots \cup \mathfrak{p}_m$ [cf. [63], Lemma 3.3], hence there exists $t \in \mathfrak{b}$, $t \notin \mathfrak{p}_1 \cup \cdots \cup \mathfrak{p}_m$. Now $A/\mathfrak{a}$ is a quasisemilocal ring with maximal ideals $\mathfrak{p}_1/\mathfrak{a}, \ldots, \mathfrak{p}_m/\mathfrak{a}$; therefore the image of t in $A/\mathfrak{a}$ is a unit of $A/\mathfrak{a}$, hence there exist

$u \in A$, $a \in \mathfrak{a}$ with $tu = 1 + a$. Since $tx \in \mathfrak{a}$, we have $x = tux - ax \in \mathfrak{a}$. Since none of the ideals $\mathfrak{a}^{(\mathfrak{p}_i)}$ contains the intersection of the other ideals $\mathfrak{a}^{(\mathfrak{p}_j)}$, $j \in \{1, \ldots, m\}$, $j \neq i$, $(*)$ is an irredundant intersection.
The uniqueness statement follows from [21], Th. 4.10.

10.11   Discriminant of a Polynomial

(10.22) (1) Let A be a ring, let $F = X^n + a_1 X^{n-1} + \cdots + a_n \in A[X]$ be a monic polynomial of positive degree n, set $E := A[X]/(F)$, and let x be the image of X in E. Note that E is a free A-algebra having $\{1, x, \ldots, x^{n-1}\}$ as a basis. We set

$$s_k := \mathrm{Tr}_{E/A}(x^k) \quad \text{for every } k \in \mathbb{N}_0.$$

Then we have [Newton's formulae, cf. [32], Ch. IV, §6, no. 5]

$$s_k + a_1 s_{k-1} + \cdots + a_{k-1} s_1 + k a_k = 0 \quad \text{for } k \in \{1, \ldots, n\},$$
$$s_k + a_1 s_{k-1} + \cdots + a_{n-1} s_{k-n+1} + a_n s_{k-n} = 0 \quad \text{for } k > n.$$

Note, in particular, that $s_k \in \mathbb{Z}[a_1, \ldots, a_n]$ for every $k \in \mathbb{N}_0$.
We define

$$\mathrm{dis}(F) := D_{E/A}(1, x, \ldots, x^{n-1}) \in A;$$

$\mathrm{dis}(F)$ is called the discriminant of F. We have

$$\mathrm{dis}(F) = \det((s_{i+j})_{0 \leq i, i < n}) \tag{$*$}$$

[cf. [32], Ch. IV, §6, no. 7, (45)]. We find that $\mathrm{dis}(F) = 1$ if $n = 1$, and

$$\mathrm{dis}(F) = \begin{cases} a_1^2 - 4a_2 & \text{if } n = 2, \\ a_1^2 a_2^2 - 4a_1^3 a_3 + 18 a_1 a_2 a_3 - 4a_2^3 - 27 a_3^2 & \text{if } n = 3. \end{cases}$$

Let $\alpha \colon A \to B$ be a homomorphism of rings, and assume that we have a factorization

$$X^n + \alpha(a_1) X^{n-1} + \cdots + \alpha(a_n) = (X_1 - x_1) \cdots (X - x_n) \text{ in } B[X].$$

Then we have [cf. [32], Ch. IV, §6, no. 7, (46)]

$$\mathrm{dis}(F) 1_B = \prod_{1 \leq i < j \leq n} (x_j - x_i)^2 = (-1)^{n(n-1)/2} \prod_{i \neq j} (x_i - x_j).$$

(2) In particular, if K is a field and if $F \in K[X]$ is a monic polynomial, then F is separable, i.e., F does not have multiple roots in a splitting field of F over K, iff $\mathrm{dis}(F) \neq 0$.
(3) Let $n \in \mathbb{N}$. There exists a unique polynomial $\Delta(X_1, \ldots, X_n) \in \mathbb{Z}[X_1, \ldots, X_n]$ with the following property: For every polynomial $F = X^n + a_1 X^{n-1} + \cdots + a_n \in A[X]$ we have $\mathrm{dis}(F) = \Delta(a_1, \ldots, a_n)$ [cf. [32], Ch. IV, §6, no. 7, Prop. 10].

Bibliography

[1] Abhyankar, S. S.: *On the ramification of algebraic functions*. Amer. J. Math., 77:575–592, 1955.

[2] Abhyankar, S. S.: *On the valuations centered in a local domain*. Amer. J. Math., 78:321–348, 1956.

[3] Abhyankar, S. S.: *Ramification Theoretic Methods in Algebraic Geometry*, volume 43 of *Annals of Mathematical Studies*. Princeton University Press, Princeton, 1959.

[4] Abhyankar, S. S.: *Local Analytic Geometry*, volume XIV of *Pure and Applied Mathematics*. Academic Press, New York, World Scientific Publishing Company reprint edition, 1964.

[5] Abhyankar, S. S.: *Resolution of singularities of arithmetical surfaces*. In *Arithmetical Algebraic Geometry (Proc. Conf. Purdue Univ., 1963)*, pages 111–152. Harper & Row, New York, 1965.

[6] Abhyankar, S. S.: *On the problem of resolution of singularities*. In *Proc. Internat. Congr. Math. (Moscow, 1966)*, pages 469–481. Izdat. "Mir", Moscow, 1968.

[7] Abhyankar, S. S.: *Resolution of singularities of algebraic surfaces*. In *Algebraic Geometry (Internat. Colloq., Tata Inst. Fund. Res., Bombay, 1968)*, pages 1–11. Oxford Univ. Press, London, 1969.

[8] Abhyankar, S. S.: *Lectures on Expansion Techniques in Algebraic Geometry*, volume 57 of *Lectures on Mathematics and Physics*. Tata Institute of Fundamental Research, Bombay, 1977.

[9] Abhyankar, S. S.: *On the semigroup of a meromorphic curve I*. In *Proc. of the Int. Symp. on Algebraic Geometry, Kyoto*, pages 249–414, 1977.

[10] Abhyankar, S. S.: *Desingularization of plane curves*. In *Singularities, Part 1 (Arcata, Calif., 1981)*, volume 40 of *Proc. Sympos. Pure Math.*, pages 1–45. Amer. Math. Soc., Providence, RI, 1983.

[11] Abhyankar, S. S.: *Square-root parametrization of plane curves*. In *Algebraic Geometry and its Applications, West Lafayette, In., 1990*, pages 19–84. Springer Verlag, Berlin Heidelberg New-York, 1994.

[12] Abhyankar, S. S.: *Resolution of Singularities of Embedded Algebraic Surfaces*. Springer Monographs in Mathematics. Springer Verlag, Berlin Heidelberg New-York, second edition, 1998.

[13] Abhyankar, S. S. and T. T. Moh: *Newton-Puiseux expansion and generalized Tschirnhausen transformation I.* J. Reine Angew. Math., 260:47–83, 1973.

[14] Abhyankar, S. S. and T. T. Moh: *Newton-Puiseux expansion and generalized Tschirnhausen transformation II.* J. Reine Angew. Math., 261:29–54, 1973.

[15] Abramovich, D. and J. Wang: *Equivariant resolution of singularities in characteristic 0.* Math. Res. Lett., 4(2-3):427–433, 1997.

[16] Alajbegović, Jusuf: *Approximation theorems for Manis valuations with the inverse property.* Comm. Algebra, 12:1399–1417, 1984.

[17] Albanese, G: *Trasformazione birazionale di una superficie algebrica qualunque in un'altra priva di punti multipli.* Rend. Circ. Mat. Palermo, 48:321–332, 1924.

[18] Angermüller, G.: *Die Wertehalbgruppe einer ebenen irreduziblen algebroiden Kurve.* Math. Z., 153:267–282, 1977.

[19] Arapović, Miroslav: *Approximation theorems for Manis valuations.* Canad. Math. Bull., 28:184–189, 1985.

[20] Aroca, José M.: *Problems of the classification of singularities of algebroid curves over fields of positive characteristic.* In *Mathematics today (Luxembourg, 1981),* pages 27–50. Gauthier-Villars, Paris, 1982.

[21] Atiyah, M. F. and I. G. Macdonald: *Introduction to Commutative Algebra.* Addison-Wesley Publishing Company, Reading, Mass., 1969.

[22] Auslander, M.: *On the purity of the branch locus.* Amer. J. Math., 84:116–125, 1962.

[23] Auslander, M. and D. A. Buchsbaum: *On ramification theory in noetherian rings.* Amer. J. Math., 81:749–765, 1959.

[24] Ban, C. and L. J. McEwan: *Canonical resolution of a quasi-ordinary surface singularity.* Canad. J. Math, 52:1149–1163, 2000.

[25] Becker, Th. and V. Weispfenning: *Gröbner Bases,* volume 141 of *Graduate Texts in Mathematics.* Springer Verlag, Berlin Heidelberg New-York, 1993.

[26] Behnke, K.: *Die Hilbertfunktion einer zweidimensionalen normalen monomialen Algebra.* Manuskript, 1991.

[27] Berger, R.: *Über verschiedene Differentenbegriffe.* S.-Ber. Heidelberger Akad. Wiss., Math-Naturw. Kl., Seiten 1–44, 1960/61.

[28] Bierstone, E. and P. D. Milman: *Canonical desingularization in characteristic zero by blowing up the maximum strata of a local invariant.* Invent. Math., 128:207–302, 1997.

[29] Bodnár, G. and J. Schicho: *Automated resolution of singularities for hypersurfaces.* J. Symbolic Comput., 30(4):401–428, 2000.

[30] Bodnár, G. and J. Schicho: *A computer program for the resolution of singularities.* In Hauser, H., J. Lipman, F. Oort, and A. Quirós (editors): *Resolution of Singularities,* volume 189 of *Progress in Mathematics,* pages 231–238. Birkhäuser, Basel Boston Berlin, 2000.

[31] Bodnár, G. and J. Schicho: *Two computational techniques for singularity resolution.* J. Symbolic Comput., 32(1-2):39–54, 2001.

[32] Bourbaki, N.: *Algèbre*. Masson, Paris, 1981.

[33] Bourbaki, N.: *Topologie Générale*. Masson, Paris, 1981.

[34] Bourbaki, N.: *Algèbre Commutative*. Masson, Paris, 1984.

[35] Brieskorn, E. und H. Knörrer: *Ebene algebraische Kurven*. Birkhäuser Verlag, Basel, 1980.

[36] Brodmann, M: *Algebraische Geometrie*, Band 1 der Reihe *Basler Lehrbücher*, *A Series of Advanced Textbooks in Mathematics*. Birkhäuser Verlag, Basel, 1989.

[37] Bruns, W. and J. Herzog: *Cohen-Macaulay Rings*, volume 39 of *Cambridge Studies in Advanced Mathematics*. Cambridge University Press, Cambridge, 1993.

[38] Brylinski, J.-L.: *Eventails et Variétés Toriques*. Dans *Séminaire sur les Singularités des Surfaces*, tome 777 de *Lecture Notes in Mathematics*, pages 247–288, Berlin Heidelberg New-York, 1980. Springer Verlag.

[39] Bump, D.: *Algebraic Geometry*. World Scientific, Singapore New Jersey London Hong Kong, 1998.

[40] Campillo, A.: *Algebroid Curves in Positive Characteristic*, volume 813 of *Lecture Notes in Mathematics*. Springer Verlag, Berlin Heidelberg New-York, 1980.

[41] Chevalley, C.: *Intersections of algebraic and algebroid varieties*. Trans. Amer. Math. Soc., 57:1–85, 1945.

[42] Chevalley, C.: *Introduction to the Theory of Algebraic Functions of One Variable*, volume 6 of *Mathematical Surveys*. American Mathematical Society, Providence, RI, 1951.

[43] Chisini, O: *La risoluzione delle singolarità di una superficie mediante transformazioni birazionali dello spazio*. Mem. Accad. Sci Bologna, VIII:1–42, 1921.

[44] Cohen, I. S.: *Commutative rings with restricted minimum condition*. Duke Math. J., 76:27–42, 1954.

[45] Cohen, I. S.: *Length of prime ideal chains*. Amer. J. Math., 76:654–668, 1954.

[46] Cohen, I. S. and O. Zariski: *A fundamental inequality in the theory of extensions of valuations*. Illinois J. Math, 1:1–8, 1957.

[47] Cohn, P. M.: *Puiseux's theorem revisited*. J. Pure Appl. Algebra, 31:1–4, 1984.

[48] Cohn, P. M.: *Correction to: "Puiseux's theorem revisited"*. J. Pure Appl. Algebra, 52:197–198, 1988.

[49] Cossart, V.: *Uniformisation et désingularisation des surfaces d'après Zariski*. Dans Hauser, H., J. Lipman, F. Oort et A. Quiròs (rédacteurs) : *Resolution of Singularities*, tome 189 de *Progress in Mathematics*, pages 239–258. Birkhäuser, Basel Boston Berlin, 2000.

[50] Cossart, V., J. Giraud, and U. Orbanz: *Resolution of Surface Singularities*, volume 1101 of *Lecture Notes in Mathematics*. Springer Verlag, Berlin Heidelberg New-York, 1984.

[51] Cossart, V., O. Piltant, and A. J. Reguera-López: *Divisorial valuations dominating rational surface singularities*. In *Valuation Theory and its Applications, Vol. I (Saskatoon, SK, 1999)*, volume 32 of *Fields Inst. Commun.*, pages 89–101. Amer. Math. Soc., Providence, RI, 2002.

[52] Cox, D.: *Toric varieties and toric resolution*. In Hauser, H., J. Lipman, F. Oort, and A. Quirós (editors): *Resolution of Singularities*, volume 189 of *Progress in Mathematics*, pages 259–284. Birkhäuser, Basel Boston Berlin, 2000.

[53] Cox, D., J. Little, and D. O'Shea: *Ideals, Varieties and Algorithms*. Undergraduate Texts in Mathematics. Springer Verlag, Berlin Heidelberg New-York, 1992.

[54] Cutkosky, S. D.: *Local monomialization and factorization of morphisms*, volume 260 of *Astérisque*. Société Mathématique de France, Paris, 1999.

[55] Cutkosky, S. D.: *Local monomialization*. In *Geometric and combinatorial aspects of commutative algebra (Messina, 1999)*, volume 217 of *Lecture Notes in Pure and Appl. Math.*, pages 165–173. Dekker, New York, 2001.

[56] Cutkosky, S. D.: *Monomialization of morphisms from 3-folds to surfaces*, volume 1786 of *Lecture Notes in Mathematics*. Springer Verlag, Berlin Heidelberg New-York, 2002.

[57] Cutkosky, S. D.: *Ramification of valuations and singularities*. In *Advances in Algebra and Geometry (Hyderabad, 2001)*, pages 147–158. Hindustan Book Agency, New Delhi, 2003.

[58] Debremaeker, R. and V. Van Lierde: *A short proof of the length formula of Hoskin and Deligne*. Arch. Math., 78:369–371, 2002.

[59] Delgado, F., C. Galindo, and A. Núñez: *Saturation for valuations on two-dimensional regular local rings*. Math. Z., 234(3):519–550, 2000.

[60] Deligne, P.: *Intersections sur les surfaces régulieres*. Dans Deligne, P. et N. Katz (rédacteurs) : *Groupes de Monodromie en Géométrie Algébrique II (SGA 7,II)*, tome 340 de *Lecture Notes in Mathematics*, pages 1–38. Springer Verlag, Berlin Heidelberg New-York, 1973.

[61] Duval, D.: *Rational Puiseux expansions*. Compositio Math., 70:119–154, 1989.

[62] Eichler, M.: *Introduction to the Theory of Algebraic Numbers and Functions*. Academic Press, New York, 1966.

[63] Eisenbud, D.: *Commutative Algebra with a View toward Algebraic Geometry*, volume 150 of *Graduate Texts in Mathematics*. Springer Verlag, Berlin Heidelberg New-York, 1994.

[64] Encinas, S. and H. Hauser: *Strong resolution of singularities in characteristic zero*. Comment. Math. Helv., 77(4):821–845, 2002.

[65] Encinas, S. and O. Villamayor: *Good points and constructive resolution of singularities*. Acta Math., 181(1):109–158, 1998.

[66] Encinas, S. and O. Villamayor: *Principalization and desingularization*. Rev. Semin. Iberoam. Mat. Singul. Tordesillas, 2(5):11–17, 1999.

[67] Encinas, S. and O. Villamayor: *A course on constructive desingularization and equivariance*. In Hauser, H., J. Lipman, F. Oort, and A. Quirós (editors): *Resolution of Singularities*, volume 189 of *Progress in Mathematics*, pages 147–227. Birkhäuser, Basel Boston Berlin, 2000.

[68] Ewald, G.: *Combinatorial Convexity and Algebraic Geometry*, volume 168 of *Graduate Texts in Mathematics*. Springer Verlag, Berlin Heidelberg New-York, 1996.

[69] Fulton, W.: *Algebraic Curves.* Mathematics Lecture Note Series. Benjamin, Reading, Mass., 1969.

[70] Fulton, W.: *Introduction to Toric Varieties*, volume 131 of *Annals of Mathematical Studies.* Princeton University Press, Princeton, 1993.

[71] Giraud, J.: *Géométrie Algébrique Élémentaire*, 1977. Cours de troisiéme Cycle.

[72] Goldin, R. and B. Teissier: *Resolving singularities of plane analytic branches with one toric morphism.* In Hauser, H., J. Lipman, F. Oort, and A. Quiròs (editors): *Resolution of Singularities*, volume 189 of *Progress in Mathematics*, pages 315–340. Birkhäuser, Basel Boston Berlin, 2000.

[73] Gräter, Joachim: *Der Approximationssatz für Manisbewertungen.* Arch. Math., 37:335–340, 1981.

[74] Gräter, Joachim: *Der allgemeine Approximationssatz für Manisbewertungen.* Monatsh. Math., 93:277–288, 1982.

[75] Gräter, Joachim: *Fortsetzungen von Manisbewertungen.* Abh. Braunschweig. Wiss. Ges., 33:147–150, 1982.

[76] Grauert, H. und R. Remmert: *Analytische Stellenalgebren*, Band 176 der Reihe *Grundlehren der Mathematischen Wissenschaften.* Springer Verlag, Berlin Heidelberg New-York, 1971.

[77] Greco, S. and K. Kiyek: *General elements of complete ideals and valuations centered at a two-dimensional regular local ring.* In Christensen, C., G. Sundaram, A. Sathaye, and C. Bajaj (editors): *Algebra, Arithmetic and Geometry with Applications*, pages 281–356. Springer Verlag, Berlin Heidelberg New-York, 2003.

[78] Griffin, F.: *Valuations and Prüfer rings.* Canad. J. Math., 26:412–429, 1974.

[79] Gröbner, W.: *Algebraische Geometrie II.* Bibliographisches Institut, Mannheim, 1970.

[80] Grothendieck, A. et J. Dieudonné: *Éléments de Géométrie Algébrique.* Inst. Hautes Études Sci. Publ. Math., 4, 8, 11, 17, 20, 24, 28, 32, 1960, 1961, 1961, 1963, 1964, 1965, 1966, 1967.

[81] Harris, J.: *Algebraic Geometry*, volume 133 of *Graduate Texts in Mathematics.* Springer Verlag, Berlin Heidelberg New-York, 1992.

[82] Hartshorne, R.: *Algebraic Geometry*, volume 52 of *Graduate Texts in Mathematics.* Springer Verlag, Berlin Heidelberg New-York, 1977.

[83] Hauser, Herwig: *The Hironaka theorem on resolution of singularities (or: A proof we always wanted to understand).* Bull. Amer. Math. Soc. (N.S.), 40(3):323–403 (electronic), 2003.

[84] Hauser, H.: *Excellent surfaces and their taut resolution.* In Hauser, H., J. Lipman, F. Oort, and A. Quiròs (editors): *Resolution of Singularities*, volume 189 of *Progress in Mathematics*, pages 341–373. Birkhäuser, Basel Boston Berlin, 2000.

[85] Hauser, H.: *Resolution of singularities 1860 - 1999.* In Hauser, H., J. Lipman, F. Oort, and A. Quiròs (editors): *Resolution of Singularities*, volume 189 of *Progress in Mathematics*, pages 5–36. Birkhäuser, Basel Boston Berlin, 2000.

[86] Heinzer, W., C. Huneke, and D. J. Sally: *A criterion for spots.* Math. J. Kyoto Univ., 26:667–671, 1986.

[87] Herrera, J.: *Series de Puiseux en Dos Variables*. Tesis de Doctorado, Universidad de Sevilla, 1985.

[88] Herrmann, M., S. Ikeda, and U. Orbanz: *Equimultiplicity and Blowing Up. An Algebraic Study with an Appendix by B. Moonen*. Springer Verlag, Berlin Heidelberg New-York, 1988.

[89] Herzog, J. und E. Kunz: *Der kanonische Modul eines Cohen-Macaulay-Rings*, Band 238 der Reihe *Lecture Notes in Mathematics*. Springer Verlag, Berlin Heidelberg New-York, 1971.

[90] Herzog, J. und E. Kunz: *Die Wertehalbgruppe eines lokalen Rings der Dimension 1*. S.-Ber. Heidelberger Akad. Wiss., Math.-Naturw. Kl., Seiten 27–67, 1971.

[91] Hironaka, H.: *Resolution of singularities of an algebraic variety over a field of characteristic zero*. Annals of Math., 79:109–203, 205–236, 1964.

[92] Hironaka, H.: *Introduction to the Theory of Infinitely Near Singular Points*, volume 28 of *Memorias de Matematica del Instituto "Jorge Juan"*. Conseja Inferior de Investigaciones Cientificas, Madrid, 1974.

[93] Hirzebruch, F.: *Über vierdimensionale Riemannsche Flächen mehrdeutiger analytischer Funktionen von zwei komplexen Veränderlichen*. Math. Ann., 126:1–22, 1953.

[94] Hoskin, M. A.: *Zero-dimensional valuation ideals associated with plane curve branches*. Proc. London Phil. Soc. (3), 6:70–99, 1956.

[95] Hübl, R. and I. Swanson: *Discrete valuations centered on local domains*. J. Pure Appl. Algebra, 161(1-2):145–166, 2001.

[96] Huckaba, J. A.: *Commutative Rings with Zero Divisors*. Marcel Dekker, New York and Basel, 1988.

[97] Hulek, K.: *Elementary Algebraic Geometry*, volume 20 of *Student Mathematical Library*. American Mathematical Society, Providence, RI, 2003.

[98] Huneke, C.: *Complete ideals in two-dimensional regular local rings*. In *Commutative Algebra, Proceedings of a Microprogram Held June 15-July 2, 1987*, pages 325–338, Berlin Heidelberg New-York, 1989.

[99] Huneke, C. and J. Sally: *Birational extensions in dimension two and integrally closed ideals*. J. Algebra, 115:481–500, 1988.

[100] Johnston, B. L. and J. Verma: *On the length formula of Hoskin and Deligne and associated graded rings of two-dimensional regular local rings*. Math. Proc. Cambridge Philos. Soc, 111:432–432, 1992.

[101] Jong, T. de and G. Pfister: *Local Analytic Geometry*. Advanced Lectures in Mathematics. Vieweg, Braunschweig Wiesbaden, 2000.

[102] Jung, H. W. E.: *Darstellung der Funktionen eines algebraischen Körpers zweier unabhängiger Veränderlichen x, y in der Umgebung einer Stelle x = a, y = b*. J. Reine Angew. Math., 133:289–314, 1908.

[103] Kaplansky, I.: *Commutative Rings*. University of Chicago Press, Chicago, revised edition, 1974.

[104] Kempf, G., F. Knudson, D. Mumford, and B. Saint-Donat: *Toroidal embeddings I*, volume 339 of *Lecture Notes in Mathematics*. Springer Verlag, Berlin Heidelberg New-York, 1973.

[105] Kiyek, K.: *Anwendung von Idealtransformationen.* Manuscripta Math., 34:327–359, 1981.

[106] Kiyek, K.: *Multiplicity sequence and value semigroup.* Manuscripta Math., 37:211–216, 1982.

[107] Kiyek, K. and M. Micus: *Semigroup of a quasiordinary singularity.* In *Topics in Algebra*, volume 26 of *Banach Center Publications*, pages 149–156, Warsaw, 1990.

[108] Kiyek, K. and J. L. Vicente: *On the Jung-Abhyankar theorem.* Arch. Math., 2004. to appear.

[109] Knebusch, M. and D. Zhang: *Manis Valuations and Prüfer Extensions I.* Number 1791 in *Lecture Notes in Mathematics.* Springer Verlag, Berlin Heidelberg New-York, 2002.

[110] Krull, W.: *Beiträge zur Arithmetik kommutativer Integritätsbereiche VI. Der allgemeine Diskriminantensatz. Unverzweigte Ringerweiterungen.* Math. Z., 45:1–19, 1939.

[111] Kuhlmann, F.-V.: *Valuation theoretic and model theoretic aspects of local uniformization.* In Hauser, H., J. Lipman, F. Oort, and A. Quirós (editors): *Resolution of Singularities*, volume 189 of *Progress in Mathematics*, pages 381–456. Birkhäuser, Basel Boston Berlin, 2000.

[112] Kunz, E.: *The value-semigroup of a one-dimensional Gorenstein ring.* Proc. Amer. Math. Soc., 25:748–751, 1970.

[113] Kunz, E.: *Einführung in die Kommutative Algebra und Algebraische Geometrie.* Vieweg, Braunschweig, 1980.

[114] Kunz, E.: *Kähler Differentials.* Vieweg, Braunschweig, 1986.

[115] Kunz, E.: *Ebene algebraische Kurven*, Band 23 der Reihe *Regensburger Trichter*. Fakultät für Mathematik der Universität Regensburg, Regensburg, 1991.

[116] Lafon, J.-P. et J. Marot: *Algèbre locale*, tome 59 de *Collection Enseignement des Sciences*. Hermann, Paris, 2002.

[117] Lang, S.: *Algebra.* Addison-Wesley Publishing Company, Reading, Mass., 1965.

[118] Larsen, M. D. and P. J. McCarthy: *Multiplicative Theory of Ideals.* Academic Press, New York and London, 1971.

[119] Laufer, H.: *Normal Two-Dimensional Singularities*, volume 71 of *Annals of Mathematical Studies*. Princeton University Press, Princeton, N. J., 1971.

[120] Lê, D. T.: *Introduction à la Théorie des Singularités. I*, tome 36 de *Travaux en Cours [Works in Progress]*. Hermann, Paris, 1988.

[121] Lê, D. T.: *Introduction à la théorie des singularités. II*, tome 37 de *Travaux en Cours [Works in Progress]*. Hermann, Paris, 1988.

[122] Lê, D. T.: *Les singularités Sandwich.* Dans Hauser, H., J. Lipman, F. Oort et A. Quirós (rédacteurs) : *Resolution of Singularities*, tome 189 de *Progress in Mathematics*, pages 457–483. Birkhäuser, Basel Boston Berlin, 2000.

[123] Lejeune-Jalabert, M.: *Linear systems with infinitely near base conditions and complete ideals in dimension two.* In Lê, D. T., K. Saito, and B. Teissier (editors): *Singularity Theory*, pages 345–369. World Scientific, Singapur, 1995.

[124] Levi, B: *Risoluzione delle singolarita puntuali delle superficie algebriche.* Atti Accad. Sci. Torino, 33:66–86, 1897.

[125] Li, H. and van Oystaeyen: *A Primer of Algebraic Geometry*, volume 227 of *Pure and Applied Mathematics. A Series of Monographs and Textbooks.* Marcel Dekker, New York Basel, 2000.

[126] Lipman, J.: *Quasi-Ordinary Singularities of Embedded Surfaces.* PhD thesis, Harvard University, Cambridge, Mass., 1965.

[127] Lipman, J.: *Rational singularities with applications to algebraic surfaces and unique factorization.* Inst. Hautes Études Sci. Publ. Math., 36:195–279, 1969.

[128] Lipman, J.: *Stable ideals and Arf rings.* Amer. J. Math., 93:449–685, 1971.

[129] Lipman, J.: *Introduction to resolution of singularities.* In *Algebraic Geometry (Proc. Sympos. Pure Math., Vol. 29, Humboldt State Univ., Arcata, Calif., 1974)*, volume 29, pages 187–230, Providence, RI, 1975. Amer. Math. Soc.

[130] Lipman, J.: *Desingularization of two-dimensional schemes.* Ann. of Math., 107:151–207, 1978.

[131] Lipman, J.: *On complete ideals in regular local rings.* In *Algebraic Geometry and Commutative Algebra, vol. I, in Honor of Masayoshi Nagata*, pages 203–231. Kinokuniya, 1987/1988.

[132] Lipman, J.: *Adjoints and polars of simple complete ideals in two-dimensional regular local rings.* Bull. Soc. Math. Belg., Sèr. A, 45:223–243, 1993.

[133] Lipman, J.: *Proximity inequalities for complete ideals in two-dimensional regular local rings.* Contemporary Mathematics, 159:293–306, 1994.

[134] Lipman, J.: *Equisingularity and simultaneous resolution of singularities.* In Hauser, H., J. Lipman, F. Oort, and A. Quiròs (editors): *Resolution of Singularities*, volume 189 of *Progress in Mathematics*, pages 485–505. Birkhäuser, Basel Boston Berlin, 2000.

[135] Lipman, J.: *Oscar Zariski 1899 - 1986.* In Hauser, H., J. Lipman, F. Oort, and A. Quiròs (editors): *Resolution of Singularities*, volume 189 of *Progress in Mathematics*, pages 1–4. Birkhäuser, Basel Boston Berlin, 2000.

[136] Luengo, I.: *A new proof of the Jung-Abhyankar theorem.* J. Algebra, 85:399–409, 1983.

[137] Marot, J.: *Anneaux de Valuations.* Manuscrit, 1973.

[138] Matlis, E.: *One-Dimensional Cohen-Macaulay-Rings*, volume 327 of *Lecture Notes in Mathematics*. Springer Verlag, Berlin Heidelberg New-York, 1973.

[139] Matsumura, H.: *Commutative Ring Theory*, volume 8 of *Cambridge Studies in Advanced Mathematics*. Cambridge University Press, Cambridge, 1986.

[140] Maurer, J.: *Puiseux expansion for space curves.* Manuscripta Math., 32:91–100, 1980.

[141] Michler, R.: *Singularities in Algebraic and Analytic Geometry*, volume 266 of *Contemporary Mathematics*. American Mathematical Society, Providence, RI, 2000.

[142] Micus, M.: *Zur formalen Äquivalenz von quasigewöhnlichen Singularitäten.* Dissertation, Universität Paderborn, Paderborn, Institut für Mathematik, 1987.

[143] Muhly, H. T. and O. Zariski: *The resolution of singularities of an algebraic curve.* Amer. J. Math., 61:107–114, 1939.

[144] Mulay, S. B.: *Abhyankar's work on desingularization.* In *Algebraic Geometry and its Applications. (West Lafayette, In., 1990)*, pages 153–160. Springer Verlag, Berlin Heidelberg New-York, 1994.

[145] Müller, G.: *Resolution of weighted homogeneous surface singularities.* In Hauser, H., J. Lipman, F. Oort, and A. Quirós (editors): *Resolution of Singularities*, volume 189 of *Progress in Mathematics*, pages 507–517. Birkhäuser, Basel Boston Berlin, 2000.

[146] Mumford, D.: *Algebraic Geometry I: Complex Projective Varieties*, volume 221 of *Grundlehren der mathematischen Wissenschaften.* Springer Verlag, Berlin Heidelberg New-York, 1976.

[147] Nagata, M.: *Local Rings.* Interscience Publishers, New York, 1962.

[148] Noether, M: *Über die algebraischen Functionen einer und zweier Variabeln.* Nachr. Königl. Akad. Wiss. Götttingen, math.-phy. Cl. II, Seiten 267–278, 1871.

[149] Noether, M.: *Über die singulären Werthsysteme einer algebraischen Function und die singulären Punkte einer algebraischen Curve.* Math. Ann., 9:166–182, 1876.

[150] Noh, S.: *The value semigroups of prime divisors of the second kind in 2-dimensional regular local rings.* Trans. Amer. Math. Soc., 336:607–619, 1993.

[151] Northcott, D. G.: *A note on the genus formula for plane curves.* J. London Math. Soc., 30:356–382, 1955.

[152] Northcott, D. G.: *Abstract dilatations and infintely near points.* Proc. Cambridge Philos. Soc., 52:178–197, 1956.

[153] Northcott, D. G.: *A general theory of one-dimensional local rings.* Proc. Glasgow Math. Assoc, 2:159–169, 1956.

[154] Northcott, D. G.: *On the notion of a first neighborhood ring with applications to the af + bφ theorem.* Proc. Cambridge Philos. Soc., 53:43–56, 1957.

[155] Northcott, D. G.: *Some contributions to the theory of one-dimensional local rings.* Proc. London Math. Soc. (3), 8:388–415, 1958.

[156] Northcott, D. G.: *Lessons on Rings, Modules and Multiplicities.* Cambridge University Press, Cambridge, 1968.

[157] Oda, T.: *Convex Bodies and Algebraic Geometry; an Introduction to the Theory of Toric Varieties*, volume 15 of *Ergebnisse der Mathematik und ihrer Grenzgebiete, 3. Folge.* Springer Verlag, Berlin Heidelberg New-York, 1988.

[158] Orbanz, U.: *Valuation semigroups and closed components.* Comm. Algebra, 7:1181–1198, 1979.

[159] Pesselhoy, J. and O. Riemenschneider: *Projective resolutions of Hodge algebras: Some examples.* In *Singularities, Part II*, volume 40 of *Proc. Sympos. Pure Math.*, pages 305–317, Providence, RI, 1983. Amer. Math. Soc.

[160] Raynaud, M.: *Anneaux Locaux Henséliens.* Lecture Notes in Mathematics. Springer Verlag, Berlin Heidelberg New-York, 1970.

[161] Reitberger, H.: *The turbulent fifties in resolution of singularities*. In Hauser, H., J. Lipman, F. Oort, and A. Quirós (editors): *Resolution of Singularities*, volume 189 of *Progress in Mathematics*, pages 533–537. Birkhäuser, Basel Boston Berlin, 2000.

[162] Ribenboim, P.: *Théorie des Valuations*. Les Presses de l'Université de Montrèal, Montrèal, 1965.

[163] Ribenboim, P.: *The Theory of Classical Valuations*. Springer Monographs in Mathematics. Springer Verlag, Berlin Heidelberg New-York, 1999.

[164] Riemenschneider, O.: *Deformationen von Quotientensingularitäten (nach zyklischen Gruppen)*. Math. Ann., 209:211–248, 1974.

[165] Riemenschneider, O.: *Zweidimensionale Quotientensingularitäten: Gleichungen und Syzygien*. Arch. Math., 37:406–417, 1981.

[166] Rockefellar, R. T.: *Convex Analysis*. Princeton University Press, Princeton, 1970.

[167] Roquette, P.: *History of valuation theory, vol. I*. In *Valuation Theory and its Applications, Vol. I (Saskatoon, SK, 1999)*, volume 32 of *Fields Inst. Commun.*, pages 291–356. Amer. Math. Soc., Providence, RI, 2002.

[168] Ruiz, J. M.: *The Basic Theory of Power Series*. Advanced Lectures in Mathematics. Vieweg, Braunschweig, 1993.

[169] Russell, P.: *Hamburger-Noether expansions and approximate roots of polynomials*. Manuscripta math., 31:25–95, 1980.

[170] Samuel, P.: *La Notion de Place Dans un Anneau*. Bull. Soc. Math. France, 85 :123–133, 1957.

[171] Scheja, G. und U. Storch: *Lokale Verzweigungstheorie*. Schriftenreihe des Mathematischen Instituts der Universität Freiburg I. Universität Freiburg I., UE. Freiburg I. UE., 1974.

[172] Schrijver, A.: *Theory of Integer and Linear Programming*. J. Wiley and sons, New York, 1986.

[173] Shafarevich, I. R.: *Lectures on Minimal Models and Birational Transformations of Two-Dimensional Schemes*. Tata Institute of Fundamental Research, Bombay, 1966.

[174] Shafarevich, I. R.: *Basic Algebraic Geometry, 2 volumes*. Springer Verlag, Berlin Heidelberg New-York, 2nd. edition, 1994.

[175] Soto, M. J. et J. L. Vicente: *Résolution de certaines équations algébriques*. rapport technique, Departamento de Álgebra, Facultad de Matemáticas, Sevilla, dec 2003.

[176] Spivakovsky, M.: *Valuations in function fields of surfaces*. Amer. J. Math., 112:107–156, 1990.

[177] Spivakovsky, M.: *On the structure of valuations centered in a local domain*. Prepublication, 1993.

[178] Tamone, G.: *A connection between blowing-up and gluings in one-dimensional rings*. Nagoya Math. J., 94:75–87, 1984.

[179] Teissier, Bernard: *Introduction to curve singularities*. In *Singularity Theory (Trieste, 1991)*, pages 866–893. World Sci. Publishing, River Edge, NJ, 1995.

[180] Tougeron, J. J.: *Idéaux de Fonctions Différentiables*, tome 71 de *Ergebnisse der Mathematik und ihrer Grenzgebiete*. Springer Verlag, Berlin Heidelberg New-York, 1972.

[181] Vaquié, M.: *Valuations*. Dans Hauser, H., J. Lipman, F. Oort et A. Quirós (rédacteurs) : *Resolution of Singularities*, tome 189 de *Progress in Mathematics*, pages 530–590. Birkhäuser, Basel Boston Berlin, 2000.

[182] Vasconcelos, W. V.: *Arithmetic of Blow-up Algebras*, volume 195 of *London Mathematical Society Lecture Notes Series*. Cambridge University Press, Cambridge, 1994.

[183] Vasconcelos, W. V.: *Computational Methods in Commutative Algebra and Algebraic Geometry*, volume 2 of *Algorithms and Computation in Mathematics*. Springer Verlag, Berlin Heidelberg New-York, 1998.

[184] Vicente, J. L.: *Singularidades de Curvas Algebroides Alabeades*. Tesis de Doctorado, Universidad de Madrid, Madrid, 1973.

[185] Villamayor, O. E .: *Patching local uniformizations*. Ann. Sci. École Norm. Sup., 25:629–677, 1992.

[186] Villamayor, O. E.: *On good points and a new canonical algorithm of reduction of singularities*. In *Real and Analytic Geometry*, pages 272–291, Berlin, 1995. De Gruyter.

[187] Villamayor, O. E.: *Introduction to the algorithm of resolution*. In *International Conference on Algebraic Geometry and Singularities, La Rabida*, volume 134 of *Progress in Mathematics*, pages 123–151, Basel, 1996. Birkhäuser Verlag.

[188] Waerden, B. L. van der: *Algebra*, Band 1. Springer Verlag, Berlin Heidelberg New York, 4. Auflage, 1955.

[189] Walker, R. J.: *Reduction of singularities of an algebraic surface*. Ann. of Math., 36:336–365, 1935.

[190] Walker, R. J.: *Algebraic Curves*. Springer Verlag, Berlin Heidelberg New-York, reprint edition, 1978.

[191] Zariski, O.: *Polynomial ideals defined by infinitely near base points*. Amer. J. Math., 60:151–204, 1938.

[192] Zariski, O.: *The reduction of the singularities of an algebraic surface*. Ann. of Math., 40:639–689, 1939.

[193] Zariski, O.: *Local uniformization of algebraic varieties*. Ann. of Math., 41:852–896, 1940.

[194] Zariski, O.: *A simplified proof for the resolution of singularities of an algebraic surface*. Ann. of Math., 43:583–593, 1942.

[195] Zariski, O.: *Analytical irreducibility of normal varieties*. Ann. of Math., 49:352–361, 1948.

[196] Zariski, O.: *A simple analytical proof of a fundamental property of birational transformations*. Proc. Nat. Acad. Sci. USA, 35:62–66, 1949.

[197] Zariski, O.: *Sur la Normalité Analytique des Variétés Normales*. Ann. Inst. Fourier, 2 :161–164, 1950.

[198] Zariski, O.: *Review to Derwidué, L.: Le Problème de la Réduction des Singularités d'une Variété Algébrique*. Math. Ann. **123** *(1951), 302–330*. Math. Rev., 13:67–70, 1952.

[199] Zariski, O.: *Le Problème de la Réduction des Singularités d'une Variété Algébrique*. Bull. Sci. Math., 78 :1–10, 1954.

[200] Zariski, O.: *On the purity of the branch locus of algebraic functions*. Proc. Nat. Acad. Sci. USA, 44:791–796, 1958.

[201] Zariski, O.: *Exceptional singularities of an algebroid surface and their reduction*. Accad. Naz. Lincei Rend. Cl. Sci. Fis. Mat. Natur. Ser. VIII, 43:135–146, 1967.

[202] Zariski, O.: *A new proof of the total embedded resolution theorem for algebraic surfaces (based on the theory of quasiordinary singularities)*. Amer. J. Math., 100:411–442, 1978.

[203] Zariski, O.: *Le Problème des Modules pour les Branches Planes*. Springer Verlag, Berlin Heidelberg New-York, Nouvelle édition, 1980.

[204] Zariski, O. and P. Samuel: *Commutative Algebra, 2 volumes*. van Nostrand, New York, 1960.

Index of Symbols

Index

Algebras and Applications

1. C.P. Millies and S.K. Sehgal: *An Introduction to Group Rings*. 2002
    ISBN Hb 1-4020-0238-6; Pb 1-4020-0239-4
2. P.A. Krylov, A.V. Mikhalev and A.A. Tuganbaev: *Endomorphism Rings of Abelian Groups*. 2003
    ISBN 1-4020-1438-4
3. J.F. Carlson, L. Townsley, L. Valero-Elizondo and M. Zhang: *Cohomology Rings of Finite Groups*. Calculations of Cohomology Rings of Groups of Order Dividing 64. 2003
    ISBN 1-4020-1525-9
4. K. Kiyek and J.L. Vicente: *Resolution of Curve and Surface Singularities*. In Characteristic Zero. 2004
    1-4020-2028-7

KLUWER ACADEMIC PUBLISHERS – DORDRECHT / BOSTON / LONDON